Program of the

Ninth Annual Conference of the Cognitive Science Society

16-18 July 1987
Seattle, Washington

Psychology Press
Taylor & Francis Group
NEW YORK AND LONDON

First published 1987 by
Lawrence Erlbaum Associates, Inc.

Published 2014 by Psychology Press
711 Third Avenue, New York, NY 10017

and by Psychology Press
27 Church Road, Hove, East Sussex, BN3 2FA

*Psychology Press is an imprint of the
Taylor & Francis Group, an informa business*

ISBN 13: 978-0-805-80166-8 (hbk)

Publisher's Note
The publisher has gone to great lengths to ensure the quality of
this reprint but points out that some imperfections in the original
may be apparent.

Officers of the Cognitive Science Society

John Anderson	Carnegie-Mellon University	1983-1988
Daniel G. Bobrow	Xerox PARC	1979-1987
Gordon H. Bower (Chair-Elect 1987-88)	Stanford University	1980-86
Joan Bresnan	Stanford University, Xerox PARC	1983-1988
Ann Brown	University of Illinois	1984-1989
Allan Collins	Bolt Beranek and Newman	1979-1987
Patrick J. Hayes	Schlumberger PA Research	1984-1989
Geoffrey E. Hinton	Carnegie-Mellon University	1986-1991
Ulric Neisser	Emory University	1985-1990
Donald A. Norman	UC San Diego	1985-1990
David Rumelhart	UC San Diego	1986-1991
Edward E. Smith (Chair 1986-87)	University of Michigan	1980-1986

Reviewers

James Anderson
Patricia Merkle-Baggett
Lee Beach
Carole R. Beal
Thomas Bever
R.C. Bolles
Alan Borning
Jeff Bradshaw
Gary Bradshaw
Greg N. Carlson
Vere C. Chappell
Patricia Cheng
John Clement
Allan Collins
Gary Cottrell
Bill Curtis
Philip Dale
Mark Derthick
Anders Ericsson
Martha Farah
Eric Fischer
Don Fisher
C.R. Fletcher
Kenneth Forbus
Lyn Frazier
Anne Gardner
Jay L. Garfield
Dedre Gentner
Donald Gentner
Philip Gough
Art Graesser
Stephen Grossberg
Goeffrey Hinton
Douglas Hintzman
Keith Holyoak
Mary Lou Hunt
Yumi Iwasaki
Aravind K. Joshi

Ninth Annual Cognitive Science Society Conference
July 16-18, 1987

Beth Kerr
Ray Kesner
Walter Kintsch
Janet L. Kolodner
Steve Kosslyn
Stan Kulikowski
Bunny Laden
Richard Ladner
Marcy Lansman
Jill Larkin
Wendy Lehnert
Alan M. Lesgold
Clayton Lewis
Jack Lockhead
Beth Loftus
Geoffrey Loftus
James Lundell
Charles Marks
Sandra Marshall
Edward Matthei
Deborah McCutcheon
Richard Mayer
Gale McKoon
Carolyn Mervis
Jose Mestre
Bonnie Meyer
James Miller
John Miyamoto
Jerome L. Myers
Donald A. Norman
Gregory Oden
John Palmer
Jim Pellegrino
David Pisoni
Peter Polson
Steve Poltrock
James Pustejovsky
Zenon Pylyshyn
Roger Ratcliffe
Keith Rayner

Ninth Annual Cognitive Science Society Conference
July 16-18, 1987

Judith Reitman-Olson
Elaine Rich
Christopher Riesbeck
Arthur Samuel
Nestor A. Schmajuk
Mallory Selfridge
Edward Smith
Elliot Soloway
Joseph Stermberger
Michael Tanenhaus
Steven Tanimoto
Alice Ter Muelen
Pamela Thibodeau-Hardiman
Kirk Thompson
R.M. Vardaris
Rajendra Wall
Jim Wise
Beverly Woolf

Table of Contents

Table of Contents

Table of Contents

Table of Contents

SATURDAY, JULY 18

Connectionism III

Linguistics II

Table of Contents

Artificial Intelligence and Simulation I

Problem Solving II

Table of Contents

POSTER PRESENTATIONS

THURSDAY, JULY 16

Table of Contents

Modifying Previously-Used Plans to Fit New Situations

Roy M. Turner
School of Information and Computer Science
Georgia Institute of Technology

Abstract

Re-using plans that were created for one situation to solve a new problem is often more efficient than creating a new plan from scratch (e.g., [Fikes *et al*, 1972] and [Carbonell, 1986]. However, a plan that was created for one problem may not exactly fit a new situation; in that case, it will have to be modified. There are two major problems with re-using plans: (1) deciding whether to modify a plan, use it as is, or discard it; and (2) modifying the plan efficiently. Our solution to these problems is to store information with plan preconditions to guide the planner during plan application. Our approach is novel in two ways. First, we have identified a type of precondition, called a *flexible precondition*, that has information associated with it that helps the planner decide whether or not to modify the plan should the precondition be violated. Second, our preconditions contain information (derived from past experience using the plan) that provides heuristics for changing the plan so that the offending precondition is either no longer violated or no longer necessary. By using this approach, our planner can quickly determine whether or not to modify a plan, then efficiently perform the modification. Our work is implemented in the Consumer-Advisor System (CAS) [Kolodner and Cullingford, 1986; Turner, 1986; Turner, in press], a common-sense advice-giving program.

1.0 Introduction

When a problem is presented to a planner, it has one of two choices: it can attempt to formulate a new plan to solve the problem by combining operators from its repertoire; or it can recall and use a plan that was formulated to solve a problem in the past.

A difficulty with re-using a plan, however, is that the problem for which the plan was originally constructed may be only similar to the current problem and not identical. Some goals in the new problem may not be met by the expected results of the plan, for example, or one or more of the plan's preconditions may be violated. In this case, the planner can do one of several things. It can discard the plan immediately. This is unattractive, however, since the plan represents previous planning experience that should not be wasted. The planner can perform additional planning in order to satisfy the violated preconditions, the traditional approach to precondition violations. This can be arbitrarily hard, however, and makes sense only when it is easier than modifying the plan. The third alternative available to a planner is to use the plan "as is". Unless the planner can predict the outcome of doing this, however, a good solution is unlikely; and predicting the outcome usually involves simulation of the plan, which is costly in terms of time. A better solution than all of these, especially when the recalled plan almost fits the current problem, is to modify the plan.

There are two major problems to be dealt with in plan modification: (1) deciding whether to modify the plan at all, or instead to use it "as is" or to discard it; and (2) modifying the plan efficiently. Our solution to both of these problems is to store planning information with a plan's preconditions. This approach is novel in two ways. First, not all of our preconditions are criteria that *must* be satisfied in order for a plan to be applied. Some preconditions, called *flexible preconditions*, have information associated with them that allows the planner to determine what the outcome of applying the plan would be should the violated precondition be ignored. Second, all of our preconditions have directives stored with them that provide heuristics to the planner to guide it during plan modification.

Our approach is implemented as part of the Consumer-Advisor System (CAS) [Kolodner and Cullingford, 1986; Turner, 1986; Turner, in press], a common-sense advice-giving program whose domain is consumer products. In this paper, we first give a brief overview of CAS, then discuss our method of plan modification.

2.0 The Consumer-Advisor System

CAS is a common-sense advice-giving program. It gives advice about acquiring consumer products, such as furniture and bookshelves. Its primary problem-solving strategy is plan instantiation: it recalls a previous plan, then attempts to fit it to the current problem.

CAS uses several concurrent processes to perform its tasks. A *dynamic memory* [Schank, 1982; Kolodner, 1984] constantly tries to remember plans and other information pertinent to the current problem situation; when it is reminded of something, it notifies the planner. The planner is a separate process that can opportunistically use information from remindings to influence its problem-solving behavior. If the reminding is of a previously-used plan, the planner attempts to apply it to the current problem, if possible. If the plan is a close fit, but the current situation violates some of its preconditions, the planner uses information stored with the plan's preconditions in an attempt to modify the plan to fit the situation.

Before describing our approach to plan modification, we present an example of CAS' problem-solving behavior to illustrate the process. The problem presented to CAS was the following:

> I want to buy some bookshelves for my study at home. I am a student, so I can't spend much. What kind of bookshelves should I buy?

We will begin at the point in CAS' problem solving when the memory returns a plan to build wooden objects, BUILD-WOODEN-OBJECTS (Figure 1); for a more complete example of CAS' behavior, see [Turner, in press].

The planner first notifies the user that it has been reminded (by the memory) of a plan for building wooden objects (Figure 1):

I am reminded of plan ⟨BUILD-WOODEN-OBJECT⟩, the plan "building wooden objects".

The planner next tries to determine, by matching goals of the problem with results of the plan, whether or not the plan is potentially applicable to the current problem:

Determining if plan is applicable...

Plan's results meet all of problem's major goals...plan is applicable.

The planner now examines the plan's preconditions in light of the current situation:

Examining preconditions of ⟨BUILD-WOODEN-OBJECT⟩...

There are several preconditions for the plan: the user should have carpentry skill, the user should know how to design the object to be built, and the user should know where to get the materials from which to build the object. CAS knows very little about the user at this point, other than that he is a student and can't spend much money. In order to determine if the preconditions are met, it must ask the user.[1]

"Do you have carpentry skill?"

The user responds negatively.

If "you do not have carpentry skill", then "object to be built will be of poor quality".

Is this acceptable?

Again, the user responds negatively. The planner then tries to modify the plan to eliminate the need for the precondition:

Modifying plan...

The planner uses the first piece of information stored with the precondition to attempt to modify the plan. This is a directive to replace the assembly step with a new step, getting someone else to assemble them for the user:

...trying directive "(:REPLACE ?ASSEMBLE ?USE-AGENT-ASSEMBLE)"...

The new step also has preconditions that must be satisfied:

"Do you know anyone who can perform plan "ASSEMBLE" for you?"

---

[1] The questions to the user are, at the moment and for the purposes of this example, canned text.

name: "building wooden objects"
actors: ?a1
props: ?object ?materials
goals: ?acquire
actions: ?design ?get-materials ?assemble
preconditions: ?pc-carpentry-skill ?pc-have-design-knowledge ?pc-know-sources-matl
results: ?have-object

——————— *Variables:*———————

a1: *isa* ↑ANIMAL
object: *isa* ↑ARTIFACT
materials: *isa* ↑SET
    type: ↑POBJ   *;;physical objects*
acquire: *isa* ↑AGOAL   *;;achievement goal [Schank and Abelson, 1977]*
       actor: ?a1
       state: *isa* ↑POSSESSION
           actor: ?a1
           object: ?object
           mode: pos
design: *isa* ↑INSTANCE
      structure: ↑DESIGN-PLAN
      bindings: ((actor . a1) (object . object))
get-materials: *isa* ↑INSTANCE
        structure: ↑GET-OBJECTS-PLAN
        bindings: ((actor . a1) (objects . materials))
assemble: *isa* ↑INSTANCE
      structure: ↑ASSEMBLE
      bindings: ((actor . a1) (object . object)
              (materials . materials))
pc-carpentry-skill: *isa* ↑PRECONDITION
           state: ↑S-CARPENTRY-SKILL
           reason: ?assemble
           if-violated: ↑S-DECREASED-QUALITY1
           fix: (or (:replace ?assemble ?use-agent-assemble)
                (:add ?learn-carpentry (?assemble before))
                (:replace ?assemble ?assemble-with-book)
                    •
                    •
                    •

ʒure 2: **A portion of the plan** BUILD-WOODEN-OBJECT. Symbols of the form "?name" are variables, referring to other slots in the frame; those of the form ↑name represent pointers to other frames.

If the user does, then the planner performs the replacement:

Replacing ASSEMBLE ("assemble object "BOOKSHELF"") with USE-AGENT-ASSEMBLE ("use agent to assemble "BOOKSHELF"").

There are other preconditions to be satisfied. Assuming that the user knows where to get the materials for building bookshelves, but does not know how to design them, the completed plan would look like that shown in

Figure 2: find a design for bookshelves in a book, get the materials required, then have someone assemble the bookshelves.

Original Plan:

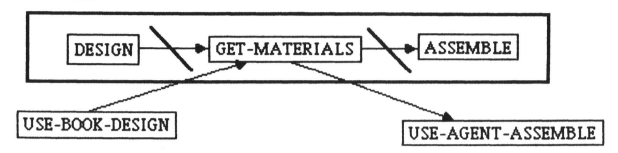

Plan after modification:

Figure 2: Example of plan modification.

In the remainder of this paper, we will discuss the knowledge CAS uses to produce this behavior and where that knowledge comes from.

3.0 Plan Structure and Origin

Before explaining the plan modification process, we need to explain the structure of plans. Recall that the plans that get modified were derived from experience. That is, small plan steps were put together in the past to successfully solve a problem. As in STRIPS [Fikes *et al*, 1972], those specific plans are generalized to create a plan skeleton.[2] It is these plan skeletons that are modified and instantiated to create a solution to a new problem. The plan skeletons (to be called plans from now on) contain plan steps (arranged hierarchically, since each step can itself be a plan) and plan preconditions. For example, Figure 1 shows a plan for building wooden objects that has three steps ("design", "get-materials", and "assemble") and three preconditions ("pc-have-carpentry-skill", "pc-have-design-knowledge", and "pc-know-sources-matl"). Each step in the plan is itself a plan, with steps and preconditions of its own. Preconditions from the steps are used, in part, to derive the preconditions of the entire plan (see [Turner, in press] for more details).

As a plan gets used several times, information about preconditions gets updated (e.g., the planner may learn that the precondition of having carpentry skill can be ignored if the user of the plan doesn't care about the quality of the object being built) and new, more specific plans are derived. Plans are indexed such that during plan retrieval the most specific applicable plan can be found. In addition, information about the modifications to the plan which yielded new plans is generalized and stored with the old plan's preconditions. This information can be used in the future to modify the plan for new problems.

The following sections discuss the content and use of the information associated with plan preconditions.

---

[2]   Explaining this process is beyond the scope of this paper. See [Turner, in press] for more details.

4

4.0 Plan Modification Directives

When a planner decides to modify a plan, it must determine the best way make the plan fit the current problem. According to our model, knowledge associated with preconditions of the plan is used to modify it for the new situation.

CAS' preconditions have information, called *directives*, associated with them that provide suggestions about how to modify the plan. For example, the precondition PC-CARPENTRY-SKILL of the plan BUILD-WOODEN-OBJECT (Figure 1) contains three heuristics: (1) replace the step "assemble" with a plan to use an agent (i.e., a plan to have someone else assemble the object); (2) add a plan to learn carpentry before the step to assemble the object; and (3) replace the assembly step with a plan to assemble the object using information from a book. The slot "fix" holds this information.[3] When a particular precondition is violated, "fix" heuristics associated with it are used by the planner as it modifies the plan.

Preconditions, when violated, can require strategic changes to a plan, i.e., changes that affect the plan's overall strategy. Such changes result in a plan that no longer requires the violated precondition, or one in which the precondition is achieved as part of the plan. An example of this can be seen in Figure 1. The precondition PC-CARPENTRY-SKILL calls for (via one of its directives) adding a step to learn some carpentry skills before applying the existing "assemble" step of the plan. Since the precondition has as its source the "assemble" step, it will be not be violated by the new plan (i.e., the user will have carpentry skill by the time he applies the "assemble" step)—indeed, it will no longer be needed as a precondition of the new plan and can be discarded. Violated preconditions can also require only tactical, or local, changes to a plan, usually replacement or modication of one step in the plan. The precondition PC-CARPENTRY-SKILL also contains an example of this: one of its directives calls for the replacement of the "assemble" step by a new plan that will use another agent to assemble the object.

We have identified three types of strategic directives and two types of tactical directives for changing a plan.

4.1 Strategic Plan Modification

A *strategic* change to a plan is one that affects the overall strategy used by the plan. Examples of this are adding a step, deleting a step, or re-ordering steps.

Adding a step to a plan. Suppose the plan BUILD-WOODEN-OBJECT is to be used by someone who doesn't know anything about carpentry; a precondition is violated. One thing that can be done is to have the user learn carpentry, then run the plan. This is not a very good idea, however, since learning all of carpentry can take a long time; the user only needs a small subset of that information to run the plan. A better approach is to add a step to the plan to learn what is needed just before the information is actually needed; that is, just before the "assemble" step of the plan. Suppose the planner decides, based on recursive planning, that this is a good modification to the plan. This information can be added to the precondition in the form of the following directive:

(:add *new-step side old-step*)

which directs the planner to add a new step described by *new-step* to the plan before or after (depending on the value of *side*) the old plan step described by *old-step*. When solving a later problem in which it was reminded of this plan and the same precondition was violated, the planner would not have to modify the plan from scratch, but rather could use this directive to quickly patch the plan for the new situation.

Deleting a step from the plan. Another strategic directive specifies that in order to eliminate the need for a precondition, a step should be deleted from the plan. This is specified by the following:

(:delete *old-step*)

where *old-step* describes a step currently called for by the plan. An example of a case in which this would be useful is the following. Suppose a plan for making chili con carne has a precondition that none of the people who will eat the chili are vegetarians. To modify the plan in response to a violation of this precondition, the step of adding meat can be eliminated from the plan. This can be represented by a :delete directive.

---

[3] This information can, in principle, come from two experiential sources (though so far in CAS it is built in). The first source, mentioned above, is previous modifications of the plan. The second source is observations of others using the plan. A program that could watch a person and learn from its observations (or that could be told about a person's behavior) could use an episode of a person modifying a plan in response to a precondition violation to learn information to add to the plan.

5

Re-ordering steps in a plan. The third type of strategic directive calls for changing the order of steps in a plan. An example of when this is appropriate is the following. Suppose, for example, that instead of having one step for designing the object to be built, our plan for building wooden objects had two steps: choose the wood and then select the configuration. It makes sense for these steps to be performed in this order if aesthetics are more important than the function of the object, since the choice of the wood constrains the configuration (e.g., thin cherry would need a great deal of structural support). However, if function is more important than aesthetics, the steps should be run in the reverse order: choose an optimal design, then select a wood that can be used in the design. One way these alternatives can be represented in a single plan is to have a precondition whose criteria for satisfaction is: "aesthetic considerations are more important than functional considerations". Associated with this precondition would be a directive to reverse the order of the two steps should the condition be violated.

4.2 Tactical Plan Modification

Tactical plan modification involves changing a single step in a plan without affecting the overall strategy of the plan. Examples of this are replacing one step with another step (that doesn't impact the remaining steps in the plan), and changing a step within a step of the plan.

Replacing a step. We saw an example of this type of tactical plan modification in our example above, when the planner substituted the plan of getting someone else to assemble the bookshelves for the step of having the user assemble them himself. The directive to specify step replacement looks like:

(:replace *old-step new-step*)

where *old-step* is a description of the step to be replaced, and *new-step* describes a plan or an action to use as the replacement. Although replacement could be accomplished by doing an addition and then a deletion, we use a single directive to specify the change. This simplifies the process of modifying the plan (i.e., only one directive, not two, needs to be followed), and is conceptually much clearer.

Modifying a step. Preconditions of a plan can also direct an internal change to be made to one of the plan's steps. For example, suppose the "get-materials" step in the plan for building a wooden object has a step of driving from store to store. If the user doesn't have a car, this step must be changed to substitute an alternate form of transportation, but the rest of the step, and the other steps in the plan, should be left the same. The form of a directive to specify this would be:

(:modify *step directive*)

where *step* is the step to be modified and *directive* is a directive, either strategic or tactical, decribing the modification to be made.

4.3 Specifying More than One Change

A violated precondition can specify more than one change to be made to a plan in order to eliminate the violation. This is done by directives that are boolean ("and" and "or" only) combinations of other directives. For example, a directive to either replace the assembly step with a plan for getting a neighbor to help or adding a step to learn carpentry would look like:

```
(:or
    (:replace ⟨description of assembly step⟩
              ⟨description of the replacement⟩)
    (:add ⟨description of learning carpentry⟩
          :before
          ⟨description of assembly step⟩))
)
```

An example of a compound directive can be seen in Figure 1.

6

5.0 Absolute and Flexible Preconditions

Researchers working on problem solving have traditionally treated preconditions as conditions that must be met before using a plan or doing a plan step. Some conditions on a plan or plan step, however, affect the degree of success of the plan rather than its absolute success. Lack of carpentry skill, for example, will preclude the complete success of the BUILD-WOODEN-OBJECT plan, but will not prevent it being applied with limited success—the quality of the object built will not be as good. We distinguish between preconditions (called *absolute preconditions*) that must be met and those (called *flexible preconditions*) that need to be met only for perfect results. CAS uses each differently during plan modification.

Flexible preconditions [4] have information associated with them that predicts the likely outcome should they be violated and the plan applied anyway. An example of a flexible precondition can be seen in Figure 1. PC-CARPENTRY-SKILL states that the user should have carpentry skill in order to apply the plan. If he doesn't, the plan can still be applied, but the result will not be as nice. Flexible preconditions allow the planner to make judgements about whether to modify a plan to fit the current problem or to ignore the violated precondition and apply the plan "as is". Absolute preconditions, on the other hand, have no such information, and no such decisions need be made about them. If they are violated, the plan simply cannot be used as is. Using it requires modifications that either make the precondition unnecessary or achieve it.

CAS recognizes a precondition as flexible if it has knowledge associated with it about the results of its violation (held in an "if-violated" slot). This information is used to decide whether or not to ignore the precondition's violation. Knowledge about precondition violation takes the form of a state that will result if the plan is applied and the precondition is violated. The knowledge associated with the precondition PC-CARPENTRY-SKILL (Figure 1), for example, indicates that if the person applying the plan lacks carpentry skill, the quality of the object built will be poor.

The information in the "if-violated" slot, like all precondition information, can be derived from past experience using the plan, though so far in CAS we have built it into our plans. When a plan is first formulated, its preconditions are derived from the preconditions of its steps and initially have no information indicating the effect of violating them. As the plan is used and its preconditions are violated, however, information is added to the preconditions describing the result of their violation. This information can come from simulating the plan, or it can come from the planner deciding to ignore the precondition, then recording the effects.

Although not yet completely implemented in CAS, flexible preconditions serve several purposes. Information in the "if-violated" slot of a precondition can be used in combination with a user model to decide when to attempt to modify a precondition and when to ignore a violation entirely. This will allow CAS to predict the outcome of applying a faulty plan without needing to simulate it. Knowledge associated with flexible preconditions also allow a planner to select and apply the best plan from a set of less than optimal plans when that is all that is available; with traditional (absolute) preconditions, no plan could be chosen.

6.0 Using Preconditions in Plan Modification

When CAS' planner is presented with a plan, it examines the expected results and compares them to the goals of the current problem. If the results satisfy all of the major goals in the problem, then the plan is deemed applicable, and the preconditions of the plan are checked to determine if there are any violations.

If a flexible precondition is violated, the planner tries to determine if the violation can be ignored. At the moment, the way it does this is to ask the user if the result of violating the precondition is acceptable. Another way to do this is to use a model of the user to help it decide whether or not a result is acceptable. If the result of ignoring the violation is acceptable, then the planner considers the next precondition; otherwise, it attempts to modify the plan, using any directives associated with the precondition. If an absolute precondition is violated, the planner assumes that the plan cannot be run if the violation occurs and immediately attempts to modify the plan.

Once the planner decides to modify the plan because of a precondition violation, it examines the directives stored with the precondition. The directives are not currently stored in any particular order. If there is only one directive, it is applied. If there are more than one, they are tried in order until one is found that is acceptable. A directive is acceptable if the change it suggests is possible and acceptable to the user.

---

[4] Called "relative preconditions" previously [Turner, 1986].

7.0 Related Work

Re-using plans is not a new idea; for example, STRIPS [Fikes *et al*, 1972] stored plans it created in structures called MACROPS for later use. Nor is the idea of plan modification itself new. MACROPS were modified by variable substitution—a process Carbonell [1983] calls "transformational analogy"—and pieces of the MACROP could be used if the entire thing was not needed. This type of modification is rather trivial and not very flexible. For instance, if a MACROP or piece of a MACROP was selected to solve a new problem, the structure was applied "as is",[5] with no changes to the structure allowed.

Carbonell [1986] proposed another form of reasoning by analogy to a previous problem, called *derivational analogy*. This work is closely related to our own, as we both rely on previous reasoning to guide plan modification. Yet there are important differences. Carbonell's approach involves stepping through the reasoning that was done in creating a plan for a previous situation, noticing problems with respect to the current problem, and fixing them. Our approach, however, relies on compiled information in the preconditions to immediately alert the planner to potential difficulty applying the plan. The planner can then evaluate the severity of the problem—again using information stored with the preconditions—and modify the plan. This is a more efficient approach, since we are not simulating planning that was done previously. Another difference between our planner and Carbonell's is that ours is hierarchical, and his is not.

Hammond [1984] proposes using information stored with *thematic organization packets*, or TOPs [Schank, 1982], to patch faulty plans. TOPs represent planning information at a high level of abstraction. For example, a TOP might suggest that in order to avoid a violation of a goal by some plan step, that plan step should be replaced or changed; however, additional reasoning would need to be done to decide how to replace or change the step. We use TOPs in CAS, too, but have found that it is advantageous to store information about modifying a specific plan with the plan itself. This allows the specific information, i.e., that which pertains specifically to the plan, to be accessible immediately upon recalling the plan, without performing any additional reasoning.

Alterman [1986] has developed an adaptive planner that can use both specific and general plans to solve a new problem. If the specific plan fails, then the failing step is generalized to find a representative category of action, then that is specialized to find a new step. Each time a step fails in a plan, this must be done. The difference between this approach and our own is that we avoid the generalization and specialization procedure by storing compiled knowledge about previous plan modifications with the preconditions. Instead of looking for an alternative step, CAS can immediately use past experience to substitute a step that worked before.

Schank and Abelson [1977] have divided the preconditions of their *planboxes* into three types: *controllable*, *uncontrollable*, and *mediating*. Controllable preconditions are those satisfiable by operators a planner knows about. Uncontrollable preconditions are those the planner doesn't know how to satisfy. Mediating preconditions are those the planner can satisfy by using other planboxes from the *persuade package*. This was a good start at categorizing preconditions by how they can be satisfied. However, the problem with this approach is that it is too static for use by a planner that learns plans over time. Preconditions that are uncontrollable now may be controllable at some future time, and preconditions that are satisfied by plans similar to those in the persuade package may at some later time be satisfiable by other plans. Their approach also was not concerned with plan modification.

8.0 Conclusion

Re-using plans often involves modifying them to fit a new situation. Plan modification is made easier by using information gained from past experience using the plan. Since this information usually relates to how the plan was modified in response to one or more precondition violations, CAS takes the approach of storing the information with the preconditions themselves. The planner can easily access this information when it needs it during the process of modifying the plan.

There are two important features of our approach. First, some preconditions are *flexible preconditions*. These contain information about the likely result of applying the plan if they are violated. This information allows a planner to quickly decide whether it should attempt to modify the plan or simply ignore the offending precondition. Since relative preconditions do not necessarily have to be satisfied in order for a plan to be applied,

---

[5]   Unless there was a precondition violation, in which case the structure was applied after additional planning to meet the preconditions.

they also have the advantage, in principle, of allowing a planner to choose the best plan from among several plans with violated preconditions, if that is all that is available.

The second feature of our approach is that preconditions contain information that can be used by the planner as heuristics during plan modification. This information is compiled from experience using the plan, and takes the form of planning directives—suggestions to the planner based on what has worked before. By making use of these directives, the planner can use previous experience to allow it to efficiently modify a plan.

Bibliography

Alterman, R. (1986). "An Adaptive Planner," in *Proceedings of the Fifth National Conference on Artificial Intelligence* (AAAI-86), Morgan Kaufmann Publishers, Inc., Los Altos, California, pp. 65–69.

Carbonell, J.G. (1983). "Learning by Analogy: Formulating and Generalizing Plans from Past Experience," in *Machine Learning: An Artificial Intelligence Approach*, eds. R.S. Michalski, J.G. Carbonell, and T.M. Mitchell, Tioga Publishing Company, Palo Alto, California.

Carbonell, J.G. (1986). "Derivational Analogy: A Theory of Reconstructive Problem Solving and Expertise Acquisition," in *Machine Learning: An Artificial Intelligence Approach, Volume II*, ed. Ryszard S. Michalski, Jaime G. Carbonell, and Tom M. Mitchell, Morgan Kaufman Publishers, Inc., Los Altos, California.

Fikes, R.E., Hart, P.E., and Nilsson, N.J. (1972). "Learning and Executing Generalized Robot Plans," *Artificial Intelligence* **3** pp. 251-288.

Hammond, C. (1984). "Indexing and Causality: The organization of plans and strategies in memory," Yale Technical Report YALEU/CSD/RR #351.

Kolodner, J.L. (1984). *Retrieval and Organizational Strategies in Conceptual Memory: A Computer Model*, Lawrence Erlbaum Associates, Publishers, Hillsdale, New Jersey.

Kolodner, J.L., and Cullingford, R.E. (1986). "Towards a Memory Architecture that Supports Reminding," *Proceedings of the Eighth Annual Conference on Cognitive Science*, Lawrence Erlbaum Associates, Publishers, Hillsdale, New Jersey.

Schank, R.C. (1982). *Dynamic Memory*, Cambridge University Press, New York.

Schank, R.C., and Abelson, R. (1977). *Scripts, Plans, Goals and Understanding*, Lawrence Erlbaum Associates, Hillsdale, NJ.

Turner, R.M. (1986). "A Derivational Approach to Plan Refinement for Advice-Giving," *Proceedings of the 1986 IEEE International Conference on Systems, Man, and Cybernetics*, pp. 858-862.

Turner, R.M. (in press). *Issues in the Design of Advisory Systems: the Consumer-Advisor System*, Technical Report #GIT-ICS-87/??, School of Information and Computer Science, Georgia Institute of Technology, Atlanta, GA 30332.

Analogical Learning:
Mapping and Integrating Partial Mental Models

Mark Burstein
BBN Laboratories

Beth Adelson
Tufts University

Abstract[1]

Descriptions of scientific and technical systems take a number of different forms. Depending upon the *purpose* of a description, it may focus on a system's behavior, causality, physical or functional topology, or structural composition. An *analogical explanation* used to teach someone about such a system is also typically geared to one or another of these purposes. In this paper we describe some research leading to the development of a theory of the role of *explanatory model types* in the generation of analogical mappings. The work is motivated by the larger question of how explanations presented as analogies are applied by students learning about new domains. Our long term goals are (1) the development of a theory of *purpose-guided analogical learning*, based on a coherent taxonomy of mental model types, and (2) the development of a theory of the integration of partial mental models during learning, using principles for relating different explanatory model types.

1. Analogical Mapping of Different Explanation Types

In recent years, researchers in artificial intelligence and cognitive psychology have begun to focus more attention on the study of analogical reasoning and its role in learning and problem solving, particularly in scientific and technical domains. A number of these researchers have independently converged on a class of models of analogical learning which stress the mapping process (Gentner, 1983, Burstein, 1986, Thagard and Holyoak, 1985, Carbonell, 1986). By this class of process model, an underlying conceptual model of a familiar source or *base* domain is abstracted and mapped to an unfamiliar *target* domain. The mapped model is then used to build a new model in the target domain. Analogical learning theories need to specify the mapping process in some detail, since mapping dictates what can be postulated from any given analogy. For example, *what* is mapped must be constrained by the *purpose* of the analogy (Kedar-Cabelli, 1985, Burstein, 1985, Thagard and Holyoak, 1985, Winston et al., 1983).

Along with specifying the mapping process, theories of analogically-based learning must also account for the use of multiple analogical models. Typically, no single analogical model can be found that completely and accurately describes a non-trivial target system. That is, when subjects generate full and correct explanations of a system, they often need to use *multiple partial models*, of varying types, and at varying levels of abstraction (Burstein, 1983, Collins and Gentner, 1983, Collins, 1985, Adelson, 1984, Sternberg and Adelson, 1978, Coulson et al., 1986). For example, Collins and Gentner found that untutored subjects used as many as three analogical models to answer novel questions about evaporation (Collins and Gentner, 1982, 1983, 1987) . Different kinds of models were used to answer different kinds of questions and frequently several models were used together.

---

[1]This work was supported in part by grants from ARI and NSF.

In another case study, Burstein (Burstein, 1985, 1986) found that three analogies were commonly used to teach students about variables and assignment for the programming language BASIC. One analogy described assignment as being like "putting things in boxes". Another analogy related assignment statements to algebraic equalities. A third analogy related the encoding and retrieval processes of human memory to analogous processes in the computer. Again, the analogies suggested partial models of different types. The box and memory analogies each describe an action that causes a result and so, as we explain in detail below, we regard these as *mechanistic causal* analogies. The analogy to algebraic equalities contributes a *behavioral* model. It explains how to infer values for variables without providing a mechanism accounting for *how* those values are derived and assigned by the computer.

2. Developing a Theory of Purpose-guided Analogical Learning

Our current research is aimed at developing a theory covering the two related issues of *mapping* and *integrating* partial mental models. Specifically, we are addressing the questions of: (1) how analogically-based models of different types are mapped to a new domain, and (2) how these different kinds of partial models are integrated in a target domain. This paper focuses on the mapping process, although we will touch on the integration issue as well.

Both parts of our theory are being developed around a taxonomy of explanatory partial model types. The research presented here is part of a series of protocol experiments being conducted to produce a detailed account of the process that maps partial models of various types. The protocols are being used to identify the kinds of relations that are (and are not) included in target models developed by mapping partial models of a given type. We have found that models of different types are distinguished by the different kinds relations that they contain, and that models of a given type map to form new models of the same type (i.e., models containing the same relations). By making the distinction between model types explicit in our theory of analogical learning, we hope to provide an account of analogical structure mapping that has clear pragmatic constraints on the amount and type of information mapped at one time from a base domain.

Future research will focus more heavily on the issue of model integration. Integrating multiple models is an important part of the learning process, since there are many situations that can only be explained by a combination of inferences from several different partial models. As we explain in section 4, we forsee that two kinds of integration processes may be used for combining partial analogical models. First, there is a reasoning process that functions to relate newly acquired partial models within a domain[2]. This process is based on principles about how, in general, partial models are inter-related. A second kind of integration process may be used when *adapting* partial models from an already active analogical *source* domain. In this case, information about how partial models are related in the source domain, can be used to avoid reasoning from first principles in the target domain. Since an understanding of the integration process is dependent on an understanding the mapping process, we have chosen to focus first on the mapping process.

---

[2]This process is general to learning in that it functions whether the models have been acquired by analogy or directly by observation or instruction in the target domain.

3. A Taxonomy of Explanatory Model Types

Our investigations of both the mapping and integration processes depend heavily on a well-defined taxonomy of model types. For our initial experiments, we have developed a working taxonomy which we expect will capture a broad set of models used in analogical learning (Figure 3-1). In developing our taxonomy of models for *analogical learning*, we have revised a taxonomy developed by colleagues at BBN initially to *categorize* textbook explanations of complex physical systems (Stevens and Collins, 1980, Stevens and Steinberg, 1981, Collins, 1985). The taxonomy was formulated during the development of the STEAMER ICAI system (Williams, Hollan and Stevens, 1981). It provides relevant background support for our work in that it has been used to show that people often use *several* different kinds of explanatory models of a single system in trying to produce a full explanation (Collins, 1985, Weld, 1983).

- **Structural models** are used to describe systems in a time-invariant manner. Structural models include:

    1. **Componential models** simply list components.

    2. **Topological models** specify configurations where the logical or functional connections between components are preserved.

    3. **Geometric models** preserve the quantitative, spatial relations between components.

- **Dynamic models** describe changes that occur in a system over time. Dynamic models include:

    1. **Functional/Behavioral models** describe a system as a "black box", in terms of inputs and outputs.

    2. **Internal Structure models** break the system down into interactions between various components. These models include:

        - **Mechanistic Causal models** describe unique behaviors for each component and break events into causal chains. These include *Action Flow* models, where some substance or energy flows through the system and more abstract *Information Flow* models, where information is described as passing between components. These models typically dictate how the outputs of individual components cause state changes in other, topologically connected components, leading to an account of the behavior of the system as a whole.

        - **Aggregate models** describe systems where the components behave in a uniform manner, subject to global constraints. In these models, components are represented prototypically, in terms of general behavioral characteristics of the group. Individual features of components are represented by distributions of values.

        - **Synchronous models** describe causal systems where events or forces occur synchronously.

Figure 3-1: A Taxomony of Explanatory Model Types

The data presented here is being used to develop a model of the analogical mapping process. Our theory suggests that *explanatory purpose*, the type of model required by the student for his ongoing problem solving or question answering activity, strongly affects the selection of base domain features and relationships for mapping. Different explanatory purposes are characterized by different types of models, which in turn can be shown to be based on different structural relations (causal relations, function/goal relations, spatial or topological relations, etc.) For example, within mechanistic causal models, temporal/causal relationships are used to relate the behaviors of a connected set of components, in order to explain a system's overall behavior. However, in aggregate models, local constraints are replaced by global constraints (e.g., conservation laws), and individual components give way to representative or prototypical entities with properties characterized as distributions.

4. Protocols of Mapping By Model Type

Our theory suggests that the type of model selected and mapped during learning is constrained by the aspect of the target situation made salient by the learning task. This prediction generates the following hypothesis: When explaining a given aspect of a target domain, subjects should map only a subset of the base domain and that subset should be coherent[3] and reflective of the purpose of the learning task.

The following situation was used to provide an initial test of this hypothesis. Subjects were provided with an analogical model of a computer programming constructs[4]. The subjects were all naive to the target construct and familiar with its base domain analog. Subjects were then asked to answer a set of questions about various aspects of the initially unfamiliar target domain concept[5]. This procedure was then repeated until subjects had been taught about the three constructs used in this study: queues, stacks and sorting. As an example, Figure 4-1 shows the texts used in teaching the concept of queues and the questions that subjects received following the texts[6].

Our hypothesis that purpose constrains selection and mapping in a way that results in a coherent and appropriate partial mental model will be supported if subjects who have a complete base domain model map only the part of the base domain model that is relevant to the question they are answering.

In order to see whether subjects were selecting and mapping partial models we constructed preliminary behavioral and causal representations for each of the concepts that the subjects had been taught (queues, stacks, and sorting). As an example, a sketch of our behavioral and causal representations for queues, in both the base and target domains, is given in figure 4-2. The representations can be thought of as vertical behavioral or causal chains of which can be read from top to bottom.

The answers to each behavioral and causal question appearing in the recorded protocols for each subject were analyzed to see how clearly they corresponded to our behavioral and causal representations of the concept. The number of times that elements from the causal or behavioral representations occurred was counted for both the behavioral and causal questions.

In answering *behavioral* questions about queues, stacks and sorting, subjects made, on the average, 2.0 references to behavioral elements and no references to portions of the causal models that were not

---

[3] In the sense suggested by Gentner's *systematicity principle* (Gentner, 1983).

[4] As described below, the subjects varied in their levels of programming experience.

[5] In this study we selected behavioral and causal models as a subset of the taxonomy presented in Figure 3-1.

[6] The full design of the study is not presented here. Text type was crossed with question type and order of presentation was counterbalanced over the full set of protocols; subjects received only one written description of the base domain situation which stressed either behavioral, causal or behavioral and causal aspects of the already familiar base domain situation. This was followed by a causal question, a behavioral question and a question about the relationship between behavior and causality. Differences in the written descriptions of the already familiar base domain did not have a discernible effect on the results described below.

Behavioral Analogy:

Frequently, at Mary Chung's (restaurant) there are people waiting to be seated. Mary keeps track of who to seat in such a way that the first person to be seated next is always the person who, among those currently waiting, came in first.

In sending files to the printer the same situation often occurs; several files need to be printed but only one can be printed at a time. In this case the computer resolves the problem in the same way that Mary does.

Causal Analogy:

Frequently at Mary Chung's, Mary has a list of people who are waiting to be seated and served. Whenever a person enters Mary puts their name at the bottom of the list. Whenever a table becomes vacant Mary calls the name at the top of the list, gives that person a table and then crosses that name off the list.

In sending files to the printer the same situation often occurs; several files need to be printed but only one can be printed at a time. In this case the computer resolves the problem in the same way that Mary does.

Behavioral Question: If Karen and then Janet and then Amy and then Karen all typed print commands one right after the other, in that order, what would the computer do?

Causal Question: Describe what you would do if you were the computer keeping track of print requests.

Behavioral/Causal Integration Question: Explain why/how the computer's method of keeping track of print requests produces the correct result? That is, what is the relationship between what gets done and how it gets accomplished?

Figure 4-1: Presented Versions of an Analogy to Queues

also part of the behavioral model. That is, *all* references to components of the causal models were references to elements that appeared in our representations of *both* the behavioral and the causal representation (Figure 4-2), because they refered to the goal and starting states in those representations. No references were made to portions of the causal model that were purely descriptions of mechanism.

In answering *causal* questions about queues, stacks and sorting, subjects made, on the average, 5.8 references to causal elements and on the average, .9 references to elements that were in both the causal and behavioral representations. Again, all of the behavioral elements that appeared in subjects' answers to causal questions were elements refering to the start and goal in those representations. This implies that a causal account of *how* something works is, in some sense, not coherent unless some goal or purpose oriented statement is included regarding *why* the mechanism is needed (Adelson, 1984).

It seems that, as our theory predicts, subjects are mapping *coherent*, purpose-oriented partial models from the target to the base domain. The models are coherent in that they provide an adequate basis for responding to questions of a given type. (This can be seen in the excerpts from the protocols that are presented below.) The models are partial in that they do not provide a complete account of the system being examined.

If our result suggests that purpose, as characterized by the *type* of model, does constrain selection and mapping, it also confirms that there are interesting and intuitively plausible interdependencies between the kinds of information mapped with each type of model. Earlier we mentioned that any learning theory needed to include an account of how partial models were integrated. Our subjects responses to

15

behavioral and causal questions suggest that the initial and goal states that occur in both types of models can be used to provide a bridge between the two. This is an example of the kind of knowledge that is required for our postulated *model integration process*. That is, a behavioral and a causal model of a given construct can be related by their shared start and goal states. Other types of models will need similar kinds of criteria to be related to or integrated with each other. Since these criteria are not domain-specific, the integration process should be applicable whether the models have been acquired through analogy or learned directly.

In the remainder of this section, we present excerpts from the protocols that motivate some central issues for theories of analogical learning.

Subject J[7] received the behavioral version of the analogy. She was then asked the following behavioral question.[8]

E: (*Behavioral Question:*) If Karen and then Janet and then Amy
      and then Karen all typed print commands, one right after the other,
      in that order, what would the computer do?

J: Give me the order again.

E: Karen, Janet, Amy, Karen.

J: All at the same moment?

E: One right after the other. So not exactly a tie.

J: Ok, well it would say that Karen is first...
      and, uh, it's just like, for me it's just like the thing
      you get when you call AAA and they tell you "Don't
      hang up, your calls are being taken in order."

J's last response, ("It's just like the thing you get when you call AAA,") is interesting because it brings up the issue of our theory's *justification* process in which newly mapped models are tested and debugged. After mapping the behavioral model, J is reminded of a recent situation where a computer-like machine was performing the kind of behavior she had just been told that computers use for print queues. That is, by our theory, when J mapped the behavioral model of first-come-first-serve from the Mary Chung (base) domain to the computer (target) domain, she appropriately placed a computer in the "agent" role that Mary had played. This new behavioral description in the target domain then triggered a *reminding* of a similar behavioral model, where the agent, a machine similar to a computer, achieved the same goal (Schank, 1982). Recalling the behavior of the AAA phone system can be seen as an attempt by J to justify or test the adequacy of her newly formed behavioral model of a queue by comparing it to a similar, already-known behavioral model.

The fact that J could have mapped a causal model when answering the behavioral question, but did

---

[7]J is a researcher in music cognition and a self-taught (LOGO) programmer.

[8]E is the experimenter.

Figure 4-2: Base and Target Domain Models of Queues

Behavioral Model

Target		**Base**
Scene 1:		*Scene 1:*
Print file command issued.	[B1]	Patron enters restaurant.
	then	
	[B2]	Mary takes person's name.
	and	
File 'queued' for printing	[B3]	Patron waits till called.
	then	
(when print request is	[B4](when Patron is first among
first among those remaining)		those waiting to arrive)
	then	
File is printed.	[B5] Patron is seated.
	= Goal State	

Causal Model

Target		**Base**
Scene 1:		*Scene 1:*
Print file command issued.	[C1]	Patron enters restaurant.
	and	
Printer not free.	[C2]	All tables full.
	initiates	
	[C3]	Mary takes person's name.
	enables	
Request put on end of queue.	[C4]	Mary puts name on bottom of list.
	results	
File in queue	[C5]	Name on list
Scene 2:		*Scene 2:*
Printer free.	[C6]	Table empty.
	and	
File is on top of Queue	[C7]	Patron name is on top of list
	enables	
Queue pointer incremented	[C8]	Mary crosses out Patron's name.
	and	
File is Removed from Queue	[C9]	Mary calls Patron's name.
	enables	
	[C10]	Patron hears name
	initiates	
	[C11]	Patron goes to Mary
	enables	
File passed to Printer	[C12]	Patron follows Mary to table
	enables	
File is printed.	[C13]	Patron sits at table.
	= Goal State	

17

not do so seems clear when her response to the behavioral question is compared to her response to the causal question that followed:

E: (*Causal Question*:) Describe what you would do if you were
   the computer keeping track of print requests.

J: Label them, I suppose, or put them if you want to think of it
   spatially.

E: If you want to draw it ...

J: OK, you've got a potential for something like this...
                                                  (*draws a row of boxes or 'slots'*)
   and you can *shove* things into these slots ...
   and Karen calls first
   so that becomes Karen,   (*writes Karen in 1st slot*)
   and Janet..
   and then Amy and then Karen   (*writes each name in a successive slot*)
   and I assume it's the same Karen. But, but, this (slot)
   is probably assigned some label like '1' and so she (Karen)
   will be '4' here and '1' there.
   So Karen is 4, or whatever,
   and when that is finished   (*points to the first slot*)
   this becomes '1'.   (*points to the second slot*)

   When that is finished   (*points to the second slot*)
   this becomes '1'.   (*points to the third slot*)

J clearly has *pieces* of the causal mechanism, rather than a coherent causal model that had been mapped during the previous behavioral question, since she is able to answer the causal question, but in a bit-by-bit rather than an all-of-a-piece manner.

Another subject, K, received the causal version of the analogy and then was asked a causal question[9]. K's answer is interesting because, for a non-programmer, she develops a surprisingly good causal model. She does this by using the *analogically related causal description* she had been given. As suggested above, in order to produce a well-motivated and coherent description, she includes the related behavioral goal in her mapping from the base domain:

E: (*Causal Question*:) If Mark and Annie and Beth had typed requests
   to the computer, one after another, how would it (the computer)
   keep track of that?

K: Can the computer work like a calculator that has a memory?

E: Yes, it can. That's right, you can have told the computer in advance.

K: It can take all of these names, put it in the memory, and then
   pull the first one out.

---

[9]K is an antique book dealer. She has used several PC-based software packages, but she does not program.

```
E:  Exactly right.

K:  So that would be virtually the same thing.  So as it's printed,
    it won't be in the memory anymore?

E:  That's right.  Let's say at this point Mark's stuff has been
    printed, and scratched off and Annies stuff and my stuff are
    waiting to be printed. Then Glenn down the hall, has something to
    be printed,  where would the computer put that request,  in order
    to keep track in just the same way Mary Chung kept track?

K:  Well, it would have to have a program that would put that at the
    end of the list.

E:  (Behavioral Question:)   So what....describes how Mark
    got to be first?

K:  First come first serve.
```

After confirming that the computer could store a list, K correctly described the mechanism of the queue, by mapping the steps that pull requests off the front of the list and add new ones at the end (statements 2 and 4 respectively). K's description of this mechanism is a clearcut example of a causal analogical mapping from an explicitly described base domain causal mechanism. It can be contrasted with subject L[10], who received the behavioral version of the analogy and was asked to describe the queue's causal mechanism.

```
E:  Frequently at Mary Chung's there are people waiting to be seated.
    Mary keeps track of who to see in such a way that the first person
    to be seated next is always the person who among those currently waiting
    came in first.  Ok, so that's the analogy is that the next guy who is
    going to get a seat is the one who ...

L:  ... came in the longest period of time ago.

E:  Yes, ok, now, in sending files to the printer the same situation
    often occurs.. if you are not working on a personal computer.  It is
    that several files need to be printed, but only one can be printed at
    a time.  And in this case, the computer resolves the problem in the
    same way that Mary does.

E:  (Causal question:) Describe what you would do if you were the
    computer keeping track of the print requests and you have a bunch
    of requests at a particular time, and you can only print one at once,
    and they came in at different times.  Ok, so can you describe to me
    how you would make this decision using the Mary Chung analogy.

L:  I would give them all a number.  And then whoever has the
    lowest number gets seated next.  Then I take that number and
    throw it out.  And so I don't have to worry about ever having to
    start over again at number 1 and seating the last person who came
```

---

[10]L is a theatre technician who does some programming as part of her technical work. L has approximately 2 years course-work experience with programming in Pascal and LISP.

19

```
          in out of order, because I am juggling my numbers as I go.
          ... I also used to work in a restaurants a lot.
```

Although we see in statement 2 that L clearly understood the *function* of a queue and her description of a mechanism satisfied that function, L's response to the causal question is interesting because it draws heavily on her knowledge of a *base* domain mechanism rather than a *target* domain mechanism. L chose to map a causal model by analogy from the base domain, despite the fact that she was *not* explicilty given a causal model of that domain and, could have generated a correct solution directly, rather than by analogy.[11]

L's generation of a causal model by analogy, using her knowledge of the restaurant (base) domain, rather than relying directly on her knowledge of programming suggests that domain-specific solutions to problems are not always preferred. Furthermore, the causal model that L retreived was from a base domain situation that was *not* the one provided by the experimenter, but was instead familiar as a result of having worked in restaurants. This suggests that the base domain retrieval process, as well as the mapping process is quite complex. We suggest that it may be guided by a combination of factors: the *type of explanation* required; the problem solving goal; and the *currently* salient attributes of the base domain model presented during learning. Holyoak (Holyoak, 1985) has also suggested a model like this, although his model would not predict that an analogical solution might *dominate* an existing target domain solution. While our model also cannot account precisely for L's preference of a base-domain solution to the problem, the protocol clearly suggests that a full theory of analogical reasoning will need to include a fairly sophisticated account of the use of base domain features in the retrieval stage of the analogical reasoning process.

5. Summary

We have presented some preliminary protocol evidence of the role of *explanatory model types* in the retrieval and mapping stages of the analogical reasoning process. Our results suggest that:

- Subjects can generate and use coherent partial models.
- Behavioral and causal models can be related by shared initial and goal states.
- Retrieval of base domain solutions can be influenced by the type of explanation required, the problem solving goal, and the salient attributes of the base domain model presented during learning.

The evidence we are collecting is being used to develop a theory of constrained, purpose-directed analogical retrieval and mapping, and a theory of the process of *integrating multiple partial models* in learning about a new domain.

---

[11]Evidence that L could describe a purely target domain soluion to the problem was gathered in a follow-up session several weeks after this protocol was collected. In that session, L was asked to describe an implementation for a queue directly rather than by analogy. She first repeated her earlier answer, but when she was asked if the numbers were necessary, she responded that the ordering of the list was sufficient, and proceeded to describe the set of steps involved in using an ordered list to implement the queue; adding new entries to the end and removing the "next" item from the front.

References

Adelson, Beth. When novices surpass experts: How the difficulty of a task may increase with expertise. *Journal of Experimental Psychology: Learning, Memory and Cognition*, July 1984.

Burstein, Mark H. Concept Formation by Incremental Analogical Reasoning and Debugging. In Michalski, R. S., Carbonell, J. G. and Mitchell, T. M. (Ed.), *Proceedings of the International Machine Learning Workshop*. Champaign-Urbana, IL: University of Illinois, 1983. Also appears in *Machine Learning: Volume II*. pp. 351-370. Morgan Kaufmann Publishers, Inc., Los Altos, CA, 1986.

Burstein, Mark H. *Learning by Reasoning from Multiple Analogies*. Doctoral dissertation, Yale University, 1985.

Burstein, Mark H. Concept Formation by Incremental Analogical Reasoning and Debugging. In Michalski, R. S., Carbonell, J. G. and Mitchell, T. M. (Ed.), *Machine Learning: Volume II*. Los Altos, CA: Morgan Kaufmann Publishers, Inc., 1986. Also appeared in the *Proceedings of the Second International Machine Learning Workshop*, Champaign-Urbana, IL., 1983.

Carbonell, Jaime G. Derivational Analogy: A Theory of Reconstructive Problem Solving and Expertise Acquisition. In Michalski, R. S., Carbonell, J. G. and Mitchell, T. M. (Ed.), *Machine Learning: Volume II*. Los Altos, CA: Morgan Kaufman Publishers, Inc., 1986.

Collins, Allan. Component Models of Physical Systems. In *Proceedings of the Seventh Annual Conference of the Cognitive Science Society*. Cognitive Science Society, 1985.

Collins, Allan and Gentner, Dedre. Constructing Runnable Mental Models. In *Proceedings of the Fourth Annual Conference of the Cognitive Science Society*. Boulder, CO: Cognitive Science Society, 1982.

Collins, Allan and Gentner, Dedre. Multiple Models of Evaporation Processes. In *Proceedings of the Fifth Annual Conference of the Cognitive Science Society*. Rochester, NY: Cognitive Science Society, 1983.

Collins, Allan and Gentner, Dedre. How People Construct Mental Models. In N. Quinn and D. Holland (Eds.), *Cultural Models in Thought and Language*. Cambridge, UK: Cambridge University Press, 1987. In press.

Coulson, R., Feltovich, P., and Spiro, R. *Foundations of a Misunderstanding of the Ultrastructural Basis of Myocardial Failure: A Reciprocating Network of Oversimplifications* (Tech. Rep. 1). Southern Illinois University School of Medicine, Conceptual Knowledge Research Project, August 1986.

Gentner, Dedre. Structure-Mapping: A theoretical framework for analogy. *Cognitive Science*, 1983, *7*(2), 155-170.

Holyoak, K. J. The pragmatics of analogical transfer. In G. H. Bower (Ed.), *The psychology of learning and motivation*. New York, NY: Academic Press, 1985.

Kedar-Cabelli, Smadar. *Analogy from a Unified Perspective* (Tech. Rep. ML-TR-3). Laboratory for Computer Science Research, Rutgers University., November 1985.

Schank, R.C. *Dynamic Memory: A theory of learning in computers and people*. Cambridge University Press, 1982.

Sternberg, R. J. and Adelson, B. Changes in Cognitive Structure via Metaphor. In *Annual Meeting of the Psychnomic Society*. San Antonio, Texas: , 1978.

Stevens, A. and Collins, A. Multiple Conceptual Models of a Complex System. In Snow, R. E., Federico, P. and Montague, W. E. (Eds.), *Aptitude, Learning, and Instruction*. Hillsdale, N.J.: Erlbaum, 1980.

Stevens, A. and Steinberg, C. *A Typology of Explanations and Its Application to Intelligent Computer Aided Instruction* (Tech. Rep. 4626). Bolt Beranek and Newman Inc., March 1981.

Thagard, Paul and Holyoak, Keith. Discovering the Wave Theory of Sound: Inductive Inference in the Context of Problem Solving. In *Proceedings of the Ninth IJCAI*. Los Altos, CA: Morgan Kaufmann Publishers, Inc., 1985.

Weld, Daniel S. *Explaining Complex Engineered Devices* (Tech. Rep. 5489). Bolt Beranek and Newman Inc., November 1983.

Williams, M., Hollan, J., and Stevens, A. Human reasoning about a simple physical system. In Gentner, D. and Stevens, A. (Eds.), *Mental Models*. Erlbaum, 1981.

Winston, P. H., Binford, T. O. Learning Physical Descriptions from Functional Definitions, Examples, and Precedents. In *Proceedings of AAAI-83*. Los Altos, CA: Morgan Kaufmann Publishers, Inc., 1983.

Analogy and Similarity: Determinants of Accessibility
and Inferential Soundness

Mary Jo Rattermann
Dedre Gentner
Department of Psychology
University of Illinois at Urbana-Champaign
603 East Daniel St.
Champaign, Illinois 61820

Abstract

Analogy and similarity are widely agreed to be
important in learning and reasoning. Yet people are
often unable to recall an analogy which would be
inferentially useful. This finding suggests that a
closer examination of the similarity factors that
promote retrieval is necessary. We approached this
problem by investigating the role of relational
commonalities (higher-order relations and first-order
relations) and common object-descriptions in the
accessiblity and inferential soundness of an analogy.

Subjects first read a large number of stories. One
week later they were given a new set of stories to
read. These new stories were designed to form matches
which shared different combinations of object-
descriptions, first-order relations and higher-order
relations with the original stories. Subjects were
asked to recall any stories from the original set that
came to mind. Afterwards they rated the matches for
subjective soundness and similarity.

The results of two experiments showed that
subjects recalled the original stories that shared
common object descriptions and first-order relations
with the new stories. These results support the idea
that similarity based access is enhanced by a
combination of surface similarity and first-order
relations. They also suggest that common higher-order
relations play a smaller part in recall. In contrast,
in both the soundness-rating and similarity-rating
tasks subjects rated the pairs that shared higher-order
relations higher than the pairs which shared surface
similarity. This suggests that those aspects of
similarity that govern recall are different than those
aspects that govern similarity-ratings and soundness-
ratings.

Analogy and similarity are widely agreed to be important in
learning and reasoning. But recent evidence has shown that
people are often unable to retreive analogies which, if
retrieved, would be inferentially useful (Gick & Holyoak, 1980,
D1983; Reed, Ernst & Banerji, 1974; Ross, 1984, in press). This
research suggests a closer examination of how similarity promotes
retrieval. Gentner and Landers (1984) approached this question

23

by comparing the retrievability and the subjective soundness of different kinds of similarity matches. Because our studies build on their method, we begin by describing this study and the theoretical issues that led up to it.

The theoretical framework for the research is Gentner's (1980, 1983) structure-mappping theory. The basic intuition of structure-mapping theory is that an analogy is a mapping of knowledge from one domain (the base) into another (the target) which conveys that a system of relations that holds among the base objects also holds among the target objects. Thus an analogy is a way of noticing relational commonalities independently of the objects in which those relations are embedded. In interpreting an analogy, people seek to put the objects of the base in 1-to-1 correspondence with the objects of the target so as to obtain maximum structural match. The corresponding objects in the base and target don't have to resemble each other at all; object correspondences are determined by roles in the matching relational structures. Central to the mapping process is the principle of **systematicity**: people prefer to map systems of predicates that contain higher-order relations with inferential import, rather than to map isolated predicates. The systematicity principle is a structural expression of our tacit preference for coherence and deductive power in interpreting analogy.

Besides analogy, other kinds of similarity matches can be distinguished in this framework, according to whether the match is one of relational structure, object descriptions, or both. Recall that *analogies* discard object descriptions and map relational structure. *Mere-appearance* matches are the opposite: they map aspects of object descriptions and discard relational structure. *Literal-similarity* matches map both relational structure and object-descriptions.

It is helpful for this discussion to decompose analogical reasoning into access and mapping-plus-inference. *Access* is the process of retrieving a base situation in memory, given a target situation that the learner is currently considering. *Mapping* occurs after a base situation has been accessed from memory. In mapping, the predicates of the base are matched with predicates of the target according to the rules given above. Further predicates are carried across and inferences may be drawn. According to structure-mapping, the subjective soundness of a possible analogy depends on the degree to which a systematic relational match can be found.

Against this background, the Gentner and Landers (1985) experiment had a two-fold purpose: (1) it tested the prediction that shared systematic structure determines the subjective soundness of a match; and (2) it asked whether the accessibility of analogy and other kinds of similarity matches mirrors their subjective soundness. The study was designed to create a situation resembling natural long-term memory access. Subjects were first given about 30 stories to read and remember. One week later, they read a new set of stories and reported any cases in

24

which they were reminded of any of the original stories. The
stories were carefully designed to embody three different kinds
of similarity matches: *mere appearance*, *true analogy* and *false
analogy*. In *mere-appearance (MA)* matches, the base and target
shared object descriptions (e.g. a hawk in the base story vs. an
eagle in the target) and first-order relations (e.g. shoot at
[x,y] vs. fire at [x',y']). In *true-analogy (TA)* matches, the
base and target shared first-order relations and higher-order
relations (i.e., relations between relations, such as CAUSE
[S(x,y), R(y,z)] vs. CAUSE [S'(x',y'), R'(y',z')] and other
constraining relations) but not object-descriptions. In *false-
analogy (FA)* matches only the first-order relations matched.
(See Table 1, below, for examples.) Note that in all three cases
the base and target shared first-order relations; the three
similarity conditions differed in which, if any, other
commonalities also existed.

     Gentner and Landers found that the proportion recalled for
the mere-appearance matches was greater than that of the true-
analogy matches, which in turn was greater than that of the
false-analogy matches. These results suggest that access to
memory is heavily influenced by surface commonalities. In
contrast, when the same subjects rated the inferential soundness
of the similarity matches, the true-analogy matches were rated as
the most sound, while the false-analogy and the mere-appearance
matches were rated significantly less sound. As predicted,
subjective soundness depended on common systematic structure.
This raises an interesting disassociation: it seems that the
kinds of similarity matches that subjects consider most sound are
not the matches that most strongly promote access. Common
surface information, such as shared object-descriptions, may have
a disproportionate affect on accessibility.

     The Gentner and Landers study revealed some interesting
points about analogical access, but also left a great many
questions unanswered. While the study suggested that surface
similarity plays a large role in remindings, it also suggested a
role for higher-order relations in remindings, since true-analogy
matches were better accessed that false-analogy matches.
However, these results do not tell us how (or whether) surface
attributes and higher-order relations combine to promote access,
nor do they tell us which aspects of surface similarity are most
effective in promoting retrieval.

     In Experiment 1 we addressed the first of these questions.
To do this we replicated the Gentner and Landers study, adding
another match type, that of literal similarity. A literal
similarity match has commonalities at all levels -- object-
attributes, first-order relations and higher-order relations.
This meant that the set of matches formed a 2 X 2 design as shown
in Figure 1 (below). Based on the results of the Gentner and
Landers study we expected that literal-similarity matches would
be recalled well in the reminding task. However, it was not
clear how much the addition of higher-order commonalities would
affect the results. They might contribute very little beyond the

effects of surface commonalities, in which case recall of the literal-similarity matches should not differ significantly from that of the mere-appearance matches. Alternately, it could be that having commonalities at all levels would lead to far greater accessibility for literal-similarity matches than would be predicted by combining the two separate effects.

One problem with the Gentner and Landers study is that it leaves open an alternative interpretation of the results: namely, that the retrievability ordering among the three types of matches was simply a function of their overall similarity. That is, it could have been the case that the mere-appearance matches were more similar to one another than the true-analogy matches, and the true-analogies more similar than the false-analogies. In this case, there would be no reason to invoke a special role for surface attributes in similarity-based access. To address this possibility, we added a similarity-rating task in order to test whether retrievability could simply be predicted from similarity ratings.

To summarize, in this study three measures were obtained for each pair of scenarios: (1) the accessibility of the first story given the second, i.e., how well the second story served as a cue for the first; (2) the inferential soundness of the analogy between the stories, as rated by the subjects; and (3) the degree of similarity between the two stories, as rated by the subjects.

Experiment 1

Method

Subjects

The subjects were 36 undergraduates who received class credit for participation in this experiment. Due to experimenter error, 18 paid subjects were used to rerun one cell in the soundness-rating task and one cell in the similarity-rating task.

Materials

There were 20 story sets, each consisting of a base story plus four different target stories designed to embody different kinds of similarity match: *literal similarity (LS)*, *mere appearance (MA)*, *true analogy (TA)*, and *false analogy (FA)*. Figure 1 shows the design of the match types; Table 1 shows example stories. In addition to the 20 story sets, there were 12

	HIGHER-ORDER RELATIONS	
	SHARED	NOT SHARED
OBJECT ATTRIBUTES SHARED	Literal Similarity	Mere Appearance
OBJECT ATTRIBUTES NOT SHARED	True Analogy	False Analogy

Note: First-order relations are roughly constant.

Figure 1: Design of Match Types

26

Table 1: Example Stories

BASE Story

Karla, an old hawk, lived at the top of a tall oak tree. One afternoon, she saw a hunter on the ground with a bow and some crude arrows that had no feathers. The hunter took aim and shot at the hawk but missed. Karla knew the hunter wanted her feathers so she glided down to the hunter and offered to give him a few. The hunter was so greatful that he pledged never to shoot at a hawk again. He went off and shot a deer instead.

Literal similarity

Once there was an eagle named Zerdia who nested on a rocky cliff. One day she saw a sportsman coming with a crossbow and some bolts that had no feathers. The sportsman attacked but the bolts missed. Zerdia realized that the sportsman wanted her tailfeathers so she flew down and donated a few of her tailfeathers to the sportsman. The sportsman was pleased. He promised never to attack eagles again.

True Analogy

Once there was a small country called Zerdia that learned to make the world's smartest computer.

One day Zerdia was attacked by it's warlike neighbor, Gagrach. But the missiles were badly aimed and the attack failed. The Zerdian government realized that Gagrach wanted Zerdian computers so it offered to sell some of it's computers to the country. The government of Gagrach was very pleased. It promised never to attack Zerdia again.

Mere Appearance

Once there was an eagle named Zerdia who donated a few of her tailfeathers to a sportsman so he would promise never to attack eagles.

One day Zerdia was nesting high on a rocky cliff when she saw the sportsman coming with a crossbow. Zerdia flew down to meet the man, but he attacked and felled her with a single bolt. As she fluttered to the ground Zerdia realized that the bolt had her own tailfeathers on it.

False Analogy

Once there was a small country called Zerdia that learned to make the world's smartest computer. Zerdia sold one of its supercomputers to its neighbor, Gagrach, so Gagrach promised never to attack Zerdia.

But one day Zerdia was overwhelmed by a surprise attack from Gagrach. As it capitulated the crippled government of Zerdia realized that the attacker's missiles had been guided by Zerdian supercomputers.

27

additional stories which Were used as fillers. Each subject
received 20 base stories: five of each of the four match types
(LS, MA, TA, and FA). Each subject saw one and only one target
story for each base story. The subjects were divided into four
groups in order to counterbalance the assignment of stories to
match types.

Procedure

Reminding Task. In the first session subjects were told to
read and remember 20 base stories and 12 filler stories. One
week later, they returned for the reminding session. They were
given booklets of 20 target stories, each of which matched one
and only one base story. For each target story, subjects were
told to notice if they were reminded of any story from the
previous session, and to write down any such stories in as much
detail as possible.

Soudness-rating Task. The soundness-rating task was given
to subjects after they had completed the reminding task.
Subjects were shown the same 20 pairs of stories they had
received in the reminding task, with each pair consisting of a
base story and a matching target story. They rated each pair for
the soundness of the match between the two stories. A sound match
was described as one in which the two situations match well
enough so that inferences in one would be likely to carry over to
the other. This description was placed in the context of what
would make a good argument. Subjects used a 1-5 scale, where 5 =
highly sound and 1 = spurious.

Similarity-rating Task. Following the soundness-rating
task subjects were again given the same pairs of stories and were
asked to rate the pairs on their overall similarity. [1] No
explicit definition of similarity was given; we simply allowed
subjects to use their own intuitions.

Scoring the Reminding Task. To score the reminding task we
had to judge whether the subjects had indeed successfully
retrieved the original stories (as opposed to just guessing, for
example). The recalls were scored in three ways, but we focus
here on one method only. Two judges scored on a 1-5 scale how
well subjects recalled each base story by comparing the subject's
recall with the correct base story. Then we computed for each
match type the proportion of stories for which a rating of 2 or
better -- indicating that the subject's description had clearly
mentioned at least a few elements of the base -- had been
assigned. This *flat match* score was designed to capture whether
any genuine retrieval of the correct base story had occurred,
without concern for whether the recall was of high quality.

----

1. As a check on whether the prior soundness-rating task had any
   influence on the subsequent similarity-rating task a later
   group of subjects was run with the reverse order of the
   soundness and similarity rating tasks. The results were
   unchanged.

Reminding task

As shown in Figure 2a, literal-similarity and mere-appearance matches led to significantly more reminding than the true-analogy or false-analogy matches.[2]

These results replicated the pattern found by Gentner and Landers in that the mere-appearance matches produced more remindings than true-analogy matches. These results support the idea that similarity-based accesss is enhanced by common surface similarity, and that common relational structure plays a smaller part in recall.

Soundness-rating task

Subjects rated the literal-similarity and the true-analogy matches as significantly more sound than the false-analogy and mere-appearance matches. (See Figure 2b.)

Again, these findings replicated those of Gentner and Landers in that true-analogy matches are considered significantly more sound than false-analogy matches and mere-appearance matches. Further, the fact that literal-similarity and true-analogy -- the two match types that utilize common higher-order structure -- were considered significanty more sound than the other two match types provides evidence for the prediction of the structure-mapping theory that soundness is governed by common relational structure.

Similarity-rating Task

Subjects rated the literal-similarity matches as significantly more similar than the true-analogy matches, and the literal-similarity and true-analogy matches were both significantly more similar than the mere-appearance and false-analogy matches. However, the mere-appearance matches and the false-analogy matches were not significantly different from each other. (See Figure 2c.)

The similarity results showed an interesting trend: the pattern of the similarity ratings mirrored that of the soundness data. Specifically, subjects rated the pairs that shared higher-order relations (the literal-similarity matches and the true-analogy matches) as more similar than the pairs which shared surface attributes (the mere-appearance matches). This suggests that in making similarity ratings subjects judge higher-order relations to be more important than surface attributes. However, it should be noted that literal-similarity matches received higher similarity ratings than true-analogy matches, showing that

----

2. One way analyses of variance were performed on the data sets in Experiment 1. In each case the analysis revealed a main effect of Match type, $p < .001$. Post hoc analyses (Tukey's, alpha = .05) were used to determine the significant differences between the means. For brevity we will omit the statistics for the remainder of the results and simply report significant differences.

Figure 2

Results from Experiment 1

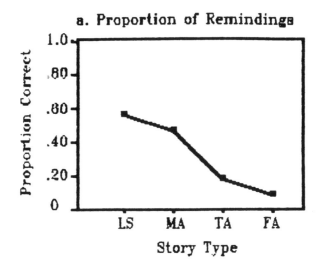

a. Proportion of Remindings

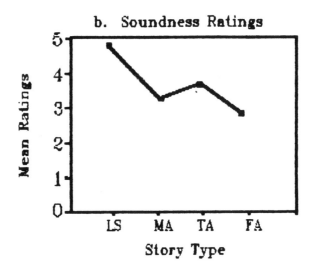

b. Soundness Ratings

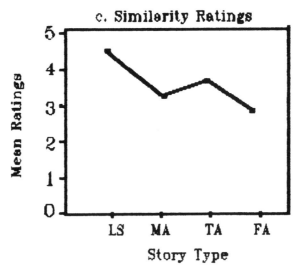

c. Similarity Ratings

(unlike soundness) similarity is also sensitive to surface features. The other important implication of the similarity findings is that, since they do not mirror the accessibility ordering, it is improbable that the accessibility results are simply an artifact of differences in similarity across match types. On the contrary, the order of accessibility match types is quite different from the order of similarity ratings.

The most important finding here lies in the comparison between the results of the recall task and the results of the similarity-rating and soundness-rating tasks. It seems that different aspects of similarity govern these different processes. In similarity-based recall, it is common surface features such as object descriptions that matter most, while in judging the soundness or similarity of two stories it is common relational structure that matters most.

Experiment 2

Experiment 2 was designed to examine more closely the issue of "surface commonalities". In the studies so far, the mere-appearance matches share both object-descriptions and first-order relations. Here we asked which aspects of mere-appearance matches led to their accessibility; and in particular, whether object-descriptions alone could promote access. In this study we created an new variant of mere-appearance matches by removing all the first-order relational commonalities from the first set, leaving the common object-descriptions (as exemplified below).

> There once was a sportsman who loved to hunt. He liked to have the animals he caught stuffed and mounted. His pride and joy was an eagle he had killed with just a crossbow and a bolt. He had been hiding in the top of an elm tree when he shot her.

We then ran the study again, pitting these new matches that shared only object-level attributes (called mere-appearance-attribute-only, or MAAO) against the mere-appearance matches that shared both object attributes and first-order relations (called mere-appearance-first-order, or MAF) against true-analogy matches and literal-similarity matches.

Method
Subjects
The subjects were 52 undergraduates who were fulfilling a course requirement.
Materials and Procedure
Subjects were divided into two groups. Each group received 14 true-analogy matches, and 14 mere-appearance matches, with half the subjects receiving mere-appearance-first-order versions and half receiving mere-appearance-attribute-only versions. In addition, each group received 6 literal-similarity matches to anchor their responses. The procedure was as in

Experiment 1. In the first session, the subjects read 20 base
stories and 12 filler stories. One week later, the subjects were
given 20 new stories: 7 TA matches, 7 of either one of the mere-
appearance matches (either MAAO or MAF) and 6 LS matches,
followed by a soundness-rating task and a similarity-rating task.
Again, as in Experiment 1, each subject saw only one target story
for each original base story.

Results

Reminding task

The results, shown in Figure 3a, were rather dramatic.
First, LS matches and MAF matches were recalled significantly
better than MAAO matches, which were recalled significantly
better than TA matches. [3]

These results show that it is not just common object-
descriptions that lead to similarity based recall; rather the
biggest gain in accessibility seems to occur for some combination
of common object-descriptions and common first-order relations.
In fact, the match type that possessed common object-
descriptions with the base, mere-appearance attributes-only, was
recalled quite poorly. What is particularly striking about these
results, however, is that TA matches were even less accessible
than MAAO matches. As discussed below, this ordering is again in
sharp contrast to the subject's own opinions concerning the
inferential soundness and similarity of the matches. Again we
find that the ordering of accessibility is different from the
ordering of soundness.

Soundness-rating task

As shown in Figure 3b, the order of soundness ratings was LS
matches followed by MAF matches followed by MAAO matches. (All
of these differences were significant.)

These findings again confirm that the most important
determinant of inferential soundness in subjects' judgements is
common systematic relational structure (which is present only in
the LS and TA matches). However, in this study, unlike the prior
two, subjects considered LS matches slightly more sound than TA
matches, indicating that for these subjects common object
descriptions also contribute to soundness (this being the only
difference between LS and TA). Not surprisingly, the MAAO
matches, which share only the cast of characters, were considered
least sound.

Similarity-rating task

Again, the similarity ranking are close to the soundness
rankings: the order of similarity ratings is LS, TA and MAF, and
then MAAO. (All differences are significant except that between
TA and MAF.) Quite reasonably, subjects consider matches with

----

3. Again to be brief we will only report significant results in
   the body of the paper. Significance was determined by Welch-
   Aspin t-tests, alpha<.05.

32

Figure 3
Results from Experiment 2

a. Proportion of Remindings

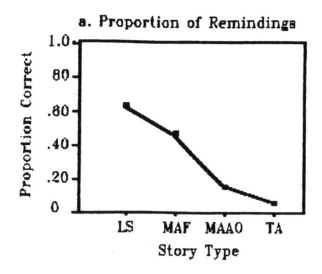

b. Soundness Ratings

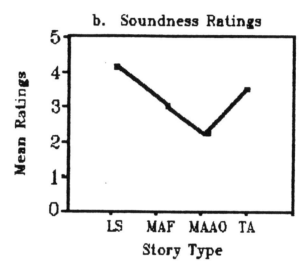

c. Similarity Ratings

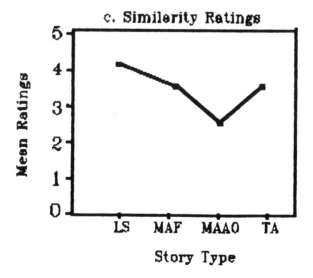

three levels of commonality (LS matches) to be the most similar, and matches with only one level of commonality (MAAO matches) to be least similar. The other two match types each share two levels (TA matches share higher-order and first-order relations, while MAF matches share first-order relations and object descriptions) and subjects here considered them roughly equally similar. (This differs slightly from Experiment 1, in which TA matches were rated as more similar than MAF matches.)

The most important point about the similarity ratings, is that they again demonstrate that the accessibility results are not due to perceived similarity. As before, the pattern of similarity is very similar to the pattern of soundness and both are quite different from the order of the accessibility.

Discussion

The most important finding here is the dissociation between the kinds of similarity that people think are inferentially reliable and the kinds of similarity that readily enable memory access. This disparity can be seen most strikingly in the comparison between true analogies and mere-appearance matches. True-analogy matches are consistently felt to be more sound than mere-appearance matches; yet mere-appearance matches are consistently better at promoting retrieval. These accessibility results cannot be attributed to differences in similarity, for the true-analogy matches are consistently felt to be as similar or more similar than the mere-appearance matches.

Thus we are left with the disturbing findings that the pattern of accessibility is very different from the patterns of subjective inferential soundness. These results are compatible with the finding that people in problem solving tasks often fail to retrieve prior analogical problems which, if retrieved, would help them solve the current problem (Gick and Holyoak, 1980, 1983; Reed, Ernst and Banerji, 1974, Ross, 1984, in press)

Such results are problematic for models of memory that assume heavy reliance on causal indexing (e.g. Schank, 1982). On the contrary, it appears that people tend to have a fairly surface-oriented default indexing scheme that emphasizes object and first-order relations.

A natural question at this point is why humans should have such as seemingly arbitrary method of memory indexing and retrieval. One part of the answer may lie in the performance of the literal similarity matches. These matches are high on every measure: they are highly accessible, they are considered extremely sound and extremely similar. One speculation is that our memory indexing is geared toward literal-similarity matches: cases in which things resemble each other on the surface and also share deeper relational structure. In many areas the strategy may work fairly well; often thing that look alike <u>are</u> fundamentally the same. But in cases where appearance is not a good predictor of underlying relational structure, our memory systems may play us false.

References

Gentner, D. (1980). The structure of analogical models in science (BBN Rpt. No. 4451). Cambridge, MA: Bolt Beranek and Newman Inc.

Gentner, D. (1983). Structure-mapping: A theoretical framework for analogy. Cognitive Science, 7(2), 155-170.

Gentner, D., & Landers, R. (1985, November). Analogical reminding: A good match is hard to find. In Proceedings of the International Conference on Systems, Man and Cybernetics (pp.607-613). Tucson, AZ.

Gick, M. L., & Holyoak, K. J. (1980). Analogical problem solving. Cognitive Psychology, 12, 306-355.

Gick, M. L., & Holyoak, K. J. (1983). Schema induction and analogical transfer. Cognitive Psychology, 15(1), 1-38.

Reed, S. K., Ernst, G. A., & Banerji, R. (1974). The role of analogy in Transfer between similar problem states. Cognitive Psychology, 6, 436-450.

Ross, B. H. (1984). Remindings and their effects in learning a cognitive skill. Cognitive Psychology, 16, 371-416.

Ross, B. H. (in press). Remindings in learning: Objects and tools. To appear in S. Vosniadou & A. Ortony (Eds.). Similarity and analogical reasoning.

Schank, R. C. (1982). Dynamic memory. New York: Cambridge University Press.

Richard Catrambone   Keith J. Holyoak

University of Michigan   University of California at Los Angeles

Transfer in Problem Solving as a Function of the Procedural Variety of Training Examples

Abstract

Students often have difficulty solving homework assignments in quantitative courses such as physics, algebra, programming, and statistics. We hypothesize that typical example problems done in class teach students a series of mathematical operations for solving certain types of problems but fail to teach the underlying subgoals and methods which remain implicit in the examples. In the studies reported here, students in probability classes studied example problems that dealt with the Poisson distribution. In Experiment 1, the four examples all used the same solution method, although for one group the examples were superficially more dissimilar than for the other group. All subjects did well on the Near Transfer target problem that used the same subgoals and methods as the training examples. However, most did poorly on two Far Transfer target problems that had different subgoal orders and different methods. These results suggest that subjects typically learn solutions as a series of non-meaningful mathematical operations rather than conceptual methods in a subgoal hierarchy. In Experiment 2, one group studied problems that demonstrated two different subgoal orders using different methods while the other group received superficially different problems which had identical subgoal orders and methods. Both groups still had difficulty with the Far Transfer problems. Subjects who received examples with varied subgoal orders and methods seemed to isolate the subgoals, however, but not the methods. This result suggests that goals and methods may be useful ways of characterizing training problems. However, students may require explicit instruction on subgoals and methods in order to successfully solve novel problems.

Introduction

A relatively consistent finding in the analogical reasoning and transfer literature is that subjects do not seem to make use of prior information to solve new problems if the new problems differ from training examples in more than minor ways (Gick & Holyoak, 1980, 1983; Reed, Dempster, & Ettinger, 1985; Spencer & Weisberg, 1986). If similarities between training examples and target problems are pointed out to subjects or if they are encouraged to consider similarities between problems or domains, then subjects have somewhat more success at noticing and applying analogies or transferring information (Gentner & Gentner, 1983; Gick & Holyoak, 1983; Tenney & Gentner, 1984).

Card, Moran, and Newell (1983) proposed the GOMS model to account for the text-editing behavior of experts performing routine tasks. In this model, the expert knowledge representation consists of four components: Goals, Operators, Methods, and Selection rules. We would like to propose that in quantitative domains such as mathematics and physics, students acquire, or should acquire, goals, methods, and selection rules for solving problems as a function of the examples they study. Operators are simple mathematical procedures which college students typically already possess, such as calculating an average. Goals are initially quite general: solve the problem. After studying several examples, a student's goal may be more refined so that it is something like "get an answer that looks like the examples' answers." This type of goal is especially likely if the examples solve for the same unknown using the same procedure. In this case, students may simply learn that in order to achieve the goal they need to string together a series of operations. However, if the examples are varied in their givens and ultimate goal (the unknown being solved for), then students are less likely simply to string together a series of operations. Rather, they may recognize and develop subgoals which correspond to the steps in the examples. In addition, students will perceive that these subgoals can be reached by particular methods, which develop after students see a set of operations used together several times to achieve some subgoal. That is, students will compartmentalize the problems into subgoals which call on particular methods to satisfy them. A method will consist of a series of mathematical operations connected together conceptually. With experience, students may develop different methods for achieving the same subgoal. The particular method chosen will depend on the particular givens in the problem. Different problems will evoke different subgoals which will in turn evoke different methods. Students will develop selection rules for choosing which method to use. If students had only studied one type of example, then they would only have one method for solving problems in that domain. In fact, the method may really be a series of operations with no clear organizing feature except order of application. Thus, varied examples may be necessary in order to demonstrate how a series of operations can be grouped as a particular method for achieving a particular subgoal.

Reed et al. (1985) conducted several experiments using college students taking an algebra course. Subjects studied word problems dealing with traditional topics like distance, mixture, and work and then solved target problems. Reed et al. manipulated the superficial similarity of the target problems to the training problems. Their general finding was that subjects exhibited little transfer of the concepts from the training problems to the target problems except in those cases where the target problems were essentially identical in solution procedure to the examples. Reed et al. (1985) concluded that subjects were relying on a syntactic approach to the problems. This suggests that in general the subjects did not understand the goals and methods being demonstrated in the problems but rather had learned a series of operations for solving the problems.

We might suppose that if students were exposed to training examples that used different solution procedures, they would be more likely to learn the underlying subgoals and methods illustrated by the examples. This might happen because they would attempt to determine the similarities (such as the goal structure) between different series of operations which produce a value for the same final goal. A resolution process could lead to the identification of subgoals, methods, and generalizations of the methods (Anderson, 1983; VanLehn, 1985). However, if the series of operations from example to example are too different, students will fail to identify subgoals or to isolate a series of operations as a method (VanLehn, 1985). Each example will be perceived as unique.

Overview of Current Studies

We suspect that the difficulty students have in grasping subgoals and methods is due to the default reasoning of students who are still relatively unsophisticated in a particular domain. By default, students focus on superficial features of problems and the operations used to achieve an end goal because the features and operations are easier to isolate than the underlying subgoals and methods (Larkin, McDermott, Simon, & Simon, 1980; Schoenfeld & Herrmann, 1982). Students have a great deal of experience with the real world objects such as decks of cards and blocks of wood which populate the world of quantitative problems. Students in quantitative courses are also quite experienced with mathematical operations such as multiplication and addition as well as somewhat more "compiled" operations such as calculating means. Thus, it is not surprising that these students would tend to focus their attention and organize their problem solving skills around the mathematical operations with which they are most familiar (Greeno, Riley, & Gelman, 1984; Hayes, Waterman, & Robinson, 1977). We would like to begin to investigate what qualities of examples can help students go beyond their default focus and help them isolate subgoals and methods in a particular domain.

We solicited paid volunteers from three upper-level probability courses at the University of Michigan. The courses are quite similar for the first third of the semester. All students learn about counting rules (e.g., ordered and unordered sampling) and are then introduced to the notion

37

of a random variable. Then students learn about certain basic discrete probability distributions such as the binomial, Poisson, and geometric. The courses introduce the binomial distribution first, followed by the Poisson distribution. Students participated in the present experiments, which dealt with the Poisson distribution, after learning the binomial distribution but before learning the Poisson distribution.

The Poisson Distribution and Some Examples

The Poisson distribution is often used to approximate binomial probabilities for events that occur in time or space with some small probability p. The Poisson equation is:

$P(X=x)=[e^{-\lambda}(\lambda)^{x}]/x!$. It can be used to calculate probabilities for various values of X. Then the predicted frequencies of various values of X can be calculated by multiplying the probabilities by the total number of events. These steps are illustrated in Figure 1.

The example in Figure 1 deals with an event occurring randomly in time. The Poisson distribution is also used to model events occurring randomly in space. For example, one could reasonably fit a Poisson distribution to the number of fossils found in each section of a partitioned quarry. This problem is presented in Figure 2. It can be solved by the same procedure as the first problem.

It seems intuitively clear that a person could learn to solve problems of this type by memorizing the series of operations without understanding the meaning of the output from the operations. Nevertheless, the two examples do differ on the surface: one is about events in time and the other is about events in space. Thus, it is possible that students who study these examples may notice that the units in the operations are different and they may be induced to consider how the units were derived and to form a generalization about the operations. On the other hand, subgoals and methods can be identified more directly by comparing procedural differences in problems. Thus it is debatable whether superficial differences are sufficient to induce students to recognize these "deeper" aspects.

The subgoals and methods (in parentheses) for the two problems described above could be listed as follows:

1) find λ (calculate λ as a weighted average)

2) find the expected probabilities for each X (plug X = x into the Poisson equation)

3) find the expected frequencies for each X (multiply each P(X = x) by the total observed frequency)

A physicist observed a radioactive substance during 2608 time intervals (each 7.5 seconds long). She recorded the number of particles reaching a geiger counter for each period. Let $\underline{x}$ be the number of particles observed in each time period. Fit a Poisson distribution to $\underline{x}$, that is, give the expected frequencies for the different values of $\underline{x}$ based on the Poisson model.

Number of Particles Observed	Observed Frequency
0	57
1	203
2	383
3	525
4	532
5	408
6	273
7	139
8	45
9	27
10	10
11 or more	6
Total	2608

Solution:

$$E(X) = [0(57) + 1(203) + 2(383) + 3(525) + 4(532) + 5(408) + 6(273) + 7(139) + 8(45) + 9(27) + 10(10) + 11(6)]/2608 = 10092/2608 = 3.87 = \lambda$$

= average number of particles that reached geiger counter each period

$$P(X=x) = [(e^{-3.87})(3.87)^x]/x! = [(.021)(3.87)^x]/x!$$

Fitted Poisson Distribution:

x	Expected Frequency
0	.021 x 2608 = 55
1	.081 x 2608 = 211
2	.157 x 2608 = 409
3	.203 x 2608 = 529
4	.196 x 2608 = 511
5	.152 x 2608 = 396
6	.098 x 2608 = 256
7	.054 x 2608 = 141
8	.026 x 2608 = 68
9	.0113 x 2608 = 29
10	.0044 x 2608 = 11
11 or more	.00153 x 2608 = 4

Figure 1: Example problem for event occurring in time.

A horizontal quarry surface was divided into 30 squares about 1 meter on a side. In each square the number of specimens of the extinct mammal Ditolestes motissimus was counted. The results are given in the table below. Fit a Poisson distribution to x, that is, give the expected frequencies for the different values of x based on the Poisson model.

Number of Specimens per Square	Observed Frequency
0	16
1	9
2	3
3	1
4 or more	1
Total	30

Figure 2: Example problem for event occurring in space.

Experiment 1 explores how well students learn methods and subgoals from examples that differ only in the superficial ways shown above. If students studied problems like those above, they should be able to solve other superficially different problems that involve the same set of operations. It is less clear would happen if they tried to solve problems that had a different subgoal order and used modified methods. Consider the problem below.

Suppose you were making a batch of raisin cookies and you did not want more than one cookie out of 100 to be without a raisin. How many raisins will a cookie contain on the average in order to achieve this result? Use the Poisson distribution to find your answer.

Solution (not presented to subject):

$$P(X=0) = .01 = [(e^{-\lambda})(\lambda^0)]/0!$$
$$.01 = e^{-\lambda}$$
$$\ln(.01) = \ln(e^{-\lambda})$$
$$-4.6 = -\lambda$$
$$4.6 = \lambda = \text{average number of raisins per cookie}$$

Figure 3: Cookie problem.

This "cookie" problem literally looks different than the prior ones. In this problem the student must realize that he or she is provided with the following piece of information: $P(X=0) = .01$ (i.e., only one cookie out of 100 should have zero raisins). He or she must also realize that the goal is to find λ—the expected value of the random variable which in this case is the average number of raisins that a cookie receives. If they recognize these two facts then the problem simply becomes a matter of inserting $P(X=0) = .01$ into the Poisson equation and solving for λ. It is unclear, however, how these realizations would follow from the types of practice problems to which the students have thus far been exposed. Students would only have learned a series of operations. They would not have learned that the calculation of λ is a subgoal which can be carried out by several different methods depending on the givens. One way to calculate λ is to find a weighted average as was done in the example problems. Another way is to find values for the other unknown in the Poisson equation (i.e., a value for some $P(X=x)$) and then solve for λ.

The subgoals and methods for this problem are listed below:

1) find the known value for some $P(X=x)$ (divide 1 by 100 to get $P(X=0)$)

2) find λ (plug $P(X=0)$ into Poisson equation and solve for λ)

Consider another problem:

Suppose you took a random sample of 500 people and found out their birthdays. A "success" is recorded each time a person's birthday turns out to be January 1st. Assume there are 365 days in a year, each equally likely to be a randomly chosen person's birthday. Fit a Poisson distribution to $\underline{x}$ (the number of people born on January 1st) and find the predicted likelihood that exactly 3 people from the sample are born on January 1st.

Solution (not presented to subject):

$\lambda = 500/365 = 1.37 =$ average number of people born on any given day

$$P(X=3) = [(e^{-1.37})(1.37^3)]/3!$$
$$= [(.254)(2.57)]/6$$
$$= .109$$
$$= \text{likelihood of exactly three people being born on}$$
$$\text{January 1st (or any other given day)}$$

Figure 4: Birthday problem.

The birthday problem requires that the student realize that λ can be calculated simply by dividing the number of days by the number of people (as opposed to being calculated as a weighted average from an observed frequency table). It also requires that the subject realize he or she was being asked to solve only for $P(X=3)$ and not to produce an expected frequency table.

The subgoals and methods for this problem are:

1) find λ (divide the number of events [birthdays] by the number of slots [days of the year])

2) find $P(X=3)$ (plug $X=3$ into the Poisson equation)

Both the cookie and birthday problems have different or modified methods compared to the training examples, yet they still have either the same subgoals (in a different order) or fewer subgoals. Students' performance on the cookie and birthday problems should indicate whether they isolated subgoals and methods during training or whether they simply learned a series of operations to achieve the single goal of producing an expected frequency table.

Experiment 1

Method

Subjects. Seventy-one students from three probability classes were recruited and were paid $7 for their participation.

Materials and Procedure. Subjects were given a booklet to study. The cover page contained a description of the relationship between the binomial and Poisson distributions and provided the Poisson equation. The next four pages contained four worked out Poisson distribution problems isomorphic to the radioactive particle and quarry problems. Subjects were told to study the problems carefully since after studying them they would be asked to solve three problems. They were also told they could refer back to the cover page but not to the examples. This was done to increase the likelihood that subjects would pay attention to the examples and how they were solved.

Subjects were randomly divided into two groups. The SAME group studied four examples which dealt with the same class of events: either four space problems or four time problems. The DIFFERENT group received problems from both classes of events: two space problems and two time problems. All problems were solved using the same procedure, which was identical to the radioactive particle and quarry problems discussed above. The example problems were picked from a pool of four space and four time problems. There was no effect in subjects' performance on the target problems as a result of the specific space or time problems a subject received, and all reported results are collapsed over this factor.

After studying the examples subjects worked on the three target problems. The first target problem is labeled the "Detroit Tiger" problem and is presented below. This problem will be called a Near Transfer problem since it embodies the same subgoals (and same subgoal order) and methods as the training examples.

In a 162-game baseball season, the Detroit Tiger infield made a total of 107 errors. The table below gives the number of games in which x errors were made. Fit a Poisson distribution to x, that is, give the expected frequencies for the different values of x based on the Poisson model.

Number of Errors x made in a game	Observed Frequency
0	85
1	52
2	20
3 or more	5
Total	162

Figure 5: Detroit Tiger Problem

The second and third target problems were the cookie and birthday problems, respectively. They will be referred to as Far Transfer problems because they involve different subgoal orders and methods than the training examples. The order of the target problems was the same for all subjects. Subjects worked at their own pace for the entire experiment. In general, subjects took about 35 minutes to complete the experiment. Subjects were asked to show all their work but could use a calculator for the basic arithmetic. The solution and error frequencies were analysed using the likelihood ratio chi-square test ($\underline{G}^2$) which is a test of equality of proportions between rows or columns.

Results

Subjects' answers to the transfer problems were first scored as correct/incorrect. Both groups did well on the Detroit Tiger (Near Transfer) problem: 91% and 94% correct for the Same and Different groups, respectively. On the cookie problem (Far Transfer) the DIFFERENT group did somewhat better than the SAME group: 42% versus 23%, $\underline{G}^2(1) = 2.9$, $\underline{p} < .09$. The DIFFERENT group also did better on the birthday problem (Far Transfer), 33% versus 23%, but this difference did not approach conventional significance levels, $\underline{G}^2(1) = .97$, $\underline{p} > .3$. Overall, 32% of the subjects solved the cookie problem and 28% solved the birthday problem.

Subjects errors were analysed separately for the cookie and birthday problems. The first type of error for the cookie problem (called ULAMBDA in Table 1) is a failure to recognize the goal of the problem, to solve for λ. That is, the subject does not realize that the average number of raisins per cookie is λ. The second error type (PX0) is a failure to recognize that $P(X=0) = .01$ is provided in the problem. The third category (FREQ1) is whether a subject attempted to make up a frequency table as a way of solving the problem (i.e., they generated hypothetical data). If a subject made up a frequency table, this would indicate that he or she was most likely trying to make the target problem appear like the examples in order to use the familiar procedure. This approach is an error since there is no way to create a useful frequency table with the information given.

There are also three error categories for the birthday problem. The first category (SLAMBDA) is a failure to recognize that λ is the average number of people that are born on any given day. This value is simply the number of people (500) divided by the number of days in the year. (A priori it seemed unlikely that a subject would understand that λ would be the average number of people born on a given day but fail to realize that this value would be 500/365. This assumption was supported by the protocols.) The second category (PX3) is a failure to realize that the problem's goal was to solve for $P(X=3)$ rather than to create a frequency table or to find only the expected value of X. The third category (FREQ2) is identical to the third category for the cookie problem; it counts how often subjects tried to make up a frequency table as an aid to solving the problem. Again, this approach will not help to solve the problem.

Sixty-eight percent (48 out of 71) of the subjects failed to solve the cookie problem and 72% (51 out of 71) failed to solve the birthday problem. Table 1 indicates the error types and their frequencies for the two far transfer problems. It also presents the frequencies collapsed across the group dimension since analyses indicated there were no differences between the groups (for subjects who got a problem wrong) with respect to the frequency of different error types.

Table 1

Percentage of Subjects Who Made Particular Types of Errors (Experiment 1)

Transfer Problem	Error Type	Group		
		SAME	DIFFERENT	Total
Cookie Problem				
		n=27	n=21	n=48
	ULAMBDA	85 (23)	67 (14)	77(37)
	PX1	96 (26)	86 (18)	92(44)
	FREQ1	41 (11)	62 (13)	50(24)
Birthday Problem				
		n=27	n=24	n=51
	SLAMBDA	96 (26)	100 (24)	98(50)
	PX3	67 (18)	62 (15)	65(33)
	FREQ2	15 (4)	21 (5)	18(9)

Note. Frequencies are given in parentheses. Percentages are based on the number of subjects who made a particular error divided by the number of subjects in each group who got the problem wrong (given at the top of each column for each of the transfer problems), not the total of number of subjects in the group.

Discussion

It was intuitively plausible to expect both groups of subjects to solve the Detroit Tiger problem equally well since it used the same series of operations as the examples. However, both groups were expected to do equally poorly on the far transfer problems because we suspected that the manipulation of SAME versus superficially DIFFERENT training examples to be unrelated to whether or not subjects learned the underlying subgoals and methods in the training examples. These expectations were largely confirmed.

It seems clear that subjects who had difficulty with the far transfer problems had difficulty because they had primarily learned a series of operations for solving problems of the training type and had not learned the underlying subgoals or formed generalizations of the methods. Sixty-eight percent of the subjects could not solve the cookie problem and for 92% of those subjects the reason seemed to be that they did not realize that they were given a piece of useful information, namely that $P(X=0) = .01$, and thus they could not figure out how to solve for λ. In addition, the fact that 77% of these unsuccessful subjects did not even realize they were solving for λ indicates that they did not recognize solving for λ as a subgoal, but rather were looking to apply the operations from the examples. This claim is further supported by the fact that half of the subjects tried to make up an observed frequency table from which to calculate λ. However, most of these subjects still went on to calculate an expected frequency table. This suggests that they did not make up the observed frequency table to calculate λ per se, but rather the table was created to help them apply the stereotyped operations so they could reach the only goal they seemed to know: to create an expected frequency table.

Experiment 2

Experiment 1 indicated that manipulations of superficial problem characteristics were not sufficient to induce subjects to isolate subgoals and methods. In Experiment 2 we manipulated the subgoals and methods used in the training problems.

Subjects were given four problems to study. The ONE-PROCEDURE group was just like the DIFFERENT group in Experiment 1: the problems used the same procedure but were different superficially. The TWO-PROCEDURE group received two problems using the same procedure as the Detroit Tiger problem and two problems using the same procedure as the cookie problem. It would not be surprising if the TWO-PROCEDURE subjects could solve the cookie problem successfully. However, the more interesting issue is whether they learned anything more than two sets of operations for solving two types of problems. That is, did they simply learn that frequency table problems require one approach and non-frequency table problems require a different approach (i.e., they learned a superficial selection rule and did not learn subgoals or methods), or did they learn that problems can have different goals, subgoal orders, and methods for obtaining those subgoals?

Subjects then attempted to solve two instances of a new problem type (the birthday problem and one isomorphic to it, the "football" problem—not illustrated here) in addition to problems whose solution procedures were already familiar to them (i.e., the Detroit Tiger problem and/or the cookie problem). Subjects' answers and errors were examined for indications that they were simply trying to apply one of two series of operations or whether they had recognized that particular subgoals existed (finding λ, then finding $P(X=x)$) and that new methods would be needed.

Method

Subjects. Fifty students from a probability class were recruited and paid $7 for their participation.

Materials and Procedure. The procedure was identical to the one in Experiment 1. The only difference was the materials. There were three groups of subjects in this experiment. The TIGER group studied four training problems that used the same solution procedure as the Detroit Tiger target problem. The COOKIE group studied four training problems which used the same solution procedure as the cookie target problem. The TWO-PROCEDURE group studied two problems which used the Detroit Tiger problem procedure and two problems which used the cookie problem procedure. All subjects then received four target problems to solve: the Detroit Tiger problem, the cookie problem, the birthday problem, and the football problem.

Results and Discussion

For some of the analyses reported below, the comparisons are between the three groups: TIGER, COOKIE, and TWO-PROCEDURE. For other analyses the TIGER and COOKIE groups are collapsed into a ONE-PROCEDURE group and thus the comparison will be between ONE-PROCEDURE and TWO-PROCEDURE subjects. In addition, the terms "near" and "far" transfer can not be used as they were in Experiment 1 since, for the COOKIE group, the cookie problem is now a near transfer problem and the Detroit Tiger problem is a far transfer problem. Thus, the target problems will be referred to by their names. Table 2 summarizes the type of transfer problem the target problems represent for each group.

Table 2

Degree of Transfer Required in Target Problems as a Function of Subject Group

Group	Near Transfer:	Far Transfer:
TIGER	Detroit Tiger	cookie, birthday, football
COOKIE	cookie	Detroit Tiger, birthday, football
TWO-PROCEDURE	Detroit Tiger, cookie	birthday, football

While all of the TIGER and TWO-PROCEDURE subjects solved the Detroit Tiger problem correctly, only 13% of the COOKIE subjects did. This difference is, of course, significant, $G^2(2) =$ 45.5, p < .0001. Similarly, while most of the COOKIE and TWO-PROCEDURE subjects solved the cookie problem correctly (87% and 86%, respectively), a much lower percentage (31%) of the TIGER subjects did, $G^2(2) = 13.9$, p < .001. There is no difference in solution rates among the three groups for the birthday or football problems which are far transfer problems for all subjects. Overall, 52% of the subjects solved the birthday problem and 50% solved the football problem. It should be noted that the 52% solution rate for the birthday problem is significantly greater than the 28% solution rate for that problem for subjects in Experiment 1, $z = 2.7$, p < .007.

Of the nine TIGER subjects who failed to solve the cookie problem, 78% failed to realize that the goal was to solve for λ, 100% did not realize that $P(X=0) = .01$ was provided in the problem, and 33% tried to make up a frequency table as an aid to solve the problem. These frequencies are similar to the ones obtained in Experiment 1.

Of the 13 COOKIE subjects who failed to solve the Detroit Tiger problem, 12 of them tried to calculate λ by taking an observed frequency for some X and plugging that into the Poisson equation and solving for λ. For those subjects who chose $X=0$, they would get an equation such as $P(X=0) = 85/162 = [e^{-\lambda}\lambda^0]/0!$. This reduces to $.52 = e^{-\lambda}$, which yields $\lambda = .65$. Given that the λ generated by the frequency table method is .66, this "cookie" approach works quite well, but in other situations it could be quite poor in comparison with the frequency table method (since it would ignore available frequency data). In addition, for the 12 subjects who took this "cookie" approach, eight of them stopped after solving for λ and did not generate the predicted frequency table. This suggests that they were performing a series of operations rather than solving for the goal of the problem. Four of the other subjects used the observed frequencies of each X in turn to solve for λ. It becomes quite messy to solve for λ when an X other than 0 is used and these subjects would set up the equations and then stop. The remaining subject who got the problem wrong calculated λ using the frequency table approach, but did not go on to generate predicted frequencies for the various values of X.

The types of errors made by subjects who were unsuccessful in solving the birthday or the football problems are presented in Table 3. The errors are presented as a function of whether subjects received examples illustrating one procedure or two (i.e., the TIGER and COOKIE groups are collapsed into the ONE-PROCEDURE group).

Table 3

Percentage of Subjects Who Made Particular Types of Errors (Experiment 2)

Transfer Problem	Error Type	Group		Total
		One Procedure	Two Procedure	
Birthday				
		n = 14	n = 10	n = 24
	SLAMBDA	93 (13)	100 (10)	96 (23)
	PX3	50 (7)	10 (1)	33 (8)
	FREQ	7 (1)	20 (2)	12 (3)
Football				
		n = 14	n = 11	n = 25
	SLAMBDA	93 (13)	100 (11)	96 (24)
	PX1	43 (6)	9 (1)	28 (7)
	FREQ	14 (2)	0 (0)	8 (2)

Note. Frequencies are given in parentheses. Percentages are based on the number of subjects who made a particular error divided by the number of subjects in each group who got the problem wrong (given at the top of each column for each of the transfer problems), not the total of number of subjects in the group.

Both ONE-PROCEDURE and TWO-PROCEDURE subjects solved the birthday and football problems about 50% of the time. These are far transfer problems for both groups. We had expected the TWO-PROCEDURE subjects to do better since we hypothesized they would have been likely to isolate subgoals such as λ and $P(X=x)$ and generalize the methods for finding them. Nevertheless, one difference did emerge in both problems. Of the ONE-PROCEDURE subjects who failed to solve the birthday problem, only 50% realized they were to solve for $P(X=3)$ while 90% of the TWO-PROCEDURE subjects realized this. This difference is significant, $\underline{G}^2(1) = 4.64$, $\underline{p} < .04$. Similarly, 50% of the ONE-PROCEDURE subjects realized they were to solve for $P(X=1)$ in the football problem while 91% of the TWO-PROCEDURE subjects realized this.

Again, the difference is significant, $\underline{G}^2(1) = 3.82$, $\underline{p} = .05$. This result suggests that TWO-PROCEDURE subjects may have at least isolated subgoals, but were unable to apply the correct method to the birthday and football problems. Most subjects did calculate λ in the birthday and football problems, but they tended to use nonsensical values such as 365/500 or 3/500 for the birthday problem. TWO-PROCEDURE subjects did not seem to learn anything about examining λ for its reasonableness, yet they did adapt to the new goal constraint (i.e., finding only a particular $P(X=x)$) while ONE-PROCEDURE subjects did not.

General Discussion

The difficulties that subjects in both experiments had with the far transfer problems suggest that procedural variety plus explicit pointing out of subgoals and methods may be required to teach students how to solve problems which have different subgoal orders and modified methods compared to training problems.

Procedural variety may mean that students should be exposed to problems that provide different givens, have different appearances, and/or which require solving for different unknowns.

These variations would presumably induce students to isolate different methods for achieving certain subgoals and to realize that there can be different goals and subgoals for solving problems in the same domain (Owen & Sweller, 1985). This induction could also be facilitated by presenting examples which give the data in different forms (such as giving λ directly rather than having it calculated from a table). The need for having students see examples which solve for different unknowns is suggested by the large number of subjects in Experiment 1 who failed to realize that they were solving for something new, namely λ, in the cookie problem. The importance of presenting similar information in different forms (e.g., tables versus text, ready-to-use values versus "low-level" values which require additional calculations before they can be used in equations) seems reasonable in light of the fact that λ was a quite simple thing to calculate in the birthday problem, yet students failed to see it or to calculate it correctly. In fact, students in Experiment 1 often used the more laborious method of making up a frequency table in order to (incorrectly) calculate λ. This problem is similar to the error Reed et al.'s (1985, Experiment 4) subjects made when they tried to use the more complex solution methods from the training examples on the simpler target problems. Both our results and Reed et al.'s indicate that students were learning series of operations rather than, or more easily than, subgoals and methods for solving problems.

We have tried to suggest that an important component of the "power" of examples is the variation that is provided in a sequence of examples. Winston's (1973) arch perceiver could only learn concepts when the examples it was presented with were given in a particular order. Negative instances of a concept were just as important (and sometimes more important) than positive instances. Failure-driven memory is an important component of Schank's (1982) model of learning. So too here, negative examples (in the form of training problems that have different subgoals and methods) are important. If a student sees several problems that are dealt with in different ways, he or she may be more likely to isolate the subgoals and methods rather than viewing the problems as a series of operations which ultimately produce some output. He or she may also form generalizations of methods. However, the student may need guidance to help him or her focus on the subgoals and methods, at least initially (Lewis & Anderson, 1985). We are currently conducting a transfer experiment using materials which provide subjects with explanatory information highlighting the subgoals and methods that are present in each training example.

It may be possible to develop a methodology for constructing examples for textbooks in quantitative domains. This methodology would involve first identifying the subgoals and methods that students need to learn (Kieras, in press; Kieras & Bovair, 1986). Then example problems and explanatory materials which highlight these subgoals and methods can be constructed. The careful procedural variation might allow students to see beyond the superficial features of examples.

Acknowledgments

This research was supported by Army Research Contract MDA903-86-K-0297. We would like to thank J.E. Keith Smith and Mary Gick for their comments on this paper.

References

Anderson, J.R. (1983). The architecture of cognition. Cambridge, MA: Harvard University Press.

Gentner, D. & Gentner, D.R. (1983). Flowing waters or teeming crowds: Mental models of electricity. In D. Gentner & A.L. Stevens (Eds.), Mental Models. Hillsdale, N.J.: Erlbaum.

Gick, M.L. & Holyoak, K.J. (1980). Analogical problem solving. Cognitive Psychology, 12, 306–55.

Gick, M.L. & Holyoak, K.J. (1983). Schema induction and analogical transfer. Cognitive Psychology, 15, 1–38.

Greeno, J.G., Riley, M.S., & Gelman, R. (1984). Conceptual competence and children's counting. Cognitive Psychology, 16, 94–143.

Hayes, J.R., Waterman, D.A., & Robinson, C.S. (1977). Identifying the relevant aspects of a problem text. Cognitive Science, 1, 297–313.

Kieras, D.E. (in press). The role of cognitive simulation models in the development of advanced training and testing systems. In N. Frederiksen, R. Glaser, A. Lesgold, & M. Shafto (Eds.), Diagnostic monitoring of skill and knowledge acquisition. Hillsdale, NJ: Erlbaum.

Kieras, D.E. & Bovair, S. (1986). The acquisition of procedures from text: A production-system analysis of transfer of training. Journal of Memory and Language, 25, 507–524.

Larkin, J., McDermott, J., Simon, D.P., & Simon, H. (1980). Expert and novice performance in solving physics problems. Science, 208, 1335–1342.

Lewis, M.W., & Anderson, J.R. (1985). Discrimination of operator schemata in problem solving: Learning from examples. Cognitive Psychology, 17, 26–65.

Owen, E., & Sweller, J. (1985). What do students learn while solving mathematics problems? Journal of Educational Psychology, 77, 272–284.

Reed, S.K., Dempster, A., & Ettinger, M. (1985). Usefulness of analogous solutions for solving algebra word problems. Journal of Experimental Psychology: Learning, Memory, and Cognition, 11, 106–125.

Schank, R.C. (1982). Dynamic memory: A theory of reminding and learning in computers and people. New York: Cambridge University Press.

Schoenfeld, A.H., & Herrmann, D.J. (1982). Problem perception and knowledge structure in expert and novice mathematical problem solvers. Journal of Experimental Psychology: Learning, Memory, and Cognition, 8, 484–494.

Spencer, R.M., & Weisberg, R.W. (1986). Context-dependent effects on analogical transfer. Memory & Cognition, 14, 442–449.

Tenney, Y.J., & Gentner, D. (1984). What makes analogies accessible: Experiments on the water-flow analogy for electricity. In Proceedings of the international conference on research concerning students' knowledge of electricity, Germany.

VanLehn, K. (1985). Arithmetic procedures are induced from examples, (Tech. Report No. ISL-12). Xerox Palo Alto Research Center.

Winston, P. (1973). Learning to identify toy block structures. In R.L. Solso (Ed.), Contemporary issues in cognitive psychology: The Loyola symposium. Washington, D.C.: V.W. Winston & Sons. Hillsdale, N.J.: Erlbaum.

Schema Acquisition from One Example:

Psychological Evidence for Explanation-Based Learning[*]

Woo-Kyoung Ahn, Department of Psychology

Raymond J. Mooney, Coordinated Science Laboratory

William F. Brewer, Department of Psychology

Gerald F. DeJong, Coordinated Science Laboratory

University of Illinois at Urbana-Champaign

Abstract

Recent explanation-based learning (EBL) models in AI allow a computer program to learn a schema by analyzing a single example. For example, GENESIS is an EBL system which learns a plan schema from a single specific instance presented in a narrative. Previous learning models in both AI and psychology have required multiple examples. This paper presents experimental evidence that people can learn a plan schema from a single narrative and that the learned schema agrees with that predicted by EBL. This evidence suggests that GENESIS, originally constructed as a machine learning system, can be interpreted as a psychological model of learning a complex schema from a single example.

Introduction

Recent explanation-based models in machine learning (DeJong & Mooney, 1986; Mitchell, Keller, & Kedar-Cabelli, 1986) allow a program to learn a concept or schema by analyzing the causal structure of a single example. Explanation-based learning (EBL) systems construct an explanation for why an instance is a member of a concept or why a particular sequence of actions achieves a goal. This explanation is then generalized, retaining only the constraints required to maintain its causal structure. (Mooney & Bennett, 1986) reviews and compares a number of similar algorithms for performing this generalization. These algorithms produce a general concept description or plan schema which can be used to improve performance on future classification, understanding, or problem solving tasks.

A major difference between EBL and other approaches to learning is the number of examples required. Similarity-based learning (Michalski, 1983; Mitchell, 1978; Quinlan, 1986) requires many examples and systems based on analogy (Carbonell, 1983; Falkenhainer, Forbus, & Gentner, 1986; Winston, 1980) require two examples, while EBL requires only a single example. Although a number of psychological experiments exist demonstrating people's ability to learn concepts or schemata from two examples using analogy (Gick & Holyoak, 1983; Spencer & Weisberg, 1986) or from many examples using similarity-based induction (Medin, Wattenmaker, & Michalski, 1986; Posner & Keele, 1968), there seems to be no experiments directed at demonstrating people's ability to learn a concept or schema from a single example. Consequently, until now, there has been no empirical evidence to support the use of EBL as a psychological model of human learning. This paper summarizes a number of recently conducted experiments which demonstrate subjects' ability to learn a new general plan schema from a single narrative describing a specific instance of the plan. The schema acquired in this way is shown to obey the variables and constraints predicted by an EBL model.

This research was supported in part by University of Illinois Cognitive Science/AI fellowships to the first two authors and in part by the Office of Naval Research under grant N-00014-86-K-0309.

Learning from One Example in GENESIS

The idea of learning a schema by analyzing the explanation of a single narrative was first presented in (DeJong, 1981). The GENESIS system (Mooney, 1985; Mooney & DeJong, 1985) is a realization of this idea which processes short English narratives and is able to acquire new plan schemata from single specific instances.

During the understanding process, GENESIS attempts to construct explanations for characters' actions in terms of the goals their actions were meant to achieve. This process involves plan-based understanding mechanisms like those employed by previous narrative processing systems (Dyer, 1983; Schank & Riesbeck, 1981; Wilensky, 1983). When the system observes that a character has achieved an interesting goal in a novel way, it generalizes the composition of actions the character used to achieve this goal into a new schema. The generalization process (described in Mooney and Bennett (1986)) consists of an analysis of the causal model of the narrative which removes unnecessary details while maintaining the validity of the explanation. The resulting generalized set of actions is then stored as a new schema and used by the system to correctly process narratives which were previously beyond its capabilities. Currently, GENESIS has learned schemata for kidnapping-for-ransom, arson-for-insurance, murder-for-inheritance, and for a police-officer impersonating a prostitute in order to entrap solicitors. In each of these cases, it demonstrates a performance improvement by using the schema it has learned to construct explanations for narratives which it previously could not explain.

The goal of the present research is to show that GENESIS can be interpreted as a psychological model. Thus, we have carried out a series of experiments to see if people can acquire a novel plan schema from a narrative describing a single specific instance of a novel action. Specifically, we predicted that people would build a causally complete representation of the text by causally connecting instantiations of existing schemata (as in Johnson-Laird's *mental model* (1983) or van Dijk & Kintsch's *situation model* (1983)). The *explanation* is the connected portion of the model which contributes to the characters achieving important goals. This explanation is immediately generalized into a schema by changing constants to variables within the constraint that the structure of the explanation remains intact. The resulting schema is characterized by a set of *variables* which are slots which can be filled by different objects or agents in each instance, and a set of *constraints* which specify necessary properties of variables and necessary relationships between variables. The *constraints* are those properties and relations required to maintain the causal validity of the explanation.

Overview of the Experiments

GENESIS was originally constructed as a machine learning system. It was not explicitly written to model an existing set of psychological data. In order to explore the validity of GENESIS as a psychological model of human learning, four psychological experiments were conducted. The basic design was to have subjects read a passage describing a specific instance of a novel plan. Each of the first three experiments used a different task to test whether or not subjects had acquired an abstract schema from this single example. The last experiment tested whether subjects generalized the narrative as a natural part of comprehension or only produced the abstract schema when they were asked general questions about the narrative.

Three passages were constructed to present situations for which the subjects presumably did not already have a pre-established schema but which they could understand using aspects of their existing knowledge. For example, one passage involves a cooperative buying scheme used in other countries. In Korea the system is called a "Kyeah" and in India it is called a "chit fund". The experimental narrative describing a single instance of this plan follows:

Tom, Sue, Jane, and Joe were all friends and each wanted to make a large purchase as soon as possible. Tom wanted a VCR, Sue wanted a microwave, Joe wanted a car stereo, and Jane wanted a compact disk player. However, after paying their expenses, they each only had $60 left at the end of every third month. Tom, Sue, Jane, and Joe all got together to solve the problem. They made four slips of paper with the numbers 1,2,3, and 4 written on them. They put them in a hat and each drew out one slip. Jane got the slip with the 4 written on it, and said, "Oh darn, I have to wait to get my CD player." Joe got the slip with the 1 written on it and said, "Great, I can get my car stereo right away!" Sue got the number 2, and Tom got number 3. In February, they each contributed the $60 they had left. Joe took the whole $240 and bought a Pioneer car stereo at K's Merchandise. In May, they each contributed their money again. This time, Sue used the $240 to buy a Sharp 600 watt 1.5 cubic foot microwave at K-mart. In August, all four again contributed $60. Tom took the money and bought a Sanyo Beta VCR with wired remote at Service Merchandise. In November, Jane got the money and bought a Technics CD player at Apple Tree Stereo.

The complexity of the experimental examples prevented a complete formal analysis and computer implementation. Therefore, constraints and variables were obtained for each schema by determining a set of roles for the schema and deciding which properties and relations of these roles were important in maintaining the underlying causal structure of the narrative. Variables and constraints were determined for each schema before conducting any of the experiments. Table 1 shows the list of variables and constraints identified for the Kyeah schema.

In addition to a group given specific narratives (Example group), experiments 2 and 3 also used a control group which was given abstract descriptions of the schemata underlying each of the example narratives (Abstract group). The description of the Kyeah schema given to the Abstract group follows:

> Suppose there are a number of people (let the number be n) each of whom wants to make a large purchase but does not have enough cash on hand. They can cooperate to solve this problem by each donating an equal small amount of money to a common fund on a regular basis. (Let the amount donated by each member be m.) They meet at regular intervals to collect everyone's money. Each time money is collected, one member of the group is given all the money collected (n X m) and then with that money he or she can purchase what he or she wants. In order to be fair, the order in which people are given the money is determined randomly. The first person in the random ordering is therefore able to purchase their desired item immediately instead of having to wait until they save the needed amount of money. Although the last person does not get to buy their item early, this individual is no worse off than they would have been if they waited until they saved the money by themselves.

Since subjects in the Abstract group had been directly told the content of the schema, they were presumed to have learned the schema. Consequently, if the Example group performed as well as the Abstract group on a task requiring knowledge of the general schema, then it is reasonable to assume that the subjects in the Example group had also acquired the schema.

In addition to the Kyeah schema, the other two situations used in the experiments included a technique for making additional money by fencing copies of a stolen collectable and a confidence game called the "phony bank-examiner ploy" (the latter was taken from Wharton (1967)). None of the experiments found appreciable differences among the three schemata; consequently, the results reported for each experiment are averaged across all three examples.

Table 1: Variables and Constraints for the Kyeah Schema

Variables	Constraints
identity of participants	participants want items of similar value
number of participants (n)	participants cannot afford items
exact time of meetings	participants trust each other
interval between meetings (t)	participants can afford m each t
amount of donation (m)	each participant donates same amount
items bought	cost of desired items $\simeq$ n X m
stores where items bought	number of meetings = number of participants
method of determining order	order must be assigned randomly

In all of the experiments, the Example group was given only a single instance of the schema. Thus any learning that occurred in this group would be outside the domain of learning theories which required multiple instances (i.e. analogy or similarity-based induction).

In order to determine whether a learned schema agreed with that predicted by EBL, subjects' learning was always judged on how well they obeyed the *constraints* and recognized the mutability of the *variables*.

Subjects in all of the experiments were undergraduate students at the University of Illinois at Urbana-Champaign and participated in the experiments to fulfill a requirement of an introductory psychology course. After each of the experiments, subjects were asked whether they had previously heard of any of the three plans described in the passages. If so, their data were discarded.

Experiment 1: General Description Generation

In Experiment 1, in order to test whether subjects had acquired a schema from a single instance, we asked them to write a general description of the schema. We predicted that in the subjects' descriptions the schema constraints would remain but the actual objects in the story would be replaced with variables.

Method

The experiment used only one group of subjects. Each subject was given the three experimental narratives and was told for each one to "write, in abstract terms, a description of the general technique illustrated in the narrative." In order to clarify the instructions, they were given a sample narrative and its corresponding general description. We included this demonstration narrative to show subjects which level of abstraction we wanted them to generate. The demonstration narrative was about skyjacking and was selected to be unrelated to the experimental passages. The demonstration passages do not reveal which portions of the narratives are constraints and which are variables. For example, an airplane is mentioned in both the demonstration narrative and its corresponding general description; however, although a VCR is mentioned in the Kyeah narrative, it is not a part of the Kyeah schema. A correct analysis can only be determined by providing an explanation for the individual example.

After reading the instructions, all of the subjects read the first narrative and wrote a general description for it and did the same thing for the second and third story at their own pace. Data were collected from 11 subjects.

Results and Discussion

In general, subjects produced good schema descriptions. The following is the description of the Kyeah schema written by one subject:

> Suppose in a group of people, each person would like to buy something expensive, but over a period of time, each person cannot earn enough to buy what he would like. By using random selection, each person could be assigned a number. when the group had saved enough money *together* to purchase an item, the person with the first number would get his item. This would continue for the rest of the group until everyone had gotten what he wished.

To provide a more objective index of the subjects' performance, we counted the number of constraints mentioned and the number of variables identified in their general descriptions. A variable was considered to have been *identified* if either an abstract term, such as "group" or "something," was used to refer to it, or if it was simply not mentioned in the description. If a particular variable is not mentioned at all, then it is reasonable to assume that the subjects believe that its particular value was not important to the overall schema.

On the average, 84% of the constraints were explicitly mentioned and 88% of the variables were identified. These percentages indicate that subjects can acquire an abstract schema from a single instance and that the characteristics of the learned schema agrees with those predicted by EBL theory.

Experiment 2: Story Generation

In Experiment 2, we tested whether the schema that subjects acquired could deal with new instances. If both Example and Abstract groups could produce another instance equally well based on what they read, it would indicate that both groups had acquired a generative schema.

Method

Subjects in the Example group were given the three narratives and told for each one to "write another story in which characters use the general method illustrated in the story but that is otherwise as different as possible." Subjects in the Abstract group were given the three abstract schema descriptions and told for each one to "write a story in which particular individuals use the technique described in the passage in a specific case." Both groups read their first passage and wrote a story and the task was repeated for the second and the third passages. There were eight subjects in the Example group and seven in the Abstract group.

Results and Discussion

In general, both groups produced equally good narratives. The following is the new Kyeah narrative written by one subject in the example condition.

> Bill, Kim, John and Mary were all business associates. Bill wanted some land in Northern Illinois, Kim wanted a new house in Switzerland, John wanted a new Porshe 928S with all all accesories, and Mary wanted to take a trip around the world. The only problem was they each only had $25,000.00 left unspent at the end of each month. They all got together and picked random variables on Bill's business computer. Mary was farthest from her variable so she would have to wait till last to get her trip around the world. John nailed his variable and jumped enthusiastically saying, "Yea, I get to get my new Porshe 928S right now." They each talked with their banker and drew the $25 Thousand dollars out and pooled it together after the first month and the next day John drove up in his new, black, 928S with all accessories. At the end of the next month they again pooled their money and Kim got her chalet in Switzerland. Again at the end of the next month they pooled their money an Bill got his land in Northern Illinois. Finally, after the fourth month they pooled their money together and Mary left for her trip around the world.

Again, to be more objective, we counted the number of constraints obeyed and the number of variables changed in the stories generated by the Example group. However, in the Abstract group, we could only count the number of constraints obeyed since there were no constants to change in their passages (e.g. number of participants, items purchased, etc.).

Overall, the Example group obeyed 90% of the constraints while the Abstract group obeyed 87%. In addition, the Example group changed 71% of the variables. These results imply that from a single specific instance subjects can acquire a schema equivalent to that acquired directly from an abstract description of the schema.

Experiment 3: Yes/No Questions

Experiment 3 was designed to test whether subjects in both groups were equally good at detecting which portions of the narratives were variables and which were constraints. Although in Experiment 2, the Example group changed only 71% of the variables, this does not necessarily indicate that they did not identify the rest of them. It is possible that the subjects simply did not take the effort to change the values of all of the variables. For example, most of the subjects did not change the number of participants in their Kyeah narratives. Experiment 3 directly tested the ability of subjects to recognize all of the variables.

Method

We developed a yes/no question for each constraint (for example, "Can some people consistently donate less than others and have the system work?") and a yes/no question for each variable (for example, "Is there any particular number of people required for this plan?"). For each question the expected answers based on EBL were sometimes "yes" and sometimes "no". Both the Example and Abstract groups read their first passage, answered the same questions with "yes" or "no", and justified their answers. Then they did the same thing for the second and the third passage at their own pace. There were ten subjects in the Example group and seven in the Abstract group.

Results and Discussion

The data supported our prediction that the Example group would perform as well as the Abstract group. The average percent correct for the Example group was 83% while that for the Abstract group was 81%. For constraints, the scores were 84% and 81% for the Example and Abstract groups, respectively. For variables, the scores were 80% and 81%, respectively.

Also, we examined the subjects' justifications for incorrect answers and found that most of their "errors" were not due to the subjects' failure to generalize in an explanation-based manner, but were due to the subjects' generating a schema slightly different from the one we had attempted to embody in the text. Some of the questions made certain underlying assumptions about the execution of the plan which could be relaxed to generate an even more general schema. Among those answers marked as incorrect, 87% of the Example group's justifications and 79% of the Abstract group's justifications presented arguments which were based on a causally consistent interpretation of the schema. For example, given the following question: "In the above plan, is it necessary that the number of meetings be the same as the number of people in the group?", one subject responded: "No, it's irrelevant. They could collect money every week and then at the end of the month the one person gets it all." An example of a causally *inconsistent* justification is when a subject was asked: "Is there any particular number of people required for this plan?" and responded: "Yes. Four is the only number of people that will make this plan work."

Experiment 4: Memory Test

Experiments 1-3 indicated that people could acquire a schema by generalizing the explanation of a single example; however, they did not indicate *when* generalization occurred. The subjects might have performed schema abstraction at the time they read the passage or only later when asked questions about it. For example, in Experiment 3, subjects in the Example group might have answered the questions by storing specific representations of the narratives in memory and then generalizing these representations after they were asked the questions. If this is the case, they should also be able to answer questions about *specific* facts in the narratives as well as questions about the *general* schemata.

Method

We tested subjects' memory for specific and general information one day after they read an experimental narrative. To test general information, we used the same questions from Experiment 3, including both questions on constraints and on variables. To test specific information, we developed new yes/no questions on each constraint and variable instantiated in the example. For example, to test the variable, "number of participants", we asked, "Were there five people in the group described?".

Subjects read only one of the three narratives and were asked to rate the quality and usefulness of the plan. They then left without knowing that there would be more tests related to what they had read. After one day, subjects returned expecting another experiment. Instead, each subject received questions on the narratives they had read previously and were asked to answer them with "yes" or "no" at their own pace. There were 17 subjects in total.

Results and Discussion

The percentage correct for general questions was 84% whereas that for specific questions was only 60%. These results indicate that subjects' responses to general questions were based on their general schemata not on specific representations of the narratives. This is because the hypothesis that the subjects were using a specific representation to answer the questions requires that the subject retrieve the specific information relevant for each abstract question. Also, since they were not aware that there would be a comprehension test one day later, these data provide evidence that the general schemata were a natural product of the comprehension process.

Conclusions

In general, the experiments described above support EBL as a viable psychological model of certain types of human learning. Specifically, they demonstrate that, like GENESIS, people can learn a schema by generalizing the explanation of a single narrative.

Previous research has generally assumed that multiple examples are required for schema or concept acquisition. For example, Rumelhart and Norman (1978) claimed that there are basically two ways in which schemata can be formed. These were *pattern generation* and *schema induction* both of which require exposure to multiple examples. Brewer and Nakamura (1984), postulating schemata as containing abstract *generic* knowledge, also assumed that schema acquisition required multiple examples.

Gick and Holyoak (1983) even specifically showed that subjects could *not* learn a particular schema (the "convergence schema") from a single example and that acquiring the schema required two analogous examples. However, the convergence schema is a very abstract concept which a standard EBL mechanism could not acquire from one example. Explanation-based generalization as performed by GENESIS involves retaining the basic structure of the plan used in a specific example and removing actions, properties, and relations which are clearly irrelevant to achieving the goal. Given a story like "The General" from Gick and Holyoak (1983), such a process would acquire a schema for capturing an enemy fortress by attacking it simultaneously from all sides; however, it could not acquire an extremely abstract concept like the convergence schema. Consequently, the results presented here do not contradict the specific results or conclusions presented in Gick and Holyoak (1983). However, our results do show that there is a large class of schemata that *can* be acquired from one example.

Although explanation-based learning can be used to learn a schema from only one example, it is unsuitable for learning certain classes of concepts. EBL is only applicable when one has sufficient knowledge of the domain and the schema to be learned is solely determined by causal constraints. In this case, people use their existing knowledge of the domain to guide the schema acquisition process and distinguish relevant from irrelevant features after given only one example. This improves the efficiency of learning and results in schemata which are free from spurious correlations. However, many schemata, such as that for a wedding ceremony or a birthday party, are determined by un-explainable social conventions as well as by necessary causal relationships among their constituent actions. Learning such schemata efficiently will require a mechanism which successfully integrates the different approaches underlying current learning mechanisms.

Nevertheless, EBL is one of the first attempts to incorporate significant amounts of causal and explanatory domain knowledge into a learning mechanism. Murphy and Medin (1985) argue persuasively for the importance of "theories" in concept formation, where important features of a "theory" include "An explanatory principle common to category members" and a "Network formed by causal and explanatory links." Investigating explanation-based learning from a single example represents an important step in understanding how "theories" can be successfully employed in concept acquisition.

Acknowledgements

The authors would like to thank Brian Ross for a number of helpful comments regarding this research.

References

Brewer, W. F. & Nakamura, G. V., (1984). The nature and functions of Schemas. In R. S. Weyer & T. K. Srull (Eds.), *Handbook of Social Cognition: Volume 1*, Lawrence Erlbaum and Associates, Hillsdale, NJ, 119-160.

Carbonell, J. G. (1983). Learning by analogy: Formulating and generalizing plans from past experience. In R. S. Michalski, J. G. Carbonell, & T. M. Mitchell (Eds.), *Machine Learning : An Artificial Intelligence Approach.* Tioga, Palo Alto, CA, 137-162.

DeJong, G. F. (1981). Generalizations based on explanations. *Proceedings of the Seventh International Joint Conference on Artificial Intelligence*, Vancouver, B.C., Canada, 67-70.

DeJong, G. F. & Mooney, R. J. (1986). Explanation-based learning: An alternative view. *Machine Learning 1*, 2.

Dyer, M. J. (1983). *In-Depth Understanding*, MIT Press, Cambridge, MA.

Falkenhainer, B., Forbus, K., & Gentner, D. (1986). The structure-mapping engine. *Proceedings of the National Conference on Artificial Intelligence*, Philadelphia, PA, 272-276.

Gick, M. L. & Holyoak, K. L. (1983). Schema induction and analogical transfer. *Cognitive Psychology, 15*, 1-38.

Johnson-Laird, P. N. (1983). *Mental Models*, Harvard University Press, Cambridge, MA.

Medin, D. L., Watenmaker, W. D., & Michalski, R. S. (1986). *Constraints and Preferences in Inductive Learning: An Experimental Study of Human and Machine Performance.* (Technical Report ISG 86-1 UIUCDCS-F-86-952). Department of Computer Science, University of Illinois, Urbana, IL.

Michalski, R. S. (1983). A theory and methodology of inductive learning. In R. S. Michalski, J. G. Carbonell, & T. M. Mitchell (Eds.), *Machine Learning: An Artificial Intelligence Approach.* Tioga Publishing Company, Palo Alto, CA, 83-134.

Mitchell, T. M. (1978). *Version Spaces: An Approach to Concept Learning.* (Technical Report STAN-CS-78-711). Stanford University, Palo Alto, CA.

Mitchell, T. M., Keller, R., & Kedar-Cabelli, S. (1986). Explanation-based generalization: A unifying view. *Machine Learning 1*, 1, January, 47-80.

Mooney, R. J. (1985). *Generalizing Explanations of Narratives into Schemata*, Masters Thesis, Department of Computer Science, University of Illinois, Urbana, IL, May.

Mooney, R. J. & Bennett, S. (1986). A domain independent explanation-based generalizer. *Proceedings of the National Conference on Artificial Intelligence.* Philadelphia, PA, 551-555.

Mooney, R. J. & DeJong, G. F. (1985). Learning schemata for natural language processing. *Proceedings of the Ninth International Joint Conference on Artificial Intelligence.* Los Angeles, CA, 681-687.

Murphy, G. L. & Medin, D. L. (1985). The role of theories in conceptual coherence. *Psychological Review, 92*, 3, July, 289-316.

Posner, M. J. & Keele, S. W. (1968). On the genesis of abstract ideas. *Journal of Experimental Psychology, 77*, 3, July, 353-363.

Quinlan, J. R. (1986). Induction of decision trees. *Machine Learning 1*, 1, 81-106.

Rumelhart, D. E. & Norman, D. A. (1978). Accretion, tuning and restructuring: Three modes of learning. In J. W. Cotton & R. L. Klatzky (Eds.), *Semantic Factors in Cognition.* Lawrence Erlbaum and Associates, Hillsdale, NJ.

Schank, R. C. & Riesbeck, C. (1981). *Inside Computer Understanding*, Lawrence Earlbaum and Associates, Hillsdale, NJ.

Spencer, R. M. & Weisberg, R. W. (1986). Context-dependent effects on analogical transger. *Memory & Cognition, 14*, 5, 442-449.

van Dijk, T. A. & Kintsch, W. (1983). *Strategies of Discourse Comprehension*, Academic Press, New York.

Wharton, D. (1967). Five common frauds, and how to avoid them. *Reader's Digest*, December.

Wilensky, R. W. (1983). *Planning and Understanding: A Computational Approach to Human Reading.* Addison-Wesley, Reading, M.

Winston, P. H. (1980). Learning and reasoning by analogy. *Communications of the Association for Computing Machinery, 23*, 12, 689-703.

Facilitation from Clustered Features: Using Correlations in Observational Learning[1]

Dorrit Billman

University of Pennsylvania

Evan Heit

Stanford University

Jennifer Dorfman

University of California, San Diego

In learning categories and rules from observation without external feedback, people must make use of the structure intrinsic to the instances observed. Success in learning complex categories from observation, as in language acquisition, suggests that learners must be equipped with procedures for efficiently using structure in input to guide learning. We propose one way that learners make use of the correlational structure available in input to facilitate observational learning: increase reliance on those features discovered to make good predictions about the values of other features. Such a mechanism predicts two types of facilitation from multiple correlations among input features. This contrasts with the effects of correlated features which have been suggested by models addressing learning with explicit feedback. Three experiments investigated learning the syntactic categories of artifical grammars, without external feedback, and tested for the predicted pattern of facilitation. Subjects did show the predicted facilitation. In addition, a simulation of the learning mechanism investigated the conditions when it would provide most benefit to learning. This research program begins investigation of procedures which might underlie efficient learning of complex, natural categories and rules from observation of examples.

Introduction

How do people learn about complex categories and rules from observation of examples? Adept performance in a variety of domains-- from linguistic to social-- suggests that people do abstract categories that reflect the organization in input. Little is known about learning from observation without benefit of explicit feedback or tutoring since research on learning, both in psychology and in artificial intelligence, has focused on explicit learning tasks with direct feedback provided. The present research explores observational learning without feedback and investigates how people use correlational structure available from examples to learn categories and rules.

Several researchers have suggested that category structure for psychologically coherent, natural catgories is rooted in the correlational structure provided by features in input (Rosch, 1978, Medin, 1983). Discovery of correlations among multiple features may be particularly important for observational learning. Maratsos and Chalkley argue that syntactic categories are acquired by extracting the system of interpredictive relations which hold among members of the same syntactic class (Maratsos & Chalkley, 1980).

Our work begins with these suggestions that correlational structure is important, and proposes how it might be used in learning without feedback. We use experimentation and simulation to test two predictions about the effects of correlational structure and to investigate learning procedures responsible for these effects.

Both predictions claim that availability of multiple, intercorrelated features in input will facilitate learning. We propose two levels of facilitation, one general, and one specific to

58

individual rules about feature covariation. The first prediction would be anticipated by many learning models: with multiple cues people should be more likely to learn at least one of the feature covariation rules, or at least something about the available categories. The odds of discovering at least one rule, or some basis for categorizing, are better when there are multiple regularities to be discovered (Trabasso & Bower, 1968).

The second prediction points to a stronger type of facilitation from multiple covarying cues and a more powerful method of capitalizing on correlational structure. This predicted pattern will be called clustered feature facilitation. It claims that subjects will be more likely to discover a target correlational rule when that rule occurs in a context which provides other rules among the same set of features. For example, applying the principle to syntactic categories would imply that learning the relations between *semantics* and *phrase structure rules* would be facilitated in a system which has, in addition, 1) covariation rules between *morphology* and *semantics* and 2) rules between *morphology* and *phrase structure rules*.

One mechanism which would produce this rule-by-rule facilitation is an attentional learning procedure, called focused sampling (Billman, 1983), which increases attention to predictive features. If a feature is predictive in one covariation rule it would become favored, or selected more often, in testing other rules. When multiple rules hold among a set of features they would provide a mutually reinforcing effect, increasing the prominence of the valuable, predictive set of features. A mechanism of this sort would result in mutual facilitation among correlated cues. This prediction contrasts with the pattern found for learning simple concepts with feedback (Trabasso & Bower, 1968): here, attending to and learning about one cue was independent of learning about a second, covarying cue.

Our research orientation and the prediction of clustered feature facilitation differ from much category learning research in two ways: we focus on learning without external feedback and we argue that successful learning here requires capitalizing on covariation among multiple cues. Specifically, we argue for a learning mechanism which uses covariation among multiple cues to facilitate rule and category learning, rather than one where learning about one cue is independent of the relations among others (Trabasso & Bower, 1968), or where learning about one cue only competes with learning about covarying cues (Zeaman & House, 1963, Lovejoy, 1966). The experiments test for facilitation from multiple covarying features.

Experiments

Rationale

Three experiments tested for general facilitation and for clustered feature facilitation. They investigated learning of syntactic categories in an artificial language. This allowed control of the regularities from which rules might be induced and maximized the chances of tapping a domain where people are naturally successful observational learners. Structures in the grammars were analogous to structures which occur in natural languages. All experiments investigated acquisition of rules distinguishing two relational classes. All experiments presented scenes with descriptive sentences which subjects observed to learn the language. All experiments compared two conditions, Structured and Isolating. All experiments compared learning of 15 college student subjects in each condition. Conditions differed in the correlational structure afforded by input. For example, Experiments 1 & 2 compared learning two relational classes (analogous to verbs and prepositions). Each class participated in a distinct set of phrase structure rules and had distinct semantics--referring to action or to relative position. In the Isolating Conditions (Experiments 1 & 2), semantics and phrase structure role were the only characteristics which correlated with one another or which could serve as a basis for learning the syntactic categories. In the Structured Conditions, additional features such as morphology, agreement rules, and syntactic marker words covaried with semantics and phrase structure rules. The two predictions were tested by comparing learning in Structured and Isolating Conditions. Some tests assessed general facilitation. Others assessed knowledge of just the target rules common to both conditions and tested clustered feature facilitation.

Table 1

Rules and Vocabulary for Structured & Isolating Conditions

Experiments 1 & 2

Examples for the Structured Condition, with English Glosses

Phrase Structure Rules with Examples of Sentence Types

1) $NP_{AG} + PP_{OB} + PP_{LC}$ (DOBOD) (BO SAFAT VULK) (BO LARAN NINK)
The DOBO-subj is beside the SAFA-obj, below the LARA-loc.

2) $NP_{AG} + PP_{OB}$ (DOBOD) (BO SAFAT VULK)
The DOBO-subj is beside the SAFA-obj.

3) $NP_{AG} + PP_{LC}$ (DOBOD) (BO LARAN NINK)
The DOBO-subj is below the LARA-loc.

4) $NP_{AG} + PP_{LOC} + VP$ (DOBOD) (BO LARAN NINK) (PIR SAFAT TOFO)
The DOBO-subj is below the LARA-loc and
exchanges places with the SAFA-obj.

5) $NP_{AG} + VP$ (DOBOD) (PIR SAFAT TOFO)
The DOBO-subj exchanges places with the SAFA-obj.

Relational Vocabulary & Characteristics

N-Verb	Structured	Isolating
Characteristics:	VP uses,	VP uses,
	actions,	actions.
	verb & marker agree,	No Other
	two syllables,	Consistent Pattern.
	PIR or TEW.	
Vocabulary:	*PIR/TEW ... TOFO/TAFA*	*BO ... GORK/GARK*
	PIR/TEW ... BOPO/BAPA	*PIR/TEW ... BOPO/BAPA*
	PIR/TEW ... JOSO/JASA	*PIR/TEW ... JUS*

N-Prep	Structured	Isolating
Characteristics	PP uses,	PP uses,
	positions,	positions.
	no agreement,	No Other
	one syllable,	Consistent Pattern.
	BO.	
Vocabulary:	*BO ... GIRK*	*PIR/TEW ... NINK*
	BO ... NINK	*BO ... TIF*
	BO ... VULK	*BO ... VALK/VOLK*

Note: / indicates two forms of a verb or marker.   ... indicates a missing shape word.

Experiments 1 & 2

Method

Stimuli

Sentences in "Neptunese" described scenes with triples of interacting objects. Three transitive actions were used, in which an agent shape moves to act on a remote object while a shape near the agent remains inactive. Sentences described the actions and relative positions among the shapes. Words referring to objects were inflected for case.

Both Structured and Isolating Conditions had two syntactic classes of relational words which had different roles in the phrase structure rules. In the Structured Condition multiple features covaried and provided a basis for grouping these words into contrasting categories. In the Isolating Condition only a subset of these features covaried, namely, semantics and distribution in phrase structure rules. In both conditions, semantics of referent and phrase structure rules covaried and distinguished two syntactic categories. Words referring to actions had one set of privileges of occurrence in phrase structure rules and will be called *N*-Verbs (for Neptunese verbs). Words referring to relative positions had contrasting (though overlapping) phrase structure rules and will be called *N*-Preps. Table 1 shows the f phrase structure rules and the five possible sentence types which can describe a scene.

Structured and Isolating Conditions differed with respect to regularities among other features, as summarized in Table 1. In the Structured Condition, additional features covaried with semantics and phrase structure distribution. All *N*-Verbs had two syllables and the vowels in the syllables changed to agree with the subject; e.g. "TOFO" was used when nouns of one (phonologically defined) gender class were used as subject and "TAFA" was used with the other gender. All *N*-Verbs used "PIR" or "TEW" as their syntactic marker at the beginning of the verb phrase, with selection between PIR and TEW again based on subject gender. In contrast, *N*-Preps were monosyllables ending in "K"; they had "I"'s and "U"'s in their stem rather than "O"'s or "A"'s; they did not change form with subject gender; and they had a single, fixed marker "BO".

In the Isolating Condition, each word had fixed phonology, agreement rules, and marker element; indeed, the values of these features here were identical to those used in the Structured Condition. However, there was no systematicity in the manner by which these features were assigned to words within a class; none of these features covaried with semantics or phrase structure distribution. *N*-Verbs had one or two syllables, fixed or changing forms, markers of "BO" or of "PIR"/"TEW"; similarly, various *N*-Preps had all possible values of these features. No pair of these features covaried with another or with semantics and phrase structure rules.

In the Structured Condition, semantics, phrase structure distribution, phonology, agreement, and marker word all covaried. In the Isolating Condition, only semantics and phrase structure distribution covaried.

Procedure

Subjects in either condition learned by watching scenes and reading out loud the accompanying descriptive sentences. The learning phase ran for two hours on each of two days and included some auxillary tasks (e.g. writing sentences). Following exposure to either the grammar of the structured or isolating condition, tests assessed what subjects had learned about the covariation patterns defining the classes. Three tests will be presented here. In all tests subjects were presented with a novel display and asked to judge whether it fit in with the learning sentence or whether it contained any error. Subjects rated sentences on a six-point scale in which one end point indicated the subject was certain the item was correct and fit in with the learning sentences and the other end point indicated certainty about an error.

The Phrase Structure Test assessed any knowledge for distinguishing between classes. Comparison between conditions tests the predicted general facilitation from multiple correlated cues. Subjects judged novel sentences using familar words but presented without any picture. They decided whether the new sentence fit into the language and was correct or whether it did not belong and contained an error. Errors were created by substituting a *N*-Verb into the context where a *N*-Prep was required or a *N*-Prep into a context for *N*-Verbs. Subjects could detect errors if they knew any of several rules, since the incorrectly used word might lead to inconsistent assignments among phonological properties, marker word, and phrase structure role. Subjects in the structured conditions might be able to detect errors based on disruption of any of several relations among predictive features, while subjects in the isolating condition would only be able to detect an error based on misplacement of the particular word in the phrase structure tree.

Two tests assessed clustered feature facilitation: the Projected Use Test and the Semantics Test. Experiment 1 used only the Projected Use Test; Experiment 2 used both. These tests evaluated knowledge of a single component of the grammar, one available to learners in both conditions.

The Projected Use Test assessed knowledge of phrase structure rules, rules which were identical in both conditions. The subject's task was to judge compatibility of a second, projected use of a novel word given an introductory use. Use of novel words means subjects cannot make judgements based on known uses of the word, but must rely on more abstract knowledge of relations among sentence forms. The task was explained by analogy to how we can make inferences about available uses of new words in English. In English, if one hears a new word, "glish", used in "She met a glish boy", one could be fairly certain that it would be grammatically okay to say "That girl is very glish." In the example, correct judgment requires use of implicit knowledge of the uses allowed for English adjectives; in the test, analogous judgments assessed knowledge of contrasting uses allowed for one of the two Neptunese categories. Subjects were told they would be given one correct, introductory use of a new word and then would be asked to judge whether a second use of the new word was also okay.

Testing clustered feature facilitiation means specifically testing knowledge of the relation among target features available to subjects in both conditions. Hence any other features which might be differentially informative for the two conditions must be eliminated. Semantic cues were eliminated by not displaying any picture. Phonological cues were removed by simply indicating the new word by the symbol *NEW*, which appeared in the sentence at the location of the target word, rather than spelling out the new word. Cues from the marker words were removed by indicating its presence by the symbol ////, without showing which marker was used. These conventions were again explained by analogy to inferences we make listening to speech; sometimes we can tell quite a bit about a new word even if we don't catch all the words in the sentence. English analogies, examples using the contrast between Neptunese relational terms versus nouns, and explaining the task back to the experimenter ensured that all subjects understood the task.

Removal of information about phonology, marker words, and agreement rules for the novel test word is critical to a fair test of clustered feature facilitation. The intent is to test whether subjects in the structured condition were more likely than those in the isolating conditions to learn an identical set of target rules (here, the phrase structure rules). While the target rules were identical in both conditions, additional cues covaried with the target rules in the Structured Condition. Clustered feature facilitation claims that the identical rule will be learned faster when it occurs as part of a related system of rules among the same features. A test of clustered feature facilitation requires that information available during learning differs between conditions, but information available at test is identical. Differential performance at test can then be attributed to differential learning about the same cues, not the availability of more cues to use in detecting errors.

The Semantics Test assessed learning the relation between semantics and the set of acceptable uses (phrase structure rules); action words are used in one set of rules, position words in another. This component of the grammar was the same in both Structured and Isolating Conditions. As in the Projected Use Test, the clustered feature prediction is tested by asking subjects to judge novel displays where correct judgments could only be made based on the target rule, common to both conditions. Subjects were told they would see scenes with some new aspect and a descriptive sentence which included a new word describing the new aspect. The new word was indicated by *NEW*. Subjects judged whether the word for the new aspect was used correctly. Subjects could judge correctly if they knew which uses were acceptable for words referring to action and which for words referring to position. Additional information about syllables, agreement, and markers was eliminated.

The Projected Use Test evaluates whether subjects are better able to learn the covariation between syntactic contexts shared by the same word (the phrase structure rules) when additional features covary with the features of the target rule. The Semantics Test evaluates whether subjects are better able to learn the covariation between contexts of use defined by phrase structure rules and semantics when additional features covary at learning. Both test the predicted clustered feature facilitation.

Results

Both experiments found a strong general advantage for learning in a grammar where multiple cues covaried to define syntactic categories. Subjects in the Structured Condition performed better than those in the Isolating Condition on the Phrase Structure Test, in both experiments. Both experiments also provided evidence for clustered feature facilitation. Subjects from the Structured Condition performed better on the Projected Use Test, in both experiments. The Semantics Test consisted of four types of items: correct action, correct position, incorrect action, and incorrect position displays. Structured Condition subjects did better than Isolating Condition subjects on two of the four measures, but did not differ on the summary measure. See Table 2.

Table 2

Scores for Experiments 1 & 2

	STRUCTURED CONDITION 1 mean(s.d.)	ISOLATION CONDITION 2 mean(s.d.)
Experiment 1		
PHRASE STRUCTURE (48 items)	4.66(.75) *	3.80(.41)
PROJECTED USE (48 items)	4.03(.74) *	3.63(.29)
Experiment 2		
PHRASE STRUCTURE (48 items)	4.59(.84) *	4.10(.50)
PROJECTED USE (48 items)	4.19(.76) *	3.83(.23)
SEMANTIC (48 items)	4.44(.29)	4.22(.21)
action/correct	5.44(.51) *	4.97(.61)
position/correct	4.17(.73)	4.57(.76)
action/incorrect	4.26(1.00)	4.14(.71)
position/incorrect	3.90(.86) *	3.19(.79)

* Structured Condition better than Isolating Condition 2 (p<.05)

Average scores on a 1 to 6 rating scale. Score of 6 means subject was certain every grammatical sentence was correct and every ungrammatical sentence was incorrect.

Experiment 3

Experiment 3 investigated the contrast between two subcategories of verbs. As in Experiments 1 & 2, learning a target rule was compared in Structured and Isolating Conditions. In Experiment 3 the relation between phrase structure rules and lexical form was preserved in both conditions, rather than the relation between the phrase structure rules and semantics, as in Experiments 1 & 2. Phrase structure forms are shown schematically here:

Class 1: $S \rightarrow NP_{AG} + NP_{OB} + (\text{Marker } NP_{LC} V_1)$

$\qquad S \rightarrow NP_{AG} + (\text{Marker } NP_{LC} V_1)$

$\qquad S \rightarrow NP_{AG} + NP_{OB} + V_1$

Class 2: $S \rightarrow NP_{AG} + NP_{LC} + (\text{Marker } NP_{OB} V_2)$

$\qquad S \rightarrow NP_{AG} + (\text{Marker } NP_{LC} V_1)$

Method

In the structured condition, Class 1 words had a distinct set of allowable uses or phrase structure rules; they referred to transitive actions; they had unchanging, single-syllable lexical forms which had "I"'s or "U"'s for vowels; and they used the marker "BO". Class 2 words had contrasting phrase structure rules; they referred to indirect causative actions; they had two syllables; they had agreement rules which changed the verb vowels to "O"'s or "A"'s to agree with the gender of the subject; and they used the markers "PIR" or "TEW", selected to agree with the subject's gender as well. In the Isolating Condition, only one of these characteristics covaried with phrase structure rules: agreement. One class had agreement rules which changed the form of the verb and the other class had fixed forms with no agreement rule. Number of syllables, marker word, and semantics were unsystematically assigned; they did not covary with each other, with uses defined by the phrase structure rules, or with fixed or changing phonological form.

Learning and test procedures were exactly analogous to those used in Experiments 1 and 2. The Phrase Structure Test assessed general facilitation in learning any rules relevant to the two syntactic categories (verb subcategories). The Projected Use Test assessed learning the dependencies among different uses (phrase structure rules) to see if learning a target rule was facilitated by the availability of additional covarying features in learning. The analog of the Semantics Test was the Lexical Test. The Lexical Test assessed subjects' knowledge of the covariation rules between lexical form (fixed or varying with agreement) of the word and allowable position in the phrase structure tree. In the Lexical Test, subjects were presented with sentences using a new word. Only the first part of the word was shown, enough to reveal whether or not it agreed with the subject but not enough to tell whether it was one or two syllables. Information about marker and semantics was also deleted. Subjects could be consistently correct only if they knew the relation between position in the sentence and whether or not the word should agree with the subject.

Results

Subjects in the Structured Condition did better than those in the Isolating Condition on all tests, the Phrase Structure Test, the Projected Use Test, and the Lexical Test. Results are summarized in Table 3.

Table 3

Scores for Experiment 3

	STRUCTURED CONDITION 1	ISOLATION CONDITION 2
	MEDIANS	MEDIANS
PHRASE STRUCTURE (48 items)	5.90 *	3.90
PROJECTED USE (48 items)	5.52	4.00
Class 1 uses	mean=5.04 **	mean=4.30
Class 2 uses	mean=5.06	mean=4.48
LEXICAL (24 items)	3.96 *	3.46

*Structured Condition median better than Isolating Condition 2 (p<.05)
  Medians used due to nonnormal distributions.

**Structured Condition mean better than Isolating Condition 2 (p<.05)

Average scores on a 1 to 6 rating scale. Score of 6 means subject was certain every grammatical sentence was correct and every ungrammatical sentence was incorrect.

Discussion of Findings

The results from the Phrase Structure tests found general benefit to learning about the system from the availability of multiple correlated features, consistent with several prior findings (Morgan & Newport, 1981, Green, 1979). The benefit was found for two pairs of syntactic classes and when semantics or morphological features covaried with distributionally defined class. In addition, the Projected Use, Semantics, and Lexical Tests suggest that when multiple features covary, learners do capitalize on the availability of one covariation pattern to discover others. Subjects did show clustered feature facilitation.

Possible Learning Mechanism

Focused sampling is part of the internal feedback approach (Billman, 1983) which specifically addresses learning without external feedback. The internal feedback approach predicts clustered feature facilitation by combining generation and testing of predictive rules with attentional learning. The learner projects hypotheses about the expected value of a second (set of) feature(s), given the value of a first. These hypotheses are evaluated by comparing the predicted with the observed value of the projected feature and using this match or mismatch as internally generated feedback. Hence the first level of learning consists of generating and testing conditional rules about predictive relations among features. Focused sampling, the attentional learning procedure, systematically alters the sampling of features used in hypothesis generation and testing. Whenever the prediction of a rule is confirmed, focused sampling increases the salience (or sampling probability) of the features which participate in that rule. Following bad predictions, salience of participant features is reduced. Focused sampling will benefit rule learning when successful participation in one rule is in fact predictive of that feature's participation in other rules. Where input does provide a system of covariation rules among an overlapping set of features, focused sampling will lead to learning the individual rules faster than when input provides only one of the individual rules by itself. Rule learning with focused sampling leads directly to the predicted clustered feature facilitation.

This account of attentional learning contrasts both with most featural and most instance-based views. Typically, models do not propose that the importance of or attention to a feature *increases* in the context of other covarying features. It may not change systematically at all (Medin & Schaffer, 1978, Trabasso & Bower, 1968), or covarying features may compete with each other (Zeaman & House, 1963, Fisher & Zeaman, 1973). Direct comparison to these models is not possible because the task changes when no feedback is available; extending these models to address learning without feedback might indeed prompt modifications in how the learner uses correlational structure. While the finding of clustered feature facilitation might be predicted by other types of models, it contrasts with a broad range of alternative types.

Simulation

Rationale

The exact effects of focused sampling are not transparent from an informal statement, so we turned to simulation. Simulation, unlike experiments with people, allows addition and deletion of a component at will; hence the contribution of that component can be directly assessed. The simulation compares two versions of a learning mechanism, with and without focused sampling. We adopted a design perspective and varied the learning problems to ask whether and when focused sampling would provide substantial facilitation (Billman & Heit, 1987).

Input for the simulation was abstract, schematically specified examples. Each example consisted of a set of features; each feature assumed one of four values. With focused sampling feature salience, or importance, changes with learning. Here each learning cycle consisted of these steps: 1) one predictor feature is sampled and, given its value, a prediction made about the value of a second feature; 2) the prediction is tested against the value specified in the input example; and 3) salience of the features as well as confidence in the prediction are both modified. Rule strength is increased or decreased depending on success. The saliences of features in the rule are also increased or decreased, changing their probablity of sampling; salience across features is normalized after each change, producing indirect, compensatory change too. The relation among features can be both supportive and competitive: supportive because of convergence on a common set of rules via focused sampling, competitive because limited attention is still allocated among the features. Without focused sampling the learning cycle does not include modifying feature salience, but is otherwise identical.

Findings

A series of experiments with varying learning stimuli were run to compare learning with and without focused sampling. Input sets varied in number of instances, number of features, and most importantly in relative number of correlated and uncorrelated features. Across a wide range of learning stimuli, focused sampling did produce benefit. Amount of benefit was assessed by comparing difference in learning with and without focused sampling at a criterion defined to match amount of exposure in the two conditions. The pattern of factors resulting in more or less benefit can be summarized in four points.

First, where comparable conditions were run varying in number of instances (4, 8, and 16), focused sampling benefit was greater for the larger number of instances.

Second, benefit was greater when input instances had more features, particularly for the first level of increase from very few (e.g. 4) to somewhat more (e.g. 6); effect of adding more features leveled off with larger numbers (e.g. 12).

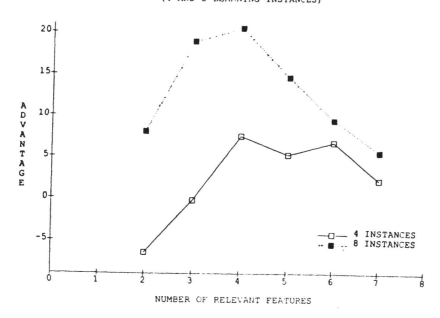

DIFFERENCES IN RULE STRENGTH
FOCUSED SAMPLING ADVANTAGE
8-FEATURE INSTANCES
(4 AND 8 LEARNING INSTANCES)

NUMBER OF RELEVANT FEATURES

Third, relative number of correlated and unpredictive features had major effects on benefit. The effects of adding more relevant or more irrelevant features interact with each other and with total number of features; only the major points are summarized here. For example, when the number of total features is fixed at 8, benefit increases as correlated features increase from 2 to 4, but benefit decreases as still more of the total are made correlated, as shown in Figure 1. Benefit increases as more features of a fixed total are made relevant. In addition, benefit increases as new relevant features are added. However, this pattern holds only for the initial increase (e.g. maximally for the increase from 2 to 3 relevant features) and attenuates or decreases with increasing additions. Benefit also increases as more *irrelevant* features are added, when irrelevant features initially are few in number. In sum, benefit is greatest when there are multiple correlated features but there are still a significant proportion of irrelevant features (from which the relevant features can steal attention). Thus, this simulation produced clustered feature facilitation over a wide range of conditions, but only where a significant number of "unsystematic" features were also present.

Finally, we found one set of conditions where focused sampling *hurt* learning the relation between the two correlated features, the target rule. This occurred when only four instances were in the learning set, and only 2 of 8 features were correlated (11111111 11222222 22333333 22444444). Here instance level structure among the *unsystematic* 6 features proved more predictive than the target rule; knowing the value of feature 3 allows perfectly reliable prediction of 5 other features, even though any such rule only applies to only one instance. The target features, overall, are less predictive and predictive in fewer rules. Thus, focused sampling is useful only when there is signficant structure available above the instance level.

These results might be glossed by the claim that focused sampling provided greater benefit for more complex (more instances, more features) and more "natural" (multiple correlated cues in a context of many uncorrelated ones, regularities beyond the instance) learning problems.

Summary

This research presents evidence and arguments for how learners use correlated structure to guide complex learning without feedback. Three experiments confirmed the predicted rule by rule facilitation from a more structured context (clustered feature facilitation). A mechanism, focused sampling, was outlined for using correlational structure to guide rule selection by altering the importance of features. The results of simulating the focused sampling procedure 1) produced clustered feature facilitation as predicted informally and as demonstrated by subjects and 2) suggested that focused sampling may be a particularly beneficial strategy for complex natural learning without feedback.

References

Billman, D.O. (1983). *Procedures for learning syntactic categories: A model and test with artificial grammars*. Doctoral dissertation, University of Michigan. Ann Arbor, Michigan: University Microfilm.

Billman, D.O. & Heit, E. (1987). Observational Learning From Internal Feedback: A simulation of an adaptive learning method. in review.

Fisher, M.A., & Zeaman, D. (1973). An attention-retention theory of retardate discrimination learning. In N.R. Ellis (Ed.), *International Review of Research in Mental Retardation*. New York: Academic Press.

Green, T.R.G. (1979). The necessity of syntax markers: Two experiments with artificial languages. *Journal of Verbal Learning and Verbal Behavior, 18*, 481-496.

Lovejoy, E. (1966). Analysis of the overlearning reversal effect. *Psychological Review, 73*, 87-103.

Maratsos, M.P. & Chalkley, M.A. (1980). The internal language of children's syntax: The ontogenesis and representation of syntactic categories. In K.E.Nelson (Ed.), *Children's Language*. New York: Gardner Press Inc.

Medin, D.L. (1983). Structural principles in categorization. In T.T.Tighe & B.E.Shepp (Eds.), *Perception, Cognition, and Development*. Hillsdale, N.J.: Erlbaum Publishers.

Medin, D.L. & Schaffer, M.M. (1978). A context theory of classification learning. *Psychological Review, 85*, 207-238.

Morgan, J.L. & Newport, E.L. (1981). The role of constituent structure in the induction of an artificial language. *Journal of Verbal Learning and Verbal Behavior, 20*, 67-85.

Rosch, E.H. (1978). Principles of categorization. In E.H. Rosch & B.B.Lloyd (Eds.), *Cognition and Categorization*. Hillsdale, N.J.: Erlbaum Publishers.

Trabasso, T., & Bower, G.H. (1968). *Attention in Learning*. New York: Wiley.

Zeaman, D. & House, B.J. (1963). The role of attention in retardate discrimination learning. In N.R. Ellis (Ed.), *Handbook of Mental Deficiency*. New York: McGraw-Hill.

Notes

[1]The research was supported by BRSG RR-07083-19, a University of Pennsylvania Research Foundation Grant, and NIMH grant R23HD20522-01A1. We would like to thank Anne D'Ulisse and Valerie Sessa for their help in running the experiments.

Address for correspondence: Department of Psychology, University of Pennsylvania, 3815 Walnut Street, Philadelphia, PA 19104.

69

A Connectionist Context-Free Parser Which is not Context-Free, But Then It is not Really Connectionist Either[1]

Eugene Charniak

Eugene Santos

Department of Computer Science
Brown University
Providence, RI 02912

ABSTRACT

We present a distributed connectionist architecture for parsing context free grammars. It improves earlier attempts in that it is not limited to parse trees of fixed width and height (i.e. fixed length sentences). The memory limitations inherent in connectionist architectures comes out in an inability to parse center-embedded sentences.

Key Words: connectionist parsing, distributed connectionism, context-free grammars.

1. Introduction

This paper describes a context-free parser designed in a "connectionist" architecture.

Connectionism has attracted considerable attention of late because it offers a new way of looking at old problems in Artificial Intelligence and Psychology - a way which also has considerable neurophysiological plausibility. Unfortunately, connectionism, or at least the "distributed" connectionism we will assume in this paper, has been hard to apply to high-level cognitive tasks. Connectionist parsing has been of interest because of parsing's intermediate role between higher and lower cognitive abilities.

Indeed, there have been several previous attempts at a connectionist context-free parser [1,2,3]. These, like the present parser, have a major failing — they are strictly speaking, only capable of parsing regular grammars. This is inevitable. Connectionism assumes a large, but bounded, number of units. Thus any connectionist scheme must be finite state.

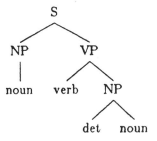

Figure 1. Parse tree on a white-board

---

[1]This research was supported in part by the Office of Naval Research under contract N00014-79-C-0592, the National Science Foundation under contracts IST-8416034 and IST-8515005, and by the Defense Advanced Research Projects Agency under ARPA Order No.4786. Thanks to James McClelland for some encouragement.

Nevertheless one can ask if such parsers could approach context-free in the limit (as the number of units is increased) and since the answer for all is "yes" it seems reasonable to call them all (including ours) "context-free."

More interesting though is to ask how the memory limitations show up in the parsing process. The failings of previous parsers have been distinctly "un-human." They are limited to fixed length sentences. Ours is more promising in this regard. While it does fine for right branching structures, it is limited in its ability to parse center embedded constructions, a point we will return to later.

Ours does, however, have a distinct failing of its own — it is not a "true" connectionist architecture. But let us say what it *is* first.

2. Representing a Parse Tree

The easiest way to visualize the parser, and how it represents a parse tree, is to imagine a parse tree drawn on a white board, as shown in Figure 1. Naturally since we are talking about a distributed connectionist scheme, we will break the white board up into many distinct units, each one of which may have one of a small number of distinct values - S, NP, VP, PP, Noun, etc. This is shown in Figure 2.

The scheme shown in Figure 2 has several representational inadequacies. For one, the representation does not indicate which constituents dominates which and as such does not distinguish the two PP attachments in Figure 3. Furthermore the same parse tree can have many distinct representations. Figure 4 denotes the same tree as Figure 2, but is quite different.

Thus we indicate a parse tree by including the entire path, from leaf to root, in every column as shown in Figure 5. This means that a single constituent, for example, the S in Figure 5, will be represented as several units. Nothing indicates that these all denote the same S constituent. There are

	S		
NP	VP		
		NP	
noun	verb	det	noun

Figure 2. Segmented parse-tree array

		S			
NP			VP		
			NP		
				PP	
					NP
noun	verb	det	noun	prep	noun

Figure 3. Ambiguous parse array

S			
NP		VP	
			NP
noun	verb	det	noun

Figure 4. Another representation of Figure 2

		S	S
S	S	VP	VP
NP	VP	NP	NP
noun	verb	det	noun

Figure 5. Tree-path parse array

different ways this could be handled. The one we have chosen is to introduce "binding" units. Each unit denoting a non-terminal (or the lack thereof, represented internally by the symbol e) has a sister unit which states which, if any, of the units in the column to the right is the same constituent. In our figures we will use lines to indicate such bindings. See Figure 6. Internally these are values from 0 to the height of the array plus a special value, b, for a "boundary" indicating that we have a right boundary of a constituent and thus there is no corresponding constituent to the right.

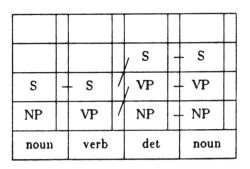

Figure 6. Actual representation of a parse

As has been implicit in our diagrams so far, the actual input to the parser is not words, but rather their parts of speech, since that is all that is relevant to our task. Thus parts of speech are the terminal symbols in the grammar. We have distinguished the bottom row of the parser, where the parts of speech appear, from the other rows, since all of the others have non-terminals only. Also the terminal row does not have binding units since we do not admit discontinuous terminal constituents. In what follows we will pretty much ignore the terminal units. So, for example, most rules which apply to non-terminals have special cases for terminals. These are ignored.

3. How the Parser Works

The basic idea is that words (really parts of speech) are read into the parser on the lower right and then shifted to the left as each new word comes in. Thus the columns are numbered from 0 to N starting from the right. See Figure 7. Initially all NT units are set to e, and all B units are set to b. At the first word the lower right unit, TR(0), then has its value "clamped" to, say, noun, and all of the NT(i,j) units, recompute their values synchronously (that is, all based upon the values in the previous iteration). Then the B(i,j)'s compute their values (using the new NT values but the old B values). This repeated is for a total of five times before the next word is read in. (Five iterations was chosen arbitrarily. In fact, it seems likely that a much lower number, like two or three, would work as well. It is one of many things we have not yet had time to test.) Figure 7 shows the starting configuration and that after three iterations. Since a unit will typically have some probability for several different values, each unit in Figure 7 has four values for it. If all four are the same it means that value has probability > .75. Three the same indicate a probability > .5, etc. This is true for lines as well, but we only indicate the most probable to reduce the jumble. In point of fact, however, all the lines in all of the examples go the same place. At this point the entire network is shifted to the left, leaving NT(0,j) = e and TR(0) is clamped to the next part of speech. Figure 8 shows a noun and verb following on the heals of the input in Figure 7. This processing, and all of the other examples, have been done using the following grammar:

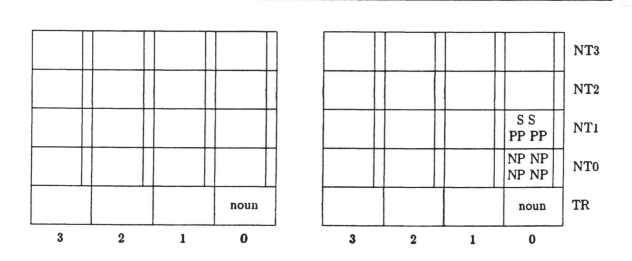

Figure 7. Start of sentence 'noun verb noun'

73

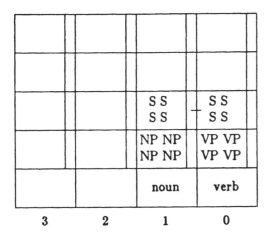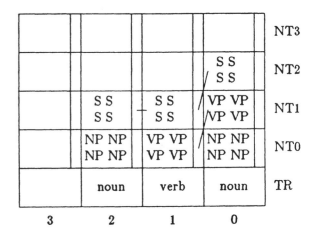

Figure 8. End of sentence 'noun verb noun'

$$S \rightarrow NP\ VP$$
$$VP \rightarrow verb\ NP\ PP*$$
$$PP \rightarrow prep\ NP$$
$$NP \rightarrow (det)\ noun\ PP*$$
$$NP \rightarrow pronoun$$
$$NP \rightarrow propnoun.$$

Shifting words left (and eventually out the left-hand side) allows the parser to handle sentences of unbounded length, at least in principle. Since such shifting is not within the traditional connectionist repertoire, it makes our architecture slightly suspect to a "traditionalist." It also means that sentences longer than the parser's width will not be completely represented at any one time. In such cases we take the total parse tree to be the "obvious" combination of the trees created during the parse.

4. Rules

Next we describe the actual rules applied at each unit. These will take us still further from connectionist orthodoxy because we allow each unit calculation to be much more complex than the typical summing plus threshold. In particular a unit decides what value to adopt by summing the influence of several sub-rules which apply to it. The sub-rules fall into two broad types, "housekeeping" rules, and grammar rules.

4.1. Housekeeping Rules

Housekeeping rules enforce the basic expectations of how the system is to work and as such remain constant over all grammars. There are three such rules.

Binding consistency. If a binding unit $B(i,j)$ has value k, then $NT(i,j)$ should denote the same constituent as $NT(i-1,k)$. More formally, the system tries to preserve the following constraint.

74

$$\{NT(i,j) = NT(i\text{-}l,k)\} + \{B(i,j) = k\}$$

This rule is applied in all ways. That is, $NT(i,j)$ tries to make itself equal to $NT(i\text{-}l,k)$, and vice versa, while $B(i,j)$ tries to pick a value which will make its left and right NT's the same.

Tree-consistency. If $B(i,j) = k$ then $B(i,j+1)$ wants to be $k+1$, and vice versa.

$$\{B(i,j) = k\} + \{B(i,j+l) = k+1\}$$

Again this is used both ways, with one exception. When a higher unit, $B(i,j+1)$, uses it to determine the value of a lower unit, $B(i,j)$, if $B(i,j+1) = k+1$, then $B(i,j) = k$ or b.

Blank-edge. If there is no constituent at a position $NT(i,j)$ then $NT(i,j) = e$, for empty. In such cases no value of $B(i,j)$ would have any real meaning. It turns out that the machine works better if such $B(i,j)$'s are assigned to be boundaries, so this rule encourages

$$\{NT(i,j) = e\} + \{B(i,j) = b\}$$

4.2. Grammar Rules

The program does not learn a grammar from examples, but must be given the grammar, which is then compiled into four rules, each of which form one aspect of the function computed it each NT and B units (though some only apply to one or the other).

Up-Down. Given a rule like $S \rightarrow NP\ VP$, the presence of an NP at position i,j should encourage the presence of an S at i, j+1, and conversely. More generally assume a rule of the form $A \rightarrow \ldots B$ ...then the following combination is encouraged:

$$\{NT(i,j) = A\} + \{NT(i,j\text{-}l) = B\}$$

Left-hand Side Start Rule. Given the rule $S \rightarrow NP\ VP$ the presence of a starting NP should encourage a starting S.

$$\{NT(i,j) = A\} + \{NT(i,j\text{-}l) = B\} + \\ forall(k)[\{B(i\text{-}l,k) \neq j\} + \{B(i\text{-}l,k\text{-}l) \neq j\text{-}l\}]$$

This rule is only applied to the B units, since it is a more restricted version of the up-down rule when applied to NT's.

Left-hand side finish. Given the rule $S \rightarrow NP\ VP$ the presence of a VP which is ending should encourage the S above it to end.

$$\{NT(i,j) = A\} + \{NT(i,j\text{-}l) = B\} + \{B(i,j) = b\} + \{B(i,j\text{-}l) = b\}$$

Right-hand side start finish. The rule $S \rightarrow NP\ VP$ encourages a finishing NP with an S above it to be followed by a starting VP with an S above it. Assuming the rule $A \rightarrow \ldots B_1\ B_2 \ldots$

$$\{NT(i,j) = A\} + \{NT(i,j\text{-}1) = B_1\} + \{B(i,j\text{-}1) = b\} + \{NT(i\text{-}1,j\text{-}1) = B_2\}$$

5. Limitations and Psychological Relevance

As already noted, the parser escapes the fixed sentence length limitation because it shifts in input into the registers, constructs a parse tree for the section in question, and eventually shifts the input out on the left.

However, parse trees are bounded not only in width, but in height. Thus long sentences will tend to have high trees, and so it is possible that some of the tree will shift up "over the top" and be lost. If we are willing to agree that the system's parse tree is the collection of the partial trees created before they shifted either left and out, or up and out, then it is nevertheless possible for the program to construct arbitrarily wide and high trees. Figure 9 shows the system parsing a modest right embedded sentence which due to the very small parser width and height still overflows the buffer in both directions. (The height and width are parameters one sets.)

75

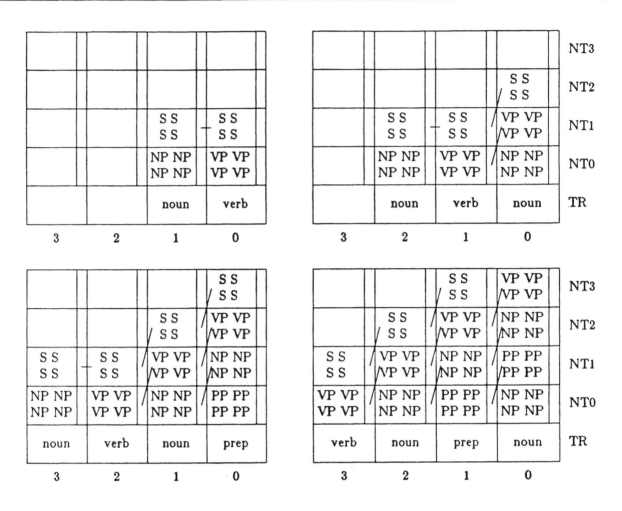

Figure 9. Overflow on 'noun verb noun prep noun'

However, the shifting has effect on what can be parsed correctly. In particular, the depth limitation means that right embedding is fine, but center embedding will be difficult. Although the front of the tree is continually being shifted left and out, and the top of the tree is going over the top, if the sentence is right embedded then the portion of the tree to which the input should attach is always present. For center embedding this will not be true, and thus the parser cannot parse such examples correctly.

6. How the Parser Really Works

We have described how the parser works in terms of the contributions of rules which we expressed as constraints to be satisfied. We have not specified exactly how the constraint satisfaction works. The actual rules are formalized in terms of array multiplication. We will consider only one of the simplest rules here, the up-down rule.

First we need to modify our notation slightly. So far we have denoted unit values with equations like $NT(1,2) = S$. However, as we have already noted, $NT(1,2)$ may be, say $.7 = S$, $.2 = NP$, and $.1 = PP$, or some such. Therefore, it would make more sense to say $NT(1,2) = (.7,.2,.1,0)$ where we use $S = 0$, $NP = 1$, $PP = 2$, and $VP = 3$ to indicated vector positions. Alternatively we could use a 3-D array and say $NT(1,2,1) = .2$. (Arrays start with position 0.) In fact, it proves to be most convenient to represent the situation as $NT_1(2,0) = .7$ $NT_1(2,1) = .2$ etc. Here the column number becomes a subscript picking out different two dimensional arrays. Next, we can represent the grammatical information needed by the up-down rule as an array itself.

$$UD(i,j) = \begin{cases} 1 \text{ iff the ith and jth nonterminals are} \\ \text{found in a rule of the form } i \rightarrow ...j... \\ 0 \text{ otherwise} \end{cases}$$

Now consider how we apply the UD rule downward. We look at each position, and consider to what degree the entity at position j suggest what should be at j-1. We can consider the NT_i array then to be this: $NT_i(k,LHS)$. We then perform the array multiplication

$$NT_i(K,LHS) \text{ X } UD(LHS,RHS) = NT_i(K,RHS)$$

The resulting array has at position K the values that the LHS at K induces on the RHS. To get this into the format we want, we then shift down K by 1, so that the values of the RHS for K-1 appear in the K-1th position. Thus the up down rule looks like this:

$$NT_i(K-1,RHS) = \text{shift-K-down-1}(NT_i(K,LHS) \text{ X } UD(LHS,RHS))$$

The other rules are similar, but often more complicated.

7. Problems and Future Research

There are many ways in which the parser could use improvement. It currently predicts that center embedding is more difficult than right or left embedding, but does not explain why some forms of center embedding (of S's) are worse than others (of PP's). If we used a grammar which represented NP's as deep trees, rather than flat ones, this issue would be even more critical.

We have only tried the parser on the simple context-free grammar given earlier, plus a few minor extensions. We have no knowledge of how it works on more complicated grammars. In particular we would like to handle extended phrase-structure grammars. For this we would expand our NT units to be many units, each for a particular feature in the extended phrase structure approach. It looks like a natural, but whether it will work is an open question.

The parser does not learn its grammars from examples. Given how far our architecture is from traditional distributed connectionism it seems unlikely that any of the standard learning algorithms will apply. Perhaps those algorithms be extended or alternatively our architecture could be made more conventional.

8. Conclusion

We have presented a distributed connectionist architecture for parsing context free grammars. It improves earlier attempts in that it is not limited to parse trees of fixed width and height (i.e. fixed length sentences). The memory limitations inherent in connectionist architectures comes out in an inability to parse center-embedded sentences, although right-embeddeding works out fine.

9. References

1. Fanty, M., "Context-free parsing in connectionist networks," TR 174, Univeristy of Rochester Computer Science Department (1985).
2. Selman, Bart, "Rule-based processing in a connectionist system for natural language understanding," Technical Report CSRI-168, Computer Systems Research Institute, University of Toronto (1985).
3. Waltz, David L. and Pollack, Jordan B., "Massively parallel parsing: a strongly interactive model of natural language interpretation," *Cognitive Science* 9 pp. 51-74 (1985).

A Principle-Based Approach To Parsing for Machine Translation

Bonnie J. Dorr

M.I.T. Artificial Intelligence Laboratory

545 Technology Square, Room 810

Cambridge, MA 02139, USA

(617) 253-7836

BONNIE@MIT-PREP.AI.MIT.EDU

Session: Paper

Keywords: Natural Language, Parsing, Principles vs. Rules, Interlingual Translation

Abstract

Many parsing strategies for machine translation systems are based entirely on context-free grammars; to try to capture all natural language phenomena, these systems require an overwhelming number of rules; thus, a translation system either has limited linguistic coverage, or poor performance (due to formidable grammar size). This paper shows how a principle-based "co-routine design" implementation improves the parsing problem for translation. The parser consists of a skeletal structure-building mechanism that operates in conjunction with a linguistically based constraint module, passing control back and forth until underspecified skeletal phrase structure is converted into a fully instantiated parse tree. The modularity of the parsing design accommodates linguistic generalization, reduces the grammar size, enables extendibility, and is compatible with studies of human language processing.[1]

1 Introduction

The problem addressed in this paper is to construct a parsing model that accommodates cross-linguistic uniform machine translation without relying on language-specific context-free rules. Typically parsing systems use grammars that describe language via complicated rules that spell out the details of their application. For example, ATN-based systems (Woods, 1970; Bates, 1978) have several hundred grammar arcs, each with detailed tests and actions; augmented phrase-structure grammars as in Diagram (Robinson, 1982) spell out the type, position, and probability of occurrence of constituents in a given phrase; and the GPSG approach (Gazdar, *et. al.*, 1985) uses a "slash-category" mechanism to incorporate long distance relations directly into the grammar rules.[2] Such systems do not work in the context of translation across several languages: the rules of a given grammar are painstakingly tailored to describe a *single* language, thus forcing a loss of linguistic generalization, limiting the addition of new languages, and inducing inefficiency (due to formidable grammar size).[3]

---

[1] Frazier 1986 provides recent psycholinguistic evidence that there is a temporal sequence of parsing consistent with the GB-based model presented here. This will be mentioned briefly in section 1, but is not the central focus of this paper.

[2] Barton (1984) describes these rule-based systems in more detail.

[3] For example, Slocum's METAL system (1984, 1985) developed at the Linguistics Research Center at the University of Texas relies on numerous language-specific context-free rules per language solely for parsing. The type of grammar formalism is allowed to vary from language to language. For example, the German

Surface Sentence

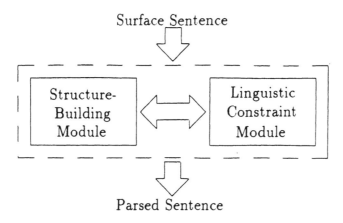

Parsed Sentence

Figure 1: Co-Routine Design of the Parser

Furthermore, these systems fail to preserve the modular organization of new theories of grammar.

In this paper I describe an implementation of a parsing model that is based on subsystems of grammatical principles and parameters.[4] The parser follows a "co-routine design:" the structure-building mechanism operates with simultaneous access to linguistic constraints of Government and Binding (GB) theory as developed by Chomsky (1981, 1982). (See figure 1.) The structure-building module assigns a skeletal syntactic structure to a sentence, and then this structure is eliminated or modified according to the principles of GB. This design is consistent with recent psycholinguistic studies (see Frazier, 1986) that indicate that the human processor initially assigns a (potentially ambiguous or underspecified) structural analysis to a sentence, leaving lexical and semantic descriptions for subsequent processing. Furthermore, the parser is designed so that it applies uniformly across all languages, allowing the user to modify the parameters of the system to accommodate additional additional languages.

The reason that parsing uniformly across languages is difficult is the parser appears to require a massive amount of "knowledge" in order to parse all possible phenomena (and their interaction effects) in any given language without allowing ill-formed sentences to also be parsed. Consider (1):

(1) Le quiere a Juan
     '(She) loves John'

Although (1) appears to be simple, it is not simple from the point of view of uniform parsing since the equivalent sentence parses differently in other languages. The Spanish and

---

parser is based on phrase-structure grammar, augmented by procedures for transformations; by contrast, the English parser employs a modified GPSG approach with no transformations. Regardless of the type of grammar formalism, each parser is nevertheless based on hundreds context-free rules of a language-specific nature. Consequently, each parser operates unilingually and has an increased running time over parsers that access smaller grammars. (As noted in Barton (1984), the Earley algorithm (1970) for context-free language parsing can quadruple its running time when the grammar size is doubled.)

[4]For example, there is a "constituent order" parameter associated with a universal principle that requires there to be a language-dependent ordering of constituents with respect to a phrase; the parameter is set by the user to be *head-initial* for a language like English, but *head-final* for a language like Japanese. This is discussed in section 2.1.

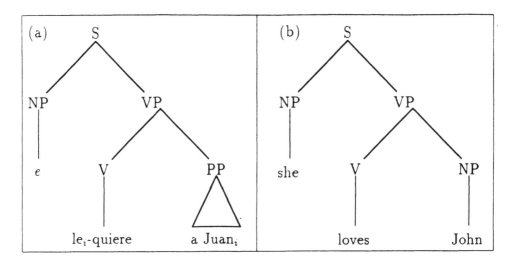

Figure 2: Spanish and English Parse Trees for an Equivalent Sentence

English parse trees for (1) are in figure 2.[5] Literally, the English translation for (1) is (2), which is ungrammatical:

(2)   him *e* loves to John

The *e* stands for a null subject that is realized as *she* in English.[6] The parsing implementation presented here rules out sentence (2) without sacrificing the ability to parse (1).

The co-routine design differs from other GB parsing/translation systems (*e.g.*, Sharp, 1985) in that the GB principles are used for "on line" verification during parsing rather than as well-formedness conditions on output. Furthermore, in Sharp's system, context-free rules (set up for English-like languages) are hardwired into the code rather than generated on the fly using principles of GB; thus, languages (like German or Japanese) that do not have the same order of constituents as English cannot be handled by the system. The primary factor that introduces this malady in Sharp's system is that the user has limited access to the principles of the system. The system described here allows the user to specify parameter values to the principles, thus modifying the effect of the principles from language to language. There are two classes of GB principles used by the system: those that are applied on line (*i.e.*, at processing time) and those that are applied off line (*i.e.*, at precompilation time).[7]

---

[5]Subscripts are used for co-referring elements. Thus *le* (= him) refers to *Juan*.

[6]Section 2.3 discusses the null subject phenomenon in Spanish.

[7]Experiments are currently underway to determine the "optimal" balance of principle clustering between the precompilation and processing phases. In order for the GB constraints to be applicable, a structure must first be created. The question under investigation is how much structure must be generated at precompilation time in order to perform on line verification of GB constraints efficiently. On the one hand, incorporating a large number of constraints into the precompilation phase causes the grammar size to become explosive, thus slowing down grammar search time; on the other hand, eliminating a large number of constraints from precompilation forces a high cost at constraint verification time. Frazier (1986) suggests that all phrase structure possibilities get multiplied out, leaving only a small subset of GB constraints to apply at processing time. In the parser presented here, a relatively small number of GB constraints (those concerning skeletal phrase structures and empty noun phrases) are accessed at precompilation time, leaving many of the GB constraints to apply at processing time. Time tests have shown this clustering of principles the most promising for efficient parsing using the co-routine design.

Both classes include parameters of variation.

The modularity imposed by the GB framework is an improvement over context-free based systems for several reasons: (a) properties common to all languages are not specified directly in rules, but are abstracted into modularized principles, thus allowing linguistic generalization to be captured; (b) multiplicative effects of linguistic constraints are not spelled out in the form of grammar rules, thus reducing grammar size (hence processing time); and (c) a separate description is not required for each language, thus the parser is easily extendible to additional languages.

2 Underlying Linguistic Theory

In order to arrive at the modules that form the basis of the structure-building and GB components of figure 1, we must separate underlying subsystems of grammar that interact to gain the effects of complicated rules systems. This section describes parameters of variation associated with the principles of three GB subtheories ($\overline{X}$-Theory, θ-Theory and Trace Theory), and discusses the relevance of these parameters within the context of the parsing model. The goal is to incorporate the parameterized principles of GB (in the form of modular subsystems of structural and well-formedness constraints) into a single, cross-linguistically uniform parsing system.

2.1 $\overline{X}$-Theory Parameters: Choice of Specifiers and Constituent Order

The central idea of $\overline{X}$-Theory is that the dictionary (henceforth *lexicon*) specifies subcategorization frames for lexical items (*e.g.*, the frame for the verb *put* includes two arguments, one that is a noun phrase, and another that is a prepositional phrase, as in *put the car in the garage*), and phrase-structures are projections of a lexical head X (= N, V, P or A).[8] $\overline{X}$-Theory assumes that phrase structures for English are derived by rules of the form:

(3) $X^{max} \Rightarrow$ (Specifier) X (Complement)

where X^{max} is the maximal projection (more commonly called XP) of the lexical head X. The Specifier of X is determined by a parameter setting associated with the $\overline{X}$ module, and the complement of X is determined by the subcategorization frame of the verb. For example, if X is a noun, X^{max} is NP, a possible Specifier is a determiner, and a possible complement is a prepositional phrase (depending on whether this is specified in the lexical entry for the noun).

English requires that specifiers of all lexical categories occur before the lexical head, and complements follow the lexical head. However, this rule does not apply to all languages (*e.g.*, Navajo, German, Japanese, *etc.*). For example, consider the following Navajo sentence:

---

[8]The lexical representation used in the parser presented here is based on the input representation required by the morphological analyzer. It includes the root forms of words and pointers to applicable affixes. Root verbs are stored with their argument structure specifications and θ-role assignment possibilities. The lexicon is discussed in Dorr (forthcoming), but will not be emphasized in this paper.

(4)   ashkii at'ééd yiyiiłtsą
      'the boy saw the girl'

This sentence literally translates as *the boy the girl saw* since Navajo requires the complement to precede the head.[9] It is assumed that the constituent order of a language is determined by a parameter of variation. Thus, before parsing begins, $\overline{X}$ rules are set up according to the constituent order of the language being parsed. This is crucial in the parsing model since many of the principles of other GB subtheories cannot apply until a valid licensed structure (with predetermined ordering restrictions) has first been built, *i.e.*, $\overline{X}$-Theory provides basic templates to which remaining parsing constraints can apply.

2.2 θ-Theory Parameters: Clitic Doubling

θ-Theory is the theory of thematic (or semantic) roles. A principle of this theory is the θ-Criterion which states that each noun phrase argument of a verb is uniquely assigned a semantic role (*e.g.*, agent, patient, *etc.*) and each semantic role is uniquely assigned to an argument. In order for a semantic role (henceforth θ-role) to be assigned, there is a principle of θ-role transmission that maps arguments in the dictionary entry of the verb to their corresponding θ-roles.

In Spanish, the phenomenon of clitic doubling is relevant to parametric variation of the θ-role transmission principle. A clitic is a pronominal constituent that is associated with a verbal object. For example, the clitic *le* in the following sentence is a clitic associated with *Juan*, the object of the verb *regalé*:

(5)   Le regalé un libro a Juan.
      'I gave a book to John.'

The phenomenon of clitic doubling is defined in terms of the pair $<$*clitic, lexical NP*$>$ where the clitic must agree in number, person and gender with the lexical NP. In (5) the clitic *le* actually stands for an NP that does not yet have a θ-role (namely, *Juan*). Thus, in order to satisfy the θ-Criterion, a parameter of variation is required for θ-role transmission. Jaeggli (1981) proposes that clitics supply θ-roles to object NPs that are doubled via a θ-role transmission rule:

(6) $[\text{CL} +case_i +\theta_j] ... [\text{NP} +case_i] \Rightarrow [\text{CL} +case_i +\theta_j] ... [\text{NP} +case_i +\theta_j]$

This rule allows a doubled NP object to receive θ-role as long as the clitic and NP must have the same case.[10] If a clitic is not present, a θ-role is assigned in the usual fashion, (*i.e.*, from the verb that contains the argument in its dictionary entry). Thus, for languages that allow clitics, clitic doubling must be available as a parameter of variation to the θ-role transmission principle of θ-Theory. The θ-Criterion can then be used as a well-formedness condition during parsing so that clitic doubling constructions will be ruled out unless (6) is allowed to fire. This is important in a parsing model since languages that allow clitics could not be analyzed uniformly without such a parameter of variation.

---

[9]Hale (1973) describes how this and several other phenomena in Navajo reveal parametric variation to GB principles.

[10]A description of Case Theory is not given here. See Chomsky (1981).

2.3 Trace Theory Parameters: Choice of Traces and Pro-drop

Trace theory is another subtheory of GB that is important for uniform parsing across languages, in particular because it provides an explanation for the distinctions between languages that allow null subjects (like Spanish) and other languages. A trace is an empty sentence position that is either base-generated or left behind when a constituent has moved. The choice of traces for a language is specified as a parameter setting to the trace module.

According to the analysis of the null subject (or pro-drop) parameter introduced by van Riemsdijk and Williams (1986), the choice of whether a language requires a sentential subject is allowed to vary from language to language. In Spanish, as in Italian, Greek and Hebrew, morphology is rich enough to make the subject pronouns redundant and recoverable. Thus, we can have the sentence:

(7) Hablé con ella.
   '(I) spoke with her.'

Since the inflection on the verb is first person singular, the subject pronoun *yo* (=I) need not be used.

The formulation of the *pro-drop parameter* by van Riemsdijk and Williams is motivated by the observation that subjects are missing in a variety of constructions, not just in cases like (7). These constructions do not appear in many other languages (*e.g.*, English, *etc.*); thus, there must be a parameter that will account for the distinction between pro-drop and non-pro-drop languages. The *pro-drop parameter*, then, is a minimal binary difference that does or does not allow empty noun phrases to occupy subject position. (For details on the pro-drop parameter, see van Riemsdijk and Williams, pp. 298-303.) The parameter setting approach is more desirable than a rule-based approach since it accounts for several types of null subject constructions without requiring several independently motivated rules.[11] The pro-drop parameter is important in the parsing model because it allows uniform analysis of pro-drop and non-pro-drop languages, ensuring that sentence without a subject are ruled out unless the pro-drop parameter is set.

2.4 Principles and Parameters

Table 1 contains a table summarizing the subsystems of principles and parameters (grouped according to subtheory) relevant to the parsing model as presented here.[12] Table 2 summarizes the parameter settings required for parsing Spanish and English.

3 Parsing Implementation

The parser is one of three translation stages in an interlingual translation system, UNI-

---

[11] A rule-based approach (*e.g.*, GPSG (1985)) would require a separate rule for every possible null subject construction allowed in a pro-drop language including free subject inversion, relative clauses, that-trace constructions, resumptive pronouns, *etc.* (These constructions are not discussed here. See van Riemsdijk and Williams (1986).) The parameter setting approach obviates the need for independent treatment of these closely related phenomena.

[12] Because of space limitations, only those parameters that are relevant to a condensed description of the parser are presented here. The actual implementation currently has 20 parameters.

Theory	Principles	Parameters
$\overline{\text{X}}$	A phrasal projection (X^{max}) has a head (X), a specifier and a complement	Constituent Order, Choice of Specifiers
θ	[CL $+case_i$ $+\theta_j$] ... [NP $+case_i$] $\Rightarrow$ [CL $+case_i$ $+\theta_j$] ... [NP $+case_i$ $+\theta_j$] if language allows clitic doubling	Clitic Doubling
Trace	Null subjects are allowed for pro-drop languages	Pro-drop
	An empty position may occur where traces are allowed	choice of traces

Table 1: Principles and Parameters of GB

Theory	Parameters	Parameter Values	
		Spanish	English
$\overline{\text{X}}$	Constituent Order	spec-head-comp	spec-head-comp
	Choice of Specifiers	V: have-aux; N: det, *etc.*	V: have-aux, do-aux; N: det, *etc.*
θ	Clitic Doubling	applicable and allowed	not applicable
Trace	Pro-drop	yes	no
	Choice of Traces	N^{max}, Wh-phrase, V, P^{max}	N^{max}, Wh-phrase, V, P^{max}

Table 2: Parameter Values for Spanish and English

TRAN (Dorr, forthcoming), which is implemented in Commonlisp and currently translates simple sentences bidirectionally between Spanish and English. In contrast to the transfer approach (*e.g.*, METAL, Slocum, 1984, 1985), the parser (and other translation modules) is uniform across all languages with respect to its theoretical and engineering basis. (See figure 3.) The transfer approach, on the other hand, requires several parsers and a third translation stage (the transfer stage) in which one language-specific representation is mapped into another. Thus, a separate parser must be supplied for each language in the transfer approach, while in the interlingual approach a single parser is used for all languages. The interlingual approach more closely approximates a true universal approach since the principles that apply across all languages are entirely separate from language-specific characteristics expressed by (user-modifiable) parameter settings.[13]

The parameters of table 2 are represented declaratively, and are subject to modification by the user. (See figure 4.) There are two types of procedures (corresponding to the two boxes of figure 1) within the system: the first type includes those procedures that perform structure-building actions (predicting, attaching and scanning), relying primarily on phrase structure templates generated at precompilation time; and the second type consists of constraint verification routines (θ-Criterion, empty NP conditions, *etc.*), performing well-formedness tests on phrase-structures built by structure building procedures.

---

[13]The approach is "universal" only to the extent that the linguistic theory is "universal." There are some residual phenomena not covered by the theory that are consequently not handled by the system in a principle-based manner. For example, the language-specific English rules of *it-insertion* and *do-insertion* cannot be accounted for by parameterized principles, but must be individually stipulated as idiosyncratic rules of English. Happily, there appear to be only a few such rules per language since the principle-based approach factors out most of the commonalities across languages.

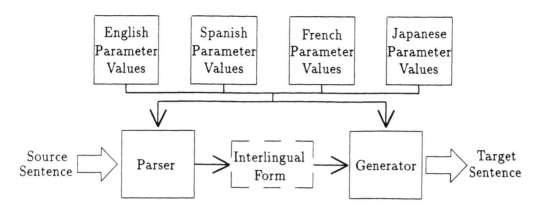

Figure 3: Interlingual Design as found in Dorr 1987

```
(DEF-PARAM CONSTITUENT-ORDER
   :SPANISH (SPEC HEAD COMP) :ENGLISH (SPEC HEAD COMP))

(DEF-PARAM CHOICE-OF-SPEC
   :SPANISH (V (HAVE-AUX) N (DET) I (N-MAX) C (WH-PHRASE))
   :ENGLISH (V (HAVE-AUX DO-AUX) N (DET N-MAX) I (N-MAX) C (WH-PHRASE)))

(DEF-PARAM CLITIC-DOUBLING :SPANISH (T T) :ENGLISH (NIL NIL))

(DEF-PARAM PRO-DROP :SPANISH T :ENGLISH NIL)

(DEF-PARAM CHOICE-OF-TRACES
   :SPANISH (N-MAX WH-PHRASE V P-MAX) :ENGLISH (N-MAX WH-PHRASE V P-MAX))
```

Figure 4: Representation of Parameter Settings for Spanish and English

Before parsing begins, the precompilation stage generates and stores a constant number of underspecified phrase-structure templates per language according to the two $\overline{X}$ parameters of figure 4: constituent order and choice of specifier. When the parser is activated, the structure-building module draws upon these templates, processing each word of input until no more structure-building actions apply. At this time, constraint verification takes place, and the last three parameters of figure 4 are accessed in order to modify or eliminate the structures derived thus far. The parse proceeds in this fashion until all sentence constituents have been successfully scanned, and all constraints have been verified. A sentence is rejected if: (a) there is a constraint violation, or (b) after consulting the constraint module no structure-building actions apply to the remaining input words. A sentence is accepted otherwise. Because the constraint component is available during parsing, the phrase-structure templates accessed by the structure-building module need not be very elaborate; consequently the grammar size need not, and should not, be as large as those found in other parsing systems.[14] Thus, the

---

[14]In fact, the number of phrase structure templates that are generated per language generally does not exceed 150 since there are a limited number of configurations per language that are allowed by the $\overline{X}$

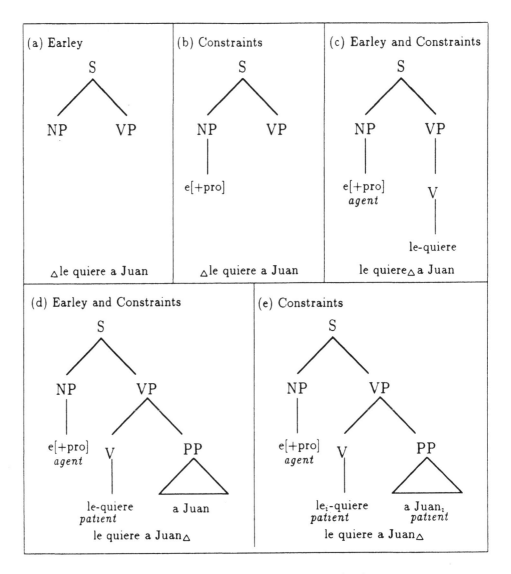

Figure 5: Snapshots of Parser in Action

system avoids high computational costs due to grammar search time.

To clarify the above description of the parsing algorithm, the next section presents and example of how the parsing modules operate.

4   An Example

Consider the problem of parsing (1) repeated here as (8):

(8)   Le quiere a Juan
       '(She) loves John'

---

principles accessed at precompilation time. Thus, the running time of the parser is not subject to the same slow-downs that are found in other systems.

Figure 5 gives snapshots of the parser in action. First the Earley structure-building component predicts that the sentence has a noun phrase (NP) and a verb phrase (VP) (see (a)), the order of which is determined by the "constituent order" parameter at precompilation time.[15] The only structures available for prediction by the Earley module are those generated at precompilation time; thus, at this point no further information about the structure is available until the linguistic constraint module takes control.

The constraint module accesses the "null subject" parameter (see section 2.3), which dictates that the empty element attached to NP is a subject. The [+pro] (pronominal) feature is associated with the node (see (b)) so the subject will accommodate both null subject source languages and overt subject source languages.[16]

In snapshot (c), the Earley module expands VP and scans the first two input words *le quiere*.[17] Now the Earley module cannot proceed any further; thus, the constraint module takes over again. First a semantic role (or θ-role, as it is called in GB Theory) of *agent* is assigned to the empty subject of the sentence. This information is determined from the dictionary entry of *quiere* which dictates that this verb requires both an agent (assigned to the subject or *external argument* of the verb) and a patient (assigned to the object or *internal argument* of the verb). The dictionary entry for *querer* (the root form of *quiere*) is encoded as follows: **(querer: [ext: agent] [int: patient] V (english: love) (french: aimer) ...)**

Now the constraint module predicts that a noun phrase (corresponding to the internal argument of *querer*) must be available. Because the clitic-doubling parameter is set to (T T), it is determined that the NP *le* can act as an object of the verb *quiere*; consequently, it receives *patient* θ-role as dictated by the lexical entry of *querer*. The constraint module then "records" the fact that a clitic has been seen, so that the NP corresponding to *le* will have a θ-role transmitted to it later if it appears in the input.[18] Once control passes back to the Earley module, the final two words are scanned, thus completing the PP. Snapshot (d) shows the parse thus far.

At this point the constraint module attempts to assign θ-role to the NP *Juan*. However, all of the θ-roles from the lexical entry of *querer* have already been assigned; thus, assigning a role from this entry would be a violation of the θ-criterion. On the other hand, leaving *Juan* without a role also violates the θ-criterion. Consequently, the constraint module determines (via the clitic-doubling parameter setting) that the θ-role transmission rule (6) is applicable, and recognizes that the NP *Juan* corresponds to the "recorded" clitic preceding the verb *quiere* (since the two match in person, number and gender). Thus, a θ-role of *patient* is transmitted to *Juan*.[19] As a result of the application of the θ-transmission rule, *le* and *Juan* are coindexed; thus, these two constituents are interpreted as coreferential during the stages following the parse. The final parse is illustrated in snapshot (e).

---

[15]Since Spanish is a *head-initial* language, NP must precede VP; however, this would not be the case for non-*head-initial* languages. (See fn. 5 for a description of the "constituent order" parameter.)

[16]For example, Italian and Hebrew do not require an overt subject, but English and French do; thus, during a later stage (generation), e[pro] will either be left as is, or lexicalized to a pronominal form (*e.g.*, *he* or *she* in English) that agrees with the main verb.

[17]Clitic adjunction is generated at precompilation time. The presence or absence of a clitic for a particular language is determined by an adjunction parameter setting associated with $\overline{X}$. This parameter will not be discussed here.

[18]Since clitic doubling is optional, the parse will not be discarded if the corresponding NP does not appear in the input; however, if it does appear (as it does in the above example), it is correctly assigned θ-role.

[19]Note that the θ-role *patient* is assigned the NP *Juan*, not to the PP *a Juan*; in general, the structural entity that is assigned semantic role is an NP, regardless of the type of phrase the containing it.

5 Conclusion

The system described here is based on modular theories of syntax that include systems of principles and parameters rather than complex, language-specific rules. The "co-routine design" allows the structure-building mechanism to operate with user-modifiable principles of current linguistic theory. The user has access to parameters associated with the system principles, thus enabling extension of the system to additional languages. The presence of linguistic constraints allows phrase structure templates to be underspecified (*i.e.*, more general), thus reducing the grammar size of a given language. In summary, the modularity imposed by the GB framework is an improvement over context-free based systems because it facilitates extensions and alterations to the system, simplifies descriptions of natural grammars, and is backed by psycholinguistic evidence (see fn. 3).

Acknowledgements

This report describes research done at the Artificial Intelligence Laboratory of the Massachusetts Institute of Technology. Support for the Laboratory's artificial intelligence research has been provided in part by the Advanced Research Projects Agency of the Department of Defense under Office of Naval Research contracts N00014-80-C-0505 and N00014-85-K-0124, and also in part by NSF Grant DCR-85552543 under a Presidential Young Investigator's Award to Professor Robert C. Berwick. Useful guidance and commentary were provided by Bob Berwick, Ed Barton, Sandiway Fong and Dave Braunegg.

References

Barton, Edward G. Jr. (1984) "Toward a Principle-Based Parser," MIT AI Memo 788.

Bates, M. (1978) "Natural Language Communication with Computers," Springer-Verlag, 191–254.

Chomsky, Noam A. (1981) *Lectures on Government and Binding*, Foris Publications, Dordrecht.

Chomsky, Noam A. (1982) "Some Concepts and Consequences of the Theory of Government and Binding," MIT Press.

Dorr, Bonnie J. (forthcoming) "UNITRAN: A Principle-Based Approach To Machine Translation," S.M. thesis, Department of Electrical Engineering and Computer Science, MIT.

Earley, Jay (1970) "An Efficient Context-Free Parsing Algorithm," *Communications of the ACM* 14, 453–460.

Frazier, Lyn (1986) "Natural Classes in Language Processing," presented at the *Cognitive Science Seminar, MIT, November*, Cambridge, MA.

Gazdar, G., E. Klein, G. Pullum, and I. Sag (1985) *Generalized Phrase Structure Grammar*, Basil Blackwell, Oxford, England.

Hale, K. (1973) "A Note on Subject-Object Inversion in Navajo," in *Issues in Linguistics: Papers in Honor of Henry and Renee Kahane*, B. Kachrue *et. al.* (eds.), University of Illinois Press, Urbana.

Jaeggli, Osvaldo Adolfo (1981) *Topics in Romance Syntax*, Foris Publications, Dordrecht, Holland/Cinnaminson, USA.

Robinson, J. J. (1982) "DIAGRAM: A Grammar for Dialogues," *Communications of the ACM* 25:1, 27–47.

Sharp, Randall M. (1985) "A Model of Grammar Based on Principles of Government and Binding," M.S. thesis, Department of Computer Science, University of British Columbia.

PARSING AND GENERATING THE PRAGMATICS
OF NATURAL LANGUAGE UTTERANCES
USING METACOMMUNICATION

David L. Sanford & J. W. Roach
Department of Computer Science
Virginia Tech
Blacksburg, Virginia 24061

Abstract

This paper reports a new theory of natural language processing and its implementation in a computer program, DIALS (for DIALogue Structures). This represents a radical departure from the paradigmatic approach to natural language processing currently dominating the fields of artificial intelligence, linguistics, and language philosophy, among others. We use the theory of metacommunication to develop a "pragmatic grammar" for the structural analysis of dialogue. We are currently able to parse and generate over 5000 surface forms of a single underlying request content. We propose using this pragmatic information to manage the communication context, including inferring some of a speakers' goals and controlling status and politeness.

Keywords: discourse analysis, pragmatics, metacommunication, requests

Introduction

Natural language processing has traditionally concentrated on syntax or semantics. Our new theory, called "Dialogue Structures," concentrates instead on pragmatic issues. This theory posits that indirect questions, emphasis, focus and speakers' goals, still problematic issues after years of research, can be determined, in part, by structural means. Every utterance consists of both a semantic, content portion and a communication management portion (Roach & Nickson, 1983, 1986). Roach and Nickson independently rediscovered the theory of metacommunication originally developed by Bateson (1951a, 1951b) and his successors (Watzlawick, Bavelas, & Jackson, 1967). Communication management, using metacommunication, guides the listener's attention, shows importance of topics, maintains a shared communication context and preserves social relationships (Sanford & Roach, in press) such as politeness, status, etc. Being able to process the pragmatic content of utterances leads, for the first time, to a system that can account for some of the tremendous range of expression in natural language. For example, the system we have built, called DIALS, can correctly generate and parse the request, "What time is it?" in more than 5000 different ways. DIALS uses a "pragmatic grammar," similar in concept to the well-known semantic grammar, to process natural language. Unlike systems that use semantic grammars, its domain is general rather than specific; i.e., DIALS' task domain is the management of communication, necessary in all dialogues. We do not claim that syntactic and semantic methods are wrong, only that one of the most important problems in parsing has been largely ignored. "Dialogue Structures" speaks, at least in part, to this problem. This paper will discuss the application of metacommunication to the understanding of the pragmatics of requests, an important cognitive task that humans learn to parse and generate correctly at a very early age.

Making requests is a common task in all types of communicative activities. Research on requests is often limited to question-answering (Lehnert, 1984) and often makes overly simplified assumptions about how communication works. Those studying language analysis from the viewpoint of linguistics, language philosophy, and related fields, seem to adopt a model of communication that says that language is explicit, that people say what they mean and mean what they say. Grice's (1975) well known Cooperative Principle, including the Manner maxims, express this "Transparent Model" of communication: speakers "avoid obscurity of expression," "avoid ambiguity," and try to "be brief."

Communication theory (Sanford & Roach, 1986) disagrees with the transparent model, arguing that people do not say exactly what they mean and seldom mean what they do say. Indeed, this "Guarded Model" of communication recognizes that speakers have hidden agendas; as Goffman (1959) says, one important goal in interactions is to save face, for which people use masks or façades. The transparent model to communication overlooks the existence of deception and face-saving. This paper will not discuss deception to any degree, but at least the guarded model recognizes its possibility. This paper will discuss how the theory of Dialogue Structures applies to requests, identifying the two major parts of a request, analyzing each part separately, and finally discussing DIALS, the computer program that implements the theory.

Theory of Dialogue Structures

According to Bateson, there are at least two subcategories of metacommunication: "the propositions about codification [i.e., communication in which the content of the utterance is the process or mechanisms of communication] and the propositions about interpersonal relationship" (1951b, p. 214). Dialogue Structures deals with the subtle expression of interpersonal relationship from the metacommunicational cues of the form in which requests are phrased. That is, a speaker's intentions cannot be directly identified but must be inferred from subtle metacommunicational cues and the interpreter's knowledge of social norms and interpersonal relationships.

Dialogue Structures states that the surface form of a request is not simply an expression of pragmatic purpose, politeness, or clarity. The main thing expressed explicitly in a request is how demanding vs. how pleading a request is, i.e., the "imperative force" of the request. A request, therefore, can be represented in the following way:

request = (content expressing a desire) +
(structure expressing imperative force)

We shall define what we mean by "structure" in the next section. Dialogue Structures, then, must explain the two parts of a request: the structure expressing imperative force and the content expressing a desire. The following will examine separately these two parts of a request.

Structure Expressing Imperative Force

"Imperative force" expresses how demanding vs. how pleading a request is. "Imperative force" was first used by Searle (1975) to express that a request is demanding; we expand it into a complete dimension of expressive power. That is, Searle said only two types of requests have imperative force: the explicit performative (e.g., "I order you to leave the room") and the flat imperative (e.g., "Leave the room"); we say that all requests have imperative force, but it ranges through many gradations from strongly demanding to strongly pleading.

Most people use the term "structure" in this context to refer only to the syntax of an utterance. Certainly syntax is a structural component of utterances, but we use "structure" in a broader sense. We identify structural aspects to the pragmatics of utterances and use this structural level as well. Structural pragmatics involve the identification not only of the syntax of an utterance, but also of key word patterns; e.g., "I was considering . . ." is not only a declarative syntactic structure, but the key words identify it as a "Claim of Deliberation" in which any desire may be embedded. (Throughout this discussion, several request categories will be mentioned. There is not enough room in this report to give the full category system. See Sanford and Roach, in press.) For example, consider trying to identify the "transparent" purpose of a request to decide how to respond. We argue that a listener cannot tell the true pragmatic purpose of a request from the surface form. However, the request may structurally appear to be a request for information, for permission, or whatever. Therefore, although a request beginning, "Do you know . . . " sounds as if it is intended as a request for a yes/no response, we cannot tell without making inferences from subtle metacommunicational cues and the interpreter's knowledge of social norms and interpersonal relationships. But we can identify it "structurally" as "Asking for Suggestion," and then investigate how demanding vs. pleading it is. A clear understanding of imperative force, therefore, is the first step in constructing an overall theory of requests.

This brings us to explaining how imperative force functions in communication. Imperative force may express politeness, status, or emphasis. Searle (1975) asserts that the choice of how to phrase a request is based solely on how polite an individual wants to be. Examining such transparent model approaches to requests would lead one to believe that politeness is most important, status is least important, and emphasis has no relation to the form of requests. Actually, the order is exactly the opposite, according to Dialogue Structures. We will first discuss the relation between politeness and status, then discuss the relation between status and emphasis, and finally present our analysis of how a hearer uses imperative force to interpret a speaker's utterances.

Status is more fundamental than politeness. Consider how the transparent model explains impolite behavior. Since people try to be clear, direct, unambiguous, and cooperative, the assumption that people would be impolite is untenable. Therefore, some other contrast to politeness is necessary. Lakoff (1973) says that the more direct the wording of a request, the more impolite but the clearer that request is; his contrast, therefore, is between politeness and clearness. Research shows this not to be the case. When Gibbs (1979) measured the time taken to interpret indirect requests embedded in a story context, it actually took longer to understand the literal than the indirect meanings. Also, research shows that children about two to three years of age have no more difficulty with indirect than with direct requests (Elrod, 1983; Shatz, 1978). Direct wordings of requests, therefore, are no more clear than indirect wordings.

Searle (1975), as mentioned, asserts that the request form is based solely on politeness. Research shows this to be an unjustified assertion:

> The politeness of the directives used by a speaker appears to be affected directly by the status relationship between the speaker and the listener. Studies with adults have shown that polite request forms such as "May I please use your phone?" are more likely to be addressed to a listener whose age or professional position places him in a superior role. (James, 1978, p. 308)

This suggests that status is fundamental to determining politeness. Hill et al. (1986) refer to two components of politeness: "discernment," which involves "conforming to the expected norm," and "volition," which allows a speaker a "more active choice." We would say that behavior under the control of norms is deference, whereas politeness is always volitional. For example, a sergeant has more status or power than a private. If a private uses a pleading request to a sergeant, the private is not being polite but is expressing deference to the power of the sergeant. If the sergeant uses a pleading request to the private, the sergeant is being polite, since the sergeant can make a demand of the private and chose to be less demanding. It is inappropriate, therefore, to apply the terms "politeness" or "clearness" as the basic dimensions of comparison of requests. This evidence strongly supports the idea that "imperative force" is a metacommunicational cue that fulfills some other function(s) in the communication of requests. Indirect requests are patterns expressing metacommunicational information in which any type of desire may be embedded. This agrees with everyday observations: if someone asks, "Do you know the time?" and receives the response, "Yes," they assume the respondent is joking or being uncooperative. Requests are demanding vs. pleading; the status of the interactants and the emphasis being expressed determine whether a request is polite.

Emphasis is more fundamental than status. To return to our example, during battle a private can address an imperative request, such as "Pass the ammunition!" to a sergeant without being considered impolite or incurring the wrath of the sergeant. The emphasis on the importance of the desire in this context precludes any consideration of status or politeness. Indeed, it might be considered "impolite" to use a pleading form, since it takes more words to express pleading, and the extra time might be life or death.

To explain more fully this issue of the emphasis expressed with a request, consider the following two dialogue examples between a customer and an airline reservation agent:

EXAMPLE 1
CUSTOMER: I need to go to L.A. Do you know if I can leave town today?
AGENT: I'm sorry, but all flights to L.A. today are booked. I could get you on a flight tomorrow.

EXAMPLE 2
CUSTOMER: I need to leave town today. Do you know if I can go to L.A.?
AGENT: I'm sorry, but all flights to L.A. today are booked. I could get you on a flight to Dallas/Ft. Worth.

Using the analysis of transparent models, the first sentence of each customer is fairly clear, has no imperative force, and is less polite, whereas the second sentence is less clear, has no imperative force, and is more polite. Most people would neither interpret these requests as mixing impolite and polite forms nor say there is no imperative force to the requests. The wording puts emphasis on the information being expressed, allowing the hearer to infer the speaker's goals. Consider how inappropriate it would be for the agent to offer the first customer a flight to Dallas/Ft. Worth; consider how equally inappropriate it would be for the agent to offer the second customer a flight on the following day.

Determining the function of imperative force in a given request. Several types of information are needed to decompose the relative importance of the three determinants of imperative force: emphasis, status, and politeness. A participant needs to know standard status levels for established societal roles, such as teacher vs. student, boss vs. worker, etc. For an ongoing relationship, one needs to know the history of status negotiation within this given relationship. For a given dialogue, one needs to know the sequence of status negotiation moves across this interaction. For example, within an interaction, status is usually the first issue addressed in a sequence of dialogue moves. This is often shown by the phrasing of the pre-requests (Jacobs & Jackson, 1983), or what we term the "empty requests," by the requestor. Two excerpts taken from transcribed tape recordings of actual interactions between airline reservation customers and agents show what we mean.

EXCERPT 1
CUSTOMER: I'm planning on a flight leaving April 5. I plan on leaving from Roanoke. I'd like to go to L. A. I was wondering if you could give me some flight information about that.

EXCERPT 2
CUSTOMER: I need some information. I would like to get some information about taking a flight leaving April 5 from Roanoke to Los Angeles.

In the first excerpt the customer emphasizes the date and airport of departure with slightly less emphasis on the airport of arrival. The last sentence is empty of any information important or relevant to the request, and would probably be considered a pre-request if it were in the first position. In the second excerpt, the customer starts with an empty request, this time more easily defined as a pre-request. Then the customer shows that all the pieces of the request are of equal emphasis, expressed with a lower level of imperative force. The speaker is using a less demanding request form than status allows, expressing politeness and the willingness to allow any of the three factors in the request to be adjusted by the agent, as circumstances require.

The process for decomposing the purpose of the imperative force of an utterance now can be clearly stated. First, the hearer determines the relative level of imperative force being used. It is well known that some people habitually express everything as if it were vitally important, while others express everything they say as if it were unimportant. That is, some peoples' range of expression stays among the strongly and moderately demanding forms, while others stay among the strongly and moderately pleading. A hearer must identify the surface level imperative force for a particular utterance and compare it to the range of imperative force used by this speaker in the past. This is one reason we choose to call our approach "Dialogue Structures," since imperative force is a structural aspect of utterances that must be examined over the course of dialogues and not just at a single point in time.

Second, the hearer checks the defined status between the speaker and hearer, originally by examining the roles of the two and the socially-defined status levels associated with those roles; then, by examining the history of status between the interactants, if there has been an on-going relationship; and finally, by examining the status negotiation, if any, during this interaction. At this stage, the speaker's referent power or the feelings of friendliness between the interactants is probably less important than the speaker's legitimate authority. Certainly there is a coloring effect from referent power and friendliness; one can easily imagine a policeman barking orders at a driver to help the driver avoid an accident (strongly demanding to express the emphasis on the importance of the information), while the driver interprets the imperative force as a status claim: "that stupid cop is yelling at me just to show how much power he has!" Here, again, we see the importance of watching this structural dimension across interactions rather than simply focusing on this particular instance of imperative force, i.e., watching dialogue structures. Policemen are trained to use their voice to achieve social control; most of their utterances will be demanding and therefore this particular utterance will not be extremely out of line with their dialogue patterns. After the hearer has determined the speaker's purpose in choosing a given level of imperative force, the speaker's referent power or the level of friendliness between the participants is much more important in helping to determine what the hearer will decide to do. One again can imagine a case where a person has made a pleading request, e.g., for the time, and the hearer correctly interprets the pleading as an attempt to be polite but decides not to answer based on the hearer's feelings of dislike for the speaker.

A hearer is ready to infer the purpose of the speaker now that the hearer has these three pieces of information: first, the imperative force of this particular request, second, the relative status of the participants, and third, the range of imperative force used by the speaker across a history of dialogue. Again, let us use an example:

STUDENT: I can't turn my paper in today.
TEACHER: Would it be convenient for you to turn it in tomorrow?

How does the student interpret the teacher's request? The surface form of the request is an "Asking about Convenience," which is moderately pleading. The student knows that teachers have higher status than students, so it cannot be a status claim. Say the teacher has a history of using the entire range of imperative force in expressing requests, so this is nothing unusual in that regard. The student can refer to the semantic domain of discussion, realize that teachers consider class assignments important, so it cannot be emphasis; therefore, it is politeness.

Take the same interaction under another circumstance. Say the student had requested to turn in an early draft of the paper for preliminary review, but the paper is not really assigned to be turned in today or tomorrow. The student could rightly assume the teacher is expressing the relative unimportance of getting it in tomorrow, expressing emphasis. Or take the same interaction with another teacher. Say this teacher has a history of using mainly demanding request forms with students; that makes this pleading form unusual. It takes on a reverse effect and becomes a highly impolite, sarcastic demand.

We are just starting to investigate the rules for combining these three types of information to make the kind of inferences exemplified above. We already have the rules needed to identify the first type of information, the imperative force for a particular utterance. This is also the raw data for the third type of information, the imperative force used by a speaker across a history of dialogue. The second type of information, the relative status of given societal roles, will be stored in the database of the pragmatic grammar for use in these inferencing procedures.

Responding to the Content that Expresses a Desire

The second issue needing explanation is the appropriate response to a request; whether to respond with information, action, confirmation, or whatever. Consider the method for making this determination using the transparent model: identify the intention of the speaker and respond to that intention. From the perspective of the guarded model, the decision is determined by the information and situation of the hearer, since one cannot trust the speaker to identify clearly the speaker's intention. Take an example from an interaction between an airline agent and a customer. A request such as, "Can you book me for the 1:45 flight?" sounds as if it is intended as a request for a yes/no answer. If there are seats available at that time, most agents would respond not with "yes" but with the action of booking a flight. Information is needed only when the circumstances prohibit action. For example, Allen (1983) grapples with the issue of whether the following is a yes/no request or a request for information:

Do you know when the Windsor train leaves?

If the one being asked knows the information, then a sensible response might be:

7:14 in the morning.

Only if the person cannot provide the information that is indirectly requested should something else be given, for example:

I don't work here.

But in neither case is a "yes" or a "no" needed. Research on people shows that they often go ahead and answer the surface question, but add the information indirectly requested. That is, they do not always include the "yes" or "no," but do so quite often. This is not ruled out by our theory; certainly, we do not lose the information that the surface form is a request for a yes/no response. We simple assume that the surface form is not the speaker's main goal and go immediately to the embedded indirect request. There are cases in which people truly want a yes/no response, but these are cases that people have difficulty identifying unless that desire has been explicitly stated. One is reminded of the many times Perry Mason had to tell a witness, "Answer only 'Yes' or 'No.'" Otherwise, the witness invariably wanted to respond to the indirect question embedded in the surface form.

Computational Results

We have constructed a parsing and generation system, called DIALS, that embodies aspects of the theory presented above. DIALS currently can handle individual sentences, separating the pragmatic from the semantic portions of a sentence and working with a caseframe of the pragmatic portion of a request. That is, it identifies the originator of the request, the proxy verbalizing the request, the receiver of the request, and the imperative force of the request, as well as the goal of the request, expressed in the semantic portion of the sentence. DIALS either starts with a surface form sentence and parses into this caseframe or starts with the caseframe and generates a surface form sentence. It handles indirect requests easily, without first determining a surface meaning then applying an inference engine to infer the indirect meaning, as suggested by Searle (1975), Allen (1983), and others assuming the transparent model. DIALS is independent of any semantic domain and therefore can be applied as a front end to any task, such as databases, operating systems, editors, etc. DIALS has been implemented in PROLOG; on a VAX 11/785, with an interpreter running at 1 klips, a pragmatic parse also producing a simplified sentence for analysis by a semantic parser requires between one and five CPU seconds.

Conclusion

We realize our approach clashes directly with the paradigmatic artificial intelligence approach introduced by Charniak (1972) and exemplified by Allen (1983); this approach says that all pragmatic inferences must be made using an immense world knowledge base. But DIALS already can parse and generate over 5000 different wordings of a single underlying request content. It has been tested by comparing its coding of 547 sentences taken from transcribed recordings of conversations between airline customers and agents against human coding of the conversations; it was able to code the imperative force of the requests correctly over 95% of the time. It is still undergoing improvement aimed at achieving

100% correctness and at being able to parse and generate tens of thousands of surface forms of single underlying request contents. Also, it is being improved to allow it to keep track of the pragmatic structure of dialogue so the rest of the theory discussed in this paper can be implemented. But our preliminary successes make us feel that the "guarded model" of communication and metacommunication may have advantages in many pragmatic issues over the "transparent model" and its attendant need for an immense world knowledge base.

References

Allen, J. (1983). Recognizing intentions from natural language utterances. In M. Brady & R. C. Berwick (Eds.), *Computational models of discourse*. Cambridge, MA: M.I.T. Press, 107-166.

Bateson, G. (1951a). Information and codification: A philosophical approach. In J. Ruesch & G. Bateson, *Communication: The social matrix of psychiatry*. New York: W. W. Norton, 168-211.

Bateson, G. (1951b). Conventions of communication: Where validity depends upon belief. In J. Ruesch & G. Bateson, *Communication: The social matrix of psychiatry*. New York: W. W. Norton, 212-227.

Charniak, E. (1972). *Toward a model of children's story comprehension* (Tech. Rep. AI TR-266). Cambridge, MA: M.I.T. AI Laboratory.

Elrod, M. M. (1983). Young children's responses to direct and indirect directives. *The Journal of Genetic Psychology, 143*, 217-227.

Gibbs, R. W., Jr. (1979). Contextual effects in understanding indirect requests. *Discourse Processes, 2*, 1-10.

Goffman, E. (1959). *The presentation of self in everyday life*. Garden City, NY: Doubleday.

Grice, H. P. (1975). Logic and conversation. In P. Cole & J. L. Morgan (Eds.), *Syntax and semantics, vol. 3: Speech acts*. New York: Academic Press, 41-58.

Hill, B., Ide, S., Ikuta, S., Kawasaki, A., & Ogino, T. (1986). Universals of linguistic politeness: Quantitative evidence from Japanese and American English. *Journal of Pragmatics, 10*, 347-371.

Jacobs, S. & Jackson, S. (1983). Strategy and structure in conversational influence attempts. *Communication Monographs, 50*, 285-304.

James, S. L. (1978). Effect of listener age and situation on the politeness of children's directives. *Journal of Psycholinguistic Research, 7*, 307-317.

Lakoff, R. (1973). The logic of politeness: Or, minding your p's and q's. *Papers from the ninth regional meeting, Chicago Linguistic Society*. Chicago: Chicago Linguistic Society, 292-305.

Lehnert, W. (1984). Problems in question answering. In L. Vaina & J. Hintikka (Eds.), *Cognitive constraints on communication: Representation and processes*. Boston: Kluwer Academic Publishers, 137-159.

Roach, J. W. & Nickson, M. M. (1983). Formal specifications for modeling and developing human/computer interfaces. In *Proceedings CHI'83 Human Factors in Computing Systems*. New York: A.C.M., 35-39.

Roach, J. W. & Nickson, M. M. (1986). Software methodology: Rule-based approach to interface design. In R. W. Ehrich & R. C. Williges (Eds.), *Human-computer dialogue design*. New York: Elsevier, 51-59.

Sanford, D. L. & Roach, J. W. (1986). The role of language in human-computer interfaces: Communication and language. In R. W. Ehrich & R. C. Williges (Eds.), *Human-computer dialogue design*. New York: Elsevier Publishers, 165-177.

Sanford, D. L. & Roach, J. W. (in press). Representing and using metacommunication to control speakers' relationships in natural language dialogue. *International Journal of Man-Machine Studies*.

Searle, J. R. (1975). Indirect speech acts. In P. Cole & J. L. Morgan (Eds.), *Syntax and semantics, vol. 3: Speech acts*. New York: Academic Press, 59-82.

Shatz, M. (1978). Children's comprehension of their mothers' question-directives. *Journal of Child Language, 5*, 39-46.

Watzlawick, P., Bavelas, J. B., & Jackson, D. D. (1967). *Pragmatics of human communication: A study of interactional patterns, pathologies, and paradoxes*. New York: W. W. Norton.

TEUCHISTIC NATURAL LANGUAGE PROCESSES

Robert K. Lindsay
The University of Michigan, Ann Arbor, Michigan 48109

Alexis Manaster-Ramer
Wayne State University, Detroit, Michigan 48202
and
IBM Thomas J. Watson Research Center
Yorktown Heights, New York 10598

Keywords: natural language, teuchistic

ABSTRACT

AI approaches to natural language specify computational processes, yet they are based on the structural concepts of language and grammar which posit necessary conditions at least for the "correct" interpretations of utterances and often also for the syntactic and/or logical representations of utterances. We argue that any such view will fail to account for a variety of important features of language behavior, which we describe. Systems based on the language/grammar model are usually accompanied by heuristic or algorithmic search/selection methods of computation. We contrast these methods with constructive (*teuchistic*) computation models based on redundant, inconsistent sets of constraints, and show that this view offers natural accounts of the phenomena described.

I. INTRODUCTION

Almost all computational work on natural languages, in and out of AI, has adopted the conception of language, derived from traditional grammar and structural linguistics, according to which there exists a body of knowledge which defines the primitives of which the language is composed, the principles by which these primitives may be combined, and the meanings associated with each primitive and with each principle of composition. This conceptual homogeneity has been obscured by the controversies surrounding virtually every other question of NL processing, such as the debate over models which postulate syntactic and/or logical levels of analysis in addition to the meaning (conceptual) representation (e.g., LUNAR (Woods *et al.* 1972), SHRDLU (Winograd 1972), PARSIFAL (Marcus 1978)) as opposed to models which relate utterances to meanings in an integrated fashion (e.g., Wilks's (1975) parser, ELI (Riesbeck and Schank 1976), and the Word Expert Parser (Small 1980)),[1] which has sometimes been viewed as involving a contrast between grammar-based and knowledge-based methodologies[2]

We are thus abstracting from the controversies about exactly how this body of knowledge is structured and whether it should be used—in parsing, say—together with world knowledge. Rather, we are concerned with the fact that almost all models purport to simulate the idealized user of a given language, conceived of as an expert on his native language (e.g., Cullingford 1986:34), and

---

[1] The issues separating these approaches have at any rate been settled to some extent at least by the compromise position of Lytinen (1986), who proposes that syntactic and semantic information form distinct knowledge bases but that these are combined dynamically during parsing.

[2] A related issue concerns the distinction, if any, between a declarative representation of knowledge of a language and a processing model. However, since in models that contain little or nothing beyond a processor, the latter still implicitly represents a description of the language, this distinction will also be ignored in this paper. After all, a context free grammar could be presented by means of a process model such as a pushdown automaton or an Earley parsing algorithm.

consequently assume that there is such a thing as a definite knowledge of *what* a native speaker does in the way of assigning some structural representations.[3]

We thus wish to contrast such **models of language** with **models of (potential) linguistic behavior**, which assume that the meanings (and any intermediate representations) are arrived at by means of *constructive* processes which take into account a variety of factors, only some of which bear a close resemblance to the linguistic knowledge as normally conceived, whereas others involve factors such as the past processing that has been done by the system, interaction with other users of the language, and even physical characteristics of the language user.

We would discard the concepts of *language* and *grammar*, replace the customary notion of linguistic competence with a concept of *potential behavior*, and view production and comprehension as nonmonotonic problem-solving processes that satisfice rather than search out an optimal choice from a pre-established theoretical set of possibilities. We further suggest that the multi-component constraint satisfaction model serve as the basis of such problem solving. In this paper we outline the approach without specifying a particular theory, and we refer to Lindsay (1971) as an example of this approach in application. We will discuss some of the principal kinds of facts of human linguistic behavior which justify and perhaps even necessitate this change of perspective. We will show that this approach is closely related to satisficing models of economic behavior and to nonmonotonic models of reasoning, while the conventional language-based models are essentially optimizing and monotonic.

II. TEUCHISTIC MODELS OF LINGUISTIC BEHAVIOR

Teuchistic Computation

Typically, AI problem solving methods involve a search through a problem space or game tree. Although there may be an exact algorithm involved in simple cases, usually the process is heuristic.[4] Our view is that NL use involves processes which are neither algorithmic nor heuristic but, to coin a new term, **teuchistic**. Teuchistic processes construct a solution rather than search for one through a set of pre-established possibilities. Trivially, many numeric computations are of this sort. For example, squaring a number (or evaluating any other closed algebraic form) is a process of constructing the answer rather than searching through the set of possibilities (the real numbers, in this case).[5]

Less trivially, at times it is not possible to define a solution space extensionally as when the space is infinite, but one can do so intensionally (for example, the real numbers). Furthermore, even when such a definition is possible, it may not be enlightening if the computation itself is not cast as a search within that solution space. Consider the case of design, say of houses or electronic circuits. Here computation is best viewed not as search but as construction, for any definition of the set of possible houses or circuits will not aid the design process as it is usually carried out by human designers. Rather the process is one of constructing a solution using available tools and meeting a set of heterogeneous constraints. Indeed if the constraints are not themselves necessary conditions, but merely design desiderata ("suggestions"), and if further the constraints may be vague and inconsistent (as design

---

[3]We will also ignore such important and controversial questions as the following: Should a computational model simulate *how* human beings represent linguistic knowledge? What level of understanding should be aimed at? Should pragmatics (knowledge of the world) be used to guide NL understanding? Will models designed for small domains of discourse scale up to whole languages?

[4]Some work in computational linguistics, in fact, views human linguistic processing as involving a highly inexact heuristic, which can handle only a restricted subset of the language allowed by the theoretical model of linguistic representation (e.g., Peters, to appears).

[5]However, some numeric computations involve search, usually systematic; for example computing a square root by Newton's method or any other algorithm that works by successive approximation may be seen as as search process.

principles generally are), the concept of a solution space is not even coherent. In such situations, as we will argue is the case for natural language use, computations must be teuchistic; further, any attempt to circumscribe the set of permissible "solutions" will be futile and a diversion from the real task.

As we will argue later, the conventional approaches to NLs appear to be optimizing in that they search for the "correct solution" to questions of syntactic or semantic representation of a particular utterance and monotonic in that the system of rules defining a correct solution is self-consistent and allows the representations to be deduced following the rules of classical logic. It should be emphasized that this is trivially true even for those models which have no syntactic or logical representations but which construct a meaning representation directly from the utterance and the knowledge of the world. The teuchistic approach, on the other hand, shares important features of satisficing and nonmonotonic models which have been developed within AI in domains other than NL.

From our point of view, the crucial distinction is between models that assume that there is a thing called language (specifically a particular language at a particular time, e.g., late 20th century American English) which can be usefully modeled computationally, usefully in the sense that such a model would behave sufficiently like a human native speaker of this language to be able to replace human beings at some significant tasks but also in the sense of casting new light on the workings of human intelligence.

Our model, if it is to be functional at all, almost necessarily has to be redundant and multipartite. This view assumes the existence of several different *sources of constraints*, each of which can be viewed as a generative component. It is not necessary that these sources be non-overlapping, consistent, nor even clearly categorized. It is only necessary that they be *redundant* in the sense that structural-interpretive information available from one set of sources may be used to infer reliably (though not necessarily with certainty) structural-interpretive information of other types. For purposes of illustration, we may assume sources of syntactic, lexical, phonetic, semantic, and pragmatic constraints (but this is an oversimplification); in addition there may be sources of logical and empirical constraints of various kinds.

It is the availability of these multiple sources of redundant constraints that drives, for example, the comprehension process. Comprehension is modelled as a constraint driven satisficing process: information is selectively attended to and used to constrain the generation of an interpretation or utterance. If all structural information is consistent, the interpretation is rapid and unproblematic. If some structural information is not consistent with the interpretation under construction, a strategy of backtracking is undertaken to revise a previous decision. If backtracking fails (outright inconsistency is encountered), some constraint must be relaxed, again according to a specific strategy.

A specific earlier proposal along these lines involved learning mechanisms based on a multiple components model which has been described and programmed by Lindsay (1971). The learning mechanism is called the *jigsaw heuristic*,[6] after the jigsaw puzzle. A jigsaw puzzle provides two sources of constraints for its solution: information from the contours (syntax?) that determines what pieces may be adjacent, and information from the design (semantics?) about the color and form of the scene depicted. These sources are redundant in the sense that the puzzle could be done if all the pieces were congruent squares *or* if all the pieces were upside down. Now suppose some of this information were incomplete – tabs were deformed, or coffee spilled on the design. Solving the puzzle would then lead to clear hypotheses about the nature of the missing information. In the NL situation, missing knowledge of word meaning or part-of-speech could be induced by a learner if sufficient information remained to permit the construction of a satisfactory interpretation.

Modeling Potential Behavior

In place of language, viewed as an object about which a definite body of information is available, we choose to focus on human linguistic behavior. The distinction is, as noted, not reducible to that

---

[6]Though we now see a crucial difference between our approach and the usual heuristic models.

between declarative grammar formalisms and processing formalisms; nor is it the same as the competence/performance distinction as formulated by Chomsky and others. Perhaps an analogy will make this clear. Human athletic ability is limited by the laws of physics; it is further limited by the specific configurations of bone and muscle and by physiological constraints; it is still further limited at any particular time and place by the exigencies of the context, the conditioning of the athlete, the weather, equipment, and so forth. To say that human athletic competence is limited only by the laws of physics, and that all other limitations (such as inability to jump over the moon) are "performance" limitations, is not a theory that in any way tells us about human beings.

On the other hand, a theory of athletic ability that specified general limitations due to human anatomy and physiology would be highly informative, even though it did not enable the prediction of a specific athlete's performance on a given day. But such a theory must tell us about the range of *potential behavior*, specifically, it must delimit what *could actually be done under the proper conditions*. Thus it would need to make reference to such factors as muscle strength, blood supply, and so forth, and how these control processes.

The theory we seek is a theory of *potential behavior* in this sense.[7] That is, we are not interested in what a specific speaker may or may not have done, nor in attempting to predict what he will or will not do in the future. Rather, we are interested in predicting what he might do under the proper conditions. It seems to us that there are a large number of claims along these lines that are essentially implicit in previous scholarship, although they have rarely been formulated in this way. There is in existence a philosophical literature (e.g. Itkonen 1978), which points out that linguistic theories are not theories of the same order as those found in other sciences precisely because they lack the ability to make these kinds of if-then claims. It seems to us that the misunderstandings about competence and performance lie at the root of the problem. One cannot make verifiable if-then claims about competence, which is by definition inaccessible, and no serious claims could be attempted for actual performance. It is only by focusing on potential, idealized *performance* (behavior) that we are able to make these kinds of claims.

III. LIMITATIONS OF LANGUAGE-BASED MODELS

What is a Language-Based Model?

Despite the diversity of fields interested in language and the diversity of approaches to its study, there is a remarkable unanimity about the existence and the boundaries of this phenomenon. The reason for this lies in the general acceptance of certain notions sometimes taken to be pretheoretical constants, but which are in fact largely the product of centuries of scholarly and other activity on language in the Western culture. Typical of this general consensus are the following: there are many languages, each of which is a separate entity; languages consist of sounds which make up words which make up sentences; hence, for each language, it is possible to set up an analysis that tells us just what it consists of; such an analysis will take the form of several chapters of a grammar plus a lexicon; this analysis represents a kind of knowledge, knowledge of one's language, that is distinct from other knowledge, such as knowledge about the physical world.[1]

The heart of the paradigm is this assumption that language can be studied as an entity in its own right. It is generally presupposed that it *is* possible to draw a boundary between knowledge of language and other knowledge. AI work which uses world knowledge nevertheless assumes that there *is* a body of knowledge about, say, modern American English which is self-contained. Thus, most AI models have been based on some conception of grammar, either explicitly by use of grammar rules plus

---

[7]We believe that a case can be made that such an approach was envisaged by the students of language known as neogrammarians at the turn of the century (e.g., Paul 1880).

[1]Even if we assume, as many AI researchers do, that knowledge of the real world guides—or assists in—the processing of natural language.

parser or implicitly by virtue of their reliance on representations of linguistic knowledge, whether in declarative or in process terms, which express the same information as a grammar.

The concept of grammar naturally suggests a model of language understanding that employs a grammar to assign structural descriptions to sentences, and then computes meanings from these structural descriptions. Research in AI, especially in the early days of NL understanding, has commonly been based on such a model. A major step in such a research program is of course to write a grammar to characterize the set of permissible sentences, and this task is formidable. However, it is not the worst of the problem. As soon as a grammar of any size is used in an analysis system, it is discovered that the "parser" produces a surprising number of structural descriptions for even very short and seemingly simple sentences. Initially, this problem was thought to be a minor inconvenience that caused very long computational times, but which could be gotten around in other ways. In fact the problem is deeper, since very large numbers of structural descriptions are the rule, not the exception. So with the absolute in-or-out criterion of a grammar, one is caught on the horns of a dilemma: either the generative capacity is severely limited, or every sentence becomes multiply ambiguous to an extreme, and no explanation is offered for the fact that humans fail to even *notice* most of the possible readings, though they can do so with effort, or at least understand explanations of them.[9]

On the other hand, more recent AI work has chosen to analyze just fragments of a language and has focused on the use of world knowledge to guide the process of arriving at an interpretation of a sentence. This approach necessarily fails in face of the same facts, though for the opposite reason. In these approaches, there must be an arbitrary dividing line drawn between those interpretations which are "normal" enough to be allowed by the world knowledge and those which will never be allowed. Yet, the observed fact is that people do arrive at readings which no AI program would allow. In fact, the reason that AI researchers have been so conscious of the need to use world knowledge in this way is precisely because, qua human users of English, we are able to arrive at all sorts of unlikely readings, which we then need to rule out.[10]

In sum, the uses to which humans put language are strongly dominated by many considerations, only some of which could be expressed as conventional linguistic knowledge (whether in terms of a grammar or of a heuristic understanding program). On the one hand, a language is more than a set of pairings of utterances and "conceptual representations."[11] On the other hand, the linguistic behavior of human beings is not restricted to some tiny set of "normal" sentences with "normal" interpretations. What interpretations will accrue to any utterance (and, likewise, how a given meaning will come to be expressed) depends not only on the supposed knowledge of the language but on the previous linguistic experiences of the language user (human or simulated), on external circumstances, and very likely on physical and other relatively low-level characteristics of the user.[12]

---

[9]While we are well aware of the proposals that human beings employ processing mechanisms which can handle only a small part of what the grammar generates, very little has actually been done to show that such a model can account for the facts. Indeed, as we point out in the next footnote, there is a strong reason for doubting that this could be the case.

[10]This argument applies with equal force to grammar-plus-processor models which claim that people can process only a subset of the language generated by the grammar, since the dividing line between what is processable ('acceptable' in the standard terminology) and what is not is equally arbitrary and empirically unmotivated.

[11]A fortiori, we reject any view of a language as a set of uninterpreted sentences.

[12]In this context, the most obvious phenomena are the properties of language which are due to the neurophysiology of articulation, the physics of air-borne sounds, and the like, as noted by Paul (1880). In the area of syntactic processing, the most striking example of a hypothesis along these lines would be Yngve's (1960) proposal that human beings use pushdown stacks of extremely limited depth.

Arguments for the Teuchistic Approach

We now turn to some general patterns of linguistic processing by human beings which seem to support a teuchistic model.

- People produce and parse rapidly a vanishingly small proportion of the theoretically possibly utterances of a given language, and moreover, intend and understand only a vanishingly small proportion of the theoretically possible interpretations of those utterances.
- Yet, given sufficient time and effort, people seem to be able to increase these proportions significantly, which is precisely why we are aware of the theoretical possibilities in the first place (in a real sense, they are not merely theoretical).[13]
- It is obvious that any written or spoken sequence can be used as a proper name of a person, place, piece of music, art, or writing, musical group, political party, etc. Such strings can moreover have the form of sentences or parts of sentences which are otherwise "impossible". Thus, *Do you know why you don't not say no sentences like this?* would be ruled out as ungrammatical by some NL systems and forced to mean something like *Do you know why you don't say any sentences like this one* by others, yet it can, and for some people is likely to, mean something like *Do you know [McCawley's (1973) article] "Why you don't not say no sentences like this one"?*. Thus, Charniak (1983: 118) is wrong in assuming that *The boys is dying* must be either left uninterpreted or else (as in his system, PARAGRAM) interpreted as *The boys are dying*. If *The boys* were the name of an avantgarde musical group, and perhaps it is, then this sentence would be interpretable without any such coercion. It follows that in a sense the task of generating all and only grammatical sentences is trivial: generate everything. What will be nontrivial will be the task of constructing appropriate interpretations for certain sentences, namely, those which can most naturally be interpreted by taking certain parts of them as (unusual) proper names.[14]
- A related difficulty, noticed for example by Robinson (1984) in work on DIAGRAM and by many others, is that common nouns (and other categories of words) can be—and are—created often enough to make reliance on any fixed dictionary an impossibility for a NL program of any scope.
- A similar but even deeper problem arises because it is possible to introduce arbitrary codes into a linguistic context. We might say, for example, that from now on *They are flying planes* is to be taken to mean that the phone is being tapped. The correct analysis of the code then depends on the interpretation of its definition at a point that is arbitrarily far removed from its use; this would require the grammar to be arbitrarily large to manage this feat. Alternatively, a grammar could trivially assign all possible interpretations to all possible strings. However, in the real-world codes are both finite and used under restricted circumstances, so the problem appears to be amenable to a teuchistic approach.

A more recent claim is that in favor of people using queues rather than pushdown stacks in linguistic processing (Manaster-Ramer, 1986).

[13]An extreme example involves expert informants, such as highly trained linguists, changing their minds over periods of years about the grammatical status or the interpretations of even short example sentences (e.g., Chomsky (1982) on parasitic gaps, Langendoen (1977; personal communication) on subject-verb agreement in *respectively* constructions). In such cases, NL users appear to operate in times often measured in years and not in milliseconds.

[14]Amsler (1987) has observed that a very large portion of naturally occurring text, such as news service reports, contains words not in dictionaries. Most of these are proper names that may be expressed in a wide variety of manners, so many that Amsler suggest the need for proper name formation rules rather than lexicons to record them; our view is that such methods will themselves be insufficient still. The same difficulty with proper names was encountered in a parser for Chinese character text developed by AMR's students as a class project.

• In real life, there are more competent readers than competent writers, more people adept at comprehension than articulate speakers. Language-based models in principle fail to allow for the possibility of such asymmetries—or else would consider it entirely outside their bounds. Models of behavior are called for.

• Equally striking differences separate different speakers, who despite different linguistic abilities, styles, dialects, and even (especially) knowledge of the topic are able to communicate to a degree, often a substantial degree. Again, language-based models offer no insight into this phenomenon, since they would have to simply recognize as many different languages as there are speakers. One argument against this is the fact that people recognize that certain individuals are better speakers of the *same* language than others (including themselves, oftentimes). All this suggests that we cannot seriously view the idealized (and idolized) native speaker as an expert whose knowledge is to be embodied in the NL programs.

• Communication can persist in the face of disruptions that cause deviations from *any* fully prescribed rules of well-formedness: typographical errors, acoustic distortions, neologisms, jargon, metaphor, halting and restarting, fractional sentences, and so forth. On the other hand, certain other types of interference, such as delaying the feedback of a speaker's own voice to his ears, is devastating. The latter fact is particularly revealing since it argues for a model that pays attention to the physical and other low-level properties of speakers.

• There are many cases where the language appears to change with use, i.e., the question of whether some utterance or utterance-interpretation pairing is licensed by a particular language cannot be answered in the abstract. Furthermore, the linguistic abilities even of a single speaker change constantly, even after puberty. Language-based theories could only deal with such facts by postulating ever newer grammars, which amounts to treating an evolving individual as a sequence of different individuals. Again, to the extent that such change is influenced by factors such as interaction with other speakers or one's physiological development, it falls outside the scope of language-based theories. The same is true, for similar reasons, of creative uses of language, such as metaphor, humor, introduction of words and terminology, and others.

• Perhaps the most striking phenomenon which argues for a teuchistic model is that of language change over generations. The existing models of language inevitably treat any language change, even a small one, as producing (or strictly speaking as being produced by) the rise of a new grammar, discontinuous from the old one. On our view (see also Paul 1880), language change is due to the fact that normal language processing involves feedback, so that the fact that particular forms were processed influences the underlying system. Since the processing is teuchistic and uses various subsystems, it follows that forms that are "incorrect" in terms of the overall system will actually occur and then in turn cause the system to be altered.[15] Moreover, since the overall system need not even be consistent, for many forms it may not even be possible to determine whether they are "correct" or not, yet they will occur, tending to restructure the system in the direction of making them "correct".[16]

All the phenomena discussed above seem to bear directly on the use of language in real-world settings, and increasingly problems due to them are becoming recognized in computational work, especially within the AI tradition. The problem of potentially unbounded ambiguity and difficulties due to the open-ended nature of names and even common words are examples which several researchers have had to deal with. As more and more ambitious system are attempted, more of these problems will be noticed to have debilitating practical effects. We believe that many of these difficulties can be solved by concerted and multi-pronged efforts of a teuchistic nature.

---

[15]The conventional view — even when supplemented with a feedback mechanism — leads to no such insights, since the it guarantees that every form produced will always be "correct" in the original grammar.

[16]For example, if a form occurs a sufficient number of times for whatever reason, it will tend to be reinforced. For example, people will often assimilate forms inconsistent with their original usage if these forms are frequently used by others in their presence (compare the use of infantile or pseudo-foreign forms by adults).

IV. PARALLELS TO OTHER WORK

There exist research paradigms in and out of AI that focus on analogous properties of nonlinguistic human behavior. Two examples are AI models of (nonmonotonic) reasoning and economic models of (satisficing) behavior. Both of these properties feature prominently in our ideas about natural language. The distinction between *optimizing* and *satisficing* processes, introduced by Simon (1947, 1955), illuminates the difference between the search-based models of NL understanding and teuchistic ones. Presumably most human problem solving involves satisficing—finding an adequate job, home, spouse—rather than optimizing—considering and comparing all jobs, houses, people—simply because information and computational resources are limited. On the view of language here proposed, the process of understanding an utterance never reaches completion in the sense of achieving *the* correct reading, nor does it ever consider all possible readings, nor is it necessary that such a set of possible readings exists. Rather, a sentence is interpreted in the context of a particular situation and purpose, so that changes in these factors could change the reading that is *constructed*. The teuchistic process continues until some adequate solution is achieved, or until one gives up the attempt for lack of time, knowledge, memory, or computational capacity. The end to be achieved may be to understand well enough to paraphrase, to answer a specific question, to give the appearance of comprehension, to determine the intelligence or political stance of the speaker, to memorize verbatim, and so on; it is not to find *the* meaning representation. An utterance is not a message to be decoded, it is a set of clues to a resolution of a problem.

The satisficing view that we have outlined leads to a potential connection between our proposals and the recent work in AI that has focused on the nonmonotonic character of human reasoning (see Turner 1984 for a survey). It would seem that most kinds of linguistic models lend themselves to a nonmonotonic manipulation. As far as the oldest attested more or less explicit grammar that has survived into our times (Pāṇini, no date, but definitely B. C.), we find that grammars contain mutually conflicting stipulations such that deductions based on two different rules (or, more generally, subgrammars) would often yield contradictory consequences. Normally, of course, deductions from a single rule of a grammar or a proper subset of the rules are not allowed. To see what a grammar generates, we have to consider the entire grammar, as befits an optimizing model. Hence, this nonmonotonic character of grammars has not had much significance. However, what we are proposing is that the normal situation in human linguistic behavior is satisficing, and this would naturally mean that, if there is such a thing as a grammar, only parts of it will be accessed at any given time in real-time processing. As a result, we would be able to explain how a speaker can consider a sentence ungrammatical for a time and then decide, upon further reflection (i.e., when more of the grammar is considered) that it is grammatical, or vice versa.

V. CONCLUSIONS

To our mind, any model of natural language which is to be useful must concern itself with the kinds of issues and phenomena we have discussed: the fluidity, imprecision, situation sensitivity, and inconstancy of language, and the robustness of human understanding. The tradition has tacitly opted for monotonic and optimizing models, not so much because of any definite arguments in favor of such models as because of a widespread failure to concern oneself with these kinds of issues in the first place. The notion that natural languages exist as ideal entities in their own right naturally leads to models that in effect are monotonic and optimizing, in the same way that considering classic logic instead of actual human reasoning or ideal economic systems instead of real ones will lead to analogous results. If we consider human linguistic behavior as our domain of study, then the satisificing and nonmonotonic models are almost inevitable. Teuchistic processes are appropriate to characterize these models.

References

Amsler, R. A. Words and Worlds. *Proceedings of TINLAP-3*. Las Cruces, NM: New Mexico State University Computing Research Laboratory, 1987, 16–19.

Charniak, E. A parser with something for everyone. In M. King (Ed.) *Parsing natural language*, New York: Academic Press, 1983, 117–149.

Chomsky, N. *Some concepts and consequences of the theory of government and binding.* Dordrecht: Foris, 1986.

Cullingford, R. E. *Natural language processing: A knowledge-engineering approach.* Totowa, NJ: Rowman & Littlefield, 1986.

Hill, A. Grammaticality. *Word, 17*, 1961, 1–10.

Itkonen, E. *Grammatical theory and metascience.* Amsterdam: John Benjamins, 1978.

Lindsay, R. K. Jigsaw Heuristics and a Language Learning Model. In N. Findler and B. Meltzer (Eds.) *Artificial intelligence and heuristic programming.* Edinburgh: Edinburgh University Press, 1971.

Lytinen, S. L. Dynamically combining syntax and semantics in natural language processing. *Proceedings of AAAI-86*, 1986, 574–578.

Manaster-Ramer, A. Copying in natural languages, context-freeness, and queue grammars. *Proceedings of Association for Computational Linguistics-86*, 1986, 85–89.

Marcus, M. *A theory of syntactic recognition for natural language.* Unpublished doctoral dissertation, MIT, 1978.

McCawley, J. D. A note on multiple negations, or, Why you don't not say no sentences like this one. *Grammar and Meaning*, 1973, 206–210.

Pāṇini. n.d. *Aṣṭādhyāyī.* Published 1887 as *Pāṇini's Grammatik* (O. Böhtlingk, ed.), Leipzig; reprinted Hildesheim: Olms, 1964.

Paul, H. *Prinzipien der Sprachgeschichte.* Halle: Niemeyer, 1880.

Peters, S. Discussion of W. C. Rounds's paper 'The relevance of computational complexity theory to natural language processing', *Proceedings of the Conference on the Processing of Linguistic Structure.* Palo Alto CA: Center for the Study of Language and Information, 1987, to appear.

Reiter, R. A logic for default reasoning. *Artificial Intelligence*, 1980, *13*, 81–132.

Riesbeck, C., and Schank, R. C. *Comprehension by computer: Expectation-based analysis of sentences in context.* Technical Report 78, Department of Computer Science. Hartford CT: Yale University, 1976.

Robinson, J. *Extending grammars to new domains.* Report RR-83-123. Los Angeles: Institute for Science and Information, 1984.

Simon, H. A. *Administrative behavior.* New York: Macmillan, 1947.

Simon, H. A. A behavioral model of rational choice. *Quarterly Journal of Economics*, 1955, *69*, 99–118.

Small, S. *Word expert parsing*. Unpublished doctoral dissertation, Department of Computer Science, University of Maryland, 1980.

Turner, R. Semantic theory of non-monotonic inference. In R. Turner (Ed.) *Logic for artificial intelligence*. Chicester: E. Horwood, New York: Halstead Press, 1984.

Wilks, Y. An intelligent analyzer and understander of English. *Communications of the Association for Computing Machinery*, 1975, *18*, 264–274.

Winograd, T. *Understanding natural language*. New York: Academic Press, 1972.

Woods, W., Kaplan, R., and Nash-Webber, B. *The Lunar Sciences Natural Language Information System: Final report*. Technical Report 2378. Cambridge MA: Bolt, Beranek, and Newman, 1972.

Yngve, V. A model and an hypothesis for language structure. *Proc. of the American Philosophical Soc.*, 1960, *104*, 444–466.

PARSNIP: A Connectionist Network that Learns Natural Language Grammar from Exposure to Natural Language Sentences

Stephen José Hanson

and

Judy Kegl

Bell Communications Research
435 South Street
Morristown, NJ 07960

Princeton University
Cognitive Science Laboratory
221 Nassau Street
Princeton, NJ 08542

Abstract

Linguists have pointed out that exposure to language is probably not sufficient for a general, domain-independent, learning mechanism to acquire natural language grammar. This "poverty of the stimulus" argument has prompted linguists to invoke a large innate component in language acquisition as well as to discourage views of a general learning device (GLD) for language acquisition. We describe a connectionist *non-supervised learning* model (PARSNIP[1]) that "learns" on the basis of exposure to natural language sentences from a million word machine-readable text corpus (Brown corpus). PARSNIP, an *auto-associator*, was shown three separate samples consisting of 10, 100 or 1000 syntactically tagged sentences, each 15 words or less. The network learned to produce correct syntactic category labels corresponding to each position of the sentence originally presented to it, and it was able to generalize to another 1000 sentences which were distinct from all three training samples. PARSNIP does sentence completion on sentence fragments, prefers syntactically correct sentences, and also recognizes novel sentence patterns absent from the presented corpus. One interesting parallel between PARSNIP and human language users is the fact that PARSNIP correctly reproduces test sentences reflecting one level deep center-embedded patterns which it has never seen before while failing to reproduce multiply center-embedded patterns.

Keyword Topics: Connectionist Models, Neural Nets, Learning, Language Acquisition

---

1. The name PARSNIP was chosen to emphasize that the present model is not a parser, but a "snippet" or precursor to a parser and is most similar to a syntactic analyzer. Our work was supported in part by a grant to Princeton University from the James S. McDonnell Foundation. We would like to thank Donald Walker, Stu Feldman and the connectionist group at Bellcore for comments on previous versions of this paper.

Introduction

Connectionist approaches to language processing (Feldman, 1985; Rumelhart & McClelland, 1986) have recently gained attention because of a need to simultaneously integrate diverse sources of information about the syntax, semantics, and pragmatics of a sentence. One important aspect of these neural-like models is their ability to combine information from various sources while at the same time allowing these sources to mutually constrain each other, reducing the need to prioritize one type of information over another during parsing.

Many questions arise concerning the computational nature of connectionist models and their potential role in natural language processing. Central to connectionist models is a *learning* process which determines how structure and rule governed behavior emerges. Unfortunately, the learning rules so far proposed (Ackley, Hinton & Sejnowski, 1985; Rumelhart, Hinton & Williams, 1986) focus primarily on the frequency of occurrence of relevant structural units within a given domain and require explicit supervision over the recognition and coding of generalizations concerning each stimulus encountered. Such constraints on learning procedures raise serious questions about the possibility of modeling natural language learning in a connectionist framework. Studies in learnability theory (Chomsky, 1957) have shown that natural language syntax cannot possibly be induced from the first-order statistics (e.g., transition matrices or conditional probabilities) available through exposure to an infinite number of examples. Children acquiring natural language are sensitive to a set of universal constraints on structural configurations and relations in language such as the A over A Condition, Subjacency, etc. (see Radford, 1981 for a description of these constraints). Even though violations of these conditions have not been explicitly corrected, discouraged, nor even experienced; children avoid violating these conditions even in their earliest linguistic utterances (Chomsky, 1965; Randall, 1982).

The acquisition of natural language poses particular problems for any learning approach. Language acquisition cannot rely on any explicit information about the grammaticality, usage, frequency, possible constituency, or any structural information about the sentence other than the linear order and cooccurrence of words in the sentence--and that information is hindered by performance errors, incomplete sentences and general noise. Under such conditions, it is hard to imagine how syntax, and natural language generally, is acquired at all. Chomsky (1972) approached this problem by assuming a nativist perspective in which the child was seen as using incoming language data in conjunction with innate linguistic knowledge to formulate hypotheses about possible grammatical rules and constraints.

Previous Work

Other computational models of language acquisition from both connectionist and rule-based approaches have tended to assume that a large amount of previous structure must be present to learn natural language syntax. A recent model (Berwick, 1985) incorporating linguistic assumptions from a Government and Binding perspective (Chomsky, 1981 and subsequent work) uses a "repair" operation on syntactic rules that are already present but need to be tuned properly. This tuning is based on incremental positive evidence in that sentences the learner hears are assumed to be grammatical and each new sentence must be incrementally accounted for. This type of acquisition where positive evidence and reactionary generalization is enforced is sometimes referred to the the "subset principle" (Berwick, 1985), Interestingly, the connectionist model proposed here can be seen as consistent with the subset principle.

Connectionist models (Feldman, 1985; Rumelhart & McClelland, 1986; Selman, 1985; McClelland & Kawamoto, 1986) have tended to provide the system with explicit rules, syntactic structure or both. They have allowed the network to learn the proper conditions under which to apply these rules or to recognize specific relations between constituent structures.

Explicit analysis of the kinds of preconditions or structure needed prior to learning natural language grammar have yet to be considered for a connectionist model. For example, no connectionist models currently exist which build up their syntactic knowledge from mere exposure to positive examples and the subsequent incremental addition of new sentences. This model begins with no assumptions about syntactic structure nor any special expectations about properties of syntactic categories other than the fact that they exist.

The Present Model

We begin with the assumption that natural language reveals to the hearer a rich set of linguistic constraints and that observable syntactic regularities serve to delimit the possible grammars that can be learned. We are not making an anti-nativist argument, in fact, the present model actually contributes to the analysis of the tradeoff between innate syntactic knowledge and previously unrecognized syntactic regularities in the data that could be used to induce grammar. Connectionist models which learn in this way can offer a new paradigm for nativist research. By filtering out those data which can be learned, we may delineate those aspects of the knowledge of language which are truly hardwired.

PARSNIP uses a variation of a *backpropagation* technique (Rumelhart, Hinton & Williams, 1986) called "auto-association" which was originally proposed by Rumelhart and Hinton. These models are multi-layer learning networks (MLL; Hanson & Burr, 1987) that have units associated with input and output as well as a modifiable set of intermediate units called "hidden units." Although backpropagation is strictly a supervised technique, auto-association is not. The difference lies in the teacher signal. Backpropagation requires a separate teacher signal for every input-output pair, whereas auto-association uses the input as the teacher signal. The auto-association network's task is to produce a veridical copy of the input with which it is presented. It must recognize this input as something it has seen before.

This seemingly straightforward task becomes difficult when the network is exposed to a large number of stimuli or when the the number of "hidden units" is small compared to the number of input/output units, forcing a compression or reduction of the information which is encoded during learning. Reducing the number of encoding units is likely to yield a new (compressed) encoding of the input information in order to adequately map it to the output, akin to chunking smaller units into higher order constituents. The auto-associator may extract regularities more general than those exhibited by the input stimuli, or it may discover features or complexes of features that are useful in predicting the output stimulus.

We are asking the following question of our network: Can it induce grammar-like behavior (rule-governed behavior) from simple exposure to a large corpus of natural language sentences. Several specific questions will also be posed: After learning on a specific set of input, can the network generalize to sentences never seen before? Does it prefer sentences that are syntactically correct? Can it recognize sentences that are more complex than those that would be predicted by simple conditional probabilities on the combinations of fragments it has previously seen? And finally, after learning on a sizeable natural language corpus, is the network resistant to learning sentences which violate syntactic well-formedness conditions purported to be universally applicable. This last question is a particularly interesting one, and is indicative of the types of questions that should be posed. If PARSNIP does not recognize such sentences or resists learning such sentences (not in this paper) after nothing more than exposure to data, this would lead us to suspect that rather than being an innate property of the learner, these constraints and conditions follow directly from regularities in the data.

A key aspect of grammar induction is the ability of the network to recognize forms that are syntactically correct but did not appear in training. Concurrently, it must not recognize syntactically incorrect forms that also never appeared in the training sample. This is a differential generalization constraint. Not only must the network generalize to new sentences, it must have a means of determining grammaticality; and worse yet, it must do so strictly on the basis of positive evidence.

Input Representation and Stimuli

PARSNIP was exposed to sentences from the Brown corpus (Francis & Kučera, 1979) consisting of one million words of running text. This corpus, compiled over a 10 year period, is composed of 500 text samples each consisting of approximately 2000 words. The texts are representative of 6 separate categories and approximately 19 subcategories, including newspaper text, religious books, technical books and novels.

President Kennedy today pushed aside other White House business to devote all his time and attention to working on the Berlin crisis address he will deliver tomorrow night to the American people over nationwide television and radio.

n-tl np nr vbd rb ap jj-tl nn-tl nn to vb abn pp$ nn cc nn in vbg in at np nn nn pps md vb nr nn in at jj nns in jj nn cc nn .

My advice , if you live long enough to continue your vocation , is that the next time you're attracted by the exotic , pass it up -- it's nothing but a headache. As you can count on me to do the same. Compassionately yours , S. J. Perelman

pp$ nn , cs ppss vb jj qlp to vb pp$ nn , bez cs at ap nn ppss+ber vbn in at jj , vb ppo rp -- pps+bez pn cc at nn . cs ppss md vb in ppo to do at ap . rb pp$$

She was a living doll and no mistake -- the blue-black bang , the wide cheekbones , olive-flushed , that betrayed the Cherokee strain in her Midwestern lineage , and the mouth whose only fault , in the novelist's carping phrase , was that the lower lip was a trifle too voluptuous.

pps bedz at vbg nn cc at nn -- at jj nn , at jj nns , jj , wps vbd at np nn in pp$ jj-tl nn , cc at nn wp$ ap nn , in at nn$ vbg nn , bedz cs at jjr nn bedz at nn ql jj .

Figure 1: Example Sentences Taken From the Brown Corpus

We chose the Brown corpus because it is one of the few sample corpora where each word of text is associated with a tag which indicates its syntactic category. The tags for each individual word were determined by linguistically informed judges. Examples of text are shown in Figure 1. The three sentences are taken from the beginning, middle and end of the corpus and provides some idea of the diversity of sentence types and topics.

Below each sentence is a string of syntactic tags. There are approximately 81 unique word class tags comprised from about 6 kinds of syntactic information including major form classes ("parts of speech"), function words, inflectional morphs and punctuation. Tags were also combined during the labeling process in order to create new codes where needed. This compounding resulted in a total of 467 unique syntactic codes over the entire corpus. We used a nine bit binary representation to code all 467 categories then input these binary representations to the auto-associator. Tags were assigned to bit pattern codes by frequency of occurrence in the corpus; most frequent were assigned to most active input codes while least frequent were assigned to less active input codes[2] (sparser). To restrict sentence diversity, the length of the sentences shown to the network was limited to 15 words or less.[3] The Brown Corpus contains approximately 35,000 sentences of 15 words or less.

Architecture

The PARSNIP network consist of a total of 585 units and 24,615 connections. Each unit's Fan Out is completely connected to units above. The unit Fan In was combined by a linear integration function over the activation states below it and over the weights connected to these states. The unit Fan Out was normalized over the interval zero to one and was compressed in the high and low ends of the scale. This type of function (e.g., logistic) transforms activation at a unit to something like "firing rate" for a neural interpretation, or "likelihood" if a probabilistic interpretation is given (Hanson & Burr, 1987).

Auto-Associator for Natural Language Syntax

585 units 24615 connections

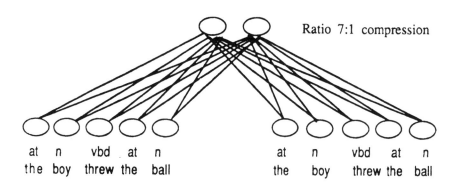

Figure 2: Schematic Version of Auto-Associator

---

2. Experiments were also attempted with random code assignments and there seemed to be little difference in the learning; although generalization performance has yet to be compared.

3. One consequence of restricting the length of sentences was the elimination of sentences containing relative clauses from the corpus. The absence of these sentence types will prove important in subsequent discussions.

A schematic version of the architecture of the auto-associator is shown in Figure 2. The input included 270 units coding 15 word positions (not including punctuation) and 14 word boundary codes. The output was identical to the input, also consisting of 270 positions. Hidden units varied in number from 10 to 60, although the data reported here is for 45 hidden units. This limitation on the number of hidden units provides a 7 to 1 compression of the data through the hidden layer.

The learning procedure was implemented with the generalized delta rule (Rumelhart, Hinton, Williams, 1986) at the output layer and was applied recursively to the layer below (between the hidden units and the input). The targets for the output layer were the input values themselves. The weights were adjusted by the following formula:

$$\Delta w_{ij}^{n+1} = \eta \, (o_{jp} * \delta_{ip}) + \alpha \, \Delta w_{ij}^{n} \qquad (1)$$

The parameter η represents the rate at which any particular sample error can affect the weights. α is a parameter that determines the effect that past deltas have had on the present delta. For α equal to 1, the present weight change and past weight change have the same effect in the weight update. An o_{jp} is the value for the jth unit and the pth pattern. And δ is the error gradient for the ith unit and the pth pattern. All experiments used an η of .1 and an α of .3.

PARSNIP Experiments

Sentences including punctuation (e.g., periods) were entered into one side of the auto-associator with padding (effectively zero or no input) after the period in order to uniformly fill 15 positions. Starting with random weights, a forward activation on the input produced activation on the output, also in 15 nine bit positions. The nine bit patterns were then compared to the input bit pattern, yielding the errors for each output value. These errors were then used to adjust the weights as specified in the delta rule.

In three separate training sessions, the PARSNIP network was separately trained[4] on three distinct sets of of sentences of sizes 10, 100 and eventually 1000.

---

4. The auto-associator/back-propagation simulator was written for a vectorizing FORTRAN compiler on a Convex C1 computer. Simulation runs, dependent on problem size, took any where from 5 hours to 3 1/2 weeks.

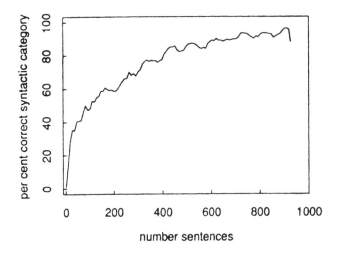

Figure 3: Learning of 10 Sentences from the Brown Corpus

The network was exposed to each set until criterion was reached (>95% correct on the entire set) or until no positive slope in the learning curve was detected. Errors were calculated from the number of missed *syntactic categories*. Thus a single bit error in the nine bit code would be counted as a miss of the entire category. Figure 3 shows percent correct ((1-error)*100) for the 10 sentence set as a function of the number of sentence presentations. Criterion was reached after about 100 cycles through the 10 sentences, namely, after about 1000 sentence presentations. In Figure 4 we show the transfer point to a new set of 100 sentences after having learned on the 10 sentence set. The first point in this graph (Figure 4) shows the last point from the 10 sentence set (Figure 3) and the next point shows the network's performance on a new sentence. Notice that performance drops dramatically from about 97% correct to 50% correct.

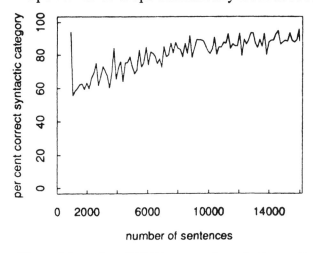

Figure 4: Learning of 100 Sentences from the Brown Corpus

Recall that the presence of word boundary information and end of sentence punctuation will allow the network to get at least 50% correct if it is able to retain just this information. In this case, after the learning on the 10 sentence set, word boundary information is all the network seems able to retain (also see below).

The network reaches criterion after about 160 cycles on the 100 sentences, namely, in about 16,000 sentence presentations. Notice that the learning curve is much more jagged in this case as compared to the 10 sentence set. Apparently, the learning of some sentence structures tends to compete with the learning of other sentence structures[5]. Finally, Figure 5 indicates the beginning of transfer to 1000 new sentences. The first point, as before, is the last point of the learning curve for the 100 sentences, and the next point shows the response of the network to a new sentence. Again, the drop is rapid. But, this time slightly more information is retained about sentence structure as can be seen by the fact that the drop only reaches about 60%. As learning proceeds, it follows a gentle positive slope, although the jaggedness of the learning curve is much greater, and learning criterion is never reached with this sentence set. To ensure that the asymptote was reached, the sentences were cycled through 180 times (180,000 sentence presentations). This time it became apparent that the network had difficulty encoding all 1000 sentences. The final performance level achieved exhibited correct recognition on about 85% of the sentences.

Acquisition by Trials. The initial output of the network involves codes that are associated with low activation. That is, the network is inhibitory in early stages of acquisition. This is attributable to the fact that error reduction drives weight changes and to the sparseness of codes. For example, if most of the codes which the network is exposed to are sparse, that is have few 1's in the target, then the network can significantly reduce error by turning off output bits and thereby making the network inhibitory. This produces a tendency for the network to retrieve codes associated with low activation. Because of the sorting of codes by frequency, these will also be low frequency categories.

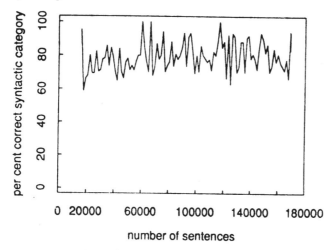

Figure 5: Learning of 1000 Sentences From the Brown Corpus

Within the first 30 trials (sentences), the network seems to pick up the first obvious regularity, that of word boundaries. Next, within the next 100 or so sentences, mass nouns and personal pronouns begin to be correctly predicted, as well as a few two sequence syntactic codes like

---

5. This type of learning curve is characteristic of learning rates that are too high for the sample. It is possible that too few hidden units are present for optimal learning. To control for the the possibility that the learning rate was too high it was dropped to half its value (.05). However, a similar amount of jaggedness was still apparent in the learning curve. In addition, experiments where 1/3 more hidden units were added in conjunction with smaller learning rates did not result in a substantial change in the texture of the curve.

article+noun and preposition+noun. As learning precedes, more complex forms begin to appear, but not with obvious predictability. Further analysis where sequences of tags are tracked through learning trials should be revealing.

Generalization Performance

The weights for all three earlier sample sizes were retained for a generalization test. With learning turned off ($\eta = 0$, $\alpha = 0$), each network was shown 1000 new sentences which were distinct from the 1110 sentences the three networks originally learned. The percent correct on the 1000 new sentences was recorded and the results are shown in Figure 6. On the x axis is the size of the sample of sentences the network had previously learned, and on the y axis is the percent correct of sentences which the network was able to predict from a novel set of 1000 sentences. Notice that the function is increasing, that is, more prior training on sentences produces greater generalization to novel sentences. As previously described, knowledge about the 10 sentence network drops to word boundary knowledge, losing about 50% of its sentence knowledge. The 100 sentence network retains about 10% more information about sentence structure (syntactic category relations), losing about 40% of what it had learned. Finally, the 1000 sentence network seems to be generalizing at about the same rate (84%) at which it had asymptotically learned. As these are log-log coordinates, it appears there is a hint that the 3 points approximate a power function of prior learning on sentence sample size.

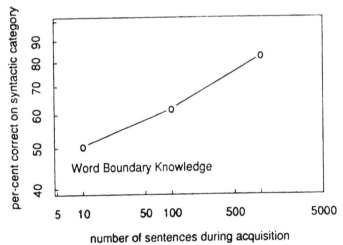

Figure 6: Generalization Performance

Recognition Performance

PARSNIP's main task is to recognize a sentence either as one it has seen before or as one that it might have seen before. This is the type of performance one might expect from any associative memory in which a large number of patterns have been stored. However, PARSNIP is much more than a pattern storer. It is in fact able to behave in a rule governed way with respect to sentence completion and recognition of sentence types it has never seen before. The question is whether the composition of sentence fragments or constituents are determined on the basis of first order statistics (conditional probabilities between sentence fragments) or whether they can be attributed to more complex generalizations arising from exposure to a large number of sentence types. All the remaining experiments were performed using the 1000 sentence network.

Pattern Completion. The first task that PARSNIP was asked to perform was sentence completion on the basis of partial input. In Figure 7 we show a sample interaction with PARSNIP. The syntactic tags representing the sentence *the boy threw a ball*, are clamped on one side of the input, Then, PARSNIP produces the same sentence, i.e. "article noun past-tense verb article noun". Suppose now that the verb is left out of the sentence and, instead, PARSNIP is shown an ambiguous code at the third position in the sentence. In this case, PARSNIP produces on the output side in the third position the tag "past-tense verb," or as a second guess "verb". That is, generalizations it has made concerning the possible structure of sentences cause PARSNIP to be reminded of the syntactic categories that best fit the empty slot in the sentence it was shown.

INPUT: ARTICLE NOUN P-VERB ARTICLE NOUN (The boy threw the ball)
PARSNIP: **ARTICLE NOUN P-VERB ARTICLE NOUN**

INPUT: ARTICLE NOUN <BLANK> ARTICLE NOUN
PARSNIP: **ARTICLE NOUN <VERB P-VERB> ARTICLE NOUN**

Figure 7: Sample Interaction with PARSNIP: Pattern Completion

Disambiguation. Another task that PARSNIP was asked to perform was one of syntactic disambiguation given a set of possible syntactic codes for a lexical item appearing in a particular sentence position. This task was similar to pattern completion except that the network was given a number of items and asked to produce the correct one. In Figure 8 we show a sample interaction where PARSNIP is given the sentence *The horse raced past the barn fell.* The word "past" could appear in a sentence as either an adverb, a preposition an adjective or a noun. The word "past" in this sentence begins the prepositional phrase "past the barn".

INPUT: ARTICLE NOUN P-VERB PREPOSITION ARTICLE NOUN P-VERB
PARSNIP: **ARTICLE NOUN P-VERB PREPOSITION ARTICLE NOUN P-VERB**

INPUT: ARTICLE NOUN P-VERB *ADVERB* ARTICLE NOUN P-VERB
PARSNIP: **ARTICLE NOUN P-VERB *PREPOSITION* ARTICLE NOUN P-VERB**

Figure 8: Sample Interaction with PARSNIP: Disambiguation

A dramatic way to demonstrate disambiguation in PARSNIP is to clamp the *incorrect* syntactic choice (adverb instead of preposition) as shown in figure 8. In response to this deliberate introduction of misinformation, PARSNIP *edits* the sentence and actually inserts the correct syntactic category. In this case, it should be noted that this particular sentence never appears in the 1000 sentence corpus to which PARSNIP was exposed.

Recursion. Sentence embedding, the ability of grammar to produce a sentence within another sentence, is considered a characteristic defining feature of natural languages. Paradoxically, it has also been shown that sentence recursion is not an unlimited feature of

115

natural language processing. Even at the next level of embedding (a sentence within a sentence within a sentence), human language users have difficulty (Miller, 1962). Therefore, general recursive rules must be filtered out somehow, and usually memory constraints are invoked in order to do this.

```
INPUT:      ARTICLE NOUN ARTICLE NOUN P-VERB P-VERB
PARSNIP:    ARTICLE NOUN ARTICLE NOUN P-VERB P-VERB


INPUT:      ARTICLE NOUN ARTICLE NOUN ARTICLE NOUN P-VERB P-VERB P-VERB
PARSNIP:    ARTICLE NOUN ARTICLE NOUN ARTICLE NOUN P-VERB NOUN VERB
```

Figure 9: Sample Interaction with PARSNIP: Recursion

In the sample interaction in Figure 9, PARSNIP is able to recognize the sentence *The rat the cat chased died.* This recognition occurs despite the lack of even a single occurrence of a center-embedded sentence within the corpus. Nonetheless, PARSNIP is able to respond to this sentence as something it recognizes. Apparently, PARSNIP is able to bind together constituents that have been used in other contexts. However, when a doubly embedded sentence, e.g., *the rat the cat the dog bit chased died* is clamped on the input side, PARSNIP produces a partial sentence but does not recognize this second level of recursion. Although constituents similar to those found in single level center-embedding are available in this more complex center-embedded, the the failure might be seen in terms of the number of and distance between constituents that must be bounded by PARSNIP's recognition rule. In other words, PARSNIP is not able to recognize constituents that it has previously recognized because they are bounded by constituents that may be unfamiliar or have not previously been useful in syntactic prediction. Note also that this effect is also independent of any memory constraints since PARSNIP is exposed to a total sentence in parallel.

Adjacency Constraints. In English, a direct object must be adjacent to a verb in order to receive case from it and thereby be allowed (licensed) to occur in object position. (This statement is phrased within the terminology of a Government-Binding approach (Chomsky, 1981).) English speaking children will probably seldom or never hear *John gave quickly the book*, where quickly intervenes between a verb and its object and blocks the assignment of accusative case by virtue of destroying the adjacency relation between the verb (the case assigner) and the direct object (the NP which must receive case). Furthermore, children acquiring English will never be explicitly discouraged from using these sentences if they should happen to hear them, e.g., "By the way, don't say this sentence". A key question in the evaluation of how language is acquired concerns the ability of the network to avoid generalizing to sentences that have adjacency violations of this type and which are not present in the training set.

INPUT: NOUN ADVERB VERB ARTICLE NOUN (men quickly steal the food)
PARSNIP: NOUN ADVERB VERB ARTICLE NOUN

INPUT: NOUN VERB ADVERB ARTICLE NOUN (men steal quickly the food)
PARSNIP: NOUN ADVERB WAS ARTICLE NOUN

Figure 10: Sample Interaction with PARSNIP: Adjacency Constraints

In Figure 10 we show PARSNIP failing to recognize an adjacency violation, in fact, it actually attempts to move what was the VERB (now retrieved as "WAS") closer to the direct object. Note that PARSNIP can also recognize *men quickly steal* or *men steal quickly* implying that the presence of the direct object is critical for this recognition failure.

Discussion

Auto-association is clearly not a plausible model for language acquisition. That is, repeated parallel exposure to a sentence with enforced production of that sentence is not a reasonable cognitive model of language acquisition, nor of a language learner's grammar production. Perhaps the closest parallel to PARSNIP's situation is that of a learner engaged in the abstract intensive study of sentences and sentence structure (similar to the activities of a linguist). It is also important to note that unlike human language learners, the network has no sense of temporal order. For PARSNIP sentences have no beginning, middle or end, but rather they exist as patterns which can be used to account for the structures it encounters. Nonetheless, there are some important parallels between the task given to PARSNIP and the task that arises for children as they learn natural language. Both PARSNIP and the child are only exposed to sentences from natural language, they both must induce general rules and larger constituents from just the regularities to which they are exposed, both one the basis of only positive evidence.

PARSNIP's ability to generalize from what it has learned to new sentences indicates that some general knowledge of constituent structure has been extracted from its experience with natural language sentences. A significant amount of coverage of sentence types occurs after training on 1000 as compared to the original 10 sentences.

It is far more interesting to us to have discovered that PARSNIP can differentially generalize to sentences that can appear in natural language (center embeddings) but cannot recognize sentences which violate natural language constraints (multiple center embeddings). As evidence that PARSNIP is using rule-like representations or possibly possesses something comparable to a grammar, we feel it is important to point out the fact no center embedded sentences appeared in the training set. In fact, even the number of adjoined relative clauses was almost nil as a result of the limitation on sentence length. Apparently, constraints from the sentences already learned allows PARSNIP to differentially generalize as though syntactic rules are in operation.

Further, the constituents that PARSNIP chooses tend neither to be predictable from first order statistics nor to be able to be generated from simple finite state grammars. PARSNIP prefers sequences of syntactic categories that often are the least likely to be predicted on the basis of the frequency with which one category follows the other in the corpus. For example, in one pattern completion interaction when PARSNIP was given the phrase *the destruction of the* *city <blank>*, it chose to fill the blank with a conjunction producing *the destruction of the city* **and**. The frequency of syntactic codes in the corpus following the syntactic codes for *the city, of the city, destruction of the city* or *the destruction of the city* were always greater (sometimes 10 times greater) for other syntactic categories (e.g. prepositions) than for conjunctions.

Although, we don't have a complete analysis of the constituents PARSNIP knows, we do have some evidence to suggest that PARSNIP recognizes noun-phrases and other higher order consituents in the hidden layer. PARSNIP was exposed to sentence fragments that were noun phrases, verb phrases, or random sentence fragments. Then, during recognition, the hidden layer values were clustered yielding groups containing either noun phrases with some random fragments or verb phrases. Much more can be done with this type of methodology in terms of isolating the constituent information that PARSNIP uses.

The acquisition strategies exhibited by PARSNIP conform to what is usually thought of as a nativist principle, the Subset Principle. This principle is usually described in terms of a child moving from one grammar to another:

> "Each step of a child's acquisition of grammar must involve movement from a smaller set to a larger set and cannot involve the reverse. The steps are motivated by pieces of input data (adult sentences) which fail to fit into the smaller set, thereby forcing an expansion of the set."
> (Roeper, in press)

This is exactly the sort of conservative generalization that one might expect from an auto-associator such as the one employed by PARSNIP. The network is lead to change its syntactic knowledge (connections/weights) based solely on single sentence violations of prior successful generalizations about a subset of sentences that it had previously constructed. This process is incremental because the entire learning process in connectionist networks is based on small incremental changes motivated the success or failure of its generalizations about the data.

References

Ackley, D. Hinton G.E. and Sejnowski, T., A Learning Algorithm for Boltzmann Machines, Cognitive Science, 9,1,1985.

Berwick, B. The acquisition of syntactic knowledge, MIT Press, 1985.

Chomsky, N. Syntactic structures, The Hague: Mouton, 1957.

Chomsky, N. Aspects of the Theory of Syntax, Cambridge, Mass.: MIT Press, 1965.

Chomsky, N. Language and Mind, Harcourt Brace Jovanovich, 1972.

Chomsky, N. Lectures on Government and Binding, Foris, 1981.

Feldman, J. Connectionist Models and Their Applications. Cognitive Science Special Issue, 9, 1, 1985.

Francis, W. N. and Kučera H., Manual of information to accompany a standard corpus of present-day edited american english for use with digital computers, Department of Linguistics, Brown University, 1979.

Hanson, S. J. and Burr, D. J., Knowledge Representation in Connectionist Networks. Submitted paper to AAAI, 1987.

Miller, G. A., Some psychological studies of grammar, American Psychologist, 17, 748-762, 1962.

McClelland J. and Kawamoto A. H., Mechanisms of sentence processing: assigning roles to constituents, in Parallel Distributed Processing, Vol II: Psychological and Biological Models, McClelland J. and Rumelhart D. (Eds.), Bradford Books/MIT Press, 1986.

Radford, A., Transformational Syntax, New York: Cambridge University Press, 1981.

Randall, J. H., Morphological Structure and Language Acquisition, Unpublished Doctoral Dissertation, University of Mass. at Amherst, Linguistics Department, 1982.

Roeper, T. Formal and substantive features of language acquisition: reflections on the subset principle and parametric variation, in Cognitive Science, S. Steele (Ed.) University of Arizona Press, in press.

Rumelhart D.E., Hinton G.E., and Williams R., Learning Internal Representations by error propagation. Nature, 1986.

Rumelhart D. E. and McClelland J. (Eds.), Parallel Distributed Processing: Explorations in the Microstructure of Cognition. Vol 1: Foundations. Bradford Books/MIT Press, Cambridge, Mass., 1986

Rumelhart D. E. and McClelland J., On learning the past tenses of english verbs, in Parallel Distributed Processing, Vol II: Psychological and Biological Models, McClelland J. and Rumelhart D. (Eds.), Bradford Books/MIT Press, 1986.

Selman, B. Rule-based processing in a connectionist system for natural language understanding (TR CSRI-168), Toronto: University of Toronto, Computer Systems Research Institute, 1985.

"Word Pronunciation as a Problem in Case-Based Reasoning"

Wendy G. Lehnert
Department of Computer and Information Science
University of Massachusetts
Amherst, MA 01003
413-545-3639
LEHNERT.UMASS@CSNET-RELAY

ABSTRACT

English word pronunciation is a challenging knowledge acquisition problem in which general rules are subject to frequent exceptions of an arbitrary nature. We have developed a supervised learning system, PRO, which learns about English pronunciation by training with words and their dictionary pronunciations. PRO organizes its knowledge in a case-based memory which preserves fragments of training items but does not remember specific training items in their entirety. After PRO has created a Case Base in response to a training set, it can pronounce novel test words with substantial degrees of success. Test items are processed by generating a search space in the form of a lateral inhibition network and embedding this search space in a larger network that reflects PRO's previous training experience with relevant fragments. Spreading activation and network relaxation are then used to arrive at a preferred pronunciation for the given test item. In this paper we report preliminary test results based on a training corpus of 750 words and a test set of 300 words.

keywords: learning, case-based reasoning, parallel processing
session preference: full paper

Introduction

The rules underlying English word pronunciation are diverse, uncertain, and frequently at odds with one another. For example, when a single vowel is followed by a consonant and a final "e" at the end of a word, a good general rule says to pronounce the vowel as a long vowel. This works for words such as "like," "rope," and "mate." Unfortunately, it doesn't work for words like "love," "move," "give," and "have." But we can't fix our rule by simply excluding the consonant "v" from the general pattern, because many words with a "v" do obey the long-vowel pattern: "save," "gave," "jive," "cove."

Most general rules for English pronunciation are subject to a large number of exceptions which appear to be largely arbitrary. It is tempting to say that one must simply learn each word on a case-by-case basis and forget about identifying a rule-base for this problem. Indeed, once one has established a "sight-vocabulary," the process of word pronunciation must be heavily aided and influenced by the process of word recognition.

Yet there is evidence that children learning to read rely heavily on a facility for phonological recoding which allows them to move from the written word to a pronunciation of that word before achieving recognition of the word [Bradley & Bryant 1983, Doctor & Coltheart 1980]. Other researchers have argued that phonological recoding is a necessary component of skilled reading as well [Gough 1972].

We have implemented a model of English word pronunciation in the form of a computer program called PRO. PRO learns to produce phonological encodings (pronunciations) for isolated input words by first training on word/pronunciation pairs. All memory structures utilized by PRO are created automatically during training, and PRO implements a model of phonological encoding that is fully independent of the word recognition problem.[1]

PRO operates by creating three types of memory structures in response to its training sessions: (1) the Hypothesis Base, (2) the Case Base, and (3) the Statistical Base. These three levels of memory are hierarchical insofar as the Case Base draws its components from the Hypothesis Base, and the Statistical Base is predicated on the existence of both the Hypothesis Base and the Case Base. As training goes on and memory expands at each level, we do see some memory interactions that shape subsequent memory expansion. For example, information from the Statistical Base is capable of influencing expansion at the level of the Hypothesis Base. But for the most part, memory grows with simple dependencies: the bigger the Hypothesis Base, the bigger the Case Base, and the bigger the Case Base, the bigger the Statistical Base.

Building the Hypothesis Base

The Hypothesis Base consists of simple associations between graphemes [Coltheart 1978] from the input word and phonemes in the target representation. A specific hypothesis in memory tells us that a particular string of letters has resulted in a particular phoneme at least once during training. The mapping defined by the Hypothesis Base from the set of set of all substrings of letters to the set of all phonemes is neither totally defined nor well-defined. Not all substrings need map to a phoneme, and when one does, it may map to any number of phonemes. For example, consonants and consonant combinations are generally less ambiguous than vowels and vowel combinations, so we will see more hypotheses associated with vowels than consonants.

To illustrate the idea of a hypothesis, consider the following segmentation of the word "action" along with its corresponding pronunciation:

A	C	TI	O	N
ă	k	sh	∂	n

When this segmentation is mapped against the target pronunciation, we can identify five underlying hypotheses (associations between graphemes and phonemes): A/a, C/k, TI/sh, O/∂,

---

[1]Another system that is very similar to PRO in its broad design is MBRtalk [Stanfill & Waltz 1986], although MBRtalk uses a very different form of memory access. MBRtalk runs on a Connection Machine whereas PRO can run reasonably in a Common Lisp environment with 4M of memory.

and N/n.

However, if this were the first word PRO encountered in a fresh training session, PRO would have no way of knowing that this is the correct segmentation of the input string. There are five ways of partitioning this six-letter word into five substrings, and PRO would be unable to decide which is best. In general, PRO consults its existing Hypothesis Base in order to determine how an input word should be segmented and matched against its target pronunciation. After generating all possible segmentations of the input word which map onto the target pronunciation, PRO checks its known hypotheses to see if any one of these segmentations results in more known hypotheses than any other segmentation. If there is a unique winner, that is the segmentation PRO picks. In the event that there is no unique winner, PRO narrows its candidates to whichever ones did maximize known hypotheses. Of these remaining segmentations, PRO checks to see if any of the new hypotheses being proposed are "close fits" against known hypotheses. A hypothesis *string1/phoneme1* is a close fit against *string2/phoneme2* if (1) *phoneme1 = phoneme2*, and (2) *string2* is a substring of *string1*. If one of the remaining segmentations supports more "close fits" than any other segmentation, then PRO identifies that one as the preferred segmentation. In most cases, PRO will be able to identify a preferred segmentation on the basis of these two filtering mechanisms. But in the event that there is still more than one segmentation which maximizes both known hypotheses and close fits to known hypotheses, then PRO consults its Statistical Base and determines which of the remaining segmentations contains hypotheses that are used with greater frequency. In this manner PRO can identify a preferred segmentation to map against the targeted phonemes.

Once a preferred segmentation is in hand, PRO consults its existing Hypothesis Base to see if all the hypotheses in the segmentation are known hypotheses. If any hypothesis is new, we index it under its grapheme and the Hypothesis Base acquires a new hypothesis. Having updated the Hypothesis Base, PRO then goes on to update the Case Base and the Statistical Base.

Building the Case Base and the Statistical Base

Given a segmentation for a training item, we can take a series of "snapshots" of the resulting hypothesis sequence where each snapshot contains exactly three consecutive hypotheses. The longer the segmentation, the more snapshots we will need to cover the full sequence. Each snapshot of three hypotheses then constitutes a case for PRO's Case Base. Cases are indexed under both the leading hypothesis and the trailing hypothesis, and all cases with a common index are organized in a tree structure. The Case Base therefore contains as many trees as there are hypotheses in the Hypothesis Base, and each tree may contain a few or a large number of cases depending on how much of PRO's training has been involved with the indexing hypothesis. All the trees have a depth of three hypothesis nodes and each complete branch in a tree corresponds to a single case in memory.

When a candidate case is taken from a "snapshot" of a training item, we check to see if that case has been recorded in the Case Base before. If it has, we do not need to alter the Case Base. If it hasn't, we grow a new branch in the appropriate case trees (one for the leading index and one for the trailing index). This dual encoding of each case allows us to maintain a more informative Statistical Base than would otherwise be possible.

The Statistical Base shadows the Case Base by maintaining frequency data for all the nodes in all the case trees. When a case is encountered during training, we add it to the Case Base if needed, and update each of the six tree nodes associated with that case in memory. For example, the case (A/a C/k TI/sh) will update the root node associated with A/a, the node that says how many times A/a has been followed by C/k, and the node that records the frequency of A/a followed by C/k followed by TI/sh. Going backwards, we update the node that records instances of TI/sh, then the node recording TI/sh preceded by C/k, and the node recording TI/sh preceded by C/k preceded by A/a. Note that the two terminal nodes should agree on the number of times a given sequence has been seen before, but the root nodes and intermediate nodes will generally maintain different frequency counts depending on whether the sequence is being traversed forwards or backwards.

Test Mode: Finding a Word Pronunciation

When PRO receives an input word in test mode, it begins by creating a search space of possible pronunciations based on all available hypotheses in memory. Each path in this space beginning with the "start" node and ending with the "end" node corresponds to a possible pronunciation for the word (see Figure 1). For example, if the search space shown in Figure 1 is generated in

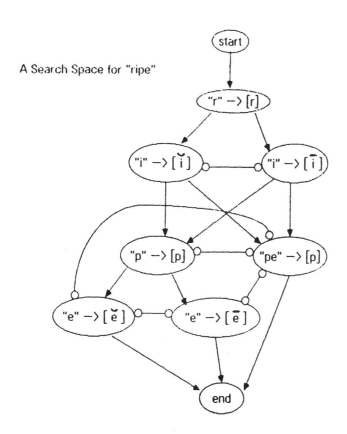

Figure 1: A Search Space using Available Hypotheses

123

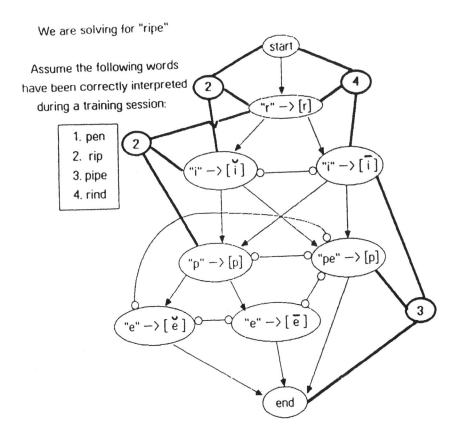

We are solving for "ripe"

Assume the following words
have been correctly interpreted
during a training session:

1. pen
2. rip
3. pipe
4. rind

Figure 2: Adding Context Nodes from the Case Base

response to the word "ripe," we have six possible pronunciations for "ripe" corresponding with the six possible paths through that space. There are two possibilities for "i" (long vs. short) and three possibilities for "e" (long vs. short vs. silent). This search space is further structured as a lateral inhibition network where instances of hypotheses which share common characters from the input word inhibit one another. In figure 1 we see that the silent e is represented by the hypothesis PE/p. Since this hypothesis contains two input characters, it is in competition with P/p as well as E/e and E/e. We therefore find three inhibition links connecting PE/p to its three competitors.

Thus far we have described a search space derived from the Hypothesis Base alone. To bring in the effects of the Case Base, we must now add context nodes representing relevant cases (see Figure 2). Each context node corresponds to a case which matches a subpath within the search space. In Figure 2 we see how four context nodes can be added to the graph on the basis of three previous training items ("rip," "pipe," and "rind.") The fourth training item ("pen") is not associated with any cases relevant for this search space. By attaching the resulting context nodes to the search space where each case applies, we see how PRO's previous training with "rip" reinforces a short "i", while exposures to "pipe" and "rind" reinforce a long "i". "pipe" also reinforces a silent "e" but no context nodes reinforce a short "e" or a long "e" at the end of the word.

The context nodes provide positive activation for a spreading activation algorithm, and the inhibition links counter this activation as it propagates through the network. Context nodes are initialized with a value based on frequency data in the Statistical Base, and a relaxation algorithm is then applied throughout the search space [Feldman & Ballard, 1982]. These networks tend to stabilize within 30 iterations, and a preferred path through the network is sought by maximizing MIN [A(N1),...,A(Nk)] where A(Ni) is the activation level associated with the ith node in a given path. In general, a unique path of maximal activation can be found and this is then the pronunciation PRO returns. If there is more than one path of maximal activation, PRO selects one of the maximal paths randomly and returns that pronunciation.

Experimental Results

We have run PRO on a training vocabulary of 750 words and tested it on a total of 300 words. 200 of the test items appear in the training set and the remaining 100 test items are "novel" words not encountered during training. We have also collected data from a modified version of PRO which substitutes a random search algorithm in place of PRO's spreading activation/relaxation algorithm. The modified version builds the same search space of possible pronunciations, but picks one randomly instead of structuring the network with additional context nodes and relaxing the net to find a pronunciation with maximal activation. The results of this random algorithm provide us with a baseline that reflects the power of the Hypothesis Base operating in the absence of the Case Base and the Statistical Base.

Test trials were run at three distinct times during training: after PRO had trained on 250 words, after 450 words, and after all 750 words. The training corpus contained words ranging from three to eight letters in length, and the test sets contained words ranging from three to five letters in length. We collected separate data for "familiar" words (the first 200 items in the training set), and "novel" words. The "familiar" items included 50 three-letter words, 100 four-letter words, and 50 five-letter words. The "novel" items included 50 four-letter words and 50 five-letter words. The modified version of PRO was run on only "novel" test items, but at the same points during training as the regular PRO runs.

Table 1 shows the results of these nine test runs. The x–axis represents the amount of training PRO has had in terms of the number of target phonemes processed. After 750 words, PRO has examined 2731 phonemes. The y-axis shows the percentage of phonemes PRO correctly identifies during test runs. The top curve shows PRO's performance on "familiar" words, the second curve shows PRO's performance on "novel" words, and the bottom curve shows the baseline performance of the random algorithm on the same set of "novel" words that we tested with PRO in tact.

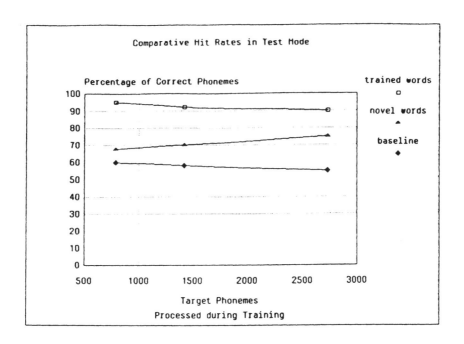

Table 1: Three Test Runs during Training

We can see that PRO is better with words it has seen during training compared to novel test words, although the distance between these two curves does diminish as PRO trains more. It makes sense for PRO's performance on familiar words to degrade somewhat because PRO's Hypothesis Base and Case Base are expanding as PRO trains. As more conflict arises between a larger number of competing cases, we might expect to see the increased error rate found in the top curve. These factors explain why PRO's performance drops off a bit (from 95% to 90%) for familiar words. This same factor must influence PRO's performance on novel words as well, but here we see a gradual increase in performance (from 66% to 75%). For novel words, the increasing search space complexity is overridden by the fact that PRO takes advantage of its growing Case Base effectively.

If we assume that these two curves eventually level out, we can estimate PRO's eventual level of competence across all test items as being somewhere in between the two asymptotes. It is safe to assume that these two curves won't cross, so they must level out somewhere between 75% and 90%. Extrapolating from our existing data, we would estimate that PRO will eventually average around 86% on all test items (this is where the two curves would intersect if they continued at their present rate of change).

To see how much of PRO's performance can be attributed to the Case Base and Statistical Base, consider the baseline curve provided by the random version of PRO. Here the hit rates drop from 60% to 55%. This is what PRO can do if it only accesses knowledge from the Hypothesis Base in order to construct a search space of possible pronunciations. Once this search space

126

has been constructed, we can pick a pronunciation from the space at random. According to our randomized simulation, any given phoneme within that pronunciation then has a 55% chance of being correct after PRO has trained on 750 words. Once again, it makes sense for this curve to diminish since the Hypothesis Base is growing over time. More hypotheses in memory will result in more possible pronunciations for each test word. If we assume that this baseline levels out at the same time our two performance curves level out, then we can estimate that the baseline will level out around 50%.

From our best estimates, we then can then conclude that PRO's Case Base and Statistical Base will result in a significant improvement over the Hypothesis Base alone (86% vs. 50% hit rates on each phoneme). Moreover, these levels of performance should stabilize once PRO has reached a saturation point in its training.

PRO's Memory: Facts and Conjectures

We have gathered additional data in conjunction with our training session to see what PRO's memory looks like after 750 words. At this point the Hypothesis Base has grown to 180 hypotheses, and the Case Base contains 1591 cases (sequences of 3 hypotheses). Some measure of English regularity might be derived from the fact that the 750 words PRO trained on could have produced roughly 4200 distinct cases if no repetitions of cases were present. (180 hypotheses are capable of generating 5,832,000 cases if all combinatorically possible combinations are considered.)

When we talk about PRO's performance stabilizing at a saturation point in training, we are assuming that the Hypothesis Base and the Case Base will eventually level out and cease to expand at any significant rate. To see how close we are to saturation after 750 words, we can examine growth curves for the Hypothesis Base (see table 2) and the Case Base (see table 3). Neither curve appears to be very level after 750 words, but it is clear that the Hypothesis Base is slowing down much faster than the Case Base. After their initial growth spurts, the Hypothesis Base acquires roughly 1 new hypothesis for every 10 training items, while the Case Base picks up 21 new cases for every 10 training items.

While it seems safe to assume that performance won't stabilize until the Hypothesis Base saturates, it is less clear that the Case Base must saturate before performance stabilizes. General correlations between memory and performance will have to be borne out by additional experiments, but if we must make some estimates, we could extrapolate from our existing growth curves and make some guesses about the state of PRO's memory when PRO's performance curves level off. Using the estimates for stabilization discussed above, it appears that PRO should acquire no more than roughly 240 hypotheses altogether, and performance will stabilize around 2800 cases. Note that we are assuming saturation of the Hypothesis Base by the time performance stabilizes, but we are less inclined to assume that the Case Base cannot continue to grow after performance stabilizes. We will only know this with certainty by expanding PRO's training corpus.

As for the estimates concerning the training vocabulary, we would guess that PRO will stabilize after about 1500 words (where a word contains an average of 3.64 phonemes - the same ratio

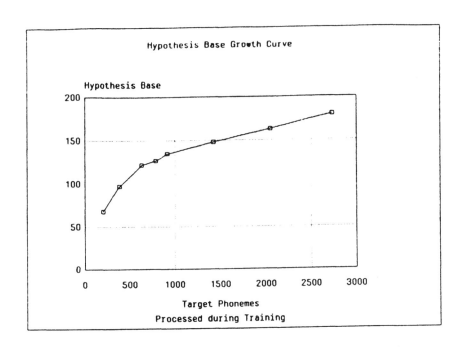

Table 2: The Acquisition of Hypotheses during Training

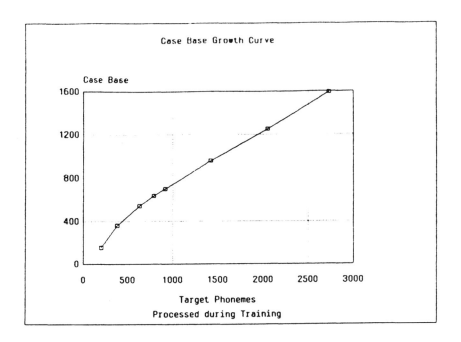

Table 3: The Acquisition of Cases during Training

128

maintained by our initial 750 word corpus). Of course all of these estimates are subject to the usual assumptions about random vocabulary and "representative" training sets. In selecting words from the dictionary we have tried to restrict ourselves to common words that any high school student would recognize. We cannot say how these restrictions relate to characterizations of English vocabulary in general.[2]

Looking inside the the Hypothesis Base, we find 47 phonemes and 101 graphemes underlying our 180 hypotheses. 69 hypotheses map from a single letter, 100 hypotheses map two-letter graphemes, 9 hypotheses map three-letter graphemes, and 2 hypotheses map four-letter graphemes. The single greatest source of ambiguity comes from the letter "a" which is associated with 10 distinct pronunciations (where "a" stands alone as a complete grapheme by itself). About two-thirds of the hypotheses have graphemes containing at least one vowel.

Conclusions

PRO shows how a large Case Base can be automatically acquired during supervised training and accessed within the framework of a spreading activation algorithm. PRO builds a three-tiered memory of hypotheses, cases, and frequency data using heuristic methods for knowledge acquisition. It is interesting to note that PRO is consistent with the concept of "timid acquisition" which is thought to be a necessary condition for success when training with a corpus of positive examples only [Berwick 1986].

A test item is processed by generating a search space based on available hypotheses and augmenting that search space with additional nodes derived from the Case Base. By structuring a network which represents both the search space and the influences of available memory, we can see in a declarative manner exactly how PRO views each test item in the context of its previous training experience.

Spreading activation and network relaxation then operate to consolidate competing information into a global consensus within the search space. While the structure of the search space is purely heuristic, the use of a relaxation algorithm to resolve the search is inspired by connectionist efforts. In this manner PRO represents a synthesis of ideas from both heuristic or symbolic methodologies and subsymbolic processing strategies.

References

Berwick, R.C. (1986) "Learning from positive-only examples: the subset principle and three case studies." in Michalski, Carbonell, and Mitchell (ed.), Machine Learning vol. 2. Morgan Kaufmann. pp. 625-645.

Bradley, L., and Bryant, P.B. (1983) "Categorizing sounds and learning to read: a causal connection." Nature, 301, pp.419-21.

Coltheart, M. (1978) "Lexical access in simple reading tasks." in Underwood, G. (ed.), Strategies of Information Processing. Academic Press: London.

---

[2]Although a claim has been made that the 1000 most common English words are also amongst the most irregular (see Sejnowski and Rosenberg 1986, p. 8).

Doctor, E., and Coltheart, M. (1980) "Phonological recoding in children's reading for meaning." Memory and Cognition, 80, pp. 195-209.

Feldman, J.A., and Ballard, D.H. (1982) "Connectionist models and their properties." Cognitive Science Vol. 6, no. 3. pp. 205-254.

Gough, P. (1972) "One second of reading." in Kavanaugh, J.P., and Mattingly, I.G. (eds.), Language by Eye and by Ear. MIT Press: Cambridge, MA.

Sejnowski, T. and Rosenberg, C. (1986) "NETtalk: A Parallel Network that Learns to Read Aloud." The Johns Hopkins University Electrical Engineering and Computer Science Technical Report JHU/EECS-86/01. Johns Hopkins University. Baltimore, MD.

Stanfill, C., and Waltz, D. (1986) "Toward memory-based reasoning." Communications of the ACM, vol. 29, no.12. pp.1213-28.

A Connectionist Architecture for Representing and Reasoning about Structured Knowledge

Mark Derthick
Carnegie-Mellon University

Abstract

μKLONE is the first sub-symbolic connectionist system for reasoning about high level knowledge to approach the representational power of current symbolic AI systems. The algorithm for building a network takes as input a knowledge base definition in a language very similar to that of KL2, which has previously been implemented only in Lisp. In μKLONE, a concept is more than a set of features: it is a complex of required and optional subparts filling well defined roles, each of which may have its own type restrictions. In addition to being able to use complex structured descriptions in its reasoning, μKLONE exhibits a facility for plausible inference due to its inherently parallel constraint satisfaction algorithm that is not shared by symbolic systems. This paper describes how the system answers a query that requires both of these characteristics. It is hoped that this is the beginnings of a response to (McDermott, 1986)'s challenge that connectionists should pay more attention to architectural issues and rely less on learning.

1. Previous Work

The semantic networks of (Quillian, 1968) were developed to model common sense reasoning using spreading activation. The degree to which concepts are related determines the influence they exert on one another. (Hinton, 1981) develops a more structured semantic network influenced by the theory of case frames in which concepts play one of three different roles in a frame.[1] In addition, the concepts themselves have a meaningful substructure: they are composed of *micro-features*. ELEPHANT is represented as the conjunction of the <big>, <gray>, and <mammal> micro-features. This gives the flexibility to reason about novel concepts that can be represented as conjunctions of existing micro-features. It also performs property inheritance automatically: since the ELEPHANT pattern contains the MAMMAL pattern, properties of mammals apply to elephants, unless they are overridden by the other micro-features of elephants.

(Shastri, 1985) is an extension of early semantic nets in a different direction. He retains the simple localist encoding of concepts, but adds a formal theory of evidential reasoning which can be encoded with connection strengths and thus efficiently executed. Structured knowledge can be represented using properties and values. Properties, like concepts, are definable in a high level language and are organized into an ISA hierarchy. In Shastri's system the formal elements, units, correspond to knowledge level entities like concepts. I will call such systems symbolic to distinguish them from systems like Hinton's in which the formal elements are at a lower level.

Although the micro-feature based representations of Hinton's system allow automatic inheritance and

---

[1] The theories of frames, schemas, and structured inheritance networks each have their own terminology. I will use "concept" uniformly instead of schema or frame and "role" instead of slot or case.

the representation of novel concepts, previously no system using them has had the expressive power of symbolic systems like Shastri's. μKLONE resembles current AI semantic networks in expressive power, particularly KL2 (Vilain, 1985), after which it was consciously modeled. It incorporates a third ontological category, individuals, in addition to concepts and roles. The language for defining a knowledge base (KB) makes it possible to specify *value restrictions* on the fillers of a role (for instance, the JOBs of MILLIONAIREs must be ARMCHAIR-ACTIVITIES) and *minimum restrictions* (for instance, a MILLIONAIRE must have at least one JOB). Appendix I gives a brief description of the KB language.

2. An Example Plausible Inference

2.1. Informal Description
The following scenario motivates the need to reason about conflicting beliefs. June has made some unwarranted conclusions about Ted from his conversation and appearance; specifically he is walking along a pier wearing a captain's hat and knowledgeably discussing the influence of independent producers on television programming. June assumes that Ted is a sailor, and that he must be deeply interested in television. The next week she sees Ted's picture in the newspaper with the caption "Millionaire Playboy Ted Turner." She now concludes that sailing is only a hobby of Ted's, since millionaires don't have jobs requiring vigorous activity but could afford an expensive sailboat. Millionaires usually have a job, so perhaps Ted is a high level television executive.

2.2. Formal Domain Description
The formal description of June's initial assumptions and general world knowledge is given in appendix I. In the knowledge base Ted is asserted to be a sailor, someone whose job is sailing, and a TV-Buff, someone who has an interest that is a television-related activity. A millionaire-playboy is defined to be a person with a hobby that is an activity requiring an expensive prop. Millionaire-playboys must have at least one job, and all their jobs must be armchair activities.

The primary query used as an example in this paper is given at the end of appendix I, and can be paraphrased "If Ted were a millionaire-playboy, what would his job and hobby be?" The system answers that sailing would be Ted's hobby and that TV-Network-Management would be his job. To demonstrate that the presupposition that Ted is a millionaire-playboy affects the reasoning process, the system can be asked "What are Ted's job and hobby?" In this case the answer is sailing is Ted's job, and he has no hobbies (TV-Watching is seen as an interest, but not necessarily either a hobby or a job).

2.3. High Level Description of Reasoning Process
The presupposition "If Ted were a millionaire-playboy" conflicts with the knowledge base because sailing is a vigorous-activity, and the jobs of millionaire-playboys must be armchair-activities. The initial impact of this conflict is that sailing is likely to be one of Ted's interests, but perhaps not his job. Since millionaire-playboys must have expensive hobbies and only two activities known to require expensive props are in the KB, either flying or sailing are most likely to be chosen. Hobbies and jobs are both interests, so scenarios in which Ted's hobby is sailing are seen to be more plausible than those where it is

flying.

Millionaire-playboys must have a job that is an armchair activity and a profitable activity. Both TV-Network-Management and Corporate-Raiding fit this category, but the former is more plausible because it is known that Ted is interested in television. The distractors TV-acting and TV-watching are in the knowledge base to demonstrate that all three factors (being TV-related, profitable, and an armchair-activity) are taken into account.

This kind of reasoning requires a structured treatment of properties. It is crucial to distinguish that vigorous-activity applies to sailing, that millionaire-playboy applies to Ted, and that Ted's relation to sailing, has-job, has its own properties, such as requiring sailing to be a profitable-activity. However, the structural properties alone are not sufficient to determine that sailing, rather than flying, is Ted's hobby. This decision is the result of residual activation of sailing as Ted's job. Default logic approaches (Reiter, 1980) are limited to binary decisions on the consistency of a theory, and have difficulty making choices like this based on relative plausibilities.

3. System Implementation

3.1. Architecture

One reason powerful non-symbolic systems are difficult to build is the problem of representing more than one concept at a time, a capability that is required to infer Ted's hobby. Since each concept is represented as a pattern over many units, it is not practical to represent very many. The best approach seems to be to build a small number of special purpose registers. This has the advantage over the localist approach that a concept can be represented in different registers in different contexts, and even in multiple places at the same time. The latter is especially difficult for localist systems.

At any one time, μKLONE can represent a single individual and information relevant to it. This includes the concepts it instantiates, any number of other individuals to which it is related, the concepts they instantiate, the relations involved, and value restrictions (VRs), value permissions (VPs), and minimum restrictions (MinRs) on the relations. The modular architecture shown in figure 1 reflects these distinctions. There are five modules primarily responsible for the representation, three auxilliary modules which mediate some of the inter-module constraints, a variable number of modules used for input and output, and a variable number of auxiliary I/O modules. There is no limit to the number of concepts, relations, or related individuals that can be represented in the five modules and used in reasoning, but the number simultaneously accessible to the user is limited by the number of I/O modules. To represent the example question of section 2.2, the pattern for Ted is clamped in the **subject** module, the pattern for MILLIONAIRE-PLAYBOY in a **subject-type-I/O** module, the pattern for HAS-JOB in one **role-I/O** module, and the pattern for HAS-HOBBY in a second **role-IO** module. After an annealing search (Kirkpatrick, 1983), the pattern for {sailing} is found in the first **role-fillers** I/O module and for {TV-Network-Management} in the second. Also represented internally are the VR that all the subject's jobs are armchair-activities, and the facts that sailing is a vigorous-activity, that TV-network-management is a profitable-activity and an armchair-activity, as well as (irrelevant) type information about the five individuals that have no known relations to Ted.

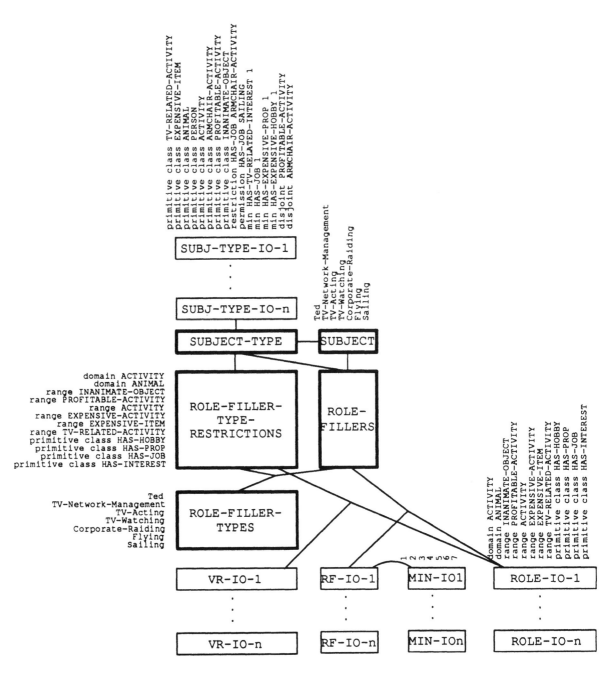

Figure 1: μKLONE has five important central modules (dark rectangles) and a variable number of subject-type, role, role-filler, minimum restriction, and value restriction IO modules. There are also auxiliary modules, which are not shown here. Those modules which directly constrain one another are connected. The meaning of each unit can be deduced from the printed descriptions. For modules with a single row of units, the meaning is the description in the unit's column. For modules with many rows, the meaning involves the conjunction of the descriptions in the row and column. For instance, the top left unit in the **role-fillers** module means that Ted is filling a role whose domain is ACTIVITY.

3.2. Representations

3.2.1. Individuals
The seven individuals in the KB are described by the concepts they instantiate and the relations they participate in, but have no internal structure. A localist representation is used, with each bit corresponding to an individual. This makes it easy to represent sets of individuals; the pattern for {Ted Flying} is seven bits long, and exactly two of the bits are on.

3.2.2. Concepts
A concept pattern has one bit for every micro-feature derived from the KB. Every primitive concept has its own micro-feature; in addition there is a micro-feature for every distinct value restriction, value permission, minimum restriction, and disjointness restriction mentioned in the KB. The active bits in the pattern for SAILOR, for instance, are those corresponding to the *<primitive class person>* and *<permission has-job sailing>* micro-features. It is apparent that a very simple algorithm is sufficient to derive the required micro-features. The concept micro-features for the example domain are listed above the **subject-type** module in figure 1.

3.2.3. Roles
Roles are also represented as sets of micro-features. In this case the possible micro-features come from primitive roles, domain restrictions, and range restrictions, and are listed to the left of the **role-filler-type-restrictions** module in figure 1.

3.2.4. Minimum Restrictions
The MinR IO groups use a unary encoding of the minimum number of fillers a role must have: zero is represented by turning all bits off, two is represented by turning the "1" and "2" bits on, etc. The maximum number that needs to be represented is the number of individuals in the KB, in this case seven.

3.2.5. Derived Representations
Value restrictions are identified by the combination of the role that is being restricted and the concept which any fillers of that role must instantiate, for example HAS-JOB ARMCHAIR-ACTIVITY. Value permissions are a combination of a role and a set of individuals filling the role, such as HAS-JOB {sailing}. Binding two entities together is not straightforward in distributed connectionist systems. μKLONE uses a kind of *coarse coding*, a technique described in (Hinton, 1986a). To bind a set of individuals whose pattern has length I with a role whose pattern has length R requires a register with R rows of I units (see figure 2). For each i and j, the unit at coordinates i,j is turned on if and only if both the ith bit in the role pattern and the jth bit in the individual-set pattern are on.

Using coarse coding a number of pairs can be stored simultaneously by superimposing their patterns. Normally there is a limit to this number because patterns begin to interfere with one another, making them difficult to reconstruct. In μKLONE the patterns are constructed so that the combination of two or more patterns always produces a pattern whose meaning is a necessary consequence of the meanings of the constituent patterns. Interference is not degrading retrieval ability, it is actually performing inferences!

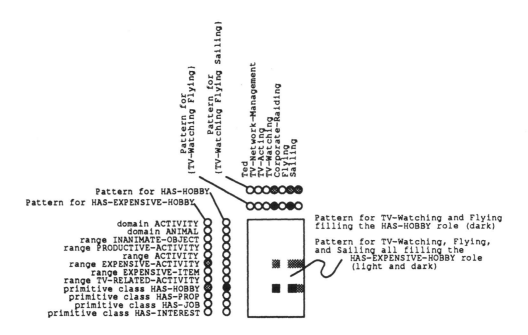

Figure 2: Two examples illustrating how pairs of patterns are conjunctively coded. The first example (black) combines the two bit pattern for (TV-Watching Flying) with the one bit pattern for HAS-HOBBY, producing a 2×1 bit pattern in the role-fillers module representing the fact that (TV-Watching Flying) is the set of fillers of the HAS-HOBBY role. The second example (gray) combines the three bit pattern for (TV-Watching Flying Sailing) with the two bit pattern for HAS-EXPENSIVE-HOBBY, producing a six bit pattern in the role-fillers module. Since the former value permission necessarily follows from the latter, its two bit pattern is contained by the six bit pattern of the latter.

The reason for this can be inferred from figure 2, which illustrates that the pattern for the value permission HAS-EXPENSIVE-HOBBY (TV-Watching Flying Sailing) contains that for HAS-HOBBY (TV-Watching Flying). Not coincidently the latter VP is a necessary consequence of the former. In the more general case where neither pattern contains the other, their superposition will yield new VPs.

The pattern for a VP, *A*, contains that for VP *B* if and only if the role and individual patterns used to form *A* contain those forming *B*. This happens if *B*'s role is more general than *A*'s, or if *B*'s individual set is contained by *A*'s, which are just the conditions under which *B* is a necessary consequence of *A*. The weaker VP is always represented by fewer bits.

In the **role-filler-type-restrictions** module, multiple concept/role pairs are represented and the same kind of inference by interference is desirable. Unfortunately a value restriction entails other VRs with more general concepts or more *specific* roles. For example if all your hobbies are profitable-activities, then surely all your expensive-hobbies are activities. If the normal role representation is used, however, the weaker VR may be represented by *more* bits than the stronger one. To fix this, patterns for VRs are calculated as in figure 2 while pretending that role patterns are complemented. This way more specific roles have fewer bits on.

136

3.3. Constraints

All the constraints on the state of the network are currently implemented as symmetric, weighted, pairwise links between units, and biases on individual units. Units linked with a positive weight tend to excite each other, and units linked by negative weights inhibit each other. The μKLONE algorithm defines a network of units and links, which can then be run as either a Boltzmann Machine (Hinton, 1986b) or a Hopfield network (Hopfield, 1984). As discussed below, using only pairwise links is proving to be a problem, and I plan to use a more powerful model in the future.

3.3.1. Constraints on Coherent Concepts and Roles

There are three places where constraints are required to ensure that coherent concepts and roles are represented: the **subject-type** module must correctly describe the subject, each column in the **role-fillers** module must correctly describe some role being filled, and each row in the **role-filler-types** module must correctly describe an individual.[2]

Two types of assertions require links to ensure a coherent pattern is represented: "specializes" clauses require the pattern for the more general concept or role to be present if the more specific one is, and "disjoint" clauses forbid two concept or role patterns to be present at the same time. In the simple case of primitive concepts or roles where each pattern has only one bit on, a single implication link (see figure 3a) or a single inhibitory link (3c) suffices to effect the constraint. Otherwise, pairwise links cannot implement the constraint and, in the Boltzmann or Hopfield networks, extra units must be used (figure 3b and d). No links are required to implement *defined* specialization relations. The pattern for MILLIONAIRE-PLAYBOY, for instance, already contains that for PERSON.

3.3.2. Other Constraints

Describing all the constraints embedded in μKLONE in detail would be exceedingly tedious. Using micro-feature based representations, all the requirements for a plausible interpretation of the query reduce to fairly local relations between unit states of the type illustrated in figure 3. These logical relations are used to implement the following constraints:

- **Role Domain and Range Constraints:** A unit in the **role-fillers** module in the "sailing" column and the "domain ANIMAL" row should only come on if the *<primitive class animal>* micro-feature is active in the **subject-type** module. A unit in the "sailing" column and the "range ACTIVITY" row should only come on if the *<sailing is an activity>* micro-feature is active in the **role-filler-types** module.

- **Individual Type Constraints:** If Ted is asserted to be a SAILOR in the KB, and Ted is the subject, then the SAILOR pattern must be active in the **subject-type** module. The SAILOR pattern must be active in the row representing Ted's type in the **role-filler-types** module no matter what the subject is.

- **Value Permission Constraints:** If SAILOR is represented in the **subject-type** module, then the HAS-JOB pattern must be active in the "sailing" column of the **role-fillers** module. There are no "instantiate-role" assertions in the example KB, but if Ted had been asserted to have an interest in flying, then the Ted unit in the **subject** module would activate the HAS-INTEREST

---

[2]Because incompatible roles may be restricted to be filled with the same type of individual, columns in the **role-filler-type-restrictions** module do not necessarily represent coherent roles. Incompatible type restrictions may be imposed on a role, so rows in this group do not necessarily represent coherent concepts.

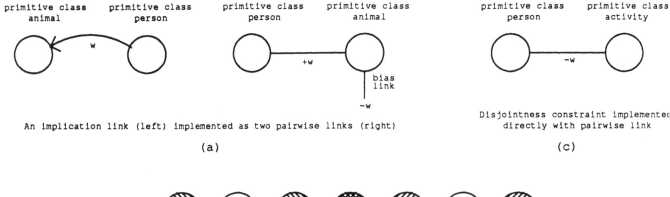

An implication link (left) implemented as two pairwise links (right)

(a)

Disjointness constraint implemented directly with pairwise link

(c)

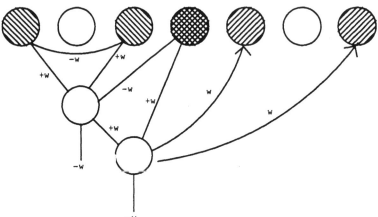

If pattern a is present (falling cross-hatching), then pattern b must be present (rising cross-hatching)

(b)

If pattern a is present (falling cross-hatching), then pattern b must not be be present (rising cross-hatching)

(d)

Figure 3: (*a*) Specializes constraints are implemented with *implication* links, which require two pairwise links. The contribution to the global energy function is +w if the *<primitive class person>* micro-feature is on and the *<primitive class animal>* micro-feature is off; otherwise the contribution is zero. Implication links are drawn with an arrowhead and labeled with an unsigned weight. (*b*) When the pattern of the more specific concept has more than one active bit, extra units are required to detect its presence; and when the pattern of the more general concept has more than one active bit not shared by the more specific concept, more than one implication link is required. (*c*) Disjoint constraints are implemented with inhibitory pairwise links if both concept patterns have a single bit on. (*d*) Otherwise a technique discovered by Steve Nowlan requiring a "winner take all" circuit attached with inhibitory links is used to make sure the union of the two patterns results in a high energy state.

138

pattern in the "flying" column of the **role-fillers** module.

- **Value Restriction Constraints:** The *<restriction has-job armchair-activity>* micro-feature in the **subject-type** module must excite the pattern for this VR in the **role-filler-type-restrictions** module. If a VR in the latter module applies to a role being filled by some individual, then the individual must be of the appropriate type, as represented in the **role-filler-types** module. Detecting whether a VR applies to a role-filler cannot be done with pairwise links alone, so there is an auxiliary module to implement these constraints. This module is actually larger than all the others put together, and so is the limiting factor on the network's speed. I plan to give up the simplicity of the current Boltzmann/Hopfield model and include more complex relations among units directly as terms in the energy equation rather than implementing them with pairwise links. This will eliminate the need for all auxiliary modules.

- **Minimum Restriction Constraints:** If the *<minimum has-job 1>* micro-feature is on in the **subject-type** module, then the HAS-JOB pattern must be active in at least one of the seven columns in the **role-fillers** module.

3.4. Input/Output Implementation

Producing and interpreting μKLONE's distributed and coarse coded representations requires rather elaborate machinery. The exception to this complexity is the **subject** module, where a local representation is used, and exactly one unit is always active. Input and output take place directly in this module. In the **subject-type** module, which uses distributed but not coarse coded representations, several concept patterns may be combined. To allow the system this freedom, constraints on the subject's type imposed by presuppositions of a query are expressed by clamping patterns into one or more **subject-type-IO** modules, rather than clamping the **subject-type** module directly. The concept represented in the **subject-type** module must be subsumed by each concept in the **subject-type-IO** modules.

Each "with" clause in a query requires a set of IO modules to represent the role, MinR, role fillers, and VR. Each **MinR-IO** module interacts solely with a **role-fillers-IO** module to ensure that a sufficient number of individuals are represented there. Using the technique of figure 2, the role and role-fillers patterns in IO modules are combined in an auxilliary module, which is constrained to pull out of the central **role-fillers** module. Pull out is a technique introduced in (Mozer, 1984) for extracting meaningful constituents from coarse-coded modules in which several patterns have been superimposed. With this arrangement, the set of individuals represented in the **Role-Fillers-IO** module filling the role represented in the **role-IO** module must be a subset of all the role fillings represented in the central **role-fillers** module. Using the same technique of having an auxiliary pull-out group, the concept represented in the **VR-IO** module as a restriction on the role represented in the **role-IO** module is forced to be one of the VRs represented in the central **role-filler-type-restrictions** module. Though described as performing input, any of these IO modules can perform output if they are not clamped. Positive biases on all the input/output units result in the retrieval of the most specific answers compatible with the scenario represented in the central modules.

3.5. Performance

The network derived from the example KB has 2531 units and 16,959 links. Using the Hopfield model for annealing is generally more efficient than using the Boltzmann model,[3] and requires an annealing

---

[3] (Marroquin, 1985) reports an order of magnitude improvement.

schedule of 500 time steps for this network. At each time step the state of each unit is updated, and the entire search takes 10 minutes on a Symbolics Lisp Machine.

4. A More Detailed Description of the Reasoning Process

Posing the example query entails clamping patterns in some of the modules. One unit clamped on is the Ted unit in the **subject** module. Since Ted is known to be a SAILOR and a TV-BUFF, this unit excites those patterns in the **subject-type** module. At the same time the pattern for MILLIONAIRE-PLAYBOY is clamped into **subject-type-IO-1**, which excites the same pattern in the **subject-type** module, where three concept patterns are now superimposed. The conflict between being a sailor and being a millionaire playboy is not felt here, though, but in the coarse coded modules. The *<permission has-job sailing>* micro-feature excites the pattern for HAS-JOB in the sailing column of the **role-fillers** module. Since sailing is asserted to be a VIGOROUS-ACTIVITY, this pattern has already become active in the sailing row of the **role-filler-types** module. The *<restriction has-job armchair-activity>* micro-feature in the **subject-type** module excites the corresponding pattern in the **role-filler-type-restrictions** module. Since sailing is filling this role, the ARMCHAIR-ACTIVITY pattern is excited in the sailing row of the **role-filler-types** module. Now two contradictory patterns, VIGOROUS-ACTIVITY and ARMCHAIR-ACTIVITY, are excited in this row. The conflict can be worked out by violating any of several constraints: that vigorous-activities and armchair-activities are contradictory, that sailing is a vigorous activity, that playboys have sedentary jobs, that sailors' jobs include sailing, that Ted is a playboy, or that Ted is a sailor. Each of these constraints has an associated cost for violation. In a learning system these costs would be chosen automatically. Here I have chosen them so the requirement that sailing must be the job of a sailor is weakest. After being excited initially, the pattern for HAS-JOB is inhibited in the sailing column of the **role-fillers** module.

The *<minimum Has-Expensive-Hobby 1>* **subject-type** micro-feature is active, so some individual must be found to fill the role. The one chosen is the one for which the HAS-EXPENSIVE-HOBBY pattern is already most active. Since flying and sailing are both known to be ACTIVITIES-REQUIRING-EXPENSIVE-PROPS, they have a head start. In addition, sailing has some residual activation of HAS-EXPENSIVE-HOBBY because of the commonality with having a job, which was formerly active, and so sailing is chosen.

The *<minimum Has-TV-Related-Interest 1>* **subject-type** micro-feature seeks a filler for its role. The leading candidates are TV-Watching, TV-Acting, and TV-Network-Management. Simultaneously, the *<minimum Has-Job 1>* **subject-type** micro-feature, part of the MILLIONAIRE-PLAYBOY pattern, seeks a filler for its role. Jobs must be profitable activities, so Corporate-Raiding, TV-Acting, and TV-Network-Management are leading candidates. TV-Acting, however, is a vigorous-activity and is inhibited by the "VR HAS-JOB ARMCHAIR-ACTIVITY" pattern in the **role-filler-type-restrictions** module. Since TV-NETWORK-MANAGEMENT is excited by both the HAS-TV-RELATED-INTEREST and HAS-JOB minimum restrictions and these roles have similar properties, it is chosen to satisfy both requirements.

Now the central modules have reached a stable configuration in which only one constraint is violated, which is the best that can be done for this query. The only remaining task is for the **role-filler-IO** modules to pull out the individuals filling the HAS-JOB and HAS-HOBBY roles.

140

5. Conclusion

μKLONE demonstrates that sophisticated knowledge representation systems can be developed which retain the advantages of micro-feature based representations. The Ted Turner example demonstrates that a sub-symbolic implementation can solve common sense reasoning problems which cause difficulties for current symbolic systems.

Acknowledgments

Geoff Hinton and Dave Touretzky have been very helpful with the design of μKLONE and the preparation of this paper. This research is supported by NSF grants IST-8520359 and IST-8516330, and an ONR Graduate Fellowship.

I. Formal Domain Definition

The following input was used by the network building algorithm to produce a Hopfield network for answering queries. The syntax derives from that of KL2's definition language. Three ontological categories are used: *concepts* are classes of *individuals*. *Roles* are classes of two-place relations between individuals. DEFCONCEPT and DEFROLE statements normally give necessary and sufficient conditions for determining whether an individual instantiates a concept or whether an ordered pair of individuals instantiates a role. Alternatively, if the language is not powerful enough to provide sufficient conditions for recognizing membership, a concept or role can be defined to be *primitive*. In this case, the extension of the concept or role must be explicitly declared using INSTANTIATE-CONCEPT or INSTANTIATE-ROLE statements. Conditions which necessarily hold of instances of concepts or roles, but are not part of the recognition criteria are asserted with ASSERT-CONCEPT or ASSERT-ROLE statements. (This is rather different from KL2, where a completely different language is used for assertions.)

```
((DEFCONCEPT Animal (PRIMITIVE))                          ;ANIMAL is a natural kind -you can't define it
 (DEFCONCEPT Person (PRIMITIVE))
 (ASSERT-CONCEPT Person (SPECIALIZES Animal))            ;PERSONs always turn out to be ANIMALs
 (DEFCONCEPT Millionaire-Playboy (SPECIALIZES Person)
   (SOME Has-Hobby Activity-Requiring-Expensive-Prop))   ;a PLAYBOY must have some HOBBY
                                                          ;which is an ACTIVITY-REQUIRING-AN-EXPENSIVE-PROP
 (DEFCONCEPT Activity-Requiring-Expensive-Prop (SPECIALIZES Activity)
  (SOME Has-Prop Expensive-Item))
 (ASSERT-CONCEPT Millionaire-Playboy (MIN Has-Job 1)     ;a PLAYBOY must have a JOB
  (RESTRICTION Has-Job Armchair-Activity))               ;a PLAYBOY's JOBs must be ARMCHAIR-ACTIVITYs
 (DEFCONCEPT Sailor (SPECIALIZES Person) (PERMISSION Has-Job Sailing))  ;sailing must be one of a SAILORs JOBs
 (DEFCONCEPT Activity (PRIMITIVE))
 (ASSERT-CONCEPT Activity (DISJOINT Inanimate-Object) (DISJOINT Animal))
 (DEFCONCEPT Armchair-Activity (PRIMITIVE))
 (ASSERT-CONCEPT Armchair-Activity (SPECIALIZES Activity))
 (DEFCONCEPT Vigorous-Activity (SPECIALIZES Activity) (DISJOINT Armchair-Activity))
 (DEFCONCEPT Profitable-Activity (PRIMITIVE))
 (ASSERT-CONCEPT Profitable-Activity (SPECIALIZES Activity))
 (DEFCONCEPT UnProfitable-Activity (DISJOINT Profitable-Activity) (SPECIALIZES Activity))
 (DEFCONCEPT Television-Related-Activity (PRIMITIVE))
 (ASSERT-CONCEPT Television-Related-Activity (SPECIALIZES Activity))
 (DEFCONCEPT Inanimate-Object (PRIMITIVE))
 (ASSERT-CONCEPT Inanimate-Object (DISJOINT Animal))
 (DEFCONCEPT Expensive-Item (PRIMITIVE))
 (ASSERT-CONCEPT Expensive-Item (SPECIALIZES Inanimate-Object))
 (DEFCONCEPT TV-Buff (SOME Has-Interest Television-Related-Activity))
 (ASSERT-CONCEPT TV-Buff (SPECIALIZES Person))
```

```
(DEFROLE Has-Interest (PRIMITIVE))
(ASSERT-ROLE Has-Interest (DOMAIN Animal)           ;only ANIMALs can have INTERESTs
 (RANGE Activity))                                    ;only ACTIVITYs can be INTERESTs
(DEFROLE Has-Job (PRIMITIVE))
(ASSERT-ROLE Has-Job (SPECIALIZES Has-Interest) (RANGE Profitable-Activity))
(DEFROLE Has-Hobby (PRIMITIVE))
(ASSERT-ROLE Has-Hobby (SPECIALIZES Has-Interest) (DISJOINT Has-Job))
(DEFROLE Has-Prop (PRIMITIVE))
(ASSERT-ROLE Has-Prop (DOMAIN Activity) (RANGE Inanimate-Object))

(INSTANTIATE-CONCEPT (Activity-Requiring-Expensive-Prop Vigorous-Activity) Sailing)
(INSTANTIATE-CONCEPT Activity-Requiring-Expensive-Prop Flying)
(INSTANTIATE-CONCEPT (Profitable-Activity Armchair-Activity) Corporate-Raiding)
(INSTANTIATE-CONCEPT (Armchair-Activity Television-Related-Activity UnProfitable-Activity) TV-Watching)
(INSTANTIATE-CONCEPT (Vigorous-Activity Television-Related-Activity Profitable-Activity) TV-Acting)
(INSTANTIATE-CONCEPT (Armchair-Activity Television-Related-Activity Profitable-Activity) TV-Network-Management)
(INSTANTIATE-CONCEPT Sailor Ted)
(INSTANTIATE-CONCEPT TV-Buff Ted))
```

The query discussed in the paper, "If Ted were a millionaire-playboy, what would his job and hobby be?" is written:

```
((SUBJECT Ted)
 (SUBJECT-TYPE Millionaire-Playboy)
 (WITH (ROLE Has-Hobby) (FILLERS ?))
 (WITH (ROLE Has-Job) (FILLERS ?)))
```

References

Hinton, G. E. Implementing semantic networks in parallel hardware. In G. E. Hinton & J. A. Anderson (Eds.), *Parallel Models of Associative Memory*. Hillsdale, NJ: Erlbaum, 1981.

Hinton, G. E., McClelland, J. L. & Rumelhart, D. E. Distributed representations. In D. E. Rumelhart, J. L. McClelland, & the PDP research group (Eds.), *Parallel distributed processing: Explorations in the microstructure of cognition.*. Cambridge, MA: Bradford Books, 1986.

Hinton, G. E., & Sejnowski, T. J. Learning and relearning in boltzman machines. In D. E. Rumelhart, J. L. McClelland, & the PDP research group (Eds.), *Parallel distributed processing: Explorations in the microstructure of cognition*. Cambridge, MA: Bradford Books, 1986.

J. J. Hopfield. Neurons with graded response have collective computational properties like those of two-state neurons. *Proceedings of the National Academy of Sciences U.S.A.*, May 1984, 81, 3088-3092.

S. Kirkpatrick, C. D. Gelatt, Jr. and M. P. Vecchi. Optimization by simulated annealing. *Science*, 1983, 220, 671-680.

Marroquin, Jose Luis. *Probabilistic Solution of Inverse Problems*. Doctoral dissertation, MIT, September 1985.

McDermott, Drew. What AI Needs from Connectionism. August 1986. Unpublished notes accompanying a lecture given at AAAI-86.

Mozer, M. C. The perception of multiple objects: A parallel, distributed processing approach. 1984. Unpublished thesis proposal.

Quillian, M. R. Semantic memory. In M. Minsky (Ed.), *Semantic information processing*. Cambridge, Mass: MIT Press, 1968.

Reiter, R. A Logic for Default Reasoning. *Artificial Intelligence*, 1980, Vol. 13(1,2).

Shastri, Lokendra. *Evidential Reasoning in Semantic Networks: A Formal Theory and its Parallel Implementation*. Doctoral dissertation, University of Rochester, September 1985. Available as TR 166.

Vilain, M.B. The Restricted Language Architecture of a Hybrid Representation System. In *IJCAI-85*. Morgan Kaufmann, 1985.

A Connectionist Encoding of Semantic Networks

Lokendra Shastri
Computer and Information Science Department
University of Pennsylvania
Philadelphia, PA 19104

Abstract

Although the connectionist approach has lead to elegant solutions to a number of problems in cognitive science and artificial intelligence, its suitability for dealing with problems in knowledge representation and inference has often been questioned. This paper partially answers this criticism by demonstrating that effective solutions to certain problems in knowledge representation and limited inference can be found by adopting a connectionist approach. The paper presents a connectionist realization of semantic networks, i.e. it describes how knowledge about concepts, their properties, and the hierarchical relationship between them may be encoded as an *interpreter-free* massively parallel network of simple processing elements that can solve an interesting class of *inheritance* and *recognition* problems extremely fast - in time proportional to the depth of the conceptual hierarchy. The connectionist realization is based on an *evidential* formulation that leads to principled solutions to the problems of *exceptions, multiple inheritance,* and *conflicting information* during inheritance, and the *best match* or *partial match* computation during recognition.

1 Introduction

Connectionist networks are playing an increasingly important role in artificial intelligence (AI) and cognitive science and have been employed successfully to deal with a variety of problems in low and intermediate level vision, word perception, associative memory, word sense disambiguation, modeling of context effects in natural language understanding, speech production, and a wide range of issues related to learning (Cognitive Science 85; McClelland & Rumelhart, 1986; Rumelhart & McClelland, 1986). However, for connectionism to be considered a scientific language of choice for expressing solutions to problems in cognitive science and AI, it must be demonstrated that it can be used to represent highly *structured* knowledge and perform inferences based on such knowledge. A common criticism leveled against connectionism is that although it is appropriate for modeling "approximate" memory processes such as semantic priming associative recall, it is unsuitable for dealing with problems related to knowledge representation and reasoning.

The work described in this paper partially answers the criticism by demonstrating that the connectionist approach is extremely effective in solving certain problems in knowledge representation and inference. This paper presents a connectionist realization of semantic networks, i.e it describes how knowledge about concepts, their properties, and the hierarchical relationship between them, may be encoded as a connectionist network that can compute principled solutions to **inheritance** and **recognition** problems with extreme efficiency. Some salient features of the system are:

i) The connectionist semantic networks use *controlled* spreading activation to solve an interesting class of *inheritance* and *recognition* problems extremely fast - in time proportional to the depth of the conceptual hierarchy.

ii) The networks compute the solutions in accordance with an evidential formalization that derives from the principle of *maximum entropy*. This formalization leads to a principled treatment of *exceptions, multiple inheritance* and *conflicting information* during inheritance, and the *best match* or *partial match* computation during recognition.

iii) The networks operate without the intervention of a central controller and do not require a distinct interpreter. The knowledge as well as mechanisms for drawing *limited inferences* on it are encoded within the network.

iv) The networks can be constructed from a high-level specification of the knowledge to be encoded and the mapping between the knowledge level and the network level is precisely specified. Furthermore, the solution scales because the design is independent of the size of the semantic memory.

1.1 Representation and retrieval: An overview

The system's conceptual knowledge is encoded in a connectionist network referred to as the *Memory network*. This network is capable of performing inheritance and recognition via controlled spreading activation. A problem is posed to the network by activating relevant nodes in it. Once activated, the network performs the required inferences *automatically* and at the end of a specified interval the answer is available implicitly as the levels of activation of a relevant set of nodes.

In keeping with the connectionist paradigm, the presentation of queries to the Memory network, and the subsequent answer extraction is also carried out by connectionist network fragments called *routines*. Routines encode canned procedures for performing specific tasks and are represented as a sequence of nodes connected so that activation can serve to sequence through the routine. Routines pose queries to the Memory network by activating appropriate nodes in it. The Memory network in turn returns the answer to the routine by activating *response* nodes in the routine. The activation returned by a node in the Memory network is a measure of the evidential support for an answer. It is assumed that all queries originating in routines are posed with respect to an explicit set of answers and there is a response node for each possible answer. Response nodes compete with one another and the node receiving the maximum activation dominates and triggers the appropriate action. Thus, computing an answer amounts to choosing the answer that receives the highest evidence *relative* to a set of *potential answers*. The actual answer extraction mechanism explicitly allows for "don't know" as a possible answer. This may happen if there is insufficient evidence for all the choices or if there is no clear cut winner. This interaction between the Memory network and routines is depicted in Figure 1.

1.2 Semantic networks, inheritance, and recognition

The term "semantic networks" has been used in a very general sense in the AI literature. We will however, only focus on the central aspects of semantic networks namely, that concepts are represented in terms of their properties and that the subsumption relationship between concepts is captured by the IS-A hierarchy. This characterization is broad enough to capture the basic organizational principles underlying *frame*-based representation langauges such as KRL (Bobrow & Winograd, 1976) and KL-ONE (Brachman & Schmolze, 1985).

The organization and structuring of information in a semantic network leads to an efficient realization of two kinds of inferences which we will refer to as *inheritance* and *recognition*. It can be argued that these two complementary forms of reasoning lie at the core of intelligent behavior and act as precursors to more complex and specialized reasoning processes.

Typically, inheritance refers to the form of reasoning that leads an agent to infer property values of a concept based on the property values of its ancestors. We define inheritance more generally to include looking up property values directly available at the concept - of course if such local information is not available then inheritance involves looking up properties attached to concepts higher up in the conceptual hierarchy. Many cognitive tasks may be shown to require inheritance as an intermediate step - word sense disambiguation, determination of case-fillers, and enforcement of selectional restrictions are some examples.

Recognition is the dual of the inheritance problem. The recognition problem may be described as follows: "Given a description consisting of a set of properties, find a concept that best matches this description". Note that during matching all the property values of a concept may not be available locally and may have to be determined via inheritance from its ancestors. For this reason, recognition may be viewed as a very general form of *pattern matching*: one in which the target patterns, i.e. the set of patterns to which an input pattern is to be matched, are organized in a hierarchy, and where matching an input pattern A with a target pattern T_i involves matching properties that appear in A with properties local to T_i as well as to properties that T_i inherits from its ancestors.

A recognition step followed by an inheritance step amounts to an important sort of reasoning namely, *pattern completion*. Using recognition a process can determine the identity of an object based on its partial description, and having determined the object's identity the process may perform an inheritance step to determine the unknown

144

properties values of the object.

In addition to their ubiquity, inheritance and recognition are also significant because in spite of operating with a large knowledge base, humans perform these inferences effortlessly and *extremely fast* - often in a few hundred milliseconds. This suggests that inheritance and recognition are perhaps *basic* and *unitary* components of symbolic reasoning - probably the smallest and simplest cognitive operations that i) produce *specific responses*, and ii) can be initiated, and have their results accessed by complex and higher-level symbolic reasoning processes. The speed with which these operations are performed also suggests that they are performed fairly automatically, and typically do not require any conscious and attentional control. Given the significance of inheritance and recognition, it appears reasonable to pursue a *computational account* of how these inferences may be drawn with the requisite efficiency.

In addition to offering computational effectiveness, the connectionist network computes solutions to the inheritance and recognition problems in accordance with a theory of *evidential reasoning* that derives from the principle of *maximum entropy*. Under the evidential formulation, inheritance and recognition are posed as problem whose answers involve choosing the *most likely* alternative from among a set of alternatives - the computation of likelihood being carried out with respect to the knowledge encoded in the conceptual hierarchy. This reformulation provides a principled way of handling *exceptions* and resolving *conflicting information* during multiple inheritance, and finding *best matches* based on *partial information* during recognition.

This paper is about the connectionist realization and a detailed discussion of the underlying evidential formulation and the motivation for adopting it, is beyond the scope of this paper. The evidential formulation, its relation to Bayes' rule, and its merits are discussed in (Shastri, 1987); a brief version that deals primarily with inheritance appears in (Shastri & Feldman, 1985). Therein, it is argued that although non-evidential treatments such as proposals based on Default Logic (Etherington & Reiter, 1983) or on the Principle of Inferential Distance Ordering (Touretzky, 1986) can handle exceptions, they do not deal with conflicting information adequately - they either make arbitrary choices or simply report an ambiguity. In contrast, the evidential approach provides a semantically justifiable way of combining all the relevant information (even though some of it may be conflicting) to obtain the most likely answer.

1.3 Related work on parallel encoding of semantic networks

Fahlman's NETL (Fahlman, 1979) was the first attempt at encoding semantic networks as a massively parallel network of simple processing elements. NETL elements communicated with one another under the control of a central controller by propagating discrete messages called *markers*. A network element could only detect the presence or absence of a marker in the input. This all or none nature of the system made it incapable of supporting "best match" or "partial match" operations. For example, in NETL recognition amounted to finding a concept that possessed *all* of a specified set of properties. Furthermore, NETL's solution to the inheritance problem was sensitive to race conditions in the presence of multiple hierarchies. These limitations of marker passing systems are discussed at length in (Fahlman, 1982; Fahlman et al., 1981; Brachman, 1985)[1]. Finally, NETL did not fully utilize the potential for parallelism because the inter-node communication critically depended on instructions issued by a central (serial) controller.

Hinton proposed a "distributed" encoding of semantic networks using parallel hardware (Hinton, 1981). The information encoded in the network was interpreted as a set of triples of the form: [relation, role1, role2]. The proposed system had several interesting properties: given two components of a triple, the network could determine the third tuple, the network could be programmed using the perceptron convergence rule, and it could perform simple property inheritance. The system however, lacked sufficient structure and control to handle general cases of inheritance and "partial matching" - especially if these occurred in a mutli-level semantic network that included multiple hierarchies and exceptional or conflicting information.

More recently, Derthik (Derthik, 1986) is implementing a variant of KL-ONE using the Boltzman machine formulation (Ackley et al., 1985). However, the representation language being implemented does not admit exceptions and conflicting information and is open to the same objections that apply to other non-evidential

---

[1]Subsequent work by Touretzky has remedied certain problems with inheritance. The use of discrete markers however, still precludes partial match and best match operations.

formalizations.

Recent work on Bayesian networks (Pearl, 1985) deals with evidential reasoning in a parallel network. Pearl's results however, apply only to singly-connected networks (networks in which there is only one underlying path between any pair of nodes). More complex networks have to be *conditioned* to render them singly-connected. This is in part due to the unstructured form of the underlying representation language employed by Pearl. The language does not make distinctions such as "concept", "property", "property-value" that we make (cf. section 2 below) and hence its ability to exploit parallellism is limited.

2 A restricted language for representing conceptual knowledge

The representation langauge may be viewed as an extension of inheritance hierarchies to include relative frequency information specifying how instances of some concepts are distributed with respect to some property values. A summary description of the language follows. The agent's knowledge consists of the quintuple:

$$\Theta = \langle C, \Phi, \#, \delta, << \rangle, \text{ where}$$

C is the set of *concepts*, Φ is the set of *properties*, # is a mapping from C to the integers I, δ the *distribution* function is a partial mapping from C X Φ to the power set of C X I, and << is a partial ordering defined on C.

For each $C \in C$, if C is a Token (instance) then #C = 1, and if C is a Type (generic concept) then #C = the number of instances of C *observed by the agent*. By extension #C[P,V] = the number of instances of C that are observed by the agent to have the value V for property P. For example, #APPLE[has-color, RED] equals the number of red apples observed by the agent. Finally, $\#C[P_1,V_1][P_2,V_2] \dots [P_n,V_n]$ = the number of instances of C, observed to have the value V_1 for property P_1, value V_2 for property P_2, ... and value V_n for property P_n.

The distribution function $\delta(C,P)$, specifies how instances of C are distributed with respect to the values of property P. Recall that a concept may have several values for the same property and hence, if C is a Type, then $\delta(C,P)$, corresponds to the summary information abstracted in C based on the instances of C. Using the # function, $\delta(C,P)$ may be expressed in terms of #C[P,V]'s. Thus, $\delta(APPLE, \text{has-color})$ may be expressed as: { #APPLE[has-color, RED] = 60, #APPLE[has-color, GREEN] = 40}. Note that δ is only a partial mapping; an agent may not know $\delta(C,P)$ for many concept-property pairs. In general, for a given C and P, an agent knows $\delta(C,P)$ only if this information may prove useful in making inferences about C.

A salient feature of the language is that either a concept is an instance of (subtype of) another concept or it is not, and the << relation specifies this unequivocally. Exceptions only apply to property values. Furthermore, both *necessary* properties as well as *default* properties may be represented. This goes a long way in assigning a clean semantics to the representation language.

In terms of the above notation, the inheritance and recognition problem may be restated as follows:

Inheritance

Given: A concept C, a property P, and a set of property values, V-SET = $\{V_1, V_2, \dots V_n\}$,

Find: $V^* \in$ V-SET, such that among members of V-SET, V^* is the *most likely value* of property P for concept C. In other words, find $V^* \in$ V-SET such that, for any $V_i \in$ V-SET, the best estimate of #C[P,V^*] $\geq$ the best estimate of #C[P,V_i]'s.

For example,the inheritance problem where C = APPLE, P = has-color, V-SET = {RED, BLUE, GREEN}, may be paraphrased as: Is the color of an apple more likely to be red, green or blue?

Recognition

Given: a set of concepts, C-SET = $\{C_1, C_2, \dots C_n\}$, and an appropriate description consisting of a set of property value pairs, i.e., a DESCR = { $[P_1,V_1], [P_2,V_2], \dots [P_m,V_m]$ }.

Find: $C^* \in$ C-SET such that *relative* to the concepts specified in C-SET, C^* is the *most likely* concept described by DESCR.

If C-SET = {APPLE, GRAPE}, DISCR = {[has-color, RED], [has-taste, SWEET]} then the recognition problem may be paraphrased as: "Is something red in color and sweet in taste more likely to be an apple or a grape"?

146

The solutions to the two problems are based on the principle of maximum entropy [Jaynes 1979] and are described in [Shastri 1987].

3 Connectionist encoding

A connectionist network (Feldman & Ballard, 1982) consists of a large number of nodes connected via links. The nodes are computational entities defined by a small number (2 or 3) of states, a real-valued potential in the range [0,1], an output value also in the range [0,1], a vector of inputs $i_1, i_2, \ldots i_n$, together with functions P, V and Q that define the values of potential, state and output at time t+1, based on the values of potential, state and inputs at time t. Nodes receive inputs via *weighted* links. A node may have multiple input *sites*, and incoming links are connected to specific sites. Each site has an associated *site functions*. These functions carry out local computations based on the inputs incident at the site, and it is the result of this computation that is processed by the functions P, V and Q.

Connectionist networks offer a natural computational model for encoding evidential formalisms because of the natural correspondence between nodes and hypotheses, activation and evidential support, and potential functions and evidence combination rules. However, in order to solve the inheritance and recognition problems, the network must perform very specific computations and it must do so *without the intervention of a central controller*. The design involves introducing explicit "control nodes" (binder and relay nodes) throughout the network to mediate and control the spread of activation.

Before describing the encoding in detail, we consider an example. Figure 2 shows a network that encodes:

"Dick is a Quaker and a Republican, most Quakers have pacifist beliefs, while most Republicans have non-pacifist beliefs"

It is assumed that one of the properties attached to persons is "has-belief", some of whose values are "pacifist" and "non-pacifist". The Figure only shows about half the connections. In particular, the connections from property values to concepts have been suppressed for better readability. The likelihoods of being pacifists and non-pacifists for Quakers and Republicans are encoded as weights of appropriate links (Cf. Section 3.1)

The question of Dick's beliefs on pacifism (or lack of it) can be posed to the network by activating the nodes DICK, has-belief, and BELIEF. The resulting potentials of the nodes PACIFIST and NON-PAC will determine whether Dick is more likely to be a pacifist or a non-pacifist. It can be shown that the potential of the node PACIFIST equals:

$$\frac{\#QUAKER[\text{has-bel},PACIFIST] \times \#REPUBLICAN[\text{has-bel},PACIFIST]}{\#BELIEF \times \#PERSON[\text{has-bel},PACIFIST]}$$

The potential of the node NON-PAC is given by an analogous expression in with NON-PAC replaces PACIFIST.

Ignoring the common factor, #BELIEF, in the above expression, the potential of PACIFIST (NON-PAC) corresponds to the best estimate of the number of persons that are both quakers and republicans and believe in pacifism (non-pacifism). Hence, a comparison of the two potentials will give the most likely answer to the question: Is Dick a pacifist or a non-pacifist.

3.1 Encoding the conceptual structure

The encoding employs five distinct unit types. These are the *concept* nodes (ξ-nodes), *property* nodes (ϕ-nodes), *binder* nodes, *relay* nodes and *enable* nodes. With reference to Figure 2, all solid boxes denote ξ-nodes, all triangular nodes denote binder nodes, and the single dashed box denotes a ϕ-node. Relay nodes are used to control directionality of spreading activation along the conceptual hierarchy, while enable nodes are used to specify the type of query (inheritance or categorization). Relay and enable nodes are not shown in Figure 2.

Each concept is represented by a ξ-node. These nodes have six sites: **QUERY, RELAY, CP, HCP, PV** and **INV**. With reference to the partial ordering <<, if concept B is a parent of concept A then there is a ↑ (bottom up) link from A to B and a ↓ (top down) link from B to A. The weight on both these links equal # A/ #B and they are incident at the site **RELAY**. As the ↑ and ↓ links always occur in pairs, they will often be represented by a single undirected arc. Arcs between DICK and QUAKER, and QUAKER and PERSON, are examples of such interconnections.

Each property is also encoded as a node. These nodes are called φ-nodes, and they has one input site: **QUERY**.

If $\delta(A,P)$ is known, then for every value V_i of P there exists a pair of binder nodes $[A,P \rightarrow V_i]$ and $[P,V_i \rightarrow A]$ that are connected to A, P and V_i as shown in Figures 3 and 4 respectively. A binder node such as $[A,P \rightarrow V_i]$ is called a i-binder node and has two sites: **ENABLE** and **EC**. The node $[A,P \rightarrow V_i]$ receives one input from node A and another from node P. Both these inputs are incident at site **ENABLE**, and the weight on these links is 1.0. The link from $[A,P \rightarrow V_i]$ to V_i is incident at site **CP** and the weight on this link is given by $\#A[P,V_i] / \#V_i$. A binder node such as the node $[P,V_i \rightarrow A]$ is called a r-binder node and has one site **ENABLE** where it receives inputs from nodes P and V_i; the weights on these links are 1.0. The output from $[P,V_i \rightarrow A]$ is incident at the site **PV** of A, and the weight on this link is given by $\#A[P,V_i] / \#A$.

If B is a parent of A such that $\delta(B,P)$ is known, and there is no concept C between A and B for which $\delta(C,P)$ is known, then there is a link from $[A,P \rightarrow V_i]$ to $[B,P \rightarrow V_i]$, incident at site **INV** with a weight of $\# A[P,V_i]/ \#B[A,V_i]$ (refer to Figure 5). Similarly, there is a link from $[P,V_i \rightarrow B]$ to A incident at site **INV** with a weight of $\#B[P,V_i] / \#B$ (refer to Figure 6). Finally, if B is such that it is the highest node for which $\delta(B,P)$ is known, then the link from $[B,P \rightarrow V_i]$ to V_i is incident at site **HCP**, instead of site **CP**.

Besides the interconnections described above, all nodes representing concepts, properties, and values (ξ-nodes and φ-nodes) have an external input incident at the site **QUERY**, with a weight of 1.0. In addition to the unit types described above, there are two other enable units: INHERIT and RECOGNIZE. These units have one input site: **QUERY**, at which they receive an external input. Each i binder node receives an input from the node INHERIT at the site **ENABLE** while each r-binder node receives an input from the node RECOGNIZE also at the site **ENABLE**.

3.2 Description of network behavior

Each unit in the network can be in one of two states: **active** or **inert**. The quiescent state of each unit is **inert**. A unit switches to an **active** state under conditions specified below, and in this state the unit transmits an output equal to its potential. The computational characteristics of various unit types are described below:

ξ-nodes:
State: Node is in **active** state if it receives one or more inputs.
Potential: If no inputs at site **HCP** then
potential = the product of inputs at sites **QUERY**, **RELAY**, **CP**, and **PV** divided by the product of inputs at site **INV**.
else potential = the product of inputs at sites **QUERY**, **RELAY**, **HCP**

i-binder nodes:
State: Node is in **active** state if and only if it receives all the three inputs at site **ENABLE**.
Potential: If state = **active** then
potential = 1.0 * the product of inputs at sites **EC**
else potential = NIL

r-binder nodes:
State: Node is in **active** state if and only if it receives all three inputs at site **ENABLE**.
Potential: If state = **active** then potential = 1.0 else potential = NIL

φ-nodes, INHERIT node, and RECOGNIZE node switch to **active** state if they receive input at site **QUERY**, and in this state their potential always equals 1.0.

The networks have the additional property that unlike other links that always transmit the output of their source node, the ↑ and ↓ normally remain disabled, and transmit activity only when they are enabled. This control is

148

affected via *relay* nodes that are associated with ϕ-nodes. The details of this mechanism are beyond the scope of this paper.

3.3 Posing queries and computing solutions

In the context of the network encoding the inheritance and categorization are posed as follows:

Inheritance

Given: A concept C, a property P, a set of possible answers, V-SET = $\{V_1, V_2, ...V_n\}$, and a concept REF where REF is an ancestor of every memebr of V-SET. (Typically, REF is a parent of V_i's. For example, if V_i's are RED, GREEN, BLUE ... then REF could be COLOR).

Find: $V^* \in$ V-SET such that relative to the values specified in V-SET, V^* is the most likely value of property P for concept C.

The inheritance query is posed by setting the external inputs, i.e. the inputs to the site **QUERY**, of nodes C, P and INHERIT to 1.0. If one or more members of V-SET reach an **active** state within three time steps, the external input to REF gets set to 1.0, and the $\downarrow$ links leaving REF are enabled. If none of the members of V-SET receive any activation, the external input to REF is set to 1.0, and the $\downarrow$ links leaving REF as well as the $\uparrow$ links leaving C are enabled. After d+3 time steps - where d is the longest path in the ordering graph defined by C and <<, the potentials of nodes will be such that for any two nodes V_i and $V_j \in$ V-SET, the following holds:

$$\frac{\text{potential of } V_i}{\text{potential of } V_j} = \frac{\#C[P,V_i]}{\#C[P,V_j]}$$

It follows that the node $V^* \in$ V-SET with the highest potential will correspond to the value that is the solution to the inheritance problem.

Recognition

Given: a set of concepts C-SET = $\{ C_1, C_2, ... C_n)$, a reference concept REF, such that REF is an ancestor of all concepts in C-SET, and a description consisting of a set of property value pairs, i.e. a set DESCR = $\{ [P_1,V_1], [P_2,V_2], ... [P_m,V_m] \}$

Find: $C^* \in$ C-SET such that relative to the concepts specified in C-SET, C^* is the most likely concept described by DESCR.

The solution to the above problem may be computed as follows: For each $[P_j,V_j] \in$ DESCR, set the inputs to the site **QUERY** of nodes P_j and V_j to 1.0. At the same time, set the input to the site **QUERY** of RECOGNIZE and REF to 1.0, and enable the $\downarrow$ links emanating from REF. Wait d + 3 time steps, where d is the longest path in the ordering graph defined by C and <<. At the end of this interval, the potential of the nodes will be such that for any two nodes C_i and $C_j \in$ C-SET, the following holds:

$$\frac{\text{potential of } C_i}{\text{potential of } C_j}$$

equals the best estimate of $\#C_i[P_1,V_1][P_2,V_2]$ ... $[P_m,V_m]$ divided by the best estimate of $\#C_j[P_1,V_1][P_2,V_2] ...[P_m,V_m]$. It follows that the node $C^* \in$ C-SET with the highest potential corresponds to the solution of the recognition problem.

4 Some examples

In order to explicate the behavior of networks and demonstrate the nature of inferences drawn by them, several examples that are often cited in the knowledge representation literature as being problematic have been simulated. The first example is an extension of the "quaker example" discussed in section 3. It demonstrates how the network performs inheritance in the presence of conflicting information arising due to "multiple inheritance". Figure 7 depicts the underlying information. There are two properties has-bel (has-belief) with values PAC (pacifist) and NON-PAC (non-pacifist), and has-eth-org (ethnic-origin) with values AFRIC (african) and EURO (european). In broad terms, the information encoded is as follows:

Most persons are non-pacifists, most quakers are pacifists, most republications are non-pacifists, most persons are of european descent, most republicans are of european descent, and most persons of african decent are democrats.

Such information is specified to a network compiler in terms of: i) the set of concepts, ii) the set of properties and their associated values, iii) a list specifying the partial ordering together with the ratios #A/#B (for all pairs of concepts A and B such that B is a parent of A), and iv) a partial mapping $\delta(C,P)$ in terms of #C[P,V]'s. The specification does not refer to any network level detail and the compiler directly translates such a specification into a connectionist network.

As our first example of inheritance, consider the query: "Is Dick a pacifist or a non-pacifist?" The normalized potentials of PAC and NON-PAC as a result of this query are 1.00 and 0.66 respectively. Thus, on the basis of the available information, Dick who is a republican and a quaker is more likely to be a pacifist than a non-pacifist, the ratio of likelihoods being 1.00 : 0.66, i.e., about 3:2. Similar simulations for RICK, PAT, and SUSAN lead to the following results: Rick who is a mormom republican is more likely to be a non-pacifist. The ratio of pacifist v/s non-pacifist for Rick being 0.39 v/s 1.00. Pat who is mormon democrat is also more likely to be a non-pacifist, but only marginally so (0.89 v/s 1.00). Finally, Susan who is a quaker democrat is likely to be a pacifist with a very high probability (1.00 v/s 0.29).

As an example of recognition, consider the query: "among Dick, Rick, Susan, and Pat, who is more likely to be a pacifist of african descent?" The resulting normalized potentials are SUSAN 1.00, PAT 0.57, DICK 0.11, and RICK 0.05. As would be expected, Susan who is a democrat and a quaker, best matches the description "person of african descent with pacifist beliefs". The least likely person turns out to be Rick (notice that Rick is neither a democrat who correlate with african origin nor is he a quaker who correlate with pacifism).

In order to illustrate how exceptions are handled, the information given in Figure 8 was encoded in a network. The information captures the following aspect of the domain: "Most Molluscs are shell-bearers, Cephalopods are Molluscs, but most Cephalopods are not shell-bearers, Nautili are Cephalopods, and all Nautili are shell-bearers".

The normalized potentials of SHELL and SKIN as a result of the inheritance of the property epidermis-type of MOLLUSC, CEPHAL, and NAUTILUS are as follows: (the potentials of FUR and FEATHER were consistently 0.0):

VALUE	MOLLUSC	CEPHAL	NAUTILUS
SHELL	1.00	0.25	1.00
SKIN	0.43	1.00	0.00

Thus, a Mollusc is more likely to be a shell-bearer. A Cephalopod is not likely to be a shell-bearer. Finally, a Nautilus is *definitely* a shell-bearer (note that the likelihood of a Nautilus having an epidermis-type other than shell computes to 0.00, this is because ALL Nautilus are shell-bearers).

5 Conclusions

This effort has lead to the design of a connectionist network that provides a computational account of how an interesting class of inheritance and recognition problems may be solved extremely fast. The networks also have a *provable behavior*; they compute solutions to the inheritance and recognition problems in accordance with a theory of evidential reasoning. The use of evidential reasoning redefines these problems so that conflicting information can be interpreted in a semantically consistent manner. The work also identifies specific constraints that must be satisfied by the conceptual structure in order to achieve an efficient connectionist realization. These are discussed at length in (Shastri 87).

Besides offering a natural way of describing the evidential interactions between pieces of knowledge, the network encoding *suggests* how a physical system may extract from its environment the information required to solve inheritance and recognition problems. An examination of the weights on the links reveals that in most cases the weights are directly related to Hebb's interpretation of synaptic weights (Hebb, 1949). The weight on these links is equal to the ratio: *"how often when the destination node was active, was the source node also active"*.

A discussion of a connectionist system often leads to the question of its biological plausibility. It may be felt that the computational characteristics of nodes described in section 3.1 are too complex to be biologically plausible. The proposed encoding is certainly not intended to be a blueprint for building "wetware". Yet it does satisfy nearly all the constraints proposed in (Feldman & Ballard, 1982). The only serious violation of biological plausibiltiy is the requirement that nodes perform high precision multiplication. One may interpret the connectionist system described here as an ideal realization of a formal model of evidential reasoning. One can try and identify more plausible "approximations" of the ideal system and study the manner in which their response deviates from the prescribed behavior. Such an exercise may be rewarding and point out further constraints that govern the organization of conceptual structure.

6 Acknowledgement

I am grateful Jerry Feldman for his guidance, insights, and support. This research was supported in part by the National Science Foundation under Grants MCS-8209971M IST-8208571, and DCR-8405720, DCR-86-07156, and U.S. Army Research Office grant ARO-DAAG29-84-K-0061.

7 References

Ackley, D.H., Hinton, G.E. & Sejnowski T.J. (1985). A learning algorithm for Boltzmann Machines, *Cognitive Science*, 9, (1), Jan. - March 1985.

Bobrow, D.G. & Winograd, T. (1976). An overview of KRL: A Knowledge Representation Language. CSL-76-4. Xerox Palo Alto Research Centre.

Brachman, R. J. (1985). I lied about the trees. *The AI Magazine*, vol.6, no. 3, Fall 1985, pp. 80-93.

Brachman R.J. and Schmolze, J. (1985). An overview of KL-ONE Knowledge Representation System. *Cognitive Science* 9(2), April, 1985.

Cognitive Science, (1985) Special Issue on Connectionism. *Cognitive Science* 9 (1), Jan. - March, 1985.

Derthik, M. (1986) A connectionist Knowledge Representation System. Thesis Proposal, CMU, June 1986.

Etherington, D. W. & Reiter, R. (1983). On inheritance hierarchies with exceptions. *Proc. AAAI-83*, Washington D.C.

Fahlman, S. E. (1982). Three flavors of parallelism. *Proc. CS-CSI-82*. Canada 1982.

Fahlman, S. E., Touretzky, D.S., and van Roggen Walter. (1981). Cancellation in a parallel semantic network. *Proc. IJCAI-81*. Vancouver, B.C. 1981.

Fahlman, S.E. (1979) *NETL: A System for Representing and Using Real-World Knowledge*. The MIT Press, 1979.

Feldman, J. A. and Ballard, D. H. (1982). Connectionist models and their properties. *Cognitive Science*, 1982, 6, pp. 205-254.

Hebb, D.O. (1949) *The Organization of Behavior*, Wiley, New York, 1949.

Hinton, G.E. (1981) Implementing Semantic Networks in Parallel Hardware. In *Parallel Models of Associative Memory*. Hinton G.E., Anderson J.A. (Eds). Lawrence Erlbaum Associates, 1981.

Jaynes, E.T. (1979) Where Do We Stand on Maximum Entropy. In *The maximum entropy formalism*, R.D. Levine and M. Tribus (Eds.) MIT Press, Cambridge Massachussets. 1979.

McClelland, J.L. & Rumelhart, D.E. (1986). (Eds) *Parallel Distributed Processing: Explorations in the Microstructure of Cognition*. Vol II. Bradford Books/MIT Press, Cambridge, MA, 1986.

Pearl, J. (1985). Bayesian Networks: A model of self-activated memory for evidential reasoning. *Proc. 7th. Cognitive Science Conference*, Irvine, CA 1985.

Rumelhart, D.E. & McClelland, J.L. (1986). (Eds.). *Parallel Distributed Processing: Explorations in the*

Microstructure of Cognition. Vol I. Bradford Books/MIT Press, Cambridge, MA, 1986.

Shastri, L. (1987) *Semantic Networks: An Evidential Formalization and its connectionist realization.* To appear Morgan Kaufman, Los Altos, CA, and Pitman, London. 1987.

Shastri L. & Feldman J.A. (1985). Evidential reasoning in semantic networks: a formal theory. In *Proc. IJCAI-85.* Los Angeles CA.

Touretzky, D. S. (1986). *The Mathematics of Inheritance Systems.* Morgan Kaufman, Los Altos, CA, Pitman London, 1986.

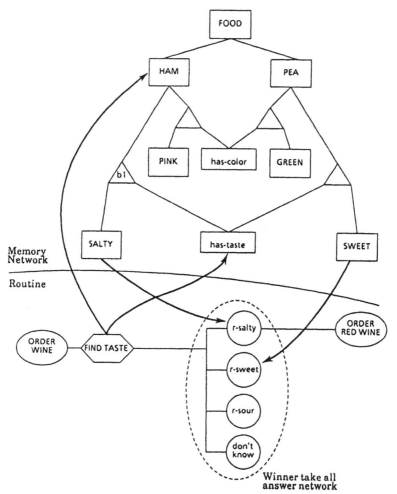

FIGURE 1: Connectionist Retrieval System

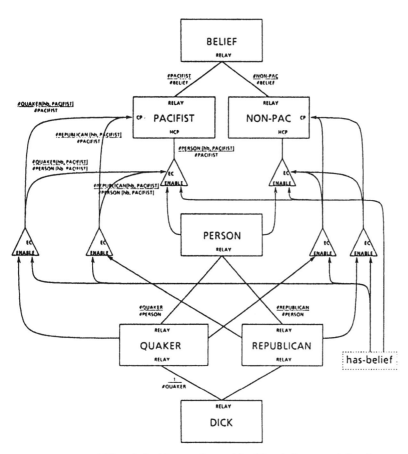

All inputs incident at site enable of δ-node have a weight of 1.0.
Not all sites and weights have been shown.
hb = has-belief

FIGURE 2: An example network

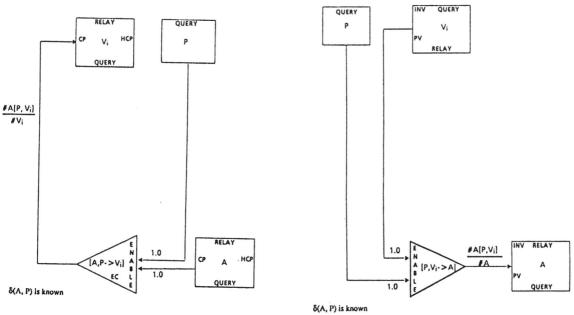

FIGURE 3: Parallel encoding for inheritance - I

FIGURE 4: Parallel encoding for recognition - I

153

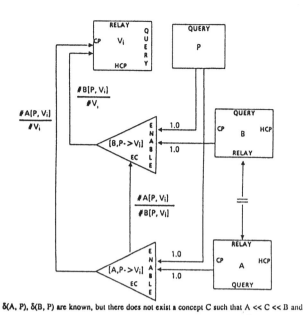

$\delta(A, P)$, $\delta(B, P)$ are known, but there does not exist a concept C such that $A \ll C \ll B$ and $\delta(C, P)$ is known

FIGURE 5: Parallel encoding for inheritance - II

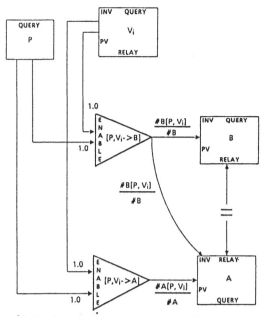

$\delta(A, P)$, $\delta(B, P)$ are known, but there does not exist a concept C such that $A \ll C \ll B$ and $\delta(C, P)$ is known

FIGURE 6: Parallel encoding for recognition - II

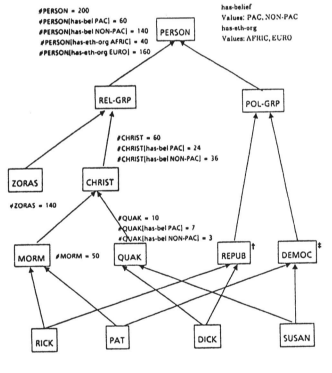

#PERSON = 200
#PERSON[has-bel PAC] = 60
#PERSON[has-bel NON-PAC] = 140
#PERSON[has-eth-org AFRIC] = 40
#PERSON[has-eth-org EURO] = 160

has-belief
Values: PAC, NON-PAC
has-eth-org
Values: AFRIC, EURO

#CHRIST = 60
#CHRIST[has-bel PAC] = 24
#CHRIST[has-bel NON-PAC] = 36

#ZORAS = 140

#QUAK = 10
#QUAK[has-bel PAC] = 7
#QUAK[has-bel NON-PAC] = 3

#MORM = 50

† #REPUB = 80
#REPUB[has-bel PAC] = 16
#REPUB[has-bel NON-PAC] = 64
#REPUB[has-eth-org AFRIC] = 5
#REPUB[has-eth-org EURO] = 75

‡ #DEMOC = 120
#DEMOC[has-bel PAC] = 44
#DEMOC[has-bel NON-PAC] = 76
#DEMOC[has-eth-org AFRIC] = 35
#DEMOC[has-eth-org EURO] = 85

FIGURE 7: The Quaker example

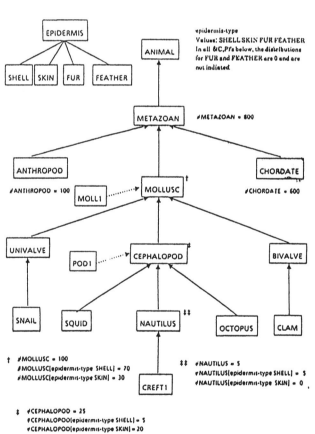

epidermis-type
Values: SHELL SKIN FUR FEATHER
In all &C,P's below, the distributions
for FUR and FEATHER are 0 and are
not indiated

#METAZOAN = 800

#ANTHROPOD = 100

#CHORDATE = 600

† #MOLLUSC = 100
#MOLLUSC[epidermis-type SHELL] = 70
#MOLLUSC[epidermis-type SKIN] = 30

‡‡ #NAUTILUS = 5
#NAUTILUS[epidermis-type SHELL] = 5
#NAUTILUS[epidermis-type SKIN] = 0

‡ #CEPHALOPOD = 25
#CEPHALOPOD[epidermis-type SHELL] = 5
#CEPHALOPOD[epidermis-type SKIN] = 20

FIGURE 8: The Mollusc example

154

A Distributed Connectionist Representation for Concept Structures

David S. Touretzky and Shai Geva
Computer Science Department
Carnegie Mellon University
Pittsburgh, Pennsylvania 15213

Abstract. We describe a representation for frame-like concept structures in a neural network called DUCS. Slot names and slot fillers are diffuse patterns of activation spread over a collection of units. Our choice of a distributed representation gives rise to certain useful properties not shared by conventional frame systems. One of these is the ability to encode fine semantic distinctions as subtle variations on the canonical pattern for a slot. DUCS typically maintains several concepts simultaneously in its concept memory; it can retrieve a concept given one or more slots as cues. We show how Hinton's notion of a "reduced description" can be used to make one concept fill a slot in another.

1. Introduction

In a typical Lisp implementation of frames, a frame is a collection of named slots with fillers (Minsky, 1975; Winston & Horn, 1984). Names are atomic symbols, and fillers are either atoms or pointers to other frames. This paper considers what a connectionist version of frames might look like. We describe a representation for frame-like structures in a neural network called DUCS. Names and fillers are diffuse patterns of activation over a collection of units. Our choice of a distributed representation (Hinton *et al.*, 1986) for frames gives rise to certain useful properties that are not shared by conventional frame systems.

One thing a connectionist frame system should be able to do is retrieve a slot given an approximation of its name. For example, suppose the frame describing Fred the cockatoo had the following slots:

$$
\left\{
\begin{array}{rl}
\text{BODY-COLOR:} & \text{PALE-PINK} \\
\text{BEAK:} & \text{GRAY-HOOKED-THING} \\
\text{CREST:} & \text{ORANGE-FEATHERED-THING} \\
\text{HABITAT:} & \text{JUNGLE} \\
\text{DIET:} & \text{SEEDS-AND-FRUIT}
\end{array}
\right\}
$$

What does Fred's nose look like? Strictly speaking, birds don't have noses; they have beaks. If the activity patterns for NOSE and BEAK are similar, which they will be if we use

155

a microfeature-based representation for symbols, then a connectionist frame system could simultaneously retrieve the description GRAY-HOOKED-THING and correct the slot name to read BEAK instead of NOSE. Ordinary frame systems could not do this unless some rule of form "look for a beak if you can't find a nose" had been explicitly established prior to the query.

A second advantage of using a distributed representation for frames is that when slots are encoded as activity patterns, instead of being limited to a small, fixed set of slot names, the names form a continuous space. Subtle nuances of meaning of the frame as a whole can be encoded as variations on the activity patterns of its slots. This is useful in the case role representation of sentences. For example, in "John sold the statue to Mary," Mary plays the role of recipient. In "John mailed the package to Boston," Boston's role would be destination. But in "John threw the ball to Mary" the role of Mary is both destination (the ball is thrown in her direction and is expected to make contact with her) and recipient (John's intention is that the ball come into her possession and be under her control.) In a system based on distributed representations, the pattern representing Mary's role could be a combination of destination and recipient, sharing microfeatures of both. Finer shadings of role names are also possible. We propose that in a connectionist version of case grammar, each combination of a verb, some case roles, and the fillers of those roles would generate slightly different role name patterns based on subtle nuances of the meaning of the sentence as a whole.

This idea was anticipated in McClelland and Kawamoto's PDP model of case role assignment (McClelland & Kawamoto, 1986). Their model provides four case slots called agent, patient, instrument, and modifier; a representation of the verb is conjunctively encoded with each slot filler. Although the names of the four slots are fixed, the fillers undergo variations from their canonical, surface forms according to the context in which they appear. For example, while there is only one pattern for representing the verb "move" in the input layer, in the case role layer the representation of "move" will be different when it has an animate agent that implicitly moves itself ("the cat moved") versus an agent that moves other things ("the boy moved the cat") versus an inanimate subject ("the rock moved") which is interpreted as a patient with the agent left unspecified.

In the following sections we describe the architecture of DUCS. The name stands for Dynamically Updatable Concept Structures. We will use the word concept rather than frame in the remainder of the paper in order to distinguish DUCS' structures from the ones used in Lisp-based reasoners. We will discuss two problems peculiar to connectionist systems. One is the problem of getting a concept to fill a slot in another concept. The other is the problem of getting concepts not recently accessed to automatically fade from working memory as new ones are created, so that the memory capacity is not exceeded.

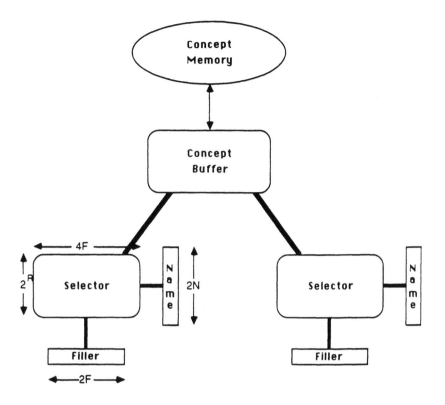

Figure 1: The DUCS Architecture.

2. The DUCS Architecture

Both slot names and slot fillers in DUCS employ distributed representations, meaning they exist as diffuse patterns of activity over a collection of units. Each slot name pattern determines a mapping of the associated slot filler pattern into an array called the concept buffer. The patterns that various slots generate in the buffer are superimposed to derive a pattern for the entire concept. See **Figure 1**.

DUCS is a two-level architecture. At the slot level, it retrieves individual slots by name, and can add, change, or delete slots by modifying the activity pattern in the concept buffer. Through the use of multiple slot mapping assemblies, each consisting of a slot name group, a slot filler group, and a selector group, several slots can be created or modified simultaneously. Thus it is possible to create a complex concept in a single operation as long as the number of slots does not exceed the available mapping hardware. Slots can also be loaded into the buffer sequentially.

At the concept level, DUCS manipulates entire concepts at a time rather than individual slots. Concepts are added to or deleted from an auto-associative concept memory via the concept buffer. DUCS retrieves a concept from concept memory using one or more slots as cues, in the following way. First the cues are loaded into slot mapping assemblies, where they generate a partial activity pattern in the concept buffer. Then the concept buffer

157

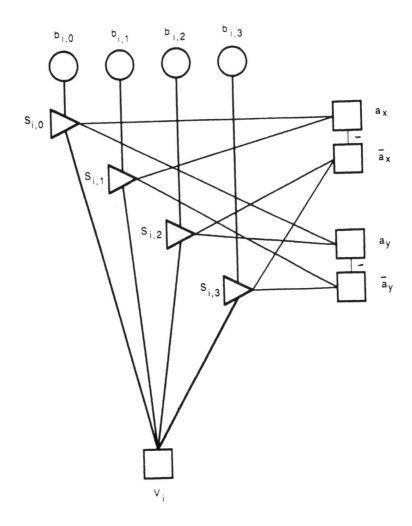

Figure 2: Mapping the slot filler bit v_i into one of 2^R concept memory bits based on an R-bit subset of the slot name. Mutually inhibitory connections among the 2^R selector units have been omitted for clarity.

supplies input to the concept memory, allowing it to complete the pattern and cause the remaining slots of the concept to materialize in the buffer.

2.1. The Slot Level

Slot names and slot fillers are N and F bit binary feature vectors, respectively, appended to their logical complements. That is, a slot name $\vec{a}$ is a $2N$-bit vector such that $a_i = \bar{a}_{(i+N)}$, $1 \le i \le N$. Similarly, a slot filler $\vec{v}$ is a $2F$-bit vector where $v_i = \bar{v}_{(i+F)}$, $1 \le i \le F$. The concept buffer and each selector group are $4F \times 2^R$ arrays, where R is a parameter between 0 and N.

Storing the filler pattern $\vec{v}$ in the slot named $\vec{a}$ generates a $4F \times 2^R$ pattern over the selector array $s_{i,j}$. Each bit v_i is copied into one of the 2^R locations in column i of the array, and independently, into one of the 2^R locations in column $i + 2F$. The location j

158

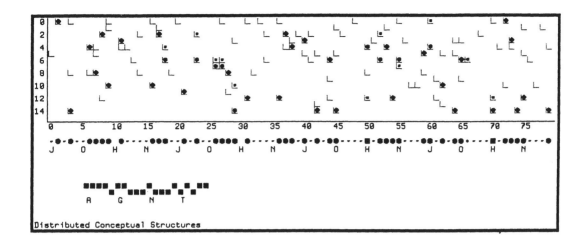

Figure 3: The state of the network part way through a retrieval of the AGENT slot.

within the column is determined by an R-bit subset of the slot name; a different subset is randomly associated with each column. After the selector group has stabilized, the active units $s_{i,j}$ are allowed to excite the corresponding units $b_{i,j}$ in the concept buffer, thereby superimposing the pattern for this slot onto the previous contents of the buffer.

Figure 2 shows the wiring for mapping the value of a single slot filler bit v_i into some $s_{i,j}$ in column i. Selector units within a column form mutually-inhibitory winner-take-all networks, so that in a stable state at most one selector per column will be active. The chosen unit will be the one with the most activation, *i.e.*, the one with the most of its R input lines from the slot name group active. In the figure, $R = 2$, so the position j is determined by two slot name bits. Note that v_i will also be copied into some position k in column $i + 2F$, where k is determined by a different pair of bits randomly chosen from the slot name. Also, v_{i+F}, which is $\bar{v}_i$, will be copied into columns $i + F$ and $i + 3F$ in positions determined by two other pairs of slot name bits. Thus the pattern developed in the selector group consists of two copies of each filler bit v_i and two copies of its complement, with each copy deposited in one of the 2^R positions in that column determined by the slot name.

The units in the slot name, slot filler, and selector groups are continuous-valued non-linear units with outputs restricted to the unit interval; all connections are symmetric. This is commonly known as a Hopfield and Tank model (Hopfield & Tank, 1985). Retrievals are accomplished by clamping the concept buffer and slot name space and setting a low gain value for selector and slot filler units. The gain then rises fairly quickly, and part way through the slot name group is unclamped. At high gain the network settles into a stable state representing the slot filler and (possibly corrected) slot name extracted from the concept buffer. Figure 3 shows the state of the network at a medium gain setting. Here, slot names and fillers are ASCII strings (using a five bit character code) rather than

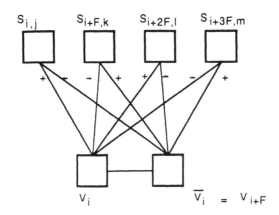

Figure 4: Error correction circuitry to exploit the redundant storage of slot fillers.

vectors of microfeatures, in order to make the model's operation more transparent. The bottom group of units represents the slot name "AGNT". The vector above that shows four copies of the slot filler "JOHN", two of which are logically inverted. The array at the top of the figure depicts the activity in the concept buffer (L-shaped symbols) and selector group (solid circles.) Three slots have been stored in the buffer; therefore each column of the array has between zero and three L-shapes, indicating active concept buffer units $b_{i,j}$. The size of the circles indicates the output level of the corresponding selector units $s_{i,j}$. Most columns have only one large circle, but some have two, indicating a pair of selectors in competition. At high gain all units will be either fully on or fully off, and there will be at most one active selector per column.

Selector units function as skeleton filters, so-called because only a skeleton subset of the units are enabled at any one time. Hinton (1981) used skeleton filters in a connectionist implementation of a semantic network. Sejnowski (1981) presents arguments for the existence of skeleton filters in the brain.

The redundant storage of fillers helps correct errors that may occur when several slots are superimposed in the concept buffer. Two slots can potentially interfere at column i when the R-bit subset of the slot name group examined by that column yields the same j value for both slots. If the first slot has a 1 in bit v_i and the second slot has a 0 (which implies the reverse situation in bit v_{i+F}), $s_{i,j}$ will be set to 1. But each filler bit is stored several times using a different R-subset each time, and the two slots are unlikely to overlap in every copy. Figure 4 shows the error correction circuitry used during slot retrieval to derive bits v_i and v_{i+F} from $s_{i,j}$, $s_{i+F,k}$, $s_{i+2F,l}$, and $s_{i+3F,m}$ via a majority voting scheme, where j, k, l, and m are determined by different R-subsets. The error correction imposes a constraint on the filler pattern that it have F bits on, and that $v_i = \bar{v}_{i+F}$, $1 \le i \le F$.

The error correction scheme is also important for associative retrieval. Suppose the network tries to retrieve a slot with a few bits of the slot name in error. Most of the $4F$ different R-subsets of the slot name will pick bits that are correct, so most filler bits will be mapped to the correct positions in their respective columns. Those columns that reference incorrect slot name bits will not be mapped properly. The error correction circuitry puts pressure on the slot filler group to settle into a pattern that meets the above mentioned constraints, and this in turn puts pressure on the selector units. If enough pressure is applied, the selector units can force the slot name units to change. This is how the network can change NOSE to BEAK during retrieval from a bird concept, assuming that the two symbols have similar activity patterns and a valid filler exists in the BEAK slot.

The parameter R determines the number of slots the concept buffer can hold. Since each filler bit maps into one of 2^R positions in its column, increasing R increases the sparseness of the concept buffer and reduces the chance that slots will interfere. Another way to increase the capacity of the concept buffer would be to to widen the array from $4F$ to, say, $6F$ units, to provide improved error correction.

2.2. The Concept Level

DUCS can memorize or retrieve entire concepts in a single operation. The activity pattern for a concept is a bit vector of length $4F \times 2^R$. The concept memory, shown in Figure 1, is a $(4F \times 2^R)$ by $(4F \times 2^R)$ matrix forming a Willshaw-style auto-associative net (Willshaw, 1981). For every pair of active concept buffer units $b_{i,j}$ and $b_{i*,j*}$ there is a concept memory unit. To store a pattern in concept memory it suffices to turn on each concept memory unit whose associated pair of concept buffer units is active.

To retrieve a concept, some subset of the pattern, generated by whatever cues were supplied, is fed into the auto-associative net. The retrieval is based on the assumption that the retrieval cues are correct. It yields a superposition of all patterns for concepts stored in the concept memory that match the given cues. For example, if two concepts with John as agent had been stored, both would be retrieved from the single cue AGENT=JOHN. A more detailed cue would be required to restrict the retrieval to a single concept. On the other hand, if no bits in addition to the original cue are retrieved, then the cue is a full specification of the desired concept, or else no such concept is stored in the memory. The network can detect whether a retrieval was successful by checking whether all pairs of bits in the concept buffer have their associated concept memory unit active. If this is the case, then the cue supplied was a valid one and only a single concept was retrieved.

3. Naming and Reduced Descriptions

In Lisp it's easy to make one structure point to another. One way to achieve a similar effect in connectionist models is to use a technique called reduced descriptions (Hinton,

1987.) For example, to represent "Bill knows that John kissed Mary" we first create and store the concept "John kissed Mary." Then we derive a small pattern for this concept, the reduced description, to fill the patient role in the concept "Bill knows that x," as shown:

$$\left\{ \begin{array}{ll} \text{AGENT:} & \text{JOHN} \\ \text{VERB:} & \text{KISS} \\ \text{PATIENT:} & \text{MARY} \end{array} \right\} \qquad \left\{ \begin{array}{ll} \text{AGENT:} & \text{BILL} \\ \text{VERB:} & \text{KNOW} \\ \text{PATIENT:} & \text{JhnKssMry} \end{array} \right\}$$

In order for this technique to work, the network must be able to retrieve the full pattern for a concept given its reduced description. In DUCS we obtain the reduced description simply by taking an F-bit slice out of the concept buffer. The exact choice of slice is unimportant; currently we use $b_{0,0}$ through $b_{F,0}$. To "follow the pointer" in the patient slot to get to the full description of what Bill knows, the F-bit filler pattern is clamped into bits $b_{0,0}$ through $b_{F,0}$ of the concept buffer, and this serves as the cue for the associative network to retrieve the rest of the concept pattern. Touretzky (1986) describes another version of pointer following in a connectionist network which does not use reduced descriptions.

4. Forgetting

The concept memory has a limited capacity. If a reasoner continually generates and stores new concepts, the memory could fill up, making further processing impossible. We view DUCS as a short term working memory for concept structures. In order to prevent its memory capacity from being exceeded, we implemented a forgetting mechanism by which concepts not recently accessed can fade and be displaced by newly stored ones.

Each unit in the concept memory has two parameters: an internal integer activation value in the interval $[0, c]$, and a binary output value that is 1 whenever the activation value is positive. To memorize a concept pattern, the first step is to decrement the activity levels of all concept memory units by one. Any unit whose activity has decayed to zero turns off. Then, for each pair of active units $b_{i,j}$ and $b_{i*,j*}$ in the concept buffer, the corresponding concept memory unit is given an initial activity level of c. With this protocol, concept memory will hold at most c distinct concepts at a time, provided that concepts are reinforced in concept memory only after non-ambiguous retrieval.

For auto-associative retrieval, concept memory units are treated as binary state units, with any non-zero activation value indicating a 1 state. Whenever a concept is retrieved into the concept buffer, it is immediately re-stored into concept memory. This refreshes the concept memory units by setting their activation levels back to c. Concepts which have not been fetched from concept memory for a while eventually decay due to lack of refresh. Non-active memory units are never reinforced, even if their associated pair of concept buffer units is active, to prevent undesired merging of multiple concepts retrieved simultaneously due to ambiguous cues.

5. Discussion

There are many architectures for building associative memories (Hinton & Anderson, 1981; Baum *et al.*, 1987). One of the unique features of DUCS is its two level structure: concepts can be recalled from concept memory into the concept buffer using slots as cues, and slots can be recalled from the concept buffer using their names as cues. The fact that slots are represented as activity patterns rather than as weights makes this possible.

Hinton's reduced description idea has great potential which we have only begun to explore. He suggests that the reduced description pattern should be meaningful (*i.e.*, interpretable) in its own right. Our current version contains too few bits to meet this criterion. An enhanced version might contain information about the type of the concept and a summary description of its slots. This would make it possible to make gross inferences about concepts (*e.g.*, that Bill knows a fact about some male person kissing some female person) without having to expand each reduced description beforehand. An ideal reduced description mechanism, rather than just taking a fixed slice out of a concept, would detect and focus on relevant features to evolve the most meaningful set of reduced descriptions possible in a given domain. We don't yet know how this might be accomplished.

Acknowledgements

Mark Derthick's μKLONE system (Derthick 1987; Touretzky & Derthick 1987), still under development, was a major source of architectural ideas for DUCS. We thank Mark Derthick and Geoffrey Hinton for frequently illuminating discussions. This work was supported in part by National Science Foundation grant IST-8516330, and by the Office of Naval Research under contract number N00014-86-K-0678.

References

[1] Baum, E. B., Moody, J., and Wilczek, F. (1987) Internal representations for associative memory. Manuscript. Institute for Theoretical Physics, University of California at Santa Barbara.

[2] Derthick, M. A. (1987) Factual and counterfactual reasoning by constructing plausible models. Manuscript. Computer Science Department, Carnegie Mellon University.

[3] Hinton, G. E. (1981) Implementing semantic networks in parallel hardware. In Hinton, G. E. and Anderson, J. A (eds.), *Parallel Models of Associative Memory*. Hillsdale, NJ: Earlbaum.

[4] Hinton, G. E. (1987) Representing part-whole hierarchies in connectionist networks. Manuscript. Computer Science Department, Carnegie Mellon University.

[5] Hinton, G. E., McClelland, J. L, and Rumelhart, D. E. (1986) Distributed representations. In D. E. Rumelhart and J. L. McClelland (eds.), *Parallel Distributed Processing: Explorations in the Microstructure of Cognition*, volume 1. Cambridge, MA: The MIT Press.

[6] Hopfield, J. J. and Tank, D. (1985) "Neural" computation of decisions in optimization problems. *Biological Cybernetics, 52,* 141-152.

[7] McClelland, J. L. and Kawamato, A. H. (1986) Mechanisms of sentence processing: assigning roles to constituents. In J. L. McClelland and D. E. Rumelhart (eds.), *Parallel Distributed Processing: Explorations in the Microstructure of Cognition*, volume 2. Cambridge, MA: The MIT Press.

[8] Minsky, M. L. (1975) A framework for representing knowledge. In P. H. Winston (ed.), *The Psychology of Computer Vision*. New York: McGraw-Hill.

[9] Sejnowski, T. J. (1981) Skeleton filters in the brain. In Hinton, G. E. and Anderson, J. A (eds.), *Parallel Models of Associative Memory*. Hillsdale, NJ: Earlbaum.

[10] Touretzky, D. S. (1986) BoltzCONS: reconciling connectionism with the recursive nature of stacks and trees. *Proceedings of the Eighth Annual Conference of the Cognitive Science Society*, Amherst, MA.

[11] Touretzky, D. S. and Derthick, M. A. (1987) Symbol processing in connectionist networks: five properties and two architectures. *Proceedings of IEEE Spring COMPCON87*, San Francisco.

[12] Willshaw, D. (1981) Holography, associative memory, and inductive generalization. In Hinton, G. E. and Anderson, J. A (eds.), *Parallel Models of Associative Memory*. Hillsdale, NJ: Earlbaum.

[13] Winston, P. H. and Horn, B. K. P. (1984) *Lisp*, second edition. Boston: Addison-Wesley.

A Dual Back-Propagation Scheme
for Scalar Reward Learning

Paul Munro
Interdisciplinary Department of Information Science
University of Pittsburgh
Pittsburgh, PA 15260

Abstract. Explicit supervised learning rules [e.g. the delta rule] require that each of the output units in a network receive a training signal indicating the "correct" response value; the unit can then adjust its parameters, so that its future response to the same stimulus is closer to the desired value. A much more realistic assumption for the nature of a supervisory signal is a single scalar "goodness-of-response" or "reward" signal. This *credit assignment problem* is handled here by a supervisory network which monitors the activities of both the sensory and effector units, and learns to predict the value of the reward signal using the generalized delta rule of Rumelhart, Hinton, and Williams (1986). The activity of a particular "predictor unit" thus comes to be associated with the expected reward. Having learned to mimic the environment's reward criteria, the supervisory network can provide each effector unit, by way of a back-propagation scheme, with an individualized correction signal that will lead to increased activity in the predictor. The actual reward is hence enhanced to the extent that the predicted reward is reliable.

The Problem

One of the most attractive features of the Parallel Distributed Processing approach to understanding cognition is its inherent framework for adaptation. Learning is implemented by mechanisms that make small adjustments to the weights such that over time the response becomes more appropriate in some sense. A particular mapping from a set of input patterns to a set of output patterns can be "programmed" in to a network using the so-called back-propagation algorithm, or generalized delta rule (Rumelhart, Hinton, and Williams, 1986). The formidable power of this modification rule as a model for learning in real-world situations is somewhat offset by the need for a "supervisor"; that is, this rule requires that learning can only take place when *each output unit in the network is given detailed error information*. This is a general problem of supervised learning rules, and various attempts have been made to exploit the power of such rules and to eliminate the need for a supervisory signal. [A notable example is the encoder network (Ackley, Hinton, and Sejnowski, 1985), which is taught to replicate its input pattern as its output; hence the input pattern provides the detailed pattern of the desired output.]

While feedback from the environment may be inadequate with regard to its detail, feedback of some kind is essential for effective learning. In order to learn, the response of an organism must be evaluated at some level. This evaluation need not be generated by a teacher *per se;* for example, if a young animal swallows food, the reward takes the form of hunger satisfaction. The problem of assigning proper credit (or blame) to the individual output units that generated the evaluative feedback is known as the *credit assignment problem.*

The Network Configuration

The network described in the present paper is built in two stages. A low-level network (henceforth LOWNET) maps the sensory input to the organism into an effective response, that is, into a response that has an effect on the external world [e.g. a "motor response"]. Another network (HIGHNET) monitors the activities of both the sensory and effector units. The probability of reward contingent on response is assumed to depend solely upon instantaneous environmental information which is *perceivable,* i.e. the sensory pattern contains information sufficient to determine a "good" response. HIGHNET has access, therefore, to all the information necessary to mimic the reward signal and hence, one of the units in HIGHNET receives reward information as a training signal, computes an errror signal, and propagates it backwards through HIGHNET using the back-propagation procedure.

Note that for a *given configuration of weights* in the network, the entire pattern of activation throughout *both* LOWNET and HIGHNET is determined by the environment through its activation of the sensory units [the environment can influence the *modification* of the weights by way of the reward signal, but the reward has no direct influence on the response of the network]. As HIGHNET learns to mimic/predict the reward signal, LOWNET begins to adapt such that the net input, consisting of the

166

combined pattern of sensory and motor activity, to HIGHNET will maximize the *mimicked* reward signal, for a given configuration of weights in HIGHNET. To accomplish this, HIGHNET translates the scalar reward signal into a more detailed pattern of error information for the motor units.

HIGHNET thus serves as a "mental model" of the environment, in that it must learn to accurately predict combinations of sensory-motor activity that yield favorable feedback from the environment.

The full network architecture, incorporating LOWNET and HIGHNET, is displayed in Figure 1. It consists of five distinct populations of units: a set **S** of sensory units is activated by some physical stimulus; a set **K** receives activation from **S**; a set **M** of motor units receives activation from **K** [the population **M** directly controls the organism's motor response]; a set **H** receives activation from both **S** and **M**; and set **R** [which consists of but a single unit] receives activation from **H**.

Thus, the network can be viewed as two overlapping three-layer feed-forward networks, each made up of an input layer, an intermediate (hidden unit) layer, and an output layer. Each of the two networks is strictly layered; that is, the hidden units receive activation from the input units only and the output units receive activation from the hidden units only. The five population sets described above map onto LOWNET and HIGHNET as outlined in Table 1. The architecture can be generalized to arbitrary numbers of intermediate layers in either network, and the input to HIGHNET may presumably be expanded to include units from the intermediate layer(s) of LOWNET.

TABLE 1. Layered Structure of LOWNET and HIGHNET

NETWORK	LAYERS		
	Input	*Intermediate*	*Output*
LOWNET	S	K	M
HIGHNET	S∪M	H	R

Network Processing

Notation. The activity levels of the various units throughout the network are denoted by the lower case letter corresponding to the population, with a subscript specifying the individual unit (see Figure 1); for example, s_2 is the activity level of unit 2 in the population set **S**. The single unit occupying population **R** is an exception; its activity level is denoted r (with no subscript). The reward, or evaluative signal, from the environment is denoted e. The weight values are specified by the symbol w with a superscript denoting the "postsynaptic" population set *followed by* a superscript denoting the "postsynaptic" population set, and subscripts denoting the individual units within those sets.

All the units in the network, with the exception of those receiving direct input from the environment, compute their responses according to a semi-linear rule, that passes a weighted linear sum of the inputs through a nonlinear nondecreasing continuous function (f). Hence, a generic unit computes its activity y_i as a function of its input values x_j and the corresponding weights w_{ij} according to the formula

$$y_i = f\left[\sum_j w_{ij}\, x_j\right] \qquad [1]$$

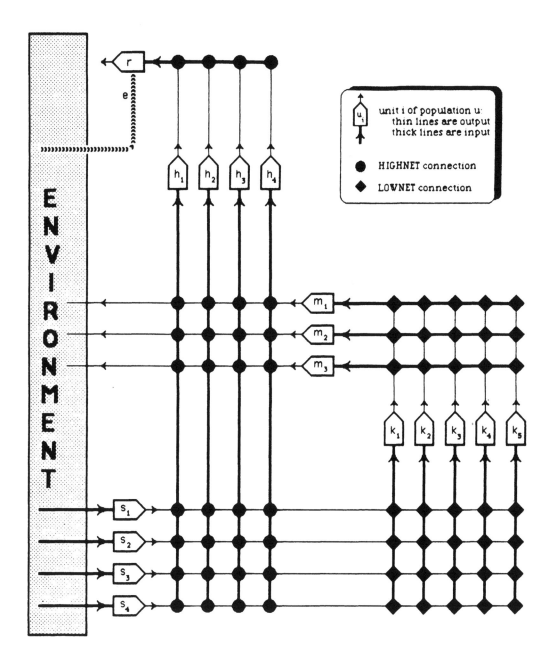

Figure 1.   *Architecture*. An example network is illustrated. Sensory input across the **S** population activates the **K** population and in turn the motor population (**M**) via the connections of LOWNET [♦]. The combined activities of the sensory and motor populations provide the input to HIGHNET. The connections of HIGHNET [●] are modified such that the unit r comes to predict the reward signal e.

Applying (1) to the network architecture described above yields the following set of equations for the activity values:

$$k_i = f\left[\sum_j w_{ij}^{KS} s_j\right] \qquad [2a]$$

$$m_i = f\left[\sum_j w_{ij}^{MK} k_j\right] \qquad [2b]$$

$$h_i = f\left[\sum_j w_{ij}^{HS} s_j + \sum_j w_{ij}^{HM} m_j\right] \qquad [2c]$$

$$r = f\left[\sum_j w_j^{RH} h_j\right] \qquad [2d]$$

Two Concurrent Learning Procedures

Rationale. A learning algorithm for minimizing a quantity Q can be obtained by the gradient descent assumption

$$\Delta w_{ij} = -\frac{\partial Q}{\partial w_{ij}} \qquad [3]$$

Removal of the minus sign from [3] will result in a learning rule that seeks *maximum* values of Q by gradient *ascent.*

If the dependence of the reward signal on the weights of LOWNET were known, the gradient descent technique could be applied directly to LOWNET, and there would be no need to introduce any more complexity to the network. However, the requisite partial derivatives depend on details of the environment that are not known to the naive organism. However, a two-stage application of gradient descent rules can be used to get at this problem. Taking both the sensory and the motor activities as input, HIGHNET generates a *predicted reward signal* r. The connections of HIGHNET are modified so as to perform a gradient descent in the squared difference between r and the actual reward e.

Since the dependence of r (but not e) on the connection strengths of LOWNET can be computed in terms of network variables; hence they follow a different learning rule, one which performs gradient ascent in r. Thus the learning dynamics are based on the following assumptions:

$$\Delta w_{ij} = -\frac{\partial (e-r)^2}{\partial w_{ij}} \qquad [w_{ij} \in \text{HIGHNET}] \qquad [4a]$$

$$\Delta w_{ij} = \frac{\partial r}{\partial w_{ij}} \qquad [w_{ij} \in \text{LOWNET}] \qquad [4b]$$

From these assumtions, learning rules are to be derived for both nets that are "factorable" in the sense that the expression for the change in any given weight is the product of a postsynaptic factor and the presynaptic activity value. For HIGHNET, the postsynaptic factor is referred to as the **effective predictor error** δ and for LOWNET it is the **effective predictor enhancement** ε; since these values depend on the response at the top level (**R**) and are propagated back down the network, the ε value must be evaluated for (or by) all units in the network, whereas computation of d is required for the HIGHNET units (sets **R** and **H**) only. The values of δ and ε for each of the population sets and the modification rules for their input connectivities are given in Table 2.

TABLE 2. Back-Propagation Formulae for the Entire Network

Network	Population	Eff. Error	Eff. Enhancement	Modification Function
HIGHNET	R	$\delta^R = (e-r)\, f'(r)$	$\varepsilon^R = f'(r)$	$\Delta w_i^{RH} = \delta^R h_i$
	H	$\delta_i^H = w_i^{RH}\delta^R f'(h_i)$	$\varepsilon_i^H = w_i^{RH}\varepsilon^R f'(h_i)$	$\Delta w_{ij}^{HX} = \delta_i^H x_j$
				[where $X \in \{M,S\}$]
LOWNET	M		$\varepsilon_i^M = \sum_j w_{ij}^{HM}\varepsilon_j^H\, f'(m_i)$	$\Delta w_{ij}^{MK} = \varepsilon_i^M k_j$
	K		$\varepsilon_i^K = \sum_j w_{ij}^{MK}\varepsilon_j^M\, f'(k_i)$	$\Delta w_{ij}^{KS} = \varepsilon_i^K s_j$

Simulation Methodology

The environmental evaluation (reward) function was evaluated as follows. Each sensory pattern was mapped onto a unique motor pattern. Deviations from this response resulted in a diminished evaluation signal e, which was computed as the negative exponential of the distance from the motor response vector **m** to a "target" response **t**.

$$e = \exp\left[-\sqrt{(t-m)\cdot(t-m)} \right]$$
[5]

In each case, the two kinds of learning were performed *separately*. HIGHNET would learn first (for several hundred thousand pattern presentations) by generating *random* motor responses for each sensory stimulus presentation. This **flailing strategy** allowed HIGHNET to explore the shape of the evaluation function over the space of possible sensory-motor combinations. It was found that learning was too slow if the random variables for the motor units were distributed uniformly over the range of the possible response values. Hence, during flailing, the units in set **M** took on either the minimum or maximum values of the function f [in this case, $\pm$ 1].

Once HIGHNET seemed to have identified the peak of the reward function for all, or nearly all, of the sensory patterns, the modification of HIGHNET connections ceased and LOWNET was permitted to function, both as the determiner of the output (no more flailing), and as a system of dynamic connection strengths. The learning in LOWNET then proceeds to drive up the value of the estimated reward, r.

Using this learning procedure, the network was guided to associate two paired pattern sets $(s \rightarrow t)$ on the basis of a reward signal alone.

Simulation Results

Experiment 1: AND & OR.

This experiment was performed on a network with $N_S=2$, $N_K=2$, $N_M=2$, and $N_H=12$, where N_X is the number of units in population **X**. The four allowed sensory stimuli were $(\pm 1, \pm 1)$. Associating -1 with "false" and +1 with "true", the desired response pattern **t** can be expressed as $(t_1, t_2) = ((s_1 \text{ AND } s_2), (s_1 \text{ OR } s_2))$. The peak is found after approximately 50000 - 60000 pattern presentations in the flailing mode. An example of the functions $e(\mathbf{m}, \mathbf{s})$ and $r(\mathbf{m}, \mathbf{s})$ plotted over **m**-space for the sensory pattern $(-1,-1)$ is shown in Figure 2 after 100000 pattern presentations. It is seen that while the details of the function do not seemed well learned, the peak (**t**) has been identified.

LOWNET learns to generate high values for r quite quickly, but due to its imperfect predictive power, high values of e do not come so quickly (Table 3). After 20000 pattern presentations, r is already quite high [note: the evaluation function was restricted to the range $0 < e < 0.8$]; but even at t=100000, e is not so high.

TABLE 3. Reward Maximization by LOWNET in Experiment 1.

	IN 1	IN 2	OUT 1	OUT 2	R	E
t = 20000	-1	-1	-0.990	-0.842	0.703	0.583
	-1	+1	-0.596	0.620	0.829	0.264
	+1	-1	-0.597	0.619	0.829	0.264
	+1	+1	0.882	0.992	0.706	0.631
t = 100000	-1	-1	-0.999	-0.868	0.706	0.614
	-1	+1	-0.825	0.853	0.850	0.507
	+1	-1	-0.827	0.852	0.850	0.507
	+1	+1	0.897	0.999	0.708	0.651

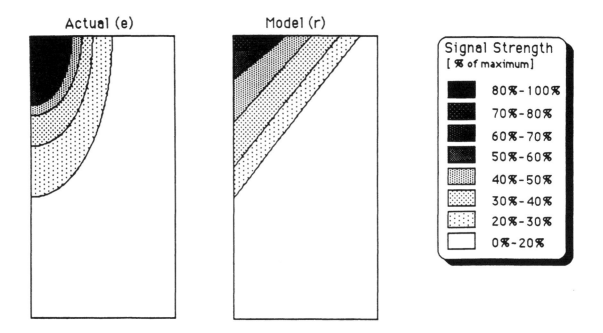

Actual and predicted rewards as a function of the motor response.

Figure 2 (above)   Contours for which the actual (left) and predicted reward values are shown for Experiment 1.

Figure 3(below)   Contours for which the actual (left) and predicted reward values are shown for Experiment 2.

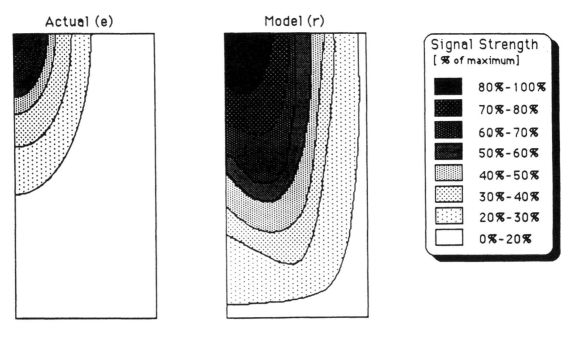

Experiment 2: NEGATION.

In the second experiment, a third "negation" unit was added to the set S. The number of possible sensory patterns was thus doubled: $s = (\pm 1, \pm 1, \pm 1)$. For the case $s_3 = -1$, the target responses depended upon s_1 and s_2 just as in the first experiment; however for the case $s_3 = +1$, the signs of the target response values were inverted. In this experiment, more flailing time is required to learn to predict the evaluation signal -- approximately 150000 trials. The functions $e(m,s)$ and $r(m,s)$ are plotted over m-space for the sensory pattern $(-1,-1,-1)$ in Figure 3. The reward maximization carried out in LOWNET learns to generate good r-values and e-values, but does so quite slowly (300000 pattern presentations) and has real trouble with the sensory pattern $(-1,-1,-1)$.

TABLE 4. Reward Maximization by LOWNET in Experiment 2.

	IN 1	IN 2	IN 3	OUT 1	OUT 2	R	E
t = 50000	-1	-1	-1	-0.292	0.992	0.061	0.012
	-1	-1	+1	0.628	0.725	0.852	0.317
	-1	+1	-1	-0.897	0.992	0.780	0.651
	-1	+1	+1	0.843	-0.831	0.781	0.504
	+1	-1	-1	-0.900	0.993	0.780	0.654
	+1	-1	+1	0.839	-0.824	0.782	0.496
	+1	+1	-1	0.135	0.308	0.775	0.087
	+1	+1	+1	-0.518	-0.758	·0.604	0.272
t = 300000	-1	-1	-1	-0.896	0.999	0.066	0.015
	-1	-1	+1	0.982	0.939	0.896	0.704
	-1	+1	-1	-0.976	0.998	0.790	0.762
	-1	+1	+1	0.994	-0.702	0.815	0.441
	+1	-1	-1	-0.976	0.998	0.790	0.763
	+1	-1	+1	0.994	-0.694	0.818	0.433
	+1	+1	-1	0.838	0.547	0.873	0.305
	+1	+1	+1	-0.969	-0.962	0.702	0.726

Discussion

The question of how instructional information is incorporated by a learning procedure appears in various incarnations at all levels of cognitive science, from expert systems to neurophysiology to network models. The recent design/discovery of the back-propagation learning procedure addresses part of this question. While there are important issues that need to be ironed out such as biological plausibility (information passes backwards across synapses!) and psychological plausibility (too much detail in the supervisory signal!), the algorithm is so powerful and elegant that it is too attractive to dismiss. Hence it seems worthwhile to try to make it fit nicely with biological and behavioral phenomenology.

The scheme described in this paper combines two forms of back-propagation learning into an architecture that requires only a single (scalar) reward value and not the explicit training information required from the environment back-propagation learning procedure by the "standard" back-propagation algorithm.

Bibliography

1. Ackley, D., Hinton, G., and Sejnowski, T. (1985) A learning algorithm for Boltzman Machines. *Cognitive Science* **9**:147-169.

2. Rumelhart, D. E., Hinton, G., and Williams, R. W. (1986) Learning internal representations by error propagation. In: *Parallel Distributed Processing: Explorations in the Microstructure of Cognition. vol 1.* D. E. Rumelhart and J. L. McClelland eds. MIT/Bradford

Using Fast Weights to Deblur Old Memories

Geoffrey E. Hinton and David C. Plaut

Computer Science Department
Carnegie-Mellon University

Abstract

Connectionist models usually have a single weight on each connection. Some interesting new properties emerge if each connection has two weights: A slowly changing, plastic weight which stores long-term knowledge and a fast-changing, elastic weight which stores temporary knowledge and spontaneously decays towards zero. If a network learns a set of associations and then these associations are "blurred" by subsequent learning, *all* the original associations can be "deblurred" by rehearsing on just a few of them. The rehearsal allows the fast weights to take on values that temporarily cancel out the changes in the slow weights caused by the subsequent learning.

1. Introduction

Most connectionist models have assumed that each connection has a single weight which is adjusted during the course of learning. Despite the emerging biological evidence that changes in synaptic efficacy at a single synapse occur at many different time-scales (Kupferman, 1979; Hartzell, 1981), there have been relatively few attempts to investigate the computational advantages of giving each connection several different weights that change at different speeds. Even for phenomena like short-term memory where fast-changing weights might seem appropriate, connectionist models have typically used the activation levels of units rather than the weights to store temporary memories (Little and Shaw, 1975; Touretzky and Hinton, 1985). We know very little about the range of potential uses of fast weights. How do they alter the way networks behave, and what extra computational properties do they provide?

In this paper we assume that each connection has both a fast, elastic weight and a slow, plastic weight. The slow weights are like the weights normally used in connectionist networks–they change slowly and they hold all the long-term knowledge of the network. The fast weights change more rapidly and they continually regress towards zero so that their magnitude is determined solely by their recent past. The effective weight on the connection is the sum of these two.

At any instant, we can think of the system's knowledge as consisting of the slow weights with a temporary overlay of fast weights. The overlay gives a temporary context–a temporary associative memory that allows networks to do more flexible information processing. Many ways of using this temporary memory have been suggested.

1. It can be used for rapid temporary learning. When presented with a new association the network can store it in one trial, provided the storage only needs to be temporary.

2. It can be used for creating temporary bindings between features. Recent work by Von der Malsburg (1981) and Feldman (1982) has shown that fast-changing weights can be used to dynamically bind together a number of different properties into a coherent whole, or to discover approximate homomorphisms between two structured domains.

3. It can be used to allow truly recursive processing. During execution of a procedure, the

177

values of local variables and the stage reached in the procedure (the program counter) can be stored in the fast weights. This allows the procedure to call a subprocedure whose execution involves different patterns of activity in the very same units as the calling procedure. When the subprocedure has finished using the units, the state of the calling procedure can be reconstructed from the temporary associative memory. In this way the state does not need to be stored in the *activity* of the units, so the very same units can be used for running the subprocedure. A working simulation of this kind is described briefly in McClelland & Kawamoto (1986).

4. It can be used to implement a learning method called "shortest descent" which is a way of minimizing the amount of interference caused by new learning. Shortest descent will be described in a separate paper and is not discussed further here.

In this paper we describe a novel use for a temporary memory stored in the fast weights: it can be used for cancelling out the interference in a set of old associations caused by more recent learning. Consider a network which has slowly and painfully learned a set of associations in its slow weights. If this network is then taught a new set of associations without any more rehearsal of the old associations, there is normally some degradation of the old associations. We show that by using fast weights it is possible to quickly restore a whole set of old associations by rehearsing on just a subset of them. The fast weights cancel out the changes in the slow weights that have occurred since the old associations were learned, so the combination of the current slow weights and the fast weights approximates the earlier slow weights. The fast weights therefore create a context in which the old associations are present again. When the fast weights decay away, the new associations return.

2. A deblurring analogy

There is an analogy between the use of fast weights to recover unretrained associations and a technique called "deblurring" which is sometimes used for cleaning up blurred images. Suppose that you are in your office and you want to take some photographs of what is on your computer screen. You set up a tripod in front of the screen and then you carefully focus the camera and take one photograph. Unfortunately, before you can take any more, your office mate moves the camera to a tripod in front of his screen and refocuses it. You now move the camera back to your tripod and it seems as if you must refocus, but there is an interesting alternative. You take another photograph of the same screen without refocussing and you compare it with your first photograph. The first photo is the "desired output" of the process that maps from screens to photos, and the difference between it and the "actual output" achieved with the out-of-focus camera can be used to estimate a deblurring operator which can be applied to the actual output to convert it to the desired output. This same deblurring operator can then be applied to any image taken with the out-of-focus camera. Focussing the camera is analogous to learning the slow weights that are required to map from input vectors (screens) to output vectors (photos). Estimating the deblurring operator is analogous to learning the fast weights that are required to compensate for the noise that has been added to the slow weights since the original learning (see figure 1).

The advantage of using fast weights rather than slow ones for deblurring is that it does not permanently interfere with the new associations. As soon as the fast weights have decayed back to zero, the new knowledge is restored.

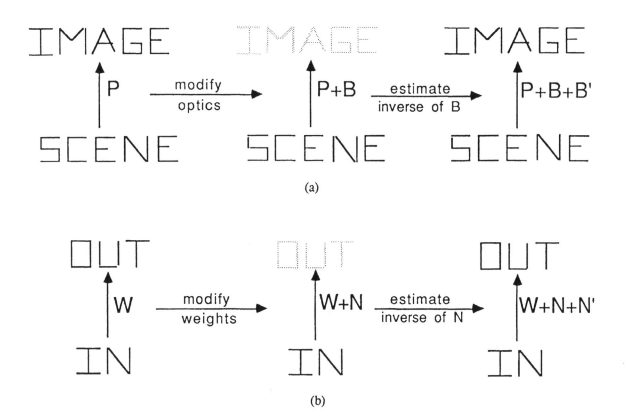

Figure 1: An illustration of the deblurring analogy between (a) images and (b) networks. The laws of projection (P) determine the mapping from the scene to the image. After applying a blurring function (B), applying the inverse of B to the blurred image restores the image. Similarly, the weights (W) define a mapping from input to output. After noise (N) is added, determining and applying the inverse of N allows the network to produce the original output.

3. The learning procedure

We used the back propagation learning procedure (Rumelhart *et al.*, 1986a; 1986b) to investigate the properties of networks that learn with both fast and slow weights. We summarize the procedure below–the full mathematical details can be found in the above references.

The procedure operates on layered, feed-forward networks of deterministic, neuron-like units. These networks have a layer of input units at the bottom, any number of intermediate layers of hidden units, and a layer of output units at the top. Connections are allowed only from lower to higher layers.

The aim of the learning procedure is to find a set of weights on the connections such that, when the network is presented with each of a set of input vectors, the output vector produced by the network is sufficiently close to the corresponding desired output vector. The error produced by the network is defined to be the squared difference between the actual and desired output vectors summed over all input-output cases. The learning procedure minimizes this error by performing gradient descent in weight space. This requires adjusting each weight in the network in proportion to the partial derivative of the error with respect to that weight. These error derivatives are calculated in two passes.

The forward pass determines the output (or *state*) of each unit in the network. An input vector is presented to the network by setting the states of the input units. Layers are then processed sequentially, working from the bottom upward, with the states of units within a layer set in parallel. The input to a unit is the scalar product of the states of units in lower layers from which it receives connections, and the weights on these connections. The output of a unit is a real-valued smooth non-linear function of its input. The forward pass ends when the states of the output units are set.

The backward pass starts with the output units at the top of the network and works downward through each successive layer, "back-propagating" error derivatives to each weight in the network. For each layer, computing the error derivatives of the weights on incoming connections involves first computing the error derivatives with respect to the outputs, and then with respect to the inputs, of the units in that layer. The simple form of the unit input-output functions makes these computations straightforward. The backward pass is complete when the derivative of the error with respect to each weight in the network has been determined.

The simplest version of gradient descent is to decrement each weight in proportion, ε, to its error derivative. A more complicated version that usually converges faster is to add to the current weight change a proportion, α, of the previous weight change. This is analogous to introducing *momentum* to movement in weight space.

Since the effective weight on a connection is the sum of the fast and slow weights, these weights experience the same error derivative. Hence they behave differently only because they are modified using different weight change parameters ε and α, and because the fast weights decay towards zero. This is achieved by reducing the magnitude of each fast weight by some fraction, h, after each weight change.

4. A simulation of the deblurring effect

To demonstrate the deblurring effect, we used a simple task and a simple network. The task was to associate random binary 10-bit input vectors with random binary 10-bit output vectors. We selected 100 input vectors at random (without replacement) from the set of all 2^{10} binary vectors of length 10, and for each input vector we chose a random output vector. The network we used had three layers: 10 input units, 100 hidden units, and 10 output units. Each input unit was connected to all the hidden units, and each hidden unit was connected to all the output units. The hidden and output units also had variable biases that were modified during the learning.

We trained the network by repeatedly sweeping through the whole set of 100 associations and changing the weights after each sweep. After prolonged training (1300 sweeps with $\varepsilon = .02$ and $\alpha = .9$) the network knew the associations perfectly, and all the knowledge was in the slow weights. The fast weights were very close to zero because the errors were very small towards the end of the training, so the tiny error derivatives were dominated by the tendency of the fast weights to decay towards zero by 1% after each weight update.

Once the 100 associations were learned, we trained the network on 5 new random associations without further rehearsal on the original 100. Again, we continued the training until the new knowledge was in the slow weights (400 sweeps). We then retrained the network on only a subset of the original associations, and compared the improvement in performance on this retrained subset with the (incidental) improvement in performance on the rest of the associations. The main result, shown in figure 2, was that in the early stages of retraining the improvements in the associations that were not retrained were very nearly as good as in the associations that were explicitly retrained. This rather surprising result held even if only 10% of the old associations were retrained.

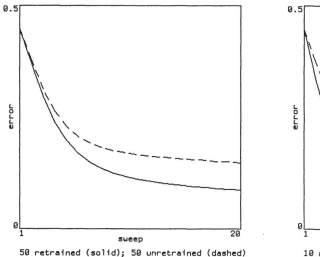 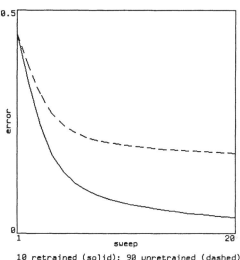

Figure 2: Performance on retrained and unretrained subsets of 100 input-output cases after (1) learning all 100 cases; and (2) interfering with the weights by learning 5 unrelated cases. The average error of an output unit is shown for the first 20 sweeps through the retrained subset.

The reason for this effect is that the knowledge about each association is distributed over many different connections, and when the network is retrained on *some* of the associations *all* the weights are pushed back towards the values that they used to have after the associations were first learned. So there is improvement in the unretrained associations even though they are only randomly related to the retrained ones. Of course, if the retrained and unretrained associations share some regularities the transfer will be even better.

4.1. A geometric explanation of the transfer effect

Consider the very simple network shown in figure 3. There are two input units which are directly connected to a single, linear output unit. The network is trained on two different associations each of which maps a two component input vector into a one component output vector. For each of the two input vectors, there will be many different combinations of the weights w_1 and w_2 that give the desired output vector, and since the output unit is linear, these combinations will form straight lines in weight space as shown is figure 4. In general, the only combination of weights that will satisfy both associations lies at the intersection of the two lines. So in this simple network, a gradient descent learning procedure will converge on the intersection point.

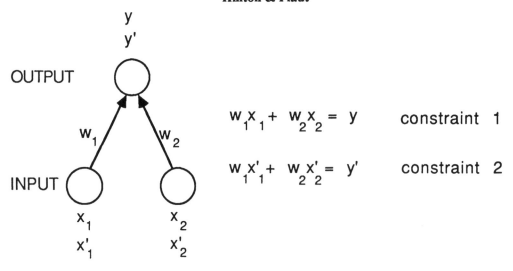

Figure 3: A simple network that learns to associate (x_1, x_2) with y and (x_1', x_2') with y'.

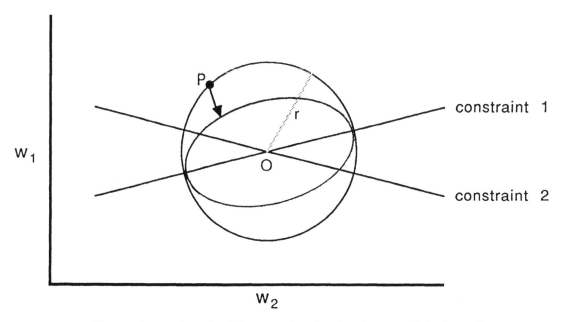

Figure 4: A plot of weight space for the simple network in figure 3.

Now, suppose we add to a network that has learned the associations a noise vector of length r that is selected at random from a distribution that is uniformly distributed over all possible orientations. The new combination of weights will be uniformly distributed over the circle shown in figure 4. If we now retrain an infinitessimal amount on association 1, we will move perpendicularly towards line 1. If we start from a point such as P that lies on one of the larger arcs of the circle, the movement towards line 1 will have a component *towards* line 2, whereas if we start from a point on one of the smaller arcs, the movement will have a component *away from* line 2. Thus, when we retrain a small amount on association 1, we are more likely than not to improve the performance on association 2, even though the associations are only randomly related to each other.

The expected positive transfer between retraining on one association and performance on the other can only be a result of starting the retraining from a point in weight space that has some special relationship to the solution point, O. We initially thought that the starting point for the retraining needed to be *near* the solution point, but the geometrical argument we have just given is independent of the magnitude, r, of the noise vector. The crucial property of the starting point is that it is selected at random from a distribution of points that are all the same distance from O, and selecting starting points in this way induces a bias towards starting points that cause positive transfer between the two associations that define the point O.

The positive transfer effect obviously disappears if the two associations are orthogonal, and so we might expect the effect to be rather small when the network is large, because randomly related high-dimensional vectors tend to be almost orthogonal. In fact, it is important to distinguish between two qualitatively different cases. If the activity levels of the input units have a mean of zero, most pairs of high-dimensional vectors will be approximately orthogonal and so the effect is small and the expected transfer from one retrained association to one unretrained one gets smaller as the dimensionality increases. If, however, the activity levels of the input units range between 0 and 1 (as in the simulation described above) random high-dimensional vectors are not close to orthogonal and we show in the next section that the expected transfer from one retrained association to one unretrained one is independent of the dimensionality of the vectors.

5. A mathematical analysis

It is hard to analyze the expected size of the transfer effect for networks with multiple layers of non-linear units or for tasks with complex regularities among the input-output associations. However, an analysis of the transfer effect in simple networks learning random associations may provide a useful guide to what can be expected in more complex cases. We therefore derive the magnitude of the expected transfer from one retrained association, A, to one unretrained association, B, in a network that has no hidden units and only one linear output unit.[1] We assume that the network has previously learned to produce the correct activity level in the output unit for two input vectors, **a** and **b**, and that it is now retraining on association A after independent gaussian noise has been added to each weight. The noise added to the ith weight is g_i and is chosen from a distribution with mean 0 and standard deviation σ. We also assume that the retraining involves changing the weights by a fixed, infinitesimal amount in the direction of steepest descent in the error function $\frac{1}{2}(y_A - d_A)^2$ where y_A is the actual output of the output unit and d_A is the desired output that was achieved before the noise was added.

A good measure of the magnitude of the transfer effect is the the ratio of two improvements: the improvement in association B caused by retraining on A and the improvement in association B that would have been caused by retraining directly on B. If the retraining involves changing the weight vector by a fixed amount in the direction of steepest descent, this ratio is simply the cosine of the angle between the direction of steepest descent for association A and the direction of steepest descent for association B.[2]

If the original learning was perfect, all of the error in the output would be caused by the noise added

---

[1] For layered, feed-forward networks with no hidden units, each output unit has its own separate set of weights. So a network with many output units can be viewed as composed of many separate networks each of which has a single output unit.

[2] If the retraining involves multiplying the gradient by a coefficient, ε, to determine the weight change, the actual ratio may differ because the magnitudes of the gradients may differ for A and B. But this will not affect the *expected* ratio, so the analysis we give for the *expected* transfer is still valid.

to the weights, so

$$\frac{\partial E_A}{\partial y_A} = y_A - d_A = \sum_i a_i g_i$$

where a_i is the activity level of the ith input unit in association A. So, for a weight, w_i, the derivatives of the error E_A for association A and E_B for association B are given by

$$\frac{\partial E_A}{\partial w_i} = \frac{\partial y_A}{\partial w_i} \cdot \frac{\partial E_A}{\partial y_A} \cdot a_i \sum_i a_i g_i, \qquad\qquad \frac{\partial E_B}{\partial w_i} = \frac{\partial y_B}{\partial w_i} \cdot \frac{\partial E_B}{\partial y_B} \cdot b_i \sum_i b_i g_i$$

Hence, the cosine of the angle, θ, between the directions of steepest descent for the two associations is given by

$$
\begin{aligned}
cos(\theta) &= \frac{\sum_i a_i b_i \cdot \sum_i a_i g_i \cdot \sum_i b_i g_i}{\left[\sum_i a_i^2 \left(\sum_i a_i g_i\right)^2 \cdot \sum_i b_i^2 \left(\sum_i b_i g_i\right)^2\right]^{1/2}} \\[2em]
&= \frac{\sum_i a_i b_i}{\left(\sum_i a_i^2 \cdot \sum_i b_i^2\right)^{1/2}} \cdot \frac{\sum_i a_i g_i \cdot \sum_i b_i g_i}{\left|\sum_i a_i g_i \cdot \sum_i b_i g_i\right|}
\end{aligned}
\tag{1}
$$

5.1. The zero-one case

The first part of equation 1 is simply the cosine of the angle between the two input vectors, and so it is independent of the weights. If the components of **a** and **b** are all 0 or 1 it can be written as

$$\frac{n_{ab}}{(n_a n_b)^{1/2}}$$

where n_a is the number of components that have value 1 in the vector **a**, and n_{ab} is the number that have value 1 in both **a** and **b**.

The second part of equation 1 depends on the weights. It always has a value of 1 or −1, depending on whether E_A and E_B have the same or opposite signs. Its numerator can be written as

$$\left(\sum_{i \in S_{ab}} g_i + \sum_{i \in S_{a\bar{b}}} g_i\right) \left(\sum_{i \in S_{ab}} g_i + \sum_{i \in S_{\bar{a}b}} g_i\right)$$

where S_{ab} is the set of input units that have value 1 in both **a** and **b**, $S_{a\bar{b}}$ is the set that have value 1 in **a** and 0 in **b**, and $S_{\bar{a}b}$ is the set that have value 0 in **a** and 1 in **b**. Since these sets are disjoint and the expected value of the products of independent zero-mean gaussians is 0, all the cross-products in the numerator of equation 1 vanish when we take expected values except for the term

$$\left\langle \sum_{i \in S_{ab}} g_i \cdot \sum_{i \in S_{ab}} g_i \right\rangle = \left\langle \sum_{i \in S_{ab}} g_i^2 \right\rangle = n_{ab} \sigma^2$$

So the expected value of equation 1 can be written as

$$\frac{n_{ab}\,\sigma^2}{\left\langle \left| \sum_i a_i g_i \cdot \sum_i b_i g_i \right| \right\rangle} = \frac{n_{ab}\,\sigma^2}{n_a^{1/2}\sigma\, n_b^{1/2}\sigma} = \frac{n_{ab}}{n_a^{1/2}n_b^{1/2}}$$

and so the expected value of $cos(\theta)$ is given by

$$\left\langle cos(\theta) \right\rangle = \frac{n_{ab}^{\,2}}{n_a n_b} \tag{2}$$

If each component of **a** and **b** has value 1 with probability 0.5 and value 0 otherwise, the expected value of $cos(\theta)$ is 0.25. So if we add random noise to the weights of a simple network that has learned 100 associations and then we retrain on 50 of them using very small weight changes, the ratio of the expected improvements on an unretrained and a retrained association at the start of the retraining will be

$$\frac{50 \times .25}{1 + 49 \times .25} = .943$$

This agrees well with simulations we have done using networks with no hidden units.

5.2. The 1, 0, -1 case

When the components of the input vectors can also take on values of -1, a modification of the derivation used above leads to

$$\left\langle cos(\theta) \right\rangle = \frac{(n_{agree} - n_{disagree})^2}{n_a n_b} \tag{3}$$

where n_{agree} is the number of input units for which $a_i b_i = 1$ and $n_{disagree}$ is the number of input units for which $a_i b_i = -1$.

For a pair of random vectors in which each component has a probability, p, of being a 1 and the same probability of being a -1, the expected value of the numerator of equation 3 is just the square of the expected length of a random walk in which each step has a probability of $2p^2$ of being to the left, $2p^2$ of being to the right, and $1 - 4p^2$ of being zero. The expected value of the numerator is therefore $4p^2n$, where n is the dimensionality of the input vector. Since the denominator has an expected value of order n^2, the expected transfer effect is inversely proportional to n.

The decrease in the expected transfer between a single pair of vectors is balanced by the fact that larger networks can hold more associations. If the number of associations learned is proportional to the dimensionality of the input vectors, and if a constant *fraction* of the associations are retrained after adding noise to the weights, the ratio of the initial improvement on the unretrained associations to the improvement on the retrained associations is independent of the dimensionality.

5.3. How the transfer effect decreases during retraining

The analyses we have presented are for the transfer between random associations during the earliest stage of retraining. As retraining proceeds, the transfer effect diminishes. Figure 4 presents a simple geometrical explanation of why this occurs. At the start of retraining, the point in weight space lies somewhere on the circle, and the probability that the transfer will be positive is simply the fraction of the circumference that lies in the two larger arcs of the circle. Retraining on association 1 moves each point on the circle perpendicularly towards the line 1 by an amount proportional to its distance from 1, so after a given amount of retraining, each point on the circle will move to the corresponding point on the ellipse. The probability of positive transfer is now equal to the fraction of the circumference of the ellipse that lies in the two larger arcs.

6. Conclusions

This paper demonstrates and analyses a powerful and surprising transfer effect that was first described (but not analysed) by Hinton and Sejnowski (1986). The analysis applies just as well if there are no fast weights and the relearning takes place in the slow weights. The advantage of fast weights is that they allow this effect to be used for temporarily deblurring old memories without causing significant interference to new memories. When the fast weights decay away the newer memories are restored.

There are many anecdotal descriptions of phenomena that could be explained by the kind of mechanism we have proposed, but we know of no well controlled studies. We predict that if a two unrelated sets of associations are learned at the same time, and if the internal representations used for different associations share the same units, then retraining on one set of associations will substantially improve performance on the other set. One problem with this prediction is that with sufficient learning, connectionist networks tend to use different sets of units for representing unrelated items. A second problem is that even if the unretrained associations are enhanced, the effect may be masked by response competition due to facilitation of the responses to the retrained items. Nevertheless, the type of effect we have described can be very large in simulated networks and so a well designed experiment should be able to detect whether or not it occurs in people.

References

Feldman J.A. Dynamic connections in neural networks. *Biological Cybernetics*, 1982, *46*, 27-39.

Hartzell H.C. Mechanisms of slow synaptic potentials. *Nature*, 1981, *291*, 539-543.

Hinton G.E. and Sejnowski T.J. Learning and relearning in Boltzmann Machines. In D.E. Rumelhart, J.L. McClelland, and the PDP research group (Eds.) *Parallel distributed processing: Explorations in the microstructure of cognition. Volume 1: Foundations,* Cambridge, MA: MIT Press, 1986.

Kupferman I. Modulatory actions of neurotransmitters. *Annual Review of Neuroscience*, 1979, *2*, 447-465.

Little W.A. and Shaw G.L. Statistical-theory of short and long-term memory. *Behavioral Biology*, 1975, *14(2)*, 115-133.

McClelland J.L. and Kawamoto A.H. Mechanisms of sentence processing: Assigning roles to constituents of sentences. In J.L. McClelland, D.E. Rumelhart, and the PDP research group (Eds.) *Parallel distributed processing: Explorations in the microstructure of cognition. Volume II: Psychological and biological models,* Cambridge, MA: MIT Press, 1986.

Rumelhart D.E., Hinton G.E. and Williams R.J. Learning internal representations by error propagation. In D.E. Rumelhart, J.L. McClelland, and the PDP research group (Eds.) *Parallel distributed processing: Explorations in the microstructure of cognition. Volume I: Foundations,* Cambridge, MA: MIT Press, 1986a.

Rumelhart D.E., Hinton G.E. and Williams R.J. Learning representations by back-propagating errors. *Nature*, 1986b, *323(9)*, 533-536.

Touretzky D.S. and Hinton G.E. Symbols among the neurons: Details of a connectionist inference architecture. *Proceedings, 9th International Joint Conference on Artificial Intelligence,* Los Angeles, August, 1985.

Von der Malsburg C. *The correlation theory of brain function.* Internal Report 81-2, Department of Neurobiology, Max-Plank-Institute for Biophysical Chemistry, P.O. Box 2841, Gottingen, F.R.G., 1981.

On the Connectionist Reduction of Conscious Rule Interpretation

Paul Smolensky

Department of Computer Science &
Institute of Cognitive Science
University of Colorado at Boulder

Abstract

Connectionist models have traditionally ignored conscious rule application in learning and performance. Conceptual problems arise in treating rule application in a connectionist framework because the level of analysis of connectionist models is lower than that which is natural for describing conscious rules. An analysis is offered of the relation between these two levels of description, and of the kind of reduction involved in connectionist modeling. From this vantagepoint an approach is formulated to the treatment of conscious rule application within a connectionist framework. The approach crucially involves connectionist language processing, and leads to a distinction between two types of knowledge that can be stored in connectionist systems.

Introduction

Connectionist models have traditionally ignored the role played by conscious application of rules in human cognition. Many tasks are learned through rules, and initially performed by consciously applying those rules; connectionist models have by and large been unable to address learning and performance in which such rule application occurs. One of the most striking phenomena in cognitive science is the shift during the acquisition of expertise from conscious processing to processing that I will simply refer to as *intuitive*. While the connectionist approach has provided successful models of purely intuitive processes, since the approach cannot now address rule application, it cannot shed light on this aspect of the transition to expertise.

Incorporating conscious rule application into a connectionist paradigm poses serious conceptual problems that must be resolved, at least provisionally, before the associated technical problems can even be recognized, let alone solved. Many of these conceptual problems stem from the fact that the *level of analysis* adopted by connectionist modeling is lower than the level at which rule interpretation is most naturally described. Conscious rules involve consciously accessible concepts, and are therefore naturally described at the level of these concepts: what I will call the *conceptual level*. In the kind of connectionist system I consider here, concepts are represented by patterns of activity over large numbers of processing units; the semantic interpretation of the individual units is considerably finer-grained than that of the consciously accessible concepts. Such a connectionist system uses *distributed representations* (eg., Anderson & Hinton, 1981; Hinton, McClelland & Rumelhart, 1986; Smolensky, 1986b). The semantics of the individual connectionist processors resides at a level lower than the conceptual level: what I will call the *subconceptual level*.

The cognitive modeling paradigm employing connectionist models with distributed representations, i.e., employing connectionist networks with subconceptual semantics, will be called the *subsymbolic paradigm*; this contrasts with the traditional *symbolic paradigm* that employs symbol manipulating models in which the symbols have conceptual semantics. The incorporation of conscious rule interpretation into the subsymbolic paradigm involves a *reduction* of the symbolic account of rule interpretation to the subconceptual level. Thus the first question to address is: *How do the symbolic and subsymbolic paradigms relate at the conceptual and subconceptual levels of analysis, and what kind of reduction is involved?*

Reduction of cognition to the subconceptual level

Imagine three physical systems: a brain that is executing some cognitive process, a massively parallel connectionist computer running a subsymbolic model of that process, and a von Neumann computer running a symbolic model of the same process. The cognitive process may involve conscious rule application, intuition, or a combination of the two. In Smolensky (1987b) I have characterized the subsymbolic paradigm as positing the following relationships between descriptions of these three physical systems at the neural, subconceptual, and conceptual levels:

(1)   a.   Describing the brain at the neural level gives a neural model.

    b.   Describing the brain approximately, at a higher level—the subconceptual level—yields, to a good approximation, the model running on the connectionist computer, when it too is described at the subconceptual level. (At this point, this is a goal for future research. It could turn out that the degree of approximation here is only rough; this would still be consistent with the subsymbolic paradigm.)

    c.   We can try to describe the connectionist computer at a higher level—the conceptual level—by using the patterns of activity that have conceptual semantics. If the cognitive process being executed is conscious rule application, we will be able to carry out this conceptual level analysis with reasonable precision, and will end up with a description that closely matches the symbolic computer program running on the von Neumann machine.

    d.   If the process being executed is an intuitive process, we will be unable to carry out the conceptual-level description of the connectionist machine precisely. Nonetheless, we will be able to produce various approximate conceptual-level descriptions that correspond in various ways to the symbolic computer program running on the von Neumann machine.

For a cognitive process involving both intuition and conscious rule application, (1c) and (1d) will each apply to certain aspects of the process.

The relationships (1a) and (1b), which are discussed at some length in Smolensky (1987b), are not relevant for the present considerations; they are mentioned here only to link the physical instantiations of the subsymbolic and symbolic models to the physical system they are in some sense models of. A number of the relations (1d) between subsymbolic and symbolic accounts of intuitive processing have been addressed elsewhere (Rumelhart, Smolensky, McClelland & Hinton, 1986; Smolensky, 1986a, 1986b; see Smolensky, 1987a, 1987b for summaries). The relationship (1c) between a subsymbolic implementation of conscious rule interpretation and a symbolic implementation is the subject of the next section.

The relationships in (1) can be more clearly understood by introducing the concept of "virtual machine." If we take one of the three physical systems and describe its processing at a certain level of analysis, we get a virtual machine that I will denote "system_{level}". Then (1) can be written:

(2)   a. brain_{neural} = neural model

    b. $\text{brain}_{subconceptual} \approx \text{connectionist}_{subconceptual}$

    c. $\text{connectionist}_{conceptual} \approx \text{von Neumann}_{conceptual}$ (conscious rule application)

    d. $\text{connectionist}_{conceptual} \sim \text{von Neumann}_{conceptual}$ (intuition)

Here, the symbol "$\approx$" means "equals to a good approximation" and "$\sim$" means "equals to a crude approximation." The two nearly equal virtual machines in (2c) both describe what I will call the *conscious rule interpreter*. The two roughly similar virtual machines in (2d) provide the two paradigms' descriptions of the *intuitive processor* at the conceptual level.

Table 1 indicates these relationships and also the degree of exactness to which each system can be described at each level—the degree of precision to which each virtual machine is defined. The levels included in Table 1 are those relevant to predicting high-level behavior. Of course each system can also be described at lower levels, all the way down to elementary particles. However, levels below an exactly describable level are ignorable from the point of view of predicting high-level behavior, since it is possible (in principle) to do the prediction at the highest level that can be exactly described (and it is presumably much harder to do the same at lower levels). This is why in the symbolic paradigm any descriptions below the symbolic level are not viewed as significant. For modeling high-level behavior, how the symbol manipulation happens to be implemented can be ignored—it is not a relevant part of the cognitive model. In a subsymbolic model, exact behavioral prediction must be performed at the subconceptual level—but how the units happen to be implemented is not relevant.

Table 1: Three cognitive systems and three levels of description

level	(process)	cognitive system		
		brain	**subsymbolic**	**symbolic**
conceptual	*(intuition)* *(conscious rule application)*	? ?	rough approximation ~ good approximation ≈	exact exact
subconceptual		good approximation ≈	exact	
neural		exact		

The relation between the conceptual level and lower levels is fundamentally different in the subsymbolic and symbolic paradigms. This leads to important differences in the kind of explanations the paradigms offer of conceptual-level behavior, and the kind of reduction used in these explanations. A symbolic model is a *system* of interacting processes, all with the same conceptual-level semantics as the task behavior being explained. Adopting the terminology of Haugeland (1978), this *systematic explanation* relies on a *systematic reduction* of the behavior that involves no shift of semantic domain or *dimension*. Thus a game-playing program is composed of subprograms that generate possible moves, evaluate them, and so on. In the symbolic paradigm, these systematic reductions play the major role in explanation. The lowest-level processes in the systematic reduction, still with the original semantics of the task domain, are then themselves reduced by *intentional instantiation:* they are implemented exactly by other processes with different semantics but the same form. Thus a move-generation subprogram with game semantics is instantiated in a system of programs with list-manipulating semantics. This intentional instantiation typically plays a minor role in the overall explanation, if indeed it is regarded as a cognitively relevant part of the model at all.

Thus cognitive explanations in the symbolic paradigm rely primarily on reductions involving no dimension shift. This feature is not shared by the subsymbolic paradigm, where accurate explanations of intuitive behavior require descending to the subconceptual level. The elements in this explanation, the units, do *not* have the semantics of the original behavior. Thus unlike symbolic explanations, subsymbolic explanations rely crucially on a semantic ("dimension") shift that accompanies the shift from the conceptual to the subconceptual levels.

The overall dispositions of cognitive systems are explained in the subsymbolic paradigm as approximate higher level regularities that emerge from quantitative laws operating at a more fundamental level with different semantics. This is the kind of reduction familiar in natural science, exemplified by the explanation of the laws of thermodynamics through a reduction to mechanics that involves shifting dimension from thermal semantics to molecular semantics. (Section discusses some explicit subsymbolic reductions of symbolic explanatory constructs.)

Indeed the subsymbolic paradigm repeals the other features that Haugeland identified as newly introduced into scientific explanation by the symbolic paradigm. The inputs and outputs of the system are not "quasilinguistic representations" but good old-fashioned numerical vectors. These inputs and outputs have semantic interpretations, but these are not constructed recursively from interpretations of imbedded constituents. And the fundamental laws are good old-fashioned numerical equations.

Haugeland went to considerable effort to legitimize the form of explanation and reduction used in the symbolic paradigm. The explanations and reductions of the subsymbolic paradigm, by contrast, are of a type well-established in natural science.

In summary, let me emphasize that in the subsymbolic paradigm the conceptual and subconceptual levels are not related as the levels of a von Neumann computer (high-level-language program, compiled low-level program, etc.). The relationship between subsymbolic and symbolic models is more like that between quantum and classical mechanics. Subsymbolic models accurately describe the microstructure of cognition, while symbolic models provide an approximate description of the macrostructure. An important job of subsymbolic theory is to delineate the situations and respects in which the symbolic approximation is valid, and to explain why.

Conscious rule application in the subsymbolic paradigm

In the symbolic paradigm, both conscious rule application and intuition are described at the conceptual level: as conscious and unconscious rule interpretation, respectively. In the subsymbolic paradigm, conscious rule application can be formalized at the conceptual level but intuition must be formalized at the subconceptual level. This suggests that a subsymbolic model of a cognitive process involving both intuition and conscious rule interpretation would consist of two components employing quite different formalisms. While this hybrid formalism might have considerable practical value, there are some theoretical problems with it. How would the two formalisms communicate? How would the hybrid system evolve with experience, reflecting the development of intuition and the subsequent remission of conscious rule application? How would the hybrid system elucidate the fallibility of actual human rule application (eg. logic)? How would the hybrid system get us closer to understanding how conscious rule application is achieved neurally?

All these problems can be addressed by adopting a unified subconceptual-level analysis of both intuition and conscious rule interpretation. The virtual machine that is the conscious rule interpreter is to be implemented in a lower-level virtual machine: the same connectionist system that models the intuitive processor. How this can, in principle, be achieved is the subject of this section. The relative advantages and disadvantages of implementing the rule interpreter in a connectionist system rather than a von Neumann machine will also be considered.

The observation is this.

> The competence to represent and process linguistic structures in a native language is a competence of the human intuitive processor, so the subsymbolic paradigm assumes that this competence can be modeled in a subconceptual connectionist system. By combining such linguistic competence with existing memory capabilities of connectionist systems, sequential rule interpretation can be implemented.

Assuming that sentences of natural language can be represented in a subconceptual connectionist system means that such sentences correspond to certain patterns of activity. Assuming that sentences can be processed means in particular that a pattern of activity representing a verbal instruction can be used to carry out that instruction. Once sentences are represented as patterns of activity, the well-known procedures of associative memories can be used to store them. These are content-addressable memories in which reinstantiation of a part of the stored item causes reinstantiation of the complete item. A collection of such memories can then be used to drive sequential behavior, as follows.

First, a set of linguistically expressed rules is presented to the connectionist system and thereby stored. For concreteness we can imagine the rules to be productions: "if *condition* holds, then do *action*." In a particular situation when the condition of a rule holds, the pattern of activity representing the condition will be instantiated in the network. This will cause the entire pattern of activity representing that rule to be reinstantiated by the memory retrieval mechanism. Now it is as if the rule had been linguistically presented to the system from an external instructor. The language processing mechanism can interpret the sentence, generating the appropriate action and leading to a new pattern of activity in the network representing the new situation. This new pattern leads to the reinstantiation of another stored rule, and the cycle repeats.

Using the stored rules the network can perform the task. The standard learning procedures of connectionist models turn this experience performing the task into a set of weights for going from inputs to outputs. Eventually, after enough experience, the task can be performed directly by these weights. The input activity generates the output activity so quickly that before the relatively slow interpretation process has a chance to reinstantiate the first rule and carry it out, the task is done. With intermediate amounts of experience, some of the weights are well enough in place to prevent some of the rules from having the chance to instantiate, while others are not, enabling other rules to be retrieved and interpreted.

Rule interpretation, consciousness, and seriality

What about the conscious aspect of rule interpretation? Since consciousness seems to be a quite high-level description of mental activity, it is reasonable to suspect that it reflects the very coarse structure of the cognitive system. Considering coarseness on the time dimension, we are lead to hypothesize:

> Patterns of activity that are stable for relatively long periods of time (on the order of 100 msec) determine the contents of consciousness.

(See Rumelhart, Smolensky, McClelland & Hinton 1986.) The rule interpretation process requires the maintenance of the retrieved linguistically coded rule while it is being carried out. Thus the pattern of activity representing the rule is stable for a relatively long time. By contrast, after connections have been developed to perform the task directly, there is no correspondingly stable pattern formed during the performance of the task. Thus the loss of conscious phenomenology with expertise can be understood naturally.

On this account, the sequentiality of the rule interpretation process is not built into the architecture; rather it is a consequence of the fact that we can follow only one instruction at a time. Connectionist memories have the capability to retrieve a single stored item, and here this is necessary to avoid asking the linguistic interpreter to simultaneously interpret more than one instruction.

It is interesting to note that the preceding analysis does not require that the "rules" be linguistic; any notational system that can be appropriately interpreted would do. Another type of "rule" is a series of musical pitches; a memorized collection of such rules allows a musician to play a tune by "conscious rule interpretation." With practice the need for conscious control goes away. Since pianists learn to interpret several notes simultaneously, the present account suggests that pianists might be able to apply more than one musical rule at a time (provided their memory for the rules can simultaneously recall more than one rule). A symbolic account of such conscious rule interpretation would involve something like a production system capable of firing multiple productions simultaneously.

Finally it should be noted that even if the memorized rules are assumed to be linguistically coded, the preceding analysis is uncommitted about the form the rules take in memory: phonological, orthographic, semantic, or whatever.

Symbolic vs. subsymbolic implementation of rule interpretation

The (approximate) implementation of the conscious rule interpreter in a subsymbolic system—a reduction of the traditional kind, as I have argued above—has both advantages and disadvantages relative to an (exact) implementation in a von Neumann machine—a reduction by intentional instantiation.

The main disadvantage is that subconceptual representation and interpretation of linguistic instructions is very difficult and we can't now actually do it. Most existing subsymbolic systems simply don't use rule interpretation.[1] They can't take advantage of rules to check the results produced by the intuitive processor. They can't bootstrap their way into a new domain using rules to generate their own experience: they must have a teacher generate it for them.[2]

There are several advantages of a subconceptually implemented rule interpreter. The intuitive processor and rule interpreter are highly integrated, with broad-band communication between them. Understanding how this communication works should allow design of efficient hybrid symbolic/subsymbolic systems with effective

---

1. A notable exception is Touretzky & Hinton 1985.

2. And when a network makes a mistake, it can be told the correct answer but it can't be told which rule it violated. Thus it must assign blame for its error in a very undirected way. It is quite plausible that the large amount of training currently required by subsymbolic systems could be significantly reduced if blame could be focussed by citing violated rules.

communication between the processors. A principled basis is provided for studying how rule-based knowledge leads to intuitive knowledge. Perhaps most interesting, in a subsymbolic rule interpreter, the process of rule selection is intuitive! Which rule is reinstantiated in memory at a given time is the result of the associative retrieval process, which has many nice properties. The "best match" to the productions' conditions is quickly computed, and even if no match is very good, a rule can be retrieved. The selection process can be quite context-sensitive.

An integrated subsymbolic rule interpreter/intuitive processor in principle offers the advantages of both kinds of processing. Imagine such a system creating a mathematical proof. The intuitive processor would suggest goals and steps, and the rule interpreter would verify the validity of proposals. The serial search through the space of possible steps that is necessary in a purely symbolic approach is replaced by intuitive generation of possibilities. Yet the precise adherence to strict inference rules that is demanded by the task can be enforced by the rule interpreter; the creativity of intuition can be exploited while its unreliability can be controlled.

Two kinds of knowledge; one medium

Most existing subsymbolic systems perform tasks without serial rule interpretation: patterns of activity representing inputs are directly transformed (possibly through multiple layers of units) to patterns of activity representing outputs. The connections that mediate this transformation represent a form of task knowledge that can be applied with massive parallelism: I will call it *P-knowledge*. For example, the P-knowledge in a native speaker encodes lexical, morphological, syntactic, semantic, and pragmatic constraints in such form that all these constraints can be satisfied in parallel during comprehension and generation.

The connectionist implementation of sequential rule interpretation described above displays a second form that knowledge can take in a subsymbolic system. The stored activity patterns that represent rules also constitute task knowledge: call it *S-knowledge*. Like P-knowledge, S-knowledge is imbedded in connections: the connections that enable part of a rule to reinstantiate the entire rule. Unlike P-knowledge, S-knowledge cannot be used massively in parallel. For example, a novice speaker of some language cannot satisfy the constraints contained in two memorized rules simultaneously; they must be serially reinstantiated as patterns of activity and separately interpreted. Of course the connections responsible for reinstantiating these memories operate in parallel, and indeed these connections contain within them the potential to reinstantiate either of the two memorized rules. But these connections are so arranged that *only one rule at a time* can be reinstantiated. The retrieval of a single rule is a parallel process, but the satisfaction of the constraints contained in the two rules is a serial process. After considerable experience, P-knowledge is created: connections that can *simultaneously satisfy* the constraints represented by the two rules.

P-knowledge is considerably more difficult to create than S-knowledge. To encode a constraint in connections so that it can be satisfied in parallel with thousands of others is no easy task. Such an encoding can only be learned through considerable experience in which that constraint has appeared in many different contexts, so that the connections enforcing the constraint can be tuned to operate in parallel with those enforcing a wide variety of other constraints. S-knowledge can be much more rapidly acquired (once the linguistic skills on which it depends have been encoded into P-knowledge, of course). Simply reciting a verbal rule over and over will usually suffice to store it in memory (at least temporarily).

That P-knowledge is so highly context-dependent while the rules of S-knowledge are essentially context-free is an important computational fact underlying many of the psychological explanations offered by subsymbolic models. Consider, for example, Rumelhart and McClelland's (1986) model of the U-shaped curve for past-tense production in children. The phenomenon is striking: a child is observed using *goed* and *wented* when at a much younger age *went* was reliably used. This is surprising because we are prone to think that such linguistic abilities rest on knowledge that is encoded in some context-free form such as "the past tense of *go* is *went*." Why should a child *lose* such a rule once acquired? A traditional answer invokes the acquisition of a different context-free rule, like "the past tense of *x* is *x+ed*" which, for one reason or another, takes precedence. The point here, however, is that *there is nothing at all surprising about the phenomenon when the underlying knowledge is assumed to be context-dependent and not context-free*. The young child has a small vocabulary of largely irregular verbs. The connections that implement this P-knowledge are capable of reliably producing the large pattern of activity representing *went*, as well as those representing a small number of other past-tense forms. Informally we can say that the connections producing *went* do so *in the context of the other vocabulary items* that are also stored in the same connections. There is no guarantee that these connections will produce *went* in the context of a different vocabulary. As the child acquires additional vocabulary items, most of which are regular, the context radically changes. Connections that were, so to speak, perfectly adequate for creating *went* in the old context now have to work in a context where very strong connections are trying to create forms ending in *–ed*; these "old connections" are not up to the new task. Only through extensive experience trying to produce *went* in the new context of many regular verbs can the "old" connections be modified to work in the new context. (In particular, strong new connections must be added that, when the input pattern is that for *go*, cancel the *–ed*; these were not needed before.)

These observations about context-dependence can also be framed in terms of inference. If we choose to regard the child as using knowledge to in some sense "infer" the correct answer *went*, then we can say that after the child has added more knowledge (about new verbs), the ability to make the (correct) inference is lost. In this sense the child's inference process is *non-monotonic*—perhaps this is why we find the phenomenon surprising. Non-monotonicity is a fundamental property of subsymbolic inference (see Smolensky, 1987b).

To summarize:

- Knowledge in subsymbolic systems can take two forms, both resident in the connections.
- The knowledge used by the conscious rule interpreter lies in connections that reinstantiate patterns encoding rules; task constraints are coded in context-free rules and satisfied serially.
- The knowledge used in intuitive processing lies in connections that constitute highly context-sensitive encodings of task constraints that can be satisfied with massive parallelism.
- Learning such encodings requires much experience.

Conclusion

The approach described above sets out a rather clear program for developing subsymbolic models of conscious rule application. The crucial technical problems to be solved involve subsymbolic models of natural language processing: the representation of linguistic structures—including procedural descriptions such as productions—and their interpretation—effecting the designated procedures as a result of activating their representation. These are hard problems, but they are problems that need to be solved anyway by a connectionist approach to language processing. With even very limited solutions to these problems, we can begin to seriously explore the interaction of rule application and intuitive processing, in both learning and performance, within a subsymbolic connectionist framework.

Acknowledgements

I am indebted to a number of people for very helpful conversations on these issues: Jerry Fodor, Zenon Pylyshyn, and especially Dave Rumelhart, Rob Cummins, and Denise Dellarosa. This research has been supported by NSF grant IST-8609599 and by the Department of Computer Science and Institute of Cognitive Science at the University of Colorado at Boulder. This paper is based on a small portion of a paper to appear in *The Behavioral and Brain Sciences*.

References

Anderson, J.A. & Hinton, G.E. (1981). Models of information processing in the brain. In G. E. Hinton and J. A. Anderson, Eds., *Parallel models of associative memory.* Hillsdale, NJ: Erlbaum.

Haugeland, J. (1978). The nature and plausibility of cognitivism. *Behavioral and Brain Sciences* 1: 215–226.

Hinton, G.E., McClelland, J.L., & Rumelhart, D.E. (1986). Distributed representations. In: *Parallel distributed processing: Explorations in the microstructure of cognition. Volume 1: Foundations,* J. L. McClelland, D. E. Rumelhart, & the PDP Research Group. Cambridge, MA: MIT Press/Bradford Books.

Rumelhart, D.E. & McClelland, J.L. (1986). On learning the past tenses of English verbs. In: *Parallel distributed processing: Explorations in the microstructure of cognition. Volume 2: Psychological and biological models,* J. L. McClelland, D. E. Rumelhart, & the PDP Research Group. Cambridge, MA: MIT Press/Bradford Books.

Rumelhart, D.E., Smolensky, P., McClelland, J.L., and Hinton, G.E. (1986). Schemata and sequential thought processes in parallel distributed processing models. In: *Parallel distributed processing: Explorations in the microstructure of cognition. Volume 2: Psychological and biological models,* J. L. McClelland, D. E. Rumelhart, & the PDP Research Group. Cambridge, MA: MIT Press/Bradford Books.

Smolensky, P. (1986a). Information processing in dynamical systems: Foundations of harmony theory. In: *Parallel distributed processing: Explorations in the microstructure of cognition. Volume 1: Foundations,* J. L. McClelland, D. E. Rumelhart, & the PDP Research Group. Cambridge, MA: MIT Press/Bradford Books.

Smolensky, P. (1986b). Neural and conceptual interpretations of parallel distributed processing models. In: *Parallel distributed processing: Explorations in the microstructure of cognition. Volume 2: Psychological and biological models,* J. L. McClelland, D. E. Rumelhart, & the PDP Research Group. Cambridge, MA: MIT Press/Bradford Books.

Smolensky, P. (1987a). Connectionist AI, symbolic AI, and the brain. *AI Review,* special issue on the foundations of AI.

Smolensky, P. (1987b). On the proper treatment of connectionism. Technical Report CU-CS-359-87, Department of Computer Science, University of Colorado at Boulder.

Touretzky, D.S. & Hinton, G.E. (1985). Symbols among the neurons: Details of a connectionist inference architecture. *Proceedings of the International Joint Conference on Artificial Intelligence.*

DOMAIN SPECIFICITY AND KNOWLEDGE UTILIZATION IN DIAGNOSTIC EXPLANATON

Vimla L. Patel
José F. Arocha
Guy J. Groen

McGill University

ABSTRACT

This paper examines the performance of cardiologists, psychiatrists, and surgeons in diagnostic explanations of cases within and outside their domain. The protocols were analyzed by techniques of transforming a propositional representation into a semantic network. Some graph-theoretic criteria for analyzing semantic networks are used for precision of analysis. The results show that the subjects interpret cases in terms of the familiar component of the problem, using specific domain knowledge. This is related to forward directed reasoning. Unfamiliar or uncertain components of the disorder are either ignored or explained using backward reasoning strategies. A tendency to move from a forward driven strategy to a backward driven strategy and viceversa is also seen in some protocols. This sequence is repeated a number of times to form a chain consisting of forward/backward reasoning sequences. This has implications for how subsequent patient information is processed in order to make decisions for treatment and management.

INTRODUCTION

Recent research on medical problem solving has shown that expert physicians solve clinical cases in a way that reflect their familiarization with specific disease types. Studies by Joseph and Patel (1986) and Patel, Arocha and Groen (1986), suggest that when experts have extensive knowledge of the disease classification in a clinical problem they are trying to solve, they work inductively, reaching a diagnostic hypothesis directly from the clinical findings. This has been documented even in situations where the physicians are forced to make tentative hypotheses due to insufficient information to diagnose the case correctly. Joseph and Patel showed that when physicians process a case serially, by reading a segment at a time, experts differ from subexperts (physicians working on a problem that falls outside their area of specialization) in the amount of information needed to reach the accurate diagnosis. That is, the experts correctly diagnose the problem early in the case. Once the experts reach the diagnosis, they interpret further findings as supporting evidence for their first hypothesis, as opposed to subexperts who produce subsequent diagnoses as they process more of the clinical information, generating multiple new hypotheses along the way.

The tendency of physicians to interpret clinical cases in terms of their specialization has also been shown by Patel, Arocha, and Groen (1986). When presented with an endocrinology case, cardiologists interpret some of the information in the text as support for cardiac involvement, even though this constituted a secondary manifestation of the primary disease of hypothyroidism. This situation is more evident when the case is at a high level of difficulty i.e., when the pattern of findings for the case overlap with that of other diseases. Precisely how this occurs was not addressed by the authors.

In a paper by Patel and Groen (1986) it was shown that in order to reach a correct diagnosis cardiologists have to make use, at least partially, of all the components in the case. However, those cardiologists who solve the case of bacterial endocarditis (see Table 1) in a more complete way, made use of the rules linking puncture wounds to intravenous drug use to bacterial infection and the rule linking early diastolic murmur to aortic valve insufficiency. Although most cardiologists recognized the infection component of the disease, all of them focused predominantly on the cardiac aspect of the problem.

The facility with which experts interpret clinical cases within their clinical specialty suggest that a fast process of pattern recognition may be taking place. This has been more clearly documented in domains involving a strong perceptual component, such as radiology (Lesgold, Robinson, Feltovich, Glaser, & Klopfer, in press). Research suggests that forward reasoning may be the counterpart of pattern recognition in more perceptually oriented domains. The issue of perception becomes a pattern of comprehension in a verbally rich domain such as medicine. The research described in Patel and Groen has pointed out that expert cardiologists process cardiology cases in a forward fashion by encoding cases in terms of hypotheses that lead to the diagnoses. Other research comparing the performance of cardiologists and endocrinologists in explanation tasks within and outside their domain specialties suggest that when a case is in the physician's own specialty and is typical of the domain, physicians use forward directed reasoning. Contrarily, if the case is outside the physicians specialty, or is at a high level of difficulty, physicians employ backward directed reasoning.

TABLE 1
ACUTE BACTERIAL ENDOCARDITIS TEXT

THIS 27-YEAR OLD UNEMPLOYED MALE WAS ADMITTED TO THE EMERGENCY ROOM WITH THE COMPLAINT OF SHAKING CHILLS AND FEVER OF FOUR DAYS DURATION. HE TOOK HIS OWN TEMPERATURE AND WAS RECORDED AT 40°C ON THE MORNING OF ADMISSION. THE FEVER AND CHILLS WERE ACCOMPANIED BY SWEATING AND A FEELING OF PROSTRATION. HE ALSO COMPLAINED OF SHORTNESS OF BREATH WHEN HE TRIED TO CLIMB THE TWO FLIGHTS OF STAIRS IN HIS APARTMENT. FUNCTIONAL ENQUIRY REVEALED A TRANSIENT LOSS OF VISION IN HIS RIGHT EYE WHICH LASTED APPROXIMATELY 45" ON THE DAY BEFORE HIS ADMISSION TO THE EMERGENCY WARD.

PHYSICAL EXAMINATION REVEALED A TOXIC LOOKING YOUNG MAN WHO WAS HAVING A RIGOR. HIS TEMPERATURE WAS 41°C, PULSE 120, BP 110/40. MUCUS MEMBRANES WERE PINK. EXAMINATION OF HIS LIMBS SHOWED PUNCTURE WOUNDS IN HIS LEFT ANTECUBITAL FOSSA. THE PATIENT VOLUNTEERED THAT HE HAD BEEN BITTEN BY A CAT AT A FRIEND'S HOUSE ABOUT A WEEK BEFORE ADMISSION. THERE WERE NO OTHER SKIN FINDINGS. EXAMINATION OF THE CARDIOVASCULAR SYSTEM SHOWED NO JUGULAR VENOUS DISTENSION, PULSE WAS 120 PER MINUTE, REGULAR, EQUAL AN DISYNCRONOUS. THE PULSE WAS ALSO NOTED TO BE COLLAPSING. THE APEX BEAT WAS NOT DISPLACED. AUSCULTATION OF HIS HEART REVEALED A 2/6 EARLY DIASTOLIC MURMUR IN THE AORTIC AREA AND FUNDOSCOPY REVEALED A FLAME-SHAPED HEMORRHAGE IN THE LEFT EYE. THERE WAS NO SPLENOMEGALY. URINALYSIS SHOWED NUMEROUS RED CELLS BUT THERE WERE NO RED CELL CASTS.

This paper focuses on the domain-specific knowledge of psychiatrists, cardiologists and surgeons which lead them to emphasize certain components of the case with which the physicians are more familiar. The problem text to be presented describes a subject with a diagnosis of acute bacterial endocarditis. The case has five main components which taken together would lead to an accurate diagnosis. These are as follows: First, the patient is admitted with signs of infection represented by fever, chills, and rigors; more specifically, the signs are an indication of bacteremia or bacterial infection. Second, he shows signs of aortic valve insufficiency indicated by low diastolic pressure and normal systolic pressure, and an early diastolic murmur. Third, the patient presents embolic phenomena suggested by a previous episode of transient blindness, hemorrhage in one eye, and the presence of red cells in his urine. The fourth component is suggestive of an acute process and it is indicated by the short duration of the illness and by the normal size of both the spleen and the heart. This is important because a failure to identify the problem as acute may change the treatment and endanger the patient's life. The fifth component, of social rather than biomedical nature, is suggested by the presence of puncture wounds on the patient's left arm, which together with the fact that he is a young, unemployed male may suggest that the patient is an intravenous drug user, which in turn, may be suggestive of bacterial endocarditis as this disease is common among intravenous drug users.

It is hypothesized that experts focus on their own domain of expertise in diagnostic reasoning because the clinical rules in these domains for solving specific problems are easily accessible. Thus expert psychiatrists explaining the clinical problem would tend to focus on the social and the "illness" aspect of the patient rather than on the biomedical explanation. A surgeon under the same condition would be expected to emphasize the infective process of the disease. A cardiologist, in contrast, would focus on the cardiac problem leading to endocarditis. This paper reports the findings from these predictions using specific and precise methods of analyses.

EXPERIMENTAL METHOD

Rationale:

The empirical paradigm used in the study involves a diagnostic explanation task in which the subject is asked to explain the underlying pathophysiology of the patient's condition. This probe has proved to be useful for assessing differences in knowledge that may underlie the problem-solving process (Patel & Groen, 1986).

We have found that physicians respond to the probe by explaining the patient's symptoms in terms of a diagnosis. The diagnosis is requested after the explanation task, giving the subjects the opportunity to provide a diagnosis during the explanation, such that the resulting protocols may reflect elements of the solution process.

It has been argued elsewhere (Groen & Patel, in press) that in highly verbal knowledge domains, problem solution requires methodologies different from those traditionally used in the literature. The attempt to combine comprehension and problem solving has led some researchers (Kintsch & Greeno, 1985; Patel & Groen, 1986) to utilize a combination of propositional analysis and production rule systems. Another way of representing the same information is through structured representations known as semantic networks, which can be rendered more precise by introducing a few concepts from graph theory (Sowa, 1984; Groen & Patel, 1987). Formally a graph is defined as a non-empty set of nodes and a set of arc leading from a node N to a node N'. A graph is connected if there exists a path, directed or undirected between any two nodes, where a directed path has every node as a source of an arrow connecting to its immediate successor. If the graph is not connected then it breaks down into disjoint components.

In terms of these notions, a semantic network is formed by nodes and connecting paths. Nodes may represent either clinical findings or hypotheses and the paths represent directed connections between nodes. Forward reasoning corresponds to an oriented path from a fact to a hypothesis. Backward reasoning corresponds to a directed path from a hypothesis to a fact. Thus forward directed rules are identified whenever the physicians attempt to generate a hypothesis from the findings in the case. Backward directed rules are identified when a physician first formulates a hypothesis and then attempts to put the data in. The reader is referred to Groen & Patel (in press) for details of the application of graph theory to semantic network.

Description of Method:

The clinical problem selected was the case of ACUTE BACTERIAL ENDOCARDITIS, previously described. It has five sub-components, infectious process, aortic insufficiency, embolic phenomena, acuteness of disease and intravenous drug use.

The subjects included cardiologists, psychiatrists and surgeons, six at each level of specialty.

The general procedure involved presentation of the clinical case, obtaining a free-recall protocol and a pathophysiological explanation protocol, and finally asking for a diagnosis.

RESULTS AND DISCUSSION

Overall summary of results in terms of number of subjects in each specialty that focus on particular components of the disease is provided on Table 2. Here focus refers to providing an explanation for the component. Most of the subjects focus on the infection component of the disease with a strong tendency for the surgeons to provide a detailed explanation for the infective process. The cardiologists have easily accessible rules for the cardiac problem, thus they focus on the aortic insufficiency aspect of the problem. In contrast, more psychiatrists than cardiologists or surgeons interpret the problem in terms of social phenomena of intravenous drug use. Surgeons identified and explained the infectious process and the acuteness of the patient's condition more than any other aspect of the disease.

The details of the results are explained with one example from each of the generated explanation protocols from a cardiologist, surgeon and a psychiatrist. Table 3 gives the explanation protocol of a psychiatrist and Figure 1 presents the semantic network or structural representation generated from the protocol.

This physician organized his protocol around the likely drug injection by the patient which was signaled in the text by the puncture wounds on the patient's arm. The psychiatrist seems to interpret the problems in terms familiar to his specialization, that is, as a drug toxicity problem. He interprets fever as a reaction to the drugs, not to the use of a contaminated needle, i.e., infective component of the disease. The subject does not mention information about the patient's aortic insufficiency or embolic phenomenon. The flame-shaped hemorrhage is also explained from the injected drug rather than from an infectious process.

TABLE 2
NUMBER OF SUBJECTS IN EACH SPECIALTY THAT EXPLAIN THE SPECIFIC COMPONENT OF THE DISEASE

SPECIALTY	NUMBER OF SUBJECTS	INTRAVENOUS DRUG USE	INFECTION	ACUTENESS	AORTIC INSUFFICIENCY	EMBOLI
CARDIOLOGISTS	6	2	4	1	6	5
PSYCHIATRISTS	6	5	5	2	2	2
SURGEONS	6	6	6	5	2	3

TABLE 3
PATHOPHYSIOLOGY PROTOCOL BY PSYCHIATRIST # 2
THE PATIENT HAS BEEN REACTING TO STRESS LIKELY BY HIS INJECTING A DRUG (OR DRUGS) WHICH HAS RESULTED IN TACHYCARDIA, A FALL IN BLOOD PRESSURE AND ELEVATED TEMPERATURE. HE IS IN OR NEAR SHOCK.
THE FLAME-SHAPED HEMORRHAGE MAY REPRESENT A SEQUELA OF AN UPSURGE IN BLOOD PRESSURE POSSIBLY AS A RESULT OF HIS INJECTION OF DRUGS.

The whole structural representation is tied to the hypothesis of drug injection. This hypothesis is causally linked to the two other hypotheses, reaction to to stress and upsurge in blood pressure and to the findings in the case i.e, tachycardia, fall in blood pressure, high temperature, toxicity and the eye hemorrhage. It is important to note that the critical information for the accurate diagnosis of bacterial endocarditis regarding the involvement of heart was completely ignored.

Figure 2 shows the four major components of the structural representation of the explanation protocol. None of the four components separately show any signs of pure forward or backward reasoning. The first component, reaction to some "stress leads to injection of drugs", has both the nodes inferred. The injection of drug is believed to be the patient's reaction to some stress and it is the drug that causes toxic problems. This component explains the injection of drugs. The second component of the network explains the diagnosis of the shock state of the patient resulting in three clinical findings and inference on the findings. The third component explains the diagnosis of the toxic state of the patient induced by the drug. The final component explains the flame-shaped hemorrhage as a result of increase in blood pressure caused by the drugs. It should be noted that everything is explained in terms of the drug induced toxicity and stress. It is the injection of drug aspect of the network that is the overlapping factor in explaining the clinical cues. Table 4 gives the pathophysiological protocol of a surgeon.

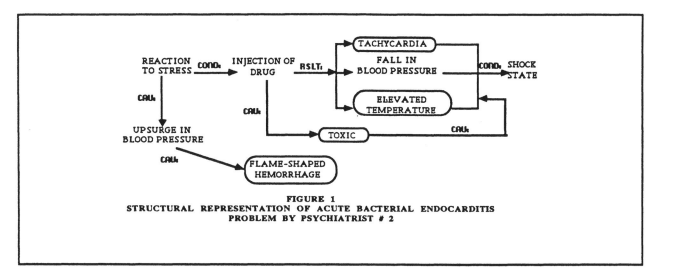

FIGURE 1
STRUCTURAL REPRESENTATION OF ACUTE BACTERIAL ENDOCARDITIS
PROBLEM BY PSYCHIATRIST # 2

KEYS FOR ALL FIGURES:

⬭ TEXT CUES

COND: conditional links
CAU: causal links
RSLT: resultive links
LOC: locative links
EQUIV: equivalence links

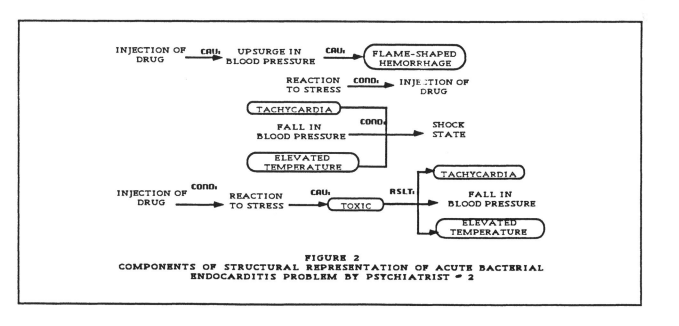

FIGURE 2
COMPONENTS OF STRUCTURAL REPRESENTATION OF ACUTE BACTERIAL
ENDOCARDITIS PROBLEM BY PSYCHIATRIST # 2

TABLE 4
PATHOPHYSIOLOGY PROTOCOL BY SURGEON # 3
ACUTE FEBRILE ILLNESS OF SOME SEVERITY, APPARENT BACTEREMIA OR VIREMIA. BACTEREMIA SEEMS MORE LIKELY BECAUSE OF THE HIGH FEVER, CHILLS, TOXICITY, AND IMPLICATION OF THE HEART I.E., POSSIBILITY OF ACUTE BACTERIAL ENDOCARDITIS, MULTIPLE ORGAN INVOLVEMENT, E.G., EYES, KIDNEYS - POSSIBLE SEPTIC EMBOLI. APPROPRIATE INVESTIGATIONS WOULD DEFINE THE ETIOLOGY. LIKELY SECONDARY TO INJECTING HIMSELF WITH DIRTY NEEDLES.

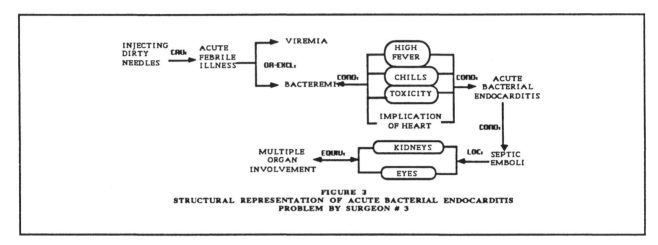

FIGURE 3
STRUCTURAL REPRESENTATION OF ACUTE BACTERIAL ENDOCARDITIS PROBLEM BY SURGEON # 3

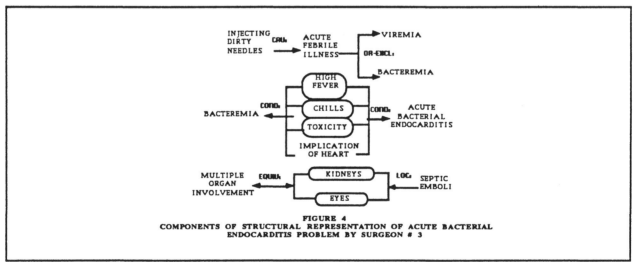

FIGURE 4
COMPONENTS OF STRUCTURAL REPRESENTATION OF ACUTE BACTERIAL ENDOCARDITIS PROBLEM BY SURGEON # 3

Figure 3 shows the structural representation of the case by a surgeon. He identifies the problem as acute febrile illness of either viral or bacterial origin. The surgeon decides that the patient has bacteremia because the patient has fever, chills, toxicity and has implications of heart involvement. His differential diagnosis is acute bacterial endocarditis. Given endocarditis, the most likely source of infection is from a dirty needle. The above description can be decomposed into three component of the representational network (Figure 4). The first component deals with acute fever due to dirty needles leading to bacteremia. The second component also deals with explaining bacteria from the given clinical cues and make the secondary diagnosis of acute bacterial endocarditis. The third component deals with septic embolism. Each of the three components deal with acute infection or sepsis, a condition that surgeons have to be very familiar with in their practice.

Figure 5 gives a structural representation of the protocol generated by a cardiologist and Figure 6 provides the four major components of this representation. The major diagnosis is of endocarditis affecting the heart valve. The mixture of forward and backward reasoning towards a diagnostic hypothesis of endocarditis affecting aortic valve as seen in component one. It should be emphasized that except for noting that infection exists, it is not explained in any details. The second component also deals with aortic valve involvement and is again a mixture of forward and backward reasoning. The use of intravenous injection by the patient is not mentioned. The third component uses a pure forward reasoning to explain the aortic valve involvement from two given textual cues. The final component also deals with the cardiac problem although it does not have any relationship to the overall coherence. All the components of the network deal with cardiac problem with little focus on the infective process.

The results show that a psychiatric emphasizes the drug toxicity and stress reaction part of the problem, a surgeon emphasizes the acute infective nature of the problem, and a cardiologist focuses on the heart valves and heart murmur that relate to the disease. The subjects make extensive use of their domain knowledge to represent the clinical problem. A mixture of forward and backward reasoning is seen in all the cases. A more forward directed reasoning strategy is used in a familiar situation and a more backward reasoning strategy is used in an unfamiliar situation.

TABLE 5

PATHOPHYSIOLOGY PROTOCOL BY CARDIOLOGIST #7

THE DIAGNOSIS APPEARS TO BE ENDOCARDITIS, THE SITE IS THE AORTIC VALVE. IT IS AORTIC RATHER THAN PULMONARY BECAUSE 1) LOW DIASTOLIC PRESSURE, 2) EVIDENCE OF EMBOLI IN EYE AND URINE, RBC CASTS, 3) HYPERDYNAMIC SYSTEMIC CIRCULATION. EXAMPLE, THE INTENSITY OF FIRST HEART SOUND. IT SHOULD BE DEMONSTRATED WITH PREMATURE CLOSING OF ATRIAL VALVE. DYSPNEA MAY POINT TO EARLY HEART FAILURE, A VERY OMINOUS SIGN WHICH MAY DEMAND EARLY SURGICAL INTERVENTION.

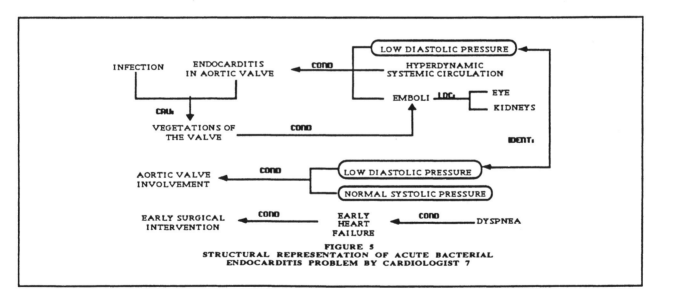

FIGURE 5
STRUCTURAL REPRESENTATION OF ACUTE BACTERIAL
ENDOCARDITIS PROBLEM BY CARDIOLOGIST 7

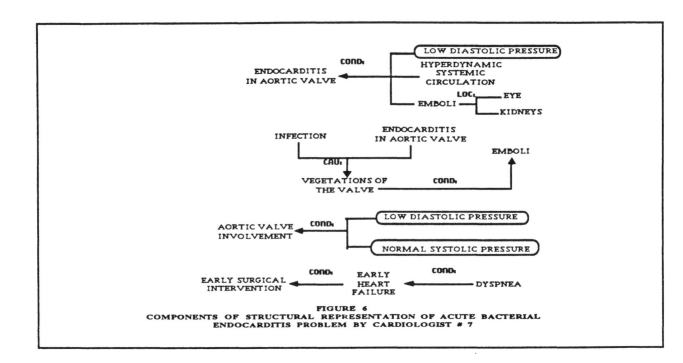

FIGURE 6
COMPONENTS OF STRUCTURAL REPRESENTATION OF ACUTE BACTERIAL ENDOCARDITIS PROBLEM BY CARDIOLOGIST # 7

ACKNOWLEDGEMENTS

This work was supported by a grant from the Josiah Macy Jr. Foundation (B8520002) and the National Science Engineering Research Council of Canada (A2598) to Vimla Patel. We would like to acknowledge the support of all the medical specialists who took part in our study and to Susan Young for typing the manuscript.

REFERENCES

Groen, G.J. and Patel, V.L. (in press) The relationship between comprehension and reasoning in medical expertise. In M. Chi, R. Glaser & M. Farr (Eds.) *The Nature of Expertise.*

Joseph, G.-M. and Patel, V.L. (1986) Specificity of expertise in medical reasoning. In *Proceedings of the Eighth Annual Conference of the Cognitive Science Society* (pp.331-343). Hillsdale, N.J.: Lawrence Erlbaum.

Kintsch, W. and Greeno, J.G. (1985) Understanding and solving word arithmetic problems. *Psychological Review, 92*: 109-129.

Lesgold, A., Robinson, H., Feltovich, P., Glaser, R. and Klopfer, D. (in press) Expertise in complex skills: Diagnosing x-ray pictures. In M.Chi, R. Glaser, & M. Farr (Eds.) *The Nature of Expertise.*

Patel, V.L. and Groen, G.J. (1986) Knowledge based solution strategies in medical reasoning. *Cognitive Science, 10,* 91-116.

Patel, V.L., Arocha, J.F. and Groen, G.J. (1986) Strategy selection and degree of expertise in medical reasoning. In *Proceedings of the Eighth Annual Conference of the Cognitive Science Society* (pp.780-791). Hillsdale, N.J.: Lawrence Erlbaum.

Sowa, J.F. (1984) *Conceptual structures..* Reading, MA.: Addison Wesley.

Measuring Change and Coherence
in Evaluating Potential Change in View

Gilbert Harman

Cognitive Science Laboratory
221 Nassau Street
Princeton University
Princeton, NJ 08542
(609) 987-2819
uucp: princeton!mind!ghh
arpanet: ghh@mind.princeton.edu

Marie A. Bienkowski

Bell Communications Research
Morristown, New Jersey

Ken Salem

Department of Computer Science
Princeton University

Ian Pratt

Department of Philosophy
Princeton University

ABSTRACT

In changing your view, you must balance the amount of change involved against the improvement in explanatory coherence resulting from the change. Even if change and improvement in coherence are measured by simply counting, there can be no general requirement that the number of modified items (added or subtracted) be no greater than the number of new explanatory and implications links. The relation between conservatism and coherence is more complex than that.

Keywords: explanation, planning, reasoning.

The purpose of this note is to discuss one aspect of the approach to the theory of reasoning described in Cullingford, Harman, Bienkowski, and Salem (1985), Harman (1973, 1986), and Harman, Cullingford, Bienkowski, Salem, and Pratt (1986). In this approach, reasoning is identified with change in view, that is, with additions and subtractions to your antecedent beliefs and plan. We suppose that among the most basic principles of change in view are (1) *conservatism*: other thing's being equal, minimize the amount of change; and (2) *coherence*: other thing's being equal, maximize the explanatory coherence in the resulting view. In order to apply these principles, you need a way of measuring how much change a suggested

The research reported here was supported in part by a research grant from the James S. McDonnell Foundation by the Defense Advanced Research Projects Agency of the Department of Defense and by the Office of Naval Research under Contracts Nos. N00014-85-C-0456 and N00014-85-K-0465; and by the National Science Foundation under Cooperative Agreement No. DCR-8420948 and under NSF grant number IST8503968. The views and conclusions contained in this document are those of the authors and should not be interpreted as necessarily representing the official policies, either expressed or implied, of the McDonnell Foundation, the Defense Advanced Research Projects Agency, or the U.S. Government.

modification would involve and how much increase in coherence. We are exploring very simple measures which simply count the number of changes and number of explanations a modification would involve. We considering exactly what it would be plausible to count and also what other principles are needed if we are to use such simple measures. In the present note, we will discuss only one aspect of the issue.

Other basic principles of reasoning include (3) *no get back*: avoid giving up things that you can infer right back; and (4) *clutter avoidance*: do not clutter your mind with trivialities (Harman 1986, chapters 2 and 6). We are not concerned with these principles in the present note. Rather, we want to consider what has to be true of reasoning if the impact of the amount of change and coherence involved can be measured simply by counting changes and explanations.

One possible outcome might be that there is no way to make this work. (The present authors disagree about whether this is the most likely outcome.) Perhaps we have to assign *weights* to propositions and explanations in such a way that some propositions and explanations count for more than others in calculating how much change in view or coherence would be involved in a particular proposed modification. It is even possible that there is no interesting way in which considerations of amount of change and coherence can play a useful role in deciding what revision to make in your plans. But let us assume as a working hypothesis that such considerations are relevant and can be measured by counting. The question then is what else might be involved in a system of revision that these considerations can be measured by counting.

Clutter avoidance requires that an acceptable change in view must promise to advance your goals. You start with an interest in whether such and such is true or an interest in having or doing something. Such interests can give you other interests, e.g., an interest in whether some other thing is true or an interest in having or doing something else. A new conclusion is acceptable only if its acceptance promises to satisfy one of your interests. For present purposes we may assume this means that a new conclusion is acceptable only if it contains a proposition P in which you are interested or want to be the case.

Comparing New Items with New Explanations

Consider an elementary logical inference. You start with a beliefs in "P" and "if P then Q" and infer "Q". We suppose that this is to infer an explanation: "Q because P and if P then Q." "Q" is the *explanandum* of this explanation and "P" and "if P then Q" are the *explainers*. (The sense in which an implication can be explanatory is discussed in Harman 1986, chapter 7.) How do we count the amount of change involved in this case and how do we count the added coherence?

We might say that this involves one new belief, namely "Q," and one new explanation, namely, "Q because P and if P then Q." In that case, the number of new beliefs is the same as the number of new explanations. The loss to conservatism involved in adding the new belief is matched by the gain in coherence involved in the new explanation. Clutter avoidance allows you to make an inference in this case only on the assumption that you are interested in whether Q.

This example may seem to suggest a requirement on acceptable changes that the number of things accepted must not be greater than the number of new explanations.

That would not mean you need an explanation of every new belief you accept. Often you accept a belief because it explains something else you already believe ("inference to the best explanation"). In that case, you do not need an explanation of the new belief. It is enough that in accepting that new belief you come to have a new explanation of something else.

The principle that the number of new things accepted must not exceed the number of new explanations would rule out arbitrarily inferring Q from any random belief P. Such an inference would involve the acceptance of one new belief with no new explanation. So, it would be ruled out by the principle as stated.

However, there is an objection to this way of counting and so to the principle that the number of new things accepted must not exceed the number of new explanations. In the first example, you come to accept not only "Q" but also that there is a certain connection between that belief and two other beliefs you already accept. That is, you come to accept that these two other beliefs imply "Q". There are really two new beliefs in this case and only one explanation. So it may seem that the principle that the number of new things accepted must not exceed the number of new explanations would prevent you from inferring "Q" from "P" and "if P then Q." In that case, the principle would be clearly unacceptable.

Now this implication is an instance of modus ponens. All such instances are immediately intelligible. In supposing that "Q" is implied by "P" and "if P then Q," you are not left wondering why there is this implication. You immediately understand the relation as implicational. You have all the explanation you need of the implication.

This is not to say that the implication is explained by the general logical principle of modus ponens. To say that would simply push the problem back one step (Carroll 1956). If we said that the implication holds because it is an instance of modus ponens, which you already accept, we would have to say that you accept three new things, (i) "Q," (ii) '"Q" is implied by "P" and "if P then Q",' and (iii) "the preceding belief (ii) is an instance of modus ponens." So, again, there would be more new beliefs than explanations unless one of the new beliefs needs no explanation. At some point you accept things that need no further explanation.

This suggests modifying the principle to say that the number of new things accepted *that require explanation* not be greater than the number of new explanations.

That would allow two methods of counting the number of changes. Either count the total number of new beliefs, including those that need no explanation, or count only the number of new beliefs that do not require explanation. Clutter avoidance favors counting all new beliefs, including those requiring no explanation. We do not want to accept even those beliefs without some special interest in whether they are true. Otherwise, there would be nothing wrong with cluttering your mind with trivialities of this sort.

This method of counting, along with our modified principle, would prevent the inference of "Q" from arbitrary "P" via the expedient of also inferring "P implies Q." If we were to count this as one new belief and one new explanation of that belief, using the initial method of counting, the inference would not conflict with our original principle. Contrary to the initial method of counting, it is obvious that two new things are accepted in this case. The inference conflicts with the modified principle, since the inference appeals to a connection between "P" and "Q" that requires an explanation. The connection needs an explanation since it is not immediately intelligible in the way in which instances of modus ponens are immediately intelligible. So the inference would involve the acceptance of two new beliefs requiring explanation and only one new explanation, which would violate the principle that the number of new things accepted that require explanation should not be greater than the number of new explanations.

It may seem that there is still a way to infer an arbitrary conclusion without violating this principle. From arbitary "P" you infer a complex e-structure containing two explanations. One explanation explains "Q" by appeal to the two explainers "P" and "P if and only if Q". The other explanation explains "P" by appeal to the two explainers "Q" and "P if and only if Q." Here there are only two new beliefs not requiring explanation, namely, "Q" and "P if and only if Q." The two implicational links are both immediately intelligible and therefore need no explanation. There are also two explanations in this case. Since the number of explanations is as great as the number of new beliefs needing explanation, this would not be ruled out by the principle that there should be a new explanation for every new item accepted that needs an explanation alone.

Notice that the complex e-structure you would accept in this case would be *circular*: Q would be taken to be explained in part by P, and P would be taken to be explained in part by Q. So, ultimately, Q would be used to explain itself. If that were allowed, you could even more simply infer an arbitrary "Q" by accepting "Q because Q." Here there would be only one new belief needing an explanation, namely, "Q" and there would be one "explanation," namely, "Q because Q." But that is no explanation; it is blatently circular.

In other words, the principle that the number of new things accepted that require an explanation should be not exceed the number of new explanations presupposes the following principle:

No circular explanations: acceptable e-structures must not contain circular explanations. If a proposition is explained in the e-structure, it must not itself serve as an explainer in the e-structure of one of its own explanatory antecedents. (It cannot be one of its own explainers, an explainer of one of its explainers, an explainer of an explainer of one of its explainers etc.)

The Flow of Acceptability from Prior Beliefs

But there is a more serious problem with our suggested principle that the number of new accepted items requiring explanation be no greater than the number of new explanations. This principle clearly fails in many cases of inference to the best explanation.

For example, Albert tells you Jack is a philosopher. You infer that Albert says this because he believes Jack is a philosopher and wants you to know whether Jack is a philosopher. Here, using the original method of counting new beliefs and ignoring new beliefs about explanatory connections, there are two new beliefs -- that Albert believes Jack is a philosopher and that Albert wants you to know whether Jack is a philosopher -- but there is only one new explanation, namely that those two things explain Albert's saying what he says. There are more new beliefs than explanations in this case, so the proposed requirement would say that this e-structure is not acceptable. But an inference to such an explanation is often quite in order even though it involves more new beliefs than explanations.

Someone might argue that the preceding example is acceptable only given something like the background belief that normally, if someone says something, then that is because the speaker believes what he or she is saying and wants the hearer to know. It might be argued that this background belief helps to provide further explanatory links between Albert's saying that Jack is a philosopher on the one hand, and, on the other hand, his belief that Jack is a philosopher and his desire that you should know that. We certainly must allow for explanations that explain why a given A is a B by noting that normally A's are B's (Harman 1986, Chapter 7). The suggestion is that you infer that a certain explanation holds on this occasion because normally, when someone says something, that sort of explanation holds. This analysis requires distinguishing two explanations. First, there is the psychological explanation of Albert's remark that Jack is a philosopher, an explanation which appeals to an assumed belief and desire of Albert's. Second, there is something like a statistical explanation of that psychological explanation's holding in this case, an explanation that appeals to what normally leads to someone saying what he or she says.

It may look as if this analysis of the inference would violate the prohibition against circular explanations, but it does not really do so. The appearance of circularity arises because you end up explaining why Albert believes that Jack is a philosopher in part by appeal to Albert's saying that Jack is a philosopher and you end up explaining Albert's saying that Jack is a philosopher in part by appeal to Albert's believing that Jack is a philosopher. But in accepting this inference you would not have to give any credit to this last explanation. The coherence in your views that it contributes would not have to counted in order to determine that your inference is acceptable. This follows from the fact that this analysis takes each one of Albert's new beliefs to be explained, so the number of new beliefs cannot exceed the number of explanations. Albert's remark and the principle about what normally explains such remarks together are taken to explain: "Albert says Jack is a philosopher because Albert believes this and wants me to know it." And the truth of the proposition just quoted obviously implies both "Albert believes that Jack is a philosopher" and "Albert wants me to know that Jack is a philosopher." So, there are at least as many explanatory and implicational connections as new beliefs requiring such connections in this analysis without counting the connection provided by explaining what Albert says via his belief and desire.

But there are problems with the analysis. It analyzes away the inference *to* an explanation, turning it into an inference *from* an explanation. Inference to an explanation seems to be a genuinely different sort of case. In many cases of inference to the best explanation there are no relevant prior beliefs about what normally or probably happens. Furthermore, if the suggested analysis were correct, there would be no real need in the present case to infer an explanation of what is said. You could, for example, infer from "Albert says that Jack is a philosopher" to "Jack is a philosopher" via the default principle, "Normally, when someone says something it is true," without having a view about how Jack's being a philosopher might be part of the explanation of Albert's saying this. This would make it difficult to account for the way in which such an inference is defeated by the discovery that Albert is insincere or that he is not in a position to know whether Jack is a philosopher (Harman, Cullingford, Bienkowski, Salem, and Pratt 1986).

If the analysis via a default rule does not work for the general case, as we are suggesting may be true, then we have to abandon the rule that there should be at least as many new explanations as new things accepted that require explanation. What might replace that principle?

Here's a proposal. Consider the case in which the relevant change in view being considered is entirely a matter of adding new things. The things added must be represented as a non-circular "e-structure," that is, as a set of "e-nodes" or simple explanations, each of which has one or more "explainers" and a single "explanandum" (the thing explained or implied). Some e-nodes themselves may be explananda in other e-nodes (for example, if a given explanatory connection is intelligible as an instance of a principle about what might explain what). We can say that an explainer or explanandum in one of these e-nodes or the e-node itself is "OK" if (but not only if) it is either already accepted or immediately intelligible (or obvious). Furthermore, if the explanandum of an e-node is OK, then all the explainers are OK; and, if all the explainers in an e-node are OK, so is the explanandum. Finally, the whole e-structure is "acceptable" only if all of its contained e-nodes and their explainers and explananda are OK. (Clearly, it is trivial to check whether this condition is satisfied.)

When you infer from "P" and "If P, Q" to "Q," the relevant e-structure has a single e-node, whose explainers are "P" and "If P, Q" and whose explanandum is "Q". Since both explainers are previously accepted, they are OK; so the explanandum is also OK. The e-node is OK since it is immediately intelligible as an instance of modus ponens. All the items in the e-structure are OK, so the e-structure as a whole is acceptable.

The simple inference from arbitrary "P" to arbitrary "Q" would be ruled out because it involves no e-nodes and so allows no way for "Q" to count as OK. This inference could not be made acceptable by adding an arbitrary e-node connecting "P" as explainer with "Q" as explanandum, since this e-node would not be OK. Nor would it help to add two e-nodes, one with explainers "P" and "P if and only if Q" with explanandum "Q" and the other with explainers "Q" and "P if and only if Q" with explanandum "P", since the resulting e-structure is circular.

Our example of inference to the best explanation is now clearly acceptable. When you infer that Albert says Jack is a philosopher because he believes it and wants you to know, the main e-node has as its explainers "Albert believes Jack is a philosopher" and "Albert wants you to kow whether Jack is a philosopher." Its explanandum is "Albert says that Jack is a philosopher." Since you already accept the explanandum, it is OK. Since the explanandum of that e-node is OK, the explainers are also OK. Finally, the e-node itself is OK, since it is intelligible as an instance of a default principle which you accept, namely, "The belief that something is so plus the desire to tell someone whether it is so can lead one to say that it is so."

When several acceptable e-structures compete (i.e. have conflicting elements), a particular e-structure can be inferred only if it is the best of the competing acceptable e-structures. We are left with the need for a way of evaluating the e-structures e.g. by counting the amount of change each involves and comparing it with the amount of coherence it brings to your overall view. We must leave for further discussion whether we should count the total number of new things accepted, the total number of new e-nodes accepted, or something else.

In any event, this last approach does not take conservatism to be just a matter of minimizing the total number of new beliefs. You also need to consider which beliefs in your projected modified view were previously accepted. In assessing an e-structure, you also need to note which of its elements are things you already accept and you need to consider whether the acceptability of those items flows over to all the elements of that e-structure in accordance with the principle just given.

Simple Plans

We conclude by considering a very simple case of practical reasoning. You want to raise your arm, where that is something that is immediately within your power. So, you decide to raise your arm. We are supposing that this involves the acceptance of the following explanation: "I will raise my arm because of my decision to raise my arm."

Here there are the following new beliefs: (1) "I will raise my arm," (2) "I decide to raise my arm," (3) (1) because of (2). (1) requires an explanation and an explanation of (1) is accepted, namely, (3). Let us suppose for the moment that we do not have to worry about (3). Either we can suppose that (3) does not require an explanation, because raising your arm is something you take to be immediately within your power. Or, we can suppose that (3) requires an explanation and is explained by something already accepted, namely that raising your arm is immediately within your power. We will come back to this below.

What about (2), "I decide to raise my arm"? Since you have a reason for deciding to raise your arm, we could suppose that your plan involves the acceptance of the idea that you decide for that reason. For example, your plan includes the thought that, because of your desire to raise your arm, you accept this very plan, which involves deciding to raise your arm, which leads to your raising your arm.

Your plan has to include a reference to the reasons for it in order to allow you to allow the plan to be abandoned if the reasons for the plan are no longer applicable. Consider a case in which you adopt a complex plan in order to obtain a goal G. You plan to do M1, which will put you in a position to do M2, which will put you in a position to do M3, and so forth, so that you are in a position to do M11, which will get you G. While in the midst of carrying out this plan, as you are doing M3, you learn that doing M11 will not get you G after all. At this point, you want to be able to abandon the whole plan so that you no longer intend to do M4, M5, and so forth. Just how you are able to do this is something we must eventually consider, but it is clear that you will be able to abandon these intermediate actions only if you keep a record as to why you are undertaking them.

Another reason to record that a plan is aimed at satisfying a particular desire, is to have a way to prevent that desire from leading to the development of other plans designed to satisfy that desire. Once you have a plan to attain a certain goal, you do not have to look for another way to attain that goal!

So, it seems that in a very simple case of deciding to raise your arm you accept a rather complex plan: "In order to satisfy my desire to raise my arm, I am led to adopt this plan, which involves my deciding to raise my arm, which leads to my raising my arm." Consider the various things this involves. First, there is the information that you desire to raise your arm. That (we may suppose) is something you already accept. (We discuss below how your acceptance of that belief might satisfy the principles of change in view.) Second, there is your recognition that you adopt the whole plan. You take your adoption of the whole plan to be explained by your desire to raise your arm. Third, there is this assignment of an explanatory link, the thought that your desire leads you to adopt the plan. We can suppose that that connection is immediately intelligible and therefore needs no further explanation. Fourth, there is the thought that you decide to raise your arm. You take that to be explained by your adopting the plan. Fifth, there is this last explanatory connection between your adopting this whole plan and the decision to raise your arm. The existence of that connection needs no explanation, since the adoption of that decision is obviously part of the plan. Sixth, there is the thought that you do raise your arm. You take your raising your arm to be explained by your decision to raise it. Finally, seventh, there is the explanatory connection between your decision to raise your arm and the fact that you raise your arm. This needs no explanation, because it is obvious to you how the decision leads to the raising.

Summary

In changing your beliefs and your plans, you accept not only simple beliefs and plans but also implications among these simple beliefs and plans, explanations of them, as well as implications among and explanations of implications and explanations. In considering whether a change in view is even minimally acceptable, it is necessary to keep track of which of its elements are already accepted and whether acceptability can flow from these elements to the whole of the proposed new e-structure. It may be that competing e-structures can be judged in part on the basis of counting the changes they involve, but there is no general requirement that the number of new items needing no explanation must be no greater than the number of new explanations.

On another occasion we will extend this analysis to include cases in which change in view involves giving up something previously accepted.

Bibliography

Carroll, L, (1956) What the Tortoise Said to Achilles. *The World of Mathematics, Volume 4,* edited by Newman, J. R. New York, Simon and Schuster. Pp. 2402-2405.

Cullingford, R. E., Harman, G., Bienkowski, M, and Salem, K. (1985). Without Logic or Justification: Realistic Belief Revision, *Proceedings of the National Academy of Sciences Workshop on Artificial*

Intelligence and Distributed Problem Solving. Washington, D. C.: National Academy of Sciences.

Harman, G. (1973) *Thought* Princeton, New Jersey. Princeton University Press.

Harman, G. (1986). *Change In View: Principles of Reasoning.* Cambridge, Massachusetts. M.I.T. Press.

Harman, G., Cullingford, R. E., Bienkowski, M. A., Salem, K., and Pratt, I. (1986) Default Defeaters in Explanation-Based Reasoning, *The Eighth Annual Conference of the Cognitive Science Society* Amherst, Masachusetts, Lawrence Erlbaum, pp. 283-291.

A Production System Model of Causality Judgment

David R.Shanks

MRC Applied Psychology Unit, Cambridge

and

Susan M.Pearson
University of Cambridge

1. Introduction

Building an internal representation of the causal structure of the world is a critically important cognitive ability. Both in order to understand the relationships between events in the external world, and in order to control events in the world, it is necessary to be able to detect causality and to be able to perform causal reasoning. In psychology this ability has traditionally been seen as of considerable importance in understanding such diverse areas as perception (e.g. Michotte, 1963), decision-making (e.g. Nisbett and Ross, 1980) and psychopathology (e.g. Seligman, 1975). Recently too, researchers studying human-computer interaction have come to recognise the importance of causal knowledge (e.g. Lewis, 1986).

In fact such has been the concentration of effort in understanding causality judgment, that it is now possible to construct a fairly complete model of causality judgment: not only can the processes involved be described, but it is also possible to specify in reasonable detail what the form of representation is likely to be. In order to do this, we have chosen to use a production system architecture to capture the principal features of the human causality judgment mechanism.

Because the empirical studies are so important in determining what features a theory of causality judgment must have, we begin by describing three experiments which illustrate the importance of certain features of causal situations in determining the formation of causal knowledge. Then a recent model of causality judgment (Shanks and Dickinson, 1987) is described, followed by some extensions of this theory which cover the way causal knowledge might be represented. Finally, an implementation of this theory in terms of a production system is presented.

2. Contiguity and Contingency

It has been recognised at least since the time of Hume that judgments of causality depend on close temporal and spatial contiguity between the target cause and the effect. Although there have been some tests of this in causality judgment (e.g. Wasserman and Neunaber, 1986), no parametric data have ever been presented. The first experiment attempts to see whether subjects' judgments of the extent to which an action causes an outcome are reduced when delays of 4, 8 or 16 sec are inserted between the action and the outcome, relative to a condition in which there is no delay.

2.1. Experiment 1

In this experiment subjects were each given eight causality judgment problems in which they were required to judge the extent to which an action (pressing the space bar on a computer keyboard) caused an outcome (the flashing of a triangle for 0.1 sec) to occur on the video screen. Each condition lasted for 2 min and was divided into 1 sec time intervals. In the four experimental conditions, if the action occurred during a particular 1 sec interval, then the outcome followed the action with probability 0.75 and never occurred independently of the action. These four experimental conditions differed in the degree of temporal contiguity between the action and the outcome. Each time an action was performed and an outcome was programmed to follow it, a delay of either 0, 4, 8 or 16 sec was inserted before the outcome. During the delay the schedule proceded normally, so that further actions could set up further outcomes. Thus for these four conditions, the probability of the outcome given the action was constant; what differed was the temporal interval before this outcome occurred.

The remaining four conditions were all control conditions. Each experimental condition was immediately followed by a control condition in which the pattern of outcomes that occurred in the experimental condition was played back to the subjects independently of their actions. Thus the temporal distribution and frequency of outcomes is matched in the experimental and control conditions. This is important because if the number of actions (and hence the number of outcomes) differed across the experimental conditions, then this would be confounded with any differences in the subjects' judgments of the extent to which the action caused the outcome. The conditions were presented in pairs consisting of an experimental condition followed by its control, but the order of pairs of conditions with respect to the action-outcome delay was random.

If people are sensitive to contiguity in their causality judgments, we would expect judgments in the experimental conditions to be reduced as the delay is increased; the control conditions provide the baseline against which the judgment in each experimental condition can be assessed.

The subjects were 16 students who were tested individually. The instructions given to the subjects at the beginning of the experiment were similar to those used in the experiments decrsibed by Wasserman, Chatlosh and Neunaber (1983). At the end of each condition the subjects were asked to make a rating of the extent to which pressing the space bar caused the triangle to flash, using a scale from 0 to 100. 100 indicated that pressing the space bar always caused the triangle to light up, and zero indicated that pressing the space bar had no effect on whether or not the triangle lit up. After typing in a number, the next problem was presented.

2.2. Results

The principal results, the judgments of causality, are shown in Table 1. As the table shows, judgments were substantially reduced by increasing the delay between the action and the outcome from 0 to 4, 8 and 16 sec. An analysis of variance found a reliable difference between the experimental and control conditions, $F(1,15) = 19.63$, a reliable overall effect of the delay, $F(3,45) = 7.75$, and a significant interaction, $F(3,45) = 7.80$. Individual tests found that there were significant differences between the experimental and control conditions at the 0 and 4 sec delays, t's$(15) = 6.37$ and 1.90, but no differences at the 8 and 16 sec delays, t's$(15) < 1.06$.

These results confirm that the occurrence of a temporal delay between an action and an outcome can decrease judgments of the extent to which the action caused the outcome. Clearly, accounting for this sensitivity to contiguity is an essential requirement for any theory of causality judgment.

3. Sensitivity to Contingency

The preceding experiment shows that the degree of contiguity between the cause and the effect does affect causality judgments. Consider the difference between the experimental and control conditions with no delay between the action and the outcome. Exactly the same pattern of outcomes occurred in these two conditions, but in the control condition this pattern was non-contingent on the pattern of actions. The notion of the contingency between the action and the outcome is a second important determinant of the formation of causal knowledge.

It is not hard to see that contiguity alone is insufficient to account for causality judgments. Imagine the two conditions illustrated in Figure 1 in which the probability of the outcome given an action, $P(O/A)$, is the same, say 0.75. If the probability of the outcome given no action, $P(O/-A)$, is zero [panel (a)], as was the case in the experimental conditions of the preceding experiment, there is little difficulty in detecting the causal relationship provided that the action-outcome delay is

211

Table 1
Results of Experiment 1

Mean judgements of causality in the Control and Experimental Conditions at each of the four action-outcome delays.

	Delay (secs)			
	0	4	8	16
Experimental	69.8	33.8	34.7	22.7
Control	22.4	19.7	26.8	19.3

Table 3
Results of Experiment 3

Mean judgements and mean actual contingency (dP x 100) for each of the conditions.

Condition	50/0	50/50	50/50(S)
Mean Judgement	49.0	20.9	32.4
Mean dP x 100	48.3	0.6	-0.1

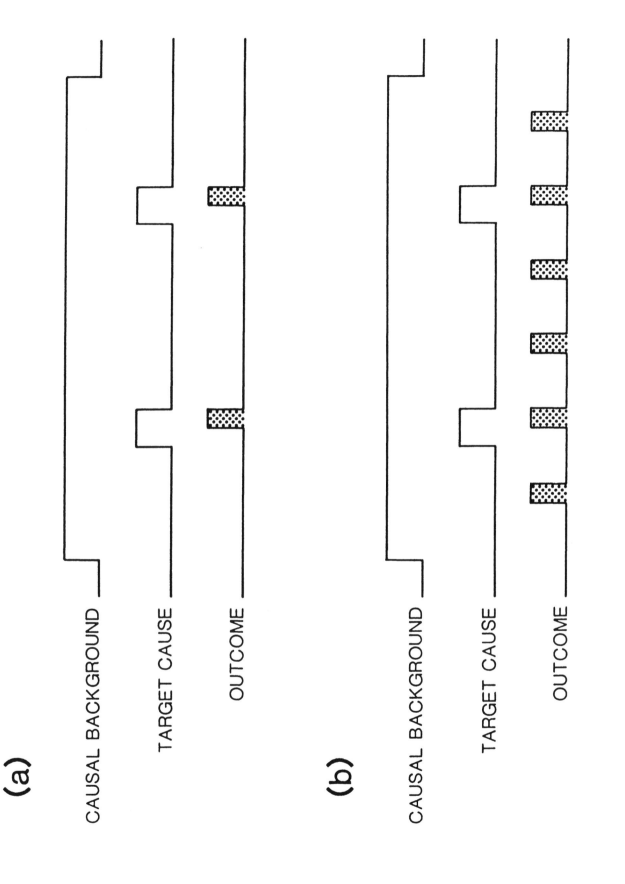

Figure 1

213

small. But suppose that in the second condition P(O/-A) is the same as P(O/A), as shown in panel (b). These two conditions have the same degree of temporal contiguity between the action and the outcome, but in the second condition the outcome is just as likely to occur independently of any action as in its presence. That is, in this second condition, the action and the outcome are noncontingent: the action does not cause the outcome, and hence subjects should judge that there is no causal relation. This is in fact exactly what happens (e.g. Wasserman et al., 1983). Clearly, sensitivity to the contingency between the action and the outcome is independent of the degree of contiguity between them. Traditionally, contingency has been defined statistically by dP, which is the difference between P(O/A) and P(O/-A).

Because this sensitivity to contingency has been a crucial factor in the formulation of theories of causality judgment, some further data will be described on this issue. It is not just important to look at the degree of sensitivity alone; for a variety of reasons which will become apparent it is also profitable to look at the learning curves for different contingencies. The following experiment therefore examines the acquisition functions for causality judgments under different contingencies, that is, the way in which judgments change as more and more information about the causal relationship is provided.

3.1. Experiment 2

In this experiment subjects were each given 4 min experience under each of six different contingencies, and in each condition they were required to make regular judgments of causality during the 4 min period. The contingencies used were as follows, where the first figure refers to P(O/A) x 1000 and the second to P(O/-A) x 1000: in three of the conditions P(O/A) was held constant as P(O/-A) was raised (conditions 875/125, 875/500 and 875/875); the last of these conditions was in fact a noncontingent one, and so another noncontingent condition was included for which the frequency of the outcome was lower, 125/125. Finally, judgments for the 875/875 condition could be compared with judgments for two other conditions for which P(O/-A) was held constant while P(O/A) was raised (conditions 125/875 and 500/875). In these latter conditions dP is in fact negative: the action actually prevents the outcome from occurring or reduces its likelihood.

There was no contiguity manipulation in this experiment. If the subject responded during a particular 1 sec interval, then the outcome occurred at the end of that interval with probability P(O/A). If there was no response during the interval, the outcome occurred at the end of the interval according to P(O/-A).

The subjects, who were 16 another members of the APU panel, were asked to make causality judgments after 10, 20, 30, 60, 90, 120, 150, 180, 210 and 240 sec in each condition. Essentially, the instructions were the same as in the preceding experiment except that here a rating scale going from -100 to +100 was used. Negative judgments were indications that the action to some extent prevented the outcome.

3.2. Results

Since the events on each trial were determined by a software random number procedure, it is important to check that the actual contingency experienced by each subject was close to the nominal contingency. The deviations of dP from the nominal contingency were very minor, and more importantly, there was no change in dP across trials, F < 1, nor was there an interaction between trials and conditions, F(45,661) = 1.07. The actual contingency was, of course, different across the conditions, F(5,75) = 586.88. The actual contingencies (dP x 100) calculated across the whole 4 min period for each condition were 76.6, 37.5, 1.4, -1.0, -34.8 and -74.5 for conditions 875/125, 875/500, 875/875, 125/125, 500/875 and 125/875, respectively.

An indication of the extent to which P(O/A) and P(O/-A) were both sampled is given by the rate of responding in each condition. Out of a possible 240 responses, the mean number of times the subjects pressed the space-bar was 106.5, indicating that the two probabilities were almost

Table 2
Results of Experiment 2

Mean actual judgements of causality across trials under each contingency. The two numbers describing each condition refer to P(O/A) x 1000 and P(O/-A) x 1000 respectively. The predicted figures are from the simulation described in the final section.

					Trial					
	10	20	30	60	90	120	150	180	210	240
Condition										
875/125 actual	**40.9**	**58.1**	**62.5**	**63.7**	**64.7**	**70.3**	**72.4**	**76.8**	**71.6**	**73.3**
875/125 predicted	34.3	52.5	60.2	68.1	68.2	68.8	69.5	67.4	66.2	66.9
875/500 actual	**26.4**	**36.9**	**33.4**	**41.9**	**39.7**	**41.3**	**34.1**	**43.8**	**48.1**	**44.6**
875/500 predicted	25.7	39.0	44.5	45.2	45.6	45.5	46.5	47.0	45.9	46.3
875/875 actual	**6.8**	**25.6**	**13.9**	**14.8**	**6.0**	**7.3**	**1.1**	**3.3**	**-0.6**	**0.6**
875/875 predicted	15.3	20.3	17.4	8.5	2.2	0.8	0.4	0.2	0.9	0.5
125/125 actual	**-6.1**	**-27.0**	**-28.9**	**-34.6**	**-39.4**	**-41.0**	**-38.4**	**-31.1**	**-37.9**	**-36.6**
125/125 predicted	1.7	0.8	0.5	0.5	-0.1	0.5	0.2	-0.3	0.0	0.1
500/875 actual	**-12.4**	**-19.0**	**-17.1**	**-13.4**	**-17.1**	**-23.0**	**-18.4**	**-24.4**	**-22.1**	**-26.4**
500/875 predicted	-2.7	-12.4	-22.2	-38.0	-42.7	-45.4	-45.9	-45.7	-44.4	-46.5
125/875 actual	**-30.6**	**-47.2**	**-55.3**	**-67.1**	**-64.1**	**-69.8**	**-77.2**	**-72.9**	**-60.0**	**-64.6**
125/875 predicted	-17.2	-33.4	-43.8	-61.3	-65.9	-67.3	-68.3	-67.9	-68.0	-68.6

equally sampled. There was no difference in the number of responses in each condition, F< 1, nor was there any trials by conditions interaction, F< 1.

Table 2 shows the mean judgments of contingency for each condition across trials. Just taking the terminal judgments in each condition, we can see that judgments were reduced from condition 875/125 to condition 875/500 to condition 875/875 as $P(O/-A)$ was increased while $P(O/A)$ was held constant. Similarly, judgments were increased (became less negative) from condition 125/875 to condition 500/875 to condition 875/875 as $P(O/A)$ was increased while $P(O/-A)$ was held constant. In addition, judgments were biased in the noncontingent conditions by the overall frequency of the outcome: judgments were substantially greater in condition 875/875 than in condition 125/125 where judgments were in fact strongly negative. This confirms previous claims (e.g. Alloy and Abramson, 1979) that in noncontingent situations judgments depend on the rate of occurrence of the outcome.

Statistically, there was a significant difference between the conditions, $F(5,75)= 39.04$, and a significant trials x conditions interaction, $F(45,675)= 2.54$. There was no overall effect of trials, F< 1. The main feature of the results, however, is the nature of the changes across trials under each contingency. When there is a positive contingency, judgments increase across trials [comparing the first and last judgments in condition 875/125, $t(15)= 3.52$], and when there is a negative contingency, judgments decrease [comparing the first and last judgments in condition 125/875, $t(15)= 3.71$]. These changes, furthermore, appear to be negatively accelerated as the judgment approaches asymptote. For the intermediate contingencies (875/500 and 500/875) these changes are less dramatic.

The experiment demonstrates two important features of causality judgments. The first is that people are highly sensitive to the actual degree of contingency as measured by the normative metric dP. This applies both to positive and to negative contingencies: people make equally reliable judgments when the action prevents the outcome as when it causes it. The second feature of the results is that these accurate judgments are derived by an incremental (or decremental) learning process. The functions represented in Table 2 are in fact simply learning curves; the subjects' judgments increase across trials under a positive contingency, towards the asymptote of the actual contingency, and decrease across trials under a negative contingency.

4. Selection Amongst Potential Causes

The two experiments described so far have illustrated two factors that have strong effects on the formation of causal knowledge, namely contingency and contiguity. The next experiment looks at a further factor, which can again have a potent effect on judgments. This is the status of other potential causes which are present at the same time as the target cause (the action in this case).

As Figure 1 illustrates, a causal sequence occurs in the context of what we might call a 'causal background', that is, a set of background stimuli which are constantly present in that situation and which represent a set of potential alternative causes of the outcome. The target cause, therefore, is not occurring in isolation, but is in competition with this background. When there is a strong contingency between the target cause and the outcome, the background is unlikely to offer much competition since the target cause is so much more informative about when the outcome will occur. But when the contingency is degraded, the causal background takes on greater significance. Referring again to Figure 1, occurrences of the outcome in the absence of the target cause in panel (b) must be attributed to the background. In terms of the associative theory of Shanks and Dickinson (1987), the background in this case will become associated with the outcome.

The significance of this association between the background and the outcome is that it suggests an explanation for sensitivity to contingency. Consider the subsequent pairings of the action and the outcome after the background has already become associated with the outcome: these outcomes can be attributed to either the background or to the action. But since the background is already associated with the outcome, surely subsequent occurrences of the outcome are now more likely to be attributed to the background than to the action, relative to the situation in which the background

216

has not become associated with the outcome. If these outcomes are less likely to be attributed to the action, then judgments of the extent to which the action caused the outcome will be reduced; this is exactly the finding when $P(O/-A)$ is increased.

Such an account of sensitivity to contingency assumes a crucial role for the causal background. It also implies that causal attribution is selective: selections will be made amongst potential causes in terms of how well established they already are as causes. The next experiment attempts to provide support for this claim.

4.1. Experiment 3

The analysis described above proposes that the impact of outcomes occurring in the absence of the target cause comes about because such outcomes are attributed to the background. But suppose this could be prevented: what would happen if such outcomes were attributed to some other event rather than to the background ? By the above analysis, if this happened then the background would not become associated with the outcome and hence would not be in strong competition with the action when subsequent outcomes occurred in the presence of the action. Thus any procedure which prevents outcomes occurring in the absence of the action from being attributed to the background should elevate judgments about the action. In this experiment a straightforward way of doing this is employed: every outcome occurring in the absence of the action is preceded by another stimulus, called the 'signal', which only occurs on those trials.

The experiment involves the same basic procedure as in the previous experiments. The subjects (24 members of the APU subject panel) were each given three conditions, presented in a random order. In one condition (50/0) there was a positive contingency between the action and the outcome: $P(O/A)$ was 0.5 while $P(O/-A)$ was zero. In the second condition (50/50) the contingency was reduced to zero by increasing $P(O/-A)$ to 0.5. The critical condition was the one in which the signal was presented [50/50 (S)]. The contingency was identical to that in the noncontingent 50/50 condition, but in this case all of the outcomes occurring in the absence of the action were preceded by the signal, which was a short tone. For half of the subjects the duration of the tone was 0.5 sec, while for the other half it was 0.75 sec. If no response occurred during a particular 1 sec interval, and the outcome was scheduled to occur at the end of that interval, then the outcome was delayed at the end of the interval for an amount of time equal to the duration of the tone; this occurred for all of the conditions. The difference in the tone's duration had no effect on the results, which are therefore collapsed across this factor.

4.2. Results

The subjects responded a mean of 55.1 times in condition 50/0, 48.2 times in condition 50/50, and 62.4 times in condition 50/50 (S). The difference between the numbers of responses in the latter two conditions was in fact reliable, Wilcoxon $W = 62.5$, $p < 0.05$. Table 3 presents the main results of the experiment.

The actual contingency, as expected, was close to 0.5 in condition 50/0 and close to zero in conditions 50/50 and 50/50 (S). Judgments were reduced in condition 50/50 relative to condition 50/0 by increasing $P(O/-A)$, $W = 50$, $p < 0.05$; this corroborates the findings of the previous two experiments that contingency is a strong determinant of causality judgments. The critical result, however, is the greater mean judgment in condition 50/50 (S) than in condition 50/50, $W = 48$, $p < 0.02$. Thus signaling those outcomes that occur in the absence of the action has the effect of elevating judgments, and hence of at least partially reversing the effect of those outcomes on causality judgments.

5. A Production System Model of Causality Judgment

If a temporal delay is interpolated between an action and an outcome, judgments of the extent to which the action is the cause of the outcome are reliably reduced, even though the delay might have no effect on the actual probability of the outcome. In Experiment 1 a short delay between the action and the outcome substantially reduced judgments. In the second experiment it was found that causality judgments follow growth functions whereby under a positive contingency judgments are incremented trial-by-trial towards an asymptote, whilst under a negative contingency they are decremented towards the asymptote. Finally, in the last experiment the importance of the causal background was illustrated in a situation where causal selection on the basis of the prior association between the background and the outcome could be prevented.

The account of causality judgment that will now be described proposes a crucial role for such associations. According to associative theories of learning, the occurrence of an outcome increments the associative strength of stimuli present at that time. A crucial aspect of such theories is the role they assign to the causal background, that set of stimuli which is always present in that context. In the model described by Shanks and Dickinson (1987), an outcome occurring in the presence of the target cause increments the associative strength of the compound of the target cause and the background according to the equation:

$$d V_{AB} = \alpha_{AB} \beta (\lambda - V_{AB}) \qquad (1)$$

where dV_{AB} is the increment in the associative strength of the compound of the action and the background, α_{AB} is a learning rate parameter for the compound, β is a learning rate parameter for the outcome, λ is the asymptote of associative strength and V_{AB} is the associative strength the compound already has. The asymptote λ is usually set to 1.0 on trials on which the outcome occurs and to zero when the outcome does not occur.

If the outcome occurs in the absence of the action, then the associative strength of the background, V_B is incremented according to the equation:

$$d V_B = \alpha_B \beta (\lambda - V_B) \qquad (2)$$

where dV_B is the increment and α_B is a rate parameter for the causal background. At the end of a series of occurrences of the action and the outcome, there will be two associative strengths, V_{AB} and V_B. The causality judgment, J_A, is then based on an inference step in which the difference between these two associative strengths is determined:

$$J_A = V_{AB} - V_B \qquad (3)$$

The model can account for sensitivity to contingency since increasing $P(O/-A)$ will increase V_B and hence reduce judgments. In fact the model can readily reproduce the pattern of acquisition functions seen in the second experiment. Table 2 gives the results of a simulation of the model run under the same contingencies as the actual experiment. Each figure represents the mean from 1000 simulated subjects with the parameters as follows: $\alpha_B = 0.1$, $\alpha_{AB} = 0.2$, β for the outcome= 0.3 and β for the nonoccurrence of the outcome= 0.8, and assuming that V_B and V_{AB} start at zero. All of the main features of the actual results are reproduced, with the exception that the negative judgments in the 125/125 condition do not emerge (see Shanks and Dickinson, 1987, for a discussion of this discrepancy).

A production system model of causality judgment which incorporates the linear operator equations of Shanks and Dickinson (1987) has recently been implemented in the computer language OPS-5 (Brownston, Farrell, Kant and Martin, 1985). It consists of two principal elements, namely the production rules and a working memory. The production rules consist of simple if...then... statements which operate when their conditions are satisfied by the contents of working memory. There will be two production rules relevant to causality judgment. The first will specify that if the causal background is represented in working memory, then the representation of the outcome should be placed in working memory, and equation (2) should be selected. This production rule has a measure of belief associated with it, which is the current value of V_B. The second rule states that if the action and the background are in working memory, then the representation of the outcome should be put in working memory and equation (1) should be selected. This rule also has a measure of belief associated with it, this time determined by V_{AB}.

Notice that the production rules do not incorporate the concept of causation. When the subject comes to make a causality judgment, the model proposes that this comes about by an inferential process which involves propostional knowledge from other sources as well as what is incorporated in the production rules. In causality judgment experiments, of course, it is assumed that any extra-experimental knowledge the subjects might bring to the situation will be minimal. In more realistic settings, though, it seems highly likely that prior knowledge about causal relationships might have a strong effect on causality judgments in particular circumstances. In terms of the model, such knowledge is brought in to the inference represented by equation (3).

How can such a model account for contiguity effects ? So far, the model does not specify what constitutes a co-occurrence of the action and the outcome [in which case equation (1) applies] and what constitutes an occurrence of the outcome in the absence of the action [in which case equation (2) applies]. But the crucial determinant is likely to be simply whether or not the action is still represented in working memory, above a certain threshold, at the time of the outcome. If we assume that the action is fully represented immediately after its occurrence but then its representation decays, then this can be captured in the model by a reduction in α_{AB}, the salience of the action-background compound, as time elapses from the occurrence of the action. Although there is no obvious reason to support one decay function over another, an exponential decay curve in the model gives predictions for contiguity effects that approximate those found experimentally. If α is being reduced, then the increment in the associative strength dV_{AB} will likewise be reduced as the action-outcome interval increases. Note that this does not affect the associative strength of the background, V_B: in accordance with the results of Shanks and Dickinson (1987); a reduction in the salience of the action brought about by an interval between the action and the outcome will reduce judgments of causality by reducing V_{AB}, but will leave V_B unaffected.

Obviously, at some point the action will no longer be represented in working memory and therefore an occurrence of the outcome must then increment V_B and not V_{AB}. A threshold below which the level of representation of the action in working memory causes equation (3) to be selected instead of equation (2) can readily be incorporated into the model.

The model can also explain the signaling effect of Experiment 3 because the introduction of the signal means that outcomes occurring in the absence of the action do not increment V_B; they increment the associative strength of the signal-background compound, which does not figure in equation (3).

In summary, the model in its present form is capable of accounting for the main features of causality judgment: first, it can explain sensitivity to contingency; secondly, it can cover contiguity effects; and thirdly, it can account for the selectional effects seen in Experiment 3. Because it is specified in the precise terms of a production system, it makes concrete predictions about a variety of causality judgment situations and can therefore readily be tested.

References

Alloy,L.B. & Abramson,L.Y. Judgment of contingency in depressed and nondepressed students: Sadder but wiser? Journal of Experimental Psychology: General, 1979, 108, 441-485.

Brownston,L., Farrell,R., Kant,E. & Martin,N. Programming Expert Systems in OPS-5: An

Introduction to Rule-Based Programming. Reading, Mass.: Addison-Wesley, 1985.

Lewis,C. Understanding what's happening in system interactions. In D.A.Norman & S.W.Draper (Eds.), User Centered System Design: New Perspectives on Human-Computer Interaction. Hillsdale, N.J.: Erlbaum, 1986.

Michotte,A. The Perception of Causality. London: Methuen, 1963.

Nisbett,R.E. & Ross,L. Human Inference: Strategies and Shortcomings of Social Judgment. Englewood Cliffs: Prentice-Hall, 1980.

Seligman, M.E.P. Helplessness. San Francisco: W.H.Freeman, 1975.

Shanks,D.R. & Dickinson,A. Associative accounts of causality judgment. In G.H.Bower (Ed.), The Psychology of Learning and Motivation (Vol.21). New York: Academic Press, 1987.

Wasserman,E.A., Chatlosh,D.L. & Neunaber,D.J. Perception of causal relations in humans: Factors affecting judgments of response-outcome contingencies under free-operant procedures. Learning and Motivation, 1983, 14, 406-432.

Wasserman,E.A. & Neunaber,D.J. College students' responding to and rating of contingency relations: The role of temporal contiguity. Journal of the Experimental Analysis of Behavior, 1986, 46, 15-35.

PREDICTIVE VERSUS DIAGNOSTIC REASONING IN THE APPLICATION OF BIOMEDICAL KNOWLEDGE

VIMLA L. PATEL
MCGILL UNIVERSITY

DAVID A. EVANS
CARNEGIE-MELLON UNIVERSITY

ANOOP CHAWLA
MCGILL UNIVERSITY

ABSTRACT: Clinical problem solving involves both diagnostic and predictive reasoning. Diagnostic reasoning is characterized by inference from observations to hypotheses; predictive reasoning, by inference from hypotheses to observations. We investigate the use of such strategies by medical students at three levels of training in explaining the underlying pathophysiology of a clinical case. Our results show that without a sound, pre-existing disease classification, the use of basic biomedical knowledge interferes with diagnostic reasoning; however, with sound classification, biomedical knowledge facilitates both diagnostic and predictive reasoning.

INTRODUCTION:
BASIC SCIENCE VERSUS CLINICAL KNOWLEDGE

It is generally believed that knowledge of basic science is required for competence in the practice of clinical medicine. Yet the precise role of basic science in *diagnostic reasoning* has remained controversial (Clancey, in press; Patil, Szolovitz & Schwartz, 1984). Recent evidence suggests that basic science and clinical problem solving may represent domains of knowledge with very limited overlap, as judged by the inability of students and clinicians to integrate basic science knowledge in the context of diagnostic explanation tasks (Patel & Groen, 1986; Patel, Arocha & Groen, 1986). Indeed, some investigators have proposed that the uses of clinical and basic science knowledge involve *qualitatively* different types of reasoning (Patel, Groen & Scott, in press; Evans, Gadd & Pople, forthcoming).

One source of evidence on the relation between basic science and clinical knowledge derives from analysis of pathophysiological explanations offered by clinicians to describe medical problems (e.g., Patil & Szolovitz, 1981). In a recent study, Patel and Groen (1986) found that expert cardiologists who accurately diagnosed a medical case manifested a pattern of inference in which clinical findings were progressively eliminated by subsumption under a diagnostic hypothesis. This pattern of monotonic reduction of uncertainty was termed *forward chaining.* Such subjects generated very few basic science descriptions in explaining the underlying pathophysiology of the problem. By contrast, cardiologists who failed to diagnose the problem generated more intermediate steps, including more basic science descriptions. While they, too, gave evidence of forward chaining, their performance was characterized by episodes of *backward chaining,* in which secondary hypotheses were produced, effectively *increasing* the number of variables that required explanations. Since the hypotheses that led to greater uncertainty were derived from basic science knowledge, it is tempting to conclude that clinical and scientific knowledge are not well integrated for problem-solving tasks.

What explanations can we offer to account for such phenomena? In routine medical problem solving, the most expeditious path towards a diagnosis leads through a chain of clinical associations, resulting in progressive elaboration of constraints (specific diagnostic hypotheses) and incremental reduction of the problem space (the clinical findings that must be explained). When a patient presents with a difficult, multi-system problem that does not conform neatly to a familiar pattern, the physician must attempt to partition the problem space by grouping findings according to their interrelations. It is in this context that the physician may rely on knowledge of basic science (Joseph & Patel, 1986). Similarly, a clinician who lacks the domain-specific knowledge or necessary clinical experience to reason diagnostically *via* clinical associations could be expected to resort to pathophysiological models of systemic disease processes to predict associations of findings.

Given the different demands of clinical practice and basic biomedical research, it is reasonable to assume that experts will show preferences in the kinds of knowledge they employ in solving a case. Patel, Arocha and Groen (1986) compared endocrine researchers and endocrine clinicians presented with pathophysiological explanation tasks. The researchers generated very detailed basic science explanations while the practitioners used principally clinical inferences to explain the underlying problem. The authors concluded that the basis of explanation was a function of the daily tasks the specialists are required to perform. Practitioners typically see a great many patients in a short period of time and are required to classify problems in order to recommend treatment. by contrast, researchers typically must attempt to understand and elucidate the mechanisms of specific biomedical phenomena.

Medical students, unlike practicing physicians or researchers, have only partially developed knowledge of basic science and clinical phenomena. In most medical schools, students spend their first two years taking basic science courses such as anatomy, biochemistry and physiology, and only in their second two years begin to focus on problems in clinical medicine. It is reasonable to suppose that, in problem-solving tasks, students' use of basic science knowledge will depend on the level of medical training, task demand, and the degree of uncertainty inherent in the problem. This paper reports on one investigation of this intuition.

HYPOTHESES ON PROBLEM SOLVING

An obvious hypothesis is that efficient causal reasoning about a disease process requires management of cognitive load -- maximizing the constraints on the problem space and minimizing the number of variables (uncertainty) that must be held in memory. Given that the task is to associate clinical manifestations to a disease process, there are *a priori* two ways in which causal networks can be build, *viz.*, by reasoning *predictively* from hypotheses to manifestations, or *diagnostically* from manifestations to hypotheses. In the sense that predictive reasoning involves the use of generalizations associated with diagnoses to identify candidates' specific details, in a case it can be regarded as *deductive*. Similarly, in the sense that diagnostic reasoning involves the use of particular details to suggest the appropriate generalized "diagnoses", it can be regarded as *inductive*.

In predictive reasoning, the problem space is controlled because one entertains a hypothesis of sufficient power and generalization to account for many possible manifestations. Uncertainty is controlled because inference is limited to what is entailed by the hypothesis. For example, if a physician *assumes* that a patient has bacterial endocarditis, he or shoe can know that an infectious process is involved and, hence, can predict (and account for) a finding of fever. Clearly, for predictive reasoning to be effective in problem solving, knowledge of diagnoses and their associated manifestations is important.

In diagnostic reasoning, it is much more difficult to constrain the problem space and eliminate uncertainty. For example, one could link fever with an infectious process, diagnostically, but it would be impossible to say which infectious process was involved without considering combinations of other findings. Furthermore, fever is not caused by infectious processes, but occurs in inflammatory disorders and certain cancers. In reasoning from specific manifestations to possible hypotheses, one introduces numerous alternatives that must be reconciled against one another. A physician who cannot quickly classify findings according to the most likely diagnoses that would account for them will be overwhelmed with the problem of managing information and inference.

We would suggest that physicians, in fact, use relatively simple classification schemata to transform the problem space of individual findings to one of a small number of diagnostic hypotheses and clustered findings. Without clinical knowledge as a basis for classification, knowledge of physiological mechanisms and scientific principles will be used to drive inferences and associate observations. Our characterization of reasoning in the domain of medicine, thus,

222

underscores the epistemological complexity of the task. This suggests that the management of uncertainty will be the principal component of effective performance. It is, therefore, important to consider structural problems that arise in reasoning under uncertainty. In particular, Henrion (1986) has identified certain phenomena associated with the representation of causal knowledge that lead to *propagation* of uncertainty in inference. We focus on two problems discussed by Henrion, *dependency effects* and *cyclical inferencing*.

Dependency effects can be illustrated with the aid of Figure 1, below. Under the first representation of the relations of observations to hypotheses, on the left, evidence from three different sources A, B, and C together strengthen the hypothesis, H_1. However, if A, B, and C are dependent on some other source, D, as represented on the right, the hypothesis is not strengthened. One method for avoiding this, suggested by Henrion, is to insure independence of evidence.

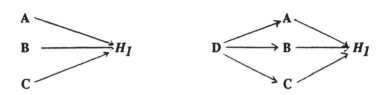

FIGURE 1: DEPENDENCY EFFECTS

Cyclical inferencing occurs when a hypothesis, based on certain specific evidence, is used to account for phenomena, including the evidence that originally gave rise to the hypothesis. Since reasoning in the medical domain involves both predictive and diagnostic strategies - i.e., reasoning from hypothesis to disease manifestations and *vice versa* - there is the danger that, without a means of keeping evidence from predictive and diagnostic sources separate, cyclical inferencing can occur. For example, consider the following reasoning, given schematically in Figure 2: *Intravenous drug use can lead to infection, which leads to fever. In a patient with fever and heart murmur, there is a high possibility of bacterial endocarditis. Bacterial endocarditis is associated with intravenous drug users, hence accounts for drug use.* The resulting network leads to propagation of uncertainty.

INTRAVENOUS DRUG USE (A) ⟶ INFECTION (B)
INFECTION (B) ⟶ FEVER (C)
FEVER (C) & HEART MURMUR (D) ⟶ BACTERIAL ENDOCARDITIS (E)
BACTERIAL ENDOCARDITIS (E) ⟶ INTRAVENOUS DRUG USE (A)

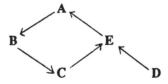

FIGURE 2: CYCLICAL INFERENCING

Henrion argues that this problem arises as a result of failure to maintain separation between the flow of predictive and diagnostic reasoning. We are particularly interested in the interaction of basic biomedical and clinical knowledge with such phenomena. Our approach has been to analyze the protocols of medical students at different levels of training who were asked to perform a diagnostic task. Our hypothesis is that clinical training provides knowledge of appropriate disease classification that enables efficient predictive reasoning, leading to explanations that manifest causal coherence. In terms of reasoning strategies, we hypothesize that classification-guided predictive reasoning will be preferred to diagnostic reasoning; and that in the absence of well-developed classification schemata, some combination of diagnostic reasoning and predictive reasoning from *naive* or *for-the-nonce* classification will be used. *Ad hoc* classification is facilitated, we suggest, by application of basic scientific knowledge edge; and in such instances, we expect to see a compounding of problems of dependency effects and cyclical inferencing. More generally, we hypothesize that any use of *induction* without an adequate basis for *deduction* - such as afforded by well-developed classification schemata - will lead to inefficient, incomplete, and incoherent diagnostic problem solving.

EXPERIMENTAL METHOD

A total of 24 subjects were selected from three levels of medical school training. Level 1 included six students just entering their first year of medical school. Level 2 included six second-year medical students who had completed all basic medical sciences, but had had no clinical work. Level 3 included twelve final-year medical students three months before graduation.

PROCEDURE: The following empirical paradigm was used: We (1) presented the subject with a clinical problem text; (2) obtained a summary protocol of the problem; (3) asked the subject to explain the underlying pathophysiology; (4) asked the subject for a diagnosis; and (5) presented the subject with three basic science texts and asked for an explanation of the clinical problem.

ANALYSIS: We use a combination of propositional and protocol analysis. Our first step is to separate the texts into segments, corresponding to syntactic units, using Winograd's system of clausal analysis (Winograd, 1972). The next step involves the representation of clauses in terms of their propositional content and structure, using the techniques of propositional analysis based on Frederiksen (1975). A detailed analysis of a clinical case identical to that used here is given in Patel & Groen (1986) and Patel & Frederiksen (1984). At a second level of analysis, propositions are linked to form a higher-ordered relational structure (frames). Such structures are representable as the causal networks discussed in the literature on medical artificial intelligence. The links in the causal networks can be converted into production rules. The directionality of such rules is important since causal rules generally lead away from a diagnosis (in a predictive direction) and only conditional rules lead toward one (in a diagnostic direction). Our analysis also uses the notion of *reference frame* of a disease, which aids us in identifying where basic science knowledge is used in reasoning in either direction.

COGNITIVE TASK: The tasks require each subject to read the case and form a representation of the problem. Subsequently they are asked to summarize the problem. This provides insight into the relations among clinical findings that they have considered. As subjects are also asked to explain the underlying pathophysiology of the problem, it is possible to assess the degree to which they can coherently account for the disease process. There are several levels of abstraction from which they can approach this task. For example, they can describe the disease process entirely in terms of clinical or anatomical features (as a static process) or they can explore the perturbations in the physiological processes that result in the patient's presenting condition (a dynamic process). After presenting a pathophysiological account, they are required to make a diagnosis. The text of the clinical problem, acute bacterial endocarditis, is given in Figure 3 and the components of the reference frames are given in Figure 4.

THIS 27-YEAR OLD UNEMPLOYED MALE WAS ADMITTED TO THE EMERGENCY ROOM WITH THE COMPLAINT OF SHAKING CHILLS AND FEVER OF FOUR DAYS DURATION. HE TOOK HIS OWN TEMPERATURE AND WAS RECORDED AT 40°C ON THE MORNING OF ADMISSION. THE FEVER AND CHILLS WERE ACCOMPANIED BY SWEATING AND A FEELING OF PROSTRATION. HE ALSO COMPLAINED OF SHORTNESS OF BREATH WHEN HE TRIED TO CLIMB THE TWO FLIGHTS OF STAIRS IN HIS APARTMENT. FUNCTIONAL ENQUIRY REVEALED A TRANSIENT LOSS OF VISION IN HIS RIGHT EYE WHICH LASTED APPROXIMATELY 45" ON THE DAY BEFORE HIS ADMISSION TO THE EMERGENCY WARD.

PHYSICAL EXAMINATION REVEALED A TOXIC LOOKING YOUNG MAN WHO WAS HAVING A RIGOR. HIS TEMPERATURE WAS 41°C, PULSE 120, BP 110/40. MUCUS MEMBRANES WERE PINK. EXAMINATION OF HIS LIMBS SHOWED PUNCTURE WOUNDS IN HIS LEFT ANTECUBITAL FOSSA. THE PATIENT VOLUNTEERED THAT HE HAD BEEN BITTEN BY A CAT AT A FRIEND'S HOUSE ABOUT A WEEK BEFORE ADMISSION. THERE WERE NO OTHER SKIN FINDINGS. EXAMINATION OF THE CARDIOVASCULAR SYSTEM SHOWED NO JUGULAR VENOUS DISTENSION, PULSE WAS 120 PER MINUTE, REGULAR, EQUAL AN DISYNCRONOUS. THE PULSE WAS ALSO NOTED TO BE COLLAPSING. THE APEX BEAT WAS NOT DISPLACED. AUSCULTATION OF HIS HEART REVEALED A 2/6 EARLY DIASTOLIC MURMUR IN THE AORTIC AREA AND FUNDOSCOPY REVEALED A FLAME-SHAPED HEMORRHAGE IN THE LEFT EYE. THERE WAS NO SPLENOMEGALY. URINALYSIS SHOWED NUMEROUS RED CELLS BUT THERE WERE NO RED CELL CASTS.

FIGURE 3: BACTERIAL ENDOCARDITIS TEXT

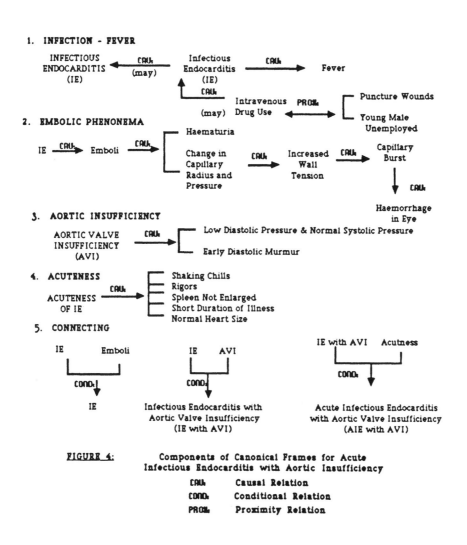

FIGURE 4:   Components of Canonical Frames for Acute
Infectious Endocarditis with Aortic Insufficiency

CAU.   Causal Relation
COND.   Conditional Relation
PROX.   Proximity Relation

225

In the second part of the experiment the subjects are asked to read three basic science texts, which they must subsequently relate to the clinical case. In effect, the subjects are provided with the opportunity at this stage to update their pathophysiological explanations using the basic science principles, abstractions, and concepts presented in the texts. There are several ways in which their explanations can change. The subjects can choose to elaborate and embellish previous characterizations of the disease process with additional detail; they may choose to reject earlier characterizations on the basis of the new information; or they may offer new pathophysiological descriptions, entirely orthogonal to their previous descriptions, resulting from a complete shift in focus. The nature of their revised representations is a function of several factors including (a) their ability to form a coherent representation of the clinical text; (b) their comprehension of the basic science material; and (c) their ability to synthesize and integrate the relevant information from the basic science texts into the context of the clinical problem.

RESULTS AND DISCUSSIONS

The performance of subjects at each level was distinct and quite uniform in terms of diagnostic and predictive reasoning using basic science knowledge in the explanation task. Students at level 1 were rated *poor* on both diagnostic and predictive reasoning; students at level 2 were rated *fair* on both; and students at level 3 were rated *good* on both. Thus, though all groups of subjects showed evidence of both types of reasoning in explaining the patient's underlying problem, reasoning improved in accuracy and coherency with level of training.

DIAGNOSTIC REASONING: It appears that reasoning in a diagnostic direction is frustrated by the application of basic science knowledge *unless* there is a strong classification of hypotheses (diagnoses). The final-year medical students, who have had some clinical experience, have partial classification schemata which assist them both in selecting an initial, accurate diagnostic hypothesis and in predicting selectively, the variables to be tested to confirm the selected hypothesis. This is illustrated in Table 1 by an analysis of a protocol, taken from a final-year student, showing the explanations produced before and after receiving basic science information. Figures 5 and 6 give the causal networks generated from these protocols. There is no evidence of cyclical inferencing or dependency effects in pre-basic science protocols but they are present to a limited extent in post-basic science protocols.

TABLE 1:

PATHOPHYSICAL EXPLANATION PROTOCOLS OF BACTERIAL; ENDOCARDITIS
BY A FINAL YEAR MEDICAL STUDENT STUDENT, WITH AND WITHOUT THE
BASIC SCIENCE TEXT PROVIDED.

CLINICAL TEXT ONLY

Bacteremia resulting in generalized symptoms of illness, i.e., fatigue, chills, fever, sweating. These bacteria accumulated in aortic valve with resultant vegetations and murmur due to ausclation of valve. Emboli from these vegations travel via carotid on left to left retinal artery resulting in temporary blindness and flame hemorrhages due to the occluded blood supply.

WITH CLINICAL AND BASIC SCIENCE TEXTS

1) Bacteria introduced into bloodstream perhaps via vena-puncture. Subsequently release endotoxins which act on bone marrow cells to produce pyrogens. These pyrogens act on the hypothalamus (pre-after) to result in fever. Chills occur due to need to increase body temperature as indicated by the hypothalamus.

2) This bacteremia results in vegetations occuring in aortic valve. The diastolic murmur is due to aortic regurgitation as these vegetations alter the closing of the valve. The murmur radiates along the blood flow tract of the aortic valve which is the left sternal border, where the left ventrical is.

3) Micro emboli are released from these vegetations and flow in the atrial circulation and eventually occlude capillaries going to the retina resulting in transient blindness. This increased capillary pressure due to occlusion resulted in retinal hemorrhage and thus are seen on fundoscopic examination.

226

It is clear that the subject inductively decides that the patient has bacteremia and the bacteremia has affected the aortic heart valve (Figure 5). In order to account for these two facets of the overall disease, certain predictions have to be made. The subject confirms that it is bacteremia because the patient has fatigue, chills, fever, and sweating, which are the findings one would expect. In order to confirm aortic valve involvement, the subject explains heart murmur and subsequent hemorrhage and blindness. The classification of the disease presumably directed his or here selection of particular variables (findings), whose presence can be taken as confirmatory evidence that the prediction was valid.

After the basic science text has been read, the subject once again uses a classification schema to provide the necessary basic science knowledge required to explain the findings (Figure 6). Here, a deeper account of the disease process is offered. There is no change in either the diagnostic explanation or in the selection of predictions in confirming the diagnosis. Basic scientific information is used to provide detailed physiological mechanisms that can account for the existence of the selected variables identified earlier. There is an effect of global coherence in both explanation protocols, due in part to the rich connectedness of details. Both causal networks resulted in only one diagnostic component each, corresponding to the diagnosis, and produced no disjoint components.

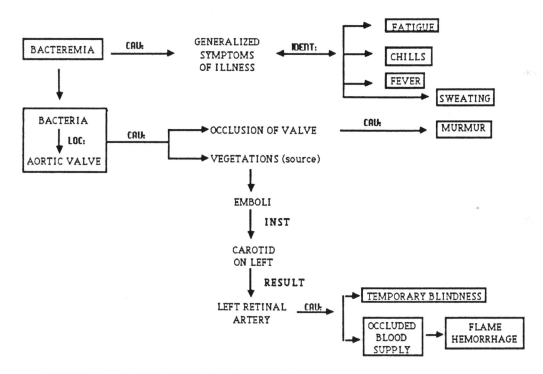

FIGURE 5:   CAUSAL NETWORK OF PATHOPHYSIOLOGICAL EXPLANATION OF ENDOCARDITIS BY A FINAL YEAR MEDICAL STUDENT
(clinical text only)

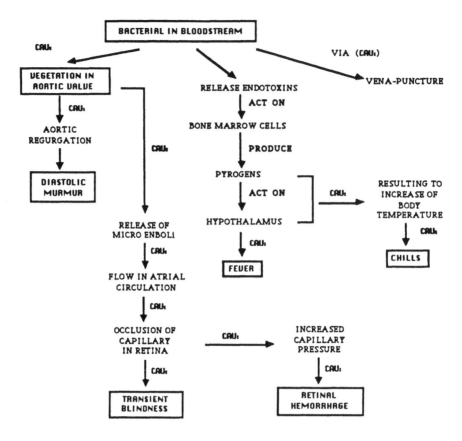

FIGURE 6: CAUSAL NETWORK OF PATHOPHYSIOLOGICAL
EXPLANATION OF ENDOCARDITIS BY A FINAL
YEAR MEDICAL STUDENT
(clinical and basic science text)

The second-year medical students, who do not have a clinical basis for disease classification, are not able to use basic science knowledge adequately in diagnostic reasoning. However, these students do have good taxonomies of some for the *components* of disease processes, such as *general infection,* which help focus hypotheses and provide local coherence in the organization of findings. This is illustrated by an explanation from a second-year student, as analyzed in the causal network in Figure 7. The analysis shows that diagnostic reasoning is facilitated by the identification of the infection component of bacterial endocarditis. Further, the subject rules out other sources of infection, *viz.*, common causes and tropical fever, and thereby increases the certainty of infection from cat bite. The subject does not use basic science information any further to reason diagnostically; rather, moves backward to explain the findings, to confirm the generated hypothesis. When more basic science information is provided, to confirm the generated hypothesis. When more basic science information is provided, it is used by the student to rule out the possibility of immune response as a differential diagnosis (Figure 8). Besides that, the additional basic science knowledge leads to a reduction in global coherence (6 hypothetical components appeared where there previously was only 1 candidate diagnosis), without a loss of local coherence. Furthermore, some use of basic science knowledge is wholly inaccurate. For example, decreases in the transfer of energy from systole to diastole do not cause hypotension.

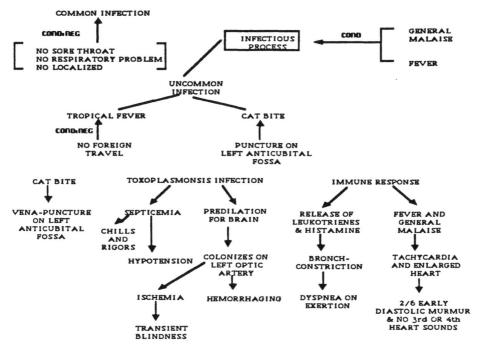

FIGURE 7:   CAUSAL NETWORK OF PATHOPHYSIOLOGICAL
             EXPLANATION BY A SECOND YEAR MEDICAL
                           STUDENT
                    (clinical text only)

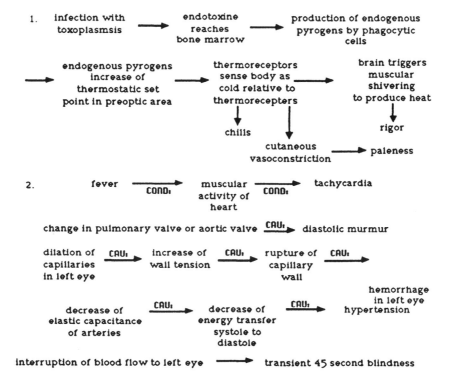

FIGURE 8:   CAUSAL NETWORK OF PATHOPHYSIOLOGICAL
             EXPLANATION BY A SECOND YEAR MEDICAL STUDENT
                    (Clinical & Basic Science Text)

229

The first-year students, who have only naive taxonomies of disease phenomena, choose a general hypothesis about the problem, which they are prepared to understand. For example, one subject diagnoses the problem as *mild shock,* then classifies the symptoms of this condition as increased heart rate, low diastolic pressure, low systolic pressure, and poor venous return (Figure 9). The balance of the student's effort is spend reflecting on connections among individual findings, to provide some coherence to the text. All subsequent explanations support blood loss and hemorrhaging, which account for the shaking chills. The sample explanations have only modest local coherence with very little global coherence (3 components remain unresolved). After the first-year students have read the basic science text, global coherence diminishes (6 unresolved components appear); though local coherence, which is still poor,improves over the first explanation. As can be seen from Figure 10, the additional basic science knowledge does not provide extra understanding affecting diagnostic reasoning.

PREDICTIVE REASONING: We might conjecture that reasoning in predictive directions would be facilitated by the application of basic science knowledge. In fact, final-year medical students, who have knowledge of relevant basic science and some disease classification schemata, have no problem in using their knowledge to predict selected variables to confirm the hypothesis. As the causal network in Figure 6 shows, additional basic science knowledge is used to add coherence to the clinical model which in turn provides the basis of the disease classification. There is a clear evidence of predictive reasoning using basic science knowledge to support the hypothesis that the patient was suffering from bacterial endocarditis.

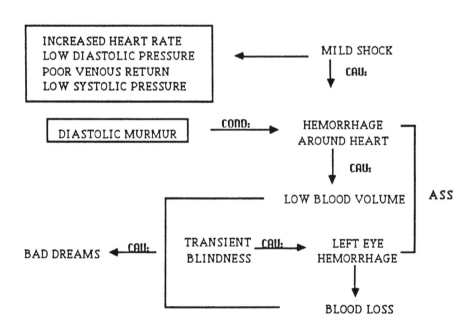

FIGURE 9: CAUSAL NETWORK OF PATHOPHYSIOLOGICAL
ENDOCARDITIS BY A FIRST YEAR STUDENT
(clinical text)

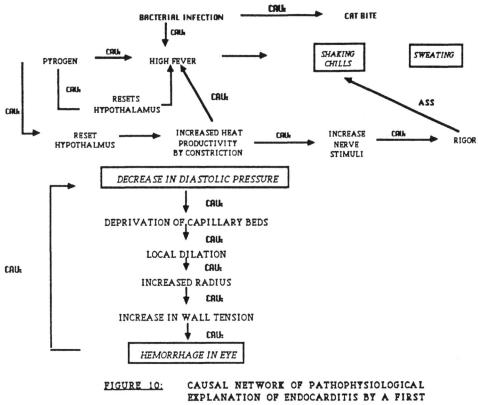

FIGURE 10: CAUSAL NETWORK OF PATHOPHYSIOLOGICAL
EXPLANATION OF ENDOCARDITIS BY A FIRST
YEAR MEDICAL STUDENT
(clinical and basic science
text

Basic science knowledge did not adequately assist the second-year students in predictive reasoning (Figure 8). It provided coherence to the local infection component of the disease, only to the extent that chills, rigor, and paleness could be predicted. The rest of the protocol shows signs of predictive reasoning using basic science knowledge, but without leading to the selection of relevant predictions.

The first-year students' protocols show poor predictive reasoning throughout. In the example case, the variables selected for confirmation were based on the diagnostic hypothesis (mild shock), which was grossly inaccurate. This hypothesis guided the selection of predictors such as blood loss and hemorrhaging. Although the use of knowledge irrelevant to the diagnosis provides some coherence to the text, it does not contribute to an explanation of the basic underlying pathophysiology of the problem. With additional basic science knowledge there is evidence of one variable predicting another, cyclically. This is seen when decreased diastolic pressure accounts for hemorrhage in the eye and *vice versa* (Figure 10). Further evidence of cyclical inference appears when fever is used to explain bacterial infection. Chills, sweating, rigor, and pink mucus membrane are caused by fever and thus provide evidence for fever. High fever is said to be caused by pyrogens which account for all the findings - fever, chills, sweating, and rigor.

The pattern of results described in detail for one subject at each level of medical training was also true for the other subjects at that level. Nine out of twelve students in their final-year of training conformed to the pattern of diagnostic and predictive reasoning described. All six of the second-year students protocols reflected the general phenomena described for one subject. However, there was somewhat greater variance seen in the protocols of the first-year medical students where, although four out of six subject protocols showed the described pattern of results, there were differences in their specific prior knowledge and in interpreting the protocols. Two subjects in this year did not follow the described pattern of results for the first-year students because of their unusual background of having a degree of Ph.D. and nursing.

CONCLUSION

For clinical problem solving, basic science knowledge serves as a powerful resource for bridging gaps during deductive reasoning; but is too powerful to be used in inductive reasoning. Inductive reasoning is successful only in the context of a well-differentiated taxonomy of goals, in particular, a sound classification schema for diagnoses and their facets. The ability to organize observations based on relevant clinical associations provides sufficient structuring of a diagnostic-problem space to exploit the details of basic science knowledge most efficiently. Once partitioned, the problem space can be constrained further by either diagnostic reasoning or predictive reasoning, with or without the addition of basic science knowledge. In sum, we find that basic biomedical knowledge interferes with diagnostic problem solving, unless there is a basis for predictive reasoning.

ACKNOWLEDGEMENTS

We wish to thank David Kaufman for his thoughtful comments during many discussions of the data in our studies. We would also like to thank Susan Young for typing the manuscript and Armar A. Archbold for his technical assistance with the figures. The authors alone, of course, are solely responsible for any errors or misrepresentations in this paper. The research reported here has been supported, in part, by grants from the Josiah Macy, Jr. Foundation to Patel and Evans.

REFERENCES

Clancey, W.J. (in press) Acquiring, representing, and evaluating a competence model of diagnostic strategy. In: M.T.H. Chi, R. Glaser & M. Farr (Eds.), *Nature of Expertise.* Hillsdale, NJ: Lawrence Erlbaum.

Evans, D.A., Gadd, C.S. & Pople, H.E., Jr. (in press) Managing coherence and context in medical problem-solving discourse. In: D. Evans & V. Patel (Eds.), *Cognitive Science in Medicine.* Cambridge, MA: MIT Press

Frederiksen, C.H. (1975) Representing logical and semantic structure of knowledge acquired from discourse. *Cognitive Psychology,*7, 371-45.

Henrion, M. (1986) Uncertainty in artificial intelligence: Is probability epistemologically and heuristically adequate? In: J. Mumpower & O. Renn (Eds.), *Expert Systems and Expert Judgement,* Proceedings of the NATO Advanced Research Workshop, in Porto, Portugal, August, 1986.

Joseph, G.-M. & Patel, V.L. (1986) Domain specificity and expertise in medicine. *Proceedings of the Cognitive Science Society Meeting, Amherst, MA,* August 1986. Hillsdale, NJ: Lawrence Erlbaum . 331-345.

Patel, V.L. & Frederiksen, C.H. (1984) Cognitive processes in comprehension and knowledge acquisition by medical students and physicians. In: H. Schmidt & H.G. deVolder (Eds.), *Tutorials in Problem Based Learning.* Assen, Holland: Van Gorcum.

Patel, V.L. & Groen, G.J. (1986) Knowledge-based solution strategies in medical reasoning. *Cognitive Science,* 10, 91-116.

Patel, V.L., Groen, G.J. & Scott, H. (in press) The uses of biomedical knowledge in clinical reasoning. In: D. Evans & V. Patel (Eds.), *Cognitive Science in Medicine.* Cambridge, MA: MIT Press.

Patel, V.L., Arocha, J.F. & Groen, G.J. (1986) Strategy selection and degree of expertise in medical reasoning. *Proceedings of the Cognitive Science Society Meeting, Amherst, MA, August 1986.* Hillsdale, NJ: Lawrence Erlbaum. 780-93.

Patil, R.S. & Szolovitz, P. (1981) Causal understanding of patient illness in medical diagnosis. *Proceedings of the 7th IJCAI.* Vancouver, B.C., Canada.

Patil, R.S., Szolovitz, P. & Schwartz, W. (1984) Causal understanding of patient illness in medical diagnosis. In: W.J. Clancey & H.E. Shortliffe (Eds.), *Readings in Medical ARtificial Intelligence.* Reading, MA: Addison-Wesley Publishing Company.

Winograd, T. (1972) *Understanding Natural Language.* New York: Academis Press.

Planning Stories

Michael Lebowitz[1]

Department of Computer Science -- Columbia University

New York, NY 10027

(212) 280-8196

4 February 1987

Full paper; Keywords: Story generation; planning

Abstract

Story generation can best be viewed as a planning task. We show here UNIVERSE, a program that generates melodrama plot outlines using hierarchical planning methods. Examples are given of the program creating story outlines using a set of characters that it also created. We indicate that story telling is open-ended, does not have to be perfect, and evaluation criteria are unclear, and contrast the sort of planning needed for story telling with other planning tasks. We suggest that certain elements of the methods used by UNIVERSE could be usefully applied to other tasks.

1 Introduction

The generation of stories is a challenging problem for Artificial Intelligence techniques. We have designed a program, UNIVERSE, that creates story outlines in the domain of interpersonal melodrama, a common form of which is soap opera. We chose this domain because it revolves around character relations, rather than action. It allows us to look at cognitive science issues such as author intention, knowledge-state assessment, character representation and eventually a range of natural language issues, in a very accessible domain. In this regard we have studied how narrative theory work such as (Barthes, 1977; Eco, 1979) applies to our task. The domain also has long-run potential in the areas of the education and interactive entertainment.

With UNIVERSE we view story telling primarily as a planning process, much in the same way that (Cohen and Perrault, 1979; Appelt, 1985) view language generation. This contrasts with, for example, making direct use of story grammars (Rumelhart, 1975; Mandler and Johnson, 1977) to generate a story.[2] In this paper we will briefly describe how UNIVERSE plans a story outline and then discuss some of the more interesting ways in which this sort of planning contrasts with planning in more traditional AI settings. Further details of UNIVERSE can be found in (Lebowitz, 1984; Lebowitz, 1985).

We have designed UNIVERSE to create plot outlines -- the major events that happen in a story -- since we are not yet ready to deal with the problems inherent in dialogue. These outlines are much like the summaries of soap operas that often appear in newspapers, or, these days, on computer services. EX1 shows a CompuServe® summary for a television melodrama. This is typical of the kinds of outlines that we have UNIVERSE produce, although UNIVERSE usually deals with broader events over a longer period of time.

---

[1]A series of projects by Paula Langer and Doron Shalmon contributed greatly to UNIVERSE.

[2]See (Black and Wilensky, 1979; Mandler and Johnson, 1980) for further discussion of story grammars.

EX1 -

Days of Our Lives-12/26/86

At her wit's end, Kim begs Shane to use the new ISA truth serum on Emma. They administer the drug and start questioning her but it is unclear as to whether Emma is really faking it or not. Poor Carrie is upset when Roman leaves for work. She is afraid that he won't be coming back. Frankie comforts her and they decide to watch the VCR together. When Frankie leaves to rent some movies, he finds her gone when he returns. He guesses that Carrie ran to the park to be with Roman. In the park, the criminal grabs her but Frankie bravely rescues her. Roman assures Carrie that he can handle his work. Finding Kayla without an assistant at the clinic, Patch offers to help out. He helps sign the patients in and keeps their spirits up much to Kayla's delight. While there, the mystery woman keeps watch on Patch. Later, Patch catches her and demands to know why she is following him.

We can see from EX1 that interpersonal melodrama involves a wide range of events and characters. The plots are quite intricate and interwoven, playing off of the personalities of the characters. The plot interconnections of EX1 are clearer if you know that Kim, Kayla and Roman are siblings, Shane is Roman's ex-partner, and Patch and Roman have a long-standing feud.

2 Story telling as planning

UNIVERSE uses a hierarchical planning algorithm much like that of NOAH (Sacerdoti, 1977). It knows about a variety of goals and uses plot fragments to achieve them. Much like NOAH's plans, or those used for story understanding in PAM (Wilensky, 1983), the plot fragments consist of a series of subgoals and have associated with them a set of constraints, mostly involving the kinds of characters that can be used as role fillers, that determine when they can be used. Currently UNIVERSE has about 65 plot fragments in its library, some of which will be seen below. This is a large enough set to convince us of the validity of our methods, and generation some interesting plot outlines, although clearly a much large set would be needed for a full-fledged program.

The key to UNIVERSE's planning algorithm is that it does not concentrate on the goals of the characters, but instead on *author* goals. In this way the planning process leads to stories that are interesting and have points. If we simply use character goals, then the characters will behave believably, but probably not interestingly. To illustrate this point, UNIVERSE uses one common goal **churn**, to keep apart two characters who are in love (by putting obstacles in their way). Clearly this is not a goal of the characters, but it leads to more interesting stories than simply having them live happily ever after. The use of author goals is the key distinction between UNIVERSE and Meehan's TALE-SPIN (Meehan, 1976) which in many ways was the inspiration for our program. TALE-SPIN would simulate characters responses to goals in Aesop's fables situations. MINSTREL (Turner and Dyer, 1985), a more recent story-telling program, does consider author goals to a point, but is more concerned with memory issues than we are here, and is still basically character oriented.

In developing UNIVERSE, we rapidly came to the conclusion that planning would be simplified if we created a set, or universe, of characters before beginning to tell stories (hence the name of the program). The rationale behind this is that the constraints needed to create believable characters -- e.g., creating all the marriages and divorces that add "color" to the story -- were unlike those for the rest of the story-telling process. UNIVERSE can create characters for a plot fragment "on the fly" when none of the existing ones are appropriate, but always doing so would decrease the coherence of the story. Each character is represented by a frame that describes parents, spouses, descriptive stereotypes (for coherence) past historical events, and, most importantly, various character traits and interpersonal relations. These values are used as constraints on the various plot fragments that can be applied. (Lebowitz, 1984) describes the character representation and creation process in detail. Figure 1 provides English summaries of a few of the characters created by UNIVERSE that will be used in the examples below.

```
Fran        -- a 25-year-old, nice, able, bureaucrat
Joshua      -- Fran's husband, a sleazy New York lawyer who is surprisingly nice
Valerie     -- another, not-so-nice, sleazy lawyer
Louis       -- Valerie's preppie husband
Gerald      -- Joshua's father, a nice fellow, who, nonetheless, recently
               had an affair with Fran
```

Figure 1: A partial cast of characters

Given a set of plot fragments, a set of characters, and a "seed goal", UNIVERSE operates much like NOAH (which turns out to be fairly similar to TALE-SPIN and even more so to micro-TALE-SPIN (Charniak et al., 1980)). It maintains a precedence graph with a partial ordering of which goals must be achieved before others. It uses a least commitment, opportunistic planning algorithm that repeatedly:

- Selects a goal with no unfulfilled goals that must precede it.

- Finds the relevant plot fragments that should achieve the goal, along with possible role bindings.

- Picks one of these plot fragments.

- Expands the selected plot fragment; this may include: 1) creating new goals, 2) modifying interpersonal relations and/or character traits, and 3) generating text (which is currently done from simple templates).

This simple algorithm appears to generate reasonable plot outlines even though our plot fragment library is still relatively small. We compare it to more traditional planning in Section 3. A few points about the algorithm are worth noting. UNIVERSE does not completely plan out a story before it generating any of it. This is because it is intended for open-ended story situations that may have no natural ending. For example, most of the plot fragments for **churn** have as their final subgoal to **churn** further. Generating as it goes leaves UNIVERSE open to the problem of running into blind alleys where it has already generated part of a story that it cannot complete. We discuss this further below, but simply note here that the program can often re-plan from the state achieved in the blind alley.

An important part of the UNIVERSE algorithm is deciding upon the plot fragment to choose when many will achieve the author goal. Since we believe in the necessity of a large plot fragment database, this is a very real issue. Indexing allows UNIVERSE to easily find the relevant fragments, but we still need a way to select one. For the moment, UNIVERSE uses two selection criteria. First it checks for plot fragments that will achieve other open goals in the precedence graph. This method, which will be illustrated below, leads to nice connections among various plot threads. The second criterion is to use interest values associated with each fragment. We feel that these interest levels should actually also depend on the characters involved (Lebowitz, 1981). If there are still multiple possible fragments, UNIVERSE selects one randomly.

Figure 2, which shows a fairly mundane beginning to a plot outline planned by UNIVERSE using characters that it created, will be used to illustrate the algorithm. Lines preceded by ">>>" are the plot outline output.

```
*(tell '(((churn JOSHUA FRAN)))

working on goal -- CHURN JOSHUA FRAN
Several possible plans with equal goal effect
LOVERS-FIGHT HIM/FRAN HER/JOSHUA
LOVERS-FIGHT HIM/JOSHUA HER/FRAN
JOB-PROBLEM P1/FRAN P2/JOSHUA JOB/BUREAUCRAT
JOB-PROBLEM P1/JOSHUA P2/FRAN JOB/SLEAZY-LAWYER
PREGNANT-AFFAIR WOMAN/FRAN HUSBAND/JOSHUA
ACCIDENT-BREAKUP P1/FRAN P2/JOSHUA
STEAL-CHILD HUSBAND/JOSHUA HER/FRAN
COLLEAGUE-AFFAIR WIFE/FRAN HUSBAND/JOSHUA
AVALANCHE-ACCIDENT HIM/FRAN HER/JOSHUA
AVALANCHE-ACCIDENT HIM/JOSHUA HER/FRAN
 -- picking one
 -- using plan LOVERS-FIGHT HIM/FRAN HER/JOSHUA

working on goal -- DO-FIGHT JOSHUA FRAN
Several possible plans with equal goal effect
PERSONAL-FIGHT A/JOSHUA B/FRAN SUBJECT/IN-LAWS
PERSONAL-FIGHT A/JOSHUA B/FRAN SUBJECT/MONEY
PERSONAL-FIGHT A/JOSHUA B/FRAN SUBJECT/SECRETS
PERSONAL-FIGHT A/JOSHUA B/FRAN SUBJECT/FLIRTING
PERSONAL-FIGHT A/JOSHUA B/FRAN SUBJECT/KIDS
 -- picking one
 -- using plan PERSONAL-FIGHT A/JOSHUA B/FRAN SUBJECT/MONEY

>>> JOSHUA and FRAN fight about: MONEY
working on goal -- DUMP-LOVER JOSHUA FRAN
 -- using plan BREAK-UP DUMPER/JOSHUA DUMPED/FRAN

>>> JOSHUA tells FRAN he doesn't love her

working on goal -- CHURN FRAN JOSHUA
Several possible plans with equal goal effect
LOVERS-FIGHT HIM/JOSHUA HER/FRAN
JOB-PROBLEM P1/JOSHUA P2/FRAN JOB/SLEAZY-LAWYER
JOB-PROBLEM P1/FRAN P2/JOSHUA JOB/BUREAUCRAT
PREGNANT-AFFAIR WOMAN/FRAN HUSBAND/JOSHUA
ACCIDENT-BREAKUP P1/FRAN P2/JOSHUA
STEAL-CHILD HUSBAND/JOSHUA HER/FRAN
COLLEAGUE-AFFAIR WIFE/FRAN HUSBAND/JOSHUA
AVALANCHE-ACCIDENT HIM/JOSHUA HER/FRAN
AVALANCHE-ACCIDENT HIM/FRAN HER/JOSHUA
 -- picking one
Multiple deferred filler -- picking one
 -- using plan COLLEAGUE-AFFAIR WIFE/FRAN HUSBAND/JOSHUA COLLEAGUE/JACK

[and the story continues with the protagonists living unhappily ever after']
```

Figure 2: The beginning of a UNIVERSE plot outline

The first action shown in Figure 2 is UNIVERSE, given the goal of churning two characters, Joshua and Fran, collecting all of the plot fragments that might satisfy that goal, filtering out those with constraints that are not met by Joshua and Fran. Since the **churn** author goal has been one of our prime examples, there are a number of possibilities. For each goal, UNIVERSE collects all the possible role bindings so that it can sensibly select among them.[3] After selecting the lovers fight fragment, essentially at random, since there are no other goals and all the fragments have the same interest value, UNIVERSE expands its subgoals. For one of these, the fight itself, we can see that UNIVERSE considered a number of possible fight topics. The topics are generated from knowledge of the characters involved and so would be different for different characters.

After UNIVERSE finishes with the lovers fight, it decides to churn the relationship further. The selection of the colleague affair fragment illustrates an important point. After its selection, UNIVERSE has to fill more roles. We discovered that there were too many combinations to allow UNIVERSE to expand out all possibilities for the various minor characters in all the relevant fragments in order to pick the best. So, instead, each plot fragment has identified the roles that we are sure it can fill with somebody (by creating a new character if need be) and are probably not relevant to the author goals of the plot, and hence are not likely to be relevant to picking the best fragment to use. The filling of these roles is deferred until after a fragment is chosen, a technique that might be applicable to other planning situations.

Figure 3 shows a second example outline that illustrates how extra goals affect the opportunistic planning of UNIVERSE. Here, as well as seeding UNIVERSE with a **churn** goal, we have also asked it to get Joshua **together** with Valerie. This goal, while given the program here, could equally well have come from its own pursuit of other goals. The notable point is that now, when looking for a fragment for **churn**, it selects accident-breakup, the only one that guarantees that the man will get together with a desired person, given appropriate bindings. We show the outline as far as where UNIVERSE begins to plan the new relationship which will, in the long run, create interest by making Joshua unavailable should Fran get better. Opportunistic planning, does an excellent job in pulling story strands together. The ability to satisfy several goals with one plan seems to be a hallmark of clever melodrama.

3 Story planning versus action planning

The planning of stories done by UNIVERSE shares much with other sorts of AI planning methods. As mentioned above, the algorithm is very similar to various hierarchical planners in its use of goals and plans (and could probably be made more similar to good effect). However, it is interesting to contrast both the task and the method with standard AI action planning. We will also indicate how it may be advantageous to apply many aspects of story planning to more standard action planning.

The key points that we see in the planning of stories, each of which we will discuss below, are:

- You can "say as you go"; it is neither necessary nor possible to fully plan out a story before beginning to tell it.

- The plan does not have to be perfect; people will accept odd combinations (and indeed fill in motivation).

---

[3]Some of the role names are a bit deceptive. For example, the *him* and *her* roles of the lovers fight plot fragment are actually genderless, which is why UNIVERSE tries both bindings.

```
*(tell '(((churn JOSHUA FRAN)) ((together JOSHUA VALERIE))))

working on goal -- CHURN JOSHUA FRAN
 -- using plan ACCIDENT-BREAKUP P1/FRAN P2/JOSHUA THIRD-PARTY/VALERIE

working on goal -- DO-DISABLE FRAN
 -- using plan DISABLE PERSON/FRAN

>>> FRAN has a spinal injury and is paralyzed

>>> FRAN doesn't want to ruin JOSHUA's life

>>> FRAN pretends to blame JOSHUA for her malady

working on goal -- DUMP-LOVER FRAN JOSHUA
 -- using plan BREAK-UP DUMPER/FRAN DUMPED/JOSHUA

>>> FRAN tells JOSHUA she doesn't love him

working on goal -- TOGETHER JOSHUA VALERIE

[again, the story continues unhappily for almost all concerned]
```

Figure 3: A multi-goal story

- Selecting the plot (plan) to use is important; heuristics that consider factors other than goal satisfaction must be used.

- Unlike most action domains, each role filler (character) is different, and they cannot be used interchangeably.

Many forms of stories, including soap opera and some children's stories, have no real end. As a result, we clearly cannot plan such stories in their entirety before generating any of the story. At the moment, UNIVERSE probably goes too far in simply generating text that appears to lead to the goal, often getting into binds that it cannot resolve. On the other hand, as we will mention below, this is not that crucial for a story. In examining television serials, one can frequently see situations where the writers have clearly headed down blind alleys of this kind. An important future research topic will be to determine just how far one should plan a story -- short of guaranteeing goal satisfaction, which is not practical -- before actually generating any of it.

We feel that more planning situations would be appropriate for an "execute as you go" paradigm than are currently approached that way. For example, if we are developing a program to plan a series of financial investments, we may want it to start and make some of the moves before waiting to plan everything. Since much of the plan is likely to be dependent on what happens at the beginning, there is no real point in planning all the way to the end. Admittedly, done more completely than by UNIVERSE this sort of planning is quite complicated. At a minimum, it involves deciding when you've planned enough, feedback from execution, and re-planning, but should be well worthwhile.

One factor that makes "say as you go" particularly feasible for story telling is that story outlines do not have to be perfect. They can go into blind alleys and backtrack in unusual ways and still be accepted by readers. This is true for at least two reasons. Obviously, there is no such thing as a single *best* story. As long as a story is coherent and does not blatantly violate rules of the world, then it is, in some sense, acceptable. Indeed, it is unlikely that at present a program could detect fine differences in story quality. A secondary reason for the lack of need to look for the best story is that our human readers will often fill in details that the program may not have had in mind (Schank and Abelson, 1977). Again, many domains beyond story telling also do not need optimal plans. For example, a program designed to come up with recipes should not try to evaluate minor differences in quality.

The next interesting point regarding story planning is that for many author goals there are typically a sizable number of different plot fragments that will achieve them, not just one or two. Further, we have seen that the program often cannot really evaluate the stories that result from different choices. Nonetheless, it is clear that some decisions will work out better than others. As a result, it is necessary that UNIVERSE pay considerable attention to which plot fragment to select, *even among those that all achieve the author goal.* In particular, it is necessary that the program use heuristics that consider factors other than simple goal satisfaction. We feel that local factors such as interest levels and patterns of character involvement can, along with the opportunistic planning that we have described, lead to intricate and interesting plot structure.

The final area of interest involving planning stories is that, unlike most action domains, every character in a fictional universe is different. In most action domains that have been studied, while there may be a number of objects in the world, they tend to fall into a few small classes (e.g., rectangular blocks) that can be treated identically by operators. In story telling, though, we have to be careful to make sure that the role fillers (characters) used by a plan are appropriate in order to retain believablility. Of course, we also have the option of creating new characters out of nowhere which is not open to other forms of planning. While the diversity of role fillers does not apply to most domains, it may be relevant in situations where we are trying to understand stories involving people or human actions.

4 Conclusion

Story telling is a form of generation that involves a number of quite interesting AI problems. Planning in terms of author goals definitely seems to be the way to think about story telling, rather than any form of story grammar. We feel that the kind of planning done for story telling reflects what people do in many different planning situations (not just stories) and could be profitably applied to AI systems dealing with planning problems. In the future, we plan to expand the basically simple methods of UNIVERSE to make use of more complex planning methods to get further intricacy into our plot outlines, including, perhaps, case-based planning of the sort described in (Kolodner, Simpson and Sycara-Cyranski, 1985; Hammond, 1986), or thematic issues such as those considered by (Turner and Dyer, 1985).

References

Appelt, D. E. (1985). *Planning Natural Language Utterances.* Cambridge, England: Cambridge University Press.

Barthes, R. (1977). *Image -- Music -- Text.* New York: Hill and Wang.

Black, J. B. and Wilensky, R. (1979). An evaluation of story grammars. *Cognitive Science, 3*, 213 - 229.

Charniak E., Riesbeck, C. K., and McDermott, D. V. (1980). *Artificial Intelligence Programming.* Hillsdale, New Jersey: Lawrence Erlbaum Associates.

Cohen, P. R. and Perrault, C. R. (1979). Elements of a plan-based theory of speech acts. *Cognitive Science, 3*, 177 - 212.

Eco, U. (1979). *The Role of the Reader.* Bloomington, Indiana: Indiana University Press.

Hammond, K. (1986). CHEF: A model of case-based planning. *Proceedings of the Tenth International Joint Conference on Artificial Intelligence* (pp. 267 - 271), Philadelphia, PA: Morgan Kaufmann.

Kolodner, J. L., Simpson, R. L. and Sycara-Cyranski, K. (1985). A process model of case-based reasoning in problem solving. *Proceedings of the Ninth International Joint Conference on Artificial Intelligence* (pp. 284 - 290), Los Angeles: Morgan Kaufmann.

Lebowitz, M. (1981). Cancelled due to lack of interest. *Proceedings of the Seventh International Joint Conference on Artificial Intelligence* (pp. 13 - 15), Vancouver, Canada: Morgan Kaufmann.

Lebowitz, M. (1984). Creating characters in a story-telling universe. *Poetics, 13*, 171 - 194.

Lebowitz, M. (1985). Story-telling as planning and learning. *Poetics, 14*, 483 - 502.

Mandler, J. M. and Johnson, N. S. (1977). Remembrance of things parsed: Story structure and recall. *Cognitive Psychology, 9*, 111 - 191.

Mandler, J. M. and Johnson, N. S. (1980). On throwing out the baby with the bathwater: A reply to Black and Wilensky's evaluation of story grammars. *Cognitive Science, 4*, 305 - 312.

Meehan, J. R. (1976). *The metanovel: Writing stories by computer.* (Tech. Rep. No. 74). Yale University Department of Computer Science.

Rumelhart, D. E. (1975). Notes on a schema for stories. In D. Bobrow and A. Collins, (Ed.), *Representation and Understanding: Studies in Cognitive Science* New York: Academic Press.

Sacerdoti, E. (1977). *A Structure for Plans and Behavior.* Amsterdam: Elsevier North-Holland.

Schank, R. C. and Abelson, R. P. (1977). *Scripts, Plans, Goals and Understanding.* Hillsdale, New Jersey: Lawrence Erlbaum Associates.

Turner, S. R and Dyer, M. G. (1985). Thematic knowledge, episodic memory and analogy in MINSTREL, a story invention system. *Proceedings of the Seventh Annual Conference of the Cognitive Science Society* (pp. 371 - 375), Irvine, CA.

Wilensky, R. (1983). *Planning and Understanding.* Reading, MA: Addison-Wesley.

PROBLEM REPRESENTATION AND HYPOTHESIS GENERATION IN DIAGNOSTIC REASONING

GUY-MARIE JOSEPH
VIMLA L. PATEL

MCGILL UNIVERSITY

ABSTRACT

In this paper we examine the role of domain knowledge in the process of hypothesis generation and problem representation during diagnostic reasoning. An on-line task environment and the combination of discourse and protocol analysis techniques were used to test the differences between two groups of experts solving a clinical problem. The groups consisted of high domain-knowledge subjects (HDK) -endocrinologists- and low domain-knowledge (LDK) subjects -cardiologists-. The results show that HDK subjects used a more efficient process of diagnostic reasoning as generated a more coherent representation of the problem. A two-stage model describing the process of hypothesis generation was proposed to explain the differences in the process of hypothesis generation.

The most frequently investigated aspect of medical problem solving is diagnostic reasoning. A central question for researchers in fields such as cognitive science, medical education, medical problem solving and decision making, is how do expert physicians go about making a diagnosis? This interest has been motivated by both theoretical and practical concerns.

Theoretically diagnosis is viewed as representative of the problem-solving processes wherein the problem solver is required to examine, evaluate and select information in order to generate an accurate solution. According to Schwartz and Griffin (1986) diagnosis is a subset of the more general skill of classification (i.e., assigning entities to different classes or categories) of diseases. Research in clinical reasoning has played a significant role in developing psychological theories of human problem solving (Wortman, 1972). In practical terms, this is directly applicable to the teaching of medicine (Wortman, 1972).

Diagnostic reasoning has been studied from two perspectives. On the one hand, computational/mechanistic models have been used to describe describe the reasoning processes of physicians while solving a problem (Johnson, 1983; Shortliffe, Buchanan, & Feigenbaum, 1979). The main goal of the computational approach is to develop computer programs that can perform (diagnose problems) as accurately as human experts. The first step in developing such programs consists in collecting think aloud protocols from experts engaged in simulated clinical tasks (e.g., the diagnosis of a patient's problem), and analyzing these transcripts to formulate models of the solution strategies and reasoning processes used. These models are then represented in a computer program (simulations) that, when executed, produces behavior that can be compared with that observed in physicians (Kassirer et al., 1982).

This is considered by many (e.g., Kassirer et al., 1978; Pople, 1982) to be a promising approach for the development of knowledge-base components in expert systems, and can contribute toward the development of theories of knowledge representation (Brachman & Levesque, 1985). The main weakness of this this approach is a lack of theories and methods for dealing with the problem of natural language in protocol analysis. Central to building a more complete account of problem solving in complex, knowledge-rich domains is the need for specific information about both how the subject's representation of the problem changes over time and how his domain knowledge is used to construct and modify these representations. "Think-aloud" protocols (e.g. Greeno & Simon, 1985; Kassirer et al., 1982) provide rich, complex data that are approximately concurrent with the subject's reasoning and therefore provide information about the subject's changing representation of the problem. However, protocol analysis methods (e.g., Ericsson & Simon, 1984) have been limited in their success at providing more than global information about subjects' processing (cf. Joseph, 1987).

From a second perspective, the psychological research on medical diagnosis has treated the physician as a processor of information. These investigations were influenced by then current models of general problem-solving ability which are summarized in Newell & Simon, (1972). Elstein and colleagues (1978) were among the first researchers to apply the information processing approach to study diagnostic reasoning. They proposed the 'hypothetico-deductive' model as the general and logical model of expert reasoning.

> "Diagnosis Diagnostic problems are solved through a process of hypothesis generation and verification. Hypotheses are consistently generated early in a workup when only a very limited data base has been obtained. While any early information may be revised or discarded if subsequent data fail to confirm it, there is a high probability that at least some of the formulations of experienced physicians will be correct. Hypotheses serve as organizing rubrics in working memory. They help to overcome limitations of memory capacity and serve to narrow the size of the problem space that must be searched for solution. Since it would be impossible to conduct an efficient inquiry without some hypothetical goal that would tell the inquirer when to stop [control structures], hypotheses serve to transform an open medical problem (What is the patient's illness?) into a set of closed problems that are much easier to solve (Is the illness X? or Y? or Z?) Elstein et al., 1978,.176".

The early generation of diagnostic hypotheses is considered as the main strategy that physicians used to reduce search in the problem space.that is most likely to yield the accurate solution. "In sum, we may propose that a set of problem formulations defines the dimensions of the fundamental problem space in which a physician's search for a diagnosis is conducted" (Elstein et al., 1978, p.176).

The basic assumptions of the use of the hypothetico-deductive method of reasoning used by expert physicians in clinical reasoning have increasingly come into question (1983; Groen & Patel, 1985; and Kassirer, et al., 1982). The assertion that expert diagnostic reasoning is characterized by a hypothetico-deductive method is questionable in view of the fact that the research in other domains have demonstrated this to be a 'weak' and inefficient method of problem solving, more characteristic of novice rather than expert performance (e.g.,

Feltovich & Barrows, 1984; and Simon & Simon, 1978). Weak methods are very general, in the sense that they can be used in almost any domain. Under realistic time constraints, however, they tend to yield either the wrong answer or no answer at all. In contrast strong methods almost always yield the correct answer, but are only applicable in a limited domain. The main problem however, is that the use of diagnostic hypotheses is not necessarily a reliable indicator of diagnostic expertise or the quality of diagnostic reasoning. In addition, further investigation on hypothesis generation parameters (e.g., timing of hypothesis generation, number of hypotheses generated), and hypothesis quality either yield inconsistent cross-situational results or are insufficient in discriminating expertise (Feltovich & Barrows, 1984).

Studies in problem solving and expertise suggest that the diagnostic hypotheses generated for a medical case constitute one component of an expert physician's representation of a patient's condition (Hasserbrock & Prietula, 1986). An expert representation should be an abstraction of commonalties existing in a patient's overall symptoms. Knowledge of pathophysiological processes provides experts with another level of abstraction. The appropriateness of the disease hypotheses being considered for a given case is likely to change if a physician shifts to a different underlying pathophysiological condition.

In the present study we are primarily interested in testing the role of domain knowledge in the process of hypothesis generation and problem representation in diagnostic reasoning. The use of an on-line task environment and the combination of discourse and protocol analysis will be used to test the differences in problem representation and hypothesis generation while solving the problem (Joseph & Patel, 1986a). Experts with specific domain-knowledge are expected to use a more efficient process of hypothesis generation as well as a more coherent representation of the problem.

METHOD

Materials: The stimulus was a text based on a real patient in a Montreal hospital and modified by an endocrinologist for the purposes of this study. The clinical findings for the case were assembled in a typed "patient file" format. That is, the clinical information

244

was arranged in the typical order of medical history, findings from physical examination, and x-ray and laboratory test results.

Subjects: Nine senior physicians associated with the Faculty of Medicine at McGill University volunteered as subjects for the study. Given that the task was the diagnosis of a patient with an endocrine disorder, the high domain knowledge (HDK) group consisted of four endocrinologists and the low domain knowledge (LDK) group consisted of five cardiologists. The physicians were all practicing physicians with from five to ten years of experience (i.e., practice) in their respective fields.

Clinical problem The endocrinology problem (Table 1) describes the case of an elderly lady who was brought to the emergency room by her daughter suffering from severe hypothyroidism. Prior to her admission into the emergency room, the patient consulted a physician complaining of difficulty in speaking and throat irritation. His diagnosis was chronic laryngitis, for which he prescribed a potassium iodide mixture as an expectorant. When taken by a healthy person, potassium iodide is quite harmless. However, when taken by a hypothyroid patient, it precipitates an acute hypothyroid crisis which leads to myxedema.

The accurate diagnosis of the case is: Hashimoto's hypothyroidism precipitated to myxedema pre-coma by the potassium iodide mixture. The diagnosis can be divided into three subcomponents varying in specificity, from general to specific. The first subcomponent is hypothyroidism. The second one is myxedema pre-coma. The third subcomponent is an autoimmune condition called Hashimoto's thyroiditis, which is the cause of the hypothyroidism.

Table 1. Endocrine problem.

Medical History

S1. A 63 year old woman with a one-week history of increasing drowsiness and shortness of breath was brought to the emergency room by her daughter.
S2. The patient had not been well for over a year.
S3. She complained of feeling tired all the time, had a loss of appetite, a 30 lb. weight gain and constipation.
S4. A month later she had been diagnosed as having " chronic laryngitis" and was prescribed a potassium iodide mixture as an expectorant.

Physical Examination

S5. Physical examination revealed a pale, drowsy, obese lady with marked periorbital edema.
S6. She had difficulty speaking, and when she did speak her voice was noted to be slow and hoarse.
S7. There were patches of vitiligo over both her legs.
S8. Her skin felt rough and scaly.

S9. Her body temperature was 36 deg. C.
S10. Pulse was 60/minute and regular.
S11. B. P. was 160/95.
S12. Examination of her neck revealed no jugular venous distention.
S13. The thyroid gland was enlarged to approximately twice the normal size.
S14. It felt firm and irregular.
S15. There was grade 1 galactorrhea.
S16. The apex beat could not be palpated.
S17. Chest examination showed decreased movements bilaterally and dullness to percussion.
S18. There was no splenomegaly.

Results From Neurology, X-Ray and Laboratory Tests

S19. Neurological testing revealed symmetrical and normal tendon reflexes but, with a delayed relaxation phase.
S20. Urinalysis was normal.
S21. Chest X-ray showed large pleural effusions bilaterally.
S22. ECG revealed sinus bradycardia, low voltage complexes and non-specific T-wave flattening.
S23. Routine biochemistry (SMA=16) showed Na=125, K=3.8, BUN=8 mg/100ml.
S24. Arterial blood gases PO_2=50 mm Hg, PCO_2=60 mm Hg.
S25. The patient was admitted to the intensive care unit for further management

Procedure: An on-line task environment (Joseph, 1987) was used to present the stimulus material to individual subjects one segment at a time on a microcomputer. The order of presentation of text segments was controlled by the experimenter. However, the rate of presentation of each segment was controlled by the subjects. Segments consisted of one or two sentences each. After the presentation of each segment subjects were asked to give a verbal report of their interpretation of the sentence with respect to a possible diagnosis. After the presentation of the entire case, subjects were asked to summarize the case and then to provide a diagnosis. Subjects' interpretations were tape recorded and later transcribed verbatim.

ANALYSES

The methods of analysis used in this study include the use of techniques from both discourse and protocol analyses as well as as the use of quantitative and qualitative measures. Three analyses were carried out on each subject's protocols.

1) Segmentation.
2) Hypothesis generation.
3) Problem representation.

Techniques were taken from discourse analysis for application in analysis 1 (functional-syntactic analysis), and analysis 3 (representing conceptual structures as frames or semantic networks). Standard techniques of protocol analysis (e.g., tabulating specific kinds of information in the protocols) were used in analysis 2.

Segmentation

At the first level of analysis, subjects' protocols were transcribed and divided into syntactic units or 'segments' (Dillinger, 1984). This division facilitates identification and further analysis of the parts of the text and of the protocols. This method is based on Winograd's system of clausal analysis (Winograd, 1983) and derived from the systemic grammar of Halliday (1967).

Process of hypothesis generation

After the segmentation, subjects' protocols were analyzed and coded for the generation of diagnostic hypotheses. In clinical reasoning the term 'diagnostic hypothesis' refers to any ideas, diagnoses, or guesses that label or propose explanations which will guide investigation of the patient's problem (Barrows, & Tamblyn, 1980). These hypotheses can refer to syndromes, specific disease entities, disorders, pathophysiological processes, and anatomical or biochemical disturbances.

> **Segment #3:** She complained of feeling tired all the time, had a loss of appetite, a 30 pound weight gain and constipation.
> **Subject's Comment:** She had a loss of appetite, a 30 pound weight gain, and constipation. OK, right now you are wondering whether she has got **hypothyroidism** when you are looking at this.

Hypotheses were coded only the first time they were generated. In the above example **hypothyroidism** was coded as a diagnostic hypothesis generated. Hypotheses which were repeated were not coded twice.

> **Segment 7:** There were patches of vitiligo over both her legs:
> **Subject's Comment:** Vitiligo over the legs, It fits with **hypothyroidism,** that you have one **autoimmune disease** that you get another autoimmune disease Hashimoto's thyroiditis and vitiligo.

In this segment only **Hashimoto's thyroiditis** was coded as a diagnostic hypothesis generated. Hypothyroidism was not coded as a diagnostic hypothesis generated because it was previously generated, and the

autoimmune disease was considered to be part of the Hashimoto's thyroiditis. When a subject ruled out the possibility of an hypothesis generated previously during the presentation of the case, it was also coded as an hypothesis generated.

Problem representation

The objective of this analysis is to generate data about the subjects' changing representation of the problem, from which it will be possible to infer some characteristics about the processes they used for constructing a problem representation and about the solution strategies they used. To do this it was necessary to develop two models: the first was a general model of the knowledge necessary for generating an accurate solution to the problem, represented as a causal network and referred to here as the "reference" or "canonical" model (Joseph, 1987; Patel & Groen, 1986). The second was a specific model of the problem (i.e., a subset of the general model) generated by each subject at each stage of solving it. This technique has proven useful in studying knowledge-based differences between experts and novices (Patel & Groen, 1986), and experts with different levels of expertise (Joseph & Patel, 1986a,b) in diagnostic reasoning.

RESULTS AND DISCUSSION

The Process of Hypothesis Generation. Analysis of the time course of the production of diagnostic hypotheses focussed on differences between HDK and LDK subjects in: a) the production of the accurate diagnostic subcomponents, b) the pattern of hypothesis generation and confirmation, and finally c) the number of diagnostic hypotheses generated by each group.

First, the HDK subjects are expected to generate the accurate diagnostic subcomponents before the LDK subjects. Second, the pattern of hypothesis generation is expected to reflect two distinct stages. The first stage consists of the generation of hypotheses, followed by a stage of hypothesis confirmation and/or ruling out inaccurate hypotheses. Third, the HDK subjects are expected to generate fewer diagnostic hypotheses than the LDK subjects. This is primarily due to the differences in domain-specific knowledge of the HDK subjects.

The overall number of diagnostic hypotheses produced by the HDK and LDK subjects is presented in Figure 1. It shows clear differences in: the pattern of hypothesis generation, the number of diagnostic hypotheses generated by the subject, and the time taken for the production of the accurate diagnostic subcomponents.

Before the presentation of Segment 7 (by which time the HDK subjects have generated all three accurate diagnostic subcomponents) the HDK subjects generated more diagnostic hypotheses than the LDK subjects. After the presentation of Segment 7 the HDKs generated fewer diagnostic hypotheses than the LDK subjects. On the other hand, the LDK subjects show a rather different pattern of hypothesis generation both before and after the presentation of Segment 7 (Figure 1). Specifically, the LDK subjects show little difference in their pattern of hypothesis generation before and after the complete diagnosis is reached.

7 (Segments 1 to 7). The letters A, B and C in Figure 1 indicate the mean segment number where the accurate diagnostic subcomponents were generated. As expected the HDK subjects generated the accurate diagnostic subcomponents earlier than the LDK subjects. Before the presentation of Segment 7 the HDK subjects generated more hypotheses than the LDK subjects. Both groups generated the first two subcomponents of the accurate diagnosis (Hypothyroidism and myxedema). However there was a time difference (i.e., number of segments) between the generation of these diagnoses in the two groups. The difference between the production of the first diagnostic subcomponent was of 1 segment (Figure 1: Diff A). The difference for the generation of the second subcomponent was of 2 segments (Figure 1: Diff B).

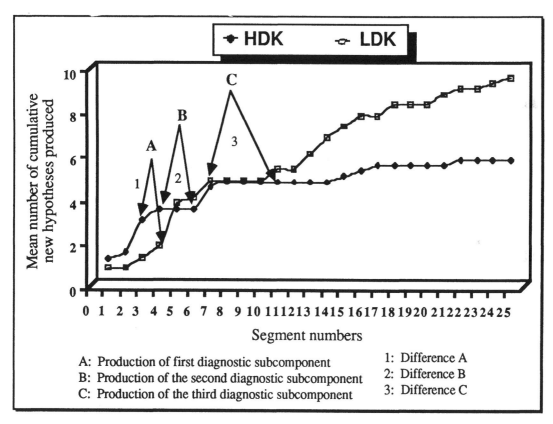

Figure 1. Cumulative number of diagnostic hypotheses generated by HDK and LDK subjects.

Hypothesis generation after Segment 7 (Segments 7 to 25): The pattern of hypothesis generation from the presentation of Segment 7 to the end of the case is also presented in Figure 1. Once again, as expected, the HDK subjects generated the third subcomponent of the accurate diagnosis earlier than the LDK subjects. The difference between the two

groups in the production of the third diagnostic subcomponent was of 4 segments (Figure 1: Diff C). Furthermore, after the production of the accurate diagnosis the HDK subjects generated very few new diagnostic hypotheses. The HDK subjects ruled out some of the hypotheses generated earlier, used the findings from the physical examination to confirm the diagnosis, and determined secondary problems.

The analysis of the time course of hypothesis generation suggests that the process of hypothesis generation differs across groups. The HDK subjects seem to distinguish two phases in their process of hypothesis generation, whereas the LDK subjects do not. The first phase consists in hypothesis generation and the second in diagnostic confirmation.

In the first stage (before the presentation of Segment 7) the HDK subjects use the information from the medical history and a few findings from the physical examination to generate the accurate diagnostic subcomponents. This explains the finding that HDK subjects generate more diagnostic hypotheses before the presentation of Segment 7. The rapidity with which the initial hypotheses are generated constitutes the most striking feature of the behavior of experienced clinicians. Often with only the age, sex and present complaint of the patient, the clinician unhesitatingly selects a single working hypothesis.

The early generation of accurate diagnostic hypotheses is also important in determining the accurate diagnosis. Work by Barrows et al. (1978) has shown that the earlier a good hypothesis set is created, the more predictive it is of the quality of the diagnosis. Differences in the time course of the generation of the accurate diagnostic subcomponents seem consistent with the findings of speed difference found in the chess-playing ability of masters vs beginners (Chase & Simon, 1973) as well as the findings from studies contrasting experts and novices solving physics problems (e.g., Larkin et al, 1980; Simon & Simon,1978). This difference is explained by the difference in domain knowledge between the HDK and LDK subjects.

The hypothesis generation process is interpreted as an indication of subjects' organization of the information provided into a more or less coherent problem representation. The results discussed so far indicate considerable differences in terms of the processes used by the two groups in generating diagnostic hypotheses and the accurate diagnostic subcomponents. This poses the question of the generative difference in how the two groups construct their problem representation over time. In the final set of analyses, the information selected from the input and the hypotheses adduced to structure that information are examined over the first six segments, i.e., the hypothesis generation phase.

Problem representation

The objective of this analysis is to generate data about the subjects' changing representation of the problem, from which it will be possible to infer some characteristics about the processes they used for constructing the problem representation and about the solution strategies they used. The method of analysis developed for this study required the use of a general model of the knowledge necessary for generating an accurate solution to the problem, represented as a causal network and referred to here as the "reference" or "canonical" model (see Patel & Groen, 1986). Using the reference model as a template, the subject's current representation of the problem was generated after each sentence presented. The subset of the reference model that the subject used was highlighted to represent the knowledge that he used to construct his current interpretation of the problem, thus yielding problem representation after sentence 1, after sentence 2, etc. for each subject.

Reference model. The reference model (Figure 2) depicts, based on techniques developed in discourse analysis and artificial intelligence (see Frederiksen, 1986; Brachman & Levesque, 1985), the conceptual relations between the cues given in the case description, some of the underlying pathophysiological processes and the components of the correct diagnosis. It was developed from protocols and interviews with two expert endocrinologists and several other physicians as well as from standard textbooks (e.g., Isselbacher et al., 1980), and is taken to represent the knowledge required for generating an expert diagnosis.

The nodes represent cues (critical in rectangles; relevant in rounded rectangles), the components of the diagnosis (filled ovals), and some of the pathophysiology linking them (text in italics). The links in the reference model represent relations between the nodes; most are causal relations (CAU) with the exception of an occasional association relation (AND), category relation (CAT), or conditional dependency relation (COND). (These relations are defined in Frederiksen, 1975.) The nodes and links are arranged to reflect the chain of ramifications which tie the disease to its manifestations: the central determinants of the patient's condition (in the middle of the diagram) cause particular conditions which eventually lead to the observed signs,

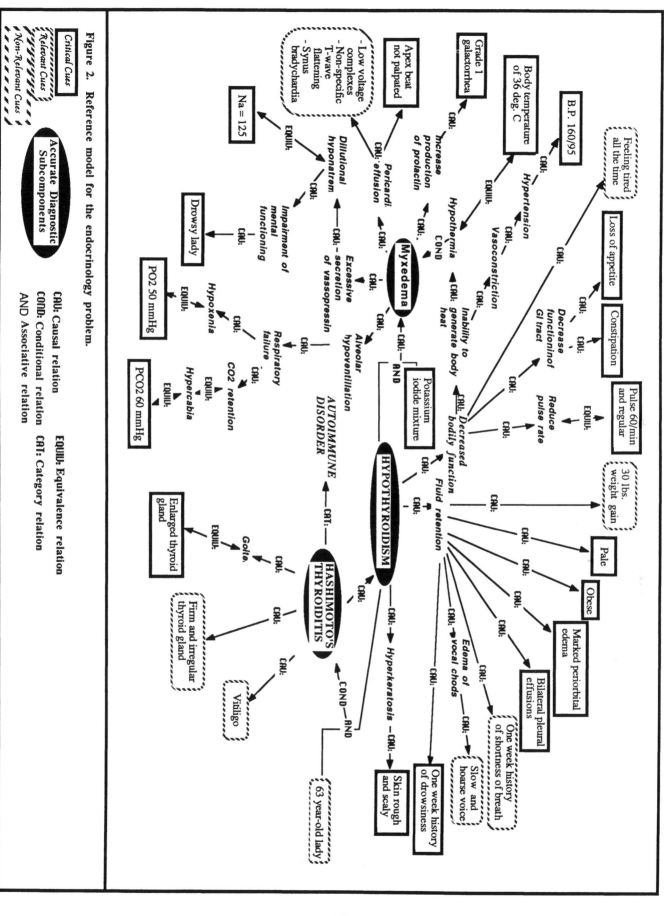

Figure 2. Reference model for the endocrinology problem.

CAU: Causal relation
EQUIV: Equivalence relation
COND: Conditional relation
CAT: Category relation
AND Associative relation

249

symptoms or laboratory results (at the periphery of the diagram). The links also represent the rules necessary to generate an accurate solution to the diagnostic problem (Patel & Groen, 1986). The reference model is used here to compare and contrast the representations of the problem which subjects generate after the presentation of each text sentence.

The HDK subjects are expected to generate a representation that is better organized and structured than the LDK subjects. Thus far the results have indicated that the most important difference between the two groups is a better organization of the information which allows a more efficient pattern of hypothesis generation and confirmation by the HDK subjects. In addition, the representation generated by the HDK subjects should be more focussed than that of the LDK subjects, primarily due to differences in domain–specific knowledge between the two groups of subjects. The results of the detailed analysis of the sample protocols will precede the results of the overall analysis of the generation of the problem representations. A detailed analysis of two sample protocols will be described in the next section.

DETAILED ANALYSIS OF TWO SAMPLE PROTOCOLS

The mapping of the subjects' representations on the reference model will be discussed for Segment nos. 3, 4, and 7. These three segments are chosen because they correspond to the segments in which the HDK subjects generated the accurate diagnostic subcomponents. The mapping consists in doing an overlay of the subjects' representation onto the reference model (Figure 2). The different types of text segments are in boxes of different types and the diagnostic hypotheses generated by the subjects are surrounded by dark circles. The pathophysiological mechanisms used by the subjects are in italics. A more detailed description of the method was given in Joseph and Patel (1986a,b). This method was originally used by Patel & Groen (1986) to study solution strategies of experts and novices in diagnostic reasoning in medicine.

The mapping of the HDK subject is expected to be more closely related to the reference model. His representation will be more organized (contain more links) and more focused (fewer types of diagnostic hypotheses generated). The HDK subject is also expected to elaborate more on the findings about the patient, i.e., generate more links from diagnosis to pathophysiological mechanisms and findings, than the LDK subject. The representation of the HDK subject is also expected to be more focused than that of the LDK subjects, due to the two-stage pattern of hypothesis generation of the HDK subjects. In addition, in the second stage the HDK subjects generated very few additional or new diagnostic hypotheses.

SEGMENT #3

The mapping of the HDK and LDK subject's representation for Segment 3 onto the reference model is illustrated in Figure 3. The findings presented in that segment activated the generation of the general diagnostic subcomponent (Hypothyroidism). After that the subject elaborated on the general condition of the patient's state, generating causal links from pathophysiological processes to other text cues (e.g., respiratory distress causing the one week history of shortness of breath and the fact that hypothyroidism caused some fluid retention which in turn caused the patient's weight gain).

The organizational pattern of the LDK's representation is also consistent with the previous segments. In this case, however, it is the evaluation of individual findings which leads to the generation of more diagnostic hypotheses (psychiatric problem, depression, and decreased glandular function). Nonetheless, the LDK subject generated a general description of the first diagnostic subcomponent (endocrine disorder or one of decreased glandular function).

The HDK subjects seem to organize the information in a way that limits the generation of multiple hypotheses, whereas the LDK subject is not able to use such efficient patterns of organization of information to limit the generation of new hypotheses. Both subjects generate the first subcomponent of the diagnosis.

SEGMENT #4

The mapping of the HDK and LDK subject's representation for this segment onto the reference model is illustrated in Figure 4. At the presentation of this segment the HDK subject generated the second subcomponent of the accurate diagnosis (myxedema), and used the same links as those illustrated in the reference model. The LDK subject also generated the second subcomponent of the diagnosis. However, as was the case in the preceding segment (#3) the subject also provided a more general explanation for myxedema: "advanced hypothyroidism".

250

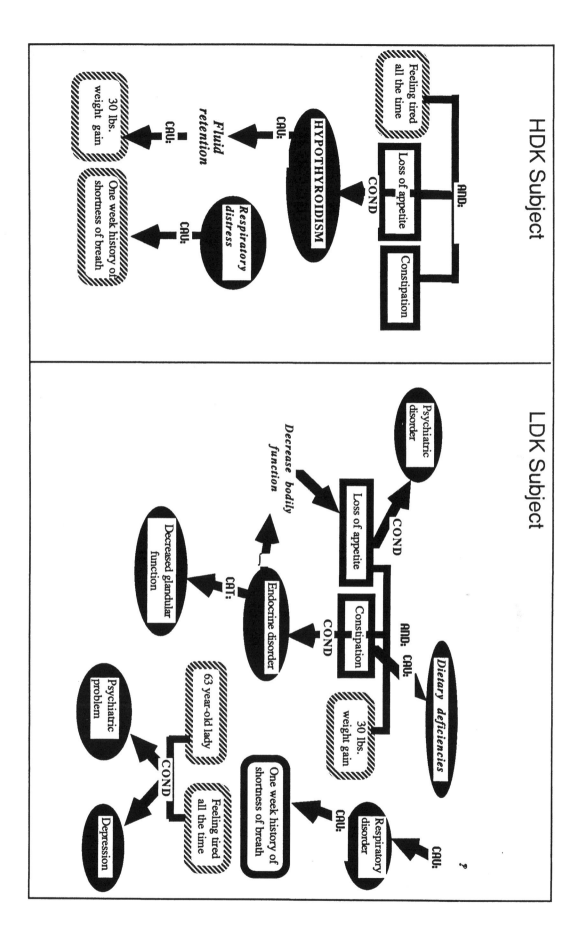

Figure 3. Parts of the problem representations generated by HDK and LDK subjects in response to sentence 3.

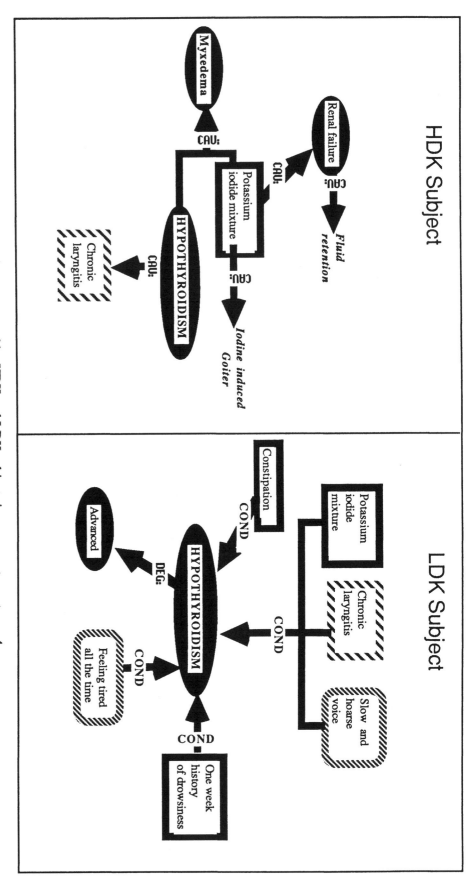

Figure 4. Parts of the problem representations generated by HDK snd LDK subjects in response to sentence 4.

252

The mapping of the HDK and LDK subjects' representation for this segment onto the reference model is illustrated in Figure 5. The HDK subject generated the specific subcomponent of the accurate diagnosis from the finding presented in that segment.

The LDK subject also recognized the autoimmune process. However, the LDK subjects had the wrong category of autoimmune process (pernicious anemia). This is interpreted as an indication of the difference in domain-specific knowledge between the HDK and LDK subject.

Overall, the results from the analysis of the problem representation generated by the HDK and the LDK subjects suggest that while there is overlap in the generation of the first two diagnostic subcomponents, on the one hand, the diagnoses of the HDK subject are more specific and precise, and on the other, the LDK subject provides a general description of the disease. Secondly, the pattern of generation of the two subcomponents is different for both subjects. The difference is mainly in the organization of the information that leads to the generation of the diagnostic subcomponents. For example, the HDK subject generates the second diagnostic subcomponent (myxedema) by associating the finding that the patient was hypothyroid and was given a potassium iodide mixture. On the other hand, the LDK subject generated a general description of the diagnosis by associating the finding of potassium iodide with the slow and hoarse voice. The HDK subject's elaboration seems to be more complete and focused than the one of the LDK subject. The representation of the HDK subject maps more directly onto the reference model than that of the LDK subject. Finally, the LDK subject re-evaluates the explanation of findings after the presentation of the last two segments, while the HDK subject does very little re-evaluation.

The pattern used by the two subjects in generating a representation seems consistent with the suggestion of Gick & Holyoak, (1983) that the HDK subject's construction of his representation is schema–driven. During the construction of the problem representation certain findings (text cues) seem to activate a disease schema from which the solver extracts the given and the goal information and connects it to existing knowledge so that an integrated representation can be formed. This explains the organizational difference (multitude of links generated from individual cues as well as the number of disease hypotheses generated from the selected cues) between the HDK and the LDK subjects during their on-line interpretation. Hence, for the expert, solving a problem begins with the identification of the right solution schema, and then the exact solution procedure involves the instantiation of the relevant pieces of information as specified in the schema. While novices also solve problems in a schema-driven way, their schema of problem types is more incomplete and incoherent than those possessed by the experts.

The results of this study raise some questions regarding the assumption of the hypothetico-deductive model as a characteristic of expert reasoning in medicine as formulated by some researchers (cf. Elstein et al , 1978). The results presented above seem to make most sense in the context of a two-step model of clinical reasoning. The first step is that of generating a coherent problem representation, and the second is that of evaluating its goodness-of-fit. (confirmation and ruling out of diagnostic hypotheses)

The processes of organizing the incoming case information are clearly the most important to the study of diagnostic reasoning (see Patel, & Frederiksen, 1984). The physician receives a list of unrelated cues (at least none of the relations are made explicit in the case description) and has to impose on that list an organization that maximizes its coherence, normally by constructing links (or adding relations) between the cues and one or more diagnoses. The resulting structure of cue-nodes and hypothesis-links is what is usually referred to as the physician's problem representation. The analysis of the process of generation of the hypotheses produced in building these representations suggests that experts divide their time assessing a case description between a first phase of active generation of hypotheses and a second in which this generation activity is absent. It seems reasonable to assume that the first phase corresponds to the organizational process of representing the problem, and the second phase corresponds to a different process, presumably one of evaluating the representation. The fact that the time course of hypothesis generation (time in which two of the accurate diagnostic subcomponents) was very similar for the two groups in the first, organizational, phase suggests a limited role of domain-specific knowledge - instead it may be determined by general problem solving skill.

The second hypothesized step is that of evaluating the problem representation(s)

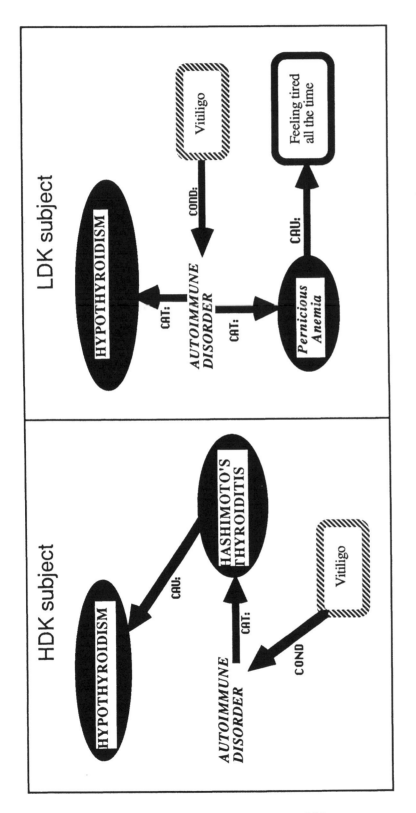

Figure 5. Parts of the problem representations generated by HDK and LDK subjects in response to sentence 7.

produced. Presumably, problem representations would have to be evaluated with respect to coverage of the cues presented, internal coherence, and relative certainty with which one hypothesis can be proposed over another. Evidence for a two-step model comes from the clear change in hypothesis-generating activity after sentence seven (for the HDK subjects), and from the fact that the LDK subjects continued to generate new hypotheses even after they had produced all of the components of the correct diagnosis. This suggests a more important role of domain-specific knowledge in evaluating hypotheses than in generating them.

This study also has implications for researchers in the field of artificial intelligence, especially for the development of expert systems (Buchanan & Shortliffe, 1984) and intelligent tutoring systems (ITS) (Anderson, Boyle & Yost, 1985; Clancey, & Letsinger, 1984; Shute & Bonnar, 1985). A central problem with the development of these systems is generating models of expert knowledge. The traditional method for developing expert models has been based on the use of protocol analysis methods to extract information from interviews with experts in a particular area. While the use of this method has not been unsuccessful, a number of researchers have become concerned about the validity and efficiency of the intuitive, non-formal methods used in protocol analysis and the limited information it provides about the expert's use of his knowledge (Kassirer et al., 1978). The use of an on-line task-environment with the combination of protocol and discourse analysis techniques seems to be a first step towards the solving this complex problem by complementing qualitative data to supplement the quantitative data normally used. This study has demonstrated the richness of the protocols thus produced.

An obvious extension of this research involves the use of an interactive on-line task environment in which subjects can control the order and content of the information that is presented by requesting what they need at a particular point in solving a problem. The main difference between such a task environment and the one used in this study is that the information here was presented in a fixed order not under the subject's control. A more flexible environment will allow subjects to explore different hypotheses freely, ask questions at any time, and backtrack to review information presented earlier. The protocols and time data thus produced, when analyzed with more refined versions of the methods used here, promise to yield much more detailed information about the intricate processes of problem solving in technical-scientific domains.

ACKNOWLEDGMENTS

We would like to thank Drs. Y. Patel, S. Magder, and D. St. Vil for their assistance with the medical aspects of the study. We particularly want to acknowledge the support of the physicians who generously volunteered their time. We also wish to thank M. Dillinger, D. Blitz, L. Coughlin and A. Cruess for their help. This work was supported in part by grants from The Josiah Macy Jr. Foundation (No. B852002) and the Natural Sciences and Engineering Council of Canada (A2598) to Vimla Patel.

REFERENCES

Anderson, J. R., Boyle, C. F., & Yost, G. (1985). The geometry tutor. In A. Joshi (Ed.), Proceedings of the Ninth International Joint Conference on Artificial Intelligence, vol. 1 (pp. 1-7). Los Altos, CA: Morgan Kaufman.

Barrows, H. S., Feltovich, J. W., Neufeld, V. R., & Norman, G. R. (1978). Analysis of the clinical methods of medical students and physicians. Final report, Ontario Department of Health. Grants ODH-PR-273 and ODH-DM-226 Hamilton Ontario, Canada, MacMaster University.

Barrows, H. S., & Tamblyn, R. M. (1980). Problem-based learning: An application to medical education, New York: Springer Verlag.

Brachman, J. R., & Levesque, J. H. (Eds.) (1985). Readings in knowledge representation, Los Altos, CA: Morgan Kauffman Publishers.

Buchanan, B. G. & Shortliffe, E. H. (1984). Rule-Based expert systems: The MYCIN experiments of the Stanford Heuristic Programming Project. Reading, MA: Addison-Wesley.

Chase, W. G., & Simon, H. A. (1973). Perception in chess. Cognitive Psychology, 1, 55-81.

Clancey, W. J. & Letsinger, R. (1984). NEOMYCIN: Reconfiguring a rule-based expert system for application to teaching. In W. J. Clancey & W. H. Shortliffe (Eds.), Readings in medical artificial intelligence: The first decade (pp. 361-381). Reading, MA: Addison Wesley.

Dillinger, M. (1984). Segmentation and Clause Analysis. Technical Report No. 3, McGill University Discourse Processing Laboratory, Montreal, Quebec.

Ericsson, K. A., & Simon, H. A. (1984). Protocol analysis: Verbal reports as data. Cambridge, MA: MIT press.

Elstein, A. S., Shulman, L. S., & Sprafka, S. A. (1978). Medical problem solving: An analysis of clinical reasoning. Cambridge, MA: Harvard University Press.

Feltovich, P. J. & Barrows, H. S. (1984). Issues in generality in medical problem solving. In H. G. Devolder & H. Schmidt (Eds.). Tutorials in problem based learning. Assen, Holland: Van Gorcum.

Frederiksen, C. H. (1975). Representing logical and semantic structure of knowledge acquired from discourse. Cognitive Psychology, 7, 371-458.

Frederiksen, C. H. (1986). Cognitive models and discourse analysis. In C. R. Cooper & S. Greenbaum (Eds.), Written communication annual, vol. 1: Linguistic approaches to the study of written discourse. Beverly Hills, CA: Sage.

Gick, M. L., & Holyoak, K. J. (1983). Schema induction and analogical transfer. Cognitive Psychology, 15 (1), 1-38.

Greeno, J. G., & Simon, H. A. (1985). Problem solving and reasoning. In R. C. Atkinson, R. J. Herrnstein, G. Lindzey, and R. D. Luce (Eds.), Stevens' handbook of experimental psychology (Revised Edition), New York: Wiley.

Groen, G. J., & Patel, V. L. (1985). Medical problem solving and cognitive psychology: Some questionable assumptions. Medical Education 19, 95-100.

Halliday, M. (1967). Notes on transitivity and theme in English, part 2. Journal of Linguistics, 3, 177-274.

Isselbacher, K. J., Adams, R. D., Brunwald, E., & Wilson, J. D. (Eds.). (1980). Harrison's Principles of Internal Medicine, Ninth Edition. New York: McGraw-Hill.

Johnson, P. E. (1983). What kind of expert should a system be? The Journal of Medicine and Philosophy, 8, 77-97.

Joseph, G.-M. (1987). The Time Course of diagnostic information processing: Levels of expertise and problem representation. Unpublished master's thesis, McGill University, Montreal Canada.

Joseph, G-M. & Patel, V. L. (1986a). Specificity of expertise in clinical reasoning. Proceedings of the Eight Annual Conference of the Cognitive Science Society, Hillsdale, NJ: Lawrence Earlbaum, 331-343.

Joseph, G.-M., & Patel, V. L. (1986b). Online analysis of expertise in medical problem solving. Technical Report # 11, Center for Medical Education, Faculty of Medicine, McGill University.

Kassirer, J. P., Kuipers, B. J., & Gorry, G. A. (1982). Toward a theory of clinical expertise. The American Journal of Medicine 73, 251-259.

Larkin, J., McDermott, J., Simon, D. P., & Simon, H. A. (1980). Expert and novice performance in solving physics problems. Science, 208, 1335-1342.

Newell, A., & Simon, H. A. (1972) Human problem solving. Englewood Cliffs: NJ, Prentice Hall.

Patel, V. L., & Frederiksen, C. H. (1984). Cognitive processes in comprehension and knowledge acquisition by medical students and physicians. In H. Schmidt & H.G. DeVolder (Eds.), Tutorials in Problem Based Learning. Assen, Holland: Van Gorcum.

Patel, V. L., & Groen, G. J. (1986). Knowledge based solution strategies in medical reasoning. Cognitive Science, 10, 91-116.

Pople, H. E. (1982). Heuristic methods for imposing structure on ill-structured problems: The structuring of medical diagnostics. In P. Szolovits (Ed.), Artificial Intelligence in Medicine (pp. 119-190). Boulder, CO: Westview Press

Schwartz, S., & Griffin, T. (1986). Medical thinking: The psychology of medical judgement and decision making. Springer-Verlag.

Shortliffe, E. H., Buchanan, B. G., & Feigenbaum, E. A. (1979). Knowledge engineering for medical decision making: A review of computer-based clinical decision aids. <u>Proceedings of IEEE</u>, <u>67</u>, 1207-1224.

Shute, V. & Bonar, J. (1986). Intelligent tutoring systems for scientific inquiry skills. In <u>Proceedings of the Eighth Annual Conference of the Cognitive Science Society</u> (pp. 353-370). Hillsdale, NJ: Lawrence Erlbaum.

Simon, D. P., & Simon, H. P. (1978). Individual differences in solving physics problems. In R. Siegler (Ed.), <u>Children's thinking: What develops?</u> Hillsdale, N.J.: Lawrence Erlbaum.

Winograd, T. (1983). <u>Language as a Cognitive Process, vol. 1: Syntax</u>. Reading, MA: Addison Wesley.

Wortman, P. M. (1972). Medical diagnosis: An information-processing approach. <u>Computers and Biomedical research</u>, 5, 315-328.

PROCESS AND CONNECTIONIST MODELS OF PATTERN RECOGNITION

DOMINIC W. MASSARO & MICHAEL M. COHEN

Program in Experimental Psychology, University of California,
Santa Cruz, California 95064 U.S.A.

Abstract

The present paper explores the relationship between a process/mathematical model and a connectionist model of pattern recognition. In both models, pattern recognition is viewed as having available multiple sources of information supporting the identification and interpretation of the input. The results from a wide variety of experiments have been described within the framework of a fuzzy logical model of perception. The assumptions central to this process model are 1) each source of information is evaluated to give the degree to which that source specifies various alternatives, 2) the sources of information are evaluated independently of one another, 3) the sources are integrated to provide an overall degree of support for each alternative, and 4) perceptual identification and interpretation follows the relative degree of support among the alternatives. Connectionist models have been successful at describing the same phenomena. These models assume interactions among input, hidden, and output units that activate and inhibit one another. Similarities between the frameworks are described, and the relationship between them explored. A specific connectionist model with input and output layers is shown to be mathematically equivalent to the fuzzy logical model. It remains to be seen which framework serves as the better heuristic for psychological inquiry.

Introduction

A growing consensus in pattern recognition is that there are multiple sources of information that the perceiver evaluates and integrates to achieve perceptual recognition. Consider recognition of the word *performance* in the spoken sentence

The actress was praised for her outstanding performance.

Recognition of the word is achieved via a variety of bottom-up and top-down sources of information. Top-down sources include semantic, syntactic, and phonological constraints and bottom-up sources include audible and visible features of the spoken word (Massaro, in press a, b).

A Fuzzy Logical Framework for Pattern Recognition

According to the this framework, well-learned patterns are recognized in accordance with a general algorithm, regardless of the modality or particular nature of the patterns (Massaro, 1979; 1984a, 1984b, in press b; Oden, 1978, 1981). The model has received support in a wide variety of domains and consists of three operations in perceptual (primary) recognition: feature evaluation, feature integration, and pattern classification. Continuously-valued features are evaluated, integrated, and matched against prototype descriptions in memory, and an identification decision is made on the basis of the relative goodness of match of the stimulus information with the relevant prototype descriptions. The model is called a fuzzy logical model of perception (abbreviated FLMP).

Central to the FLMP are summary descriptions of the perceptual units (Oden & Massaro, 1978). These summary descriptions are called prototypes and they contain a conjunction of various properties called features. A prototype is a category and the features of the prototype correspond to the ideal values that an exemplar should have if it is a member of that category. The exact form of the representation of these properties is not known and may never be known. However, the memory representation must be compatible with the sensory representation resulting from the transduction of the input. Compatibility is necessary because the two representations must be related to one another. To recognize an object, the perceiver must be able to relate the information provided by the object itself to some memory of the object category.

Prototypes are generated for the task at hand. The sensory systems transduce the physical event and make available various sources of information called features. During the first operation in the model, the features are evaluated in terms of the prototypes in memory. For each feature and for each prototype, feature evaluation provides information about the degree to which the feature in the speech signal matches the corresponding feature value of the prototype.

Given the necessarily large variety of features, it is necessary to have a common metric representing the degree of match of each feature. Two features must share a common metric if they eventually are going to be related to one another. To serve this purpose, fuzzy truth values (Zadeh, 1965) are used because they provide a natural representation of the degree of match. Fuzzy truth values lie between zero and one, corresponding to a proposition being completely false and completely true. The value .5 corresponds to a completely ambiguous situation whereas .7 would be more true than false and so on. Fuzzy truth values, therefore, not only can represent continuous rather than just categorical information, they also can represent different kinds of information. Another advantage of fuzzy truth values is that they couch information in mathematical terms (or at least in a quantitative form). This allows the natural development of a quantitative description of the phenomenon of interest.

Feature evaluation provides the degree to which each feature in the stimulus matches the corresponding feature in each prototype in memory. The goal, of course, is to determine the overall goodness of match of each prototype with the stimulus. All of the features are capable of contributing to this process and the second operation of the model is called feature integration. That is, the features (actually the degrees of matches) corresponding to each prototype are combined (or conjoined in logical terms). The outcome of feature integration consists of the degree to which each prototype matches the stimulus. In the model, all features contribute to the final value, but with the property that the least ambiguous features have the most impact on the outcome.

The third operation during recognition processing is pattern classification. During this stage, the merit of each relevant prototype is evaluated relative to the sum of the merits of the other relevant prototypes. This relative goodness of match gives the proportion of times the stimulus is identified as an instance of the prototype. The relative goodness of match could also be determined from a rating judgment indicating the degree to which the stimulus matches the category. The pattern classification operation is modeled after Luce's (1959) choice rule. In pandemonium-like terms (Selfridge, 1959), we might say that it is not how loud some demon is shouting but rather the relative loudness of that demon in the crowd of relevant demons. Two important predictions of the model are 1) two features can be more informative than just one and 2) a given feature has a greater effect to the extent a second feature is ambiguous.

Relationship to Connectionist Models

The framework provided by the FLMP anticipated many of the distinguishing properties of new connectionism (Massaro, 1986a, 1986b; Oden & Rueckl, 1986). A connectionist model of perception (CMP) also is an information-processing system having and manipulating information (McClelland & Rumelhart, 1986). The information is represented in terms of the activations and inhibitions of neural-like units. The units are assumed to exist at different levels; for example, the TRACE model of speech perception (McClelland & Elman, 1986) consists of units at the feature, phoneme, and word levels. The units interact with one another via connections among the units. The connectivity is implemented by positive and negative weights that are either specified in advance or learned through feedback.

A prototypical connectionist framework shares several fundamental properties with the current theoretical framework as instantiated in the FLMP. First, both frameworks assume continuous rather than discrete representations; the fuzzy truth values of the FLMP are analogous to the continuous levels of activation and inhibition of connectionist models. Second, both frameworks acknowledge the existence of multiple simultaneous constraints on human performance. Both frameworks provide an account of the evaluation and integration of multiple sources of information in pattern recognition. Third, there is the parallel assessment of multiple candidates or hypotheses at multiple levels in both models. Fourth, both frameworks provide a common metric for relating qualitatively different sources of information. In the FLMP, each source of information is represented by fuzzy truth values representing the degree to which alternative hypotheses are supported. Activation level plays the analogous role in connectionist models. Fifth, the automatic categorization of a novel instance can be accomplished in both frameworks. Finally, both frameworks conceptualize pattern recognition as finding the best fit between the relevant constraints and the pattern that is perceived.

The close fit between the present framework and connectionism dictates an exploration of their similarities and differences. Although the two frameworks appear to agree on important theoretical criteria, the specific models to date differ in terms of the amount of connectivity in the system. The FLMP assumes no top-down influences of a higher-level unit on activation of a lower-level unit and no inhibition among units at a given level. Connectionist models, such as the interactive activation models of written word recognition and speech perception, usually make both of these assumptions. As presently formulated, many of the connectionist models with two-way connections among different levels of units and connectivity among units at a given level are too powerful. They are capable of predicting not only observed results but also results that do not occur (Massaro, 1986a). That is, some connectionist models can simulate results that have not been observed in psychological investigations and results generated by incorrect process models of performance (Massaro, in preparation).

Mathematical Equivalence of Two Models

It can be shown that the FLMP makes mathematically equivalent predictions to those made by a two-layer CMP, with input and output units. As in all instantiations of a theory, particular assumptions must be made about the description of the results of interest. Different assumptions would probably change the relationship between the two models. The models are compared in an expanded factorial designs in which two or more dimensions of information are varied independently of one another in a pattern recognition task. Each of the dimensions is also presented alone. Labeling the dimensions as X and Y, X_i would correspond to the ith level of the X dimension. Similarly, Y_j would correspond to the jth level of the Y dimension. A given stimulus composed of a single dimension would be labeled X_i or Y_j, and a given combination would be represented by $X_i Y_j$.

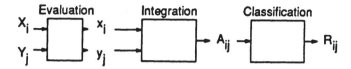

Figure 1. Schematic representation of the three operations involved in perceptual recognition, according to the fuzzy logical model of perception.

Figure 1 illustrates the three stages involved in pattern recognition. The sources of information are represented by uppercase letters. The evaluation process transforms these into psychological values (indicated by lowercase letters) that are then integrated to give an overall value. The classification operation maps this value into some response, such as a discrete decision or a rating.

The FLMP assumes three operations between presentation of a pattern and its categorization, as illustrated in Figure 1. Feature evaluation gives the degree to which a given dimension supports each test alternative. The physical input is transformed to a psychological value, and is represented in lowercase. For a given response alternative A_{ij}, X_i would be transformed to x_i, and analogously for dimension Y_j. Each dimension provides a feature value at feature evaluation. Feature integration consists of a multiplicative combination of feature values supporting a given alternative A_{ij}. If x_i and y_j are the values supporting alternative A_{ij}, then the total support for the alternative A_{ij} would be given by the product $x_i y_j$.

$$A_{ij} \quad : \quad x_i y_j.$$

The third operation is pattern classification, which gives the relative degree of support for each of the test alternatives. In this case, the probability of an A_{ij} response given X_iY_j is

$$P(A_{ij} | X_i Y_j) = \frac{x_i y_j}{\sum} \tag{1}$$

where $\sum$ is equal to the sum of the merit of all relevant alternatives, derived in the same manner as illustrated for alternative A_{ij}.

The CMP is assumed to have an input layer and an output layer, with all input units connected to all output units. It is assumed that each level of each dimension is represented by a unique unit at the input layer. Each response alternative is represented by a unique unit at the output layer. Figure 2 gives a schematic representation of two input units connected to a single output unit.

An input unit has zero input, unless its corresponding level of the stimulus dimension is presented. Presentation of an input unit's target stimulus gives an input of one. The activation of an output unit by an input unit is given by the multiplicative combination of the input activation and a weight w. With two active inputs X_i and Y_j, the activation entering output unit A would be

$$A_{ij} \quad : \quad x_i + y_j$$

where $x_i = w_i X_i$ and $y_j = w_j Y_j$. The total activation leaving an output unit is given by the sum of the input activations, passed through a sigmoid squashing function (McClelland & Rumelhart, 1986).

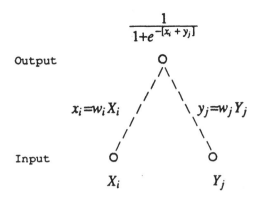

$$\frac{1}{1+e^{-[x_i + y_j]}}$$

Output   O

$x_i = w_i X_i$ $y_j = w_j Y_j$

Input   O   O

X_i Y_j

Figure 2. Illustration of connectionist model with two input units and one output unit.

$$A_{ij} \quad : \quad \frac{1}{1+e^{-[x_i + y_j]}}$$

A connectionist model does not specify completely the input-output relationship. The output activations have to be mapped into a response, and Luce's choice rule is usually assumed to describe this mapping (McClelland & Rumelhart, 1985). Taking this tack, the activation A_{ij} transformed into a response probability by Luce's choice rule gives

$$P(A_{ij} \mid X_i \ Y_j) = \frac{\dfrac{1}{1+e^{-[x_i + y_j]}}}{\Sigma} \tag{2}$$

where Σ is equal to the sum of the activations of all relevant outputs, derived in the manner illustrated for alternative A_{ij}.

The FLMP does not specify the psychophysical relationship between the physical stimulus and its sensory transformation. Neither does a connectionist model; both models require free parameters to specify this relationship. The free parameters are weights in the connectionist model and truth values in the FLMP. A unique weight is assumed for each level of each dimension in the CMP, and a unique truth value is required for each level of each dimension in the FLMP. Thus, the same number of free parameters is required by the two models. The number of free parameters is equal to the number of levels of the X dimension plus the number of levels of the Y dimension. Although a threshold unit is sometimes assumed in connectionist models, no such unit is assumed here. We also have Luce's choice rule operating for both the CMP and the FLMP. In this case, a formal equivalence between the two models exists if adding the weighted activations at input and transformed by the sigmoid squashing function is mathematically equivalent to multiplying fuzzy truth values. Given that the CMP's activated X_i and Y_j input units are equal to one, the activations entering an output unit are equal to w_i and w_j. It follows that the activation of an output unit in the CMP is predicted to be $\frac{1}{1+e^{-[w_i + w_j]}}$. The degree of support for a given test alternative for the FLMP is equal to $x_i \ y_j$.

The truth values in the FLMP are constrained between zero and one, following the assumption of fuzzy logic (Zadeh, 1965). Accordingly, $x_i \ y_j$ must lie between zero and one. The sigmoid squashing function also takes on values only between zero and one, even though the weights are unbounded. It follows that the models can make mathematically identical predictions because 1) for every x_i, there is a w_i, and 2) for every y_j, there is a w_j such that $\frac{1}{1+e^{-[w_i + w_j]}}$ equals $x_i \ y_j$. It can be shown that there exists a correspondence between these predictions such that equivalent predictions can be made by the two models.

$$x_i \, y_j = \frac{1}{1+e^{-[w_i + w_j]}} \tag{3}$$

The proof of the above equivalence is most obvious for the single-dimension conditions of the expanded factorial design. There exists a unique relationship between the weights in the CMP and the truth values in the FLMP if an expanded factorial design is used. In this case, it can be proved that $\frac{1}{1+e^{-[w_i]}}$ equals x_i and $\frac{1}{1+e^{-[w_j]}}$ equals y_j. Equivalently, weight w_i equals $-\ln(\frac{1}{x_i} - 1)$. Data from an expanded factorial design always give only one set of parameters for the FLMP, and also force the CMP to come up with a unique set of mathematically equivalent weights. Given this equivalence, we can translate directly between the two kinds of parameters. We might argue also that the truth values are more informative in the FLMP analysis because it is easy to conceptualize values between 0 and 1, and the truth value gives the contribution of a source of information uncontaminated by other sources. This latter feature is another value of independence models relative to models with high interconnectivity, in which the contribution of one source can not be pulled apart from the contribution of other sources.

Similar predictions exist for three or more stimulus dimensions and three or more response alternatives. Increasing the number of response alternatives does not change the relationship between the two models because this increase only affects the number of outputs, and these are handled equivalently by Luce's choice rule in both models. Increasing the number of dimensions adds the same number of terms to both models, preserving the equivalence shown in Equation 3. In the FLMP, the three dimensions of support for alternative A_{ij} would be

$$A_{ij} \quad : \quad x_i \, y_j \, z_k$$

In the CMP, the activation of three input units would give

$$A_{ij} \quad : \quad \frac{1}{1+e^{-[x_i + y_j + z_k]}}$$

where $x_i = w_i \, X_i$, $y_j = w_j \, Y_j$, and $z_k = w_k \, Z_k$. The total activation of an output unit is given by the sum of the three input activations passed through the sigmoid squashing function, and so on for a larger number of inputs.

The FLMP specifies mathematically evaluation and integration processes. The CMP implements evaluation and integration by activations and inhibitions between input units and output units. Evaluation corresponds to the activation along a single connection between an input unit and an output unit. Integration in the CMP corresponds to the sum of all the activations entering a given output unit, and transformed by the sigmoid squashing function. The correspondence between the FLMP and CMP reveals that the two models, couched in different theoretical frameworks, can make identical predictions in practice. A remaining issue is how process and connectionist models differ from one another, and whether there is an advantage of one over the other.

References

Luce, R. D. (1959). *Individual choice behavior*. New York: Wiley.

Massaro, D. W. (1979). Reading and listening (Tutorial paper). In P. A. Kolers, M. Wrolstad, & H. Bouma (Eds.), *Processing of visible language: Vol. 1* (pp. 331-354). New York: Plenum.

Massaro, D. W. (1984a). Building and testing models of reading processes. In P. D. Pearson (Ed.), *Handbook of reading research* (pp. 111-146). New York: Longman.

Massaro, D. W. (1984b). Time's role for information, processing, and normalization. *Annals of the New York Academy of Sciences, Timing and Time Perception, 423*, 372-384.

Massaro, D. W. (1986a, November). *Connectionist models of mind.* Paper given at the twenty-seventh annual meeting of the Psychonomic Society, New Orleans.

Massaro, D. W. (1986b). The computer as a metaphor for psychological inquiry: Considerations and recommendations. *Behavior Research Methods, Instruments, & Computers, 18*, 73-92.

Massaro, D. W. (in press a). Integrating multiple sources of information in listening and reading. In D. A. Allport, D. G. MacKay, W. Prinz, & E. Scheerer (Eds.), *Language Perception and Production: Shared Mechanisms in Listening, Speaking, Reading and Writing.* London: Academic Press.

Massaro, D. W. (in press b). *Speech perception by ear and eye: A paradigm for psychological inquiry.* Hillsdale, NJ: Lawrence Erlbaum Associates.

Massaro, D. W., & Cohen, M. M. (1983). Evaluation and integration of visual and auditory information in speech perception. *Journal of Experimental Psychology: Human Perception and Performance, 9*, 753-771.

Massaro, D. W., & Oden, G. C. (1980). Speech perception: A framework for research and theory. In N. J. Lass (Ed.), *Speech and language: Advances in basic research and practice: Vol. 3* (pp. 129-165). New York: Academic Press.

McClelland, J. L., & Elman, J. L. (1986). The TRACE model of speech perception. *Cognitive Psychology, 18*, 1-86.

McClelland, J. L., & Rumelhart, D. E. (1981). An interactive activation model of context effects in letter perception: Part I. An account of basic findings. *Psychological Review, 88*, 375-407.

McClelland, J. L., & Rumelhart, D. E. (1985). Distributed memory and the representation of general and specific information. *Journal of Experimental Psychology: General, 114* 159-188.

McClelland, J. L., & Rumelhart, D. E. (1986). *Parallel distributed processing* Cambridge: MIT press.

Oden, G. C. (1978). Semantic constraints and judged preference for interpretations of ambiguous sentences. *Memory & Cognition, 6*, 26-37.

Oden, G. C. (1981). A fuzzy propositional model of concept structure and use: A case study in object identification. In G. W. Lasker (Ed.), *Applied systems and cybernetics: Vol. VI* (pp. 2890-2897). Elmsford, NY: Pergamon Press.

Oden, G. C., & Massaro, D. W. (1978). Integration of featural information in speech perception. *Psychological Review, 85*, 172-191.

Oden, G. C., & Rueckl, J. G. (1986, November). *Taking language by the hand: Reading handwritten words.* Paper given at the twenty-seventh annual meeting of the Psychonomic Society, New Orleans.

Selfridge, O. G. (1959). Pandemonium: A paradigm for learning. In *Mechanization of thought processes* (pp. 511-526). London: Her Majesty's Stationery Office.

Zadeh, L. A. (1965). Fuzzy sets. *Information and Control, 8*, 338-353.

Acknowledgements

The research reported in this paper and the writing of the paper were supported, in part, by grants from the National Science Foundation (BNS-83-15192) and the graduate division of the University of California, Santa Cruz. Bruce Bridgeman, Brian Fisher, Ray Gibbs, Alan Kawamoto and other members of the PEP Program made helpful comments on this enterprise.

PROPERTIES OF CONNECTIONIST VARIABLE REPRESENTATIONS[1]

Deborah Walters

Department of Computer Science, State University of New York at Buffalo, Buffalo, NY

Abstract

A theoretical classification of the types of representations possible for variable in connectionist networks has been developed [1]. This paper discusses the properties of some classes of connectionist representations. In particular, the representation of variables in value-unit, variable-unit and intermediate unit representations are analyzed, and a course-fine concept of representation developed. In addition, the relation between the measurement of a feature and it's representation is discussed.

1. Connectionist Networks

Connectionist networks consist of a large number of very simple processing elements, which are highly interconnected, with each processor receiving input from and sending output to many other processors. In a broad sense of connectionism there are various types of connectionist networks: in cellular automata networks, the connections are generally limited to those between nearest neighbors, and the computations are generally deterministic [2]; in cooperative and competitive networks it is the dynamical analysis of feedback, shunting etc. within and between layers of processing units that is generally studied [3]; hile in the "connectionist school" complete interconnectivity is permitted, and either local or distributed representations of features are used [4,5]. This paper is concerned with connectionist networks in this broad sense: it concerns data-parallel processing where each processing element within a group has the same program, and each processing element is connected to other processors which lie within its local neighborhood.

Connectionist networks have been studied for a variety of reasons. One motivation comes from cognitive science, where the desire is to understand the brain as a computational device. As the brain consists of large numbers of massively interconnected simple processing elements, the study of connectionist networks may ultimately aid us in understanding neural computations. However, this goal must be treated with caution; the analogy to neurons must be made at the proper level.

A more recent motivation for studying connectionism has arisen from the computer science emphasis on parallel processing, as connectionist networks are an example of fine-grain parallelism. The style of computing that is possible with such parallelism is very different from that possible with uniprocessor systems, or with parallel systems containing a small number of processors, and thus represent a distinct class of parallel computations. With the advent of the Connection Machine [6], a fine-grain parallel machine, there is increasing motivation to understand the types of computations possible with such hardware.

1.1. The Representation of Variables in Connectionist Networks

Just as a different style of computation is possible in a connectionist network, the styles of representation of variables that are natural for a connectionist network may be very different from the types of representation natural for serial, or coarse grain parallel processing. For example, Feldman and Ballard use explicit local representations of image features, which they refer to as parameter spaces or feature spaces. Shapes or objects are represented by a set of particular values of certain features, and the connectionist computations involved are basically indexing operations into the feature space, and constraint satisfaction between sets of features. Grossberg, however, does not use such a feature representation, but rather investigates the more global patterns of activity in a field of connectionist units. Similarly, Hinton, McClelland and Rumelhart [4], and Kohonen [7] use distributed

---

[1] This research is funded by NSF Grant IST 8409827 awarded to the author.

representations.

A general theoretical classification of the types of variable representations possible in connectionist networks, in which the different existing variable representations can be expressed, and new types of representations can be indicated, has been created [8]. The theoretical framework enables the properties of different connectionist representations to be formally analyzed which allows the principled choice of the optimal representation for a particular application. As the framework has been treated in detail elsewhere [1], only a brief description occurs here.

2. Preliminary Assumptions

The theoretical framework for connectionist representation of variables is based on certain general assumptions about connectionist networks, the processing elements which participate in connectionist networks, and the nature of variables that are to be represented in such networks. These assumptions are enumerated below.

2.1. Connectionist Units

Connectionist networks are to contain a large number of simple, identical processing units, each of which are capable of signaling a limited number of values. For example, if a connectionist unit consisted of an 8-bit memory word, it could represent or signal only 256 separate values. There are two means by which connectionist units can represent information: explicitly through it's level of activity or the value it stores; and implicitly, as each unit in a multi-unit system can represent a different value or range of vales. For example, in representing color, three units could be used; one each to implicitly represent red, blue and green. Each of these units could then explicitly represent the intensity value of it's implied color component.

Notice that it is not the connections between units which encode the information here, as we are primarily concerned with short-term rather than long-term encoding of visual information. Other researchers have used the connections between units to encode longer term information.

2.2. Representation of Variables

The basic property of a variable is that it can take on a range of possible values. For each variable there is an n, and an injection g, such that g maps the set of values of that variable into euclidean n space, and only k distinct values lie along each dimension of a variable.

3. Definitions of Types of Variable Representations

The general theory of variable representations for connectionist networks assumes that representations can be classified in terms of the following three properties.

3.1. Conjunctive versus Disjunctive Representations

For an n dimensional variable, it is possible to have a completely conjunctive, a completely disjunctive or a partially conjunctive / partially disjunctive representation. A completely disjunctive representation could be thought of as containing n separate one-dimensional representations, one for each dimension of the variable. This contrasts with the completely conjunctive representation which would contain only a single n dimensional representation. A partially conjunctive / partially disjunctiove representation would consist of at least one one-dimensional representation, and at least one m-dimensional representation, with $m < n$.

One way to classify representations in terms of this property is to express the number of disjunctive dimensions present in a given representation. With no disjunctive dimensions, the representation would be completely conjunctive, and with n disjunctive dimensions a representation would be completely disjunctive, and intermediate values would be partial.

266

3.2. Variable-unit versus Value-unit Representations

Connectionist representations of variables can also vary in terms of the number of distinct values that each constituent unit can signal. One way to express this is in terms of the "memory size" of the connectionist units. It will be assumed here, without loss of generality, that connectionist units use a binary representation, and the memory size will be expressed in bits.

There are two extremes to the possible variable representations in terms of the memory size of connectionist units required to represent one value of a variable. In the variable-unit representation each unit has $\log_2 k'$ memory size, where k' is the number of distinct values of the variable to be represented. In the value-unit representation each unit has a memory size of 1. A single variable-unit can be used to signal the presence of any of the k' distinct values, while a single value-unit can only represent one of the k' values, and k' value-units would be required to represent any arbitrary value along a dimension.

The two types of coding referred to are the logical extremes of codings well known in the various disciplines which are concerned with connectionist computations. Neurophysiologists refer to the first type as a frequency code, and to the second type a a labeled-line code [9]. Ballard has discussed the implications of each type of coding, which he refers to as variable and value coding respectively [5].

The value unit and variable-unit encodings are the two possible extremes of variable representation in terms of the memory-size of units, and the number of units required to represent a variable. It is also useful to consider a representation which is intermediate between these extremes. In an intermediate representation the memory size of the units is b, $(1 < b < \log_2 k)$. In both biological and machine systems, the intermediate-unit representation is often used.

3.3. Response Overlap

Representations can also vary in terms of response overlap. In a no response overlap representation a particular value of a variable along one dimension is represented by the activity of a single unit. In response overlap representations the activity of each processing unit represents a range of d discrete values of a variable along each dimension, and each particular value is encoded by the activity of a number of overlapping units. Connectionist units participating in a response-overlap representation are said to have a diameter of d.

4. Theoretical Classification of Variable Representations

Variable representations can be classified along the three dimensions described above, which form the connectionist variable representation space (VRS) illustrated in Figure 1 [1]. VRS provides a theoretical framework for describing the connectionist encodings used in previous research. For example, Hinton's coarse coding scheme is a conjunctive, overlapping, value-unit representation [10], while Ballard uses a conjunctive, non-overlapping value representation in his connectionist shape perception algorithm [5]. The neurons in the mammalian striate cortex that are selectively sensitive to a small range of spatial frequencies (or edge widths) and a small range of edge orientations are using a conjunctive, overlapping, intermediate-unit representation. In the MT region of visual processing, there is one set of neurons which are selectively sensitive to velocity, and another set which are selectively sensitive to the direction of motion. These neurons appear to be using a disjunctive, overlapping intermediate representation. In computer vision programs where edge information is represented as an intensity map, and an orientation map, the disjunctive, nonoverlapping variable representation is being used.

Each point in VRS represents a class of variable representations. In order to completely define a specific representation, more than it's location in VRS must be known; in addition the response mapping function must be specified. The response mapping function, f, defines the response of a given connectionist unit to the k values along each dimension of a variable.

267

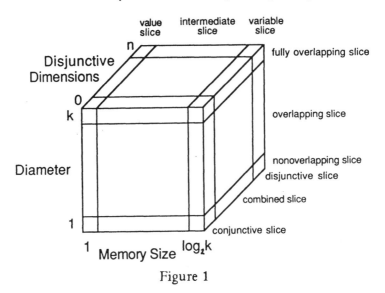

Variable Representation Space (VRS)

Figure 1

5. Measurement Issues: Response Function Analysis

By analyzing the properties of each region of VRS, the optimal class of representations for a given problem can be determined. The chosen representation can then be completely specified by defining the response mapping function, f. However, in an actual implementation such as computer vision, the real problem is to find a method for <u>measuring</u> the feature values present in an image. Thus it is not a question of choosing a variable representation and defining a mapping, because the measurement process defines the response mapping, and therefore constrains the choice of representation. In order to choose the correct representation, the response mapping function for a particular measurement process must be determined. This is an important point that has been previously neglected because much connectionist research has primarily emphasized intermediate and high-level visual processing and thus assumed that the input to the network had already been processed into the appropriate form.

5.1. Determining Response Mapping Functions

In many cases the methods for measuring a feature value directly from an image do not yield a simple, single dimensional mapping. The response of a detector can be a function not only of the value of the desired feature, but also of many related features. Thus the first step in representing variables measured from images is to determine just what is being measured by plotting the multidimensional response function for the measurement process, as a function of a priori image properties. For example, convolving an image with a template is one means of measuring the orientation and amplitude of an edge in an image. The question arises as to how well the output of the convolution correlates with the presence of an edge. To study this, the response function for an edge detector might be plotted against the following properties of edges: location, length, width, amplitude, orientation, curvature, image sampling, signal-to-noise ratio, orientation and/or curvature discontinuities, and edge profile. An example of orientation response functions for one set of oriented edge operators is shown in Figure 2a. Each curve is the response, as a function of the edge orientation, of one operator when convolved with a step edge. From these response functions it is clear that the edge operators create an overlapping, intermediate unit representation. But the response of an edge operator is not a single dimensional function: the response can vary as a function of edge amplitude, width, profile, orientation, and distance from an edge. For example, Fig. 2b shows the responses of one operator to step edges of different amplitudes. The response is obviously a function of both edge orientation and amplitude, which suggests a conjunctive representation of orientation and amplitude. Similar results would be obtained if the other dimensions were explored.

268

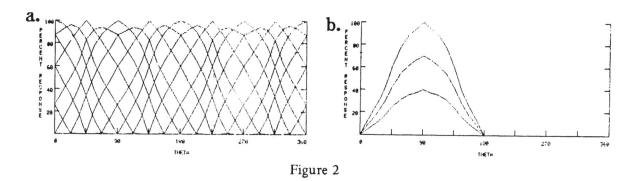

Figure 2

If a particular type of representation is required, then the measurement process must be designed to give the appropriate response mapping function. Thus response function analysis may provide useful constraints in the design of image based operators.

6. Properties of Feature Space Representations

At present only the partial analyzes of particular representations exist: Hinton [10] and Sullins [11] have both analyzed the distributed coarse coding representation; Ballard [5] has investigated some of the implications of value-unit versus variable-unit representations; and, Saund discusses a representation useful for dimensionality reduction [12].

In analyzing representations, several properties should be considered, including the following:

1) Representation of multiple values
The representations differ in terms of the number of distinct values of a feature that can be represented by one copy of the representation. For example, the variable code requires i copies to represent i values, while the value code requires only one copy, as long as there is a one-to-one mapping between values and units. Such differences effect overall coding efficiency.

2) Match between representation and implementation architecture
Connectionist units may have limited memory, which would influence the choice of a representation. For example, value-unit or intermediate-unit encoding is useful for units with small memory size, while variable unit representations are possible when units have enough memory.

3) Total Representation and Generalization
How many copies of a representation are required to represent all possible feature values? In general, the larger the diameter of a unit, the more copies required to simultaneously represent all possible feature values, but this also depends upon the internal code of the units. Thus a system with narrow diameter units has the advantage of being able to simultaneously represent multiple values. But in terms of generalization, the opposite is true. For example, if the mapping is not ordered, then a metric other than the simple distance metric must be used to determine the similarity between feature values. When the unit diameter is broad, a suitable simple metric exists: feature values which both activate the same unit are similar. Intermediate representations provide a useful compromise between total representation and generalizations.

4) Item density
The distribution of the feature values that will be encountered in a given situation is important, as representations differ in the density and distributions which can be handled.

5) Required degree of accuracy
Another property that is influenced by the characteristics of the variable to be represented is the required degree of accuracy, and related sampling issues, as are discussed in Section 6.1.

6) Diameter of response ranges and degree of overlap
These properties influence the generalization capabilities, the efficiency and the suitability for implementation in a particular architecture.

269

7) Probabalistic Representation

An efficient probabalistic representation is suitable for many applications.

These properties can be used to determine the type of representation best suited to represent particular types of information. For example, there are a variety of questions that could be asked about the represented information: "Is value x of feature y present?"; "How many instances of value x of feature y are present?"; "What is the value of feature y at a given spatial position?"; etc. Which types of information should be available thus depends on the nature of the features being represented, and on the types of computations in which the features will be involved.

6.1. Resolution of A Representation

Another property of representations is the resolution to which values of a variable can be encoded. This accuracy will obviously depend upon the sampling resolution of the representation. For example, in the variable-unit representation each unit can distinctly signal the k values of a variable, thus the resolution is k values/dimension, which is the best possible resolution. But even representations with coarse sampling, such as conjunctive, overlapping value units, can have a resolution equal to the variable representation.

Another way of stating the resolution issue is to discuss the degree of accuracy to which the value of a variable can be determined when it is encoded in a particular representation. In the remainder of this section the resolution of the various representations are discussed.

6.1.1. Estimation of the Value of a Variable from a Variable-unit Representation

In all types of variable-unit representations, the value of a variable is represented by the activity of a single processing unit. Thus if f is the response mapping function of a processing unit, then $f(x)$ is the representation of the value x of the variable. If f is a single-valued function, and if $f(x_1) = f(x_2)$, then $x_1 = x_2$, then x can be uniquely determined from $f(x)$ if f is known, and the resolution is k values/dimension.

6.1.2. Estimation of the Value of a Variable from a Value-unit Representation

In the non-overlapping value unit representation there is a separate processing unit, P_x to represent each value x of a variable, thus the activity in a unit uniquely represents the value of a variable, and the maximal resolution is achieved.

In the overlapping value-unit representation, the value of a variable is represented by the activity of a collection of processing units, and the resolution of each unit is coarse. Hinton has shown that the value of a variable represented in a coarse coding scheme (an overlapping, value-unit representation) can be uniquely determined if the number of values being represented is $\leq ((k/d)-1)^n$ [10].

6.1.3. Estimation of a Value of a Variable from an Intermediate-unit Representation

In an intermediate-unit representation with no overlap, each dimension can be broken down into $k/(2^b)$ sections, with a variable unit representation in each section. In this case, the resolution is maximal, and the value of a variable is represented both by the activity-level of a given processing unit, and which unit is active.

In terms of the estimation of a value, the most interesting representation is the overlapping, intermediate-unit encoding. The following analysis applies to the disjunctive class of this type of representation for sparse data.

Assume each dimension of a variable is periodic, with period P, thus $x + P = x$. Further assume response functions are strictly monotonically decreasing for $x \geq r$, and are strictly monotonically increasing for $x \leq r$, where r is the peak value of f. Figure 3a shows an example response function, f_{r_1}, which satisfies this assumption. Assume a given representation consists of m identical response functions, thus for all r_1, r_2, x; $f_{r_1}(x) = f_{r_2}(x + (r_2 - r_1))$. Figure 3b shows a protion of a

270

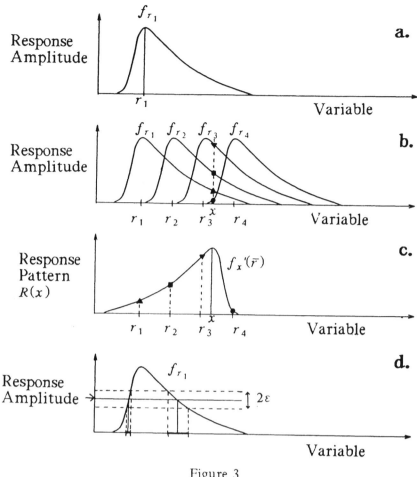

Figure 3

representation, with four response functions, one for each of the four units r_1, r_2, r_3 and r_4.

The total response pattern, $R(x)$, is the set of responses of the m different response functions, thus $R(x) = f_{r_1}(x), f_{r_2}(x), \cdots, f_{r_m}(x)$. The small shapes in Figure 3c are the response pattern which represents x, the value of the variable indicated by the arrow in Figure 3b. Note the response pattern is simply the response of each unit to the value x, replotted at the maximum of the response function (r_i) for each unit.

The mirror-image response of response function f_x to the peak values of the set of response functions is $f_x'(\bar{r}) = f_x'(r_1), f_x'(r_2), ..., f_x'(r_m)$, where $f_x'(r) = f_x(2x - r)$.

From these assumptions, it is possible to show that: $R(x) = f_x'(\bar{r})$, because by assumption 3, with $x = r_i$, it is seen that each element of $R(x)$ is equal to an element of $f_x'(\bar{r})$. This means that $R(x)$ contains all the information needed to obtain x, and that x is found by taking a mirror-copy of the response function, and sliding it along $R(x)$ until the best match is found. Then the maximum value of the mirror-copy will occur at x. Figure 3c show the best fit of $R(x)$, and it's subsequent indication of the value of x.

This analysis shows that only two distinct response units are required, and that they need not be orthogonally spaced over the variable space. However, the minimum permissible distance between r_1 and r_2 is a function of the measurement error, ε, and thus of γ (as defined below), with $|r_1 - r_2| \geq 2\gamma$.

The second assumption can also be relaxed, with the same general results still holding. It is only necessary that the f_i's be strictly monotonically decreasing for $r_i \leq x < y$, for y such that $f_i(y)=0$. The only additional requirement is that there is still a minimum of two nonzero responses for each value of x.

6.1.4. Accuracy of Estimate of x

The accuracy to which x can be found is a function of the accuracy to which the f_i's are defined, and the accuracy with which the $f_i(x)$'s are measured. So if it is assumed that all $f_i(x)$'s are continuously defined, there is no error in the definition of f_i. So assume the error in the measurement of $f_i(x)$ is ε.

Given the above, it is possible to determine x to within $\pm\gamma$, where $\gamma = \min_{i=0,r} (\max(y_{(i,1)}, y_{(i,2)}))$, for $y_{(i,1)}$ such that $f_i(y_1) = f_i(x) + \varepsilon$, and $y_{(i,2)}$ such that $f_i(y_2) = f_i(x) - \varepsilon$. Figure 3d shows the disjoint range of possible estimates of x that results from one particular measured response of f_{r_1}. Note that where the slope of f_{r_1} is steep, γ is small, but the shallow slope yeilds a large γ. This suggests that in order to give the most accurate estimates in the region of it's maxima, the response function should be steep in the region of it's maxima. Thus gaussian shaped response functions [12] are not desirable.

Alternatively, if f_i is sampled at intervals of δ, and $\varepsilon = 0$, then it is possible to determine x to within $\pm\delta$. In other words, under these assumptions, $\gamma = \delta$.

6.1.5. Coarse versus Fine Estimation of x

The previous discussion assumes that the goal of the computation is the accurate estimation of x, given the transfer function f_i and the total response pattern $R(x)$. However, another goal might be the rapid estimation of x, given only the total response pattern $R(x)$. One possible method for estimating x from $R(x)$ alone is: $x = r_i$, such that $f_{r_i}(x) \geq f_{r_j}(x)$, for all $j = 1, m$. This method yields only a coarse estimate of x, there being only m possible values of the estimate. The original method for estimating x yields a much more accurate or fine estimate, but at the cost of requiring more information, and a more complex computation.

When variables are represented by overlapping encodings, there are two modes in which the information can be used: the explicit, coarse representation of a variable can be used to yield a rapid estimate of the value of a variable at a resolution of k/d; while more accurate estimates must be based on the implicit fine representation, which requires more intensive processing, and yields estimates at a resolution of k. Coarse estimates of the values of a variable can be made in parallel across an image, allowing the next stages of processing to proceed in parallel. Fine estimates may require serial processing, and thus not occur automatically over all regions of an image.

As biological systems appear to use overlapping representations of feature variables, they may use the coarse mode to make rapid judgements, such as those in the preattentive, parallel stage of visual perception, while using the more complex, and perhaps serial mode to make fine judgements such as those involved in hyperacuity. For example, humans appear to be able to perform certain visual tasks in parallel, such as the discrimination of two texture regions which differ in terms of the orientation of line elements [13]. Humans can also make very fine discrimination judgements of the orientation of line segments [14]. It may be that such fine judgements require more complex processing and cannot necessarily be made in parallel.

References

1. D. Walters, "The Representation of Variables in Connectionist Networks," Proceedings of First International Conference on Computer Vision, London, England (June, 1987).

2. K. Preston and M.J.B. Duff, *Modern Cellular Automata: Theory and Applications*, Plenum, New York (1984).

3. S. Amari and M.A. Arbib, *Competition and Cooperation in Neural Nets*, Springer-Verlag, New York, NY (1982).

4. G.E. Hinton, J.L. McClelland, and D.E. Rumelhart, "Distributed Representations," pp. 77-109 in *Parallel Distributed Processing*, ed. J.L. McClelland, MIT Press, Cambridge. MA (1986).

5. Dana H. Ballard, "Cortical connections and parallel processing: Structure and function," *The Behavioral and Brain Sciences* **9** pp. 67-120 (1986).

6. W.D. Hillis, *The Connection Machine,* MIT Press, Boston (1985).

7. T. Kohonen, E. Oja, and P. Lehtio, "Storage and Porcessing of Information in Distributed Associative Memory Systems," in *Parallel Models of Associative Memory*, ed. J.A.Anderson,Earlbaum, Hillsdale, NJ (1981).

8. D.K.W. Walters, "Variable Representation in Neural Networks," Proceedings of the First International Conference on Neural Networks, San Diego, CA (June, 1987).

9. H.B. Barlow, "Single units and sensation: A neuron doctrine for perceptual psychology? ," *Perception* **1** pp. 371-394 (1972).

10. G.E. Hinton, "Shape representations in parallel systems''," Proceedings 7th IJCAI, Vancouver, BC (1981).

11. J. Sullins, "Value cell encoding strategies," U. Roch. TR 165 (1985).

12. Eric Saund, "Abtraction and representation of continuous variables in connectionist networks," Proceedings of AAAI (1986).

13. J.Beck, "Textural Segmentation," in *Organization and Representation in Perception*, ed. J.Beck,Erlbaum, Hillsdale, NJ (1982).

14. J.P. Thomas and J. Gille, "Bandwidths of orientation channels in human vision," *J. Opt. Soc. Am.* **69** pp. 652 - 660 (1979).

Individualism and Theories of Action

Steven Davis
Philosophy Department
Simon Fraser University

Abstract

In a recent series of articles Tyler Burge has presented arguments which cut against individualist theories of intentional states. In this paper I shall try to show what consequences Burge's arguments have for individualist theories of behavior. I shall take Jerry Fodor, who is one of the leading exponents of individualism in psychology, as representative of this view. First, I shall lay out one of Burge's arguments against individualist theories of intentional states; second, I shall describe the leading principles of Fodor's individualist metatheory for psychology; and lastly, I shall draw some of the consequences that Burge's arguments have for Fodor's theory of behavior.

* * *

Suppose that I believe that the sun sets over Vancouver Island and suppose that my report of the belief is *de dicto*.[1] It seems that I can have the belief, even if it were false and even if there were no sun nor Vancouver Island. That I have a belief and what belief that I have does not seem to depend upon the existence of anything except me. It appears that for me to have the beliefs that I do only requires that I have certain concepts or notions and the ability to combine them in certain ways. We might think that it follows from this that the conditions of individuation of beliefs depend on what is internal to an individual and not on the external objects to which an individual is related or on his social or linguistic community. One might further hold that a theory of beliefs, and intentional states in general, can be adequate without making reference to anything external to an individual. Borrowing a term from Tyler Burge, we shall call the views about the individuation and explanation of intentional states 'individualist' and theories which presuppose this view 'individualist theories of intentional states'(1979, p. 73).[2]

There are many things which I do which can be described in such a way that what I do does not seem to depend on anything except my existence, for example, my rubbing my thigh or my performing an action with the intention to relieve the pain in my thigh. We can extend Burge's notion to apply, as well, to the individuation of behavior and to theories of behavior. We can say that a criterion of individuation or a theory of behavior is individualist, if it is committed to the view that an adequate criterion or theory can be given without presupposing the existence of anything external to the individual. There is an obvious connection between individualist criterion of individuation and individualist explanatory theories; the latter presupposes the former. Hence, if it can be shown that the individuation of intentional states or of behavior is not individualist, then it follows that the explanatory theories of intentional states and of behavior cannot be individualist.

---

1. In what follows all the reports of intentional states are *de dicto*.

2. There are some states which are regarded to be intentional, such as knowing, which seem to depend for their existence on objects external to the individual. These states do not fall within the domain of individualist theories (Fodor, 1981, p. 228).

In a recent series of articles Tyler Burge has presented arguments which cut against individualist theories of intentional states. In this paper I shall try to show what consequences Burge's arguments have for individualist theories of behavior. I shall take Jerry Fodor, who is one of the leading exponents of individualism in psychology, as representative of this view. First, I shall lay out one of Burge's arguments against individualist theories of intentional states; second, I shall describe the leading principles of Fodor's individualist metatheory for psychology; and lastly, I shall draw some of the consequences that Burge's arguments have for Fodor's theory of behavior.

Let us begin then with Burge's argument against individualism. In this argument Burge presents a thought experiment in which he keeps constant the objects to which a subject is causally related and his internal states and changes only the linguistic practices of his surrounding community. In the actual situation a person, whom I shall call 'Oscar,' has the thought that he has arthritis in his thigh.[3] That is, he uses 'arthritis' to refer to a rheumotoid condition in his joints and a similar pain which he has in his thigh. This use is contrary to the use of those in Oscar's speech community to whom he defers on matters about English, to his doctor's for example. Burge, then, describes a counterfactual situation in which Oscar's internal states remain the same. There is no change in the history of his stimulations, in his internal physical states, in his dispositions to behavior, when his behavior is described non-intentionally, and in the causal relations among them. The only difference is that those in his speech community to whom Oscar defers on linguistic matters use 'arthritis' in the way in which Oscar uses it mistakenly in the actual situation. In the counterfactual situation Oscar does not have the thought that he has arthritis in his thigh, for no one in the counterfactual situation has any notion of arthritis. They have a notion of a disease which can occur in the joints and in thighs which is not a notion of arthritis. Burge suggests that we could introduce the term 'tharthritis' into English as it is actually spoken which would express the notion that 'arthritis' expresses in the counterfactual situation. We, then, could describe the thought that Oscar has in this situation, namely the thought that he has tharthritis in his thigh. But the thought that Oscar has in the actual situation, the thought that he has arthritis in his thigh is not the same as the thought that he has in the counterfactual situation, namely the thought that he has tharthritis in his thigh, for the two thought events do not have the same content, since arthritis is not the same as tharthritis. So in the counterfactual situation Oscar lacks a thought he actually has and he has a thought that he actually lacks. The conclusion of this argument is that there are cases in which a person's internal states do not individuate his intentional states; reference must be made to the practices of the linguistic community of which he is a part. And the conclusion Burge draws from the thought experiments is that a criterion of individuation of intentional states cannot be adequate and be individualist. A necessary condition for their adequacy is that they make reference to objects and linguistic practices which are external to the subject.

Let us now turn to the consequences I think that Burge's thought experiment has for individualist theories of action. I shall take as representative of these views a theory of Jerry Fodor who is one of the leading exponents of individualism in psychology. Fodor holds that our ordinary ways of talking about and explaining actions, when made rigorous and systematic, form the basis of cognitive psychology. What I wish to show is that our ordinary views about actions do not presuppose an individualist criterion of the individuation of action and in so far as psychology adopts our ordinary views about actions its theories cannot be individualist. Before

---

3. Following Burge, I shall use 'actual situation' and 'counterfactual situation' where others might use 'actual world' and 'possible world'.

turning to my arguments for this, I shall lay out the principle doctrines of Fodor's views about how psychology should be done.

Fodor holds then that one way of giving an ordinary explanation of a particular human action is by appealing to a subject's intentional states which cause the action. And one of the goals of cognitive psychology is to utilize this form of explanation. Hence, Fodorian cognitive psychology takes as one of its goals the explanation of individual human actions and presupposes that some of these actions are caused by the subject's intentional states (1982, p. 100). These intentional states are representations which relate a subject to a content and a psychological theory which is committed to this is a representational theory of mind. In addition Fodor holds that representations, instantiated internally in a subject, are in a "language" of thought where only the formal properties, that is, the syntactic properties, of the "sentences" of the language play a causal role. Semantic properties of the representations, such as being true, having a referent or having a meaning, play no role in internal mental processes (1981, pp. 231). It is, then, the formal properties of a subject's internal representations which cause the subject's behavior and by appealing to these internal representations we can explain human behavior (1981, p. 239). On Fodor's view this is tantamount to the hypothesis that mental states and processes are computational.

Fodor maintains that the computational theory of mind entails a version of what Putnam has called "methodological solipsism" (1975, p. 136). On Fodor's view an adequate theory of human behavior which attributes intentional states to subjects need make no appeal to the subject's external environment, including the actual objects with which he is causally related or the social relations in which he is embedded. What explains the subjects behavior are the internal causal relations among formal properties of representational states (1981, p. 244). Fodor's commitment to methodological solipsism is a commitment to what I have called 'individualist' theories of intentional states and of behavior.

The identity conditions for representational states entailed by Fodor's methodological solipsism are formal: if a and b are representational state tokens of a subject who bears the same relation to a and b and a and b have the same syntactic properties, then $a=b$. Moreover, sameness of formal properties entails identity of causal powers, since the causal powers of representational states are contained in their formal properties.

Fodor, further, holds that particular representational states, including wants and beliefs, are token identical to internal physical states of the individuals which have them (1981, p. 9 and p. 145). In the case of humans intentional states are token identical to particular brain states. Consequently, if there is no change in the internal physical states of a subject, then there are no changes in his intentional states. It follows on Fodor's view that if there are no changes in a subject's internal physical states, then there is no change in his behavior, since his behavior is caused by his internal representational states.

The last principle is not one Fodor adopts. But there is nothing in Fodor's work which suggests that he would reject it. If a and b are rigid designators for bits of behavior and there is a cause of a which is possibly not a cause of b, then a is not identical to b. This principle is similar to part of Davidson's criterion for the individuation of event tokens (1969, p. 179). However, he would not accept the modal addition to the antecedent. Despite this, I shall call it 'Davidson's principle'. I take this principle to be plausible, since the cause of an event is what brings about an event. That is, the cause of an event is the origin of an event and the origin of an event is essential to it. This is parallel to the essential origin which Kripke claims for material objects (1980, p. 114). Kripke offers something like a proof for his claim about material objects which, I believe, can be applied to my claim about event tokens. And an event's causes being essential to an event entail what I have called Davidson's principle.

It follows immediately from Fodor's commitment to methodological solipsism that in an adequate psychological theory a subject's behavior cannot be described in such a way that it presupposes the existence of any object other than the subject. We cannot describe Oscar as taking aspirin to relieve the arthritic pain in his thigh, for that presupposes the existence of aspirin and arthritis, but we could describe Oscar as either performing an action which he intends to be a taking of aspirin which relieves the arthritis in his thigh or moving his body in a certain way. Neither seems to contravene methodological solipsism. The former does not, since correctly describing someone as having the intention to take aspirin to relieve arthritis in his thigh does not presuppose the existence of aspirin or arthritis. It might seem that the latter description is contrary to methodological solipsism, since it presupposes the existence of Oscar's thigh which on some views could be taken to be external to him. For the moment I shall suppose that Oscar's thigh is not external to him and that it is not ruled out by methodological solipsism. Methodological solipsism, then, limits psychological theories to descriptions of Oscar's movements of his body or to descriptions of his actions which are described intentionally.

Fodor suggests that his representational theory of mind accords with current theories in cognitive psychology (1981, p. 226). But limiting psychological theories to descriptions of actions which make reference only to a person's moving his body or to the intentional states which bring about the action is to propose a radical revision of current psychological theorizing. Moreover, I believe that Burge's thought experiment creates problems for Fodor's proposals for a cognitive theory of behavior even where the theory is limited in the descriptions it permits for behavior. I shall present two arguments against Fodor's theory which cut against the two sorts of descriptions which methodological solipsism seems to allow for actions. I shall begin with a problem for his theory where the descriptions of actions are descriptions of an agent's moving his body.

Let us suppose that Oscar rubs his thigh because of his beliefs that he has arthritis in his thigh and that rubbing a part of his body which suffers from arthritis will reduce the pain and his desire to lessen the pain in his thigh. We can say, then, that Oscar's action of rubbing his thigh is caused by his beliefs and his desire. In the counterfactual situation nothing changes about Oscar's internal states, that is, about his stimulation patterns, brain states, dispositions to behavior, and the causal relations among them. And on Fodor's theory since representational states are internal physical states, these, too, do not change. If we take Oscar's language of thought to be the language which he speaks, described non-semantically, then there is no difference between the actual and counterfactual situations in the sentences of his internal language. Hence, nothing should change about his behavior, since it is caused by his internal states. Oscar's rubbing his thigh in the actual situation should be identical to his rubbing his thigh in the counterfactual situation. However, in the counterfactual situation Oscar does not have the belief that he has arthritis in his thigh. He cannot have this belief, because he has no notion of arthritis. Hence, in this situation this belief cannot be a causal factor in his rubbing his thigh. But if we suppose that his rubbing his thigh is the same event token in the actual and counterfactual situation and in the counterfactual situation it is not caused by the belief that he has arthritis in his thigh, then this belief cannot play a causal role in the actual situation. Let us suppose that it does. Then, his rubbing his thigh in the actual situation has a cause which it fails to have in the counterfactual situation. But it follows from Davidson's principle that his rubbing his thigh in the actual and counterfactual situations cannot be the same act token. However, if we maintain that in the actual and the counterfactual situations we have the same act token, then we must give up its having a cause in one situation which it does not have in the other. Consequently, it seems that Fodor must give up the causal efficacy of beliefs and with it any hope of explaining actions by generalizing over intentional states. But this dooms the representational/computational theory of mind, since the purpose of the theory is to allow for such generalizations, while providing an account of causal mental states and processes.

Fodor can hold onto the causal efficacy of beliefs and their explanatory role by denying that in the actual and counterfactual situations in Burge's thought experiments the subject performs the same actions. But it would follow that there can be changes in a subject's behavior, where this is not described by making reference to objects external to the subject, even though there is no change in the subject's internal states. It would not, then, be only the formal properties of the internal representational states which cause actions, since they do not change from the actual to the counterfactual situation. Hence, the representational/computational theory of mind which, Fodor argues, entails methodological solipsism must be abandoned, since beliefs which are not identical to any internal states can cause actions. It follows that to have a full account of these beliefs appeal must be made to the linguistic practices of the linguistic community of which Oscar is a part. Thus, reference must be made to objects which are external to the subject. As a consequence, on the assumption that behavior is caused by intentional states a theory which entails methodological solipsism cannot be an adequate theory of behavior.

In the example above Oscar's action was described as an action of his moving his body. We also obtain consequences which are unacceptable to individualist theories of behavior, if we consider an action of Oscar which is intentionally described. Let us suppose that Oscar performs an action which he intends to be an act of taking two aspirins to relieve the arthritic pain in his thigh. And let us suppose further that he performs this action, because he believes that he has arthritis in his thigh and that taking aspirin relieves the pain of arthritis and he desires to relieve the pain in his thigh. In this case, as in the example above, we can say that Oscar's action is caused by his beliefs and desires. In the actual and counterfactual situations in Burge's thought experiment there is no change in Oscar's internal states and hence, on Fodor's theory there should be no change in Oscar's behavior. But in the counterfactual situation Oscar does not have the belief that he has arthritis in his thigh, nor can he have the intention of performing an action which relieves an arthritic pain in his thigh. He can have no such belief or intention in the counterfactual situation, because he and everyone in his speech community lack the notion of arthritis. Hence, in the counterfactual situation Oscar does not perform an action which he intends to be a taking of two aspirin to relieve the arthritic pain in his thigh. Thus, Oscar does not perform the same action in the actual and the counterfactual situations, even though there is no change in Oscar's internal states. Since the same causes should have the same effects, it follows that either in the actual or the counterfactual situation the internal representational states which Fodor's theory attributes to Oscar cannot be the sole causal factor which brings about his actions and cannot, therefore, be used to give a complete explanation of his behavior. Once again appeal must be made to the linguistic community of which Oscar is a part to have an adequate account of his behavior.

I have considered two sorts of descriptions of Oscar's behavior: his moving his body and his performing an action with a certain intention. Perhaps, a way out of the difficulties that I have raised for Fodorian cognitive psychology is to describe Oscar's behavior as bodily movements, rather than either his moving his body or his performing an action with a certain intention. If we consider again the example in the previous paragraph we could take it that there is a bodily movement of Oscar, or a series of them, which is caused by his intentional states which are internal representational states of Oscar and the bodily movement or a series of them is actually contingently identical to his action which he performs with the intention that it be a taking of two aspirin to relieve arthritic pain. However, in the counterfactual situation we would have the same bodily movement, but not the same action, for Oscar, lacking any notion of arthritis, could not have an intention involving this notion and thus, could not perform an action with this intention.

It would, then, be the movements of Oscar's body, so described, which are caused by his intentional states. This does not mean, of course, that actions are not caused by intentional states. Rather, it has as a consequence that what action is caused by a particular set of intentional states can vary depending on changes in the non-internal environment of a subject. In the actual situation Oscar's beliefs and desire cause his performing an action which he does with the intention that it be the taking of two aspirin to relieve arthritic pain in his thigh, since this action is identical to movements of his body.

This will not do. In the actual and the counterfactual situations by hypothesis the representational states are the same, but the actions performed are not. In the actual situation the action performed is an action done with the intention that it be a taking of two aspirin to relieve arthritic pain. This action is not performed in the counterfactual situation. Since it is appeal to representational states which is supposed to explain actions and the representational states in the counterfactual and the actual situations are the same, there is no explanation for the difference in the actions between the actual and the counterfactual situations. Hence, the representational/computational theory of mind cannot be a complete theory of human action.

I have assumed that methodological solipsism allows descriptions of an agent's moving his body and movements of his body which make reference to parts of an agent's body. But I think that a variant of Burge's thought experiment can be used to show that methodological solipsism rules out such descriptions. Let us imagine that Oscar is as we described him in the actual situation, but that in the counterfactual situation he has no legs to rub and that his brain is hooked up to electrodes which produce in him the visual experiences about his legs which are identical to the visual experiences which he has in the actual situation. We suppose further that he has exactly the same beliefs and other intentional states and the same dispositions to behavior which he has in the actual situation. That is, his internal states are identical in the actual and the counterfactual situations. On Fodor's theory cognitive psychology has within its domain the formal properties of internal representational states which can be specified without making any reference to external objects. As Fodor puts it, "...[Representational states] have no access to the *semantic* properties of such representations, including the property of being true, of having referents, or, indeed the property of being representations *of the environment*." (1981, p. 231) But, then, a theory of such representational states cannot explain the difference between Oscar's actually rubbing his thigh and counterfactually believing falsely that he does, since there is no difference in his representational states. Hence, the theory cannot explain Oscar's rubbing his thigh and thus has no place for descriptions of actions which make reference to parts of Oscar's body.

Let me conclude by summarizing what I think that I have shown. On Fodor's view if psychology is successfully to provide a theory of behavior, then it must make reference to the beliefs, desires, and wants, that is, to the intentional states of a subject. Further, Fodor claims that this goal can only be achieved, if the working cognitive psychologist is committed to methodological solipsism. Burge's thought experiments can be taken to show that if Fodor is construing psychological theories to be more rigorous and precise versions of our ordinary want and belief explanations, then these theories must violate methodological solipsism by making reference to objects and linguistic practices external to an individual. My arguments show that if intentional states are the cause of our behavior, then theories of behavior cannot account for it fully and be individualist. That is, they contain descriptions of behavior which violate methodological solipsism. Moreover, and perhaps more importantly if the computational theory of mind entails methodological solipsism, as Fodor contends, then no computational theory can give us a complete account of behavior.

Burge, Tyler. (1979). Individualism and the Mental. <u>Midwest Studies in Philosophy: Studies in Metaphysics</u>, Vol IV, eds. Peter A. French, *et al*. Minneapolis: University of Minnesota Press, pp. 73-122.

——————————. (1982a). Other Bodies. <u>Thought and Object</u>, ed. Andrew Woodfield. New York: Oxford University Press, pp. 97-120.

——————————. (1982b). Two Thought Experiments Revisited. <u>Notre Dame Journal of Formal Logic</u>, XXII, 3, pp. 284-293.

——————————. (1986a). Individualism and Psychology. <u>Philosophical Review</u>, XLV, 1, pp. 3-45.

——————————. (1986b). Intellectual Norms and Foundations of Mind. <u>Journal of Philosophy</u>, LXXXIII, 2, pp. 697-720.

Davidson, Donald. (1969). The Individuation of Events. <u>Essays on Actions and Events.</u> Oxford: Clarendon Press, pp. 163-181.

Fodor, Jerry A. (1981). <u>Representations.</u> Cambridge, Mass.: MIT Press.

——————————.(1982). Cognitive Science and the Twin-Earth Problem. <u>Notre Dame Journal of Formal Logic</u>, Vol.23, 2, pp. 98-118.

Kripke, Saul. (1975). <u>Naming and Necessity.</u> Cambridge, Mass.: Harvard University Press.

Putnam, Hilary. (1975). The Meaning of Meaning. <u>Minnesota Studies in the Philosophy of Science</u>, ed. Keith Gunderson. Minneapolis: University of Minnesota Press, pp. 131-193.

EXTERNAL REPRESENTATIONS AND

THE ADVANTAGES OF EXTERNALIZING ONE'S THOUGHTS

DANIEL REISBERG

REED COLLEGE

We consider some of the functional differences between internal and external representations. In particular, we argue that different knowledge and skills can be brought to bear on external representations than on internal ones; that external representations, as non-intensional entities, are open to reinterpretation in ways that internal representations are not; and that one can discover omissions from external representations that cannot be found in internal representations, suggesting that attention is deployed differently in these two cases. We consider several implications of this view, including the advantages to be gained by externalizing one's thoughts, i.e. by writing out the content of mental propositions or by sketching the content of mental images.

A vast amount of cognitive science is concerned with the formation and use of mental representations. These representations may be "distributed" or "symbolic;" they may be "frames" or "prototypes" and so on. However we conceive them, though, internal representations indisputably play an essential role in cognition. In the present paper, I will argue that there is also an immensely important function served by external representations. This function is not limited to the obvious role that external representations play in communication: To convey my thoughts to you, I must translate them into words or pictures; you must decode these to understand me. Beyond this, though, external representations also serve an important non-communicative function, in the examination and development of one's own ideas. As we will see, the function of external representations derives from the particular nature of internal representations, and, in fact, from the limits of internal representations.

External representations are often used an an _aide memoire_ or as an _aide pensèe_. The scientist writes the hypothesized equations on the blackboard; the artist sketches the lay-out in a notebook. Or, to take a rather different case, I spent some months contemplating the contents of this paper before sitting down to write it. Yet much of the discovery and development happened only when I began to write. I believe this is a common experience -- that translating one's thoughts into words can be extremely instructive; indeed, I will try to argue that some of the instruction is only possible once the thoughts are externalized.

Before proceeding, it is worth noting the diversity of what we are calling external representations. The examples just listed include pictures, equations on a blackboard, and long papers; some of these are symbolic, some are depictive. These are different from each other in many regards, and, for many purposes, can not be grouped together. Nonetheless, I will treat them as interchangeable for present purposes. Indeed, much of the evidence which informs the present discussion concerns a single paradigm case -- the contrast between mental images as internal representations, and pictures as external ones. The issue of whether we can generalize from this case will be a central theme in what follows.

There are of course some trivial reasons why it is helpful to externalize a thought. Working memory is limited in capacity, setting boundaries on how much information can be kept ready-at-hand. Thus one can mentally multiply 2-digit numbers; for 3- and 4-digit numbers, one uses an external aid in retaining the interim results. This is obviously important when contemplating the long series of steps in a complex argument, or the many facets of a complicated problem. Additionally, this is exacerbated when the mere maintenance of the internal representation itself requires effort. Kosslyn (1980, 1983), for example, has argued that maintaining a mental image requires constant effort, as parts of the image "fade" and must be refreshed. This potentially sets limits on both the maximum complexity of mental images and the kinds of effortful operations which can be applied to an image.

These considerations of memory load and of effort are no doubt correct, but there is no reason to believe they exhaust the limitations of internal representations. The following sections describe three other, more interesting, ways that external representations can enhance cognitive performance. To anticipate the argument, I consider some of the ways in which external representations are functionally different from internal ones, by virtue of being open to different operations, and by tapping different skills. Said differently, we will consider the functional limitations of internal representations, and so, by contrast, the advantages of external representations. While our stress will be on the consequent benefits of externalizing one's thoughts, this perspective also has many other implications, and I discuss some of these in the final section.

Perceptual knowledge

There is a growing body of evidence that the knowledge to which one has conscious access may be interestingly different from the knowledge used in perceiving or identifying objects in the world. To start with a trivial case, a baseball outfielder's perceptual anticipation of a flyball's trajectory may (implicitly) require some speedy calculus; the player's explicit, conscious grasp of calculus may be far less sophisticated. Figure 1 illustrates the converse case.

AS SMART AS HE WAS, ALBERT EINSTEIN COULD NOT FIGURE OUT HOW TO HANDLE THOSE TRICKY BOUNCES AT THIRD BASE.

This important distinction between implicitly available and explicitly available knowledge has recently surfaced in discussions of "intuitive physics" (e.g. McCloskey et al., 1980). College students' explicit judgments about inertia and momentum and the like seem systematically incorrect. Even after a year of training in physics, subjects seem to hold an "impetus theory" similar to the one rejected by modern science over 400 years ago. Remarkably, subjects make predictions which are clearly contrary to their two decades of interaction with the world, predicting (for example) anomalous trajectories different from any they have seen in their lifetimes.

Subjects fare far better, though, if we request an implicit judgment rather than an explicit one: Rather than asking subjects to predict a motion, Proffitt and Kaiser (1986) showed subjects videotapes of actual motions and (simulated) anomalous motions, and asked subjects which looked correct. In this mode of test, subjects' intuitive grasp of physics seems quite good, as they reliably reject anomalous motions in this perceptual task. Apparently, then, subjects have perceptual knowledge about motions which exceeds their conscious knowledge: they cannot answer simple problems, but they do know what "looks right."

In a different vein, Jacoby and others have repeatedly shown that conscious memories can easily be dissociated from perceptual learning. For example, Jacoby and Dallas (1981) had subjects study a word-list. After a 24-hour delay, subjects' conscious recognition of these words was at near-chance levels. However, in a tachistoscopic-recognition paradigm, subjects were twice as likely to identify the previously viewed words (compared to control words). More, different variables seem to influence conscious recognition and perceptual identification, with depth of processing (for example) being critical for the former but irrelevant to the latter.

In describing these data, Jacoby argues that prior exposure leads to "perceptual fluency." The fluency by itself does not ensure that the item will be "recognized" as familiar. The recognition will occur only if the perceiver attributes the fluency to the relevant prior encounter. In the absence of this attribution, the item will not be acknowledged as (say) being from the previously viewed list, but, even in this case, the perceptual fluency still has demonstrable effects: the item will be recognized more readily (Jacoby and Dallas, 1981), may be judged as more attractive in appearance (Seamon, 1981), may be judged as being more "prominent" (e.g. in a judgment of relative "fame" - Jacoby, 1985) and so on. Thus we are influenced by perceptual fluency -- things "look right" or "look familiar" -- even when we cannot recall what is right or attribute the familiarity. It remains an open question whether this perceptual knowledge simply provides a more sensitive assessment of what we know, or whether, more strongly, it is a different species of knowledge. One way or the other, though, these identification processes seem to be privileged with regard to various procedures, and it is therefore to one's advantage to be able to tap this knowledge.

The proposal, then, is that one benefit of external representations is that they give us access to knowledge and skills which are otherwise unavailable to us. Subjects in the "intuitive physics" experiments could, at least potentially, improve their performance by actualizing the motions they are attempting to judge; by creating the requisite input for accessing perceptual knowledge, that knowledge could be employed. Subjects in Jacoby and Dallas' memory experiments could likewise employ a species of the "generate and test" strategy -- generating candidate memory items, and then responding on the basis of the fluency with which these are processed. If these cases seem far-fetched, consider a more familiar case: In trying to spell a rarely encountered word, one writes the two possible spellings out on a piece of paper, then judges which "looks right." The suggestion is that this is a sensible strategy, tapping the same skills in perceptual identification which are evident in Jacoby's procedures. It is interesting that one seems unable to derive the same benefit by merely thinking about or imaging the alternate spellings; in these latter cases, one has no access to the perceptual skill. One gains access by externalizing the representation, by creating the relevant input for the perceptual process.

This kind of claim is entirely consistent with (although does not rest on) current claims about the "modularity" of cognitive functioning. The thesis of modularity is that separate and independent processes are employed by various input and output modules; these are inaccessible to and impenetrable by the central processes. Perceptual processes are one clear candidate for "modular" status; the impenetrability, for example, is evident in the fact that illusions remain effective even when one knows one is looking at an illusion. The inaccessibility of these processes is evident in the "unconscious," automatic status of much perceptual "inference." The notion here, however, is that one can in a sense defeat the modularity by externalizing one's thoughts. To put this concretely, the perception module "knows" the correct spelling of a word, but one cannot directly access this knowledge. However, one can offer the perception module a candidate spelling and so to speak learn how the module reacts. In this way, by creating inputs for the module, and by monitoring the module's output, one indirectly gains access to the specific knowledge or skills inhering in the module.

The non-intensionality of external representations

It is extremely important that external representations are both representations and also things in themselves -- ink marks on paper, acoustic signals in the air. Thus the representations exist after we no longer are thinking about them, and they are open to interpretation by others, even if the others do not know the intent with which the representation was created. Indeed, even the creator of the representation has the option of setting his or her understanding of it aside, returning so to speak to the "raw material," and interpreting it anew. This opens the possibility that a different interpretation may be reached, rendering the external representation at least potentially ambiguous.

This potential for ambiguity is not present for internal representations, providing an important functional contrast between these and external representations. This point can be illustrated with an example drawn from our work with mental images (Chambers and Reisberg, 1985; Reisberg and Chambers, 1987). Many have argued that mental images are picture-like depictions, inherently ambiguous, and in need of interpretation (presumably via processes related to perception). In contrast, our claim has been that mental images, as mental representations, are inherently meaningful, representing some particular thing or state of affairs. Mental images are thus intensional entities, embedded in a particular context of understanding, and so necessarily unambiguous.

To test this claim, Chambers and Reisberg led subjects to encode some of the classical "ambiguous figures" (e.g. the duck/rabbit, the Necker cube, etc.).

Subjects were then asked to form mental images of these figures, and to examine the image for an alternative construal of the form. Subjects had previously received practice and instruction with other ambiguous figures, but, critically, were naïve to the test figures. In addition, subjects were not given enough time at encoding to find both construals of the test figures. In these ways, any reconstruals of the image will be bonafide "discoveries," indicating that images can support reinterpretation and so are ambiguous.

Across a variety of procedures, exactly zero subjects succeeded in reconstruing their images. An important control, though, makes clear the contrast between internal and external representations. After subjects had tried (and failed) to reconstrue their image, they were asked to draw a picture from the image, and then to attempt reconstrual of their own drawing. In sharp contrast to the imagery data, 100% of the subjects were able to reinterpret their own drawings. (These drawings were in fact ambiguous, as a new group of subjects was also able to find both construals in them.) Once there was a stimulus on the scene (even a self-created one), subjects could set aside the understanding they had in mind in creating the stimulus, and interpret it anew. In imagery, the understanding is inherent in the representation, so that there simply is no representation separate from the understanding. With no freestanding icon to interpret, no reinterpretation is possible. Consistent with this, our broad pattern of imagery data repeatedly show clear boundaries on what subjects can learn from or discover in their mental images. Discoveries easily occur when these are consistent with the way an image is understood. Discoveries which require a change in the image's understanding seem rigidly impossible. When subjects draw pictures of their images, though, there is a stimulus to be interpreted, and reinterpretations are routinely possible.

While the empirical content of this work has focused on imagery, I believe that the lesson is a general one. As long as ideas are internally represented, they exist only via a certain context of understanding, so there can be neither doubt nor ambiguity about what is intended. If I believe I am thinking about (say) crocodiles, then I am. It may turn out that I know very little about crocodiles, or I may later on discover that my knowledge about crocodiles is instead true of alligators. But, at the moment of having the thought, if I understand it as being about crocodiles, then it is about crocodiles, because it is only through my understanding of it that the thought has any definition at all. This is obviously of functional importance, since it entails (among other things) that "lapses of communication" cannot happen within thought. At the same time, though, it may place limits on thought, especially if one is looking for a novel interpretation of a prior conception.

I believe that this is what is at stake when one tries to evaluate one's own ideas by "distancing" or "detaching" oneself from them. One needs to learn how well the idea fares outside of the context in which it was created, away from the implicit assumptions which may have accompanied it, divorced from the specific construals or emphases one initially had in mind. One likewise needs to discover if understandings which accompanied the idea were essential corollaries, or gratuitous associations. To achieve all this, one needs to preserve what is central in the idea while changing as much as possible about the mental context. One important way to achieve this is by externalizing the idea, so that there is something to be interpreted or reinterpreted once one's perspective has changed.

I do not wish to imply that this reinterpretation will be easily achieved. It is often difficult to change one's initial understanding of an external representation. In the case of ambiguous geometric figures, for example, reinterpretation requires foreknowledge that the figure is in fact ambiguous (Girgus et al., 1977); reinterpretations are delayed if one is prevented from concentrating on the figure (Reisberg, 1983; Reisberg and O'Shaughnessy, 1984). Reisberg and O'Shaughnessy argue that these difficulties lie, not in finding a new construal, but in ousting the old, in subverting the processes which maintain the stimulus's interpretation. Be that as it may, interpretation is possible: stimuli can be ambiguous even if mental representations cannot. Hence, one can learn by turning one's representations into stimuli -- i.e. by externalizing them.

Discovering omissions

Our final distinction between internal and external representations is the one, I believe, with broadest applicability. Internal representations often strike us as being rich and detailed, but the detail is uneven: Aspects of the representation which are attended will be elaborated; aspects which are not attended will be undetailed and vague. It is unclear whether the attention creates the detail or vice versa, but, in either case, this co-occurrence often causes us to be surprised when the areas of vagueness are discovered. As long as they are unattended, the omissions from the representation can remain unnoticed.

Once again, our principal example comes from imagery. Even the most vivid of images omits myriad details, often details that seem central. There has been considerable discussion in the philosophy literature of the fact that the imaged tiger may have an indeterminate number of stripes (e.g. Dennett, 1981; Fodor, 1981). One can, if one chooses, image three or eight or however many stripes. If, however, one simply images a tiger, the number of stripes may be unspecified, and this indeterminacy can easily be overlooked.

This omission from the image can be discovered in various ways. As a simple extreme, one can ask the imager to count the tiger's stripes; the imager will typically register surprise at his/her inability to do this. As a more sophisticated variation of this, one can challenge subjects to form an image of a word (e.g. "pumpkin") and then to read off the letters of this word backward (n-i-k-p...). (Cf. Morris and Hampson, 1983, pp. 221-223.) To their surprise, subjects find this reasonably difficult, even when the imaged word was subjectively quite clear and vivid. Apparently, then, the image is appreciably less clear than the corresponding picture (from which the letters can easily be read), and, importantly, appreciably less complete than the subjects themselves had believed.

A final example brings us back to our theme: It is a common experience that even if one has a very clear image in mind, it is nonetheless difficult to "copy" the image onto a drawing. In part, it seems difficult simply to reproduce the contours "seen clearly" by the mind's eye. More important, one discovers in the process of drawing that some aspects were simply absent from the image. For example, one first images a horse, then attempts to draw this image. For most of us, it is very hard to draw the horse's knees; to use John Kennedy's description, the pictured horse often ends up looking as if it were wearing pajamas, and the picture looks clearly anomalous. The initial image, though, did not look at all anomalous -- one may have omitted the knees from the image (if, for example, one has no idea about how these look), but, as long as attention is elsewhere, the omission is consistently overlooked.

There are probably several reasons why externalizing the image leads to the discovery of omissions. After all, one could in principle apply the same selective filter to the drawing as to the image (attend only to the nose of the pictured horse, not its knees), and so discover nothing new. One reason this does not occur is the production process itself: At some point, one's pen must near the horse's knees, and then the contour must be dealt with, not ignored. At this point, one will immediately realize the absence of any clear idea about just how the horse's knees should look. In addition, it seems possible that the control processes for attention to stimuli may not be the same as those for mental representations. In the former case, one can easily allocate attention spatially: I will attend to the drawing's lower-right corner, or to the global contour, etc. This may be more difficult or perhaps just less likely with internal representations. In this case, attention seems governed by something like a content-addressing system, so that one focuses more easily on specified targets or aspects, rather than on whatever-is-in-region-X. This certainly would fit with our claim above that images are intensional entities, and not depictions to be effortlessly reparsed into arbitrary spatial regions.

As before, while our illustrations for this point have been drawn from imagery, I believe the phenomenon to be a general one. In thinking through a verbal argument, for example, one attends to what is included, and not to the gaps; thus the gaps are overlooked. It is only in explicitly and externally spelling out the argument that the gaps are discovered and the vaguenesses unveiled.

Conclusions and implications

Our prior data have pointed to clear differences between images and pictures; we have discussed some of the consequent advantages of externalizing images. The attempt to generalize beyond these data, at least for now, obviously rests more on conjecture than on hard evidence. One point of this paper, then, has been to argue that this evidence is worth gathering. In addition, this research is obviously relevant to the psychology of instruction and the psychology of discovery. Given current interest in these topics (e.g. Glaser and Takanishi, 1986; Gruber, 1974; Tweney et al., 1981), this by itself motivates this kind of inquiry. Finally, this attempt to contrast internal and external representations will inevitably sharpen our understanding of each, again an aim of some importance within cognitive science.

For the moment, therefore, the worth of this perspective rises or falls on the richness of its set of implications, and not on the strength of its empirical base. In addition to the implications we have mentioned, consider the fact that much work in artificial intelligence has sought to segregate the problems of intelligence and machine vision. This is obviously consistent with current claims about modularity, but, if I am right, it is in some ways an unwise strategy. To take just one salient case, we know that Darwin sketched a great deal in his notebooks (cf. Gruber, 1974); our suggestion is that these sketches may have been causes of his insights as much as expressions of them. Current "unseeing" models of discovery, therefore, (e.g. Bradshaw et al., 1983) might for this reason exemplify the wrong way of conceiving this achievement.

Or, to indulge in one last conjecture, we know that humans have created decorative arts since the beginnings of our evolution as a species. Many have speculated about the cultural or psychological causes which may underlie this urge to decorate (e.g. Arnheim, 1971; Festinger, 1984; Gombrich, 1960). Our perspective suggests at least one possible factor: Humans may have discovered early on the advantages of drawing rather than imaging, of thinking out loud, so to speak, rather than thinking to oneself. These advantages of externalizing for various forms of problem-solving may be one element which encouraged and maintained the development of decorative art.

Much of this paper has aimed at motivating questions about external representations as aids to thought. It seems important to close, therefore, with one of the critical and early questions that has to be asked, namely, what is an external representation? By this I do not mean to ask what is a representation at all. Instead, our concern lies with what it is that distinguishes internal from external representations. I have employed straightforward examples here (e.g. mental images vs. pictures) but there are many less clear cases. An example will illustrate this point: Reisberg et al. (1987) asked subjects either to listen to or (in other conditions) to image an ambiguous auditory stimulus. That is, the stimulus (rapid repetitions of the word "stress") could be parsed as either repetitions of "stress" or of "dress" or of "rest" or "tress." When subjects heard this stimulus, they easily discovered the various construals. When subjects imaged the soundstream, the results depended on whether additional steps were taken to block subvocalization.

Without subvocalization (e.g. if subjects imaged while chewing a large piece of candy), the results follow the pattern of our data with visual imagery: subjects reliably fail to reconstrue the auditory image. With subvocalization (e.g. no candy), though, subjects do discover the alternative interpretations of the image. Reisberg et al. argue that this is because subvocalization creates a stimulus, in this case an articulatory event. This articulatory stimulus, as an "external" representation, is then subject to interpretation and will support reinterpretation. Note, therefore, the possibility of <u>covert</u> external representations. In fact, while I have used the term "external representation" throughout this paper, the relevant, non-intensional, stimulus event need not be literally external. I am nonetheless inclined to retain the "internal"/"external" terminology (for expository ease), but now the problem of definition is in plain view. In addition, in this paper I have suggested several empirical properties of external representation, with regard to perception, or to ambiguity, or to controlling attention, but these properties may not be co-extensive. Given all this, I am reluctant to stipulate a definition. Instead, these properties of representations need to be assessed empirically. This is exactly what we have tried to do in our work with imagery; a major purpose of this paper has been to recommend a broadening of this effort.

References

Arnheim, R. (1971) <u>Art and visual perception</u>. Berkeley: University of California Press.

Bradshaw, G, Langley, P, and Simon, H. (1983) Studying scientific discovery by computer simulation. <u>Science</u>, 222, 971-975.

Chambers, D. and Reisberg, D. (1985) Can mental images be ambiguous? <u>Journal of Experimental Psychology: Human Perception and Performance</u>. 11, 317-328.

Dennett, D. (1981) The nature of images and the introspective trap. In N. Block (Ed.), <u>Imagery</u> (pp. 51-61). Cambridge, MA: MIT Press.

Festinger, L. (1984) <u>Reflections on Human Nature</u>. New York: Columbia University Press.

Fodor, J. (1981) Imagistic representation. In N. Block (Ed.), <u>Imagery</u> (pp. 63-86). Cambridge, MA: MIT Press.

Girgus, J., Rock, I. and Egatz, R. (1977) The effect of knowledge of reversibility on the reversibility of ambiguous figures. <u>Perception and Psychophysics</u>, 22, 550-556.

Glaser, R. and Takanishi, R. (Eds.) (1986) Psychological science and education. American Psychologist, 41, 1025-1168.

Gombrich, E. (1960) Art and illusion. New York: Pantheon.

Gruber, H. (1974) Darwin on Man. London: Wildwood House.

Jacoby, L. (1985) The relationship between learning and recollection: Memory attributes vs. Memory attributions. Paper presented at the Annual Meeting of the Psychonomic Society, Boston, Mass.

Jacoby, L. and Dallas, M. (1981) On the relationship between autobiographical memory and perceptual learning. Journal of Experimental Psychology: General, 3, 306-340.

Kosslyn, S. (1980) Image and mind. Cambridge, MA: Harvard University Press.

Kosslyn, S. (1983) Ghosts in the mind's machine: Creating and using images in the brain. N.Y.: Norton.

McLoskey, M., Caramazza, A., and Green, B. (1980) Curvilinear motion in the absence of external forces: Naïve beliefs about the motion of objects. Science, 210, 1139-1141.

Morris, P. and Hampson, P. (1983) Imagery and consciousness. New York: Academic Press.

Proffitt, D. and Kaiser, M. (1986) Why you cannot see what holds a gyroscope up. Paper presented at the Annual Meetings of the Psychonomic Society, New Orleans.

Reisberg, D. (1983) General mental resources and perceptual judgments. Journal of Experimental Psychology: Human Perception and Performance, 9, 966-979.

Reisberg, D. and Chambers, D. (1987) Neither pictures nor propositions: What can we learn from a mental image? Manuscript under review.

Reisberg, D. and O'Shaughnessy, M. (1984) Diverting subjects' attention slows figural reversals. Perception, 13, 461-468.

Reisberg, D., Smith, J.D. and Sonenshine, M. (1987) "Enacted" auditory images are ambiguous; "pure" auditory images are not. Manuscript under review.

Tweney, R., Doherty, M. and Mynatt, C. (1981) On scientific thinking. N.Y.: Columbia University.

The Content of Event Knowledge Structures

Brian J. Reiser

Cognitive Science Laboratory
Department of Psychology
Princeton University

Abstract

Autobiographical retrieval has been modeled as a predictive retrieval process, in which strategies elaborate the original retrieval cue relying on information accessed in knowledge structures to direct the search. Previous studies have demonstrated that event concepts differ in their utility in this process. The present study examines the type of information made available by accessing two such event concepts, activities and general actions. Activity structures are shown to enable more concrete predictions about included objects, people, and setting information, while general actions tend to be associated with internal mental states. These differences in available features are consistent with previously observed retrieval time differences between these types of concepts and support a general underlying mechanism of predictive inferencing in retrieval. The results suggest the types of information that computer models of memory organization should utilize in their representations of event structures and the reasoning mechanisms that depend on those structures.

Recent proposals concerning the organization of individual events in memory have suggested that experiences are stored in memory associated with knowledge structures, indexed by their differentiating features (Kolodner 1983, 1984; Reiser, 1986a; Reiser, Black, & Abelson, 1985; Reiser, Black, & Kalamarides, 1986; Schank, 1982). In the context-plus-index model, experiences are retrieved from memory by first selecting a context in which the target experience was likely to have occurred, then elaborating the original description until enough features are assembled to discriminate an event with the target properties (Reiser, 1986a, 1986b; Reiser & Black, 1983; Reiser et al., 1985; Reiser et al., 1986). This elaboration relies on information contained in knowledge structures that have been accessed to direct the search. Computer models of retrieval have demonstrated that knowledge structures used to make inferences during comprehension provide the type of information needed for such a strategic elaborative retrieval process.

In order to model the inferencing process, it is necessary to determine both the types of knowledge structures and the features they make available that would be useful in retrieval. Experiments examining the effectiveness of event descriptions have demonstrated that event knowledge structures differ in their utility in autobiographical memory retrieval (Reiser, 1986a; Reiser, Black, & Abelson, 1985). For example, phrases describing common activities (e.g., *saw a movie*) are more effective retrieval cues than phrases describing actions generalized across contexts (e.g., *paying*). We have argued that these retrieval time differences arise because knowledge structures differ in their ability to provide features useful for predictive inferencing during retrieval. Consider the example from Reiser et al. (1985) of trying to predict and interpret an event involving "paying in a restaurant". While knowledge about paying can be used to infer that a financial transaction occurs, most of the details in the event, such as the appearances of the people involved, the physical characteristics of the surroundings (tables, chairs, food smells, wine, music), the social roles involved (waiter, maitre d'), and motivations for the event (hunger, socializing) are drawn from knowledge about restaurants.

The argument based on these retrieval results suggests that event structures that are predictive of more features, particularly those features useful in elaboration of events, will be more effective retrieval structures. The present experiment examines this hypothesis by investigating the types of features encoded in representations of common events. Our hypothesis is that the number and variety of features encoded in an event structure could be used to predict its utility in autobiographical retrieval.

We investigated two types of event descriptions used in previous experiments of autobiographical memory search. Common activities are represented as a stereotypical sequence of deliberate actions undertaken to achieve one or more goals, e.g., *Grocery Shopping* and *Going to the Movies* (Schank, 1982; Schank & Abelson, 1977). General actions represent the common aspects of a component of several different activities. For example, *Make Reservations* is a component of several activities such as *Eating in Restaurants, Going on vacation*, and *Playing racquetball*. Reiser et al. (1985) found activities to be more effective retrieval cues than general actions, suggesting that these structures provide more of the information needed for inferencing during retrieval. Within the general actions, those that described the failure of a goal and were therefore more unusual also led to slower retrievals. We argued that this difference arises because more careful inferencing is required to assemble a set of features in which such a non-normative event would have occurred.

The goals of this experiment were to determine whether the amount of features these structures made available could account for their effectiveness in retrieval. In addition, examing the types of features these structures elicit will suggest the types of concepts that must be encorporated in models of general and specific event knowledge.

Method

In order to determine the amount and type of information encoded in these structures, we asked subjects to list the common features for each of twenty activities and twenty general actions. We followed Rifkin's (1985) modification for event categories of Rosch, Mervis, Gray, Johnson, Boyes-Braem's (1976) original procedure for eliciting features of object categories. Subjects were told to think about what the given type of situation is "usually like", then to list the characteristics common to that situation. They were told their characteristics should be described in five words or less, and were warned not to be redundant. They were also instructed that they should list general characteristics that come to mind when they imagine the situation, and not to respond in terms of associations, such as reporting that the item *you are food shopping* "reminds you of 'my friend who works in the vegetable section of the A & P.'"

The stimulus phrases were taken from the activities and general actions used in the Reiser et al. (1985) experiments, and converted to the present tense for use in this experiment. Of the general action phrases, ten were regular general actions, describing normative components of events, such as *you are paying at a ticket booth, you are waiting for your turn*. Ten phrases described the failure of a goal, e.g., *you don't get what you asked for* or *you want to leave early*. The forty phrases were divided into two stimulus sets, each containing ten activities, five regular actions, and five failure actions. An equal number of subjects were assigned each of the two stimulus sets. Subjects were given a booklet containing the twenty stimuli in randomized order, and were given two minutes for each event to list the characteristics that came to mind. Twenty Princeton University undergraduates were paid for their participation in the experiment.

Results

We first considered the number of responses listed for each event. The mean number of responses per event for each event type are listed in the first column of Table 1. The number of responses listed for each item was significantly different for the three types of stimuli ($F(2, 37) = 36.2$, $p < .001$). Subjects listed more features for activity cues than for general action cues. Among the general action cues, more features were listed for the regular actions than for the failure action cues. This pattern of results mirrors the retrieval time effects found by Reiser et al. (1985), also shown in Table 1. As expected, those items previously found to elicit faster retrieval of autobiographical experiences also elicited retrieval of more event features in the present experiment.

Table 1: Mean Number of Features Generated
and Retrieval Time for Activities and Actions

Event Type	Mean Number Responses Listed	Mean Number Features Listed	Retrieval Time (sec)
Activities	14.3	15.8	2.100
Regular General Actions	11.5	12.6	5.139
Failure General Actions	10.1	11.0	5.306

In order to determine whether the three event types elicited different types of information, we categorized all the responses generated by subjects into *feature types*. Each response was coded according to the type of information it contained. We developed 16 categories that captured the various types of information mentioned by subjects. The categories included internal states (Mental States, Physical States, Evaluation), actions and action elaborations (Action, Action Condition, Action Modification), taxonomic relations (Event Subcategory), relations to other events or states (Goal, Event Outcome, Reason, Other Event), physical and abstract referents in the event (Physical Object, Abstract Object, Setting), people (Role), and references to time (Time). Some typical examples are shown in Table 2.

Table 2
Examples of Categorized Features

Feature Type	Response	Stimulus Item
Emotion/Mental State	nervous	you sit down and wait
Physical State	cramp in hand	you are taking an exam
Action	fill out form	you go to a bank
Action Condition	admission price	you go to a movie
Action Modification	leisurely	you visit a museum
Event Subcategory	cast parties	you go to a party
Physical Object	counter	you pay at the cash register
Abstract Object	excuse	you want to leave early
Event Outcome	wrong turn	you get directions to find something
Reason	celebration	you go out drinking
Goal	want to sit down	you are waiting for your turn
Time	weekend	you go to a nightclub
Setting	big auditorium	you attend a concert
Role	go up to cashier	you pick out what you want
Evaluation	unfriendly	you take a ride on a train
Other Event	restaurants	you go on vacation

Categorizing the responses revealed that a number of responses (approximately 10%) included more than one feature. For example, the phrase "look at menu" was mentioned by a subject in response to the activity *you go to a restaurant*. This cue was categorized as both an Action and a Physical Object. Similarly, the phrase "go up to cashier" in Table 2 includes both Action and Role features. The three stimulus types do not differ in the mean number of features listed per response; thus the number of features listed for activities and actions show the same pattern as the number of responses. The second column in Table 1 displays the mean number of features listed for each event type.

We next considered whether the patterns of the types of features generated were different for the three types of events. Table 3 summarizes the mean number of features of each type per item listed by subjects. We used a discriminant analysis to determine whether the types of features generated would significantly discriminate the three feature types. The analysis revealed that membership in the three groups could be significantly predicted by the features. Those feature types that significantly discriminated the three types of stimuli are indicated in the table.

Table 3: Mean Number of Features Generated for Each Feature Type

Feature Type	Activities	Regular Actions	Failure Actions
Emotion/Mental State*	.46	1.13	2.15
Physical State	.35	.20	.44
Action	3.34	3.24	3.97
Action Condition	.03	.17	.01
Action Modification	.22	.15	.10
Event Subcategory	.23	.11	.01
Physical Object*	5.82	3.69	1.14
Abstract Object	.23	.49	.25
Event Outcome*	.10	.34	.39
Reason*	.10	.07	.35
Goal	.17	.21	.24
Time	.29	.28	.30
Setting*	2.14	.80	.33
Role*	1.34	.82	.64
Evaluation	.67	.30	.28
Other Event	.30	.55	.31
Unclassified	.02	.05	.05

*Significantly discriminates activities, regular actions, and failure actions.

The two strongest predictors are the Emotions and Physical Objects feature types. Activities elicit more Physical Objects than do the general actions, while the general actions elicit more Emotion features than activities. In addition, activities elicit more Setting and Role information, while general actions elicit more Reason and Event Outcome features. These findings are very consistent with the notion of activity structures as they are used in models of comprehension (e.g., Graesser & Clark, 1985; Schank & Abelson, 1977). Retrieval of an activity structure not only provides a stereotypical sequence of actions that achieve a goal, but includes a description of the various social roles involved in the interaction, and physical objects and other setting information characteristic of the event. The concrete predictions provided by activity structures are used in planning and comprehension of events, and protocol studies of autobiographical retrieval suggest that predictions based on these features can be used to elaborate a retrieval description during autobiographical retrieval (Reiser et al., 1986). The general actions provide less of this information, and appear to provide more information based on internal mental states and emotions. Thus, the differences in features elicited by these structures appear to be quite consistent with the observed differences in their efficacy during retrieval.

To summarize our findings to this point, we have demonstrated that event structures that are more effective in retrieval also elicit a greater number of features. Analyses of the specific types of features elicited indicated that activities are associated with more concrete predictions about included objects, people, and setting information, while general actions are more associated with internal mental and states. The next question to address concerns whether the differences in features can account reliably for the retrieval time differences found by Reiser et al. (1985). To investigate this, we performed a multiple regression using the number of features for each type as independent variables to predict retrieval time. The two largest predictors of retrieval time were Mental States and Physical Objects. A larger number of Physical Object features were associated with faster retrieval times ($r = -.55$), while a larger number of Mental State features were associated with slower retrieval times ($r = .62$). Not surprisingly, then, the feature types that best discriminate between the three stimulus types also predict retrieval time.

Discussion

We have found that presenting activity cues to subjects enables them to generate more features of a typical event than do general action cues. Analyses of the types of features listed indicate that in addition to the total number of features retrieved, these event structures also differ in the type of features they access. Activities retrieve more information about physical objects, characteristics of the physical setting, and types of people present in the event. General actions retrieve more information about mental states and emotions, and connections to other events (reasons and outcomes). Finally, two of these factors also predicted retrieval time -- an increased number of physical objects is associated with faster retrievals, and an increased number of mental states is associated with slower retrievals.

These findings are consistent with a model of autobiographical retrieval as a predictive process. Retrieval requires elaborating the cue in order to assemble features sufficient to discriminate a target experience from others of its general type. First, accessing activity structures appears to provide more information than does accessing general actions. This additional information appears to be useful in retrieval, as indicated by the retrieval time effects of previous studies. Furthermore, studies of retrieval protocols reveal that subject given information more abstracted from context, such as general actions or emotions, tend to focus their reasoning on finding a plausible activity context in which the event could have taken place.

The findings concerning the nature of retrieved features also suggest the types of information required in the retrieval process. Activities structures appear to provide more concrete information that is generally useful for search strategies. General actions on the other hand retrieve less of this type of information, and instead seem to elicit more features referencing internal states. It is important to consider how the types of features retrieved by activities and actions differ in their utility in retrieval.

Memory retrieval is an iterative process, in which each search retrieves partial information that can be used in further probes of memory (Kolodner, 1984; Norman & Bobrow, 1979; Reiser, 1986a, 1986b). Thus, event structures that access features useful for directing further probes of memory will be more effective retrieval structures. Consider the types of features that activities make available that were the most highly associated with retrieval time. These features seem to be themselves better predictors than those that general actions retrieve. For example, the activity *riding a subway* elicits physical object features such as maps, graffiti, tokens, signs, and so on. These features are themselves highly correlated, so that any of them might be used to predict other features. Similarly, physical object features such as a chalkboard, a blue exam book or social roles such as a waiter, conductor, and bartender are all predictive of particular events. Such features would appear to be quite useful by inferential retrieval mechanisms. On the other hand, emotions and mental states are generally not diagnostic of particular activities. One may feel excited, bored, or tired in a great variety of situations. Inferring a mental state for an event would not appear to be as useful for further inferences. Thus, it appears that the information activities access can be iteratively used to direct search, while the information accessed by general actions is of less utility to the reasoning mechanisms in retrieval. These results are consistent with the approach taken by Kolodner (1984) and Lebowitz (1983) in their computer models of knowledge structure organization. In these models, knowledge structures are continually developed by building structures to capture new patterns of features that discriminate a number of events within a category. The features selected to build these structures and index them within their more general category are those that are *predictive* of other characteristics of the event.

These findings are also quite consistent with the memory processes of *reality monitoring* discussed by Johnson and Raye (Johnson, 1985; Johnson & Raye, 1981). Johnson and Raye describe the mechanisms by which subjects distinguish between memories of external origin (records of real experiences) and internally derived memories, such as thoughts, imagined events, and dreams. They found that externally derived memories have more contextual attributes and sensory attributes, while internally derived memories have more traces of mental operations such as decisions and reactions. Our results concerning the type of information provided by activities and general actions appear to map well into Johnson and Raye's reality monitoring mechanism. The physical objects, setting, and person information provided by activities are contextual information of the type useful for reality monitoring. On the other hand, the information provided by general actions references internal states. Such information cannot be used to discriminate real autobiographical events from internally generated events such as thoughts, imagined events, or dreams. Thus, the key difference between activities and general actions may be summarized as the retrieval of external information by activities and internal information by general actions. The same information used as a basis for reasoning about events by activity structures can also be used by reality monitoring mechanisms to discriminate real from imagined events. The precise relation of strategic elaboration of retrieval descriptions and reality monitoring remains to be explored in future research. At this point, it seems clear that the differences in information content between activities and general action event structures can play a role in two ways. Our retrieval time and protocol studies results suggest that activities provide information of greater use to strategic retrieval mechanisms. In addition, activities appear to trigger retrieval of features that are of greater utility in distinguishing externally derived from internally generated events.

In summary, these results support the notion that retrieval is a predictive process. We have shown that structures that enable prediction of more features, particular externally derived features, lead to faster retrieval of individual experiences. Of course, it must be stressed that our analyses of the role of the features in these event structures in influencing retrieval time is purely correlational. That is, we have demonstrated that two types of event structures, activities and general actions, differ in the amount and type of information they make available. This difference in features is highly predictive of the retrieval time difference. However, the possibility remains that there are one or more underlying factors that account for both the differences in feature elicitation and retrieval time.

It is also important to note that these studies have concerned only activities and general actions. We have focused on these structures in our investigations because they are of great interest in studies of comprehension, and have been a focus of computational models of reasoning and comprehension. However, we do not mean to suggest that all events can be encoded as activities or actions, nor do we mean to suggest that these are the only useful retrieval structures in autobiographical memory. Indeed, our work suggests that in part activities are useful in that they provide additional information about physical objects, settings, and people. In fact, Barsalou (in press) has demonstrated that cues of this sort may be as effective in retrieving experiences as event cues.

Our results concerning activities and general actions suggest a general underlying mechanism of predictive inferencing in retrieval. The present experiment suggests the type of information that computer models of memory organization should utilize in their representations of event structures and the reasoning mechanisms that depend on those structures.

Acknowledgements

I am greatly indebted to Shari Landes for her assistance in designing, executing, and analyzing the study. I also thank Stephen Hanson and Marcia Johnson for helpful discussions of this work. The research reported here was supported in part by a research grant from the James S. McDonnell Foundation to Princeton University. The views and conclusions contained in this document are those of the authors and should not be interpreted as necessarily representing the official policies, either expressed or implied, of the McDonnell Foundation.

References

Barsalou, L. W. (in press). The content and organization of autobiographical memories. In U. Neisser & E. Winograd (Eds.), *Real events remembered: Ecological approaches to the study of memory.* Cambridge: Cambridge University Press.

Graesser, A. C., & Clark, L. F. (1985). *Structures and procedures of implicit knowledge.* Norwood, NJ: Ablex.

Kolodner, J. L. (1983). Reconstructive memory: A computer model. *Cognitive Science, 7,* 281-328.

Kolodner, J. L. (1984). *Retrieval and organizational strategies in conceptual memory: A computer model.* Hillsdale, NJ: Erlbaum.

Johnson, M. K. (1985). The origin of memories. In P. C. Kendal (Ed.), *Advances in cognitive-behavioral research and therapy, Volume 4,* New York: Academic Press.

Johnson, M. K., & Raye, C. L. (1981). Reality monitoring. *Psychological Review, 88,* 67-85.

Lebowitz, M. (1983). Generalization from natural language text. *Cognitive Science, 7,* 1-40.

Norman, D. A., & Bobrow, D. G. (1979). Descriptions: An intermediate stage in memory retrieval. *Cognitive Psychology, 11,* 107-123.

Reiser, B. J. (1986a). The encoding and retrieval of memories of real-world experiences. In J. A. Galambos, R. P. Abelson, & J. B. Black (Eds.), *Knowledge structures,* Hillsdale, NJ: Erlbaum.

Reiser, B. J. (1986b). Knowledge-directed retrieval of autobiographical memories. In J. L. Kolodner & C. K. Riesbeck (Eds.), *Experience and reasoning,* Hillsdale, NJ: Erlbaum.

Reiser, B. J., & Black, J. B. (1983). The roles of interference and inference in the retrieval of autobiographical memories. *Proceedings of the Fifth Annual Conference of the Cognitive Science Society,* Rochester, NY.

Reiser, B. J., Black, J. B., & Abelson, R. P. (1985). Knowledge structures in the organization and retrieval of autobiographical memories. *Cognitive Psychology, 17,* 89-137.

Reiser, B. J., Black, J. B., & Kalamarides, P. (1986). Strategic memory search processes. In D. C. Rubin (Ed.), *Autobiographical memory,* New York: Cambridge University Press.

Rifkin, A. (1985). Evidence for a basic level in event taxonomies. *Memory and Cognition, 13,* 538-556.

Rosch, E., Mervis, C. B., Gray, W. D., Johnson, D. M., & Boyes-Braem, P. (1976). Basic objects in natural categories. *Cognitive Psychology, 8,* 382-439.

Schank, R. C. (1982). *Dynamic memory: A theory of reminding and learning in computers and people.* New York: Cambridge University Press.

Schank, R. C., & Abelson, R. P. (1977). *Scripts, plans, goals, and understanding.* Hillsdale, NJ: Erlbaum.

A clausal form of logic of belief

— A logic programming model

Y. J. Jiang

Computer Science Department,
University of Essex,
Wivenhoe Park,
Colchester, C04 3SQ

Abstract

In this paper, we present a clausal logic of belief which formalizes beliefs in an *extended clausal form* of logic. Our aims are to solve the representational problem of quantified beliefs and to allow an efficient resolution-like proof procedure with controlled granularity to be developed. A *levelled* intensional scheme that enables the clausalization of beliefs is proposed. An inferential power bounded resolution rule of beliefs for the formalism is introduced. The formal semantics of the formalism is defined. A general circumscriptive non-monotonic reasoning system for belief revision is described. Finally, a scheme for handling the consistency of beliefs under Tarski's *truth definition theorem* is developed.

Keywords: logic programming, computational models of belief, intension, imputation, non-monotonic reasoning and semantic paradoxes.

1 Introduction

There are currently three approaches to formalization of beliefs. In the semantic approach (eg. [Moore 85], [Halpern&Moses 85]), beliefs are characterized by accessibility relations between possible worlds. In the partition approach (eg. [Kobsa 85]), beliefs are identified with the presence of representation structures in specific nested belief spaces reserved for the respective agent. In the syntactic approach (eg. [Konolige 85]), belief of an agent is equated with derivability of a first order theory of the agent. Despite the fundamental differences of the three approaches, one common characteristic is that they all assume the use of *arbitrary forms* of logic in their representation of beliefs.

Although the form of representation can play a part in the *meaning* of a sentence from a strict cognitive sense (for example, representing the belief 'there does not exist a person who does not like Mary' as 'Everybody likes Mary' could be misleading because an agent of the belief may not know that these two sentences are equivalent), however several problems can arise from the use of arbitrary forms of logic. One problem is the difficulty of developing an *efficient* proof procedure for reasoning about beliefs. The solution is usually based on a natural deduction approach which does not seem to have much success in AI applications.

Another problem is the difficulty of formalizing the intensionality of arbitrary quantifications. (In fact, many schemes, in particular the semantic ones such as [Levelsque 84] and [Halpern&Moses 85] are careful to avoid this problem by sticking with propositional beliefs.) For example, it is easy to see Konolige's [85] semantics of the statement $BEL(Jim, \exists x Unicorn(x))$ as: $\exists x Unicorn(x)$ is in the belief space of Jim; however it is difficult to see how the semantics of the *quantifying-in* belief $\exists x BEL(Jim, Unicorn(x))$ can be formalized in a similar way because the quantification is outside the scope of BEL predicate. Konolige's solution is to treat the second sentence in a similar way as $\exists x P(x)$ where P is any ordinary predicate. However this seems to obscure the semantics of the BEL predicate. For example, given the following universally qualified belief

$\forall x BEL(Jim, Unicorn(x) \rightarrow Likes(Jim, x))$.

and the following quantifying-in belief, $\exists x BEL(Jim, Unicorn(x))$, following Konolige's semantics, it is difficult to see how we can derive:

$\exists x BEL(Jim, Likes(Jim, x))$

which we should do. The problem of intensional quantification gets more serious when we try to represent nested beliefs, eg. $BEL(Jim, \exists x BEL(Tom, Unicorn(x)))$ in which Jim believes that there is a particular individual in Tom's mind whom Tom believes to be a unicorn although Jim has no idea himself who this individual could be.

A third problem is the granularity of implied beliefs of an agent as discussed in [Levelsque 84]. In the possible world approach, beliefs of an agent are represented by a set of possible worlds that are compatible with what the agent believes. A recuring problem in this approach is *logical ominiscience*, ie. the set of beliefs is closed under logical implication. This means that anyone who can be persuaded of the truth of Peano's postulates knows everything about number theory that anyone else knows. In addition, every valid sentence must be believed by every agent and contradictory beliefs of an agent imply that the agent believes anything. The possible world approach is thus *too coarse-grained* in the sense that it cannot distinguish same logical sets of beliefs. The syntactic and partition approaches on the other hand are *too-fine grained* in the sense that it distinguishes too much on same logical sets of beliefs. For example, in these approaches, an agent may believe A and B but not $A \wedge B$. To avoid these spurious syntactic distinctions, the obvious axioms must be present which could complicate the proof theory of belief.

Thus we propose an alternative approach which formalizes beliefs in a quantifier-free *canonical form* of logic. As argued in [Moore & Hendrix 79], beliefs should be represented in an internal language from a computational perspective rather than an external language of thought although different external languages of the same content may not have the same meaning. Here we further argue that beliefs should be represented in a canonical form of an internal language from a *logic programming* perspective although different forms of the language of the same content may not have the same meaning. We have chosen an extended clausal form of logic as such a canonical form for three reasons. The first is that a clausal form of logic is quantifier-free. This would simplify the task of formalizing the semantics of quantified beliefs. The second is that clausal forms of logic form the basic foundation of current logic programming systems and Japanese Fifth Generation Computer Kernel Languages([Kowalski 83]). In other words, there are well-developed and efficient proof procedures for clausal forms of logic. The third reason is that it seems easier to control the granularity of implied beliefs in a clausal form of logic. However it should be understood that clausal forms of logic are *by no means* the only forms used in logic programming. In fact, the concept of logic programming can be applied to *any forms of logic*. The reason we call our approach a "logic programming model", is simply because the *current most promising* logic programming approach is to operate on a clausal form of logic.

It is worth noting that a recent attempt has been made by [Cerro 86] to allow modal reasoning in Prolog [Clocksin&Mellish 81]. However no concern is given to the problem of formulating clausal forms of logic of abitrarily quantified beliefs. In particular, the problem of intensionality of terms is not addressed. In addition, the semantics of the scheme is based on the possible world approach, hence also suffers the granularity problem of possible world approach. However, Cerro's approach does have many practical applications, especially those involving knowledge base systems for which the granularity problem of modal approach is less serious.

The paper is organized as follows. Section 2 introduces the basic preliminaries. Section 3 describes the problem and solution of clausalizing beliefs. Section 4 discusses the inference mechanism and the formal semantics of the proposed formalism. Section 5 presents an explicit circumscriptive non-monotonic reasoning scheme for handling belief revision. Section 6 proposes a scheme to allow a modified Tarski's *truth definition theorem* to be consistent with our formalism.

2 Clausal form of Logic, Skolemization and Resolution

Every standard first order formula can be transformed into a prenex-normal form which is logically equivalent. A Prenex-normal form can be written as QM, where Q consists of all the quantifiers in the formula ($\exists$ and $\forall$),

while M is a quantifier-free well-formed formula (wff)[1].

By introducing Skolem functions, one can eliminate existential quantifiers, and hence the universal quantifiers since they are then implied, i.e. indicated by leaving variables free. Skolemization is achieved by replacing each of the existentially quantified variables by a skolem function f whose arguments are all of those universally quantified variables that precede the existential variable. For example, the formula $\forall x \exists y Likes(x, y)$ can be skolemized into $Likes(x, f(x))$ where f is a skolem function. If there is no universal quantifier preceding an existential quantifier, a skolem function with no arguments called *a skolem constant* will be produced.

Every prenex-normal formula can be transformed into a quantifier-free clausal form of logic which is logically equivalent. To do this, a prenex-normal formula must be skolemized if it contains any existential quantifiers. A clausal form of logic can be one of the following two equivalent forms:

- Implication form

  $P_1 \wedge ... \wedge P_n \rightarrow Q_1 \vee ... \vee Q_m$

- Disjunctive form

  $\neg P_1 \vee ... \vee \neg P_n \vee Q_1 \vee ... \vee Q_m$

where P_i and Q_j are (positive) literals. When m=1, the clause is called a *Horn clause*. We will mix the use of these two forms throughout this paper.

The resolution principle [Robinson 65] is a rule of inference that can be applied to clausal forms of logic. The principle can be defined as follows.

Given two clauses,

$S_1 \vee p1 \vee S_2$

---

[1] This section is mainly extracted from [Chang & Lee 73].

$S_3 \vee \neg p2 \vee S_4$

where S denotes disjunctive literal set and p1a=p2a (ie. p1 and p2 are *unifiable* where "a" is a substitution called the Most General Unifier (MGU)),

we can deduce the following resolvent,

$S_{1,2,3,4}a$

Resolution is the only rule of inference that is necessary in order to find proofs to all theorems. Although resolution is complete [Robinson 65], unlimited applications of resolution may cause many irrelevant and redundant clauses to be generated. Thus many restricted forms of resolutions have been developed. One such an example is the linear resolution strategy in which one of the two resolved clauses is always the most recent resolvent. A more restricted linear resolution strategy called *linear input resolution* on Horn-clauses in which linear resolution is begun with the input goal is used in Prolog.

3 Clausalizing Beliefs

Standard first order logic is concerned with *extensional objects* or things that exist. However in a belief logic, namely, a logic supplemented with a BEL (or B) predicate, we are additionally concerned with *intensional objects* or concepts which may not have extensions (eg. unicorns). Thus unlike predicates of standard first order logic, a B predicate introduces intensional scopes that quantifiers cannot be moved outside. This can be illustrated by the following two sentences:

1. $\exists x B(Simon, Unicorn(x))$

2. $B(Simon, \exists x Unicorn(x))$.

In the first sentence, there is a particular individual in Simon's mind whom Simon believes to be a unicorn; while in the second sentence, Simon does not know which individual

is a unicorn but he believes a unicorn exists. This means that we cannot have a prenex-normal form, hence a clausal form of logic to represent quantified beliefs.

To solve this problem, we propose a *levelled* intensional scheme that allows an extended clausalization to be achieved. In this scheme, each logical term is associated with a number denoting the level (or depth or nestness) of intensional scope it is *meant* to be in (the default level is 0). This is represented by using a built-in predicate structure called $LEVEL$ so that $LEVEL(x,i)$ denotes a i levelled intensional term x. For clarity reason, we write $LEVEL(x,i)$ in a subscript notation: x_i. To unify two terms, we additionally require the levels of the two terms to be unifiable.

Because quantified terms of different intensional scopes are distinguished/levelled, the intensional scheme allows quantifiers to be moved outside the scope of BEL predicates so as to produce prenex-normal forms which can then be skolemized (if there is any existential quantifier) into clauses. For example, the formula

$$\forall x B(Simon. \exists y B(John, likes(x,y)))$$

after introducing levels of intension can be transformed into the following prenex-normal form

$$\forall x_0 \exists y_1 B(Simon, B(John, Likes(x_0, y_1)))$$

which can then be skolemized into the following clause

$$B(Simon, B(John, Likes(x_0, f(x_0)_1)))$$

where f is a skolem function.

We elaborate more on the levelled intensional scheme with the representation of the four intensional interpretation of the statement 'Jim believes every unicorn likes something' (omitting the logical form before clausalization):

304

1. 'There is a particular thing Jim believes every unicorn likes'

$$B(Jim, Unicorn(x_1) \rightarrow Likes(x_1, c1_0))$$

2. 'Jim believes every unicorn likes a particular type of thing'

$$B(Jim, Unicorn(x_1) \rightarrow Likes(x_1, c2_1))$$

3. 'Jim believes every unicorn likes something of his own type'

$$B(Jim, Unicorn(x_1) \rightarrow Likes(x_1, f(x_1)_1))$$

4. 'Jim believes every unicorn likes his own particular thing '

$$B(Jim, Unicorn(x_1) \rightarrow Likes(x_1, g(x_1)_0))$$

where c1, c2 are skolem constants and f, g are skolem functions.

We adopt the following notations in our formalism:

- A built-predicate is denoted by an uppercase string, eg. BEL/B.

- A non-skolem term is denoted by a lowercase string preceded with a upper-case letter.

- A skolem constant is denoted by a lowcase string preceded with a letter of the set {a,b,c,d,e}.

- A skolem function is denoted by a lowercase string preceded with a letter of the set {f,g,h}.

- A variable is denoted by a lower-case string preceded with a letter of the set {x,y,z}.

To demonstrate the expressiveness of our formalism, we describe its representations of some more examples.

The following sentence attributed to Russel is discussed by McCarthy [79]: "I thought that your yacht was longer than it is." It can be expressed in the clausal approach as (omitting tense and pronouns):

$Len(Yt, c1_0)$.
$B(I, Len(Yt, c2_1))$.
$B(I, c2_1 > c1_0)$.

where c1 and c2 are skolem constants. Here B is better understood as Believed rather than *Believe*.

From the example, it can be seen that conjunctive beliefs are modelled as separate clauses in our formalism, eg. the formula $B(I, A \wedge B)$ is represented as two clauses $B(I, A)$ and $B(I, B)$. Such modelling helps to remove the spurious syntactic distinction problem that often persist in the syntatic formalization of beliefs.

To express "Your yacht is longer than Peter thinks it is", we have the following formulae:

$Len(Yt, c1_0)$.
$B(Peter, Len(Yt, c2_1))$.
$c1_0 > c2_1$.

where c1 and c2 are all skolem constants.

Quine [56] discusses an example in which Ralph sees a person skulking about and concludes that he is a spy, and also sees him on the beach, but doesn't recognize him as the same person. The facts can be expressed in our formalism as:

$See(Ralph, Sk(b_1))$.
$B(Ralph, Spy(b_1))$.
$See(Ralph, onbeach(c_1))$.
$B(Ralph, Spy(c_1))$.
$b = c$.

where b and c are skolem constants.

Note that a non-skolem constant can also have a level of intension other than zero. This can be illustrated by the following two sentences:

1. $B(Simon, Loves(Mars_0, Venus_0))$

2. $B(Simon, Loves(Mars_1, Venus_1))$.

In the first sentence, Mars and Venus are extensional objects of the real world; while they are intensional concepts of Simon which may not have extensions in the second sentence.

To see how nested beliefs can be modelled with levels of intension, we illustrate with the following example:

1. $\exists x B(Simon, B(John, Unicorn(x)))$

2. $B(Simon, \exists x B(John, Unicorn(x)))$

3. $B(Simon, B(John, \exists x Unicorn(x)))$.

The intensional differences of these sentences can be explained as follows. In the first sentence, Simon has a particular individual in his mind whom he thinks John believes to be a unicorn. In the second sentence, Simon believes there is a particular individual in John's mind (though Simon may not know which one it is) whom he thinks John believes to be a unicorn. In the third sentence, Simon believes that John believes that there exists a unicorn though Simon has no idea himself or no idea about John's idea about who the unicorn is.

These sentences can be represented in our formalism as follows:

1. $B(Simon, B(John, Unicorn(c1_0)))$

2. $B(Simon, B(John, Unicorn(c2_1)))$

3. $B(Simon, B(John, Unicorn(c3_2)))$

where c1, c2 and c3 are all skolem constants.

We have so far only shown that beliefs can be represented in an extended clausal form of logic. Our formalism also allows beliefs to be nested within clauses. This can be illustrated with the following example: "Every Person (P) believes that every ET (E) believes that Unicorns (U) exist". One intensional representation of the statement in an arbitrary form of logic could be:

$$\forall x(P(x) \rightarrow B(x, \forall y(E(y) \rightarrow B(y, \exists z U(z)))))$$

This can be represented in our formalism as:

$$P(x_0) \rightarrow B(x_0, E(y_1) \rightarrow B(y_1, U(f(x_0, y_1)_2))).$$

From the above examples, it can be seen that unlike McCarthy [79] which appears to have only two levels of intension, our approach

allows theoretically an infinite level of intension depending on the nestness of beliefs. Thus instead of using terms like "de-dicto" and 'de-re" to describe intensions and extensions, we talk about levels of intension in our formalism. In addition, McCarthy uses different notations to express intensions and extensions of same concepts which could complicate the first order deductive calculus, we use same notations for them whose intensions are distinguished by the levels of intensional scope they are in.

Barnden [1986] has criticized the existing models of beliefs including the modal approach (eg. [Helpern&Moses 85]), the quotation scheme (eg. [Perlis 85]) and the concept formation scheme (eg. [Creary 79]), for introducing unwarranted inferences which he call the *imputation* problem. In these models, the belief of Jim "Sue is smart" would be represented in the same way as the nested belief of Tom "Jim believes that Sue is smart". They introduce the unwarranted inference (or "opacity violation") in that Jim's mental state of smartness of Sue is the same as Tom's or as anyone else's. Undesirable imputations of a similar sort arise in the belief models described in Barwise&Perry [83], if they are extended in the natural way to deal with nested beliefs. The extension causes imputations, to ordinary agents, of beliefs about their models' "situation types'.

Barnden then proposed an alternative scheme based on Creary's concept formation scheme [Creary 79]. In Barnden's scheme, the mental state of each agent is *explicitly* denoted as a concept formation. For example, one intensional interpretation of the nested belief of Tom "Jim believes that Sue is smart" in Barnden's scheme is:

$$B(Tom, \$_{Tom}(B(\$_{Tom}Jim, \$_{Tom}\$_{Jim}(S(Sue)))))$$

where $\$_{agent}$ denotes the mental state of the agent. Although Barnden's scheme may be theoretically sound, it does not seem to be computationally viable.

306

It is felt that the cause of imputation of existing models of belief lies on their basis of only two levels of intension. Thus in contrast with Barnden's scheme, the intensional scheme in our formalism can distinguish the mental states of nested agents *implicitly* by their intensionally levelled terms, eg.:

$$B(Tom, B(Jim_1, S(Sue_2)_2))$$

This approach is more viable computationally than Barden's explicit scheme since intensional structures can be regarded as functional structures – a similar analogy to a Prolog *structure* [Clocksin & Mellish 81].

It should be noted that our intensional scheme and clausalization mechanism can be generalized to intensional predicates other than BEL such as WANTS, SEEKS, AWARE etc. This can be illustrated by the following example discussed in [Hobbs etal 77]: "Everyone seeks a frog". We can have the following four interpretations of the statement represented in our clausal forms as shown below.

- "There is a particular frog everyone seeks"

  $$Person(x) \rightarrow SEEKS(x, c1_0).$$

- "Everyone seeks a particular type of frog"

  $$Person(x) \rightarrow SEEKS(x, c2_1).$$

- "Everyone seeks his own particular frog"

  $$Person(x) \rightarrow SEEKS(x, f(x)_0).$$

- "Everyone seeks his own particular type of frog"

  $$Person(x) \rightarrow SEEKS(x, g(x)_1).$$

where c1, c2 are skolem constants and f,g are skolem functions.

4 Inference and Semantics

In this section, we discuss the inference mechanism, semantics, soundness, completeness, consistency and recursiveness of our formalism.

4.1 Inference

In the previous section, we have described the syntactic logical form of our formalization of beliefs, namely, an extended clausal form of logic supplemented with a BEL/B predicate. However it is pointless to talk about beliefs outside the context of a world in which the beliefs may be true or false. This means that we need rules of inference to reason about beliefs of an agent to obtain his implied beliefs.

The only inference rule in our approach is a linear resolution-like principle (such as Linear Input Resolution or , SL resolution [Hayes&Kowalski 79]) for all agents of beliefs. Exactly what this principle is, is not the concern of this paper.

To distinguish the different reasoning capability of different agents of beliefs, we assign an Inferential Power (IP) to each agent. We could have designed a more clever and complicated measurement (such as deduction, learning and memory abilities of an agent plus resource-bound factors) as the inferential power of an agent, for simplicity and illustration reason however, we have chosen the maximum inferential depth of resolutions allowed to an agent as the inferential power of the agent. So given IP(Simon,3), Simon can only invoke at most, a depth of three resolution inferences in a deductive process. By *inferential depth of resolution*, it means the depth of a AND/OR-search tree of a linear resolution proof.

For example, given the following beliefs of Simon and ID(Simon,3),

$$P \leftarrow Q$$
$$Q \leftarrow R$$

$$R \leftarrow S$$
$$S$$

we can infer that Simon also believes S, Q, R but not P.

To answer a query about an agent's belief, we negate the belief and prove refutationally that the negated belief is inconsistent with the belief space of the agent. For example, to prove $BEL(Simon, p)$, is to show $BEL(Simon, \neg P)$ to be inconsistent with Simon's beliefs. This means that in addition to the normal resolution rule, we need to define the following resolution rule regarding beliefs.

Given

$S_1 \vee B(agent, S_2 \vee P \vee S_3) \vee S_4$

$S_5 \vee B(agent, S_6 \vee \neg Q \vee S_7) \vee S_8$

where $Pa = Qa$ (where a is the MGU between the two literals) and S denotes disjunctive literal set.

we can obtain the following resolvent (within the inferential power of the agent),

$(S_{1,5} \vee B(agent, S_{2,6,3,7}) \vee S_{4,8})a$.

Our inference mechanism can be compared and contrasted with Konolige's approach [85] which uses multiple rules of inference for arbitrary forms of a first order logic supplemented with a B predicate. In his approach, the inferential power of each agent is determined by the set of rules of inference (or sequents) he has. As argued in the introduction, multiple rules of inference for arbitrary forms of logic may not be supported efficiently.

In addition, unlike the too-coarse-grained possible-world approach, our formalism represents beliefs as syntactic clausal structures to be manipulated and the consequential closure of an agent's beliefs is controlled by his inferential power. However unlike other systactic schemes (eg. [Konolige 85]) which suffers the problem of too-fine granularity, we use a canonical clausal form which helps to remove many spurious syntactic distinctions. In this sense, our formalism can be seen as a balance between the fine-grained syntactic approach and the coarse-grained semantic approach.

Finally, it may be noted that our inference mechanism does not have any axioms. However certain axioms may be useful. Two such ones are the positive introspection $B(a, p) \rightarrow B(a, B(a, p))$ and the negative introspection $\neg B(a, p) \rightarrow B(a, \neg B(a, p))$. Belief introspection [Konolige 85a] is useful because it allows an agent to reflect upon the workings of his own cognitive function. This can be done in our formalism by issuing recursive queries to the inference system with perhaps reduced inferential power for each agent. However this discussion is outside the scope of this paper.

4.2 Semantics

The basic syntax of our belief logic is a first order clausal form of logic supplemented with a BEL/B predicate which can take a clause as argument in the form of $B(agent, clause)$ where *clause* can be a variable or a clause instance. In addition to the normal semantics of a standard first order clausal form of logic which we will not describe, the B predicate presents an additional semantics which allows $B(agent, clause)$ to be logically implied by a set of clauses believed by the agent. It is this type of semantics we will describe in this paper.

Because our intensional mechanism have replaced all the quantifiers by intensionally levelled variables and skolem functions, the semantics of our formalism can *uniformly* be defined as follows:

> $B(agent, clause)$ is true iff the clause is a member of the belief space of the agent.

A clause c is a member of the belief space of an agent a iff it is a IP-controlled consequence

308

of the members of the belief space of a. This effectively determines the soundness and completeness results of the formalism which can be stated as follows.

The soundness result of the formalism is:

> if a clause c follows from the IP-controlled inference of an agent a, then $B(a, c)$ is true.

The completeness result of the formalism is:

> if $B(a, c)$ is true, then the clause c follows from the IP-controlled inference of the agent a.

Because the belief space of an agent is consequentially-closed under the inferential power of the agent, the consistency of a belief space need to be defined differently from that of a standard first order logic. In particular, we can have contradictory beliefs without causing an agent to believe anything (as is the case in the possible world approach). In our formalism, a belief space of an agent (closed under the inferential power of the agent) is consistent iff it does not contain two literals $p1$ and $\neg p2$ such that $p1$ and $p2$ are unifiable. This means that an inconsistent set of clauses in a standard first order logic, can be consistent in a belief space of an agent provided the agent's inferential power will not allow contradictions to be deduced. For example, a person may believe the following statements "Every Professsor has a PhD", "Every MD does not have a PhD" and "John is a Professor and has a MD" without realizing that they are inconsistent because he has a limited inferential power or he may have chosen relevant theories for the statements in such a way that he will not establish inconsistent beliefs. The latter will be discussed in the next section.

All the above results can be similarly applied to recursive/nested beliefs. To do this, each belief space of an agent is organized hierarchically with each sub-space of a space denoting the next level of agents in a nested belief. Thus $B(a1, B(a2, c))$ is true iff c is a member of the belief space a2 which is a sub-space of a1. However the inferential power of all the sub-agents in a nested belief is determined by the outmost agent. A more detailed description of nested beliefs can be found in [Jiang 86].

It may be noted that the basic semantics of our formalism is similar to the semantics of Konolige's [85] deductive model of belief except that we use a canonical clausal form of logic and a single rule of inference as the computational structures of an agent. From a *logic programming* perspective, our formalism may be treated as a more-viable computational model of belief than Konolige's.

5 Circumscriptive Nonmonotonic Reasoning

Unlike a piece of knowledge, a belief may not be true in the real world. This means that with new beliefs in hand, an agent can retract his old beliefs. This form of reasoning is sometimes called belief revision. One possible approach of revising beliefs is to build inference paths of beliefs so that a retracted belief can be traced back from such inference paths. Its purpose is to detect contradictions, identify their causes and try to resolve the contradictions by revising beliefs. Two examples of this approach are London's dependency networks [78] and Doyle's truth maintenance system [79]. Another possible approach of revising beliefs is the explicit non-monotonic reasoning. Its purpose is not to detect the cause of inconsistency but to ensure that the belief system is always consistent. In this approach, non-monotonic beliefs are *explicitly* represented as non-monotonic rules and can only be derived if they are consistent within a certain belief space. Although it is felt that inference paths

309

in linear resolution systems on clausal forms of logic are easier to find than arbitrary forms of logic, in this paper however, we are only concerned with the problems addressed in the latter approach.

There are currently two important types of non-monotonic reasoning. One is McCarthy's circumscription rule [80]. Another is Reiter's default rule [Reiter 80][2]. Both can be specified in the following meta-axiom:

$$x \wedge C(y) \rightarrow y$$

where $C(y)$ is true if $\neg y$ cannot be proved.

The difference between these two approaches lies in the area of consistency checking. In McCarthy's approach, it is the whole belief space including all the non-monotonic rules; while it is only the area of the belief space that excludes default rules in Reiter's approach.

The problem with McCarthy's approach is that it is *too fine-grained* in the sense that it tends to fail to conclude anything. This can be seen from the following two default beliefs:

1. if x is a professor and there is no proof that x is a Mr then we may infer that x is a Dr (Ph.D):

$$\begin{aligned} Dr(x) \quad &\leftarrow \quad Prof(x) \\ &\wedge \quad C(\neg Mr(x)) \end{aligned}$$

2. if x has a MD and there is no proof that x is a Dr then we may infer x is a Mr:

$$\begin{aligned} Mr(x) \quad &\leftarrow \quad MD(x) \\ &\wedge \quad C(\neg Dr(x)) \end{aligned}$$

In addition, we assume the belief that no one can both be a Dr and a Mr:

3. $\leftarrow Dr(x) \wedge Mr(x).$

---

[2]Moore [85a] may argue that this is best called auto-epistemic reasoning.

If we assume that John has MD, and is a professor, then in proving $Dr(John)$, we need to prove $C(\neg Mr(John))$ or $Mr(John)$ which lands us in the proving of $C(\neg Dr(John))$, or $Dr(John)$; hence back to the original goal, ie. looping.

Reiter's approach on the other hand is *too coarse-grained* in the sense that it tends to conclude everything. For instance, though it does not loop for the above example, Reiter's approach would give us both $Dr(John)$ and $Mr(John)$, ie. a contradiction (to belief (3)).

To solve these problems, we propose a neutral approach (based on [Bowen&Kowalski 82]) that covers a range or varing granularities of non-monotonic reasoning systems. Instead of being restricted in one single predefined theory/area of belief space, an agent can perform consistency checking within an explicitly defined theory. This means that in our formalism, a belief space can be divided into various overlapping areas/theories. Allowing various explicit specification of theories is cognitively feasible because an agent may only use a subset of his beliefs which he thinks is relevant to achieve a certain process of reasoning (eg. to establish the proof that John is guilty) and use another subset of his beliefs (maybe overlapping with the former set) to achieve another process of reasoning. This relevance of beliefs to an agent is sometimes called *circumscriptive relevance* [Konolige 85]. For this reason, we call our belief revision approach *circumscriptive non-monotonic reasoning*.

To define the explicit theory of a clause, we further extend the syntax of our basic logic of belief by allowing each clause to have a distinguishing label or number. The idea of attaching a distinguishing number to a wff was initiated by Godel [Enderton 72], thus born the name of *godelization* of a wff. Godel showed that the godelizations of wffs are representable. This means that we can use a number to denote a wff. In other words, by making assertions on godelizations of wffs, we can indirectly express assertions about other assertions. Thus based on the idea of godelization, we can express the explicit theory of a clause by qualifying the godelization of the clause with the theory. For example, if the clause c is in the theory $t1$, we can represent this as:

$$THEORY(\#c, t1)$$

where $\#c$ is the godelization of clause c.

However unlike [Bowen&Kowalski 82], these theories are organized in hierarchical structures in our formalism so that a clause at a higher-level theory can be inherited by a lower-level theory unless it is false there. Making the theories explicit has the advantage that subtle differences in meanings can be expressed by appropriate organization of theories.

Thus to solve the above looping and contradiction problems, we can represent the theory hierarchies in such a way that T2 is one level below T1 and T2 has belief #2 but T1 does not. This effectively assigns a higher priority to belief #1 over #2. This can be shown in our formalism as follows:

$\#1 : Dr(x) \leftarrow Prof(x) \wedge C(\neg Mr(x), T1).$
$\#2 : Mr(x) \leftarrow MD(x) \wedge C(\neg Dr(x), T2).$
$SUB - THEORY(T2, T1).$
$THEORY(\#2, T2).$
$\neg THEORY(\#2, T1).$
$\leftarrow Dr(x) \wedge Mr(x)$

Finally, it should be noted that our explicit control over the area of belief space for handling belief revision can be generalized to ordinary proofs in the spirit of *relevance logics* [Anderson&Belnap 75]. In this case, every proof must be associated with an area of a belief space. To achieve this, we introduce another meta-predicate PROVE so that $PROVE(p, t)$ stands for "p can be proved in the theory t". Relevance proofs are useful in knowledge representation. For example, it distinguishes $Guilty(x)$ from $PROVE(Guilty(x), t)$ as we should do because we may believe a person to be guilty, but we may not think we can *prove*

it given the evidences we think are *relevant* to the case.

6 The consistency of self-referential paradoxes

Another expressive feature of our formalism is its ability in representing self-referential beliefs, eg. "John believes that his belief is false" and mutual-referential beliefs, eg. "Simon believes that Tom's belief is true and Tom believes that Simon's belief is false". For example, the above example of mutual belief can be represented in our godelized clausal form of belief logic as follows:

$B(Simon, \#2 : TRUE(\#3))$
$B(Tom, \#3 : \neg TRUE(\#2))$

As argued before, it is pointless to talk about people's beliefs outside the context of a world in which the beliefs may be true or false. However this could introduce inconsistencies to belief spaces. This is shown by Tarski [36] in his No Truth Definition Theorem, which states that

$$TRUE(\#a) \leftrightarrow a$$

is inconsistent. This can be seen from the example if we assume Simon's belief to be true and Simon's inferential power is capable of making the following inferences:

$$
\begin{aligned}
TRUE(\#2) \quad &\rightarrow \quad TRUE(TRUE(\#3)) \\
&\rightarrow \quad TRUE(\#3) \\
&\rightarrow \quad TRUE(\neg TRUE(\#2)) \\
&\rightarrow \quad \neg TRUE(\#2)
\end{aligned}
$$

ie. a paradox inconsistency has arisen.

Tarksi's solution is attaching numerical subscripts or levels to 'true'. In Tarksi's approach, a truth $true_m$ is restricted to apply to sentences containing no predicate $true_n, n \geq m$. This requirement effectively blocks the derivation of contradiction. However Tarski's approach suffers the problems of inefficiency due to different operations at different levels [Warren 81], and limited expressiveness in representing beliefs [Perlis 85] such as "I have a false belief".

Kripke's solution [75] thus introduces truth gaps to account for paradoxes. However the solution incurs the invalidity of the excluded-middle principle, ie. $(p \lor \neg p)$. This could make the design of an efficient proving system difficult. Thus we take an alternative approach in the spirit of Perlis [85].

In our approach, the Tarski's Truth Definition Theorem is modified so that the axiom

$$TRUE(p) \leftrightarrow p$$

(called the *positive axiom*) holds for all positive p but may not be true for negative p. For the negative p, we adopt Gilmore's reduction rule [74] in our approach in such a way that the axiom

$$TRUE(\neg TRUE(p)) \rightarrow TRUE(\neg p)$$

(called the *negative reduction axiom*) holds for all p. To allow further reductions, we introduce another axiom (called the *restricted negative axiom*):

$$TRUE(\neg p) \rightarrow \neg TRUE(p)$$

which holds if the dereference of p involves no TRUE predicate. By dereference of p, we mean that if p is a label, then the dereference of p is the wff named by p; otherwise, the dereference of p is p itself. The restricted negative axiom allows us to deduce $\neg TRUE(EQ(1,2))$ from $TRUE(\neg EQ(1,2))$ because the dereference of $EQ(1,2)$ contains no $TRUE$ predicate; while from $TRUE(\#3)$ in the above example, we cannot deduce $\neg TRUE(\#2)$ because the dereference of $\#3$, ie. $\neg TRUE(\#2)$, contains a TRUE predicate. Note that for the above axioms to work correctly, we need to represent a clause in the disjunctive clausal form mentioned in Section 2.

Using the *restricted negative axiom*, we can *preserve the excluded-middle principle*

$$TRUE(p) \lor \neg TRUE(p)$$

but not

$$TRUE(p) \lor TRUE(\neg p).$$

In other words, we cannot have both $TRUE(p)$ and $\neg TRUE(p)$ in a consistent belief space, but we can have both $TRUE(\neg p)$ and $TRUE(p)$ in a consistent belief space. The fact that $TRUE(p) \land TRUE(\neg p)$ holds, helps to reveal a paradox without letting this create an inconsistency to our formalism. The price is simply that we stick literally with what the statements express, and this inconvenience will be as rare as are these sentences in typical discourse situations. A consistency proof of the modified truth definition theorem can be found in [Jiang 86].

To see how we have solved the paradox inconsistency, we use the earlier example as an illustration.

Suppose we assume that Simon is right, then we have the following inference chain,

$$
\begin{aligned}
TRUE(\#2) \quad &\rightarrow \quad TRUE(TRUE(\#3)) \\
&\rightarrow \quad TRUE(\#3) \\
&\rightarrow \quad TRUE(\neg TRUE(\#2)) \\
&\rightarrow \quad TRUE(\neg \#2)
\end{aligned}
$$

which is consistent.

Suppose we assume that Tom is right, then we have the following inference chain,

$$
\begin{aligned}
TRUE(\#3) \quad &\rightarrow \quad TRUE(\neg TRUE(\#2)) \\
&\rightarrow \quad TRUE(\neg \#2) \\
&\rightarrow \quad TRUE(\neg TRUE(\#3)) \\
&\rightarrow \quad TRUE(\neg \#3)
\end{aligned}
$$

which is still consistent.

In both cases, paradoxes are *revealed* and at the same time consistencies are *preserved*.

7 Conclusions

In this paper, we have presented a scheme from a logic programming perspective which

312

formalizes beliefs in *an extended clausal form of logic*. We have shown that our formalism is free-from the quantification problem that often persists in existing formalisms. In particular, we have indicated that our formalism allows an efficient resolution-like proof procedure to be developed. A levelled intensional scheme which enables the clausalization of beliefs has been proposed. It has been argued that the intensional scheme is free from the imputation problem. An inferential power bound resolution rule of belief has been introduced. A general circumscriptive non-monotonic reasoning system for handling belief revision has been described. The concept of godelization which increases the expressiveness of our formalism has been introduced. In particular, a modified Tarski's Truth Theorem has been shown to be consistent with our formalism.

There are issues such as common beliefs and implied beliefs of *a group of agents* which have not beed discussed in this paper due to the space limit. In addition, we have neither addressed belief introspection nor *implicit* belief revision. These problems will be sbjected to further research.

As regards to implementation of our formalism, it is felt that it can be done quite easily in Prolog. In particular, an intensionally-levelled term can be represented as a structure of the form (term,level); and a godelized clause can be represented as a structure of the form (label,clause). However this discussion is outside the scope of this paper.

Acknowledgements

This work is supported by the Alvey Directorate and ICL under grant GR/D/45468-IKBS 129. It is part of a collaborative project between ICL and the Univ. of Essex and Manchester. It is a pleasure to acknowledge the many discussions I have had with my collegues in Essex, Manchester and other institutes. In particular, I like to thank Peter Aczel, Alan Rector, David Warren (Manchester), Simon Lavington, Raymond Turner, Sam Steel (Essex), John Barnden (Indiana) and Richard Frost (Glassgow) for their helpful discussions.

References

A. Anderson & N. Belnap (1975) *Entailment:The logic of relevance and necessity.* Vol.1 Princeton Univ. Press.

J. Barnden (1986) *Interpreting propositional attitude reports: towards greater freedom and control* Proc. 7th European Conf. on AI, July, 1986.

J. Barwise & J. Perry (1983) "Situations and attitudes". MIT Press.

K. A. Bowen and R. A. Kowalski (1982) *Amalgamating language and meta-language in logic programming..* In Logic Programming, ed. Clark and Tarnlund, pages 153-172, Academic Press, 1982.

L.F. Cerro (1986) *MOLOG: A system thay extends Prolog with modal logic.* New Gen. Computing 4 pp.35-50.

C.L. Chang & R.C.T. Lee (1973) *Symbolic logic and mechanical theorem proving.* Academic Press.

W.F. Clocksin and C.S Mellish (1981) *Programming in Prolog.* Springer Verlag.

L.G. Creary (1979) *Propositional attitude: Fregean representation and simulative reasoning.* Proc. of 6th IJCAI. Tokyo, Japan, August.

J. Doyle (1979) *A truth maintenance system.*, AI. Vol. 12, No.3, pp.231-272.

H. Enderton (1972) *A mathematical introduction to logic.* Academic Press. 1972.

P. Gilmore (1974) *The consistency of partial set theory without extensionality.* ed. T. Jech, Axiomatic set theory, Amer. Math. Soc. Providence RI 1974,

J. Hintikka (1962) *Knowledge and belief.* Ithaca, New York, Cornell Univ. Press.

Y.J. Halpern, Y. Moses (1985) *A guide to the modal logics of knowledge and belief: preliminary draft.* IJCAI 85, Vol.1

P.J. Hayes & R. Kowalski (1969) *Semantic trees in automatic theorem proving.* Machine Intel. 4, Edinburgh Univ. Press, pp.87-101.

J.R. Hobbs & S.J. Rosenschein (1977) *Making computational sense of Montague's intensional logic.* AI 9 pp.287-307.

Y.J. Jiang (1986) *A formalism for representing qualified knowledge and its implementation for large knowledge bases* Ph.D thesis, Comp. Sci. Dept., Univ. of Manchester, 1986.

A. Kobsa (1985) *Using situations and Russellian attitudes for representing beliefs and wants.* IJCAI 85, Vol.1.

K. Konolige (1985) *Belief and incompleteness.* in Formal theories of the commonsense world, Ed. J.R. Hobbes, R.C.Moore. Ablex Pub. Corp. 1985. pp.359-403.

K. Konolige (1985a) *A computational theory of belief introspection.* IJCAI 85 Vol. 1.

R. Kowalski (1983) *Logic Programming..* IFIP 83, pp.133-145.

S. Kripke (1975) *Outline of a theory of truth.* J. Philosophy 72, pp.690-716.

H.J. Levelsque (1984) *A logic of implicit and explicit belief.* Proc. National Conf. on Artificial Intelligence. pp.198-202.

H.J. Levelsque (1984a) *The logic of incomplete knowledge bases.* On Conceptual Modelling, eds M.Brodie, J.Mylopouos&J.Schmidt, Springer-Verlag, 1984, pp.165-187.

P. London (1978) *Dependency networks as representation for modelling general problem solver.* Ph.D. Dissertation, Tech. Report, Dept. of Comp. Science, Univ. of Maryland, 1978.

J. McCarthy (1979) *First order theories of individual concepts and propositions.* Machine Intel. 9, Eds, J.Hayes, D.Michie, L. Mikulich, Ellis Horwood.

J.McCarthy (1980) *Circumscription-A form of non-monotonic reasoning.* AI 13, pp.27-40, 1980.

R.C. Moore & G. Hendrix (1979) *Computational models of belief and the semantics of belief sentences.* SRI Technical Note 187, Menlo Park CA.

R. C. Moore (1985) *A formal theory of knowledge and action.* in Formal theories of the commonsense world, Ed. J.R. Hobbes, R.C.Moore. Ablex Pub. Corp. 1984.

R.C. Moore (1985a) *Semantic considerations on non-monotonic logic.* AI. Vol. 25, pp.75-94.

D. Perlis (1985) *Languages with self-reference I: foundations..* Artificial Intel. 25, pp.301-322, 1985.

W.V. Quine (1956) *Quantifiers and propositional attitudes..* Journal of Phil. 53.

R.Reiter (1980) *A logic for default reasoning* AI 13, pp.81-132,1980.

J.A. Robinson (1965) *A machine oriented logic based on the resolution principle.* in JACM, Vol 12, pp.23-41.

A. Tarski (1952) *The semantic conception of truth.* Semantics and Philosophy. (L. Linsky ed) Univ. of Illinois.

D.H. Warren (1981) *Higher order extensions to Prolog- are they needed?.* in Machine Intel. 10, Ellis Horwood.

Using Cognitive Models of Learning in Instructional Design

Dana S. Kay
Graduate School of Education
University of California, Berkeley

The primary goal of this research was to facilitate learning by applying a model of the changes in knowledge representations that occur with increased computer experience to instructional design. The model proposes that users' knowledge representations of computer systems evolve from being organized according to preconceptions to being organized according to the goals, plans and rules for using the system. The model was implemented in an on-line training system. It was proposed that using a model of the natural evolution of knowledge to organize the presentation of information in the training would create a better fit between learning process and learning materials and thus, facilitate learning. The model-based training system was compared to four other training systems that implemented other methods of organization that are currently used in research and industry. The results illustrated only partial validation for the model. That is, the training based directly on the model did not lead to the best peformance, but also did not lead to the worst. The best performance was observed for instructions that were organized according to tasks and the worst performance was observed for instructions that were organized according to categories of commands. It is proposed that (1) using models of learning in instructional design requires more than mapping the learning assumptions onto an instructional setting and (2) instructional design provides an important method for testing the assumptions of the learning model.

There has been a substantial amount of attention given to the investigation of the acquisition of knowledge. From this research has grown an interest in the application of cognitive models of the student to instructional design. That is, researchers have begun to examine the benefits of using knowledge-based models of learning to guide the organization of instructional information. In a recent study, Kay and Black (1985) presented a model that traced the changes in the content and organization of knowledge that occur with the acquisition of text-editing knowledge. The present study compares the performance that results from instruction that is based upon this model to the performance that results from instruction that is based upon other methods of organizing computer knowledge.

By combining the results of three studies that examined the changes in text-editing knowledge representations, Kay and Black (1985) proposed a four phase model of the evolution of knowledge as one proceeds from naive to expert computer user. In this model, phases represent snapshots of the mental structures that exist at varying levels of ability. Phase One of this model describes the knowledge representations that exist before the learning process begins. In this phase, knowledge is organized according to preconceptions about the terminology that will later refer to text-editing commands. In Phase Two, users' knowledge representations begin to reflect computer-related information. Because one of the first things that users learn is the goals that the commands accomplish, the new organization of the commands is based upon the relationships between the commands and the relevant goals that they accomplish.

Having acquired the basic text-editing goals and commands, users begin to note that there are certain commands that are frequently used together in accomplishing a goal. It is this realization that leads to another reorganization of knowledge in which actions that were organized separately in Phase Two are now combined into the plan-based representations of Phase Three. The final phase of the model describes the knowledge representations for users who are experts in text-editing. The representation proposed for this phase completes the acquisition process (aside from minor forms of tuning that will continually take place) and is similar to the GOMS (Card, Moran & Newell, 1983) account of text-editing experts. In this phase, knowledge is reorganized to incorporate (1) compound plans that are composed of the simple plans of Phase Three and (2) selection conditions for choosing the most appropriate plan for accomplishing a given goal.

Computer Instruction

Scenario Machines. There are many forms of computer instruction. One form, that has been given attention by Carroll and his colleagues, is training systems. These systems are designed to encourage users to learn by exploring the functions available in the system. Thus, users are actively engaged in the learning process. The "training wheels" interface (Carroll & Carrithers, 1984a) is a training system in which new users are blocked from accessing advanced system functions and from committing typical user errors. The motivation behind this design is to encourage "discovery learning," while sparing users the confusion and frustration that often result when new users wander into the depths of complex computer systems. By blocking off certain system states, the set of possible actions and consequences of these actions is constrained. Thus, users are free to actively participate in and control their learning.

The "Scenario Machine" (Carroll & Kay, 1985; McKendree, Schorno, & Carroll, 1985) is a more restricting version of the "training wheels" interface. In this design, not only are advanced functions and user errors blocked, but the user is directed through a single scenario that represents the basic functions of the system. The use of this type of system design provides researchers with knowledge of what the user will be required to do and thus, we can better control our investigation of user learning. The current experiment uses the Scenario Machine design as a vehicle for examining the application of cognitive models of student knowledge representations to the design of computer instruction.

Current methods of computer instruction. If we survey the commercial forms of computer instruction that are currently used, there are two prominent methods of organization that are used -- command-oriented and task-oriented. Command-oriented instruction is designed by organizing the computer information according to categories of commands that can be used in the system. The problem with this type of instructional organization is that the order of presentation is guided by categories of commands, rather than by the order in which the commands are encountered as part of an editing session. Task-oriented instruction illustrates an attempt to find a better match between order of presentation and order of use in an editing session. A task-orientation presents the editing commands according to plans for accomplishing editing tasks. Recent research has improved upon the commercial task-oriented instruction and proven it to be quite effective (Carroll, 1984; Carroll & Carrithers, 1984).

Example-based instruction is a method of instruction that has not been implemented in commercial systems. However, this type of instruction has been examined within the context of research on teaching people how to use computers and has been shown to be an important part of learning (Rissland, Valvarce, & Ashley, 1984). Example-based instruction is similar to the task-oriented instruction in that the command information is presented in accordance with its use in computer sessions or tasks. However, using only examples does not provide the user with conceptual knowledge of the system and thus, the user may have difficulty recovering from errors due to the use of incorrect commands.

Current Experiment Closer examination of the current methods of organization shows that none of these methods is based directly upon the changes in the content and organization of user representations as presented by the Kay and Black four phase model. Therefore, an on-line training system was designed for this experiment using the four phase model to guide the organization of information. The information presented in the training system is for a command-based, experimental database system. In addition to the four phase model training system, training systems implementing the other methods of instruction were also designed for the database system. By comparing the performance (i.e. time to complete tasks and errors committed) that results from using each of these training systems, we can assess each of these methods of instruction. The basic hypothesis is that for the model-based training, there will exist a better fit between the learning process and learning materials. As a result, it should be easier for users to acquire computer information. If using this model facilitates learning, then in addition to supporting the use of cognitive models in instructional design, there will also be further validation of the four phase model.

Method

Subjects

Forty Yale undergraduates participated in this study. All participants had little or no computer experience. That is, they had either never used a computer or had used the computer a few times for word processing purposes only.

Materials

Database System. A simple experimental database system was designed for the study. The system was a command-based, interactive system that contained information about popular songs. There were fifteen commands that could be used in the system and five basic tasks that sequences of these commands could accomplish. The system kept track of the keystrokes entered and the reaction time between keystrokes.

Training Systems. In addition to the basic database system, five training systems were also designed. These systems taught subjects how to use the database system by explaining the commands used to accomplish the database system tasks. The presentation of the commands within the training systems was dependent upon the method of instruction implemented in that training system. Four of the training systems were based upon the methods of organization previously described -- command-oriented, task-oriented, example-based and four phase model. The fifth training system was used as a control condition for the four phase model. Because it is believed that the ordering of the phases is important to the learning process, a training system was designed that organized the four phases in reverse order.

There were two parts to the design of these systems. In the first part, each of the five methods of organization was mapped onto the fifteen commands used in the database system. In the second part, these mappings were used to design a method of organization that generalized across the mappings and resulted in an ordering for the commands that could be applied in all the training systems. The latter part of this design process was used so that the same exercises could be used regardless of the method of instruction.

There were five parts to the training systems. The first part of the training was the same for all five training system conditions and consisted of two parts, an introduction to the database system and a description of the Quit command. This command was described separately because it is used throughout the system, regardless of the task.

In Part 2 of training, the commands used for listing, printing, and sorting the database were presented together. In Part 3, the commands for adding songs to the database were presented. In Part 4, the commands for searching in the database were presented and finally, in Part 5, the commands for changing the database were presented together. Because of the similarity in command relationships, the systems could be designed so that regardless of instructional organization, the same exercises were presented after each section.

The exercises were the same for all five training systems. In performing these exercises, subjects were taken into a modified version of the database system that had the same screen design as the full database system. However, the exercise environment differed in its response to subject input. As described previously, the system used a Scenario Machine design (Carroll and Kay, 1985) in which the system stalls until the correct command (that required for the task at hand) is input by the subject. That is, for each exercise, there is a fixed scenario for accomplishing the task and unless the subject inputs the correct command for the scenario, the system beeps and prompts the subject to input a command. This design was used to keep subjects from entering incorrect commands and getting lost in the system.

Procedure

Learning Phase. The learning phase took place on the first day of the experiment. Subjects were told that they would be following an on-line description of how to use the database system and then after each part of the training, they would demonstrate their understanding of the system by working through a series of exercises related to what they had just learned. They were also told that for each of the exercises, they would not be required to use any command that they had not been previously introduced to and that if they input a wrong command, the system would beep and then prompt them to input another command.

Subjects were also given a manual that contained the same information that was presented on the computer. They were informed that they could use this manual while working on the exercises, but that they should let the on-line system guide their training. The experimenter informed the subjects that they were to work on their own, but that if they felt that they were in trouble, they could call for help. In addition, subjects were informed that their interaction with the computer would be video-taped.

When the subjects completed the training tasks, a computer concepts test was administered to assess conceptual understanding of the database system. This test consisted of nine questions. Some of these questions asked users "how " to accomplish certain database tasks. Other questions asked subjects about the functions associated with certain database commands. Two versions of this test were used. One version was administered half way through the study (after training) and the other was given at the end of the study (after transfer).

Transfer Phase. The transfer phase took place on the second day of the study. Subjects were told that they would be using the database system again. However, in this part of the study, they would not be using the training system that guarded them against errors, but rather, they were using the complete database system. Subjects were given the manual that they had used the day before and a booklet of 12 transfer exercises of varying levels of difficulty. Four of the exercises were direct mappings from the training exercises that were part of the training phase. Four of the exercises were simple extensions from the training exercises (i.e. they combined two of the exercises presented in training). The last four were more complex exercises (i.e. they combined more than two of the exercises presented in training). Again subjects were told that their interaction with the computer would be video-taped and that they were to work on their own, but could call for help if they were in trouble.

After completing the transfer tasks, subjects were given another computer concepts test about the system. In this test, some of the questions were repeated from the training test and some of the questions were new. Regardless, the questions asked "how" to accomplish tasks in the system and "what" tasks certain combinations of commands accomplished.

Results

Performance Measures

Each time subjects used the database, five performance measures were recorded. The training systems and the database system recorded the time to complete the exercises performed while using the systems. An overall time measure for each phase was also collected by totalling across the exercise times. In addition to the time measures, two error measures were also recorded. One type of error is a "related" error in which the subjects uses an incorrect command that is related to the correct command. The second type of error is an "unrelated" error in which the command input is not related to the correct command. In most cases, the "unrelated" errors result when the user appears to be inputting random commands.

The fourth performance measure is the number of times that the subject requested aid from the experimenter (experimenter interventions). The range for this measure was zero to four interventions. The fifth performance measure is the score on the computer concepts test. This score is a number (0-9) representing the number of correct answers to the questions asked.

Performance during learning

Table 1 presents the relevant data for the learning phase. The first row of the table presents the mean time to complete training. The results of this analysis showed that the five training systems differed in their time to complete the learning exercises $F_{(4,35)} = 4.4, p < .01$). A follow-up analysis showed that (1) the task-oriented training training system took significantly less time to complete the exercises than the four phase model, the reverse phase model and the command-oriented training systems at the .05 level and (2) the example-based training system was significantly faster than the command-oriented training system.

As mentioned previously, in addition to the reaction time measures, other performance measures were also collected -- "related" errors, "unrelated" errors, experimenter interventions, and score on the computer concepts test. The results for these measures during training are presented in the second through fifth rows in Table 1. A multivariate analysis of variance showed a significant main effect for the type of instruction used (Hotellings trace criterion, $F_{(16}[1], 102) = 2.2, p < .01$).

To localize these effects, univariate analyses of variance for the individual measures were used. A significant effect for type of training system was present for the number of "related" errors ($F_{(4,30)} = 3.17, p < .05$) and for the score on the computer concepts test $F_{(4,30)} = 3.58, p < .025$). Further analysis of these differences showed that for the "related" errors, the command-oriented training system lead to significantly more errors that the other four training systems. It appears that subjects using the command-oriented training system had to try more commands before finally using the correct command. In addition, the analysis of score on the computer concepts test showed that the example-based training system had significantly poorer performance on the test than the task-oriented, four phase and reverse phase training systems. Thus, although this training system lead to faster performance than some of the other training systems, the conceptual knowledge of the subjects using this training system was not as good as the knowledge extracted from the other training systems.

Table 1

Performance for Learning Phase

	Training System Treatments				
	Command	Task	Example	Four Phase	Reverse Phase
Total Time (in mins)	37.462	20.376	26.054	27.541	30.265
Related Errors	60.375	11.125	29.000	30.000	31.375
Unrelated Errors	50.250	9.875	13.875	14.625	23.875
Inter-ventions	1.250	0.500	1.125	0.750	1.625
Concepts Test	6.563	7.188	6.125	7.438	7.188

---

[1]The degrees of freedom in this analysis are smaller than expectedbecause an analysis of covariance was used. The covariates used and the results of these covariates will not be discussed in this paper due to lack of significant effects and space constraints.

Table 2

Performance for Transfer Phase

	Training System Treatments				
	Command	Task	Example	Four Phase	Reverse Phase
Total Time (in mins)	41.474	35.841	38.857	42.641	44.923
Related Errors	66.875	11.125	24.750	21.500	41.250
Unrelated Errors	17.000	3.000	7.375	5.125	10.250
Inter-ventions	0.375	0.000	0.125	0.103	0.250
Concepts Test	7.313	7.750	7.625	7.500	7.563

The results for the learning phase show that the command-oriented treatment took longer to complete the learning exercises and committed more errors while doing these exercises. On the other end of the continuum, the task-oriented treatment took the least amount of time. Although the example-based training system had relatively fast times and few errors, this treatment seemed to lack in conceptual knowledge of the system as is shown by the poor performance on the concepts test. It appears that using the examples to train users, eases use of the system, but does not provide users with an understanding of the system beyond the tasks at hand.

Transfer Phase

Table 2 presents the performance measures for the transfer phase. As in the learning analyses, an analysis of total task time was used first. However, there was no significant effect for training system types $(F(4,30) = .63, ns)$. As in the learning phase, there were four additional performance measures recorded in addition to time to complete the transfer exercises. As can be seen from the table, the mean for experimenter interventions for the task-oriented training system was zero. Because the value for this variable can only be positive, the variance for this variable for the task-oriented training system can only be zero. Since a zero variance disturbs the homogeneity of variance for this variable, it was analyzed separately. The results of this analysis showed no significant difference between training system treatments in the number of experimenter interventions $(X^2 = 2.927, ns)$.

A multivariate analysis of variance was performed on the remaining three measures. The results of this analysis showed a significant difference between the training system treatments (Hotellings trace criterion, $F(12,95) = 1.95, p < .05$). This main effect was localized to a significant difference in the number of "related" errors committed $(F(4,35) = 5.68, p < .01)$. A follow-up analysis revealed that the command-oriented training system treatments committed significantly more errors that the four phase, reverse phase, and task-oriented treatments. In addition, the reverse phase treatment committed significantly more errors than the task-oriented treatment.

In the transfer phase, the results show that all subjects learned to use the system and were able to complete the exercises in roughly the same amount of time. However, when we look at the results for the errors committed, the command-oriented and reverse phase treatments committed the most errors. The system used in this study was a simple system. However, if a more complex system was used, we can imagine that committing a great deal of errors would lead to the users getting in trouble and in the end, taking longer to complete the tasks. This is an empirical question to be addressed in the future.

Discussion

The primary goal of this study was to facilitate learning and improve performance by designing instruction that is based upon an analysis of the changes in knowledge representation with the acquisition of computer knowledge. The results from the training phase of the study showed that the task-oriented and for some measures the example-based organization led to the best performance, while the command-oriented organization led to the worst. The instructions organized according to the four phase model seemed to fall in the middle with the four phase model being somewhat better than the reverse four phase organization (at least as measured by the number of errors committed during transfer). Thus, the original hypothesis that the four phase treatment would lead to the best performance is not totally supported.

Four Phase Facilitation of Learning

Utility of Phase Two. One possible explanation for the lack of support of the original hypothesis refers back to the original proposal of the model. In this presentation, Kay and Black (1985) provide several reasons for the existence of the four phases that they observed. They argued that the first phase was necessary because the only information that new users have to use in their initial attempts with a computer system are the prior knowledge associations that they bring to the computer domain. In addition, the fourth phase is necessary because past research (albeit minimal) has suggested that complex plans and selection rules are necessary for skilled computer performance (Card, Moran & Newell, 1983). The question then becomes are both the intermediate phases also necessary?

It would be difficult for users to progress directly from Phase One to Phase Four because this transition requires the use of more knowledge than human working memory is capable of holding at one time. Thus, the two intermediate phases provide a way of moving from Phase One to Phase Four within the constraints of working memory capacity. That is, these phases allow for the progression from Phase One to Phase Four to follow steps that can be handled by working memory. However, Kay and Black also suggested that it might not be necessary to have two intermediate phases. They proposed that Phase Three was a necessary precondition for Phase Four, but that Phase Two might be an artifact of the way that users currently learn to use a computer. That is, until recently, most commercial computer training materials have used a command-oriented organization and therefore, it is not surprising that novices incorporate this organization into their representation of the domain in the form of goal/action relationships.

The training system results suggest that it might indeed be the case that Phase Two is an artifact of learning materials. If we examine the task-oriented training system, we see that this system takes subjects directly from Phase One (preconceptions) to Phase Three (plans). Because of the simplicity of the database system and the plans necessary for using the system, the presentation of these plans can easily be handled within the limits of working memory. Therefore, subjects can learn these plans and begin using them immediately. In the four phase model training system, subjects are first guided through the Phase Two representation and then presented with the Phase Three plans. It is possible that because the Phase Two representation was unnecessary, this information interfered with subjects' ability to learn the plans presented later in the instructions and performance was hampered.

Process + Representation. Another possible explanation pertains to the method by which the four phase model was applied to the training system design. Recall that one of the original goals of the four phase model was to describe the changes in the content and organization of users' knowledge representations. However, in applying the four phase model to instructional design, the organization was focussed on the changes in knowledge representation and not on the processes that lead to these changes. That is, the instructions guided users through the four phases, but did not describe the acquisition processes. It is possible and probable that for the model to be used successfully, instruction needs to include process and representation information.

Mapping Learning onto Instruction An implicit assumption in the design of the four phase training system is that by mapping knowledge representations and learning processes directly onto instruction, we can design better instruction that will facilitate learning. It is possible that this assumption is not totally correct and therefore, the four phase model did not lead to the best performance. Although understanding the user may be important to instructional design, it may be the case that what we know about knowledge representation and learning needs to be reorganized into a framework that is organized around the goals of the instruction. One source of the need for this reorganization is that learning analyses work from the means to the end (i.e., they look at how people progress to a certain level of expertise), whereas, instruction works from the end (knowledge to be learned) to the means. If we consider instructional design as a problem solving task, we can propose that the problem specification should reflect the goals of the instruction and the domain knowledge that is used to solve the problems and accomplish these goals is our knowledge of the "how" and "what" of learning.

Example-based Learning

Although the example-based subjects were able to successfully perform the training tasks without an extreme number of errors, their performance on the concepts test was significantly poorer than the other training treatments. Thus, although these subjects could use the database system, they did not have a good conceptual representation of how the system worked. One possible explanation for this result is that these subjects were performing the tasks by analogically mapping them onto the examples presented in the training system. That is, these subjects were applying the commands in the examples without understanding how the commands interact.

Learning by analogy has been shown to be a prominent method used by novices in a domain (Anderson, Farrell, & Sauers, 1984). However, as suggested by the current results, once the problems from which students analogize are taken away, the students are no longer able to perform successfully. In addition, past research has shown that using analogies often leads to misconceptions in students' thinking (Douglas & Moran, 1983) because students have difficulty noting how the analogy and the problem are similar/different (Halasz & Moran, 1982).

One of the motivations behind using the example-based training system was to examine examples as a sole source of instruction since it is more often the case that examples are included as part of more detailed instruction. It appears that examples when presented by themselves provide only performance information and do not provide information at a more conceptual level. Therefore, one might propose that examples should be included in addition to other forms of information that do provide conceptual information.

Task-oriented vs. Command-oriented

Schema representations. One possible explanation for the observed results is that the task-oriented method of instruction allows the subjects to see the use of the commands as a coherent whole. That is, presenting the commands in the context of the plans that they are part of provides subjects with enough information to form at least a sketchy schema for the plan that can later be easily applied. On the other hand, presenting the commands as separate fragments of information as in the command-oriented organization forces subjects to develop the plans on their own. The error results for subjects in the command-oriented treatment suggest that this instruction leads to a strategy in which learners decide on a goal to accomplish and then try all the commands that are related to this goal until they get their desired result. As a result of this trial and error strategy, subjects obscured the coherence of the inter-command relationships. Thus, when required to perform a similar task, they were not able to remember how they had previously performed the task. The results for the transfer phase in which the command-oriented subjects were still committing significantly more errors than the other subjects lends support to this claim.

Procedural/declarative representation. Another way of explaining the training system results pertains to the procedural/declarative distinction and more specifically, its influence on the acquisition of knowledge as proposed in the ACT* theory (Anderson, 1983). Applying the procedural/declarative representation to the methods of organization used in the current study, one can propose that the task-oriented organization provides users with procedural information whereas, the command-oriented organization provides only declarative information. When we present subjects with a procedural organization of information, we are showing them the how to accomplish domain-specific tasks and thus, facilitating the learning process. When we present subjects with a declarative organization of information, we are requiring them to first use their general problem solving procedures to interpret the declarative knowledge and then use this interpretation process to develop domain-specific procedures. Thus, we are making the the learning process more difficult.

Thus, combining this distinction and the results of the current experiment suggests that using a procedural organization of knowledge can lead to better performance. There are, however, two caveats to this proposal. First, it is not necessarily the case that procedural information alone will lead to the best performance. In the current experiment, the example-based training also provided some procedural information in that it provided users with information about how the commands are used together in a sequence for a specific example. However, as was discussed previously, this organization did not provide users with a complete conceptual understanding of the computer system. One possible explanation for the difference between the task-oriented and example-based training systems is that the task-oriented organization provides more abstract procedural information that includes information about the general goals that drive the procedures and thus, drive performance.

The other caveat for interpreting the use of procedural information to facilitate learning pertains to the system used in the current experiment. The database system used is a simple computer system (i.e. requires learning for only a small set of commands that can accomplish a small number of tasks) and only requires learning simple procedures. It is not clear that procedural instruction would afford the same benefits in a more procedurally complex domain.

The goals of this research were to improve instruction and learning by using a knowledge-based analysis of learning to guide the design of the instruction and at the same time use instructional design as a method of validation for the Kay and Black four phase model. The results of the study provide only partial evidence in support of these goals. As previously described, the four phase model training system was designed by mapping the learning model directly onto instruction. The results from the training system comparisons suggest that it is important to consider instructional goals in addition to learning goals and thus, instructional design should be viewed as more than a mapping process. In addition, the results from the model-based training suggests that instructional design can and should be used as an important test of the assumptions made in proposed models of learning. Future research will address the relationship between learning models and instructional design as it pertains to these issues.

References

Anderson, J.R. (1983). The architecture of cognition. Cambridge, MA: Harvard Press.

Anderson, J.R., Farrell, R. & Sauers, R. (1984). Learning to program in LISP. Cognitive Science, 8, 87-129.

Card, S.K., Moran, T.P., & Newell, A. (1983). The psychology of human-computer interaction. Hillsdale, NJ: Lawrence Erlbaum Assoc.

Carroll, J.M. (1984). MINIMALIST training. Datamation, 30(18), 125-136.

Carroll, J.M. & Carrithers, C. (1984). Training wheels in a user interface. Communications of the ACM, 27, 800-806.

Carroll, J.M. & Kay, D.S. (1985). Prompting, feedback and error correction in the design of a Scenario Machine. Proceedings of CHI '85, San Francisco, CA, 149-153.

Douglas, S.A. & Moran, T.P. (1983). Learning text editor semantics by analogy. Proceeding of the CHI '83 Conference on Human Factors in Computing Systems 207-211.

Halasz, F. & Moran, T. (1982). Analogy considered harmful. Proceedings of Human Factors in Computer Systems Conference, National Bureau of Standards, Gaithersburg, MD.

Kay, D.S. & Black, J.B. (1985). The Evolution of knowledge representations with increasing expertise in using a system. Proceedings of the Seventh Annual Conference of the Cognitive Science Society. Irvine, CA, August.

McKendree, J., Schorno, S., & Carroll, J.M. (1985). Personal planner: The Scenario Machine as a research tool. Videotape demonstration presented at CHI '85.

Rissland, E.L., Valcarce, E.M., & Ashley, K.D. (1984). Explaining and arguing with examples. Proceeding of the AAAI. 288-294.

Question Asking During Procedural Learning: Strategies for Acquiring Knowledge in Several Domains

Scott P. Robertson
Psychology Department - Rutgers University

Merryanna Swartz
Psychology Department - Catholic University
Army Research Institute - Arlington, VA

Abstract

Questions asked during acquisition of a complex skill reflect the types of knowledge that learners require at different stages. Questions that learners ask themselves may serve to generate incomplete conceptual frames that can be used to guide explanation of future events. Question-asking data collected from students learning to use a spreadsheet program suggest that learners initially require knowledge about plans and the structure of the skill domain. Next they require knowledge about the structure of tasks that they will be performing. Finally they concentrate on plan refinement. Models of skill acquisition and explanation-based learning should incorporate mechanisms for monitoring levels of knowledge in several distinct domains and dynamically altering strategies for knowledge acquisition within these domains.

Introduction

The acquisition of new skills is difficult for both humans and machines. Traditional psychological learning theories stress the role of practice and reinforcement in learning. Current models of learning and skill acquisition stress cognitive components. Change in the structure of goal hierarchies and the clustering of actions with practice has been studied extensively by researchers in cognitive science (Anderson, 1982, 1983a, 1983b, 1986; Neves & Anderson, 1981; Newell & Rosenbloom, 1981; Robertson & Black, 1986; Rosenbloom & Newell, 1986; Rumelhart & Norman, 1978). Recently, researchers have begun to concentrate on the ability to generalize and reason from single instances or examples. Researchers in this area have emphasized analogy (Burstein, 1986; Carbonell, 1983; Forbus & Gentner,

----

This research is supported by the National Science Foundation under grant no. IST-8696141, "Errors, explanation, and plan modification in cognitive skill learning," to the first author.

1986; Gentner, 1983), schema-based reasoning (DeJong, 1986; Schank, 1982), and more formal rules of induction and generalization (Hayes-Roth, 1983; Mitchell, Utgoff, & Banerji, 1983; Sammut & Banerji, 1986; Stepp & Michalski, 1986).

Reasoning from examples, especially during learning of complex interactive behaviors, requires utilization of knowledge from many different domains. What types of knowledge are typically required during acquisition of a new skill? How is domain search constrained during learning? Here these issues are addressed by examining the questions that people ask during complex skill acquisition.

Question Asking

In contrast to question answering, question asking is not a widely studied phenomenon. In one of the few AI implementations of a question asking system (Sammut & Banerji, 1986), questions to a teacher are used to test hypotheses about a problem. Miyake and Norman (1979) conducted one of the few studies of question asking in the psychological literature. They found that the number of questions asked depended on a combination of the learner's level of knowledge and the difficulty of the task. Novices asked more questions while doing easy tasks, but experts asked more questions while doing hard tasks. Miyake and Norman made the point that more questions are asked when the task difficulty is appropriate for the level of knowledge because question askers must have enough knowledge to formulate questions. While this study suggests how level of knowledge might affect question asking, it does not address how questions might be used to shape the learning process.

A question presents a concept or proposition to a listener along with information about unknown information related to the proposition. The question "How do I get to Newark from here?", for example, presents the fact that the question asker has a goal but no plan for achieving that goal. A cooperative answerer will provide a plan for achieving the goal as an answer to the question. The steps of the plan will become associated with the goal if they succeed so that the question asker can achieve the goal in the future without asking a question. We argue that self-directed questions are posed in order to generate incomplete knowledge structures (like a plan-less goal) that can be embellished by information gained from the learning situation. Thus self-directed question generation can be viewed as a strategy for acquiring knowledge by generating incomplete concepts and using them to guide exploration and constrain interpretation and explanation of new information.

As an example, consider a learner trying to learn about a spreadsheet program (this example is used because the upcoming data was collected from learners in this situation). At some point the question "How do I get rid of data in this cell?" might arise. A conceptual representation of the question shows a set of relationships among known and unknown aspects of the concept:

```
(CAUSE
    (?ACT          (ACTOR learner)
                   (OBJECT ?system-obj))

    (STATE-CHANGE  (ACTOR system)
                   (INITIAL-STATE (CONTAIN cell-x data-obj))
                   (FINAL-STATE (CONTAIN cell-x nil))))
```

In question answering systems, the specified parts of the representation would serve as templates for simple memory search. When learners ask questions like this, however, they know that the answer is not in their memory and so the representation serves another purpose. If other types of reasoning processes (e.g. induction, analogy) can not solve the problem, the question representation can serve as an explanation daemon which will match any subsequent occurence of the desired outcome of the unknown action (i.e. the STATE-CHANGE). When the desired system action occurs, by design or by accident, the daemon will recognize a previously unknown relationship between learner and system actions and be able to fill in the unknown action (?ACT) and system object (?system-obj) slots. This constrains the process of explanation required when an unexpected system action is encountered and allows earlier lapses in understanding to be turned into opportunities for learning.

Some Data on Question Asking During Skill Acquisition

The previous diccussion suggests that questions can be used stategically to instantiate incomplete knowledge structures which will be useful for learning. If this is so then the questions that people ask during learning should fall into categories that reflect distinct and useful knowledge domains. As knowledge is acquired in certain domains, questions in that domain should decrease and questions in other domains should increase proportionately. In this section such data is discussed.

Six students were asked to learn the use of a subset of commands for a popular spreadsheet program and enter four sets of data. The students were given brief instructions about relevant keys, the nature of a spreadsheet and the spreadsheet display. The students were instructed to talk

```
-----------------------------------------------------------------
```
Table 1. Examples of plan, system operation, task, and act
         questions asked during the learning trails.

Plan questions:

    "How do I erase it?"
    "Can I go back and erase and move it out?"
    "Do I hit 'escape' to get out of this thing?"

System operation questions:

    "Where did that go?"
    "Is there a certain amount of spaces it will leave?"
    "Why did it do that?"

Task questions:

    "Would I do those lines underneath there?"
    "Now I use the adding form?"

Act questions:

    "What did I do wrong?"
    "What was I doing before?"
    "I don't even know what I did."
```
-----------------------------------------------------------------
```

aloud as they completed the four data-entry tasks. We
stressed that questions were of greatest interest, although
the students were told they would not be answered.

The students asked a total of 166 questions, an average
of 27.7 questions per subject. Students asked fewer
questions as time went on, however this overall trend was
not evident for all types of questions.

Questions were categorized by the feature of the
learning situation that they referenced. Relevant features
were the system being learned, human plans, human actions,
and the spreadsheet tasks. The surface forms of questions
were not used for categorization purposes. For example, the
questions "How do I erase a cell?" and "Can I erase a cell
with the DEL-key?" are procedural and verification questions
respectively in terms of structure (Lehnert, 1978). Both
questions refer to human plans for performing a system
operation, however, so they would both be categorized as
"plan questions."

Questions about plans, actions, system operation, and
spreadsheet tasks accounted for 79% of all observed
questions. Table 1 shows examples of each type of question.
Figure 1 shows changes in the proportions of questions in
each of these categories as the learning trials progressed.

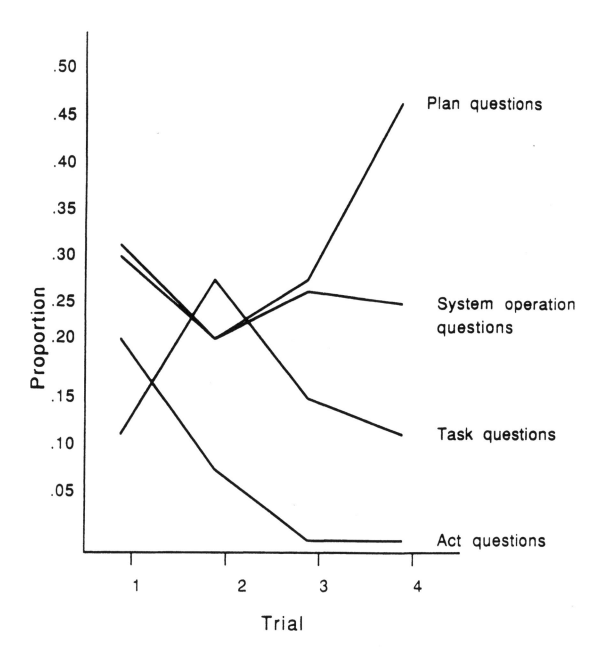

Figure 1. Changes in the proportions of different question types over trials.

Plan questions reflected lack of knowledge about a procedure, a step in a plan, or a goal. Plan questions often specified a goal or included a proposed procedure. As Figure 1 shows, plan questions were among the most consistent and important of all question types. They accounted for for 30% of all questions on the first trial and their frequency increased to 45% by the last trial.

System operation questions were the second most frequent type of question overall. These questions were about states of the system, reasons for system behavior, and causal links between system actions. System operation questions were as frequent as plan questions initially, but their frequency dropped to 24% of all questions by the last trial (Figure 1). Even so, in the last trial plan and system operation questions combined accounted for 56% of all questions asked.

Task questions were about the nature of the task itself or what was required of the learner. Specific procedures or characteristics of the system were not mentioned in these questions. Rather, they sought to clarify intentions of the experimenter or demands of the task. Figure 1 shows that the proportion of task questions increased in the second trials but remained constant and moderate otherwise.

Finally, act questions were about the learners' own actions. They indicated a lack of understanding of the consequences of an action or even a failure to remember what action was just performed. They generally arose in response to an unexpected system behavior. Even though act questions were rare overall, they were very frequent initially, accounting for 20% of all questions in the first trial (Figure 1). By the last two trials, act questions had disappeared completely.

Questions in each of the four conceptual categories either referred to something that had just happened (e.g. "What did I press?", "Did I fill in the right cell?") or to something that might happen (e.g. "What should I press?", "Could I press the delete key to remove that entry?"). The former are "reactive" questions and the latter are "predictive" questions. Novice subjects have no prior knowledge on which to base predictions or generate explanations, so few predictive questions and many reactive questions would be expected initially. As expertise develops, however, reactive questions should decrease and predictive questions should become more common. Figure 2 shows that both types of questions decreased over trials, but reactive ones decreased more rapidly. Also, reactive questions were much more frequent in the first trial than predictive ones, but this trend had reversed by the last trial.

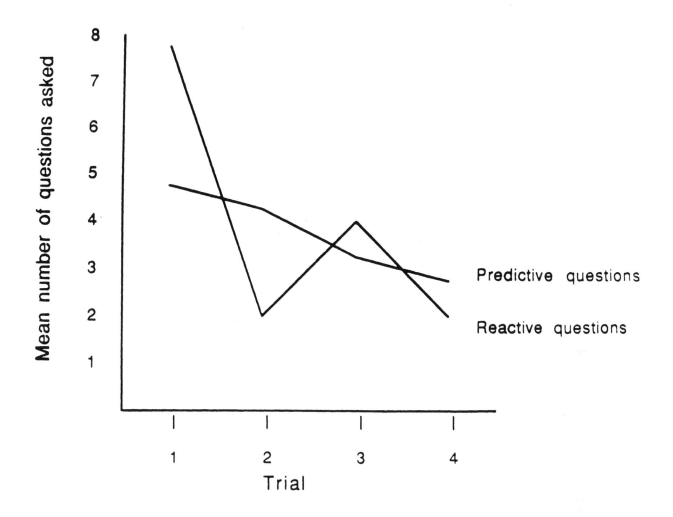

Figure 2. Mean number of predictive and reactive questions
over trials.

Summary and Interpretation of the Question Asking Data

Several different types of questions that are relevant to distinct knowledge domains arose during learning. The distribution of question types changed dramatically as learning progressed. The most frequent questions were about plans and the operation of the system being learned. This suggests that priority is given initially to building a "user model" (Norman, 1986) of the system and its operation and to acquiring rules for its use. When basic concepts and rules have been acquired, the efficient achievement of specific tasks becomes more important and an increase in the number of task questions was observed. Unlike other question types, the proportion of plan questions increased in later trials suggesting that plan refinement begins after basic skills and an understanding of the tasks have been acquired.

The initial tendency for reactive questions to dominate predictive questions suggests that learners were trying to explain observed phenomena in the absence of well understood goals, plans, or system knowledge. Later, as knowledge develops it can be used to generate hypotheses about plans and their effects. Thus, predictive questions become more prevalent.

Implications for Models of Knowledge Acquisition

The questions that people ask during learning demonstrate that different learning strategies arise at different stages of skill acquisition. Novice learners must reason from very general plans, often from other skill domains, and without much understanding of the system they are using. A novice's strategy, therefore, is to test alternative plans, fill in procedures for achieving goals, and notice any behavior of the system that will contribute to problem solving. In order to do this, novices construct "question frames" that represent the relationships between human actions and system operations. These questions arise in reaction to unexpected system actions and serve to guide problem solving and constrain explanation during early, failure-driven learning.

Novice learners restrict themselves to elementary operations. Only when they understand basic concepts about the system and acquire procedures for performing elementary actions do they begin to worry about plans for achieving complex tasks. At this stage, learners begin to predict system actions based on a developing model of the system. Questions become more abstract and relate more to tasks and refined plans. This shift in strategy is an important consideration for models of skill acquisition.

General mechanisms for learning, whether they are automatic like plan compilation and chunking, or more strategic, like structure mapping and analogy, may only apply at one stage of learning. An exploratory learner will use different strategies that are determined by the learner's level of knowledge in several distinct domains. It is important in future work to understand how learners determine what they need to know and how they select exploration strategies based on their level of knowledge. Viewing self-generated questions as strategic components of knowledge construction during exploratory learning provides a productive perspective for this effort.

References

Anderson, J.R. (1982). Acquisition of cognitive skill. Psychological Review, 89, 369-406.

Anderson, J.R. (1983a). The architecture of cognition. Cambridge, MA: Harvard University Press.

Anderson, J.R. (1983b). Acquisition of proof skills in geometry. In R.S. Michalski, J.G. Carbonell, & T.M. Mitchell (Eds.), Machine learning: An artificial intelligence approach. Palo Alto, CA: Tioga.

Anderson, J.R. (1986). Knowledge compilation: The general learning mechanism. In R.S. Michalski, J.G. Carbonell, & T.M. Mitchell (Eds.), Machine learning: An artificial intelligence approach. Vol II. Palo Alto, CA: Tioga.

Burstein, M.H. (1986). Concept formation by incrimental analogical reasoning. In R.S. Michalski, J.G. Carbonell, & T.M. Mitchell (Eds.), Machine learning: An artificial intelligence approach. Vol II. Palo Alto, CA: Tioga.

Carbonell, J.G. (1983). Learning by analogy: Formulating and generalizing plans from past experience. In R.S. Michalski, J.G. Carbonell, & T.M. Mitchell (Eds.), Machine learning: An artificial intelligence approach. Palo Alto, CA: Tioga.

DeJong, G. (1986). An approach to learning from observation. In R.S. Michalski, J.G. Carbonell, & T.M. Mitchell (Eds.), Machine learning: An artificial intelligence approach. Vol II. Palo Alto, CA: Tioga.

Forbus, K.D. & Gentner, D. (1986). Learning physical domains: Toward a theoretical framework. In R.S. Michalski, J.G. Carbonell, & T.M. Mitchell (Eds.), Machine learning: An artificial intelligence approach. Vol II. Palo Alto, CA: Tioga.

Gentner, D. (1983). Structure mapping: A theoretical framework for analogy. Cognitive Science, 7, 155-170.

Hayes-Roth F. (1983). Using proofs and refutations to learn from experience. In R.S. Michalski, J.G. Carbonell, & T.M. Mitchell (Eds.), Machine learning: An artificial intelligence approach. Palo Alto, CA: Tioga.

Lehnert, W.G. (1978). The process of question answering. Hillsdale, NJ: Erlbaum.

Mitchell, T.M., Utgoff, P.E., & Banerji R. (1983). Learning by experimentation: Acquiring and refining problem-solving heuristics. In R.S. Michalski, J.G. Carbonell, & T.M. Mitchell (Eds.), Machine learning: An artificial intelligence approach. Palo Alto, CA: Tioga.

Miyake, N. & Norman, D.A. (1979). To ask a question one must know enough to know what is not known. Journal of Verbal Learning and Verbal Behavior, 18, 357-364.

Neves, D.M., & Anderson, J.R. (1981). Knowledge compilation: Mechanisms for the automatization of a cognitive skill. In J.R. Anderson (Ed.). Cognitive skills and their acquisition. Hillsdale, N.J.: Erlbaum.

Newell, A. & Rosenbloom, P.S. (1981). Mechanisms of skill acquisition and the law of practice. In J.R. Anderson (Ed.). Cognitive skills and their acquisition. Hillsdale, N.J.: Erlbaum.

Norman, D.A. (1986). Cognitive engineering. In D.A. Norman & S.W. Draper (Eds.). User centered system design: New perspectives on human-computer interaction. Hillsdale, NJ: Erlbaum.

Robertson, S.P. & Black, J.B. (1986). Structure and development of plans in computer text editing. Human-Computer Interaction, 2, 201-226.

Rosenbloom, P.S. & Newell, A. (1986). The chunking of goal hierarchies: A generalized model of practice. In R.S. Michalski, J.G. Carbonell, & T.M. Mitchell (Eds.), Machine learning: An artificial intelligence approach. Vol II. Palo Alto, CA: Tioga.

Rumelhart, D.E. & Norman, D.A. (1978). Accretion, tuning, & restructuring: Three modes of learning. In J.W. Cotton & R. Klatzky (Eds.), Semantic factors in cognition. Hillsdale, N.J.: Erlbaum.

Sammut, C. & Banerji, R.B. (1986). Learning concepts by asking questions. In R.S. Michalski, J.G. Carbonell, & T.M. Mitchell (Eds.), <u>Machine learning: An artificial intelligence approach.</u> Vol <u>II</u>. Palo Alto, CA: Tioga.

Schank, R.C. (1982). <u>Dynamic memory: A theory of reminding and learning in computers and people.</u> Cambridge, MA: Cambridge University Press.

Stepp, R.E. & Michalski, R.S. (1986) Conceptual clustering: Inventing goal-oriented classifications of structured objects. In R.S. Michalski, J.G. Carbonell, & T.M. Mitchell (Eds.), <u>Machine learning: An artificial intelligence approach.</u> Vol <u>II</u>. Palo Alto, CA: Tioga.

ThinkerTools: Enabling Children
to Understand Physical Laws

Barbara Y. White and Paul Horwitz

BBN Laboratories

Abstract

This project[1] is developing an approach to science education that enables sixth graders to learn principles underlying Newtonian mechanics, and to apply them in unfamiliar problem solving contexts. The students' learning is centered around problem solving and experimentation within a set of computer microworlds (i.e., interactive simulations). The objective is for students to gradually acquire an increasingly sophisticated causal model for reasoning about how forces affect the motion of objects. To facilitate the evolution of such a mental model, the microworlds incorporate a variety of linked alternative representations for force and motion, and a set of game-like problem solving activities designed to focus the students' inductive learning processes. As part of the pedagogical approach, students formalize what they learn into a set of laws, and critically examine these laws, using criteria such as correctness, generality, and parsimony. They then go on to apply their laws to a variety of real world problems. The idea is to synthesize the learning of the subject matter with learning about the nature of scientific knowledge -- its form, its evolution, and its application. Instructional trials found that the curriculum is equally effective for males and females, and for students of different ability levels. Further, sixth graders taught with this approach do better on classic force and motion problems than high school students taught using traditional methods.

1 Introduction

Research has demonstrated that students can succeed in high school and even college physics courses, while still maintaining many of their misconceptions and without acquiring an understanding the physical principles addressed in the course (Caramazza et al. (1981), Clement (1982), diSessa (1982), Larkin et al. (1980), McDermott (1984), Trowbridge & McDermott (1981), Viennot (1979), White (1983), and many others). For example, they make incorrect predictions about what will happen to the motion of a ball when it emerges from passing through a spiral tube (McClosky et al., (1980)). Such questions do not call for computation or the algebraic manipulation of formulas; rather, they require understanding the implications of the fundamental

---

[1]This research was sponsored by the National Science Foundation, under award number DPE-8400280 with the Directorate for Science and Engineering Education.

tenets of Newtonian mechanics. Students' failure to correctly answer such questions reveals a deficiency in their knowledge of the causal principles that underly the formulas they have been taught.

We believe that students need, at an early age, experiences that will enable them to acquire accurate causal models (Bobrow (Ed.) (1985), Gentner & Stevens (Eds.) (1983), White & Frederiksen (1986(a)) & (1986(b))) for how forces affect the motions of objects. This will inhibit the development of misconceptions and foster the type of understanding that older students appear to lack. In contrast to our view, many cognitive and educational theorists believe that attempts to teach children physics will inevitably fail (see, for example, Shafer & Adey (1981)). They argue that understanding physical principles requires formal operational thinking (Piaget & Garcia, 1964), and that many students have not reached this stage of cognitive development at the high school or even the college level, let alone at the elementary school level. Consequently, such students cannot be expected to master physical principles. We have found this not to be the case. This paper will describe an instructional approach that enabled sixth graders to understand important aspects of Newtonian mechanics. Further, it will illustrate how they also began to learn about the nature of science -- what are scientific laws, how do they evolve, and why are they useful?

2 The Progression of Microworlds & Subject Matter

The objective was for students to evolve a mental model of sufficient sophistication to enable them to analyze projectile motion problems (i.e., problems involving motion under a constant, uniform gravitational force). The desired model would incorporate such fundamental concepts of Newtonian mechanics as force, velocity, and acceleration, as well as causal principles, such as *forces cause changes in velocity*. In order to enable the students to acquire such a causal model, we created a progression of increasingly complex microworlds. Associated with each microworld is a set of problem solving activities and experiments designed to help the students discover the laws governing the microworld. These microworlds gradually introduce the full set of principles needed to analyze projectile motion problems. All of them require students to control the motion of a computer-generated graphic object via the application of forces. Within these simulations, the complications introduced by friction and gravity can be selectively eliminated, allowing students to encounter first simpler situations obeying Newton's first law (objects do not change their velocity unless a force is applied to them), and later to analyze more complex situations in terms of such basic laws. In addition, the microworlds incorporate a number of different representations for the application of forces and for the motions of objects. For example, there is the datacross (see Figure 1), which is essentially a pair of crossed "thermometers" that register the horizontal and vertical velocity components of an object via the amount of "mercury" in them. Also, there are wakes (also shown in Figure 1) that provide a record of an object's past speed and direction

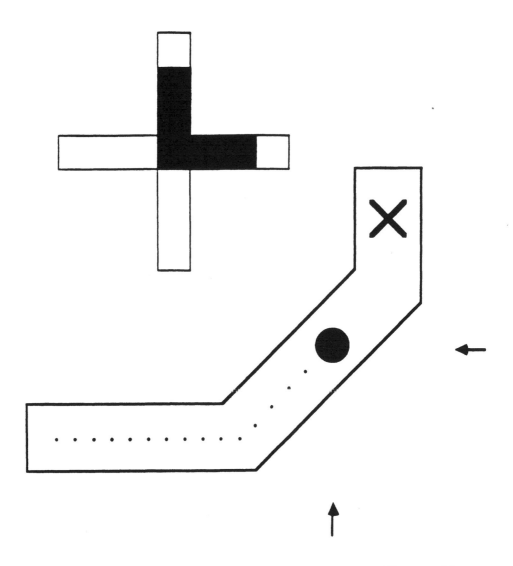

Figure 1: The Representation of Motion within the Microworlds

In this game, one must control the motion of the dot so that it navigates the track and stops on the target X. The shaded circle in the middle of the angled path is the dot. It represents a physical object which may be given fixed sized impulses in the left—right or up—down directions. In this figure the dot has been given the (optional) property of leaving "wakes" in the form of little dots laid down at regular time intervals. These denote, by their position and relative separation, the past history of the motion. The large cross in the middle of the figure is the "datacross" —— a device for displaying the instantaneous values of the X and Y velocity components. Here the datacross is depicting a velocity inclined at +45 degrees to the horizontal. The arrows at the bottom and right side of the figure continually point to the dot, unless it leaves the screen, in which case they "get stuck" at the edge of the screen. They represent the X and Y coordinates of the dot's position, and are useful for determining its location to within a quadrant when it is off the screen. Their motions, while the dot is on the screen, dynamically illustrate the X and Y components of the dot's velocity.

338

of motion by leaving a mark on the screen at fixed time intervals. These representations allow the effects of forces on an object's motion and velocity components to be directly observed, and thereby facilitate students' attempts to formulate the principles governing these causes and effects.

Microworld #1: At the beginning of the curriculum, we introduce a simple one-dimensional microworld which has no friction or gravity. In this world, students try to control the motion of an object (referred to as the "dot") by applying fixed sized impulses to the right (-->) or to the left (<--). The students observe that whenever they apply an impulse to the dot, it causes a change in its speed: If the impulse is applied in the same direction that the dot is moving, it adds to its speed; applied in the opposite direction, it subtracts from the speed. In this way, the students discover that a formalism they learned in second grade, scalar arithmetic (e.g., 3 - 2 = 1), will enable them to make predictions about the effect that a particular impulse, or sequence of impulses, will have on the motion of the dot. As part of this process, the students have discovered a corollary of Newton's first law -- Whenever you apply an impulse to an object, you change its velocity.

Microworld #2: Next, the students are given a two-dimensional microworld. Again, there is no friction or gravity. In this world, the students can apply impulses up or down, as well as to the left or right. Through carefully designed problem solving activities (for example, see White, 1984), the students discover that the law they developed for the horizontal dimension applies equally well to the vertical dimension of the dot's velocity. Further, they learn that these two components of the dot's motion are independent of one another -- for instance, if you apply an upwards or downwards impulse, it has no effect on the horizontal velocity of the dot[2]. Finally, they acquire the foundation for an understanding of vector addition -- they learn how the vertical and horizontal velocity components combine to determine the speed and direction of the dot's motion.

Microworld #3: The next step is to provide students with a microworld where the rate at which they can apply impulses can be varied. The purpose of this microworld is to introduce students to continuous forces via a limit process: The students can repeatedly double the frequency with which they can apply impulses, while at the same time, the size of each impulse is halved. At the end of this process, the students are applying very small impulses closely spaced in time. In this way, the students learn to think of continuous forces, like gravity, as a lot of small impulses applied one after another. This enables students to apply their causal model, learned in the simpler

---

[2]This is learned by giving one student the capability to apply only horizontal impulses to the dot and another the capability to apply only vertical impulses. The first student is then given a task that requires controlling the arrow whose motion represents the dot's horizontal velocity, and the second student has to control the arrow whose motion represents its vertical velocity. They discover that the other student's impulses have no effect on the motion of their arrow.

microworlds, to understand the effects of continuous forces. For example, they are asked to think about what happens when you throw a ball up into the air. They discover that gravity is constantly applying an impulse that continually adds a small amount to the vertical velocity of the ball in the downwards direction. This causes the ball to go upwards at a slower and slower rate until it finally stops, turns around, and accelerates downwards. By analyzing such problems, the students develop an understanding of acceleration and of F=ma for the case where the mass and the force are constant. In other words, they learn a simpler causal form of F=ma, that is, F --> a.

Microworld #4: Finally, the students are presented with a microworld in which gravity is acting, and they can again apply impulses to the left and right, as well as up and down. In this world, in the absence of other forces, motion is characterized by constant acceleration in the vertical dimension and constant speed in the horizontal dimension. The students are given problems of the form: "Imagine that you give two balls a horizontal push off the side of a table. One ball gets a soft push and the other ball gets a hard push. Both balls are pushed at the same time. Which ball hits the floor first?" Working in this microworld thus enables students to make connections to some interesting real world situations. Solving such problems requires the students to apply the causal model that they learned from interacting with the prior microworlds.

3 The Instructional Approach

The set of microworlds we have created focuses on, simplifies, and makes concrete certain aspects of Newtonian mechanics. The challenge was to devise instructional techniques, centered around the microworlds, that would facilitate students' acquisition of the desired mental model. A central aspect of the approach that we developed is to synthesize the teaching of subject matter with teaching about the form, evolution, and application of scientific knowledge. Students are given a variety of laws that have been proposed for a given microworld, and are asked to determine which laws are correct and which laws are incorrect (i.e., demonstrably false). Then, for the correct laws, they have to devise criteria for deciding which laws are better than others. Finally, they have to apply their laws to real world contexts and thereby discover that the laws are general and enable one to make predictions about a wide range of physical phenomena.

Within each of the existing four modules of the curriculum (corresponding to the four microworlds described above), instruction was divided into four distinct phases:

The Motivation Phase. In the first phase, students are asked to make predictions about what they think will happen in simple real world contexts. For example, in the first module of the curriculum, the teacher asks the students: "Imagine that we have a ball resting on a frictionless surface and we blow on the ball. Then, as the ball is moving along, we give it a blow, the same size as the first, in the opposite direction. What will be the effect of this second blow on the motion of the ball?" The teacher simply tabulates the different answers and reasons for these answers without commenting on their correctness. This process demonstrates to the students that not everyone holds the same beliefs. For instance, some think that the second blow will cause the ball to turn around and go in the opposite direction, others think that it will make the ball stop, and yet others believe that it will simply cause the ball to slow down. Since not everyone can be right, students are motivated to find out who has correct explanations and who has misconceptions. Further, since the predictions are about the behaviors of real world objects, this phase sets up a potential link between what happens in the computer microworld and the real world.

The Evolution Phase. In this phase students solve problems and perform experiments in the context of the computer microworld. For instance, one of the problems in the first module requires the students to make the dot hit a stationary target while moving at a specified speed. Once the student succeeds, the dot returns to its starting location and a new target speed is specified. By attempting to solve this problem, the students learn how impulses affect the speed of the dot. As described in the previous section, the computer microworld increases in complexity with each new module of the curriculum. The problems and experiments are designed and sequenced to build upon the students' prior knowledge and to enable them to induce increasingly sophisticated concepts and laws relevant to understanding the implications of Newton's laws of motion.

The Formalization Phase. In this phase the students are asked to evaluate a set of laws formulated to describe the behavior of objects within the microworld. Examples of such laws are (1) *whenever you give the dot an impulse to the left, it slows down*, and (2) *if the dot is moving, and you do not apply an impulse, it will keep moving at the same speed*. Initially students are asked to sort the laws into two piles -- those that they can prove wrong and those for which they cannot find a counterexample. Then, for the subset of "true" laws, they are asked to pick the rule they like best: "if you could only have one of these rules in your head to base predictions on, which one would you pick and why?" This activity typically engenders discussions of (1) the precision of a rule's predictions, (2) the range of situations to which it applies, and (3) its simplicity and memorability.

The Transfer Phase. In this phase the objective is to get students to appreciate how the rule they have selected applies to real world contexts. In the first stage of this process, students apply their rule to the predictive question that they were asked at the beginning of the instructional cycle. They then compare the answer that the rule generates to the set of answers generated by the class. For answers that differ, students go to the microworld and experiment, for example, by putting friction and/or gravity into the microworld, to see which of their "wrong" predictions can become correct predictions under these more complex circumstances. In the second stage of the transfer process, students conduct experiments with real world objects, or design their own experiments to illustrate how the rule they evolved in the microworld context holds in these real world contexts.

4 Experimental Results

The curriculum was implemented by a teacher who taught five science classes in the sixth grade of a middle school located in a middle class Boston suburb. One of the five classes was used for a pilot trial of the first two modules of the curriculum. Of the remaining classes, two were used as a peer control group (containing 37 students), and the other two were given the ThinkerTools curriculum (containing 41 students). The curriculum took two months to complete. During this period, the students had a science class every school day for 45 minutes. The ThinkerTools curriculum occupied the entire class period. Students in the control classes received the standard curriculum which, at this point in the year, was devoted to a unit on inventions. All students had completed a physics unit earlier in the year which included material on Newton's laws.

In addition to the control group of sixth grade students, a second control group was employed, consisting of two classes of high school physics students (containing 41 students) drawn from the same school system as the sixth graders. These students had just completed two and one half months studying Newtonian mechanics using the text book "Concepts in Physics" (Miller, Dillon, & Smith (1980)).

We utilized a variety of evaluation instruments to help us determine the effectiveness of the ThinkerTools curriculum. During the course we observed numerous classroom sessions and kept videotape and audiotape records of certain sessions. At the end of the course we administered three written tests measuring: (1) ability to translate between the alternative representations of motion (datacrosses and wakes -- see Figure 1) employed within the curriculum, (2) subject matter knowledge in the computer microworld context, and (3) transfer of the underlying principles to real-world contexts. This third test was also given to the two control groups. In addition, following the administration of the written tests, seven of the ThinkerTools students were interviewed on an individual basis. These protocols allowed us to explore in depth the nature of the mental models they had acquired.

Since the first two tests were given only to the experimental subjects, we will focus in this limited space on the results of the third test. The findings will be discussed with respect to the two primary objectives of the course: (1) understanding the principles underlying Newtonian mechanics, and (2) learning about the form and evolution of scientific knowledge.

4.1 Understanding Newtonian Mechanics

The experimental students did well on the first two tests, with a third of them getting more than ninety percent of the questions correct. Protocols of high scoring students reveal that their pattern of correct answers was produced by the consistent application of the desired mental model for reasoning about force and motion problems. This is in marked contrast to the inconsistent and misconception fraught reasoning that sixth graders display prior to instruction (White & Horwitz, in preparation), and that high school physics students exhibit following a traditional physics course (White, 1983).

The third test, administered to the experimental group and two control groups, measures students' understanding of Newtonian mechanics in real world problem solving contexts. It is composed of questions used by other researchers in studying misconceptions among physics students (Clement, diSessa, McClosky, McDermott, Minstrell, White). The particular questions used are simple predictive questions to which high school and college students frequently give wrong answers. They all require reasoning from basic principles, rather than constraint-based, algebraic problem solving.

In the first analysis of this transfer test, we compared sixth graders who had the ThinkerTools curriculum with those who did not. A three way between subjects analysis of variance was carried out with (1) treatment (experimental versus control), (2) gender, and (3) ability (low, middle, and high, based upon California Achievement Test (CAT) total scores) as the three factors. With this design we could assess the effectiveness of the experimental curriculum for subjects of each gender and ability level. There was a highly significant main effect of instructional treatment ($F_{1,62}=62.9$, $p<.0001$). The average number of questions correct for the experimental subjects was 11.15 out of 17, while the average for the control subjects was 7.56. In addition, there was no significant interaction of gender with treatment ($F_{1,62}=.219$, $p=.64$), or ability with treatment ($F_{2,62}=.834$, $p=.44$). Thus, the ThinkerTools curriculum was equally effective for girls and boys as well as for students of different ability levels as measured by the CAT.

With respect to the ThinkerTools students, the questions on the test can be classified into two categories:

1. Those that involve the application of a principle taught in the course and that the students have applied in a context similar to the one presented in the problem; and

2. Those that involve a principle addressed in the course but that the students have never applied to the particular context presented in the problem.

An item analysis revealed that the experimental students did better than the control students on both types of problems. This suggests that the ThinkerTools students not only learned the principles focused on in the course, but also could apply them to unfamiliar contexts.

Finally, it is noteworthy that the Thinkertools students also did significantly better on this transfer test than the high school physics students ($t_{80} = 1.7$, $p < .05$), who were on the average six years older and had been taught about force and motion using traditional methods. An item analysis revealed interesting differences between these groups. The ThinkerTools students performed better (in some cases dramatically better) than the high school physics students on problems that involved analyzing the effects of forces in terms of velocity components. The high school students, however, performed better on problems that involved constraint forces (such as a fixed length string constraining the motion of a pendulum bob). This latter result is not too surprising, since constraint forces were not dealt with in the ThinkerTools curriculum.

4.2 Acquiring Scientific Inquiry Skills

In addition to teaching students principles underlying Newtonian mechanics, we had the symbiotic goal of helping them learn about the form, evolution, and application of scientific knowledge. In evaluating our success with respect to this second major objective of the curriculum, we relied partly upon observations of students' classroom performance. For instance, we examined the quality of the laws and experiments that they formulated for themselves, as well as the sophistication of the discussions they held when they were attempting to select the best law. In addition, we looked at the results of the written tests, particularly the transfer test, to aid in this aspect of the evaluation.

Understanding the Form of Scientific Knowledge. Knowing the characteristics of a useful scientific law is an important aspect of understanding the form of scientific knowledge. The instructional technique we developed was to present students with alternative laws for each microworld, and have them select the best law. We observed that when students were evaluating these sets of laws, they spontaneously engaged in discussions concerning the simplicity of a law, the precision of its predictions, and its range of applicability. The set of laws was carefully constructed to elicit such discussions and this approach thus appears to have been highly successful.

Developing Scientific Inquiry Skills. It is important to understand that falsification is part of the process by which scientific knowledge evolves. For a rule to be a potential scientific law, it must be capable of being proven wrong. Being able to develop and reason from counter evidence is an important scientific inquiry skill. We observed that when the students were evaluating the sets of laws given to them, they were adept at designing experiments that would falsify a particular law.

When we went on to look at what the students did when they formulated laws for themselves, there were clear limits to their scientific inquiry skills. For example, one group of students discovered the "linear friction law": in the microworld the effect of friction is linearly proportional to the speed with which the object is moving. The consequence is that when you apply a sequence of impulses to the dot, it does not matter whether you apply them one right after the other or whether you separate them further in time, the dot will come to rest at the same point. The students discovered this fact, but they did not fully explore its implications, nor did they go on to investigate whether it was true for the kind of real world friction that affects, for instance, rolling balls or sliding hockey pucks.

If one looks at our instructional approach, this limitation in their inquiry skills is understandable. We gave the students activities to help them induce the laws, as well as sets of possible laws to evaluate, and real world activities that enabled them to see that the laws generalized. Toward the end of the course they were asked to formulate their own laws and to design real world activities that illustrated the laws. This was clearly too abrupt a transition, and an important area of our future research will be the development of bridging activities that enable a more gradual transition to independent scientific discovery.

Acquiring Scientific Problem Solving Skills. Students need to understand that the laws they are evolving are of increasing general applicability, and need to be able to apply them in new contexts. Based upon classroom observations and the results of the transfer test, we see that the students were indeed able to generalize principles derived in the microworld contexts to a variety of simple real world contexts. This was achieved by a process of abstracting what they learned from the computer microworld into a set of laws and then learning how to map the laws onto different real world problem solving situations.

The general conclusion is that the ThinkerTools students learned that a useful scientific law is a concise principle that enables predictions across different contexts. In addition, they developed skill at designing experiments to falsify or show the limitations of a law, and applying a given law to a variety of different domains. This view of scientific knowledge and these inquiry skills are an important component of understanding what science is all about.

5 Discussion

The design of the curriculum was based upon extensive protocol studies of sixth graders' reasoning about force and motion problems. Based upon this research, we determined which aspects of their prior knowledge we could build upon, and which misconceptions we could use to motivate their learning about Newtonian mechanics. The progression of increasingly complex computer microworlds was then designed to correspond to the desired evolution of the students' understanding of the phenomena.

Further, the design of the microworld made abstractions, such as Cartesian components of displacement and velocity, into concrete observable data-objects, and introduced simplifications, such as quantized impulses, that enabled students to learn and make concrete what are normally regarded as abstract and difficult concepts.

Another aspect of the instructional approach that we believe was crucial to its success is the process of reification -- students were asked to consider alternative descriptions of what they learned from the computer microworld in the form of a set of laws, and had to evaluate the properties of the various laws. This enabled the students to develop a concept of what it was they were trying to learn -- for example, rather than learning a set of facts, they were trying to induce a set of laws and learn about the properties of scientific laws. Further, the process of getting students to apply the laws they induced from the microworld to real world contexts was important both for their understanding of Newtonian mechanics and for their perception of the nature of scientific knowledge. They learned that their laws apply in a wide range of contexts and they gained experience in transferring what they learned in one context (i.e., the computer microworld) to another context (i.e., a particular real world situation). We conjecture that these formalization and transfer phases of our curriculum are responsible for the ThinkerTools students being able to apply their knowledge to unfamiliar contexts -- a result which is rarely obtained in educational research.

6 References

Bobrow, D.G. (Ed.) (1985). Qualitative Reasoning about Physical Systems. Cambridge, MA: MIT Press.

Caramazza, A., McCloskey, M., & Green B. (1981). Naive beliefs in "sophisticated" subjects: Misconceptions about trajectories of objects. Cognition, 9, 117-123.

Clement, J. (1982). Students' preconceptions in elementary mechanics. American Journal of Physics, 50, 66-71.

diSessa, A. (1982). Unlearning Aristotelian physics: a study of knowledge-based learning, Cognitive Science, 6(1), 37-75.

Gentner, D., & Stevens, A. (Eds.) (1983). Mental Models. Hillsdale, NJ: Lawrence Erlbaum Associates.

Larkin, J.H., McDermott, J., Simon, D.P., & Simon, H.A. (1980). Expert and novice performance in solving physics problems. Science, 208, 1335-1342.

McCloskey, M., Caramazza, A., & Green, B. (1980). Curvilinear motion in the absence of external forces: Naive beliefs about the motion of objects. Science, 210, 1139-1141.

McDermott, L.C. (1984). Research on conceptual understanding in mechanics. Physics Today, 37, 24-32.

Piaget, J. & Garcia, R. (1964). Understanding causality. New York: Norton.

Shayer, M. & Adey, P. (1981). Towards a Science of Science Teaching. London, England: Heinemann Educational Books.

Trowbridge, D.E., & McDermott, L.C. (1981). Investigation of student understanding of the concept of acceleration in one dimension. American Journal of Physics, 49, 242-253.

Viennot, L. (1979). Spontaneous reasoning in elementary dynamics. European Journal of Science Education, 1, 205-221.

White, B. & Horwitz, P. (in preparation). Multiple Muddles: Novice Reasoning about Force and Motion.

White, B. & Frederiksen, J. (1986 (a)). Progressions of qualitative models as a foundation for intelligent learning environments. BBN Report No. 6277. Cambridge, MA. (To appear in Artificial Intelligence.)

White, B. & Frederiksen, J. (1986 (b)). Intelligent tutoring systems based upon qualitative model evolutions. In Proceedings of the Fifth National Conference on Artificial Intelligence. Philadelphia, PA.

White, B. (1984). Designing computer activities to help physics students understand Newton's laws of motion. In. Cognition and Instruction, 1, 69-108.

White, B. (1983). Sources of difficulty in understanding Newtonian dynamics. In Cognitive Science, 7(1), 41-65.

FROM CHILDREN'S ARITHMETIC TO MEDICAL PROBLEM SOLVING: AN EXTENSION OF THE KINTSCH-GREENO MODEL

Guy J. Groen & Janos P. Jerney
McGill University

ABSTRACT

It has been found that expert physicians use forward reasoning in diagnostic explanations of clinical cases. This paper shows that the Kintsch-Greeno model for solving arithmetic word problems, which assumes a forward chaining process, can be extended to explain this phenomena. The basic approach is to modify the lexicon and the schema structure of the existing simulation program while retaining the basic control structure. The principle modifications are in the structure of the schemata which make use of three slots: indicator, abnormality and consequence. As with the Kintsch-Greeno theory, the model proceeds by using these schemata to build super-schemata from the propositional representation of the problem text.

From Children's Arithmetic to Medical Problem Solving: An Extension of the Kintsch-Greeno Model

Guy J. Groen & Janos P. Jerney
McGill University

Recently, two approaches have evolved that specifically combine the methodologies of propositional analysis and protocol analysis. The first is the theory by Kintsch and Greeno (1985) of the processes used in solving algebraic and arithmetic word problems and the series of computer simulation programs that are based upon it (Dellarosa, 1986). Such problems pose a simple arithmetic problem as a story. The simulation begins with a propositional representation of the story which it transforms into a set of frame structures called proposition frames. On this basis, it builds "set frames" that represent each number specified in the story as a set of objects. It then builds a "superschema" that specifies the relational structure betweeen these sets. The presence of a satisfactory superschema triggers an appropriate algorithm that generates the answer. If an incomplete superschema has been created, then the program produces "intelligent guesses" about the answer. Crucial to this theory is the distinction (van Dijk & Kintsch, 1983) between a textbase and a situation model. The textbase is the semantic representation of the input text. The situation model is the representation of the knowledge required to solve the problem. In the simulation program, the textbase is the proposition frames. The situation model is the final superschema.

The second is an approach developed by Patel and Groen (1986) to study clinical reasoning in medicine. This involves transforming a propositional representation of a reasoning protocol into a semantic network, and deriving from this a set of production rules that are adequate to simulate the reasoning task. This was applied to study reasoning in a task involving the diagnosis of a case of acute bacterial endocarditis. The subjects were seven specialists in cardiology. It was shown that the diagnostic explanations of subjects making an accurate diagnosis could be accounted for in terms of a model consisting of pure forward reasoning (i.e. from data to diagnosis) through a network of causal-conditional rules, actuated by relevant propositions in the stimulus text. In contrast, subjects with inaccurate diagnoses tended to make use of a mixture of forward and backward reasoning, beginning with a high level hypothesis and proceeding in a top-down fashion to the propositions embedded in the stimulus text or to the generation of irrelevant rules.

It may seem that childrens' arithmetic word problems are a world apart from clinical cases in medicine. However, both involve situations in which the task is to make inferences from narrative discourse. The simulation program, with appropriate parameters, performs like an expert in its domain and has one feature that is extremely important for our purposes. It builds the superschema from the initial frames by means of a simple, well-known forward chaining procedure (Winston & Horn, 1981) on the basis of a set of production rules. Since the strongest finding by Patel and Groen was the use of forward reasoning by experts with accurate diagnoses, it seems reasonable to explore at a more precise level how the two approaches are related.

Groen and Patel (in press) have proposed that the van Dijk-Kintsch (1983) theory of comprehension can be combined with a generalization of the Kintsch-Greeno model and Dellarosa's program to yield a simulation model of expert diagnostic reasoning in medicine. Its important aspects would be as follows:

. The reading of a text results in the creation of a propositional textbase and a frame based situation model.

. Reasoning tasks result in an elaboration of the situation model.

. Expertise resides in the rules that develop the situation model. More satisfactory rules result in a more satisfactory situation model.

. An individual will apply these rules by a process of forward chaining until he or she is aware of an inaccurate or incomplete situation model. This accounts for the phenomenon of forward chaining by experts, but does not preclude forward chaining by the less than expert.

. An individual's awareness of an inaccurate or incomplete situation model will result in some form of backward chaining.

. The situation model (in a preliminary form) will also affect the structure of the textbase by means of various strategies specified by van Dijk and Kintsch, which would also affect the nature of the macropropositions. An expert situation model would result in higher level macropropositions.

The purpose of the research reported in this paper was to use these notions to develop a program that is capable of accounting for the results concerning forward reasoning obtained by Patel and Groen (1986). The basic strategy was to use ARITHPRO, Dellarosa's instantiation of the Kintsch and Greeno model as a shell with which to develop it. This simulation runs on a XEROX 1108/1186 computer. The program includes a rule base written in Interlisp and a lexicon written in LOOPS. Within the rule base, seperate rules exist for 1) building the proposition frames of the textbase, 2) building the set frames and superschemata of the situation model, 3) encoding the procedural knowledge for solving arithmetic problems, 4) encoding the procedural knowledge for converting one type of problem to another and 5) encoding the procedural knowledge for default solution strategies enabling "good" guesses (consistent with guesses young children would make) given incomplete information. The lexicon, on the other hand, relies on LOOPS' inheritance system to create a taxonomy of the words likely to be encountered in the problem text.

Since ARITHPRO is currently the most complete and accurate instantiation of the Kintsch-Greeno model, every attempt was made to work within the existing framework of the program. As such, nearly all of the simulation's control structure was taken directly from it. In a similar fashion, most of the rules for constructing the proposition frames of the textbase were retained though slightly modified. The similarity ends here, however, as none of the set building or arithmetic counting procedures present in the Kintsch-Greeno model are applicable to the domain of medical problem solving. Therefore, a completely new set of rules had to be constructed to build the schemata necessary for diagnosing cases of acute bacterial endocarditis. In addition to this, the lexicon was modified to incorporate the new words appropriate for this domain.

In constructing the situation model, the simulation relies on a single general data structure called an Abnormality schema which contains three slots: 1) the abnormality slot which is usually a physiological disorder (such as an emboli) or disease category, 2) the indicator slot which is generally a primary clinical indicator of the abnormality (such as transient blindness) and 3) the consequence slot which can either be a clinical or physiological consequence of the abnormality. In the current implementation, these slots appear to be adequate for encapsulating the information the program needs to use in order to arrive at an acceptable solution. However, considerations are being made to expand the schema structure in future implementations to allow for the explicit storage of additional information. This is facilitated by the existing rule structure which allows for a graceful incrementation. The structure of an Abnormality schema along with an example is given in Table 1.

Table 1.
Structure of Abnormality Schema with example

```
(<schema name>    (INDICATOR (: <indicator list>))
                  (CONSEQUENCE (: <consequence list>))
                  (ABNORMALITY (: <abnormality>)))


(EMBOLI     (INDICATOR (: BLINDNESS))
            (CONSEQUENCE (: HEMORRHAGE))
            (ABNORMALITY (: EMBOLI)))
```

Our simulation, like ARITHPRO, does not have a natural language parser. Instead, a propositional representation of the input text is presented to the program. An example of a (simplified) problem case is presented in Table 2a along with a corresponding propositional representation in Table 2b which would be input to the simulation. In order to use ARITHPRO's existing rules to build the proposition frames, the general form of the input propositional representation had to be retained. In particular, the propositions with the head element DURATION are modelled after the quantity propositions in ARITHPRO. This proposition causes the goal MakeAbnormSchema to be placed on the goal list in much the same way that a quantity proposition creates a MakeSet goal in ARITHPRO. Subsequent rules then look into STM to find the abnormality. If the other elements of the premise are present in either short term or long term memory, the appropriate abnormality schema building rule is fired. An example of an abnormality schema rule (paraphrased in English) is given in Table 3.

The simulation proceeds by attempting to build an abnormality schema for each cue in the problem text. Only one schema is build for each abnormality and subsequent cues leading to an existing abnormality schema are appended to the indicator slot. When one or more of the indicators of an abnormality are themselves schemas, a superschema is created. The BACTERIAL-INFECTION superschema rule is shown in Table 4a while the actual superschema that it builds is illustrated in Table 4b. In this example, one of the indicators is a cue from the text, PUNCTURE-WOUNDS, while the other is the INFECTION schema. These superschemata are similar in function to the ones used by Kintsch and Greeno in that a logical relationship is established between the indicators the superschema subsumes. As with Dellarosa's program, key information such as the contents of STM (the textbase and the situation model), the goal list and the next rule to fire is constantly updated and displayed on the screen. The process continues until the end of the input text has been reached and no more schema building rules can fire. When this happens, the top-level superschema is returned as the most likely diagnosis and the simulation ends. The final result, for the case discussed in this paper, is a superschema for acute bacterial endocarditis.

The role of the consequence slots have not been completely defined or implemented yet. In the most general case, the consequences will be used to direct the focus of the schema building rules by placing various goals on the goal list. This process may be use to alert the simulation to follow up on certain cues and will probably be used extensively in a future implementation that includes backward chaining. This will enable a more complete account of the empirical phenomena found by Patel and Groen (1986) and Patel, Arocha and Groen (1986).

Table 2a.
English Text: Acute Bacterial Endocarditis

This unemployed young male was admitted to the emergency room complaining of a fever of four days duration. Functional inquiry revealed a transient loss of vision in his right eye which lasted approximately 45 s on the day before admission to the emergency ward. Funduscopic examination revealed a flame shaped hemorrhage. Examination of his limbs showed puncture wounds on his arm. Auscultation of his heart revealed a 2/6 early diastolic murmur in the aortic area. There was no splenomegaly. Urinalysis showed numerous red cells.

Table 2b.
Propositional Representation of Text

```
(((P1 (EQUAL X MALE))
 (P2 (ATT X P3))
 (P3 (DURATION UNEMPLOYED-YOUNG UNKNOWN)))
((P4 (COMPLAIN X P5))
 (P5 (DURATION FEVER 4-DAYS)))
((P6 (COMPLAIN X P7))
 (P7 (DURATION BLINDNESS 45-SEC)))
((P8 (HAVE X P9))
 (P9 (DURATION HEMORRHAGE UNKNOWN)))
((P10 (HAVE X P11))
 (P11 (DURATION PUNCTURE-WOUNDS UNKNOWN)))
((P12 (HAVE X P13))
 (P13 (DURATION EARLY-DIASTOLIC-MURMUR UNKNOWN)))
((P14 (HAVE X P15))
 (P15 (DURATION RED-BLOOD-CELLS-IN-URINE UNKNOWN)))
((P16 (HAVE X P17))
 (P17 (DURATION NORMAL-SPLEEN UNKNOWN))))
```

Table 3.
Emboli Schema rule

Rule Make-Emboli-Schema

IF: 1) MakeAbnormSchema is on the goal list, and
      2) The indicator BLINDNESS is present in STM

THEN: 1) Create and add the schema EMBOLI to STM, and
      2) Bind BLINDNESS to the indicator slot of the EMBOLI schema, and
      3) Bind HEMORRHAGE to the consequence slot of the EMBOLI schema, and
      4) Bind EMBOLI to the abnormality slot of the EMBOLI schema, and
      5) Remove the MakeAbnormSchema goal from the goal list.

Table 4a.
Bacterial Infection superschema rule

Rule Make-Bacterial-Infection-Schema

IF: 1) MakeAbnormSchema is on the goal list, and
     2) The indicator PUNCTURE-WOUNDS is present in STM, and
     3) The INFECTION schema is present either in STM or LTM

THEN: 1) Bring the INFECTION schema into STM from LTM if necessary, and
       2) Create and add the superschema BACTERIAL-INFECTION to STM
       3) Bind PUNCTURE-WOUNDS and the INFECTION schema to the indicator slot of the
           BACTERIAL-INFECTION superschema, and
       4) Bind BACTERIAL-INFECTION to the abnormality slot of the BACTERIAL-INFECTION
           superschema, and
       5) Remove the MakeAbnormSchema goal from the goal list.

Table 4b.
Bacterial Infection Superschema

```
[BACTERIAL-INFECTION
      (INDICATOR
            (INDICATOR1: (PUNCTURE-WOUNDS))
            (INDICATOR2: ((INFECTION
                              (INDICATOR (: (FEVER)))
                              (ABNORMALITY (: (INFECTION)))))))
      (ABNORMALITY (: BACTERIAL-INFECTION))]
```

It is important to note that this program is not designed to provide a direct simulation of subjects' behavior in the diagnostic explanation task examined by Patel and Groen (1986). Rather, it is designed to provide a detailed model of the diagnostic process in which the detailed rules, frames and schemata can be mapped onto propositions in diagnostic expanation protocols. This may enable a detailed examination of the hypothesis that the diagnostic explanation task reflects elements of the diagnostic process, which underlies much of our previous research. It may also enable the use more complex diagnostic problems, since it will remove the necessity of selecting clinical cases for which the logical validity of models of diagnostic reasoning can be verified by hand simulation.

It also begins to throw light on a far more general issue. There is currently a widespread belief both in Artificial Intelligence and Cognitive Psychology that problem solving is highly domain specific. Our attempt to generalize the Kintsch-Greeno model is a thrust in the opposite direction. As the program develops, it should become possible to make a far more precise differentiation between those aspects of problem solving that are domain specific and those that are common to many domains, regardless of their complexity.

ACKNOWLEDGEMENTS

This research was supported by grants from the Josiah Macy Jr. Foundation (No. B8520002) and the Province of Quebec.

REFERENCES

Dellarosa, D. (1986) A computer simulation of childrens' arithmetic word-problem solving. Behavior Research Methods, Instruments, and Computers, 18:142-154.

van Dijk, T.A. & Kintsch, W. (1983) Strategies of Discourse Comprehension. New York:Academic Press

Groen, G.J. & Patel, V.L. (in press) The relationship between comprehension and reasoning in medical expertise. In M. Chi, R. Glaser, & M. Farr (Eds.), The Nature of Expertise. Lawrence Erlbaum Associates.

Kintsch, W. & Greeno, J.G. (1985) Understanding and solving word arithmetic problems. Psychological Review, 92:109-129.

Patel, V.L., Arocha, J.F., & Groen, G.J. (1986) Strategy selection and degree of expertise in medical reasoning. In Proceedings of the Eight Annual Conference of the Cognitive Science Society, Lawrence Erlbaum Associates:NJ, 780-791.

Patel, V.L. & Groen, G.J. (1986) Knowledge based solution strategies in medical reasoning. Cognitive Science, 10:91-116.

Winston, P.H. & Horn B. (1981) LISP First Edition. Addison-Wesley.

A Temporal-Difference Model of Classical Conditioning

Richard S. Sutton

GTE Laboratories Incorporated

Andrew G. Barto

University of Massachusetts

Abstract—Rescorla and Wagner's model of classical conditioning has been one of the most influential and successful theories of this fundamental learning process. The learning rule of their theory was first described as a learning procedure for connectionist networks by Widrow and Hoff. In this paper we propose a similar confluence of psychological and engineering constraints. Sutton has recently argued that adaptive prediction methods called *temporal-difference methods* have advantages over other prediction methods for certain types of problems. Here we argue that temporal-difference methods can provide detailed accounts of aspects of classical conditioning behavior. We present a model of classical conditioning behavior that takes the form of a temporal-difference prediction method. We argue that it is an improvement over the Rescorla-Wagner model in its handling of within-trial temporal effects such as the ISI dependency, primacy effects, and the facilitation of remote associations in serial-compound conditioning. The new model is closely related to the model of classical conditioning that we proposed in 1981, but avoids some of the problems with that model recently identified by Moore et al. We suggest that the theory of adaptive prediction on which our model is based provides insight into the functionality of classical conditioning behavior.

Introduction

The increasing interest in connectionist or parallel distributed processing models of cognitive behavior provides a new rationale for examining animal conditioning behavior. Many of the rules used for adjusting connection weights in connectionist models are the result of postulating that single neuron-like units exhibit simplified analogs of animal behavior in conditioning experiments. Connectionist theories of higher functions therefore provide vehicles for integrating insights from animal learning research into more comprehensive theories of behavior. At the same time, the mathematical theories associated with connectionist learning provide new theoretical perspectives on conditioning behavior.

This research was supported in part by the Air Force Office of Scientific Research through grant AFOSR-87-0030. The authors wish to thank Harry Klopf, Jim Morgan, Jim Kehoe, John Moore, John Desmond, Diane Blazis, and Neil Berthier for sharing their ideas and simulation results with us. We also particularly thank John Moore for reading and providing valuable comments on an earlier draft of this paper.

Viewed at the trial level, classical, or Pavlovian, conditioning is related to supervised associative learning as studied by engineers and computer scientists and embodied in many connectionist learning systems. The system is repeatedly presented with an input pattern, corresponding to a conditioned stimulus (CS), together with a specification of a desired response, which corresponds to the presentation of an unconditioned stimulus (US) and the unconditioned response (UR) that it reflexively elicits. After a number of such CS–US pairings, the CS comes to elicit a conditioned response (CR) that closely resembles the UR or some part of it.* When details occurring within trials are considered, classical conditioning is seen to involve the extraction of predictive relationships among stimuli as if causal rules are being learned.

In a previous paper (Sutton and Barto, 1981), we pointed out that the Rescorla-Wagner model of classical conditioning (Rescorla and Wagner, 1972) is nearly identical to the learning algorithm introduced earlier by engineers Widrow and Hoff (1960), which is used in practical engineering applications (Duda and Hart, 1973; Widrow and Stearns, 1985) as well as in recent connectionist models (e.g., see Rumelhart and McClelland, 1986). That there is this degree of correspondence between psychological models and engineering methods should not be surprising given the similarity of the functional demands made in each case. In this paper, we propose a refinement of this correspondence. We propose a new model of classical conditioning based on a new theory of engineering methods called *temporal-difference methods* (Sutton, 1987). Temporal-difference methods have been shown to be superior in certain respects to the Widrow-Hoff algorithm and to other engineering algorithms for adaptive prediction. Here, we argue that the new model of classical conditioning, which we call the Temporal-Difference, or TD, model, also provides a better account of animal learning data than the Rescorla-Wagner model. In addition, the TD model and the theory of temporal-difference methods provides specific new suggestions about the functional nature of classical conditioning.

The TD model is a minor variant of the Adaptive Heuristic Critic (AHC) algorithm developed by Sutton for temporal credit assignment (Sutton, 1984; Barto, Sutton, and Anderson, 1983) and combined with the error back-propagation method of Rumelhart, Hinton, and Williams (1985) by Anderson (1986). The AHC algorithm itself is closely

---

* For example, a human subject is repeatedly presented with the sound of a bell (CS) followed by a puff of air to his eye (US), which causes him to blink (UR). After several such pairings, the subject blinks immediately (CR) in response to the bell alone.

related to the model of classical conditioning that we proposed in 1981 (Sutton and Barto, 1981; Barto and Sutton, 1982), which we here call the Sutton-Barto, or SB, model, and which was strongly influenced by the work of Klopf (1972, 1982). In this paper, we present the TD model as a substantially modified version of the SB model that solves some of the problems with that model identified by Moore et al. (1986). We show how the TD model performs in simulations of single-CS acquisition and extinction, trace and delay conditioning, blocking, conditioned inhibition, second-order conditioning, and several serial-compound conditioning paradigms. We also discuss what the theoretical basis of the TD model suggests about what animals are doing in classical conditioning. Finally, we briefly mention some of the limitations of the TD model.

Real-Time Models of Classical Conditioning

Whereas many models of classical conditioning (e.g., Rescorla and Wagner, 1972; Mackintosh, 1975; Pearce and Hall, 1980) specify changes in associative strength only as the result of a trial as a whole, the TD and SB models specify changes in associative strengths from moment to moment within trials. We will call models with this property *real-time* models (after Moore and Stickney, 1980; Blazis et al., 1986). Real-time models have also been proposed by, e.g., Gelperin, Hopfield, and Tank (1985), Gluck and Thompson (in press), Hawkins and Kandel (1984), Klopf (1986), Moore et al. (1986) Tesauro (1986), and Wagner (1981).

Real-time models have several kinds of advantages over trial-level models. First, since real-time models distinguish between times within a trial, they can make predictions about the effects of varying the temporal relationships among stimuli within a trial, whereas trial-level models can't. The trial-level Rescorla-Wagner model, for example, does not make predictions about the effect of the inter-stimulus interval between CS and US, even though this is well-known to have a strong effect on conditioning. A second advantage of real-time models is that they are more mechanistic and thus it is easier to see how they might be implemented by physical mechanisms. In particular, they are a step closer to neural models since their behavior can be compared more directly with electrophysiological correlates of learning.

Some real-time models, including the SB model, have been presented in the form of rules for altering the connection weights of a neuron-like adaptive element, and we follow this tradition with our description of the TD model. Although this form of presentation suggests possible relationships to the cellular basis of learning and makes it clear how the model can be used as a learning rule for connectionist networks, it is not essential to the TD model as a model of conditioning behavior. Nor is the realization of the model suggested by this adaptive-element the only way the model could be implemented in a nervous system.

The SB Model

We first describe the SB model and then discuss several of its shortcomings. Following our 1981 paper (Sutton and Barto, 1981) we present it as a set of rules for adjusting the connection weights of a neuron-like element, but we use a slightly different notation. Figure 1 shows a neuron-like adaptive element with $n+1$ input pathways, labeled $x(0), \ldots, x(n)$, and a single output pathway labeled y. For each i, $i = 0, \ldots, n$, $x_t(i)$ denotes the strength of the signal on pathway i at time t; y_t denotes the strength of the output signal at step t. Associated with each input pathway $x(i)$ is a weight $w(i)$ that specifies the efficacy of that pathway; $w_t(i)$ denotes the weight's value at time t. Pathway $x(0)$ is the US pathway and its weight $w(0)$ is positive and constant over time. Patterns of activity over the remaining input pathways represent stimuli that can be associated with the US—the CSs.* Changes in the weights of the CS pathways over time represent changes in the the associative strengths of the CSs with respect to the US. We denote by x_t the input vector at time t consisting of the n components of the CS vector, i.e., $x_t = (x_t(1), \ldots, x_t(n))$. Similarly, w_t denotes the n-component vector of weights of the CS pathways at time t. The element output, y, is assumed to contribute to both the UR and the CR.

---

* Tesauro (1986) correctly points out that the original description of the SB model suggests that the model is applicable only when a CS is represented locally by activity on a single input pathway. However, the model obviously also applies to the case of distributed CSs, and we wish to allow that possiblity here. This is also true of the TD model, but in the simulations presented here, locally represented CSs are used for simplicity.

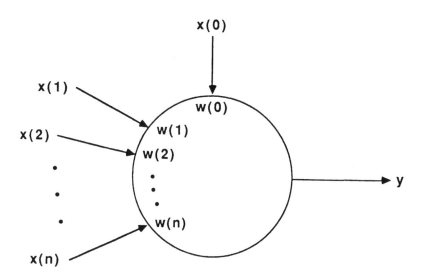

Figure 1. A neuron-like adaptive element used in the SB model. There are n modifiable CS input pathways, $x(1), \ldots, x(n)$, and a pathway $x(0)$ with fixed weight $w(0)$ that corresponds to the US. The element output y corresponds to both the UR and the CR.

The element output at time t is a function of the weighted sum of the inputs at time t:

$$y_t = f\left\{ \sum_{i=0}^{n} w_t(i) x_t(i) \right\}, \tag{1}$$

where $f\{\cdot\}$ is some S-shaped function; in our earlier simulations we assumed it was the identity function. We assume that this input/output relationship is instantaneous because the model does not address intrinsic response latencies, which vary across response systems.

The connection weights of the CS pathways are updated at each time step as follows:

$$w_{t+1} = w_t + c(y_t - y_{t-1})\bar{x}_t, \tag{2}$$

where $c > 0$ and $\bar{x}_t$ is the vector of *eligibility traces*, each component of which is a weighted sum of past values of the corresponding input signal.[*] We compute these traces using the following recursion:

$$\bar{x}_t = \beta \bar{x}_{t-1} + (1 - \beta) x_{t-1}, \tag{3}$$

---

[*] In Sutton and Barto (1981) and Barto and Sutton (1982), the model used an output trace $\bar{y}_t$ in place of y_{t-1} in Equation 2. However, in all the simulations described there we used only $\bar{y}_t = y_{t-1}$, which is the special case of a trace resulting from letting $\beta = 0$ in Equation 3. Because we now believe that this special case is best for reasons made clear in the theory underlying the TD model, we explicitly specify this case in our restatement of the SB model.

where $0 \leq \beta < 1$.[†]

Equations 1, 2 and 3 constitute the SB model. We can describe the learning process as follows: Activity on any input pathway i, $i = 1, \ldots, n$, can immediately influence the element's output, y, if $w(i) \neq 0$, but also causes that pathway to become "tagged" by the stimulus trace $\bar{x}(i)$ as being eligible for modification in the future (for as long as the trace is nonzero). A connection weight changes only if the pathway is eligible and reinforcement occurs, where reinforcement is defined as a deviation of the current output from the immediately preceeding output (for continuous time, reinforcement is the rate-of-change of the output). Figure 2 shows the time courses of the relevant signals for a single trial with an initially neutral CS.

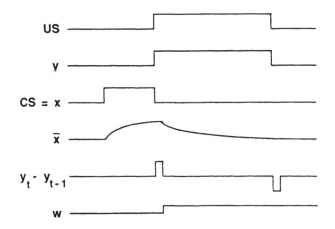

Figure 2. Time courses of element variables in the SB model for a trial in which an initially neutral ($w = 0$) CS is followed by the US.

In Sutton and Barto (1981) and Barto and Sutton (1982) we showed that this model is closely related to the Rescorla-Wagner model and could similarly account for phenomena in classical conditioning such as blocking and conditioned inhibition. Additionally, we showed how it could account for inter-stimulus interval (ISI) effects, anticipatory CRs, and aspects of higher-order and serial-compound conditioning. Recently, a novel prediction of the model concerning blocking and serial-compound conditioning has been tested and

[†] In Sutton and Barto (1981), $\bar{x}_t$ was defined as in Equation 3 except that the factor of $(1 - \beta)$ was absent. This factor, which was used in our presentation in Barto and Sutton (1982), simply normalizes the trace in such a way as to ensure that the trace of input that is constant over time will converge to that constant value as $t \to \infty$.

confirmed by Kehoe, Schreurs, and Graham (in press). That result is discussed further in the section on serial-compound results.

Despite these successes, the SB model suffers from several major problems. In our original presentation of the SB model, we avoided many of the problems by using a US that was very long, which ensured that all CS traces had fallen to zero by the time of US offset. Moore et al. (1986) have since found that if shorter USs are used, the SB model does not generate appropriate conditioning behavior as a function of the CS-US inter-stimulus interval (ISI). For example, if CS onset is simultaneous with or shortly after US onset, then the SB model incorrectly predicts strong inhibitory conditioning to the CS. Even worse, if CS offset is simultaneous with US offset, as in standard delay conditioning, then the unmodified SB model predicts that the CS will fail to acquire a positive association at any ISI.

Moore et al. succeeded in producing a modified version of the SB model, called the Sutton-Barto-Desmond, or SBD, model, that largely solves these and other problems, and also reproduces key features of response topography and CR-related neuronal firing (Moore et al., 1986; Blazis et al., 1986). The primary modifications to the SB model were 1) allowing the effect of the US to vary as a function of current weight values, 2) specifying a particular lagged relationship between CSs and their corresponding signals $x(i)$, and 3) making the trace decay rate β depend on CS duration. Together, these modifications constitute a substantial increase in the complexity of the model. With the TD model, we are attempting to solve the ISI problems of the SB model in a simpler way. The modifications made by Moore et al. to give the SB model a more realistic reponse topography and to relate it to neuronal firings may also be applicable to the TD model, but this has not yet been explored. Space limitations prevent us from making a full comparison of the TD model with the SBD model and with other competing real-time models (e.g., Klopf, 1986, in prep.; Tesauro, 1986; Gluck and Thompson, in press).

The Temporal-Difference (TD) Model

A key desirable feature of the SB model and some other models (Gelperin, Hopfield, Tank, 1985; Hawkins and Kandel, 1984; Klopf, 1986, in prep.; Moore et al., 1986, Tesauro,

1986) is that reinforcement is caused by the onsets and offsets of previously conditioned CSs. Since the US is treated exactly like a previously conditioned CS in the SB model, the US's reinforcing effects also occur at its onset and offset. Experimentally, however, it seems as if simply the presence of the US is reinforcing rather than changes in its presence. This is the basic difference between the SB model and the TD model—in the TD model, US presence itself is directly reinforcing, not its initiation and termination.

We define the TD model by referring to the adaptive element shown in Figure 1. The element's output at time t is

$$y_t = r_t + P(w_t, x_t),$$

where r_t denotes the value at time t of a signal indicating the presence and strength of the US (i.e., r_t is the same as $w_t(0)x_t(0)$ of the SB model) and $P(w_t, x_t)$ is defined by

$$P(w, x) = \begin{cases} \sum_{i=1}^{n} w(i)x(i), & \text{if } \sum_{i=1}^{n} w(i)x(i) > 0; \\ 0, & \text{otherwise.} \end{cases} \tag{4}$$

The weights are updated according to the rule

$$w_{t+1} = w_t + c\Big(r_t + \gamma P(w_t, x_t) - P(w_t, x_{t-1})\Big)\bar{x}_t, \tag{5}$$

where $c > 0$, $0 < \gamma < 1$, and $\bar{x}_t$ is as defined by Equation 3.

This model is similar to the SB model and basically works in the same manner, but it differs from that model in several crucial ways. First, note that the sum P plays a role in the weight update equation similar to the role the output y plays in the SB model (Equation 2): Changes in P over time are critical determinants of weight changes. But here the sum P does not include a contribution from the US as the sum y does in the SB model (Equation 1). The US directly contributes to weight changes through the term r_t in Equation 5. Consequently, in the TD model, the presence of the US (signaled by a nonzero value of r_t), rather than its onset and offset, acts as reinforcement. This is accomplished while retaining the feature of the SB model whereby a CS with an existing association generates reinforcement at its onset and offset (through the CS's contribution to P).

A second major feature distinguishing the TD model from the SB model concerns the parameter γ. The theoretical interpretation of this parameter is discussed in a later section. Here it suffices to point out that this parameter causes a CS with an existing associative strength to generate reinforcement throughout its presence and not just at its onset and offset. In Equation 5, if P is constant over time, then to the extent that γ is less than 1, reinforcement is still generated. The strength of this reinforcement is proportional to the strength of the CS's existing association, but of opposite sign. The choice of γ determines the relative importance of reinforcement generated by CSs with existing associations due to their constant presence, and due to their onsets and offsets. γ is usually chosen to be near 1 (e.g., $\gamma = .95$ in all simulations described here), so that the presence of a CS generates much less reinforcement than does its onset or offset.

Basic Results

In this section we present simulation results showing the behavior of the TD model in a range of basic conditioning paradigms—single-CS acquisition and extinction, ISI curves for trace and delay conditioning, blocking, conditioned inhibition, and the lack of extinction of conditioned inhibitors. We regard such results as basic because they do not involve complicated temporal relationships between CSs and because previous models have demonstrated each of these abilities. Nevertheless, to our knowledge only the SBD model (Moore et al., 1986; Blazis, 1986) has previously demonstrated all of these abilities.

The parameter values used in all simulations were $c = .01$, $\beta = .8$, and $\gamma = .95$. These values were chosen so as to approximately match ISI data for the rabbit nictitating membrane response (as discussed below), under the interpretation that each time step corresponds to .05 seconds. When a stimulus was present, the corresponding input signal (x or r) was set to 1, and when the stimulus was absent, the signal was set to 0. The time interval between trials was long enough for all traces to fall to zero. Since no stimuli were presented during the inter-trial interval, it is clear from Equation 5 that no weight changes will occur during the bulk of this time. Thus, most of the inter-trial interval was simulated simply by setting the traces to zero.

Figure 3 shows the behavior of the TD model in a single-CS acquisition and extinction paradigm. The temporal relationships among stimuli during the acquisition phase of the experiment are shown in Figure 3A. During extinction, only the CS was presented. Over acquisition trials, the CS gains associative strength in a negatively accelerated way, asymptotically approaching a fixed value. During extinction, associative strength is lost

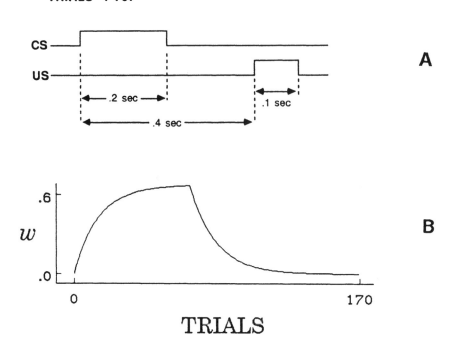

TRIALS 1-70:

Figure 3. Simulation of Single-CS Acquisition and Extinction in the TD Model. A) Timing relationships between stimuli during acquisition. **B)** The behavior of the weight corresponding to the CS during acquisition (trials 1-70) and extinction (trials 71-170). During extinction, the CS is presented not followed by a US. The time intervals are given in seconds under the interpretation that each time step corresponds to .05 seconds.

in a similar manner.

Figure 4 shows the ISI curves produced by the TD model in trace and delay conditioning experiments. These curves show the final associative strength generated by the TD model after 80 CS-US pairings as a function of the inter-stimulus interval between CS and US. The general shape of these curves is independent of parameter settings, but not important details such as how rapidly associative strength declines as the ISI increases. Roughly speaking, β determines the rate of decline in trace conditioning, and, for fixed β, γ determines the rate of decline in delay conditioning. The parameter values given above were selected to approximate the ISI data for rabbit NMR conditioning shown in Figure 5.

The TD model exhibits complete blocking if first-stage training is conducted until asymptotic associative strength is achieved and if the CS added in the second stage has

TRACE CONDITIONING:

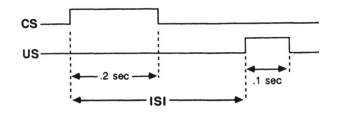

DELAY CONDITIONING:

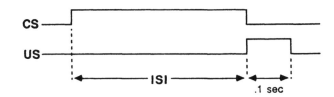

A

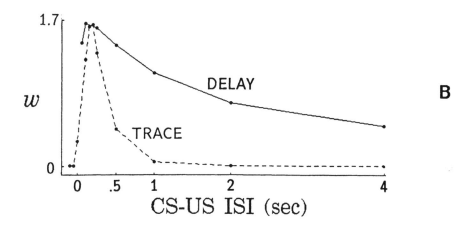

B

Figure 4. Effect of the CS-US Inter-Stimulus Interval in Trace and Delay Conditioning of the TD Model. A) Timing relationships between stimuli in trace and delay acquisition trials. **B)** Resultant CS weight after 80 acquisition trials as a function of ISI.

exactly the same time course as the first CS. This is apparent from inspection of Equation 5—the weights for the two CSs experience exactly the same increments during a second-stage trial; if the weight of the first CS no longer experiences any net change, then neither will the weight of the added CS.

Figure 6 shows the behavior of the TD model in a conditioned inhibition (CI) training regime. In CI, reinforced and unreinforced trials of the two types shown in Figure 6A are

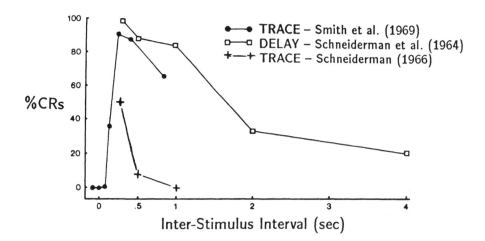

Figure 5. Effect of the CS-US ISI in Trace and Delay Conditioning of the Rabbit Nictitating Membrane Response (NMR). The time course of the ISI dependency varies widely between species and response systems. The parameter values used here in the TD model were chosen so that the model's ISI dependency, shown in Figure 4, approximately matches this rabbit NMR data.

intermixed. CS^+ is followed by the US except in the presence of CS^-. CS^+ is found experimentally to become positively conditioned whereas CS^- becomes a conditioned inhibitor, that is, it tends to inhibit CRs. This result is also found in the simulation. In the extinction phase of the CI experiment shown in Figure 6, both stimuli were presented individually without the US. The result shown is also the same as that found experimentally: The association to the excitor extinguishes, but the association to the inhibitor does not (Zimmer-Hart and Rescorla, 1974). Moore et al. (1986) showed that the SB model will reproduce the desired behavior if the output y is prevented from becoming negative (this corresponds to a particular choice for f in Equation 1), and this is essentially what we have done in the TD model by using a threshold operation in Equation 4.

Serial-Compound Results

Real-time conditioning models are interesting primarily because they make predictions for a wide range of situations that cannot be represented by trial-level models. These situations involve conditionable stimuli that occur together but not strictly simultaneously.

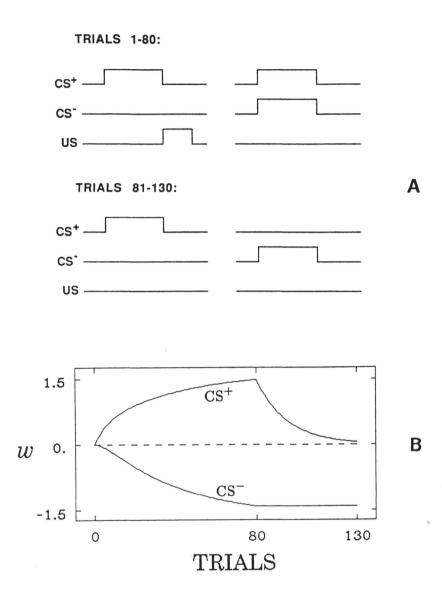

Figure 6. **Conditioned Inhibition and its Extinction in the TD Model. A)** Time traces showing the two kinds of trials presented alternately in a conditioned inhibition experiment (trials 1-80) and in a subsequent attempt to extinguish the resultant associations (trials 81-130). **B)** Behavior over trials of the weights associated with CS^+ and CS^-. During acquisition, the weight for CS^+ becomes positive, while the weight for CS^- becomes negative. The association to CS^+, but not to CS^-, is extinguished by nonreinforcement. Both CSs are .2 seconds in duration and the US is .1 second in duration.

Any such compound stimulus whose components do not both begin and end at the same time is called a serial-compound stimulus. It should be recognized that almost all learning involves serial-compound stimuli, either because the animal distinguishes earlier and later portions of a stimulus that may be viewed as a single stimulus by the experimenter, or because the animal's behavior gives rise to a predictable sequence of situations leading to reinforcement, as in maze running. Kehoe (1982) surveys the theoretical issues and empirical results relevant to serial-compound conditioning.

One of the theoretical issues arising in serial-compound conditioning concerns the facilitation of remote associations. It has been found that if an empty trace interval between the CS and the US is filled with a second CS to form a serial compound stimulus, then conditioning to the first CS is facilitated. Figure 7B shows the behavior of the TD model in a simulation of such an experiment, the timing details of which are shown in Figure 7A. Consistent with the experimental results, the model shows facilitation of both the rate of conditioning and the asymptotic level of conditioning of the first CS due of the presence of the second CS.

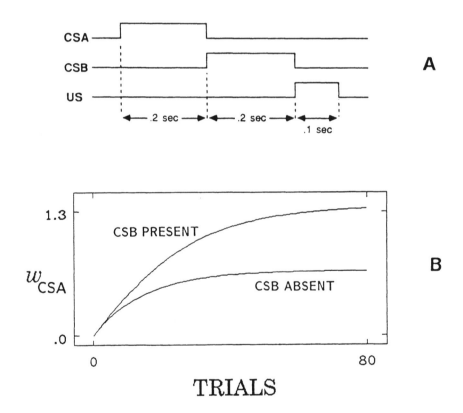

Figure 7. Facilitation of a Remote Association by an Intervening Stimulus in the TD Model. A) Temporal relationships among stimuli within a trial. **B)** The behavior over trials of CSA's weight when CSA is presented in a serial compound, as in **A**, and when presented in an identical temporal relationship to the US, only without the presence of CSB.

The stimulus context effects such as blocking and conditioned inhibition that the Rescorla-Wagner model is so successful at reproducing involve effects on the conditioning of one CS due to the presence of others. However, since it is a trial-level model, the Rescorla-Wagner model does not take into account the temporal relationships between the CSs, which are known to be capable of producing dramatic behavioral consequences. One of the best-known early demonstrations of this is the Egger-Miller (1962) experiment that involved two overlapping CSs in a delay configuration as shown in Figure 8A. Although CSB is in a better temporal relationship with the US, the presence of CSA reduces conditioning to CSB substantially as compared to controls in which CSA is absent. Figure 8B shows the same result being generated by the TD model in a simulation of this experiment.

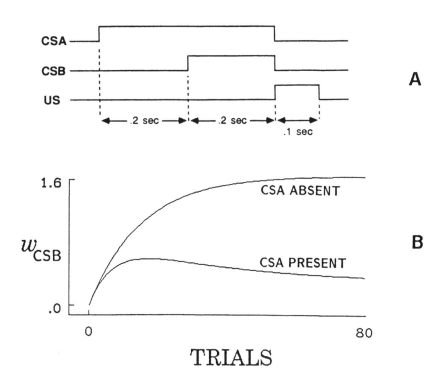

Figure 8. The Egger-Miller or Primacy Effect in the TD Model. A) Temporal relationships among stimuli within a trial. B) The behavior over trials of CSB's weight when CSB is presented with and without CSA.

In Sutton and Barto (1981), we presented simulation results with the SB model for an experiment similar to the Egger-Miller experiment discussed above. The experiment we simulated differed from the Egger-Miller experiment in that CSB was given prior training until it was fully associated with the US. When CSA was subsequently introduced, the pre-established association to CSB decreased to zero as training continued. Although we did not realize it at the time, this is a novel and surprising prediction of the SB model. Why should a well-trained CS that continues to be paired with the US in a good temporal relationship lose associative strength just because a new CS is introduced with no initial association and in a poor temporal relationship? This is a situation in which one might expect the original CS to block and limit association to the new CS. However, the SB model predicts a decrement in the other direction. Recently, Kehoe, Schreurs, and Graham (in press) have tested and confirmed the prediction that CSB will lose associative strength under these conditions. They also note that alternative theories do not make this prediction and have considerable difficulty in explaining this result. The behavior of the TD model under these conditions is shown in Figure 9. This behavior is in slightly better accord with the data than is the SB model's behavior, in that the association to CSB is reduced after the introduction of CSA, but not completely eliminated.

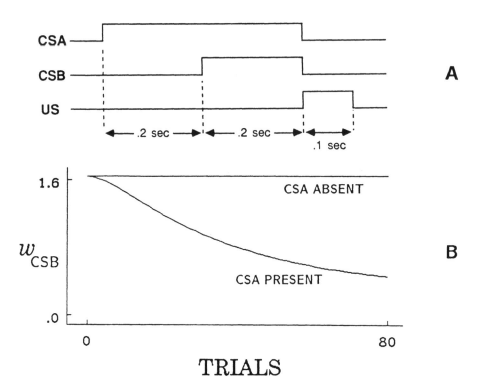

Figure 9. Temporal Primacy Overriding Blocking in the TD Model. A) Temporal relationships between stimuli. B) The behavior over trials of CSB's weight when CSB is presented with and without CSA. The only difference between this simulation and that shown in Figure 8 was that here CSB started out fully conditioned—CSB's weight was initially set to 1.653, the final level reached when CSB was presented alone for 80 trials, as in the "CSA-absent" case in Figure 8.

Figure 10 shows the behavior of the TD model in a second-order conditioning experiment. In the first phase (not shown in the figure), CSB is pretrained with the US. In the second phase, CSA is paired with CSB in the sequential arangement shown in Figure 10A, in the absense of the US. Experimentally, CSA is found to acquire associative strength even though it is never paired with the US. In the TD model, CSA first acquires a substantial association and then this association and the original one to CSB are extinguished. This is the same pattern seen experimentally.

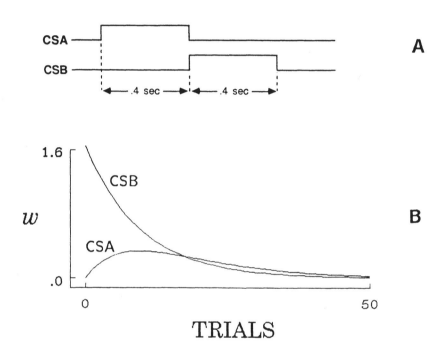

Figure 10. Second-Order Conditioning of the TD Model. A) Temporal relationships between stimuli. **B)** The behavior of the weights associated with CSA and CSB over trials. The second stimulus, CSB, has an initial weight of 1.653 at the beginning of the simulation.

Figure 11 shows the ISI curve for the TD model in second-order conditioning. It plots the associative strength after 100 trials as a function of the CSA–CSB ISI. This ISI curve differs significantly from the CS–US ISI curve shown in Figure 4 primarily in that here simultaneous presentation results in the formation of a large negative association instead of a small positive one. Recall that the TD model treats the reinforcement due to USs and previously conditioned CSs differently: US signals directly cause reinforcement whereas *changes* in the signals of previously conditioned CSs cause reinforcement. Thus, in simultaneous presentation, a US's reinforcement is delivered thoughout the presentation,

371

whereas a previously conditioned CS delivers reinforcement only at its onset, and negative reinforcement at its offset, so that a simultaneously paired CS will be much more affected by the negative reinforcement than by the positive reinforcement.

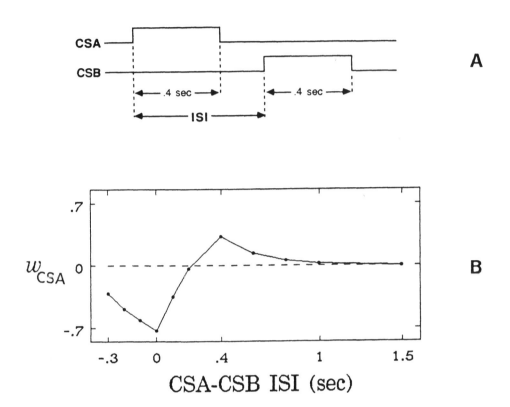

Figure 11. Effect of the CSA-CSB ISI on Second-Order Conditioning of TD Model. A) Temporal relationships between stimuli. **B)** Resultant value of CSA's weight after 10 trials as a function of CSA-CSB ISI.

Experimentally, second-order conditioning is observed to occur with both simultaneous and sequential CSA–CSB pairings. To explain this observation in terms of the TD model we must appeal to indirect associations, which are outside the scope of the model per se. That is, the model clearly predicts that no direct CSA → US association will develop, but does not preclude the development of both CSA → CSB and CSB → US associations, which together could have the same effect. This explanation of second-order conditioning is in fact partially confirmed experimentally. One observed difference between simultaneous and sequential second-order conditioning is that the association to CSA is eliminated by extinguishing CSB in simultaneous second-order conditioning, but not in sequential second-order conditioning (Rescorla, 1980). This suggests that simultaneous second-order conditioning in fact does not result in a direct CSA → US association.

Theoretical Basis of the TD Model

In addition to providing an account of the range of classical conditioning phenomena described above, the TD model has a theoretical basis that suggests an account of the functionality of these phenomena. Sutton (1987) has developed a class of methods for adaptive prediction called temporal-difference (TD) methods and has shown that they have certain advantages over other prediction methods for problems having a certain structure. The advantages of TD methods include reductions in memory requirements, a more even distribution of computation over time, and better generalization from past experience to new situations. If classical conditioning involves prediction, as many believe it does, then TD methods are likely candidates for the underlying learning procedure. Here we provide a brief introduction to the theory as it relates to the TD model.

At each time step t, the subject receives a pattern of CSs represented by the stimulus vector x_t, from which it forms a prediction $P(w, x_t)$, using its current weight vector w. But what does $P(w, x_t)$ predict? Clearly, $P(w, x_t)$ should tell the subject something about the values of the US signal r in the near future. For example, $P(w, x_t)$ might predict something like

$$E\left\{\sum_{k=1}^{N} r_{t+k}\right\},$$

where N is the number of steps remaining in the current trial. The sum is a natural way to have the ideal prediction vary with the intensity, duration, and number of USs occurring on the trial, and the expected value provides a principled way to deal with statistical variation from trial to trial.

However, the particular sum given above, in which all the r_{t+k} values in the rest of the trial are given equal weight, is problematic for two reasons. First, trials and trial boundaries are generally in the mind of the experimenter and unknown to the subject. Second, experimentally the association formed to a CS depends strongly on the time elapsing between it and the US—the more closely the US follows the CS, the stronger the association it will support. This last observation suggests that subjects are predicting a sum in which greater weight is given to r_{t+k} values for smaller values of k. Although there are many ways of varying the weighting with time, the TD model is based on an exponential weighting in which the weight of each r_{t+k}, $k \geq 1$, is γ^{k-1}, for $0 < \gamma < 1$. That is, the TD model is based on the hypothesis that the subject attempts to adjust w

so that, at each time t:

$$P(w, x_t) \approx E\left\{\sum_{k=0}^{\infty} \gamma^k r_{t+k+1}\right\}. \tag{6}$$

The parameter γ is called the *discount rate* because is determines the rate at which later values of r are discounted.

Although the theorems so far obtained for TD methods (Sutton, 1987) do not apply to predicting the quantity given by Equation 6, TD theory nevertheless provides a methodology for constructing a TD learning method specialized for predicting this quantity. The distinguishing feature of TD methods is that the error term they use is the difference between temporally successive predictions. $P(w, x_{t-1})$ and $P(w, x_t)$ are temporally successive predictions, but it is not appropriate to use their difference directly as an error because they are predictions of two different quantities, $P(w, x_{t-1})$ of $E\left\{\sum_{k=0}^{\infty} \gamma^k r_{t+k}\right\}$, and $P(w, x_t)$ of $E\left\{\sum_{k=0}^{\infty} \gamma^k r_{t+k+1}\right\}$. However, these two predictions are closely related as follows:

$$\begin{aligned}
P(w, x_{t-1}) &\approx E\left\{\sum_{k=0}^{\infty} \gamma^k r_{t+k}\right\} \\
&= E\left\{r_t + \sum_{k=1}^{\infty} \gamma^k r_{t+k}\right\} \\
&\approx r_t + \gamma E\left\{\sum_{k=0}^{\infty} \gamma^k r_{t+k+1}\right\} \\
&\approx r_t + \gamma P(w, x_t).
\end{aligned}$$

Thus, $r_t + \gamma P(w, x_t)$ is a prediction of the same quantity predicted by $P(w, x_{t-1})$, but it is available one time step later and is based on slightly better information—on the newly-available actual value of r_t and on the new stimulus vector x_t. It is thus the difference between these two predictions, that is, $(r_t + \gamma P(w, x_t)) - P(w, x_{t-1})$, that is used as a reinforcement or error in the TD model's update rule (Equation 5).

The TD model proposed here is not the first model of classical conditioning to be based on changes or temporal differences in net associative strength. This mechanism is a key part of the SB model, and also of the models proposed by Hawkins and Kandel (1984), Gelperin, Hopfield and Tank (1985), Klopf (1986, in prep.), Moore et al. (1986), and Tesauro (1986). What is different about the TD model is that the precise way temporal differences are used is based on a formal, engineering theory of prediction, coupled with a specific proposal for the quantity being predicted.

Limitations and Conclusion

Neither the SB model nor the TD model are complete models of classical conditioning. Among the major classes of phenomena that are beyond the scope of these models and which have been treated by other models are configuration and patterning phenomena (e.g., Kehoe, 1986, and Granger and Schlimmer, 1986), attentional and stimulus selection effects, learning to learn, and learned salience/associability changes (e.g., Moore and Stickney, 1980; Schmajuk and Moore, 1986; Kehoe, 1986), sensory preconditioning and other effects of indirect associations (e.g., Schmajuk and Moore, 1986), CR topography (e.g., Moore et al., 1986; Frey and Sears, 1978), and stimulus preprocessing issues (e.g., Gelperin, Hopfield, and Tank, 1985). Some of these phenomena may be addressable with connectionist mechanisms such as backpropagation (Rumelhart, Hinton, and Williams, 1985) learning-rate adjustment rules (e.g., Frey and Sears, 1978; Sutton, 1986; Barto and Sutton, 1981, Appendix C), and recurrent networks (e.g., Sutton and Barto, 1981a; Sutton and Pinette, 1985).

Although animal learning is complex and subtle, with different processes operating at different levels and time scales, its regularities are far more striking than its variations. Although one theory that explains all animal learning remains a goal, most progress in this area has been made by focussing on identifiable component processes of animal learning. Against this background, the TD model actually represents a substantial integration, since its behavior subsumes nearly all the behavior of the trial-level Rescorla-Wagner model but additionally generates predictions and explanations for within-trial phenomena. The simulations of the TD model described in this paper, together with the theoretical basis of the TD model, suggest that these phenomena might be regarded as consequences of an adaptive process for predicting a discounted sum of future values of the US signal.

References

Anderson, C.W. 1986. Learning and problem solving with multilayer connectionist systems. Ph.D. dissertation, Dept. of Computer and Information Science, University of Massachusetts.

Barto, A.G., Sutton, R.S. 1981. Goal seeking components for adaptive intelligence: An initial assessment. *Air Force Wright Aeronautical Laboratories/Avionics Laboratory Technical Report AFWAL-TR-81-1070*, Wright-Patterson AFB, Ohio.

Barto, A.G., Sutton, R.S. 1982. Simulation of anticipatory responses in classical conditioning by a neuron-like adaptive element. *Behavioral Brain Research 4*: 221–235.

Barto, A.G., Sutton R.S., Anderson, C.W. 1983. Neuronlike elements that can solve difficult learning control problems. *IEEE Trans. on Systems, Man, and Cybernetics, SMC-13*, No. 5, 834–846.

Blazis, D.E.J., Desmond, J.E., Moore, J.W., Berthier, N.E. 1986. Simulation of the classically conditioned nictitating response by a neuron-like adaptive element: A real-time variant of the Sutton-Barto model. *Proceedings of the Eighth Annual Conf. of the Cognitive Science Society*, 176-186.

Duda, R.O., Hart, P.E. 1973. *Pattern Classification and Scene Analysis*. New York: Wiley.

Egger, D.M., Miller, N.E. 1962. Secondary reinforcement in rats as a function of information value and reliability of the stimulus. *Journal of Experimental Psychology 64*: 97–104.

Frey, P.W., Sears, R.J. 1978. Model of conditioning incorporating the Rescorla-Wagner associative axiom, a dynamic attention process, and a catastrophe rule. *Psychological Review 85*: 321–348.

Gelperin, A., Hopfield, J.J., Tank, D.W. 1985. The logic of *Limax* learning. In: *Model Neural Networks and Behavior*, A. Selverston, Ed. New York: Plenum Press.

Gluck, M.A., Thompson, R.F. In press. Modeling the neural substrates of associative learning and memory: A computational approach. *Psychological Review*.

Hawkins R.D., Kandel, E.R. 1984. Is there a cell-biological alphabet for simple forms of learning? *Psychological Review 91*: 375–391.

Kehoe, E.J. 1982. Conditioning with serial compound stimuli: Theoretical and empirical issues. *Experimental Animal Behavior 1*: 30–65.

Kehoe, E.J. 1986. A layered network model for learning-to-learn and configuration in classical conditioning. *Proceedings of the Eighth Annual Conf. of the Cognitive Science Society*, 154–175.

Kehoe, E.J., Schreurs, B.G., Graham, P. In press. Temporal primacy overrides prior training in serial compound conditioning of the rabbit's nictitating membrane response.

Klopf, A.H. 1972. Brain function and adaptive systems—A heterostatic theory. Air Force Cambridge Research Laboratories Special Report No. 133 (AFCRL-72-0164). Also DTIC

Report AD 742259 available from the Defense Technical Information Center, Cameron Station, Alexandria, VA 22304.

Klopf, A.H. 1982. *The Hedonistic Neuron: A Theory of Memory, Learning, and Intelligence.* New York: Harper & Row / Hemisphere.

Klopf, A.H. 1986. A drive reinforcement model of single neuron function: An alternative to the Hebbian neural model. In J.S. Denker (Ed.) *Neural Networks for Computing*, AIP Conference Proceedings *151*, New York: American Institute of Physics, 265–270.

Klopf, A.H. In preparation. A neuronal model of classical conditioning.

Mackintosh, N.J. 1975. A theory of attention: Variation in the associability of stimuli with reinforcement. *Psychological Review 82*: 276–298.

Moore, J.W., Desmond, J.E., Berthier, N.E., Blazis, D.E.J., Sutton, R.S., Barto, A.G. 1986. Simulation of the classically conditioned nictitating membrane response by a neuron-like adaptive element: Response topography, neuronal firing and interstimulus intervals. *Behavioral Brain Research 21*: 143–154.

Moore, J.W., Stickney, K.J. 1980. Formation of attentional-associative networks in real time: Role of the hippocampus and implications for conditioning. *Physiological Psychology 8*: 207–217.

Pearce, J.M., Hall, G. 1980. A model for Pavlovian conditioning: Variations in the effectiveness of conditioned but not of unconditioned stimuli. *Psychological Review 87*: 532–552.

Rescorla, R.A. 1980. Simultaneous and successive associations in sensory preconditioning. *Journal of Experimental Psychology: Animal Behavioral Processes 6*: 339–351.

Rescorla, R.A., Wagner, A.R. 1972. A theory of Pavlovian conditioning: Variations in the effectiveness of reinforcement and nonreinforcement. In: *Classical Conditioning II*, A.H. Black and W.F Prokasy, Eds., 64–99. New York: Appleton-Century-Crofts.

Rumelhart, D.E., Hinton, G.E., Williams, R.J. 1985. Learning internal representations by error propagation. Institute for Cognitive Science Technical Report 8506, UCSD, La Jolla, CA 92093. Also in Rumelhart and McClelland (1986), 318–362.

Rumelhart, D.E., McClelland, J.L. 1986. *Parallel Distributed Processing: Explorations in the Microstructure of Cognition, Volume 1: Foundations.* Cambridge, MA: MIT Press.

Schlimmer, J.C., Granger, R.H. 1986. Simultaneous configural classical conditioning. *Proceedings of the Eighth Annual Conf. of the Cognitive Science Society*, 141-153.

Schmajuk, N.A., Moore, J.W. 1986. A real-time attentional-associative network for classical conditioning of the rabbit's NMR. *Proceedings of the Eighth Annual Conf. of the Cognitive Science Society*, 794–807.

Schneiderman, N. 1966. Interstimulus interval function of the nictitating membrane response of the rabbit under delay and trace conditioning. *Journal of Comparative and Physiological Psychology 62*: 397–402.

Schneiderman, N., Gormezano, I. 1964. Conditioning of the nictitating membrane of the rabbit as a function of the CS-US interval. *Journal of Comparative and Physiological Psychology 57*: 188–195.

Smith, M.C., Coleman, S.R., Gormezano, I. 1969. Classical conditioning of the rabbit's nictitating membrane response at backward, simultaneous and forward CS-US intervals. *Journal of Comparative and Physiological Psychology 69*: 226–231.

Sutton, R.S. 1984. Temporal credit assignment in reinforcement learning. Ph.D. dissertation, Dept. of Computer and Information Science, University of Massachusetts. Available from the author or as Technical Report #84-2.

Sutton, R.S. 1987. Learning to predict by the methods of temporal differences. Technical Report TR87-509.1, GTE Labs, Waltham, MA.

Sutton, R.S., Barto, A.G. 1981. Toward a modern theory of adaptive networks: Expectation and prediction. *Psychological Review 88*: 135–171.

Sutton, R.S., Barto, A.G. 1981a. An adaptive network that constructs and uses an internal model of its environment. *Cognition and Brain Theory Quarterly 4*: 217–246.

Sutton, R.S., Pinette, B. 1985. The learning of world models by connectionist networks. *Proceedings of the Seventh Annual Conf. of the Cognitive Science Society*, 54–64.

Tesauro, G. 1986. Simple neural models of classical conditioning. *Biological Cybernetics 55*: 187–200.

Wagner, A.R. 1981. SOP: A model of automatic memory processing in animal behavior. In: *Information Processing in Animals: Memory Mechanisms*, N.E. Spear and R.R. Miller, Eds., 5–48. Hillsdale, NJ: Erlbaum.

Widrow B., Hoff, M.E. 1960. Adaptive switching circuits. *1960 WESCON Convention Record Part IV*, 96–104.

Widrow, B., Stearns, S.D. 1985. *Adaptive Signal Processing*. Englewood Cliffs, NJ: Prentice-Hall.

Zimmer-Hart, C.L., Rescorla, R.A. 1974. Extinction of Pavlovian conditioned inhibition. *Journal of Comparative and Physiological Psychology 86*: 837–845.

A REPRESENTATION FOR NATURAL CATEGORY SYSTEMS

Sandra L. Peters and Stuart C. Shapiro
Department of Computer Science
State University of New York at Buffalo
Buffalo, NY 14260
peters@buffalo.csnet, shapiro@buffalo.csnet

ABSTRACT

Most AI systems model and represent natural concepts and categories using uniform taxonomies, in which no level in the taxonomy is distinguished. We present a representation of natural taxonomies based on the theory that human category systems are non-uniform. That is, not all levels of abstraction are equally important or useful; there is a basic level which forms the core of a taxonomy. Empirical evidence for this theory is discussed, as are the linguistic and processing implications of this theory for an artificial intelligence/natural language processing system. We present our implementation of this theory in SNePS, a semantic network processing system which includes an ATN parser-generator, demonstrating how this design allows our system to model human performance in the natural language generation of the most appropriate category name for an object. The internal structure of categories is also discussed, and a representation for natural concepts using a prototype model is presented and discussed.

1. INTRODUCTION.

Knowledge-base systems typically model and represent natural concepts and categories using *uniform* inheritance networks [Quillian 1967, 1968, 1969; Collins & Quillian 1970; Fahlman 1979] or frame systems [Brachman 1983; Brachman & Schmolze 1984]. We will present a representation of natural taxonomies based on the theory that human category systems are non-uniform, i.e., that not all levels of abstraction are equally important or useful. This theory is supported by a substantial body of empirical evidence from the fields of psychology, anthropology, and linguistics [Rosch et al. 1976, 1978; Mervis & Rosch 1981; Berlin 1978; C. H. Brown et al. 1976; Tversky 1978; Hunn 1976; Cantor et al. 1979; Smith & Medin 1981]. We will discuss some of the evidence for this theory, as well as some of the linguistic and processing implications of this theory for an AI system modeling human cognitive behavior.

The need for a non-uniform representation will be extended as we consider the internal structure of natural concepts. We will present and discuss some of the empirical evidence that supports the use of a prototype model for these concepts, and present a representation using this model. We will argue, however, that certain levels of abstraction exhibit more of a prototypicality structure than others, and that, therefore, distinct representations are again needed in modeling the internal structure of concepts at different levels of abstraction in a taxonomy.

Our implementation uses the SNePS semantic network processing system which includes an ATN parser-generator [Shapiro, 1978, 1979, 1982, 1986].

2. THEORY - THE VERTICAL DIMENSION OF CATEGORY SYSTEMS - A BASIC LEVEL.

Our representation is based on the following principles of human categorization set forth by Eleanor Rosch. Categories within taxonomies are structured such that there is one level of abstraction at which the most basic category cuts can be made. This level of abstraction forms the "core" [Berlin 1978, p. 24] of a taxonomy, and is called the basic level. Basic categories are: (1) those which carry the most information; (2) those whose members have the most attributes in common; and (3) the categories

379

most differentiated from one another. Basic level categories are, in fact, disjoint. Chair, car, and dog are examples of basic level objects.

Levels of a taxonomy above the basic level are called superordinate categories (e.g., furniture, vehicle, mammal). Fewer attributes are shared among members of superordinate categories, i.e., there is less category resemblance. Categories below the basic level are called subordinate categories (e.g., kitchen chair, station wagon, collie). Subordinate categories contain many attributes which overlap with those of other subordinate categories, i.e., there is less contrast between categories across a subordinate level.

2.1. Empirical Evidence.

The following summarizes some of Rosch's empirical evidence supporting the existence of a basic level which forms the core of a taxonomy. [Rosch et al. 1976, 1978; Mervis & Rosch 1981].

2.1.1. Attributes of Objects.

When subjects were asked to list attributes of basic, superordinate, and subordinate level objects, very few attributes were listed for superordinate categories, a great number of attributes were listed for basic categories, and an insignificant number of additional attributes were listed for subordinate level categories. This result supports the theory that the basic level is the most inclusive or general level at which the objects of a category possess a large number of attributes in common. Attributes appear to be clustered at the basic level.

2.1.2. Object Recognition.

Experiments using averaged shapes, obtained by superimposing outlines of objects to form normalized shapes, showed that the basic level is the most inclusive level at which the averaged shape of an object can be recognized. That is, basic objects (e.g., chairs, dogs) were the most general objects that could be identified from these shapes; superordinate objects (e.g., furniture, animals) could not be identified from averaged shapes. This suggests that basic level objects, are the most inclusive categories for which a concrete mental image of the category as a whole can be formed. We can form an image of a cat or dog which reflects the average members of the class, however, we cannot form an image of a mammal that reflects the appearance of the class as a whole.

2.1.3. Object Names - Categorization.

Studies of picture verification have demonstrated that objects are first recognized as members of their basic level category. When subjects were shown pictures of objects, the basic level name was the name chosen for an object. With additional processing time, subjects were able to categorize objects at their subordinate and superordinate levels. Thus, subjects knew the subordinate and superordinate names of objects, but categorized objects first at the basic level. Rosch further states that basic level objects are the first categorizations made during perception of the environment, as well as the categories most named, and most necessary in language.

2.1.4. Development of Categories.

Basic level objects are not only the first categories learned by children, they appear to be formed differently from categories at other levels. That is, basic categories are not learned explicitly by acquiring a definition or deductive rule, but rather are learned implicitly by exposure to multiple instances of the category; i.e., they are formed inductively. This is often called the acquisition of types through ostensive definitions [Jackendoff 1983]. Categories subordinate and superordinate to this level are often formed by the acquisition of a deductive rule [Berlin 1978]. For example, the concept mammal might be learned in terms of a rule which lists attributes such as: warm-blooded; body usually covered with hair; female gives milk to young.

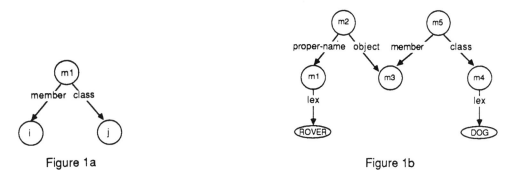

Figure 1a

Figure 1b

2.1.5. Summary of Empirical Evidence.

Thus, recent categorization research provides a great deal of empirical evidence supporting the importance of basic level categories in a taxonomy, and the non-uniformity of human category systems. Basic level categories are the first categories developed, they are formed differently than non-basic categories, they are the most used and useful categories, and therefore, they must be distinguished from non-basic categories in some way.

3. REPRESENTATION AND USE OF CATEGORIES IN AN AI/NLP SYSTEM.

If an artificial intelligence/natural language processing (AI/NLP) system modeling human category systems must be able to distinguish basic level categories from non-basic categories, an important issue to be considered is how and where to make the distinction. Basic level objects are used in two kinds of categorization: "ordinary" categorization, i.e., the classification of an individual in a class, and generic categorization, i.e., categorization involving two classes or types. It seems clear that since basic level categories are formed early in life, they are formed via ordinary categorization. The teaching of these names is limited to the presentation of examples and counter-examples. Thus, a child may learn the basic level name 'dog', as someone points to Rover and says 'dog'. Therefore, our system makes the distinction between basic and non-basic levels in the representations for ordinary categorization, i.e., in the individual/class relations.

Figure 1a shows the case frame used for ordinary categorization of a basic level object. Here **m1** represents the proposition that the individual represented by **i** is a member of the basic level category represented by **j**. Figure 1b shows the representation for "Rover is a dog". [See Shapiro & Rapaport 1986 for the syntax and semantics of other constructs.]

Since non-basic categories are formed later than basic categories, and are formed in the course of the investigation of underlying principles rather than ostensive features, we use a slightly more complex case frame to represent membership in a non-basic level category. Thus, in Figure 2a **m1** represents the proposition that the individual represented by **i** is a member of the non-basic category represented by **j**. The representation of "Rover is a mammal" is shown in Figure 2b.

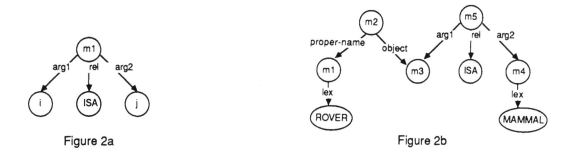

Figure 2a

Figure 2b

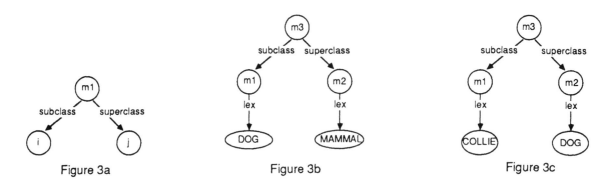

Figure 3a   Figure 3b   Figure 3c

These case frames in SNePS are the built-in syntactic structures of our modeled minds. The use of the *member/class* case frame reflects the basic or primitive nature of categorization in basic categories, whereas the use of the *arg1/rel/arg2* case frame treats membership in non-basic categories as an ordinary binary relation. Thus, our system distinguishes two cases of ordinary categorization: one representation is used when the class membership involves a basic level category, another representation, when the class membership involves a non-basic category.

In addition to this ordinary categorization, a system must, of course, be able to represent generic categorization, i.e., class/class relations, such as "Dogs are mammals". These relations are represented using a subclass/superclass case frame. See Figure 3a. Here **m1** represents the proposition that the class of **i**'s are a subclass of the class of **j**'s. The representations of "Dogs are mammals" and "Collies are dogs" are shown in Figures 3b and 3c respectively.

Thus, we build a traditional uniform type hierarchy of class/class relations. We see no reason to distinguish any relations in the hierarchy, since we find no evidence that generic categorization sentences such as "collies are dogs", "dogs are mammals", "mammals are vertebrates", require different underlying representations. (It is noteworthy that there are no class/class relations between two basic level categories.)

Since an ability to form abstract concepts is required for generic categorization, this categorization occurs at a later stage of development than does ordinary categorization of basic level objects. Therefore, the type hierarchy, which is formed after basic level concepts are formed, is not the appropriate place to make the distinction between basic and non-basic categories. In summary, a single representation is used for class/class relations, but two distinct representations are used for individual/class relations.

KRL-0 [Bobrow & Winograd 1977a, 1977b, 1979] is, to our knowledge, the only other AI system to distinguish basic and non-basic levels in the representation of taxonomies. KRL-0 used *units* to represent both classes and individuals. Three distinct levels of abstraction were used in the representation of classes or types in *units*: a basic level, an abstract level, and a specialization level. Bobrow and Winograd stated that they did not, however, find an appropriate way to use these unit categorization levels for classes, and removed unit categorization from KRL-1 [Bobrow & Winograd 1979, p. 41]. Although not precisely specified in their papers, Bobrow and Winograd appear to have made distinctions among the levels of abstraction in the type hierarchy of frames only, not in the individual/class relationships. We could not find any evidence that distinctions were made in the units representing individuals [Bobrow & Winograd 1977a p. 23].

3.1. INHERITANCE AND LINGUISTIC IMPLICATIONS.

3.1.1. Inheritance.

One of the organizational principles to which most semantic networks and frame systems adhere is that of storing properties in the hierarchy at the place covering the maximal subset of nodes sharing them. This is an efficient organizational scheme in which properties do not have to be replicated at different places in the network, for they are inherited by nodes below the ones in which they are stored. This principle fits in well with the theory of cognitive economy, for one can gain a great deal of information from a category system organized in this way, while conserving resources.

Categorization research studies, however, do not support this principle of organization. As stated above, properties appear to be clustered at the basic level, not at the level covering the maximal subset of nodes. This means that there is not a great deal of inheritance of properties taking place in the type hierarchy. Instead most inheritance occurs at the individual level, i.e., from the basic level category to the individual. Thus, Rover inherits attributes from the basic level category dog.

3.1.2. Linguistic Implications.

Perhaps the most dramatic enhancement to our system resulting from our distinguishing basic and non-basic level categories is our ability to model human performance by choosing the most appropriate category name for an object. Systems using uniform taxonomies have to make arbitrary word choice decisions. For example, the NIGEL generator [Sondheimer et al. 1986] generates as specific a term as possible. However, we know from human categorization research that in the absence of a specific context that would lead one to use a non-basic level name for an object, the basic level name should be used.

The dialog shown in Figure 4 illustrates our system's ability to model human performance in this respect. Since the basic level name is the most useful and most used name, the most appropriate answer to the question "What did Lucy pet?" is not the specialization "collie" or the superordinate level name "mammal, but the basic level name "dog".

The additional dialog shown in Figure 5 demonstrates that the basic level name is chosen regardless of the order in which categories are mentioned, and also demonstrates the natural language input of classificational information.

4. THEORY - THE INTERNAL STRUCTURE OF CATEGORIES.

Although categories have been viewed traditionally as concepts established by necessary and sufficient criteria, and many AI systems have modeled natural concepts in this way [Brachman 1983; Brachman & Schmolze 1984; Fahlman 1979], recent categorization research on naturally occurring concepts does not support this view [Rosch 1976, 1978; Mervis & Rosch 1981; Smith & Medin, 1981]. Rosch has suggested that another way to achieve separateness and clarity of categories is by "conceiving of each category in terms of its clear cases" [Rosch 1978, p. 36], that is, in terms of prototypes. Thus, all members of a category are not equally representative of the category, but, rather, some exemplars are more representative than others. Prototypical category members are those members which have the most attributes in common with other members of the category and the fewest attributes in common with members of contrasting categories.

4.1. Empirical Evidence Supporting Prototype Theory. A large body of empirical evidence exists which supports the theory that categories are conceived of in terms of prototypes, and that, therefore, category members vary in representativeness [Rosch 1978; Mervis & Rosch 1981; Smith & Medin 1981]. A brief overview of some supporting evidence follows.

4.1.1. Production of Exemplars. When subjects are asked to list exemplars of superordinate categories, they list the most representative exemplars. Similarly, when subjects are asked to sketch an exemplar for a particular category, they sketch the most representative exemplars [Mervis & Rosch, 1981].

4.1.2. Categorization: Speed of Processing. Reaction times studies show that the verification of a robin as a bird is performed faster than the verification of an ostrich or penguin as a bird. Thus, subjects are able to verify category membership faster for those objects rated as being prototypical, i.e., for representative exemplars, than for those rated as non-prototypical [Mervis & Rosch, 1981]. In addition, atn parser initialization

: Lucy petted a yellow animal
I understand that Lucy petted a yellow animal

: The animal was a dog
I understand that the yellow animal is a dog

: The dog was a collie
I understand that the yellow dog is a collie

: What did Lucy pet
Lucy petted a yellow dog

: ^end
(end atn parser)

Figure 4

atn parser initialization

: Mary petted a dog
I understand that Mary petted a dog

: The dog is a mammal
I understand that the dog is a mammal

: The dog was a labrador
I understand that the dog is a labrador

: What did Mary pet
Mary petted a dog

: Jane petted a manx
I understand that Jane petted a manx

: The manx is a cat
I understand that the manx is a cat

: A cat is a mammal
I understand that cats are mammals

: Mammals are animals
I understand that mammals are animals

: Who petted an animal
Mary petted a dog
and
Jane petted a cat

: ^end
(end atn parser)

Figure 5

when a prime, i.e., the prior mention of the category name, is provided, the response time to verify category membership of representative exemplars decreases. Priming, however, increases the response time necessary to verify the membership of non-representative exemplars [Mervis & Rosch, 1981].

4.1.3. Learning and Development of Categories.

Category membership is established first for those exemplars that are most representative of the category; membership for non-representative exemplars vacillates for some time. Thus, the formation of category prototypes is related to the initial formation of categories: the most representative members of categories are the ones first established as category members. In addition, categories are learned more easily if initial exposure is confined to representative exemplars [Mervis & Rosch 1981].

4.1.4. Categorization: Indeterminacy of Category Membership.
Category boundaries are not well defined. Many experiments have revealed that subjects disagree concerning the category to which poor exemplars belong [Berlin & Kay 1969; Labov 1973; McCloskey & Glucksberg 1978]. Both Sokal [1974] and Rosch & Mervis [1975] have demonstrated that poor exemplars contain attributes that overlap with those of contrast categories.

4.1.5. Perception of Typicality Differences.
There is general agreement among subjects when they are asked to rate how good an example is of its category, or to choose the exemplar most representative of its category. [Rosch 1978; Mervis & Rosch 1981].

4.2. Summary of Empirical Evidence.
If natural categories are determined by necessary and sufficient criteria, then category members should be equivalent, i.e., any member of the category should be as good an example of the concept as any other one. The empirical evidence outlined above does not support this view. Reaction time studies showing the speed of categorization, studies of the learning and development of categories, studies of typicality ratings, and studies involving the production of exemplars, demonstrate that all members are not equally good examples of their category, but rather that some members of a category are more representative than others.

We are not proposing that a prototype model is an appropriate one for all concepts, but are considering only natural concepts in this paper. Some classes of entities may indeed be determined by necessary and sufficient criteria.

4.3. Representation.

As Rosch points out, use of a prototype model does not specify a representation of categories, it merely constrains our choice of representation [Rosch 1978]. Different theories of semantic memory can accomodate this view. However, it seems clear that a representation based on the classical model, i.e., on necessary and sufficient criteria, cannot accomodate the evidence discussed above.

4.3.1. Additional Constraints.

We agree with the proposal [Rosch 1978] that any representation for a natural concept should satisfy the requirements of (1) mirroring the structure in the perceived world and (2) cognitive economy. Cognitive economy dictates that a representation for the concept *dog* should include only the essential information for the categorization of novel instances of this concept, i.e., information about how dogs resemble one another and differ from other concepts. In addition, we believe that cognitive economy further dictates that we mentally represent categories in terms of an *abstraction* which is an amalgamation of the most salient and most modal features of category members. An alternative to this featural model is the mental image model of prototypes. However, as Johnson-Laird [1983] points out, images are highly specific. Thus, we do not form an image of trees in general or of fruit in general, rather we form an image of a specific tree or a specific fruit. We cannot form an image that is isomorphic to the class as a whole for these categories. The image model would require mental images for each of the clear-cut cases or typical category members, rather than a composite of the features of these typical members, and yet seems no more useful in categorizing novel instances. Thus, the featural model is a more economical one. The featural model is also able to capture functional attributes which cannot easily be captured in a mental image or holistic model.

Our representation makes use of one further constraint: our belief that types or categories are non-projectable or non-referring, i.e, types do not exist. As Jackendoff [1983] points out, we cannot point to a type, but only to instances of types. We do not consider a prototype to be an additional member of its category, but rather consider it to be an abstraction: a list of features. Therefore, although our representation for the prototypical *dog* may contain the feature *four-leggedness*, we do not concern ourselves with structural information such as this prototypical dog's typical left front leg. We are only interested in the abstract feature or property *four-leggedness*, for its facilitation of categorization. This constraint allows us to avoid some of the inheritance problems with which Fahlman's NETL [1979] must deal, since he must concern himself that an individual elephant does not inherit the typical elephant's typical left front leg. (See also Shapiro [1980] for a review of NETL.)

4.4. Implementation of Prototype Model in SNePS.

Our implementation represents a category as a collection of abstract features or properties. A separate proposition is used to capture each abstract feature. Thus, the category *bird* is represented in terms of the most salient and modal features of the members of the concept *bird*, e.g., *flies*, *feathered*, *winged*, *has a beak*, *sings*, etc. Each of these properties is represented with a default generalization, e.g., *flies* is represented with a default generalization that may be paraphrased as: For all x if x is a bird, then presumably x flies. Figure 6 shows a partial representation for the prototype *bird*, using two of the features mentioned above. Thus, a prototype in our system consists of a bundle of default generalizations, all of which share the same antecedent, x is a bird, in our example, and the same variable.

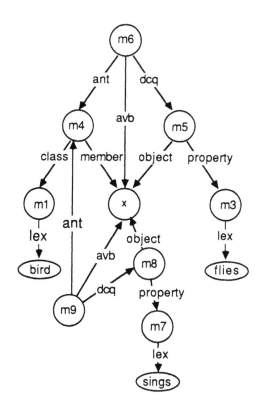

Figure 6

386

4.5. Effect of the Level of Abstraction on the Representation.

While superordinate level categories are formed in the course of the investigation of underlying principles, basic level categories are formed by exposure to multiple instances, on the basis of ostensive features. Therefore, it seems that basic level level categories should exhibit more of a protypicality structure than superordinate level categories. The basic level is also the level of abstraction at which attributes are clustered; very few attributes cluster at higher levels in the taxonomy. Thus, it seems appropriate to use distinct representations for the internal representation of basic and non-basic level categories: a prototype representation for basic level categories and perhaps, a mixed representation that combines deductive rules which are true universals, e.g., mammals are warm-blooded, with an exemplar component, that is, with a listing of prominent exemplars, for superordinate level concepts.

4.6. Understanding Generic Sentences.

This work grew out of an interest in understanding generic sentences, i.e., sentences containing generic concepts. We are now considering whether the prototype model is of use in understanding these sentences when they involve basic level or subordinate level concepts. One of the problems these sentences pose for an AI/NLU system is that a multiplicity of quantifiers can be posited for sentences of this type. Consider the following sentences:

(1) Cardinals are red. (The mature males are red.)

(2) Mosquitos carry malaria. (Very few actually carry malaria.)

(3) Dogs are four-legged. (Most are four-legged.)

(4) Shamrocks are green. (Most are green.)

(5) Men are mortal. (Mortality is a necessary criterion in this case.)

It may be more appropriate to examine the properties or features expressed in terms of saliency, rather than to try to determine the sense of quantification in these sentences. Features which we consider to be especially salient, such as *transmits dangerous disease*, may be part of the abstraction that form the prototypical mosquito, regardless of the sense of quantification in the sentence or utterance.

4.7. PROCESSING IMPLICATIONS.

The non-uniformity of human category systems also has implications for a processing model for categorization. Category research has established that objects can be identified as members of their basic level category more rapidly than as members of their superordinate or subordinate categories. A possible processing model for our implementation, compatible with Rosch's empirical evidence and the current general processing assumptions about categorization involving featural models [Smith & Medin 1981] such as ours is the following. An object is first identified or recognized as a member of its basic class, since properties or attributes are clustered at the basic level. Because of this bundling of attributes at the basic level, this processing involving feature matching can be performed quickly. Categorization of an object as a member of its subordinate classes requires additional processing time, because additional features must be matched, some of which are much less salient than the features for categorizing an object at the basic level. Categorization of an object as a member of its superordinate classes requires inferencing using the type hierarchy. We use path-based inference to accomplish this [Shapiro 1986, 1978]. Performing this inferencing, of course, requires additional processing time.

5. CONCLUSIONS.

We have incorporated principles of categorization derived from several years of research in our AI/NLP system. Principles of categorization, taken seriously, affect the design and representation of taxonomies for natural concepts. We distinguish one level, the basic level, as the core of our taxonomies, using a representation for membership in basic categories distinct from that used for membership in non-basic categories. This representation allows our system to model human performance in the generation of appropriate names for objects: when there is no context effect, the basic level name is used.

The use of a prototype representation and storing of attributes at the basic level will allow us to model human performance in categorization tasks involving basic and non-basic level objects.

REFERENCES

(1)   Ashcraft, M. H. (1978), "Property Norms for Typical and Atypical Items from 17 Categories: A Description and Discussion," *Memory and Cognition*, vol. 6, pp. 227-232.

(2)   Berlin, B. (1978), "Ethnobiological Classification," *Cognition and Categorization* (Hillsdale, NJ: Lawrence Erlbaum Associates) pp. 9-27.

(3)   Berlin, B., Kay, P. (1969), *Basic Color Terms: Their Universality and Evolution* (Berkeley: Univ. Calif. Press).

(4)   Bobrow, D. G., Winograd, T. (1977a), "Experience with KRL-0, One Cycle of a Knowledge Representation Language," *IJCAI-77*, vol. 1, pp. 213-222.

(5)   Bobrow, D. G., Winograd, T. (1977b), "An Overview of KRL, a Knowledge Representation Language," *Cognitive Science*, vol. 1:1, pp. 3-46.

(6)   Bobrow, D. G., Winograd, T. (1979), "KRL: Another Perspective," *Cognitive Science*, vol. 3, pp. 29-42.

(7)   Brachman, R. J. (1979), "On the Epistemological Status of Semantic Networks," *Associative Networks: Representation and Use of Knowledge by Computers* (New York: Academic Press), Edited by N. V. Findler, pp. 3-50.

(8)   Brachman, R. J., Fikes R. E., Levesque, H. J. (1983), "KRYPTON: A Functional Approach to Knowledge Representation," *IEEE Computer*, vol. 16, pp. 67-73.

(9)   Brachman, R. J., Schmolze, J., (1984), "An Overview of the KL-ONE Knowledge Representation System," Fairchild Technical Report Number 655, (Palo Alto, CA: Fairchild Laboratory for Artificial Intelligence Research), September 1984.

(10)   Brachman, R. J., (1985), "'I Lied About the Trees" Or, Defaults and Definitions in Knowledge Representation," *The AI Magazine*, (Fall 1985), pp. 80-93.

(11)   Brown, C. H., Kolar, J., Torrey, B. J., Truong-Quang, T., Volkman, P. (1976), "Some General Principles of Biological and Non-Biological Folk Classification," *Am. Ethnol.*, vol. 3, pp. 73-85.

(12)   Brown, R. (1958), "How Shall a Thing Be Called?," *Psychol. Rev.*, vol. 65, pp. 14-21.

(13)   Cantor, N., Mischel, W. (1979), "Prototypes in Person Perception," *Adv. Exp. Soc. Psychol.*, vol. 12, pp. 3-52.

(14)   Carlson, G. N. (1982), "Generic Terms and Generic Sentences," *Journal of Philosophical Logic*, vol. 11, pp. 145-181.

(15)   Collins, A. M., Quillian M. R., (1970), "Does Category Size Affect Categorization Time?," *Journal of Verbal Learning and Verbal Behavior*, vol. 9, pp. 432-438.

(16)   Delgrande, J. P., (1986), "A Propositional Logic for Natural Kinds," *Proc. Canadian Conf. on AI-86*, pp. 44-48.

(17)   Fahlman, S. E., (1979), *NETL: A System for Representing and Using Real-World Knowledge* (Cambridge, MA: MIT Press).

(18)   Gaschnig, J., (1979), "Preliminary Performance Analysis of the PROSPECTOR Consultant System for Mineral Exploration," *IJCAI*, vol. 6, pp. 308-310.

(19) Hunn, E., (1976), "Toward a Perceptual Model of Folk Biological Classification," *Am. Ethnol.*, vol. 3, pp. 508-524.

(20) Jackendoff, R., (1983), *Semantics and Cognition* (Cambridge, MA: The MIT Press).

(21) Johnson-Laird, P.N., (1983), *Mental Models* (Cambridge, MA: Harvard U. Press).

(22) Labov, W., (1973), "The Boundaries of Words and Their Meanings," *New Ways of Analyzing Variations in English* (Washington: Georgetown Univ. Press) Editors: C. J. Bailey, R. Shuy, pp. 340-373.

(23) Kitts, D. B., Kitts, D. J., (1979), "Biological Species as Natural Kinds," *Philosophy of Science*, vol. 46, pp. 613-622.

(24) Lehnert, W., Wilks, Y., (1979), "A Critical Perspective on KRL," *Cognitive Science*, vol. 3, pp. 1-28.

(25) McCloskey, M. E., Glucksberg, S., (1978), "Natural Categories: Well Defined or Fuzzy Sets?" *Mem. Cognit.*, vol. 6, pp. 462-472.

(26) Melle, W. van, Shortliffe, E. H., Buchanan, B. G., (1981), "EMYCIN: A Domain-independent System that Aids in Constructing Knowlege-Based Consultation Programs," *Machine Intelligence, Infotech State of the Art Report*, vol. 9:3.

(27) Mervis, C. B., Rosch, E., (1981), "Categorization of Natural Objects," *Ann. Rev. Psychol.*, vol. 32, pp. 89-115.

(28) Quillian, M. R., (1967), "Word Concepts: A Theory and Simulation of Some Basic Semantic Capabilities," *Behavioral Science*, vol. 12, pp. 410-430.

(29) Quillian, M. R., (1968), "Semantic Memory," *Semantic Information Processing* (Cambridge, MA: MIT Press), Editor, M. Minsky.

(30) Quillian, M. R., (1969), "The Teachable Language Comprehender: A Simulation Program and Theory of Language," *Communications of the ACM*, vol. 12, pp. 459-476.

(31) Rosch, E., Mervis, C. B., Gray, W. D., Johnson, D. M., Boyes-Braem, P., (1976), "Basic Objects in Natural Categories," *Cognitive Psychology*, vol. 8, pp. 382-439.

(32) Rosch, E., Lloyd, B. B., (1978), *Cognition and Categorization* (Hillsdale, NJ: Lawrence Erlbaum Associates).

(33) Shapiro, S. C., (1978), "Path-Based and Node-Based Inference in Semantic Networks," *TINLAP-2: Theoretical Issues in Natural Language Processing* (New York: ACM), Editor, D. Waltz, pp. 219-225.

(34) Shapiro, S. C. (1979), "The SNePS Semantic Network Processing System," *Associative Networks* (New York: Academic Press), Editor, N.V. Findler, pp. 179-203.

(35) Shapiro, S. C. (1980), "Book Review: NETL: A System for Representing and Using Real-World Knowledge," *Amer. Journal of Comp. Ling.*, vol. 6, pp. 183-186.

(36) Shapiro, S. C., (1981), "COCCI: A Deductive Semantic Network Program for Solving Microbiology Unknowns," Technical Report Number 173, (Buffalo: SUNY Buffalo Dept. of Computer Science).

(37) Shapiro, S. C., (1982), "Generalized Augmented Transition Networks Grammars for Generation from Semantic Networks," *Amer. Journal of Comp. Ling.*, vol. 8, pp. 12-25.

(38) Shapiro, S. C., Rapaport, W. J., (1986), "SNePS Considered as a Fully Intensional Propositional Semantic Network," *Proc. AAAI-86*, vol. 1, pp. 278-283.

(39) Shapiro, S. C., (1986), "Symmetric Relations, Intensional Individuals, and Variable Binding," *Proc. of the IEEE*, vol. 74, no. 10, pp. 1354-1363.

(40) Smith, E. E., Medin, D. L., (1981), *Categories and Concepts* (Cambridge, MA: Harvard University Press).

(41) Sondheimer, N. K., Nebel, B. (1986), "A Logical-Form and Knowledge-Base Design for Natural Language Generation," *AAAI-86*, vol. 1, pp. 612-618.

(42) Tversky, A., Gati, I., (1978), "Studies of Similarity," *Cognition and Categorization* (Hillsdale, NJ: Lawrence Erlbaum Associates), pp. 81-95.

(43) Wegner, P., (1986), "The Object-Oriented Classification Paradigm," Technical Report No. CS-86-11, (Providence, RI: Brown University Dept. of Computer Science).

(44) Woods, W., (1978), "Semantics and Quantification in Natural Language Question Answering," *Advances in Computers* (New York: Academic Press), Editor, M. C. Youvits, vol. 17, pp. 2-87.

Cascaded Back-Propagation on Dynamic Connectionist Networks

Jordan B. Pollack
Computing Research Laboratory
New Mexico State University

ABSTRACT

The Back Propagation algorithm of Rumelhart, Hinton, and Williams (1986) is a powerful learning technique which can adjust weights in connectionist networks composed of multiple layers of perceptron-like units. This paper describes a variation of this technique which is applied to networks with constrained multiplicative connections. Instead of learning the weights to compute a single function, it learns the weights for a network whose outputs are the weights for a network which can then compute multiple functions.

The technique is elucidated by example, and then extended into the realm of sequence learning, as prelude to work on connectionist induction of grammars. Finally, a host of issues regarding this form of computation are raised.

1. Introduction

Most "Connectionist" (Feldman & Ballard, 1982) or "Parallel Distributed Processing" (Rumelhart et. al., 1986b) models use fixed-structure networks, in which the weights are set programmatically or are adjusted slowly by some iterative learning algorithm. The resultant networks are essentially "hard-wired" special-purpose computers that perform some application, like a 10-city traveling Salesman problem (Hopfield & Tank, 1985), past-tense verb conjugation (Rumelhart & McClelland, 1986), text-to-speech processing (Sejnowski & Rosenberg, 1986), or context-free parsing of bounded-length sentences (Fanty, 1985; Selman, 1985). This last application is particularly disturbing because a bounded-length context-free grammar is simply a regular grammar, recognizable by a simple finite-state machine. If connectionism entails a return to pre-Chomskian theories of linguistic capabilities, then it will be in trouble.

One of the major differences between our work in connectionist language processing (Pollack & Waltz, 1982; Waltz & Pollack, 1985) and others (Cottrell, 1985; Fanty, 1985; Selman, 1985) is our use of dynamically changing network structure, i.e., weights that are modified during a computation. Various researchers have seen the need for dynamic connections, including (Feldman, 1982) and (McClelland, 1985), but the resulting systems are very difficult to manage. In the most unconstrained case of a system using multiplicative connections, each weight in a system of n nodes can be a function of the activities of all n nodes, leading to a system with n^3 "parameters" instead of n^2.

But without some form of dynamic connections the generative capacity of connectionist models is suspect. In the past, we have used "normal" computer programs such as a chart parser (Kay, 1973) to dynamically connect our network. This paper outlines steps towards a better way.

A modified form of back-propagation is applied to networks with constrained structures of multiplicative connections and feedback to build systems capable of learning to sequentially process inputs. Using multiplicative connections allows all weights in the system to be dynamically modified for each input.

The technique, called *Cascaded Back-Propagation*, is introduced by comparison to normal back-propagation on feed-forward networks.

2. Back-Propagation

The basic form of the (Rumelhart et. al., 1986a) learning algorithm is as follows. A non-iterative feed-forward network of several layers computes input/output relationships using a continuous version of a perceptron.

Each unit i has an output bounded between 0 and 1. This bounded output is computed by "squashing" its input, x, (a linear combination of weights and other outputs) with the sigmoid function:

$$\Gamma(x) = \frac{1}{1+e^{-x}}$$

which has a derivative (after some algebra) of:

$$\Gamma'(x) = \Gamma(x)(1 - \Gamma(x))$$

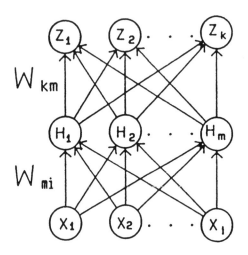

Figure 1:
    A simple feed-forward network. Each layer is completely connected to the next. The weights, therefore, are representable by rectangular arrays.

For a simple layered network as shown in Figure 1, this feed-forward computation is as follows:

$$\vec{H}_m = \Gamma(W_{mi} \cdot \vec{X}_i)$$

$$\vec{Z}_k = \Gamma(W_{km} \cdot \vec{H}_m)$$

Where $\vec{X}_i$ is the set of inputs, $\vec{H}_m$ are the outputs of the hidden units, and $\vec{Z}_k$ are the outputs. Back-propagation is given a set of cases consisting of matched pairs of input and desired output vectors. The overall error, E, can be computed as the distance between all desired output vectors $\vec{D}_k^c$ and the actual output vectors computed by the forward-pass, $\vec{Z}_k^c$:

$$E = \sum_c \sum_k (D_k^c - Z_k^c)^2$$

The backward pass works by distributing this error to all the weights in the system. For a particular input/output case, c, this computation is as follows:

$$\frac{\partial E}{\partial \vec{Z}_k} = (\vec{Z}_k - \vec{D}_k)\vec{Z}_k(1 - \vec{Z}_k)$$

$$\frac{\partial E}{\partial \vec{H}_m} = \frac{\partial E}{\partial \vec{Z}_k} \cdot W_{km}\vec{H}_m(1 - \vec{H}_m)$$

$$\frac{\partial E}{\partial W_{km}} = \frac{\partial E}{\partial \vec{Z}_k} \times \vec{H}_m$$

$$\frac{\partial E}{\partial W_{mi}} = \frac{\partial E}{\partial \vec{H}_M} \times \vec{X}_i$$

And by summing the weight errors over all cases and updating each weight by a fraction of its error, μ, plus a fraction, α, of its previous error, ΔW, the algorithm can find a set of weights by gradient descent:

$$W' = W - \mu\frac{\partial E}{\partial W} + \alpha\Delta W$$

$$\Delta W = -\mu\frac{\partial E}{\partial W}$$

A couple of details complete the algorithm. First, initial weights need to be chosen; if all weights are initially 0, there is no way for the system to allocate error, so usually weights are chosen as very small random numbers, say, between ± 0.5. Secondly, the network needs to be "grounded" by adding a "bias" to each hidden and output unit. This amounts to adding another input unit whose output is always 1; the adjustment of the biases then co-occurs with adjustment of all the other weights in the system.

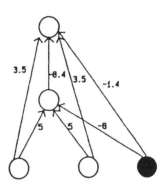

Figure 2:
A simple feedforward network for the exclusive-or problem. Instead of putting biases inside the circles, they are shown as links from a unit with an output of 1.

The simplest test of the algorithm is to learn to compute "Exclusive-or", a function which cannot be learned in a single layer of perceptrons. Figure 2 shows a standard feedforward network on which this algorithm is capable of learning XOR. Using $\mu=0.5$ and $\alpha=0.9$, back propagation can find these weights usually in a one to two hundred iterations.

3. Cascaded Networks

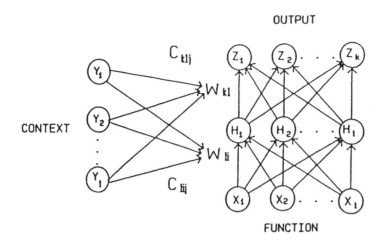

Figure 3:
> *Cascaded network. A context network with fixed weights runs first and sets variable weights on the function network.*

Instead of learning a fixed set of weights for one group of input-output relationships, one network is used to compute some input-output function (the "function network"), and another network (the "context network") is used to compute the weights for the function network (figure 3). By varying the inputs to the context network, the function network can be used to compute various functions.

The forward-pass consists of a forward pass on the context network, which sets the weights on the function network, then a forward pass on the function network:

$$W_{li} = C_{lij} \cdot \vec{Y}_j$$

$$W_{kl} = C_{klj} \cdot \vec{Y}_j$$

$$\vec{H}_l = \Gamma(W_{li} \cdot \vec{X}_i)$$

$$\vec{Z}_k = \Gamma(W_{kl} \cdot \vec{H}_l)$$

Where C_{lij} and C_{klj} represent the fixed weights of the context network, W_{li} and W_{kl} are the varying weights of the function network, $\vec{Y}_j$ are the inputs to the context network, $\vec{X}_i$ are the inputs to the function network, $\vec{H}_l$ are the outputs of the hidden layer, and $\vec{Z}_k$ are the outputs of the function network.

The backward pass consists of computing the errors for the variable weights of the function network, and then using them to compute the errors for the fixed weights:

$$\frac{\partial E}{\partial \vec{Z}_k} = (\vec{Z}_k - \vec{D}_k)\vec{Z}_k(1 - \vec{Z}_k)$$

$$\frac{\partial E}{\partial \vec{H}_l} = \frac{\partial E}{\partial \vec{Z}_k} \cdot W_{kl}\vec{H}_l(1 - \vec{H}_l)$$

$$\frac{\partial E}{\partial W_{kl}} = \frac{\partial E}{\partial \vec{Z}_k} \times \vec{H}_l$$

$$\frac{\partial E}{\partial W_{li}} = \frac{\partial E}{\partial \vec{H}_l} \times \vec{X}_i$$

$$\frac{\partial E}{\partial C_{lij}} = \frac{\partial E}{\partial W_{li}} \times \vec{Y}_j$$

$$\frac{\partial E}{\partial C_{klj}} = \frac{\partial E}{\partial W_{kl}} \times \vec{Y}_j$$

3.1. Exclusive-or Problem

This approach can build networks which runs multiple functions over the same set of units. To compute the exclusive-or function, for example, this amounts to learning the two functions:

$$f(y) = \begin{cases} y & \text{if } x = 0 \\ \neg y & \text{if } x = 1 \end{cases}$$

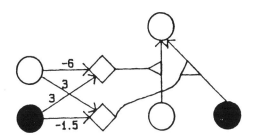

Figure 4:
Cascaded network for the XOR problem. The function network acts as either an inverter or non-inverting buffer depending on the context bit.

Figure 4 shows the cascaded network for the XOR problem. This network needs to learn only 4 weights instead of 7, and, with the learning parameters $\mu = 0.5$ and $\alpha = 0.9$, cascaded back propagation only needs about 30 cycles to learn exclusive-or.

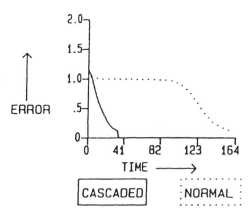

Figure 5:

Typical learning curves for normal versus cascaded XOR network. The horizontal axis represents time as iterations of error propagation, and the vertical represents the global error, E, over all 4 test cases. The algorithms halt when all outputs are within .2 of their desired values.

Figure 5 shows the learning curves for typical runs of back-propagation and cascaded back-propagation for the exclusive-or networks. The number of iterations is represented along the horizontal axis and the global error, E, is represented on the vertical.

3.2. The 4-1 Multiplexor

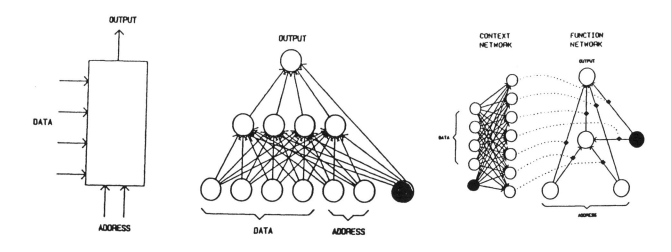

Figure 6:

Three versions of a 4-1 multiplexor. First is the standard block diagram used for logic design; next is a 6-4-1 layered network; and finally, a cascaded network.

Another example is a 4-to-1 multiplexor. A classic logical functional unit used in computer design, it can be thought of as a programmable 2-1 logic function. Figure 6 shows three views of a multiplexor. One of the 4 "data" lines is selected by the values on the 2 address lines. In a normal feed-forward network, there is no distinction between these six inputs, and 4 hidden units are needed to learn all 64 input/output cases. The cascaded network, composed of a single-layer 4-7 network connected to the standard XOR network, essentially learns 5 interacting sets of 7 weights, which produce 16 different

396

sets, one for each logic function.

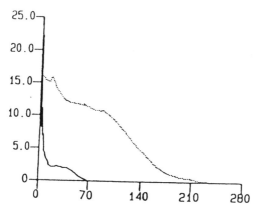

Figure 7:
Typical learning curves for runs of normal versus cascaded back-propagation on the multiplexor problem.

In general, the cascaded solution for the multiplexor problem converges much quicker than the feed-forward solution. Figure 7 shows typical behavior for both solutions with $\mu=0.5$ and $\alpha=0.9$.

4. Sequential Cascaded Networks

When the outputs of the function network are used as inputs to context network, a system can be learned which sequentially processes inputs by dynamically changing the weights in the function network after each input. In parsing terms, it could be said that each word is processed in the context of all the preceding words. And although the number of possible intermediate states are, of course, finite,[1] this system can learn grammars which are bounded in depth, but unbounded in length. The intermediate states must encode various up/down counters. Figure 8 shows a block diagram of a simple sequential cascaded network. Given an initial context, $\vec{Y}_j(0)$, and a sequence of inputs, $\vec{X}_i(t), t=1...n$, the network can compute a sequence of function output/context input vectors, $\vec{Y}_j(t), t=1...n$ by dynamically changing the set of weights, $W_{kl}(t)$ and $W_{jl}(t)$:

---

1. Unless it is assume that the outputs are true analog values or rational numbers.

397

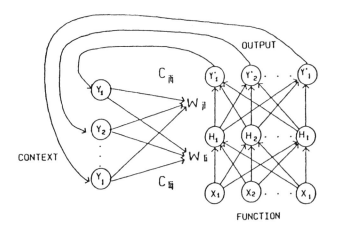

Figure 8:

The outputs of the function network are used as the next inputs to the context network, yielding a system whose function varies over time.

$$W_{li}(t) = C_{lij} \cdot \vec{Y}_j(t-1)$$

$$W_{jl}(t) = C_{jlj} \cdot Y_j^{\flat}(t-1)$$

$$\vec{H}_l(t) = \Gamma(W_{li}(t) \cdot \vec{X}_i(t))$$

$$\vec{Y}_j(t) = \Gamma(W_{jl}(t) \cdot \vec{H}_l(t))$$

The error correction phase can be applied to just the final input:

$$\frac{\partial E}{\partial \vec{Y}_j(n)} = (\vec{Y}_j(n) - \vec{D}_j)\,\vec{Y}_j(n)(1 - \vec{Y}_j(n))$$

$$\frac{\partial E}{\partial \vec{H}_l(n)} = \frac{\partial E}{\partial \vec{Y}_j(n)} \cdot W_{kl}(n)\vec{H}_l(n)(1 - \vec{H}_l(n))$$

$$\frac{\partial E}{\partial W_{kl}(n)} = \frac{\partial E}{\partial \vec{Y}_j(n)} \times \vec{H}_l(n)$$

$$\frac{\partial E}{\partial W_{li}(n)} = \frac{\partial E}{\partial \vec{H}_l(n)} \times \vec{X}_i(n)$$

$$\frac{\partial E}{\partial C_{lij}} = \frac{\partial E}{\partial W_{li}(n)} \times \vec{Y}_j(n-1)$$

$$\frac{\partial E}{\partial C_{klj}} = \frac{\partial E}{\partial W_{kl}(n)} \times \vec{Y}_j(n-1)$$

where $\vec{D}_j$ is the desired output for a particular sequence.

4.1. Learning Parity

When Exclusive-or is generalized to more than 2 inputs, it becomes the parity problem, to determine whether a boolean string has an odd or even number of 1's in it. This problem was discussed at length, both by (Minsky & Papert, 1969), as a hard problem for perceptrons and by (Rumelhart et. al., 1986a) as a test case for back-propagation.

Figure 9:
A simple 2-state machine is shown on the left, and a sequential cascaded network is shown on the right.

A problem for normal back-propagation is that the parity problem of size K requires K hidden units to work, so a system which learned to determine parity of 5 bits would not work for 6. This problem can be overcome by "going sequential", using the cascaded exclusive-or network with feedback between the output of the function network and the input to the context network. This network, and the corresponding small finite-state machine are shown in figure 9.

One problem with the cascaded network approach is that if the system is trained to within .2 of the solution for each training case, the weights "fuzz out" for longer tests than the ones given. There are several solutions to this. The simplest one is to put a truncating filter between the function output and context input which converts outputs above 0.8 to 1 and below 0.2 to 0. Other possible solutions include more complicated filters such as an auto-associative memory or other relaxation system which corrects fuzzy states..

4.2. Parenthesis Balancing

Unfortunately, parity is very unnatural and extremely finite state "language". A real solution to a temporal credit assignment problem with application to language processing is not served by learning such finite systems as parity or even 6-letter sequence completion as used by (Rumelhart et. al., 1986a) to demonstrate recurrent networks.

A connectionist network at least should be able to learn a context-free language from example in order to claim any service to language processing. Accordingly, experiments have been performed in learning the second simplest context-free language known to man: Parenthesis balancing.[2] We have successfully used a sequential cascaded network for parenthesis balancing consisting of of a layered 1-3-2 function network and a 2-14 context net. The input is either 1 or 0 for left and right parentheses and one output signifies grammaticality of the prefix and the other works as a stack, by shifting outputs from 1 to .5 to .25 as more left parentheses are input.

4.3. Other grammars

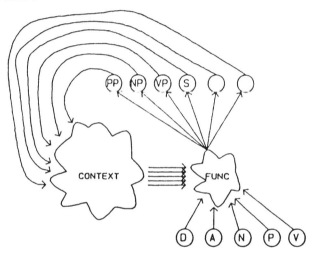

Figure 10:
  A sequential cascaded network for parsing. The current state is tied back into the context for the next input, and the unlabeled units would develop necessary features.

Using a sequential cascaded network engenders a time-distributed representation of a parse-tree. For example if a network were used as shown in figure 10, with lexical categories for input and phrase markers for output and context, a system could be developed that, for a simple declarative sentence like "The fat man ate the spaghetti with sauce", could have time-varying outputs which implicitly code its parse tree.

The following table shows the outputs of the phrase-marker units over time. Each contiguous group of "on" states could be interpreted as a single node of a tree, with interval-inclusion determining dominance.

---

2. The simplest context free grammar is described by the regular expression $a^n b^n$.

400

S	*	*	*	*	*	*	*	*
VP				*	*	*	*	*
PP							*	*
NP	*	*	*		*	*		*

There are several problems with this representation, the main one being that self-recursive categories are not recoverable. One possibility which is currently being examined is the combination of this learning technique with the representational assumption that outputs can have arbitrary fractional resolution used as stack. Experimentation with representing and learning larger grammars of this type has just begun; clearly more work need be done.[3]

5. Discussion

Cascaded networks do not solve everything, unfortunately. They are basically a very constrained type of network with multiplicative links. This algorithm, being a variation of back-propagation, does not solve its inherent problems. For example, it can still get stuck in local minima, and the exact topology needed to solve a particular problem must be defined beforehand.

But the combination of faster convergence and the computational power engendered by multiplicative connections make cascaded back-propagation a useful technique for connectionist modeling and worthy of further study. Some issues for further discussion are presented below.

5.1. Why is it Faster?

For both the exclusive-or and multiplexor problems given above, as well as various other problems we have experimented with, cascaded back-propagation converges on solutions significantly faster than normal back-propagation. We think this is due, essentially, to the well-known algorithmic technique of "divide-and-conquer". By breaking the solution to exclusive-or into two simpler problems (i.e. an inverting and non-inverting buffer) or the multiplexor into 16 smaller problems (i.e. each 2-1 boolean function), we reduce the amount of work involved tremendously.

Consider running normal back-propagation multiple times, once for each simple problem on the function net, saving the discovered weights, and then once for mapping the context inputs to to these weights. If both nets are capable of learning their functions, then this scheme will work, and the number of iterations needed will be the sum of all the smaller cases.

But when we learn all these subproblems at the same time, the number of iterations will be related to the hardest subproblem to learn. Thus for the exclusive-or problem, the number of iterations we need is related to how hard it is to learn to invert (i.e. even a *perceptron* can do this), and for the multiplexor problem the number of iterations needed will be related to how hard it is to learn normal exclusive-or or equivalence.

Furthermore, a particular solution to a subproblem found by back-propagation is a discrete point in "weight space", somewhere along the edge of a region of good solutions. Running all subproblems first makes the context mapping problem more difficult by adding this unnecessary edge constraint; merging the subproblems and mapping problem removes this constraint and

---

3. One learning trial may be considered a success, however: We inadvertently used a training set with a simple principle of grammaticality -- the system discovered that all grammatical sentences were of length $0 \bmod 3$!

allows each subproblem solution to be anywhere in its region.

5.2. Relation to Sigma-Pi Units

Williams (1986) has classified various activation functions for connectionist models. The ultimate function for combining inputs to a unit are called "Sigma-Pi" functions, which linearly combined multiplied subsets of inputs. So for n inputs, $x_1, \ldots, x_n$, a unit with j weights may provide as output:

$$\sum_{S_j \in P} \prod_{i \in S_j} x_i$$

Where P is the set of all subsets of $\{1, ..., n\}$. This is the ultimate in combining functions because when $j = 2^n$ a single unit can implement a general polynomial. The down side is that having 2^n weights associated with a single unit is the combinatorial *brut existant.*

As far as classification, however, almost any multiplicative connection system is a special case of Sigma-Pi. For example, the gating activation function described by (Hinton, 1981) uses $j = \dfrac{n}{2}$ weights, where n inputs are separated into $\dfrac{n}{2}$ pairs which are multiplied and combined. A cascaded network can be also seen as a special case, where the n inputs are broken into 2 sets whose elements are multipled in pairs with $j = \dfrac{n^2}{4}$ weights.

5.3. Single-layered Context Networks

In the examples given in the paper, single-layered context networks were used. This construction will work only if the good regions in weight space for each subproblem can be linearly composed with respect to the context inputs. That a single-layer context works with the exclusive-or problem is obvious; that it worked for the multiplexor is surprising. For harder problems, it may turn out that more hidden units are needed in the function network to provide the flexibility for this kind of context network. On the other hand, if back-propagation works, there is no constraint that the context network has to be single-layered.

6. Conclusion: A Universal Neural Network?

Consider the notion of a Universal Turing Machine. A very simple construction which, when presented with a description of any other Turing Machine and its initial state, runs a simulation of that TM to completion. This is similar to a virtual machine emulator or programming language interpreter running on a normal computer. One difference is that because of the random-access property of a computer versus the serial tape access of a UTM, the normal computer runs simulations much faster. For each simulated operation of the TM, the UTM may have to step from one end of its tape to the other. For a program and tape of size n this amounts to about an n^2, or *polynomial* simulation time. A programmed interpreter, on the other hand, has to look up an operation in a table and call a simple routine which updates the state of the interpreted system. Assuming each simulated operation takes about k machine instructions, then the simulation takes kn, or *linear* time. This efficiency advantage is why no modern computers are built like Turing machines.

Consider, finally, an extended cascaded network where the context network is presented a description of a machine and produces the set of weights for the function network. The function network can then run at full "neural speed". If the context network takes a constant time, k, to do its computation, then this simulation runs in $k+n$, or *constant* time.

The point of all this is to solve a conundrum for connectionists: When attempting to model a high-level cognitive domain one quickly realizes the folly of equating a neuron with an element of that domain. Neurons are arrayed in a fixed network, and only die as time goes by. If the memory of your grandmother were localized to a single neuron, and that neuron failed, you would forget her.

One backup position (which this author has resorted to occasionally) is something like the following: The units in **my** system are not really neurons, but a elements of a higher level system which are somehow simulated by neurons. The problem for this position has been that simulation takes time, and given the finite number of cycles available for "real-time" cognition, i.e. 10-100 cycles, there isn't any time for the simulation to take place.

If a universal neural network could really run a simulation of a neural network in constant time, then the backup position becomes viable -- the units in the higher-level system are temporarily run on neurons at neural speed. Cascaded networks are a first step in this direction.

7. References

Cottrell, G. W. (1985). Connectionist Parsing. In *Proceedings of the Seventh Annual Conference of the Cognitive Science Society*. Irvine, CA.

Fanty, M. (1985). Context-free parsing in Connectionist Networks. TR174, Rochester, N.Y.: University of Rochester, Computer Science Department.

Feldman, J. A. (1982). Dynamic Connections in Neural Networks. *Biological Cybernetics, 46*, 27-39.

Feldman, J. A. & Ballard, D. H. (1982). Connectionist models and their properties. *Cognitive Science, 6*, 205-254.

Hinton, G. E. (1981). A Parallel Computation that Assigns Canonical Object-Based Frames of Reference. In *Proceedings of the Seventh International Joint Conference on Artificial Intelligence*. Vancouver, B.C., 683-685.

Hopfield, J. J. & Tank, D. W. (1985). 'Neural' computation of decisions in optimization problems. *Biological Cybernetics, 52*.

Kay, M. (1973). The MIND System. In Rustin, (Ed.), *Natural Language Processing*. New York: Algorithmics Press.

McClelland, J. L. (1985). Putting Knowledge in its Place. *Cognitive Science, 9*, 113-146.

Minsky, M. & Papert, S. (1969). *Perceptrons*. Cambridge, MA: MIT Press.

Pollack, J. B. & Waltz, D. L. (1982). Natural Language Processing Using Spreading Activation and Lateral Inhibition. In *Proceedings of the Fourth Annual Cognitive Science Conference*. Ann Arbor, MI, 50-53.

Rumelhart, D. E. & McClelland, J. L. (1986). On Learning the Past Tenses of English Verbs. In J. L. McClelland, D. E. Rumelhart & the PDP research Group, (Eds.), *Parallel Distributed Processing: Experiments in the Microstructure of Cognition*, Vol. 2. Cambridge: MIT Press.

Rumelhart, D. E., Hinton, G. & Williams, R. (1986). Learning Internal Representations through Error Propagation. In D. E. Rumelhart, J. L. McClelland & the PDP research Group, (Eds.), *Parallel Distributed Processing: Experiments in the Microstructure of Cognition*, Vol. 1. Cambridge: MIT Press.

Rumelhart, D. E., McClelland, J. L. & Group, the PDP research (1986). In *Parallel Distributed Processing: Experiments in the Microstructure of Cognition*, Vol. 1. Cambridge: MIT Press.

Sejnowski, T. J. & Rosenberg, C. R. (1986). NETtalk: A parallel network that learns to read aloud. JHU/EECS-86/01: The Johns Hopkins University, Electrical Engineering and Computer Science Department.

Selman, B. (1985). Rule-Based Processing in a Connectionist System for Natural Language Understanding. CSRI-168, Toronto, Canada: University of Toronto, Computer Systems Research Institute.

Waltz, D. L. & Pollack, J. B. (1985). Massively Parallel Parsing: A strongly interactive model of Natural Language Interpretation. *Cognitive Science, 9*, 51-74.

IMPLEMENTING STAGES OF MOTION ANALYSIS IN NEURAL NETWORKS

Margaret E. Sereno
Psychology Department
Brown University

Abstract

A neural model is proposed for human motion perception. The goal of the model is to calculate the two-dimensional velocity of elements in an image. Unlike most earlier approaches, the present model is structured in accord with known neurophysiological data. Three distinct stages are proposed. At the first level, units are sensitive to the components of motion that are perpendicular to the orientation of a moving contour. The second level integrates these initial motion measurements to obtain translational motion. The third level uses translational motion measurements to compute general three-dimensional motion such as rotation and expansion. The model shows a high level of performance in solving the measurement of two-dimensional translational motion from local motion information. Most importantly, the present model uses nervous system structure as a natural way to formulate constraints. The psychological implications of staged motion processing are discussed.

Visual motion perception serves many important functions, including the segregation of objects, the estimation of object motion, the control of eye movements, and the estimation of the three-dimensional structure of objects & the environment. The operations responsible for the perception of motion, however, are not well known.

As three-dimensional surfaces move in space, they project light onto the eye, forming a two-dimensional image of the world that changes with time. The visual system must reconstruct a three-dimensional world from this two-dimensional image. This reconstruction can be accomplished by using information about the organization of movement in the changing image. However, the motion of elements in the two-dimensional image (i.e., their speed and direction) is not an inherent property of the image but must be inferred from the varying intensities of the image. Thus, motion analysis is often considered a two-stage process (Hildreth, 1983).

The goal of the first stage is the measurement of two-dimensional motion of elements in an image (i.e., extracting the velocity--speed and direction--of moving elements). To accomplish this goal there must be initial motion detection and measurement by motion sensors, an integration of the initial motion measurements to compute an instantaneous two-dimensional velocity field (the so-called "aperture" problem), and the detection of motion discontinuities. The second stage consists of an interpretation of the three-dimensional structure of surfaces from two-dimensional motion.

I present a neural network model of part of the first stage of motion analysis (i.e., the integration of initial, local measurements to compute a two-dimensional velocity field). The model extracts the true two-dimensional motion of an entire pattern from ambiguous local motion information available at the pattern's component contours. In other words, it solves the "aperture problem" for rigid two-dimensional motion in the plane. Local motion detectors provide ambiguous information because they only measure the component of motion perpendicular to the orientation of a moving contour. A family of possible motions exists that can give rise to the locally detected motion. The aperture problem, then, reduces to the assignment of a unique velocity to an object given only local motion measurements (See Figure 1).

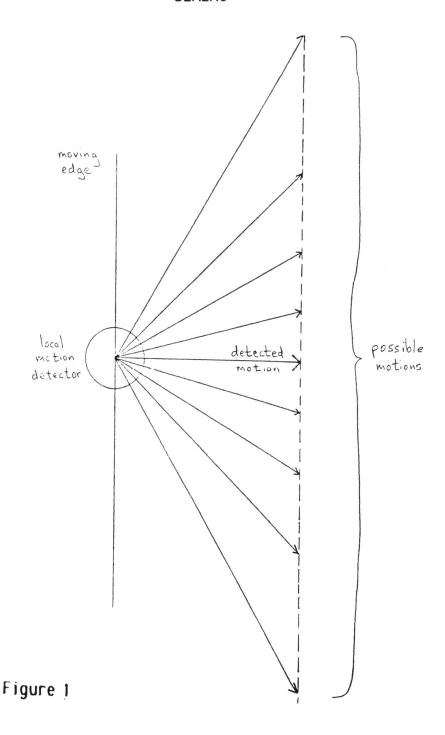

Figure 1

Hildreth (1983) has proposed a computational model for the measurement of two-dimensional motion. In her model, local measurements are obtained from the image and are then combined to compute a unique two-dimensional velocity field by applying constraints to limit the solution. For example, the "smoothness constraint" is based on the observation that objects usually have smooth surfaces. This constraint is implemented by finding the velocity field of least variation. The model works well on simple figures for planar and general three-dimensional motion (e.g., rotation and expansion).

The basic motivation for formulating the present model is to build more structure into the model to enable it to perform transformations on the input data leading to a *unique* solution. This is done by closely adhering to both neurophysiological and psychological data on motion analysis. The model is structured in accord with neurophysiological data because I assume that the nature of the hardware profoundly affects how the problem is solved. The ultimate goal is to integrate the neurophysiological and psychological information to form a more coherent theory of motion perception.

Two ideas about the basic operations involved in motion analysis emerge from the psychological, psychophysical, neurophysiological, and mathematical work on motion. One is that there are primitives of optic flow that are analyzed by specialized neural mechanisms. Work on the mathematics of optic flows demonstrates that any flow field can be decomposed into a linear vector combination of several basic types: translation, rotation, shear, and dilation (Koenderink & Van Doorn, 1976; Longuet-Higgins & Prazdny, 1980). Psychophysical data from adaptation studies have provided evidence for translation, rotation, and expansion sensitive mechanisms (Regan & Beverly, 1978; Regan, 1986). Also, neurophysiological studies in macaque visual cortex (area MST) demonstrate that neurons are sensitive to linear, rotational, and dilational motion (Saito et al., 1986).

The second idea is that the integration of local one-dimensional motion measurements into a full two-dimensional velocity field occurs in several stages. Psychophysical studies demonstrate that one-dimensional motion measurements are combined to compute two-dimensional translational motion (Adelson & Movshon, 1982; Nakayama & Silverman, 1983). Neurophysiological data suggests that the computation of all types of motion in the nervous system does not occur in a single step. A pervasive aspect of the cortical architecture of sensory systems is the presence of multiple topographic representations or maps of sensory surfaces projecting to each other. Several areas involved in motion analysis in the macaque visual cortex include Areas V1, MT, and MST. Area V1 neurons are involved in the analysis of component motion while some MT neuons respond to linear pattern motion (Movshon, Adelson, Gizzi, & Newsome, 1985). As previously noted, a recent study of cells in a visual area (MST) upstream to area MT has discovered neurons that respond selectively to translating, expanding, contracting, and rotating patterns (Saito et al., 1986).

As a first step, a model is constructed to solve the aperture problem for rigid motion in the plane (i.e., translation). This is accomplished, first, by using some formal observations on how to uniquely limit the solution and, second, by structuring the model in accord with neurophysiological organization. It is then proposed that this two-dimensional translation information is combined to compute other general motions.

Adelson and Movshon (1982) discuss a solution to unambiguously determine the two-dimensional motion of a pattern given the motion of its local components (See Figure 2). The dashed lines indicate the family of global pattern velocities which are consistent with the locally measured component velocity vector. They note that when at least two nonparallel moving contours belonging to the same pattern are compared, only one vector is common to both one-dimensional families, and it describes the motion of the entire pattern. This vector is the point in velocity space at which the two dashed lines intersect.

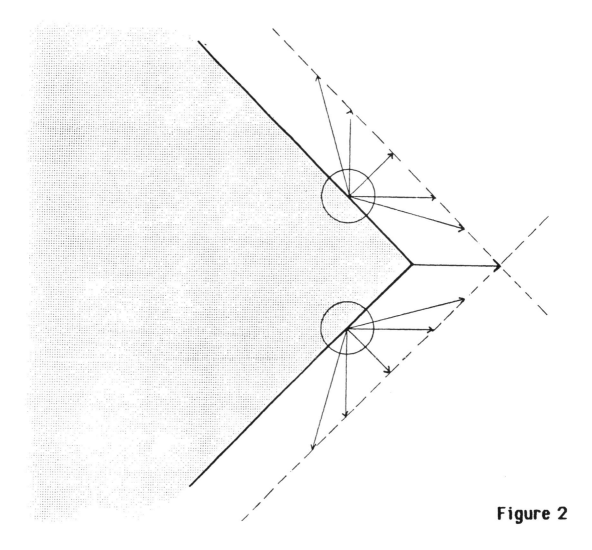

Figure 2

This constraint was implemented in a model (Sereno, 1986) that was structured in accord with the following neurophysiological facts. Some neurons in striate cortex (Area V1) are selective for orientation, speed and direction of edges. However, they only respond to the perpendicular component of motion. Area MT, an area involved in motion analysis, receives a direct topographic projection from V1, is selective for the direction and speed of motion of a stimulus while having little selectivity for spatial structure, and possesses larger receptive fields, indicating spatial summation of its inputs. Moreover, 25% of MT neurons exhibit "pattern" direction selectivity, that is, they are selective for the motion of the pattern as a whole (Movshon, Adelson, Gizzi, & Newsome, 1985).

A "Boltzmann Machine" (Ackley, Hinton, & Sejnowski, 1985) was constructed with an input layer of units representing V1 and output layer of units representing area MT. Each unit is selective for a specific speed and direction of motion (See Figure 3). Specifically, layer V1 contains 32 units (8 directions, 2 speeds and 2 locations) while layer MT contains 24 units (8 directions, 3 speeds and 1 location). V1 units respond only to the component of motion perpendicular to the orientation they are sensitive to; MT units respond to two-dimensional motion.

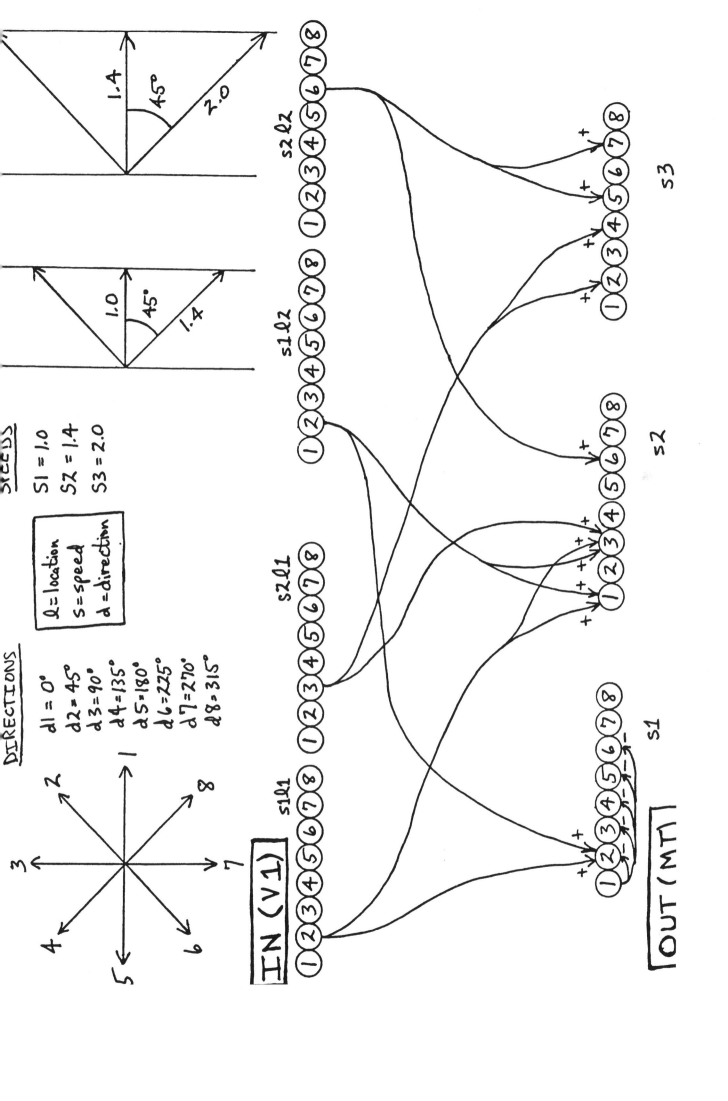

The formal solution described above was hardwired into the system by having each V1 unit project to the family of pattern velocities in the output layer that could describe the true motion underlying its response. With this predefined connectivity, when a number of differently oriented line segments belonging to the same moving pattern are input to the system, a gradient descent algorithm results in the system changing to a configuration in which the activity of the output unit describing the pattern motion is selectively enhanced. Figure 4 presents an example of a pattern of line segments moving across the two sets of input units (See Figure 4). After 20,000 iterations, the output unit describing the pattern velocity is driven to an "on" state 100% of the time. In addition, a motion illusion (the Split Herringbone Illusion) is presented to the model. The alternating columns of lines actually move in opposite directions while the perceived motion is perpendicular to these directions, consistent with an "intersection of constraints" solution. After 20,000 iterations, the perceived direction is selectively enhanced.

These results demonstrate that the intersection of constraints described above can be realized in a two-layered neural network. The specific implementation makes a testable neural prediction about how the first layer of neurons (area V1) projects to the second layer of neurons (MT) to transform the neural response from selectivity for one-dimensional motion to selectivity for two-dimensional motion. The projection, consequently, produces MT units with a wider range and higher cut-off of preferred speeds than V1 units, a finding consistent with existing neurophysiological data (Van Essen, 1985). Another important aspect of the model is that it predicts that two-dimensional motion measurements result from the integration of one-dimensional motion measurements from nearby spatial locations.

To summarize, a positive aspect of the model is that it is neurally-based with the result that it produces one solution to a given input. No post hoc assumptions or constraints are needed to limit the solution. However, a major limitation of the model is limited to the discrete values of speed and direction of movement to which the input units are sensitive. A neurally plausible solution to this problem of representing intermediate values of speed and direction is to let the information be carried by an ensemble code. This requires that individual units have continuous valued activities. For example, a speed or direction that lies exactly in between the values of 2 units can be represented by activity in each unit that is 1/2 the maximum activity. Such a representation, however, cannot be implemented on a Boltzmann Machine because the units cannot have continuous valued activity. However, it is not difficult to show that the intersection of constraints illustrated in Figure 2 amounts to a solution of a set of linear equations and hence, it can be solved with linear methods that permit continuous-valued output. Therefore, a second model was constructed using a simple linear associator with error correction, such as that used by Anderson (1983) (See Figure 5).

In the simple linear associative model, the same neurophysiological assumptions hold, except that learning can occur. This means that the connection weights are modifiable. The matrix, A, of modifiable synaptic weights describes the projection of the input layer of neurons to the output layer. The vectors f and g represent the activities across the input and output layers, respectively. Learning occurs when pairs of these vectors, one pair per pattern, are associated to form the connectivity matrix. To do this, two assumptions are made: The first assumption is that a neuron's activity results from the linear summation of its input. That is, the activity of each neuron, in the second layer, is determined by the activity of its inputs weighted by their connection strengths. Second, the matrix of connection strengths is constructed according to the generalized Hebbian rule for connectivity modification which asserts that synaptic strength is porportional to pre- and postsynaptic cell activity. This learning rule is used with error correction in which the difference between the true association,

Diamond

% "on" after 20,000 iterations	Speed and Direction
35%	S1 D2
43%	S1 D8
100%	S2 D1
38%	S2 D3
42%	S2 D7
0%	other units

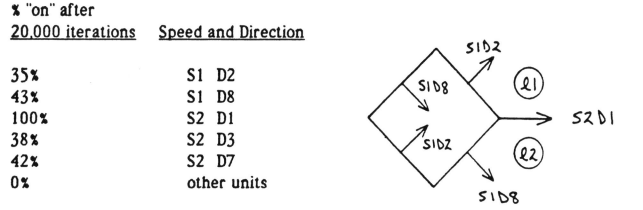

Split Herringbone

% "on" after 20,000 iterations	Speed and Direction
20%	S1 D2
20%	S1 D4
20%	S2 D1
100%	S2 D3
21%	S2 D5
0%	other units

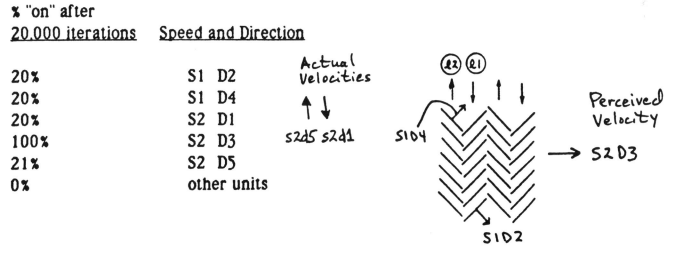

Figure 4

Two assumptions of the linear associative model:

f_i = vector of input layer neuron activities representing component velocities for the ith pattern

g_i = vector of output layer neuron activities representing pattern velocities for the ith pattern

g_i' = vector of output layer neuron activities that results when a pattern, f_i, is input to the system

1) Neurons take a linear summation of their input:

$$g_i' = A \quad f_i$$

2) Learning Rule: Synaptic strength is proportional to the product of pre-synaptic and post-synaptic activities:

$$\Delta A = \sum_{i=1}^{n} g_i \quad f_i^{T}$$

Error Correction Procedure:

$$\Delta A = k \quad (g_i - g_i') \quad f_i^{T}$$

ΔA is learned and added to the developing A connectivity matrix:

$$A_{t+1} = A_t + \Delta A$$

Figure 5

g, and the actual association, g', is learned and added to the developing A connectivity matrix.

To teach the model, different patterns moving at different velocities are input to the system. For each pattern, a vector, f, describing component velocities and a vector, g, describing pattern velocities are associated using error correction.

After learning is completed, the matrix is tested. The output of each stored input is computed. That is, each f is input to the system to get an output g'. The output g' is then compared to the true association g by taking the cosine between them. If the vectors are the same, the cosine will equal 1. The system is then tested with nonassociated vector pairs to see how well the system generalizes to new stimuli.

One simulation will be described to illustrate the performance of the system. For this simulation, direction sensitive units are placed every 15 degrees and have bandwidths of 90 degrees (peak response tapers off to 0, 45 degrees on either side of the peak direction). There are 17 peak directions (spanning 180 degrees) and 8 peak speeds (spanning 30 degrees/sec). Since each unit is sensitive to both a speed and a direction, a total of 136 units (136 speed/direction combinations) are available at each location. In this simulation, the system learns on 50 patterns and is then tested on these 50 patterns and on 50 new patterns. The patterns are composed of 1 to 3 line segments positioned at different angles relative to each other. Some example patterns are shown in Figure 6 (See Figure 6). Each pattern is moved at a different velocity.

Linear Associative Model

Example Patterns:

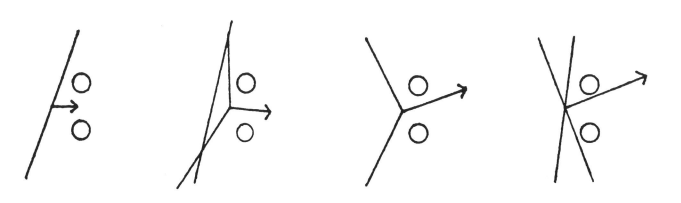

Figure 6

After 15 associations per vector pair, the system reaches stable performance and is tested. The mean cosine between the true association that the system learns, g, and the actual association that the system produces, g', is equal to .98. This represents very good performance. Moreover, the mean cosine for the new, nonassociated vectors is equal to .97. This also represents very good performance and demonstrates that the system is able to generalize quite well to stimuli it has never seen before.

To obtain a finer performance measure, a calculation was made to determine one value of speed and one value of direction for each pattern. A weighted average was taken in which each unit's preferred speed or direction was weighted by its activation level (See Figure 7). The mean difference between the weighted average for the real direction (g) and the reconstructed direction (g') for old patterns was 3.0 degrees while the mean difference for new patterns was 4.2 degrees. The mean difference between weighted averages for real and reconstructed speeds for old patterns was 1.1 degrees per second compared to 1.6 degrees per second for new patterns.

Weighted Average Calculation

$$\text{pattern speed} = \Sigma_i \, (r_i * s_i) \, / \, \Sigma_i \, r_i$$

$$\text{pattern direction} = \Sigma_i \, (r_i * d_i) \, / \, \Sigma_i \, r_i$$

where i = unit number

r = activation level of unit

s = speed to which unit is most sensitive

d = direction to which unit is most sensitive

Figure 7

In sum, the model shows excellent performance for extracting two-dimensional translational motion from one-dimensional motion information.

The present model is then extended to handle the two-dimensional projected velocity of objects moving in depth (e.g., in rotating and expanding objects). Again, the model is constructed taking into account the relevant neurophysiological data. Saito et al. (1986), for example, describe three classes of directionally selective cells with large receptive fields (about 35 degrees compared to a mean of about 6 degrees for MT cells) in area MST, an area which receives a direct projection from MT. One class of cells is sensitive to translation in the plane, a second class (size-change cells) is selective for expanding or contracting patterns, and a final class (rotation cells) is selective for rotating patterns (clockwise or counterclockwise) in the frontoparallel plane, or rotating patterns in depth. A common feature of these neurons is that they respond to appropriate patterns anywhere in their large receptive fields at the expense of being able to precisely signal information about location. Saito et al. (1986) argue that these cells are sensitive to "whole events" of visual motion because they integrate elemental motion signals from MT cells.

These data suggest that the visual system utilizes several distinct stages for motion analysis. In an analogous fashion, the present model takes the output of a second layer that responds to two-dimentional linear motion and feeds it into a third layer that responds to motion of rotation, dilation, or contraction.

The proposed model will be tested using complex motion (the combination of simpler motions). Moreover, the model will be introduced to moving patterns which give rise to illusory perception such as the rotating spiral illusion. In this illusion, a rotating spiral appears to expand or contract. The three layers of the present model result in the extraction of elemental motion which can then be combined in an ensemble code to compute the perceived two-dimensional motion.

The obvious advantage of such a model is that it makes use of the structure of the nervous system as a natural way to constrain the model. Consequently, it can provide insight into the sequential processes involved in motion analysis.

References

Adelson, E.H. & Movshon, J.A. (1982). Phenomenal coherence of moving visual patterns. *Nature*, 300, 523-525.

Ackley, D.H., Hinton, G.E., & Sejnowski, T.J. (1985). A learning algorithm for Boltzmann machines. *Cognitive Science*, 9, 147-169.

Anderson, J.A. (1983). Cognitive and psychological computation with neural models. *IEEE Transactions on Systems, Man, and Cybernetics*, SMC-13, 799-815.

Hildreth, E.C. (1983). *The measurement of visual motion.* Cambridge: MIT Press.

Koenderink, J.J. & Van Doorn, A.J. (1976). Local strructure of movement parallax of the plane. *Journal of the Optical Society of America*, 66, 717-723.

Longuet-Higgens, H.C. & Prazdny, K. (1980). The interpretation of a moving retinal image. *Proceedings of the Royal Society of London B*, 208, 385-397.

Movshon, J.A., Adelson, E. H., Gizzi, M.S., & Newsome, W.T. (1985). The analysis of moving visual patterns. In C. Chagas, R. Gattas, and C.G. Gross (Eds.), *Pattern Recognition Mechaisms.* Rome: Vatican Press.

Nakayama, K. & Silverman, G.H. (1983). Perception of moving sinusoidal lines. *Journal of the Optical Society of America*, 72.

Regan, D. (1986). Visual processing of four kinds of relative motion. *Vision Research*, 26, 127-145.

Regan, D. & Beverly, K.I. (1978). Looming detectors in the human visual pathway. *Vision Research*, 18, 415-421.

Saito, H., Yukie, M., Tanaka, K., Hikosaka, K. Fukada, Y., & Iwai, E. (1986). Integration of direction signals of image motion in the superior temporal sulcus of the macaque monkey, *Journal of Neuroscience*, 6, 145-157.

Sereno, M.E. (1986). A neural model for the measurement of visual motion. *Journal of the Optical Scociety of America*, 3, p. 72.

Van Essen, D.C. (1985). Functional organization of primate visual cortex. In A. Peters & E.G. Jones (Eds.), *Cerebral Cortex: Vol. 3: Visual cortex.* New York: Plenum Publishing Corporation.

Consistency and Variation in Spatial Reference [1]

Sarah A. Douglas
David G. Novick
Computer and Information Science Department
University of Oregon

Russell S. Tomlin
Linguistics Department
University of Oregon

Abstract

Modeling the meaning and use of linguistic expressions describing spatial relationships holding between a target object and a landmark object requires an understanding of both the consistency and variation in human performance in this area. Previous research [Herskovits 1985] attempts to account for some of this variation in terms of the angular deviation holding among objects in the visual display. This approach is shown to fail to account for the full range of human variation in performance, and a specific alternative algorithm is offered which is grounded in task variability and the notions of corridor and centroid. The significance to this algorithm of task variation, of the separation of semantic from pragmatic issues, and of the role of function and structure is discussed.

Keywords: spatial relations, natural language, reference.

1 Introduction

There is a growing body of literature in cognitive science which deals with the cognition of spatial relations. The aim of this research is to discover the principles which underlie the appropriate use and comprehension of expressions concerning spatial relationships among objects. However, current approaches to the problem of the language of spatial relations typically face difficulties in explaining why there seem to be so many different spatial relations with similar descriptions. Why do minor variations in physical relationships seem to give rise to major differences in language used to describe them, when in other cases major physical variations result in consistent linguistic descriptions? The answer to this question comes from the recognition that current theories of spatial relations tend to oversimplify the range of human linguistic performance that must be accounted for, and tend to conflate semantic concerns with pragmatic ones. In this paper, we show that accounting for these factors can explain one of the problems explicitly recognized as unaccountable in earlier work. We demonstrate that a more complex theoretical treatment of the grammar of spatial relations permits a good account of cases problematic for Herskovits [1985], and provides a practical algorithm for use in computational systems.

Herskovits [1985] looked at spatial relations in terms of "ideal relations"–what she might have called prototypes, except for the baggage of controversy over meaning which follows prototypes. Herskovits built on the work of, among others, Talmy [1983], who observed that in the enormous set of spatial relations among objects, only a relatively small number were lexicalized. Herskovits sought to explain lexical choice in locative descriptions. This paper expands on one aspect of her work–the ability of speakers and hearers to produce and accept apt locative descriptions where the relation between the referents deviates from the "prototype." Herskovits described this variation in angular terms:

---

[1] This research is supported by FIPSE grant #84.116C.

417

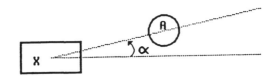

Figure 1: Is **A** *directly to the right* of X?

> There is a certain tolerance for deviation from truth of the ideal meaning, (or from the truth of the transformed ideal meaning when such a transformation has taken place.). I am concerned here with gradual deviations measurable in terms of an angle at a distance. For example, I want to ask: how far apart can two objects be so that one can say one is *at* the other? . . . Or consider [Figure 1]. How close to the right axis must an object A be, for:
>
> > *A is directly to the right of X.*
>
> to be true? [Herskovits, 1985: 366]

Herskovits identifies indeterminacy arising from the nature of the objects and accuracy of perception, but argues that allowable variation (which she calls "tolerance"), depends chiefly on relevance of the relation to the discourse. In assessing the potential effects of these factors, she relies on angular deviation as the measure of variation. She then adds

> But often, the tolerance reflects an accumulation of practices, of interactions with the objects, making prediction impossible. In our constant intercourse with the objects in our world, we have integrated into our knowledge strategies that allow us to count or discount some fact according to context; what those strategies are is still very much a mystery. Yet, tolerance is one direction in which the search for systematicity could proceed. [Herskovits, 1985: 367]

We believe that the mystery of acceptance of deviation from prototypical relations can be explained. The solution is in two parts. First, we show that angular differences are not the most useful measure of deviation. Second, we develop a plausible algorithm for description of spatial relations which inherently accounts for consistency and variation.

2 Acceptable Sources of Variation

Given the need to establish the basic semantics of spatial relations for *right of, left of, above, below,* and *between,* for the purpose of building an ICAI tutor for beginning second language instruction, we looked originally at the linguistics and artificial intelligence literature on spatial relations, but were unable to find descriptions for these relations which were adequate for the explanation of even the most simple cases of semantic distinction. Accordingly, we began a systematic inquiry into the ranges and boundaries of acceptable variation in spatial relations from which we could infer an algorithmic description.

Beginning with the simplest cases, we note that the example posed by Herskovits quickly leads to difficulties. In Figures 2 (a) and (b), circles A and B have a common angular deviation α from the horizontal axis, yet A seems describable as "directly to the right of" the square X, while B clearly is not prototypically "directly to the right of" X.

The problem shown in Figure 2 is not simply attributable to increasing distance. First, increasing distance should not be a factor in the angular deviation model, particularly at the relatively small change in distance in Figure 2. Second, a similar problem occurs in the case of decreasing distance. In Figure 3, why does rectangle C seem to be *below* square X while rectangle D is not prototypically *below* X? In both cases the angular deviation is identical, yet the prototypical spatial relation holds only for rectangle C.

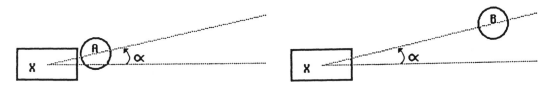

(a) A is directly to the right of X, although offset by α.

(b) B is not directly to the right of X, although again offset by α.

Figure 2: With constant angular deviation, change in distance affects judgments about directional spatial relations.

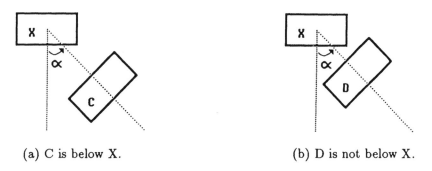

(a) C is below X.

(b) D is not below X.

Figure 3: With constant angular deviation, change in distance affects judgments about directional spatial relations, yet the effect of the change in distance on the judgment is opposite that in Figure 2.

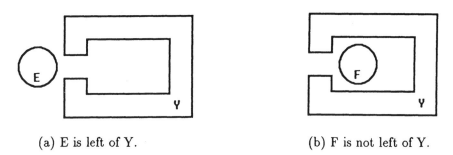

(a) E is left of Y.

(b) F is not left of Y.

Figure 4: Despite constant angular deviation, qualitative nature of the spatial relations changes markedly.

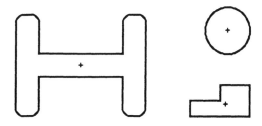

Figure 5: Centroids of various shapes.

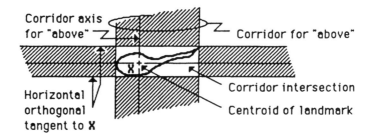

Figure 6: Boundaries and regions defined by a landmark shape.

Another problem situation for angular deviation is shown in Figure 4. In Figure 4 (a), the circle E is clearly directly to the left of the object Y. Yet in Figure 4 (b), with the same null angular deviation from the x-axis, circle F cannot reasonably be considered directly to the left of Y. One might say that circle E is *inside* object Y, but that is not a result predicted by constant angular deviation.

3 Fundamental Loci

To overcome these problems, we propose an algorithm which uses the following concepts:

1. Using terminology similar that of Langacker [1986], a spatial relation exists between a target and a landmark. The target is the object whose location an utterance seeks to declare. The landmark is a different object which is used to locate the target. Thus in the sentence "The circle is directly to the left of the square," the circle is the target and the square is the landmark.

2. The centroid of an object is the center of its area. That is, the centroid is like a flat object's "center of gravity," except that the object is assumed to have uniform density [Rosenfeld, 1976]. Thus the centroid of a circle is its center, the centroid of a symmetrical dogbone would be in the center of its shank, and the centroid of an asymmetrical shape is proportionately offset. See Figure 5.

3. The corridors of a landmark are the horizontal and vertical "shadows" of the landmark defined by its orthogonal tangents. As shown in Figure 6, the corridors of the landmark object X are indicated in gray. The corridor intersection is the rectangular region bounded by the landmark's orthogonal tangents. The centroid axis of a corridor is a line extending from the landmark's centroid in a direction parallel to the corridor.

With these concepts, we now propose an algorithm for deciding which term in the set *left of*, *right of*, *above*, and *below* best fits a given spatial relation (if at all).

IF the target does not overlap the landmark's corridor intersection
THEN   IF   (1) the centroid of the target is in a corridor,
       OR   (2) exactly one of the landmark's corridor
                 contains any point of the target,
       OR   (3) the vector from the centroid of the landmark
                 to the centroid of the target
                 is closer to the centroid axis of a corridor
                 than to the centroid axis of any other corridor,
       THEN   the relation between the target and the landmark
                corresponds to the direction of the corridor,
ELSE no particular spatial relation from this set is apparent.

The spatial relations algorithm is illustrated in Figure 7. Square A shows case (1); it is clearly to left of the landmark. Square B shows case (2); it is below the landmark. Square C shows case (3); it is more above than right of the landmark because the vector from the centroid of the landmark to the centroid of the target is closer to the *above* centroid axis than to the *right of* centroid axis.

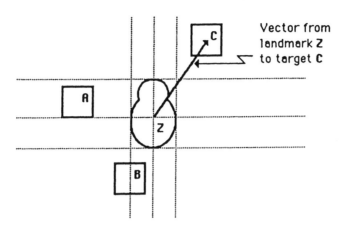

Figure 7: Cases of the spatial relations algorithm.

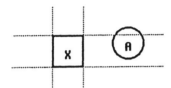

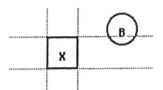

(a) The centroid of A is in the *right of* corridor of X.

(b) The centroid of A is not in the *right of* corridor of X.

Figure 8: Cases (1) and (3) of the spatial relations algorithm distinguished.

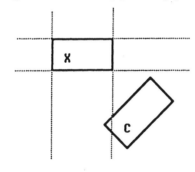

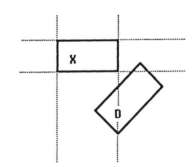

(a) Only one corridor of X contains any points of C.

(b) More than one corridor of X contains points of D.

Figure 9: Cases (2) and (3) of the spatial relations algorithm distinguished.

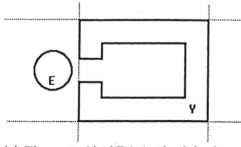

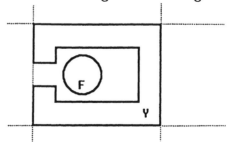

(a) The centroid of E is in the *left of* corridor of Y.

(b) The centroid of F is in the corridor intersection of Y.

Figure 10: Applicability of the non-overlapping cases of the spatial relations algorithm.

Figures 8 through 10 show the algorithm applied to the examples of Figures 2 through 4. In Figure 8 (a), the centroid of target circle A is in the *right of* corridor of landmark square X, so case (1) applies. In Figure 8 (b), the centroid of target circle B is not in any corridor of landmark square X; at best, case (3) applies. In Figure 9 (a), exactly one corridor of landmark square X contains points of target rectangle C; this is thus case (2). Figure 9 (b) displays the corresponding case (3) situation. In Figure 10 (a) target circle E is clearly to the left of landmark object Y; E's centroid is in Y's *left of* corridor. In Figure 9 (b), however, circle F is in Y's corridor intersection, so none of cases (1), (2), or (3) applies.

This approach allows for variation from the prototypical loci, yet avoids the anomalous results of angular deviation as the standard. For example, as long as the centroid of the target is in a corridor of the landmark, the target's location can vary freely and yet the perceived spatial relationship will continue to hold. An important aspect of this approach is that it is not reducible to boolean evaluation of simple relational predicates. Rather, the approach seeks the best choice from the set of possible linguistically determined spatial relations based on the physically determined spatial relations in their semantic context.

4 Discussion

4.1 Theoretical Premises

Four theoretical or methodological premises have shaped both the criticisms and the alternatives we offer above.

4.1.1 Task Variation

First, the range of human performance in dealing with the language of spatial relations is broader than typically addressed. In particular, it is important to recognize that there are at least four different kinds of tasks involving spatial relations which subjects routinely perform. These tasks are illustrated in sentences (1) - (4). One can ask subjects to *verify* the truth of some sentence, as in (1). One can ask subjects to *choose* the most appropriate expression describing some state of affairs from a set of alternative, semantically "correct" expressions, as in (2). One can ask subjects to *describe*, either simply or in great detail, the relation holding between some target object and a landmark, as in (3). And, one can ask subjects to *manipulate* some part of the world in response to an utterance using a spatial relation expression, as in (4).

> (1) The circle is above the square.
>
> (2) Is the circle above or to the right of the square?
>
> (3) Where is the circle located?
>
> (4) Put the circle above the square.

Analysis of (i) seven protocols of two language teachers spontaneously teaching these spatial relations with geometric tiles and (ii) five protocols of subjects describing a videotape of moving circles and squares discloses no task outside this set.

Previous work typically focused almost exclusively (and implicitly) on the first kind of task, relying principally on introspective consideration of the grammaticality or felicity of sentences like (1). But while a subject will agree to the truth of (1) when shown Figure 8 (b), no subject ever places a circle in that location in response to the command in (4). A complete accounting, then, of the grammar of spatial relations requires that we account for this wider range of human behavior.

4.1.2 Lexical Alternatives

Second, the problem one faces in modeling spatial relations is not one simply of specifying a precise meaning or rules of use for individual lexical expressions in isolation from each other. Rather, within a given semantic domain, such as space, the speaker makes choices among acceptable linguistic expressions, identifying the most appropriate expression for the given context. So, for example, Figure 8 (b) can be described either by (6) or by (7),

(6) The circle is to the right of the square.

(7) The circle is above the square.

for each correctly describes the state of affairs represented by Figure 8 (b), and subjects asked to verify the truth of either sentence alone will accept each as true. But of the two, (7) is recognizable as more appropriate. The use of linguistic expressions, then, is not wholly determined by the meaning of a given expression alone but in concert with semantically associated items in the same domain.

4.1.3 Semantics or Pragmatics

Third, previous approaches to spatial relations tend to conflate semantic facts concerning the intrinsic meaning of expressions with pragmatic issues concerning their use. In the approach taken here, the semantics of an expression represents the full range of cases for which the expression is true. The inherent semantic meaning of a lexical expression for a spatial relation like *left* or *above* is represented by the region whose boundaries are defined by the cases for which the expression is true versus those for which the expression is false. Methodologically, it is the verification task that permits one to discover exactly what that region is. This approach motivates the basic notions used in the algorithm.

The pragmatics of linguistic expressions represents the principles governing the choice among alternative true expressions. In our approach, one selects the best-fitting expression arising from the domain. In pragmatics, the other three kinds of linguistic task described above are also important. Choice tasks help determine the boundaries of alternatives in conflict. Manipulation tasks help determine optimal choices. Description tasks help determine the variables relevant to pragmatic principles of use.

Overall, we have found that a strict separation of semantic and pragmatic issues allows one to account for the wide range of cases represented by the four task types. Further, there appears to be no need to formulate prototype meanings for any expression, for the prototype effects fall out from this treatment.

4.2   Task Variation and the Proposed Algorithm

If variation in the use and comprehension of linguistic expression of spatial relations depends in part on task, then task serves as one predictor for acceptance of deviation from any prototypical locus. The algorithm presented above addresses some of the variability in use of expressions due to task variation.

For manipulation tasks, the locus of acceptable points for a spatial relation corresponds to a special case of case (1) in the algorithm. This reflects the case, for example, in which one tells the subject "Put the circle to the left of square." We expect that the subject will typically put the target relative to the landmark so that the centroid of the target is on the centroid axis of the landmark's left corridor, and so that the target is near the landmark without overlapping. The meaning of *near* is the subject of related work [cf. Denofsky, 1976].

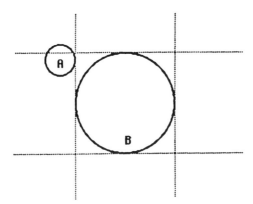

Figure 11: Spatial relations can be non-commutative.

For the description and choice tasks, a subject spontaneously uses the best term to describe a spatial relation between a target and landmark. This task corresponds to case (1) of the algorithm. If the centroid of the target is anywhere in the left corridor of the landmark, then we expect a subject to describe the relation between the objects as "The circle is to the left of the square." (This assumes that the objects don't overlap.) Anywhere else, the subject chooses the closest vector.

For the verification task, the subject chooses a truth value for a statement about the spatial relation between objects. This task corresponds to cases (2) and (3) of the algorithm. Subjects will find case (2) clear, and will find case (3) less clear as the vector from the landmark to the target becomes equidistant from two neighboring centroid axes of the landmark. In Figure 7, with respect to the landmark object Z, square A is in a case (1) relation, square B is in a case (2) relation, and square C is in a case (3) relation. Note that the location of square A corresponds to a prototypically correct response to the command "Put the square to the left of Z." Likewise, the locations of squares B and C will typically elicit an affirmative response to the the question "Is the square above (or below) Z?"

Note that these tasks require an explicit identification of target and landmark. If these roles are reversed for a given display, the perceived spatial relation between objects may not be maintained. This lack of commutativity is reflected in the algorithm. For example, in Figure 11, a subject would probably describe circle A as being to the left of circle B, but would not describe circle B as being to the right of A.

4.3   Function and Structure

The spatial relation algorithm presented above, on its terms, does not apply to many possible cases, especially where the target lies wholly or in part in the landmark's corridor intersection. Evidence we have collected suggests that many of these cases can be resolved by reference to world knowledge of function and structure. By function, we mean uses of objects, such as containment. By structure, we mean the constituent parts of an object which make up its shape (and the relation of these parts to each other, of course).

Just as the task in which objects are involved influences the nature of their perceived spatial relations, so too may the function of objects affect spatial relations. In Figure 4 (b), the target circle would be best described as *in* the landmark object. However, simple geometric definitions of *in* and *out* are not adequate for domains which purport to model real-world semantics. While the landmark object in Figure 4 (b) has no apparent semantics aside from its polygonal structure (and perhaps the quality of being a reversed letter "C"), we believe that the real-world semantics of objects such as those in Figure 12 would overwhelm the naive geometrical approach. At this point, then, we represent the function of objects by explicit reference to their functions. For example, a thing is or is not a *container*.

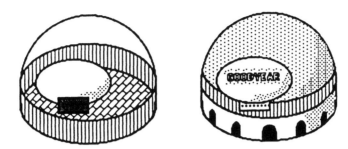

Figure 12: The domain semantics of shapes affect their spatial relations. Thus identical shapes (in outline) in the same physical relation in two-dimensional representation will create different apparent linguistic relations depending on the domain information imparted by their internal representations.

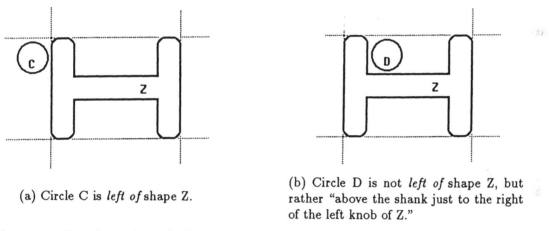

(a) Circle C is *left of* shape Z.

(b) Circle D is not *left of* shape Z, but rather "above the shank just to the right of the left knob of Z."

Figure 13: Local spatial relations used where the target overlaps the corridor intersection of the landmark.

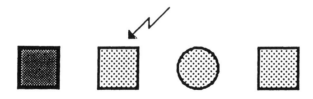

Figure 14: Spatial relation disambiguated by combining shapes in an implied global structure. The arrow indicates for the shape to which the experimenter pointed; the arrow was not present in the figure actually presented to subjects.

With respect to structure, we noticed that where subjects had difficulty describing a relation (typically because the target lay within the landmark's corridor intersection), the subject would resolve the problem by identifying a part or parts of the landmark and then using the usual spatial relations expressions relative to those parts. A typical example is shown in Figure 13. In Figure 13 (a) the target circle C is unambiguously *left of* the landmark object Z, and subjects will give the expected response. But when faced with Figure 13 (b), subjects will produce a description like "The circle is above the shank just to the right of the left knob." That is, the subject has broken down the landmark into parts from which the normal discourse of spatial relations can be constructed. Interestingly, we also observed a converse effect in which ambiguity among spatial referents was avoided by ideating a structure comprised of the referents and then constructing an utterance based on a spatial relation involving the structure. For example, when Figure 14 was presented and the indicated object pointed out, subjects typically identified the object as "the second square from the left," impliedly creating a horizontal structure encompassing all four objects. Subjects typically did not identify the object as "the square to the left of the circle."

4.4 Future Work

Our future work in this area involves, among other things, systematic experimental verification of the spatial relations algorithm. We plan to present subjects with a variety of tasks and relations from which we can extract data on manipulation, description, choice, and verification, with reaction times where appropriate. This work will be conducted on a Symbolics 3645 Lisp machine using testing software specially developed for these series of experiments.

Other work will include continued research into the meaning of *near*. We expect to apply similar techniques to this term. We are also continuing to analyze tasking in the discourse of spatial relations. The algorithm that we have presented in this paper, while inherently accommodating the semantics if *left of, right of, above,* and *below,* nevertheless lacks explicit procedures for generating the pragmatic aspects of task, function, and structure that we have discussed. Our future work will focus on that elaboration and its empirical validation.

5 References

Denofsky, Murray Elias (1976). *How Near is Near?*, AI Memo No. 344, Massachusetts Institute of Technology Artificial Intelligence Laboratory, February 1976

Ehrich, V. (1985) The Linguistics and Psycholinguistics of Secondary Spatial Deixis. In G.A.J. Hopenbrouwers, P.A.M. Seuren, and A.J.M.M. Weijters, *Meaning and the Lexicon*. Foris Publications: Dordrecht, Holland.

Hayes, P. (1985b). Naive Physics I: Ontology for liquids. In J.R. Hobbs and R.C. Moore (Eds.), *Formal Theories of the Commonsense World*. Norwood, NJ: Ablex Publishing Corp.

Herskovits, A. (1985). Semantics and pragmatics of locative expressions. *Cognitive Science, 9*, 341-378.

Langacker, R. W. (1986). An Introduction to Cognitive Grammar, *Cognitive Science 10*, 1-40.

Rosenfeld, A., & Kak, A. C. (1976). *Digital picture processing*. New York: Academic Press.

Talmy, L. (1983). How language structures space. In H. Pick and L. Acredolo (Eds.) *Spatial Orientation: Theory, Research, and Application*. New York: Plenum Press.

SIMULATION OF TASK PERFORMANCE AS GUIDE TO THE IDENTIFICATION OF SOLUTION STRATEGY FROM EYE-MOVEMENT RECORDINGS

Gerhard Deffner

Department of Psychology, University of Hamburg

Von-Melle-Park 5, 2000 Hamburg 20, West Germany

phone: 0049 40 4123 5471

Summary

An approach is presented which was developed for the goal of comparing subjects with respect to their use of solution strategies. They were given a series of eight tasks of a type which can well be solved by two distinctly different methods: n-term series tasks in the form of statements about spatial relations. One of these methods relies on mental imagery (Method of Series Formation) while the other is more of an abstract, analytical method (Elimination). For the purposes of comparison, eye-movements were recorded during task performance. The sequence of gazes recorded during each individual solution attempt was subsequently matched to patterns derived from information processing models of task performance which simulate the different strategies. The percentage of successful matches to one or the other pattern was used in further analyses comparing the two groups of subjects.

Keywords: cognitive simulation, eye-movement data, strategy identification

427

Identification of Strategies

1. Introduction

The methodology described here was developed for an experiment which continued a series of studies (Deffner, 1984) in which silent control subjects were compared to subjects who were required to think aloud during their attempt to solve experimental tasks. In the new experiment, the main variable of interest was strategy use, especially shifts in the selection of strategies over time.

2. Material

The experimental tasks were derived from Ohlsson's (1980) collection of tasks. They consist of n-term series tasks where propositions refer to the relative positions of persons sitting on a bench, thus the 'terms' in these tasks are names, and the propositions use spatial relations. In contrast to Ohlsson, who for the purposes of studying the solution of more complex tasks used a larger number of relational terms (immediately to the left/right of, left-/rightmost, left-/rightmost but one, left/ right, between) only one type of relation was used in the present context. An example is given in Figure 1, where line numbers are included to aid in the understanding of subsequent examples.

```
SEVERAL  BOYS  ARE  SITTING  ON  A  BENCH:        (line 0)

        KEITH  IS  LEFT  OF  CHUCK               (line 1)
        CHUCK  IS  LEFT  OF  ROY                 (line 2)
        MARK   IS  RIGHT OF  ROY                 (line 3)
        MARK   IS  LEFT  OF  PHIL                (line 4)
```

Figure 1: n-term series task

In each task, subjects are required to answer a specific question with respect to the the position of one particular person, for example:

"Who is immediately to the right of Mark?"

Eight experimental tasks were used (four 4-term and four 5-term series tasks). Eye-movements were recorded using an Applied Sciences 1996 system.[1] This equipment measures x- and y-coordinates for gaze direction at a sampling rate of 60

---

[1] I want to thank Richard Ohlsson from the University of Colorado for the privilege to use his laboratory and for his generous help.

428

Hz, and recordings are made on the level of fixations (Kliegl & Ohlsson, 1981). This means that successive samples are aggregated as long as they lie on the same character position of the display (Kliegl, 1981). This aggregation is very fine and it results in a resolution of distinct fixation points within individual words. For the present analysis, a more natural unit of analysis is the 'gaze' rather than 'fixation'. Gazes incorporate all fixations on the same word. This resolution was chosen because in the present context we are not concerned with an analysis of the reading process during task performance but want to use units which are relevant entities in information processing models of task performance (for a discussion of this principle, see: Just & Carpenter, 1976). For this reason, only three gaze positions were identified per line:

    1) the term (name) on the left,
    2) the relational information in the middle
            ("is to the left/right of"),
    3) the term (name) on the right.

Figure 2 shows the beginning of a recorded sequence of gazes where the first number in each parenthesis stands for gaze position relative to the layout of the task (the first digit refers to line number and the second digit stands for first, second or third position in a proposition); the second number stands for the duration of the gaze on that location.

    (03 38)(02 7)(01 13)(03 25)(11 5)(13 15)(12 7)(11 7)
    (13 70)(21 8)(13 36)(12 7)(11 12)(13 51)(12 13)(11 6)
    (13 61)(22 14)(21 17)(23 15)(21 13)(23 11)(22 7)(21 9)
    (23 7)(22 6)(21 10)(13 12)(22 8)(21 8) ...

Figure 2: Sequence of gazes recorded during performance
          of an n-term series task

Data of this kind were recorded for 18 think-aloud subjects and 18 silent controls.

3. Task analysis

There has been much research on how subjects solve n-term series problems (see Sternberg, 1980 for an overview). This resulted in a large variety of models for the solution process. These on the one extreme are based upon the idea of subjects building a mental image of the sequential arrangements of terms in the propositions presented to them. When they answer the final question they "read off" from their mental image (DeSoto, London & Handel, 1965; Huttenlocher, 1968). On the other extreme is that position which claims

that subjects extract information for each term separately
and then answer the final question on the basis of integrating all individual pieces of information (Clark, 1969). More
recently, it has been argued that we should not try to
determine which of these models is the right one, but rather
pay attention to the time course of strategy choice during
repeated trials (Johnson-Laird, 1972). Wood, Shotter & Godden
(1974) have shown that subjects tend to begin with an imagery
strategy and during succesive trials change over to an
analytical strategy.

Most research concentrated upon tasks using the left/right
dimension, asking subjects questions with respect to the
extreme ends of the series. This leaves room for quite a
number of strategies. By choosing Ohlsson's way of concluding
the experimental tasks (asking for a specific position within
the series), the field can be narrowed down to two strategies. These in one case go back to the original (DeSoto,
London & Handel, 1965; Huttenlocher, 1968) imagery model:
Series Formation, and in the other are a combination of
Quinton and Fellows' (1975) and Clark's (1969) model:
Elimination. Both strategies at their top level can easily be
described as LISP programs.

First, this is the definition of the Series Formation
strategy:

```
(DEFUN SF (PROP-LIST QUESTION)

  (READ-OFF (SERIATE PROP-LIST) (CADR_QUESTION) (CAR QUESTION)))
```

where PROP-LIST contains the propositions of a task (terms
symbolized by "T1", "T2" ...; relations by "R" or "L"), and
QUESTION is a list contining a relation and a term. READ-OFF
builds a series using SERIATE and returns that element of the
series which is connected to the target (CADR QUESTION)
through the target relation (CAR QUESTION). SERIATE is
defined as:

```
(DEFUN SERIATE (PROP-LIST)

  (DO ((SERIES (PREFERRED-DIRECTION (CAR PROP-LIST)))
       (P-LIST (CDR PROP-LIST)))
     ((NULL P-LIST) SERIES)
     (SETQ SERIES (INTEGRATE SERIES (PREFERRED-DIRECTION (CAR P-LIST))))
     (SETQ P-LIST (CDR P-LIST))))
```

where PREFERRED-DIRECTION returns the two terms of a proposition in the (preferred) left-to-right order and INTEGRATE returns a new series from an old one combined with a new pair of terms in left-to-right order.

Next is the definition for the Elimination strategy:

```
(DEFUN EL (PROP-LIST QUESTION)

   (DO ((ANSWER NIL)
        (ANSWER-TEMP NIL)
        (TARGET-RELATION (CAR QUESTION))
        (TARGET (CADR QUESTION)))
      ((NULL PROP-LIST) ANSWER)
      (COND ((MEMBER TARGET (CAR PROP-LIST))
              (SETQ ANSWER-TEMP (GET-ANSWER TARGET (CAR PROP-LIST)
                        TARGET-RELATION))
               (SETQ ANSWER (COND ((NULL ANSWER-TEMP) ANSWER)
                            (T ANSWER-TEMP)))))
      (SETQ PROP-LIST (CDR PROP-LIST))))


(DEFUN GET-ANSWER (TARGET PROP RELATION)

   (COND ((NULL PROP) NIL)
          ((EQUAL (CADR PROP) RELATION)
            (COND ((EQUAL (CADDR PROP) TARGET)(RETURN (CAR PROP)))
                (T NIL)))
          (T (COND ((EQUAL (CADDR (CONVERT PROP)) TARGET)(CADDR PROP))
                (T NIL)))))


(DEFUN CONVERT (PROP)
    (LIST (CADDR PROP)
         (COND ((EQUAL (CADR PROP) 'R) 'L)
            (T 'R))
         (CAR PROP)))
```

These two strategies have to be extended so that they can cope with an increase in task difficulty: propositions need not be presented in the order of sequential overlap of their terms. As an example, the task in Figure 1 can be presented as shown in Figure 3:

                SEVERAL   BOYS   ARE   SITTING   ON   A   BENCH:

                    KEITH   IS   LEFT   OF   CHUCK
                    MARK   IS   LEFT   OF   PHIL
                    CHUCK   IS   LEFT   OF   ROY
                    MARK   IS   RIGHT   OF   ROY

Figure 3: n-term series task without sequential overlap of
          terms

In order to process such tasks, an additonal function is
required for Series Formation which checks whether propositions overlap (function: OVERLAP?). Using this function,
there are two alternatives for extending the definition of
SERIATE: 1) SERIATE-DISCARD: if a second premise does not
overlap, then the old series is discarded and a new series is
built; the old proposition will have to processed again
later, 2) SERIATE-2ND-SERIES: if there is no overlap, a
second series is built which has to integrated into the first
one after all propositions have been processed.


```
(DEFUN SERIATE-DISCARD (PROP-LIST)
  (DO ((SERIES (PREFERRED-DIRECTION (CAR PROP-LIST)))
       (P-LIST (CDR PROP-LIST)))
  ((NULL P-LIST) SERIES)
  (COND ((OVERLAP? SERIES (CAR P-LIST))
           (SETQ SERIES (INTEGRATE SERIES
                             (PREFERRED-DIRECTION (CAR P-LIST))))
           (SETQ P-LIST (CDR P-LIST)))
        (T (SETQ SERIES (PREFERRED-DIRECTION (CAR P-LIST)))
           (SETQ P-LIST
                 (APPEND (CDR P-LIST)(LIST (CAR PROP-LIST)))))))))


(DEFUN SERIATE-2ND-SERIES (PROP-LIST)

  (DO ((SERIES1 (PREFERRED-DIRECTION (CAR PROP-LIST)))
       (P-LIST (CDR PROP-LIST))
       (SERIES2)(PAIR))
  ((NULL P-LIST)(COND (SERIES2 (INTEGRATE SERIES1 SERIES2))
                      (T SERIES1)))
  (SETQ PAIR (PREFERRED-DIRECTION (CAR P-LIST)))
  (COND ((OVERLAP? PAIR SERIES1)
           (SETQ SERIES1 (INTEGRATE SERIES1 PAIR)))
        (T (SETQ SERIES2 (INTEGRATE SERIES2 PAIR))))
  (SETQ P-LIST (CDR P-LIST))))
```

For Elimination, the situation is more simple, because checking for the occurrence of target terms is already part of the strategy, and overlap therefore does not have any importance.

Before these models can be used any further, they have to be judged with respect to their plausibility. One important qestion is, whether they use storage mechanisms that are compatible with our knowledge of human memory. In the case of Series Formation, an analog representation is postulated which holds up to five items in sequential order that can be rehearsed and recalled over a short period of time. In the case of Elimination, three items have to be kept available for immediate access: TARGET, TARGET- RELATION, and ANSWER; during intermediate steps, one additional temporary variable is required: ANSWER-TEMP. No matter what limitations and mechanisms of short term memory we assume: both these assumptions are extremely plausible.

Another question is related to the complexity of the operations used in the models. This is not critical in the present case: the most difficult operations are those concerning the conversion of relational information, i.e. from "T1 is left of T2" to "T2 is right of T1", and we do consider humans capable of this.

4. Strategy identification

In contrast to earlier approaches (Deffner, 1985) where verbal descriptions of information processing models of task performance served as a basis for strategy identification, the programs Series Formation and Elimination can be used in the present case. Input and output to and from operators in these models can be seen in analogy to attentional proccesses during task performance. But not all attentional processes are observable, since reference to internally represented information need not be related to observable behaviour. Access to items in the external display nevertheless is accompanied by overt behaviour: gazes on these items can be understood to be indicators of such attentional processes, and the overall gaze sequence is a sequence of items attended to.

One note of caution is necessary, though. Gaze direction is not a definite indicator of subjects attending to the item looked at. At least on the level of fixations, perceptual processes determine some eye-movements (c.f. Groner 1978), and also there is the possibility of 'empty stares' where a gaze is not at all directed at the item on the display. For the present purposes, gaze direction nevertheless remains the richest source of data on attentional processes as they are related to visually displayed tasks.

The basic rationale of the present strategy identification is that of matching observed gaze sequences against sequences predicted by models which stand for different strategies. For each model, a program trace can provide this prediction. The degree to which an observed sequence of gazes resembles traces from these models will be used as a basis for quantitative evaluation.

4.1. Derivation of ideal sequences

In order to establish these ideal sequences, both simulation programs were run on all experimental tasks and traced. Tasks 3,4,7, and 8 were of the non-sequential-overlap type, of which SERIATE-DISCARD could solve tasks 3 and 4 and only SERIATE-2ND-SERIES could solve the last two tasks. These more complex versions of SERIATE were used for the task which required them. Only those functions were traced, which can without doubt be considered information processing stages involving visual input. This rules out processes which have to be considered perceptual, and also it rules out any processes involving symbolic processing without immediate or clear reference to the displayed task (c.f. storage/rehearsal mechanisms).

For the SF-program, functions PREFERRED-DIRECTION, OVERLAP?, INTEGRATE, and READ-OFF were traced. Figure 4 shows a trace for the experimental task from figure 3 where terms are symbolized by "T1" through "T5" and the relations left and right by "L" and "R" respectively.

In the case of EL, only GET-ANSWER and CONVERT were traced. MEMBER also is an important function in the program, but it was considered too close to fast perceptual processes to be relevant in the present context of slower information processing stages. Figure 5 presents a trace of EL on task 7.

In a next step, elements in the trace were matched to gaze positions and segmented into stages. Gaze positions within each stage were treated as having equal probability of being looked at, and were consequently joined into sets of possible gaze locations for a given stage. The sequences of sets made up ideal patterns of items attended to under the one or the other strategy. Figure 6 presents an example for task 7:

```
* (SF TASK7 '(R T4))
Entering: PREFERRED-DIRECTION, Argument list: ((T1 L T2))
Exiting: PREFERRED-DIRECTION, Value: (T1 T2)
Entering: PREFERRED-DIRECTION, Argument list: ((T4 L T5))
Exiting: PREFERRED-DIRECTION, Value: (T4 T5)
Entering: OVERLAP?, Argument list: ((T4 T5) (T1 T2))
Exiting: OVERLAP?, Value: NIL
Entering: INTEGRATE, Argument list: (NIL (T4 T5))
Exiting: INTEGRATE, Value: (T4 T5)
Entering: PREFERRED-DIRECTION, Argument list: ((T2 L T3))
Exiting: PREFERRED-DIRECTION, Value: (T2 T3)
Entering: OVERLAP?, Argument list: ((T2 T3) (T1 T2))
Exiting: OVERLAP?, Value: T
Entering: INTEGRATE, Argument list: ((T1 T2) (T2 T3))
Exiting: INTEGRATE, Value: (T1 T2 T3)
Entering: PREFERRED-DIRECTION, Argument list: ((T4 R T3))
Exiting: PREFERRED-DIRECTION, Value: (T3 T4)
Entering: OVERLAP?, Argument list: ((T3 T4) (T1 T2 T3))
Exiting: OVERLAP?, Value: T
Entering: INTEGRATE, Argument list: ((T1 T2 T3) (T3 T4))
Exiting: INTEGRATE, Value: (T1 T2 T3 T4)
Entering: INTEGRATE, Argument list: ((T1 T2 T3 T4) (T4 T5))
Exiting: INTEGRATE, Value: (T1 T2 T3 T4 T5)
Entering: READ-OFF, Argument list: ((T1 T2 T3 T4 T5) T4 R)
Exiting: READ-OFF, Value: T5

T5
```

Figure 4: Trace of SF on task 7

```
* (EL TASK7 '(R T4))
Entering: GET-ANSWER, Argument list: (T4 (T4 L T5) R)
 Entering: CONVERT, Argument list: ((T4 L T5))
 Exiting: CONVERT, Value: (T5 R T4)
Exiting: GET-ANSWER, Value: T5
Entering: GET-ANSWER, Argument list: (T4 (T4 R T3) R)
Exiting: GET-ANSWER, Value: NIL

T5
```

Figure 5: Trace of EL on task 7

```
SF: ((11 12 13)(21 22 23)(31 32 33)(41 42 43)(52 53 23))
EL: ((52 53)(21 22 52 53 23)(42 43 52 53 23))
```

Figure 6: ideal patterns for SF and EL on task 7

There was one more complication, however: with the Elimination method, there is no need to use only one set order in
which the propositions are processed. Though it does not seem
reasonable to assume that the order is random, it must be
conceded that there are two plausible orders. In one,
subjects work their way down the list of three or four
propositions from the top to bottom, whereas in the other
case, they start from the bottom line and work upwards. For
this reason, there have to be two ideal sequences for EL,
thus a third pattern has to be added to Figure 6:   EL-2: ((52
53)(42 43 52 53)(21 22 52 53 23)).

4.2. Matching gaze sequences to program traces

Matching was straightforward: Starting with the first list in
the pattern, its elements were checked against successive
elements of the observed gaze sequence. A match started with
the first element from that list and was continued until more
than one successive element in the gaze sequence was extraneous to the list from the pattern, or the duration of an
individual extraneous gaze was longer than the average gaze
duration in the total gaze sequence. When a match was
discontinued, the next list from the pattern was used; if the
pattern was exhausted, matching started from the beginning of
the pattern. Because of this strict sequential order in which
lists from the pattern were matched, only such backup in task
performance could be identified which would start from the
very beginning. All attempts to allow for partial backup
resulted in substantial loss of clearness of strategy
identification, and were not included in the final version of
the algorithm.

This algorithm was used three times for each gaze sequence:
once for the Series Formation strategy and twice for the
Elimination strategy (using the forward and the backward
pattern). The numerical information used for further data
analyses consisted of the percentage of total gaze sequence
which could be matched to each pattern. For Elimination, only
the higher of the two percentages was used, so that differences between backward or forward task performance could be
ignored in the overall comparison of Elimination to Series
Formation. Another extension was that these analyses were
performed for both gaze duration and gaze frequency as a
basis of percentages. The reason for using both these
measures was that no plausible argument could be found to
favor either one or the other on theoretical grounds. Table 1
presents an example of the resulting output.

436

Table 1: Sample output of percent matches

Subj.	%SF/dur	%EL/dur	%SF/freq	%EL/freq
S-EL	0.00	59.38	0.00	46.50
10	86.89	25.40	86.17	14.17
11	0.00	68.75	0.00	38.02
13	95.45	40.91	99.08	48.31
14	38.03	8.22	44.18	9.35
15	29.41	17.07	28.64	14.07
17	60.87	10.20	59.68	9.64
18	46.51	10.20	36.63	5.55

As can be seen from Table 1, using duration or frequency as a basis of percentages did not result in great differences, the two measures are highly correlated. Also, the percentages for EL tended to be smaller. This argues for less appropriateness of the patterns and/or pattern matching used for EL. These figures nevertheless could be used well, the only consequence was that they should not be treated as variables on the same dimensions.

Two methods were used to extract information from the four variables per person and task: principal components analysis and cluster analysis. The former resulted in very clear factor structures with one factor explaining a large proportion of the variance. It was identified as a bipolar SF - EL factor and scores on this factor were used for analyses involving continously scaled numercial information.

Cluster analysis was used as a basis of a binary categorization of individual task performances with respect to their predominant strategy. Using the k-means method, it was possible to obtain clear two-cluster solutions for all eight tasks. In all eight cases, these could easily be identified as an Elimination cluster and a Series Formation cluster. Cluster membership was then used for the categorization of task performance.

5. Validity

The validity of these measures was checked in two ways. Firstly, the analysis was performed on additional data recorded from subjects who had received prior training in one of the strategies. Out of a group of six, there were only two subjects who in their subsequent judgement were positive of having used nothing but the trained strategy - one subject trained to use Elimination (called S-EL) and one trained to use Series Formation (S-SF). Table 2 presents the percentage of successful matches for these two sets of data.

Table 2: percent matches for two trained subjects

	Subj.	%SF/dur	%EL/dur	%SF/freq	%EL/freq
Task1	S-EL	0.00	59.38	0.00	46.50
	S-SF	75.00	29.17	83.71	11.62
Task2	S-EL	0.00	47.06	0.00	58.24
	S-SF	80.95	4.76	85.97	2.20
Task3	S-EL	0.00	100.00	0.00	100.00
	S-SF	59.46	27.03	65.11	14.74
Task4	S-EL	0.00	91.67	0.00	98.02
	S-SF	67.86	7.14	83.65	2.71
Task5	S-EL	0.00	60.00	0.00	76.95
	S-SF	42.86	6.98	49.47	3.35
Task6	S-EL	0.00	63.64	0.00	54.96
	S-SF	62.07	6.90	75.61	1.39
Task7	S-EL	0.00	35.29	0.00	30.32
	S-SF	76.20	7.94	69.89	9.16
Task8	S-EL	0.00	25.00	0.00	23.63
	S-SF	59.25	9.09	52.02	6.44

As can be seen from Table 2, there is very good separation between S-SF and S-EL.

The other line of approach was based on an idea used by Wood et al. (1974). These authors surprised subjects with a repeted presentation of a task where in the repetition they did not ask for the position of one specific person, but required subjects to give the total arrangement instead. The difference in solution time between the first and the second solution of the task was used to estimate whether subjects had been using the Series Formation strategy - long times on the repetition standing for prior use of the Elimination strategy. In the present case, solution time on the repetion of task 8 (expressed as a factor of solution time for task 8) was compared for subjects whose solution attempt had been classified as SF or EL on the basis of cluster analyses. Table 3 shows means scores:

Table 3: Means scores for relative solution time on task 9

	$\overline{x}$	s	
SF (N = 18)	2.72	3.05	t = 3.908 p < .01
EL (N = 17)	7.04	4.65	

There is a clear difference in the expected direction:
subjects whose solution of task 8 was categorized as Series
Formation required significantly less time when asked for the
total arrangement.

Thus, both approaches to testing the validity of strategy
identification gave very clear and reassuring results.

6. Conclusion

Instead of arguing for the appropriateness of the assumptions
underlying the approach to the identification of strategies
presented here, I shall present a brief glimpse at the
results.

The comparison of mean factor scores (Factor SF-EL) revealed
significant differences in the case of task 3 and 4: mean
scores were higher in the silent group (indicating more use
of the Elimination strategy). This is borne out by the
categorizations on the basis of on the basis of cluster
analyses, which lend themselves more readily for graphical
presentation: Figure 7 shows the frequency of Elimination in
the two groups.

These differences can be interpreted as follows: There is no
big difference between the two gropus with respect to
strategy use. What is different, is the speed at which they
discover the Elimination strategy: Thinking-aloud subjects
are slower.

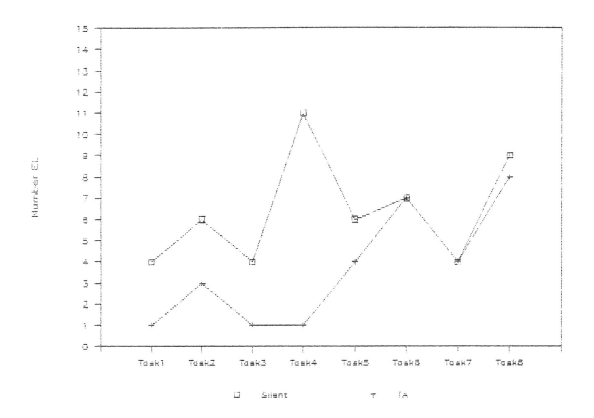

Figure 7: Use of the Elimination strategy over 8 tasks

References

Clark, H.H. (1969) Linguistic Processes in deductive reasoning. <u>Psychological Review</u>, <u>76</u>, 387-404.

Deffner, G. (1984) <u>Lautes Denken - Untersuchung zur Qualität eines Datenerhebungsverfahrens</u>. Frankfurt: Lang.

Deffner, G. (1985) Identification of solution strategies on the basis of eye-movement data. In R. Groner, G.W. McConkie & C. Menz (Eds.), <u>Eye movements and human information processing</u>. Amsterdam: North Holland.

DeSoto, C.B., London, M. & Handel, S. (1965) Social reasoning and spatial paralogic. <u>Journal of Personality and Social Psychology</u>, <u>2</u>, 513-521.

Groner, R. (1978) <u>Hypothesen im Denkprozess</u>. Bern: Huber.

Huttenlocher, J. (1968) Constructing spatial images: A strategy in reasoning. <u>Psychological Review</u>, 7, 351-357.

Johnson-Laird, P.N. (1972) The tree-term series problem. <u>Cognition</u>, <u>1</u>, 57-82.

Just, M.A. & Carpenter, P.A. (1976) Eye fixation and cognitive processes. <u>Cognitive Psychology</u>, <u>8</u>, 441-480.

Kliegl, R. (1981) Automated and interactive analysis of eye fixation data in reading. <u>Research Methods & Instrumentation</u>, <u>13</u>, 115-120.

Kliegl, R. & Ohlsson, R.K. (1981) Reduction and calibration of eye monitor data. <u>Behavior Research Methods & Instrumentation</u>, <u>13</u>, 107-111.

Ohlsson, S. (1982) <u>Competence and strategy in reasoning with common spatial concepts</u> (Working Papers from the Cognitive Seminar, No 6) Stockholm: Department of Psychology.

Quinton, G. & Fellows, B.J. (1975) 'Perceptual' strategies in the solving of three-term series problems. <u>British Journal of Psychology</u>, <u>66</u>, 69-78.

Sternberg, R.J. (1980) Representation and process in linear syllogistic reasoning. <u>Journal of Experimental Psychology: General</u>, <u>109</u>, 119-159.

Wood, D. Shotter, J. & Godden, D. (1974) An investigation of the relationships between problem solving strategies, representation and memory. <u>Quarterly Journal of Experimental Psychology</u>, <u>26</u>, 252-257.

Linguistic Descriptions of Visual Event Perceptions

Anthony B. Maddox
James Pustejovsky

Department of Computer Science
Brandeis University
Waltham, MA 02254
617-736-2700
tony@brandeis.csnet-relay

jamesp@brandeis.csnet-relay

Abstract

In this paper we address the problem of constructing a computational device that is able to describe in natural language its own conceptualization of visual input. This addresses the basic issues of event perception from raw data, as well as what connnection a language with a limited vocabulary has to this event construction. We outline a model of how the perceptual primitives in a system act to both constrain the possible conceptualizations and naturally limit the language used to describe events.

Topic: Visual Perception, Natural Language, Lexical Semantics

1. Introduction

In order for an artificially intelligent system to interact with humans, it is desirable that it be able to communicate with them. Characterizing this interaction will, in part, require considering the impact of language and perception on the communication process. This paper will address the use of language in describing visually perceived events. The focus will be on a theoretical but practical description of the interface between a limited vocabulary linguistic system which supports both tense and aspect and a perceptual representation for visual events. Two major issues discussed are visuo-linguistic temporal granularity and the effect of the interaction between "hard-wired" and learned *focus of attention* on event conceptualization. The paper begins with a discussion of vision-language research and the problems associated with integrating vision and language. In section 3, we present our linguistic and visual concept structures. Section 4 follows with a description of the visuo-linguistic interface illustrated by 4 examples. Section 6 concludes the paper with a summary and directions for future research.

442

2. Language-Vision Research

There has been little research concerning the interface of visual and linguistic processes. One reason for this is that they each currently appear to involve very different and difficult processes. Considerable energy has been focused on low-level or early vision. Marr's [17] primal sketch includes several low-level primitives from which scenes can be constructed. Many others have developed formalisms that relate low-level visual information to the analysis of polyhedral scenes in the blocks world [1]. Their work points out the difficulty in analyzing even the most simple scenes. There has also been some research concerning high-level vision. [10] implement a global blackboard memory in a scene interpretation system, generating scene descriptions by sharing the blackboard at several abstract levels of visual interpretation. [5] uses geometric models to identify aircraft objects in aerial images of an airport scene. [2]

There has been some interest in the use of language to describe events and spatial relationships. [4] develops an event calculus which uses some low-level visual primitives to guide the interpretation of events in a robot assembly environment. [3] describes the use of spatial prepositions for generating descriptions to scenes from the viewpoint of a scene observer. [11] analyzes locative prepositions and points out that the use of such locatives establishes "ideal" relationships which must be made to fit to each particular instance of its usage. She has also pointed out that there is an implied "geometric conceptualizaton" when locatives are interpreted. [16] develops a cognitive grammar which helps to formalize the use of spatial and perceptual relationships through the use of referents and trajectors as keys which relate a linguistic grammar to the conceptualization of the objects which are spatially related. [23] explore verb-driven event processing in the observation of traffic scenes for the generation of natural language descriptions. [29] has contributed to the research with explorations of the relationship between language and spatial relations. [24] has implemented a system using visual predicates for early language development. While research has been accomplished toward understanding verbal scene description, there has not been enough work on describing the visuo-linguistic interface in terms of how vision and language influence and constrain each other to determine visual and linguistic conceptualizations.

The perceptual activities and structures associated with visual perception are not well-defined. From an apparently small set of "hard-wired" visual percepts, people seem to eventually build a relatively large set of complex visual concepts. While language helps people communicate, it is often required to also efficiently convey a large amount of perceptual information. A complete analysis of the verbal description of visual concepts would require considering the verbal communication process from perceptually low-levels through the generation of linguistic responses. This paper will concentrate however, on outlining how linguistic concepts of tense and aspect can be generated from mostly intermediate-level visual percepts.

3. Conceptualizating the Event

To further discuss the model of a visuo-linguistic interface, it is important to define what concepts and conceptualizations are. A *concept* is an association of object, state, and event (object and state changes) representations which have perceptual, linguistic, physical, and cognitive foundations. *Conceptualization* is the process of associating those representations under a common conceptual theme as concepts. There is no default structure for concepts since they are representations of distributed knowledge sources and may be associated with several other concepts. Conceptual association is constrained by the memory and processing capability of the conceptualizing agent. In this paper, we are concerned with the *visuo-linguistic conceptualization of events:* the process of associating sequential visual object, state, and event changes with

---

[1] Space does not permit us to review the low-level vision research but Cf. [2], [6], [13].

[2] An excellent collection of papers concerning computer vision systems may be found in [10].

language and vice versa. To address this concern, the description of linguistic and visual concepts must be presented. The following sections will outline linguistic and visual concepts and discuss their properties.

3.1 Lexical Semantics for Verbs

In this section we outline the framework that defines our domain for linguistic and lexical conceptualization. We will adopt an interval-based semantics, the *Extended Aspect Calculus* ([25]), which provides a semantics for lexical items and constrains what word meanings are possible for lexicalization in a language. The thesis of this approach is to decompose the events denoted by verbs into the subintervals that compose them (cf. [7]).

Our model is a first-order logic that employs special symbols acting as operators over the standard logical vocabulary. These are taken from three distinct semantic fields. They are: *causal*, *spatial*, and *aspectual*. The predicates associated with the causal field are: $Causer(C_1)$, $Causee(C_2)$, and $Instrument(I)$. The spatial field has two predicate types: *Locative* and *Theme*. Finally, the aspectual field has three predicates, representing three temporal intervals: t_1, beginning, t_2, middle, and t_3, end. From the interaction of these predicates all thematic types can be derived.[3]

Let us illustrate the workings of the calculus with a few examples. For each lexical item, we specify information relating to the argument structure and mappings that exist to each semantic field; we term this information the *Thematic Mapping Index (TMI)*.

Part of the semantic information specified lexically will include some classification into one of the following event-types (cf. [1], [7], [15], [26], [30]).

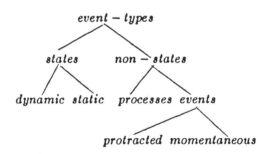

For example, the distinction between state, activity (or process), and accomplishment can be captured in the following way. A state can be thought of as reference to an unbounded interval, which we will simply call t_2; that is, the state spans this interval.[4] An activity or process can be thought of as referring to a designated initial point and the ensuing process; in other words, the situation spans the two intervals t_1 and t_2. Finally, an event can be viewed as referring to both an activity and a designated terminating interval; that is, the event spans all three intervals, t_1, t_2, and t_3.

We assume that part of the lexical information specified for a predicate in the dictionary is a classification into some event-type as well as the number and type of arguments it takes [18], [31]. For example, consider the verb *run* in sentence (1), and *give* in sentence (2).

(1) John ran yesterday.

(2) John gave the book to Mary.

---

[3] The presentation of the theory is simplified here, as we do not have the space for a complete discussion. See [25] for discussion.

[4] This is a simplication of our model, but for our purposes the difference is moot. A state is actually interpreted as a primitive homogeneous event-sequence, with downward closure. Cf. [26],

We associate with the verb *run* an aspect structure P (for process) and an argument structure of simply $run(x)$. For *give* we associate the aspect structure A (for accomplishment), and the argument structure $give(x, y, z)$. The Thematic Mapping Index for each is given below in (3) and (4).

(3)

$$
run = \begin{pmatrix}
& C_1 & \\
& | & \\
& x & \\
& | & \\
& Th & \\
& \diagup\; | & \\
t_1 & t_2 &
\end{pmatrix}
$$

(4)

$$
give = \begin{pmatrix}
& C_1 & C_2 & \\
& \diagup & \diagup & \\
x & y & z & \\
| & | & | & \\
L & Th & L & \\
| & | & | & \\
t_1 & t_2 & t_3 &
\end{pmatrix}
$$

The sentence in (1) represents a process with no logical culmination, and the one argument is linked to the named thematic (or case) role, *Theme* [14], [27]. The entire process is associated with both the initial interval t_1 and the middle interval t_2. The argument x is linked to C_1 as well, indicating that it is an *Actor* as well as a moving object (i.e. *Theme*). This represents one TMI for an activity verb.

The structure in (2) specifies that the meaning of *give* carries with it the supposition that there is a logical culmination to the process of giving. This is captured by reference to the final subinterval, t_3. The linking between x and the L associated with t_1 is interpreted as the thematic role *Source*, while the other linked arguments, y and z are *Theme* (the book) and *Goal*, respectively. Furthermore, x is specified as a *Causer* and the object which is marked *Theme* is also an affected object (i.e. *Patient*). This will be one of the TMIs for an accomplishment.

Finally let us consider how this lexical information is actually used when we form sentences in the language. In particular, let us examine the distinction between the *simple past* forms of a sentence (5) and the *progressive* forms in (6).

(5) a. The plane landed.

b. The plane descended.

(6) a. The plane is landing.

b. The plane is descending.

Notice that (6b) entails (5b) but it is not true that (6a) entails (5a). That is, although we can say that the plane has descended if we say that it is currently descending, it is not the case that the plane has completed its landing if we say that it is landing. If we classify *descend* as an activity and *land* as an accomplishment, however, we are able to capture this distinction in entailments.

Let us say that the progressive acts as an operator over an event sequence, and picks out the middle interval t_2 as the one being referred to.

This means that the subevent being referred to by use of the progressive is inside the event t_2, and does not entail the completion of a landing, since there is no culminating event associated with the progressive. Given this analysis for the progressive, we now can explain why process verbs allow the inference *if x is V-ing, then x has V-ed*.

3.2 Visual Event Concepts

Visual event concepts (or visual events) are associations of percepts which are "hard-wired" low-level visual primitives (motion, location, intensity, size, color, etc.) and spatio-temporal relations defined by those primitives (e.g. under, near, between, etc.). A majority of percepts represent object states and relations, while only a few percepts represent events (e.g. motion). Our definitions for motion are found in Figure 1.

```
[motion
 (object (motion-right motion-left motion-forward
                       motion-backward motion-up motion-down 0)))]

[motion-right
 (object (location 0) (location (x (increase 1)))))]
                          ⋮
[motion-down
 (object (location 0) (location (y (decrease 1)))))]
```

[Figure 1]

We assume a viewer-oriented coordinate system with the origin at the center of the field of view. The z-axis is the line of sight of the observer (+z) through the origin. The x-axis corresponds to the observers' right (+x) and left while the y-axis is up (+y) and down. The numbers are visual sampling indices which suggest the expected sequence of location states which determine motion. We present these definitions to illustrate that though motion and location can be decomposed into more primitive elements (coordinates), we will define location and motion (change of location) as our most primitive state and event, respectively. Our theory is concerned with percepts which are sufficient for verbal descriptions using tense and aspect, therefore percepts which are determined by low-level location and motion are considered intermediate-level visual percepts.

Since events are more salient than states, percepts which denote events have greater control of an observer's visual focus of attention than percepts which simply denote states. Visual events which include such percept changes can influence (support, interrupt, suspend, or terminate) the observer's attention. Furthermore, short-term memory constraints force the observer to attend to perceptual changes during an observation. We define object, state, and event changes as simple events so that a visual event may be defined as a sequence of one or more simple events. This sequence may be a sub-sequence (sub-event) of any number of other distinct visual events. For the remainder of this paper, visual events will be referred to simply as events and simple events which bound visual events will be termed initial and final events (a simple event is the initial and final event of itself).

Events are largely developed through observation which is the concurrent processing of identifying objects and their behavior, predicting and matching event-schemata, and evoking linguistic, cognitive, and physical descriptive procedures. These procedures are evoked at some level of abstraction which is appropriate for the description, which by default is the highest level. Description complexity may vary from simple perceptual recognition to combinations of linguistic, cognitive, and physical procedures. Generally, events are percept changes defined in terms of object combinations and the polyadicity (number of object arguments) of the percepts. We have identified percepts which require one, two, or three objects similar to those in [19]. Each visual sampling interval (scene) is represented by the simple event which is the set of monadic, dyadic, and triadic percepts using each object, object pair, and object triple as arguments, respectively.

446

The purpose of observation is to describe known event-schemata and to define new event-schemata. Generally, new schemata are constructed from a visual activity history by: retaining the history as it was observed; defining sub-events from changing "hard-wired" percepts; defining sub-events from any changing percepts or simple events; or by matching predicted simple events from known event-schemata. New event-schemata are named through interaction with a critic or by concatenating the names of previously defined events and recognized percepts. The probability of event recognition is measured by the degree to which event-schemata are matched.

4. A Visuo-linguistic Interface Model

At each scene, matched percepts, simple events, sub-events, and event-schemata will determine the generation of verbal descriptions where objects assume thematic roles. The description process is guided, in part, by recognizing whether the objects, the initial event, and the final event can be identified, partially identified, or unidentified. This will in turn determine whether the event is a definite, probable, or possible past, present, or future process, achievement, or accomplishment. The *past* is considered when all events have occurred prior to the present scene; the *present* when all simple events occur within the present scene; and the *future* is used when all events will occur after the present. The following is our algorithm for the visuo-linguistic interface:

1. OBSERVE *scene_n*.

When an observation begins, the observer creates a visual history in intermediate-term blackboard memory. At each scene the observer confirms the recognition of objects and the spatio-temporal relations between objects.

2. PREDICT event-schemata.

Predictions of long-term memory event-schemata are goal-based when selected by the observer through non-visual (e.g. verbal) input, object-based when selected by visibly identifying objects which are event-schemata agents, and event-based when selected by identifying spatio-temporal relation sequences of visibly unidentified objects and plausibly inferring event-schemata agents.

3. DETERMINE changed percepts.

The observer's attention is driven by percept changes during each scene. The degree of attention is roughly proportional to the number of changed percepts: the larger the number of changes, the greater the need for attention. Goal-based prediction will evoke expectation-driven attention while object-based and event-based predictions will evoke data-driven attention. Putting these together, we can define the total attention to be the cooperative and/or competitive interaction between the data-driven and expectation-driven mechanisms.

4. MATCH observed sub-events with predicted event-schemata.

Sub-events are identified by focusing attention on default "hard-wired" and/or learned percepts. Identified sub-events are matched with sub-events of predicted event-schemata.

5. CLASSIFY predicted event-schemata using matched sub-events.

Matched sub-events are compared with the structure of predicted event-schemata by verifying object agents, and determining the state (past, present, or future) of predicted event-schemata and their matched sub-events. Predicted event-schemata are subsequently classified as processes, achievements, or accomplishments .

6. PRIORITIZE classified predicted event-schemata.

Predicted event-schemata are ordered based on the percept salience of the sub-events by which they were classified. For example, a nearby object quickly moving toward the observer may be more salient than a distant object moving slowly away from the observer.

7. DESCRIBE successfully predicted event-schemata.

Verbal descriptions are generated based on the salience, state, and classification of the predicted event-schemata. This will include tense, aspectual, and causal references. The fine-grained temporal granularity of visual perceptions will be mapped into medium-grained perceptual changes which are mapped into coarse-grained linguistic descriptions.

8. COMMENT and generate QUERIES about unsuccessful predictions.

Verbal comments are generated for predicted event-schemata whose descriptions suggest improbable occurrence and minimal salience. The observer will direct questions to an interactive critic in an attempt to relate successful predictions to unsuccessful predictions ([6], [21]).

9. REFINE, UPDATE, and CREATE event-schemata.

Through dialogue the observer will attempt to assign credit to percepts and sub-events in an effort to create new event-schemata and revise known event-schemata.

10. REPEAT UNTIL $scene_f$.

The process continues until the observation is terminated by a minimal amount of salience in the scene for an extended period of time, or through the volition of the observer.

It should be pointed out that event-schemata predictions are made in order to reduce the search problem of a large number of event-schemata with a very large number of percepts. Goal-based predictions are specific and require the less of the observer's attention resources than object-based predictions while event-based predictions are general and require more of the observer's attention than object-based predictions. This attention disparity exists at the beginning of the observation, but it is expected that by the end of the observation a small number of event-schemata will have actually been described.

5. Examples

To illustrate our theory of the integration of language and perception, consider that an observer and a critic witness air show events at an airport. There are two objects at the show: a plane and a runway. The observer is a novice and can identify planes and runways and the critic is an aviation expert. Assume that for every scene in the observation the observer perceives the location and motion of both objects. From these "hard-wired" percepts the observer determines other percepts: on, over, above, velocity, and altitude. Let us say that the observer focuses on percept changes between each scene and represents them in a visual activity history. If an observation yields the following history of percept changes:

```
(visual-history
 (runway (location 0))
 (plane  (location 0) (motion 3)
         (velocity (zero 0) (increase 4) (constant 15) (decrease 19)
                   (constant 27))
         (altitude (constant 0) (increase 10) (constant 16) (decrease 20)
                   (constant 25))
         (runway (on 0) (over above 10) (above 13) (over above 20)
                 (on 25))))
```

and the observer can verbally describe percepts, it could describe the activity in any scene in terms of the percepts:

```
(scene-0
 (runway (location 0))
 (plane  (location 0)
         (velocity (zero 0))
         (altitude (constant 0))
         (runway (on 0))))
```

$$ING(sit\ x)$$
$$ON(Th, L)$$
$$Theme \rightarrow plane/[-motion]$$
$$L \rightarrow runway$$

"A plane is sitting on a runway."

Scene 0 suggests that a motionless plane is the direct agent of sitting on a runway location. This would be the case until scene 4:

```
(scene-4
 (runway (location 0))
 (plane  (location 0) (motion 3)
         (velocity (increase 4))
         (altitude (constant 0))
         (runway (on 0))))
```

$$ING(move\ x)$$
$$ON(Th, L)$$
$$Theme \rightarrow plane/[+motion]$$
$$L \rightarrow runway$$

"The plane is moving faster on the runway."

Scene 4 shows that the plane had been on the runway since scene 0, moved since scene 3, and increased velocity in scene 4. The observer can also generate sub-events based on any particular changing percept. For instance, the observer can define a simple sub-event by focusing on the change in velocity of the plane at scene 15 and include all percept changes which occurred between the last two successive velocity changes in scenes 4 through 15 and call it a "foo":

```
(sub-event-foo
 (runway (location 0))
 (plane  (location 0) (motion 3)
         (velocity (increase 4) (constant 15))
         (altitude (increase 10))
         (runway (over above 10) (above 13))))
```

$$ING(foo\ x)$$
$$Theme \rightarrow plane/[+motion]$$
$$L \rightarrow above\ runway/[+location]$$

"The plane is increasing altitude above the runway at constant speed."
or
"The plane has fooed."

The observer could continue to generate descriptions of this visual activity in such terms, but for long and complex events there could be a very large number of percepts and sub-events making verbal descriptions too detailed, awkward, lengthy, or ridiculous. For these reasons, it is sometimes desirable that sub-events have more concise and meaningful descriptions. Sub-events could be identified by an interactive critic who can recognize and label them linguistically. Consider that the following dialogue takes place after witnessing the visual activity:

```
Critic: "The plane takes-off when it accelerates on the runway
         and then ascends."
Observer: "What is ascending?".
Critic: "The plane ascends when it increases altitude."
```

This verbal exchange causes the observer to focus attention on "increasing altitude" at scene 10. The observer now constructs an "ascend" sub-event schema:

```
(ascend
 (runway (location 0))
 (plane  (location 0) (motion 3)
         (velocity (zero 0) (increase 4))
         (altitude (constant 0) (increase 10))
         (runway (on 0) (over above 10))))
```

From the observation and the dialogue, the role of the runway in the plane's ascending is not clear. Furthermore, the critic has not given any definite indication as to when an ascend begins and ends. If the dialogue continues:

```
Observer: "When does an ascend begin?"
  Critic: "The plane begins to ascend when it increases
           altitude."
Observer: "When does it end?"
  Critic: "When the plane stops increasing altitude."
Observer: "Does a plane need a runway to ascend?"
  Critic: "No."
Observer: "Does it need velocity to ascend?"
  Critic: "Yes."
```

450

and scene indices are normalized, the observer may generate a more refined schema for "ascend":

```
(ascend
  (plane (location 0) (motion 1)
         (velocity (increase 2) (constant 4))
         (altitude (increase 3) (constant 5))))
```

Careful guidance by the critic could result in other refined event-schema definitions such as take-off, descend, and landing. The observer could now describe the same visual activity at a higher level of abstraction (? indicates unobserved percept):

```
(scene-4
  (runway (location 0))
  (plane  (location 0) (motion 3)
          (velocity (increase 4))
          (altitude (constant 0))
          (runway (on 0))))
```

```
(take-off
  (runway (location 0))
  (plane  (location 0) (motion 3)
          (velocity (increase 4)) (altitude (increase ?)))
          (runway (on 0) (over above ?))))
```

```
                "The plane is taking-off."
```

```
(scene-23
  (runway (location 0))
  (plane  (location 0) (motion 3)
          (velocity (decrease 19))
          (altitude (decrease 20))
          (runway (over above 20))))
```

```
(descend
  (plane (location 0) (motion 3)
         (velocity (decrease 19) (constant ?))
         (altitude (decrease 20) (constant ?))))
```

```
(land
  (runway (location 0))
  (plane  (location 0) (motion 3)
          (velocity (decrease 19)) (altitude (decrease 20)))
          (runway (over above 20) (on ?))))
```

```
        "The plane is descending and has almost landed."
```

In these cases, the observer is guided to define sub-events by focusing attention on suggested percepts rather than focus attention on "hard-wired" or motion-related percepts though all percepts remain building blocks for sub-events. Partial event-schemata matches were found to be helpful in generating descriptions with the use of words such as "almost" and "partially" though the events never completely occurred. The observer may now describe new visual activity in terms of events that it can recognize.

Without the benefit of instruction, it would take our observer several observations outside the proximity of an airport to notice that planes often ascend without runways and sometimes ascend due to increased wind velocity. [5] While the plane's velocity is not essential for visually recognizing ascent or descent, such percepts can be included in event-schemata to help the observer make causal inferences in verbal descriptions.

Our examples show that our visual event definitions are hierarchical (since sub-events are constructed from events) and concurrent. We are quick to point out that without the benefit of language, event boundaries may be determined by percept salience alone, however, language can help to determine and label visual events on non-salient or non-visual bases. Thus the default temporal granularity and focus of attention during event processing can be altered by using language.

6. Summary and Future Work

Our theory relates the thematic roles of objects in events to lexical and perceptual semantics. It presents a plausible mapping from visual percepts to linguistic descriptions and the inverse transformation from linguistic descriptions to visual event-schemata. We have suggested the role that language may play in describing perceptions and provide an algorithm which describes this mapping process. We introduce goal-based, object-based, and event-based prediction and show how such predictions are integrated to focus attention on input which may be linguistic as well as perceptual.

The authors would like to point out several significant directions that our research in perceptual-linguistic interfacing and related issues can be explored. First, though we are directly concerned with vision and language in this paper, such work should lead towards investigations in perceptual modality and descriptive integration. For example, the next step in defining formalisms could be to select another perceptual modality (e.g taction) and another descriptive mechanism (e.g. motor-control) and develop formalisms which describe how an intelligent, observing entity may physically move as a result of how it is physically touched. Along with the theory outlined in this paper a more complete characterization of perceptual description may result.

Another interesting avenue to explore would be how modal and descriptive integration can be controlled. One idea is that the lexicon, percepts, and event-schemata can be nodal processors in a massively parallel fine-grained computational network similar to [12] and more sophisticated memory and inference and search reduction mechanisms such as [28] may be employed. We are exploring such implementation details and find that a "Society of Mind" [20] architecture may be most promising.

---

[5] This is the same problem as learning the necessary conditions for an event or concept. The more general notion of the concept will arise with the right training instances. See [21], [22].

Bibliography

[1] Bach, Emmon, "The Algebra of Events", in *Linguistics and Philosophy*, 1986.

[2] Ballard, Dana H, Brown, C., *Computer Vision*, Prentice-Hall, New Jersey, 1982.

[3] Boggess, L. C., "Computational Interpretation of English Spatial Prepositions," Report T-75, Coordinated Science Laboratory, University of Illinois at Urbana-Champaign, February 1979.

[4] Borchardt, G. C., "Event Calculus," in *Proceedings, Ninth International Joint Conference on Artificial Intelligence*, Los Angeles, August 1985, 524-27.

[5] Brooks, R. A., "Symbolic Reasoning Among 3-D Models and 2-D Images," Report STAN-CS-81-861, Department of Computer Science, Stanford University, June 1981.

[6] Cohen, P. R., and Feigenbaum, E. A., *The Handbook of Artificial Intelligence*, 3, William Kaufmann, Los Altos, 1982.

[7] Dowty, David R., *Word Meaning and Montague Grammar*, D. Reidel, Dordrecht, Holland, 1979.

[8] Fillmore, Charles, "The Case for Case", in *Universals in Linguistic Theory*, E. Bach and R. Harms (eds.). New York, Holt, Rinehart, and Winston, 1968

[9] Gruber, Jeffrey, "Studies in Lexical Relations" unpublished PhD, MIT, 1965

[10] Hanson, A. R., and Riseman, E. M., (Eds.), *Computer Vision Systems*, Academic Press, New York, 1978.

[11] Herskovits, A., "Semantics and Pragmatics of Locative Expressions," *Cognitive Science*, 9, 3, 1985, 341-78.

[12] Hillis, D., *The Connection Machine*, MIT Press, Cambridge, 1985.

[13] Horn, B.K.P., *Robot Vision*, MIT Press, Cambridge, 1986.

[14] Jackendoff, Ray, *Semanic Interpretation in Generative Grammar*, MIT Press, Cambridge, MA. 1972

[15] Kenny, Arthur, *Actions, Emotions, and Will*, Humanities Press, New York. 1963

[16] Langacker, R. A., "An Introduction to Cognitive Grammar," *Cognitive Science*, 10, 1, 1986, 1-40.

[17] Marr, D., *Vision*, W. H. Freeman, San Francisco, 1982.

[18] Miller, George, "Dictionaries of the Mind" in Proceedings of the 23rd Annual Meeting of the Association for Computational Linguistics, Chicago, 1985.

[19] Miller, G. A., and Johnson-Laird, P. N., *Language and Perception*, Belknap/Harvard University Press, Cambridge, 1976.

[20] Minsky, M. L., *The Society of Mind*, Simon and Schuster, New York, 1987.

[21] Mitchell, Tom, "Version Spaces: A Candidate Elimination Approach to Rule Learning," in *Proceedings, Fifth International Joint Conference on Artificial Intelligence*, August 1977.

[22] Michalski, R.S. "A Theory and Methodology of Inductive Learning," in Michalski et al (eds.), *Machine Learning I*, Tioga Press, 1983

[23] Neumann, B., and Novak, H.-J., "Event Models for Recognition and Natural Language Description of Events in Real-World Image Sequences," in *Proceedings, Eighth International Joint Conference on Artificial Intelligence*, Karlsruhe, W. Germany, August 1983, 724-26.

[24] Pustejovsky, James, "The Acquisition of Lexical Entries: The Perceptual Origin of Thematic Relations," to appear in *Proceedings of the 25th Meeting of the Association of Computational Linguistics*, Seattle, 1987.

[25] Pustejovsky, James, "The Extended Aspect Calculus", Submission to special issue of *Computational Linguistics*, 1987

[26] Ryle, Gilbert, *The Concept of Mind*, Barnes and Noble, London, 1949

[27] Schank, Roger, *Conceptual Information Processing*, North-Holland, Amsterdam, 1975.

[28] Stanfill, C., and Waltz, D. L., "The Memory-Based Reasoning Paradigm," Thinking Machines Corporation, Cambridge, 1987.

[29] Talmy, L., "How Language Structures Space," in *Spatial Orientation: Theory, Research, and Application*, Acredolo, L., and Pick, H., (Eds.), Plenum Press, 1983.

[30] Vendler, Zeno, *Linguistics and Philosophy*, Cornell University Press, Ithaca, 1967

[31] Wilks, Yorick "Preference Semantics," *Aritficial Intelligence*, 1975.

RHO-SPACE:
A NEURAL NETWORK FOR THE DETECTION AND REPRESENTATION OF ORIENTED EDGES[1]

D.K.W.Walters, Computer Science Dept., State University of New York at Buffalo, NY 14260

Abstract

This paper describes a neural network for the detection and representation of oriented edges. It was motivated both by the inherent ambiguity of convolution-style edge operators, and the processing of oriented edge information in biological vision systems.

The input to the network is the output of oriented edge operators. The computations within the network are based on orientation dependent, three-dimensional, excitatory and inhibitory neighborhoods in which computations such as lateral inhibition and linear excitation can occur.

Rho-space has a variety of interesting properties, which have been investigated. These include:
1) Both coarse and fine representation of the orientation information is possible.
2) No global thresholding is required, and the local adaptive thresholding is localized in orientation, as well as in spatial position.
3) The filling-in of dotted and dashed lines readily occurs.
4) There is a natural representation of connectivity, which agrees with human perception.
5) Illusory contours, of one type produced by the human visual system are produced.
6) All processing is completely data-driven, and no domain dependent knowledge or model based processing is used.

1. Introduction

The most universally applied stage of low-level visual processing is the detection of image edges, be they intensity edges, motion edges or texture edges. Yet there are some basic theoretical problems which complicate the detection and representation of image edges. One such problem is the inherent ambiguity in the response of any single convolution-style edge operator; the operator responds to the conjunction of edge location, orientation, amplitude, etc., and the values of these contributing factors can not be untangled from a single response. In this paper it is argued that the ambiguity problem lies not with edge operators themselves, but with how the operator responses are being interpreted in current computer vision systems. For example, neural-based biological vision systems use convolution-style, oriented edge operators and appear to have solved the ambiguity problem. This suggests that a solution exists using the style of representation and computation possible in neural networks.

This paper explores the use of a neural network for the detection and representation of oriented edges. Grossberg and Mingolla [1] have also addressed the edge detector ambiguity problem through the use of networks which model neural computations at the level of the dynamic, cooperative and competitive interactions of feedback, shunting, etc. Our research differs in two main ways: first, the neural networks studied perform static, noniterative, discrete computations of the type that could be easily implemented in a clocked, discrete, parallel digital architecture; and second, emphasis is placed on the type of distributed representation that is possible in such networks.

2. Representations for Oriented Edges

Although there is considerable debate amongst vision researchers as to the existence of an optimal edge detector [2, 3, 4, 5], there is general agreement about how oriented edges should be represented. The standard representation consists of an amplitude or gradient image, $A(x,y)$, in which each point represents the amplitude or gradient of the edge at spatial position (x,y); and an orientation image

---

[1] This research is funded by NSF Grant IST 8409827 awarded to the author.

$B(x,y)$, in which the value of each point represents the orientation or gradient direction at spatial position (x,y). The implicit assumption is that the local edge amplitude and orientation at each image point can be measured. But the ambiguity problem invalidates this assumption for the response of individual edge operators. One potential solution is to look at the response of a set of edge operators at each image location, such as set of oriented operators. It will be shown that the response of the set as a whole contains significant structural information. But most techniques using oriented edge operators use either thresholding or local averaging to produce a single estimate of edge amplitude and orientation from the set, and this loses the structural information. This point has been made by Zucker [6], who proposes a model-matching scheme for reconstructing the edge information from the responses. A model-free approach is taken here which uses a distributed representation of oriented edge information, ρ-space, in which the responses of oriented edge operators for a single image point are not combined through thresholding or other techniques, which allows the subsequent computations to disambiguate the orientation, position and amplitude information.

3. Rho-Space Representation

Rho-space is a three dimensional space, where the x and y dimensions represent the spatial dimensions of an image, and the third dimension, ρ, represents the orientation of image contours (intensity edges, texture boundaries, lines, etc.). The space is discretized in all dimensions.

The input to ρ-space is currently produced by convolving an image with a set of Canny-type oriented edge operators [2], of either 8 or 18 separate orientations. Figure 1 shows a diagram of the ρ-space. Each orientation plane shown in Fig. 1 can be thought of as the result of convolving the image with an edge operator of a given orientation. The value at each location of a single orientation plane is then the amplitude of the output of that particular operator at that image location. Figure 1 shows 6 orientation planes, but in the computer implementation either 8 or 18 orientation planes were used.

The algorithms developed for the ρ-space representation assume that there is a simple processor associated with each point or pixel in the space, and each processor is locally connected only to those processors in its three-dimensional neighborhood (as defined below). Each simple processor is actually a network of neural-type units, but can be simulated as single processor which runs a simple internally stored program. The ρ-space processors perform local, noniterative computations such as discrete forms of lateral inhibition, short and long range linear excitation, and short and long range linear inhibition. These computations put the oriented edge information into a form that is usable by a wide range of

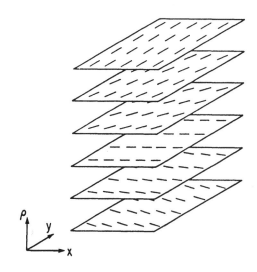

Figure 1

low and intermediate level visual computations such as perceptually based enhancement, segmentation and grouping [7]. Further descriptions of the ρ-space representation can be found in [8,9].

3.1. Definitions

Assume there are k processors in ρ-space, and define x_j to be the activity of processing unit j.

3.2. Excitatory and Inhibitory Neighborhoods in Rho-Space

All ρ-space computations, and the definitions of oriented lines in ρ-space are based on the concept of the local neighborhood of each point in ρ-space. The excitatory neighborhood, E_j, of point j includes all points which are directly connected to j, and which participate in excitatory operations. The inhibitory neighborhood, I_j, of point j consists of all points which are directly connected to j, and which participate in inhibitory computations.

All E_j and I_j are defined as functions of the edge operators used to generate the input to ρ-space. The two spatial dimensions of the neighborhoods, n by m, are the spatial dimensions of the convolution kernels of the edge operators. The orientation dimension, d, of the neighborhoods are determined by the number of separate orientations represented in the convolution kernels. For example, for sixteen oriented 7 by 7 operators of the type illustrated in Figure 2a, the excitatory neighborhood for the central solid horizontal pixel is that shown in Figure 2b, where the non-empty circles represent the locations in the excitatory neighborhood. Figure 2b shows just a small portion of ρ-space, part of each of three consecutive orientation planes. The middle plane corresponds to horizontal edges (0 degrees), while the top and bottom planes contain information about orientations of +22.5 and -22.5 degrees respectively. The excitatory neighborhood of each point in ρ-space lies within a small rectangular box shaped region of ρ-space.

The inhibitory neighborhood, I_j, of point j is the compliment of the excitatory neighborhood, E_j, defined over the n by m by d space centered on the point j.

Each neighborhood can be divided into two halves by a plane which passes through the central point, and which is orthogonal to the orientation direction of the central point. Let $E_{j,1}$ and $E_{j,2}$ refer to the two halves of E_j.

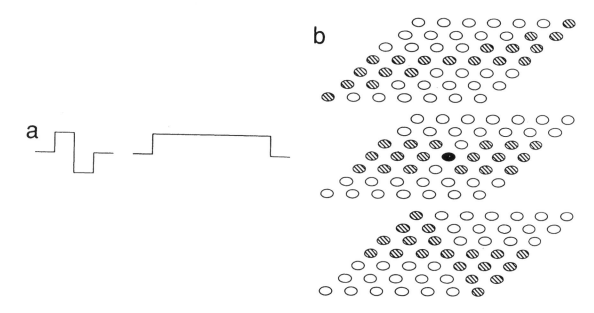

Figure 2

457

4. Computations Using the Rho-Space Representation

The computations in ρ-space are all local computations which involve only a single processor and it's local neighbors. Four such computations are lateral inhibition (LI), short-range linear inhibition (SRLI), short-range linear excitation (SRLE), and mid-range linear inhibition (MRLI). Each processor, x_j, is composed of 5 neural units: O_j, is the output of the oriented edge operators; T_j is the output of LI, L_j is the output of SRLI; F_j is the output of SRLE; and R_j is the output of MRLI. For example, LI is defined :

$$T_j = O_j \left(\max \left((O_j - M_{O,I_j}), 0 \right)/(O_j - M_{O,I_j}) \right)$$

where $M_{O,I_j} = \max_i (O_i \ st \ O_i \ is \ an \ element \ of \ I_j)$.

SRLI is defined as :

$$L_j = T_j \left(\max \left(-M_{T,E_j}, 0 \right)/-M_{T,E_j} \right)$$

where, $M_{T,E_j} = \max_i (T_i \ st \ T_i \ is \ an \ element \ of \ E_j)$.

All four computations are illustrated in Figure 3 where (a) shows a black-white checkerboard image in which uniform noise has been added to each pixel, and part (b) shows the nonzero pixels present in the input image. In part (c) each white pixel indicates that at least one of the oriented operators had a non-zero response at that image point. Thus, the operators are indicating that all but one of the points in this image could be an edge point. A common method for interpreting such responses is to use either a global or a local threshold to remove unwanted responses, but it is generally not possible to find a threshold which removes all of the noise-generated responses, and none of the edge-generated responses. Part (d) shows the results after LI, which is a kind of local adaptive thresholding, but which has the added advantage that it is orientation selective. Thus a high amplitude horizontal edge does not inhibit a neighboring low amplitude vertical edge. Part (e) shows the results after SRLI which has the function of removing potential edge points which are not connected to other edge potential points, and thus could not have arisen from true edges. Part (f) shows the small gaps in the lines representing edges being filled in by SRLE, while part (g) shows the short, unconnected lines removed after MRLI, yielding the connected edges of the checkerboard. (More details of these excitatory and inhibitory interactions can be found in [10]).

5. Definitions of Lines in Rho-Space

As one of the goals for computations in ρ-space is to group local edges into a form which represents more global image edges, a means of defining an image edge is required. If we were working in euclidean space, then we might define an image edge as a connected line of local edges, but in ρ-

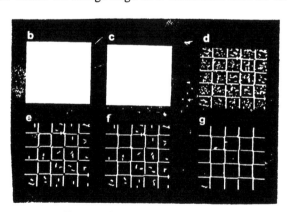

Figure 3

space the euclidean definitions of connectivity and of lines do not hold. Thus the following ρ-space definitions are required.

A point j is <u>connected</u> to another point j' if and only if both are nonzero, and j lies within $E_{j'}$ and j' lies within E_j.

(Note that two processors in rho-space can be connected, without the edge points associated with them being connected.)

A <u>line</u> is a set of connected points which lack neighbors in their inhibitory neighborhoods.

A <u>line end</u> is a line point, j, which has neighbors in either $E_{j,1}$ or $E_{j,2}$, but not in both.

6. An Example of Grouping and Segmentation

From the above definitions, it is possible to group local edge points in ρ-space into more global image edges or lines, and from structural information about line ends, it is also possible segment an image into sets of lines which are likely to have arisen from a single object [7]. An example of this process is seen in Figure 4, where parts (a) through (f) show the results of the ρ-space computations described above, while parts (g) and (h) each show one of the two segments generated by the segmentation algorithm. Note that although the circle and the rectangle are connected in the image space, they are not connected in ρ-space, which aids in their segmentation.

7. Rho Space Properties

Rho-space has a variety of interesting properties, which have been investigated. These include:
1) Both coarse and fine representation of the orientation information is possible.
2) No global thresholding is required, and the local adaptive thresholding is localized in orientation, as well as in spatial position.
3) The filling-in of dotted and dashed lines readily occurs.
4) There is a natural representation of connectivity, which agrees with human perception.
5) Illusory contours, of one type produced by the human visual system are produced.
6) All processing is completely data-driven, and no domain dependent knowledge or model based processing is used.

As an example, consider the illusory contour property. One of the properties of the current implementation of ρ-space is that the response to a single line, is the line itself, and two orthogonal end lines. This behavior was pointed out by Marr and Hildreth as being an undesirable side effect of oriented edge operators [11]. However, the orthogonal end lines might play a positive role in perception - the formation of illusory contours. For example, when the pattern in Fig.5a is viewed from the

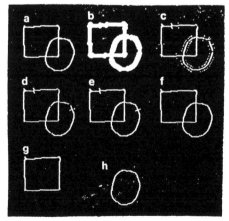

Figure 4

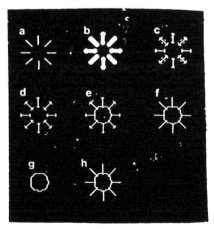

Figure 5

459

appropriate distance, the central region may appear "darker" than the rest. This can be explained in part in terms of an illusory circular contour being formed, and the orthogonal end lines can provide such a contour. To illustrate this, an example ρ-space computation is shown in Figure 5. Part (a) shows the input pattern. Part (b) shows the nonzero convolution responses, and part (c) shows the nonzero responses after LI. Part (d) shows the nonzero responses after SRLI, and the orthogonal end lines are apparent. Part (e) shows the nonzero responses after the SRLE, and a roughly circular contour which may correspond to the illusory contour is formed. In part (f) the short unconnected line segments have been removed by MRLI to form the final percept. This is an interesting result because an illusory contour has been formed in a completely data-driven manner, without reference to models, or inferring depth.

Grossberg has shown that the illusory contour formed by this pattern disappears when the individual lines are rotated by 45 degrees about their interior end points. Under such rotation the orthogonal end lines of the lines would not join to form a closed contour, and no illusory contour would be formed, thus supporting the orthogonal end line hypothesis for this illusion.

References

1. S. Grossberg and E. Mingolla, "Neural Dynamics of surface perception: Boundary-webs, illuminants and shape from shading," *Comp. Vision, Graph. & Image Proc.* **37**(1) pp. 116-165 (1987).

2. John F. Canny, "Finding Edges and Lines in Images," MIT Master's thesis (1983).

3. R.M. Haralick, "Digital step edges from zero crossing of second directional derivatives," *IEEE Trans. Pattern Anal. Machine Intell.* **PAMI-6** pp. 58-68 (1984).

4. S.E.L. Grimson and E.C. Hildreth, " "Comments on "Digital tesp edges from zero crossings of second directional dirivatives," *IEEE Trans. Pattern Anal. Machine Intell.* **PAMI-7** pp. 121-127 (January, 1985).

5. V. Torre and T.A. Poggio, "On edge detection," *IEEE T-PAMI*, pp. 147-163 (March, 1986).

6. S. Zucker, "Early Process for Orientation Selection and Grouping," pp. 170-200 in *From Pixels to Predicates*, ed. A.P. Pentland,Ablex, Norwood, NJ (1986).

7. D.K.W.Walters, "Selection of image primitives for general-purpose visual processing," *Computer Vision Graphics & Image Processing*, ((in press)).

8. D.K.W.Walters, "Selection and use of image features for segmentation of boundary images," Proceedings IEEE CVPR (1986).

9. D.K.W.Walters, "A Computer Vision Model Based on Psychophysical Experiments," pp. 88-120 in *Pattern Recognition by Humans and Machines*, ed. H.C.Nusbaum,Academic Press, New York, NY (1986).

10. D.K.W.Walters, *Parallel Computations in Rho-Space*. In preparation.

11. D. Marr and E. Hildreth, "Theory of edge detection," *Proceedings of the Royal Society of London B* **207** pp. 187 - 217 (1980).

LEARNING INTERNAL REPRESENTATIONS FROM GRAY-SCALE IMAGES:
AN EXAMPLE OF EXTENSIONAL PROGRAMMING

Garrison W. Cottrell
Institute for Cogntive Science
University of California, San Diego

Paul Munro
Department of Information Science
University of Pittsburgh

David Zipser
Institute for Cogntive Science
University of California, San Diego

ABSTRACT

The recent development of powerful learning algorithms for parallel distributed networks has made it possible to program computation in a new way. These new techniques allow us to program massively parallel networks by example rather than by algorithm. This kind of *extensional programming* is especially useful when there are no known techniques for solving a problem. This is often the case with the computations associated with basic cognitive processes such as vision and audition. In this paper we apply the technique to the problem of learning an efficient internal representation of image information directly from a gray-scale image. We compare the results of this to the engineering version of this problem, i.e., image compression. Our results demonstrate that a very simple learning method learns internal representations that are nearly as efficient as those developed by the best known techniques in image compression. Thus we have a technique whereby neuron-like networks can self-organize to form a compact representation of a visual environment.

LEARNING INTERNAL REPRESENTATIONS FROM GRAY-SCALE IMAGES: AN EXAMPLE OF EXTENSIONAL PROGRAMMING

Garrison W. Cottrell
Institute for Cogntive Science
University of California, San Diego

Paul Munro
Department of Information Science
University of Pittsburgh

David Zipser
Institute for Cogntive Science
University of California, San Diego

INTRODUCTION

The recent development of powerful learning algorithms for parallel distributed networks has made it possible to program computation in a new way. These new techniques allow us to program massively parallel networks by example rather than by algorithm. This kind of *extensional programming* is especially useful when there are no known techniques for solving a problem. This is often the case with the computations associated with basic cognitive processes such as vision and audition. In this paper we apply the technique to the problem of learning an efficient internal representation of image information directly from a gray-scale image. We compare the results of this to the engineering version of this problem, i.e., image compression. Our results demonstrate that a very simple learning method learns internal representations that are nearly as efficient as those developed by the best known techniques in image compression. Thus we have a technique whereby neuron-like networks can self-organize to form a compact representation of a visual environment.

The technique we employ is known as *back propagation*, developed by Rumelhart, Hinton, and Williams (1986). While we will not go into the details of it here, back propagation can be considered a generalization of the perceptron learning procedure for multilayer nonlinear networks of neuron-like computing elements. Training the network consists of repeated presentations of input-output pairs representing the function to be learned. The learning algorithm operates by adjusting the weights between the elements of the network in such a way as to reduce the overall error in the output. In many cases, the network finds a solution to the problem that was unknown in advance to the user. In doing so, it develops its own *internal representation* of the input that is useful for solving the problem. It is often difficult to analyze this representation because many units are involved and the representations are highly *distributed* over the set of internal units. A subgoal of the present research is to make a first step towards unraveling the nature of these representations by applying the learning mechanism to a domain where the types of useful representations have been well studied.

Another aspect of this work is that the representation of images in an efficient format by neuron-like computing elements may give us clues to the way such information is represented in actual neural tissue. The learning procedure itself is not particularly biologically plausible, but the mechanisms it discovers for solving problems are (Zipser, in press). Whether or not there is anything like back propagation in the brain, we learn something about how the brain *could* solve problems from the "neural" solutions it discovers. Such information could be useful in guiding neurobiologists in their observations of cell firings during cognitive tasks.

Encoder Networks

The problem of finding an efficient internal representation of an environment is called the *encoder problem*.[1] In PDP networks using back propagation, this problem is solved by giving a network the problem of performing an identity mapping over some set of inputs. The network is constrained to perform this mapping through a narrow channel of the network, forcing it to develop an efficient encoding in that channel. There are two interesting aspects to this: (a) the network is developing a compact representation of its "environment"; and (b) although the algorithm used was developed as a supervised learning scheme, in this case the learning can be regarded as unsupervised—since the training signal is the same as the input, the system self-organizes to encode the environment.

A network appropriate for performing this task in the image domain is shown in Figure 1. It consists of an 8×8 input patch, corresponding to a two-dimensional patch of an image, that is completely connected to sixteen *hidden units*, the "narrow channel" through which the patch of image must be transmitted. These hidden units are completely connected to an 8×8 output patch, where the image is reconstructed.

PROCEDURE

We trained the above network with a digitized image of the Intelligent Systems Group (ISG) at UCSD (Figure 2). A digitized image is an $M{\times}N$ light intensity function $f(x, y)$, where x and y correspond to the spatial coordinates within the image, and $f(x, y)$ is a light intensity value from 0 to 255. One element of $f(x, y)$ is often referred to as a *pixel*, for *picture element*. Thus the original image has eight bits of information for each pixel. However, there is a great deal of redundancy in this information. Neighboring pixel values will tend to be highly correlated. If the network can capture this redundancy, it can represent the image more compactly.

We trained our network by randomly sampling 8×8 patches of this image, converting the gray level value linearly to the range [0,1].[2] These values form the input to the network. Activation passes through the net, and

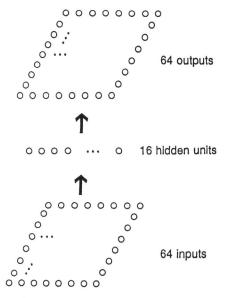

Figure 1. The network used in most of our examples.

---

[1] Ackley, Hinton, and Sejnowski (1985) were the first to demonstrate a learning algorithm for PDP networks that could solve the encoder problem.

[2] We used the usual sigmoidal activation function, with the output range scaled to [-1,1]. Since this function only asymptotically achieves the end values, it is easier for a unit to achieve values in the middle of the range. Hence converting the gray scale values to the range [0,.85] works better on this problem. We show results from the [0,1] conversion for historical reasons.

Figure 2. The original image of the Intelligent Systems Group (ISG) at UCSD.

activation of the output patch is obtained. This is compared with the input value, error is propagated back through the network, and the weights are updated according to the back propagation algorithm. We used an initial learning rate of .25 (no momentum), and trained the network on 100,000 patches of the image. Then the learning rate was lowered to .01 and the network was trained for an additional 50,000 iterations.

The result of this training is a "patch compressor." A reproduction of the image can be obtained by systematically applying this patch compressor across the original image, reconstructing a (nonoverlapping) patch at a time. In this way, the entire image is passed through the narrow channel of the hidden units, and we can view the reconstructed image to get an idea of the fidelity of the representation obtained by the hidden units.

In order to compare our results to that of image compression techniques, it is necessary to obtain a comparable measure of the number of bits used to represent the image. Image compression is measured by the number of bits transmitted per pixel of the reproduced image. In our case this corresponds to:

$$bits/pixel = \frac{(bits/hidden\ unit\ output) \times (\#\ hidden\ units)}{\#\ of\ pixels\ reproduced}$$

We must *quantize* (round off to a fixed number of values) the outputs of the hidden units in order to use this formula. For example, if we round off to 32 different output values, then this corresponds to five bits per hidden unit output. In the following examples, we used a uniform quantizer—the rounded-off values are equally spaced between [-1,1]. We could have done better (in terms of resulting error) by quantizing in ranges where the hidden unit outputs spend most of their time.

Finally, we need an objective *fidelity criterion* to measure how close the reconstructed image is to the original. The standard measure used is the mean square error normalized with respect to the squared intensity of the image. If $g(x, y)$ is the reproduced image, then the error is given by

$$e(x, y) = g(x, y) - f(x, y) ,$$

464

the mean-square error is given by

$$MSE = \frac{1}{M'N} \sum_{x=0}^{M-1} \sum_{y=0}^{N-1} e^2(x, y) \, ,$$

and the normalized MSE with respect to the average squared intensity of the image is given by

$$NMSE = \frac{MSE}{\frac{1}{M'N} \sum_{x=0}^{M-1} \sum_{y=0}^{N-1} f^2(x, y)} \, .$$

This is what we will use, expressed in percent.

RESULTS

The problem is to develop an efficient representation of the information in a digitized image. The network of Figure 1 does this by processing repeated presentations of samples of the visual environment (the image in Figure 2), using back propagation to correct the internal representation. The result is that the image can be represented with very little loss of information with 1 bit/pixel, representing an eight-fold compression of the information in the image. Also, the same representation does a good job of reproducing several images the network was not trained on.

Some reconstructions of the ISG image are shown in Figure 3. In Figure 3A five bits of hidden unit output were used, representing 1.25 bits/pixel. The most noticeable degradation from the original image is that the stripes on the shirt of the seated gentleman (Don Norman) are gone. This is not noticeably different from the result if we do not quantize the hidden units. On the other hand, reducing the output levels by another bit (16 values) more noticeably degrades the image (Figure 3B).

It turns out we can recover the shirt stripes if we use more hidden units, but compression suffers. Figure 4A is a reconstruction using 32 hidden units, with 16 output values each. This represents a compression of 2 bits/pixel. Higher compression can be obtained by using fewer hidden units, but the result is less satisfying. Figure 4B shows the results of using a network with 8 hidden units and 32 output values, resulting in .625 bits/pixel. More examples exploring the space of numbers of hidden units vs. numbers of quantization levels can be found in (Cottrell, Munro, & Zipser, in press).

How good a representation is this for images other than the training image? We naively expected that perhaps a network could be trained that would work well for all images, justifying the expense of the initial training. This is a somewhat misplaced dream, given that our network learns, in some sense, the statistics of the image it is trained on, and different images have different statistics. However, it may work well for a class of images. It turns out that it does a good job of reproducing some images that it wasn't trained on. Two of the images we tested it on and their reproductions are shown in Figure 5. We expect that it would not work well for images with very different statistics, such as text, but have not had a chance to try it on such images yet.

The Internal Representation

What is the internal representation at the hidden unit layer? Figure 6 shows the internal representation for eight hidden units. Each row corresponds to one hidden unit. Figure 6A shows the weight matrix for each of eight hidden units thresholded at various levels, one hidden unit per row. The center column, representing a threshold of 0, identifies which weights are negative and which positive. This gives an idea of the kind of pattern that excites each hidden unit the most. Figure 6B shows the output patch driven by each hidden unit alone. Again, each row corresponds to one hidden unit, and the columns correspond to different levels of activation from the hidden unit. The right-hand column thus corresponds to the output weights from that hidden unit. One obvious thing to note here is that the hidden units try to reproduce what they "see." Figure 6C shows the same information as 6B, in a gray scale image (6B is a thresholded version of 6C).

Figure 3. Quantization effects. A: 5 bits, 1.25 bits/pixel, NMSE 0.474%. B: 4 bits, 1 bit/pixel, NMSE 0.676%.

Figure 4. A: The reproduced image using 32 hidden units, four bits of quantized levels, 2 bits/pixel, NMSE 0.625%. B: The reproduced image using eight hidden units, 32 quantizer levels, resulting in .625 bits/pixel, NMSE 1.182%.

A

B

Figure 5. Two images (on this page and next) reproduced by the network trained on the image in Figure 2. A: The Symbolics Graphics group. B: Reproduced image, using six bits of quantized values, 1.5 bits/pixel, NMSE 1.267%.

Figure 5. C: Cadillac. D: Reproduced Cadillac, 1.5 bits/pixel, NMSE 0.764%.

What do these weights represent? We don't have an analytic answer to this question. However, we can compare the network's solution to a standard technique, the Principal Components Transform (PCT), to get an idea of what it does.

First we set up some correspondences between our network and the usual image compression system. The first step in a transform encoding system is to multiply the patch vector by a matrix to obtain less correlated coefficients:

$$\mathbf{y} = \mathbf{A}\mathbf{x}.$$

The y_i's are sent through a channel in a coded form, and at the other end they are transformed back into image space. The reconstructed image is the inverse transform

$$\hat{\mathbf{x}} = \mathbf{A}^{-1}\mathbf{y} \ .$$

It is the form of $\mathbf{A}$ that determines the type of transform. In the principal components transform, the rows of $\mathbf{A}$ are the eigenvectors of the covariance matrix of the $\mathbf{x}$ patch vector. This corresponds to setting up a new coordinate system with axes along the directions of maximum variance, and sending the coordinates in this new system. Then the inverse matrix converts back into image coordinates. For a principal components transform, this inverse matrix is just the transpose of $\mathbf{A}$. What is often done in this case is to just send the coordinates along the first k dimensions—the ones with highest variance. What this means is that the coefficients themselves (the coordinates along these high-variance axes) also have variance that is high for the first coordinate and that monotonically decreases.

The analog in our network is that $\mathbf{A}$ is the weight matrix between the input and hidden unit layers, with each row of $\mathbf{A}$ corresponding to the input weights on one hidden unit, and each hidden unit output a semilinear version of y_i. Similarly, the weight matrix between the hidden units and the output patch corresponds to $\mathbf{A}^{-1}$.

Now, we can begin to understand what the network does. First, observation has shown that during

469

(a) (b)

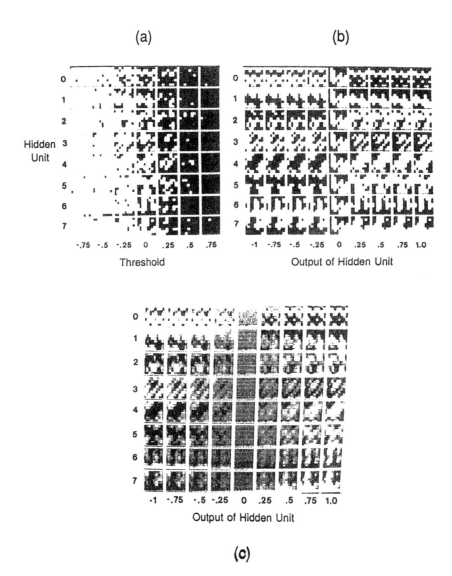

(c)

Figure 6. The internal representation. A: The weight matrices from the input patch to eight hidden units, thresholded from −.75 to +.75. The middle column (zero threshold) shows the "canonical" feature responded to by that hidden unit. B: The output patch driven by each hidden unit at different output values from −1 to 1. C: The same picture as (B) on a color monitor.

reconstruction of an image, the hidden unit outputs are mostly in the linear range of the activation function. So the network makes little use of the nonlinearity. Second, note that Figure 6 shows that the network also uses the transpose of the input weights as the output weights.

Finally, notice that the final image can be regarded as a linear combination of *basis images*: one for each coefficient (or hidden unit output). For comparison purposes, Figure 7 shows the basis images from the principal components transform for a picture of a cameraman (from Gonzales & Wintz, 1977). Figure 6B, the last column, shows the same thing for eight units of our network. Unlike the principal components transform, there is no obvious way to order the basis images.

This is reflected in the variances of the hidden units' outputs: They are all about equal (to 0.1) and the amount of error in the output accounted for by each one is of comparable size. Back propagation has spread the error relatively evenly across the hidden units. In the principal components transform, the "hidden units" would

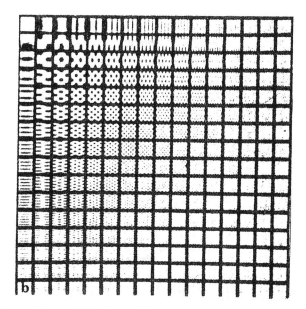

Figure 7. The set of Hotelling basis images for a particular image. (From Gonzales & Wintz, 1977. Reprinted by permission.)

have monotonically decreasing variance, and the variance typically falls off very quickly, so that they differ by orders of magnitude. Our conjecture at this point is that the hidden units span the space of the first several principal components, but are rotated so that each can have about equal variance.

DISCUSSION

This study has produced results that have implications for both connectionist networks and image compression. T⸋ ⸋se are discussed and summarized below.

Implications for Connectionist Networks

Extensional Programming

The major result of this study is that a relatively straightforward application of the back-propagation learning procedure to a problem that has been studied for many years results in near state-of-the-art performance. The key point is that this performance was obtained not by programming a connectionist solution to the problem, but by the process of *extensional programming*. In this procedure, many examples of the desired behavior are presented and the network must program itself to achieve the behavior. This suggests that other problems, where solutions are not known in advance, may be solved by back propagation.

A major problem with this technique is determining post hoc how the network solved the problem. In our case, we have some pieces of the answer, mainly because image compression is a well-studied problem. Hence we have some idea what to look for, if not an analytical solution. By comparison of our network to the techniques of image compression, we can gain insight into the solution found. However, this will not be the case in general. The importance of back propagation is that whether we know how to solve the problem or not, whether we know of an algorithm for the solution or not, back propagation will in most cases find a solution to the mapping simply from examples of the input-output patterns.

Another key point is that the network self-organizes to represent its environment. This is discussed elsewhere with relation to answering the question of how meaning might be grounded in perception (Chauvin, 1986; Cottrell, 1987).

Linear Networks

There is currently a bias in the connectionist community, shared by the authors of this article, against linear networks. This is partly due to the assumption that "interesting" problems must require nonlinearity for their solution. While the results were not reported here, we found that a linear version of the network produced results compatible with the nonlinear version. Since identity mapping is a linear problem this is not too surprising. However, it is useful to check whether nonlinearity is necessary for a particular problem. If not, the elimination of evaluating the logistic function can lead to more efficient solutions. If both approaches appear viable, comparison of the two can lead to a better understanding of nonlinear solutions, since the linear network lends itself to analysis much more readily than the nonlinear one (Williams, 1985). This approach needs to be carried further in future work.

Internal Representations

One of the typical ways to speak of the solutions discovered by back propagation learning is to say that the network discovers regularities in the input. This paper adds at least a new vocabulary for discussing the kinds of regularities discovered in the case of autocoding. We can look at the *variance* of the hidden units as indicative of their usefulness in the resulting solution. It may not be the case that hidden units span the principal subspace of the covariance matrix, that is, the space spanned by a PCT solution, but it is possible that the hidden units are finding the best approximation to this within the constraint that the logistic function imposes of a limited range on the coefficients. If this turns out to be true, then we may speak of the hidden units as finding at least an analog of the principal subspace and as discovering useful *axes of covariance* of the input.

Implications for Image Compression

A major result of this work is the application of a new way of minizing mean square error to a real-world problem that shows it is competitive with PCT. This new technique has several possible advantages over PCT and other current techniques. These need confirmation by further investigation.

One advantage is the relatively equal distribution of error among all of the coefficients. This should lead to a reduction in the effects of channel errors. In PCT and other techniques that approximate it, channel errors that affect the coefficients with high variance can result in a patch that is dominated by the corresponding basis image. The relatively equal contributions of each basis image in the network solution should mitigate these effects. In particular, we know in advance what range the value should be in, and if a coefficient is suspected of being in error, an acceptable restoration of the patch can probably be effected by simply eliminating that coefficient or replacing it with its average value.

Second, because of the fixed range of the coefficients, problems with "tracking" the coefficients by an adaptive quantizer is mitigated. Adaptive quantizers try to follow coefficients as they change, changing the quantization as the coefficients shift. They can "lose track." In our system, we know in advance the range of the coefficients, which should make this less of a problem.

Third, the ability of our network to generalize to novel images is striking. The performance of the linear network is especially encouraging in this regard. This requires some qualification. First, it is likely that this generalization does not apply to images with very different statistics, such as text. Second, we are not aware of work in this area investigating the ability of PCT to generalize to images other than the "training" image. Further work should compare these techniques.

CONCLUSIONS

The major result of this work is in demonstrating the efficacy of current connnectionist techniques for programming by example, rather than algorithm. We have termed this *extensional programming*. The results here suggest that this technique is a powerful one. Its naive application to a problem of current interest among engineers resulted in respectable performance compared to current methods.

However, back propagation is not a panacea—it brings new problems of its own. Designing a connectionist representation of the input (and output for nonautocoding problems) is itself an art. The representation must contain enough information to license solution of the problem, without providing so much that the solution is trivial. However, Hinton (1986) has shown that at least in some domains, back propagation can even design the input representation simply from the occurrence of a token in context.

The results of this study suggest that one useful approach to problems for which no algorithm is known, or for which no parallel algorithm is known, is to use connectionist representations of the problem and allow the network to discover the program itself. Analysis of the the programs thus discovered may aid in our understanding of the problem and lead to methods for doing the programming ourselves. A variety of problems that cognitive science is concerned with are of this character—the input-output behavior is known, but the algorithm is not. With extensional programming we can begin to investigate algorithms that we did not invent ourselves.

REFERENCES

Ackley, D. H., Hinton, G. E., & Sejnowski, T. J. (1985). A learning algorithm for Boltzmann machines. *Cognitive Science, 9,*147-169.

Chauvin, Y. (1986). Hypermnesia, back propagation, categorization, and semantics. Preprints of the Connectionist Models Summer School, Carnegie-Mellon University, Pittsburgh, PA, June 21-29, 1986.

Cottrell, G. (1987). Toward connectionist semantics. In Y. Wilks (Ed.), *Theoretical issues in natural language processing: Position papers* (conference proceedings from TINLAP-3). Association for Computational Linguistics.

Cottrell, G., Munro, P., & Zipser, D. (in press). Image compression by back propagation: An exmaple of extensional programming. In N. E. Sharkey (Ed.), *Advances in cognitive science* (Vol. 3). Norwood, NJ: Ablex.

Gonzales, R. C., & Wintz, P. (1977). *Digital image processing.* Reading, MA: Addison-Wesley.

Hinton, G. E. (1986). Learning distributed representations of concepts. In *Proceedings of the Eighth Annual Conference of the Cognitive Science Society*, August 15-17, 1986, Amherst MA.

Rumelhart, D. E., Hinton, G. E., & Williams, R. J. (1986). Learning internal representations by error propagation. In D. E. Rumelhart, J. L. McClelland, & the PDP Research Group, *Parallel distributed processing: Explorations in the microstructure of cognition. Vol. 1. Foundations.* Cambridge, MA: MIT Press/Bradford Books.

Williams, R. J. (1985). *Feature discovery through error correction learning* (Tech. Rep. 8501). La Jolla: University of California, San Diego, Institute for Cognitive Science.

Zipser, D. (in press). Programming neural nets to do spatial computations. In N. E. Sharkey (Ed.), *Advances in cognitive science* (Vol. 2). Norwood, NJ: Ablex.

Order Information and Distributed Memory Models

Roger Ratcliff

Psychology Department
Northwestern University
Evanston, IL 60201

Abstract

Current versions of distributed models have difficulty in
accounting for the representation of order information in
matching tasks. In this article, experiments are presented
that allow discrimination between physical and ordinal
representations of ordinal information, discrimination between
position-dependent codes and context-sensitive codes, and
generalization of the results of matching tasks from strings of
letters to long-term memory for triples of words. Data from
these experiments constrain the kinds of models that can be
developed to account for matching and order, and present
problems for several current memory models, including
connectionist models. Suggestions are made for modifications
of these models to account for the results from matching tasks.

The representation of order information is important to normal
functioning in many cognitive domains. In speech, both perception and
production involve the processing of a continuous temporal stream of
information that requires the maintenance of order information. In
the perception of visual patterns, the processing system is usually
required to maintain either the absolute or relative positions of
objects in the visual scene. The study of the representation of order
information was of importance over 10 years ago as a topic in its own
right (Lee & Estes, 1977; Murdock, 1974, pp. 157-174), but more
recently it has become a subtopic within different processing domains.
The domain studied in this paper concerns the maintenance of order
within a simultaneously presented string of letters or words.

The task used in the experiments of this article is a matching
task in which subjects study a string of items (letters in Experiments
1 and 2, words in Experiments 3 and 4) and then must decide whether a
test string matches the study string (see Murdock, 1984, for a related
recall task). When strings of letters are the studied items, a test
string is presented immediately after each study string. When words
are used, the test is delayed by presenting study strings in blocks,
and then presenting test strings for all the study strings in the test
block. The primary experimental manipulation considered in this paper
is one in which the order of the studied items is rearranged at test
(Angiolillio-Bent & Rips, 1982; Proctor & Healy, 1985; Ratcliff, 1981;
Ratcliff & Hacker, 1981). When items adjacent in a study string are
interchanged in the test string, subjects find it difficult to respond
that the test string is "different;" accuracy is poor and reaction
time slow relative to "different" conditions in which new items in the
test string replace old items in the study string. Results also show
that the greater the displacement in the switch (e.g., adjacent
letters switched versus remote letters switched), the easier it is for
subjects to respond "different." Ratcliff (1981) developed a model to

account for these effects that assumes that the representations of items are distributed across position so that when items are interchanged in the test string, there is a contribution from close positions to the match between study and test strings.

This article continues this work with three major aims. First, new experimental results on matching tasks are presented that test some specific hypotheses about the nature of the representation of order and also generalize the results from letter strings to word strings. Second, the results allow discrimination between two main hypotheses about the representation of order, position dependent codes versus context sensitive or associative codes. Third, it is argued that memory models designed to account for memory for single words, pairs of words, and so on, should be capable of representing order, and in the last part of the paper, some of these models are evaluated. Specifically, there are important implications of the experimental data for distributed memory models and connectionist models. The argument is simple: models in which items are represented as vectors of features assume that elements within the vectors are independent. Thus, in their present form, they are incapable of dealing with the transposition data presented here and in Ratcliff (1981).

The empirical part of the paper will present four new experiments. The first demonstrates that manipulations of the order of the items in the study string are not sensitive to exact physical position. The second experiment examines performance on different permutations of the study string to contrast the hypotheses of position dependent versus context sensitive memory codes. The third experiment replicates the letter matching task in the memory domain using triples of words, and Experiment 4 uses a response signal procedure to examine the time course of processing in this memory task.

Experiment 1

The letter matching task has been traditionally called the "perceptual" matching task. This label comes from another class of perceptual letter recognition tasks, in which letters are displayed briefly for identification, and physical variables such as the spacing of letters in the string to be identified affect performance (e.g., Bjork & Murray, 1979; Estes, 1982). A similar effect in the matching task would point to a mental representation based closely on physical features of the stimuli.

Method

To examine whether performance in matching depends on physical location, the spacing of the letters in a string was altered between study and test. Subjects studied three letters presented in the center of the display for 500 ms (e.g., _ABC_). There were two manipulations at test. Spacing was tested by altering the positions of the letters in 5 slots e.g., ABC__, AB_C_, AB__C, A_BC_, (all these would require a positive response) and so on. Spacing was crossed with positive and negative conditions: One third of the trials were positive trials in which the three study letters were presented at test in the same order as at study. For negative trials, there were 5 permutation conditions, and three conditions in which a single letter was replaced by a new letter. The test string was displayed immediately after the study string for 200 ms and then removed to eliminate possible eye movement effects. Subjects were presented with 10 blocks of 120 trials. Eighteen Northwestern undergraduate subjects participated in a one-hour session for course credit. (See Ratcliff, 1981, for further details of the experimental procedure.)

Results

Results are shown in Tables 1 and 2. Table 1 shows accuracy and reaction time for the positive conditions as a function of spacing. The effect of spacing is significant (reaction time: $F(9,153)=3.82$, $p<.05$, mse=1720; error rate: $F(9,153)=2.45$, $p<.05$, mse=.00226), but inspection of Table 1 shows that the effects are quite small. Tukey's HSD = 45 ms, so that differences between pairs of reaction times in Table 1 larger than 45 ms are significant. For error rates, Tukey's HSD is 0.051. Inspection of Table 1 shows that only condition 6 (test A__BC) differs from some of the other conditions in accuracy, and only conditions 3 and 6 in reaction time. The power of these contrasts is high because there are around 600 observations per condition.

For negative conditions, there was no effect of spacing but large effects of permutation, replicating Ratcliff and Hacker (1981) and Ratcliff (1981) (see Table 2). For reaction time: spacing effect, $F(9,153)=1.3$, not significant, negative condition, $F(7,119)=31.5$, $p<.05$, and the interaction between spacing and negative condition, $F(63,1071)=1.02$, not significant. For accuracy: spacing, $F(9,153)=1.7$, not significant, negative condition, $F(7,119)=26.6$, $p<.05$, and the interaction, $F(63,1071)=1.02$, not significant.

Table 1
Reaction Time and Accuracy for <u>Same</u> Conditions in Experiment 4

Condition	Number	Accuracy	Reaction Time (ms)
1 ABC__	607	.916	624.2
2 AB_C_	609	.917	645.3
3 AB__C	603	.910	674.3
4 A_BC_	599	.903	627.1
5 A_B_C	611	.919	629.2
6 A__BC	576	.869	670.8
7 _ABC_	626	.939	612.7
8 _AB_C	603	.904	657.2
9 _A_BC	605	.917	635.8
10 __ABC	612	.911	629.5

Note. The study string was presented as _ABC_ where the symbol _ refers to a blank.

Table 2
Reaction Time and Accuracy for <u>Different</u> Conditions in Experiment 4 averaged over Spacing

Negative Condition	Number of Responses	Accuracy	Reaction Time (ms)
ACB	1289	.777	766.9
BAC	1574	.952	666.6
BCA	1602	.965	630.8
CAB	1608	.972	619.1
CBA	1627	.978	610.4
XBC	1627	.973	597.0
AXC	1589	.958	645.3
ABX	1567	.943	666.2

Note. The study string is denoted ABC and X is a letter other than A, B, or C.

We can conclude that spacing differences of the letters between study and test have small effects that are detectable only with experiments with high power. Thus, it is wise to view the word "perceptual" in the term perceptual matching as a name for the task and not as a description of what kinds of variables are likely to affect performance.

Experiment 2

The second experiment was designed to provide data to distinguish between models in which an item is encoded in terms of its absolute position and models in which relative position is encoded. The idea is that certain test conditions allow these models to be contrasted. If the string ABCDE is studied, then a test string BCDEA has four letters in their correct adjacent order (BCDE) and none in their correct absolute position. In contrast, the test string AECDB has three letters in their correct absolute positions but only one pair in the correct adjacent order (CD). The relative difficulties of such test strings can be used to discriminate the two kinds of models.

To perform this experiment, all permutations of the final four letters were the main conditions studied (Ratcliff, 1981, found that performance when the first letter was changed was near ceiling). There were also conditions in which one letter was replaced by a new letter and some fillers in which the first letter was permuted.

Method

The method was similar to that of Experiment 1, except 5 Dartmouth undergraduates were volunteer subjects (paid at $3/hr) for 7 one-hour sessions. The study string was presented for 1.2 s and the test string was presented for 250 ms.

Results

Results are shown in Table 3. To compare the various negative conditions, Tukey's HSD test was used, and differences in accuracy greater than .05 and in reaction time greater than 41 ms are significant.

Table 3
Reaction Time and Accuracy for the Letter Matching Experiment 2

Condition	Number of Observations	Accuracy	Reaction Time (ms)
Same	6520	.926	650
ABCED	197	.430	734
ABDCE	282	.609	667
ABDEC	353	.762	645
ABECD	349	.746	659
ABEDC	360	.763	643
ACBDE	311	.670	656
ACBED	360	.776	632
ACDBE	347	.732	609
ACDEB	388	.822	606
ACEBD	393	.836	590
ACEDB	396	.843	596
ADBCE	347	.737	626
ADBEC	397	.838	603
ADCBE	351	.756	618
ADCEB	391	.827	582
ADEBC	404	.861	588
ADECB	408	.863	591
AEBCD	392	.838	604
AEBDC	390	.826	608
AECBD	397	.841	625
AECDB	369	.790	597
AEDBC	402	.843	589
AEDCB	401	.851	584
AXCDE	351	.750	614
ABXDE	369	.782	607
ABCXE	307	.667	691
ABCDX	286	.620	722
BACDE	409	.887	607
CBADE	427	.910	578
DBCAE	423	.910	581
EBCDA	426	.908	595
XBCDE	397	.871	601
CABDE	425	.900	574
BCADE	419	.895	579

Note. It is assumed that the study string is ABCDE and X is a letter other than A, B, C, D, or E.

Several comparisons can be made that address the issue of relative order versus absolute location. Assuming that ABCDE is the studied string, the string AEBCD with items BCD in correct order but four items in incorrect location can be compared with ACEBD and ADBEC which also have 4 items in incorrect location but no pairs in correct relative positions. Results in Table 3 show that these conditions produce the same values of reaction time and accuracy. Thus, the position of an item with respect to other studied items is not a dimension that is of importance in matching. In contrast to this test of pairwise order, correct location of an item in the string does have a large effect. For example, ABCED, ABDCE, and ACBDE are hard to reject (accuracy less than 0.7 and reaction time greater than 680 ms).

These results argue for a model with a position dependent code (Ratcliff, 1981) or a position dependent retrieval process (Proctor & Healy, 1985). This does not mean that a position dependent code is always used or that a context dependent code is never used, only that in matching procedures, a position dependent code is used. But it does mean that models must have the capability of using a position dependent code (see McNicol & Heathcote, 1986, for similar arguments and experimental support).

Experiment 3

Ratcliff (1981) presented two perceptual matching experiments that used letters strings as stimuli. The model developed to account for the results assumed that the difficulty in responding <u>different</u> to a reordered letter string was located in the distributed representation of letters in memory. If this is correct, then the results should generalize across paradigms and materials and provide contact with both theoretical and empirical work in memory research (e.g., Gillund & Shiffrin, 1984; Murdock, 1982). Experiments 3 and 4 are analogs of Experiments 1 and 2 in Ratcliff (1981) using word triples instead of 5-letter strings, and a study-test procedure in which 8 triples are studied and then tested. Also, a large pool of different words was used instead of repetitions of the same set of letters.

Method

Subjects were 17 Northwestern University Undergraduates participating for course credit. Thirty-two lists were presented, each made up of 8 study triples (6 seconds per triple) followed by 8 test triples. Subjects were encouraged to be fast and accurate; feedback ("TOO SLOW!!") was given for responses slower than 2500 ms.

Results

Results are shown in Table 4. The results are similar to those found in letter matching (Ratcliff, 1981): adjacent switches (test strings ACB and BAC for the study string ABC, where each letter represents a word) are difficult with low accuracy, other permutations are less difficult, and single replace conditions (ABX, AXC, and XBC) are least difficult. Angiolillio-Bent and Rips (1982) defined dis-

479

placement count as a measure of the size of the permutation, so ACB has a displacement count of 2 and BCA has a displacement count of 4. Displacement count seems to be the main determiner of accuracy for the permutations in this study. Reaction time results differ from the normal pattern (which is low accuracy, slow responses) because accuracy and reaction time are correlated only when accuracy scores are all above (or below) accuracy .5 will be slowest. In general, the results are qualitatively similar to those found in Ratcliff (1981), namely, the smaller the displacement, the more difficult is a negative response.

Table 4
Reaction Time and Accuracy for Word Triple Matching Experiment 3

Condition	Number of Observations	Accuracy	Reaction Time (ms)
Same	1464	0.717	1346
ACB	119	0.465	1499
BAC	109	0.427	1489
BCA	147	0.576	1511
CAB	154	0.609	1505
CBA	146	0.575	1479
ABX	180	0.711	1424
AXC	178	0.698	1486
XBC	197	0.770	1401

Note. The studied word triple was ABC and X refers to a word other than A, B, or C.

Experiment 4

This experiment uses a response signal procedure to examine the growth of accuracy in the same experimental conditions in Experiment 3. The experiment is designed to provide the kind of evidence obtained in Experiment 2 in Ratcliff (1981) which examined the growth of accuracy in the letter matching task. In that procedure, subjects were required to respond when a signal to respond was presented. Results showed that accuracy grew rapidly, at the same rate for each of the different negative conditions. This provided strong evidence for a parallel holistic matching process.

Method

The method was the same as that of Experiment 3 except for one main difference: Subjects were required to respond upon presentation of a signal (within 200 to 300 ms) and the signal was presented at lags of 150, 300, 600, 900, and 2000 ms after the test string. The signal was a row of asterisks presented directly under the test triple. Eight Northwestern undergraduates served as subjects for course credit and participated in 5 experimental sessions preceded by one practice session.

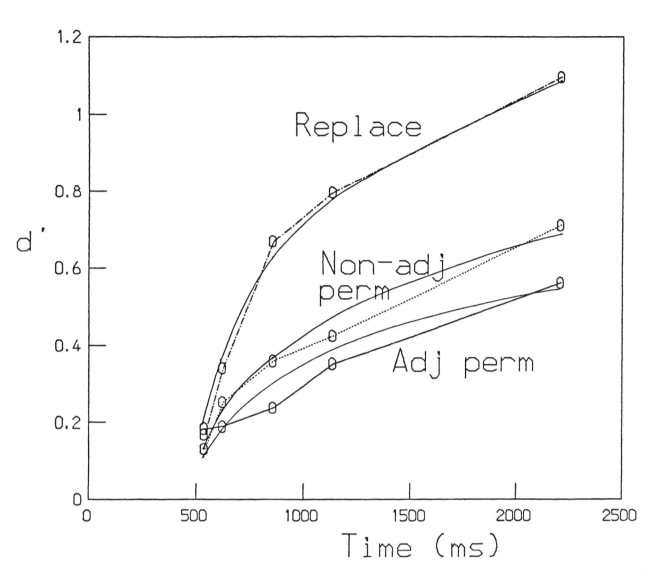

Figure 1
Growth of accuracy as a function of time for Experiment 4. The continuous curves are fits of the diffusion model (Ratcliff, 1978; 1981)
to the data. Parameters of the model are T_{er}=501, v=2910, and d'_a
for the three conditions (top to bottom)=1.84, 1.12, 0.88. The diffusion equation is: $d'(t)=d'_a/sqrt(1+v/(t-T_{er}))$.

Results

        The main results are shown in Figure 1. The two test conditions
ACB and BAC are collapsed to give the "adjacent permutation"

condition, the other permutations are collapsed to give the "non-adjacent permutation" condition, and the three single replace conditions are collapsed. Accuracy grows for all the conditions essentially in parallel to different asymptotes. This means that information about goodness-of-match is available for the three different classes of negatives at the same rate. Thus any model that explains order effects by a systematic, serial, element-by-element comparison process is not supported by this data.

The results of this experiment are qualitatively similar to those found in Experiment 2, Ratcliff (1981). All the functions begin at the same point in time and grow in parallel. Fits of the diffusion retrieval model (Ratcliff, 1978; 1981) to the response signal results are shown in Figure 1. The fits are excellent and support the claim for parallel growth of these functions, so that information about all the different kinds of negative conditions is available to the same extent at any time during the course of matching. These results show that the memory procedure provides equivalent results to the letter matching procedure, and so suggests that we should give special attention to models that account for both sets of phenomena with similar representations and/or processes.

Theory

In this section, several memory models are considered with respect to their abilities to account for position dependent effects (Experiment 2), and the immediate availability of position codes for use in letter-matching and word triple matching in long-term memory (Experiment 4).

Vector models. This class of models includes both distributed memory models (e.g., Anderson, 1972; 1973; Murdock, 1982) and connectionist models (Ackley et al, 1985; McClelland & Rumelhart, 1985). The first point to note about current theories is that many have no built-in capability for dealing with position codes and distance effects. In vector models, the first assumption that might be made is that groups of elements within a vector represent letters within a letter string. Thus the memory vector for a five letter string consists of 5 groups of n elements. The main problem with this idea is that each element is assumed to be independent of each other element so switching elements within the vector, the analog of studying ABCDE and testing ABDCE, would produce results identical to testing ABXYE. The data in Ratcliff (1981) and Experiments 1-4 here show that the former is much more difficult, so this approach will not work.

One way of implementing position dependence in the framework of a vector model is to assume that each letter string is a vector (with letters being sub-vectors within that vector) and that the letter string vector is added to a longer memory vector many times at randomly varying positions. I have implemented this model and found that 100 presentations of the letter vector are needed with a memory vector of length 200, a letter vector of length 10, and storage position normally distributed with standard deviation 15. For this model to produce the correct distance effects, more copies of the letter string vector must be stored in the central position (position at which the test vector matched) and the number of copies stored at more distant positions must decrease with distance from the center (this is why the normal distribution of storage positions is used). To illustrate this, suppose that ABCDE is stored and ABDCE is tested. During encoding, some copies of ABCDE are stored in positions BCDE_, i.e., the string is shifted one letter position to the right, so that the C in the test string matches the stored C in position D.

Another possibility for implementing position dependence is to assume that at test time, a cross correlation is computed between the study and test strings. For a vector model, this means that not only the match between the two strings is assessed (e.g., if f and g are the study and test vectors, then the correlation is SIf_i*g_i), but also the match between the study and test strings with the vector shifted in position (SIf_i*g_{i+k} where k varies from 1 to n-k; note some weighting may be needed). Thus, if two items were interchanged, there would be a component of match from the cross-correlation.

I have implemented both these schemes and they both mimic the model presented in Ratcliff (1981) that assumed that letters are distributed across position. The first vector model is a specific implementation of the distributed memory scheme and the cross-correlation model is a retrieval implementation. The critical issue for this class of models is not whether a model can be developed, but whether a unified model with this kind of mechanism could also account for the same range of data as a more traditional model.

Context sensitive models. Instead of the vector scheme outlined above, a different representation could be used in which an item is encoded with respect to its neighboring items (a context sensitive code). Several such models have been developed (e.g., Cohen & Grossberg, 1986; Rumelhart & McClelland, 1986; Wickelgren, 1969). For the domain of speech processing, the context sensitive representation is appropriate, but the results from Experiments 2 and 3 demonstrate that a context sensitive code will not work for the matching task.

Another representation that might be used is one in which both item and position information are represented. However, extra assumptions would be necessary to account for distance effects when letters or words were switched in position. The model of Ratcliff (1981) and the cross-correlation scheme noted above would both provide that metric.

Gillund and Shiffrin (1984) model. This model assumes an associative code for the representation of information. At test time in recognition, the familiarity of the test probe is assessed from pairwise associations between the test probe and memory. There is no position dependence in this code. In recall, the associative code is used in a sampling scheme for recall. Again, the associative code will not explain the results of Experiments 2 and 3. Thus some kind of position dependent code is required for this model to account for the experimental results presented above.

Distributed memory models. For Anderson's (1973) and Murdock's (1982) vector model, we could assume that each letter (or word) is represented as a subvector within the vector representing the whole item string and then strings are entered into the memory vector at different positions. The issue is whether multiple representations will affect the signal to noise ratio and thus make recognition performance too low. The Anderson and Murdock models also have an associative component that stores pairwise associations (Anderson, 1972, a matrix; Murdock, a convolution). This associative component would not account for position dependent effects because an associative code is context sensitive and so would not deal with the results from Experiment 2.

There are two connectionist models that are relevant here. The auto-associative distributed memory model of McClelland and Rumelhart (1985) assumes a vector representation and assumes that memory is represented in a matrix of connections between each element and each other element. The multilayer connectionist model (e.g., Ackley, et al, 1985) assumes that there are three layers of units, e.g., an input layer, a hidden layer, and an output layer. For the encoder problem (essentially learning to produce an output pattern that is the same as the input pattern), memory for the pattern resides in the two matrices of connections between the input and hidden layer and hidden and output layers. Both these models as they stand do not allow for distance effects as noted earlier because they assume that elements of the vectors are independent so that switches of items within the string would be the same as replacements of new letters. One might think that one of the schemes noted above could be used to encode position dependent information, i.e., multiple copies or cross-correlation at retrieval. But multiple copies of the studied materials shifted in position would not help because patterns are stored as units (partly as a result of nonlinear processing) and permutations of letters across position would not produce the required crosstalk. Cross-correlation at test time may be a better candidate, but if cross-correlation were used routinely in all retrieval processes, the signal to noise ratio at test would be severely reduced (because of all the contributions from nonmatching cross-correlations). Perhaps some of the notions of Zipser (1986) could be incorporated into these schemes. The main issue for these distributed models is to come up with a consistent and coherent account that applies across a range of paradigms including tests of order information as well as other phenomena (e.g., those in Gillund & Shiffrin, 1984).

The final issue to be considered is whether positional information is routinely encoded into memory or whether it is only encoded when it is needed. McNicol and Heathcote (1986) argue that within the domain of short-term memory, their results are most compatible with a theory that allows different subsystems in short-term memory each with its own format for preserving order (see Ratcliff & McKoon, 1987 for data showing no position dependent code and availability of order information late in processing). I endorse this and argue that the results of Experiments 3 and 4 extend the use of a position dependent code into the domain of long-term memory. However, the critical point is that a model must have the capability of representing position dependent codes without the invocation of a completely new model just for that task.

References

Ackley, D.H., Hinton, G.E., & Sejnowski, T.J. (1985). A learning algorithm for Boltzmann machines. Cognitive Science, 9, 147-169.

Anderson, J.A. (1973). A simple neural network generating an interactive memory. Mathematical Biosciences, 14, 197-220.

Anderson, J.A. (1973). A theory for the recognition of items from short memorized lists. Psychological Review, 80, 417-438.

Angiolillio-Bent, J.S., & Rips, L.J. (1982). Order information in multiple element comparison. Journal of Experimental Psychology: Human perception and Performance, 8, 392-406.

Bjork, E.L., & Murray, J.T. (1977). On the nature of input channels in visual processing. Psychological Review, 84, 472-484.

Cohen, M., & Grossberg, S. (1986). Neural dynamics of speech and language coding: Developmental programs, perceptual grouping, and competition for short-term memory. Human Neurobiology, 5, 1- 22.

Estes, W.K. (1982). Similarity-related channel interactions in visual processing. Journal of Experimental Psychology: Human Perception and Performance, 8, 353-382.

Gillund, G. & Shiffrin, R.M. (1984). A retrieval model for both recognition and recall. Psychological Review, 19, 1-65.

Lee, C. & Estes, W. K. (1977) Order and position in primary memory for letter strings. Journal of Verbal Learning and Verbal Behavior, 16, 395-418.

McClelland, J.L., & Rumelhart, D.E. (1985). Distributed memory and the representation of general and specific information. Journal of Experimental Psychology: General, 114, 159-188.

McNicol, D., & Heathcote, A. (1986). Representation of order information: An analysis of grouping effects in short-term memory. Journal of Experimental Psychology: General, 115, 76-95.

Murdock, B.B. (1974). Human memory: Theory and data. Potomac, Md.: Erlbaum.

Murdock, B.B. (1982). A theory for the storage and retrieval of item and associative information. Psychological Review, 89, 609- 626.

Proctor, R.W., & Healy, A.F. (1985). Order-relevant and order-irrelevant decision rules in multiletter matching. Journal of Experimental Psychology: Learning, Memory, and Cognition, 11, 519-537.

Ratcliff, R. (1978). A theory of memory retrieval. Psychological Review, 85, 59-108.

Ratcliff, R. (1981). A theory of order relations in perceptual matching. Psychological Review, 88, 552-572.

Ratcliff, R., & Hacker, M.J. (1981). Speed and accuracy of same and different responses in perceptual matching. Perception and Psychophysics, 30, 303-307.

Ratcliff, R., & McKoon, G. (1987). Item familiarity versus relational information: The time course of retrieval of sentences. In preparation.

Rumelhart, D.E., & McClelland, J.L. (1986). On learning the past tense of English verbs. In J.L. McClelland & D.E. Rumelhart, Parallel distributed processing: Explorations in the microstructures of cognition. Vol. 2: Psychological and biological models. Cambridge, Mass.: MIT Press.

Wickelgren, W.A. (1969). Context-sensitive coding, associative memory, and serial order in (speech) behavior. Psychological Review, 76, 1-15.

Zipser, D. (1986). Biologically plausible models of place recognition and goal location. In J.L. McClelland and D.E. Rumelhart, Parallel distributed processing: Explorations in the microstructures of cognition. Vol. 2: Psychological and biological models. Cambridge, Mass.: MIT Press.

Footnote

This research was supported by NSF grant BNS 82 03061 to Roger Ratcliff and NSF grant BNS 85 16350 to Gail McKoon. I would like to thank Gail McKoon and Scott Gronlund for their comments on this article.

A Parallel Natural Language Processing Architecture with Distributed Control

George Berg
Computer Science Department
Northwestern University
Evanston, IL 60201

Abstract

This paper describes work on the autonomous semantic network (ASN) knowledge representation and natural language processing architecture and its implementation - the NO HANS simulator. An ASN is an enhanced spreading activation semantic network, but one without a centralized controller. Rather, in addition to the semantic network are types of nodes which have the ability to change links or add nodes in the network. These nodes are activated by the energy spreading through the underlying semantic network. Thus, the same spreading activation which infuses the knowledge representation also drives the control mechanism as well. Because of this, ASN's offer a compromise between the distributed but restrictive connectionist model and the powerful but heretofore essentially serial conceptual natural language processing models. Spreading activation is also the basis for the search capability, which is loosely based on the connectionist winner-take-all idea. We construct a simple conceptual analyzer in this model and indicate how it works.

1. Introduction

The goal of this work is to be able to use the conceptual style of natural language processing in a massively-parallel form. Conceptual natural language processing merges syntactic, semantic and world knowledge and facilitates processing by allowing high-level inferences (cf Schank and Abelson, 1977; Dyer, 1983). One of the problems with this approach is that the number of alternatives that needs to be searched becomes prohibitive using a serial model of computer control. Massively-parallel architectures (cf Hillis, 1985; Fahlman, 1979) have tens of thousands of nodes, each of which has a limited processing capability. Their forte is extremely fast search. By casting the conceptual techniques into a form where we can utilize parallelism, we hope to gain insights into how to increase the size and complexity of the domains which can be handled by such systems, eventually transcending the micro-world limitiations which have plagued the field.

In recent years there has been much work on distributed natural language processing. These efforts fall into two main categories - connectionist and symbolic. The connectionist systems (Cottrell, 1985; Waltz and Pollack, 1985; Selman and Hirst, 1985) are primarily parsers - they produce a pattern of activation corresponding to a parse tree. They do not add a representation of the sentence to their networks. Because of this, they are not good candidates for conceptual analyzers - an important feature of conceptual systems is that they can build and manipulate a representation of the input text. The symbolic systems, such as the Word Expert Parser (Small and Rieger, 1982) do produce a representation of the new information. Unfortunately, they use a high-level model of parallelism. They use powerful processing units which engage in complex communication with one-another. We would like to be able to use the simpler processing units and communications abilities which are suited to massive-parallelism.

What we propose in this paper is a new model - the autonomous semantic network (ASN). ASN's offer parallelism comparable to the connectionist models while potentially allowing the sophisticated processing of the symbolic, conceptual systems.

2. Autonomous Semantic Networks

The idea underlying an ASN is a spreading-activation semantic network (cf Collins and Loftus, 1975). These systems have used a central, controlling program to direct their activity. Our approach differs because, to take advantage of the parallelism inherent in such an architecture, it is necessary to distribute the controller among nodes connected to the network. To do this, new kinds of nodes were added which alter the network in response to new information. In addition to the normal nodes and links of the spreading activation network, we add *construction nodes* ("c-nodes") which can manipulate the nodes and links. These c-nodes are connected to the semantic network and are enabled by the same activation which spreads through the network. The activation energy which provides focus in the knowledge representation now drives the system's control mechanisms. In this way we distribute the controller. There is also a type of node, the *Winner-take-all* ("wta"), which provides a search ability by comparing the activation of network nodes.

Although the semantic network which underlies this model is essentially connectionist, it was felt that it was necessary to use the c-nodes, with their relatively sophisticated abilities. Current efforts to get connectionist models to do high-level tasks still require large numbers of nodes (cf Touretzky and Hinton, 1985). So, to retain the distributed qualities of connectionism, but to be able to do the relatively sophisticated operations necessary for conceptual natural language processing, the model adds the c-nodes, each with a limited ability to change the network. This way the system is not bound by the computational limits of connectionism, but still retains its amenability to massive-parallelism. For a detailed treatment of the ASN model see Berg (1987).

The behavior of nodes in the spreading activation portion of an ASN is like the units in a local connectionist model (cf Feldman and Ballard,1982). They receive excitatory and inhibitory inputs and, based on a simple activation and decay function, output activation. They are connected by three kinds of links: *activation, hierarchical-activation* ("h-act") and *inhibition*. Activation links are one-way connections which spread positive activation between nodes. Inhibition links are one-way connections which spread negative activation.

The h-act links are where we begin to see the interface between the semantic network and the c-nodes. As far as the semantic network is concerned, one of these is a pair of activation links, one in each direction. Their importance, providing a backbone for search, will be discussed in the next section. The input and output to the system will be through the semanitc network. Certain nodes, called *lexical-input* nodes will be activated when the system is to "read" a word. When a node designated as a *lexical-output* node is activated, it prints a word on the system's output. We will identify four types of c-nodes: *link-h-act, link-act , de-link* and *make*. Link-h-act and link-act nodes will construct a node of the appropriate type between the two nodes to which they point. Since links are directed, these kinds of nodes use two specialized links: *from* and *to* (figure 1). Similarly, a de-link node will remove a link between two nodes. C-nodes can also attach links

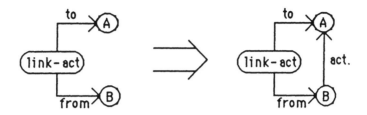

Figure 1. A Link-act node making a link

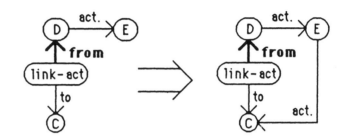

Figure 2. A link-act node using indirection in making a link

indirectly. When a c-node manipulates a link indirectly, it changes the link not on the node to which it points, but to the node to which the second node points (figure 2). Indirect operations by construction nodes will be indicated by bold-face links in the figures. If there is no node connected appropriately to one of its links, a c-node will wait until there is one (however c-nodes, like normal nodes, are subject to becoming inactive through decay). Indirection allows us to use intermediate nodes, or *structural markers,* as pointers to nodes. Structural markers serve an analogous role to slots in a frame system; and the nodes connected to them by activation links can be seen as the slot fillers (see section 4). A make node can introduce new nodes into the network by creating a new node and placing an activation link from itself to the new node.

3. Distributed Search

A conceptual natural language processing system needs a search capability. We provide that by establishing a parallel network which shadows the network of normal nodes and their h-act links. For every normal node which has incoming h-act links there is a *wta* node, and corresponding to every h-act link is a *search* link. Normal nodes which have no incoming h-act links are represented by themselves in this network. The resulting network represents the inheritance hierarchy implicit in the semantic network. The notion of wta comes from Feldman and Ballard's *winner-take-all* networks. When activated, the wta node will select whichever of the nodes connected to it by incoming search links is activated. If a wta node is connected to another by a search link, the subordinate wta node is activated, and the winner of its search is presented to the superordinate wta node. If more than one node is activated, the wta will inhibit all of them and defer a decision until only a single node is active. This way, nodes which will be reactivated by other nodes in the network will be selected in preference to those with spurious or poorly supported activation. In effect, a search starting at a wta node in this network says, "find the

489

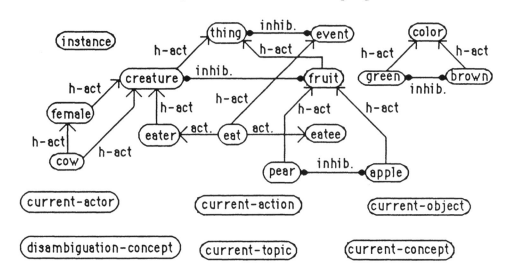

Figure 3. The semantic network before the text is read.

instance (leaf node) in the network below you which is in focus (has the highest supported activation)." The wta node will construct an activation link between itself and the "winner". That way c-nodes can access the winning node indirectly through the wta node.

C-nodes can use their indirection to find the wta node associated with a normal node in the semantic network. This is necessary because, in many cases, the wta node to be activated will not be known *a priori*, and it is necessary to access it through its corresponding normal node. For example if we want to know which instance of "cow" is currently active (or "in focus") we can connect the node for "cow" to a c-node through a structural marker. The c-node will find the wta node associated with the normal node (here **wta-cow**). An example of this is shown in section 4. C-node links which initiate a wta search are identified as *wta-from* and *wta-to* links.

4. Example: A Simple Analyzer

To give an indication of how spreading activation can direct the analysis of a sentence, we give an extremely oversimplified trace of how the sentence "A cow ate an apple" is processed. When a lexical-input node is activated it activates c-nodes which change the network to reflect the word's meaning in the context of the network. Figure 3 shows the semantic network before the sentence is read. To simplify the presentation, the initial network has no nodes which are instances of its concepts and the structural markers are initially unconnected. In this and subsequent figures, h-act links will be shown as one-way nodes to emphasize their role in wta searches, although they spread activation both ways in the semantic network.

Most of the nodes in this figure represent concepts in our taxonomy of things and events. However, we use several structural markers. A structural marker points with an activation link to the node which fills the indicated category. Their purpose is similar to the *binders* of Selman and Hirst (1985) and Cottrell (1985). **Current-concept** is used to indicate which concept is being processed. By connecting **current-concept** to a node with an activation link, the node becomes accesible to c-nodes. In a similar fashion, **current-actor, current-action** and **current-object** are used to point to those nodes which represent those categories in our

490

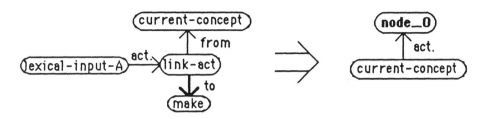

Figure 4. Reading "A"

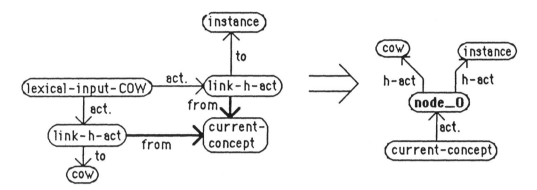

Figure 5. Reading "COW"

sentence. **Disambiguation-concept** is used to point to a node when we want a c-node to access its wta node. **Current-topic** points to a node indicating the overall topic of the current text.

Figure 4 shows the results of activating the lexical node for "A". That node spreads activation to a link-act node. That node constructs an activation link from **current-concept** indirectly through **make**. The indirect reference through **make** causes it to construct a new node, which is the node the link-isa wants. The result of activating those nodes creates a new node (**node_0**) which is the focus of the system's attention by virtue of being connected to **current-concept**.

When **lexical-input-COW** is activated (figure 5) it spreads activation to two link-h-act nodes. When these nodes are activated they connect the node connected to **current-concept** to **instance** and **cow**. In our sentence, **node_0** is identified as an instance of a cow.

When the node for "ATE" is activated, it activates c-nodes which assemble the structure of the sentence (figure 6). **Lexical-input-EAT** (we ignore tense) activates several construction nodes. The three boxes represent groups of c-nodes. The c-nodes in **sub-actor** change **node_0** from having a link from **current-concept** to having one from **current-actor**. **Sub-action** has c-nodes which make a new node (**node_1**) and link it to **current-action** and establish an activation link from the new node to the current-actor (**node_0**). **Sub-object's** c-nodes link the node attached to **current-concept** (which is the as yet unread second noun phrase in the sentence - hence figure 6 shows the c-nodes waiting) to **current-object** and creates a link from **node_1** to the node attached to **current-object**. The three link-h-act nodes in figure 6 identify the fillers of the current-actor, current-action and current-object roles as **eater**, **eat** and **eatee**, respectively, when activated.

Activating the nodes for the second noun phrase ("a pear") works the same as the first. It will

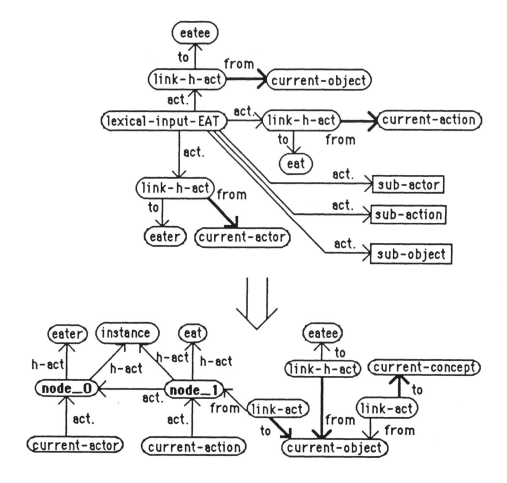

Figure 6. Reading "ATE"

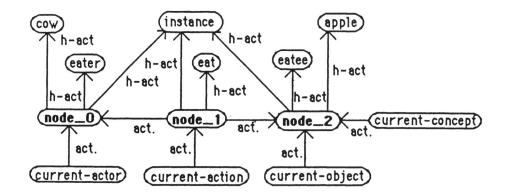

Figure 7. The result of the sentence.

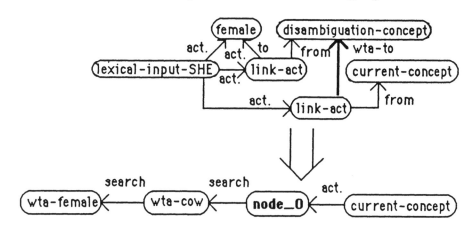

Figure 8. Reading "SHE"

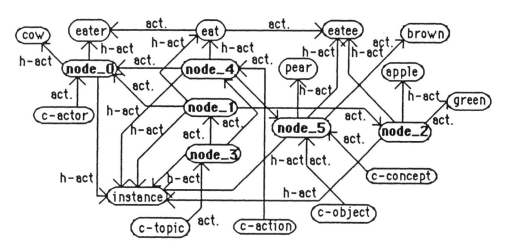

Figure 9. The semantic network after the text has been read.

create a node (**node_2**) and link it to **instance** and **pear**. **Node_2** will be identified as the current-concept, and thus be acted on by the c-nodes activated by "ATE" and which were stalled waiting for the second noun phrase to be read. Figure 7 shows the semantic network after the entire sentence has been processed.

In addition to our sentence let us consider what the activation of one other lexical-input node does. When the system reads "SHE" (figure 8) two c-nodes are activated. The first links **female** to **disambiguation-concept**. This allows the second to access **wta-female** through an indirect wta-to link. What this does is activate **wta-female**. The wta node will initiate a search. Here its only link is to **wta-cow**. **Wta-cow** is activated and, in turn does a wta search. In this example, there is only one candidate - **node_0**. Had there been more than one active node along the search paths, they would have been inhibited until only one (presumably reactivated by its neighbors in the semantic network) was active. A pointer to this "winning" node is passed up from **wta-cow** to **wta-female** and is connected by the c-node to the **current-concept**, identifying it as the

493

pronoun's referent.

Our presentation omits many important details of the actual analyzer. It handles definite noun phrases, adjectives and subtle problems with using structural markers which are beyond the scope of this short discussion. To give some flavor of how the system would analyze a text, figure 9 shows the network after it has read "A cow ate an apple. The apple was green. She also ate a brown pear." In addition to the nodes from our sentence, **node_4** represents the "eating event" of the third sentence, **node_5** is the brown pear, and **node_3** is a node representing the entire text.

This analyzer has been implemented on the NO HANS system. NO HANS is a compiled Franz Lisp program which simulates the spreading activation semantic network and the more powerful construction and wta nodes which the network activates. NO HANS supports all of the features of the ASN model, simulating them as a generic, massively-parallel machine. On NO HANS, the analyzer runs the example text in approximately five minutes of real-time on a VAX 780.

6. Future Work

We currently have several directions for our research on ASN's. Our short-term goal is to use them to address many of the issues in conceptual natural language processing. These include handling complex sentence structure, word sense disambiguation, conflict detection, inference and classification. A simple *conflict* would be produced by introducing two contradictory pieces of information such as "Spot is a dog. He is also a pear" or introducing information which violates knowledge already in the network. Our prototypical *inference* problem is "Janet hit Sue. She felt guilty." The inference we need to make is that Janet might feel guilty for hitting Sue, thus being the referent of the pronoun. An example of *classification* would be to have the system read "Clyde is large, has four legs, big ears and a trunk" and have the system's internal representation indicate that Clyde is an elephant.

For the long term, this research is aimed at understanding longer and more complex texts - taking advantage of the distributed control to be able to consider more inferences during processing. As massively-parallel machines come to accomodate this, and more powerful, models, the size and breadth of natural language processing systems can increase. The parallelism will reduce the time necessary to analyze texts. But, ultimately, for any large-scale system of this type to be effective, progress must be made on the ability to make the abstractions and complex inferences necessary for learning. Otherwise we become bogged down creating the representations for every new word and concept our systems will use.

Acknowledgements

I would like to thank Gilbert Krulee, Cathie Love and Betty Modlin, for each helping in their own way. The work presented here benefitted greatly from discussions with John Rager and Jim Roberge´.

References

[1] Berg, George (1987) "Autonomous Semantic Networks for Natural Language Processing', Technical Report in preparation.

[2] Collins, Allan and Loftus, Elizabeth (1975) "A Spreading-Activation Theory of Semantic Processing", *Psychological Review*, 82: 407-428.

[3] Cottrell, Garrison (1985) *A Connectionist Approach to Word Sense Disambiguation*, Ph.D. thesis, University of Rochester, Department of Computer Science, Rochester, NY.

[4] Dyer, Michael (1983) *In Depth Understanding*, MIT Press, Cambridge, MA.

[5] Fahlman, Scott (1979) *NETL: A System for Representing and Using Real-World Knowledge*, MIT Press, Cambridge MA.

[6] Feldman, Jerome and Ballard, Dana (1982) "Connectionist Models and their Properties", *Cognitive Science*, 6: 205-254.

[7] Hillis, W. Daniel (1985) *The Connection Machine*, MIT Press, Cambridge, MA.

[8] Schank, Roger and Abelson, Robert (1977) *Scripts, Plans, Goals and Understanding*, Lawrence Erlbaum, Hillsdale, NJ.

[9] Selman, Bart and Hirst, Graeme (1985) "A Rule-Based Connectionist Parsing System", *Proceedings of the Seventh Annual Conference of the Cognitive Science Society*, Irvine, CA, pp. 212-219.

[10] Small, Steve and Rieger, Chuck (1982) "Parsing and Comprehending with Word Experts", In: Lehnert, W. and Ringle, M. (Eds.), *Strategies for Natural Language Processing*, Lawrence Erlbaum, Hillsdale, NJ.

[11] Touretzky, David and Hinton, Geoffrey (1985) "Symbols among the Neurons: Details of a Connectionist Inference Architecture", *Proceedings of the Ninth International Joint Conference on Artificial Intelligence*, Los Angeles, CA, pp. 238-243.

[12] Waltz, David and Pollack, Jordan (1985) "Massively Parallel Parsing: A Strongly Interactive Model of Natural Language Interpretation", *Cognitive Science*, 9: 51-74.

Organization of Action Sequences in Motor Learning:

A Connectionist Approach

Yoshiro Miyata
Institute for Cognitive Science
University of California, San Diego

Abstract

This paper presents a connectionist model of motor learning in which performance becomes more and more efficient by "chunking" output sequences, organizing small action components into increasingly large structures. The model consists of two sequential networks: one that maps a stationary representation of an intention to a sequence of action specifications or action plans, and one that maps an action plan to a sequence of action components. As the network is trained to produce output sequences faster and faster, the units that represent the action plans gradually discover representational formats that can encode larger and larger chunks of subsequences. The model also shows digraph frequency effects similar to that observed in typewriting, and it generates capture errors similar to that observed in human actions.

Organization of Action Sequences

The idea that the size of perceptual and motor units increases with experience is not new. Various models have proposed different ways in which the units are organized: in a hierarchical manner (Bryan & Harter 1897, Lashley 1951); as associative chains of small elements (Wickelgren 1969); or elements linked together by inhibitory connections (Rumelhart & Norman 1982), for example. Recently, Grudin and Larochelle (1982) and Jordan (1986) have argued that when a complex motor skill such as typing is learned, the learner develops representations of motor sequences at levels higher than individual action components, such as digraphs in typing. Representation of "chunks" of motor sequences seems to be formed.

The connectionist framework (Feldman & Ballard 1982, Rumelhart & McClelland 1986) has been successfully applied to the domain of motor control in a number of studies (Hinton & Smolensky 1984, Rumelhart & Norman 1982, Jordan 1986). These studies stressed the role of parallel computation in solving the problems of many degrees of freedom and of multiple and complex constraints. However, the issues of how representation of action sequences is developed and how performance becomes increasingly efficient have not been addressed directly by these models. In general, these models tended to focus on the performance of an expert rather than the process of learning itself. This paper presents a model of motor learning, of the shift from a novice to an expert performance.

496

The Architecture of The Model

In the model, there are three basic levels of representation. The first is a conceptual representation of the sequence to be produced, such as a word to be typed, that is more or less independent of motoric components or physical requirements of the task. The second is a representation in which actions are specified to the action system. The third is the output of the action system that represents each component to be executed. The execution of an action sequence in the model thus involves two mappings: the mapping from the conceptual representation to the action specification, and the mapping from the action specification to the actual action components. In both mappings, an input vector is mapped to a sequence of output vectors.

Jordan's Network

One important requirement for a model of action control is that it is capable of generating sequences of output vectors. The simulations described in this paper used a technique developed by Jordan (1985) to generate sequences. Figure 1 shows the architecture of a Jordan network. It receives an input to a layer of *plan* units, called the *plan vector*. For example, a plan vector might specify the sequence "ABCD.." and the task of the network is to produce first output vector A, then B, C, and so on. Once a plan vector is

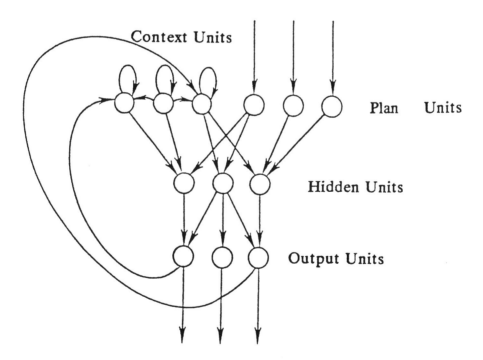

Figure 1. The basic architecture of a Jordan network. Output is determined by a stationary plan and temporal context of past outputs.

given, it produces the first output vector by sending activation forward through one layer of hidden units. Then the output vector is fed back to a layer of *context* units, which store the history of output vectors by recurrent connections to themselves. At each time step, the next output vector is determined both by the plan vector which does not change during the sequence and by the context vector which changes at each time step and thereby specifies place in the sequence.

The Hybrid Model

The model was constructed as a hybrid of two Jordan networks, one for each mapping (the left half of Figure 2). The upper network, called *Plan-net* supplies plan vectors for the lower network called *Action-net*. The right half of Figure 2 illustrates the time course of updating the state of the network. At time 0, a specification of the sequence to be produced is given at the layer of units labeled "Intention" which serves as the input to Plan-net. Plan-net then generates its output as a sequence of three vectors at the layer of units labeled "Plan" at time 0, 3, and 6. The plan units actually represent the input for Action-net. In response to each plan vector, Action-net generates a sequence of three output vectors, one at each time step, at the layer labeled "Output". Then each output vector is converted to another vector at the layer labeled "Action" by choosing the most active output unit. Thus, Action-net updates its state more often than (in this particular simulation, three times as often as) Plan-net. There is a connection from the context units of Action-net to the

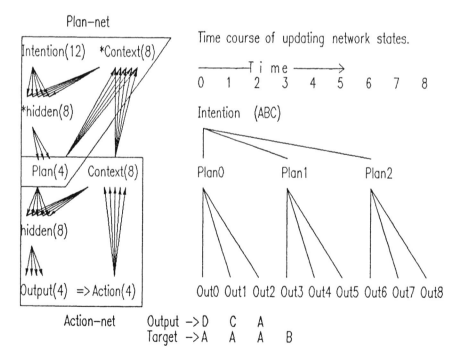

Figure 2. The architecture of the model (left) and the time course of updating the state of the network (right). Plan-net maps from an *Intention* to a sequence of three *Plans*. Action-net maps from each *Plan* to a sequence of three *Outputs*.

context units of Plan-net: this allows Plan-net to get some information about what Action-net has done so far.

Training The Network

The entire network was trained using the back-propagation algorithm (Rumelhart, Hinton & Williams 1986). The training procedure for a network with recurrent connections is slightly more complicated than the training of a straight feedforward network and is described in detail in Rumelhart et al., (1986).

Each output vector and each action vector were compared with a target vector. Initially the target vector represented the first component to be produced in the sequence. For example, suppose the target sequence is *ABC*. Then the target to compare with the output and action vectors is initially *A*. If the action vector does not match the target then the target stays the same so that the same target is used at the next time step. If the action vector does match the target then the target for the next time step is changed to the next component, *B* in this case. At each time step, the error between the output vector and the target is propagated back through the network and all the weights in the network[1] are modified so as to reduce the error.

Pre-training

In any learning situation for humans, the learner usually has a fair amount of a priori or background knowledge before the learning starts. This prior knowledge was modeled by dividing training into two parts: one to put background knowledge and one to train the task itself. Consider the typing task : Even a novice typist can type correctly by "hunt and peck." The problem is that the novice is very slow. The network was pre-trained so that it can perform the task analogous to the performance of a novice typist. Thus the Plan-net was trained to generate a sequence of plan vectors, each plan vector representing only one action component. The Action-net was pre-trained so that in response to each plan vector it could generate the action represented by the plan vector.

Figure 3 illustrates the performance of the network after the pre-training phase. An activity pattern in a layer of units (a vector) is shown as a row of vertical bars, the height of a bar representing the activation level of a unit. A sequence of vectors is shown as a matrix composed of several rows of vertical bars, each row representing the vector at each time step. The bottom row represents the first vector produced and the top row the last vector. The 12-dimensional vector labeled "Intention", is the input pattern to Plan-net at Intention layer. This pattern represents the output sequence *ABC*. In response to this input, Plan-net produces a sequence of three 4-dimensional plan vectors, shown to the right as three rows of vectors, representing the actions *A*, *B*, and *C*, respectively. In response to the first plan

---

[1] The recurrent connection from the units in Context layer to themselves had fixed set of weights which formed a diagonal matrix: i.e., each unit in Context layer was connected only to itself.

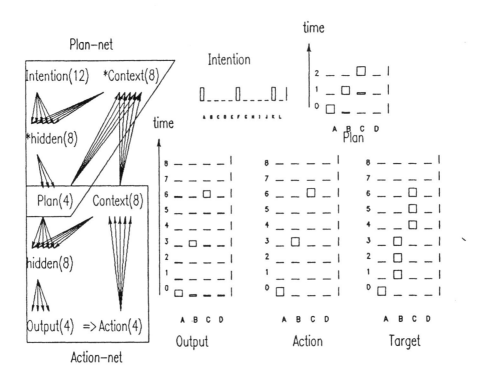

Figure 3. Response of the network after the pre-training phase: In response to an input (Intention) specifying the sequence *ABC*, Plan-net produces a sequence of three plan vectors (Plan) specifying the actions *A*, *B* and *C*. In response to each plan vector, Action-net produces the action specified by the plan. It takes seven time steps to complete the sequence.

vector, Action-net then produces the action *A*, and in response to the second plan vector it produces the action *B*, and so on. The three matrices at the bottom show sequences of nine output vectors, action vectors, and target vectors, at time step 0 through 8. In this simulation the network was trained on all the possible sequences of three outputs, each output representing one of four actions, *A*, *B*, *C* and *D*. There are 64 such sequences. After the pre-training the network can perform all the sequences but only very slowly.

Results - New Plan Representations

This situation changes as the result of training. Notice that as soon as the first action is made correctly the target shifts to the second component and tries to turn on the unit that corresponds to that component. In effect, this kind of training would be expected to speed up the execution of the entire sequence. That is in fact what happens. Figure 4 shows the response of the network to the same input sequence as shown in Figure 3 after some 1600 presentations of all 64 patterns. Before training, it took seven time steps to complete each sequence. After training, all the sequences are completed in three steps, which is the maximum rate. The plan units which, before the training, could represent only one action component at a time, now represent the entire sequence. These units have developed representational formats that can encode all 64 sequences.

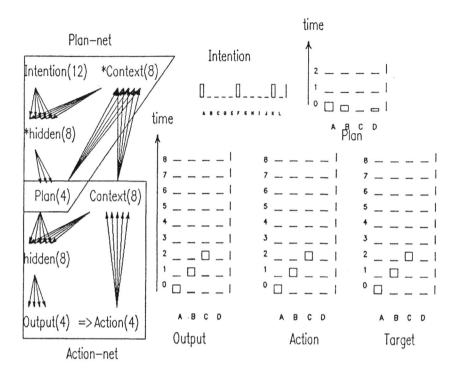

Figure 4. Response of the network after the training phase to the same input as in Figure 3. Only one plan vector is needed to specify the sequence *ABC*. The sequence is completed in three time steps.

The network, as the result of training, showed a transition from an inefficient performance analogous to that of a novice to a more efficient performance. The next section demonstrates that during the process of training, the performance of the network exhibits a certain characteristic also observed in human experts.

Digraph Frequency Effects

One strong source of evidence for the existence of higher-order (multi-character) representational units in typewriting comes from the digraph frequency effects. Grudin and Larochelle (1982) found that the inter-keystroke-intervals for higher frequency digraphs were reliably shorter than for lower frequency digraphs. The present model speeds up its output by developing higher-order representation of the output sequences. It is interesting to see if the model exhibits the same kind of effect: Does a transition from one action to another become faster if that particular transition is experienced more frequently by the network? To test this possibility, a set of twelve sequences of three actions were constructed using six digraphs (Table 1). Each sequence was presented with different frequency, as shown in the table, so that as a result three of the six digraphs (*AC*, *BA*, and *CB*) were presented twice as often as the other three (*AB*, *BC* and *CA*). A frequency means the number of times an item is presented to the network during each cycle of training. The network used in this simulation was identical to the one described in the previous section except that, because there were only three possible actions, there were only three units each in the output,

Table 1

Twelve sequences were presented to the network with different frequencies so that
three of the digraphs contained were presented twice as often as the other three.

Frequency	Sequences presented to the network	Frequency	Digraphs contained in the sequence
4	ACB, BAC, CBA	High (8)	AC, BA, CB
2	ABA, ACA, BAB, BCB, CAC, CBC	Low (4)	AB, BC, CA
1	ABC, BCA, CAB		

action, and plan layers and nine units in the intention layer. The training procedure was
identical to the one used before.

Results

Intervals (number of time steps) between two successive actions were recorded during
the training phase. Three pairs of digraphs were compared: *AB* vs. *AC*, *BC* vs. *BA*, and *CA* vs.
CB, the second digraph in each pair having the higher frequency. These pairs were chosen
because two digraphs in each pair appeared in very similar contexts during the training: for
example, *AB* appeared in the sequences *ABA, ABC, BAB* and *CAB* while *AC* appeared in *ACA,
ACB, BAC* and *CAC*.

The network was faster with the higher-frequency digraph than with the lower-
frequency digraph within every pair of digraphs. Table 2 shows the digraph frequency
effects in each pair of digraphs during five blocks of 100 learning cycles. Following Grudin
and Larochelle (1982), the "digraph frequency effect" (DFE) is defined as the average interval
for the lower-frequency digraph minus the average interval for the higher-frequency digraph.
This represents the time saved producing the second action of a higher-frequency digraph.
The interval for a digraph is defined as the time steps required between the first and the

Table 2

Digraph Frequency Effects (time saved producing
second action of higher-frequency digraph)

| Blocks of learning cycles | Digraph Pairs Compared | | |
	AB—AC	BC—BA	CA—CB
500-599	1.27	2.18	1.37
600-699	1.36	1.80	1.43
700-799	0.81	2.20	1.43
800-899	1.00	2.00	0.86
900-999	1.04	2.04	0.51

second actions in the digraph. If the action B in the digraph AB is produced at time step immediately following the time step the action A was produced, then the interval for the digraph is 1. The overall digraph frequency effect was significant (i.e., DFE > 0), $F(1, 2) = 20.72$, $p < 0.05$, using the pairs as the random variable.

Capture Errors

One interesting aspect of action control is that errors seem to exhibit certain general patterns. This has been shown in controlled experiments in speech (Motley, Camden & Baars 1982, Dell 1984) and in typing (Grudin 1983, Sellen 1986) as well as observations in natural settings (Norman 1982, Reason 1979). One class of errors are called *capture errors*. "A capture error occurs when a familiar habit substitutes itself for an intended action sequence. .. Pass too near a well-formed habit and it will capture your behavior (Norman 1982)." The present model seems to have a potential to generate this type of errors. This is because the mappings in the network have the characteristic that similar inputs tend to be mapped to similar outputs. If two output sequences have similar subsequences, this results in similar contextual information in the context vector that might lead to an error.

This possibility was tested by simulating Sellen's psychological experiment (1986) using a Jordan network as the subject. The design of the experiment is summarized in Table 3. The network was given four plan-target pairs to learn. In response to each of the four plan vectors $a, b, c,$ and d, the network was to generate four different output sequences $ABC, ABD,$ $EFG,$ and EHI. The first two sequences — ABC and ABD — are very similar to each other: these are called "high similarity sequences". The last two sequences — EFG and EHI — have only one common component: these are called "low similarity sequences". The sequences ABC and EFG are called "high familiarity sequences" because they are presented to the network three times more often than the "low familiarity sequences" ABD and EHI. There were nine possible actions, $A, B, C, ..., H, I$, and thus the network had nine output units. The action produced by the network was decided by choosing the most active output unit.

Table 3

Similarity and familiarity of the four
plan-target sequence pairs learned by the network.

Plan -->	Target Sequence	Similarity	Familiarity
a	ABC	High	High
b	ABD	High	Low
c	EFG	Low	High
d	EHI	Low	Low

Table 4 shows the four possible "capture errors" that can occur. A capture error can occur either between two high similarity sequences ABC and ABC, or between two low similarity sequences *EFG* and *EHI*. It can occur either from a high familiarity sequence to a low familiarity sequence or vice versa. For example, if the network generates the sequence *ABD* in response to the plan 'a', it is a capture error between high similarity sequences and it is a high-to-low familiarity error.

The network used was a Jordan network with four plan units, six hidden units, four context units and nine output units. Each output unit represented one of the nine possible actions: *A, B, C,..., I*. The network was trained on the four pairs of plans and output sequences described above. Gaussian noise was added to each plan unit so that the network continued to make some errors even after it learned the sequences. Testing was done by presenting each plan, with noise added, one thousand times after the network had reached stable state. Seven networks were run that were identical except for initial random weights.

Results

Figure 5 shows the results. Both similarity and familiarity affected the probability of capture errors. There were more capture errors between the high similarity sequences *ABC* and *ABD* than between the low similarity sequences *EFG* and *EHI*, $F(1, 6) = 141.9, p < .001$. And there were more capture errors from a low-familiarity sequence to a high-familiarity sequence than from high to low familiarity, $F(1, 6) = 293.9, p < .001$.

In Sellen's (1986) experiment the familiarity factor had a significant effect on the probability of capture errors, but the effect of similarity was small and not significant. The discrepancy between her experimental results and this simulation can be attributed to a number of factors: The difference can simply be because the experiment did not have enough power to detect the effect (fewer trials, smaller overall error rate). Another possibility is that the temporal context stored in the context units did not correspond to the experiment. A finer grained manipulation of similarity in the experiment as well as manipulation of temporal characteristic of the context units in the simulation might be needed to make them

Table 4

Four possible capture errors

Plan ->	Target	Capture Error	Similarity	Familiarity
a	ABC	ABD	High	High -> Low
b	ABD	ABC	High	Low -> High
c	EFG	EHI	Low	High -> Low
d	EHI	EFG	Low	Low -> High

CAPTURE ERRORS BY NETWORKS

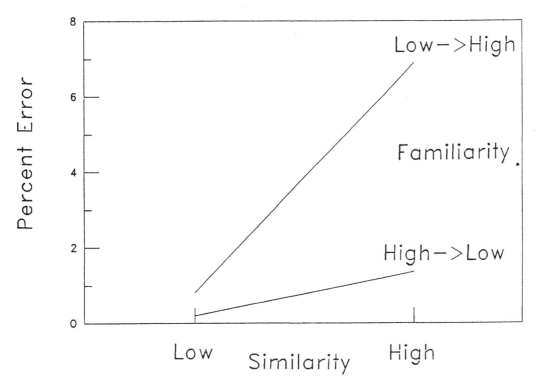

Figure 5. The effect of similarity and familiarity on capture errors made by seven networks: Capture errors were more likely from low-familiarity sequence to high-familiarity sequence than the reverse direction, and more likely between more similar sequences.

more comparable.

This simulation used a single Jordan network with noise to generate capture errors. Interestingly, capture errors were also observed without noise when the hybrid network learned many sequences. When some of the sequences were presented more often than other sequences, it was observed that a low-frequency sequence that shared a subsequence with a high-frequency sequence was sometimes captured by the high-frequency sequence, but a capture error in the reverse direction was rare. This suggests that some action errors can occur as the result of interaction among many different mappings that a single network is trying to learn.

Summary

The model presented in this paper has a number of interesting properties. First, it was able to model the shift from a serial, novice performance to a highly parallel, expert performance. Training of the network started from an initial state where Action-net did only a simple mapping of one plan to one action, and much of the work was done by Plan-net that mapped an intention to a sequence of plans. The training reversed the situation:

505

the mapping from intention to plan became one-to-one, and the mapping from plan to action became one-to-sequence. For Plan-net, the task changed from a serial one to a highly parallel one.

Second, this model seems to show a way "chunking" can be done in a connectionist network. The network was able to discover new representation of action plans so that each plan represents a increasingly large chunk of output sequence. The process of chunking in the model is gradual rather than discrete, and no new representational entity needs be created when chunks are formed. It is not necessary to presuppose what "level" of information each component in the model should represent: rather such a "level" is a dynamic property of each component that can change through learning.

Third, the model exhibits digraph frequency effects similar to those observed in studies of typing. The model also simulated capture errors and has the property that capture errors are more likely to be made from a low frequency sequence to a higher frequency sequence than the reverse direction.

Acknowledgement

This work was supported by a Grant from Nippon Telegraph & Telephone Corporation to the Institute for Cognitive Science. I would like to thank Don Norman, Dave Rumelhart, and Mike Jordan for useful discussions.

References

Bryan, W. L., & Harter, N. (1897). Studies on the physiology and psychology of the telegraph language: The acquisition of a hierarchy of habits. *Psychological Review, 4*, 27-53.

Dell, G. S. (1984). Representation of serial order in speech: Evidence from the repeated phoneme effect in speech errors. *Journal of Experimental Psychology: Learning, Memory, and Cognition, 10*, 222-233.

Feldman, J. A., & Ballard, D. H. (1982). Connectionist models and their properties. *Cognitive Science, 6*, 205-254.

Grudin, J. T., & Larochelle, S. (1982). *Digraph frequency effects in skilled typing.* (Tech. Rep. No. 110). University of California, San Diego, Center for Human Information Processing.

Grudin, J. T. (1983). Error patterns in skilled and novice transcription typing. In W. E. Cooper (Ed.), *Cognitive aspects of skilled typewriting* (pp. 95-120). New York: Springer-Verlag.

Hinton, G. E., & Smolensky, P. (1984). *Parallel computation and the mass-sprint model of motor control.* (Tech. Rep. No. 123). University of California, San Diego, Center for Human

Information Processing.

Jordan, M. I. (1985). *The learning of representations for sequential performance*. Unpublished doctoral dissertation, University of California, San Diego.

Jordan, M. I. (1986). Attractor dynamics and parallelism in a connectionist sequential machine. In *Proceedings of the eighth annual conference of the Cognitive Science Society* (pp. 531-546).

Lashley, K. S. (1951). The problem of serial order in behavior. In L. A. Jeffress (Ed.), *Cerebral mechanisms in behavior*. New York: Wiley.

McClelland, J. L., & Rumelhart, D. E. (1986). *Parallel distributed processing: Explorations in the microstructure of cognition. Vol. 2. Psychological and biological models*. Cambridge, MA: MIT Press/Bradford Books.

Motley, M. T., Camden, C. T., & Baars, B. J. (1982). Covert formulation and editing of anomalies in speech production: Evidence from experimentally elicited slips of the tongue. *Journal of Verbal Learning and Verbal Behavior, 21*, 578-594.

Norman, D. A. (1981). Categorization of action slips. *Psychological Review, 88*, 1-15.

Reason, J. T. (1979). Actions not as planned. In G. Underwood & R. Stevens (Eds.), *Aspects of consciousness*. London: Academic Press.

Rumelhart, D. E., & Norman, D. A. (1982). Simulating a skilled typist: A study of skilled cognitive-motor performance. *Cognitive Science, 6*, 1-36.

Rumelhart, D. E., Hinton, G. E., & Williams, R. J. (1986). Learning internal representations by error propagation. In D. E. Rumelhart & J. L. McClelland (Eds.), *Parallel distributed processing: Explorations in the microstructure of cognition. Vol. 1. Foundations* (pp. 318-362). Cambridge, MA: MIT Press/Bradford Books.

Rumelhart, D. E., & McClelland, J. L. (1986). *Parallel distributed processing: Explorations in the microstructure of cognition. Vol. 1. Foundations*. Cambridge, MA: MIT Press/Bradford Books.

Sellen, A. J. (1986). *An experimental and theoretical investigation of human error in a typing task*. Unpublished Master's thesis, University of Toronto.

Wickelgren, W. A. (1969). Context-sensitive coding, associative memory, and serial order in (speech) behavior. *Psychological Review, 76*, 1-15.

MUSACT: A Connectionist Model of Musical Harmony

JAMSHED J. BHARUCHA
Dartmouth College

A connectionist model of musical harmony is proposed to account for intuitions of schematic expectancy and sequential consonance. The model distinguishes between schematic and episodic structures, and the current version focusses on the former. Units representing tones, chords and keys are linked in a network. Activation received by tone units from a musical event spreads through the network phasically, reverberating until the network settles into a state of equilibrium. The settled state represents the expectancies for events to follow and the perceived consonance of these events should they occur. The model accounts for psychological data on rating judgments and recognition memory. Evidence for the spread of activation comes from experiments on priming of chords.

Several features of connectionist models--in which simple processing units, connected by weighted links in a network, activate each other in parallel--recommend them as models of music cognition. First, they are highly interactive, so that low level processes can influence higher level processes and vice versa. Thus, a sequence of tones can imply a chord or a key, and the implied chord or key can in turn influence the perception of tones that follow. Second, the architecture of a connectionist system hooks up naturally with a sensory front end that codes frequency as a (spatial or temporal) pattern of activations of neurons in the inner ear. Third, pattern matching of a currently heard sequence can be achieved in parallel by content-addressing all similar memory traces, enabling recognition of sequences and variations thereof. Finally, connectionist networks can internalize persistent regularities, such as occur in our musical environment, without explicit instruction.

A network's ability to internalize regularities--by altering the strengths of connections between units (McClelland & Rumelhart, 1986)--enables the parsimonious view that music cognition is a consequence of general principles of cognition operating on structural regularities in the environment. The average listener has little, if any, explicit knowledge of musical structure, yet shows evidence of considerable tacit knowledge. For example, even listeners with no formal training in music show systematic differences in processing time for chords as a function of the prior harmonic context (Bharucha & Stoeckig, 1986, 1987).

The model proposed here, MUSACT, is designed to capture musical intuitions and psychological data concerning expectancy, sequential consonance and short term memory for musical harmony. Schenker (1906/1954) observed that one of the qualities of the dominant chord is "to indicate that the tonic is yet to come" (p.219). Expectancies of this sort are driven by cognitive structures and processes that have internalized regularities in the musical environment in order to facilitate subsequent perception. Hand in hand with expectancies are intuitions about sequential consonance. The greater an event's expectancy, the greater its context-dependent consonance. A composer may choose to satisfy or violate these expectancies to varying degrees, thereby evoking varying degrees of sequential consonance or dissonance. The aesthetic value of subtle departures from the expected has figured in numerous theoretical writings about music and about emotion (e.g. Mandler, 1984; Meyer, 1956). A new piece is heard as culturally anomalous to the extent that expectancies are violated. Indeed, the connection strengths between units are assumed to be trained by minimizing, over the history of one's exposure, the discrepancy between expectancies generated by the network and transitional probabilities in the music of one's culture.

People exposed to Western music are assumed to acquire a network representation of chord functions (hereafter referred to as chords) and their organization in the form of keys, which serves to schematize subsequent perception. Constraints on the combining of tones into chords and constraints on the sequencing of these chords are among the more obstinate regularities in Western music. Every amateur musician knows that a mastery of only six chords enables you to accompany a vast majority of popular songs. These basic harmonic regularities have even begun to permeate much of the popular music of countries in the East. Given the pervasiveness of these regularities, and given the evidence of tacit knowledge of them (Bharucha & Stoeckig, 1986, 1987), it is reasonable to conclude that they have been internalized as cognitive structures that facilitate and bias subsequent perception.

Schematic & Episodic Structures: Schematic & Veridical Expectancy

Two broad classes of cognitive structures for music are envisioned: schematic structures, which represent abstract structural regularities (sometimes formalized as grammars) of the music of one's culture, and episodic structures, which represent particular musical sequences (Bharucha, 1984b). The former embody typical relationships between types or classes of events, and the latter embody relationships between particular event tokens. The expectation generated by a dominant chord for the tonic to follow arises from schematic structures that encode the typicality of relationships, whereas the expectation for a VI chord to follow a particular dominant chord in a particular familiar piece arises from episodic structures that encode the particular events in that piece. The former generate schematic expectancies and the latter veridical expectancies. The two are usually in agreement but often in conflict, giving rise to the peculiar effect known as the deceptive cadence (a dominant chord followed by a VI chord). The unavoidable effect of schematic expectancies, even when listening to a piece that violates them, provides a resolution of Wittgenstein's puzzle (see Dowling & Harwood, 1985) concerning the possibility of violating expectancies when listening to a familiar piece.

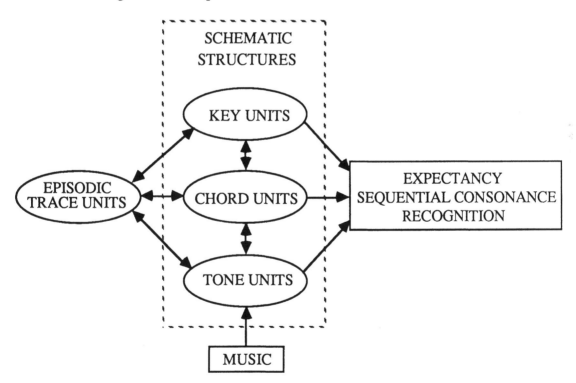

FIGURE 1. A sketch of the model

Although schematic expectancies can in principle be generated from episodic structures by adding veridical expectancies over episodic traces activated in parallel, a model of this sort would require documenting all the episodic memory traces in a typical person's brain. MUSACT focusses on schematic structures, employing episodic structures only to the extent of implementing experiments on short term memory for sequences of chords.

The Model

Architecture

We have the sense that certain typical tone clusters, such as major and minor chords, are heard as unitary. We also have the sense of an even more abstract, meaning-like, state induced by music, called the key. Sequences or clusters of tones may unavoidably suggest a chord or several alternative chords, and combinations of chords establish a sense of key. One of the advantages of the model is that it enables chord and key instantiations to be graded and ambiguous. These ambiguities are an important feature of music, exploited during modulations or other transitions, and used to create graded degrees of expectancy violation.

The schematic network consists of three layers of units, representing tones, chords and keys (see Figures 1 and 2). There are symmetric links between units of adjacent layers (i.e. between tone units and chord units, and between chord units and key units) but no links between units within a layer. The links between tone and chord units reflect the relationships between tones and the chords of which they are components. The links between chord and key units reflect the relationships between chords and their parent keys. In the current version of the model, only major and minor chords are implemented, and only major keys.

The input to the network is a sequence of events, each event being a simultaneous cluster of tones. Input is received via the tone units, which represent the twelve octave-equivalent pitch categories. This layer constitutes a discrete pitch schema to which the pitch continuum is assimilated (see Shepard & Jordan, 1984). The frequency responses of these units are equally spaced along a logarithmic scale of frequency, and are fixed only relative to each other, underlying the relational nature of pitch memory. The sensory front end that provides the input to these units is beyond the present scope, but neural net models that extract octave-equivalent pitch categories (see Deutsch, 1969) can be adapted quite naturally to a connectionist model of more abstract phenomena as proposed here.

The output of the model is the pattern of activation of the chord and key units. A chord unit is activated either by the explicit sounding of some or all of its component tones, or by indirect influences, via its parent keys, from related chords. When only some of the chord's component tones are sounded, the context may help disambiguate the chord by top-down activation from parent key units. A key unit is activated by some or all of its daughter chords, or by indirect influences, via its daughter chords, from related keys. Indirect activation of chord units permits smooth excursions (such as secondary dominants and modulations) from the focus of activation.

Phasic Activation

After an event is heard, activation spreads through the network, via the weighted links, reverberating back to units that were previously activated. In this model, activation is phasic, meaning that units respond only to changes in activation of neighboring units. Phasic activation was selected because of the salience of event onsets in music. On each cycle, units are synchronously updated on the basis of activation levels, from the previous cycle, of neighboring units. Phasic activation eventually dissipates until the network settles into a state of equilibrium. The network will settle if no unit transmits more phasic activation than it received on the previous cycle. This requirement is easily satisfied if the weights are small relative to the fan-in or fan-out.

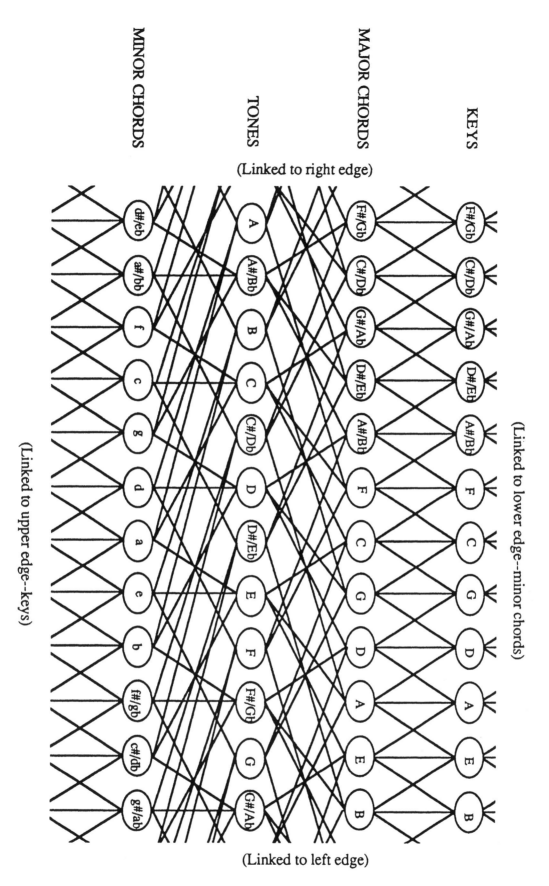

FIGURE 3. The network of tone, chord and key units.

511

Activation, Tonality, and Expectation

The pattern of activation of key units represents the degree to which keys are established. Tonal music will tend to build up activation in one region of the network, such that one key unit is most highly activated, with activation tapering off with increasing distance from the focal key. Atonal music will typically induce a less focussed pattern, and polytonal music would result in multiple, though not very strong, foci. The model thus allows for gradations of key, and for multiple keys, consistent with the findings of Krumhansl (Krumhansl & Kessler, 1982; Krumhansl & Schmuckler, 1986).

The pattern of activation of chord units represents the pattern of expectancies for chords to follow. A chord whose unit is highly activated is strongly expected, and a strongly expected chord is heard as consonant. The pattern of activation of chord units thus underlies our harmonic expectations as well as our intuitions of context-dependent sequential consonance.

In future versions of the model that incorporate additional pitch structure, such as pitch proximity (see Bharucha 1984a; Deutsch, 1978) for voice leading, tone units will also serve as output units responding to top-down and lateral influences. At present, although the tone units do propogate top-down activation, the model is intended to be tested only for its harmonic, not melodic, adequacy.

Weights

All links between tone units and chord units are assumed to have equal weights. There are six classes of weights between chord units and key units, corresponding to the six chords in each key. All the links of the same class must have the same weight, resulting in a repeating pattern of weights over the network. (For example, the G major chord has the same relationship--the dominant--to the key of C as the C# major chord has to the key of F#.) In music-theoretic terms, a chord's relationships to its parent keys are its functions in those keys. The six classes of weights thus correspond to the six chord functions. Since some chord functions are stronger instantiators of key than others (e.g. the key of C is more strongly instantiated by the C major chord--the tonic--than by the F major chord--the subdominant--even though both are daughter chords), it would be reasonable to assume that links for stronger functions have higher weights. However, it turns out that the model can exhibit all the essential qualitative patterns of behavior even when the weights are not differentiated according to function. Thus, the typical hierarchy of strengths of the three major chord functions (tonic, dominant, subdominant) emerges even if their weights are equal. The pattern of connectivity alone is sufficient to bring about functional differentiation. The model thus generates the desired functional hierarchy of chords simply by knowing which chords are members of which keys and which tones are components of which chords. This is a remarkable and unanticipated property of the model, and points to its power.

The weights are assumed to be higher for major chord units than for minor chord units, since major chords are stronger instantiators of key. As discussed below, with only these elementary constraints on links and their weights, based on fundamental tenets of music theory, a set of weights can be found that enable the model to account for a complex array of psychological data on the perception of harmony as well as some subtle aspects of music theory. Indeed, one of the advantages of a connectionist model is that complexities and interactions may emerge naturally from simple constraints on architecture.

The Spread of Activation

The activation, $a_{i,e}$, of the ith unit after the network has reverberated to equilibrium following the eth event is:

$$a_{i,e} = a_{i,e-1}(1-d)^t + sA + \sum_{c=1}^{equil} \Delta a_{i,e,c},$$

where **d** is the rate of decay $(0 < d_i < 1)$ for one time unit, **t** is the time elapsed since the offset of the last event, **A** is the source activation due to the stimulus, and **s** is 1 if the unit is receiving environmental input and 0 otherwise. (For simplicity, the present version of the model assumes that all events are of equal duration; duration is varied simply by repeating events.) $\Delta a_{i,e,c}$ is the change in activation after reverberation cycle **c**, and is the output of the unit on cycle **c+1**. $\Delta a_{i,e,c}$ is the sum of the outputs of its **n** neighboring units, weighted by their links, w_{ij}, $(0 < w_{ij} < 1)$. Thus,

$$\Delta a_{i,e,c} = \sum_{j=1}^{n} w_{ij} \Delta a_{j,e,c-1}.$$

Simulations

The simulation results reported below were obtained with a weight assignment in which all chord units of the same mode (major or minor) have equally weighted links with their parent keys. This enables us to observe the differentiation of chord functions without forcing their differentiation via the weights. Major chord units were assigned more strongly weighted links (0.244) with their parent key units than were minor chord units (0.22, 90% of 0.244), since major chords more strongly establish their parent keys. Links between tone units and chord units were also uniform (0.0122, 5% of 0.244), with no preference given to the root of the chord. This weight assignment yields an initial strong influence of the input tones on their local regions of the network, followed by an extended percolation during which the two more abstract layers exert their influence. Thus, if the input tone cluster is {C,E,G}, i.e., a C major chord, the chord units linked to these tone units show an initial prominence, so that, for example, the A major chord unit is more highly activated than the D major chord unit. However, after the activation has had a chance to reverberate for a number of cycles, the D major chord unit overtakes the A major chord unit, by virtue of its greater proximity to the eventual activation peak, which is at C major. Within 40 cycles, all constraints inherent in the network are satisfied, and the pattern of activation does not change qualitatively. Within 50 cycles, the phasic activation has dissipated to the point at which the ratio $\Delta a_{i,e,c}/A$ does not exceed 0.005 for any unit. For the results discussed below, the equilibrium state is stipulated to be the state of the network when this 0.005 criterion is reached.

The tone cluster {C,E,G}, i.e., a C major chord, played without any prior context, generates the following pattern of activation (see Figure 3). The most highly activated key is C, of which the source chord is the tonic. Activations of other key units decrease monotically with distance from C, the lowest being F#. The pattern of activation of the chord units mirrors that of the key units with the same alphabetic name.

Even though the weights between the source chord, C major, and its parent keys, F, C, and G, are equal, the parent keys are not activated to the same degree. In decreasing order, the activations of these three keys are C, F, and G, which are the keys in which the source chord is the tonic, dominant, and subdominant, respectively. This is exactly the hierarchy of harmonic functions to be expected from music theory.

There seems at first to be a paradox here. On the one hand, a chord more strongly instantiates the key of which it is the dominant than the key of which it is the subdominant. On the other hand,

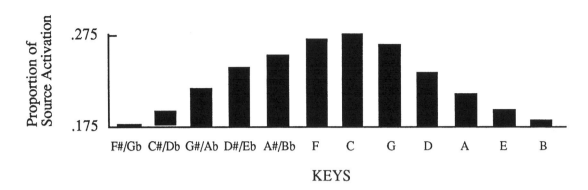

FIGURE 3. The pattern of activation of key units after activating the tone units C, E, and G (heard as a C major chord).

the key of the dominant is generally thought to be closer to the current key than is the key of the subdominant, and should therefore have a higher activation. This paradox is not peculiar to the present model; the model simply forces us to confront it. (Interestingly, the model conforms to the latter alternative if the link between a tone unit and a chord unit is weighted more heavily for the root of the chord than for the other tones, illustrating the unanticipated interdependence of apparently disparate factors.) The resolution of the paradox lies in the fact that dominant chords typically occur much more often than subdominant chords in a piece of music, presumably because dominant chords contribute to a more stabe key. Given the high frequency of dominant chords, the dominant builds up more activation than the subdominant.

For an input consisting of a sequence of events, the rate at which activation decays between events was set at 0.3. If the input sequence is the tone cluster {F,A,C} followed by {G,B,D}, i.e., an F major chord followed by a G major chord, the model shows the most highly activated key unit to be C, even though the tonic chord of that key, the C major chord, has not occurred. This, again, is consistent with what would be expected from music theory.

Psychological Data

The model's performance in experiments eliciting rating judgments and memory confusions is qualitatively equivalent to human performance on these tasks. In a series of experiments on the perception of harmony (Bharucha & Krumhansl, 1983; Krumhansl, Bharucha, & Castellano, 1982), the perceived relationship between chords as a function of context was studied using these tasks.

Rating Judgments

In the rating task, subjects were presented with two chords in succession and rated, on a scale from 1 to 7, how well the second followed the first. For the simulation, the rating judgment consisted of reading off the level of activation of the last chord. On this assumption, the observed rating judgments are equivalent to judgments of expectancy or sequential consonance.

When both chords of the pair shared a parent key, subjects gave higher judgments when the last chord was major than when it was minor. This result was particularly interesting for a pair consisting of one major chord and one minor chord, because the same two chords elicited different judgments depending upon their temporal order. In the model, a major chord activates its unit slightly more than does a minor chord, even though the source activation, **A**, is the same, because of more reverberatory activation from its parent keys. This asymmetry doesn't require asymmetric links, since it follows directly from the fact that major chords establish their parent keys more strongly than do minor chords. In music, this translates into a tendency to end with a major chord, often even if the key is minor.

When the two chords were preceded by a context that established a key, subjects gave higher judgments the more closely related the last chord was to the key of the context. In the model, this occurs because the context activates closely related chords more than distantly related chords. An interesting asymmetry was contained in this result as well: two chords of the same mode (major or minor) elicited a higher judgment if the one closer to the key of the context was played last. Thus, in the key of C, an F major chord followed by a C major chord is a more stable ending than the reverse, even though the reverse may be true in the absence of a prior context.

Recognition Memory

Memory for a sequence does not take the form of an explicit representation, but rather is encoded by weight increases of links between the units in the network and episodic units that are temporally organized. Each event in a sequence increases the weights of links between a proprietary episodic unit and the schematic units, in proportion to the activations of these latter units. The episodic units are then activated every time the tone, chord or key units to which they have strong links are highly activated. Once activated, the episodic units in turn activate the schematic units. This architecture can accomplish pattern completion (recalling a piece given only a few notes or a sketchy rendition), recognition of variations, and retrieval of past musical memories in a parallel, content-addressable fashion rather than through serial search.

In the present implementation, episodic units serve only to simulate short term recognition of a sequence. Consider a to-be-remembered sequence of chords. Each chord token in the sequence has its proprietary episodic unit. Links between this unit and the schematic units have their weights increased in proportion to the phasic activations of the schematic units after that chord is heard. In this way, the pattern of activation of the network after hearing a particular chord token can be recovered simply by activating its episodic unit.

Data from experiments on recognition memory for chord sequences show trends similar to those observed for rating judgments (Bharucha & Krumhansl, 1983; Krumhansl, Bharucha, & Castellano, 1982). In these experiments, subjects judged whether two presentations of a sequence of chords were identical or had different chords (the target chords) in one serial position. If the sequence as a whole established a key, a change was less likely to be detected the more closely related the second target was to the key. This manifest itself in two ways. First, a change was less likely to be detected if both targets were closely related to the key than if both were distantly related to the key. Second, if one target was more closely related to the key than the other, a change was less likely to be detected when the more closely related one occurred in the second presentation rather than the first. In general, if a change is made to a sequence, it is less likely to be detected if it renders the sequence more coherent than more anomalous. Coherence is a consequence of the strong activation of a subsuming unit, such as a key unit in music, or a subsuming semantic unit in language (Bharucha, Olney, & Schnurr, 1985), relative to the other subsuming units in the network.

In the model, the probability with which the second target is judged to be the same as the first is monotonically related to the activation of the unit representing the second target, relative to the activations of the other chord units, when the first target was heard. Since the more closely related the second target is to the key of the context, false alarm rates increase with closeness of the second target to the key of the context.

This short term memory architecture also predicts that if the second presentation is transposed to a different key, false alarm rates should decrease with the distance (along the network) between the two keys. This would be consistent with recognition memory results for sequences of tones (Cuddy, Cohen, & Miller, 1979).

Evidence of Spreading Activation: Priming

Evidence that chords indirectly activate representations of related chords comes from experiments on priming. Bharucha and Stoeckig (1986) presented subjects with two major chords in succession, the first called the prime and the second the target. On half the trials, the chord in the target position was a mistuned foil. Subjects were instructed to judge, as fast as possible, whether the chord in the target position was in-tune or out-of-tune. Subjects first practiced without the prime until a criterion level of accuracy was reached. In the main task, response times were significantly faster when the prime and target were close together along the network than when they were distant. This demonstrates that the prime activates units corresponding to closely related targets, as would be predicted by the model.

Error rates mirrored response times, so that the target was more likely to be judged in-tune the closer it was (along the network) to the prime. The response time and error rate data measure the prime's influence on the target's expectancy and sequential consonance, respectively.

An alternative explanation of the above results is that the priming is due to overlapping harmonic spectra between closely related prime and target chords. In a subsequent study (Bharucha & Stoeckig, 1987), harmonics that overlapped were removed from the stimuli and priming was still observed. This demonstrates that there must be a spread of activation at a fairly abstract cognitive level.

References

Bharucha, J.J. (1984a). Anchoring effects in music: The resolution of dissonance. Cognitive Psychology, 16, 485-518.

Bharucha, J.J. (1984b). Event hierarchies, tonal hierarchies, and assimilation: A reply to Deutsch and Dowling. Journal of Experimental Psychology: General, 113, 421-425.

Bharucha, J., & Krumhansl, C.L. (1983). The representation of harmonic structure in music: Hierarchies of stability as a function of context. Cognition, 13, 63-102.

Bharucha, J.J., Olney, K.L, & Schnurr, P.P. (1985). Coherence-disrupting and coherence-conferring disruptions in text. Memory and Cognition, 13, 573-578.

Bharucha, J.J., & Stoeckig, K. (1986). Reaction time and musical expectancy: Priming of chords. Journal of Experimental Psychology: Human Perception & Performance, 12, 403-410.

Bharucha, J.J., & Stoeckig, K. (1987). Priming of chords: Spreading activation or overlapping harmonic spectra? Manuscript under review.

Deutsch, D. (1969). Music recognition. Psychological Review, 76, 300-307.

Deutsch, D. (1978). Delayed pitch comparisons and the principle of proximity. Perception & Psychophysics, 23, 227-230.

Dowling, W.J., & Harwood, D.L. (1985). Music cognition. New York: Academic Press.

Cuddy, L.L., Cohen, A.J., & Miller, J.(1979). Melody recognition: The experimental application of musical rules. Canadian Journal of Psychology, 28, 148-157.

Krumhansl, C. L., Bharucha, J., & Castellano, M. A. (1982). Key distance effects on perceived harmonic structure in music. <u>Perception & Psychophysics</u>, <u>32</u>, 96-108.

Krumhansl, C.L., & Kessler, E.J. (1982). Tracing the dynamic changes in perceived tonal organization in a spatial representation of musical keys. <u>Psychological Review</u>, <u>89</u>, 334-368.

Krumhansl, C.L., & Schmuckler, M.A. (1986). The <u>Petroushka</u> chord: A perceptual investigation. <u>Music Perception</u>, <u>4</u>, 153-184.

Mandler, G. (1984). <u>Mind and body: Psychology of emotion and stress</u>. New York: Norton.

McClelland, J.L., & Rumelhart, D. (1986). <u>Parallel distributed processing: Explorations in the microstructure of cognition</u>. M.I.T. Press.

Meyer, L. (1956). <u>Emotion and meaning in music</u>. Chicago: University of Chicago Press.

Schenker, H. (1954). <u>Harmony</u> (O. Jones, Ed., E.M. Borgese, Trans.). Cambridge: MIT Press. (Original work published 1906).

Shepard, R.N, & Jordan, D.S. (1984). Auditory illusions demonstrating that tones are assimilated to an internalized musical scale. <u>Science</u>, <u>226</u>, 1333-1334.

Acknowledgements

This paper was written while the author was at Carnegie-Mellon University. The author thanks James McClelland and Herbert Simon for helpful discussions. Katherine Olney contributed to early implementations of the model. The author also thanks Paul Cohen, Diana Deutsch, William Estes, Leonard Meyer, Caroline Palmer, Mark Schmuckler and, especially, Carol Krumhansl, for suggestions and discussions.

Learning Acoustic Features From Speech Data Using Connectionist Networks

Raymond L. Watrous [1]
Department of Computer and Information Science
University of Pennsylvania
Philadelphia, PA 19104
1-215-898-8542
Siemens Research and Technology Laboratories
Princeton, NJ

Lokendra Shastri
Department of Computer and Information Science
University of Pennsylvania

Keywords: connectionist networks, machine learning,
speech recognition

March 13, 1987

Abstract

A method for learning phonetic features from speech data using connectionist networks is described. A *temporal flow model* is introduced in which sampled speech data flows through a parallel network from input to output units. The network uses hidden units with recurrent links to capture spectral/temporal characteristics of phonetic features. A supervised learning algorithm is presented which performs gradient descent in weight space using a coarse approximation of the desired output as an target function.

A simple connectionist network with recurrent links was trained on a single instance of the word pair "no" and "go" represented as fine time-scale filterbank channel energies, and successfully learned to discriminate the word pair. The trained network also correctly separated 98% of 25 other tokens of each word by the same speaker. The same experiment for a second speaker resulted in 100% correct discrimination. The discrimination task was performed without segmentation of the input, and *without a direct comparison of the two items*.

A second experiment designed to extended the use of this model to discrimination of voiced stop consonants in various vowel contexts is described. Preliminary results are described in which the network was optimized using a second-order method and learned to correctly classify the voiced stops. The results of these experiments show that connectionist networks can be designed and trained to learn phonetic features from minimal word pairs.

[1]Thanks to Wolfgang Feix, Alex Waibel, Max Mintz and Bruce Ladendorf for helpful discussion.

1 Introduction

Connectionist networks offer significant advantages in addressing problems of machine perception because of their inherently parallel structure, which is well matched to the biological architecture that has served as their paradigm. Their learning capabilities, robust behavior, noise tolerance and graceful degradation are all capabilities which are becoming increasingly well understood and documented [SR86].

The solution of certain perceptual problems requires that the temporal relationships among stimulus characteristics be properly represented. This is especially true in speech recognition, where the relationship between time and frequency is wonderfully complex. One major result from the past thirty years of speech recognition and synthesis research is that it is generally impossible to define speech as a sequence of events with static spectral characteristics. Instead, speech is produced and perceived as a continuous flow of sound, with constantly changing spectral properties. In the production of speech, basic speech units (phonemes) are integrated into a smooth sequence, so that the acoustic boundaries can be very difficult to specify. Moreover, phonemes are often co-produced (coarticulated), so that the phonemes exert a strongly context-dependent interaction. The effect of context is seen in the changes in formant trajectory, duration and energy contours. Thus, the perception of speech depends on the correct analysis of dynamic temporal/spectral relationships.

Many solutions to the problem of speech recognition have been advanced, including signal processing, feature extraction, pattern matching with dynamic non-linear time alignment, linear predictive coding, stochastic modeling, segmentation and labeling, syntactic grammars and expert systems, with explicit rule-based knowledge representation. These approaches share the goal of capturing the regular structure inherent in the speech signal in the presence of tremendous variability. Although these techniques have all succeeded to some extent, a general shortcoming has been that minor irregularities in the input, whether from signal noise, background acoustic noise, or speaker variability, have major negative effects on performance. This lack of robustness has been very frustrating, because it is contrary to our experience of speech communication, in which minor irregularities are easily overcome, if consciously perceived at all.

The connectionist network approach is attractive because it offers a computational model which has inherently robust properties. The networks consist of simple processing elements which integrate their inputs and broadcast the results to the units to which they are connected. Thus, the network response to input is the aggregate response of many interconnected units. It is the mutual interaction of many simple components that is the basis for robustness.

Connectionist networks also provide a fundamentally different language for knowledge representation. This is important in establishing a conceptual framework in which solutions can be conceived and investigated. In addition, the computational speed of parallel networks is a requirement for real-time performance in non-trivial speech systems.

The problem of designing connectionist networks which can learn the dynamic spectral/temporal characteristics of speech has not yet been widely studied. Most work in connectionist networks so far has focussed on the static relationship between input/output pairs, such as associative memories [KL81,Hop82], various encoding, decoding, parity and addition problems [RHW86], and mapping from word spelling to phoneme labels [SR86].

The TRACE model [EM86,ME86] is the first well-developed model for studying speech recognition using connectionist networks. As discussed below, this model represents temporal sequence directly using sets of network units allocated to subsequent time slices. The approach developed in this paper is different in that temporal sequence is represented implicitly.

Learning to associate static input/output pairs can be accomplished with layered connectionist networks with feedforward links alone. But recurrent, or feedback, links are required to provide the network with state sequence information, in order to capture sequential behavior. [Jor86,Sut85,RHW86].

The experiments reported here were designed to explore the capabilities of parallel networks to learn *dynamic properties of time-varying data*.

We first choose a moderately difficult speech recognition problem to test the extent to which a connectionist network could form an internal representation of the temporal/spectral characteristics which distinguish two similar words. A simple network with recurrent links was trained on a single instance of the word pair "no" and "go", and successfully discriminated 98% of 25 other tokens of each word for the same speaker. The experiment was repeated for a second speaker and resulted in 100% discrimination performance.

This research is being extended to more difficult problems of speech recognition. Preliminary results are reported which show that connectionist networks can be used to successfully discriminate the voiced stop consonants, /b,d,g/, in various vowel contexts.

The results of the preliminary experiments show that connectionist networks can indeed be designed and trained to successfully discriminate similar word pairs by learning acoustic-phonetic features.

2   Experiment I

The experiment selected for this first examination of connectionist networks in speech recognition was the discrimination between the minimal pair "no" and "go". This is a typical speech recognition problem, which is included in a standard database for evaluation of speech recognizers [DS81]. The utterances "no" and "go" share for the major and final portion the voiced phoneme /o/. The "no" utterance is characterized by a lower energy nasal murmur preceding the transition to the back vowel /o/. The "go" is distinguished by a very low energy voicing interval during the lingua-palatal closure, a brief burst as the closure is released, and a voiced transition to the full vowel.

The distinction between "no " and "go", therefore, is concentrated in the brief interval of relatively low energy at the beginning of the word. These differences consist in the relative voicing energy, burst spectrum, and formant value and transition pattern.

2.1   Network Architecture

For this first experiment, a three-layer connectionist network consisting of an input layer, one hidden layer and an output layer was implemented, as shown in Figure 1. The sampled speech data flowed through the network in time sequential order. Thus, the 16 channel energies were applied to 16 input units, from which activation spread toward the output units simultaneously as the input units were updated by subsequent speech samples. This design will be referred to as the *temporal flow model*, or, more simply as the *flow model*.

Other approaches have used an array of input units, and represented the time axis along one index of input unit array [PNH86,EM86,ME86]. In these cases, time is spatialized across units. The temporal flow model was chosen because it does not require 'chunking' of variable length utterances onto a fixed size network, it avoids the problem of temporal symmetry, and the temporal flow model seems to be closer to the biological model of speech processing.

Integration over time of spectral characteristics is accomplished by the recurrent unit links. A positive recurrent link weight will feed back as input some of the unit output to reinforce and integrate the unit response.

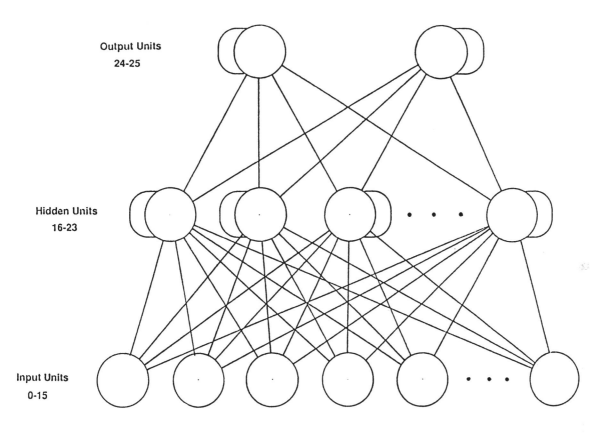

Figure 1: "Temporal Flow Model showing input, hidden and output layers"

2.1.1 Unit Functions

The functions which define the unit behavior were chosen from ones in common use in connectionist networks [SR86,RHW86]. These functions approximate the computational properties of neural cells, and have convenient mathematical properties for the learning algorithm used in this experiment.

The unit output, $o_j(t)$, is given by the sigmoid function:

$$o_j(t) = \frac{1}{1 + e^{-p_j(t)}}$$

where $p_j(t)$, the potential function is given by:

$$p_j(t) = \sum_{i,d} w_{ijd} o_i(t - d)$$

where d is the time delay along the link between units u_i and u_j.

2.2 Back-Propagation Learning Algorithm

For this experiment, an extended form of the back-propagation learning algorithm was chosen to accommodate networks with recurrent links [RHW86]. The derivation of a more general form of the algorithm for variable delay and recurrent links is given in [WS86].

The error-propagation algorithm modifies the unit connection weights in order to minimize the mean squared error between the actual and desired output values. The weight change rule can be written as:

$$\Delta w_{ijd} = \eta \sum_{\tau} \delta_j(t - \tau) o_i(t - \tau - d)$$

where $\delta_j(t - \tau)$ is the error signal at unit j at time $t - \tau$, with respect to the target values at the output units at time t. This error is given by:

$$\delta_j(t - \tau) = \sum_{\alpha,k} w_{jk\alpha} \delta_k(t - \tau + \alpha) \frac{\partial o_j}{\partial p_j}(t - \tau)$$

for $\alpha \leq \tau$.

The error signal for an output unit is defined by the difference between the actual and target values, times the unit function slope at time t:

$$\delta_j(t) = (o_j(t) - targ_j(t)) \frac{\partial o_j}{\partial p_j}(t)$$

The target function for the output units consists of a simple ramp. For the output unit which corresponded to the utterance being trained, the ramp increased from a value of 0.5 to 1.00 over the duration of the utterance. The other unit was correspondingly decreased from 0.5 to 0. This represented the intuition that evidence for or against a particular word accumulates over its duration, and reaches a level of confidence after the utterance is completed.

2.3 Data

The data used for this experiment consisted of speech data for one male (GD) and one female speaker (CP) from the Texas Instruments standard isolated word recognition database [DS81]. The digitized data was played through an A/D converter (Digital Sound Corporation DSC 2000) into a commercial speech recognition device (Siemens CSE 1200), where it was passed through a 16-channel filter bank, full-wave rectified, log compressed and sampled every 2.5 milliseconds. The filters were low Q bandpass filters, with linear-log spaced center frequencies [Mar70]. Twenty-six repetitions of each word comprise the corpus, for a total of fifty-two utterances (26 "no" and 26 "go") for each speaker. The filter bank response to the training utterances is shown in Figure 2.

2.4 Results

The connectionist network experiments were conducted on a sequential machine using a network simulator, written specifically for this experiment. The experiments were carried out on a VAX 8650 and a SUN 3/165 workstation. The network described previously was initialized with small random link weights and trained on a single pair of no/go utterances for 6000 training iterations. Each speaker's data was used to optimize separate networks.

The results of the optimization are shown in Figure 3 for the first speaker. The value of the squared-error term is neither monotonic decreasing nor a smooth function of the number of optimization iterations. This is thought to be due to the local nature of the weight change algorithm, and the limited extent of back-propagation in time. The network coinciding with the sharp notch in the squared error value was chosen for further study.

2.4.1 Output Unit Response to Training Data

The response of the output units for the network at the selected critical point in the learning process was recorded, and can be seen in Figure 4. The output units respond in equal and opposite ways to the input stimuli; in addition, their time response roughly approximates a ramp. Since the learned response closely fits the training function, the network shows very good discrimination between the single pair of the training set.

The significance of this result should not be overlooked. First, the application locally of global optimization metric provided a successful optimization path to a desired network response pattern. Second, although no segmentation decisions were made, the network was able to form discriminating spectral features independently. Third, the approximations of constant weight value, and restrictions to maximum τ value in the extended back-propagation algorithm did not prevent convergence to a good solution. Fourth, although the shape of the error contour is unknown, it is almost certainly not smooth; consequently, the learning path apparently avoided local minima in arriving at a solution.

2.4.2 Extension to Test Set

In order to test the generality and robustness of the internal representations obtained from the training word pair, and to further investigate the characteristics and behavior of the hidden units, the network of least squared error value was tested on a set of 25 additional pairs of no/go utterances by the same speaker. Using a simple deterministic decision algorithm, the input word could be clearly categorized by the network response. Under these conditions, the trained network successfully discriminated all but one of the test cases (98% accuracy). The results for the second speaker were similar (100% accuracy).

The responses of the hidden units were analyzed for the 50 test utterances as well as the 2 training utterances for each speaker. In nearly every respect, the hidden unit responses of the test utterances were isomorphic to the response to the training data. A single hidden unit provided the discriminatory response. In the single error case, this single unit failed to respond to the input

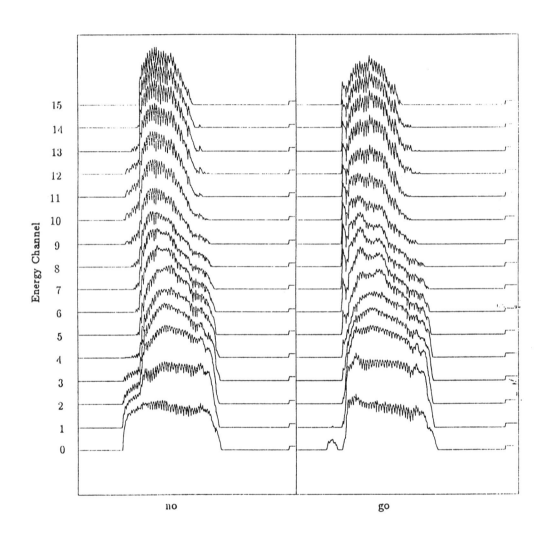

Figure 2: "Channel Energies for no/go pair"

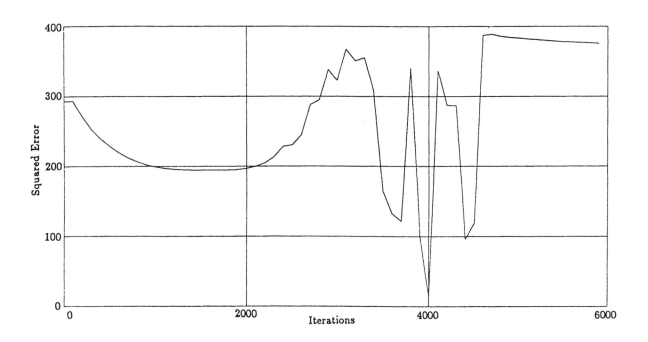

Figure 3: "Squared Error for Training Set vs. Iteration"

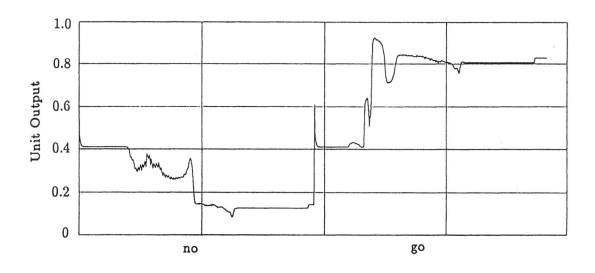

Figure 4: "Output Unit 24 Response to No/Go Pair"

data. The channel energy input data for this utterance is extremely low level, especially in the mid to upper channels.

Based on these encouraging results, a second set of experiments was designed to explore extensions of the use of connectionist networks for more difficult problems in speech recognition.

3 Experiment II

The next set of experiments were designed to learn acoustic-phonetic properties of the phonemes in the category of the voiced stop consonants. The method, however, is completely general and will be extended to other classes of speech sounds. The experimental design is presented, followed by the network architecture and learning algorithm.

3.1 Design of Experiments

The plan of the experiments to learn acoustic properties of phonemes uses a principle of incremental optimization. An initial network is optimized to discriminate the phonemes /b,d,g/ in a particular vowel context, say /i/. When the network has been modified to correctly discriminate the CV words /bi,di,gi/, the training data is expanded to include another vowel context.

The subsequent vowel context for optimization was chosen by two methods. In the first, the subsequent vowel was chosen to be phonetically close to the first. This is designed to test the extent to which the network is able to generalize the consonant discrimination across vowel contexts. The second method was to choose a subsequent vowel phonetically far from the first vowel. This is designed to test the extent to which the network could make context-specific consonant discriminations.

The choice of subsequent vowels was done to increase the likelihood of successful optimization by minimizing the incremental learning required. A series of incremental experiments is defined in this way, by which increasingly dissimilar contexts are added to the network for invariant discrimination of /b,d,g/ and increasing similar contexts are subtracted from the network design to respond selectively to /bi,di,gi/.

An alternative design would involve optimizing the respective networks over the complete data set of positive and negative samples in a single step. This approach has the advantage that the network is forced from the start to attend to the desired goal; a weakness of the incremental approach is that the network could be optimized for one context using an internal representation which is inappropriate for the larger task. In this case, the network would be required to "unlearn" the original solution, which could prove a difficult optimization problem. Nevertheless, the incremental approach was selected for these initial experiments because the progress of optimization could be more closely controlled and evaluated.

It is clear either the subtractive or the additive case would be sufficient for speech recognition. Either context-dependent or context-independent recognition would achieve the goal of consonant discrimination. There is, however, an important theoretical question at issue here of acoustical invariance [Blu86,BS79,BS80,LGB84], which will be explored in a subsequent paper. The results obtained to this point do not warrant a full discussion of the issue of invariance.

3.2 Network Architecture

For these preliminary experiments, a three-layer temporal flow model was implemented, as shown in Figure 1, with a third output unit to accommodate the three voiced consonants.

3.3 Network Learning Algorithm

For these experiments, a second-order optimization algorithm was selected called the Broyden-Fletcher-Goldfarb-Shanno algorithm (BFGS) [Fle80]. This algorithm combines a linear search along a minimizing vector with an approximation of the second-derivative of the objective function f. In this way, knowledge about the structure of the error surface is used to select optimal search directions and achieve much more rapid convergence, especially in the neighborhood of the minima of the objective function. Although such second-order methods do not share the locality property of first-order methods, the BFGS algorithm was employed for the purposes of more rapid optimization using a sequential machine.

The BFGS update formula is given as:

$$H_{(k+1)} = H_k + \left(1 + \frac{\gamma^T H_k \gamma}{\delta^T \gamma}\right) \frac{\delta \delta^T}{\delta^T \gamma} - \left(\frac{\delta \gamma^T H_k + H_k \gamma \delta^T}{\delta^T \gamma}\right)$$

where H is the approximate inverse of $G = \nabla^2 f(\vec{w})$, and:

$$\gamma = \vec{g}_{(k+1)} - \vec{g}_{(k)}$$
$$\delta = \vec{w}_{(k+1)} - \vec{w}_{(k)}$$

The algorithm basically iterates through three steps, as follows:

1. compute the search direction as $\vec{s} = -H\vec{g}$.

2. execute a linear search along $\vec{s}$; that is, minimize $f(\vec{w} + \alpha\vec{s})$ over α, a scalar, $\alpha > 0$.

3. update H according to the BFGS formula above.

The computation of the gradient vector $\vec{g}$ was accomplished by an extended form of the back-propagation learning algorithm for networks with recurrent links as described above.

For these experiments, the initial value of H was chosen to be I. In cases where the linear search failed, H was reset to I, and the search continued.

The linear target function described in the previous experiments was also used for the consonant discrimination experiments.

3.4 Data

The speech data used for the second set of experiments was taken from a small database of isolated consonant-vowel (CV) utterances for a single speaker (RW) consisting of the stop consonants (/p,t,k,b,d,g/) in combination with ten vowels (/i,I,e,ae,a,ˆ ,o,u,U,3/). Five repetitions of each CV word for a total of three hundred utterances were recorded on a Nakamichi Model 480 tape recorder using the Digital Sound Corporation Model 240 preamplifier. The recorded speech was played into a commercial speech recognition device (Siemens CSE 1200) where it was filtered and sampled as described above. Additional data from the original and other speakers will be collected for further experiments.

The data files were segmented by hand to extract the transition portion of the CV word. The initial segmentation boundary was set at a point of silence at least 50 ms prior to the consonant release and the final segment boundary at the point of maximum vocalic energy, approximately in the center of the vowel nucleus. This segmentation was performed without difficulty and did not involve an attempt to identify the consonant-vowel boundary. The segmentation was done primarily to decrease the computational load on the optimization algorithm. It is certain that sufficient if not complete discriminatory information remained in the segmented data.

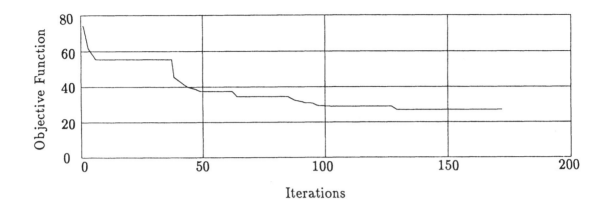

Figure 5: "Objective Function During Optimization"

4 Results

The network described previously was trained on a single set of /bi,di,gi/ using the back-propagation method for networks with recurrent links used in experiment I. The network converged after approximately 3000 iterations.

The resulting network was then trained in two experiments, an additive experiment incorporating /bI,dI,gI/ and a subtractive experiment reducing by /bu,du,gu/. These experiments were conducted with the BFGS algorithm described above. In both cases, the networks converged for proper discrimination in both contexts.

The objective function value during optimization for the subtractive experiment is shown in Figure 5. The response of the output units to the positive training sample for the optimized network can be seen in Figure 6.

The network makes an unambiguous discrimination between the voiced stop consonants in the response of the output units. The unit responses roughly approximate the target function and clearly begin discriminating responses at the initial word boundary.

5 Discussion

The results of the "no/go" experiment have been discussed at length elsewhere [WS86]. In summary, the network formed an internal representation of an acoustic-phonetic feature characteristic of the burst-release of the velar consonant /g/. In addition, the network formed a similar discriminatory mechanism for both speakers. This discriminatory feature was quite robust across repetitions by the same speaker of the same word. Taken together, these two facts strengthen the conclusion that connectionist networks can be used to infer significant acoustic-phonetic features directly from real speech data. This suggests that connectionist learning may be used to uncover acoustical characteristics of the speech waveform in which computational optimality may be found to have perceptual significance.

The results of the consonant discrimination task are preliminary; it is significant, however, that the temporal flow model was correctly optimized to make this discrimination in the context of a single vowel, and that this discrimination was extended additively and subtractively to the nearest corresponding phonetic contexts. Further testing of the invariance hypothesis in the context of

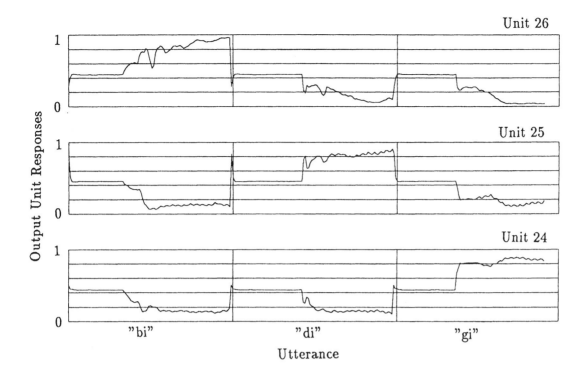

Figure 6: "Output Unit Responses to Training Data"

connectionist networks is currently in progress.

References

[And86]   Charles William Anderson. *Learning and Problem Solving with Multilayer Connectionist Systems.* PhD thesis, University of Massachusetts, September 1986.

[Blu86]   Sheila E. Blumstein. On acoustic invariance in speech. In Joseph S. Perkell and Dennis H. Klatt, editors, *Invariance and Variability in Speech Processes*, chapter 9, pages 178–201, Lawrence Erlbaum Associates, Hillsdale, NJ, 1986.

[BS79]   Sheila E. Blumstein and Kenneth N. Stevens. Acoustic invariance in speech production: evidence from measurements of the spectral characteristics of stop consonants. *Journal of the Acoustical Society of America*, 66(4):1001–1017, October 1979.

[BS80]   Sheila E. Blumstein and Kenneth N. Stevens. Perceptual invariance and onset spectra for stop consonants in different vowel environments. *Journal of the Acoustical Society of America*, 67(2):648–662, February 1980.

[DS81]   George R. Doddington and Thomas B. Schalk. Speech recognition: turning theory into practice. *IEEE Spectrum*, 26–32, September 1981.

[EM86]   Jeffrey Elman and John McClelland. Exploiting lawful variability in the speech wave. In Joseph S. Perkell and Dennis H. Klatt, editors, *Invariance and Variability in Speech Processes*, chapter 17, pages 360–380, Lawrence Erlbaum Associates, Hillsdale, NJ, 1986.

[Fle80]   Roger Fletcher. *Practical Methods of Optimization.* Volume 1 Unconstrained Optimization, John Wiley, New York, 1980.

[Hop82] John J. Hopfield. Neural networks and physical systems with emergent collective computational abilities. *Proceedings of the Natural Academy of Sciences USA*, 79:2554–2558, 1982.

[JFH63] R. Jakobson, Gunnar Fant, and Morris Halle. *Preliminaries to Speech Analysis*. MIT Press, Cambridge, MA, 1963.

[Jor86] Michael I. Jordan. Attractor dynamics and parallelism in a connectionist sequential machine. In *Proceedings of the Eighth Annual Conference of the Cognitive Science Society*, Lawrence Erlbaum, Hillsdale, NJ, 1986.

[KL81] Tuevo Kohonen and Pekka Lehtio. Storage and processing of information in distributed associative memory systems. In G.E. Hinton and J.A. Anderson, editors, *Parallel Models of Associative Memory*, pages 105–143, Lawrence Earlbaum Associates, Hillsdale, N.J., 1981.

[LGB84] Aditi Lahiri, Letitia Gewirth, and Sheila E. Blumstein. A reconsideration of acoustic invariance for place of articulation in diffuse stop consonants: evidence from a cross-linguistic study. *Journal of the Acoustical Society of America*, 76(2):391–404, 1984.

[Mar70] Thomas B. Martin. *Acoustic Recognition of a Limited Vocabulary in Continuous Speech*. PhD thesis, University of Pennsylvania, 1970.

[ME86] John L. McClelland and Jeffrey L. Elman. Interactive processes in speech perception: the trace model. In J.L.McClelland D.E.Rumelhart and the PDP research group, editors, *Parallel Distributed Processing: Explorations in the Microstructure of Cognition: Volume II Psychological and Biological Models*, chapter 15, MIT Press, Cambridge, MA, 1986.

[PNH86] David C. Plaut, Steven Nowlan, and Geoffrey Hinton. *Experiments on Learning by Back Propagation*. Technical Report CMU-CS-86-126, Carnegie-Mellon University, 1986.

[RHW86] David E. Rumelhart, Goeffrey Hinton, and Ronald Williams. Learning internal representations by error propagation. In J.L.McClelland D.E.Rumelhart and the PDP research group, editors, *Parallel Distributed Processing: Explorations in the Microstructure of Cognition: Volume I Foundations*, chapter 8, MIT Press, Cambridge, MA, 1986.

[SB78] Kenneth N. Stevens and Sheila E. Blumstein. Invariant cues for place of articulation in stop consonants. *Journal of the Acoustical Society of America*, 64(5):1358–1368, 1978.

[SR86] Terrence J. Sejnowski and Charles R. Rosenberg. *NETtalk: A Parallel Network that Learns to Read Aloud*. Technical Report JHU/EECS-86/01, Johns Hopkins University, 1986.

[Sut85] Richard S. Sutton. The learning of world models by connectionist networks. In *Proceedings of the Seventh Annual Conference of the Cognitive Science Society*, Erlbaum, Hillsdale, NJ, 1985.

[WS86] Raymond L. Watrous and Lokendra Shastri. *Learning Phonetic Features Using Connectionist Networks: An Experiment in Speech Recognition*. Technical Report MS-CIS-86-78, University of Pennsylvania, October 1986.

Teaching a Minimally Structured Back-Propagation Network to Recognise Speech Sounds

T. K. Landauer
C. A. Kamm
S. Singhal
Bell Communications Research, Morristown, N.J. 07960

Abstract: An associative network was trained on a speech recognition task using continuous speech. The input speech was processed to produce a spectral representation incorporating some of the transformations introduced by the peripheral auditory system before the signal reaches the brain. Input nodes to the network represented a 150-millisecond time window through which the transformed speech passed in 2-millisecond steps. Output nodes represented elemental speech sounds (demisyllables) whose target values were specified based on a human listener's ability to identify the sounds in the same input segment. The work reported here focuses on the experience and training conditions needed to produce natural generalisations between training and test utterances.

The primary goal of this work is to explore the use of learning networks as an analytical tool for gaining insight into complex human pattern recognition processes. This goal motivates the choice of input representation, architecture, and training regimen. The strategy is to give a minimally structured network the experience and training it needs to perform a complex pattern recognition task of a kind that humans readily master, and then to use the way in which it learns, generalises, and fails, and the internal weight organisation which it adopts, as a means of studying ways in which the pattern classification in question can be accomplished. To the extent that the features of the input stimuli and the information processing that a successful network uses resemble those used by a human, these results may suggest new hypotheses about how humans accomplish the pattern recognition task.[1]

The input to our network model of speech recognition was processed to mimic several transforms that the peripheral auditory system imposes on the acoustic signal, but that are not specific to recognising speech. Thus, the information in the input might be considered grossly analogous to information the brain receives from the inner ear. The network was required to learn to extract speech-relevant information from a continuous signal. Error feedback consisted only of information about which speech elements a human could detect in the signal, and no knowledge or theory about the process or mechanism of speech perception as such was embedded in the internal architecture of the learning network or the coding of its output.

The network readily learned to pick a small number of speech sounds, for example, the initial portions of the syllables do, re, mi, fa, so, la, and ti, out of continuous "sentences" consisting of only these syllables, and generalised successfully to other sentences composed of the same elements spoken in different orders. The success of generalisation and the naturalness of the errors and partial recognitions evinced by the network appear to depend in interesting ways on the training set to which it was exposed and the discriminations required of its output. We will present both systematic data and impressions gained from experiments with these networks. First, however, we give necessary details on the input transformation, network configuration, learning rules, and training procedures.

1. Input Speech Transformation

The motivation for the signal processing of the input speech was to approximately simulate several of the modifications that the inner ear imposes on an acoustic signal. The processing consisted of five steps. First, the input speech was digitised and low-pass filtered to 5 kHz bandwidth. Spectral estimates were

---

1. While this is the primary goal, the research can also be viewed as an attempt to apply learning network methods to the problem of automatic speech recognition. In this respect, it asks whether the high-dimensional non-linear representation of such nets and the solution-optimisation procedure of back-propagation will produce the same, or perhaps interestingly different, results from those obtained using more traditional pattern-matching or pattern-classification algorithms.

obtained using 128-point FFTs. To mimic the sensitivity of the human listener to rapid changes in high frequency components and to fine frequency distinctions at low frequencies, several FFTs were computed, using temporal windows of 4 to 20 ms, with a 2-ms frame shift. For each 2-ms frame, a composite amplitude spectrum was obtained by extracting low frequency components from the FFT with 20-ms window, the highest components from the FFT with 4-ms window, and intermediate components from FFTs computed with temporal windows between 4 and 20 ms. Second, the frequency scale was transformed to a Bark scale to reflect the frequency spacing along the basilar membrane of the inner ear (Schroeder, Atal & Hall, 1979). Third, the Bark-scaled spectra were convolved with an asymmetric filter simulating the spread of excitation along the basilar membrane (Schroeder & Hall, 1974), which results in a highly smoothed output spectrum. The combination of steps two and three serves to simulate the filtering known to occur in the peripheral auditory system. Fourth, to model the short-term adaptation of the peripheral auditory system, changes in the amplitude of each component were modified by applying a multiplicative function of the difference between successive frames, with exponential decay, producing an enhancement of spectrotemporal "edges". Fifth, the amplitude of each component was scaled relative to the overall minimum amplitude by a power function with exponent 0.6 to simulate the transform from acoustic pressure to relative loudness. Finally, 15 of the 128 transformed amplitudes, spaced at one Bark (approximately one critical band) intervals from 3 to 17 Barks (287 to 4884 Hz), were selected for input to the network.

Clearly this transformation does not perfectly represent the signal sent from ear to brain - for one thing it carries much less information - but it does have several of the important characteristics of that signal. Thus, the information the network learns to extract from this input to recognise speech sounds should have a fair chance of being information the human brain could also extract.

2. Network Architecture and Training Procedure

There were 1,125 input nodes representing the 15 transformed amplitudes for each of 75 successive 2-millisecond frames of a 150-millisecond window of speech. On each training cycle a real value between 0 and 1, proportional to the amplitude of each spectral component, was applied to each input node. Various numbers of hidden nodes, usually 20, were fully interconnected with the input nodes and with a varying number of output nodes depending on the number of speech-sound elements that were to be detected. Training proceeded by stepping the 150-millisecond speech window across an utterance in 2-millisecond steps, at each one calculating the results of forward activation to the output nodes, calculating an error signal, and updating weights by the standard back-propagation procedure (Rumelhart, Hinton and Williams, 1986).

To specify targets, judgments of whether each of the speech sounds to be trained was or was not present in the input window were made by listening to 150-millisecond segments of the signal and by visual inspection of the speech waveform and a speech spectrogram of the signal. Judgments were made on a nine-point confidence scale, from "possibly" to "definitely" present. The error signal at each output node was adjusted in proportion to this confidence value. A separate judgment was made as to whether the system should be trained on a particular sound element for that window, or allowed to produce an output without error being propagated back from the corresponding output unit (this strategy is related to the "don't care" procedure of Jordan, 1986).

3. Results, Observations, Modifications, Comparisons and Lessons.

In early trials with a small number of artificially synthesised speech-like stimuli, it quickly became apparent that the system would learn a training set readily, but did not generalise well to the same nominal "speech" sounds in other contexts, and that the errors it made were not always sensibly related to the confusions between sounds that a human listener would make. Two features of the procedure seemed at fault. First, we thought that the system was not being exposed to enough different speech sounds in enough different contexts to be able to extract the important features and structure of the stimuli. Second, the error calculation procedure was forcing the system to discriminate strongly between sounds that would be perceived as similar by a human, such as "fa" and "va". Therefore, for our main investigation, we used natural speech input with a greater variety of speech sounds in many different contexts, although still a very tiny subset of English. A professional announcer spoke 14-syllable "sentences" composed of two tokens of each of seven syllables (do, re, mi, fa, so, la, ti) strung together in a Latin square design such that over the sentences studied each syllable followed and preceded every other syllable, including itself, equally often. The speaker did not sing the "sentences", but spoke them with wide variations in intonation, duration and phrasing. (Examples of the recordings will be played.) The total set contained 39 demisyllables (7 initial and 32 final demisyllables) of the roughly 800 to 1,000 demisyllables needed to transcribe all of English speech (Fujimura, Macchi & Lovins, 1977). Two representative sentences are given in Table 1. We typically trained the system to recognize the subset of initial demisyllables.

Table 1. Sample Input "Sentences"

1. do do re so mi re fa la so mi la ti ti fa.

2. mi mi fa ti so fa la do ti so do re re la.

In addition to using this richer training set in the main experiments, we also altered the rule for error calculation to encourage more generalization. In particular, the error for output nodes corresponding to demisyllables not present in the window was multiplied by a constant between 0 and 1, so that a high value on a node corresponding to an absent element was not as strongly corrected as a low value on a node corresponding to a present element. For a given residual criterion training error, this causes the system to move towards a solution in which it adopts high output values for all positive patterns but allows itself moderate output values for other output nodes that tend to be excited by the same inputs.

Results of this training procedure were quite encouraging. The seven plots in Figure 1 show the output values for the seven units corresponding to the initial demisyllables as a function of time for a test sentence after training on just one other sentence. The syllables at the top of the figure, and the lines at the top of each panel indicate the positions in the utterance where each target syllable occurred. The figure shows that, in all instances, the output unit corresponding to the target syllable had the highest output of the seven units. There are several instances where a second output unit also had relatively high activation (e.g., the "so" unit shows activation of about 0.8 when the "do" syllable is in the input window). Such evidence of similarity (or confusability) typically occurred for output units sharing the same vowel sound (e.g., "do" and "so", "fa" and "la").

We are also studying the effect of variation in the training set by testing on one set after having trained on one or more others. Preliminary data suggest that more varied training produces somewhat but not dramatically better generalization. A more detailed characterization of these effects awaits the completion of additional experiments.

Another technique that we are exploring in an attempt to improve the generalization of the network involves training the system to simultaneously recognize the set of speech elements and perform an encoder or auto-association function. In this network, the output layer consists of output nodes of the set of speech elements and 1,125 additional output nodes, fully interconnected to the hidden layer. The error on each of the latter nodes is calculated as the difference between its output and the activation applied to a corresponding input node. The rationale for this combination of directed training and auto-association is that this architecture should supply a much greater degree of constraint on the representations adopted by the hidden nodes, and that these additional constraints may result in a solution that demonstrates better generalization for identifying speech sounds. The system is required to maintain its ability to represent faithfully the raw spectral information at the same time it is learning to recognize a particular speech element, and should thus be less likely to generate an idiosyncratic representation which distorts its overall representation of sound. Preliminary results using this network suggest that the performance and the apparent naturalness of the generalization are improved. Figure 2 shows time functions displaying the activation of each inital demisyllable output node for a training run of this combined direct-training/auto-associative network. Notable in Figure 2 is that demisyllables with common vowel portions appear to be slightly better differentiated than the same demisyllable pairs in Figure 1, and that the silent

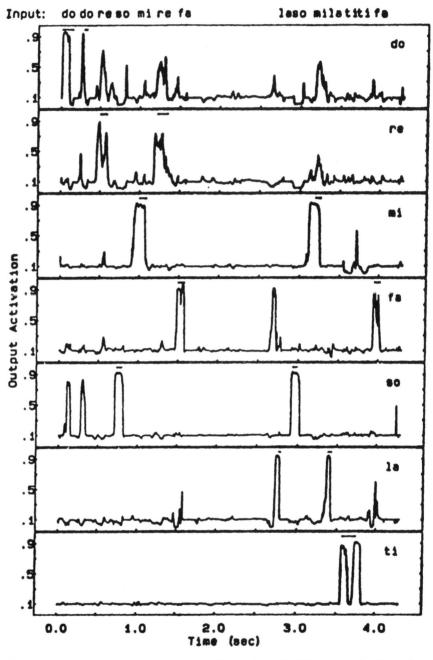

Figure 1. Generalization Test following Directed Training

Input: do do re so mi re fa   laso milatitife

Fig 1. The network whose recognition performance is shown here had been trained on just one other "sentence" consisting of the same syllables in a different order. Shown in each horizontal frame is the activation level of an output node trained to respond to the presence of a particular demisyllable, as a function of time as the continuous test sentence was stepped through the 150 msec input window. Labels across the top of the figure indicate the demisyllables judged present by a human observer; the small horizontal lines indicate intervals in which the highest confidence of presence was assigned.

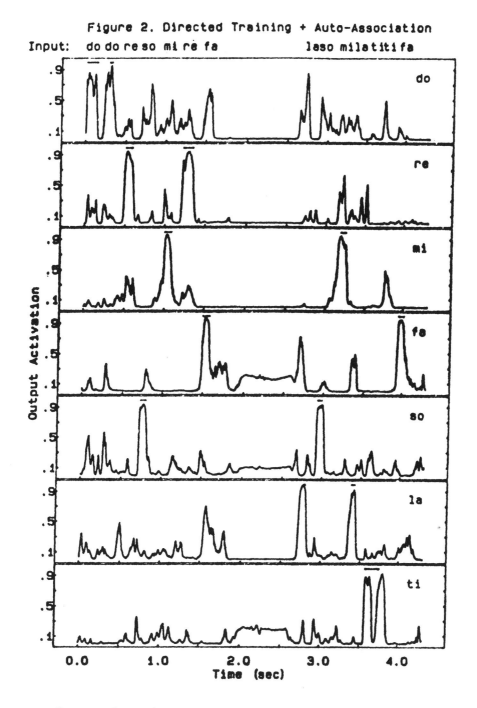

Figure 2. Directed Training + Auto-Association

Input: do do re so mi re fa   laso milatitifa

Fig 2. Performance after a moderate degree of training for a network that was simultaneously learning to recognize seven demisyllables and to reconstruct the activation levels of its input nodes. Only the demisyllable recognition node activities are shown. The training (and test) sentence is the same as the test sentence of fig 1.

period (at approximately 1.7 - 2.6 seconds) contains only low levels of activation, the highest of which are from output units corresponding to demisyllables with voiceless initial consonants containing frication noise or noise bursts (/f/, /t/, /s/).

This work clearly represents only the first few steps toward the goal of being able to use the network to analyze the important constituents of speech sounds and their processing. It does appear, however, that, with sufficient experience and proper training techniques, such a complex yet minimally-structured system can learn to use approximately natural input to recognize speech sounds in a way that leads to fairly natural generalization, as evidenced by the recognition performance for non-training tokens of the syllables and the representation of perceptual "similarities" in the output activation patterns. This work also encourages the development of methods for studying what the network is extracting from the transformed speech signal. Experiments along this line may well resemble ones that might be performed using human observers, for example making various systematic modifications of the speech signal and observing the result, or trying various systematically arranged generalization tests. Among the advantages of using networks as opposed to using human listeners would be that such experiments could be run very rapidly and that the internal states of the network, including the continuous activation functions on all of the different output nodes, as well as the auto-association function, are available for analysis.

4. References

Fujimura, O., Macchi, M. J. and Lovins, J. B. Demisyllables and affixes for speech synthesis. Paper presented the 9th International Congress on Acoustics, Madrid, Spain, July 4-9, 1977.

Jordan, M. I. Serial Order: A Parallel Distributed Processing Approach. ICS Report No. 8604, Institute for Cognitive Studies, UCSD, La Jolla, California, 1986.

Rumelhart, D. E., Hinton, G. E. and Williams, R. J. Learning internal representations by error propagation. In D. E. Rumelhart, J. L. McClelland, and the PDP Research Group, *Parallel Distributed Processing*, Vol. *1*, 318-362, 1986.

Schroeder, M. R., Atal, B. and Hall, J. L. Optimizing digital speech coders by exploiting masking properties of the human ear. *J. Acoust. Soc. Am., 66*, 1647-1652, 1979.

Schroeder, M. R. and Hall, J. L. Model for mechanical to neural transduction in the auditory receptor. *J. Acoust. Soc. Am., 55*, 1055-1060, 1974.

Revealing the Structure of NETtalk's Internal Representations

Charles R. Rosenberg

Cognitive Science Laboratory, Princeton University,
Princeton, New Jersey 08542, USA

(April 3, 1987)

Abstract

NETtalk is a connectionist network model that learns to convert English text into phonemes. While the network performs the task with considerable accuracy and can generalize to novel texts, little has been known about what regularities the network discovers about English pronunciation. In this paper, the structure of the internal representation learned by NETtalk is analyzed using two varieties of multivariate analysis, hierarchical clustering and factor analysis. These procedures reveal a great deal of internal structure in the pattern of hidden unit activations. The major distinction revealed by this analysis of hidden units is vowel/consonant. A great deal of substructure is also apparent. For vowels, the network appears to construct an articulatory model of vowel height and place of articulation even though no articulatory features were used in the encoding of the phonemes. This interpretation is corroborated by an analysis of the errors or confusions produced by the network; The network makes substitution errors that reflect these posited vowel articulatory features. These observations subsequently led to the discovery that articulatory features of place of articulation and, to some extent, vowel height, are largely present in first-order correspondences between vowel phonemes are their spellings. This work demonstrates how the study of language may be profitably augmented by models provided by connectionist networks.

Introduction

Speech synthesis is the translation of written text into an acoustic speech signal. In most speech synthesis systems, two distinct knowledge sources are used in the determination of the pronunciations of words: rules which encode regularities, and a dictionary of exceptions to those rules which handle those cases where the regularities break down. When a correspondence can be predicted on the basis of regularities operating at some level, it can be encoded more efficiently as a generative rule, such as the rule that when the letter "c" occurs before a high, front vowel, typically spelled with the letters "i", "e", or "y", it is pronounced like an "s" as in "icy" or "center" but like a "k" in other contexts. But even the best letter-to-sound rules have exceptions, and so the rules must be augmented by a dictionary which is checked before any rules are applied. This dictionary is generally used with great frequency because the most frequent words in English are also the most irregular.

NETtalk is a connectionist network model which learns the pronunciation of English text, represented as phonemes, or distinctive speech sounds (Sejnowski & Rosenberg, 1986; 1987a; 1987b). No rules of pronunciation were provided to NETtalk. Rather, the network reaches a reasonable level of performance by being presented with a number of training examples and being incrementally corrected using the back-propagation rule (Rumelhart, Hinton & Williams, 1986).

Is NETtalk's knowledge of pronunciations divided up in this way, between dictionary-like knowledge and rule-like knowledge? Since all knowledge shares the same architectural/representational space in NETtalk, an answer to that question is not immediately obvious. However, there is some indication that both exceptions and regularities are learned. For example, it learns correctly that "of" is pronounced /xv/ (see Appendix), even though the letter "f" is not pronounced this way in any other case. On the other hand, when trained on a 16,000 word selection from the entire 20,000 word Webster's Pocket dictionary, the network is able to generalize to the rest of the corpus with an accuracy exceeding 90% correct phonemes. Moreover, the confusions are usually between phonemes that are phonologically similar. This ability to generalize to novel words indicates that the network does much more than memorize specific input-output pairs as found in a dictionary. But what exactly *are* the regularities that it discovers in English pronunciation?

Many network models to date have been simple enough so that the functionality of many of the hidden units could be discovered through direct visual inspection of the unit activations and weight values. In larger networks, where one may have hundreds of hidden units and tens or hundreds of thousands of weights, it becomes difficult or impossible to detect underlying structures by such direct methods. We must turn to more sophisticated techniques.

Analytical tools

Factor analysis

Factor analysis (eg. Harman, 1976; Rummel, 1967) is a well-known technique for attempting to account for the variance in a large number of variables in terms of a much smaller number of relatively independent underlying factors. Factor analysis is based on the assumption of a linear model, meaning that the observed variables must be predicted by a linear, weighted, combination of underlying factors. Specifically, a factor is defined as

$$F = \sum_{i=1}^{k} w_i X_i,$$

where the w's are the factor weights (to be estimated from the data) and the X's are the k original variables. The factor loadings are the correlations between the final factors and the original variables.

Factors are determined in two stages. The initial factor extraction is based on the method of principle-components analysis. The first principle-component is that weighted combination of variables that accounts for the greatest amount of the total variance in the data. The second principle-component accounts for the greatest amount of variance not accounted for by the first principle-component, and so on. The principle-components are chosen so as to be mutually uncorrelated or statistically independent, but are typically hard to interpret. Consequently, a second process is generally performed where these initial factors are rotated.

Cluster analysis

A technique which makes fewer assumptions about the underlying form of the data is cluster analysis (eg. Everitt, 1974). Here, items are progressively grouped or clustered together based on relative similarity within a cluster, and relative dissimilarity between clusters. The method is very simple: Given some matrix of similarities, cluster analysis iteratively merges the two most similar clusters. Of course, there are many ways to determine which groups are most similar. The three most common methods are "centroid", where the distance between two groups is defined as the average of

TEACHER:

/t/	/r/	/æ/	/n/	/z/	/l/	/e/	/S/

GUESS:

/t/	/r/	/æ/	/m/				

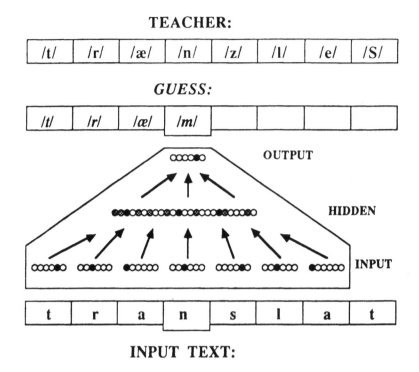

t	r	a	n	s	l	a	t

INPUT TEXT:

Figure 1: The Architecture of NETtalk.

distances between all pairs of members in each group, "complete linkage" or "farthest neighbor", where the two groups are merged that have the nearest most remote members, and "single linkage" or "nearest neighbor", where groups are merged on the basis of their nearest members. The resulting clusters can be graphed as a dendogram, and cuts through the dendogram yields the groups formed at that particular depth or distance.

A Brief Overview of NETtalk

NETtalk is composed of simple processing units arranged into layers. In the experiments reported here, I used an architecture with three layers of units and two layers of modifiable weights completely connecting successive layers of units (see Fig. 1). The activation of a unit in the network can take any value from zero to one. The connection strengths or "weights" can have any value, positive or negative. Each unit computes a weighted sum of unit activations times the weights from units in the layer below and determines its output value according to a nonlinear function that is zero for large negative inputs and monotonically approaches one for large positive inputs. (Hopfield, 1984; Rumelhart et al., 1986). Thus, information travels through the network, from the input layer to the intermediate layer of "hidden" units and finally to the output layer. The input layer receives information from an English text and the network produces a pattern of activation at the output layer of units that is a representation of how the input text should be pronounced. The representations on the input and output layers are fixed, but those of the hidden units is constructed by the learning procedure. The representations formed at this hidden layer are the focus of the present work.

The pronunciation of a given letter is generally influenced by the surrounding letters. Letter context was represented in the network by extending the input layer out over seven letters, where each of the seven letter positions was represented by separate groups of input units (see Fig. 1). At each sequential step, the network received input simultaneously from all the letters that fell within the fixed-sized input window. Based on this information, a guess was made as to the pronunciation of the middle letter of the window in two steps: 1. the values of the output units were determined by forward-propagating the activation of the input units through the network to the output layer, then 2. the sum of the squares of the differences was computed between the output units and each of the possible phoneme targets. The phoneme which minimized this distance was chosen as the guess made by the network. [1]

Changes were computed for the weights following each forward-propagation step, but these changes were actually incorporated into the weights only between words in order to reduce computation time. The output units were compared, unit-by-unit, with the correct phoneme supplied to the network, and the weight changes for internal connections were computed by recursively applying back-propagation from the output layer to the input layer. The seven-letter input window was then moved down one letter position in the input text, and the process repeated. When the end of the corpus was reached, the network continued at the beginning. (See Sejnowski & Rosenberg, 1986, 1987b for details on NETtalk and Rumelhart et. al, 1986 for details on the error back-propagation algorithm.)

---

[1]This procedure for selecting the phoneme is slightly different from that previously reported (Sejnowski & Rosenberg, 1986, 1987b).

The Analyses

These multivariate techniques were applied to the analysis of the patterns of hidden unit activities observed in NETtalk, where phonemes constituted the variables of the analysis. For each phoneme, average hidden unit activations were collected over all letter contexts. Since there were 80 hidden units in this version of NETtalk, the representation of each phoneme at the hidden layer was cast as an 80-dimensional vector. Pair-wise correlations between these vectors were computed and this correlation matrix was submitted as similarity data to cluster analysis and factor analysis. If phonemes were encoded as orthogonal dimensions in this space, then this correlation matrix would have one's along the diagonal (phonemes are correlated with themselves), but zero's everywhere else. On the other hand, if there is structure in the way the network represents phonemes, the phonemes might be found to cluster together in some way.

Initial Training

A network with a single hidden layer of 80 units was trained on an eleven thousand word selection from the full 20,000 word Websters Pocket Dictionary. Phonemes and letters were both encoded using a local encoding, where a single unit in each input and output group encoded a single letter or phoneme. There were seven groups of 29 input units per group and a single group of 55 output units with complete connectivity between layers. The network made ten complete passes through this corpus, thus being trained on a total of 160,000 words, of which 16,000 were different words. The order of the words was randomized. Performance at this point in learning was 92% correct phonemes and 56% fully correct words on the training corpus. From the remainder of the 20,000 word corpus, 1000 words were selected at random. These words were not included in the training set. Learning was turned off and the network was tested on this new corpus. 90% of the phonemes and 49% of the words were completely correct.

Data Collection

With learning turned off, data was collected from this network as it went through this 1000-word corpus of novel words. For each of 48 phonemes (the three "special" symbols, /./, /-/, and /_/, and four other phonemes /!/, /K/, /q/, and /I/ were dropped, the latter because they did not occur in the 1000 word corpus), hidden unit activations were averaged over all input windows that had that phoneme as it's pronunciation. Trials where the network failed to guess the phoneme correctly were not included in the means. Since the network guessed 90% of the phonemes correctly, roughly ten percent of the trials were thrown-out for this reason. [2] Thus, an 80-dimensional vector of mean activation values was constructed for each phoneme. The sample correlation coefficient, r, then was computed for each pair of phonemes, resulting in a 48 x 48 correlation matrix. Each cell in this matrix represented the correlation of the pattern of hidden unit activity that resulted for the ith and jth phoneme.

Results from Hierarchical Clustering

A hierarchical clustering of all the phonemes using the centroid clustering method is shown in Fig. 2, where the dissimilarities, were defined as $1 - r$. This analysis revealed two major clusters, corresponding to vowels and consonants. All phonemes on the left major branch are vowels, and nearly all of the phonemes on the right-side are consonants, with the exceptions of /*/ and /y/. Centroid clusterings are shown, though similar results were obtained using complete linkage clustering.

Different organizational schemes are apparent within the two major groups. Within the vowel group, the next major division is between the "front" vowels, /i/, /I/, /E/, /e/, and /@/, where the sounds are produced with the tongue towards the front of the mouth, and the "back" vowels, /u/, /U/, /a/, /c/, and /o/, produced more towards the rear of the mouth. Next, the central vowels, /x/ and /^/ split off from the "back" group. Thus, the major organizational principle within the vowel cluster appears to be place of articulation. The next major division is based on the *height* of the tongue when the sound is produced ("vowel height"), as the "high" phonemes, /i/ and /I/ split off from the "low" phonemes /e/ and /@/, with the mid-vowel, /E/, falling between the two groups. The same phenomena occurs for the back vowels, where the low /o/ and /a/ group are distinguished from the high /U/ and /u/.

The consonant cluster appears to quickly divide into anywhere from five to ten major groups. Unlike the vowels, the consonant clusters correspond to place of articulation only weakly. Rather, they seem to cluster around typical input letters. For example, one major grouping consists of the phonemes /T/, /D/, /C/, /S/ and /t/: these are the phonemes that the letter "t"

---

[2] However, the pattern of results does not change significantly if this requirement is relaxed.

All Phonemes

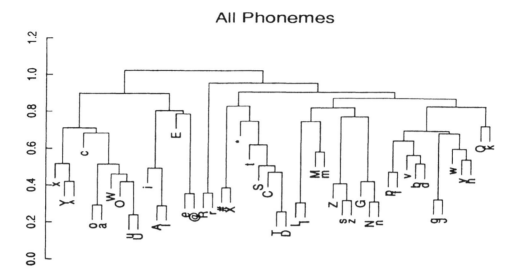

Figure 2: Hierarchical clustering results for 48 phonemes.

typically corresponds to. Another grouping are the phonemes /#/ and /X/. These are the phonemes for the letter "x". With few exceptions, the same pattern follows for the "m" group (/M/, /m/), the "s" group (/s/, /z/, /Z/), the "n" group (/n/, /N/, /G/), the "p" group (/f/, /p/), and the "g" group (/g/, /J/).

Results from Factor Analysis

A set of ten non-dipthong vowels were selected for factor analysis. Since dipthongs, such as /A/ in the word "bite", involve a change in the place of articulation and height during the course of their pronunciation, it was not desired to complicate matters by including these more ambiguous vowels in the data set. A 10 x 10 correlation matrix was prepared and submitted to a Varimax rotated orthogonal factor analysis. Three factors were extracted, which together accounted for 68% of the total variance. The rotated factors are presented in Table 1. Factor loadings less than 0.55 are generally considered unreliable (Comrey, cited in Kim & Mueller, 1978). Loadings less than 0.25 were deleted in Table 1.

The results from the factor analysis in general confirmed the analysis based on clustering, that the vowels organized according to place of articu-

```
                    SORTED ROTATED FACTOR LOADINGS
            ------------------------------------------
```

PHONEME	FACTOR 1	FACTOR 2	FACTOR 3
c /cAUght/	0.805		
a /fAther/	0.771	0.345	
@ /bAt/	0.746	-0.266	0.372
u /bOOt/		0.909	
U /bOOk/		0.901	
^ /bUn/	0.469	0.516	
i /bEAt/			0.808
I /bIt/			0.788
x /womEn/	0.411	0.278	0.676
E /bEt/			0.499
Var explained	23.04%	22.42%	22.37%
Cumm var	23.04%	45.46%	67.83%

Table 1: Three factors and accompanying factor loadings extracted from a Varimax rotated orthogonal factor analysis employing 10 vowels as variables. Zero factor loading was 0.25.

lation and vowel height. The three vowels that loaded highest on the first factor are the low vowels, /c/, /a/ and /@/, which suggests the interpretation of the first factor as a "lowness" factor. /u/ and /U/ both loaded high on Factor 2, leading to the interpretation that this factor relates to "backness". /I/ and /i/ both load high on the third factor, suggesting that this factor describes "frontness". A bit of an anomaly is the high loading of the schwa /x/ on this "frontness" dimension. Thus, Factor 1 was interpreted as representing vowel height, but two factors, Factors 2 and 3, represented place of articulation.

It was desired to collapse these three factors to two dimensions so that they could be compared to known relationships between place and height. Standard values for place of articulation and vowel height can be found in any introductory linguistics textbook. The set of values used for the purposes of this analysis are plotted in Fig. 3.

Using least-squares linear regression, weights were found to predict these true values of place of articulation and vowel height based on the factor loadings on the three factors for each vowel. Excellent fits were found in both cases. For "place", $r = 0.96$, $F(3,6) = 25.37$, $p < 0.001$, and for

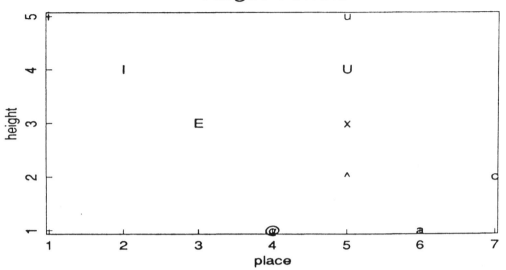

Figure 3: "Place of articulation" and "vowel height" attribute values for the set of ten vowels used the analysis.

"height", $r = 0.96$, $F(3,6) = 23.41$, $p < 0.005$. These two dimensions were not orthogonal, but were negatively correlated with each other, $r = -.079$. The factor loadings were then substituted back into the regression equations and the derived estimates for "place" and "height" are plotted in Fig. 4.

Thus, the pattern of results from the factor analysis largely agreed with the results based on hierarchical clustering: the representation of vowels at the hidden layer of NETtalk appears to be organized around two articulatory features of vowels, place of articulation and vowel height. In addition, the clustering of all of the phonemes suggests that the vowels and consonants are represented by distinctly different patterns of hidden unit activation.

Results from Confusion Data

If the internal representations of vowels and consonants are structured in the way suggested by the previous analysis of the hidden units, then these patterns ought to be reflected in the overt behavior of the network. For example, if vowels and consonants are as distinguishable as they appear to be, then the network should rarely confuse them, by guessing a consonant when the target was a vowel, and visa versa. We might also expect the structure of

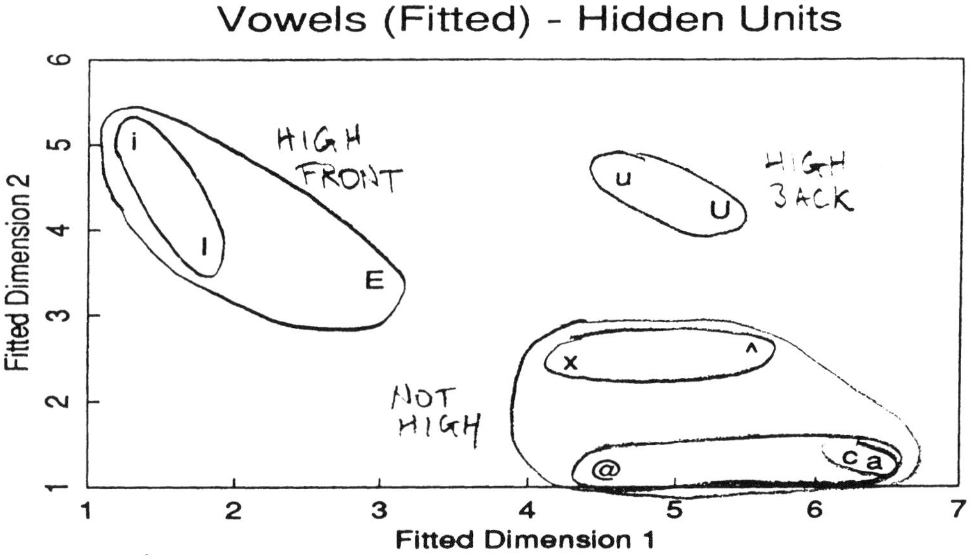

Vowels (Fitted) - Hidden Units

Figure 4: Fits based on linear regression of the three factor solution on the attribute dimensions "place of articulation" and "vowel height", plotted in Fig. 3. Data based on hidden unit activations. The groupings are based on a centroid hierarchical cluster analysis of the original correlation matrix.

the internal representation of vowels to be reflected in the pattern of confusions displayed, so that more confusions should be apparent between nearby vowels in articulatory space, than between more distant vowels. Of course, this would improve the intelligibility of NETtalk, since mistakes would be limited to similar phonemes. The following experiment was designed to test this prediction.

With learning turned off, the network went through the same 1000-word corpus used in the previous experiment. For each of the 48 phonemes examined previously, the guesses made by the network were recorded. These responses were cast as a 48 x 48 confusion matrix, where the rows were the target phonemes and the columns were guesses or responses made by the network to that target. The fraction of the time the network guessed phoneme j in response to target i was recorded in the ith x jth cell of the matrix. Correlations were computed between each of the target phonemes (rows) of the confusion matrix, and the resulting correlation matrix was clustered using the "centroid" or "average" clustering procedure. The results are shown in Fig. 5. Similar results were found using the complete

Confusions: All Phonemes

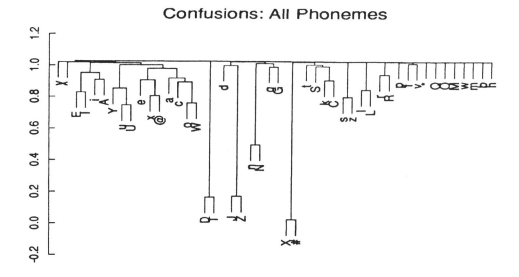

Figure 5: Hierarchical cluster analysis of ten vowels, based on the pattern of confusions produced by NETtalk.

linkage clustering procedure.

Besides four pairs of consonants that were consistently confused, /D/ and /T/, /J/ and /Z/, /n/ and /N/, and /X/ and /#/, nearly all of the dissimilarities are quite high - the network simply did not confuse very many of the phonemes. This lack of structure makes any analysis difficult. However, more confusions were made on the vowels than the consonants, and some structure is apparent. Three main clusters form, conforming fairly closely to "front" (/i/, /E/, /I/), "back" (/u/, /U/), and "not-high" (/x/, /@/, /a/, /c/, /o/). The dipthongs, /W/, /e/, /Y/, /A/ and /O/, are more difficult to interpret for reasons given previously. The "not-high" group then appears to split further into what might be called a "back, not-high" group (/a/, /c/, /o/) and a "not-back, not-high" group (/@/, /x/).

A factor analysis extracted a set of four factors which are very similar to the groups observed in the cluster analysis. As before, least-squares linear regression analysis was used to predict vowel height and place of articulation from the factor loadings. The fitted dimensions are plotted in Fig. 6. Fits to the attributes were not quite as good as in the hidden unit analysis. The correlation, r, between the fitted dimension for "place" and the attribute vector was 0.87, $F(4,5) = 3.92$. For "height", $r = 0.85$, $F(4,5) = 3.35$. Both

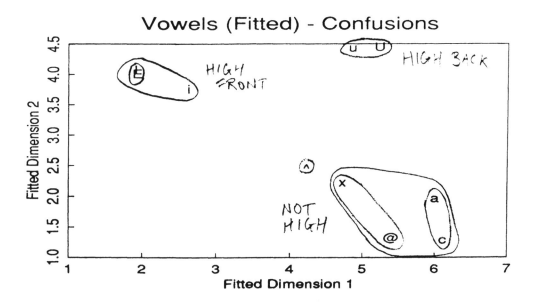

Figure 6: Fits of "place of articulation" and "vowel height", based on linear regression of the four factor solution on the attribute dimensions plotted in Fig. 3. Data was based on confusions. The groupings are based on a centroid hierarchical cluster analysis of the original correlation matrix.

F-tests narrowly failed to reach significant levels.

The overall pattern of results from the confusion data is similar to that found previously based on hidden unit activations: 1. vowels tend to be confused with vowels, and consonants with consonants, and 2. vowels that were similar to each other in terms of place of articulation and height were more likely to be confused.

Letter-to-Phoneme Correspondences

Why should vowels organize around these articulatory dimensions? Apparently, there is a "map" of articulatory features hidden in letter-to-phoneme correspondences for vowels, a map which does not exist (at least not to the same extent) for consonants. Upon closer inspection of first-order letter-to-phoneme correspondences, this was found to be true.

First-order correspondences between vowel letters and their pronunciations in the Websters corpus were investigated in much the same way as the hidden activations and confusions were analyzed previously. The full

	a	e	i	o	u	y	other
a	32.5	0.6	0.2	66.7	0.0	0.0	0.0
c	42.4	0.0	0.0	57.6	0.0	0.0	0.0
i	0.0	35.4	25.4	0.0	0.0	39.1	0.0
u	0.0	3.7	0.0	45.7	50.5	0.0	0.0
x	24.8	15.0	20.6	25.7	13.5	0.5	0.0
E	6.2	93.6	0.0	0.0	0.1	0.0	0.1
I	2.1	17.9	76.9	0.1	0.1	3.0	0.0
U	0.0	0.0	0.0	50.0	50.0	0.0	0.0
@	99.9	0.0	0.1	0.0	0.0	0.0	0.0
^	0.0	0.0	0.0	6.3	93.8	0.0	0.0

Table 2: Frequency with which the ten vowel phonemes are spelled by the letters "a", "e", "i", "o", "u", and "y".

Websters corpus of 20,000 words was searched for the percentage of times each of the ten vowel phonemes was spelled with each letter. The result was cast as a 10 x 7 matrix, where the rows correspond to the phonemes and the columns to their spelling. These percentages are presented in Table 2. As can be seen by inspection, there are obvious correlations between the phonemes in their spelling, and that these correlations agree, to some extent, with the articulatory dimensions of place and height. For example, the high-front vowels /I/ and /i/ both frequently correspond to the letters "e" and "i". The low-back vowels, /u/ and /U/ both frequently correspond to "o" and "u". Performing a factor analysis on this data, again using Varimax orthogonal rotation, four factors were found to account for 89% of the variance. As before, linear least-squares regression was performed in order to predict the attribute dimensions of "place of articulation" and "vowel height". "Place" was predicted well, the correlation between the attribute values and the predicted values being 0.95, $F(4,5) = 12.98$, $p < 0.001$. "Height" was predicted less well, $r = 0.79$, $F(4,5) = 2.06$, $p < 0.25$. Estimates for "height" and "place" based on the regression of the factor loadings on the attribute dimensions are plotted in Fig. 7.

These regularities inherent in these first-order letter-to-phoneme correspondences go far in explaining the previous hidden unit and confusion results: Similar phonemes, where similarity is based on place of articulation and vowel height, overlap in the letters with which they are typically spelled. This "map" of articulatory features is hidden within letter-to-phoneme correspondences for vowels. However, "height" was not as well represented in this first-order data as it was in the previous analyses using NETtalk. This suggests that that vowel height, unlike place of articulation, may require information from several letters to be adequately predicted. If true, then

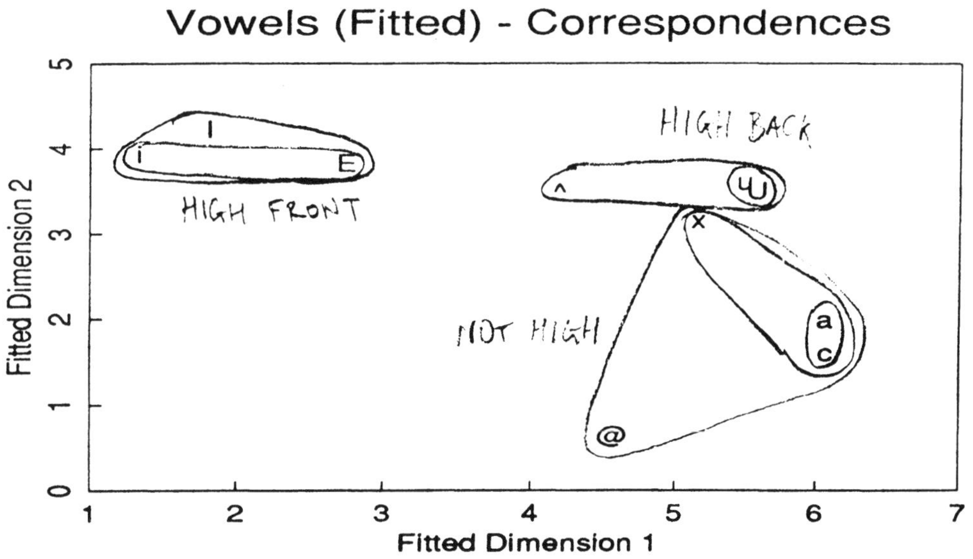

Vowels (Fitted) - Correspondences

Figure 7: Fits of "place of articulation" and "vowel height", based on linear regression of the four factor solution on the attribute dimensions plotted in Fig. 3. Data was based on correspondences between phonemes and letters. The groupings are based on a centroid hierarchical cluster analysis of the original correlation matrix.

networks with only a single input group and output group ought to exhibit a similar disadvantage in the acquisition of vowel height.

Conclusions and Directions for Future Research

Summing up, vowels and consonants are encoded at the hidden layer of NETtalk as distinctly different patterns of activation. Within these two categories, vowels organize according to place of articulation and vowel height, whereas the consonants seem to organize around input letter. The pattern of confusions or errors produced by the network are similar to those observed at the hidden layer.

These observations led me to go directly to the training corpus, the Webster's Pocket dictionary, to investigate letter-to-phoneme correspondences for similar regularities. Surprisingly, articulatory features were found to be largely present in first-order correspondences between phonemes and letters; It is possible to reconstruct a fairly accurate map of vowel height and place of articulation using distances derived from the dissimilarities between the vowels in terms of their spellings. In short, similar vowel phonemes are spelled in similar ways. To my knowledge, this regularity has not been described previously but can be seen as a useful property of language since it makes it possible to make a good guess about the spelling of a word by "sounding it out" phonetically.

All of the hidden unit analyses reported here were based on averaging activity levels over all inputs for a given phoneme. A similar analysis has been performed for letter-to-phoneme correspondences (Sejnowksi & Rosenberg, 1987b), and even more could be learned by examining graphemes and other letter combinations at the hidden layer. For example, we have clustered hidden unit activation patterns in contexts in which the letter "c" is at the center, and have found regularities and subregularities in the coding scheme even for irregular cases (Sejnowski & Rosenberg, 1987a).

There is a great deal that could be learned about language and about representation from analyzing larger connectionist networks, once analytical tools have been appropriately adapted for this purpose. Judging from previous studies of small networks for problems such as XOR and the encoder problem, it is possible that NETtalk solves certain problems in letter-to-phoneme translation in elegant and perhaps novel ways. It also seems possible that similar kinds of techniques to those found useful in analyzing model systems such as NETtalk, may eventually be applicable to the understanding of how knowledge is organized in real neural systems.

Acknowledgments

I would especially like to thank Dr. Terrence Sejnowski, with whom I collaborated on the construction of NETtalk. I would also like to thank Drs. Stephen Hanson and George Miller, who helped me in thinking about many of the issues reported here, Drs. Ken Church, Mitch Marcus, Mark Liberman, Judy Kegl, and Chris Tancredi. Dr. Jerome Feldman suggested the use of local input and output representations, which simplified the analysis and interpretation of the hidden representations. Bell Communications Research generously provided computational support. Support was provided by a grant from the James S. McDonnell foundation to the Human Information Processing Group at Princeton University.

Appendix

Table of Phonemes

Phoneme	Sound	Phoneme	Sound
/a/	father	/C/	chin
/b/	bet	/D/	this
/c/	bought	/E/	bet
/d/	deb	/G/	sing
/e/	bake	/I/	bit
/f/	fin	/J/	gin
/g/	guess	/K/	sexual
/h/	head	/L/	bottle
/i/	Pete	/M/	absym
/k/	Ken	/N/	button
/l/	let	/O/	boy
/m/	met	/Q/	quest
/n/	net	/R/	bird
/o/	boat	/S/	shin
/p/	pet	/T/	thin
/q/	uh-oh	/U/	book
/r/	red	/W/	bout
/s/	sit	/X/	excess
/t/	test	/Y/	cute
/u/	lute	/Z/	leisure
/v/	vest	/@/	bat
/w/	wet	/!/	Nazi
/x/	about	/#/	examine
/y/	yet	/*/	one
/z/	zoo	/|/	logic
/A/	bite	/^/	but

Output representations for phonemes and punctuations. The symbols for phonemes in the first column are a superset of ARPAbet and are associated with the sound of the italicized part of the adjacent word. Compound phonemes were introduced when a single letter was associated with more than one primary phoneme. The continuation symbol, /-/, not shown here, was used when a letter was silent.

553

References

[1] Everitt, B., (1974). *Cluster Analysis*. Heinemann: London.

[2] Harman, H. H. (1976). *Modern Factor Analysis*. University of Chicago Press: Chicago.

[3] Hopfield, J. J. (1984). Neurons with graded response have collective computation abilities, *Proceedings of the National Academy of Sciences USA*, **81**, 3088-3092.

[4] Kim, J. & Mueller, C. W. (1978). *Introduction to Factor Analysis*, Sage University Paper series on Quantitative Appliciations in the Social Sciences, 07-013: Beverly Hills.

[5] Rumel, R. J. (1967). Understanding factor analysis, *Conflict Resolution*, **11**, 444-480.

[6] Rumelhart, D. E., Hinton, G. E. & Williams, R, J. (1986). Learning internal representations by error propagation. In D. E. Rumelhart & J. L. McClelland, (Eds.), *Parallel Distributed Processing: Explorations in the Microstructure of Cognition. Vol. 1: Foundations*, Cambridge: MIT Press.

[7] Sejnowski, T. J. and Rosenberg, C. R. (1986). *NETtalk: A parallel network that learns to read aloud* (Technical Report 86/01). Baltimore: Johns Hopkins University, Department of Electrical Engineering and Computer Science.

[8] Sejnowski, T. J., and Rosenberg, C. R. (1987a). Connectionist Models of Learning. In M. S. Gazzaniga (Ed.), *Perspectives in Memory Research and Training*. Cambridge: MIT Press.

[9] Sejnowski, T. J. and Rosenberg, C. R. (1987b). Parallel networks that learn to pronounce English text, *Complex Systems*, **1**, 145-168.

Generation of Simple Sentences in English Using the Connectionist Model of Computation

Jugal Kalita and Lokendra Shastri
Department of Computer and Information Sciences
University of Pennsylvania, Philadelphia, PA 19104

March 13, 1987

Abstract

This paper discusses the design and implementation of a connectionist system for generation of well-formed English sentences of limited length and syntactic variability. The design employs several levels of interacting units for making appropriate decisions. It uses a simple technique for specifying assignment of input concepts to roles in a sentence and also has a *reusable* subnetwork for the expansion of noun phrases. The same NP-subnetwork is used for the expansion of noun phrases corresponding to the subject as well as the object phrases of the generated sentences. The input to the system consists of parallel activation of a cluster of nodes representing conceptual specification of the sentence whereas the output is in the form of sequential activation of nodes corresponding to the words constituting the sentence. The system can produce simple sentences in both active and passive voices, and in several tenses. Results of a simulation experiment performed are also included.

1 Introduction

It is generally accepted that text generation involves three distinct sequential phases – content determination, text planning and surface generation. Content determination involves identifying information that needs to be included in the generated text. Text planning is concerned with appropriately sequencing the intended contents in order to achieve coherence in the generated text. The final phase — *surface generation* produces the actual sentences given their internal representations. In this paper, we are not involved with the first two phases. Our emphasis is on the generation of well-formed sentences assuming that the initial two phases have been successfully performed.

Currently, there are several approaches to surface generation. These include the approach based on functional grammar as used by McKeown in her TEXT system [McKeown 85], the propositional logic based system employed by Appelt [Appelt 83] and the stepwise refinement approach taken by McDonald in the MUMBLE system [McDonald 83]. The functional grammar approach is non-deterministic; the generation of a sentence from a conceptual specification involves the process of unification which is slow and inefficient taking potentially exponential computation time. Appelt assumes homogeneity of various decision processes; his approach which requires implementation of theorem proving techniques is also not easily amenable to a parallelism. MUMBLE employs several levels of processing to achieve its goal. Consequently, processing requirements are relatively simple at each level enhancing efficiency and modifiability. It is deterministic in the nature of processing involved.

Our approach to surface language generation is similar to that of McDonald. We have modeled the generation process as a hierarchy of processing levels. Decisions have to be made at each of these levels until nodes corresponding to the words constituting the sentence are activated in an appropriate order. We attempt to implement the various processing levels by using the connectionist model of computation in the spirit of [Feldman et al 82]. The techniques employed in our implementation have been influenced by Cottrell's system [Cottrell 85] for word-sense disambiguation and parsing. We successfully generate English sentences of limited length and limited syntactic variability given a non-linguistic specification of their contents.

2 The Levels of Processing

As indicated earlier, there are several levels of computation in the generation of a well-formed sentence from its conceptual specification in our generational paradigm. In this respect, our chosen approach is similar to the approaches to perceptual processing reported in [McClelland et al 81,Sabbah 85] where each level of processing is concerned with forming representation of the input at a different level of abstraction. In our system, levels represent various decision steps that need to be taken in order to generate a sentence.

The main levels of processing employed in the system are

- input level

- realization-class level

- choice level

- constituent level

- morphology level

The relationships among the various levels are shown in figure 1.

Each level of processing is carried out by one or more specially designed cluster of nodes. The *input* level nodes represent a communicative goal (along with some constraints to be discussed later). This information is in the form of a non-linguistic conceptual specifications of what needs to be conveyed through an intended utterance. Examples of concept level nodes are *concept-eat*, *concept-monkey-1*, *concept-banana-1* etc.

Each concept node is linked to a *realization-class* node. Several concept nodes may be linked to the same realization-class node, but each concept node is connected to a unique realization-class node. The realization-class node which is connected to a concept node determines how that concept will be translated into English. The translation of all concepts whose corresponding nodes are connected to a realization-class node proceed in an identical fashion. Examples of realization-class nodes are *svo* and *individual-item*. The *svo* node requires concept nodes linked to it be translated into a sentence where the subject, verb and the object are all present. Concept units corresponding to transitive verbs such as *eat, give* etc., are connected to the *svo* node. Concept nodes such as *concept-monkey-1*, *concept-baboon-1*, *concept-banana-1*, which denote individual objects that can be expressed in English in terms of noun phrases, are each connected by an excitatory link to the *individual-item* node in the realization-class level.

Each realization-class node has activation links to one or more *choice* level nodes. The choice nodes connected to a realization-class node form a *winner-take-all* network. In other words, only one of them can be active at any instant of time. Each choice node connected to a realization-class

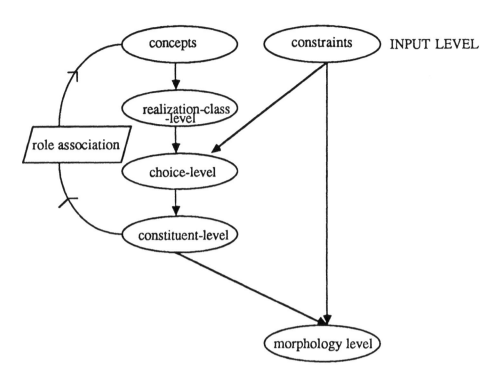

Figure 1: **Various levels and their interactions**

node represents a different grammatical way of realizing a concept linked to that realization-class node. For example, the realization-class node *svo* is connected to two choice nodes: *active-svo-choice* and *passive-svo-choice*. This allows a concept such as *concept-eat* (whose corresponding node is connected to the *svo* node) to be translated into either an active or a passive sentence after its subject and object roles are filled.

The choice nodes have other activation links from *constraint-specification* nodes (which are input nodes) incident upon them. Activation on each input link is necessary for a choice node to become active. The constraint-specification nodes must be activated by the text planner before the process of surface generation is started. Examples of constraint-specification nodes include *current-voice-is-active*, *current-aspect-is-perfect* etc. In order for the *active-svo-choice* node t⸱ be active, it must receive activation inputs both from the *svo* node (a realization-class node) and the *current-voice-is-active* node (a constraint-specification node).

A choice node has activation links to one or more *constituent* level nodes. The constituent nodes specify the roles that need to be filled in order to achieve a particular choice of realization of an input concept. Examples of constituent nodes connected to the choice node *active-svo-choice* are *subject-slot*, *active-svo-verb* and *object-slot* nodes. The activation of the constituent nodes is sequentialized. Thus, the translation of a concept node connected to the *svo* node (a realization-class node), assuming the *current-voice-is-active* node is active, involves filling in three roles in sequence: subject, verb and object. Sequencing is accomplished using *sequentialization* nodes discussed later.

Each constituent node is either connected to one or more nodes that handle morphology or a cluster of nodes that specify role association. Activation of a constituent node may result in excitation of one or more word nodes in a predetermined sequence. If a constituent node is connected to a node that specifies role assignment, then its activation starts expansion of another realization-specification node corresponding to the conceptual specification of a part of the sentence (e.g. the subject or object role of the sentence).

The morphology handling level contains two types of nodes: *generic-word* and *word* nodes. It

557

is assumed that activation of a word node leads to the word being spoken aloud or written out. A generic-word node represents a word whose lexicalization is dependent upon the grammatical context. For example, the node *generic-word-eat* is connected to the word nodes *word-eat* and *word-eats*. It should be noted that complex forms of verbs such as *has been eating* (i.e., three word nodes: *word-has, word-been* and *word-eating* whose activation is sequenced) are also connected to the *generic-word-eat* node. Depending on constraints such as current tense, current voice and current aspect, one of these sets of word nodes will be activated.

In addition to activation links from binder and constraint-specification nodes, the morphology nodes may have activation links from concept nodes via the cluster of nodes specifying role association. The realization-specification node to which a concept node is connected, determines the structure of the phrase or clause that results from the concept's translation through its links to choice and constituent nodes. One or more of the word nodes which need to be activated to fill in the slots in the translated phrase or clause may be determined directly by the concept node itself by having activation links onto specific generic-word or word nodes.

3 Implementation Details

Each of the levels of processing discussed in the previous section has one or more unit types associated with it. Thus, we have the following unit types: concept, constraint-specification, realization-class, choice, constituent, generic-word and word.

Input nodes which include concept and constraint-specification nodes are activated by the text planner. They are activated simultaneously and thereafter they drive the process of generation.

Generation of a well-formed sentence involves activating the nodes which correspond to words in the sentence. However, parallel activation of word nodes cannot be considered as generating a sentence. Sequentialization of activation of word nodes is imperative in order to generate a meaningful sentence. It should be noted at this point that a connectionist model is inherently parallel, and hence additional efforts are required to achieve the sequentialization needed by the nature of the problem under consideration. The sequencing technique discussed next is similar to the one employed in [Cottrell 85].

In order to achieve appropriate word sequencing, a new unit type has been defined: the *sequentialization* type. Sequentialization nodes also assist in meeting the crucial requirement that only one word node remain active at any instant of time. Feedback from a corresponding sequentialization node turns off a word node a few cycles after it has been activated. Sequentialization nodes also facilitate the deactivation of units which have already participated in the generation process and are no longer required.

Each unit in the network has several sites. The nature of these sites is dependent on the unit type. Among the sites are: *or, and, expansion-completed* and *reuse-site*. An *or* site computes its value to be the highest among all its weighted input signals, whereas the computation carried out in at the *and* sites involves selecting the minimum of its weighted inputs as its value. The function of the other two sites are discussed later.

The units can be in three possible states. They are *initial-inactivation, activation* and *final-inactivation*. All nodes are initially in the initial-inactivation state. A node is switched to the activation state on receiving appropriate excitation signals at its *or* or *and* sites. Once in the activation state, a unit remains so until such time when a signal is received at its *expansion-completed* site from a sequencer unit. This forces a unit to *final-inactivation* state. Thus, when all activity ceases, the nodes which participated in the processing are left in *final-inactivation* state.

The unit behavior discussed thus far is sufficient to generate simple sentences, but in order to

558

achieve better resource utilization, it is necessary to introduce another site called the *reuse-site*. A positive input at this site of a unit in *final-inactivation* state causes the unit to change its state to *initial-inactivation* state so that the unit can be reused. Reuse-sites are used only for the nodes in the subnetwork for noun phrase expansion. This enables us to use the same noun phrase expansion subnetwork for expanding the subject as well as the object of a sentence. Using two separate subnetworks for noun phrase expansion – one for the subject and the other for the object would lead to a wastage of resource. Since the task is essentially the same, we have decided to design a reusable NP-subnetwork. When a noun phrase expansion is complete, a specially designated unit in the NP expansion subnetwork gets activated. This unit, then, sends a positive activation to all units in the noun phrase subnetwork at the *reuse-site* as a result of which the units become ready to resume their activity.

This, in our view, is a significant feature of our implementation. Supposing we are generating an active sentence, the subject NP needs to be expanded first. In order to achieve this, the *subject-slot* constituent node is activated. This initiates the expansion of the subject NP. Once this expansion is complete, all nodes that participated in this expansion in the NP subnetwork are reset by a signal at the *reuse-site* as explained earlier and are available for reuse. After subject NP expansion, the active verb phrase is generated. Following this, the same NP expansion subnetwork is used to generate the noun phrase corresponding to the object.

In order to generate active as well as passive sentences, we have a *passive-svo* choice unit connected to the realization class unit *svo*. The passive choice unit gets activated only when it receives activation from the realization class unit *svo* and the constraint specification unit *current-voice-is-passive*. The subnetwork employed for generation of subject and object noun phrases for active sentences is also used for passive sentences without any modification. New sequencing nodes have been introduced so that the expansion of the object phrase precedes that of the verb and agent phrases for the passive case. Units for handling passive forms of verbs have also been implemented.

The distinction between the two inactive states is introduced to prevent oscillation of a node between active and inactive states. This requirement is extremely important, in particular, for word nodes in order to achieve proper sequentialization of utterance. In our implementation, it was required of a unit to remain in inactive state once there is a transition from active to inactive state to prevent unexpected potential and state oscillations. A node which is inactive to start with plays its appropriate role in the process of generation by becoming active on receiving excitatory inputs. Consequently, it activates other units, and in turn, it becomes inactive again (final-inactivation). However, the arrival of a positive input at the *reuse-site* causes a unit in final-inactivation state to make a transition to initial-inactivation state enabling its participation in repetitive processing.

The mechanism by which the input to the network is specified is described below with the aid of figure 2. We should be able to associate concepts with roles such as subject and object in a sentence. Corresponding to each object that can fill in the subject/object role of a sentence, we have a concept node such as *concept-monkey-1, concept-banana-1*, etc as discussed earlier. Each of these concept nodes is connected to the realization-class unit *individual-item*. Each of these concept units is also connected to two binder units for the purpose of role association. Consider the concept unit *concept-monkey-1*. It is connected to two units labeled *e-s-1* and *e-o-1* in the diagram. *e-s-1* is called a *subject-binder* unit; *e-o-1* is an *object-binder* unit. There are two such units for each concept that can fill the role of subject/object. In this example, *e-s-1* plays a role in associating *concept-monkey-1* unit to the subject role; and *e-o-1* helps associate *concept-monkey-1* to the object role. It should be noted that each *subject-binder* node (for each concept) has an activation link from the *subject-slot* unit (which is a constituent unit driven by the svo-active/svo-passive choice units for the realization-class unit svo). Similarly, the object-binder node for each concept has an activation link from the *object-slot* constituent unit.

559

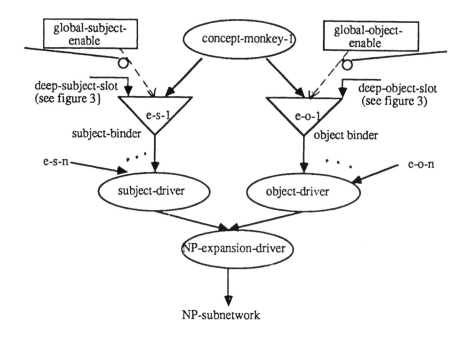

Dashed lines indicate links some of which must be enabled at
input time for role assignment

Figure 2: **Binding of Concepts to Roles in Input Specification**

Several other units are also used in this process. Two such units are the *global-subject-enable* and *global-object-enable* nodes. These nodes always kept active. There is a link from *global-subject-enable* node to the *subject-binder* node for each concept node. There is a similar link from *global-object-enable* to the *object-binder* node for that concept. Initially, each such link is in a *disabled* state. Appropriate links are enabled before processing a sentence to specify role assignments. Enabling of role assignment links is done at input time before processing is initiated. This is a required in order to correctly specify at a conceptual level which concept plays what role in the sentence to be generated. If, however, our system needs to be interfaced with higher level generation systems that perform text planning, such enabling of links will have to be performed automatically. This is a problem which future research must address.

In order to assign the *concept-monkey-1* unit to the subject role, the link from the *global-subject-enable* to the subject-binder unit for *concept-monkey-1* (here, unit *e-s-1*) is enabled at input time. (Please note that all other links from *global-subject-enable* unit to all other subject-binder nodes are still disabled.) And, similarly to assign another unit (say, *concept-banana-1*) to the object role, the link from *global-object-enable* unit to the object-binder unit for *concept-banana-1* is also enabled. So, the crucial step in specifying the input properly constitutes enabling the linkages appropriately before the processing of a sentence starts.

The subject-binder and the object-binder nodes are such that they need all their inputs to get activated. A subject-binder node provides excitatory input to the *subject-driver* node. Excitation from any subject-binder node excites the *subject-driver* node. Similarly, there is an *object-driver* node which receives excitatory input from the object-binder nodes. The *subject-driver* and the *object-driver* nodes have excitatory inputs to an *np-expansion-driver* node. This node drives a node called the *binding-completion-signaler* which initiates the expansion of the *choice* nodes connected to the *individual-item* realization-class node. Thus, excitation of the *binding-completion-signaler*

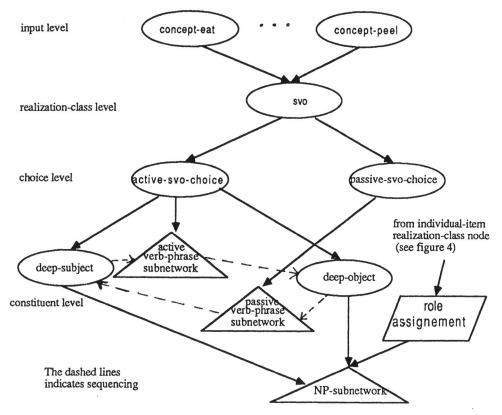

Figure 3: **A section of the relevant portion of the network** (also see figure 4)

unit which sits between the realization-class node *individual-item* and its choice units initiates the excitation of units that leads to the expansion of the noun phrase corresponding to the subject or object of a sentence.

The choice unit *active-svo* has excitatory inputs to *subject-slot, active-svo-verb* and *object-slot* units – the activation of these units is sequentialized. Thus, during the expansion of the subject, only the *subject-slot* unit is active; as a result, the subject-driver node is activated (object-driver node is off). This leads to the generation of a noun phrase corresponding to the subject of the sentence. Similarly, when the *object-slot* unit is active, the *object-binder* unit corresponding to the object concept gets activated. This activates the *object-driver* node which finally results in the generation of a noun phrase corresponding to the object concept.

4 An Example Simulation

We will now briefly run through an example simulation showing the various steps involved in generating a sentence given the conceptual specification of its contents. The network for the simulation was built using the ISCON simulator [Fanty 85]. The network has about one hundred units. The relevant portions of the network are shown in figures 3 and 4. Sequentialization nodes are not shown here.

Initially the input units corresponding to the conceptual level description of the sentence to be generated are activated. For example, in order to generate the sentence *A monkey is eating a banana*, we activate the units corresponding to *concept-eat, concept-monkey-1* and *concept-banana-1*. We also activate the constraint-specification units – *current-voice-is-active,* and *current-aspect-*

561

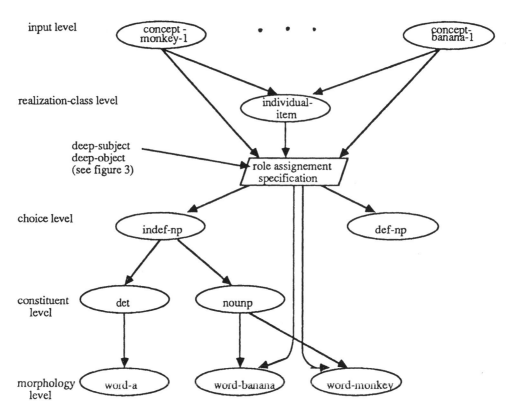

Figure 4: **Section of the network handling noun phrases**

is-continuous. The units *global-subject-enable* and *global-object-enable* are also activated and the links from *concept-monkey-1* to its subject-binder (viz., *e-s-1*) and from *concept-banana-1* to its object-binder (viz., *e-o-2*) are enabled.

Activation of the *concept-eat* unit activates the realization-class unit *svo.* This along with the fact that the *current-voice-is-active* unit is on, turns on the choice unit *active-svo-choice* which, in turn, sequentially activates the three constituent units connected to it, viz., *deep-subject, active-svo-verb* and *deep-object.*

Meanwhile, activation of the concept-level unit *concept-monkey-1* had turned on the realization-class unit *individual-item* which feeds an activation link to the *binding-completion-signaler* node which however, remains inactive. Among the constituent units, the *deep-subject* unit is activated first. Now, the set of units which participate in the binding process comes to play its role. Since *deep-subject* unit is active, the unit *e-s-1* receives activation from all its inputs, and this results in the binding of the *concept-monkey-1* unit to the subject role. Finally, as a result of binding, the *np-expansion-driver* unit provides activation to the *binding-completion-signaler* unit which becomes active. It feeds the activation link to the choice unit *indefinite-np* which is turned on. It provides excitatory input to the constituent units *determiner* and *nounp* which are activated resulting in the sequential activation of the units *word-a* and *word-monkey.* Expansion of the subject is now complete. A special sequencer node is activated; it resets the various nodes involved in the expansion of the subject making the NP subnetwork available for reuse for the object phrase expansion, and also starts the expansion of the verb phrase by sending an excitatory signal to the *active-svo-verb* constituent unit.

The expansion of the verb phrase of the sentence leads to the utterance of the word units *word-is*

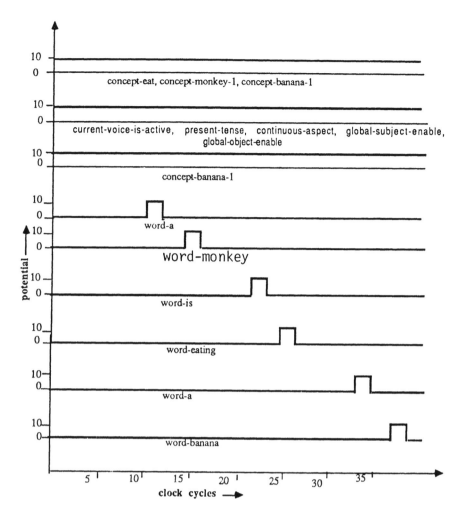

Figure 5: **Graph of potentials of relevant units**

and *word-eating* in sequence. All units involved in the expansion of the verb phrase are turned off, and expansion of the noun phrase corresponding to the object of the sentence is initiated resulting in activation of the word units *word-a* and *word-banana* in sequence. This completes generation of the whole sentence and is followed by deactivation of all units that took part in the process of generation and are still active. Finally, a specialized sequentialization unit is turned on. Activation of this unit can be used to generate an appropriate pause (in case of speech) or the punctuation symbol '.' in case of written text.

A graph showing the potentials of the concept, restriction-specification and the word units for this example are shown in figure 5. This clearly shows that the words that constitute the sentence are activated in proper sequence. Each word unit is active for a short period of time during which it is assumed to be spoken or written. Examples of other sentences that can be generated by our system include *A monkey has been eating a banana, A banana is being eaten by a monkey, A baboon was peeling a banana*, etc.

5   Discussions

The sentences generated by the current network are all of the form *svo* where a sentence contains a subject, a verb and an object. They can be in active or passive voice and in several different tenses. All noun phrases generated have a determiner followed by a noun. All verbs used so far are transitive. Also, all passive sentences currently generated have the *by*-phrase following the verb.

563

We need to incorporate intransitive verbs and allow the agent phrase to be optional in passive sentences. Future research is also necessary in many other fronts, such as, incorporation of relative clauses, repeated constituents such as prepositional phrases and adjectives, anaphoric pronominal phrases, etc.

An open problem that needs to be addressed is regarding mechanisms for interfacing our surface generation system with other systems that perform content determination and text planning in natural language generation. Our approach to specifying this interface is simple, but inefficient. A more sophisticated technique might involve using *exploded case frames* introduced in [Cottrell 85]. This will allow us to prevent the generation of such absurd sentences such as *A banana is eating a monkey* by imposing selectional constraints on the fillers of roles of the verbs. This will necessitate the incorporation of an elaborate knowledge base describing the concepts that are available in our domain of interest. The specification and organization of such a knowledge base such that responses to pertinent queries can be elicited efficiently can be modeled after [Shastri 85]. Additionally, in the current implementation, we need two binder units – a subject-binder and an object-binder for each concept that can fill the subject/object role of a sentence. Since, the number of such concepts is usually very large, it will be worthwhile to investigate other approaches to the problem of binding. In general, this problem is an instance of *the binding problem* in connectionist systems. This problem is not specific to our work, and any elegant solution to it will be applicable here.

Currently, we are working on automating the process of building up the network from a declarative specification of the realization-specifications, choices and constituents and the bindings. This will help in adding new sentence types and component structures easily for the purposes of testing and development. It will also involve automatically identifying units which can be shared among the various constituent schemata and how sequentialization nodes are to be interconnected.

Another direction in which we intend to pursue further research is regarding the choice of lexical tokens in the generated sentences. Lexical choice is governed by a large number of factors – syntactic, semantic and pragmatic. These include considerations such as relevant characteristics of the speaker and the hearer, their interpersonal goals, the conversational setting, etc. Identification of such factors, and a careful analysis of the nature of their interplay is essential in order to be able to develop a systematic theory of lexical choice. It is believed that different factors provide different levels of evidence for the selection of competing words, and these have to be appropriately combined in order to be able to select a particular word over its competitors.

References

[Appelt 83]   Appelt, D.E., Planning Natural Language Utterances, *Proceedings of the Eighth International Joint Conference on Artificial Intelligence*, 1983, pp. 59-62.

[Cottrell 85]   Cottrell, G.W., *A Connectionist Approach to Word Sense Disambiguation*, TR-154, Department of Computer Science, University of Rochester, 1985.

[Fanty 85]   Fanty, M., *Connectionist Network Simulation Tools*, Technical Report, Department of Computer Science, University of Rochester, 1985.

[Feldman et al 82]   Feldman, J.A., and Ballard D., Connectionist Models and their Properties, *Cognitive Science*, Volume 6, 1982, pp. 205-254.

[McClelland et al 81] McClelland, J.L., and Rumelhart, D.E., An Interactive Activation Model of Context Effects in Letter Perception: Part 1, An Account of Basic Findings,

Psychological Review, Volume 88, Number 5, September 1981, pp. 375-405.

[McDonald 83] McDonald, D.D., Description Directed Control: Its Implications for Natural Language Generation, *Computers and Mathematics with Applications*, Volume 9, No. 1, 1983, pp. 111-129.

[McKeown 85] McKeown, K., *Text Generation: Using Discourse Strategies and Focus Constraints to Generate Natural Language Text*, Cambridge University Press, 1985.

[Sabbah 85] Sabbah, D., Computing with Connections in Visual Recognition of Origami Objects, *Cognitive Science*, 9, 1985.

[Shastri 85] Shastri, L., *Evidential Reasoning in Semantic Networks: A Formal Theory and its Parallel Implementation*, Ph.D. Thesis, Department of Computer Science, University of Rochester, 1985.

Inferences in Sentence Processing: The Role of Constructed Representations

Margery M. Lucas
Wellesley College

Michael K. Tanenhaus
University of Rochester

Greg N. Carlson
University of Iowa

Abstract

Recent studies have revealed interesting differences between lexical decision and naming tasks. Naming responses seem to be primarily sensitive to lexical processes and lexical decisions to both lexical and message-level processes. This differential sensitivity to level of representation was used to investigate the following questions: 1) Are probable instruments for an action routinely inferred during sentence comprehension? Previous work may have failed to show that instruments are inferred, in part, because processing measures were used that were relatively insensitive to the level of representation involved in the inference and 2) If instruments are inferred, does this process require accessing elements of the linguistic or the constructed representation? Four experiments were performed that used cross-modal lexical decision and naming tasks as measures of instrument priming in sentences that implied the use of an instrument. No priming was found for sentences with no context, replicating Dosher and Corbett (1982). When sentences were preceded by a context that explicitly mentioned the instrument, however, priming was found with the lexical decision task. In combination with the result of the first two experiments, this suggests that instruments are inferred when the instrument implied by a sentence is available from the context but not when sentences are presented without contexts. Priming was not found with the naming task, however. The lexical decision/naming data together suggest that making an instrument inference involves accessing elements of a constructed representation of the discourse.

In addition, in sentences that contained pronouns that referred to the instruments, priming was found for appropriate referents with the lexical decision task but not with naming. This suggests that locating antecedents for pronouns also involves a constructed representation.

Introduction

The lexical decision and naming tasks have both been widely used in the study of word recognition. Recently, interesting differences between the two tasks have surfaced, indicating that they are sensitive to different types of information. Seidenberg, Waters, Sanders, and Langer (1984) compared the two tasks and found that naming responses were sensitive primarily to linguistic (particularly lexical) -level processes and lexical decisions were sensitive to both lexical processes and post-lexical message-level processes. Seidenberg et al. suggested that this was because the lexical decision task requires an explicit yes-or-no discrimination that benefits from a post-access check to determine if the target is compatible with its context. Naming, on the other hand, requires only procedural knowledge about pronunciation and is, therefore, more likely to be influenced solely by information tied directly to the lexical representation of the word.

This difference is potentially important because it may provide a methodological tool for discriminating the different levels of representation used in sentence comprehension. For example, based on the linguistic evidence provided by Hankamer and Sag (1976), Tanenhaus, Carlson, and Seidenberg (1985) suggested that antecedent assignment for different types of anaphors involves linking the anaphor to elements in either a linguistic representation of the surface or logical form of a sentence or a conceptual representation. The latter is a model of the discourse constructed by using the outputs of the linguistic system. An earlier study used the lexical decision/naming task difference to test this hypothesis for definite noun phrases (Lucas, Tanenhaus, Carlson, and Senytka, 1985). Knowledge-level processing is generally required to determine the antecedents of definite noun phrases, suggesting that a constructed representation is more likely to be involved in antecent assignment for that kind of anaphor than a linguistic representation. In the study, subjects listened to sentence pairs that contained definite noun phrases at the end of the second sentence. Subjects then made lexical decisions or naming responses to targets that were appropriate or inappropriate antecedents. Naming responses did not show an appropriateness effect indicating that linguistic representations were not involved in antecedent assignment. There was a sizeable appropriateness effect for lexical decision responses, however, suggesting that the definite noun phrases were linked to elements of the constructed representation.

One of the purposes of the present studies was to extend this methodology to other forms of inference, in particular to inferences involving knowledge of the tools or instruments typically used to accomplish some action. For example, upon hearing the sentence, "Mary cut into a steak" listeners might infer that Mary used a knife. Presumably, making such inferences requires accessing some form of representation of the instrument. One issue investigated in the studies reported here concerns the level of representation involved. This is a particularly interesting test of the lexical decision/naming difference because it is not obvious whether the linguistic or conceptual level is involved. Schema theories that maintain that words are mapped onto well-formed conceptual structures which form the basis for inference (Schank, 1974; Bobrow and Norman, 1975) would predict that instruments commonly used in certain actions would be inferred at the level of the schema or conceptualization for that action. On the other hand, Fillmore's case theory, in which case roles like instruments are part of the representation of the verb in the lexicon, is compatible with the view that typical instruments for actions can be derived from the lexical representation of the verb (Fillmore, 1971).

Before this issue can be investigated, however, the issue of whether instrument inferences are drawn must be addressed. The empirical literature on this issue is mixed. There is some evidence that instrument inferences are encoded and stored as part of the memory representation of a sentence (Johnson, Bransford, and Solomon, 1973; McKoon and Ratcliff, 1981; Paris and Lindauer, 1976), but other evidence contradicts this (Corbett and Dosher, 1978;Dosher and Corbett, 1982; Singer, 1979). One possible reason for the discrepancy is that some studies used on-line measures of sentence processing, but many involved recall or recognition. These latter studies, therefore, provide more information about memory processes than about the process of computing inferences on-line. Also, recognition tasks require the subject to explicitly consider the instrument in relation to the discourse and therefore may not reflect normal sentence processing. The one study that used an on-line processing measure that did not also require a recognition decision was Dosher and Corbett (1982). Using stroop interference, Dosher and Corbett found no evidence that implied instruments were activated following simple sentences. They argued that encoding implicit instruments is not necessary for discourse processing. Given the evidence that different priming tasks can tap different levels of representation, however, it is possible that the Stroop interference task is not sensitive to the representational level involved in instrument inferences. A different priming task may work where Stroop failed.

Also, Dosher and Corbett presented sentences with no context, but some previous research suggests that instrument inferences may occur when there is a context that explicitly mentions the instrument (McKoon and Ratcliff, 1981). This suggests that studies of instrument inferences must also be sensitive to the fact that comprehension strategies depend on the availability of context.

The first two studies reported here attempted to replicate the Dosher and Corbett findings with isolated sentences using cross-modal lexical decision and naming tasks. If Dosher and Corbett failed to find evidence for instrument priming because the Stroop interference task does not tap processes at the appropriate level of representation, then it is possible that the lexical decision or naming tasks will tap the appropriate processes. In any case, the results of these "no context" experiments will form a basis for comparison for the next two studies which examine comprehension strategies when the same stimuli are preceded by context.

Experiments 1 and 2

Method

Subjects. Thirty-two University of Rochester undergraduates participated in the first two experiments, sixteen in the lexical decision version and sixteen in the naming version.

Materials. The experimental sentences were generated from twenty-eight verb-instrument pairs. Sentences for each verb-instrument pair were composed whose verb phrases contained one of the verbs but which did not explicitly mention the instrument. An example of an experimental sentence is given in (1).

(1) He swept the floor every week on Saturday.

Each sentence was paired with either a target that was a plausible instrument or a control target that was an implausible instrument for the action specified by the verb. For example, targets for the sentence in (1) were "broom", the appropriate instrument and "closet" the inappropriate control. Instruments and controls were similar in length and frequency. Two presentation conditions were produced by combining the sentence list with two target lists that counterbalanced type of target (instrument or control). The experimental sentences were intermixed with 88 filler sentences and the first six sentences in the list were fillers. In the lexical decision version, 60 of the filler trials were paired with non-word targets and the remaining 32 with word targets. Materials for the naming version were identical to the lexical decision version except that nonword targets were replaced by word targets in the filler sentences.

Procedure. In each experiment, eight subjects were assigned to each of the two presentation conditions. Each presentation version was preceded by 10 practice trials. Subjects heard the sentences binaurally over stereo headphones. A timing tone inaudible to the subject was placed at the end of the second phrase following the verb but never at the end of a sentence. For example, the tone in (1) appeared at the end of "week". Subjects, therefore, had plenty of time to make the inference before the target was presented if they were going to do so. The tone initiated presentation of a target stimulus midscreen on an Apple IIe computer monitor. In the lexical decision version, subjects pressed a button to indicate their response. Lexical decision times were recorded by a Digitry millisecond timer from the onset of the target to the subject's buttonpress. In the naming version subjects said the target word out loud as soon as it appeared. Responses were spoken into a microphone and naming times were measured from the onset of the target to the onset of the subject's spoken response. In order to ensure that subjects were attending to the sentences, comprehension questions were asked following one third of the trials.

Results and Discussion

Analyses of variance (ANOVA's) with subjects and items as random factors and Appropriateness (plausible instrument or implausible control) as a fixed factor were run separately on the lexical decision and naming data. For each subject outlier scores greater than or less than 2.0 standard deviations from the subject mean were replaced by the 2.0 standard deviation cutoff score. Condition means for both the lexical decision and naming experiments are presented in Table 1. There was no evidence for an Appropriateness effect in either the lexical decision version, $F(1,15) = .01$ by subjects and $F(1,27) = .001$ by items, or the naming version, $F(1,15) = 1.91$ by subjects and $F(1,27) = 1.64$ by items. This is true even when the results of both experiments are combined. An ANOVA on the combined data with Task as an additional factor revealed only a main effect of Task, $F(1,30) = 25.2$, $p < .0001$ by subjects and $F(1,54) = 160.8$, $p < .0001$ by items. This was due to lexical decision times being about 300 ms slower than naming times on the average. But there was still no main effect of Appropriateness, $F(1,30) = .39$ by subjects and $F(1,54) = .34$ by items, and no interaction between Task and Appropriateness, $F(1,30) = .61$ and $F(1,54) = .36$.

The fact that there was no appropriateness effect in either experiment indicates that the implied instrument was no more accessible than an inappropriate control. The results, therefore, replicate Dosher and Corbett in finding no evidence for instrument priming in sentences without context, using lexical decision and naming tasks instead of Stroop interference. In the next two experiments we sought evidence that providing a context that explicitly mentions the instrument would cause subjects to infer the use of the instrument.

One problem addressed in the next set of studies concerned the fact that concluding that instrument inferences are not routinely made depends on getting a null result. In such cases it is unclear whether the effect did not occur because the null hypothesis is correct or because there is a defect in the materials or the technique. To counter this possibility, in the following studies instrument inferences were directly compared with another form of inference - antecedent assignment for pronouns. Although instrument inferences may not be necessary for comprehension, antecedents must be assigned to pronouns if sentences are to be understood. Antecedent assignment, then, can serve as a check on the experimental design, to insure that the technique is sensitive to inferences drawn using our materials.

In addition, using the lexical decision/naming difference as a diagnostic, the levels of representation involved in both instrument inferences and antecedent assignment for pronouns could be determined. Earlier, it was argued that instrument inferences could require access to either a linguistic or a constructed representation. Likewise, Tanenhaus, Carlson, and Seidenberg (1985) have argued that finding the referent of a pronoun could require linking the anaphor to elements in either a linguistic or a constructed representation.

The next two experiments, then, were undertaken with several goals in mind. One was to determine if instrument inferences would be made when sentences with implied instruments were presented following contexts that explicitly mentioned the instrument. Sentences that referred to the instrument by using pronouns would serve as a check on the materials and design of the experiment should there be another null result for the instrument inferences. Also, the lexical decision/naming difference would reveal whether representations for implied instruments and pronoun antecedents were found at a linguistic or a conceptual level.

Experiments 3 and 4

Method

Subjects. Forty-eight subjects from the University of Rochester participated in the next two experiments, twenty-four in the lexical decision version and twenty-four in the naming version.

Materials. The sentences from the first two experiments were also used in these experiments. There were two modifications to the experimental design. First, a context sentence was constructed for each of the experimental sentences. This context sentence contained both the instrument and control target words from the first study. For example, the sentence "He swept the floor every week on Saturday" from the first experiments was given the context, "John took the broom out of the closet". In half the context sentences the implied instrument occurred earlier than the control. The reverse was true for the remaining half. A second modification involved the introduction of an additional set of sentences, contructed from the same sentences that implied the instruments, in which the second phrase following the verb was replaced by a phrase that contained a pronoun. The pronoun always referred to the instrument explicitly mentioned in the preceding

570

sentence. In the example sentence, the sentence in the pronoun condition would become "He drove to work in it every morning".

Procedure. Subjects all heard the same context sentences but heard final sentences in either their pronoun or non-pronoun form. Four presentation conditions were formed combining two sentence lists which counterbalanced pronoun and non-pronoun versions of the final sentence with the two target lists from the first two experiments. For both the lexical decision and naming versions, six subjects were assigned to each of the four presentation conditions. The procedure for these experiments was identical to that for the first two experiments with one exception. In sentences containing pronouns, targets were presented at the end of the pronoun. Since phrases with pronouns replaced phrases from the original sentences, targets were the same distance from the verb in the pronoun versions as they were in the non-pronoun versions.

Results and Discussion

ANOVA's with subjects and items as random factors and Appropriateness (plausible instrument or implausible control) and Type (pronoun or non-pronoun version) as fixed factors were run separately on the lexical decision and naming data. Condition means for both experiments are presented in Table 2. There was a main effect of Appropriateness in the lexical decision data, $F(1,23) = 10.15$, $p < .005$ by subjects and $F(1, 26) = 6.31$, $p < .02$ by items, but not in the naming data, $F(1,23) = .52$ by subjects and $F(1,26) = .63$ by items. The main effect of Type and the interaction of Type by Appropriateness was not significant in either the lexical decision or the naming data.

The fact that there was a significant Appropriateness effect in the lexical decision data and no significant interaction of Type by Appropriateness indicates that there was instrument priming not only in the pronoun version of the sentences, where listeners needed to locate the instrument antecedent if they were to understand the sentence, but also in the non-pronoun versions where the use of the instrument was merely implied. In combination with the results of the previous experiments, these results support the hypothesis that, although instrument inferences are not drawn for sentences out of context, they will be drawn when sentences are presented in context, particularly if the sentences explicitly mention the instruments.

The results just discussed suggest that certain forms of inference, instrument inferences and antecedent assignment for pronouns, are made when the right context conditions are in place. What can be said about the form of representation accessed in both types of inference? Because the Appropriateness effect occurred only in the lexical decision data and not in the naming data it seems that the constructed representation of the discourse is the sole level accessed for both types of inference. An ANOVA with Task, Type, and Appropriateness as factors was run on the combined lexical decision and naming data for further confirmation of this. Although there was a main effect of Task, $F(1,46) = 20.04$, $p < .0001$ by subjects and $F(1,27) = 332.26$, $p < .00001$ by items and a main effect of Appropriateness, $F(1,46) = 7.51$, $p < .01$ by subjects and $F(1,27) = 5.39$, $p < .05$ by items, no other main effects or interactions were significant. If the conceptual level (and not the linguistic level) of representation had been accessed, we would expect an interaction of Task by Appropriateness. But this was only marginally significant by subjects, $F(1,46) = 2.92$, $p < .10$ and not by items, $F(1,27) = 1.84$. An inspection of the means shows that this may be due to a small Appropriateness effect for instrument inferences in the naming version. This suggests that listeners accessed linguistic representations when these inferences were made but this is not confirmed elsewhere in the analysis (either by a main effect of Appropriateness in the naming data alone or in any interactions involving type in either the naming or the combined data). The tentative conclusion remains, therefore, that both antecedent assignment for pronouns and instrument inferences require access to a constructed representation.

Conclusion

The studies reported here provide further evidence that the lexical decision/naming task difference can be used as a diagnostic test for different levels of representation in sentence comprehension. Comparing results in the two tasks suggests that drawing an instrument inference and finding an antecedent for a pronoun both involve making links to elements in a constructed representation of the discourse. In addition, the evidence showed that inferences are not routinely made under all conditions of sentence processing. There was no instrument priming in sentences out of context, but, when sentences were presented in contexts that explicitly mentioned the potential instrument, instrument priming did occur.

This research provides a demonstration of how the lexical decision and naming paradigms can be used together to reveal more about the systems involved in sentence comprehension than either task could reveal alone. This methodology should prove valuable in future investigations of different forms of inferential processing.

Table 1

Mean reaction times in ms for experiments 1 (lexical decision) and 2 (naming)

Type of Target

Type of Task	Appropriate (Plausible Instrument)	Inappropriate (Inplausible control)
Lexical Decision	929	926
Naming	604	630
Combined	766	778

Table 2

Mean reaction times in ms for experiments 3 (lexical decision) and 4 (naming)

Type of Target

Pronoun

Type of Task	Appropriate (Plausible Instrument)	Inappropriate (Inplausible control)
Lexical decision	987	1032
Naming	735	739
Combined	861	886

Non-Pronoun

Lexical decision	982	1040
Naming	718	738
Combined	850	889

Pronoun/Non-Pronoun refers to whether or not the sentence contained an anaphor that referred to the instrument.

References

Bobrow, D.G. and Norman, D.A. (1975). Some principles of memory schemata. In D.G. Bobrow and A. Collins (Eds.), Representation and understanding (pp. 131-150). New York: Academic Press.

Corbett, A.T. and Dosher, B.A. (1978). Instrument Inferences in Sentence Encoding. Journal of Verbal Learning and Verbal Behavior, 17, 479-491.

Dosher, B.A. and Corbett, A.T. (1982). Instrument Inferences and Verb Schemata. Memory and Cognition, 10, 531-539.

Fillmore, C.J. (1971). Types of lexical information. In D.D. Steinberg and L.A. Jakobovits (Eds.), Semantics: An Interdisciplinary Reader in Philosophy, Linguistics & Psychology (pp. 370-392). New York: Cambridge University Press.

Hankamer, J. and Sag, I. (1976). Deep and surface anaphors. Linguistic Inquiry, 7, 391-426.

Johnson, M.K., Bransford, J.D. and Solomon, S. (1973). Memory for tacit implications of sentences. Journal of Experimental Psychology, 98, 203-205.

Lucas, M.L., Tanenhaus, M.K., Carlson, G.N., and Senytka, D. (1985, March). The level of representation involved in the processing of anaphoric reference during sentence comprehension. Paper presented at the meeting of the Eastern Psychological Association, Boston, MA.

McKoon, G. and Ratcliff, R. (1981). The comprehension processes and memory structures involved in instrumental inference. Journal of Verbal Learning and Verbal Behavior, 20, 671-682.

Paris, S.G. and Lindauer, B.K. (1976). The role of inference in children's comprehension and memory for sentences. Cognitive Psychology, 8, 217-227.

Schank, R. (1974). Conceptual dependency: A theory of natural language understanding. Cognitive Psychology, 3, 552-632.

Seidenberg, M.S., Waters, G.S., Sanders, M. and Langer, P. (1984). Pre- and postlexical loci of contextual effects on word recognition. Memory and Cognition, 12, 315-328.

Tanenhaus, M.K, Carlson, G.N., and Seidenberg, M.S. (1985). Do listeners compute linguistic representations? In D.R. Dowty, L. Karttunen, and A.M. Zwicky (Eds.), Natural Language parsing: Psychological, Computational, and Theoretical Perspectives, New York: Cambridge University Press.

Semantic Relations, Metonymy, and Lexical Ambiguity Resolution : A Coherence-Based Account

Dan Fass

Rio Grande Research Corridor
Computing Research Laboratory
New Mexico State University
Box 30001, Las Cruces, NM 88003.
CSNET: dan@nmsu

Abstract

An account of coherence is proposed which tries to clarify the relationship between semantic relations, metonymy, and the resolution of lexical ambiguity. Coherence is the synergism of knowledge (synergism is the interaction of two or more discrete agencies to achieve an effect of which none is individually capable) and plays a substantial role in cognition. In the account of coherence, semantic relations and metonymy are instances of coherence and coherence is used for lexical ambiguity resolution. This account of coherence, semantic relations, metonymy and lexical ambiguity resolution is embodied in Collative Semantics, which is a domain-independent semantics for natural language processing. A natural language program called meta5 uses CS; an example of how it discriminates a metaphorical relation is given.

1 Introduction

An account of coherence is proposed which attempts to unpick the relationship between semantic relations, metonymy, and the resolution of lexical ambiguity. Coherence is defined as the synergism of knowledge, where synergism is the interaction of two or more discrete agencies to achieve an effect of which none is individually capable. In the account, semantic relations and metonymy are instances of coherence and coherence is also used in resolving lexical ambiguity. This account of coherence, semantic relations, metonymy and lexical ambiguity resolution is embodied in Collative Semantics, hereafter CS. CS is a semantics for natural language processing which extends the ideas of Preference Semantics (Wilks 1973, 1975a, 1975b, 1978; Fass and Wilks 1983).

To explain our account of coherence, we establish two sets of relationships that involve coherence and then unify those relationships. Section 2 establishes the first relationship which is between coherence, semantic relations, and metonymy. We take the general conception of coherence used in theories of discourse and extend it downwards from the discourse level to the sentence level to argue that semantic relations and metonymies in sentences are instances of coherence.

Section 3 establishes a second relationship which is between coherence and the resolution of lexical ambiguity. We develop a conception of coherence that is grounded in properties of semantic networks which are a common kind of knowledge representation scheme. Two basic kinds of coherence are distinguished that are termed "inclusion" and "distance." This general conception of coherence is extended upwards and it is argued that inclusion and distance underlie two of the main

575

approaches to lexical ambiguity resolution.

The last part of section 3 integrates the two sets of relationships to produce the skeleton of our account of coherence, which is that [1] basic notions of coherence are founded on principles of knowledge representation, [2] semantic relations and metonymy within sentences are instances of coherence, [3] coherence is used for lexical ambiguity resolution, and [4] discourse phenomena are instances of coherence.

Section 4 connects [1] to [2] and [2] to [3] by describing the four components of CS and their interrelationships. The four components are "sense-frames," "collation," "semantic vectors," and "screening." CS is embodied in a natural language program called meta5. An example sentence is given that meta5 analyses. The sentence contains a metaphorical relation and illustrates the interactions between the components of CS. Section 5 provides a brief summary.

2 Coherence, Semantic Relations, and Metonymy

This section selectively surveys the literatures on coherence, semantic relations, and metonymy, and argues that semantic relations and metonymy are cases of coherence.

Coherence is a central notion in theories of discourse (e.g., Van Dijk 1977, Chapter 4; de Beaugrande and Dressler 1981, Chapter V; Van Dijk and Kintsch 1983, Chapter 5; Myers et al, pp.6-8) and truth (e.g., Rescher 1973). In discourse theories, coherence refers to how a discourse "hangs together," "makes sense," or is "meaningful." Discourse theories view the coherence of a discourse as amalgamated from the coherence relations between sentences in that discourse (e.g., Van Dijk, 1977). Little attention is paid by discourse theories to coherence relations that exist within parts of sentences. The coherence relations within a sentence determine the coherence of that sentence. These coherence relations include semantic relations and metonymy.

In our view, semantic relations between terms are complex systems of mappings between descriptions of those terms within some context. This view of semantic relations draws from definitions of metaphor as systems of relationships or "implicative complexes" (Black 1979), mappings (Carbonell 1981), correspondences between domains (Tourangeau and Sternberg 1982), and selective inferences (Hobbs 1983).

Next, we develop some terminology for describing semantic relations. The two terms in a semantic relation are called the **"source"** and the **"target"** (Martin 1985). The source initiates and directs the mapping process, the target has mappings laid upon it, and there is direction from the source towards the target.

We distinguish six types of semantic relation. These are termed literal, metaphorical, anomalous, redundant, inconsistent, and novel relations. Brief definitions of the six semantic relations are now given, together with an example sentence for each relation. These sentences assume a null context in which there are no complicating effects from prior sentences or the pre-existing beliefs of producers or understanders. The definitions of literal, metaphorical, and anomalous semantic relations develop the observation by Katz, Wilks and others that a satisfied selection restriction or preference indicates a literal semantic relation whereas a violated restriction indicates a metaphorical or anomalous semantic relation. The description of redundant, inconsistent, and novel semantic relations is an expansion of Katz and Fodor's ideas on "attribution" (1964, pp.508-509), which were a development of some ideas by Lees (1960). The meta5 program analyses all six sentences.

(1) "The man drank beer."

There is a literal relation between 'man' and 'drink' in (1) because 'drink' prefers an animal as its agent (it is animals that drink) and a man is a type of animal so the preference is satisfied.

(2) "The car drank gasoline." (adapted from Wilks [1978])

By contrast, the semantic relation between 'car' and 'drink' in (2) is metaphorical because 'drink' preferred an animal as its agent but a car is not a type of animal so the preference is violated. However, there is an analogy between animals and cars that is relevant in the context of a sentence about drinking, such as (2). The relevant analogy is that animals drink potable liquids as cars use gasoline.[1]

(3) "The idea drank the heart."

In (3), the semantic relation between 'idea' and 'drink' is anomalous because 'idea' is not an preferred agent of 'drink' and no relevant analogy can be found.

(4) "His wife is married."

The semantic relation between 'wife' and 'married' is semantically redundant in (4) because a wife is by definition a married women so the information that a wife is married is not new information.

(5) "His wife is unmarried."

In (5), the semantic relation between 'wife' and 'married' is inconsistent because the information added by 'unmarried' is incompatible with 'married' in the definition of a wife as a married woman. In out terminology, inconsistent semantic relations include contradictory and contrary ones.[2]

(6) "His wife is young."

Finally, (6) contains a novel semantic relation between 'wife' and 'young' because the information that a wife is young is new information.

Semantic relations are manifestations of coherence because they are the synergism of knowledge. Synergism, remember, is the interaction of two or more discrete agencies to achieve an effect of which none is individually capable. Consider, for example, the metaphorical relation observed in (2) and the relevant analogy that is central to its recognition. The analogy arises from the interaction of three agencies that are 'car' (the surface subject), 'animal' (the expected agent), and 'drink' (the relevant context; also the main sentence verb). The analogy is an effect achieved of which none of the three agencies is individually capable. The analogy is a synergism of knowledge and the metaphorical relation as a whole is a more complex synergism of knowledge.

Another form of coherence apart from semantic relations is metonymy. Metonymy is a nonliteral figure of speech in which the name of one thing is substituted for

---

[1] It has been frequently claimed that the critical match in a metaphorical relation is some analogy or correspondence between two properties (e.g., Wilks 1978; Ortony 1979, p.167; Tourangeau and Sternberg 1982, pp.218-221; Gentner 1983). The importance of relevance has been argued by Tversky (1977) and Hobbs (1983).

[2] Contrary relations exist between terms gradable on some scale, e.g., hot/cold and big/small whereas contradictory relations exist between ungradable terms, e.g., female/male, single/married (Lyons 1963, pp.460-469; Lehrer 1974, p.26). This difference is compatible with the standard philosophical distinction between contraries and contradictories (Lyons 1977, p.272).

that of another related to it (Lakoff and Johnson 1980, pp.35-40). Lakoff and Johnson group individual cases of metonymy into seven general "metonymic concepts" (1980, pp.38-9). One of those metonymic concepts, with example sentences, is :

PRODUCER FOR PRODUCT
> "He bought a *Ford*."
> "I hate to read *Heidegger*."

Our treatment of metonymy distinguishes it from semantic relations but that treatment can only be described very briefly here. Metonymy is treated as a type of inference and metonymic concepts are encoded as **"metonymic inference rules."** Four types of metonymic concepts are currently represented. These are **Part for Whole, Container for Contents, Artist for Artform,** and **Co-Agent for Activity**

This section has developed a conception of coherence that includes semantic relations and metonymy. The next section develops another conception of coherence that is used as a means of resolving lexical ambiguity.

3 Coherence and Lexical Ambiguity Resolution

This section discusses two well known approaches to the resolution of lexical ambiguity and attempts to show how they utilise coherence. We call these approaches the **"inclusion-based"** and **"distance-based"** approaches. Inclusion-based approaches include Katz's semantic theory (Katz and Postal 1964; Katz 1972), Preference Semantics, message passing (Rieger and Small 1979; Small and Rieger 1982), and CS. Distance-based approaches include schemes for spreading activation (Quillian 1968) and marker passing (Charniak 1983; Hirst 1983). Both approaches use semantic networks. Our contention is that each approach is founded on two different basic notions of coherence that exist in semantic networks, that we call **"inclusion"** and **"distance."**

A semantic network is a network with nodes organised as a taxonomy of genus and species terms linked by arcs that have labels denoting class inclusion.[3] In a semantic network, a path between any pair of nodes has two intrinsic properties that are the two basic notions of coherence.

One intrinsic property is the semantic distance, or number of arcs traversed, between the two nodes. For example, 'vehicle' and 'car' have a small distance between them whereas 'animal' and 'car' have a much greater distance between them.

The other intrinsic property is the inclusion relation between the two nodes. For example, the path between network nodes for 'vehicle' and 'car' denotes inclusion because a car is a type of vehicle; on the other hand, the path between 'animal' and 'car' denotes exclusion because a car is not a type of animal.

Both distance and inclusion describe kinds of conceptual relatedness. A short distance indicates close conceptual relatedness (i.e., 'vehicle' and 'car') whereas a long distance indicates remote conceptual relatedness (i.e., 'animal' and 'car'). Inclusion

---

[3] The genus is the name of a class that includes subordinates called the species. A species is distinguished from other species of the same genus by its differentia. Take for example

Car : a vehicle that carries passengers.

The word 'vehicle' serves as the genus term while "that carries passengers" differentiates cars from other species such as buses and motorbikes.

signifies conceptual relatedness (i.e., a 'vehicle' is a 'car') whereas exclusion signifies conceptual unrelatedness (i.e., a 'car' is not an 'animal').

Note that these basic notions of coherence (distance and inclusion) are not explicit in a semantic network but instead that they are a by-product of path building between nodes in that network. In other words, this new conception of coherence is the synergism of knowledge, as was the conception of coherence in section 2. Synergism, once again, is the interaction of two or more discrete agencies to achieve an effect of which none is individually capable. The agencies here are network nodes, the interaction is path building, and the effect achieved is the basic kinds of coherence, i.e., distance and inclusion. Next, it is shown how these two basic kinds of coherence underlie inclusion-based and distance-based approaches to lexical ambiguity resolution.

Inclusion-based approaches to lexical ambiguity resolution are based on the satisfaction and violation of selection restrictions which are called expectations in message passing and preferences in Preference Semantics and CS.[4] The notions of "satisfaction" and "violation" are based on inclusion, which is one of the two basic kinds of coherence. Satisfied selection restrictions, preferences and expectations are all paths denoting inclusion through a semantic network. Conversely, violated selection restrictions, preferences and expectations are all paths denoting exclusion. For example, if a selection restriction was for 'vehicle' then 'car' would satisfy the restriction because a car is a type of vehicle; but if the restriction were for 'animal' then 'car' would cause a violation because a car is not a type of animal.

Distance-based approaches all seek the path with the shortest distance between two nodes in a semantic network, i.e., the second basic kind of coherence. Search is unconstrained except for the ruling out of certain path sequences (Charniak 1983, 1986) and the use of mathematical functions for limiting the length of network paths (Hirst 1983; Charniak 1986).

The conception of coherence developed in this section is grounded in properties of semantic networks. It is argued that distance and inclusion are basic kinds of coherence that are emergent from network paths and underpin two of the main approaches to lexical ambiguity resolution. If this conception of coherence is combined with the conception of coherence from section 2 then our account of coherence is that

[1] basic notions of coherence are founded on principles of knowledge representation (from section 3),

[2] semantic relations and metonymy in sentences are instances of coherence (from section 2),

[3] coherence is used for lexical ambiguity resolution (section 3), and

[4] discourse phenomena are instances of coherence (section 2).

What is missing is the links between [1], [2], [3], and [4]. Section 4, which is on CS (Collative Semantics), attempts to supply the missing links between [1], [2], and [3].

---

[4] The terms 'preference' (Wilks 1975a) and 'expectation' (Schank 1975) highlight different aspects of the use of selection restrictions. Wilks emphasises that selection restrictions may or may not be satisfied, hence the word 'preference'. Schank stresses that selection restrictions are used for top-down prediction, hence the term 'expectation'.

4 Collative Semantics

CS has four components which are **"sense-frames," "collation," "semantic vectors,"** and **"screening."** Sense-frames are the knowledge representation scheme and represent individual word-senses. Collation matches the sense-frames of two word-senses, finds any metonymies, and discriminates the semantic relations between the word-senses as a complex system of mappings between their sense-frames. Semantic vectors represent such systems of mappings produced by collation and hence the semantic relations encoded in those mappings. Screening chooses between two semantic vectors by applying rank orderings among semantic relations and a measure of conceptual similarity, thereby resolving lexical ambiguity.

CS has been implemented in the meta5 program. The meta5 program is written in Quintus Prolog and consists of a lexicon containing the sense-frames of 460 word-senses, a small grammar, and semantic routines that embody collation and screening, the two processes of CS.[5] An example sentence shows how semantic relations are discriminated by meta5, and hence by CS.

(2) "The car drank gasoline."

There is a metaphorical relation between 'car' and 'drink' in (2). Figure 1 shows the sense-frames for car1, drink1, and animal1 which is the agent preference of drink1. Sense-frames are composed of other word-senses that have their own sense-frames, much like Quillian's (1968) planes. There are no semantic primitives in the sense of Schank's (1975) Conceptual Dependency or Wilks' Preference Semantics.[6]

```
                    sf(drink1,
                        [[arcs,
                            [[supertype, ingest1]]],
                        [node2,
                            [[agent,
                                [preference, animal1 ]],
                            [object,
                                [preference, drink1]]]]]).


sf(car1,                                sf(animal1,
    [[arcs,                                 [[arcs,
        [[supertype, motor_vehicle1]]],         [[supertype, organism1]]],
    [node0,                                 [node0,
        [[it1, carry1, passenger1]]]]).         [[biology1, animal1],
                                            [it1, drink1, drink1],
                                            [it1, eat1, food1]]]]).
```

Figure 1. Sense-frames of car1, animal1, and drink1 (verb).

---

[5] Meta5's grammar is adapted from the grammar of XTRA (Huang 1985), an English-Chinese machine translation program also written in Prolog. XTRA is the latest in a succession of programs that originate from Boguraev's (1979) natural language analyser that was written in LISP. Huang's and Boguraev's programs use versions of Preference Semantics.

[6] One of the main claims of CS to be a semantics is its treatment of semantic primitives. See Fass (1986) for details.

The node part of a sense-frame is the differentia that provides a description of the word-sense represented by the sense-frame that differentiates it from other word-senses. Sense-frame nodes for nouns (node-type 0) resemble Wilks' (1978) pseudo-texts. They contain lists of two-element and three-element lists called **"cells."** Each cell expresses a piece of functional or structural information and can be thought of as a complex semantic feature or property of a noun.

The arcs part of a sense-frame contains a labelled arc to its genus term (a word-sense with its own sense-frame). The most common arc labels describe types of class inclusion such as 'supertype' that denotes membership of a class of individuals by a class of individuals and 'superinstance' that denotes membership of an individual within a class of individuals. Together, the arcs of all the sense-frames comprise a densely structured semantic network of word-senses called the **"sense-network."** This general architecture of a semantic network with frame-like structures as nodes is similar to many frame-based and semantic network-based systems, such as Quillian's (1968) memory model, schema theory (Norman and Rumelhart 1975), KRL (Bobrow and Winograd 1977), FRL (Roberts and Goldstein 1977), KLONE (Brachman 1979), and frail (Wong 1981).

Collation is the second component of CS. In (2), collation matches what was expected (animal1) against what is there in the sentence (car1) so the sense-frames of animal1 and car1 are matched together. Collation finds a system of multiple mappings between those sense-frames, thereby discriminating the metaphorical relation between animal1 and car1. Collation contains a graph search algorithm and a frame-matching algorithm. The graph search algorithm distinguishes five types of sense-network path. Two path-types denote inclusion; three denote exclusion. These path-types are the basic mappings produced by collation. The frame-matching algorithm matches the sets of cells from two sense-frames. Seven types of cell match are distinguished. These cell matches are more complex mappings that are built from sense-network paths and hence also embody inclusion.

First, collation finds a preference or expectation violation : a car is not a kind of animal. Next, collation matches the inherited cells of animal1 and car1. What is "relevant" in the present context is the action of drinking because that is what (2) is about. Collation then inherits the cells of animal1 down the sense-network and searches those cells for one that refers to drinking.

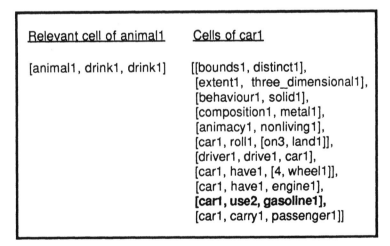

Relevant cell of animal1	Cells of car1
[animal1, drink1, drink1]	[[bounds1, distinct1], [extent1, three_dimensional1], [behaviour1, solid1], [composition1, metal1], [animacy1, nonliving1], [car1, roll1, [on3, land1]], [driver1, drive1, car1], [car1, have1, [4, wheel1]], [car1, have1, engine1], **[car1, use2, gasoline1],** [car1, carry1, passenger1]]

Figure 2. Match of relevant cell from animal1 with cells from car1

It finds a relevant cell [animal1, drink1, drink1] and seeks a match for that cell against the list of inherited cells for car1 (see figure 2). It finds a match with [car1, use2, gasoline1] (highlighted in figure 2) which is the relevant analogy that animals drink potable liquids as cars use gasoline. This relevant analogy is crucial to recognising the semantic relation between car1 and the drink1 as metaphorical.

Non-relevant cells of animal1	Non-relevant cells of car1	Cell matches
[[bounds1, distinct1], [extent1, three_dimensional1], [behaviour1, solid1],	[[bounds1, distinct1], [extent1, three_dimensional1], [behaviour1, solid1],	3 identical cell matches
[composition1, flesh1], [animacy1, living1],	[composition1, metal1], [animacy1, nonliving1],	2 sister cell matches
[animal1, eat1, food1], [biology1, animal1]]		2 distinctive cells of animal1
	[car1, roll1, [on3, land1]], [driver1, drive1, car1], [car1, have1, [4, wheel1]], [car1, have1, engine1], [car1, carry1, passenger1]]	5 distinctive cells of car1

Figure 3. Matches of non-relevant cells from animal1 and car1.

Finally, collation matches together the remaining non-relevant cells of animal1 and car1 (see figure 3) because such matches may figure in the aptness of a metaphor.[7] The cell [car1, use2, gasoline1] has been removed to prevent it from being used a second time. All of these matches made by collation are recorded in a semantic vector which figure 4 shows.

```
[preference,
    [[path_type,
        [0, 0, 0, 0, 1]],        First array :
                                  exclusive sense-network path
    [cell_matches,
        [[relevant,
            [0, 0, 1, 0, 0, 0, 10]],    Second array :
                                         analogical match of relevant cell
        [non_relevant,
            [0, 3, 2, 0, 0, 2, 5]]]]]]    Third array :
                                           matches of non-relevant cells
```

Figure 4. Semantic vector for a metaphorical semantic relation.

Semantic vectors are the third component of CS. Semantic vectors are a form of representation, along with sense-frames; but sense-frames represent knowledge, whereas semantic vectors represent coherence. Semantic vectors are therefore a kind of **"coherence representation."** A semantic vector is a data structure that contains

---

[7] Tourangeau and Sternberg (1982) have claimed that the more distance between the conceptual domains of the source and target, the better the metaphor. We have developed a measure that tests this claim.

nested labels and ordered arrays structured by a simple dependency syntax. The columns of the arrays record different kinds of mapping between sense-frames.

The crucial elements of the metaphorical relation in (2) were the preference violation and the relevant analogy. In figure 4, the preference violation has been recorded as the 1 in the first array (fifth column) and the relevant analogy is the 1 in the second array (third column). The aptness of a metaphor may be determined by the matches of non-relevant cells. In figure 4, those matches are recorded in the third array (compare with figure 3). Other semantic relations have different semantic vectors.

Together, the labels and arrays of a semantic vector specify the synergistic interaction of sources of knowledge in a semantic relation; in other words, the labels and arrays represent coherence. To see this, recall once again that we define coherence as the synergism of knowledge, and that synergism is the interaction of two or more discrete agencies to achieve an effect of which none is individually capable. In the semantic vector of figure 4, the discrete agencies are three knowledge sources ([1] the surface subject car1; [2] the agent preference animal1; and [3] drink1, the relevant context and also the sense of the main sentence verb), the interaction of those sources is the systems of mappings, and the effect achieved is the metaphorical semantic relation in (2).

The fourth component of CS is the process of screening. During analysis of a sentence constituent, meta5 computes a semantic vector for pairwise combinations of word-senses. These word-sense combinations are called **"semantic readings"** or simply "readings." Each reading has an associated semantic vector. Screening chooses between two semantic vectors and hence their attached semantic readings. Rank orderings among semantic relations are applied. In the event of a tie, a measure of conceptual similarity is applied.

The detail can now be supplied for the missing links from our skeleton account of coherence in section 3. The first missing link was from [1] to [2], namely how inclusion is used in the recognition of semantic relations and metonymy (distance is not used in CS). Our explanation is that collation discriminates semantic relations and performs metonymic inferencing by finding multiple sense-network paths between two sense-frames.

The second missing link was from [2] to [3], which is how semantic relations and metonymy are used in the resolution of lexical ambiguity. Our explanation is that semantic relations are represented in semantic vectors as systems of mappings and that screening uses those mappings to apply rank orderings of semantic relations and, if necessary, a measure of conceptual similarity to choose between semantic readings and thereby resolve lexical ambiguity (metonymy helps to establish semantic relations and does not figure directly in lexical ambiguity resolution).

The missing links between [1] and [3] are filled by the four components of CS. What unifies our account of coherence is the treatment of semantic relations that collation discriminates ([1] to [2]) and semantic vectors represent ([2] to [3]).

5 Summary

This paper has attempted to describe the relationship between semantic relations, metonymy, and lexical ambiguity resolution. Coherence was used as an explanatory concept that organised that relationship. CS was introduced as a theoretical framework in which the role of coherence in the relationship between semantic relations, metonymy, and lexical disambiguation was made more concrete. Finally, an example from meta5 was used to make the description of CS and the relationship between all four phenomena (coherence, semantic relations, metonymy, and lexical disambiguation) more concrete still.

Coherence is the main theoretical focus of CS, together with semantic primitives. Collation produces coherence, semantic vectors represent coherence, and screening uses coherence. In CS, the representation and processing of knowledge (sense-frames and collation) are distinguished from the representation and processing of coherence (semantic vectors and screening).

There are many phenomena that a coherence-based approach such as CS can explore. We have argued that semantic relations and metonymy are manifestations of coherence. Coherence exists between linguistic structures of all sizes. We have argued that coherence exists within sentences and is prominent in approaches to lexical ambiguity resolution. Coherence also exists between sentences -- it is basic to theories of discourse. Coherence merits thorough investigation as it appears to play a substantial role in cognition, not just semantic relations, metonymy and lexical ambiguity resolution.

6 Acknowledgements

The work reported here is funded by the New Mexico State Legislature, administered by its Scientific Technical Commission as part of the Rio Grande Research Corridor.

7 References

de Beaugrande, Robert, and Dressler, Robert (1981) *Introduction to Text Linguistics*, New York, Longman.

Black, Max (1979) More about Metaphor. Andrew Ortony (Ed.) *Metaphor and Thought*, London : Cambridge University Press, pp. 19-43.

Bobrow, Daniel G., & Winograd, Terry (1977) An Overview of KRL, A Knowledge Representation Language. *Cognitive Science*, 1, pp. 3-46.

Boguraev, Branimir K. (1979) Automatic Resolution of Linguistic Ambiguities. Technical Report No.11, University of Cambridge Computer Laboratory, Cambridge, England.

Brachman, Ronald J. (1979) On The Epistemological Status of Semantic Networks. In Nicholas V. Findler (Ed.) *Associative Networks : Representation and Use of Knowledge By Computers*, New York : Academic Press, pp. 3-50.

Carbonell, Jaime G. (1981) Metaphor : An Inescapable Phenomenon in Natural Language Comprehension. Research Report CMU-CS-81-115, Dept. of Computer Science, Carnegie-Mellon University.

Charniak, Eugene (1983) Passing Markers : A Theory of Contextual Influence in Language Comprehension. *Cognitive Science*, 7, pp.171-190.

Charniak, Eugene (1986) A Neat Theory of Marker Passing. *Proceedings of the 5th National Conference on Artificial Intelligence (AAAI-86)*, Philadelphia, Pa., pp. 584-588.

Van Dijk, Teun A. (1977) *Text and Context : Explorations in the Semantics and Pragmatics of Discourse*, New York : Longman.

Van Dijk, Teun A., and Kintsch, Walter (1983) *Strategies of Discourse Comprehension*, New York : Academic Press.

Fass, Dan C. (1986) Collative Semantics : An Approach to Coherence. Memorandum MCCS-86-56, Computing Research Lab., New Mexico State University, New Mexico.

Fass, Dan C. & Wilks, Yorick A. (1983) Preference Semantics, Ill-Formedness and Metaphor. *American Journal of Computational Linguistics*, 9, pp. 178-187.

Gentner, Dedre (1983) Structure Mapping : A Theoretical Framework for Analogy. *Cognitive Science*, 7, pp. 155-170

Hirst, Graeme John (1983) Semantic Interpretation against Ambiguity. Technical Report CS-83-25, Dept. of Computer Science, Brown University.

Hobbs, Jerry R. (1983) Metaphor Interpretation as Selective Inferencing : Cognitive Processes in Understanding Metaphor (Part 1). *Empirical Studies of the Arts*, 1, pp. 17-33.

Huang, Xiuming (1985) Machine Translation in the SDCG (Semantic Definite Clause Grammars) Formalism. *Proceedings of the Conference on Theoretical & Methodological Issues in Machine Translation of Natural Languages*, Colgate University, New York, USA, pp. 135-144.

Katz, Jerrold J. (1964) Analyticity and Contradiction in Natural Language. In Jerry A. Fodor & Jerrold J. Katz (Eds.) *The Structure of Language : Readings in the Philosophy of Language*, Englewood Cliffs, NJ : Prentice-Hall, pp. 519-543.

Katz, Jerrold J. (1972) *Semantic Theory*, New York : Harper International Edition.

Katz, Jerrold J., & Fodor, Jerry A. (1964) The Structure of A Semantic Theory. In Jerry A. Fodor & Jerrold J. Katz (Eds.) *The Structure of Language : Readings in the Philosophy of Language*, Englewood Cliffs, NJ : Prentice-Hall, pp. 479-518.

Katz, Jerrold J., & Postal, Paul (1964) *An Integrated Theory of Linguistic Description*, Cambridge, Mass. : MIT Press.

Lakoff, George, & Johnson, Mark (1980) *Metaphors We Live By*, London : Chicago University Press.

Lees, R.B. (1960) The Grammar of English Nominalizations. *International Journal of American Linguistics*, 26, 3.

Lehrer, Adrienne (1974) *Semantic Fields and Lexical Structure*, North Holland : Amsterdam.

Lyons, John (1963) *Structural Semantics*, Blackwells : Oxford.

Lyons, John (1977) *Semantics Volume 2*, Cambridge : Cambridge University Press.

Martin, James H. (1985) Knowledge Acquisition though Natural Language Dialogue. *Proceedings of the 2nd Annual Conference on Artificial Intelligence Applications*, Miami, Fl.

Myers, Terry, Brown, Keith, & McGonigle, Brendan (1986) Introduction : Representation and Inference in Reasoning and Discourse. In Terry Myers, Keith Brown, & Brendan McGonigle (Eds) *Reasoning and Discourse Processes*, Academic Press : London, pp.1-12.

Norman, Donald, A., Rumelhart, David, E., and the LNR Research Group (1975) *Explorations in Cognition*, San Francisco : W.H. Freeman.

Ortony, Andrew (1979) Beyond Literal Similarity. *Psychological Review*, 86, pp. 161-180.

Quillian, M. Ross (1968) Semantic Memory. In Marvin Minsky (Ed.) *Semantic Information Processing*, Cambridge, Mass : MIT Press, pp. 216-270.

Rescher, Nicholas (1973) *The Coherence Theory of Truth*, Clarendon Press : Oxford.

Rieger, Chuck, & Small, Steve (1979) Word Expert Parsing. Technical Report TR-734, Dept. of Computer Science, University of Maryland.

Roberts, R. Bruce, & Goldstein, Ira P. (1977) The FRL Manual. MIT AI Memo 409.

Schank, Roger C. (1975) The Structure of Episodes in Memory. In Daniel G. Bobrow & Allan Collins (Eds.) *Representation and Understanding*, New York : Academic Press, pp. 237-272.

Small, Steve, & Rieger, Chuck (1982) Parsing and Comprehending with Word Experts (A Theory and its Realization). In Wendy G. Lehnert & Martin H. Ringle (Eds.) *Strategies for Natural Language Processing*, Hillsdale, NJ : Erlbaum Assocs., pp. 89-147.

Tourangeau, Roger, & Sternberg, Robert J. (1982) Understanding and Appreciating Metaphors. *Cognition*, 11, pp. 203-244.

Wilks, Yorick A. (1973) An Artificial Intelligence Approach to Machine Translation. In Roger C. Schank & Kenneth M. Colby (Eds.) *Computer Models of Thought and Language*, San Francisco : W.H. Freeman, pp. 114-151.

Wilks, Yorick A. (1975a) A Preferential Pattern-Seeking Semantics for Natural Language Inference. *Artificial Intelligence*, 6, pp. 53-74.

Wilks, Yorick A. (1975b) An Intelligent Analyser and Understander for English. *Communications of the ACM*, 18, pp. 264-274.

Wilks, Yorick A. (1978) Making Preferences More Active. *Artificial Intelligence*, 10, pp. 1-11.

Wong, Douglas (1981) On the Unification of Language Comprehension with Problem Solving. Technical Report CS-78, Department of Computer Science, Brown University.

Thematic Roles in Language Processing[1]

Michael K. Tanenhaus
Curt Burgess
Susan Hudson D'Zmura
University of Rochester
Greg Carlson
University of Iowa

1 Abstract

We present some ideas about how thematic roles (case roles) associated with verbs are used during on-line language comprehension along with some supporting experimental evidence. The basic idea, following Cottrell (1985), is that all of the thematic roles associated with a verb are activated in parallel when the verb is encountered. In addition, we propose that thematic roles are provisionally assigned to arguments of the verbs as soon as possible, with any thematic roles incompatible with such an assignment becoming inactive. Active thematic roles that are not assigned arguments within the sentence are entered into the discourse model as unspecified entities or addresses. In our first experiment we show that temporary garden-paths arise when subjects initially assign the wrong sense to a verb as in *Bill passed the test to his friend*, but not when subjects initially assign the wrong role to the noun phrase, as in *Bill loaded the car onto the platform*. This prediction follows directly from our assumptions. In our second experiment we show that definite noun phrases without explicit antecedents in the preceding discourse can be more readily integrated into a preceding discourse when they can be indexed to an address created by an open thematic role.

2 Introduction

Although case roles have played an important part in linguistic theory since the seminal work of Fillmore (1968), and they are frequently incorporated into AI models of natural language understanding, there has been little experimental work that examines their role in human language processing. In this paper we present experimental evidence suggesting that thematic roles (case roles) associated with verbs, play an important role in language comprehension. The idea that we will be exploring is that thematic roles can provide a mechanism whereby the parser can make early semantic commitments, yet quickly recover from the inevitable missasignments that occur as a consequence of the local indeterminacy that is characteristic of natural language. We further suggest that thematic roles provide a mechanism for interaction among the parser, the discourse model, and real-world knowledge.

---

[1]This research was supported in part by NSF grant BNS-8217378.

3 Motivation

Our motivation for exploring these ideas comes from a confluence of findings from the language comprehension and word recognition literature. First, research on language processing suggests that the language processor makes extremely early commitments, with each word being fully interpreted and integrated with preceding context as it is processed (Marslen-Wilson, 1975). Secondly, the processor appears to compute structures serially (Frazier, 1978; Ford et al, 1983; Frazier & Rayner, 1982). Evidence comes from studies demonstrating local increases in processing complexity when the parser pursues an analysis that turns out to be inconsistent with the remainder of the sentence or resolves an ambiguity in a manner that is contextually inappropriate. Yet feedback from context clearly enables the parser to rapidly recover from and perhaps occasionally avoid these local garden-paths. This picture suggests that the parser computes structures serially but also has rapid access to alternative structures.

4 Multiple Codes

Our central assumption is that thematic roles associated with a verb are activated in parallel. Placing the parallelism in the lexicon is attractive because there is a large body of research on lexical processing demonstrating that multiple codes associated with a word become activated regardless of processing context. For example, multiple senses of ambiguous words are initially accessed even in the presence of contextual bias (Seidenberg et al, 1982; Swinney, 1979; Tanenhaus et al, 1979). Moreover, a number of lexical and sublexical effects such as the word superiority effect, effects of spelling-sound regularity, and of orthographic regularity in visual word recognition can be explained elegantly by the assumption that there is a great deal of bottom-up parallel activation, with incompatible representations inhibiting one another (McClelland & Rumelhart, 1981; Seidenberg, 1985). When representations are compatible, however, multiple codes remain active (Seidenberg & Tanenhaus, 1979; Tanenhaus, Flanigan & Seidenberg, 1980).

Frazier and colleagues (Frazier, 1986; Rayner, Carlson & Frazier, 1984) have argued that the vocabulary of thematic relations is shared by the parser, discourse model, and world knowledge. They have proposed a thematic processor which provides a channel of communication among these domains. In light of the growing evidence for multiple code activation in lexical processing, for strong lexical effects in parsing, and for on-line serial commitment and rapid local garden-path recovery, it seems reasonable to seek a mechanism whereby lexical structures can help to organize a parse, guide local garden-path recovery, and communicate with the discourse model. Thematic roles provide a promising candidate for such structures.

5 Representational and Processing Assumptions

In order to motivate our experiments it will be necessary to briefly outline some representational and processing assumptions. These assumptions are presented in more detail in Carlson and Tanenhaus (1987).

5.1 Representational Assumptions

We assume that a verb meaning is represented in terms of a *core meaning* (sense) and an associated set of thematic roles. Following Carlson (1984) we assume that the main function of thematic roles is to relate the arguments of a verb to the core meaning in semantic interpretation. We also assume that an integral part of the interpretation of a discourse is a mental model, or discourse model, which represents an ongoing record of the discourse (Heim, 1982; Johnson-Laird, 1983; Kamp, 1979). We also make the following standard assumptions about the mapping between thematic roles and arguments.[2]

1. Every argument of a given verb is assigned a thematic role.

2. No argument is assigned more than one thematic role.

3. Every argument of a verb is assigned a unique thematic role.

We finally assume that the set of arguments that are assigned thematic roles by a verb are the subject of the sentence and the subcategorized phrases in the verb phrase.[3]

5.2 Processing Assumptions

In addition, we make the following processing assumptions:

1. Lexical access makes available the core meaning (sense) of a verb and the thematic roles associated with the sense. For an ambiguous verb, all the senses will be activated in parallel, along with the set of thematic roles associated with each sense (see Cottrell, 1985, for a similar proposal).

2. Only the sense of the verb that is contextually appropriate (or, in the absence of biasing context, the most frequent sense) remains active, along with its thematic roles.

3. Thematic roles are provisionally assigned to arguments of the verb as soon as possible; any active thematic roles incompatible with such an assignment becomes progressively less active.

4. Any active thematic roles not assigned to an argument remain as open thematic roles in the discourse model, appearing as free variables or unspecified *addresses* in the model. Thus, we do not assume that every active thematic role assigned by a verb ends up assigned to the meaning of some syntactic constituent.

6 Experiment 1: Sense and Thematic Role Ambiguities

The model we have sketched above predicts a processing difference for sentences exhibiting temorary ambiguities such as those illustrated in (1) and (2).

---

[2]See Carlson and Tanenhaus (1987) for some qualifications.

[3]The phrasing here is a matter of convenience. It is actually the meanings themselves that are assigned thematic roles. See Carlson (1984) and Carlson and Tanenhaus (1987) for discussion.

1. Bill *passed* the test to his friend.
2. Bill *loaded* the car onto the platform.

In sentence (1), *passed* is ambiguous between two senses, roughly, earning a *passing grade*, and *hand over*. In sentence (1), the phrase which follows the verb, *the test*, biases the *grade* sense. Readers should experience a small garden-path when this sense later turns out to be incorrect. This follows from the assumption that lexical access will make available multiple senses of such a word as *pass*, but only the contextually most appropriate (or, in absense of context, most frequent (Simpson & Burgess, 1985)) sense will remain active and the others become unavailable (see Simpson, 1984, for a review of relevant literature). When a reader or hearer initially selects the wrong sense of an ambiguous verb, reinterpretation would require retrieving the alternative sense. This should take time and processing resources. In sentence (2), the noun phrase *the truck* could either be the location of the loading, or what is being loaded. When the wrong thematic assignment is initially made, thematic reassignment should be relatively cost-free because: (a) the core meaning of the verb remains constant, and hence the verb's lexical entry need not be reopened; (b) the alternative thematic roles are generally active and available; and, (c) the syntactic-thematic mappings provide explicit information about how roles are to be assigned so only a limited domain of information needs to be reexamined. Thus, thematic roles allow the processor to make early commitments without undue cost. The null hypothesis is that both ambiguities are really just sense ambiguities, and hence are not fundamentally distinct.

6.1 Experimental Methodology

6.1.1 Stimuli

The experimental materials are illustrated in (3) and (4), for sense and thematic role ambiguities, respectively.

3. a. Bill *passed* the test to his complete surprise.
3. b. Bill *failed* the test to his complete surprise.
3. c. Bill *passed* the test to the person behind him.
3. d. Bill *handed* the test to the person behind him.
4. a. Bill *loaded* the truck with bricks.
4. b. Bill *filled* the truck with bricks.
4. c. Bill *loaded* the truck onto the ship.
4. d. Bill *drove* the truck onto the ship.

In examples (3a) and (3c), different senses of *pass* are selected by the final disambiguating phrase; disambiguation does not take place until after presentation of the direct object noun phrase. Examples (3b) and (3d) are control sentences using unambiguous verbs that have core meanings related to the appropriate sense in the ambiguous version of the sentence. The sentences of (4) repeat that same pattern for the thematic ambiguities; (4a) and (4c) involve temporary ambiguity of thematic assignment to the direct object, to be disambiguated by the final constituent; (4b) and (4d) serve as unambiguous controls. Sets of sentences similar to those in (3) and (4) were constructed for 16 verbs with different senses and 16 verbs for which the thematic assignment of the following noun phrase is ambiguous. The four sentences for each verb were counterbalanced across four lists, with each subject seeing only one list.

6.1.2 Procedure

The sentences were displayed on a CRT and the subjects' task was to decide as quickly as possible whether or not the sentence *made sense*. We assumed that subjects would initially select the incorrect verb sense or thematic assignment on approximately half the trials where temporary ambiguity is possible. If incorrect sense selection results in a garden-path once disambiguating information to the contrary arrives, this should be reflected either in fewer sentences with sense ambiguities being judged to make sense or in longer reaction times to comprehend these sentences, all relative to control sentences. In contrast, thematic role ambiguities should result in much weaker garden paths. Filler trials included some sentences that were incongruous, such as: *Several people borrowed ideas under the bed.*

6.2 Results and Discussion

Data from 28 subjects are presented in Table 1, which displays reaction time (in msec) to the sentences judged to make sense, and the percentage of sentences judged to make sense. Separate ANOVAs were conducted on the judgment data and on the reaction time data. The ANOVA on the judgment data revealed a significant interaction between Verb Type (Sense versus Thematic) and Ambiguity (Ambiguous or Control conditions) with subjects as a random factor, ($F(1, 24) = 6.17$, $p = .02$). The interaction obtained because sentences with sense ambiguities were less likely to be judged to make sense than their controls ($F(1, 13) = 6.07$, $p = .029$), whereas sentences with thematic role ambiguities were not ($F(1, 15) = .02$, $p = .88$).

The reaction time results were less clear cut. Both sense and thematic role ambi-

Ambiguity Type	Verb Type	Control
SENSE	2445 (77)	2290 (94)
THEMATIC	2239 (92)	2168 (93)

Table 1: Latencies (in msec) for sentences judged to make sense. Percent judged to make sense in parentheses.

guities took longer to comprehend when they were judged to make sense than their controls ($F(1, 24) = 14.69$, $p = .0008$), and although the effect was numerically larger for the sense ambiguities, the interaction between type of verb and ambiguity was not significant. However, the effect of ambiguity was significant only for the sense ambiguities in the item analysis ($F(1, 13) = 6.73$, $p = .02$). The reaction time results were more consistent with the judgment results, when using our intuitions,

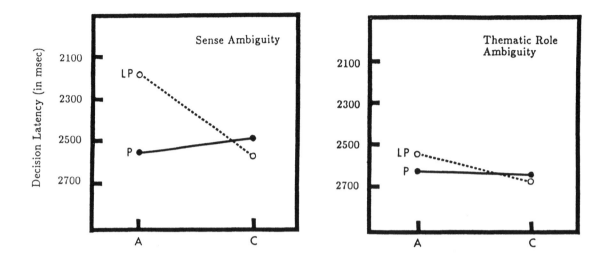

Figure 1: Latencies for sentences judged to make sense when preferred (P) or less preferred (LP) sense or thematic assignment turned out to be correct (A = ambiguous verb; C = unambiguous control verb).

we divided the ambiguous sentences into those in which the preferred and less preferred initial sense or thematic assignment is correct and incorrect (see Figure 1). We see that the sense ambiguities are more difficult than their controls only when the less preferred sense turns out to be correct. This is reflected in the interaction between Sense (preferred and less preferred) and Ambiguity ($F(1, 13) = 6.79$, $p = .02$). This interaction is not present with the thematic role ambiguities ($F(1, 15) = 1.19$, $p = .29$).

7 Experiment 2: Open Thematic Roles

The results of Experiment 1 demonstrate a processing difference between thematic role ambiguties and sense ambiguties, thus lending support to our assumption that thematic roles can be distinguished from core meanings. Experiment 2 tested the idea that open thematic roles are entered into the discourse model as unspecified discourse entities. Open thematic roles obtain when an active thematic role is not filled by an argument. For example, the verb *load* has three thematic roles associated with it: Agent, Theme and Location. In a sentence such as (5)

5. Bill *loaded* the car.

one of the thematic roles, most likely the Theme, would be left open. If this idea is correct, then listeners should not have difficulty interpreting a subsequent sentence that begins with a definite noun phrase as long as the noun phrase could plausibly fill the open role. Thus sentence (6)

6. *The suitcases* were heavy.

should be relatively natural when it follows (5) in a discourse because *the suitcases* can fill the open Theme role. In contrast the same sentence should be more difficult to understand when there is not an open thematic role, because the reader will have to make an inference to build a bridge from the noun phrase to the context (Haviland & Clark, 1974).

7.1 Experimental Methodology

7.1.1 Stimuli

We constructed sixteen sets of two sentence discourses in which a target sentence beginning with a definite noun phrase (8) was preceded by either a context sentence that introduced an open thematic role that the NP could fill as in (7b) or a sentence that created a plausible context for the target sentence but that did not leave an open role, as in (7a).

7. a. John had difficulty running fast to catch his plane
7. b. John had difficulty loading his car.
8. *The suitcases* were very heavy.

The two context sentences for each set were counterbalanced across two presentation lists.

7.1.2 Procedure

Subjects were presented with the context sentence on a CRT. When they pressed a button indicating that they had read and understood the context sentence, they were presented with the target sentence. Their task was to decide whether or not the target sentence made sense given the context.

7.2 Results

Target sentences were judged to make sense more often when the context sentence introduced an open thematic role than when it did not, (97% vs 84%; $F(1, 22) = 5.76$, $p = .025$). Latencies to target sentences judged to make sense were faster when the context introduced an open thematic role than when it did not, (1628 msec vs 1847 msec; $F(1, 22) = 6.32$, $p = .019$). Thus the results strongly supported the hypothesis.

8 Discussion

The studies that we have presented provide initial encouragement for the framework that we have been developing. Two interesting, and to our knowledge, previously unobserved phenomena were predicted and confirmed experimentally. It is important to acknowledge, however, that thematic roles are not the only possible explanation for our results. It is difficult, for example, to rule out the hypothesis that thematic roles are not distinct grammatical entities, but rather just aspects of verb meanings as Ladusaw and Dowty (1987) have argued. A great deal of future research will be necessary before it becomes clear which of these frameworks will provide the more interesting insights about comprehension processes.

References

Carlson, G. (1984). Thematic roles and their role in semantic interpretation. *Linguistics, 22*, 259-279.

Carlson, G., & Tanenhaus, M. K. (1987). Thematic roles and language comprehension. To appear in W. Wilkens (Ed.), *Thematic relations*. New York: Academic Press.

Cottrell, G.W. (1985). *A connectionist approach to word sense disambiguation.* (Tech. Rep. No. TR154). Rochester, NY: University of Rochester, Department of Computer Science.

Fillmore, C. J. (1968). The case for case. In E. Bach & R. T. Harms (Eds.), *Universals in Linguistic Theory.* New York: Holt, Rinehart & Winston.

Ford, M., Bresnan, J., & Kaplan, R. (1983). A competence-based theory of syntactic closure. In J. Bresnan (Ed.), *The mental representation of grammatical relations* (pp. 727-796). Cambridge, MA: MIT Press.

Frazier, L. (1978). *On comprehending sentences: Syntactic parsing strategies.* University of Connecticut doctoral dissertation.

Frazier, L. (1986). Theories of sentence processing. To appear in J. Garfield (Ed.), *Modularity in knowledge representation and natural language.* Cambridge, MA: MIT Press.

Frazier, L., & Rayner, K. (1982). Making and correcting errors during sentence comprehension: Eye movements in the analysis of structurally ambiguous sentences. *Cognitive Psychology, 14*, 178-210.

Haviland, S. E.,& Clark, H. H. (1974). What's new? Acquiring new information as a process in comprehension. *Journal of Verbal Learning and Verbal Behavior, 13*, 512-521.

Heim, I. (1982). *The semantics of definite and indefinite noun phrases.* University of Massachusetts doctoral dissertation.

Johnson-Laird, P. (1983). *Mental models.* Cambridge, MA: Harvard University Press.

Kamp, H. (1979). Events, instants, and temporal reference. In R. Bauerle, U. Egli & A. von Stechow (Eds.), *Semantics from different points of view* (pp. 376-417). Berlin: Springer-Verlag.

Ladusaw, W., & Dowty, D. (1987). *Toward a non-grammatical account of thematic roles.* To appear in W. Wilkens (Ed.), *Thematic relations.* New York: Academic Press.

McClelland, J., & Rumalhart, D. (1981). An interactive activation model of context effects in letter perception: Part 1. An account of basic findings. *Psychological Review, 88*, 375-405.

Rayner, K., Carlson, M., & Frazier, L. (1984). The interaction of syntax and semantics in sentence processing: Eye movements in the analysis of semantically biased sentences. *Journal of Verbal Learning and Verbal Behavior, 22*, 358-374.

Seidenberg, M. S. (1985). Constraining models of word recognition. *Cognition, 14*, 169-190.

Seidenberg, M. S., & Tanenhaus, M. K. (1979). Orthographic effects in rhyme monitoring. *Journal of Experimental Psychology: Human Learning and Memory, 5*, 546-554.

Seidenberg, M. S., Tanenhaus, M. K., Leiman, J. M., & Bienkowski, M. (1982). Automatic access of the meanings of ambiguous words in context: Some limitations of knowledge-based processing. *Cognitive Psychology, 14,* 489-537.

Simpson, G. B. (1984). Lexical ambiguity and its role in models of word recognition. *Psychological Bulletin, 96,* 316-340.

Simpson, G. B., & Burgess, C. (1985). Activation and selection processes in the recognition of ambiguous words. *Journal of Experimental Psychology: Human Perception and Performance, 11,* 28-39.

Swinney, D. A. (1979). Lexical access during sentence comprehension: (Re)Consideration of context effects. *Journal of Verbal Learning and Verbal Behavior, 18,* 645-659.

Tanenhaus, M. K., Flanigan, H., & Seidenberg, M.S. (1980). Orthographic and phonological code activation in auditory and visual word recognition. *Memory & Cognition, 8,* 513-520.

Tanenhaus, M. K., Leiman, J. M., & Seidenberg, M. S. (1979). Evidence for multiple stages in the processing of ambiguous words in syntactic contexts. *Journal of Verbal Learning and Verbal Behavior, 18,* 427-440.

Can Synchronization Deficits Explain Aphasic Comprehension Errors?

Helen M. Gigley

Department of Computer Science
University of New Hampshire
Durham, NH 03824

Henk J. Haarmann[1]

ABSTRACT

The context dependent nature of language processing requires the synchronization of several subprocesses over time. One claim which follows is that de-synchronization is likely to disturb language processing.

Aphasia is a language disorder which arises following certain types of brain lesion. Many theories of aphasia are competence based theories and do not address aspects of performance which can be affected under conditions of processing degradation. The discussion in this paper will focus on de-synchronization as a possible explanation for aphasic language comprehension problems.

HOPE, a computer model for single sentence comprehension, provides a tool to systematically study the effects of various hypothesized de-synchronization problems on different language processing levels and on overall comprehension performance. HOPE includes a neural-like architecture that incorporates a grammar which functions in a predict/feedback manner. It illustrates one way in which serial-order input can map into synchronous, parallel subprocesses that can effectively produce normal sentence comprehension performance.

Using HOPE, the study of explicit de-synchronization effects on a cover set of stimuli sentences suggests error patterns to be sought in neurolinguistic evidence. Within a subset of a cover set of stimuli, simulation results from a slowed propagation lesion experiment will demonstrate how timing problems can result in observed aphasic comprehension performance.

1. INTRODUCTION

The context dependent nature of language processing requires synchronization of several subprocesses over time. One claim which follows is that de-synchronization as an example of a processing degradation is likely to cause problems with language processing.

Following brain lesion, usually on the left side, there is often a disturbance of language facility called aphasia. While it is known that brain process subserves language, until recently, theories of aphasia have been chiefly defined within competence theories of language instead of processing ones. (For some exceptions see. Von Monakow, 1914; Kolk, Van Grunsven, and Keyser 1985; Kolk and Van Grunsven, 1985.)

Neurolinguistic studies of aphasia, focusing on linguistic competence theories, analyze observed language difficulties in aphasia as the result of knowledge dissolution and/or rule degradation. In contrast, a processing based theory, the focus of this paper, attends to competence issues as studied in behavioral research, but also employs constraints developed within architectural considerations of the processing mechanism underlying the behavior.

There are several possible reasons for the general emphasis on competence theory motivated studies in neurolinguistics as opposed to processing degradation motivated ones.

(1)   It is difficult to keep track of the on-line effect that a particular de-synchronization problem has on different levels of processing by just using paper and pencil. To give an example of a particular de-synchronization problem: How does an hypothesized slowing of propagation of activity after phonetic recognition affect the processing on different levels, such as the syntactic and the semantic level?

---

[1] Department of Psychology. University of Nijmegen. Montessorilaan 3. 6525 HE Nijmegen. The Netherlands.

The initial development of the reported research was supported though an Alfred P. Sloan Foundation Grant entitled "A Training Program in Cognitive Science" at the University of Massachusetts at Amherst. Continuing development is supported through a Biomedical Research Support Grant at the University of New Hampshire and through a grant from the Netherlands American Fulbright Committee and from IBM Netherlands NV.

(2) Human performance studies can only induce synchronization problems by manipulating the input to the language system. Internal manipulations, such as the slowing of activity propagation, as previously mentioned, are impossible. But such slowing might explain part of the differences in the results found in comprehension studies of aphasia which manipulate the rate of sentence input. (Brookshire, 1971; Blumstein, Katz, Goodglass, Shrier and Dworetski, 1985; Laskey, Weidner, and Johnson, 1976; Liles and Brookshire, 1975).

HOPE, a model of single sentence comprehension, was explicitly designed to provide a conceptual tool that enables on-line study of the effects of various hypothesized de-synchronization problems (Gigley, 1982; 1983; 1985b; 1986). Previous models of pathological language comprehension (Baron, 1975; Lavorel, 1982, Marcus, 1982) implemented to varying degrees, are rule based and do not include facility to manipulate the synchronization of information flow in the manner included in HOPE.

Using generally observed, independently affected competence knowledge as the basis of representations, HOPE was designed to be able to study how processing changes could affect overt behavior. It was not designed to specifically model, ''aphasia type x'' from the literature. In direct contrast to the role computational models assume in research in general,, the nature of the approach in HOPE is to study processing degradations, forming hypothesized patient performance profiles and to go to the literature or to design specific studies to determine the best fit to the behavioral data.

The best fit of simulation results that has been found to date appears in the data of studies of agrammatic patients. Kean (1985) describes, *Agrammatism is typically defined as a disorder of sentence production involving the selective omission of function words and some grammatical endings on words.* An extensive discussion of the clinical picture can be found in Kean (1985) and issues of knowledge loss theories related to it in Kolk and Van Grunsven (1985). Various difficulties which completely define any classification are discussed by Caplan (1985.)

HOPE's unique role as a model of language processing that includes issues related to neurolinguistic and neural-like processes will be discussed. To illustrate the specific features of HOPE's design which address the processing motivated lesions, we will describe how HOPE's lesionability is dependent on its neural-like architecture and on the time-coordinating function of its grammar. Then we will proceed by describing various possible ways to disturb the synchronization of HOPE's sub-processes. While HOPE also includes ways to lesion its competence factors, such as grammatical knowledge, or interpretation ability, these will not be discussed in this paper.

A brief description of a simulation run of HOPE will be used to demonstrate how de-synchronization can produce results which can be attributed to competence-based analyses. While other de-synchronization simulations can be shown to produce distinctly different results (Gigley, 1982; 1983; 1986) their scope cannot be described in detail here. Furthermore, within the simulations, we will discuss the need for a cover set of sentence stimuli and will discuss the implications of it on the design of suitable test stimuli during validation.

Finally, based on an hypothesized patient profile, the role of such modelling in providing a mechanism for exploring processing motivated theories of language performance will be shown. The final claim is that through the evolution of such modelling attempts a better understanding of the neural mechanisms subserving language will be gained.

2. THE LEVELS OF REPRESENTATION IN HOPE

The basic unit of computation is based on a lexical item. An example of an open class item is given in Figure . (Closed class items have a less complex distributed form.) Each lexical item includes a phonetic representation, all meaning representations associated with it, an interpretation representation called pragmatic that reflects the correct meaning for the sentential context, and morphological information appropriate to the phonetic form.

Further competence knowledge includes a meaning for each syntactic category type that defines a down-line predicted category type and an associated semantic category type that occurs following correct interpretation and composition. Interpretation of a syntactic type is ''computed'' within an interpretation function that is defined for each syntactic type.

The lexical item is represented in a connected hypergraph formalism (Berge, 1970.) Spaces or hypergraphs are shown as enclosed areas in Figure 1. The neural-like computations over these units is described

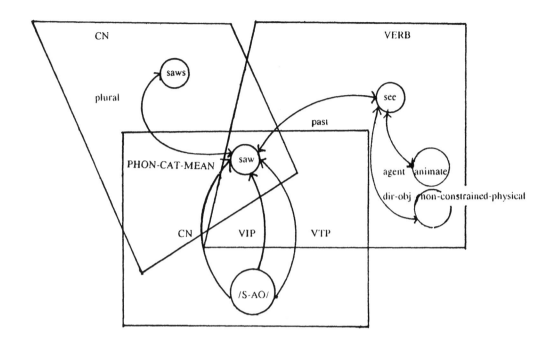

Figure 1: Example open-class word representation.

Open-class word representations are distributed across spaces, PHONETIC, PHON-CAT-MEAN, and morphologically related VERB and CN spaces. The order of activation of this information across time during processing is tuned to produce normal performance results.

below.

Briefly, sentences are input as phonetic word strings one word at a time. Then all meanings for each homophone are activated simultaneously (Seidenberg, Tanenhaus, Leiman and Bienkowski, 1982; Swinney, 1979) through spreading activation. These meanings interact with subsequent input producing the final state of simulated sentence comprehension. This is represented as the binding determinations for agent and object relationships on the interpreted verb form of the sentence and is activated as a subgraph in a pragmatic space.

At this point, one should note that each of units in the representations as illuustrated in Figure 1 receive activity in a neural-like manner. They are not all activated simultaneously, but over time as part of the process of syntactically motivated meaning disambiguation. Furthermore, each unit may lie in more than one space in the graph. Spaces currently denote, phonetic, meaning(PHON-CAT-MEAN), grammar, and pragmatic interpretations of the units. The hypergraph representation captures the redundancy supported in neurophysiology of multiple meaning/ multiple effect as each unit when active can affect different connected information in different ways depending on the spaces in which it lies. The details of this are relevant to processing but not critical to the focus of this paper. (For more detail see Gigley, 1982; to appear; and Gigley and Boulicaut, 1985.)

3. HOPE'S NEURAL-LIKE ARCHITECTURE

This section will describe HOPE's neural-like architecture as far as is necessary to acquaint the reader with the general notion of using a grammar as a coordinator of activation among levels of representation. For a complete understanding the interested reader is referred to Gigley (1982; 1983)

HOPE's architecture is neural-like. It is connectionist in the sense that all knowledge is represented as a distributed connected network of units that are activated over time. Each unit is intended to correlate with a pattern of activation over more abstract neural-like units (Hinton, 1981; Hinton, McClelland, and

Rumelhart, 1986; Smolensky, 1986.) A unit is **not** a grandmother cell. At an AI level, each lexical item is a collection of units that together are the "lexical item."(See Figure 1.)

HOPE's basic information unit is defined as a threshold unit with memory whose activity value is being manipulated by several automatically applied processes:

(1) Decay: All active units that do not receive additional input have their activity value exponentially reduced.

(2) Change of State computations: A unit can be in four different states and changes state automatically after it reaches threshold and fires.

    a) Short-term-memory state. The initial state of activation for a unit that has not reached threshold.

    b) Firing state. The state when the unit has reached or surpassed threshold and fires.

    c) Refractory state. When a unit has just fired, it is set to a state in which it cannot affect the computation.

    d) Post-refractory state. The state the unit is in after having been in the refractory state. Currently, the unit is subject to a different decay rate than before firing.

(3) Firing-information-propagation. Firing state propagates activity to other units. The propagated information is either excitatory or inhibitory. Excitatory information serves two different functions through feedback or feedforward (meaning spread) (Collins and Loftus, 1975; Hinton, 1981; Quillian, 1968/1980).

Activity input to units occurs over time across lexical item representations. Units are passive receivers of input; there is no self-modulation. The units that comprise each lexical item set are activated in a fixed-order over time using a spreading activation paradigm. The next section will illustrate this spreading activation. In Contrast, each time a unit fires, it dynamically computes where its information is sent. The effect of the activity on the units which receive inhibitory or excitatory input may differ from context to context.

An assumption shared with the connectionist approach is that units do not have the ability to look around to see if certain preconditions necessary for their action are fulfilled (Feldman and Ballard, 1982; Gigley and Lavorel, 1985; McClelland and Rumelhart, 1986.) Such preconditions viewed within an AI sense, can be thought of as patterns to be matched such as "DET followed by a CN."

HOPE does not compute rule interpretations of the occurrence of a rule pattern as suggested in competence theories. Any results in HOPE instead depend on activities arriving at the appropriate information within certain time-bounds. Propagation results are not all or non in effect but must have concurrency with other related activity propagation from other inputs down-line to attain a result similar to rule application. Evidence that a rule or pattern has occurred arises in the time-sequence trace of activity propagations during a simulation. A rule as stated above, becomes a prediction of the anticipated down-line information in a subsequent time interval.

Because of the predictions and feedback when they are confirmed, there is no formal syntactic structure built during the processing. As each unit fires it affects subsequent represented aspects of the lexical item in multiple ways. These generate a compositional interpretation for the input string (a predicate formalism) which captures the dynamic bindings of agent and object for the simple S-V-O sentences over which HOPE is currently being run.

This is in direct contrast to other connectionist-type models dealing with any of the normal processing issues in HOPE. These models include syntactic preprocessing or filtering (Cottrell, 1985; Waltz and Pollack, 1985; McClelland and Kawamoto, 1986.) HOPE claims this is not necessary and furthermore, that syntactic and semantic processing must occur in a fully integrated on-line fashion. Simultaneous on-line syntactic disambiguation and semantic interpretation is coordinated through the categorial grammar formalism, described later.

Computations are made over the entire graph representation of the current state of the solution (all active nodes of information) in a time-synchronized, lock-step fashion. A process interval is defined as one application of all processes to all information units. There is a separate interval driven process that determines the time-course of the introduction of the phonetically encoded words of a sentence. After each process time-interval, one can study the state of the entire graph (the process trace.)

The processes are governed by a set of modifiable parameters that determine rate of decay, amount of inhibition and excitation, height of the firing threshold, time lapse of activity propagation, and the initial activity levels of the short-term-memory, refractory and post-refractory states. Furthermore, a word-time-interval parameter determines when a new word is introduced. The processes are synchronized or externally tuned by the model user to define the **normal process model.** The modifiable parameters must be set in such a way that for a complete cover set of sentences, each will be correctly interpreted. A complete **cover set** of sentences is the minimal set of sentences which capture one example of each correct syntactic sentence the model is defined to process and include all combinations of syntactic form-class combinations for each position of the input. The parameter values are set relative to each other.

Once tuned, no computations such as waiting for signals, are necessary for synchronization. Furthermore, relative changes in the parameter values from their synchronized settings define processing lesion conditions. Note that usually only one modification of a parameter is made per simulation experiment although multiple "lesions" are possible in the model. Of critical import is the synchronization role of the grammar discussed next.

4. THE COORDINATING ROLE OF HOPE's GRAMMAR

The function of HOPE's grammar is to coordinate the on-line interactions of the compositional meaning representations of a sentence with the incoming words (Gigley, 1982, 1983, 1985a). This is achieved by using the activity states of the grammar to control feedforward and feedback relations which are encoded as part of the meaning of syntactic categories. Furthermore, the grammar currently serves as a filter of the amount of activity propagated in these relations. The goal of this section is to introduce this general notion of a predict-feedback cycle by means of a specific example. Later on we will see how de-synchronization can cause comprehension problems by disrupting this predict-feedback cycle and the events that happen as a consequence of it.

To give an example, the meaning of the lexical category, determiner, is encoded as: **DET** := **TERM/CN.** This is called a derived specification and is to be read as follows: A determiner (DET) PREDICTS a common noun (CN) and FORMS a TERM after successful compositional interpretation with the CN meaning. The interpretation computation is part of the category meaning as previously described.

The definition is drawn from categorial grammar (Ajdukiewicz, 1935, Lewis, 1972)), which makes the assumption that correct meaning composition can be assumed simultaneously with syntactic well-formedness. In categorial grammar, for example, a DET followed by a CN is semantically equivalent to a TERM, which can also be a syntactic type. Thus, the categories of both phrases THE SAW (DET CN) and PAUL (TERM) are syntactically and semantically TERMs which denote specific instances.

As an example, Figure 2 shows the time course of the semantic composition of the phrase /TH-UH S-AO/ (THE SAW) relative to the incoming phonetically encoded words using the determiner meaning DET := TERM/CN. The numbered E's in Figure 2 refer to events that will now be described. For clarity, decay and state computations (except firing) will not be discussed.

In the first time-interval: /TH-UH/ fires representing PHONETIC recognition (E-1). This has two effects:

SPACE	time1	time2	time3	time4	time5
PRAGMATIC				$E-7$	$E-12$
GRAMMAR		$E-3$		$E-10$	
PHON-CAT-MEAN	$E-2$		$E-5, E-6$	$E-8, E-9, E-11$	
PHONETIC	$E-1$		$E-4$		

Figure 2: Example of processing over time.

Numbered events (E-n) which occur during a simulation are described in the text. Time intervals are labelled, timeN

(1) Spreading activation to PHON-CAT-MEAN (lexical meaning) occurs in the same time-interval. This activates the category-meaning entries of the recognized phonetic word. /TH-UH/ has only one meaning entry: TH-UH <-- DET - THE. DET is its category and THE is a spelling representation for its meaning in PHON-CAT-MEAN (E-2).

(2) The predictive meaning of the category aspect of the recognized phonetic word's meaning, if it has one, will be activated in the GRAMMAR space the next time-interval. This is an example of fixed-time spreading activation. In this example the predictive meaning of DET in the grammar, CN, will be activated at time-interval 2 (E-3).

In the third time-interval, /S-AO/ fires at the PHONETIC level(E-4). Again, in the same time-interval spreading activation activates the category-meaning representations of /S-AO/: S-AO <-- VTP-SAW; <-- VIP SAW and <-- CN - SAW, where VTP stands for a transitive verb phrase and VIP for an intransitive verb phrase (E-5). The category-meaning <-- CN - SAW that is associated with /S-AO/, receives, like all other CN-meaning pairs, additional activity, because it is predicted in the grammar. This causes it to fire (E-6). Firing has three effects that will be visible in the fourth time-interval.

(1) The second aspect of each category meaning, the category specific interpretation function, is applied. The interpretation function for CN will interpret the lexico-graphically encoded meaning "SAW" which is associated with the firing CN as a generic NOUN on the PRAGMATIC level (E-7 in time4.)

(2) The activity of categories and meanings competing with <-- CN - SAW will be inhibited (E-8).

(3) Because CN is a predicted category and active in the GRAMMAR, all lexical entries whose associated category predict a CN (here only DET) will receive additional activation by feedback (E-9) and the activity of the predicted category CN will be dampened in the grammar (E-10). Note that the CN itself does not predict another category in the simulation defined model.

Also, in the fourth time-interval, the category-meaning pair associated with /TH-UH/ <-- DET - THE fires (E-11) after having received additional activation due to feedback. Because it does not have competing meanings or categories, no inhibition occurs. Its only effect will be the application of the interpretation function associated with the category DET, which checks to see if there is only one CN available for composition on the PRAGMATIC level and attaches it to a TERM indicating that it has been interpreted as a specific instance (E-12 in time5.)

HOPE allows a user to define his own specific model within the constraints of HOPE computations, his lexicon, category-meaning set, grammatical interactions, and associated interpretation computations for each category. In the 1982 model the following grammar was defined over a lexicon of items having different degrees of syntactic form-class ambiguity: 1) DET := TERM/CN; 2) VIP := SENTENCE/ENDCONTOUR; 3) VTP := VIP/TERM.

The derived specifications for DET and VTP (transitive verb) have been adopted from a categorial grammar of English, whereas the ENDCONTOUR, a symbol for the end of sentence intonation contour, in VIP's derived specification (intransitive verb), triggers the completion of a sentence and the stopping of processing. Among other things, firing of VTP results in the interpretation of the object case relation of the verb (dir-obj) and firing of VIP in the interpretation of the agent case relation of the verb. It is at this time, after firing, during interpretation, that the morphological and tense constraints between the verb and the lexical items are activated. The composition of a VTP with a TERM phrase is semantically equivalent to a VIP category as can be concluded from the derived specification.

Because of the time-sequence activation of parallel predictions and their affect on down-line input, the semantic composition of a phrase will only succeed if the individual units involved in its composition receive a sufficient amount of activity (inhibitory or excitatory) at the right moment in time. Here moment of time really is a time bounded window of one or more computed time-intervals.

Some of the modifiable set of parameters determine the relevant time-course employed for any given specific model defined. Other parameters affect the degree of interaction due to firing and decay.

5. DE-SYNCHRONIZATION OF THE PREDICT-FEEDBACK CYCLE

Any deviation from normal fine-tuning can be viewed as de-synchronization. Studying the effect of de-synchronization across the complete cover set of sentences provides the basis for defining hypothesized aphasic patient profiles. These can be used to better define and understand aspects of performance problems in aphasia.

Examples of de-synchronization effects are:

(1) Increasing the decay rate of all units may result in an activity level that is insufficient for threshold firing. The result of non-firing may be incomplete or erroneous interpretation.

(2) Increasing the rate of new word introduction. can cause incorrect down-line syntactic disambiguation. If a word or phrase associated with a predicted category comes in before the predict category is active in GRAMMAR it will not fire.

A more explicit encoding of the second lesion, achieved by slowing the effective propagation of spreading activation to grammatical meaning(s) (Haarmann, 1987), will be used here to illustrate how timing issues can be a factor of aphasic performance. Then the observed simulated performance will be compared to several studies of agrammatic aphasic performance to show where there is support for the hypotheses and to point out where additional studies must be done.

The effects of different de-synchronization lesions can be distinguished by running simulations using an entire cover set of sentences under particular ``lesion`` conditions. For example, for the correct (S-V-O) syntactic sentence form of; DET CN VTP DET CN, one must include sentences with lexical items having form-class combinations that match exactly, ie that are unambiguously of the correct form class, and also:

DET	CN	CN	DET	CN
	VIP	VTP		VIP
DET	CN	CN	DET	CN
	VIP	VTP		VIP
	VTP			
DET	CN	CN	DET	CN
	VIP	VIP		VIP
	VTP	VTP		...etc.

Table 1 shows the syntactic form of several of a cover set of sentences which are currently defined in HOPE. At least two examples for each form are given. The first one contains unambiguous lexical items. The second and subsequent sentences (4c and 4d) contain different degrees of lexical form-class ambiguities.

For each ``lesion`` simulation, one can define a patient profile that summarizes the performance pattern on the whole cover set of sentences. Even though only a very limited domain of linguistic constructions is covered (simple declaratives, S-V and S-V-O sentence types) the patient profiles appear to be suitably fine-grained to allow distinctions between various ``lesion`` conditions (Gigley, 1982; 1983; 1985b; 1986.) The patient profiles of the increased decay and the slowed propagation ``lesions`` are described in Gigley (1985c; 1986). Here we want to draw the attention to an interesting difference we found between the effect of slowed propagation on a cover set of sentences with and without syntactic ambiguities under the ``lesion`` condition of slowed propagation only.

6. AMBIGUITY AND SLOWED PROPAGATION

During simulation runs where the length of time to propagate activity from a syntactic category to its meaning is increased, with cover sets of sentences with and without syntactic lexical ambiguities, we have found differences that suggest how syntactic lexical ambiguity can affect language understanding. Instead of describing the differences for the whole cover set of sentences, this section will contrast the final processing states for the set of sentences of Table 1 in the ``slowed propagation lesion`` state. The final states of interpretation for each are summarized in Tables 2 and 3. Table 2 contains the results of the unambiguous sentences while Table 3 shows the results of the ambiguous ones. All syntactic form class sets are derived from definitions in Webster (1981).

Because of the levels of representation and the remaining activity of active units in them, HOPE provides an hypothesized patient profile across each level. Here, the discussion will focus on only the PRAGMATIC or interpretation level results. Be cautioned that an hypothesized patient profile requires a complete cover set of sentences for analysis and we are only dealing with a subset here. A full cover set is necessary to mathematically define all possible outcomes based on syntactic interactions and note specific patterns that

603

(1)	TERM	VTP	TERM	
	a. Paul	slapped	John.	
	b. Paul	saw	John.	

	DET	CN	VTP	TERM
(2)	DET	CN	VTP	TERM
	a. The	girl	slapped	John.
	b. The	girl	saw	Paul.

	TERM	VTP	DET	CN
(3)	TERM	VTP	DET	CN
	a. Paul	slapped	the	girl.
	b. Paul	saw	the	girl.

	DET	CN	VTP	DET	CN
(4)	DET	CN	VTP	DET	CN
	a. The	girl	slapped	the	boy.
	b. The	girl	saw	the	boy.
	c. The	girl	saw	the	seal.
	d. The	girl	saw	the	building.

Table 1: Examples from a syntactic cover set of sentences.

Note: (a) sentences are unambiguous in each lexical item; (b) sentences contain lexical ambiguities. Sentences (4c) and (4d) contain more complex combinations of lexical ambiguities than their syntactic counterparts in sentences (4a) and (4b).

		CORRECTLY INTERPRETED CONSTITUENTS AND BINDINGS			
	|AGENT	|OBJECT	|VERB	|CONSTITUENT|	
SENTENCE					
(1a)	|PAUL	|JOHN	|SLAPPED			correct
(2a)	|*absent	|PAUL	|SLAPPED			
(3a)	|*absent	|*absent	|*absent	|TERM-PAUL		
(4a)	|*absent	|THE BOY	|SLAPPED			

Table 2: Summary final states of interpretation for unambiguous sentences.

In the table, *absent indicates there was no interpreted bound referent for the given role as intended from the input. Sentence 3a shows that while the proper name referent is understood, there is no verb understood and hence no bindings occur.

produce them.

The patient profile based on the information in Tables 2 and 3 will necessarily omit aspects of the profile definition due to its incompleteness. To give an idea of such a profile in this limited context, one can state that comprehension ability is inconsistent across an S-V-O analysis of sentence structure. Based on the simulation, one expects some sentences to be interpreted correctly, but not all. Evidence that this occurs in agrammatic aphasics has been reported in Schwartz, Saffran, and Marin, (1980) and related affects in agrammatic production have been reported for Dutch patients, in Kolk and Van Grunsven (1985).

A more fine-grained observation is that sentences which have agents and objects that are both proper names can be correctly understood and bound in some syntactic contexts, but not all (Gigley, 1985b; 1986). Compare sentences (1a) and (1b) where the verb syntactic-form-class ambiguity differs.

The simulation also suggests that certain error types will occur within certain syntactically ambiguous contexts. For one error type, where a noun meaning for an intended verb occurs as the final interpretation, recently presented evidence has been reported for French in the performance of a French aphasic (Hannequin, Deloche, Branchereau, and Nespoulous, 1986.) Such findings have not been clinically studied for

	CORRECTLY INTERPRETED CONSTITUENTS AND BINDINGS			
	AGENT	OBJECT	VERB	CONSTITUENT	
SENTENCE					
(1b)	JOHN	*absent	SAW-VIP	TERM-PAUL	reversal?
(2b)	*absent	PAUL			
		THE SAW	SAW-VTP		
(3b)	*absent	*absent	*absent	TERM-PAUL	
(4b)	*absent	*absent	*absent	TERM-THE SAW	
(4c)	*absent	*absent	*absent	TERM-THE SAW	
(4d)	non-morpho-logical agreement	*absent	BUILDING-VIP	TERM-THE SAW	

Table 3: Summary final states of interpretation for lexically ambiguous sentences.

Lexical ambiguity affects the ability of the hypothesized patient to understand. Often, only a noun will be understood. Often, it is not a noun which was intended by the ''speaker.'' Sentence, (4d), shows two misinterpretations, one with respect to a noun referent and the other, a misinterpretation of an intended noun as a verb. Of additional interest is the fact that these meanings are semantically related.

English to date. It is through the enumeration of these contexts and corresponding error hypotheses and the ability to manipulate and study them that HOPE contributes a new dimension in the study of natural language processing.

In addition, within this simple lesion context, one sentence, (1b), demonstrates an incomplete interpretation which includes reversal of agent binding from the order presented in the sentence. These last two aspects of performance have been noted in clinical evaluations without any supported explanation. We suggest, based on the simulation results, that timing difficulties not directly observable nor manipulable, can provide an explanation of the agrammatic patient performance problems. For several misinterpreted sentences, on analysis of the over-time trace of the simulations (not provided due to space constraints) shows the cause of the misinterpretation to be that the predict category in GRAMMAR, CN, is activated too late for the CN. (See sentences 4b,c,d.) Slowed propagation can thus lead to the non-interpretation of certain sentence elements as well as misinterpretation of the intended syntactic sense of others. These ''lesion'' results suggest that neurolinguists, aphasiologists and clinicians should be careful in considering syntactic lexical ambiguities which might cause non-obvious difficulties in processing. It also demonstrates the import of the selectional constraints or the set of possible answers for the task and the recorded observations.

7. HOPE's UNIQUE PLACE IN ARTIFICIAL APHASIA

This section will discuss several characteristics that give HOPE a unique place in developing a processing motivated theory of normal natural language comprehension and comprehension in aphasia: dynamic lesionability and time-driven parallelism.

7.1. Dynamic Lesionability

Dynamic lesionability refers to the ability in HOPE to manipulate the dynamics of the process without rewriting or redesigning the system. It makes lesionability a factor which is separable from stored knowledge issues. The point is that one does not need to assume that a lesion causes a loss of some sort of knowledge or rule. It is not necessary that lesions affect competence defined in the linguistic sense.

The results of the de-synchronization lesions that were induced by changing the fine-tuning of HOPE's control process parameters seem to offer a conceptual justification of the importance that neuropsychologists attach to temporal constraints in brain processing (Von Monakow, 1914; Goldstein, 1948; Lenneberg, 1967; Luria, 1970; and Ojemann, 1983.) Jackson (1965/1884) was among one of the first neurologists to point out that brain lesions do not necessarily lead to a loss of knowledge. Almost a century later Wood (1978/82) and Gigley (1982,1983) provided computer simulations of this idea, although in different clinical contexts. Recently, within adaptation theory the importance of temporal constraints has been discussed by Kolk and van Grunsven (1985).

7.2. Time-driven Parallelism

A number of arguments have been put forward that emphasize the need and existence of parallel processing in intelligent systems (see for instance: Fahlman, 1982; Feldman and Ballard, 1982; Marslen-Wilson and Tyler, 1980; McClelland and Rumelhart, 1981; Scholes, 1978; Waltz and Pollack, 1985). But after choosing parallel processing one still needs to decide how to conceptualize it. The inclusion of lesionability directly affects the conceptualization of HOPE's parallelism.

HOPE's conceptualization of parallelism is depicted in Figure 3. The Figure shows a simplified abstraction from HOPE's real architecture, which was defined earlier. Spaces can be thought of as perspectives that provide a context for the interpretation of the knowledge they contain. They do not imply a rigid hierarchy of processing levels, but represent different viewpoints on representations activated over time.

We have previously discussed the lock-step parallel application of the basic computations of the HOPE architecture which "cut" through spaces. (See Figure 3). Simultaneously, HOPE's time-driven parallelism includes a third dimension.

One can study the events which occur within a space over time for all spaces simultaneously. The events which occur within a space over time take the appearance of separate independent processes within one type of knowledge.

Contrary to some other parallel approaches to natural language processing (Cottrell, 1983; 1985; McClelland and Kawamoto, 1986; Scholes, 1978; Waltz and Pollack, 1985), HOPE does not implement the processes within an individual space over time as separate independent processes. HOPE assumes no level specific processes. Instead, the processes are unit specific and homogeneous across all levels of representation. Level specific processes are not directly programmed but depend on the time-coordination of the results of all processes applied over the entire graph of information, and can be observed within one space representation over time.

This section has described dynamic lesionability and its role in processing models of natural language comprehension. It has also illustrated the HOPE conceptualization of parallelism. It shows that there may be many levels of parallelism in parallel processing models of cognition and that some apparent parallel effects may be the result of computations which are non-specific to those effects.

8. CONCLUSION

By virtue of its neural-like parallel architecture, which employs internal synchronization and time-driven parallelism, HOPE provides a tool to systematically study the effects of various de-synchronization problems on processing over time. De-synchronization problems occur as a consequence of a disturbance of the time-coordinating role of HOPE's grammar. By changing the values of the process parameters several lesions can be defined although only one is discussed in detail here. Lesions that imply the violation of temporal constraints in language processing appear to affect sentences differently depending on whether the sentences contain syntactic lexical ambiguities or not and to what degree the ambiguities exist.

The problem of comprehension may be characterized in terms of Karl Lashley's problem of serial order of behavior (1967): From a serially encoded input that is coming in on-line, a simultaneous meaning representation has to be constructed. Using HOPE, as a model of how this "construction" may be viewed, this paper demonstrates the vital role that synchronization plays in such a process. Furthermore, it demonstrates that synchronization problems, especially problems with temporal constraints, can affect this construction and thus be a cause of comprehension performance problems of aphasics.

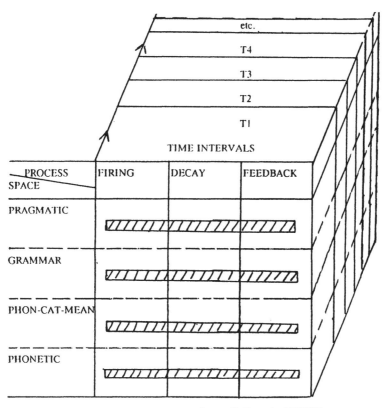

Figure 3: Overview of parallelism in HOPE.

To each unit in each space, each process is applied at the same time. Not all control processes of HOPE are shown: feedforward, inhibition, excitation, meaning spread, and change-of-state are not depicted.

Note: ———— = process boundary.

  ----- = space boundary.

  ⬛ = one unit.

9. REFERENCES

Ajdukiewicz, D. Die syntactische konnexitat. 1935. Translated as "Syntactic Connection" in **Polish Logic**, S. McCall, Oxford, 1967, 207-231.

Baron, R.J. Brain architecture and mechanisms that underlie language: an information processing analysis. In S. Harnad, H. Skelis, and J. Lancaster (Eds.), **Origin and evolution of language and speech**, The New York Academy of Sciences, 1975, 280, 240-256.

Berge, C. **Graphes et Hypergraphes**. Paris: DUNOD, 1970.

Blumstein, S.; Katz, B.; Goodglass, H.; Shrier, R. and Dworetsky, B. The Effects of Slowed Speech on Auditory Comprehension in Aphasia. **Brain and Language**, 24, 1985, pp.246-265.

Brookshire, R. Effects of trial time and inter-trial interval on naming by aphasic subjects. **Journal of Communication Disorders**, 1971, 3, 289-301.

Caplan, D. Syntactic and Semantic Structures in Agrammatism, in Kean, M.L. (Ed.) **Agrammatism**. Academic Press, 1985.

Collins, A.M., and Loftus, E.A. A spreading activation theory of semantic processing. **Psychological Review**, 1975, 82(6), 407-428.

Cottrell, G. W. **A Connectionist Approach to Word Sense Disambiguation**. Unpublished Ph.D. Dissertation, Department of Computer Science, University of Rochester, 1985.

Fahlman, S.E. Three flavors of parallelism. **Proceedings of the 4th National Conference of the Canadian Society of Computer Studies of Intelligence**, 1982.

Feldman, J.A., and Ballard, D.H. Connectionist models and their properties. **Cognitive Science**, 1982, 6, 205-254.

Gigley, H.M. **Neurolinguistically Constrained Simulation of Sentence Comprehension: Integrating Artificial Intelligence and Brain Theory**. Unpublished Ph.D. Dissertation, University of Massachusetts, Amherst, 1982.

Gigley, H.M. HOPE--AI and the dynamic process of language behavior. **Cognition and Brain Theory**, 1983, 6(1), 39-88.

Gigley, H.M. Grammar viewed as a functioning part of a cognitive system. **Proceedings of the 23rd Annual Meeting of the ACL**,

607

Chicago, July, 1985a.

Gigley, H.M. Computational Neurolinguistics -- What is it all about? **IJCAI-85 Proceedings**, Los Angeles, August, 1985b.

Gigley, H.M. Studies in Artificial Aphasia -- Experiments in Processing change. Baltimore, November, 1985c. **Proceedings, Ninth Annual Symposium on Computer Applications in Medical Care; Selected from conference to appear in IEEE Journal of Computer Methods and Programs in Biomedicine, 1986.**

Gigley, H. M. Lexical Ambiguity Resolution in Aphasia, Tech. Rep. 86-37, Department of Computer Science, University of New Hampshire, Durham, NH 03824; Draft of chapter for Cottrell, G.W., Small, S.L., and Tanenhaus, M.K. (Eds.) **Lexical Ambiguity Resolution in the Comprehension of Human Language**, to appear

Gigley, H.M. and Boulicaut, J.-F. Grasper: A Grpah Processing Tool for Knowledge Engineering in Cognitive Modelling, **Proceedings COGNITIVA 85**, Paris, 1985.

Gigley, H.M., and Lavorel, P.M. Computational Neurolinguistic Modelling Integrating "Natural Computation" Control with Performance Defined Presentations. CS-85(c)-26, Technical Report University of New Hampshire, Durham, NH.

Goldstein, K. **Language and Language Disturbances.** New York, Grune and Stratton, 1948.

Haarmann, H. J. **A computer simulation of syntactic understanding in normals and aphasics: slowed activity propagation.** MS Thesis, Cognitive Science, Psychological Laboratory, University of Nijmegen, The Netherlands, 1987.

Hannequin, D.; Deloche, G.; Branchereau, L. and Nespoulous, J.-L. Noun-Verb Disambiguation in French and Sentence Comprehension in Broca's Aphasics. Paper presented at the Academy of Aphasia, Nashville, Tennessee, 1986.

Hinton, G.E. Implementing semantic nets in parallel hardware. In G.E. Hinton and J.A. Anderson (eds.), **Parallel models of Associative Memory**. Lawrence Erlbaum Associates Publishers, 1981.

Hinton, G.E.; McClelland, J. L.; and Rumelhart, D. E. Distributed Representations, in J.L. McClelland and D. E. Rumelhart (eds.) **Parallel Distributed Processing**, MIT Press, 1986.

Jackson, J.H. The Croonian Lectures on the Evolution and Dissolution of the Nervous System. Reprinted in: R.J. Herrnstein and E.G. Boring (Eds.), **A Source Book in the History of Psychology**, Cambridge: Harvard University Press, 1965/1884.

Kean, M.L. (Ed.) **Agrammatism.** Academic Press, 1985.

Kolk, H.H.J., Van Grunsven, M.J.F. Agrammatism as a variable phenomenon. **Cognitive Neuropsychology**, 1985, 2(4), 347-384.

Kolk, H.H.J., Van Grunsven, M.J.F., and Keyser, A. On parallelism between production and comprehension in agrammatism. In M.L. Kean (Ed.), **Agrammatism**. New York: Academic Press, 1985.

Lashley, K.S. 1967. The problem of serial order of behavior. In L. Jeffress (Ed.), **Cerebral Mechanisms of Behavior**, Hafner Publishing Company, 1967, pp 112-146.

Laskey, E.Z., Weidner, W.E., and Johnson, J.P., Influence of linguistic complexity, rate of presentation, and interphrase pause time on auditory verbal comprehension of adult aphasic patients, **Brain and Language**, 3, 1976, pp. 386-396.

Lavorel, P.M. Production Strategies: A systems approach to Wernicke's aphasia. In M.A. Arbib, D. Caplan, and J. Marshall (Eds.), **Neural Models of Language Processes**. Academic Press, 1982.

Lenneberg, E.H **Biological foundations of language.** New York: Wiley and Sons, 1967.

Lewis, D. General semantics. In Davidson and Harmon (eds.), **Semantics of Natural Language**, 1972, 169-218.

Liles, B.Z., and Brookshire, R.H., The effects of pause time on auditory comprehension of aphasic subjects, **Journal of Communication Disorders**, 8, 1975, pp. 221-235.

Luria, A.R. **Traumatic Aphasia.** The Hague: Mouton, 1970.

Marcus, M.P. Consequences of functional deficits in a parsing model: Implications for Broca's aphasia. In M.A. Arbib, D. Caplan, and J.C. Marshall (Eds.), **Neural Models of Language Processes.** Academic Press, 1982.

Marslen-Wilson, W., and Tyler, L. The temporal structure of spoken language understanding, **Cognition**, 8, 1980, pp. 1-71.

McClelland, J. L. and Kawamoto, A. H. Parallel Mechanisms of Sentence Processing: Assigning Roles to Constituents of Sentences; in J.L. McClelland and D. E. Rumelhart (eds.) **Parallel Distributed Processing**, MIT Press, 1986.

McClelland, J.L. and Rumelhart, D.E. An interactive activation model for context effects in letter perception: Part 1. An account of basic findings. **Psychological Review**, 1981, 88, 5, 375-407.

McClelland, J.L. and Rumelhart, D. E. A General Framework for Parallel Distributed Processing, in J.L. McClelland and D. E. Rumelhart (eds.) **Parallel Distributed Processing.** MIT Press, 1986.

Ojemann, G.A. Brain organisation for language from the perspective of electrical stimulation mapping. **The Behavioral and Brain Sciences**, 1983, 4, 189-230.

Quillian, M.R. Semantic memory. In M. Minsky (Ed.), **Semantic Information Processing.** Cambridge, Ma: MIT Press, 1980.

Schwartz, M.F., Saffran, E., and Marin, O.S.M. The word order problem in agrammatism. I. Production. **Brain and Language**, 10, 1980.

Seidenberg, M., Tanenhaus, M., Leiman, J., and Bienkowksi, M. Automatic access of the meanings of ambiguous words in context: Some limitations of knowledge based processing. **Cognitive Psychology**, 1982.

Scholes, R. Syntactic and lexical components of sentence comprehension. **The Aquisition and Breakdown of Language**, Johns Hopkins Press, Baltimore, 1978, pp. 163-194.

Smolensky, P. Neural and Conceptual Interpretation of PDP Models, in J.L. McClelland and D. E. Rumelhart (eds.) **Parallel Distributed Processing.** MIT Press, 1986.

Swinney, D.A. Lexical Access during Sentence Comprehension: (Re)Consideration of Context Effects. **Journal of Verbal Learning and Verbal Behavior**, 1979, 18, 645-660.

Von Monakow, C. **Die Lokalisation im Grosshirn.** Wiesbaden: Bergmann, 1914.

Waltz, D. and Pollack, J. Massively parallel parsing: A strongly interactive model of natural language interpretation. **Cognitive Science**, 1985, 9, 1, 54-75.

Webster's New Collegiate Dictionary, G. & C. Merriam Co. 1981.

Wood, C.C., Implications of simulated lesion experiments for the interpretation of lesion in real nervous systems. In M.A. Arbib, D. Caplan, and J. Marshall (Eds.) **Neural Models of Language Processes**, Academic Press, 1982.

608

A Model of Purpose-driven Analogy and Skill Acquisition in Programming

Peter Pirolli
University of California, Berkeley

Abstract

X is a production system model of the acquisition of programming skill. Skilled programming is modelled by the goal-driven application of production rules (productions). *Knowledge compilation* mechanisms produce new productions that summarize successful problem solving experiences. *Analogical problem solving* mechanisms use representations of example solutions to overcome problem solving impasses. The interaction of these two mechanisms yields productions that generalize over example and target problem solutions. Simulations of subjects learning to program recursive functions are presented to illustrate the operation of X.

Introduction

Theoretical progress in cognitive science hinges crucially on the ability of theories to address issues of knowledge acquisition. In turn, theories of knowledge acquisition have direct bearing on theoretical and practical issues in instruction. I present a model of analogical problem solving and skill acquisition, called X, developed as an extension of the ACT* theory (Anderson, 1983) for the domain of learning to program recursive functions. This model was developed to explore the notion insightful conceptual understandings of example solutions are instrumental in skill acquisition. Analogical problem solving makes use of goal-relevant and plan-relevant information that is encoded in a mental representation of an example solution. The content and detail of this information has an impact on the success of analogical problem solving. In turn, skill acquisition mechanisms transform these analogical problem solving experiences into skills. Ultimately, the quality of the skills acquired from analogy rests upon the content and detail of the goal-relevant and plan-relevant information in the example representation. Analogy thus serves as a means towards effecting *repairs* (Brown & Van Lehn, 1980) to domain-specific problem solving procedures.

Analogy is a term with many denotations and connotations in cognitive science as well as everday language. The particular analogical problem solving mechanisms implemented in X work from representations of example solutions that might be presented by a textbook or teacher. Basically, these representations constitute an explanation of the structure and functionality of components of the solution structure along with constraints on the conditions under which such components apply. In this respect, the X analogical problem solving mechanisms are similar to machine learning work on explanation-based learning (DeJong & Mooney, 1986; Mitchell, Kellar, & Kedar-Cabelli, 1986). The invocation of analogical problem solving and the selection of relevant analogs is driven by active problem solving goals. This aspect of X is similar to machine learning work on purpose-driven analogy (Kedar-Cabelli, 1985). Although similar in spirit to these clusters of machine learning research, X evolved from the ACT* production system theory of problem solving and is a minor variant of another descendent of ACT*, the PUPS production system (Anderson & Thompson, 1986). Like other members of the ACT* family, accounting for phenomena of human cognition is one major impetus for developing X. Another motivation for developing X, is the notion that richer theories of learning will lead to richer, more effective, and more efficient means of instruction.

Programming Recursion

An Ideal Model

The X model was initially developed to address observations of people learning to program recursive functions. Recursion is usually a novel concept for introductory programming students and consequently serves as a useful domain for studying knowledge acquisition. A typical example of a recursive function in LISP is the definition of the function FACT to compute factorials presented in Table 1. Like all recursive functions, FACT is defined in terms of itself. The body of the definition for FACT consists of a conditional structure containing a series of conditional clauses. The conditional structure is implemented by the LISP form COND and each conditional clause is represented as a list within the COND structure (a list in LISP is anything enclosed in parentheses; e.g., (A B C) is a list). Each conditional clause contains two parts. The first part is a list that specifies a condition and the second part specifies an action to take if the condition does not evaluate to NIL, which stands for "false." The first clause in FACT states that if the input N is equal to zero, then the result of FACT will be 1. The second clause, states that in all other cases, the result of FACT will be N times the result of FACT applied to N - 1.

Table 1

A recursive LISP function to compute factorials

_____

Definition   Comments

_____

```
(DEFUN FACT (N)                    ; define n!
    (COND ((ZEROP N) 1)            ;If n = 0 then return 1
          (T (TIMES N (FACT (SUB1 N))))))    ;Else return n x (n - 1)!
```

_____

In previous studies of expert and novice programming (Pirolli & Anderson, 1985) we identified concepts that appeared to be crucial to efficiently learning general skills for programming recursion. A prescriptive model of programming skills for recursion, called an *ideal model*, has been implemented in the GRAPES production system and used as the basis for instruction on recursion in an intelligent tutoring system (Pirolli, in press). GRAPES is a system that emulates a subset of the ACT* theory (Anderson, 1983).

Recursive functions, in general, have a conditional structure consisting of two types of cases. The *recursive cases* compute a result by using the results of one or more *recursive calls* to the function. For example, the last conditional clause of the FACT function presented in Table 1 is a recursive case that involves computing the result of the recursive call, (FACT (SUB1 N)), and uses that result to compute the result of (FACT N). The *terminating cases* terminate the recursive process. In Table 1 the terminating case is the first conditional clause, which returns the value 1 when the input N = 0.

610

The *recursive relation* holds between the value of the function and the value of a recursive call to the function. In FACT, the recursive relation is the relation between between the result of $n!$ and $(n - 1)!$--that is, $n! = n \times (n - 1)!$. Students often have a great deal of difficulty identifying and planning out the recursive cases of recursive functions. Based on our analyses of expert problem solving with recursive functions (Pirolli & Anderson, 1985) the general method for determining the recursive relation is to: (a) assume that the result of a recursive call can be achieved, then (b) determine how to achieve the result of the function using the result of the recursive call.

GRAPES productions, in general, decompose programming goals into subgoals until some action can be achieved, forming a hierarchical goal tree. A typical production for programming LISP might be:

```
P1:  IF   the goal is to write a function definition in LISP
          and the name of the function is given
          and the arguments to the function are given
     THEN  write "(DEFUN <name> <arguments> <body>)"
          where <name> is the name of the function
          and <arguments> are the arguments to the function
          and set a subgoal to code the <body> of the function
          which implements a process that achieves the function
```

Some observations on analogical problem solving

Instruction in problem solving domains usually includes descriptions of the entities in the domain, general rules for problem solving, and example solutions. Protocol studies (Pirolli & Anderson, 1985) have indicated that subjects have a tough time deriving solution methods from general definitions or rules of thumb for a domain and turn to example solutions for guidance in early problem solving attempts . These subjects are usually successful in producing some solution by analogy to an example once the example has been selected (see also Gick & Hoyoak, 1980; Reed, Dempster, and Ettinger, 1985; Reed, Ernst, and Banerji, 1974). We also observed (Pirolli & Anderson, 1985) that problem solving procedures are learned from such analogical problem solving (see also Gick & Holyoak, 1983; Sweller & Cooper, 1985). However, subjects sometimes conceptualize examples in a shallow and literal manner and sometimes conceptualize examples in a deep and insightful manner (Pirolli & Anderson, 1985). These variations have an impact on the problem solving behavior generated by analogy to examples. In some cases subjects basically copy program code that they do not understand and in other cases they produce solutions based on the more principled methods they see embodied in the example programs. These variations in early problem solving experiences involving the use of examples in turn have an impact on early skill development (Pirolli & Anderson, 1985).

The pervasiveness of analogical problem solving in the early stages of skill acquisition is borne out by an experiment in which subjects had access to an online example of a recursive function, that had been part of their instruction. While coding their first recursive function, subjects spent approximately 30% of their time looking at the example. Analysis of target problem solutions revealed that portions that were similar to the example accounted for fewer errors. Later, when subjects were coding their fourth recursive function, looking at the example solution only accounted for about 5% of their problem solving time.

The impact on skill acquisition of the particular manner in which an example is encoded is illustrated by another experiment in which the same example program was presented to one group of subjects (structure group) in the context of an explanation of how recursive functions are written (based on the GRAPES ideal model) and to another group (process group) in the context of how recursive functions execute (the typical pedagogical approach). The structure group took less training time than the process group in achieving the same level of proficiency indicating that knowing how an example recursive function is written yields more efficient learning than knowing how a recursive function works. A simulation of a structure group subject is presented later in this paper.

The X Model

The X model of analogical problem solving was implemented as an extension of the GRAPES production system. The X model is also a subset of a production system architecture for analogical problem solving and skill acquisition, called PUPS, that is being developed by Anderson and Thompson (1986). Basically, X takes several ideas developed in PUPS about analogical problem solving and instantiates them in the GRAPES architecture.

Like several other proposals for problem solving by analogy (e.g., Carbonell, 1986; Gick & Holyoak, 1980, 1983) the X analogy mechanisms supply a method for problem solving when domain-specific methods are lacking or inadequate. The general notion is that the learner has some declarative knowledge of how the structure, S_e, of an example achieves various functions, F_e, under certain preconditions, C_e, and is faced with achieving goals G_t, under conditions C_t in a target problem. The task in analogical problem solving is to come up with a target solution S_t by solving the analogy $S_e:F_e::S_t:G_t$, subject to the constraint that the mapping of S_e onto S_t transforms C_e, into a set of preconditions that are in C_t or satisfiable in the target solution.

Representation

The X system makes use of a representation scheme for declarative knowledge that captures the functionality, structure, and conditionality, of concepts or actions in a problem solving domain. This knowledge representation scheme is crucial to the working of analogy and is an important addition to the ACT* theory. The principle components of this scheme are schematic knowledge structures called *units* that have slots that are filled or instantiated by particular values. Although arbitrary slots are allowed, there are three types of slots that have preeminence in the representation: (a) *functionality* which describes the purpose or goals achieved by a unit, (b) *structure* which describes the composition of a unit from other units, and (c) *conditionality* which describe constraints on the unit.

Table 2

Examples of X representations

_____

The POWER program.

power-definition
    functionality: defines(power-function args power-result)
    preconditions: implemented-in(power-function LISP)
    structure: steps(defun power-name args body)

The Factorial Problem.

fact-definition
    functionality: defines(fact-function x-arg fact-result)
    preconditions: implemented-in(fact-function LISP)

_____

Some examples of the representation can be seen in Table 2, which presents a declarative description of part of an example program and a target problem in LISP. The example is a recursive definition of a function, POWER, that computes m^n. The target program is the factorial function presented in Table 1. Units thus provide a knowledge representation scheme that captures important goal-relevant and plan-relevant information for use in problem solving. A unit can be thought of as rule of the form

$$conditionality \quad \wedge structure \quad \Rightarrow functionality$$

So, *power-definition* in Table 2 can be translated into the rule

$$steps(defun\ power\text{-}name\ args\ body)$$
$$\wedge implemented\text{-}in(power\text{-}function\ LISP)$$
$$\Rightarrow defines(power\text{-}function\ args\ power\text{-}result)$$

Problem solving

Goals in the X system are to-be-achieved units that have functionality but no structure. An example of a to-be-achieved goal in the Table 2 is *fact-definition*. The X system considers one goal at a time and considers only productions that are applicable to that goal. The propositions on the structure slots represent orderings, partial orderings, and hierarchical relationships among the actions represented by units. The agenda for goal processing is achieved by productions that search through the structural links from a current active goal. Units encountered in this search that have a specified functionality but no structure are placed on the goal agenda. Analogical problem solving is invoked by X at problem solving impasses--in other

613

words, when a goal is activated and no production matches, analogical problem solving is invoked. X selects an analog for further processing based on a specificity principle. In theory, this is an associative memory retrieval achieved by the spreading activation mechanisms of ACT* (Anderson, 1983; Anderson & Pirolli, 1985). In the computer implementation of X, the effects of spreading activation are approximated by a specificity principle based on the number of correspondences and identical elements that hold between the the example and target. Since the goal of analogical problem solving is to map an existing solution structure from an analog unit to a target unit, one constraint on the selection of an analog unit is that it must have a filled-in structure slot. There are three major subprocesses involved in solving a target problem by analogy:

- *Function matching.* the first step taken by X is to place the target goal unit into correspondence with the functionality of potential analogs and to select the best analog. Two function propositions can be placed into correspondence if the predicates in both functions are identical. The arguments of one function proposition are placed into correspondence with a another function proposition by virtue of the slots they fill within the propositions. These correspondences are used to map information about the analog onto new information in the target. Function matching also checks that the conditions on the analog unit are not violated in the target problem. If there is a violation then there is no match.

- *Structure mapping.* This involves mapping an analog structure onto a new target structure. However, there is no guarantee that the correspondence set will be elaborate enough to permit such a mapping.

- *Function elaboration.* This occurs when an element of an analog's structure has no correspondence. There are a number of ways that function elaboration can be carried out in order to map a particular structural element e_a of an analog onto a new structural element e_t in a target. First, the functionality of e_a may match the functionality of some existing target unit e_t. The correspondence set can be elaborated with this correspondence plus the correspondences resulting from the function match of e_a and e_t. Second, a new target unit can be created and assigned a functionality mapped from e_a and this may recursively invoke further function elaboration. Third, additional correspondences can be found by elaborating the match of an analog unit to a target unit. This is achieved by recursively matching the functions of elements already placed into correspondence. This may lead to an elaborated set of correspondences that permits e_a to be mapped onto a new e_t.

Learning from Analogy

One major outcome of analogical problem solving is the induction of new production rules by a set of *knowledge compilation* mechanisms that generalize over information present in the declarative units representing analog and target problems and their solutions. Knowledge compilation mechanisms create new productions that summarize the problem solving involved in analogy (for the details of knowledge compilation see Anderson, 1983). These new productions apply in situations similar to those that invoked analogical problem solving in the first place. The compilation of solutions produced by analogy yield general problem solving operators and thus the interaction of analogy and knowledge compilation offers an alternative procedure for the generalization of cognitive skills in ACT* (Anderson, 1986).

To see how skills are acquired from analogy, consider that analogical problem solving in X consists of two things: (a) matching conditions and functionalities of the analog and target, and (b) creating target conditions, functionalities, and structures based on the analog. Knowledge compilation creates new productions with conditions that specify the target functions and conditions that were matched in the analogy process and that have actions that specify the structures, functions, and conditions that were created in the target by analogy. The conditions created by compilation retain the components of the target that matched exactly to the analog and variablizes over target information that mismatched. Thus, a compilation of the analogical problem solving involved in achieving the goal unit *fact-definition* based on the example *power-definition* produces the production:

```
L1:  IF   the goal is to achieve
          =definition
                functionality: defines(=function =arguments =result)
                precondition: implemented-in(=function LISP)
          and
          =name
                functionality: name-of(=function)
     THEN the structure of =definition is
                steps(defun =name =arguments =body)
          and the functionality of =body is
                implements(=function)
```

where the items preceded by the equal sign denote variables. Production L1 applies when the goal is to achieve a function definition in LISP when the name of the function has been decided on. The action specified by L1 lays out a template for the code to define the function. Production L1 has the same semantics as production P1 presented earlier.

Example Simulations

Previous protocol analyses and experiments on learning recursion (Pirolli, 1985; Pirolli & Anderson, 1985) indicate that subjects with richer representations of how recursive functions are written learn more efficiently and effectively than subjects who either just understand how recursive functions operate or who have a encoded a rather literal representation of examples. With X it is possible to explore in greater detail what knowledge promotes efficient and effective learning. Two simulations of X are presented here. These simulations address verbal protocol data analysed and modelled previously (Pirolli & Anderson, 1985) in a more general manner. The first simulation illustrates how a rather literal representation of an example can lead to a successful solution by analogical problem solving, with very little gain in skill acquisition. The second simulation illustrates a case of analogical problem solving that lead to effective problem solving skills for programming recursion.

Literal Analogy

The first simulation addressed the data gathered from subject JP, an eight year old learning recursion in LOGO. Her first programming problem was to write a function that would recursively draw a set of squares of increasing size. JP's final solution, called TUNNLE [sic] was

```
TO TUNNLE :X
SQUARE :X
IF :X = 42 THEN STOP
TUNNLE :X +10
END
```

In coding TUNNLE, JP used an example program, CIRCLES, to guide her analogical problem solving. After writing TUNNLE (which works), JP was unable to code even slight variants of TUNNLE (e.g., drawing more squares).

The analysis of protocol data and interview data suggested that JP has a very literal representation of the CIRCLES program and largely copied that solution onto the TUNNLE [sic] solution. Figure 1 presents the representation of CIRCLES that is encoded initially in X for the simulation of JP. In Figure 1, literal code from the example is in uppercase. The structure of the CIRCLES code is not represented at any deeper level (e.g., as a tree structure representing the different LOGO structures, terminating cases, recursive cases, etc.). The functionality of only some of the program symbols are elaborated (e.g., CIRCLES is the name of the function, "50" is the maximum size of the circles).

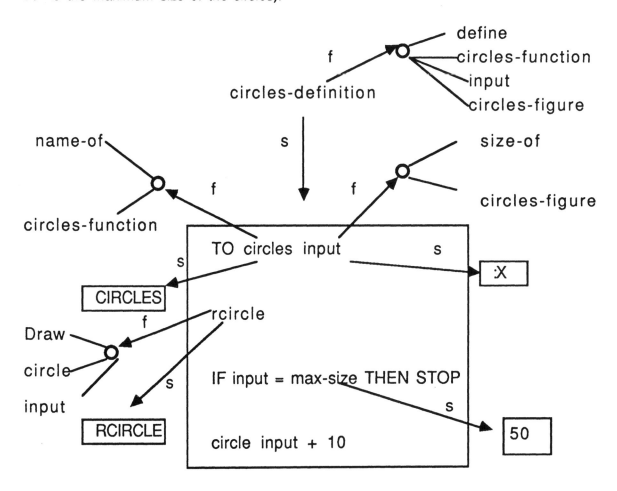

Figure 1: JP's representation of the CIRCLES example. Arrows labelled *s* indicate structure slots; *f* indicates function slots. Boxes indicate structures that fill structure slots.

Analogical problem solving works in this case because the target problem is similar enough to the example representation to permit successful mappings. The following matches are made in the simulation:

- Both definitions define a function taking an input and produce a composite figure
- The process implemented by both functions is to repeatedly draw a figure of increasing size
- RCIRCLE draws an element of the composite circles figure and SQUARE draws an element of the composite squares figure
- 50 is the maximum size of the circles figure and 42 is the maximum size of the square figure
- The functionality of the inputs match

Knowledge compilation of this analogical problem solving yields the following production

```
L2:  IF  the goal is to define a function =name with input =x
         that draws a =composite-figure
         by repeatedly drawing a figure with a function =figure-drawer
         up to maximum size =number
     THEN  write
               TO =name =x
               =figure-drawer
               IF =x = =number THEN STOP
               =name =x + 10
```

Production L2 will basically code other fuctions that draw composite figures of increasing size like CIRCLES, but is not effective for coding recursive functions in general.

Insightful Analogy

The second simulation addressed data gathered from subject AD, a college student learning recursion in SIMPLE (Shrager & Pirolli, 1983). AD's programming tasks centered on writing functions that searched and gathered selections from a database of library entries. These library entries could be identified by a number or title, and SIMPLE predicates were available to test whether entries belonged in one of three categories (science, religion, or fiction). The recursion problems assigned to AD involved collecting library entries of various categories into lists with different orderings placed on the list items. AD's instruction on recursion was identical to that given to the structure group in the experiment discussed earlier--that is, it emphasized how recursive functions are written. The example discussed in this instruction was SORT, a function that sorted a list of entries such that science books were at the beginning of the list result of SORT.

From AD's protocol gathered as she read instructions out loud and wrote her first recursive function, it was clear that she had encoded a rich representation of the SORT example. After writing her first recursive function, AD coded an additional 19 recursive functions without error. Part of the encoding of SORT given to X in simulating AD is presented in Figure 2, which depicts the representation of a recursive case of SORT. The representation includes the notion that the recursive case is a conditional structure and that the action in this case involves a recursive relation. The recursive relation is achieved by determining the result of a recursive call to SORT and then comparing it to the result of SORT.

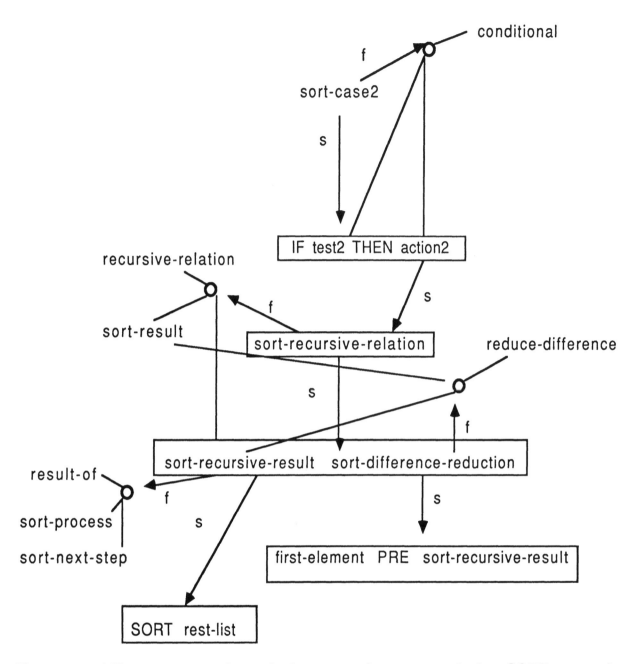

Figure 2: AD's representation of the recursive case of the SORT example. Arrows labelled *s* indicate structure slots; *f* indicates function slots. Boxes indicate structures that fill structure slots.

Knowledge compilation of AD's analogical problem solving yield the following productions

L3: IF   the goal is to code the action of a conditional clause
          of a function that places all elements
          of a specified type into a result in a specified order
     THEN set a goal to code the recursive relation

L4: IF the goal is to code a recursive relation
   THEN set subgoals to
      1. refine the recursive call
      2. achieve the result of the function using the recursive call

L5: IF the goal is to code a recursive call to a function
      and a value that is one step closer
      to the terminating case than the current function input is known
   THEN write the name of the function followed by that value

L6: IF the goal is to achieve a value that is one step closer
      to the terminating case than the current function input
      and the function input is a number
   THEN set a goal to write code that will subtract 1 from the input

Production L3 specifies that the action of a conditional clause that is supposed to place all elements of a certain type in an output list can be solved by coding the recursive relation. Production L4 lays out a plan for inducing a recursive relation to satisfy the constraints of the program. Production L5 codes a recursive call. Production L6 specifies that the argument to a recursive call in a function that has a numeric input should be one less than the input. The key point to be made in this simulation of subject AD is that having an abstract representation of the underlying structure and functionality of a recursion example that encoded how recursive functions are written facilitated the learning of productions that are similar to those of the ideal model implemented in GRAPES. The generality of the productions acquired by X in simulating subject AD accounts for the ease with which AD coded her subsequent recursion problems.

Summary

The X model of analogical problem solving and skill acquisition was developed as an extension of ACT* (Anderson, 1983) to deal with the pervasive phenomena of analogical problem solving. A major difficulty with ACT* has been with its ability to deal with the structuring of problem solving performance when encountering a novel domain (e.g., Anderson, 1983, ch. 6). The analogical problem solving mechanisms in X (and PUPS, Anderson & Thompson, 1986) comprise a weak problem solving method that appears to serve this function in a number of domains such as programming, geometry (Anderson, Greeno, Kline, & Neeves, 1981), and algebra (Reed et al., 1985; Sweller & Cooper, 1985). The interaction of analogy and knowledge compilation yields generalized productions from a single problem solving episode. This appears to more accurately fit the phenomena of early skill acquisition (e.g., Kieras & Bovair, 1986) better than the ACT* mechanism of production generalization which requires two similar production applications.

The major gap in X is that it does not address the process of comprehending example solutions to form the representation that serves as the basis for later analogical problem solving. Current work is focused on filling this gap. It is assumed that understanding an example is driven by a process of attempting to explain (by instantiating declarative schematic knowledge about plans and goals) how an example solution achieves various goals (cf. DeJong & Mooney, 1986). The goal of this effort is to work towards a model that addresses some aspects of how variations in prior knoweledge about plans and goals interact with variations in instruction using examples.

References

Anderson, J.R. (1983). *The architecture of cognition*. Cambridge, MA: Harvard University Press.

Anderson, J.R. (1986). Knowledge compilation: The general learning mechanism. In R.S. Mickalski, J.G. Carbonell, and T.M. Mitchell (Eds.) *Machine Learning, Volume 2*. (pp. 289-310). Los Altos, CA: Morgan Kaufmann.

Anderson, J.R., Greeno, J.G., Kline, P.J., and Neves, D.M. (1981). Acquisition of problem-solving skill. In J.R. Anderson (Ed.), *Cognitive skills and their acquisition* (pp. 191-230). Hillsdale, NJ: Lawrence Erlbaum.

Anderson, J.R. and Pirolli, P.L. (1984). Spread of activation. *Journal of Experimental Psychology: Learning, Memory, and Cognition, 10*, 791-798.

Anderson, J.R. and Thompson, R. (1986). *Use of analogy in a production system architecture*. Unpublished manuscript, Carnegie-Mellon University, Department of Psychology, Pittsburgh, PA.

Brown, J.S. and VanLehn, K. (1980). Repair theory: A generative theory of bugs in procedural skills. *Cognitive Science, 4*, 379-426.

Carbonell, J.G. (1986). Derivational analogy: A theory of reconstructive problem solving and expertise acquisition. In R.S. Mickalski, J.G. Carbonell, and T.M. Mitchell (Eds.) *Machine Learning, Volume 2*. (pp. 371-392). Los Altos, CA: Morgan Kaufmann.

DeJong, G. and Mooney, R. (1986). Explanation-based learning: An alternative view. *Machine Learning, 1*, 145-176.

Gick, M. L. and Holyoak, K. J. (1980). Analogical problem solving. *Cognitive Psychology, 12*, 306-355.

Gick, M. L. and Holyoak, K. J. (1983). Schema induction and analogical transfer. *Cognitive Psychology, 15*, 1-38.

Kedar-Cabelli, S. (1985). *Purpose-directed analogy*. In Proceedings of the Cognitive Science Society. Irvine, CA: CSS.

Kieras, D.E. and Bovair, S. (1986). The acquisition of procedures from text: A production-system analysis of transfer of training. *Journal of Memory and Language, 25*, 507-524.

Mitchell, T.M., Kellar, R.M., and Kedar-Cabelli, S.T. (1986). Explanation-based generalization: A unifying view. *Machine Learning, 1*, 47-80.

Pirolli, P.L. (1985). *Problem solving by analogy and skill acquisition in the domain of programming*. Unpublished doctoral dissertation, Carnegie-Mellon University.

Pirolli, P. (in press). A cognitive model and computer tutor for programming recursion. *Human-Computer Interaction*.

Reed, S.K., Dempster, A., and Ettinger, M. (1985). Usefulness of analogous solutions for solving algebra word problems. *Journal of Experimental Psychology: Learning, Memory, and Cognition, 11*, 106-125.

Reed, S.K., Ernst, G.W., and Banerji, R. (1974). The role of analogy in transfer between similar problem states. *Cognitive Psychology, 6*, 436-450.

Pirolli, P.L., and Anderson, J.R. (1985). The role of learning from examples in the acquisition of recursive programming skills. *Canadian Journal of Psychology, 39*, 240-272.

Shrager, J. & Pirolli, P. L. (1983). *SIMPLE: A simple language for research in programmer psychology* [Computer program]. Pittsburgh: Carnegie-Mellon University, Department of Psychology.

Sweller, J, and Cooper, G.A. (1985). The use of worked examples as a substitute for problem solving in learning algebra. *Cognition and Instruction, 2*, 59-89.

The Operational Level of a Commonsense Planner

Richard Alterman
Department of Computer Science
Brandeis University

ABSTRACT

This paper characterizes the operational level of a commonsense planner. In particular it will compare two approaches on the problem of level of operation. A 'plan instantiater' planner has access to a sort of summarization of the activities involved in each of the subkinds of a given category of plans, and by a process of instantiation it selectively adds details to match the current planning situation. Such planners abandon the details of the lower plan and dynamically recreate them under the exigencies of a given situation. A 'reference point' planner selects a subordinate plan, from a given category of plans, to represent the category as a whole. The reference point planner assumes a level of operation which can directly access a greater number of functional details than its plan instantiation counterpart, but perhaps at the loss of flexibility.

The thrust of this paper is that reference point planners are the appropriate model of planning for the commonsense domain.

1. Introduction

The term commonsense planning refers to the mundane day-to-day activities of human planners [21]. This includes the planning of such activities as fetching a newspaper, making a deposit at the bank, taking a child to daycare, fixing breakfast, riding a bus to work, going to a movie, making an airline reservation, shopping at a supermarket, eating at a restaurant - in short - planning for the quotidian concerns of human existence as it varies under the flux of daily circumstances. In commonsense planning it is less the case that the activities of the planner vary a great deal, and more the case that the circumstances underwhich the plans are applied vary.

Adaptive Planning [1,2] is an approach to planning in the commonsense domain. An adaptive planner takes advantage of the habitual nature of many of the planning situations for which it plans by basing its activities on pre-existing plans. A critical issue, and the subject of previous papers, is the question of flexibility: How does an adaptive planner refit an old plan in order to meet the demands of some new planning situation?

One of the important themes of adaptive planning is the way it characterizes the level of operation. It takes the following position: specific plans are foregrounded (in order to take advantage of the mass of details associated with the more specific plan), and general plans are in the background (serving more to organize old plans into categories). This paper is intented to act as a clarification of this position. In particular it will compare two approaches on the problem of level of operation. Approach one is that commonsense planners work from plans that are basically underspecified and then add details as the situation demands. Approach two is that the planner works from plans that, if anything, are overspecified and then removes details and refits the plan only as indicated by the situation. A critical point in this discussion will be the habitual nature of the commonsense planning domain. The paper is fairly tendentious (in favor of the second approach).

622

Consider the category of plans concerning TELEPHONING (see figure 1).

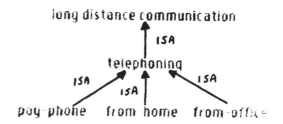

Figure 1: Two Approaches.

I will refer to a planner that works at the level of TELEPHONING as a planner that works via a process of **plan instantiation**. The notion here is that the planner has access to a sort of summarization [17] of the activities involved in each of the subkinds of TELEPHONING, and by a process of instantiation it selectively adds details to match the current planning situation. Such planners abandon the details of the lower plan and dynamically recreate them under the exigencies of a given situation.

I will refer to a planner that works from one of the subordinate level plans (i.e. from-office, from-home, or pay-phone) as a **reference point** [10,14] planner. Each of the subordinate plans listed for TELEPHONING can serve as a reference point for the category depending on the planning context. The general idea here is to use a subordinate plan to represent the category as a whole. The reference point planner assumes a level of operation which can directly accesses a greater number of functional details than its plan instantiation counterpart, but perhaps at the loss of flexibility.

This paper will articulate the trade-off between a planner that works operationally as a plan instantiator and one that works operationally as a reference point planner.

2. Generalizing the Information in a Subordinate Plan

I will distinguish between two sorts of generalization: object generalization and sequence of action generalization. Object generalization is concerned with generalizing the objects, or props, associated with a given set of plans: an object that can be generalized out of the three telephone plans is the 'telephone'. A sequence of action generalization is concerned with generalizing over the steps of the subordinate plans. The focus of this discussion will be on sequence of action generalizations: I will give a detailed accounting of the kinds of action knowledge that is lost as a result of summarizing a given category of plans. This analysis should support two sorts of conclusions. First that making a useful generalization for a given set of subordinate level plans is by-and-large a difficult proposition. Second that even in the event that a useful generalization can be made, the problem of re-instantiating details on the fly for a given habitual situation is doubly complicated since the generalization has lost not only the details associated with individual steps, but also the details that correspond to the inter-relationships amongst those steps.

2.1 Delete Steps

Below are shown the sequence of actions associated with the two kinds of TELE-PHONING:

from-home - 1) lift receiver 2) hear dial tone 3) remember number 4) dial 5) ringing sound 6) ask-for 7) talk

pay-telephone - 1)get change from pocket 2) lift receiver 3) hear dial tone 4) insert coins 5) dial number 6) ringing sound 7) ask-for 8) talk

It would be possible to form a generalization over these two kinds of telephone calling by removing two steps from the pay-telephone plan: getting change from one's pocket and inserting coins into the telephone box. There is an implicit loss of default information when individual steps are deleted. The subordinate pay telephone plan encodes the fact that the planner's preferred plan for getting change is to reach into his/her pocket. For another plan, say doing the laundry at the laundromat, the preferred plan for getting change is to use a change machine.

2.2 Generalize Individual Actions

Consider the following two plans:

theatre-plan - 1) buy theatre ticket 2) enter playhouse 3) view play 4) leave
movie-plan - 1) buy movie ticket 2) enter movie 3) view movie 4) leave

In this case it would be possible to form a generalization by generalizing three of the individual steps, i.e.,

show-plan - 1) buy ticket 2) enter show 3) view show 4) leave show

There is a cost involved in generalizing an individual step. Consider the first steps of the theatre and movie plans. They both involve buying tickets, but one usually requires a reservation and for the other a reservation is not possible. In the theatre plan situation the planner calls and makes a reservation, thus saving a trip if the show is sold out. Similar sorts of loss of information occur for step two (in one case refreshments are bought before the show) and step three (in one case there is an intermission). Where a plan instantiator must re-discover these details in a piecemeal fashion, a reference point planner, by selecting a subordinate level plan, accesses all of the details, of one or the other plan, all at once.

2.3 Relax Sequence Constraints

There is another kind of information that can be lost when generalizing over subcategories of a plan and that is sequencing information. Consider three subcategories of a restaurant plan:

sitdown restaurant - 1) seating 2) ordering 3) serving 4) eating 5) paying
fastfood plan - 1) order 2) pay 3) serve 4) sit 5) eat
cafeteria plan - 1) select food 2) pay 3) sit 4) eat

These three plans include many of the same actions, but they are performed in a differing orders. In this case, it would be possible to factor out many of the actions, with some generalizations, but sequencing information would be lost.

Relevant to this issue is the problem of sequencing. There are three sorts of positions taken on the sequencing problem in the AI literature. One approach (linear assumption [19]) is that the planner commit itself to any ordering of actions, and if difficulties arise, due to subgoal interactions, make the appropriate changes. The second approach (least commitment [15]) is to make a commitment to an ordering only when necessary. The third approach (opportunism [6]) is to perform a given step when convenient. Working from reference points offers a fourth approach: Select an old ordering. The set of reference points, in effect, offers a short list of possible orderings. Encoded in each reference point are sequencing considerations, many of which are dictated for causal sorts of reasons. A good heuristic for domains of habitual activity is to select, or reselect, orderings from the short list.

3. Does a good reference point always exist?

How do we know that there is always a reasonable reference point? Suppose the category of plan concerned with telephoning consists of the following two subordinate plans:

from-home - 1) lift receiver 2) hear dial tone 3) remember number 4) dial 5) ringing sound 6) ask-for 7) talk

pay-telephone - 1)get change from pocket 2) lift receiver 3) hear dial tone 4) insert coins 5) dial number 6) ringing sound 7) ask-for 8) talk

Furthermore suppose the new situation is that the planner intends to make a telephone call from work and must dial a nine in order to get an outside line. Claim: if there exists a generalization over these two plans that is applicable and reasonable for the planning situation above, then one of the two reference points is reasonable for the same planning situation. Or, more generally, if for a given planning situation there is a reasonable preexisting general plan, then there always exists a reasonable reference point plan for the same situation. Below is shown an argument as to why this must be the case.

[1] If a generalization gets to be too abstract then it stops being useful. Consider the following example: Suppose a graduate student comes to you and asks you for a plan to get a Phd. One answer is: pass your generals, pass your orals, write a dissertation, defend your dissertation. Such a plan is a description of what the student must do, but it is much too abstract to be of any use.

[2] If a plan P is instantiated from a generalization G, G is a generalization of P.

[3] Given a set of plans the amount of abstraction that is required over that set in order to form a generalization works as a function of the difference between any pair of plans in that set. The greater the difference between any two plans in the set, the more abstract the general plan must be. Given any number of plans $P_1...P_n$ there exists a reasonable generalization over these plans iff the differences between any two plans, $\Delta(P_i, P_j)$, is reasonable as defined by some constant C, i.e.

$$\Delta(P_i, P_j) < C \quad for\ 1 \leq i, j \leq n$$

[4] Any plan P which is derived from a generalization G is part of the set of plans which G generalizes over (2), and consequently, by our notion of a more general plan, must not be too different from the other plans in the class (3), i.e.

$$\Delta(P_i, P) < C \quad for\ 1 \le i \le n.$$

The above shows that if there is a reasonable generalization which applies to a given situation then there is a reasonable subordinate plan which also applies. That does not mean that there is always a reasonable pre-existing general plan. The previous section demonstrates some of the difficulties in forming a reasonable general plan. In general, it is not hard to find cases where there exists a category of plans where it is difficult to find a useful summary, but is nevertheless a useful category of plans. Consider the following two plans for telephoning.

from-home 1) lift receiver 2) hear dial tone 3) dial 5) ringing sound 6) ask for 7) talk
from-work 1) lift receiver 2) buzz secretary 3) request call 4) hang-up and wait 5) hear buzz and pick-up receiver 6) greet and talk

It is hard to see what a useful generalization would be over these two plans, yet it is clear that they represent two kinds of telephoning. Moreover, there are any number of planning situations where one of these two subordinate level plans would be an appropriate starting point.

Perhaps it would be possible to get around this kind of problem by splitting the category or allowing for multiple generalizations for a given category of plans. These kinds of techniques would work - the generalizations are making finer and finer discriminations - but dividing up the category is just another way of converging on reference points.

There is an assumption here that the planner is working from prestored plans. Perhaps if the more general plans are dynamically created it would be possible to work from more general plans. A whole generation of AI planners, beginning with [15], has been constructed along these lines. But there are difficulties for this sort of approach in the commonsense domain. Largely these difficulties arise from the fact that the planner fails to take advantage of the habitual nature of most of his/her activities. The constraint of habituality decidedly favors using pre-existing plans.

Notice that there is no claim here that there always exists a useful pre-existing plan. There will be planning situations that require the planner to plan for a non-habitual activity. The point is that for planning and acting with regards to habitual activies there is nothing to lose - and a lot to gain - in selecting and working from a reference point.

Finally, the claim is about the existence of a reasonable reference point plan. There is a question as to how hard, or easy, it is to find a reference point plan. The associative techniques suggested by [18,7] are good candidates for the task of finding a reference point for a given planning situation. Since the activities we are concerned with are habitual, they should be relatively easy to find.

4. Flexibility and Levels

How can a commonsense planner maintain the advantages of both flexibility and habituality? One candidate answer is to have the commonsense planner work from specific plans and when things go awry, it can retreat to weak methods [3,4]. Under certain circumstances this could be the appropriate choice, but there are many instances where the planner can succeed without a full retreat to weak methods. In this section I will describe a reference point planner PLEXUS that can perform flexibly without retreating to weak methods. For further details and traces see [1,2].

Assume that the reference point planner has access to the content and organization of background knowledge associated with the old plan. In [1,2] the importance of background knowledge is described, as well as some details about four kinds of background knowledge. Here, it will suffice to draw attention to one of the kinds of background knowledge, distributed categorization knowledge. Distributed categorization knowledge refers to the categorization hierarchies associated with the various steps and subplans of the reference point plans (as opposed the categorization hierarchy associated with the overall plan).

Assume the commonsense planner selects a reference point from the category of plans that it intends to accomplish. In a given situation, if a problem arises, the planner's refitting behavior is directed towards finding an alternate reference point. Alternate reference points can take one of three forms:

[1] *Select a different reference point from the overall category of plans.* Suppose the planner intends to make a call from work, and its usual plan for the office situation is with-secretary (see figure 2). Furthermore, suppose the secretary was out of the office. In this case planner is directed towards choosing an alternate reference point for the overall category of plan, i.e. from-home.

[2] *Select a different reference point for the individual step.* Suppose the planner decides to use a pay-phone and discovers, as it attempts the first step of the plan, that it has insufficient pocket-change (see figure 3). Adapting to this situation does not require that the overall plan be changed - instead an alternate version of the individual step must be used. Within the distributed categorization hierarchy associated with getting-change there exists several alternate reference points: getting change from a merchant, asking somebody else for change, or using a change machine. Depending on the current planning circumstance one or another of the reference points is appropriate.

[3] *Select a different reference point for the implicit subplan.* A given plan can be construed as a collection of interleaved subplans. Under certain circumstances the planner must find an alternate reference point for one of the subplans. Suppose the planner intends to make a telephone call from a pay-phone and discovers that it has no money at all (see figure 4). In this case it is the implicit pay-direct subplan for which an alternate reference point needs to be found, i.e. call-collect

The commonality for each of these methods is that they all involve **abstraction** and **specialization** within a categorization hierarchy. By a process of abstraction the planner removes details that are not appropriate to the planner's current circumstances. By a process of specialization the planner adds details, *en masse*, which are appropriate for the current circumstances. See [1,2] for heuristics that apply to each of these processes and a description of four kinds of situation difference that can occur between an old plan and a new situation.

Each of the above methods of adaptation require that the planner return to the reference point level of operation. In a sense the higher levels of the categorization hierarchies act in an organizational capacity, while the more special plans capture the habitual qualities of the commonsense domain.

5. Discussion and Summary

5.1 Discussion

Until recently, Wilensky's work on commonsense planning virtually stood alone in the AI literature as a computational model of a commonsense planner [21]. As oppposed to weak method reasoners [12] that apply to knowledge poor domains, Wilensky contended that commonsense planning was knowledge intensive. Although the emphasis of his work was on developing goal and plan meta-structures that were shared by both planning and understanding activities, Wilensky did propose an architecture for commonsense planning based on a cycle of plan proposal, plan projection (or simulation), and goal detection. Wilensky assumed that his planner was working from the most specific plan that was appropriate to a given situation, but never really developed or defended this idea.

The notion of cognitive reference points is attributable to Rosch [14]. This paper has been most directly influenced by the recent work of Lakoff on reference points (Lakoff refers to reference point reasoning as metonymic reasoning) [10]. Both Rosch and Lakoff were interest in using reference points to account for prototype effects. This work complements these previous studies by looking at domains other than natural kinds and natural language (i.e. plans). Here the role of the reference point is important because it encapsulates many of the important details that are not available at higher levels.

This work is also related to the recent work on case-based reasoners [8,9,13,5]. For the purposes of this paper I would like to differentiate between two types of knowledge that case-base reasoners can access: semantic knowledge and episodic knowledge. Roughly, an episode is a particular experience in the biography of an individual, and semantic knowledge refers to the conceptual thesaurus of the individual [20]. Semantic knowledge can be used to index and interpret the episodes of an individual [16]. Episodes can be interpreted by more than one piece of semantic knowledge. When a case-base reasoner works from an episode, it must first extract the relevant plan structure and then adapt it. When a case-base reasoner works from semantic structures, it need only adapt the structure. By-and-large the habitual nature of the commonsense domain suggests that the planner is working from semantic structures - in all likelihood the correct interpetation of an episode has already been made.

5.2 Summary

A plan instantiator planner works at the level of the general plan and recreates the details of a more specific plan as the situation demands. A reference point planner selects one of the subordinate plans as a basis for planning activity, thus it can access a greater number of functional details. This paper presents several analyses which suggest that commonsense planners are reference point planners.

One analysis demonstrated the difficulties associated with forming generalizations over sets of plans. In order to create a general plan from a subordinate plan, three kinds of changes can be necessary: the deletion of steps, the generalization of the individual actions of the plan, and the relaxation of sequence constraints. Each kind of change represents a loss of information and has an associated cost.

Whenever there exists a reasonable pre-formed general plan for a given task, there exists a reasonable reference point plan. In many cases the subordinate plans differ sufficiently to make the formation of a useful generalization difficult, if not impossible - nevertheless the subordinate level plans of such categories can prove to be quite useful. Planners that dynamically create general plans, before instantiating them, are, on the whole, inappropriate for commonsense planning because they fail to take advantage of the habitual nature of the commonsense domain.

A reference point planner can achieve flexibility by taking advantage of the distributed categorization hierarchies that exist in the background knowledge. Achieving flexibility is largely a function of finding an alternate reference point. PLEXUS is an example of a commonsense planner that adapts a plan, without resorting to weak methods, via the selection of alternate reference points. In a given situation PLEXUS can adapt a plan in one of three ways: it can select a different reference point from the overall category of the plan; it can select a different reference point for an individual step; it can select a different reference point for the implicit subplan. Each of these three adaptive methods can be characterized as a process of abstraction and specialization.

In some ways this argument is analogous to the following choice of wagers:

Heads you win. Tails no bet.

Heads you win. Tails you lose.

For habitual activites reference point planning is equivalent to the first wager and plan instantiation is equivalent to the second wager. Under the best of circumstances the reference point plan is tailor-made for a given situation, and under the worst of conditions it is no worse than working from a general plan via a process of plan instantiation. Thus in domains of habitual activity - that is domains where habitual activities are the rule and not the exception - gearing the planner, and its representation of planning knowledge, towards reference points can be a saving heuristic.

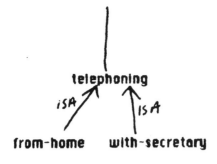

Figure 2: Overall Plan

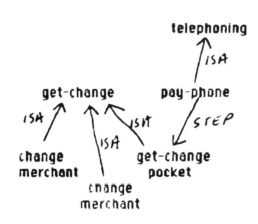

Figure 3: Individual Step.

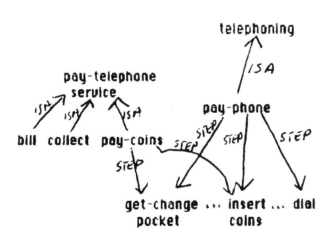

Figure 4: Implicit Subplan

REFERENCES

[1] Alterman, R. "An Adaptive Planner" in AAAI-86.

[2] Alterman, R. "Issues in Adaptive Planning." Technical Report No. UCB/CSD 87/304, Computer Science Division, UC Berkeley, 1986.

[3] Carbonell, J. "A computational model of analogical problem solving," in IJCAI 7, 1981.

[4] Carbonell, J. "Derivational analogy and its role in problem solving," AAAI-83.

[5] Hammond, K. "Chef: A model of case-based planning" in AAAI-86.

[6] Hayes-Roth, B and Hayes-Roth, F. "A cognitive model of planning." *Cognitive Science Journal,* 1979,275-310.

[7] Hendler, J. "Integrating marker-passing and problem solving" in COGSCI-85.

[8] Kolodner, J., and Simpson, R, "Experience and problem solving: a framework" in COGSCI-84.

[9] Kolodner, J., Simpson, R., and Sycara-Cranski, K, "A process model of case-base reasoning in problem solving" in IJCAI-1985.

[10] Lakoff, G. *Women, Fire, and Dangerous Things.* University of Chicago Press, 1987.

[11] McCarthy, J. and Hayes, P.J. "Some philosophical problems from the standpoint of Artificial Intelligence" in Meltzer and Michie (Eds.) *Machine Intelligence 4.* Edinburgh University Press, 1969.

[12] Newell, A. "Artificial Intelligence and the Concept of Mind" in R. Schank and K. Colby (Eds.) *Computer Models of Thought and Language.* W.H. Freeman and Co., 1973.

[13] Rissland, E. and Ashley, K. "Hypotheticals as heuristic device" in AAAI-86.

[14] Rosch, E. "Prototype Classification and Logical Classification: The Two Systems." Paper presented at a meeting of the Jean Piaget Society, Philadelphia, 1981.

[15] Sacerdoti, E. *A structure for plans and behavior.* Elsevier North-Holland, 1977.

[16] Schank, R.C. "Language and Memory." *Cognitive Science Journal,* 1980,4,243-284.

[17] Smith, E. and Medin, D. *Categories and Concepts.* Havard University Press, 1981.

[18] Stanfill, C. and Waltz, D. "Toward memory-based reasoning." *Communications of the ACM,* 1986,29,1213-1239.

[19] Sussman, G. *A computer model of skill acquisition.* Elsevier North-Holland, 1975.

[20] Tulving, E. "Episodic and Semantic Memory" in E. Tulving and W. Donaldson (Eds.), *Organization of memory.* Academic Press, 1972. Compiler."

[21] Wilensky, R. *Planning and Understanding.* Addison- Wesley, 1983.

Conditioning and Learning Within Rule-Based Default Hierarchies

Keith J. Holyoak
University of California, Los Angeles
Kyunghee Koh   Richard E. Nisbett
University of Michigan

Abstract

We present a theory of classical conditioning based on a parallel, rule-based performance system integrated with mechanisms for inductive learning. Inferential heuristics are used to add new rules to the system in response to the relationship between the system's predictions and environmental input. A major heuristic is based on "unusualness": novel cues are favored as candidates to predict important, unexpected events. Rules have strength values that are revised on the basis of feedback. The performance system allows rules to operate in parallel, competing to control behavior and to obtain reward for successful prediction of important events. Sets of rules can form default hierarchies, in which exception rules "censor" useful but imperfect default rules, protecting them from loss of strength. The theory, implemented as a computer simulation, accounts for a variety of phenomena (e.g., rapid learning in certain inhibition paradigms, and failure to extinguish non-reinforced inhibitory cues) that previous associationist accounts have not dealt with successfully.

Introduction

Intelligence manifests itself in the adaptation of goal-directed systems to complex and potentially dangerous environments. The kinds of learning that underlie such adaptation fall under the rubric of *induction*, broadly defined as those inferential processes that expand knowledge in the face of uncertainty (Holland, Holyoak, Nisbett, & Thagard, 1986). Holland *et al.* presented a framework for induction that encompasses phenomena ranging from animal learning to human categorization, analogical reasoning, and scientific discovery. In the present paper we present a theory, derived from the general Holland *et al.* framework, that applies to some of the simplest forms of inductive learning: those observed in studies of classical conditioning in animals. The theory is implemented in a computer simulation that creates and revises rule-based default hierarchies. We will first describe some limitations of previous conditioning models.

CER Paradigm and Earlier Models

In a typical conditioning experiment, a rat is first trained to press a lever to get food. After many sessions of pressing the lever for food, a distinctive tone (the conditioned stimulus, or CS) is presented for several seconds. Just as the tone goes off, a shock (the unconditioned stimulus or US, also termed the "reinforcer") is delivered to the rat's feet. As this sequence of events is repeated, the rat soon begins to show signs of fear when the tone is heard. While the tone is on, the animal suppresses its routine lever pressing and eating, and displays a collection of behaviors, such as crouching, that constitute a *conditioned emotional response* (CER). The tone now signals shock and the rat exhibits fear in response to it.

Early theories of conditioning generally assumed that temporal contiguity of the CS and the US was necessary and sufficient to establish a conditioned response (CR). In fact, however, an event may be paired with another event and still not result in conditioning, as in the "blocking" phenomenon observed by Kamin (1968) (for recent reviews, see Mackintosh, 1983; Rescorla & Holland, 1982). Further difficulties for conditioning theories based simply on association arise in studies of *conditioned inhibition*. Kamin (1968) trained rats to associate noise with shock. He then paired the noise with light. This compound was presented for several trials, but was never paired with shock. Traditional associationist assumptions would predict that the light would take on the acquired fear conditioned to the noise. In fact, however, the effect of the light was to *inhibit* fear, which it did from the very first trial. The phenomena associated with conditioned inhibition indicate that whether conditioning is excitatory or inhibitory depends on the information that the cue provides about occurrence of the US, and cannot be predicted by simple temporal contiguity.

Rescorla and Wagner (1972) proposed an associationistic model of conditioning, in which the strength V_{CS} of an association is revised in accord with a linear model,

$$\Delta V_{CS} = \alpha_{CS}(\lambda_{US} - \sum_{j=1}^{n} V_j) \tag{1}$$

where α_{CS} is a constant that determines how fast conditioning can occur for a given CS, λ_{US} denotes the asymptotic limit of conditioning that can be supported by the US, and ΣV_j represents the sum of the current strengths of associations to the US from the stimuli present (the particular CS plus all other concurrent cues).

Equation 1 is essentially equivalent to the Widrow-Hoff rule familiar in adaptive-systems theory, and is a generalization of the perceptron convergence rule (see Sutton & Barto, 1981). The Rescorla-Wagner model is thus closely related to a major class of strength-revision procedures used to model associative learning within adaptive networks of the sort currently being explored by connectionist theorists (Rumelhart, Hinton, & Williams, 1986). Within the connectionist framework, all learning is viewed as the product of incremental changes in connection strengths with experience.

Problems with the Rescorla-Wagner Model

Despite notable empirical successes, the Rescorla-Wagner model does not account well for a wide variety of other phenomena, some of which antedate their treatment and some of which are more recent. These inadequacies have spurred development of numerous other models that provide refinements and alternatives (Mackintosh, 1975; Pearce & Hall, 1980; Wagner, 1978, 1981). All of the models subsequent to that of Rescorla and Wagner (1972) adopt variations of the standard associationist framework in which conditioning is treated solely in terms of strength revision. In our view, none solves all of the empirical difficulties that beset the Rescorla-Wagner theory while preserving its successes. Here we describe some of these difficulties.

Learned irrelevance. A major problem for the Rescorla-Wagner formulation is that it does not account for changes in the processing of CSs with experience. Equation 1 predicts that a stimulus that is uncorrelated with a reinforcer will begin and end with zero associative strength. No distinction is drawn between stimuli that the animal has encountered before and those that it has not. But in fact conditioning is severely retarded if the to-be-conditioned stimulus has been presented previously (Baker & Mackintosh, 1977).

Rapid learning effects. None of the associationist models provides a satisfactory account of extremely rapid learning effects that are sometimes observed in conditioning studies. Kamin

(1968) and Rescorla (1972) have both shown that conditioning can take place after one trial, or indeed, even during an initial trial. For example, rats in the Kamin study showed less fear on the very first trial on which the excitatory stimulus was preceded by the inhibitory stimulus, despite the fact that the animal had no way of knowing whether this lessened fearfulness was justified. Such one-trial and no-trial effects look much more like inferences than like products of traditional trial-and-error learning.

Of course, learning is often much less rapid, and associationist models all include free parameters that govern rate of learning. The problem, however, is that the models provide no principled specification of when to expect rapid one-trial learning and when to expect slow learning requiring scores or even hundreds of trials. The rapid acquisition of conditioned inhibition is particularly problematic for all connectionist models of learning, because it involves the rapid alteration of a well-learned response (fear elicited by the excitatory CS), which has been gradually strengthened by reinforcement over many trials. There has been no demonstration that any connectionist model, using a consistent set of parameters for strength revision, can simultaneously account for the slow acquisition of a strong response with reinforcement, its typical pattern of slow extinction with nonreinforcement, and its immediate displacement by an incompatible response under certain specifiable conditions.

Conditioned inhibition. The Rescorla-Wagner model can account for the basic phenomenon of conditioned inhibition, as established in an A+, AX- paradigm (that is, stimulus A is always followed by the US, whereas the compound stimulus AX never is). The model assumes that strength values can be negative as well as positive. Cue A will reach an asymptotic strength equal to λ_{US}, whereas X will reach an asymptote equal to $-\lambda_{US}$; thus the net strength of the AX compound will be 0.

However, this account incorrectly predicts that inhibitory conditioning effects, once established, should be extinguishable by presenting the inhibitory CS alone, in the absence of either the US or of excitatory CSs. As Zimmer-Hart and Rescorla (1974) put it, "Assuming that nonreinforcement supports a zero asymptote ... a simple nonreinforcement of a previously established inhibitor should produce a change. If V_X is negative, then the quantity $(0 - V_X)$ is positive and consequently V_X should be incremented toward zero when it is separately nonreinforced. That is, the theory predicts that repeated nonreinforced presentation of an inhibitor should attenuate that inhibition" (pp. 837–838). One might expect, however, that when a cue predicts the nonoccurrence of a US, then repeated presentations of the cue in the absence of the US will provide additional confirmations of the expectation of nonoccurrence, and hence would *enhance* its inhibitory properties. In fact (when ceiling effects on inhibition were controlled), this is the result obtained by Zimmer-Hart and Rescorla (1974).

Adaptation Within a Rule-Based Default Hierarchy

In view of the above and other limitations of associationist models of conditioning, it seems worthwhile to investigate an alternative approach. The theory of conditioning we will present is derived from the framework for induction proposed by Holland et al. (1986). We will first sketch the general framework, and then outline the specific theory of conditioning and describe its embodiment in a computer simulation.

Rule-Based Mental Models

Holland *et al.* proposed that representations of the environment take the form of sets of rules, with associated strength values, that comprise mental models. A mental model is an internal representation that encodes the world into categories, and uses these categories to define an internal transition function that mimics the state changes that unfold in the world. In the relatively simple world of a rat's conditioning chamber, for example, the animal may learn that an occurrence of an instance of the category "loud tone" signals a transition to the environmental state "painful shock". The rat's knowledge about the relationship between tones and shocks might be informally represented by a rule such as, "If a tone sounds in the chamber, then a shock will occur, so stop other activities and crouch." Rules at different levels of generality form default hierarchies, as in situations that give rise to conditioned inhibition. For example, if tones are typically followed by shock, unless paired with a light, the rule "If tone, then expect shock" can serve as a useful but fallible default, to be overridden by the exception rule "If light and tone, then do not expect shock" when both rules are matched.

The Holland *et al.* (1986) framework provides a number of specific principles that can be applied to animal conditioning:

(1) The probability that a set of rules will control behavior increases monotonically with the strengths of the rules in the set relative to the strengths of competing rules that are matched. The function relating strength and response probability must be probabilistic in order to allow opportunities for weaker rules to be tested, and to gain strength if they prove more useful than their competitors.

(2) Rules form default hierarchies in which useful but imperfect default rules are "protected" from strength reduction by more specific exception rules that can override the defaults in particular circumstances.

(3) Only rules that succeed in controlling responses are subject to strength revision. In terms of an economic analogy, rules that control behavior "pay" for the privilege by a reduction in their strength, and must "earn" at least as much reward in the form of a subsequent strength increase in order to make the transaction worthwhile. Rules that do not gain control over responses produce no consequences for the system, and therefore neither gain nor lose strength.

(4) When multiple rules operate as a set to control behavior, they divide any attendant reward. This competition for reward implies that rules accrue greater reward when they uniquely make a correct prediction than when other rules make the same prediction (cf. Kelley's, 1973, · "discounting principle"). Reward competition provides an inductive pressure that tends to favor general rules over redundant rules that are more specific, and that impairs learning of new rules that serve the same function as existing strong rules.

(5) New rules are generated in response to particular states of the system that suggest a new rule might be useful. In the current implementation of our conditioning theory, three triggering situations are identified: (a) the occurrence of an unexpected and important event; (b) the failure of a prediction based on a rule that had previously been highly successful; and (c) the occurrence of an unusual feature in temporal contiguity with a known predictor of an important event.

(6) Inferential heuristics favor certain features over others as building blocks for new rules. In particular, unusual features of the environment are favored as candidates to build the conditions of new rules.

A Rule-Based Theory of Conditioning

We have constructed a simulation model of conditioning based on the above principles. Figure 1 depicts the basic components of the processing system. In general terms, the system matches the conditions of rules against a "message" representing the current state of the environment (simulating perceptual input), and uses the matched rules to select an effector action and to generate a message describing the predicted next state of the environment. The predicted message is compared to that observed on the next time step, and the result governs the reward given to rules that generated the prediction, storage of unusual events in a short-term buffer, and triggering of the generation of new rules.

Knowledge representation. As Figure 1 indicates, the model includes four types of information, which for simplicity we will describe in terms of four memory stores. Three of these store information of a declarative nature: a long-term store of all the cues that have appeared in the environment, tagged with a measure of degree of familiarity; a short-term store for recent unusual events; and currently active messages. The fourth store contains the rules in the system. Each of these stores is dynamically updated as the model operates in a simulated environment.

In the simulation, rules are represented in "classifier" notation (Holland, 1986). For purposes of exposition, we will use a more mnemonic notation. Thus the rule, "If tone occurs, then expect shock and crouch," will be represented simply as T$\Rightarrow$ S. We will use the classifier symbol "#" to represent a maximally general condition. Thus the rule, "If in the conditioning chamber, expect shock and crouch," will be represented as # $\Rightarrow$ S. A rule with the condition "#" will be matched on every processing cycle.

The three declarative stores are each extremely simple. At any time, the message buffer contains three active messages: one describing the current environment, one describing the environment of the preceding cycle, and one, created on the preceding cycle, that predicts the current environment.

The short-term event buffer holds recent "unusual" events that occur in the environment. In the simulation, an unusual event is defined as an observed message that includes the onset of an "unfamiliar" feature, where familiarity of a feature is a function of its number of occurrences. This familiarity count for each feature is maintained in the long-term feature store. When an unusual event occurs, the message representing it, with the unfamiliar features tagged, is placed in the short-term event store and held for a few cycles, during which period it may be used by the rule-generation heuristics described below to form conditions of new rules.

Performance system. Each cycle of the processing system begins when a message describing the current state of the environment is received. The feature portion is compared with the corresponding portion of the prediction message posted on the previous cycle. The result determines how much reward, R (if any), is given to the rules that acted on the previous cycle. Specifically, reward is given in three circumstances: (a) if shock was predicted and occurred (a large positive reward); if absence of shock was predicted and shock did not occur (a lesser positive reward); and (c) if absence of shock was predicted but shock occurred (a negative reward, i.e., a punishment). Otherwise the reward is zero. The reward, if any, is added to the strength values of the relevant rules. The comparison is also used to trigger the generation of new rules (see below).

The message representing the current environment is compared to the message representing the previous environmental state to determine whether an unusual event has occurred. If a feature-onset occurs, a check of the long-term feature list is made to determine if the feature is unfamiliar, in which case the current event is defined as unusual and entered in the short-term event store.

Rule matching and response selection. Conditions of all rules are then matched against the message describing the current environment. Each matched rule posts a *bid*, which is a proportion of its strength. That is, the bid b made by Rule i will be

$$b_i = k * s_i, \tag{2}$$

636

where k is a constant between 0 and 1. Rules that make the same prediction sum their bids and act together as a set.

The rules that will govern the system's response are then selected in accord with Principles 1 and 2 above. Principle 1, the assumption of a probabilistic relationship between relative strengths of competing rules and response selection, is realized by applying a simple version of the Luce (1963) choice model to the strengths of matched rules. Principle 2, the assumption that default rules are protected by exception rules, is realized by allowing exception rules to "censor" their corresponding default rules (cf. Winston, 1986). When exception rules censor a default, the exception rules substitute for the default on that cycle. The rules that are selected to determine the system's response on a cycle will be termed the *winning set*, W. The action called for by the winning set is performed, creating a new predicted message, and the indicated effector action is taken. The next cycle then begins.

Strength revision. Strength revision takes place in two steps. In accord with Principle 3 above, only the rules in W have their strengths changed. The rules in W effectively compete for reward, as called for by Principle 4, in accord with the following scheme. First, when the winning set is selected, each rule in the set has its strength *reduced* by an equal portion of the summed bid made by the rules in W, $\sum_{j=1}^{n} b_j$, where n is the number of rules in W. Second, when the reward R is assigned to the winning set on the subsequent cycle, an equal portion of R (i.e., R/n) is added to the strength of each rule in W. The net change in the strength of Rule i, then, is

$$\Delta s_i = (R - \sum_{j=1}^{n} b_j)/n \text{ for } i \in W, \ 0 \text{ otherwise.} \tag{3}$$

Rule generation. The program contains three inferential heuristics for generating new rules, triggered by particular states of the system as specified by Principle 5. All three heuristics are specific instantiations of the unusualness heuristic (Principle 6).

(1) *Covariation Detection.* If a shock occurs unexpectedly, and is preceded by or concurrent with an unusual event stored in the short-term buffer, then a new rule will be constructed. The new rule will include the unfamiliar feature of the unusual event in its condition, and will have an action specifying expectation of a shock and crouching. For example, if an unfamiliar tone begins prior to an unexpected shock, the rule $T \Rightarrow S$ will be generated. If no unusual event is stored, and no other heuristic applies, then with some probability less than 1 a general rule is constructed with a maximally general condition and the same action as above. Thus if an unexpected shock occurs, and no unusual event is stored, the rule $\# \Rightarrow S$ may be generated.

(2) *Exception Formation.* If a strong rule makes an erroneous prediction about the presence or absence of a shock, and an unusual event occurred prior to or concurrent with the prior cycle (when the failed rule was matched), then exception rules are formed by (a) adding the unusual feature to the condition of the failed rule, and substituting the appropriate action, and (b) using the unusual features alone to form the condition. The failed rule is preserved as a default, tagged with the newly created exception rules. For example, suppose a tone occurs paired with an unfamiliar light, and the strong rule $T \Rightarrow S$ is a member of the winning set and creates an expectation of shock, which fails to occur. The rules $L + T \Rightarrow \bar{S}$ and $L \Rightarrow \bar{S}$ will then be generated. The former exception rule corresponds to the hypothesis that the unusual cue (L) signals nonoccurence of shock only in the presence of the known predictor (T); the latter exception rule captures the more general possibility that the unusual cue might signal absence of shock regardless of whether the known predictor occurs.

(3) *Chaining.* If an unusual event is stored in the buffer, and a strong rule is included in the current winning set, then a new rule may be formed that uses the unfamiliar features of the

unusual event in the condition, and which has the same action as the parent rule. The initial strength is set to a proportion of the strength of the strong rule from which the new rule was constructed. Chaining tacitly seeks earlier predictors of the US; hence the strength of the new rule is set higher if the onset of the unusual event preceded the old CS (that is, preceded the cycle on which the parent rule was matched) than if it was concurrent with it. As an example, suppose a tone occurs concurrently with an unfamiliar light, and the strong rule $T \Rightarrow S$ is a member of the winning set. Then the rule $L \Rightarrow S$ may be generated (at the lesser initial strength value).

The three heuristics serve related but distinct functions. Covariation Detection provides initial rules to explain unexpected occurrences, whereas Exception Formation and Chaining build on existing partial knowledge. Exception Formation creates exception rules that may censor (and hence protect from further strength reduction) strong default rules that err under identifiable circumstances. Exception Formation is thus a reaction to an erroneous prediction. In contrast, Chaining represents an opportunistic attempt to identify an earlier predictor of a US that is already predicted by a CS.

Simulations of Conditioned Inhibition

We have simulated several variations of the CER paradigm, including blocking (Kamin, 1968) and the effects of statistical predictability on learning (Rescorla, 1972). The unusualness heuristic provides an explanation of the fact that unfamiliar cues are maximally conditionable. Here we will present simulations of two studies of conditioned inhibition described earlier, which pose difficulties for the Rescorla-Wagner model.

Rapid learning effects. An experiment by Kamin (1968) provides a dramatic demonstration of rapid inhibitory conditioning. Kamin trained rats for 16 trials to associate white noise with shock. He then created two different groups. Group LN received eight trials of a compound, simultaneous light-plus-noise stimulus that was never reinforced by shock, followed by four trials of the original noise stimulus which was again nonreinforced. Group N simply received 12 standard extinction trials, during which the noise was presented but never with shock. All animals received four trials per day.

Let us analyze the predictions our model makes for this study, based on rule generation and subsequent competition. Group N is of course expected to show just the customary gradual extinction as the rule $N \Rightarrow S$ dies a slow death due to nonreinforcement. The situation is much more complex for Group LN. Given that a trial in Kamin's experiment spanned a fairly long interval (several minutes), this time period would correspond to several cycles of matching and firing rules. Since no shock was presented on the first trial in which the light occurred along with the tone, the strong rule $N \Rightarrow S$ would repeatedly fail. Given the availability of an unusual event—the occurrence of the light—heuristics for rule generation will be triggered. In particular, Exception Formation should on the initial extinction trial generate new exception rules, $L + N \Rightarrow \bar{S}$ and $L \Rightarrow \bar{S}$. Both new rules will have an immediate inhibitory influence.

When the light is first paired with the excitatory noise, Chaining may create the excitatory rule $L \Rightarrow S$, which will compete with the other new rules. However, because the new cue does not occur prior to the original CS, the initial strength of the rule generated by Chaining will be low. Consequently, the influence of the new inhibitory rules will outweigh the influence of the excitatory ones, so that Group LN might be expected to show some inhibition of suppression even on the very first trial. Furthermore, the new inhibitory rules will of course be confirmed, and so the animal should show rapid development of inhibition over trials—much faster than Group N, which has no cue to suggest that the initial learning situation has now changed.

What should happen when, after the first eight trials, the noise alone is presented? For Group N, nothing interesting. This is merely a continuation of the slow competition between the rule N $\Rightarrow$ S and its original competitor # $\Rightarrow$ Press. For Group LN, however, we expect a reversion to substantial suppression effects, because for rats in this condition the rule N $\Rightarrow$ S will have been protected to some extent due to the censoring effect of the successful exception rules.

Figure 2 presents the the simulation results, which capture the major qualitative aspects of Kamin's findings. The data are presented as *suppression ratios*—the ratio of bar-pressing rate during CS presentations to the rate during the CS plus during its absence. A ratio of .5 indicates no excitatory conditioning to the CS, and a ratio of 0 indicates maximal conditioning. The results for Group N, presented with the noise alone, may be seen at the bottom of Figure 2. These animals showed the customary slow extinction process. The results for group LN are utterly different. The very first trial shows a substantially reduced suppression effect. The next trial, the first that confirms the new inhibitory rules, shows a further reduced suppression effect. By the fourth experience of the nonreinforced compound, the suppression ratio has become asymptotic.

Then, four trials after that, the single stimulus N is introduced. For Group N, this is by now simply the standard occurrence, but for Group LN it is an event not encountered since the original conditioning trials, during which N alone was always accompanied by shock. Because the rule N $\Rightarrow$ S has been partially protected from strength reduction by the exception rule L + N $\Rightarrow$ $\bar{S}$, Group LN rats show considerable suppression on the very first presentation of N alone.

Increased inhibition due to an "extinction" procedure. We earlier discussed a phenomenon that is especially problematic for the Rescorla-Wagner formulation—failure to demonstrate extinction of the inhibitory power of an inhibitory cue that is presented in the absence of either the excitatory cue or reinforcement (Zimmer-Hart & Rescorla, 1974). In one experiment all rats were first given training in bar pressing to obtain food, followed by an initial session in which a 30-second tone was presented four times, ending each time with a shock. This would establish the rule T $\Rightarrow$ S.

The animals then were divided into two groups, which received different procedures for inhibitory conditioning. For both groups, each subsequent session involved four presentations of the tone paired with shock, intermixed with four presentations of the tone in combination with a flashing light without shock. These events would generate the inhibitory rules L + T $\Rightarrow$ $\bar{S}$ and L $\Rightarrow$ $\bar{S}$ by Exception Formation, establishing the light as an inhibitory cue.

Group 1 received no other presentations of CSs. However, Group 2 also received four intermixed presentations of the light alone without shock. From the perspective of the Rescorla-Wagner theory, these were "extinction" trials that should have diminished the inhibitory power of the light, thus slowing down the acquisition of inhibition to the light-tone compound for Group 2 relative to Group 1. In contrast, from the point of view of our theory these are additional occasions for strengthening of the L $\Rightarrow$ $\bar{S}$ rule. Since this rule contributes an inhibitory influence when the light-tone compound is presented, its strengthening when the light is presented alone should actually accelerate early acquisition of inhibition to the compound. However, the "light" rule will share reward with the other "light plus tone" exception rule when the compound is presented. Greater strength of the former rule will eventually lead to diminished strength of the latter. Accordingly, our model predicts that Group 2 will show less suppression than Group 1 to the compound early in acquisition, but that the two groups will show comparable suppression later in training.

Figure 3 presents the data from our simulation of this experiment. These data were obtained on test trials involving three reinforced presentations of the tone and two nonreinforced presentations of the light-tone compound. Suppression to the tone presented alone was asymptotic for both groups over the entire test period. The simulation model correctly indicates that animals in Group 2, which experienced separate presentations of the inhibitory light CS without reinforcement, should exhibit *increased* inhibition to the light-tone compound during early trials.

Conclusions and Future Directions

We are optimistic that the present theory can provide insights into forms of learning more complex than classical conditioning. The model can be extended to the acquisition of rule sequences by adding the "bucket brigade" algorithm for back-chaining strength to early rules in a sequence that eventually achieved a goal (Holland, 1986). We also hope that the theory will at some level prove relevant to understanding higher-level human cognition. Some recent work has begun to apply theoretical models derived from studies of animal learning to human categorization and decision making. Gluck and Bower (1986) found evidence that people weight cues to category membership in terms of their relative predictive power: the degree to which a cue is used to predict category membership is decreased by the presence of other more valid cues. This phenomenon can be accounted for by models such as that of Rescorla and Wagner (1972) and our own in which redundant cues compete to acquire strength. Interestingly, as Gluck and Bower point out, most current theories of human categorization do not display this property. Given that the present model exhibits both competitive learning and the capacity to represent nonindependent cues (an important aspect of human categorization), it may prove applicable to the analysis of performance in categorization tasks.

Our general aim has been to develop a model that integrates mechanisms of hypothesis formation and strength revision within a comprehensive performance system. Theories that invoke the notion of hypothesis generation often have been unduly restrictive, in our view, in assuming that hypotheses are entertained and tested serially, and ultimately rejected if any exceptions are found. By representing hypotheses as rules that can operate in parallel, take on continuous strength values, interact as defaults and exceptions, and actively compete to control behavior and to gain reward for predictive successes, a great deal of theoretical power is gained. The present theory makes extensive use of mechanisms for strength revision; however, the introduction of heuristics for rule generation is crucial in allowing us to account for phenomena, such as rapid acquisition of exception rules, that purely associationist mechanisms have not dealt with successfully. We suspect that the explanatory power of learning models based entirely on strength revision will prove to have limits, and that those limits will be found to lie short of a full account of classical conditioning in rats, far less of human cognition. A major limitation of current connectionist models of learning is that the proposed algorithms for adjusting connection weights become computationally intractable when the number of interconnected units grows large. Heuristics that propose plausible candidate rules can function to drastically reduce the effective size of the search space in which strength-revision procedures operate.

640

Acknowledgements

Preparation of this paper was supported by NSF Grants BNS-8615316 to K. Holyoak and SES-8507342 to R. Nisbett. Requests for reprints may be sent to Keith J. Holyoak, Department of Psychology, University of California, Los Angeles, California 90024.

References

Baker, A. G., & Mackintosh, N. J. (1977). Excitatory and inhibitory conditioning following uncorrelated presentations of CS and UCS. *Animal Learning and Behavior, 5*, 315-319.

Gluck, M. A., & Bower, G. H. (1986). Conditioning and categorization: Some common effects of informational variables in animal and human learning. In *Proceedings of the Cognitive Science Society Conference.*

Holland, J. H. (1986). Escaping brittleness: The possibilities of general purpose machine learning algorithms applied to parallel rule-based systems. In R. S. Michalski, J. G. Carbonell, & T. M. Mitchell (Eds.), *Machine learning: An artificial intelligence approach*, Vol. 2. Los Altos, Calif. Kaufmann.

Holland, J. H., Holyoak, K. J., Nisbett, R. E., & Thagard, P. R. (1986) *Induction: Processes of inference, learning and discovery.* Cambridge, Mass.: Bradford Books/MIT Press.

Kamin, L. J. (1968). "Attention-like" processes in classical conditioning. In M. R. Jones (Ed.), *Miami symposium on the prediction of behavior: Aversive stimulation.* Miami, Florida: University of Miami Press.

Kelley, H. H. (1973). The processes of causal attribution. *American Psychologist, 28*, 107-128.

Mackintosh, N. J. (1975). A theory of attention: Variations in the associability of stimuli with reinforcement. *Psychological Review, 82*, 276-298.

Mackintosh, N. J. (1983). *Conditioning and associative learning.* New York: Oxford University Press.

Pearce, J. M, & Hall, G. (1980). A model for Pavlovian learning: Variations in the effectiveness of conditioned but not of unconditioned stimuli. *Psychological Review, 87*, 532-552.

Rescorla, R. A. (1972). Informational variables in Pavlovian conditioning. In G. H. Bower (Ed.), *The psychology of learning and motivation*, Vol. 6. New York: Academic Press.

Rescorla, R. A., & Holland, P. C. (1982). Behavioral studies of associative learning in animals. *Annual Review of Psychology, 33*, 265-308.

Rescorla, R. A., & Wagner, A. R. (1972). A theory of Pavlovian conditioning: Variations in the effectiveness of reinforcement and nonreinforcement. In A. H. Black & W. F. Prokasy (Eds.), *Classical conditioning II: Current theory and research.* New York: Appleton-Century-Crofts.

Rumelhart, D. E., Hinton, G. E., & Williams, R. J. (1986). Learning internal representations by error propagation. In Rumelhart, D. E., McClelland, J. L., & the PDP Research Group (1986). *Parallel distributed processing: Explorations in the microstructure of cognition*, Vol. 1. Cambridge, Mass.: Bradford Books/MIT Press.

Sutton, R. S., & Barto, A. G. (1981). Toward a modern theory of adaptive networks: Expectation and prediction. *Psychological Review, 88*, 135-170.

Wagner, A. R. (1978). Expectancies and the priming to STM. In S. H. Hulse, H. Fowler, & W. K. Honig (Eds.), *Cognitive processes in animal behavior.* Hillsdale, NJ: Erlbaum.

Wagner, A. R. (1981). SOP: A model of automatic memory processing in animal behavior. In N E. Spear & R. R. Miller (Eds.), *Information processing in animals: Memory mechanisms.* Hillsdale, NJ: Erlbaum.

Winston, P H. (1986). Learning by augmenting rules and accumulating censors. In R. S. Michalski, J. G. Carbonell, & T. M. Mitchell (Eds.), *Machine learning: An artificial intelligence approach*, Vol. 2. Los Altos, Calif. Kaufmann.

Zimmer-Hart, C. L., & Rescorla, R. A. (1974). Extinction of Pavlovian conditioned inhibition. *Journal of Comparative and Physiological Psychology, 86*, 837-845.

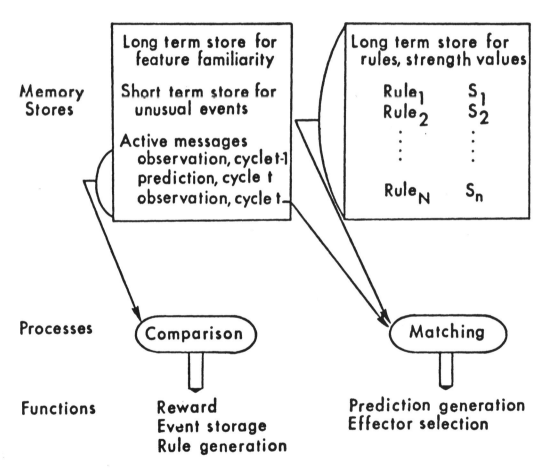

Figure 1. The memory stores and major processes involved in the process model of conditioning.

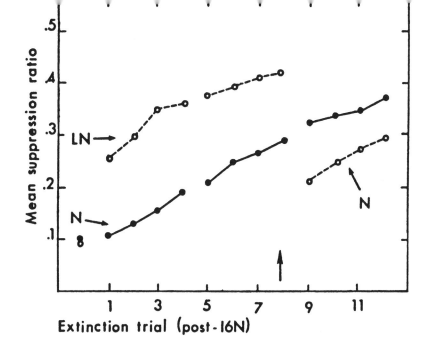

Figure 2. Simulation of experiment by Kamin (1968), showing predicted extinction of suppresion, by trial, following 16 sessions of conditioning to noise (N). The groups were extinguished either to noise alone or to a light-plus-noise (LN) compound. The arrow in the abscissa indicates point at which group extinguished to compound was switched to noise alone.

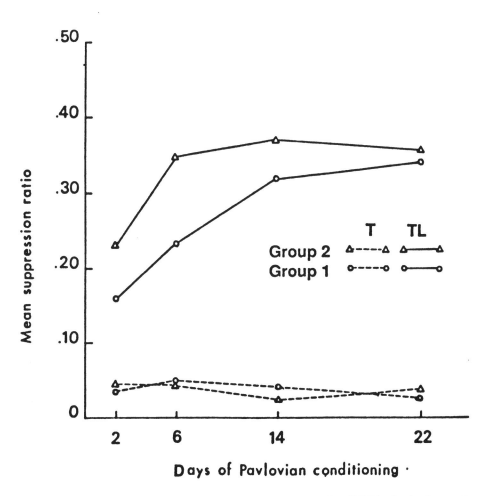

Figure 3. Simulation of experiment by Zimer-Hart and Rescorla (1974), depicting mean suppression ratios during tone-alone (T) and light-tone (LT) trials in single test sessions following various amounts of training. Both groups received T+, LT- presentations; Group 2 also received intermixed L- presentations.

SEAS: A DUAL MEMORY ARCHITECTURE FOR COMPUTATIONAL COGNITIVE MAPPING

Nestor A. Schmajuk
Center for Adaptive Systems
Department of Mathematics
Boston University

1987

ABSTRACT

We introduce a **dual memory architecture** that, by way of computing conditioned-conditioned stimulus (CS-CS) associations and conditioned-unconditioned stimulus (CS-US) associations, is capable of **computational cognitive mapping**. The network is able to describe complex classical conditioning paradigms in which cognitive mapping is presumably involved such as blocking, overshadowing, sensory preconditioning, second-order conditioning, compound conditioning, serial compound conditioning, and sensory preconditioning. By assuming that limbic-cortical regions of the brain are involved in CS-CS associations, the network is able to describe several cognitive impairments that have been reported after limbic-cortical lesions.

INTRODUCTION

Two major approaches characterize the study of the neurobiological basis of memory. One approach considers that memory is a unitary process that involves the whole brain. Another approach regards memory as a multiple process that involves different areas of the brain, each area being involved in a different type of memory (Kesner, 1984). For example, Squire (1982) suggested that hippocampal and amygdalar regions of the brain are participated in the acquisition of new information about the world (declarative memory) but not in the acquisition of new perceptual-motor skills (procedural memory). In the same vein, other authors proposed that the limbic-cortical regions of the brain would be involved in processes such as off-line associations (Hirsh, 1974), stimulus configuration (Mishkin and Petri, 1984), vertical associative memory (Wickelgren, 1979), or representational memory (Thomas and Spafford, 1984). Striatal and cerebellar regions of the brain would be involved in processes such as on-line associations (Hirsh, 1974), habit formation (Mishkin and Petri, 1984), horizontal associative memory (Wickelgren, 1979), or dispositional memory (Thomas and Spafford, 1984).

In line with the approach that regards memory as a multiple process, we have introduced a **dual memory architecture** that, by way of computing conditioned-conditioned stimulus (CS-CS) associations and conditioned-unconditioned stimulus (CS-US) associations, allows to build **computational cognitive maps** (Schmajuk, 1986a; Schmajuk, 1986b; Schmajuk and Moore, 1986). In the context of the multiple memory process approach, CS-CS associations might be regarded as components of off-line associations, declarative memory, stimulus configuration, vertical associative memory, or representational memory. CS-US associations might be regarded as components of on-line associations, procedural memory, habit formation, horizontal associative memory, or dispositional memory. Limbic-cortical areas would be involved in CS-CS associations, whereas striatal and cerebellar regions would be involved in CS-US associations.

The present paper presents a second-order associative network, designated the SEAS network (as a mnemonic for **SE**cond-order **AS**sociative), and illustrates its behavior in complex classical conditioning paradigms. The SEAS network is able to describe conditioning paradigms such as conditioned inhibition,blocking, overshadowing, sensory preconditioning, second-order conditioning, compound conditioning, serial compound conditioning, and sensory preconditioning. The network is also able to describe some very well known effects of limbic-cortical and striatal-cerebellar lesions.

THE SEAS NETWORK

First-order associations. Consider the case of one CS, CS_i, that predicts event k. Net associative value, $V_i{}^k$, represents the first-order prediction of event k by CS_i. When the CS_i is accompanied or followed by event k, the associative value between CS_i and event k, $V_i{}^k$, increases by

$$\triangle V_i{}^k = S_i \ \beta_i{}^r \ \tau_i \ (\ \Gamma^k - B^k \ ), \ [\ 1 \ ]$$

where S_i is the salience of CS_i, $\beta_i{}^r$ is $\beta_i{}^r = \theta_i{}^r$ $(0 < \theta_i{}^r < 1)$ when $\Gamma^k > B^k$, and $\beta_i{}^r = \theta_i{}^r{}'$ $(0 < \theta_i{}^r{}' < \theta_i{}^r)$ when $\Gamma^k < B^k$, τ_i is the trace of CS_i, Γ^k the intensity of event k, and B^k the aggregate prediction of event k.

Second-order associations and cognitive mapping. Consider now the case of two CSs, CS_i and CS_r, that predict event k. It is assumed that CS_i predicts k directly by $V_i{}^k$ and indirectly by predicting CS_r, by $V_i{}^r$. In turn CS_r predicts k by $V_r{}^k$. The second-order net prediction of event k by CS_i, is expressed as the product $V_i{}^r V_r{}^k$. The product $V_i{}^r V_r{}^k$ can express - quantitatively -four logical inferences. For example, if CS_i predicts the absence of CS_r (negative $V_i{}^r$), and CS_r predicts the presence of event k (positive $V_r{}^k$), CS_i will predict the absence of event k (negative $V_i{}^r V_r{}^k$)

$B_i{}^k$, the first- and second-order prediction of event k by CS_i, is

$$B_i{}^k = (\ V_i{}^k + \Sigma_r \ w_i{}^r \ V_i{}^r \ V_r{}^k) \ \tau_i . \qquad [\ 2 \ ]$$

$V_i{}^k$ is the net associative value of CS_i with event k. The sum over the index r involves all CSs with index r = k. $V_i{}^r$ is the net associative value of CS_i with all CSs with index r = k. $V_i{}^r$ is the net associative value of all CS with event k. τ_i is the trace of CS_i . The mathematical expression for τ_i is given below. Coefficient $w_i{}^r$ serves to adjust the relative weights of first- and second- order predictions in paradigms such as conditioned inhibition. In order to avoid redundant CS_i -US and CS_i -CS_i - US associations, $w_i{}^r$ = 0 when i = r, and $w_i{}^r$ > 0 when i $\neq$ r. B^k , the aggregate prediction of event k made upon all CSs (including the context) with τ > 0 at a given moment, is

$$B^k = \Sigma_i \ B_i{}^k . \qquad\qquad [\ 3\ ]$$

Variable B^k participates in the rules governing the computation of $V_i{}^k$. In addition, B^{US} determines the topography of the NM response, as described below.

The integration of different predictions, $V_i{}^r \ V_r{}^k$, into a larger and new prediction, $B_i{}^k$, is similar to the process Tolman (1932) called <u>inference.</u> For Tolman, expectancies can be combined in order to form new expectancies and organized in a "cognitive map". Up to the present, models for classical conditioning did not have any mechanism to account for "inference" processes. The introduction of second-order associations allows to build "computational cognitive maps" in which CS-CS predictions can be combined among them, and with CS-US associations. By the introduction of second-order associations the SEAS model is capable of describing sensory preconditioning and secondary reinforcement.

Figure 1 shows how SEAS explains sensory preconditioning. Sensory preconditioning is predicted by allowing CS_B to be associated to CS_A in a first phase, denoted by the solid circle $V_B{}^A$, and CS_A to be associated to the US in a second phase, denoted by the solid circle $V_A{}^{US}$. When CS_B is presented alone in a test trial, it activates the A representation through node $V_B{}^A$, and this A representation activates the node $V_A{}^{US}$, generating a conditioned response (CR).

<u>Trace</u> <u>function.</u> It is assumed that a CS_i generates a trace, τ_i , that increases over time to a maximum, stays at this level for a period of time independent of the CS duration, and then gradually decays back to zero. Formally, trace τ is defined for t <= 200 msec by

$$\tau(t) = CS_{max} (\ 1 - e^{-(\ k1\ t\ )}\ ), \qquad [\ 4\ ]$$

where CS_{max} is the maximum intensity of the CS and k1 is a constant, 0 < k1 < 1. Parameter k1 is selected so that the ISI for optimal conditioning is 200 msec.

$\tau(t)$ remains equal to CS_{max} as long as the CS does

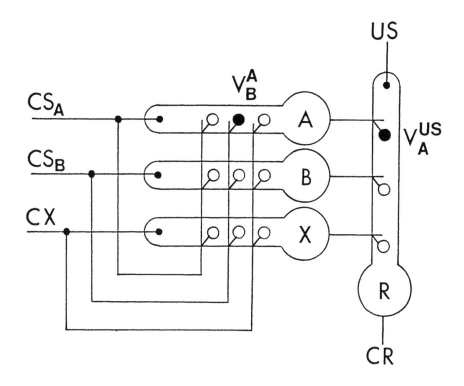

Figure 1. **SEAS neural network**. CSA and CSB : conditioned stimuli . CX: Context. US: unconditioned stimulus. CR: condioned response. VAB: CSA-CSB associative value. VAUS: CSA-US associative value. For explanation see text.

not decay. If the CS = 0 and t > 200 msec, τ (t) decays by

$$\tau(t) = CS_{max} (e^{-(k_1 t)}), \qquad [5]$$

If CS_i is not present 200 msec after its onset, the trace decays to zero.

Performance Rules. The SEAS network incorporates performance rules that permit realistic descriptions of rabbit's classically conditioned nictitating membrane (NM) responses in real time (Gormezano, Kehoe, and Marshall,

1983). Performance rules relate variable B^{US} to the topography of NM responses.

Time of CR onset is the earliest time t such that

$$\Sigma^{t}_{t'=ti} \; \Sigma_j \; B_j^{US}(t') \; >= \; L1 \; , \qquad\qquad [\; 6 \;]$$

where t_i denotes the time step at which CS_i onset occurs. The sum over the index j involves B_j^{US} of all CSs with $\tau_j >$ 0 , excluding the context. Sum over index t involves all time steps for which $\tau_j > 0$, starting at the time step when the amplitude of the NM response as defined by Equation 7 equals zero. L1 is a threshold greater than zero. Equation 6 implies that as B_j^{US} increases over trials, CR onset moves progressively to an asymptote determined by L1.

During the CS period, for time steps $t > t_i$, the amplitude of the NM response, NMR(t), is changed by

$$\triangle NMR \; (t) \; = \; k2 \; (\; B^{US}(t) \; - \; NMR(t)), \qquad [\; 7 \;]$$

where k2 is a constant ($0 < k2 < 1$).

During the US period, while $B^{US}(t) > \Gamma^{US}(t)$, is given by Equation 7. However, when $B^{US}(t) < \Gamma^{US}(t)$, NMR (t) increases by

$$\triangle NMR \; (t) \; = \; k2 \; (\; \Gamma^{US}(t) \; - \; NMR(t)), \qquad [\; 8 \;]$$

When $B^{US}(t)$ and $\Gamma^{US}(t)$ equal zero, NMR(t) decays to baseline by

$$\triangle NMR \; (t) \; = \; - \; k2 \; NMR(t). \qquad\qquad [\; 9 \;]$$

Effects of cerebellar lesions. A description of the effect of cerebellar lesions (CL) in agreement with Lincoln, McCormick, and Thompson's (1982) results, is obtained by assuming that lesions of this limbic structure impair CS-US associations but not the computation of CS-CS associations. Mathematically, after CL it is $V_i^{US} = 0$.

Effects of hippocampal lesions. A description of the effect of hippocampal lesions (HL) in agreement with experimental data (see Schmajuk, 1984, for a review) is obtained by assuming that lesions of this limbic structure impair CS-CS associations but not the computation of CS-US associations. Mathematically, after HL it is $V_i^{\Gamma} = 0$.

Impairments in CS-CS associations imply impairment in **cognitive mapping**. Since $V_i^{\Gamma} = 0$, B_i^k is given by

$$B_i^k \; = \; V_i^k \; \tau_i \; . \qquad\qquad [\; 10 \;]$$

Because B_i^k for HL animals computed with Equation 10 is larger than B_i^k for normal animals given by Equation 2, use

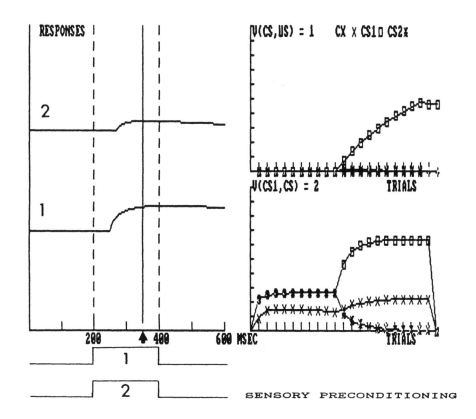

Figure 2. **Sensory preconditioning.** [1] : CS(1). [2] :
CS(2). [X] : Context. Left Panels: NM response topography
in 1- and 2- trials, after 10 CS(1)-CS(2) nonreinforced
trials and 10 CS(1)reinforced trials. Upper-Right Panels:
CS-US associative values, V(CS,US), at the end of each trial,
as a function of trials. Lower-Right Panels: CS1-CS
associative values, V(CS1,CS), at the end of each trial, as a
function of trials.

of Equation 10 implies impairments in several classical
conditioning paradigms, including blocking and sensory
preconditioning.

COMPUTER SIMULATIONS

In the simulations, continuous time was converted to
discrete time steps or bins of 10 msec in duration. Each
trial consisted of 60 bins. Otherwise specified, the
simulations assumed 200 msec CSs, the last 50 msec of which
overlaps the US.

Initial values of Vs were zero for all i's. Parameters
values for variations of associative values were : S_1 = 1, S_2
= 1, S_X = .1 . $\theta_i{}^r$ = 0.3 and $\theta_i{}^{r'}$ = 0.03 for r = US . $\theta_i{}^r$ =

0.015 and $\theta_i r' = 0.0015$ for $r = US$. For computations of $B_i k$: $w_i k = 2$ when $i \neq r$; and $w_i k = 0$ when $i = r$. For computations of the NM CR : $L1 = 2$. For computation of the trace: $k1 = 0.1$, and for the NM response topography : $k2 = 0.5$.

Simulation results.

Sensory preconditioning. Figure 2 shows simulations of a sensory preconditioning paradigm. In the first phase, 10 nonreinforced trials with a compound CS(1 and 2). During the second phase, one of the nonreinforced CSs (1) was reinforced for 10 trials. A test trial assessed the CR to CS(2) never paired with the US. Simulations showed that context associability decreases during preconditioning. In the nonreinforced test trial CS(2) acquired inhibitory associative value because it was presented in a context with excitatory associative value. CS(2) generated a CR. Simulation results are in agreement with data reported by Port and Patterson (1984) for normal animals. After HL, CS-CS associations are absent and therefore sensory preconditioning is also absent, a result in agreement with Port and Patterson (1984), who found that fibrial (hippocampal output) lesions in rabbits impaires sensory preconditioning.

Serial Compound Conditioning. Figure 3 shows simulations of a serial compound conditioning paradigm, in which two conditioned stimuli (CS1 and CS2) are followed by the US. The temporal primacy of CS1 over CS2 determines CS1 to become more strongly associated with the US than CS2, in spite of the contiguity of CS2 and the US. As shown in Figure 3, CS1 generates a CR larger than that generated by CS2.

Our results are in agreement with Wickens, et al (1973), who found that, after a CS1-CS2 serial compound had been paired with a US, associations acquired by CS1 and CS2 were functions of the CS1-CS2 interval. With a long CS1-CS2 interval, each CS-US association was inversely proportional the respective CS-US interval, and therefore, the CR generated by CS2 was larger than the CR elicited by CS1. With an intermediate CS1-CS2 interval, as in the case of our simulation, the CR elicited by CS1 was larger than that elicited by CS2. Finally, with a short CS1-CS2 interval, the CR generated by CS2 was larger than that produced by CS1. According to the SEAS model, associations acquired by CS1 and CS2 are functions of the CS1-CS2 interval, because the CS1-CS2 interval establishes the degree of association between CS1 and CS2 ($V_1 2$ by Equation 1), and this degree of association between CS1 and CS2 controls the associative value of CS1 and CS2 with the US (B^{us} by Equation 2).

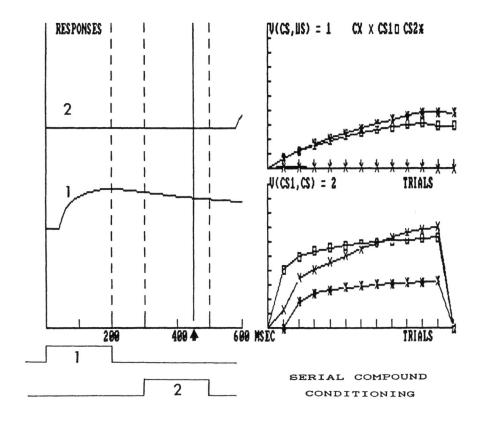

Figure 3. **Serial compound conditioning.** [1] : CS(1). [2] : CS(2). [X] : Context. Left Panels: NM response topography in 1- and 2- trials, after 10 CS(1)-CS(2) nonreinforced trials and 10 CS(1)reinforced trials. Upper-Right Panels: CS-US associative values, V(CS,US), at the end of each trial, as a function of trials. Lower-Right Panels: CS1-CS associative values, V(CS1,CS), at the end of each trial, as a function of trials.

The SEAS network predicts that serial compound conditioning is impaired after HL, each CS being able to acquire associations inversely proportional to their contiguity with the US. This predictions awaits experimental testing.

Blocking. Figure 4 shows simulations of a blocking paradigm. Experimentals received 10 trials with CS (1) (blocker) paired with the US followed by 10 trials with CS (1) and CS(2) (blocked CS) paired with the US. The network showed simulated blocking in the normal case (N) because the designated blocked CS(2) does not generate a CR. After HL the network predicts that the blocked CS (2) will show a larger CR than it does in the normal case. The results agree with blocking data in the normal rabbit NM response preparation as reported by Marchant and Moore (1973), and in the HL rabbit as reported by Solomon (1977).

DISCUSSION

The present paper introduce SEAS, a dual memory architecture that is capable of generating computational cognitive maps. When applied to classical conditioning, the network describes several complex classical conditioning paradigms in real time.

The SEAS network is able to describe paradigms, such as serial compound conditioning, that had been succesfully explained by attentional theories of conditioning (see Kehoe, 1983). This fact points out to a degree of equivalence between attentional approaches and higher-order associative approaches such as that presented here. This equivalence between attentional and higher-order associative approaches might be based on the fact that both are dual memory systems that rely on the existence of a second memory for storing information not inmediately connected to
open responses, such as CS-US associations.

In addition to the description of normal behavior, the SEAS network can describe some of the effects of cerebellar and hippocampal lesions on the classically conditioned NM response in the rabbit.

This study was supported in part by NSF Grant IST8417756.

REFERENCES

Gormezano, I., Kehoe, E.J., & Marshall, B.S. (1983). Twenty years of classical conditioning research with the rabbit. **Progress in Psychobiology and Physiological Psychology**, 10, 197- 275.

Hirsh, R. (1974) The hippocampus and contextual retrieval of information from memory: A theory. **Behavioral Biology**, 12, 421-444.

Kehoe, E.J. (1983). CS-US contiguity and CS intensity in conditioning of the rabbit's nictitating membrane response to serial compound conditioning. **Journal of Experimental Psychology**, 9, 307-319.

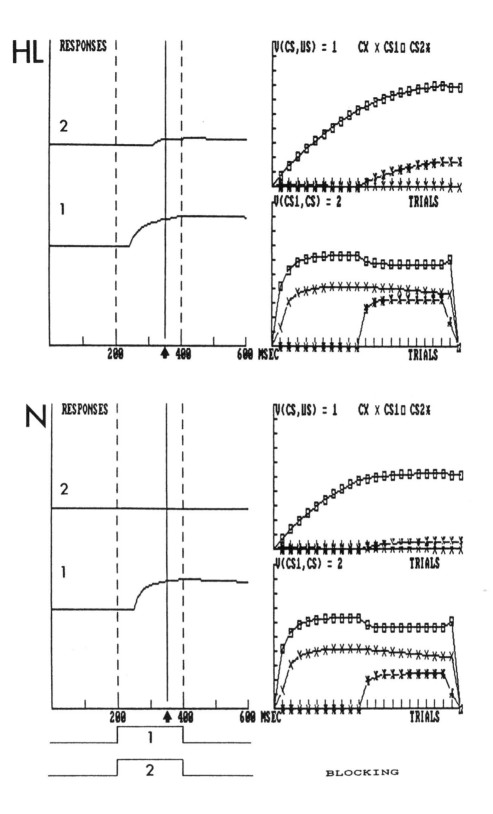

Figure 4. **Blocking.** N: normal case. HL: hippocampal
lesioned case. [1] : CS(1). [2] : CS(2). [X] : Context.
Left Panels: NM response topography in 1- and 2- test trials
after 10 CS(1) reinforced trials and 10 CS(1) and CS(2)
reinforced trials. Upper-Right Panels: CS-US associative
values, V(CS,US), at the end of each trial, as a function of
trials. Lower-Right Panels: CS1-CS associative values,
V(CS1,CS), at the end of each **trial, as** a function of trials.

Kesner, R.P. (1984). The neurobiology of memory: implicit and explicit assumptions. In **"The Neurobiology of Learning and Memory"**, G.Lynch, J.L.McGaugh, and N.M.Weinberger (Eds.), New York: Guilford Press.

Lincoln, J.S., McCormick, D.A., & Thompson, R.F. (1982). Ipsilateral cerebellar lesions prevent learning of the classically conditioned nictitating membrane/eyelid response. **Brain Research, 242**, 190-193.

Marchant, H.G., & Moore, J.W. (1973). Blocking of the rabbit's conditioned nictitating membrane response in Kamin's two-stage paradigm. **Journal of Experimental Psychology, 101**, 155-158.

Mishkin, M., & Petri., H.L. (1984). Memories and habits: Some implications for the analysis of learning and retention. In L.R. Squire and N. Butters (Eds.), **Neuropsychology of Memory.** New York: Guilford Press.

Port, R.L., & Patterson, M.M. (1984). Fimbrial lesions and sensory preconditioning. **Behavioral Neuroscience, 98**, 584-589.

Schmajuk, N.A. (1984). Psychological theories of hippocampal function. **Physiological Psychology, 12**, 13-22.

Schmajuk, N.A. (1986a). Real-time attentional associative models of classical conditioning and the hippocampus. Unpublished Doctoral Dissertation. University of Massachusetts.

Schmajuk, N.A. (1986b). A real-time model for classical conditioning. 19th Annual Meeting of the Society of Mathematical Psychology. Cambridge, Massachusetts.

Schmajuk, N.A., & Moore, J.W. (1985). Real-time attentional models for classical conditioning and the hippocampus. **Physiological Psychology, 13**, 278-290.

Schmajuk, N.A., & Moore, J.W. (1986). A real-time attentional-associative network for classical conditioning of the rabbit's NMR. Proceedings of the 8th Annual Conference of the Cognitive Society, Amherst, Massachusetts.

Solomon, P.R. (1977). Role of the hippocampus in blocking and conditioned inhiition of rait's nictitating response. **Journal of Comparative and Physiological Psychology, 91**, 407-417.

Squire, L.R. (1982). The neuropsychology of human memory. **Annual Review of Neuroscience, 5**, 241-273.

Thomas, G.J., & Spafford, P.S. (1984). Deficits for representational memory induced by septal and cortical lesions (Singly and combined) in rats. **Behavioral Neuroscience, 98**, 394-404.

Tolman, E.C. (1932). Cognitive maps in rats and men. **Psychological Review, 55**, 189-208.

Wickelgren, W.A. (1979). Chunking and consolidation: A theoretical synthesis of semantic networks, configuring in conditioning, S-R versus cognitive learnig, normal forgetting, the amnestic syndrome, and the hippocampal arousal system. **Psychological Review, 86**, 44-60.

Wickens, D.D., Nield, A.F., Tuber, D.S., & Wickens, C.D. (1973). Stimulus selection as a function of CS1-CS2 interval in compound conditioning of cats. **Journal of Comparative and Physiological Psychology, 85**, 295-303.

JANUS: An Architecture for Integrating Automatic and Controlled Problem Solving

David S. Day
Experimental Knowledge Systems Laboratory
Department of Computer and Information Science
University of Massachusetts
Amherst, Massachusetts 01003

Abstract: This paper attempts to unify two problems in cognitive science: the relationship between "controlled" and "automatic" processing and the competing computational models of intelligence proposed by symbolic Artificial Intelligence and the connectionist school. An architecture is proposed in which symbolic and connectionist problem solving systems interact and take advantage of their different strengths. It is argued that the resulting system can account for much of the problem solving behavior associated with automatic and controlled processing as well as their complex interplay. Thus, the architecture can account for how expertise can be transformed from "explicit" to "compiled" forms via automatization, and how the opacity of the resulting automatic behavior can be counterbalanced in a cognitively plausible manner by explanations generated *ex post facto*.

1. Introduction.

This paper proposes a model for integrating *controlled* and *automatic* processing. This model assumes that these two problem solving styles are distinct not only in behavior but in implementation as well, and so an architecture has been designed in which the two competing mechanisms can also communicate and cooperate with each other. It is believed that this will result in system-wide behavior that accords well with psychological accounts of controlled and automatic problem solving and related behaviors.

The assumption of the distinct implementation of controlled and automatic problem solving grows out of the current progress in connectionist models of cognition and the resulting tension between this view and that of the symbolic Artificial Intelligence (AI) approach to these same problems. A debate has formed that pits these two views against each other as competing models. The present work joins a small but growing corpus of research devoted to establishing that a synthesis of these two models is both possible and desirable. (See, for example, Touretsky & Hinton [17], Touretsky [18], Derthick [3], Rumelhart, Smolensky, McClelland & Hinton [11], Anderson [1], Derthick & Plaut [4], among others.) Whereas other work has concentrated on establishing the theoretical possibility of incorporating one model within another—usually by simulating symbolic AI techniques in connectionist systems—we advocate the pragmatic incorporation of both models into a single system to study the complex interplay between these types of computation. Specifically, this paper claims that the proper computational model for controlled problem solving is derivable from the symbolic AI view of rule-based, categorical reasoning, while the proper computational model for automatic problem solving can be found in the connectionist view of parallel distributed processing.

This claim requires demonstrating how such distinct computational models can interact so as to present a plausible account of the rich interplay characteristic of the corresponding cognitive phenomena. The design of an architecture to support this interplay is the first step in this direction; an implementation of the system described here has not yet been completed.

The *ultimate* goal of this research is to eliminate the knowledge- or rule-based techniques used to implement this system. That is to say, we believe that the categorical, multi-step reasoning that is currently best exhibited by symbolic AI programs can eventually be incorporated into a wholly connectionist framework. However, we feel that attempting to accomplish this directly delays addressing important questions whose answers can help direct the eventual development of the totally connectionist systems. In addition, some of the issues brought out by this attempt to integrate what can be called the "propositional" and the "experiential" are interesting in their own right. For example, we believe that it is necessary for connectionist models to adopt some form of propositional representation to successfully model the important cognitive behaviors described later in this paper.

2. Competition and Cooperation Between the Controlled and the Automatic.

The concepts of controlled and automatic processing have been a topic of research in psychology for some time. (See, for example, Schneider & Shiffrin [16], Treisman [19], and Schneider, Dumais & Shiffrin, among others.) Shiffrin & Dumais [13] characterize automatic processes as highly parallel, exhibiting a marked ability to improve with practice, a limited propensity to "transfer" this expertise to dissimilar problem solving situations, making minimal demands on processing resources (other than those on which the processing is being directly carried out), and being outside of the explicit control of the problem solver. Controlled processing, on the other hand, is characterized as serial, exhibiting little improvement with practice[1], being under the direct and explicit control of the problem solver, and being much more amenable to its application in "unfamiliar" situations.

While these behaviors seem quite distinct, there is considerable interplay between the two types of processes. First, virtually all automatic skills (we will concentrate on cognitive

skills in this paper) are originally played out under the direct control of the problem solver. Thus, a complex chess opening requires considerable analysis by a player when it is first encountered, but given that this opening appears a large number of times in subsequent play, it is likely that recognizing and reacting to the defense will tend to become automatic and stylized. How is this transformation, or "automatization," effected? Some problem solving is of a "mixed" character, where some steps are automated, and others require "strategic" intervention. Even when some behavior has been automated it may be possible to override its "suggestions" and solve the problem again, "from first principles." This might be done in situations that call for extreme care for one reason or another. In addition, fully automated behaviors tend to be opaque with respect to introspection and explanation, and yet sometimes explanations (or "justifications") are nonetheless provided for what clearly seem automatic cognitive skills.[2] All of this suggests that an explanation of this behavior requires a model not just of the independent mechanisms but of their interaction as well.

While the psychological literature on these issues has grown, Schneider [14] points out that the treatment of many of the particular phenomena associated with this distinction has tended to remain at the level of only vague verbal descriptions of the underlying mechanisms. Schneider's paper proposes a four phase model of the development of automatic processing. While Schneider's work concentrates on a particular problem and provides a very detailed account of this example, the present paper advocates a general computational architecture. Schneider describes a particular algorithm for variably-mapped category search with two categories. The imple-

---

[1] This excludes the process of automatization, of course, in which the skill becomes automatic over time.

[2] This has been illustrated in many places. For example, in Expert System construction it has appeared when interviewing experts about why they performed certain actions in some task. Often their explanations seem either to disagree with the rapidity of the subject's choice of action, or else the explanation is insufficient to account for actions chosen in other similar situations.

656

mentation involves the comparison, modification and combination of activation values of threshold units. As a controlled process, these steps are carried out by an algorithm that is "hard-coded" into the system, which manipulates the values and gates in the system of units.

The JANUS architecture described in this paper might be seen as a generalisation of this process, since this and any other algorithm can be encoded within a system of productions.[3] This allows the researcher to further study effects of attention and problem solving strategy within the production system paradigm—at least so far as *controlled* cognitive skills are concerned. The relationship between these productions and the values of units that are involved in the subsequent automatization is much more complex than that in Schneider's model due to the translation of these rules into a distributed representation. Despite these differences, we believe a set of phases similar to Schneider's four phase model will fall out of the system described here as well (see Section 3).

2. A Description of the JANUS Architecture.

The proposed system contains three modules (see Figure 2-1). One contains a simple rule-based problem solving system. This is called the P-module (for "propositional"). As with any rule-based system, the P-module consists of a long term memory in which are stored the rules, and a working memory (PWM in Figure 2-1) in which are stored what is currently asserted (or "believed") by the P-module.[4] The PWM of the P-module contains the description of the current problem state. As the steps in solving a problem are followed,

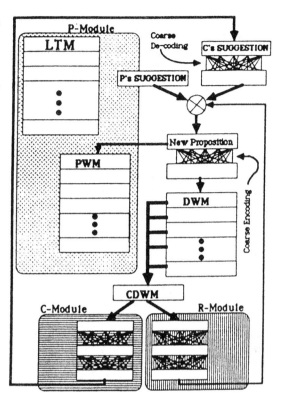

Figure 2-1.

the intermediate results, hypotheses and final solution proposed are also placed in this working memory.

The other two modules are both three layer feedforward networks (where two layers have modifiable weights) utilizing the backpropagation learning rule (Rumelhart, Hinton & Williams [10]). These two modules are the C-module (for "connectionist") and the R-module (for "resolution"—though it, too, is a connectionist module). The input layer of both connectionist modules is obtained in the following way. As each new proposition is asserted by the P-module and thereby placed in its local propositional working memory (PWM), it gets coarse-coded (in a manner to be described shortly), and this distributed representation of the proposition is placed at the front of a relatively small queue of such encoded propositions—it can hold only ten such

---

[3]Schneider makes note of the logical connection between the controlled-processing operations in his model and rules in a production system.

[4]One atypical restriction on this rule-based system is that the size of the working-memory is limited to a fairly small number of propositions—approximately twenty or so. This is to keep the P-module's available working memory close to the same size as that of the C- and R-modules, discussed below.

propositions. This queue is called the DWM (Distributed Working Memory) and as each new distributed proposition is added to the top-most cell, all the other proposition are pushed down, removing the proposition in the bottom-most cell altogether. All ten of these distributed representations of propositions are then coalesced into one distributed representation of all of them by simply adding the vectors together. It is this coalesced vector (the CDWM in the figure) which is the input layer for both the C- and R-modules.

It might at first appear that information is being hopelessly confused when the separate propositions in the DWM are "coalesced" into the single CDWM. However, the loss of information can be significantly reduced by using an adequate coarse-coding scheme. This is because each feature in the coarse-coded representation has the ability to distinguish the role being played by some domain object (or concept) in the proposition from which it is derived. This allows the C-module and the R-module to include a more complete "description" of the current problem state by viewing up to ten propositions at one time. (McClelland's CID mechanism [8] utilizes a similar technique.)

The C- and R-modules differ, however, in their output layers and the effects of different values on these output layers. The output of the C-module is a distributed representation of a proposition. The output of the R-module is a single binary unit. The C-module's output layer is placed in a vector ("C's Suggestion") which is then translated into a local (propositional) representation of the module's output. That is, there is a single layer network which takes a distributed representation of a proposition and generates on its output layer a "local" representation of that proposition. This process is simply computing the inverse of the coarse-coding process mentioned earlier. If the value of the R-module's output unit is *greater* than zero, it has the effect of placing this local representation of the C-module's output layer in the "New Proposition" buffer (see Figure 2-1). If the R-module

has instead produced an output *less* than zero, then the C-module's output layer would be ignored, and the contents of the local working memory buffer within the P-module ("P's Suggestion") would be the source of the next addition to the propositional working memory ("New Proposition") within the P-module.

Thus, the basic structure of the system is that the P- and C-modules are both generating propositions based on the contents of the P-module's working memory (though the C-module's version is somewhat removed from the original because of the coarse-coding and coalescing processes), and both are "competing" to have their proposed propositions "asserted" by the system as a whole by having them placed in the "New Proposition" buffer, from whence it is forwarded to the working memories of all three modules. The arbiter of this competition is the R-module. The actions that the C- and P-modules would carry out on the P-module's propositional working memory (PWM) are 'gated,' and this gate is controlled by the R-module.

Training the Connectionist Modules.

The R-Module. The R-module is a back-propagation network, and the basis of its training schedule is to prefer the suggestions from that module which lead to a solution. However, the R-module also has a built-in bias towards letting the C-module set the PWM (since we want to encourage this more efficient form of problem solving). The R-module is thus computing a confidence rating of the C-module's abilities. This means that in the early stages of performance, since the C-module has had too little time to learn anything, the R-module will quickly learn to favor the P-module's suggestions for how to change the state of PWM in order to solve a given problem. Over time, as the C-module begins to learn to mimic the P-module's actions, the R-module's bias towards automatic processing will lead to the C-module's taking over some or all of the problem solving. The R-module's input layer is provided with the whole CDWM in order to allow it to distinguish those problem states for which the suggestions from either the C- or P-module should be selected.[5]

The C-Module. The C-module is only trained (it's error backpropagated and weights modified) if the *P-module's* "suggestion" is placed in the PWM (via the "New Proposition" buffer). In this event, the C-module is trained to match the P-module's suggestion. If the C-module's *own* suggestion has been chosen by the R-module to be placed in the PWM, then no training of the C-module is done. If the resulting solution leads to an error, then only the R-module is penalized for relying on the C-module's suggestion. (This is discussed further in the next section.)

Distributed Representation of Propositions. The propositional working memory (PWM) of the P-module requires that propositions be used to represent the problems and the intermediate states leading to solution in order to allow rules to be invoked using pattern matching. But symbolic propositions cannot be presented directly to the connectionist modules C- and R-; they require feature vectors. Space precludes a detailed treatment of this important representational issue, but a transformation of the P-module's local propositional WM can be performed by a coarse-coding technique adapted from the work of Hinton [6], McClelland & Kawamoto [7], and Saund [12]. The basic idea here is that symbols in the propositional patterns of WM should be identifiable independent of the different roles they might play in different propositions. At the same time, the role they play is also important to express the meaning of the proposition. The method adopted by Hinton [6] is particularly useful in this regard.

The distributed representation is generated by a training schedule which must be per-

formed before the various modules are put together. This network would be trained to map the local feature vector representation of the proposition to another local feature vector representation of this same proposition through the intermediate layer of units, which would consist of a smaller number of units than either of the other two layers (see Saund [12] for a similar setup). It is this process which derives the coarse coding representation used in the C- and R-modules of the JANUS architecture. Once this set of weights has been developed, the weights between the first and intermediate layer of this network (coarse *en*coding) are used to translate between the most recent propositional addition to the P-module's PWM (via the "New Proposition") and the distributed representation of that proposition in the top-most queue entry of the DWM (see Figure 2-1). The weights between the intermediate and output layers (coarse *de*-coding) are used to map the C-module's propositional "suggestion" before being placed in the "New Proposition" buffer.

3. Using the Architecture to Model Cognitive Behavior.

The architecture just outlined was developed as an attempt to model the following types of behavior.

Categorical Reasoning. The basis of connectionist problem solving is, so far, via problem *recognition*; a solution to even a novel situation is based on having generated a mapping from problem states to solutions which, through generalization, will capture situations sufficiently similar to those in the training set. However, there are classes of problems—addition is a good example, as is trying to determine the voltage on some wire in a fairly complex circuit diagram—where it seems that a multiple step solution is called for based on what has been called a 'model' of the domain. Similarity to previous problems tends to be a weak indicator. So far, such models have been able to be most easily expressed and used in 'categorical,' symbolic AI systems. As the P-module is a relatively standard rule-based

[5]It is probable that the R-module could actually be incorporated into the C-module by adding an extra output unit to the C-module with the above interpretation. However, this aspect is separated out for ease in testing alternative formulations of the R-module's behavior.

system, such knowledge will be able to be encoded and used in this powerful way by the P-module. As mentioned in the previous section, the P-module will very soon tend to be favored over the C-module as the module to be "trusted" in novel situations—that is, in novel states of the PWM.)

Automaticity. The C-module is not idle, however, when the P-module is being deferred to. While the P-module is holding sway,[6] the C-module is being trained to produce the solutions generated by the P-module, which is the source of training examples for the C-module. As a growing number of problems are presented to the system as a whole, it is presumed that the C-module will be able to generate a mapping such that the solutions can be suggested on the basis of the similarity of the problem state to previously seen problems.[7] This is simply the standard way in which connectionist networks learn to perform mappings.

Automatization. This refers to the transfer of knowledge from an explicit, categorical (or propositional) form (in the P-module) to the distributed and more efficient recognition-based form in the C-module. It is the main goal of this research to demonstrate that the knowledge that is initially encoded explicitly and that can be used only by chaining through some number of reasoning steps will be subject to incorporation into the C-module after sufficient instances of such problem solving are presented. This process would proceed in stages not dissimilar to those described in Schneider [14] and the "compilation" process modelled in

Anderson, et al. [2]. As inferences become automated by the C-module, this allows the corresponding propositions in the P-module to be skipped. This, in turn, allows more propositions of the PWM to "fit into" the distributed working memory (DWM), which would allow further automatization to occur. In this way, more and more multi-step inferences are slowly encompassed by single step inferences carried out by the C-module. The advantage of this scheme over the rule-based scheme adopted by Anderson, et al. [2] is that the C-module can generalize and thereby generate plausible suggestions for changing the PWM for problems never seen before by the system as a whole, as long as there is sufficient similiarity to previously seen problems.

Explanation. A major difficulty of connectionist systems has been their opacity. When a network presents some 'solution' on its output units, it is not at all clear the 'reasons' that support this computation. Of course, the only truly *valid* explanation would involve a complete listing of all the net's previous training examples which led it to have the weights it now has; and then these weights, and their ramifications, might also be explicated in some fashion. Except for trivial problems, this approach seems ludicrous.

On the other hand, following a plausible model of how we generate some explanations ourselves, we can imagine that the explanations of fully (or partially) automatic behaviors are generated *post facto* by utilizing any relevant categorical knowledge within the P-module. For example, in solving for the voltage on a wire in some complex circuit diagram a system as described here might generate an opinion based on the similarity of the network layout to a large number of previously seen layouts. Asked for an explanation ('justification' might be a more accurate term), the system might use the proposed solution and the known initial problem state to generate a solution using the *P-module's* knowledge alone. This *post facto* (and more expensively derived) solution has the advantage of perspicuousness of the knowledge used to generate the solution.

---

[6]This will tend to happen for relatively long periods of time, since the PWM changes only slowly at each cycle, and it is on the basis of recognizing states of the PWM that the R-module assigns priority to either module.

[7]This relies, of course, on the R-module *also* noting this similarity and being prepared to 'trust' the solution 'proposed' by the C-module. This suggests that the "natural degree" to which the R-module has preferential bias for the C-module might be out of line—either too much or too little—with the rate at which the C-module actually learns; this should make itself readily apparent when testing the system.

Of course, this is hardly the same way in which the original solution was in fact generated. On the other hand, to the extent that the C-module's knowledge is the result of automatization of previous explicitly encoded knowledge in the P-module, such an 'explanation' might not be so inappropriate or unrelated to the automatically derived solution.

It should be noted that with this method it is perfectly possible that the "explanation" would not agree with the original solution. This would hardly seem to be undesirable, however, since if we are to make assumptions on the basis of generalizing from past experience, it is very useful to know when those assumptions disagree with the knowledge acquired directly in the form of rules or other propositions.

Other Types of 'Mixed Reasoning'. The previous description of explanation introduces other possibilities for intermixing the two types of knowledge to obtain satisfactory performance. Such techniques would take the form of the R-module's alternating between the two 'knowledge sources' to modify working memory in the intermediate steps of solving some larger problem. For example, in the circuit analysis example, an initial set of solutions might be 'proposed' by the C-module on the basis of the circuit's similarity to other circuits, which might be followed by an explanation-like process in which the rough possible solutions are checked and/or refined by more expensive (but possibly better supported by reference to 'first principles') categorical reasoning of the P-module.

4. Summary.

This paper proposes a method of integrating *automatic* and *controlled* forms of problem solving by building an architecture in which connectionist and standard symbolic AI implementation techniques complement each other in a single system. This is seen as an important step towards eventually incorporating the propositional (or rule-based) form of knowledge found in AI systems into completely connectionist networks. This architecture is described structurally. The paper describes various cognitive behaviors which derive from the interplay between the two problem solving styles and that this architecture would be able to model. The emphasis of this research is on the way in which expertise can be transformed from "explicit" to "compiled" via automatization, and how the opacity of the resulting automatic behavior can be counterbalanced in a cognitively plausible manner.

5. Bibliography.

1. Anderson, J. A., (1983) "Cognitive and Psychological Computation with Neural Models," *IEEE Transactions on Systems, Man and Cybernetics*, SMC-13, 799-815.

2. Anderson, John R., J. G. Greeno, P. J. Kline & D. M. Neves, (1981) "Acquisition of Problem Solving Skill," in J. R. Anderson (Ed.), *Cognitive Skills and Their Acquisition*, Hillsdale, N.J.: Erlbaum.

3. Derthick, Mark (1986) "A Connectionist Knowledge Representation System", CMU Thesis Proposal.

4. Derthick, Mark & David Plaut, (1986) "Is Distributed Connectionism Compatible with the Physical Symbol System Hypothesis?," Proceedings of the Cognitive Science Society.

5. Gallant, Steven, (1985) "The Automatic Generation of Expert Systems From Examples," Proceedings of Second International Conference on AI Applications, Piscataway, NJ, 313-319.

6. Hinton, Geoffrey, (1986) "Learning Distributed Representations of Concepts", Proceedings of the Cognitive Science Society.

7. McClelland & Kawamoto, (1986) "Mechanisms of Sentence Processing: Assigning Roles to Constituents," *Parallel Distributed Processing, Vol. II*, McClelland & Rumelhart, eds., Cambridge, MA: MIT Press.

8. McClelland, J., (1986) "The Programmable Blackboard Model of Reading", *Parallel Distributed Processing, Volume II*, McClelland & Rumelhart, (Eds.), Cambridge: MIT Press.

9. McClelland & Kawamoto, (1986) "Mechanisms of Sentence Processing: Assigning Roles to Constituents," *Parallel Distributed Processing, Volume II*, McClelland & Rumelhart, (Eds.), Cambridge: MIT Press.

10. Rumelhart, D., G. Hinton & R. Williams, (1986) "Learning Internal Representations by Error Propagation," *Parallel Distributed Processing, Volume I*, Rumelhart & McClelland, (Eds.), Cambridge: MIT Press.

11. Rumelhart, D., P. Smolensky, J. McClelland & G. Hinton, (1986) "Schemata and Sequential Though Processes in PDP Models," *Parallel Distributed Processing, Volume I*, Rumelhart & McClelland, (Eds.), Cambridge: MIT Press.

12. Saund, Eric, (1986) "Abstraction and Representation of Continuous Variables in Connectionist Networks," Proceedings of the AAAI.

13. Shiffrin, Richard M. & Susan T. Dumais, (1981) "The Development of Automatism," in *Cognitive Skills and Their Acquisition*, edited by John Anderson, Hillsdale, NJ: Erlbaum.

14. Schneider, Walter, (1985) "Toward a Model of Attention and the Development of Automatic Processing," in Posner & Marin (Eds.), "Attention and Performance XI," Hillsdale, N.J.: Erlbaum.

15. Schneider, W., S. Dumais & R Shiffrin, (1984) "Automatic and Control Processing and Attention," in *Varieties of Attention*, Academic Press.

16. Schneider & Shiffrin, (1984) "Controlled and Automatic Human Information Processing: I. Detection, Search and Attention," *Psychological Review*, 84.

17. Touretzky and Hinton, (1985) "Symbols Among the Neurons", Proceedings of the International Joint Conference on AI.

18. Touretzky, (1986) "BoltzCONS: Reconciling Connectionism with the Recursive nature of Stacks and Trees," *Proceedings of the Cognitive Science Society*.

19. Treismann, Anne, (1969) "Strategies and Models of Selective Attention." *Psychological Review* 76, 282-299.

A Dynamic Connectionist Model Of Problem Solving

James Lundell and Bernice Laden
University of Washington

Connectionist computing offers a flexible approach to modelling automatic cognitive processes, while production systems provide reasonable models of controlled processes involved in problem solving. Our initial motivation for the development of this problem solving simulation was to build a model which could predict the time required to solve a problem and the types of errors people are likely to make under time pressure. Aside from its predictive ability, we believe the simulation offers a unique approach to modelling cognitive tasks. We refer to the model as a Dynamic Connectionist Model of Problem Solving (DCOMPS). It is connectionist because working memory consists of propositions which are organized into a connected activational network, and it is dynamic because this network changes its organization over time as a result of the firing of production rules. Its architecture is designed to handle a variety of problem solving tasks, including arithmetic word problems, two-term series and a stroop task.

Our purpose in this project is to develop a model of problem solving that predicts the time required to solve a problem and the errors which people are likely to make under time pressure. We assume that problem solving occurs as a series of mental events, each of which consists of the recognition of a pattern and a response. The time required to respond to a pattern is a function of the interaction between automatic and controlled processes (Neely, 1977). The two most commonly used architectures in modelling cognitive processes, connectionist networks and production systems, closely parallel automatic and controlled processes, respectively. Connectionist networks are used to simulate the pattern-matching capabilities of the cognitive system - they are particularly good at resolving incomplete or ambiguous patterns. These networks, however, do not easily accomodate the goal-driven, serial nature of human cognitive processing. Production systems, on the other hand, are goal-driven and serial in nature. Connectionist networks do not easily deal with instantiation of several concurrent concepts; production systems can accommodate instantiation and variable binding.

We would like to thank Earl Hunt, Aura Hanna, Simon Farr and Penny Yee for their comments on earlier drafts. This research was supported by a grant from the Office of Naval Research Grant N0014-84-K-003 to the University of Washington, Earl Hunt Principal Investigator.

Taken together, production systems and connectionist models compensate for each others' weaknesses. The human cognitive system may have evolved these two types of processing primarily because they compensate for each others' deficiencies. Thus there are theoretical as well as pragmatic reasons for attempting to integrate the two approaches to cognitive modelling.

Our work builds on two attempts to combine models of automatic and controlled processing: Anderson's ACT* (1983) and Hunt and Lansman's (1986) production-activation model. In the ACT* approach, production rules are embedded in a connected network, and the rules fire when they attain a sufficient level of activation.

Hunt and Lansman's theory is similar to ACT* but focuses upon simple cognitive tasks such as the Stroop task, four and eight-choice reaction time tasks, and has been extended to simple arithmetic tasks (Richardson and Hunt, 1986). Production rules are matched to features in working memory. A production rule is fired when activation strength exceeds that of all competing productions by a parameter *delta*. The time required for a given production to fire is a decreasing function of the degree of match between the production pattern and the contents of working memory, and an increasing function of the number of competing productions.

Our simulation is based upon the Hunt and Lansman system, but has several basic changes. Hunt and Lansman represented objects as vectors of features; we chose a propositional representation. In the Hunt and Lansman model, activation spreads through a network of production rules. In our model, activation spreads through a network of propositions in which the links between the propositions are determined by a few general principles, and can be modified by production rules.

DCOMPS: An Overview

The program is called DCOMPS, which stands for Dynamic Connectionist Model of Problem Solving. DCOMPS is based upon three assertions: 1) automatic processes can be modelled by the parallel activation of related nodes in a propositional network, 2) controlled processes can be modelled by a production system and, 3) the time required to solve a problem is a function of the interaction between the two processes.

The following points describe some of the more salient features of DCOMPS:

First, we have loosely adapted the system described by Kintsch (1974), for the propositional representation of problems. Complex problem solving tasks are represented by simple propositions, where a proposition consists of a predicate and one or more arguments.

Second, following Hunt and Lansman we distinguish between internal and external channels. Working memory consists of a semantic (internal) channel and one or more external channels. Each channel may contain several linked propositions. Each proposition has an activation level that is determined by the activity in the network. Representations in the external channel have been processed to a certain extent, but have not been semantically encoded. In the current version, there is only one external channel which receives propositional representations of visually presented words or sentences. Other external channels, such as auditory and tactile channels, could be added. Connections across channels are allowed, although they are generally weaker than within-channel connections.

Third, control of the problem solving process is governed by the pattern of activation of propositions resident in working memory over time. This in turn is governed by simple rules which dictate how propositions may be linked together, and by the incorporation and transformation of propositions in working memory.

The incorporation of propositional information is based on Kintsch and van Dijk's (1978) model of discourse understanding. Kintsch and van Dijk distinguish between a microstructure and a macrostructure representation of text. The microstructure consists of the part of the discourse which is currently being comprehended. The macrostructure consists of propositions which have been extracted from the microstructure and which are retained in working memory as an aid to understanding incoming propositions. Because we eventually intend to simulate many different types of visual and auditory tasks, we have chosen to call the macrostructure and microstructure the semantic and external channels, respectively. Propositions are extracted from the external channel and incorporated into the semantic channel based on the concept of *referential coherence*. That is, the propositions which contain arguments that are most often referred to are extracted first, while other propositions tend to be discarded. Kintsch (personal communication, July, 1986) has suggested how this theory may be implemented in a connectionist framework.

Propositions are linked to themselves by a connection strength of 1.0. Propositions that share arguments are linked by a connection strength of 0.5. Propositions that are both referenced by a third proposition are linked by a strength of 0.3. For example, P1(agent (John)) and P3 (Object (ball)) would be linked to each other by a strength of 0.3 if proposition P2(has (P1 P3)) were also present. Propositions that are related more distantly are linked by a connection strength of 0.2. For example, if P4 were (Color (red P3)), it would be linked to P1 by 0.2. Thus for the propositions P1-P4 we have the following connection matrix:

	P1	P2	P3	P4
P1	1.0	0.5	0.3	0.2
P2	0.5	1.0	0.5	0.3
P3	0.3	0.5	1.0	0.5
P4	0.2	0.3	0.5	1.0

We have chosen these link strengths arbitrarily. The actual values themselves are not as important as the idea that the connection strengths decrease with the distance from related propositions. Propositions placed in the external channel are given an initial activation value of 1.0. Each proposition's activation value at time t is computed synchronously by the following formula:

$$A_{nt} = \sum A_{i(t-1)} C_i / N$$

where A_{nt} is the activation of proposition n at time t, $A_{i(t-1)}$ is the activation of proposition i at time t-1, and C_i is the connection strength between proposition n and proposition i. N is a normalizing factor which prevents activation in the network from becoming unacceptably high. The activation level of each node in the network is iteratively computed until the network becomes sufficiently "stable", i.e. when the change in activation levels for each node drops below a criterion amount. Once the network reaches stability, the propositions which are most highly referenced are the most active. In this way, a hierarchy of activation levels is formed, and the least active propositions may be dropped from the network. The most active propositions will then be transformed into a semantic code and placed into the semantic channel. The macrostructure will eventually contain a parsimonious representation of the topic of discourse, based on the concept of referential coherence.

As Kintsch and van Dijk (1978) have noted, referential coherence only partially accounts for discourse comprehension. Since comprehension requires the additional use of information resident in long term memory, the system contains two other types of nodes: schema nodes and rule nodes. These can be conceived of as knowledge the system already has. Schema nodes reside in the semantic channel of working memory where they differentially activate certain types of propositions, and thus help control the activation of the important aspects of a statement. For example, if we were to read *John has three dollars* in the context of an arithmetic problem, we would probably attend to the quantity *three*. This is because our schema for arithmetic problems activates all propositions which pertain to numbers. If we were to read the same sentence at a time when we happen to need money, we would attend more strongly to the object of the sentence, *dollars*.

Rules are part of the knowledge base of the system, and are linked to classes of propositions or schema nodes which appear in working memory. The activation algorithm for rules is simple: rules inherit the activation of the nodes to which they are connected. All rules are negatively linked to all other rules by a value of -0.5. Each time the network stabilizes, the most active rule fires. When a rule fires, it may change the connection strength between propositions or it may insert new propositions into working memory. This method of inserting new propositions or 'nodes' into the network constitutes a break with traditional connectionist modeling.

A Specific Example: The Two-term Series Problem

The DCOMPS' architecture will be illustrated by showing how the simulation works on two-term series problems.The two-term series problem is a deductive reasoning task containing a relational statement and a question. An example is: *John is better than Pete. Who is best?* In this type of problem people make inferences based upon linguistic, rather than logical interpretations. Performance has been shown to be influenced by three factors: 1) lexical marking, 2) use of negations and 3) statement-question congruence (Clark,1969). We have incorporated these aspects into the simulation.

TABLE 1

Eight Versions of the Two-Term Series Problem

	Proposition	Question
Unmarked	John is better than Pete.	Who is best?
	John is better than Pete.	Who is worst?
	Pete is not as good as John.	Who is best?
	Pete is not as good as John.	Who is worst?
Marked	Pete is worse than John.	Who is best?
	Pete is worse than John.	Who is worst?
	John is not as bad as Pete.	Who is best?
	John is not as bad as Pete.	Who is worst?

There are eight versions of the two-term series problem if one takes into consideration lexical marking, negation and congruence. These are summarized in Table 1. We have selected one of them to illustrate the details of our model. In text form the problem is:

John is worse than Pete. Who is best? In propositional form the text is represented as follows:

```
P1(Agent John)
P2(Worse_than P1 P3)
P3(Agent Pete)
P4(Question P5 P6)
P5(Agent unknown)
P6(Best P5)
```

Initially, working memory contains a reasoning schema and an instruction node, I1, in the semantic channel. The semantic channel is organized as depicted in Figure 1. I1 is the initial instruction to read the problem. S1 is linked to the rule R1. When R1 is fired the system will read the first sentence of the problem. S2 is linked to rule R2 which when fired will attempt to solve the problem. S4 is linked to R5. R5 will be fired if the text contains a negation, such as *John is not as good as Pete*, and will transform those propositions to a positive interpretation, *Pete is better than John.* S3 is a special node which will look for any relational propositions, such as "Better_than" or "Worse_than" in the external channel and link itself to that proposition. Thus any relational propositions will tend to be more active, and so will receive a greater amount of 'attention'.

S0 is initialized with an activation level of 1.0, and the system iterates until the network stabilizes. All positive links in the superschema are given a value of 0.5, and all negative links are given a value of -0.5. When the system settles, rule R1, read_sentence, is highly active. The system reads in the first sentence, *John is worse than Pete* by placing the first three propositions in the external channel. The worse_than relationship is lexically marked relative to the better_than relationship. In the current example, lexical marking is simulated through the addition of inferences that are made by the system whenever a worse_than relationship is encountered. DCOMPS infers that Pete is bad, and that John is very bad. These inferences are constructed as propositions and linked to P1(Agent John) and P3(Agent Pete) respectively. All inference propositions are linked to their related propositions by a connection strength of 0.2. Inferred propositions are given an initial activation level of 0. According to Clark, lexically marked adjectives take longer to recover from memory. In our interpretation, lexically marked adjectives are adjectives which cause a number of inferences to be made. This means that a sentence containing a lexically marked adjective will link to more (inferred) propositions. It will take longer, on average, for the larger network to stabilize.

At this point, node S3 in the term schema is linked to proposition P2, (Worse_than P1 P3). In order to keep the computational overhead down, rules from the rulebase are only added

Semantic Channel

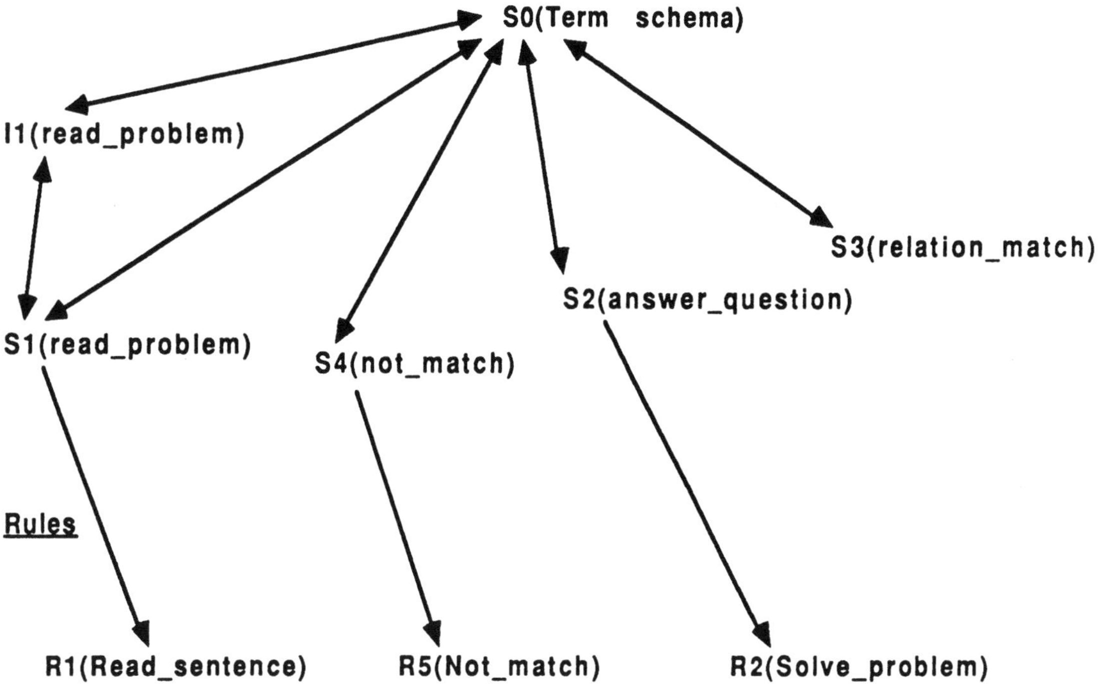

S0(Term schema)
I1(read_problem)
S3(relation_match)
S2(answer_question)
S1(read_problem)
S4(not_match)
Rules
R1(Read_sentence)
R5(Not_match)
R2(Solve_problem)

External Channel

Empty

P - Propsitional Node
S - Schema Node
R - Rule Node
I - Instruction Node

Figure 1. Initially, working memory contains a schema organized in the semantic channel.

to the network if there is a corresponding proposition match in working memory. Rule R4, worse_than_relation, is added to the network because it now has a corresponding match in P2. The state of the network at this point is described in Figure 2.

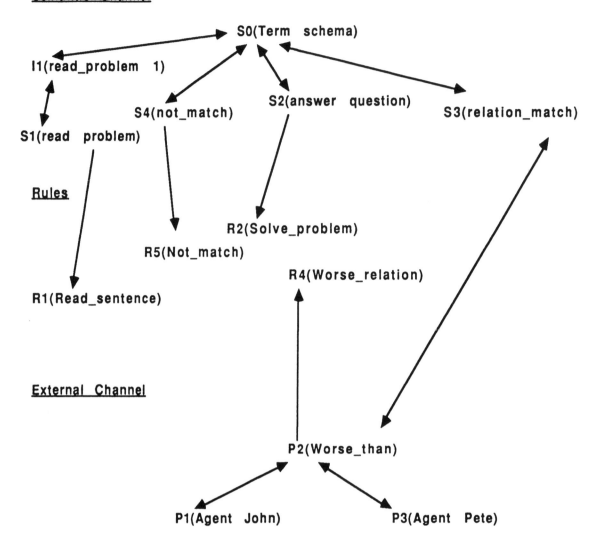

Figure 2. The first sentence is read in. The relational
node of the schema, S3 is linked to the proposition P2.
Rule R4 is added to the network.

The system again cycles through the activation updating function
until the network stabilizes. This time R4 fires, which does two
things: it adds a proposition to the semantic channel asserting
that this first sentence contains a Worse_than relation, and it
incorporates the propositions P1-P3 into the schema by changing the
link strengths between the propositions to 0.3 if one proposition
is contained in another's argument list, and 0.2 if two
propositions share a common argument. Now the schema looks like
Figure 3.

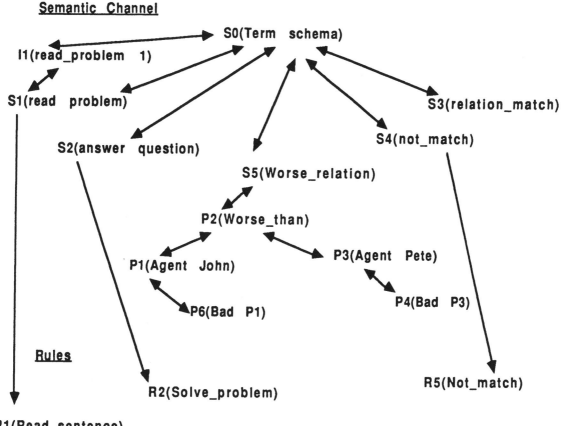

R2(Solve_problem)

R1(Read_sentence)

External Channel

Empty

Figure 3. Both text and inferred propositions are incorporated into the schema.

The presence of the term schema was necessary for proper understanding of the first sentence. The schema node S3 directed attention towards the relation stated in the text. As a result, R4 fired and created a new proposition (S5) linking P2 and S0. Without the schema the model would not have been able to extract information from the sentence which is crucial to solving the two-term series problem.

The connection strengths between the propositions are lowered once the propositions are incorporated into the semantic channel.

671

This is a decay mechanism used in the simulation which we found was necessary to allow the system to attend to new propositions placed on the external channel.

During the next cycle the system proceeds as before which eventually results in the rule R1 firing. The system places the next sentence, *Who is best?*, into the visual channel, setting all activation values for P7-P9 to 1.0. Rule R6(Quest_rule) is linked into the network. Rule R6(Quest_rule) wins out, which creates proposition S6(Find_agent Best). I1, the instruction to read, is removed from the network because the Quest_rule proposition is a signal that this is the last sentence in the problem.

The system cycles through again, now firing R2 because proposition I1 has been removed from the network. Rule R2 is a complex rule which forms various hypotheses about the answer. At this point three hypotheses are possible: *John is best*, *Pete is best* or the statement is incongruent with the question and must be converted to the proper representation. These hypotheses are linked into the network as S7, S8 and S9. DCOMPS gathers evidence for each hypothesis by linking each one to the propositions in the semantic channel which support it. In this case there are no propositions which support the John and Pete hypotheses, because the proposition terms such as P4(Bad P1) are incongruent with the question, *Who is best*. Hypothesis S9 is most active because there is a worse_than relation and a best question present in the current representation, signalling that a conversion is necessary.

Three rules are linked to the three newly formed hypotheses. As with all other rules which are present in the network, inhibitory links are created between them. When the network stabilizes, rule R9(Convert) fires. This rule will reformat the propositions created by the first sentence so that the relation is congruent with the question. The convert rule changes *worse_than* to *better_than* in P2, as well as S5. It also switches the agents, John and Pete, so that the initial sentence now has the interpretation *Pete is better than John*. The "bad" inferences drop out of working memory as well as the convert hypothesis, S9.

Now that the problem is represented correctly, the conversion rule no longer applies. Two hypotheses remain: John is best or Pete is best. These hypotheses can now gather evidence because the worse_than proposition has been converted to better_than. Figure 4 depicts how these hypothese are now linked to working memory. Two rules, R7(Agent1) and R8(Agent2) are added to the rule list. R8 is most active, since it is linked to P1 (Agent Pete), and P2(better_than P1 P3). Thus, *Pete is best* is determined to be the answer.

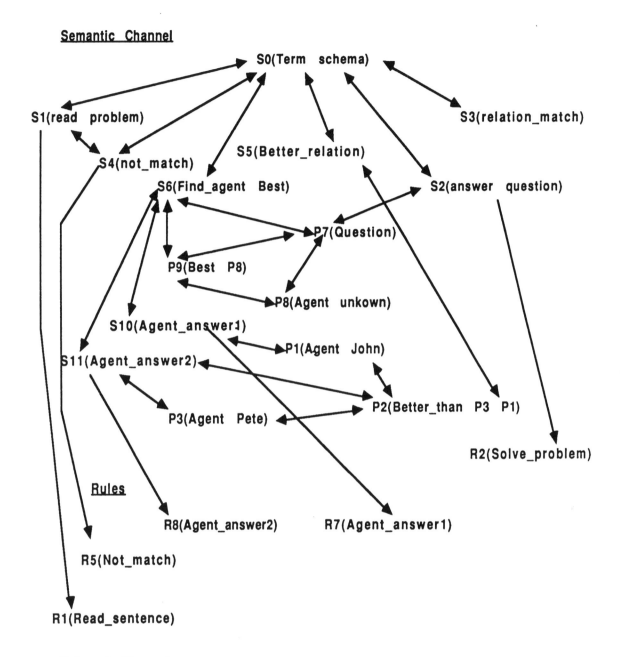

S0(Term schema)

S1(read problem)

S3(relation_match)

S4(not_match)

S5(Better_relation)

S6(Find_agent Best)

S2(answer question)

P7(Question)

P9(Best P8)

P8(Agent unkown)

S10(Agent_answer1)

P1(Agent John)

S11(Agent_answer2)

P2(Better_than P3 P1)

P3(Agent Pete)

R2(Solve_problem)

Rules

R8(Agent_answer2)

R7(Agent_answer1)

R5(Not_match)

R1(Read_sentence)

External Channel

Empty

Figure 4. Hypotheses linked into working memory

673

How does DCOMPS compare to empircal results?

Table 2 presents a comparison of the original Clark data and the number of cycles required for DCOMPS to solve the same problem. There is a Spearman correlation of .994 between the rank order of the times required for solution by human subjects and by DCOMPS. A tie occurs between the times required to solve *John is worse than Pete. Who is best?* and *John is not as good as Pete. Who is best?* We attribute this to the fact that DCOMPS takes less time to process *not* statements than do people. Because *John is not as good as Pete. Who is best?* is the quickest of the negatively framed problems and *John is worse than Pete. Who is best?* is the longest of the positively framed problems, a tie occurs between the times required to solve these two problems by DCOMPS.

TABLE 2

A comparison between data from Clark (1969) and DCOMPS for performance on two-term series problems.

Problem	Question	Clark secs.	DCOMPS cycles
John is better than Pete.	Who is best?	.61	91
John is worse than Pete.	Who is worst?	.62	95
John is better than Pete.	Who is worst?	.68	108
John is worse than Pete.	Who is best?	1.00	113
John is not as good as Pete.	Who is best?	1.17	113
John is not as good as Pete.	Who is worst?	1.47	128
John is not as bad as Pete.	Who is worst?	1.58	130
John is not as bad as Pete.	Who is best?	1.73	147

The time required to solve these problems is a function both of controlled, serial processing and automatic, parallel processing. That is, problems involving incongruent questions or negative assertions require the firing of additional productions which transform the appropriate propositions. The time required for these factors is a result of serial operations, and can be predicted by a simple production system model. However, the difference between solving problems with marked and unmarked adjectives is a result of the parallel activation of the inferred propositions which were incorporated into the network. During some steps of the problem solving process, it takes longer for the network to "settle" when these additional propositions are present.

Conclusion

DCOMPS currently performs two other tasks: arithmetic word problems and Stroop tasks. The system solves word arithmetic problems in a manner similar to two-term series problems. DCOMPS begins with an arithmetic schema which aids in interpreting sentences which appear on the external channel. In particular, the arithmetic schema differentially activates any *Quantity* propositions. As each sentence is read, rules fire which incorporate the propositions into the schema, and classify the propositions according to the type of set which is indicated. For example, the sentence *John has five more marbles than Sarah* is classified as a more_than set. When a question is encountered, such as *How many marbles does John have?*, DCOMPS generates hypotheses about the operations required to solve the problem based on the sets which have been identified.

Like the two-term series problems, a theory of the particular problem solving process already exists (Kintsch and Greeno, 1985). Our simulation is an attempt to implement the theory within the DCOMPS architecture, and to compare the times required by the system to solve various problems with the times required by humans.

The time to respond to various types of Stroop conditions compares favorably with human data in three ways. First, color words presented in contrasting colors are not identified as quickly as color words presented in the same color. Second, it takes DCOMPS longer to identify the color of the word than the meaning of the word. Third, DCOMPS displays between-trial effects. It takes the system longer to identify a feature which was inhibited on the previous trial. For example, if the task is to identify the color of the word and the first word presented is the word *blue* printed in the color *red*, it would take longer to identify the color of the next word if its color is *blue*, since that feature was inhibited on the previous trial.

We are currently working on simulating two additional types of tasks using DCOMPS: factual and counterfactual reasoning, and musical chord priming. Future plans for DCOMPS include: 1) pursuing several approaches to relate DCOMPS' performance with human data, 2) studying individual differences in problem solving by manipulating the parameters of the model, and 3) simulating several additional cognitive tasks.

DCOMPS offers a flexible compromise between production systems and connectionist models. Although some researchers have suggested a way in which a static connectionist model might simulate sequential processess (e.g, Rumelhart, Smolensky, McClelland, & Hinton, 1986) this method is very likely to be computationally expensive and, more importantly, to be so complex as to be intractable in simulating problem solving tasks. DCOMPS explores the possibility that a simple system such as the one proposed can solve problems which involve both sequences of actions and parallel processing.

References

Anderson, J.R.(1983). *The Architecture of Cognition*. Cambridge, MA: Harvard University Press.

Clark, H.H. (1969). Linguistic processes in deductive reasoning. *Psychological Review*, Vol. 76, No. 4, 387-404.

Hunt, E. and Lansman, M. (1986). Unified model of attention and problem solving. *Psychological Review*, 93(4), 445-461.

Kintsch, W. (1974). *The Representation of Meaning in Memory*. Hillsdale, N.J.: Erlbaum.

Kintsch, W. and Greeno, J.G. (1985). Understanding and solving word arithmetic problems. *Psychological Review*, Vol. 92, No. 1, 109-129.

Kintsch, W. and van Dijk, T.A. (1978). Toward a model of text comprehension and production. *Psychological Review*, 85, 363-394.

Neely, J.H. (1977). Semantic priming and retrieval from lexical memory: Roles of inhibitionless spreading activation and limited-capacity attention. *Journal of Experimental Psychology: General*, Vol. 106, No. 3, 226-254.

Richardson and Hunt, E. (1986). Problem solving under time pressure. University of Washington Technical Report.

Rumelhart, D.E., Smolensky, P., McClelland, J.L., and Hinton, G.E. (1986). Schemata and sequential thought processes in PDP models. In J.L.McClelland and D.E.Rumelhart, Eds, *Parallel Distributed Processing: Explorations in the Microstructure of Cognition, Vol. 2: Psychological and Biological Models*. Cambridge: The MIT Press, 7-58.

UNDERSTANDING THE MICROSTRUCTURE OF SCIENCE: AN EXAMPLE[1]

by
Ryan D. Tweney & Catherine E. Hoffner
Department of Psychology
Bowling Green State University
Bowling Green, Ohio   43403
(419) 372-2301

The present paper is intended to demonstrate the feasibility of a cognitive approach toward understanding real-world science through the intensive analysis of a scientific diary. The approach draws on many concepts of contemporary cognitive science, especially protocol analytic (PA) techniques. As Ericsson & Simon (1984) emphasize, not every verbal record of thought is amenable to such analysis. Generally, PA is applied to "think aloud" protocols gathered during the solution of a problem. In most cases, the problem subjected to analysis has been a carefully designed task with a clear goal state, a minimum of background information needed for solution, and a level of difficulty which can be solved by most subjects within one hour or less. The classic examples meet these criteria; for example, the DONALD & GERALD problem and proof problems in symbolic logic.

In applying PA to the analysis of a working scientific diary, there is frequently no clear problem space, more than one problem space is often under exploration, and the background knowledge brought to the task is extensive and often not easy to characterize. Specific diary entries (unlike utterances in a think aloud protocol) may reflect either the contents of working memory or retrieved information brought from long term memory. The temporal sequence of a diary is generally clear, but the fullness of the record is very suspect. Finally, it is very likely that keeping a diary is reactive.

In spite of these problems, it is possible to modify conventional PA to permit meaningful analysis of the microstructure of a scientific diary. Our basis for this claim is a modified PA of a portion of the scientific diary of Michael Faraday (1791-1867), the eminent English physicist who discovered electromagnetic induction in 1831. Part of Faraday's diaries have been published (Faraday, 1932-1936); a fragment of the published portion served as the source protocol for the present analysis.

Goal states in the present paper are taken to be empirical observations of phenomena which had not previously been observed by Faraday. States and operators were defined as corresponding roughly to "ideas" and to "actions," respectively. Throughout the series, Faraday was acting upon some mental representation in the hope of transforming it into another mental representation. Thus, a state can refer to a highly abstract construal of a theoretical notion or to a very specific, concrete record of an empirical observation. The operators which apply to these states possess a similar range from highly abstract mental transformations (analogies, metaphors, instances of formal and informal reasoning, etc.) to very specific manual operations upon apparatus.

Analysis focused on a sequence of over 100 experiments conducted by Faraday between August 29, 1831 and November 4, 1831. The first experiment in the series recorded his discovery of induction. Taking an iron ring wound with two coils, he found that a brief transient current was generated in one coil whenever a battery current was turned on or off in the second coil. Over the next several months, Faraday explored the properties of the new phenomena, finally reading a paper on November 4 to the Royal Society. Two previous papers have dealt with the problem of what constitutes an experiment in this context (Tweney, 1984) and with the general problem of where the idea for the experiment came from (Tweney, 1985). The present account demonstrates that the microstructure of experimental exploration of the new discovery is also amenable to analysis.

Each record was first segmented into unitary propositions. Segmentation was highly reliable (in excess of 90% agreement for two independent coders). Each segment was classified as representing a state alone, an operator alone, or a combination of one or more states and one operator. Distinguishing states and operators was highly reliable among independent coders. However, classifying operators into specific categories proved only moderately reliable (about 60% agreement between the two authors working independently). Repeated attempts to refine the classification system resulted in no gains in reliability, probably because many of the judgements depend upon contextual knowledge of the relevant physics, Faraday's overall goals, and the heuristics used by Faraday (Tweney, 1985). Disagreements were resolved by discussion and reference to prior historical accounts of Faraday's experimentation. The most frequent operators were DO, OBSERVE, INFER, COMPARE and USE ANALOGY.

DO operators were generally followed by OBSERVE operators. In the early portion of the record, non-DO operators were much less frequent than DO operators, all of which involve some form of manipulation of physical entities; in later portions, Faraday sometimes used very few DO operators. A "Problem Behavior Graph," PBG, for the first two days of the record is shown in the figure. Each geometrical shape corresponds to an operator located in the numbered segment printed within the shape. The graph explicitly shows only those operators which culminate a sequence of manipulations having some sort of consequence, either an observation (shown by a square box) or some other operator (shown by a circle). Time is shown in the diagram by movement either rightward or downward. Rightward movement was used whenever a new empirical observation was made; downward movement was used in all other cases. If an observation was made which Faraday judged to be spurious (either then or later), then the temporal sequence was jumped backward to the last non-spurious observation, and moved downward to the next operator. Thus the overall shape of the graph is moving rightwards when Faraday was learning new things, downwards when he was not observing new things, and backwards when he was temporarily "tricked" into believing something which later turned out to be false.

Similar graphs were prepared covering the entire sequence of experiments, and inspected. The graph moves rightward at a generally high rate initially, at a low or zero rate during the middle, rightwards again toward the end of the series, and straight downward at the very end. Very little "branching" exists; this PBG is far less "foliated" than PBGs developed from laboratory studies of problem solving. In general, Faraday does not look as if he were blindly searching a problem space, gradually tracing a path to a solution by eliminating blind alleys. Instead, he appears to have had very little patience with unproductive results. If a particular set of manipulations failed to produce new results very quickly, then Faraday abandoned the line of inquiry and turned to another. This appears to be a kind of "working forward" generally not observed in laboratory studies of human problem solving.

Tweney (1985) argued that Faraday's 1831 researches could be understood as the application of specific scripts applied to specific schemata and guided "in the large" by powerful heuristics that regulated search for confirmatory and for disconfirmatory results. The present analysis extends such a view by tying it to lower-level states and operators. DO operators reflect

678

instantiations of specific scripts; during the analyzed series there is reason to believe that Faraday is in the process of developing a new script, "Produce an induced current," which figured extensively in his subsequent research. States in our analysis most often instantiate perceptual information, but on a few occasions represent instantiations of a slowly developing schema concerning the nature of forces as field-like phenomena. Faraday's greatest theoretical contribution to science is implicit here, since he is generally considered to have developed the first truly non-Newtonian conception of field forces (Miller, 1984; Nersessian, 1984). Gooding (1985) argued that Faraday typically proceeded from fairly loose "construals" to tightly defined scientific "concepts," via a series of studies that culminated in clear-cut, simple demonstration experiments. The present analysis displays such a sequence.

The most striking implications for cognitive science concern the differences between the present analysis and other recent accounts of science. Klahr & Dunbar (in press), for example, demonstrated that the process of discovering how an electronic device worked could be represented as a dual search through two problem spaces, an hypothesis space and an experimental space. Such a view is inadequate as a description of Faraday's work because it is not helpful to construe Faraday as searching through an experimental and an hypothesis space; no finite list of hypotheses or experiments can capture the boundless possibilities facing him after the initial discovery in 1831. Instead, Faraday's activity is better construed as a multi-level search in which large numbers of promising lines of exploration are abandoned in favor of lines which coincide with higher level goals (to elaborate a field-like theory of force, say). In contrast to Klahr & Dunbar's problem, there is not one definable goal state but many overlapping goals.

Langley, et al. (1986) focus upon the pursuit of intermediate goals via selective search of intermediate situation trees. Such a representation proves to be extremely cumbersome for Faraday, however, because one needs to postulate a new intermediate tree every few experiments. In effect, each new observation opens a new situation tree. We prefer a schematic approach via the redefined notion of goal, and we believe that such an approach holds out more promise for the successful analysis of the kind of science which Faraday conducted.

FOOTNOTE

Presented at the Ninth Annual Conference of the Cognitive Science Society, July 16-18, 1987, Seattle, Washington. Grateful acknowledgement is made to the Faculty Research Committee of Bowling Green State University for its support of this project, to the Royal Institution of Great Britain (London, England) for access to archival materials, to Rena Corbin and Kim Schaller for assistance in coding, and to Karin G. Hubert for assistance in preparation of graphical analyses.

REFERENCES

Ericsson, K.A. & Simon, H.A. (1984). Protocol analysis: Verbal reports as data. Cambridge: MIT Press.

Faraday, M. (1932-36). Faraday's Diary ... during the years 1820-62. (7 volumes). London: Bell. (edited by T. Martin).

Gooding, D. (1985). "In nature's school:" Faraday as an experimentalist. In D. Gooding & F.A.J.L. James (eds.) Faraday rediscovered: Essays on the life and work of Michael Faraday, 1791-1867. London/New York: Macmillan/Stockton.

Klahr, D. & Dunbar, K. (in submission). The psychology of scientific discovery: Search in two problem spaces.

Langley, P.W., Simon, H.A., Bradshaw, G.L. & Zytkow, J.M. (1986). Scientific discovery: An account of the creative processes. Cambridge: MIT Press.

Miller, A.I. (1984). Imagery in scientific thought: Creating 20th-century physics. Boston: Birkhauser.

Nersessian, N.J. (1984). Faraday to Einstein: Constructing meaning in scientific theories. Dordrecht: Martinus Nijhoff.

Newell, A. & Simon, H.A. (1972). Human problem solving. Englewood Cliffs, NJ: Prentice-Hall.

Tweney, R.D. (1984). Cognitive psychology and the history of science: A new look at Michael Faraday. In S. Bem, H. Rappard & W. van Hoorn (eds.) Studies in the history of psychology and the social sciences 2. Leiden: Psychologisch Instituut van de Rijksuniversiteit Leiden.

Tweney, R.D. (1985). How Faraday discovered induction: A cognitive approach. In D. Gooding & F.A.J.L. James (eds.) Faraday rediscovered: Essays on the life and work of Michael Faraday, 1791-1867. London:New York: Macmillan/Stockton Press.

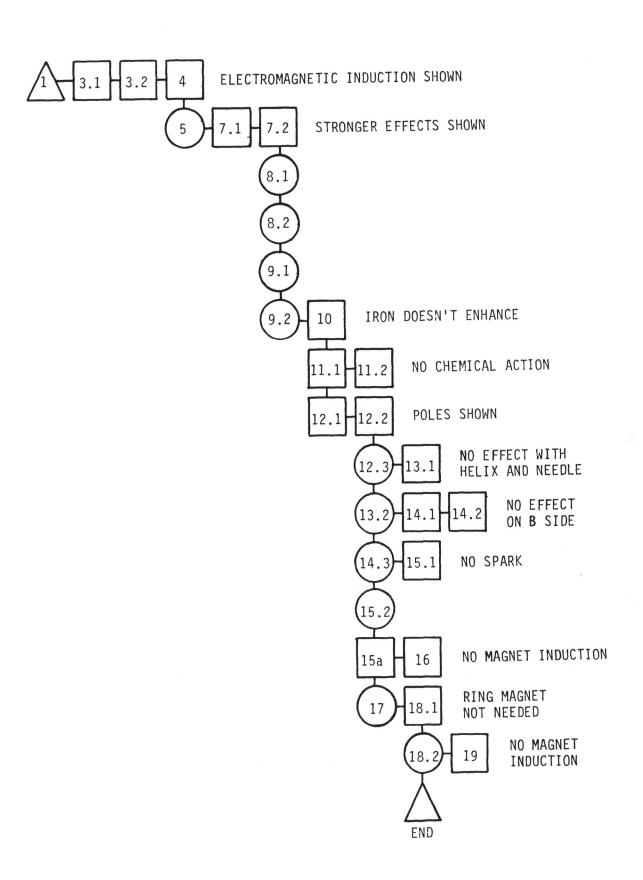

PROBLEM BEHAVIOR GRAPH FOR AUGUST 29 AND 30, 1831

Explanation-Based Decision Making

Nancy Pennington and Reid Hastie

University of Chicago   Northwestern University

Abstract

In complex decision tasks the decision maker frequently constructs a summary representation of the relevant evidence in the form of a causal explanation and relies on that representation, rather than the "raw" evidence base, to select a course of action from a choice set of decision alternatives. We introduce a general model for this form of decision making, called explanation-based decision making, because of the central role played by the intervening evidence summary. Several original empirical studies of judicial decision making, a prototype of the class of explanation-based decision tasks, are reviewed and the findings are adduced in support of the explanation-based decision model. In legal decision making tasks subjects spontaneously construct evidence summaries in the form of stories comprising the perceived underlying causal relationships among decision relevant events. These explanations are primary mediators (i.e., causes) of the subjects' decisions.

Many important decisions in engineering, medical, legal, policy, and diplomatic domains are made under conditions where a large base of implication-rich, conditionally-dependent pieces of evidence must be evaluated as a preliminary to choosing an alternative from a set of prospective courses of action. We propose that a general model of explanation-based decision making describes behavior under these conditions (Pennington & Hastie, 1986). According to the explanation-based decision model, decision makers begin their decision process by constructing a causal model to explain the available facts. Concommitant with or subsequent to the construction of a causal model of the evidence, the decision maker is engaged in a separate activity to learn or create a set of alternatives from which an action will be chosen. A decision is made when the causal model of the evidence is successfully matched to an altenative in the choice set. The three processing stages in the explanation-based decision model are shown in Figure 1.

The distinctive assumption in our explanation-based approach to decision making is the hypothesis that decision makers construct an intermediate summary representation of the evidence and that this representation, rather than the original "raw" evidence, is the basis of the final decision. Interposition of this organization facilitates evidence comprehension, directs inferencing, enables the decision maker to reach a decision, and determines the confidence assigned to the accuracy or success of the decision. This means that the locus of theoretical accounts for differences in decisions rendered by different individuals, systematic biases shared by many individuals, and the effects of most variations in decision task characteristics will usually lie in the evidence evaluation stage of the decision process.

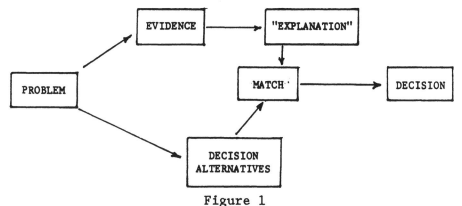

Figure 1

Overview of the Processing Stages of the Explanation-based Model

The structure of the causal model constructed to explain the evidence will be specific to the decision domain. For example, we have proposed that a juror uses narrative story structures to organize and interpret evidence in criminal trials. Different causal rules and structures will underlie an internist's causal model of a patient's condition and its precedents (Pople, 1982), an engineer's mental model of an electrical circuit (de Kleer & Brown, 1983), a merchant's image of the economic factors in a resort town (Hogarth, Michaud, & Mery, 1980), or a diplomat's causal map of the political forces in the Middle East (Axelrod, 1976). Thus, a primary task in research on explanation-based decision making is the identification of the type of intermediate summary structure that is imposed on evidence by decision makers in a specific domain of decision making. This is in contrast with earlier process-oriented calculational models where the theoretical focus was on attentional processes and the computations whereby separate sources of information were integrated into a unitary value or utility (Anderson, 1981; Edwards, 1954; Kahneman & Tversky, 1979).

Explanation-based Decision Making in Judicial Decisions

In the present paper we concentrate on the example of juror decision making. The juror's decision task is a prototype of the tasks to which the explanation-based model should apply: First, a massive "database" of evidence is input at trial, frequently requiring several days to present. Second, the evidence comes in a scrambled sequence, usually several witnesses and exhibits convey pieces of a historical puzzle in a jumbled temporal sequence. Third, the evidence is piecemeal and gappy in its depiction of the historical events that are the focus of reconstruction: event descriptions are incomplete, usually some critical events were not observed by the available witnesses, and information about personal reactions and motivations is not presented (often because of the rules of evidence). Finally, subparts of the evidence (e.g., individual sentences or statements) are interdependent in their probative implications for the verdict. The meaning of once statement cannot be assessed in isolation bacause it depends on the meanings of several related statements.

Evidence Summary. Empirical research has demonstrated that the juror's "explanation" of legal evidence takes the form of a "story" in which causal and intentional relations among events are prominent (Bennett & Feldman, 1981; Hutchins, 1980; Pennington, 1981; Pennington & Hastie, 1986). The story is constructed from information explicitly presented at trial and knowledge possessed by the juror. Two kinds of knowledge are critical: (a) expectations about what makes a complete story and (b) knowledge about events similar in content to those that are the topic of dispute.

General knowledge about the structure of human purposive action sequences, characterized as an episode schema, serves to organize events according to the causal and intentional relations among them as perceived by the juror. An episode schema specifies that a story should contain initiating events, goals, actions, consequences, and accompanying states, in a particular causal configuration (Mandler, 1980; Pennington & Hastie, 1986; Rumelhart, 1977; Stein & Glenn, 1979; Trabasso & van den Broek, 1985). Each component of an episode may consist of an episode so that the story the juror constructs can be represented as a hierarchy of embedded episodes. The highest level episode characterizes the most important features of "what happened." Components of the highest level episode are elaborated in terms of more detailed event sequences in which causal and intentional relations among subordinate story events are represented. Expectations about the kinds of information necessary to make a story tell the juror when important pieces of the explanation structure are missing and when inferences must be made. Knowledge about the structure of stories allows the juror to form an opinion concerning the completeness of the evidence, the extent to which a story has all its parts.

More than one story may be constructed by the juror, however one story will ususally be accepted as more coherent than the others. Coherence combines judgments of completeness, consistency, and plausibility. Consistency concerns the extent to which the story does not contain contradictions and the plausibility of alternative stories may be assessed by comparing story sequences to known or imagined events in the real world. If more than one story is judged to be coherent, then the story will lack uniqueness and great uncertainty will result. If there is one coherent story, this story will be accepted as the explanation of the evidence and will be instrumental in reaching a decision.

Choice Set. The decision maker's second major task is to learn or to create a set of potential solutions or action alternatives that constitute the choice set. In some decision tasks the potential actions are given to the decision maker (instructions from the trial judge on verdict alternatives) or known beforehand (treatment options available to a physician). In others, creation of alternatives is a major activity of the decision maker (for example, drafting alternate regulations for industrial waste disposal, planning alternate marketing strategies, or negotiating alternate acceptable trade contracts). These solution design tasks may invoke their own (embedded) decision tasks.

In criminal trials the information for this processing stage is given to jurors at the end of the trial in the judge's instructions on the law. The process of learning the verdict categories is a one-trial learning task in which the material to be learned is very abstract. Interference may occur from jurors' prior knowledge of concepts such as first degree murder, manslaughter, armed robbery, etc. The juror attempts to learn the defining features (elements of the crime) of each verdict alternative and a decision rule specifying their appropriate combination. We hypothesize that the conceptual unit is a category (frame) defined by a list of criterial features referring to identity, mental state, circumstances, and actions linked conjunctively or disjunctively to the verdict alternative (Kaplan, 1978; Pennington & Hastie, 1981).

Match Process. The final stage in the global decision process involves matching solution alternatives to the summary evidence representation to find the most successful pairing. Confidence in the final decision will be partly determined by the goodness-of-fit of the evidence-solution pairing selected and the uniqueness of the winning combination when compared to alternative pairings. Because verdict categories are unfamiliar concepts, the classification of a story into an appropriate verdict category is likely to be a deliberate process. For example, a juror may have to decide whether a circumstance in the story such as "pinned against a wall" constitutes a good match to a required circumstance, "unable to escape," for a verdict of Not Guilty by Reason of Self Defense.

The classification process is aided by relatively direct relations between attributes of the decision categories and the components of the episode schema. The criminal law has evolved so that the main attributes of the decision categories suggested by legal experts (Kaplan, 1978)-- identify, mental state, circumstances, and actions -- correspond closely to the central features of human action sequences represented as episodes-- initiating events, goals, actions, and accompanying states.

The story classification stage also involves the application of the judge's procedural instructions on the presumption of innocence and the standard of proof. That is, if not all of the verdict attributes for a given verdict category are satisfied "beyond a reasonable doubt," by events in the accepted story, then the juror should presume innocence and return a default verdict of not guilty.

Confidence in Decisions. Several aspects of the decision process influence the juror's level of certainty about the final decision. First, the accepted story is judged to be the most coherent but the level of coherence will affect confidence. Thus, if the story lacks completeness, consistency, or plausibility, confidence in the story and therefore in the verdict will be diminished. Second, if a story lacks uniqueness, that is, there is more than one coherent story, then certainty concerning the accuracy of any one explanation will be lowered (Einhorn & Hogarth, 1986). Finally, the goodness-of-fit between the accepted story and the best-fitting verdict category will influence confidence in the verdict decision.

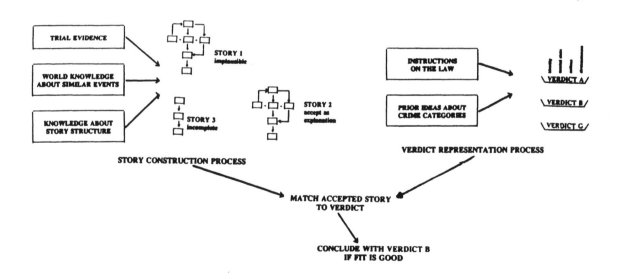

Figure 2
The "Story Model" for Juror Decision Making

685

In summary, our application of the general explanation-based decision model to legal decisions is based on the hypothesis that jurors impose a narrative story organization on trial information, in which causal and intentional relations between events are central (Bennett & Feldman, 1981; Pennington, 1981; Pennington & Hastie, 1986). Meaning is assigned to trial evidence through the incorporation of that evidence into one or more plausible accounts or stories describing "what happened" during events testified to at the trial. The story organization facilitates evidence comprehension and enables jurors to reach a predeliberation verdict decision. We call our application The Story Model because of the central role played by narrative, story-like evidence summaries in the decision process. The Story Model includes three components: (a) evidence evaluation through story construction, (b) representation of the decision alternatives by learning verdict category attributes, and (c) reaching a decision through the classification of the story into the best fitting verdict category (see Figure 2).

Previous Research

Our previous research on the Story Model provided descriptions of mental representations of evidentiary information and verdict information at one point in time during the decision process (Pennington & Hastie, 1986). In that research we established that the evidence summaries constructed by jurors had story structure (and not other plausible structures); verdict representations looked like feature lists (or simple frames); and that jurors who chose different verdicts had constructed different stories such that there was a distinct causal configuration of events that constituted a story corresponding to each verdict category. Moreover, jurors choosing different verdicts did not have systematically different verdict representations, nor did they apply different classification criteria. Thus verdict decisions covary with story structures but do not covary with verdict learning or story classification. However, the interview method used in this research precluded strong inferences concerning the spontaneity of story construction the functional role of stories in the decision phase.

In a second empirical study we established that decision makers spontaneously constructed causal accounts of the evidence in the legal decision task (Pennington & Hastie, 1987). In this study, subjects' responses to sentences presented in a recognition memory task were used to draw conclusions about subjects' post-decision representations of evidence. Subjects were expected to "recognize" as having been presented as trial evidence sentences from the story associated with their decision, with a higher probability than to recognize sentences from stories associated with other (rejected) decisions. This implies that hit rates (correct recognitions) and false alarm rates (false recognitions) for sentences from each story can be predicted from subjects' verdicts. These predictions were confirmed; verdict decisions predicted the high hit and false alarm rates found for sentences in the subjects' stories. Thus, a different method, subject population, and stimulus materials yielded results converging with the interview study conclusions about the correlation between memory structure and decision outcome. Even though we can conclude that story representations were constructed <u>spontaneously</u>, the causal role of stories in decisions is still not established because subjects could decide on a verdict and then (spontaneously) justify it to themselves by constructing a coherent story.

The Effect of Evidence Structure on Decisions

An experiment was conducted to study the effects of variations in the order of evidence presentation on judgments. Our primary goal was to test the claim that the construction of stories in evidence evaluation causes decisions. A secondary goal was to determine whether story coherence and uniqueness influence judgments of confidence in the correctness of verdicts. We used legal case materials based on the transcript of an actual murder trial, titled Commonwealth v. Johnson, and varied presentation order to influence the ease with which a prosecution (First Degree Murder) or defense (Not Guilty by Reason of Self Defense) story could be constructed (see Pennington & Hastie, 1986 for a summary of the trial). The "logic" of the experiment was summarized in our hypothesis that (manipulated) ease of story construction would influence verdict decisions; easy-to-construct stories would result in more decisions in favor of the corresponding verdicts.

Stories were considered easy to construct when the evidence was ordered in a temporal and causal sequence that matched the occurrence of the original events (Story Order; Baker, 1978). Stories were considered difficult to construct when the presentation order did not match the sequence of the original events. We based the non-story order on the sequence of evidence as conveyed by witnesses in the original trial (Witness Order). One-hundred and thirty college student mock-jurors listened to a tape recording of a 100-sentence summary of the trial evidence (50 prosecution statements and 50 defense statements), followed by a judge's charge to choose between a Murder verdict and a Not Guilty verdict. The 50 prosecution statements, constituting the First Degree Murder story identified in our initial interview study (Pennington & Hastie, 1986), were presented either in a Story Order or a Witness Order. Similarly, the defense statements, the Not Guilty story, were presented in one of the two orders creating a four-cell factorial design. In all four order conditions the prosecution evidence preceded the defense evidence as per standard legal procedure. After listening to the tape recorded trial materials, the subjects completed a questionnaire indicating their verdict, confidence in the verdict, and their perceptions of the strengths of the prosecution and defense cases.

As predicted, subjects were likeliest to convict the defendant when the prosecution evidence was presented in Story Order and the defense evidence was presented in Witness Order (78% chose guilty) and they were least likely to convict when the prosection evidence was in Witness Order and defense was in Story Order (31% chose guilty, see Table 1). Conviction rates were intermediate in conditions where both sides of the case were in Story Order (59% convictions) or both were in Witness Order (63% convictions). Statistically, the best summary of the effects of evidence order on verdict choice was two main effects, one for defense side order (Story versus Witness) and one for prosecution side order (log-linear model analysis, chi-squared "badness-of-fit" statistic (2 df) = .42, p > .80.

Analyses were conducted on the ratings of strength of the defense and prosecution cases and these ratings were influenced by presentation order, with Story Order evidence rated as stronger than Witness Order. Furthermore, the perceived strength of one side of the case depended on both the order of evidence for that side and for the other side of the case. This finding supports our claim that the uniqueness of the best-fitting story is one important basis for confidence in the decision. We also examined the verdict confidence ratings and found that, regardless of verdict chosen, jurors who heard both sides of the case in Story Order were verdict chosen, jurors who heard <u>both</u> sides of the case in Story Order were more confident than jurors who heard one or neither side in Story Order. This result reinforces our conclusion that alternate story strength is important, although the finding was not predicted.

	Defense Case Presented In	
	STORY FORM	**WITNESS FORM**
STORY FORM	59%	78%
WITNESS FORM	31%	63%

Prosecution Case Presented In (row labels at left: STORY FORM, WITNESS FORM)

Table 1
Percentages of Subjects Choosing the "Guilty" Verdict

Conclusions

We have introduced a model of decision making that describes human behavior in tasks where a large, implication-rich, conditionally-dependent set of propositions constitute an evidence base for selection of an option from a limited set of decision alternatives. We propose that considerable processing occurs to understand the evidence base by constructing a summary explanation or causal model of the decision-relevant facts, assumptions, and premises. Once a satisfactory explanation has been constructed, the decision maker attempts to select an option by matching features of the summary explanation to corresponding characteristics of solutions or courses of action in the decision set.

The decision process is divided into three stages: construction of a summary explanation; determination of decision alternatives; mapping the explanation onto a best-fitting decision alternative. This subtask framework is in contrast to the uniform on-line updating computation or the unitary memory-based calculation hypothesized in most alternative approaches (cf. Hastie & Park, 1986). Furthermore, we diverge sharply from traditional approaches with our emphasis on the structure of memory representations as the key determinant of decisions. We also depart from the common assumption that, when causal reasoning is involved in judgment, it can be described by algebraic, stochastic, or logical computations that lead directly to a decision (e.g., Anderson, 1974; Einhorn & Hogarth, 1985; Kelley, 1973). In our model causal reasoning plays a subordinate but critical role by guiding inferences in evidence evaluation and construction of the intermediate explanation.

References

Anderson, N.H. (1974). Cognitive algebra: Integration theory applied to social attribution. In L. Berkowitz (Ed.), <u>Advances in Experimental Social Psychology</u>, Volume 7. New York: Academic.

Anderson, N.H. (1981). <u>Foundations of Information Integration Theory</u>. New York: Academic.

Axelrod, R. (Ed.) (1976). <u>Structure of Decision: The Cognitive Maps of Political Elites</u>. Princeton: Princeton University Press.

Baker, L. (1978). Processing temporal relationships in simple stories: Effects of input sequences. <u>Journal of Verbal Learning and Verbal Behavior</u>, <u>17</u>, 559-572.

Bennett, W.L., & Feldman, M. (1981). <u>Reconstructing Reality in the Courtroom</u>. New Brunswick, N.J.: Rutgers University Press.

deKleer, J., & Brown, J.S. (1983). Assumptions and ambiguities in mechanistic mental models. In D. Gentner and A.L. Stevens (Eds.), <u>Mental Models</u>. Hillsdale, N.J.: Erlbaum.

Edwards, W. (1954). The theory of decision making. <u>Psychological Review</u>, <u>51</u>, 380-417.

Einhorn, H.J., & Hogarth, R.M. (1986). Judging probable cause. <u>Psychological Bulletin</u>, <u>99</u>, 3-19.

Hastie, R., & Park, B. (1986). The relationship between memory and judgment depends on whether the judgment task is memory-based or on-line. <u>Psychological Review</u>, <u>93</u>, 258-268.

Hogarth, R.M. Michaud, C., & Mery, J.L. (1980). Decision behavior in urban development: A methodological approach and substantive considerations. <u>Acta Psychologica</u>, <u>45</u>, 95-117.

Hutchins, E. (1980). <u>Culture and Inference</u>. Cambridge, Mass.: Harvard University Press.

Kahneman, D., & Tversky, A. (1979). Prospect theory: An analysis of decision under risk. <u>Econometrica</u>, <u>47</u>, 263-291.

Kaplan, J. (1978). <u>Criminal Justice: Introductory Cases and Materials</u> (2nd Ed.). Mineola, N.Y.: Foundation Press.

Kelley, H.H. (1973). The processes of causal attribution. <u>American Psychologist</u>, <u>28</u>, 107-128.

Mandler, J.M. (1980). Categorical and schematic organization in memory. In C.R. Puff (Ed.), <u>Memory Organization and Structure</u>. New York: Academic.

Pennington, N. (1981). <u>Causal Reasoning and Decision Making: The Case of Juror Decisions</u>. Unpublished doctoral dissertation, Harvard University.

Pennington, N., & Hastie, R. (1981). Juror decision making models: The generalization gap. _Psychological Bulletin_, _89_, 246-287.

Pennington, N., & Hastie, R. (1986). Evidence evaluation in complex decision making. _Journal of Personality and Social Psychology_, _51_, 242-258.

Pennington, N., & Hastie, R. (1987). Explanation-based decision making: The effects of memory structure on judgment. Unpublished manuscript, University of Chicago.

Pople, H.E., Jr. (1982). Heuristic methods for imposing structure on ill-structured problems: The structuring of medical diagnostics. In P. Szolovits (Ed.), _Artificial Intelligence in Medicine_. Boulder, Colo.: Westview Press.

Rumelhart, D.E. (1977). Understanding and summarizing brief stories. In D. LaBerge and S.J. Samuels (Eds.), _Basic Processes in Reading: Perception and Comprehension_. Hillsdale, N.J.: Erlbaum.

Stein, N.L., & Glenn, C.G. (1979). An analysis of story comprehension in elementary school children. In R.O. Freedle (Ed.), _New Directions in Discourse Processing_, Volume 2. Norwood, N.J.: Ablex.

Trabasso, T., & van den Broek, P. (1985). Causal thinking and the representation of narrative events. _Journal of Memory and Language_, _24_, 612-630.

Diagnosing Errors in Statistical Problem-Solving: Associative Problem Recognition and Plan-Based Error Detection

Marc M. Sebrechts
Lael J. Schooler
Cognitive Science Program & Department of Psychology
Wesleyan University

Abstract

This paper describes our model for diagnosis of student errors in statistical problem-solving. A simulation of that diagnosis, GIDE, is presented together with empirical validation on student solutions. The model consists of two components. An "intention-based" diagnostic component analyzes solutions and locates errors by trying to synthesize student solutions from knowledge about the goal structure of the problem and related knowledge about planning errors. This approach can account for about 82% of the lines and over 95% of the goals in a set of 60 student t-tests. When solutions contain errors in procedural implementation such plan-based analysis is quite effective. In many cases, however, students do not pursue an "appropriate" solution path. The diagnostic model, therefore, includes a second component which is used to determine which type of problem the student is using; it is modeled by a spreading activation network of statistical knowledge. On a sample of 38 student solutions, the simulation correctly identified 86% of the problem types. The model appears to account for a wide range of problem-solving behavior within the domain studied. The preliminary performance data suggest that our model may serve as a useful part of an intelligent tutoring system.

Introduction

An important part of problem-solving is the ability to detect and explain errors in proposed solutions. Such "diagnostic" ability has gained particular prominence from attempts to analyze faulty problem-solving during nuclear power plant failures. However, a similar type of skill constitutes an important part of the daily activity in almost any classroom. In both settings, the assumption is that we can learn by our failures.

Although the behavior is common, it is not simple. Diagnosis includes a wide range of general strategies and domain specific behaviors. In this paper, we report our initial attempts to analyze one common type of instructional diagnosis: locating and explaining errors in solutions to statistics problems.

We will first describe the general principles that guide our conception of the task. They are based on both our informal analysis of expert protocols in statistics and on reported strategies in the literature. These principles then serve as the basis for our model of diagnosis. That model is made explicit in a simulation. Finally, we report tests of our model on data collected from students. These analyses are meant to provide a test of sufficiency only. Additional data will be needed to verify a more detailed mapping between the "process" of our implementation and that of experts.

General Principles

Before describing the diagnostic system, it should be noted that "diagnosis" covers a wide range of problems. Its most familiar use is in medicine, and substantial advances have been made in developing a model of medical diagnosis (Clancey, 1985; Clancey and Lestinger, 1984; Shortliffe, 1976). Although this research borrows heavily from the medical domain, the "diagnostic" goals are rather different. In medicine, diagnosis is used to determine the underlying cause of a set of symptoms. In the case of statistics (or other situations where we are trying to instruct the student), the ultimate diagnostic goal is not only to understand the source of the error in the solution, but to specify the source of the error *with respect to the student's conception of the problem.*

Our model of this type of diagnosis is guided by a series of general principles about the nature of errors. First, it is recognized that errors in problem-solving are frequently systematic rather than random. Even for an apparently simple domain such as subtraction, there are large numbers of potential systematic errors. Friend and Burton (1981), for example, listed 110 'primitive' subtraction errors which can lead to a much larger number of compound errors (Brown & Burton, 1978; Burton, 1982). Treating error diagnosis as a problem of detecting misdirected systematicity makes it possible to develop useful strategies for isolating and correcting the mistakes. The model of the expert diagnostician must therefore include systematic deviations from correct procedures as part of its knowledge base.

Second, an error is, at least in part, generated by its context. It is often difficult to say whether or not a particular action was appropriate or inappropriate without knowing the reason for its inclusion. The prototype case for such reasoning is programming. There are an infinite number of syntactically correct steps that are possible in even a moderately complex program. However, the number of legitimate steps for a particular program are far more constrained.

In the more general case of problem-solving, it is often difficult to judge the correctness of a component procedure in isolation. It is therefore necessary to embed knowledge of errors within a context that characterizes the normal structure of problem solutions; and the knowledge of the diagnostician must somehow capture that embedded structure. Below it is argued that such a context must include a problem description and associated goal structure in order to capture the "intention" of the student.

692

The third general principle is closely related to the second: solution strategies share many concepts in common. This means that solution components can be viewed as part of complex heterarchy. Detecting individual concepts or formulae will not be sufficient to analyze a solution. With respect to our model, this principle argues for a process of search that is goal and plan directed. It is the goal/plan structure that disambiguates the role of a procedure. It is important to recognize that this model attempts to capture "diagnosis" by an instructor; that process need not, and usually does not, mimic student problem solving. There is evidence that an explicit plan structure can improve students' performance in programming (Miller, 1978), but that data actually suggests that students do not usually follow such systematic planning in their unconstrained problem-solving.

The Intention-Based Approach

These principles concerning the relationship of a knowledge base and error diagnosis served as the basis for our model of diagnosis in statistics (Sebrechts, Schooler & LaClaire, 1986; Sebrechts, LaClaire, Schooler & Soloway, 1986). This portion of the model is based on a strategy developed by Johnson and Soloway (1985) in the realm of programming (PROUST). This approach is called "intention-based" diagnosis, since it uses a goal structure that represents the problem the student apparently "intended" to solve. In this approach a problem is described in terms of a series of necessary and sufficient goals. The goals in turn have a variety of plan implementations, both correct and "buggy" that can be used to describe the student's solution. The instructors' knowledge is thus described as consisting of knowledge about the domain and knowledge about common errors. In addition, errors are thought to be understood within a context of goals and plans. A statement's "meaning" is determined relative to the broader solution context.

The simulation for statistics, called GIDE, analyzes a solution by attempting to describe how the student solution matches, or fails to match, some realization of an intended solution. The procedure consists of analysis-by-synthesis: an attempt is made to analyze a solution by building a plausible description of how that solution satisfies the goals defined by the problem.

For each problem, GIDE has a problem description that indicates those goals that must be satisfied by any correct solution. GIDE attempts to find some satisfaction of those goals in the student's solution by testing a series of plans. If no correct solution is located, GIDE tries to explain the discrepancy. This can be done by trying "buggy" plans that represent common conceptual errors in a plan (such as confusing standard-error and standard-deviation), or by examining "buggy" rules representing common mistakes across different types of plans (such as reversing a sign in a computation).

GIDE attempts to extend the model used for programming in PROUST to different domains. Student solutions in statistics differ in numerous ways from programming solutions. First, students frequently leave out steps. Components that are "obvious" or can be calculated mentally are left out. In addition, there is a very loose syntax, which allows frequent changes in construction. In contrast to most programming languages, students use assignment to values, they include free (unbound) expressions in their solutions, and they occasionally change the symbols they are using. These are all components that can be analyzed successfully by teaching assistants.

In order to handle the wide range of expressions used, GIDE can match a procedure in several ways. The first strategy is to look for an appropriate symbolic structure. If a student writes "StdE = 20 / sqrt n", GIDE will match this to its internal representation of one form for standard error (?Se = ?Sd / sqrt ?count). Another way in which the student may indicate the standard error is by using the appropriate values. The form "5 = 20 / 4" ,would also be recognized as standard error *if* ?Sd had been previously bound to 20 and ?count had been bound to 16. Finally, if a plan is not satisfied symbolically or by value, GIDE will attempt to find a free expression. Thus, for example, the "-14" on line 9 of Figure 2, is recognized as a deviation score.

In order to handle steps that are left out, GIDE uses "implicit matching". GIDE contains a dependency tree that shows the relationship among knowledge of concepts: for example, determining the variance requires the sum-of-squares. If the student includes in their solution the correct variance, but not the sum-of-squares, GIDE infers that the student understands the sum-of-squares. The details of these techniques are described in Sebrechts, Schooler and LaClaire (1986). Here we will focus on the empirical outcome of this general approach.

Empirical Tests of Plan Analysis

In order to evaluate our diagnostic model, we examined GIDE's performance on a series of 43 solutions collected from students in an introductory statistics course. In order to examine the types of plans needed for different solution strategies we used two testing conditions. In the first condition (called "Directed") the problem was accompanied by a set of directions indicating the appropriate steps for solution, and was similar to a homework assignment. In the second condition ("Undirected") another problem was given as an exam problem without any accompanying materials.

The specific statistical problem we have examined is a repeated-measures t-test. This type of problem attempts to determine whether or not there is a reliable difference between two measurements on the same group. For example, you might use this test to determine if a training program has improved the efficiency of workers by measuring efficiency before and after training (Figure 1).

Student solutions were written on paper and were coded directly into GIDE. An example solution from each of the two conditions is shown in Figures 2 and 3. The only information added by GIDE are line numbers for reference. GIDE's output for the two solutions is included with each solution.

Buggy Solutions. Not surprisingly, a greater percentage of solutions had bugs in the "undirected" condition. Of those solutions with bugs, however, the greatest percentage in both conditions consisted of missing and implicit goals. An implicit goal is one that is not explicitly stated but is obvious from other concepts. For example, in order to compute an average, the student must know the sum. GIDE catches over 90% of such implicit goals based on its knowledge of

694

Problem Statement:

An employer at an automotive plant is interested in determining whether or not a training program can improve efficiency of his employees. He has tried it out tentatively for four work groups in one of the plants. Below are the results of the trial: efficiency is measured by a standardized test ranging from 1 (extremely efficient) to 25 (extremely inefficient). Determine whether or not there is a reliable change in efficiency.

Work Group	Pre-Training	Post-Training	Post Minus Pre
A	14	9	-5
B	19	7	-12
C	16	10	-6
D	18	5	-13

Figure 1. An example of a repeated-measures t-test.

---

dependencies. Missing goals are those that never appear in the solution and are not implicit. GIDE catches a moderately high number of such goals (71% and 86% for directed and undirected problems respectively). The main reason why performance is not higher is that the mechanism for implicit matching is too powerful. It occasionally gives credit for knowing a concept that is actually not evident in the student solution.

GIDE was able to detect all of the relatively small number of specific plan errors. There were, however, important differences between the types of errors generated in the two conditions, as can be seen in Figure 4. For Directed problems, the errors tended to be more low level, involving failures in computational plans. In the Undirected case, students tended to make higher level mistakes. For example, in Figure 3, the student confused the standard deviation with the standard error.

Problem-Recognition Through an Associative Network

GIDE performed extremely well on the set of problems initially provided. As we began to extend the domain of GIDE's capabilities, however, we discovered that the system could become involved in extensive searches that were not reflected in our protocols of experts. When we tested experts on these problems, they would say, "this student is using the wrong kind of test", before doing a careful analysis. In brief, there was a problem recognition phase that was not captured by the systematic approach of GIDE.

Of course, GIDE can manage to get around this difficulty by an exhaustive search. However, such a search would eliminate the heuristic value of intention-based diagnosis. As a consequence it

Directed Student Solution (id #48)

```
 1:        -5
 2:       -12
 3:        -6
 4:       -13
 5:         n
 6:         4
 7:     -6 + -13 + -5 + -12 = -36
 8:     d_dbar
 9:       -14
10:       -21
11:       -15
12:       -22
13:     d_dbar2
14:       196
15:       441
16:       225
17:       484
18:      1346
19:     1346/3 = 448 = s2
20:     sqrt 448 = 21 = s
21:     sdbar = 21/2 = 11
22:      -9/11 = t
```

The average was implicitly included in your solution.

Your calculation of the deviations was incorrect. A useful way to avoid this mistake is to note that the sum of the deviations should always equal 0. The sum of your calculated deviations equals 72.0

In the process of calculating the standard deviation, you added the individual deviation scores to the mean of the deviations: instead, you should have subtracted the mean of the deviations from the individual deviation scores. For example, where you have -14, you should have 4.0.

These erroneous values were used to complete the analysis.

___

Figure 2: Student Solution (id #48) and GIDE's associated comments.

Undirected Student Solution (id #2)

```
 1:        d
 2:       -4
 3:       -8
 4:       -4
 5:       -4
 6:     -20 = sum d
 7:     x_dbar
 8:        1
 9:        3
10:        1
11:        1
12:     x_dbar2
13:        1
14:        9
15:        1
16:        1
17:     ss = 12
18:     dbar = -5
19:     t = (dbar - Uo)/ sdbar
20:     Sd = sqrt (12/3) = sqrt 4 = 2
21:     sdbar = sd/sqrt n = 2/sqrt 4 = 1
22:     t = -5/2 = -2.5
23:     alpha = 0.05
24:     tcrit = 3.18
```

The count of the observations was implicitly included in your solution.

The variance was implicitly included in your solution.

On line 22, when calculating the t statistic, you substituted the standard deviation for the standard error.

This erroneous value was used to complete the analysis.

___

Figure 3: Student solution (id #2) and GIDE's associated comments.

GIDE's Performance

Criteria	Directed		Undirected	
Solutions				
Total number of solutions	43		17	
Total receiving complete analysis	43	(100.0%)	16	(94.0%)
Total receiving partial analysis	0	(0.0%)	1	(6.0%)
Goals				
Total number of goals in solutions	445		216	
Total number of goals correctly analyzed	443	(99.5%)	207	(96.0%)
Total number of goals incorrectly analyzed	2	(0.5%)	9	(4.0%)
Lines				
Total number of lines included in solutions	1203		403	
Total number of lines correctly interpreted	986	(82.0%)	329	(81.6%)
Total number of lines misinterpreted	1	(0.0%)	9	(2.2%)
Total number of lines unaccounted for	216	(18.0%)	65	(16.1%)
Bugs				
In solutions:				
total number of solutions with bugs	27	(62.8%)	17	(100.0%)
total number of bugs	69		91	
missing goals	28	(40.5%)	37	(40.6%)
implicit goals	35	(50.7%)	48	(52.7%)
plan errors	6	(8.6%)	5	(5.5%)
errorneous solutions	0	(0.0%)	1	(1.1%)
Detected:				
total number of bugs detected	58	(84.0%)	84	(92.0%)
missing goals	20	(71.4%)	32	(86.5%)
implicit goals	32	(91.4%)	47	(97.9%)
plan errors	6	(100.0%)	5	(100.0%)
erroneous solutions	0		1	(100.0%)
false alarms	8		4	

Figure 4: Summary of GIDE's performance on directed and undirected student solutions.

would require a knowledge base and search times that are unreasonable as parts of a psychological model of the task.

In many cases, although the problem description does suggest what the student is trying to do, it does not capture the strategy that the student selects toward that end. Experts appear to be able to recognize how statements are related to more global problems. This was demonstrated in the case of programming by Adelson (1981). She found that although novices grouped program statements syntactically, experts grouped statements according to the programs from which they were derived. In terms of our diagnostic problem, this implies that experts should be able to recognize the type of solution being presented by examining constituent statements.

This is consistent with the problem analysis we have observed in statistics. The instructor tends to classify solutions fairly quickly, frequently with reference to specific aspects of the solution. Our model of how this occurs is that the diagnostician scans through the problem, identifying the component terms of the solution. Each term provides some degree of activation for associated concepts. The problem type that receives the highest level of activation during scanning is selected as the appropriate response.

We have simulated these processes by a network structure which consists of letters, symbols ("words"), goals, and problem types. The basic layer structure is similar in concept to that described by McClelland (1979) in his cascade model. Nodes at each level have excitatory links to nodes at the next higher level. In addition, there are reciprocal links from symbols ("words") to individual letters as in the McClelland and Rumelhart (1981) model, but in contrast to their model, there are no inhibitory links. (Likewise our model does not include any phonetic or morphemic information.)

The model is constrained by the fact that its only recognition capability is the particular statistical domain of interest. There are at least two plausible ways to characterize this network. It can be viewed as a subset of a more complete "reading" network in which the domain-relevant elements have higher initial activation levels; as such, only those initially activated nodes are relevant. Alternatively, it can be viewed as a domain-specific network for statistics which is activated more or less as a unit. The model does not make any differentiation between those options.

At the beginning of a particular problem type, GIDE's goal structure is used to construct the appropriate network. This process can be thought of as potentiating the portions of the network that are relevant to the statistics problems. Based on the problem description, each goal is linked to its subgoals. Thus, for example, the goal for Standard-Deviation would be linked to the subgoals of Mean, Count, Sum-of-Squares. The goals and subgoals are likewise linked to associated symbols. Standard-Deviation is linked to the symbol 'Std', a common abbreviation. Finally, each symbol is linked to its constituent letter nodes, which indicate both the character and the position of each letter. So, the letter node "s-" would indicate an initial 's' and would activate the "Std" symbol, whereas the letter node "-s" indicates a terminal 's' and would not directly activate 'Std'.

Once the network is established in this way, each of the symbols in the solution is "read" and decomposed into its constituent letters. The related symbols and goals are then activated through

698

spreading activation following a reduced version of the general model described by Anderson (1983a; 1983b) for ACT*. There are no direct connections between letters, but one letter can indirectly activate another letter through a symbol or goal. The network is run as though activation occurred in parallel. Following the spread on each cycle, all elements are decayed. This results in a dynamic set of activation levels which are updated as the lines are "scanned."

This network approach to recognition makes sense in light of behaviors we have observed. Statistics includes a set of fairly common symbols. However, unlike the case of reading English in which words and non-words are differentiated, in statistics there are numerous deviations from those symbols, depending on idiosyncracies of individual students. Spreading activation helps to capture such deviations as well as to locate minor errors without having to anticipate all possible forms of each symbol. Thus, for example, "d-dbar" is an appropriate form for finding a deviation score in a repeated measures t-test. Some students use "x-dbar", which may indicate either a different notation or an error in the procedure. In either case, it does suggest that the student is getting deviation scores base on the mean of deviations ("dbar"). When the network is presented with "x-dbar" it will activate "d-dbar" given the strong similarity of the two symbolic expressions.

Empirical Tests of Network Activation Analysis

In order to evaluate this portion of the model, we conducted network activations on 38 student solutions. The network included four standard statistical tests that served as problem types: repeated-t, independent-t, repeated ANOVA, and independent ANOVA. We compared the highest goal activation level in the network with categorization by the two authors. On average, the system was able to identify 86% of the problems correctly. It was more successful at identifying repeated-measures t-tests (95%) than independent sample t-tests (75%). This is due to the fact that repeated measures t-tests in our sample are usually conducted using deviation scores: the terms associated with those deviations are more distinct that the terms in the independent t-test.

Informal analysis suggests that this performance is roughly comparable to that of a teaching assistant. More importantly, the problems that tend to create the greatest difficulty for instructors are also those that provide the least differentiation for the simulation. Likewise the simulation's microbehavior reflects sensible changes in the relative activations of different goals as lines are scanned. Figure 5 shows a trace of activation levels for four types of statistical test. Through symbol line 14, the system indicates roughly equal activation for repeated and independent t-tests. This is because most of the symbols are common to both tests. The presence of SUMX2 and XBAR2 has provided slightly greater activation for independent-t, since repeated-t is usually constructed with deviation scores rather than a second set of sums. DBAR (lines 20 and 25 in Figure 5), however, is the mean deviation score and is used only in repeated-t. As a consequence the final solution is judged to be a repeated t based on overall activation. Since these tests are in fact quite similar, and the student has used symbols common to both, as we would expect, the activation differences are not very large.

Activation Level Analysis of Student Solution # 16

Student Solution

1 d
2 4
3 8
4 4
5 4
6 sum x1 = 42
7 n1 = 4
8 xbar1 = 10.5
9 ss = 29
10 alpha 0.05
11 sumx2
12 n2 = 4

13 xbar2 = 5.5
14 ss = 5
15 n = 4
16 k = 2
17 alpha = .05
18 sum d = 20
19 nd = 4
20 dbar = 5
21 ss = 12
22 sd = sqrt (12/(n-1))
23 sdbar = sqrt (12/3) = 2
24 sdbar = 2/sqrt 4 = 1
25 t = (dbar-ud)/sdbar = (5-0)/1 = 5

Goal Activation Levels

Symbols in Solution	Repeated t-test	Independent t-test	Repeated ANOVA	Independent ANOVA
D	0.0	0.0	0.0	0.0
X1	0.036	0.035	0.0	0.0
N1	0.012	0.012	0.0	0.0
XBAR1	0.363	0.554	0.100	0.028
SS	0.188	0.378	0.021	0.405
ALPHA	0.112	0.144	0.004	0.338
SUMX2	0.029	0.760	0.000	0.048
N2	0.006	0.461	0.000	0.010
XBAR2	0.157	0.887	0.047	0.014
SS	0.112	0.471	0.010	0.351
N	0.072	0.302	0.006	0.224
K	0.491	0.638	0.004	0.144
ALPHA	0.230	0.254	0.001	0.291
D	0.230	0.254	0.001	0.291
ND	0.975	0.104	0.000	0.119
DBAR	0.873	0.604	0.092	0.041
SS	0.309	0.204	0.019	0.350
SD	0.989	0.123	0.004	0.297
N	0.633	0.079	0.003	0.190
SDBAR	0.955	0.484	0.090	0.065
SDBAR	1.640	0.877	0.143	0.066
T	3.471	2.982	0.092	0.042
DBAR	2.612	1.855	0.208	0.069

Figure 5: A trace of goal activation levels for a line-by-line evaluation of a student problem.

Initial problem recognition is good, but it should be noted that there are several assumptions implicit in these results. In this analysis, we have assumed that the highest activation level is the appropriate characterization of the goal selection. The reasoning behind this approach is that an expert actually scans a problem in a non-linear fashion. The problem-type is defined by focusing on the most salient features in the problem. In the simulation, salience is described by peak activation. Our analysis suggests that the simulation can handle student solutions in ways similar to that of an instructor. However, additional data will be needed to confirm the model at a process level.

Summary and Conclusions

We have presented two components of what we believe to be a reasonable characterization of diagnosis for statistics. The combination of "automatic" problem recognition and more deliberate goal-directed search is consistent with other theories in psychology. The general distinction has been part of psychological models for a number of years (Shiffrin & Shneider, 1977). Only recently, however, have these components been integrated into a model that attempts to account for problem solving (Hunt & Lansman, 1986). Despite substantial advances in the development of low level models of processing (McClelland, Rumelhart, and the PDP Group, 1986), there is still a case to be made for separating automatic associative processing from goal-based knowledge (Norman, 1986, calls this "deliberate conscious control").

This two component model also exhibits behavior that is comparable in many ways to that of instructors. The goal-direction, "intention-based" aspect provides a way to account for a range of errors. This approach provides analysis of over 80% of the individual lines and almost all of the goals in a set of t-tests that we have collected from students. Observation of instructors, however, indicates that they are able to select individual types of procedures by scanning the problem. We have modelled this behavior as the spreading activation of an associative network of letters, symbols ("words"), goals and problem types. Using highest level of activation as the criterion for goal selection, this method can correctly identify 86% of the problems tested. In order to serve as a good psychological model, difficulties encountered by the simulation should mirror those of instructors. Our preliminary observations confirm that match, although it will be necessary to verify the behavior with a larger sample of independent judges.

Although the model has met reasonable performance criteria, it is not an exhaustive model of diagnosis, and there are at least two important qualifications on the results. First, the two components of the model, automatic recognition and goal-based reasoning, are currently only weakly linked in the simulation. Our model of these components would suggest greater interaction between the components, in a manner similar to Hunt and Lansman's (1986) production-activation model. The activation levels available during problem-scanning should be used to provide more specific direction to the deliberate search. In the current simulation those levels are only used to select among problem types at the global level. Second, the data we have reported deal with a relatively circumscribed problem space in statistics; additional data will be needed to demonstrate the generality of the model.

701

The proposed model seems to provide a reasonable approximation to several important aspects of diagnosis. The "intention-based" component of the model has now been validated in both programming (Soloway & Ehrlich, 1984; Sack et al., in prep.) and in statistics. The spreading activation component has not been explicitly validated outside of statistics in its current form, but it has been shown to be of general utility as a model for different kinds of psychological processes (Anderson, 1983b). Extending the model we have described here should prove useful for the development of intelligent tutoring systems.

References

Adelson, B. (1981). Problem-solving and the development of abstract categories in programming languages. **Memory and Cognition**, *9*, 422-433.

Anderson, J.R. (1983a). **The architecture of cognition.** Cambridge, MA: Harvard University Press.

Anderson, J.R. (1983b). A spreading activation theory of memory. **Journal of Verbal Learning and Verbal Behavior**, *22*, 261-295.

Brown, J.S. & Burton, R.R. (1978). Diagnostic models for procedural bugs in basic mathematical skills. **Cognitive Science**, *2*, 155-192.

Burton, R.R. (1982). Diagnosing bugs in a simple procedural skill. In D. Sleeman & J.S. Brown, (Eds), **Intelligent tutoring systems.** pp.157-183.

Clancey, W.J. (1985). **Acquiring, representing, and evaluating a competence model of diagnostic strategy.** Report No. STAN-CS-85- 1067, Department of Computer Science, Stanford University.

Clancey, W.J. and Lestinger, R. (1984). NEOMYCIN: Reconfiguring a rule-based expert system for application to teaching. In W.J. Clancey, and E.H. Shortliffe, (Eds.), **Readings in medical artificial intelligence: The first decade.** pp. 361-381.

Friend, J. & Burton, R.R. (1981). **Teacher's manual of subtraction bugs.** CIS Working Paper. Xerox Palo Alto Science Center.

Hunt, E. & Lansman, M. (1986). Unified model of attention and problem solving. **Psychological Review**, *93(4)*, 446-461.

Johnson, W.L. (1985). **Intention-based diagnosis of errors in novice programs.** Research Report #395. Department of Computer Science, Yale University, New Haven, CT.

Johnson, W.L. & Soloway, E. (1985).   PROUST: An automatic debugger for Pascal programs. **Byte**. April, 179-190.

McClelland, J.L. (1979).   On the time relations of mental processes: An examination of processes in cascade. **Psychological Review**, 86, 287-290.

McClelland, J.L. & Rumelhart, D.E. (1981).   An interactive model of context effects in letter perception: Part I. An account of basic findings. **Psychological Review**, 88, 375-407.

McClelland, J.L., Rumelhart, D.E., & the PDP Research Group (1986).   **Parallel distributed processing**. Cambridge: MIT Press.

Miller, M.L. (1978).   A structured planning and debugging environment for elementary programming. **International Journal of Man-Machine Studies**, 11, 79-95.

Norman, D.A. (1986). Reflections on cognition and parallel distributed processing. In J.L. McClelland, D.E. Rumelhart, and the PDP Group (Eds.), **Parallel Distributed Processing**. 531-546.

Sack, W., Littman, D., Spohrer, J.C., Liles, A. & Soloway, E. (in prep.) **Empirical evaluation of PROUST**.

Sebrechts, M.M., LaClaire, L., Schooler, L.J., & Soloway, E. (1986).   Towards Generalized Intention-based diagnosis: GIDE.   In **Proceedings of the 7th National Educational Computing Conference**. San Diego, CA.

Sebrechts, M.M., Schooler, L.J., & LaClaire, L. (1986).   Matching strategies for error diagnosis: A statistics tutoring aid.   In **Proceedings of the International Conference on Systems, Man and Cybernetics**. Atlanta.

Shiffrin, R.M. & Schneider, W. (1977).   Controlled and automatic human information processing: II. Perceptual learning, automatic attending, and a general theory. **Psychological Review**, 84, 127-190.

Shortliffe, E.H. (1976). **Computer-based medical consultations: MYCIN**. New York: Elsevier.

Soloway, E. & Ehrlich, K. (1984).   Empirical investigations of programming knowledge.   **IEEE Transactions of Software Engineering**.   SE-10(5).

A Time-Dependent Distributed Processing Model of Strategy-Driven Inference Behavior

Kurt P. Eiselt

Richard H. Granger, Jr.

Irvine Computational Intelligence Project
Department of Information and Computer Science
University of California
Irvine, California 92717

Abstract

Experimental evidence suggests that some readers make inference decisions early on in text understanding and mold the inferences from later text to fit with the earlier inferences, while other readers postpone inference decisions until later in the text and then base their final interpretation of the text on those postponed inferences. This behavior has been called *strategy-driven inference behavior* because it was originally ascribed to different strategies used by readers to guide the course of their inference decisions. This paper presents a new theory of how this behavior comes about, attributing the observed differences in behavior not to different strategies but to very small differences in the underlying cognitive architecture. This theory is illustrated by a simple model of inference processing during text understanding. The inference processing model employs a hybrid connectionist network whose behavior is extremely sensitive to the order of activation of nodes in the network, which in turn corresponds to the order of presentation of events in the story.

1 Introduction

One of the more intriguing mysteries of natural language understanding is that of *strategy-driven inference behavior*. According to the theory of strategy-driven inference behavior, different readers consistently employ different strategies to guide their choice of an interpretation for a text. These strategies are time-dependent; that is, they are sensitive to the order of presentation of events in the text. Experimental evidence suggests that some readers make inference decisions early on in text understanding and mold the inferences from later text to fit with the earlier inferences, while other readers postpone inference decisions until later in the text and then base their final interpretation of the text on those postponed inferences. These differences in inference behavior can be elicited with the use of specially constructed texts that have two equally likely interpretations. Upon reading these texts, some readers will arrive at one interpretation while the other readers will find the alternative interpretation, and their choices appear to be entirely dependent upon the sequence of events in the story. If the order of presentation is reversed, the different sets of readers will reverse their interpretations.

Previously, we have built two computational models of text understanding in attempts to shed light on the processes underlying strategy-driven inference behavior. Though the explanations provided by these models were satisfying at the time the models were constructed, we have since found weaknesses in these explanations. This paper describes our new model of text understanding which explains observed differences in inference behavior as the result of a connectionist network in which the interpretation settled upon is determined by the order of activation of the nodes in the network.

2 Old Problems, Old Solutions

Granger and Holbrook (1983) reported the results of a psychological experiment that investigated the processes people use in making pragmatic inferences while reading text. These results provided support for Granger and Holbrook's theory that different readers employed different strategies for selecting from alternative interpretations of a single text. In this experiment, subjects read a number of short texts that had two equally plausible interpretations, such as the following text:

Text 1: Wilma began to cry.
Fred had just asked her to marry him.

Interpreting this text requires that a causal relationship between Fred's proposal and Wilma's tears be inferred. The experimental results showed that many subjects inferred that Wilma was happy about Fred's proposal and was crying "tears of joy," while approximately the same number of subjects inferred that Wilma was crying because she was saddened or upset by the proposal. However, the subjects' interpretations were not based on their predispositions toward a particular interpretation; for example, tests run on another set of subjects drawn from the same subject pool confirmed that the subjects almost unanimously associated crying with sadness and marriage with happiness. Instead, the results indicated that the subjects' interpretations were based on the *order of presentation* of the events in the story. In other words, when some subjects read Text 1, they determined that Wilma was sad based on inferences generated from the first story event, the fact that Wilma was crying, while other readers determined that she was happy based on the second story event, Fred's marriage proposal.

Granger and Holbrook theorized that some readers consistently make inference decisions as early as possible in reading and try to make inferences from later text that agree with the earlier inferences; these readers were called *perseverers*. Other readers, called *recencies*, postpone making inference decisions. When recencies eventually do make decisions, they are based on the most recently read text. A computational model of the processes suggested by this theory was developed soon thereafter (Granger, Eiselt, & Holbrook, 1983). This model, called STRATEGIST, arrives at either of two interpretations of an input text using the same component inference processes but different rules for deciding when the processes are invoked, resulting in different interpretations of the same text.

STRATEGIST was later subsumed by another computational model of inference processing. This model, ATLAST, attempts to unify lexical and pragmatic inference decision processes and offers an explanation of how readers are able to correct erroneous inference decisions (Eiselt, 1985; Granger, Eiselt, & Holbrook, 1986; Eiselt, 1987). In addition, ATLAST also provides a framework in which to further study strategy-based inference behavior.

ATLAST uses marker-passing to search a relational network for paths which connect meanings of open-class words from the input text. A single path is a chain of nodes, representing objects or events, connected by links that correspond to relationships between the nodes. Any nodes in a path that are not explicitly mentioned in the text are events or objects that are inferred; therefore, these paths are called inference paths. A set of inference paths which joins all of the words in the text into a connected graph represents one possible interpretation of the text. In this respect ATLAST resembles a number of other models of text understanding that utilize marker-passing or spreading activation (e.g., Charniak, 1983; Cottrell & Small, 1983; Hirst, 1984; Quillian, 1969; Riesbeck & Martin, 1986; Waltz & Pollack, 1985).

For any given text, however, there may be a great number of possible interpretations, many of which are nonsensical. The problem then is determining which of the possible interpretations provides the best explanation of the text. ATLAST deals with this problem by applying inference evaluation metrics. These metrics are used to compare two competing inference paths and select the more appropriate one. Two inference paths compete when they connect the same two nodes in the relational network via different combinations of links and nodes. The path that fits better with the existing interpretation is then added to the interpretation. The choice of one inference path over another is made as soon as ATLAST discovers that the two paths compete; unlike STRATEGIST, ATLAST does not postpone inference decisions. As the marker-passing search mechanism finds more paths, ATLAST constructs an interpretation consisting of those paths that survive the evaluation process. When the marker-passing and evaluation processes end, the surviving inference paths make up the final interpretation of the text.

Most of the evaluation metrics attempt to make either a quantitative or qualitative judgment of the relative merits of competing inference paths in order to provide the most parsimonious interpretation: one metric favors the shorter of two paths, another favors the path that shares more nodes with the current interpretation, and still another metric favors the more specific path as determined by the relationships represented by the links in the path. If these metrics fail to yield a decision the last remaining metric is invoked; this metric alone determines the difference between perseverer and recency behavior. When the programmer wants ATLAST to model perseverer behavior, a rule that chooses the inference path found earlier is used; recency behavior is obtained by using a rule that selects the inference path found later. Thus the theory of strategy-driven inference behavior embodied in ATLAST differs significantly from that of its predecessor, STRATEGIST. In STRATEGIST, the different inference behaviors were caused by different orderings of invocation of the same inference decision processes. In ATLAST, the inference decision processes (i.e., the evaluation metrics) are invoked in the same order for both types of readers, and the difference in inference behavior is attributed to the use of a different "tie-breaker" metric that is applied only when the other metrics are unable to choose one inference path over another.

There are at least two problems with this explanation of strategy-based inference behavior. The first problem is with the nature of the evaluation metrics. In ATLAST, the perseverer/recency metric has the same status as the other metrics in that they are all rules to be applied after competing inference paths have been discovered. However, human understanders exhibit substantial differences in the behavior that would be determined by perseverer or recency metrics, but they do not appear to exhibit significant differences in the behavior that would be determined by the other metrics. In other words, while we see equally large proportions of perseverers and recencies in the laboratory, there is nothing that suggests that readers who favor parsimonious interpretations and those who do not exist in these same proportions. Yet, if the difference between perseverer and recency behavior is in fact rule-based, and similar rules are used to determine other inference decision behavior as well, we would expect to see differences in behavior along those lines. We simply do not see these other differences; in fact it is difficult to imagine how anyone could function in this world while always pursuing the least plausible interpretation of everything read or heard.[1] Thus, a model that could explain why readers exhibit differences in some types of inference behavior but not in others without resorting to arbitrary differences in rules would be superior to ATLAST.

The second problem is that perseverer/recency behavior in ATLAST depends upon a serial ordering of the competing inference paths as determined by the relative times at which they were discovered. Although ATLAST is intended to be a parallel model, the parallelism is only simulated and the inference paths are discovered and evaluated serially; ATLAST takes advantage of this latter fact. In a truly parallel system, two competing inference paths may be discovered at exactly the same time. In this case, ATLAST's perseverer and recency metrics would not be able to make a decision. A more desirable explanation of the difference between perseverers and recencies would be one that could function in a true parallel processing environment.

3 The New Solution

The new model, called CATLAST (for Connectionist ATLAST), is directly inspired by the work of Cottrell and Small (1983) and Waltz and Pollack (1985). These systems do not use the marker-passing style of spreading activation that ATLAST employs. Instead, these systems spread continuously variable numeric quantities representing activation energies through a network of nodes connected by excitation and inhibition links. An interpretation of input text is chosen not by application of rules, but by the iterative adjustment of activation energies at the nodes. While this method of making inference decisions is clearly an example of distributed processing, the representation scheme is just as clearly not distributed: these networks all adhere to the "one node equals one concept" principle (cf. Collins & Loftus, 1975; Quillian, 1968). Thus, these models share features of rule-based symbol-manipulation systems and parallel distributed processing systems (Rumelhart & McClelland, 1986).

---

[1]There may be rare exceptions to this rule, but we are unaware of them. Though we have not run experiments to test for this sort of anomalous behavior in human subjects, we have run ATLAST with the metrics modified to favor the least parsimonious interpretation of a story. The resulting interpretations bore no similarity to any of the interpretations offered by human readers.

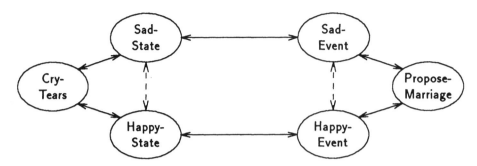

Figure 1: CATLAST's memory network.

The excitation/inhibition network used in CATLAST is simply a modified version of the relational network used in ATLAST (see Figure 1). Two uni-directional inhibitory links, represented by single dashed lines in Figure 1, have been added between pairs of incompatible nodes. The inhibition weights can take on values from 0 to −1. A solid line represents two uni-directional excitatory links. The values of excitation weights can range from 0 to 1. Some of the excitatory links are *preferred* links in that their excitation weights are higher than the norm. For the examples discussed in this paper, the preferred links are from Cry-Tears to Sad-State (but not from Sad-State to Cry-Tears) and from Propose-Marriage to Happy-Event (but not from Happy-Event to Propose-Marriage).

The activation function used to compute activation levels at the individual nodes or computing units is loosely based on the interactive activation function described by McClelland and Rumelhart (1981):

$$a_j(t+1) = min(1, max(0, C_1 a_j(t) + C_2(\sum w_{ij} a_i(t))/n_j))$$

where $a_i(t)$ is the activation energy at node i at time t, w_{ij} is the excitation or inhibition weight on the link from node i to node j, n_j is the number of inputs to node j, and C_1 and C_2 are damping factors that represent degrees of confidence in their respective terms.

With CATLAST we are trying only to find a better explanation for strategy-driven inference behavior; we are not, at this time, attempting to build a robust language understanding system. Therefore, we have taken the liberty of simplifying CATLAST's input texts down to the events explicitly stated in those texts. When CATLAST processes a text, the first event is read, the corresponding node is raised to its maximum activation level (i.e., 1), and the network is allowed to settle through iterative application of the function given above. Then the next event is read, its corresponding node is activated, and the network settles again. CATLAST's interpretation of the text is determined by comparing the final activation levels at the nodes, with high levels favored over low levels.

By adjusting the parameters, CATLAST can be tuned so that it either always selects the "sad" interpretation or it always selects the "happy" interpretation, regardless of which event is presented first. Much more interesting is the fact that CATLAST can also be tuned so that it selects the interpretation preferred by the first event presented. That is, it exhibits perseverer behavior: if Cry-Tears is activated first CATLAST chooses the "sad" interpretation, but it chooses the "happy" interpretation if Propose-Marriage is activated first.

However, equipped with only the features described so far, CATLAST cannot be tuned so that it consistently exhibits recency behavior. In order to correct this deficiency, CATLAST must be endowed with the ability to postpone making a decision until it reads the second event. This is accomplished by the addition of an inhibition threshold for the inhibitory links and an excitation threshold for the nodes.

The inhibition threshold prevents the inhibitory link from feeding inhibition energy into a destination node until the activation energy at the source node exceeds a prescribed value. Increasing this threshold has the effect of delaying the onset of inhibition, thus helping to postpone the inference decision. The excitation threshold holds the activation of a node at the minimum value (i.e., 0) until the total activation energy at that node exceeds a prescribed value. Increasing this threshold effectively slows the spread of activation energy through the network. Neither of these features alone can postpone the decision long enough to cause CATLAST to exhibit recency behavior, as far as we have been able to determine. On the other hand, the two thresholds combined do provide the necessary delay when the appropriate values have been assigned.

4 The New Solution in Action

Figure 2 shows the time course of the activation of the individual nodes while CATLAST exhibits perseverer behavior on an abbreviated version of Text 1. In this example, the first event is read and Cry-Tears is raised to its maximum activation level.[2] As activation energy feeds into Sad-State and Happy-State, the activation level at Cry-Tears dips briefly and then recovers. Sad-State and Happy-State follow almost the same course of activation for awhile, though the level at Sad-State is always slightly higher because the weight on the excitation link from Cry-Tears to Sad-State is greater than the weight on the link from Cry-Tears to Happy-State.

Activation energy then spreads to neighboring nodes. Because Sad-State is always higher than Happy-State, the excitation threshold at Sad-Event is exceeded before that threshold is reached at Happy-Event. The activation level at Sad-Event begins to rise, which further encourages the increase at Sad-State. As the level at Sad-State approaches the maximum, the inhibition threshold on the inhibitory link from Sad-State to Happy-State is exceeded and the activation level at Happy-State begins to plummet. Meanwhile, activation energy from Sad-Event spreads to Propose-Marriage which in turn spreads to Happy-Event. Receiving support from both Propose-Marriage and the still active Happy-State, Happy-Event exhibits a brief increase in activation which quickly dies as Happy-State drags Happy-Event down. The network settles into a stable state in which all nodes are at maximum activation levels except for Happy-State and Happy-Event, which are at minimum levels; this occurs before the second event is read. Activating Propose-Marriage does not change the distribution of energy in the network as Propose-Marriage is already at the maximum level and the network is stable, so CATLAST maintains its "sad" interpretation of Text 1. Had CATLAST been presented with the same events but

---

[2] For those who wish to try this at home, the parameters are set as follows: C_1 is 0.6, C_2 is 0.9, the inhibition weight is -1.0, the standard excitation weight is 0.9, the preferred excitation weight is 0.95, the inhibition threshold is 0.999, and the excitation threshold is 0.192.

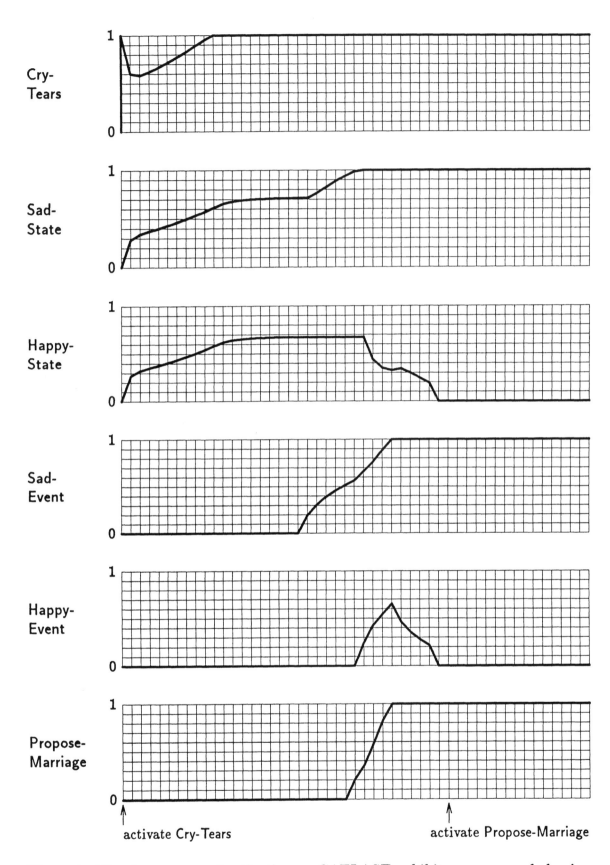

Figure 2: The time course of activation as CATLAST exhibits perseverer behavior.

in the reverse order, it would have settled upon the "happy" interpretation. Because of the symmetry of the network, the resulting time course of activation would have been the same as that shown in Figure 2, with the exception that the order of the labels assigned to the individual graphs would be inverted. Thus, Cry-Tears would exhibit the time course that Figure 2 now shows for Propose-Marriage, Sad-State would display the time course now shown for Happy-Event, and so on.

Figure 3 shows the time course of node activation as CATLAST processes Text 1 as a recency.[3] After reading the first event, CATLAST's behavior as a recency parallels its behavior as a perseverer, but raising the excitation threshold has prevented activation energy from spreading beyond Sad-State and Happy-State; the network stabilizes with a slightly higher activation level at Sad-State than at Happy-State, again because of the increased weight on the excitatory link from Cry-Tears to Sad-State.

CATLAST reads the second story event and raises Propose-Marriage to the maximum activation level. Because both Sad-Event and Happy-Event are now receiving activation energy from two sources, their activation levels rise sharply, though Happy-Event rises a bit more quickly because it is preferred by Propose-Marriage. At the exact time that Happy-Event reaches maximum activation, Sad-Event is just far enough behind to make all the difference in CATLAST's behavior because the energy at Happy-Event has exceeded the inhibition threshold but the energy at Sad-Event has not. Happy-Event begins to inhibit Sad-Event but not vice-versa, and Sad-Event begins to fall. At the same time, Sad-State and Happy-State have reached maximum activation and inhibit each other. They then fall below the inhibition threshold, their activation levels rise to maximum, and they inhibit each other again, thus displaying oscillatory behavior. Sad-State, though, loses reinforcement as Sad-Event declines, activation levels at both nodes decrease to the minimum, and CATLAST settles upon the "happy" interpretation. Again, if the story events had been presented in the reverse order, CATLAST would have settled on the alternate interpretation. The resulting time course of activation can be constructed by inverting the order of the labels of Figure 3 as described previously.

5 Conclusion

The explanation of strategy-driven inference behavior offered by CATLAST is that such behavior is not strategy-driven after all. CATLAST explains differences in inference behavior as resulting from exactly the same architecture with only very slight differences in the computing units' sensitivity to activation energy. (In the examples given above, the abrupt change from perseverer to recency behavior was caused by an increase in the excitation threshold of only 0.001.) Whether CATLAST behaves as a recency or a perseverer, the model's preference for the most parsimonious interpretation is preserved, as such preferences are inherent in connectionist architectures (e.g., given two competing inference paths in a connectionist network, the nodes in the shorter path will have higher levels of activation than the nodes in the longer path and the shorter path will be favored). In addition, the model does not rely on any serial ordering which arises through simulation

---

[3]The values of the parameters for this example differ from the previous example only in that the excitation threshold is now 0.193 instead of 0.192.

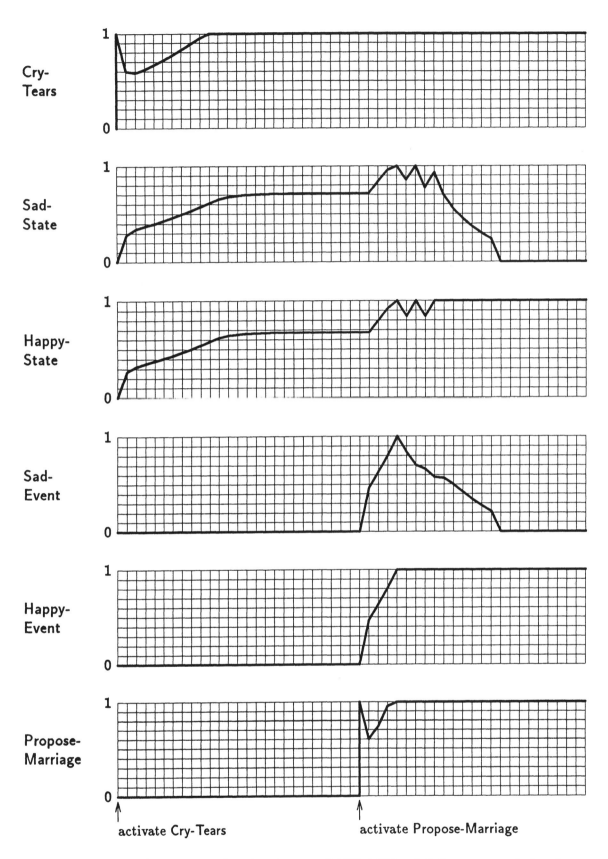

Figure 3: The time course of activation as CATLAST exhibits recency behavior.

of a parallel process on a serial processor. Thus, CATLAST provides solutions to the two problems discussed earlier that were left unsolved by ATLAST.

Not all inference decision behavior is automatic or unconscious, but we believe the perseverer/recency behavior discussed in this paper certainly is. Anecdotal evidence from the experiment described in Section 2 provides some support for this claim: several experimental subjects not only were consistent in the interpretations they assigned to test stories, they protested that other interpretations were entirely implausible (Granger et al., 1983). If attentional or conscious processes had been involved in interpreting a given story, one would expect that the subjects would have had some memory of different possible interpretations they must have entertained while reading the text; in at least these cases they did not. Thus it seems reasonable to try to explain this aspect of the human language understanding mechanism as the result of subtle differences in the underlying cognitive architecture, though it is not necessarily the case that all aspects of language understanding can be modeled adequately at this level.

Finally, some might conclude that the question of how perseverer/recency behavior comes about does not warrant the attention we have given it. After all, this behavior is usually only elicited by specially constructed texts; everyday texts are seldom if ever constructed in such a way as to leave the reader with two equally likely interpretations and no clue as to which interpretation is correct. However, the fact that people do consistently exhibit these differences, even if it occurs only in the laboratory, informs us about the human language understanding mechanism—it provides constraints which must guide the development of a comprehensive model of language understanding, however far in the future that may be.

6 Acknowledgments

This research was supported by the National Science Foundation under grant IST-85-12419. We wish to thank Jeff Schlimmer for the insights he provided during the development of CATLAST. We also wish to thank Kathleen Delling for helping to beat a large body of data into submission.

7 References

Charniak, E. (1983). Passing markers: A theory of contextual influence in language comprehension. *Cognitive Science*, *7*(3), 171-190.

Collins, A.M., & Loftus, E.F. (1975). A spreading-activation theory of semantic processing. *Psychological Review*, *82*(6), 407-428.

Cottrell, G.W., & Small, S.L. (1983). A connectionist scheme for modelling word sense disambiguation. *Cognition and Brain Theory*, *6*, 89-120.

Eiselt, K.P. (1985). A parallel-process model of on-line inference processing. *Proceedings of the Ninth International Joint Conference on Artificial Intelligence*, Los Angeles, CA.

Eiselt, K.P. (1987). Recovering from erroneous inferences. *Proceedings of the National Conference on Artificial Intelligence*, Seattle, WA (to appear).

Granger, R.H., Eiselt, K.P., & Holbrook, J.K. (1983). STRATEGIST: A program that models strategy-driven and content-driven inference behavior. *Proceedings of the National Conference on Artificial Intelligence*, Washington, DC.

Granger, R.H., Eiselt, K.P., & Holbrook, J.K. (1986). Parsing with parallelism: A spreading-activation model of inference processing during text understanding. In J.L. Kolodner & C.K. Riesbeck (Eds.), *Experience, memory, and reasoning*. Hillsdale, NJ: Lawrence Erlbaum Associates.

Granger, R.H., & Holbrook, J.K. (1983). Perseverers, recencies, and deferrers: New experimental evidence for multiple inference strategies in understanding. *Proceedings of the Fifth Annual Conference of the Cognitive Science Society*, Rochester, NY.

Hirst, G. (1984). Jumping to conclusions: Psychological reality and unreality in a word disambiguation program. *Proceedings of the Sixth Annual Conference of the Cognitive Science Society*, Boulder, CO.

McClelland, J.L., & Rumelhart, D.E. (1981). An interactive activation model of context effects in letter perception: Part 1. An account of basic findings. *Psychological Review*, *88*, 375-407.

Quillian, M.R. (1968). Semantic memory. In M. Minsky (Ed.), *Semantic information processing*. Cambridge, MA: MIT Press.

Quillian, M.R. (1969). The teachable language comprehender: A simulation program and theory of language. *Communications of the ACM*, *12*(8), 459-476.

Riesbeck, C.K., & Martin, C.E. (1986). Direct memory access parsing. In J.L. Kolodner & C.K. Riesbeck (Eds.), *Experience, memory, and reasoning*. Hillsdale, NJ: Lawrence Erlbaum Associates.

Rumelhart, D.E., & McClelland, J.L. (Eds). (1986). *Parallel distributed processing: Explorations in the microstructure of cognition. Volume 1: Foundations*. Cambridge, MA: MIT Press.

Waltz, D.L., & Pollack, J.B. (1985). Massively parallel parsing: A strongly interactive model of natural language interpretation. *Cognitive Science*, *9*(1), 51-74.

Capitalizing on Failure Through Case-Based Inference*

Janet L. Kolodner
School of Information and Computer Science
Georgia Institute of Technology
Atlanta, Georgia 30332

Abstract

In case-based reasoning, previous reasoning experiences are used directly to solve new problems and make inferences, rather than doing those tasks from scratch using generalized methods. One major advantage of a case-based approach is that it can help a reasoner avoid repeating previously-made mistakes. When the case-based reasoner is reminded of a case in which a mistake was made, it provides a warning of the potential for a mistake. If the previous case was finally solved successfully, it can provide a suggestion of what to do instead. In this paper, we describe the process by which a case-based reasoner can take advantage of previous failures. We illustrate with cases from the domains of common-sense mediation and menu planning and show a program called JULIA reasoning in the domain of menu planning.

1. Introduction

Over the past several years, there has begun to be a great deal of interest in case-based and analogical reasoning (e.g., Alterman, 1986, Ashley, 1986, Carbonell, 1983, 1986, Hammond, 1986, Holyoak, 1984, Kolodner, et al., 1984, 1985, Rissland, 1986, Simpson, 1985). Case-based reasoning is a problem solving method in which previous reasoning experiences are used directly to solve a new problem, rather than solving the problem from scratch using generalized methods. The major advantages of a case-based approach are that it can provide shortcuts in problem solving and that it can help a reasoner avoid repeating previously-made mistakes.

We shall see that previous failures serve several purposes during problem solving. They can provide warnings of the potential for failure in the current case, and they may also provide suggestions of what to do instead. Analyzing the potential for failure in a new case, a necessary part of capitalizing on an old failure, may require the problem solver to gather additional information, thus causing the problem solver to change its focus of attention. A previous failed case that was finally solved correctly can help the problem solver to change its point of view in interpreting a situation if that is what is necessary to avoid potential failure.

We shall illustrate the processes involved in capitalizing on failure using examples from two domains: common-sense mediation of everyday disputes and menu planning. Case-based resolution of common-sense disputes is implemented in the MEDIATOR (Kolodner, et al., 1985, Simpson, 1985), an early case-based reasoning program. JULIA (Cullingford & Kolodner, 1986) interactively solves problems in the catering domain. The processes that capitalize on failure are implemented in JULIA.

2. Background

In the simplest case, making a case-based inference involves the following steps:*

---

* This work is supported in part by NSF under Grant No. IST-8317711 and Grant No. IST-8608362, by ARO under Contract No. DAAG29-85-K-0023, and by ARI under Contract No. MDA-903-86-C-173. Programming of the examples, and much work on analogical reasoning that is incorporated into JULIA's case-based reasoner was provided by Hong Shinn. Discussions with other members of the AI Group, past and present, have also been useful.
* Each of these steps, of course, is a complicated process. For more information about step 1, see Kolodner (1983), Hammond (1986), Holyoak (1984), Schank (1982); about step 2, see Kolodner, et al. (1985), Simpson (1985); for step 3, see Alterman (1986), Ashley (1986), Carbonell (1983, 1986), Hammond (1986), Kolodner (1985, 1986), Kolodner et al., (1985), Rissland (1986), Simpson (1985); for step 4, see Simpson (1985).

1. Recall a relevant case from memory

2. Determine which parts of that case are appropriate to make the necessary problem solving decision for the new case (i.e., focus on appropriate parts of the previous case)

3. Achieve the targetted problem solving goal for the new case by making an inference based on the old case

4. Check the consistency of what is derived in step 3 to the new case

Consider, for example, the following case:

Avocado Dispute 1

A problem solver is attempting to resolve a dispute over possession of an avocado. Two people want it. The problem solver is attempting to fill in the underlying goals of the disputants (i.e., why does each want the avocado?). It is reminded of a dispute in which two kids wanted the same candy bar. They both wanted to eat the candy bar, and the reasoner compromised by dividing the candy bar equally between them, having one divide it and the other choose his half first.

The problem solver has already been reminded of another case (step 1). Because the problem solver's goal is to infer the underlying goals of the disputants in the avocado case, it focuses on the underlying goals of the disputants in the candy dispute (step 2). They both had the goal of eating the whole candy bar. This goal was inferred through a *default-use* inference. The reasoner makes the case-based inference that the disputants in the avocado dispute also want to eat the disputed object (i.e., the avocado) (step 3). Because this hypothesis is consistent with what is already known about the case (step 4), the representation of the case is updated to include this inferred knowledge.

When a recalled case resulted in failure, however, reasoning is not as straightforward. Consider, for example, the following:

Avocado Dispute 2

A problem solver is attempting to resolve a dispute over possession of an avocado. Two people both want it. The problem solver is trying to infer the underlying goals of the disputants. This time it is reminded of a case where two sisters both wanted the same orange. The problem solver in that case inferred the sisters' goals by using a *default-use* inference to infer that both disputants wanted to eat the orange. It turned out, however, that the goal of one of the disputants was to use the peel of the orange to bake a cake. The *default-use* inference applied to the orange as a whole led to selection of the wrong plan for resolution of the conflict, and the plan failed. We shall call this part of the case *orange-dispute-f*.

The problem solver reinterpreted the dispute and solved it. The goals of the sisters were amended: one wanted possession of the fruit of the orange, the other its peel. Their underlying goals were also amended: one wanted to satisfy hunger by eating the fruit, the other wanted to bake with the peel. It finally resolved the problem by dividing the orange in a better way. One sister was given the fruit and the other was given the peel. We shall call this part of the case *orange-dispute-s*.

The problem solver also analyzed its failure in *orange-dispute-f*, and added its analysis to its memory of that case: Failure was due to a *wrong-goal inference*. *Default use* applied to the entire disputed object (orange) resulted in failure, while *default use* applied to parts of the orange (the peel and the fruit) would have resulted in success.

Suppose now that the problem solver is reminded of *orange-dispute-f*, the case that resulted in failure. This case acts as a warning to the problem solver of the potential to make a faulty inference in the current case. It must check to see if the inference used previously would also result in error in the current case. The question that must be asked of the avocado dispute based on analysis of the orange dispute is whether an avacado also has parts used for different purposes that might predict the goals of the current disputants better than if they were computed by applying *default-use* to the whole avacado. In other words, based on its reminding of *orange-dispute-f*, which failed, case-based

716

reasoning alerts the reasoner to the fact that if the disputed object has several parts, the goals of the disputants may have something to do with the parts and not necessarily with the avocado as a whole.* The potential for failure is flagged and two alternative solutions are presented.

Errors in reasoning can happen during any problem solving step. The problem might have been misunderstood initially, resulting in incorrect classification of the problem or incorrect inferences during the problem elaboration phase. Since problem understanding is an early part of the problem solving cycle, such misunderstandings and incorrect inferences propagate through to the planning phase, resulting in a poor plan. A problem might be understood correctly and all the necessary details known about it, but might still be solved incorrectly because poor decisions were made while planning a solution. In general, such errors are due to faulty problem solving knowledge. The problem solver might not have complete knowledge, for example, about under what circumstances a particular planning policy or plan step is appropriate. Finally, a problem might be solved correctly but carried out incorrectly by the agent carrying out the plan, or unexpected circumstances might cause execution to fail. Reminding of a case where any of these things happened warns the the problem solver of the potential for the same type of error in the new case. If the previous case was finally resolved correctly, details of its correct resolution suggest correct decisions for the current case.

3. Some Problem Solving Assumptions

Before presenting the set of processes that capitalizes on previously-failed cases, we briefly present the relevant parts of our problem solving paradigm. First, when we refer to problem solving, we include the entire cycle of understanding a problem and elaborating its features, coming up with a plan for its solution, executing that plan, analyzing the results, and if necessary, going back to the beginning and trying again. Our own previous work (Kolodner, et al., 1985, Simpson, 1985) and that of others (e.g., Hammond, 1986) has shown that case-based inference can be used for a variety of tasks during any of these problem solving phases.

The second important assumption of our paradigm is that memory access and problem solving are happening in parallel (Kolodner, 1985, Kolodner & Cullingford, 1986). The memory's job is to integrate the case that is currently being reasoned about into the memory that already exists (Schank, 1982), resulting in remindings. Memory can return generalized knowledge (e.g., knowledge structures or rules) for the problem solver to use or a previous case that is similar to what the problem solver is currently dealing with. As the problem and its solution are further elaborated, memory is able to recall both more relevant general knowledge and better related cases for the problem solver to use.

Our third important assumption is that case-based reasoning is happening in the context of a set of reasoning goals and that, in addition to the case-based reasoner, other reasoners are also keeping track of those goals and making any suggestions they can. Thus, in addition to the case-based reasoner, a problem reduction problem solver might be available to break the problem into smaller parts, while a constraint propagator might do forward chaining inferences, and a truth maintenance system might be checking for inconsistencies and constraint violations. Something we'll call the overall problem solver keeps track of reasoning goals and subgoals as they come up, and each of the reasoners watches the goal network and attempts to achieve any goal it can.

Finally, the processes we present below assume that reminding has been of the failed part of a case that might have been resolved correctly later. In the case of the orange dispute, for example, we assume reminding has been of the episode that failed, *orange-dispute-f*. Reminding during problem solving may be of either the successful or the failed version of any case. When reminding is of the successful instance of solving it, the faulty reasoning that preceeded the successful solution is bypassed and a good solution is suggested immediately. The problem solver is never alerted to possible problems. Only when reminding is of a failed attempt at resolving a case is the problem solver alerted and the analysis described below done.

---

* It may be judged in this case that inference based on the parts is inappropriate (since one rarely plants avocado seeds).

4. The Process

Given this set of assumptions, we see that the problem solver might be reminded of a previous case that resulted in failure any time during problem solving. Because of this, the processes that capitalize on previous failures must be applicable during any part of the problem solving cycle. The following set of steps are executed any time during problem solving that a failed case is recalled.

1. Determine whether the failed case was ever followed up on, and, if so, recall the entire reasoning sequence that followed it.

This step makes alternatives that were attempted previously to solve the recalled problem available to the problem solver.

In the representation we are currently using, each full analysis of a problem is kept separately with pointers between them. Thus, the representation for a case that failed and was reanalyzed, such as the orange dispute, is actually represented as two cases. The first is the one that failed (*orange-dispute-f*), where one set of assumptions was made about the goals of the disputants. That one includes the mistaken problem description, the suggested plan (cut it in half), feedback after suggesting or carrying out that plan (after suggesting that the orange be cut in half), and the analysis of what went wrong (a *wrong-goal-inference*). The first (failed episode) also includes a pointer to the next problem solving episode, i.e., the reasoning that is carried out to solve the problem after the failures of the first episode have been diagnosed and repaired. Thus, *orange-dispute-f* points to *orange-dispute-s*, where the problem is described as one where the disputants have the second set of goals, and the solution plan that goes with that (divide agreeably) is recorded.

2. Recall or determine what was responsible for the previous failure.

In some instances, responsibility for failure will already have been attributed during previous reasoning. In that case, this step is an easy step of retrieving the error attribution from the representation of the case. In other instances, there might not have been any analysis of why the previous problem occured. When this happens, it is appropriate for the problem solver to try to figure out why the previous error happened. We do not go into that process in this paper.[*]

In general, failures happen because some inference was made incorrectly or not made at all. This might be due to faulty or missing information about the problem itself, or faulty or incomplete problem solving knowledge. An analysis of a failure may record only which inference was made incorrectly or was not made, or it may record the reasons why the inference was made incorrectly. As we shall see, the better an analysis of a previous failure is, the more the problem solver will be able to capitalize on the failure. The best analysis of a failure will record reasons for faulty reasoning all the way back to a point in the reasoning where it could have been corrected, i.e., where the missing or faulty information can be obtained or fixed. For example, failure in *orange-dispute-f* can be traced to a *wrong-goal inference*. The goals were inferred incorrectly. The reason for this is that *default-use* was applied to the wrong object (i.e., to orange as a whole rather than the parts of the orange). The reason for this is that the problem solver was viewing the orange in the wrong way: as a whole rather than as a thing with functional parts. If the reasons for this inference error are recorded to this level, then by using this case and following the set of steps to be presented, the problem solver will be able to consider whether some other object might be better viewed as a thing with functional parts. If only the fact that the goal was inferred incorrectly were recorded, it would not have as much to go on, but would only be able to consider if there is another goal associated with the object.

3. Determine the relationship of the decision currently being focussed on to the previous failure and refocus as required:

---

[*] If responsibility for failure is not known at the end of this step, it is still possible to capitalize on the failure.

(a) Was the decision analogous to the one the problem solver is currently trying to make responsible for the failure? If so, maintain current problem solving focus.

(b) If not, was the decision analogous to the one the problem solver is currently trying to make dependent on the one responsible for the failure, or alternatively, did the value the problem solver is currently attempting to derive change in the final solution to the problem? If so, refocus the problem solver on the decision analogous to the one that was responsible for the previous failure.

(c) If not, then refocus as in (b) to be careful or maintain current focus to be fast.

When the decision the problem solver is currently trying to make was responsible for the previous failure (i.e, the answer to 3(a) is yes), then more effort must go into making that decision. This is the case in *avocado dispute 2*. The problem solver has the goal of inferring the goals of the disputants, and it was this decision that was responsible for the failure in *orange-dispute-f*.

The more interesting cases, however, is when the answer to 3(b) is yes. In these cases, some decision other than the one currently being attempted was responsible for the previous failure. The problem solver will have to *refocus* itself on that decision, and (re)make it for the current case before continuing. Consider, for example, the following:

Panama Canal Dispute

Both Panama and the United States want possession of the Panama Canal Zone. The problem solver is attempting to figure out how to classify the dispute. The problem solver is reminded of the dispute between Israel and Egypt over the Sinai. Both wanted the Sinai, and the problem solver had originally classified it as a *physical dispute* over possession of the land. It had therefore suggested that they cut it down the middle and share it. Both Israel and Egypt baked. On further analysis, the failure of this suggestion was tracked down to a set of *missing-goal inferences*. The goals of Israel and Egypt with respect to the Sinai had not been inferred. Israel wanted military control of the area for security reasons, while Egypt wanted possession of the land itself for reasons of national integrety. This interpretation makes the dispute into a *political dispute* rather than a physical one, i.e., one for which political alternatives are suggested rather than alternatives having to do with the physical object itself.

Responsibility for the failure in the previous case (the Sinai Dispute) had already been tracked down to missing goal inferences. The problem solver is currently attempting to decide what kind of dispute it is (e.g., physical or political?). The original classification of the Sinai Dispute as a physical dispute was not per se the reason that solution failed. Rather that decision was based on the goals of the disputants, which had been inferred incorrectly previous to attempting classification. The physical classification, however, changed to political in the final analysis, and was dependent on what was responsible for the failure in reasoning. Reminding of the Sinai Dispute should *refocus* the problem solver on the set of decisions that were responsible for its failure, namely inference of disputant goals.

If the decision being focussed on at the beginning of this set of steps was a correct one for the previous case and if it did not change when the case was reanalyzed (case c), there is no reason why the problem solver must consider the previous failure at all. However, a careful problem solver will also consider whether that failure is possible in the current environment, thus refocusing itself on whatever caused the failure previously before going on.

In cases where the problem solver changes its focus, it continues by trying to redo the task that could have been made in error, following the set of steps below. If the problem solver changes a decision it had made previously, then it must also remake any decisions that depended on it before going on. After this set of steps is complete, the problem solver must refocus appropriately to finish solving the problem. Processing that happens in the course of recomputing already-made decisions may direct the problem solver in different directions than it had been planning when it was interrupted by the failed case. On the other hand, if there are no other recomputations to be made or if no other problem solving directions are suggested, the problem solver continues after this step as it had been planning originally. That is, it goes back to the goal it was working on when it reached this step and continues from there.

719

The processing that happens after step 3 depends on whether or not a successful solution was ever found in the previous case and whether or not analysis can be or has been done of the previous failure. If there was neither a solution found to the previous problem nor an explanation of the previous failure, then only an analysis of the potential for failure can be contributed by the the previous case. And, if there is no explanation of the failure, then less can be contributed than if there is an explanation. With an explanation, we know what features of the previous case were responsible for the failure and we can check for the presence of those in the new case. Without that explanation, we can use the justifications for previously made inferences and see if they hold in the new case, but such analysis is in a sense "superstitious" since no causal explanation available.

4. Recall the inference rules and justifying conditions used to infer the focused-on portion of the failed case. IF there was followup, THEN also recall the inference rules and justifying conditions used to infer the focused-on portion of each of the followup cases.*

The inference rules and justifying conditions of any failed cases will be used to check for the potential for failure in the current case. Those from the successfully-resolved case will be used to guide the problem solver to a correct decision.

In the case of *orange-dispute-f*, the inference rule used to infer the goals of the disputants was *default-use* applied to the disputed object. It is justified by its preconditions, i.e., there is an object of current interest (the orange) that has a default use (eating). It might also have been justified by its use previously in the candy dispute, where it worked fine. For *orange-dispute-s*, there were two inference rules used to infer the goals of the disputants. In one case, *default-use* was applied to the fruit of the orange, in the other it was applied to the peel of the orange. The fruit and peel of the orange are its major parts and each are used for different purposes.

5. Check to see if there is the same potential for failure in the new case. This is done by a variety of methods. We list two here.

(a) Check the reason why the reasoning error was made in the first case. An error can be made because of incomplete information, because of faulty information, because of a faulty inference rule, or because of faulty focus (which might itself be tracked down to one of these causes).

(b) Determine if the justifying inference rules and conditions from the failed and successful cases also hold in the new case.

Let us consider (a) first. This is the way we determine potential for failure in a new case if we know why the previously-made decision failed. If a previous reasoning error was made because of lack of knowledge, the appropriate knowledge is now sought for the current case. If it was because of faulty information, this step will require clarification of the analogous knowledge in the new case. If it was because of a faulty inference rule, that rule will be ruled out in this case. And if it was because of faulty focus (probably due to one of the other types of error), a suggestion will be made from the previous case of where to focus in the new case. Analyzing the orange dispute using this step, we find that the reason for the *wrong-goal-inference* was faulty focus. Focus had been on the orange as a whole while it should have been on its functional parts. The suggestion is thus made to focus on the functional parts of the avacado, rather than the avacado as a whole in inferring the goals of the disputants with respect to the avacado. As in the analysis of the orange dispute from above, in the next steps, the reasoner will either ask the disputants which parts of the avacado they are interested in or will decide that the only functional part that is worth considering is the fruit.

When there is no knowledge about why a previously-made decision was in error, the best that can be done is to evaluate whether conditions that led to that decision are also present in the current case. This is case (b). These

---
* Recall that the problem solver might have refocused its goals in the last step, so the portion of the case being focused on now might not be the one originally considered.

conditions can be found in the justifications for the value that was computed previously. If justifications of both the failed and the successful decision are applicable in the new case, an evaluation must be done of which is best. In *orange-dispute-f*, for example, the goals of each disputant were computed using a *default-use* inference applied to the disputed object. Justification for the *default-use* inference comes from its antecedent clause, which asks whether there is some major default use for the object in question that has an "obvious" goal associated with it. An orange and an avocado, of course, both have the same default use (eating) and "obvious" goal (satisfy hunger). In *orange-dispute-s*, the goals of each disputant were computed using a *default-use* inference applied to the functional parts of the disputed object. Justification for this application of this inference rule is a combination of the justification for choosing the objects to be focussed on (the disputed object has functional parts) and the antecedent clause of *default-use* applied to each of those parts.

Using the orange dispute as a model for the avocado dispute, justifications for each of the goal decisions made in resolving that dispute are evaluated with respect to the avocado dispute. Since the avacado has a default use (eating), the inference from *orange-dispute-f* can be made. Since it also has parts with default uses (the fruit is eaten while the seed can be planted), the inferences from *orange-dispute-s* can also be made. In this case, further evaluation is needed to determine which way to make the inference. While case-based reasoning, in this case, does not provide an answer, it does warn of the potential for misinterpreting the case and it also provides suggestions of alternate interpretations. It thus acts as a *preventive measure* to aid in avoiding failure.

It is interesting to note that the knowledge necessary to do the computations just described may not yet have been considered (e.g., the problem solver may not have considered if an avacado has parts used for different purposes). Sometimes, gathering appropriate knowledge consists of just an easy question to the user. In some cases, however, answering the questions posed in this set of steps may require significant reasoning. This extra computation, while significant, is done only when a previous case points to the need to look out for a problem. As we stated previously, it is a preventive aid in avoiding failure.

The output of this step is an evaluation of whether the previous failure could happen in the new case, and if the previous case was solved successfully, an evaluation of whether the previous successful solution is applicable to the new case. Based on these two evaluations, the reasoning continues.

6.

(a)   If the previous failure will not repeat itself in the new case, go on with the problem solving. The (failed) suggestion from the previous case can be transferred to the new case if there is some independent reason that it can be supported or or a decision can be made independent of the recalled case.

(b)   If the previous failure could repeat itself in the new case, rule out the inference rule or value used previously for the new case.

(c)   If the previous successful solution is judged applicable to the new case, use it and apply case-based reasoning methods to derive a value for the new case based on it.

(d)   If the previous successful solution or any of the interim solutions from the previous problem are judged inapplicable to the new problem, rule them out for the new case.

(e)   If both the failed and successful solutions to the previous problem are judged applicable to the new one, use some decision-making procedure to decide between them.

5. Case-Based Inference in JULIA

In the following problem solving session, we see JULIA following the set of steps above to capitalize on a previous problem solving failure. JULIA (Cullingford & Kolodner, 1986) is designed to be an automated colleague whose task is to help a caterer design a meal. JULIA's problem solving components include a case-based reasoner, a problem reduction planner, a constraint propagator, and a reason-maintenance system. It also has a memory for events. Each decision JULIA makes when it is solving a problem is recorded along with the justifications for the decisions. Thus, in later problem solving, those justifications are available to use in case-based reasoning.[*] In the first case-

---

[*] See Kolodner (1986) for a description of the representational support for these processes, especially the content and structure of the justifications JULIA maintains.

based reasoning example, we see JULIA using a previous case to avoid serving a spicy Mexican meal to people who don't eat spicy food. The previously-failed case, which failed because of a lack of this information, causes JULIA to gather the appropriate information in the new case, thus letting it plan for those people immediately. This makes JULIA retract two previously-made decisions (the entree and the cuisine). JULIA refocuses itself on the cuisine, retracts and remakes that decision, and continues with its problem solving.

The dialog begins with JULIA introducing itself and the user stating her problem. We state the problem in English here. Actual input for this problem is a conceptual representation.

<JULIA> Hello. I'm JULIA, a Caterer's Advisor. May I help you?
USER =>
  I'm having a dinner next week for my research group.
  There is no room for all of them in the dining room.
<JULIA>
*** The initial problem is:
***** Frame #<FRAME 34502032> *****
Name: M-MEAL774
   Isa: (<M-MEAL>)
   Category: INDIVIDUAL
   Slots:
      ACTUAL-RESULT: NIL
      CHARACTERS: (?HOST ?GUESTS ?PARTICIPANTS)
      CONSTRAINTS: (<C-LIMITED-SPACE778>)
      DEFINED-SLOTS: NIL
      DESCRIPTOR: NIL
      EXPECTED-RESULT: NIL
      FOLLOW-UP: NIL
      GOALS: (<E-EAT776> <S-HUNGER777>)
      GUESTS: <*JLK*S-GROUP>
      HOST: <*JLK*>
      ORDER: NIL
      PARTICIPANTS: (?HOST ?GUESTS)
      SETTING: <*JLK*S-HOUSE>
      STEPS: NIL
      TIME: NIL

JULIA's problem reduction problem solver starts out. It immediately reduces the instantiate-a-meal task to its component parts and begins working on the first of them, the meal descriptor task. Meal descriptors include, among other things, the meal's cuisine.

***** Planning for the task MEAL-TASK775 ....
  TASK-TREE = ((<MEAL-TASK775>))
  The active goal chain is (M-MEAL)
  Reducing the task MEAL-TASK775 to subtasks (<MEAL-DESCRIPTOR-TASK>
    <MAIN-COURSE-TASK> <APPETIZER-TASK> <SALAD-TASK> <DESSERT-TASK>)
***** Planning for the task MEAL-DESCRIPTOR-TASK ....
  TASK-TREE = ((<MEAL-DESCRIPTOR-TASK> <MAIN-COURSE-TASK> <APPETIZER-TASK>
          <SALAD-TASK> <DESSERT-TASK>) (<MEAL-TASK775>))
  The active goal chain is (MEAL-DESCRIPTOR <= M-MEAL)

722

As part of this task, JULIA attempts to formulate a cuisine, but can't do it with the information supplied so far by the user, so it asks.

Constraint Satisfaction ......
   Checking constraints for slot CUISINE of MEAL-DESCRIPTOR783 ...
   Instantiating slot CUISINE of frame MEAL-DESCRIPTOR783 ...
<JULIA> What cuisine would you like?
USER => mexican
<JULIA>
Constraint Satisfaction ......
   Checking constraints for slot CUISINE of MEAL-DESCRIPTOR783 ...
   Instantiating slot CUISINE of frame MEAL-DESCRIPTOR783 ...
   --> The value is MEXICAN-CUISINE

JULIA goes on to its next task, choosing a main course, and suggests chili for the main course (we leave out the way it did that). At that point, it is reminded of a meal in which chili was served. If this meal had been a successful one, the reminding would provide suggestions about other parts of the meal. In this case, however, some guests, who did not eat hot food, were left unsatisfied.

***** Planning for the task MAIN-COURSE-TASK ....
   TASK-TREE = ((<MAIN-COURSE-TASK> <APPETIZER-TASK> <SALAD-TASK>
            <DESSERT-TASK>) (<MEAL-TASK775>))
   The active goal chain is (SC-MAIN-COURSE <= M-MEAL)
<JULIA> What about CHILI791 for the main course?
USER => ok
<JULIA>
***** Reminded of MEAL80, where chili was the main course
***** Case-based reasoning with the case MEAL80
*** Trying to do analogy-mapping with the case MEAL80 ...
Checking if the previous plan for goals S-HUNGER80 E-EAT80 was successful ......
Previous plan execution failure found
The set of goals failed was S-HUNGER80 E-EAT80
It was because ((NOT EVERY ONE ATE SPICY DISH))

JULIA will try to avoid making this mistake again. It finds (through looking at the representation of the previous case) that the previous failure was because of a missing constraint about spices and seeks to find out if this constraint should be taken into account in the current case. After asking, it finds that to be so and creates a a "non-spicy-food" constraint for the current case. It propagates that constraint and checks it against what it has already decided. It finds out that chili and Mexican food are spicy, and rules both out. Because choosing a main course is dependent on having a value for cuisine, it deletes the choose-a-main-course task from the task network, reschedules the meal-descriptor task and the choose-a-main-course task, and attempts the meal-descriptor task again in an effort to choose a cuisine.

*** Attempting to avoid the previous plan failure......
The assigned blame was that C-NON-SPICY-PREF80 had not been considered.
To avoid previous plan failure ...
Asking the user of a missing constraint C-NON-SPICY-PREF
<JULIA> Is there anyone who doesn't like spicy food? (How many?)
USER => 3
<JULIA>

Trying to propagate the constraint C-NON-SPICY-PREF793 ....
   --> Generating a new constraint C-NON-SPICY-CUISINE794
   --> Generating a new constraint C-NON-SPICY-DISH795
Applying constraint C-NON-SPICY-DISH795 to CHILI791
      --> Aborting CHILI791
Applying constraint C-NON-SPICY-CUISINE794 to MEXICAN-CUISINE
      --> Aborting MEXICAN-CUISINE
   --> Killing the current task MAIN-COURSE-TASK ......
   --> Rescheduling MEAL-DESCRIPTOR-TASK MAIN-COURSE-TASK into the task network ......
***** Planning for the task MEAL-DESCRIPTOR-TASK ....
   TASK-TREE = ((<MEAL-DESCRIPTOR-TASK> <MAIN-COURSE-TASK> <APPETIZER-TASK>
         <SALAD-TASK> <DESSERT-TASK>) (<MEAL-TASK775>))
   The active goal chain is   (MEAL-DESCRIPTOR <= M-MEAL)
Constraint Satisfaction ......
   Checking constraints for slot CUISINE of MEAL-DESCRIPTOR783 ...
   --> Applying constraint C-NON-SPICY-CUISINE794 to slot CUISINE
      --> The slot CUISINE is not yet filled in
Instantiating slot CUISINE of frame MEAL-DESCRIPTOR783 ...

Because there has been little in the way of preferences offered by the user up to now, JULIA cannot suggest a new cuisine by itself at this point. It asks the user again for a cuisine preference, this time telling the user constraints on the preference. The user suggests Italian, and JULIA goes on. To complete the menu, JULIA continues its reasoning, choosing lasagne for the main course and is reminded of a case in which vegetarians were at a lasagne dinner and could not eat. JULIA knows that in the previous case, they could have eaten if the meatless version of the dish had been served, and proposes the same in this case. The meal JULIA finally comes up with includes vegetarian antipasto as the appetizer, veggie lasagne and Italian bread for the main course, mixed green salad as the salad, and ice cream for dessert.

6. Discussion

In our scheme, potential failures can be encountered and thus need to be dealt with during any step of the problem solving. Any time the problem solver encounters a case with a previous problem, it considers whether there is the potential for that problem in the new case. This may cause it to refocus itself until the potential for failure is determined, and if such potential is determined and the problem solver has to retract decisions made previous to the current one, then it must remake any decisions dependent on those decisions. Such processing, of course, requires that the problem solver be integrated with a reason-maintenance system that keeps track of the dependencies among its decisions. Other steps require that the reasoner record justifications for each of the decisions it makes. We have not done a great deal of work in these areas, but our experience so far leads us to believe that a standard truth maintenance system (Doyle, 1979, McAllester, 1980, DeKleer, 1986) is not adequate to do all of the work we need such a system to do. In particular, in addition to its standard bookkeeping functions, such a system will need strategies or policies to follow in making decisions about how to make the world consistent when a condition check fails, or will need to interact with a reasoner that can make such decisions. While it is standard for a truth maintenance system to retract decisions that are inconsistent and to propagate those retractions as far as it needs to, in the problem solving situation we are looking at, it is often more advantageous to try to satisfy constraints in a different way (e.g., to replace a retracted value with another that satisfies the necessary constraints).

Hammond (1986) takes the complexity out of this issue by having the reasoner explicitly try to avoid mistakes in one of its early planning steps. The advantage of this, of course, is that after potential mistakes are discovered, the problem solver need only keep them in mind during the remainder of problem solving rather than having to deal with new issues and possible change of focus part of the way through. There is thus no need for the complexity of a truth maintenance system. On the other hand, the reasoner can only avoid those mistakes that can be foreseen at the onset of problem solving, but cannot avoid mistakes that the problem solver might not be able to anticipate until late in the problem solving.

Carbonell (1986) deals with this issue in yet another way. His work assumes that each old problem is stored as a sequence of reasoning steps, and that any time two problems are similar in their set of steps, the second is stored on with the first, branching from it at the place they begin to be different. Thus, once the case-based reasoner is reminded of a previous case, it has available to it all of the cases that have been solved by the same initial set of steps as the one it is currently trying to solve. This means that at each decision point in the problem solving, each of the previous decisions that have been made are available along with their justifications. Reasoning similar to that described in this paper happens to evaluate which of the possibilities is appropriate for the new case. The advantages of this method are similar to the advantages in Hammond's method: the problem solver, in general, never needs to refocus itself, and there is no need for a truth maintenance system. The major disadvantage, however, is that once Carbonell's problem solver finds a set of previous cases that are similar to its current one, it is wedded to that set, and no other cases that might be similar along a different set of dimensions can contribute to the problem solving.

7. Summary

Previous problem solving failures can be a powerful aid in helping a problem solver to become better over time. When a previous case in which an error was made is recalled, it flags the potential for a similar mistake and the reasoner considers whether the same potential for error exists in the new case. The direct result of this is that reasoning is directed to that part of the current problem that was responsible for the previous error, sometimes changing the problem solver's focus. Evaluation of the potential for error in the current case may require the problem solver to gather knowledge it doesn't already have, another way focus might be redirected. A case with an error may also suggest a correct solution for the new case. The combination of these helps the problem solver to avoid repeating mistakes and suggests shortcuts in reasoning that avoid the trial and error of previous cases.

8. References

Alterman, R. (1986). An Adaptive Planner. *Proceedings of AAAI-86*, pp. 65-69.

Ashley, K. (1986). Knowing What to Ask Next and Why: Asking Pertinent Questions Using Cases and Hypotheticals. *Proceedings of the Eighth Annual Conference of the Cognitive Science Society.*

Carbonell, J. G. (1983). Learning by Analogy: Formulating and Generalizing Plans from Past Experience. In Michalski, R. S., Carbonell, J. G, & Mitchell, T. M., *Machine Learning*, Tioga.

Carbonell, J. G. (1986). Derivational Analogy: A Theory of Reconstructive Problem Solving and Expertise Acquisition. In Michalski, R. S., Carbonell, J. G., & Mitchell, T. M., *Machine Learning II*, Morgan Kaufmann Publishers, Inc.

Cullingford, R. E. & Kolodner, J. L. (1986). Interactive Advice Giving. In *Proceedings of the 1986 IEEE International Conference on Systems, Man, and Cybernetics.*

de Kleer, J. (1986). An Assumption-Based TMS. *Artificial Intelligence*, vol 28, pp. 127-162.

Doyle, J. A truth maintenance system, *Artificial Intelligence*, vol 12, pp. 231-272.

Hammond, K. (1986). Learning to Anticipate and Avoid Planning Problems through the Explanation of Failures. *Proceedings of IJCAI-86.*

Holyoak, K. J. (1984). The Pragmatics of Analogical Transfer. In Bower, G. (Ed.), *The Psychology of Learning and Motivation*, Academic Press.

Kolodner, J. L. (1983). Reconstructive Memory: A Computer Model. *Cognitive Science*, vol 7.

Kolodner, J. L. (1985). Experiential Processes in Natural Problem Solving. Technical Report No. GIT-ICS-85/23. School of Information and Computer Science. Georgia Institute of Technology. Atlanta, GA 30332.

Kolodner, J. L. (1986). Some Little-Known Complexities of Case-Based Inference. *Proceedings of TICIP-86.*

Kolodner, J. L. & Cullingford, R. E. (1986). Towards a Memory Architecture that Supports Reminding. In *Proceedings of the 1986 Conference of the Cognitive Science Society.*

Kolodner, J. L. & Simpson. R. L. (1984). Experience and Problem Solving: A Framework. *Proceedings of the Sixth Annual Conference of the Cognitive Science Society.*

Kolodner, J. L., Simpson, R. L., & Sycara, K. (1985). A Process Model of Case-Based Reasoning in Problem Solving. In *Proceedings of IJCAI-85.*

McAllister, D. (1980). An Outlook on Truth Maintenance. Artificial Intelligence Laboratory, AIM-551, MIT, Cambridge, MA.

Rissland, E. L. & Collins, R. T. (1986). The Law as a Learning System. *Proceedings of the Eighth Annual Conference of the Cognitive Science Society.*

Schank, R. C. (1982). *Dynamic Memory.* Cambridge University Press.

Simpson, R. L. (1985). A Computer Model of Case-Based Reasoning in Problem Solving: An Investigation in the Domain of Dispute Mediation. Ph.D. Thesis. Technical Report No. GIT-ICS-85/18. School of Information and Computer Science, Georgia Institute of Technology, Atlanta, GA.

Problem Solving in a Natural Task as a Function of Experience*

Juliana S. Lancaster
Janet L. Kolodner

School of Information and Computer Science
Georgia Institute of Technology
Atlanta, GA 30332

Abstract

Problem solving is known to vary in some predictable ways as a function of experience. In this study, we have investigated the effects of experience on the problem solving behavior and knowledge base of workers in an applied setting: automobile mechanics. The automobile itself is a highly complex system with many interconnected subsystems. Problem descriptions (i.e., symptoms) presented to a mechanic who needs to diagnose a car, however, are usually quite sketchy, requiring the collection of more information before solution. Novices are less able than experts to diagnose any but the obvious problems, and we are interested in identifying the qualitative differences between mechanics at different levels of expertise. In the study reported, we observed three student mechanics in a post-secondary technical school, each at a different level of expertise, diagnose six problems introduced into cars in the school. We then analyzed the protocols we collected to find the knowledge and strategies used in solving each problem. We also analyzed the series of protocols for each student to find the changes in knowledge and strategies used in solving later problems as compared to earlier problems. Differences were seen in both the knowledge used by the subjects and in their general approach to diagnosis. As a result of experience, the student mechanics seemed to improve in three areas: (1) their knowledge of the relationships between symptoms and possible failures was augmented, (2) their causal models of the car's systems were augmented, and (3) their general troubleshooting procedures and decision rules were much improved.

1. Introduction

Problem solving is known to vary in some predictable ways as a function of expertise. When the process of problem solving first came under scrutiny by psychology and computer science researchers, the problems studied were in knowledge-lean domains in which well-defined situations have known solutions (Reed, Ernst, & Banerji, 1974; Reed & Johnson, 1977; Reitman, 1976; Simon, 1975). In that work, the behavior of interest was generally a variable such as number of steps to completion or number of correct solutions. Recently however, interest in problem solving has leaned more toward problems in knowledge-rich domains such as physics (Chi, Glaser,& Rees, 1982; Simon & Simon, 1978), thermodynamics (Bhaskar & Simon, 1977), architecture (Akin, 1980), and political science (Voss, Greene, Post, & Penner, 1983; Voss & Tyler, 1981). Within these domains, researchers have continued to look at the steps and plans generated in coming to a solution, but they have also developed a further interest in the nature or organization of the knowledge used in the process of problem solving. A major question regarding the nature or organization of knowledge has been how that knowledge and its changes influence performance.

Our knowledge of the differences between novices and experts has reached the point where several general statements can be made. First, experts in any field are more able to recognize and remember typical conditions within their area of expertise. Second, experts generally organize their knowledge by functional characteristics of problems while novices are more likely to use surface features to characterize problems.** There have not been a lot of explicit conclusions, however, about the particular knowledge structures used by experts and novices. Nor has there been work describing the particular changes in knowledge and processing behavior that happen as a result of a single experience.

* This research is supported in part by the Army Research Institute for the Behavioral and Social Sciences under Contract No. MDA-903-86-C-173. Thanks to Ken Allison and Gita Rangarajan, who provided representations for the paper and ideas about analyzing the protocols.
**See Chi, et al (1982) and Glaser (1985) for more discussion of novice/expert differences.

727

Our primary goal is to discover the changes that individual experiences have on a problem solver. In order to achieve that goal, we first have to find out what knowledge the problem solver starts with before solving any problem and what knowledge he has later to solve a similar problem. While earlier work has indicated that "good" diagnostic ability is a function more of knowledge about the problem area being diagnosed than of general diagnostic skills (Miller, 1975), we find that the diagnostic skills of novices and experts also differ, and therefore also observe initial strategies of problem solvers and those used after a particular experience.

In the particular experiment to be discussed, we had two goals. Our first was to find out what knowledge subjects at different levels of expertise had and to be able to state the problem solving strategies used by subjects at varying levels of expertise. This, we felt, would give us a good idea of what things experience teaches. Based on our previous work on memory and problem solving (Kolodner, 1985; Kolodner & Simpson, 1984; Kolodner & Kolodner, 1987), we expected that differences would be in both the amount known and accessibility (or organization) of known knowledge. Our second goal was to identify particular changes over time in each individual's handling of specific problems and types of problems. The sequence of problems presented to the subjects was derived such that this would be possible.

The task domain we have chosen to look at, diagnosis of automotive problems, is interesting for several reasons. The automobile engine is a highly complex entity. It consists of a number of interacting systems acting to produce the car's motion. Failures in any component or system of the engine usually produce noticable symptoms or changes in the car's performance, but the failures themselves are seldom obvious to the amateur. In addition, a given symptom can indicate numerous possible failures within the engine. The person who comes to the shop with a problem describes a symptom or set of symptoms to the mechanic, and it is the mechanic's job to further investigate the car to find out which of the many possible problems that could cause the reported symtom(s) is in fact responsible for it. Experts are much better than novices at determining the causes of automotive problems. (As the old story goes; it's ten cents for the screw and twenty dollars for knowing which one to replace.)

The domain is knowledge-rich, and the depth of knowledge and ability to use it are both important in making a good diagnosis. Schools teach about cars in general, but since there are so many different kinds of cars, each of which have their own peculiarities, textbooks and schools can't teach everything. Diagnosing a car with a given set of symptoms may depend as much on the age and type of engine as on the symptoms presented. A given failure can be a common cause of a particular symptom in one engine and not possible in another. Experience with different types of cars and different types of problems is thus essential in gaining expertise. Furthermore, there are too many types of cars (most models change at least a little every year) and too much in the sets of manuals for individual cars for a mechanic to know everything about every car. Thus, it is essential for the expert mechanic to draw his own generalizations about cars that allow him to organize and access knowledge appropriate to any particular car and problem he is looking at.

In the work reported here, three student mechanics were observed while diagnosing car failures. Six problems were presented at weekly intervals and think-aloud protocols were collected while the students worked and were transcribed and coded for later analysis. Each week the instructor demonstrated the correct or optimum troubleshooting sequence for diagnosis of the failure after all subjects were finished. Thus, each student had an opportunity for feedback and an explanation of the car's problem whether or not he had diagnosed it correctly. Each failure was introduced into the car deliberately and each problem was caused by only one failed part. Analysis of the data focussed on the knowledge and strategies used by students at different levels of training, how their knowledge was organized, and how their knowledge and strategies changed with experience.

We expected that the more experienced student would solve more problems and would give evidence of having a more organized knowledge base than the less experienced students. In addition, we expected that individuals would show evidence over the series of problems of acquiring new diagnostic skills and new knowledge and connections within their knowledge.

2. Method

2.1. Subjects

Three students at a post secondary technical school volunteered to participate in the project. The technical program is a two-year, eight-quarter program. During much of the second year, the students work in a shop setting within the school. Cars belonging to school personnel and friends of the students and instructors are diagnosed and repaired by students. In addition, the school owns several cars that can be used in teaching students to teach about specific problems.

Each of the three student volunteers was at a different point in the program. The novice student was in his first quarter of the program and had no prior training or experience. The intermediate student was at the beginning of his second year in the program. The advanced student was near the end of the second year and held a part-time job as a mechanic outside of school. Each student worked on at least four of six problems.

2.2. Procedure

Subjects were observed once a week while diagnosing an actual problem in a car. The problems used were selected by an instructor in the program in consultation with the experimenter. The problems and the information given as the customer's complaint are described in Table 1. Each fault was introduced into a car by the instructor or by a student not in the study under the direction of the instructor. The cars used were all owned by the school with one exception: a new car brought in by a school official that had symptoms we had been presenting to the students in previous weeks. In every case, a single complaint was given and a single fault could be traced to account for the complaint. Students were told to track down the fault, but not to fix it unless repair was necessary to confirm the diagnosis.

In each session, the student was led to the car and, with the experimenter posing as a customer, told that the car was exhibiting a particular symptom. The student was then allowed to perform any tests desired on the car and its engine, with the exception of a driving road test, prohibited primarily by the symptoms presented by the car. The student was instructed to think aloud as he worked to find the failed component in the car. His comments were tape recorded by the experimenter, who also served as an assistant to the student when necessary.

Table 1		
Faults and their complaints as presented to subjects		
Problem	Complaint (Symptom)	Fault
1	cranks but will not start	sediment or other blockage in gas line
2	cranks slowly when starting	bad cell in battery-will not hold charge
3	cranks but will not start	bad connection behind fuse panel and fuel pump fuse
4	cranks but will not start	loose ground wires from Electronic Control Module (computer)
5	cranks but will not start	open tach circuit
6	detonation on acceleration	poorly adjusted timing

2.3. Coding

After all protocols were transcribed, each statement was coded into one of six categories, shown in Table 2 with examples. Statements coded as *hypotheses* were those in which a specific system or component was first named as a possible source of the failure or in which the system or component was accepted or rejected as the

729

Table 2		
Coding Categories for Protocols		
Category	Additional Specifications	Examples
Hypotheses	Number and Status	Could be starved for gas (N-P1) It could be, could be the starter (N-P2)
Rules	Topic(Failure, normal functioning, or troubleshooting)	Fuel Pump should come on for 3 seconds (I-P4) First of all, I have to locate the connector to the back of the fuel pump (A-P3)
Information Gathering	Source of information obtained	Before I look in the book, I'm going to check the fuse (A-P3)
Observation	Topic (hypothesis(number) or complaint)	What we don't have is fuel to the throttle body (A-P3) I don't believe I hear it running (A-P3)
Restatements	Topic (complaint or summary of observations)	to rephrase that-the throttle body is not injecting fuel (A-P3)

source of the failure. Hypotheses were numbered in order of appearance and, each time one was mentioned, its status was noted. Its status could be *open, accepted, confirmed, or rejected. Rules* were statements giving known, constant information about an engine or about the process of diagnosis. Statements coded as *information gathering* were generally descriptions of the actions being taken by the subject at the time. Such actions could elicit or obtain information from the customer, from a book, or via a procedure or test applied to the engine. *Observations* were statements giving the information obtained from the action taken. *Restatements* were repetitions of previously stated or collected information rather than new information. Each statement falling into one of the last three categories was identified with a specific hypothesis by its number if possible. All other statements were uncodable and were marked as such.

3. Results and Discussion

As expected, the ability of the students to correctly diagnose the problems changed substantially between the novice level and the intermediate and advanced levels. The diagnoses given by each subject and the number of hypotheses considered are shown in Table 3. The novice correctly diagnosed only one of four problems attempted, while the intermediate student correctly diagnosed three of six and the advanced student three of four. In addition, the number of hypotheses considered increased with expertise. The novice generated a mean of 3.0 hypotheses per problem and the intermediate and advanced students generated 6.8 and 5.0 hypotheses per problem respectively.

Table 3			
Final Diagnoses and Number of Hypotheses Considered by Each Subject			
Problem	Novice	Intermediate	Advanced
1	not getting fuel(4)	clogged fuel line(3)	--------
2	dead battery cell(5)	starter(4)	dead battery cell(6)
3	--------	fuel pump relay(5)	fuel pump fuse(5)
4	fuel pump(3)	no diagnosis(6)	injector solenoid(9)
5	no diagnosis(0)	open tach circuit(9)	--------
6	--------	bad timing(10)	bad timing(2)

3.1. Knowledge Structures and Knowledge Organization

In general, the diagnostic behavior we saw was similar to that reported by other researchers (Hunt, 1981; Rasmussen, 1978; 1979; Rasmussen & Jensen, 1974). Students generated one or more possible hypotheses for the failure immediately after observing the symptom(s). These hypotheses were then tested in a fairly systematic (albeit sometimes idiosyncratic) way either by observation of the inputs to and outputs from specific components and systems or by performance of specific diagnostic tests. In successful cases, a single diagnosis ultimately was given, accompanied by an explanation of how or why that failure would generate the observed symptom(s).

We interpret this process as being indicative of an interaction between two types of knowledge structures. The first, a *causal model* of the car's engine, contains knowledge about individual components and their inputs, outputs, and normal behavior; relates components within a system to one another; and describes the relationships and connections between systems. It is used to evaluate hypotheses in light of the evidence obtained from the failed engine and to lead the mechanic through the engine to the source of the problem in a systematic way. The causal model is generally quite large, and the second type of knowledge structure, *symptom-fault sets*, is used to index into the causal model at appropriate places. Symptom-fault sets represent the relationships between particular symptoms or sets of symptoms and failures. For example, given the symptom "the car cranks but will not start", the symptom-fault sets will identify three systems as possible locations for the failure: the fuel system, the air intake system, and the ignition system. Within each of these systems, additional symptom-fault sets will identify individual components that may cause the symptom(s). For the fuel system, these would be a failed fuel pump, an empty gas tank, or a blocked fuel line. For the ignition system, these would be a bad distributor, bad spark plug wires, or bad spark plugs. These symptom-fault sets are used to derive initial hypotheses, directing the mechanic to look at only appropriate places in the causal model.

If, in fact, mechanics are using these two types of knowledge structures during troubleshooting, then we can predict several changes we should expect to see in these structures as a result of experience, and from those, we can predict the processing differences that would result from these changes. First, we predict that through experience, a mechanic's set of symptom-fault sets increases and that the sets he already knows become more accurate. As a result of these changes, the mechanic should have better ways to index into the causal model, leading to more efficient searches for the correct failure. Second, the causal model should become more filled out with experience, both through addition of components and/or systems that were previously unknown and through addition of relationships and dependencies between the known components. The causal model, like symptom-fault sets, should also become more accurate. As a result of having a better causal model, a mechanic should be better able to systematically reason about the way the car works, allowing him to find engine failures more systematically and in more cases.

We did, in fact, see clear differences between students at different levels of experience reflecting exactly these changes in their knowledge structures. First, we saw evidence that both the organization and number of symptom-fault sets increased with experience. The advanced student seemed to know more symptom-fault sets than the novice, as evidenced by the larger number of hypotheses he was able to generate for each problem. In addition, the advanced student seemed to organize his symptom-fault sets differently than the novice, evidenced by the more systematic procedure he used for generating and testing hypotheses. The advanced student's procedure was to zero in

731

on one of the engine's subsystems and then to consider which component of that system was faulty, while the novice did not differentiate between systems and components of systems in diagnosis. While for the novice, all faults are equal and an hypothesis at the component level was as likely to be selected as the first to investigate as an hypothesis at the system level, the more advanced troubleshooter seemed to organize his symptom-fault sets into two categories, each used for different purposes. One set pointed to faulty *subsystems* within the car (e.g., fuel system, electrical systems) and was used early in diagnosis to zero in on the faulty subsystem, while the second set pointed to faulty *components of these systems* (e.g., the fuel pump, the battery) and was used to diagnose the problem within that system. Such a change requires that the mechanic also reorganize his knowledge about the car's engine in a more hierarchical way that differentiates between systems and components of systems. Figure 1 shows a portion of the novice and advanced student's organizations of the causal model of the engine.

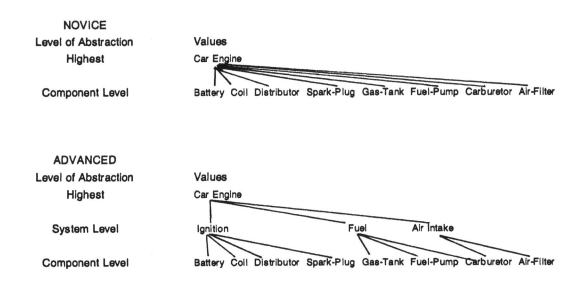

Figure 1

Novice and Advanced Student Representations of the Car's Engine

We also saw evidence that content of the causal model changed with experience. The causal model of the more advanced students contained not only more knowledge about individual components, but also more knowledge about the interconnected nature of the engine's systems. The behavior of the students during troubleshooting illustrates these findings. Consider, for example, the behavior of the advanced student in Problem 4. His reasoning went as follows:[*]

> The first thing you want to do, which is the easiest thing to do, is look and see if we have any fuel, because you gotta have fuel, air, and heat... Dont have fuel..The first thing I want to do is check the fuse...they're OK... hook this jumper lead to the bypass to the fuel pump...the fuel pump is running... check and see our connection up here to the energizer...going from the ECM up to the injector is OK...try to energize this solenoid by hand...check to see if we got any gas...all the lines are alright...got gas to the throttle body... my diagnosis is the solenoid is bad because everything else checks out.

The hypotheses generated by this student are in an order that reflects the multi-level and highly integrated

---

[*] For a full protocol of the session, write to the first author.

organization of both his causal model and his symptom-fault sets. He first determined which of three possible systems of the engine was affected and then investigated its components and others that could impinge on the behavior of the system under focus. In fact, his primary focus was on the electronic (or computer controlled) influences on the behavior of the fuel pump and fuel injectors. This reasoning showed an awareness (reflected in the student's causal model) of the interdependencies between subsystems. His reasoning shows that he knows that systems (such as the fuel and electronic systems) may intersect at several points and that an apparently or possibly failed component in one system may reflect an action, or lack of action in another system.

In contrast, the novice generated relatively few hypotheses for any given problem. His protocols indicate that this is because he has little knowledge about the relationships between given symptoms and their causes and also because his causal model is inadequate. In solving the same problem the advanced student was working on above, the novice reasoned:

> This problem could be in the fuel system, ignition system...we know it's not in the starting system because the car will crank over...One small drop of fuel...in that bowl...so it's in the fuel system...the fuel pump's... supposed to turn for 10 to 15 seconds...I can't hear it...It might just be a bad fuel pump.

We can see little evidence of an integrated hierarchy of levels in his organization of symptom-fault sets. While his hypotheses were sometimes at the system level (i.e., fuel system) and sometimes at the component level (i.e., fuel pump is bad), in only one problem (this one) did he clearly consider first a system and then a component within that system. More commonly, he generated hypotheses at both levels and then investigated only specific components. Furthermore, he showed a similar lack of integration in his causal model. Specifically, he never considered the possibility that one system could affect the behavior of another. His knowledge appeared to stop at the individual component's behavior and did not include the possibility that the actions of another system (the electronic system) could be affecting the behavior of the component he was considering (the fuel pump).

While the novice knew about many of the components of the car's engine and about what their connections were within a single system, he did not know how the systems and the components in different systems were interrelated. The advanced student, on the other hand, knew both the connections between components and the connections between systems. Thus the advanced student had a more integrated and complete understanding of the car's engine, while the novice's understanding seemed to be highly disjoint. Figure 2 shows our interpretation of what the novice and advanced students knew about the fuel pump, for example.

PUMP   Source: a container
                Substance: a substance in the container
                Conduit: a pipe
                Destination: a container
                Energy-Source: an energy device

NOVICE		ADVANCED	
FUEL PUMP	ISA PUMP	FUEL PUMP	ISA PUMP
	Source: gas tank		Source: gas tank
	Substance: gasoline		Substance: gasoline
	Conduit: hose		Conduit: hose
	Destination: carburetor		Destination: carburetor
			Energy-Source: electrical system

Figure 2

Novice and Advanced Student Representations of a Fuel Pump

Note that the general information about pumps is available to both the novice and the advanced student. However, the information that the fuel pump requires an energy source which is the electrical system of the car is not part of the novice's representation of the fuel pump. If asked "What makes the fuel pump run?", the novice is able to construct the appropriate answer by using the more general information about pumps, but he does not use this knowledge during problem solving. The same pattern is probably true of knowledge about systems and components. The novice can undoubtedly tell an inquirer what system of the engine a particular component resides in, but he does not maintain this information where it is readily usable during problem solving.

We also saw within-subject changes in these knowledge structures over the course of the experiment. These changes were most evident in the intermediate student. Two examples will serve to demonstrate changes across problems. In working on problem three, the intermediate student made a long and protracted search for the fuel pump relay using both written reference materials and extended visual examination of the engine. While working on problem four, he was able to immediately locate and check the same part. This component, and its physical relationship to others, had been incorporated into the causal model during or following problem three. Similarly, the symptom-fault sets changed as new information was acquired. For example, the first hypothesis the intermediate student checked at the component level for problem four was the fuel pump fuse, which was the correct diagnosis for problem three. He made the point as he worked that he was checking this possibility out first because of the previous case. ("I'm gonna check the fuel pump fuse *first* [this time].")

3.2. Diagnostic Strategies

In addition to the changes experience makes in knowledge structures and organization, we also saw differences in diagnostic style. Diagnostic strategies seemed to be used differently by subjects at different levels of expertise and evaluation criteria changed significantly with experience. Some of these changes are due to the development of better strategies for testing and confirming hypotheses with experience while others appear to result from the differences in the knowledge available for diagnosis as a mechanic gets more experienced.

The change in how the mechanics tested and confirmed hypotheses was striking. As the example above showed, the novice student was willing to accept an hypothesis when preliminary evidence could be interpreted as congruent with that hypothesis and not pursuing the task any further (i.e. "can't hear the fuel pump"). In contrast, the advanced student sought, for each hypothesis, specifically confirming or disconfirming evidence that was part of a causal explanation. While he was willing to select an hypothesis to pursue on the basis of preliminary evidence, he would not accept or reject it without causally based information (i.e. "the fuel pump's not running, now we have to find out why").

The changes in diagnostic strategies that resulted from changes in the knowledge structures were more apparent in the efficiency of diagnosis. As the causal model gets filled out, it should allow the mechanic to pursue a longer systematic search through the engine and also allow him to evaluate information in more detail and with more concern for the real effects of the behavior observed. At the same time, as the number and complexity of symptom-fault sets increases, long searches should become less necessary, because the mechanic is able to index into his model in more, and more effective, locations.

These two types of changes in the mechanic's diagnostic strategies work together to produce the results we saw. As the mechanic gains experience with making correct and incorrect diagnoses, he gains a sense of what kind and how much information is "enough" to be sure of his opinions. In addition, as his causal model and symptom-fault sets become more complete and accurate, he is more able to select hypotheses for investigation appropriately and to continue invgestigating a problem to the point that only one hypothesis remains as a possible diagnosis. Consequently, the conditions under which he will accept an hypothesis as a final diagnosis will become more accurate and the path by which he reaches his diagnosis will become more efficient.

This result is clearly evident in protocols of the novice and advanced students. When the novice's working hypothesis was a that a particular component was faulty, he either accepted it or rejected it as the cause of the symptom. He never investigated other effects on or inputs to that component. For example, in problem 2, the failure was a dead battery cell which caused the car to crank very slowly. The novice based his diagnosis on the following information:

> First of all, we'll have to check this battery...it could be the starter... it could be the alternator...it
> could be a voltage loss...could be a dead cell in the battery...we've only got 10 volts in the battery-
> -each battery cell is 2 volts and there's 6 cells in the battery, so dead battery cell.

Here we see the novice generating both system and component level hypotheses but, because his knowledge is not hierarchically organized, not pursuing them in that order. Rather, he looks first at the battery charge. Because it is low, he accepts the hypothesis of a dead cell. His diagnostic strategy does not require that he consider any hypotheses relating to why the battery might be low, such as a malfunction in another system.

In contrast, the advanced student generally collected more information before giving a diagnosis. If possible, he confirmed his diagnosis by visually finding the condition that created the symptom (i.e., the disabled fuse panel connection in Problem 3). When that was not possible, he justified his diagnosis within his causal model. For example, in Problem 2, the failure could not be confirmed by visual evidence. Instead, the advanced student reaches his diagnosis with the following information:

> ...check the starter draw...it's pulling enough down to get the starter to go alright...We put the bat-
> tery under load, you can see the amps rising and it's charging the battery...So the alternator's
> working OK...what I believe we have is the cell is dead in the battery...Try the test on the VAT...As
> you see on the indicator is also showing that it needs charging for the battery is bad...So what we
> have here is a battery with a couple of cells dead, and it's a sealed battery and you cannot check
> the specific gravity with a hydrometer to check and see which one's dead.

He reached and justified his diagnosis by eliminating all other possibilities from his symptom-fault sets and the causal model. In other words, he tested and verified normal functioning of both the starting system ("it's pulling enough down to get the starter to go alright") and the charging system ("So the alternator's working OK"). These are the only two systems, other than accessories such as headlights and radio, that affect the level of charge in the battery. Consequently, according to the student's causal model, if the battery's charge is low and the starting and charging systems are functioning correctly, the only remaining component in which the failure can be located is the battery itself. In some types of batteries, this conclusion can be tested directly, but in the car used in this problem, the battery is sealed. Therefore, the mechanic must stop with his explanation rather than attempt to verify the diagnosis any further. In comparison to the novice, he selected his hypotheses more efficiently, first eliminating competing systems from consideration. In addition, he based his acceptance of the diagnosis on a full causal explanation rather than on superficial evidence.

4. Conclusions

The results are as predicted by our interpretation of the diagnostic behavior as an interaction between several knowledge structures. Both the causal model and the symptom-fault sets change with experience, and we have seen some examples of exactly what changes occur. In the causal model, the most notable change is the increasing complexity of the model, reflected in the growing awareness of the interconnectedness of systems within the engine. The novice is clearly unaware of the possibility that electronic failures can affect things like fuel delivery, since he knows little about the dependencies between the fuel system and the electrical system, while the more advanced mechanic not only knows that such relationships exist, he considers them a highly common source of failures. Similarly, the number, organization, and accuracy of the symptom-fault sets changes with increasing experience. Ultimately, they are able to represent a complex, hierarchical system of relationships. The data suggest that components are organized hierarchically under their respective systems and are never directly considered unless their system is determined to house the failure, or at least to be the source of information crucial to locating the failure.

Building partly on these changes in the knowledge structures, and partly on independent effects of experience on decision processes, the mechanic's procedures and guidelines for accepting hypotheses as diagnoses also change. The processes or procedures used become increasingly focussed on information that will allow a causal interpretation of the behavior observed. At the same time, the developing knowledge structures allow the mechanic to search for and aquire more, and more accurate, information from his symptom-fault sets and his causal model. The interaction of these changes in both knowledge and process lead to the more accurate and efficient problem solving seen in experts.

Thus, we see that experience is providing the mechanic with three things. His overall level of knowledge is increasing; the organization and integration of his knowledge structures, both the symptom-fault sets and the causal model, are increasing; and his processes and criteria for reaching diagnoses are becoming more accurate, more efficient, and more focussed on causal information.

5. References

Akin, O. (1980) *Models of architectural knowledge* London: Pion.

Bhaskar, R. & Simon, H.A. (1977) Problem solving in semantically rich domains: An example from engineering thermodynamics. *Cognitive Science, 1,* 193-215.

Chi, M.T.H., Glaser, R., & Rees, E. (1982) Expertise in Problem Solving. In R.J. Sternberg (ed) *Advances in the Psychology of Human Intelligence* Hillsdale: Lawrence Erlbaum.

Glaser, R. (1985) Thoughts on expertise. Technical Report #8. Learning Research and Development Center, University of Pittsburgh, Pittsburgh, PA 15260.

Hunt, R.M. (1981) Human pattern recognition and information seeking in simualated thought diagnosis tasks. Report #T-110, Coordinated Science Laboratory, University of Illinois at Urbana-Champaign.

Kolodner,J.L. (1985) Experiential processes in natural problem solving. Technical Report #GIT-ICS-85/23. School of Information and Computer Science, Georgia Institute of Technology, Atlanta, GA 30332.

Kolodner,J.L. & Kolodner,R.M. (1987) Using experience in clinical problem solving: Introduction and framework. *IEEE Transactions on Systems, Man, and Cybernetics.*

Kolodner,J.L. & Simpson,R.L. (1984) Experience and Problem Solving: A Framework. *Proceedings of the Sixth Annual Conference of the Cognitive Science Society.*

Miller, P.B. (1975) Strategy selection in medical diagnosis, Project MAC. Report #TR-153, Massachusetts Institute of Technology, Cambridge, MA.

Rasmussen, J. (1978) Notes on diagnostic strategies in process plant environment Riso National Laboratory Report #RISO-M-1983, Roskilde, Denmark.

Rasmussen, J. (1979) On the structure of knowledge-A morphology of mental models in a man-machine context. Riso National Laboratory Report #RISO-M-2192, Roskilde, Denmark.

Rasmussen, J. and Jensen, A. (1974) Mental procedures in real life tasks: A case study of electronic troubleshooting. *Ergonomics, 17(3),* 293-307.

Reed, S.K., Ernst, G.W., & Banerji, R. (1974) The role of analogy in transfer between similar problem states. *Cognitive Psychology, 6,* 436-450.

Reed, S.K. & Johnson, J.A. (1977) Memory for problem solutions. In G.H. Bower (ed) *The Psychology of Learning and Motivationk* New York: Academic Press.

Reitman, J.S. (1976) Skilled perception in go: Deducing memory structures from inter-response times. *Cognitive Psychology, 8,* 336-356.

Simon, H.A. (1975) The functional equivalence of problem solving skills. *Cognitive Psychology, 7,* 268-288.

Simon, D.P. & Simon, H.A. (1978) Individual differences in solving physics problems. In R. Siegler (Ed.) *Children's Thinking: What Develops?* Hillsdale, N.J.:Lawrence Erlbaum

Voss, J.F., Greene, T.R., Post, T.A., & Penner, B.C. (1983) Problem solving skill in the social sciences. In G. Bower (Ed.), *The psychology of learning and motivation: Advances in research theory.* New York: Academic Press.

Voss, J.F. & Tyler, S. (1981) *Problem solving in an social science domain.* Unpublished manuscript, University of Pittsburgh, Learning Research and Development Center.

Distinguishing – A Reasoner's Wedge [1]

Kevin D. Ashley [2]
Department of Computer and Information Science
University of Massachusetts
Amherst, Massachusetts 01003

Telephone: (413) 545-0332
CSNET address: ASHLEY@UMass

Abstract

In this paper we focus on the *Distinguisher's Wedge*, an intellectual tool for responding to an argument that two cases are alike by asserting reasons why they are different and why the differences matter. We characterize the wedge as involving a search for *distinctions*, factual differences between the cases that tie into justifications for treating them differently. We show how the wedge can be modelled computationally in a Case-Based Reasoning ("CBR") system using *precedential* justifications and describe how the model is realized in our *HYPO* program which performs legal reasoning in the domain of trade secret law. Legal argument, with its emphasis on citing and distinguishing precedents and lack of a strong domain model, is an excellent domain for studying the wedge. We show how HYPO uses "dimensions", "case-analysis-record" and "claim lattice" mechanisms to cite and distinguish real cases and suggest how the model may be extended to cover more sophisticated kinds of distinguishing.

1. Introduction

Distinguishing is making explicit what is different about two cases and why that difference justifies *not* treating them the same way. In everyday argument, we are often called upon to distinguish cases. For example, a twelve year old, whose birthday is next month, demands that his over-protective parents tell him why *he* can't go to the movies to see "Little Shop of Horrors" but his fifteen year old sister can. The parents' task is to *distinguish* the brother's case from the sister's. The short answer, as if it were enough, is, "Because your sister is three years older." But, one might ask, or – if one is twelve, demand – why *that* difference makes a difference, a question to which there are many dubious responses, among them: "Your sister is more mature"; "The movie gets out too late – it's past your bedtime."; "The movie is rated PG-13 – it's for teenagers.".., to which there are at least as many indubitable retorts: "But I *am* a teenager! I'm going to be 13 next month."; or "I could see it if it was on a VCR but not at the movies!"; or "Noah's parents

---

[1] This work was supported (in part) by: the Advanced Research Projects Agency of the Department of Defense, monitored by the Office of Naval Research under contract no. N00014-84-K-0017, and an IBM Graduate Student Fellowship.

let *him* go see it." Dubious as the responses are, there are plenty of other differences that meet the issue even less: "Your sister is a girl and you're a boy."; "Your sister has more money than you do." (One can imagine the issues for which *these* differences would make a difference.); and some differences that appear to favor the opposite conclusion: "Your sister has more homework than you do."

Distinguishing is so fundamental it is like one of mechanics' simple machines, like a reasoner's *wedge*, a tool that uses the point of the factual difference to pry the two cases apart. One hears this kind of argument in many diverse domains: in arguments about mathematics (If figure 6 is a polyhedron then why isn't figure 7? [Lakatos, 1976]); in scientific research (If two persons were exposed to the same virus why did only one get the disease?); in historical political analysis (America's Nicaragua policy is leading to another Vietnam.); And, of course, in legal arguments [Levi, 1949]. In each domain, the distinctions (or lack of them) elucidated in the responses are the bedrock of effective analysis.

2. The Distinguisher's Wedge

Distinguishing involves searching for *distinctions*, factual differences between the two cases that tie into *justifications* for treating them differently. In its most general form, the wedge works like this:

(1) A reasoner *must* respond to an assertion in an argument by analogy that since case α is just like case β, α should be treated in the same way as β;

(2) Among the tools in the reasoner's kit is the wedge. In searching for distinctions, the would-be distinguisher has two alternatives that he pursues in parallel:

1. Search for factual differences with which to take advantage of known justifications. If possible, the reasoner will try to find out or infer new facts about α or β that constitute such differences.

2. Search for possible justifications with which to take advantage of known or possible factual differences between α and β.

(3) Assuming that the search for distinctions is successful, the reasoner must still evaluate how useful the distinctions are as responses. By citing the distinction, the reasoner changes the state of the argument and may introduce new facts or justifications that will have to be defended.

The need to distinguish drives both factual inquiry and theory building. It prompts the distinguisher to investigate for new factual differences and to find or create theories that assign the desired significance to the differences. The parents' responses, for example, point out factual differences, in the childrens' ages, maturity and bedtimes, as well as theories why the differences matter, ranging from invoking the rule that movies are only for their rated age group to an implied theory about movies' effect on children. The factual difference and the justification are, of course, intimately related. The sex difference, for example, may not appear relevant except, arguably, in light of a theory that girls mature faster than boys.

A purported distinction can be disputed at a range of levels, including disagreements about the asserted factual differences, their significance, or the validity of the justifications. Among the ways to respond to a distinction are: (1) to cite another case that minimizes the difference asserted by the distinguisher, in effect granting that the difference does matter (e.g., citing the example of twelve year old Noah with the permissive parents. One can almost hear the parent's next distinction: "Just because Noah's parents allow him to do certain things does not mean ..."); (2) to assert that the justification applies to α just as much as it applies to β, (e.g., "I am a teenager!"), in effect granting that the justification is valid but questioning the meaning of the justification. This is an example of how the wedge puts pressure on a definition; (3) to grant the difference but show that it actually leads to the opposite of the distinguisher's conclusion. For example, the supposed difference in maturity may imply that the sister has greater responsibility – like homework – and drive a factual query whether sister does not have a big report due tomorrow and shouldn't be going to the movies after all. In other words, the asserted distinction backfires; or (4) to cite a counter example to the justification (e.g., Noah saw the PG-13 movie despite being less than thirteen.)

There are many ways to evaluate who wins the battle over whether there is a distinction between the cases. Good ways to lose such arguments, at least in some domains, are to lead off with an example α for which the opponent can cite numerous equally analogous counter examples, or to assert a distinction that can be turned against the distinguisher's position. Good arguers take these evaluation criteria into account in strategically planning their arguments.

3. Computationally Modelling the Wedge – a Case-Based Approach

One major problem for computationally modelling the wedge is controlling the inferencing necessary to find a distinction between cases α and β. The justification part of a distinction may be an explanatory chain of inferences in terms of some causal theory, for example the "maturity" theory of the initial example, or more simply, invocation of the rule that PG-13 movies are only for teenagers, whose conclusion applies to β but not, because of some factual difference, to α. In light of the theory, that difference is crucial and worthy of a distinction. *But how does a system find the difference and the theory?* There may be many differences between α and β, not all of which are relevant to the issue posed. The relevance of even a crucial difference may not become apparent except after a possibly long chain of inferences, and it may "cut" the wrong way, that is, lead to a chain of inferences that hurts the distinguisher's position. There may also be many theories that would lead to the conclusion that that α and β should be treated differently. Which one should the system try? How far should it backchain before deciding that there is no crucial difference between α and β along the lines of that theory?

Another problem is in dealing with the "open texture" of the predicates in the rules representing the theory. It is not possible simply to define the predicates and hope that there's an end to it. The boy's response is quite reasonable. Depending on the context, "Teenager" sometimes means "≥ 13" but sometimes it may also mean "of the level of maturity of at least an average 13 year old." Add this wrench into the machinery of controlling inferencing and things get really messy.

A case-based approach to modelling the wedge takes a short cut across the problem of controlling inferencing. A central element of a CBR approach is the use of *precedential justifications* instead of chains of inferences to simplify the control problem. A CBR system assumes that the facts that a prior case had a certain cluster of features, and that the decision of the prior case was made because of some of those features and inspite of others, are a basis for a precedential justification for coming to the same conclusion in a future case with a similar combination of features.

Consistently with its use of precedential justifications, CBR assumes that when a decision is made about a case, the decider assigns credit or blame to some of the case's factual features as either contributing in favor of or against the decision. In effect, the decision of a case: (1) *Selects* certain features that are important enough for purposes of credit assignment; (2) *Clusters* the selected features; and (3) "Weights" them, ranking the features in the cluster that favor the decision higher than those against it, at least in that case. The factual similarities and differences among cases that are important in the domain are the ones that previous cases have found to be important; they are predefined for the system and used as the basis for *indexing* the cases in the CKB. In other words, the cases in the CKB are indexed by the same features that are involved in credit assignment and precedential justifications.

In essence, the CBR approach controls the inferencing problem by flattening out the depth of inferencing needed to come up with precedential justifications. Although the depth of inferencing is shallow, the depth of analysis is not by virtue of the breadth of the index, the size and diversity of the CKB, and the mechanism for selecting the best cases to use as precedents.

4. The HYPO Program

HYPO is a CBR program that reasons about a fact situation by critically comparing it to precedent cases. Its domain is trade secret law [Rissland & Ashley, 1986; Rissland & Ashley, 1987; Ashley & Rissland, 1987]. The law is an excellent domain for studying the wedge, since distinguishing and the use of precedential justifications are primary components of legal argument [Levi, 1949].

The main sources of legal knowlege in HYPO are contained in HYPO's CKB and its library of dimensions. Dimensions represent the legal relationship between various clusters of operative facts and the legal conclusions they support or undermine. Dimensions provide not only indices into lines of cases but a scale for comparing cases in terms of important factual differences that affect the strength, or weakness, of a fact situation with respect to that line of reasoning. For instance, one line of trade secret cases focusses on the degree to which the "cat (i.e., secret) has been let out of the bag", even by the complaining plaintiff, himself: that is, how many disclosures of the putative secret were there and of what kind? This way of looking at a trade secret case (captured by the *Disclose-Secrets* dimension) provides one approach to resolving a misappropriation dispute and was used in the *Data General* and *Yokana* cases discussed below. Another approach might emphasize the competitive advantage gained by the defendant at the plaintiff's expense or the switching of a key employee from the plaintiff to the defendant [Rissland & Ashley, 1986]. Each dimension has: *prerequisites*, expressed in terms of *factual predicates*, that tell whether a dimension applies to a case or not; *focal slots* that single out the particular facts making a case stronger or weaker along the dimension and range information that tells how a change in the focal slot affects that strength (e.g., for *Disclose-Secrets*, the focal slot is the number of disclosees. Increasing that number weakens the plaintiff's position.) See generally [Ashley, 1986].

5. HYPO's Model of Distinguishing

HYPO's model of distinguishing uses *dimensions, case-analysis-records* and *claim lattices* to find both relevant factual differences and precedential justifications. Dimensions provide HYPO's handle on factual differences and a means for extrapolating from prior cases. By virtue of dimensions' definitions, HYPO knows that a particular difference is significant and which side, π or δ, the difference *favors*. Dimensions also provide a precedential justification for why the difference matters – because the difference represents either a strengthening or weakening of features that mattered in prior cases indexed by the dimension. The case-analysis-record and claim lattices enable HYPO to compare cases in terms of multiple dimensions' cumulative effects as well as along individual dimensions. The differences they deal with represent either a strengthening or weakening of the *closeness of the analogy* between a prior case and the cfs.

Here is how the model works. *First*, in analyzing a new cfs, HYPO runs through the library of dimensions and produces a **case-analysis-record** that contains: (1) applicable factual predicates; (2) applicable dimensions; (3) near-miss dimensions; (4) potential claims and (5) relevant cases from the CKB. **Near-miss** dimensions are those for which some, but not all, of the prerequisites are satisfied. The combined list of applicable and near-miss dimensions is called the **D-list**. Figure 1 describes a cfs based, for purposes of illustration, on *Data General v. Digital Computer Controls, Inc.*, a real case in the CKB (but with one difference: the cfs involves 12000 disclosures of plaintiff's secret while *Data General* involved only 6000.) Figure 2 shows the case-analysis-record for the cfs.

Second, HYPO uses the case-analysis-record to construct the **claim lattice**, which is a lattice such that: (1) the root is the cfs together with its D-list; and (2) successor nodes contain pointers to cases that share a subset, usually proper, of the dimensions in the cfs's D-list. Figure 3 shows the claim lattice actually generated by the HYPO program for analyzing the cfs of Figures 1 from the viewpoint of a trade secrets misappropriation claim [Ashley & Rissland, 1987]. (There is a separate claim lattice for each possible claim.)

The ordering scheme enables claim lattices to capture a sense of closeness to the cfs of cases in the CKB. Those sharing more dimensions are nearer to the cfs. Those nodes closest to the root whose subsets of the cfs's D-list do not contain near-miss dimensions can be considered most-on-point-cases "**mopc's**" to the cfs; leaf nodes are the least-on-point. All of the cases displayed are relevant to the cfs because they all share some legally important strengths or weaknesses with the fact situation as represented by the dimensions shared with the cfs.

Third, HYPO uses the cases in the claim lattice to make and respond to arguments by analogy about the cfs citing those cases as precedents. Different major branches of the lattice indicate different ways to argue the case, effectively one way for each group of mopc's. HYPO can argue the case for *side 1*, let us say the plaintiff ("π") in the cfs, by *citing* a pro-π mopc. As it happens, in Figure 3, the only mopc is the pro-π *Data General* case, so HYPO cites it in favor of the plaintiff:

[a] ↪ For *Side 1*: (point) π wins.
      Case Cited: *Data General*
      (π in cfs and cited case disclosed secrets
      but disclosures subject to restriction.)

Since mopc's share the most legally important strengths and weaknesses with the cfs (i.e. mopc's are the closest analogies to the cfs), *Data General* is the most persuasive case HYPO could cite for the defendant as *side 1*. (For purposes of illustration *Data General* is also the basis of the cfs in Figure 1. It is encouraging that after analyzing the cfs, HYPO has found the nearly indentical case to be most on point!) *Telex v. IBM*, for example, is not a mopc because, although it is very close to the root, the *Competitive-Advantage* dimension which applies to *Telex*, and which would help π if it applied to the cfs, is only a near-miss for the cfs. (Note that *Competitive-Advantage* is *'d in Figure 3.) In support of the citation, HYPO draws the analogy between the mopc and the cfs. The relevantly similar facts are just those summarized by the dimension[s] that apply to both

(i.e., *Disclosed-Secrets* and *Restricted-Disclose*, the latter capturing the idea that if the disclosees agree to maintain confidentiality, then the secret is safe.)

HYPO responds to points like [a] by *distinguishing* the cited case using three basic methods: (1) Comparing the strengths of cfs and cited case along the dimensions they share in common; (2) Finding strengths or weaknesses, represented by dimensions, that cfs and cited case do *not* share. (3) Finding other cases that are more on point than the cited case.

Responses [b], [d] and [f] below illustrate these methods. As an example of the **first** method, consider how HYPO distinguishes *Data-General* from the cfs, on behalf of *side 2*, the defendant, by comparing values of the focal slots of the shared dimension:

> [b] ↩ For *Side 2*: (response to [a])
> Case Distinguished by δ: *Data General*
> (π in case disclosed to 6000 outsiders;
> $6000 \ll 12000$ disclosures in cfs.)

HYPO knows from the claim lattice and the range information about the *Disclose-Secrets* dimension, that *Data General* presents a stronger case because π disclosed the confidential information to fewer outsiders.

The **second** method involves focussing on facts associated with *unshared* dimensions that helped a party (π or δ) in the cited case or hurt the corresponding party in the cfs. Suppose the *Telex* case were cited on behalf of plaintiff as *Side 1*:

> [c] ↪ For *Side 1*: (point) π wins.
> Case Cited: *Telex*.

In Figure 3, *Telex* is a *potential mopc*, that is, a case close to the root, some of whose applicable dimensions are only near-misses with respect to the cfs. HYPO distinguishes *Telex* on behalf of defendant as *Side 2* by pointing out the facts that help π in *Telex* and that are associated with dimensions that are either near-misses (*Competitive-Advantage*) or inapplicable (*Bribe-Employee*) to the cfs:

> [d] ↩ For *Side 2*: (response to [c])
> Case Distinguished by δ: *Telex*
> (π in case gained competitive advantage;
> δ in case bribed π's employees.)

The **third** method of distinguishing involves finding a case, favorable to the responding side, whose *overall factual difference* from the cfs is less than that of the cited case. A mopc, for example, distinguishes all pro-opponent cases in any successor nodes of the claim lattice because it is more on point. (A mopc does not distinguish opponent's cases in other branches of the lattice – that would be like comparing apples and oranges.) If one were starting the argument from the other side and cited the *Midland-Ross* and *Yokana* cases on behalf of defendant as *Side 1*:

> [e] ↪ For *Side 1*: (point) δ wins.
> Cases Cited: *Midland-Ross, Yokana*
> (π in cfs and cited cases disclosed secrets to outsiders.),

HYPO responds by pointing out the pro-π strength that *Data General* shares with the cfs but which is missing from *Midland-Ross* and *Yokana*, namely that the disclosures were subject to restrictions to maintain confidentiality (a feature captured by the *Restricted-Disclose* dimension that applies to the cfs and *Data General* but not to *Midland-Ross* or *Yokana*). In other words, HYPO's response for plaintiff as *side 2* is:

[f] $\hookrightarrow$ For *Side 2*: (response to [e])
Cases Distinguished by π: *Midland-Ross*, *Yokana*
(Although π in cases and cfs made disclosures,
disclosures in cases were not on confidential basis;
Data General is more on point.)

HYPO uses the knowledge of how a case may be distinguished to prompt the user to find new factual differences and new justifications that would strengthen or weaken the argument. For example, HYPO uses the distinction between the *Telex* case and the cfs elucidated in response [d] to pose a hypothetical variant of the cfs in which the factual difference between the two is reduced [Rissland & Ashley, 1986; Ashley & Rissland, 1987]. The hypothetical prompts the user to investigate whether π in the cfs gained a competitive advantage or δ in the cfs bribed π's employees, in which case the *Telex* case would become a powerful precedent for the π that could not be so readily distinguished.

6. Evaluating HYPO's Model of the Distinguisher's Wedge

HYPO's model allows it to distinguish cases in a manner similar to what is actually done in court opinions in cases involving similar issues to our cfs. In *Mixing Equipment Co. v. Philadelphia Gear, Inc.*, 436 F.2d 1308, 1315 (3d Cir., 1971), the court distinguished the *Yokana* case as follows:

[Another case] and *Midland-Ross Corp. v. Yokana*, 293 F. 2d 411 (3 Cir. 1961) cited by appellants are inapposite. They involve situations in which restrictive covenants had not been utilized by the former employer.

In *Data General Corp. v. Digital Computer Controls, Inc.*, 357 A.2d 105, 109 (Del. Ch., 1975), the real case on which the cfs is based, the court took pains to point out, with respect to the drawings that had been disclosed to customers:

Such drawings bore a proprietary notice or legend, and the machine itself was accompanied by a ...confidentiality agreement limiting the use of such drawings to maintenance, as opposed to manufacture, which was stated to be forbidden without plaintiff's consent in writing.

In *National Rejectors, Inc. v. Trieman*, 409 S.W. 2d 1, 40-42 (Sup. Ct. Mo., 1966) the court said:

[W]e do find some significant parallels between the facts of this case and those of *Midland-Ross Corporation v. Yokana*, (D.C. N.J.), 185 F Supp. 594. ...What was lacking in *Yokana* as in this case, was any evidence that, prior to defendant's competition,

743

plaintiff considered the information which Yokana sought to use trade secrets. The court pointed out that plaintiff's blueprints in *Midland-Ross* were furnished plaintiff's suppliers and customers and potential customers. The court found an absence of precautions on the part of plaintiff to keep secret information regarding its machines.

7. Extending HYPO's Model of the Wedge

HYPO's model of distinguishing does not yet support all of the kinds of disputes that attorneys have about justifications. Attorneys argue about the significance of a court's holding in a prior case in more abstract terms, using predicates that obtain wide currency in the analysis of particular claims (e.g., what attorneys would call the *elements* of a claim, generalized statements that purport to define the necessary requirements of a claim.) Consider, for example, alternative ways for defendant's attorney to state the response in [b]:

> *Data General* does not help the plaintiff. With 12000 disclosures, plaintiff doesn't *have* a secret to protect anymore, regardless of whether the disclosees agreed to maintain confidentiality or not.

More interestingly, the attorney might address the court.

> If you hold in favor of a plaintiff who has disclosed its "secret" to 12000 outsiders then, regardless of whether they have agreed to maintain confidentiality or not, you are effectively doing away with the requirement imposed in *Midland Ross* and *Yokana* that trade secrets be secret.

This kind of response (reminiscent of the "I *am* a teenager!" response) is common enough in the law. HYPO avoids representing detailed definitions of predicates used in justifications like "secret" so as not to compound the inference control problem. Compare this to the classic rule-based approach of Waterman and Peterson [1981] who would use ever more refined rules to define legal predicates. As Gardner points out, there is no way in jurisprudence to specify logical definitions from which it is possible to deduce whether the predicate is satisfied [Gardner, 1984]. Gardner proposed to use cases to resolve "hard" issues about the meanings of predicates, but her main approach and her "cases" were very rule-like.

HYPO's model does, however, point to a middle ground for providing predicates with a kind of operational meaning, not in terms of a logical definition, but in terms of the boundaries of fact situations to which the predicate has been held to apply or not and knowledge of how those boundaries may be stretched. HYPO does know, for example, that *Data General* is an extreme case for plaintiffs in some sense. It is a *boundary* case for plaintiff along the *Disclose-Secrets* dimension, the weakest case along the dimension that the plaintiff still won. A boundary case may or may not be as on point as a mopc, but it is still useful as a precedent. A boundary case may be used to convince a court that it is not making new law by showing that, even if the court *were* to rule for the plaintiff in the cfs, the cfs would still not be the *worst* case, at least in some sense, that a plaintiff has won.

Of course, the meaning of a predicate in a justification like "secret" is not one-dimensional but then neither is the meaning of the claim of trade secrets misappropriation. HYPO represents the latter by associating various claims with clusters of dimensions. In a sense, the cases indexed by the dimensions scope out the boundaries of fact situations that have or have not been deemed to present winning trade secrets claims. In a similar way, HYPO could "tag" certain important predicates such as the elements of a claim, with cases that scope out the boundaries of fact situations that have or have not been deemed to satisfy the predicate. HYPO would then know what kinds of factual changes effect not only the strength of the claim, but also the meaning of the predicates. Whatever a "secret" is, HYPO would know what factual circumstances would make something more or less of one.

8. Conclusion

In this paper we have described the Distinguisher's Wedge, a tool for responding to an argument that two cases are alike by asserting distinctions, that is, factual differences and justifications why the differences matter. We have shown how a computational model of the wedge can be incorporated into a Case-Based Reasoning (CBR) system that makes and responds to arguments using precedential justifications. We have described how the HYPO program realizes this model and have suggested an extension of the model useful in generating more sophisticated disputes about the terms of justifications.

9. Figures

Plaintiff Data General (π), who developed and marketed the Nova 1200 minicomputer, complained that defendant Digitial (δ) developed a competing minicomputer, the D–116, by misappropriating π's trade secrets. Specifically π complained that δ copied π's drawings of the design of the Nova 1200 and used them to design the substantially identical D–116. The drawings appeared in a maintenance manual that π distributed to 12,000 customers who purchased the Nova 1200. The drawings and manual contained a legend that prohibited their copying except by written permission of π.

Figure 1: Current Fact Situation (cfs) based on *Data General Corp. v. Digital Computer Controls, Inc.*

Applicable Factual Predicates:
exists-corporate-claimant, exists-confidential-info, exists-disclosures ...

Applicable Dimensions: *Disclose-Secrets, Restricted-Disclose*

Near-Miss Dimensions:
Competitive-Advantage, Vertical-Knowledge

Potential Claims: Trade Secrets Misappropriation

Relevant CKB cites: See claim lattice, Figure 3

Figure 2: Case-Analysis-Record for CFS

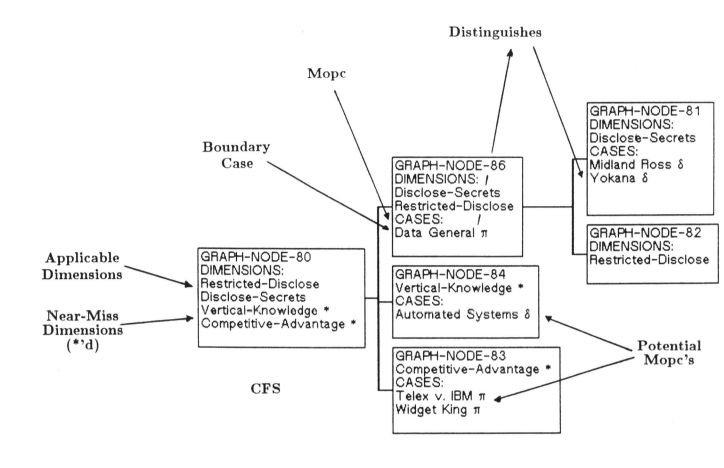

The root node represents the cfs and its D-list. (Dimensions that are near-misses for *cfs* have *'s.) Successor nodes contain pro-plaintiff (π) or pro-defendant (δ) cases, involving trade secrets misappropiation claims, that are on point to cfs. Nodes closest to root that do not have near-miss dimensions contain mopc's; otherwise they may contain **potential mopc**'s. Leaf nodes are least-on-point. Each major branch of lattice that contains mopc's represents one way of arguing the cfs. Mopc's **distinguish** cases in successor nodes. **Boundary** cases are examples of extremes along particular dimensions. Hypothetical **hybrid mopc**'s combine features of different mopc's that hold for π and δ. Potential mopc's suggest fruitful hypothetical variants of cfs.

Figure 3: A Claim Lattice.

References

[1] Kevin D. Ashley. *Modelling Legal Argument: Reasoning with Cases and Hypotheticals – A Thesis Proposal*. Project Memo 10, The COUNSELOR Project, Department of Computer and Information Science, University of Massachusetts, 1986.

[2] Kevin D. Ashley and Edwina L. Rissland. Creating Neighborhoods of Cases, Projections Through a Case Space. Submitted: IJCAI-87, 1987.

[3] A. vdL. Gardner. *An Artificial Intelligence Approach to Legal Reasoning*. PhD thesis, Department of Computer Science, Stanford University, 1984.

[4] I. Lakatos. *Proofs and Refutations*. Cambridge University Press, London, 1976.

[5] Edward H. Levi. *An Introduction to Legal Reasoning*. University of Chicago Press, 1949.

[6] Edwina L. Rissland and Kevin D. Ashley. HYPO: A Case-Based Reasoning System. Submitted: IJCAI-87, 1987.

[7] Edwina L. Rissland and Kevin D. Ashley. Hypotheticals as Heuristic Device. In *Proceedings of the Fifth National Conference on Artificial Intelligence*, American Association for Artificial Intelligence, August 1986. Philadelphia, PA.

[8] D. A. Waterman and M. Peterson. *Models of Legal Decisionmaking*. Technical Report R-2717-1CJ, The Rand Corporation, Santa Monica, CA, 1981.

Belief Revision and Induction

Donald Rose
Pat Langley

Irvine Computational Intelligence Project
Department of Information & Computer Science
University of California, Irvine CA 92717 USA

Abstract

This paper describes how inductively produced generalizations can influence the process of belief revision, drawing examples from a computational model of scientific discovery called REVOLVER. This system constructs componential models in chemistry, using techniques from truth maintenance systems to resolve inconsistencies that arise in the course of model formulation. The latter process involves reinterpreting observations (premises) given to the system and selecting the best of several plausible revisions to make. We will see how generalizations aid in such decisions. The choice is made by considering three main factors: the number of models each premise supports, the number of premises supporting the generalized reaction, and whether a proposed revision to that premise matches any predictions made by any generalizations. Based on these factors, a cost is assigned to each premise being considered for revision; the hypothesis (set of revisions) having the lowest cost is chosen as best, and its revisions are carried out. By viewing generalized premise reactions as a paradigm, we will argue that the revision process of REVOLVER models how scientific paradigms shift over time.

Introduction

In this paper, we discuss three main topics: how to form simple scientific theories, how to revise theories in order to account for new information, and how empirical generalizations can help to direct that revision process. In earlier papers, we have discussed STAHLp (Rose & Langley, 1986a, 1986b), a computer program that discovers explanatory models of chemical substances. In this respect it was similar to the STAHL system (Zytkow & Simon, 1986), which also constructed such models. However, unlike its predecessor, STAHLp featured a unified mechanism for reinterpreting its given observations when inconsistencies arose.

Yet STAHLp itself had a number of limitations. One of the most important is the need to take generalizations into account during the belief revision process. For instance, one should recognize when revising a premise will affect strongly held generalizations – i.e., those supported by many observations. Taken together, a set of generalizations can be viewed as a paradigm in the sense of Kuhn (1970). For example, many of STAHLp's runs involved observations associated with two main paradigms of 18th century chemistry: phlogiston theory and oxygen theory. The phlogiston framework, which came first historically, was based on the assumption that burning substances emitted a substance (phlogiston) during combustion. Oxygen theory took an opposing view of this process, stating that when a substance burns, it gains another substance (oxygen) in the process.

Historically, scientists like Lavoisier used generalizations to argue for their paradigm (e.g., oxygen theory) and to reinterpret observations made by supporters of competing

748

paradigms (e.g., phlogiston theory). Gradually, predictions made from the general reactions summarizing the oxygen paradigm were confirmed by new experiments, many of which were proposed after the observations of the phlogiston paradigm were reinterpreted. Later in the paper, we describe REVOLVER, a model of scientific theory formation that takes such inductive generalizations into account during its belief revision process. But first, let us recount the earlier work on STAHLp.

STAHLp: Scientific Discovery and Belief Revision

As we have mentioned, STAHLp constructed componential models based on many kinds of observations. For example, suppose the system is given initial beliefs from phlogiston theory: charcoal and calx-of-iron (known as iron oxide today) react to form iron and ash, and charcoal decomposes into phlogiston and ash. In shorter notation, the premises are {CI Ch} → {I Ash} and {Ch} → {Ph Ash}. The program would first infer the components of charcoal (Ch = {Ph Ash}), then substitute its components into the first reaction, yielding {CI Ph Ash} → {I Ash}. Cancelling ash from both sides yields {CI Ph} → {I}, and the system would now infer a model for iron (I = {CI Ph}).

Using this method, STAHLp constructed many componential models, replicating several episodes from the history of science. However, in the process of discovering such componential models, inconsistencies can arise. This occurs when the premises leading to these beliefs are themselves mutually inconsistent; either specific observations may be faulty, or groups of premises cannot be believed simultaneously. In both cases, some observations must be reinterpreted in order to arrive at a consensus. In an attempt to model how scientists reinterpret their observations when confronted with inconsistencies, STAHLp used belief revision techniques based on those of truth maintenance systems (Doyle, 1979; de Kleer, 1984).

Let us look at an example of how STAHLp handles the task of reinterpreting (i.e., revising) its premises. The first premise given to the system is again from phlogiston theory: the belief that mercury decomposes into calx-of-mercury and phlogiston ({M} → {CM Ph}). The model M = {CM Ph} is then inferred. Next STAHLp is given a second premise, one which embodies oxygen theory: {M O} → {CM}. Substituting mercury's components into the second premise yields {CM Ph O} → {CM}, and cancelling CM from both sides of this transformed reaction results in {Ph O} → {}. This is an inconsistent reaction, because it has inputs but no outputs.

At this point STAHLp invokes belief revision to find the premises that caused this error, propose revisions to those premises, and implement the best set of revisions. Each proposed revision to a premise can be viewed as a reinterpretation of the observation encapsulated by that premise. In our example, the system proposes four sets of revisions (hypotheses), each of which would remove the inconsistent reaction:

(1) Premise 2: outputs really had Ph and O;
(2) Premise 2: outputs really had Ph, inputs really had no O;
(3) Premise 1: inputs really had O, outputs really had no Ph;
(4) Premise 1: outputs really had no Ph; Premise 2: inputs really had no O.

Now the system must evaluate each hypothesis. STAHLp used one heuristic to drive its evaluation: *prefer the revision of premises that support the least number of models.* That is, the system tried to reach a consensus by altering the current theory (set of models plus the inconsistency) in the least drastic way. In our example, the cost of revising premise 1 is **2** because that premise supports the model of mercury and the inconsistency. In contrast, premise 2 has a cost of **1** because it supports only the inconsistency. Since the cost of each hypothesis is the cost of its suggested revisions, the four hypotheses have a cost of **1, 1, 2** and **2**, respectively. The system then selects the best (lowest cost) hypothesis – in this case, either the first or second hypothesis. Either set of revisions results in removal of the inconsistency when inferencing begins again, and thus the premises will be mutually consistent.

REVOLVER: Using Generalizations to Influence Belief Revision

We have seen that the STAHLp program modelled an important aspect of scientific discovery: the need to reinterpret one's observations when conflicts cause an inconsistent theory to be formed. While STAHLp's successes were significant, its belief revision process left no place for generalizations like those used by Lavoisier. In response to this limitation, we are integrating inductive reasoning into REVOLVER, a new model of scientific theory formation. This system will be able to use generalizations as part of the belief revision process and, ultimately, to formulate these generalizations on its own initiative.

The use of generalization in REVOLVER takes the form of two new heuristics, incorporated into the evaluation function that decides which revisions to make during belief revision. New heuristic (1) is used to *prefer revision to premises that support relatively weak generalized beliefs.* For example, when considering which of a set of premises to revise, REVOLVER would change the premise that led to the generalization having the least number of supporting premises, all other factors being equal. When selecting among candidate revisions, new heuristic (2) is used to *prefer revisions that confirm predictions made by strong generalized beliefs.* For example, if plausible revisions have been generated, and only one of them matches a prediction made by some generalization, then REVOLVER would select that revision, all other factors being equal. If there are several such matches, the system would select the revision matching the prediction that is part of the most heavily supported generalization.

Let us take a closer look at how these rules will be incorporated into REVOLVER. The method of detecting inconsistencies and generating plausible revisions will remain the same; only the function used to evaluate the revisions will change. The new evaluation function selects revisions of premises which support few models and generalizations, and which match a prediction if possible. Ignoring prediction matching for the moment, the new evaluation function computes the cost for each proposed revision by adding the number of models supported by the premise to be revised, plus the *total* number of premises supporting each generalization to which that premise lends support. The higher this number for a given premise, the more damage would be done to the belief system by revising that premise. However, if revising a premise would match a prediction made by a generalization, then the cost would be decreased, indicating that the revision is more desirable.

To illustrate how these new rules will be used in REVOLVER, let us reanalyze our previous example, taking into account the above changes. The previous example only involved two premises: {M} → {CM Ph} (premise 1) and {M O} → {CM} (premise 2).

Suppose two new premises are added to the system: $\{I\} \rightarrow \{CI \ Ph\}$ (premise 3) and $\{I \ O\} \rightarrow \{CI\}$ (premise 4), where I represents iron and CI represents calx-of-iron. At this point, the system can form two general reactions: premises 1 and 3 lead to $\{X\} \rightarrow \{CX \ Ph\}$, while premises 2 and 4 lead to $\{Y \ O\} \rightarrow \{CY\}$. The new classes X and Y represent those substances obeying the general reactions described (in this example, both classes contain M and I). The first generalization represents phlogiston theory: when any combustible burns, the calx of that substance (an element) remains and phlogiston is emitted. The second generalization represents oxygen theory: when any combustible burns, oxygen is gained and the calx of that substance (a compound) remains.

After this inductive step ends, standard deduction takes place. REVOLVER follows the same inference path as before, arriving at $M = \{CM \ Ph\}$ and then at the inconsistent reaction $\{Ph \ O\} \rightarrow \{\}$. The sets of revisions (hypotheses) proposed are also the same. However, the cost assigned to each is now different. Let us reexamine the four hypotheses proposed by both STAHLp and REVOLVER, along with the costs assigned by each system. Each of the first two hypotheses involved changing only premise 2 ($\{M \ O\} \rightarrow \{CM\}$). While STAHLp would assign each a cost of 1 (because premise 2 only supports the inconsistency), the new REVOLVER would assign each hypothesis a cost of **3**, since premise 2 also supports the generalization $\{X \ O\} \rightarrow \{CX\}$, which has two premises as support. The third hypothesis involves changing premise 1 ($\{M\} \rightarrow \{CM \ Ph\}$). While STAHLp would assign a cost of 2 (because premise 1 supports one model plus the inconsistency), REVOLVER would assign a cost of 4, since premise 1 also supports the generalization $\{X\} \rightarrow \{CX \ Ph\}$, which has two premises as support. The fourth hypothesis involves changing both premise 1 and premise 2. STAHLp would again assign a cost of 2, since premise 1 and 2 together support one model plus the inconsistency; in contrast, REVOLVER would assign a cost of **6**, since each premise supports a generalization that has two premises as support.

To summarize, STAHLp's hypotheses had costs of 1, 1, 2 and 2, respectively; those of REVOLVER had costs of **3, 3, 4** and **6**. Note that the first two hypotheses will be considered the best by both systems, but that the last hypothesis is clearly the worst in the view of REVOLVER, since it involves premise changes that would affect two generalizations (all others would impact only one generalization). If we continue altering this example by adding more premises, the choice of best hypothesis will also become different between the two systems. In particular, consider the addition of another premise that fits oxygen theory. This would mean that its generalization ($\{X \ O\} \rightarrow \{CX\}$) would now have three supporting premises, and thus the hypothesis costs would now become 4, 4, 4 and 7, respectively. Note that three hypotheses now tie for best; the new entry is the third hypothesis, which suggests revising only premise 1 – a belief from phlogiston theory.

In other words, adding more support to the oxygen theory generalization makes revision of the phlogiston theory premise more plausible. This trend continues further if we add yet another oxygen theory premise; this increases the support of its associated generalization to four premises. The new hypothesis costs would thus be **5, 5, 4** and **8**, respectively; note that the third hypothesis is now the sole best choice. In short, as the general belief embodying oxygen theory gained in strength, it became less desirable to revise premise 2 (which supports it), and thus more desirable to revise premise 1 (which supports the general belief embodying phlogiston theory).

Discussion

While the previous example involved induction on a relatively small number of premises, one can envision examples involving large systems of beliefs, where the generalizations embody laws supported by substantial observational evidence. If we look at each law (plus its supporting premises) as a paradigm and note that competing paradigms may result from different subsets of the premises, we feel that our new heuristics relating induction and belief revision bring us closer to modelling how paradigms shift over time. In this view, older paradigms would usually consist of well-supported generalizations (i.e., laws that summarize many observations), while new paradigms would consist of generalizations having only a few observations as supporting evidence. Initially, new heuristic (1) – preferring revisions to premises supporting weak generalizations – would tend to protect older paradigms. We claim that this is a plausible model of how science normally proceeds; old paradigms tend to become entrenched and require a steady accumulation of negative evidence to overthrow them.

This negative evidence comes as more observations are gathered that fit the generalization for the competing paradigm. In this manner, the competing paradigm gains strength and each of its premises becomes less vulnerable to revision. Coupled with this effect are the effects of new heuristic (2), where predictions made by the new paradigm are confirmed by revised premises. Each confirmed prediction can now add its support to the new paradigm, which in turn makes each supporting premise less vulnerable to revision. In short, our two induction heuristics reflect two directions in which paradigms can shift; heuristic (2) tends to build support for newer generalizations having little confirmed support but many predictions, while heuristic (1) tends to retain support for older generalizations having firm support. In our future research, we plan to use this general approach to model historical shifts in scientific paradigms, in particular the shift from phlogiston theory to oxygen theory.

Acknowledgments

This research was supported by Contract N00014–84–K–0345 from the Information Sciences Division, Office of Naval Research. We would like to thank Randy Jones, Bernd Nordhausen, and Paul O'Rorke for discussions that helped us form the ideas presented in this paper.

References

de Kleer, J. (1984). Choices without backtracking. *Proceedings of the Fourth National Conference on Artificial Intelligence* (pp. 79–85). Austin, TX: Morgan Kaufmann.

Doyle, J. (1979). A truth maintenance system. *Artificial Intelligence, 12*, 231–272.

Kuhn, T. S., (1970). *The structure of scientific revolutions.* Chicago, IL: University of Chicago Press.

Rose, D., & Langley, P. (1986a). STAHLp: Belief revision in scientific discovery. *Proceedings of the Fifth National Conference on Artificial Intelligence* (pp. 528–532). Philadelphia, PA: Morgan Kaufmann.

Rose, D., & Langley, P. (1986b). Chemical discovery as belief revision. *Machine Learning, 1*, 423–451.

Zytkow, J. M., & Simon, H. A. (1986). A theory of historical discovery: The construction of componential models. *Machine Learning, 1*, 107–136.

HOT COGNITION:

MECHANISMS FOR MOTIVATED INFERENCE

Paul Thagard and Ziva Kunda,
Psychology Department,
Princeton University,
Princeton, NJ 08544.

Keywords: motivation, inference, computation, generalization.

Abstract

We present an implemented computational theory of motivated inference intended to account for a variety of experimental results. People make motivated inferences when their conclusions are biased by their general motives or goals. Our theory postulates four elements to account for such biasing. (1) A representation of the self, including attributes and motives. (2) A mechanism for evaluating the relevance of a potential conclusion to the motives of the self. (3) Mechanisms for motivated memory search to retrieve desired conceptions of the self and evidence supporting desired conclusions. (4) Inference rules with parameters that can be adjusted to encourage desired inferences and impede undesired ones.

1. INTRODUCTION

Pascal (1966) said that the heart has its reasons that reason does not know. But relatively little attention in cognitive science has been paid to "hot" cognition involving motivation and affect, as opposed to "cold" cognition involving problem solving, learning, and so on. Phenomena of motivated inference have been discussed by philosophers such as Fingarette (1969) and Haight (1980), and have been investigated experimentally by social psychologists (for reviews see Greenwald, 1980; Wicklund & Brehm, 1976). But aside from the path-breaking study by Abelson (1963), there has been little investigation of *how* motivated inference takes place, of the mechanisms by which motivations of the self influence the conclusions that it reaches (cf. Hastie, 1983; Sorrentino & Higgins, 1986).

Here we propose a computational model of motivated inference that accounts for a variety of phenomena that have been investigated empirically. Our account builds on the PI model of cold cognition developed by Thagard and Holyoak (1985; Holland, Holyoak, Nisbett, and Thagard, 1986). PI is a computational model of problem solving and learning.[1] We will describe an extension of PI, Motiv-PI, in which inferences such as generalization can be biased by the motivations of the system.

2. ELEMENTS OF A THEORY OF MOTIVATED INFERENCE

2.1. Kinds of Motivated Inference.

Our model is designed to provide an integrated account of several kinds of motivated inference.

(a) *Motivated changes of self-conceptions.* How people see themselves may be influenced by how they would like to see themselves. Thus people led to believe that extraversion is predictive of academic success come to view themselves as more extraverted (Kunda & Santioso, 1986).

(b) *Motivated changes of theories about the world.* People tend to generate those theories about the causal determinants of events that are most likely to support their goals. Thus people tend to believe that their own attributes are more predictive of happy marriage than are other people's attributes (Kunda, 1987). This allows them to

---

[1] "PI" is short for "processes of induction" and is pronounced "pie".

maintain the belief that they will achieve a happy marriage.

Our model views both these phenomena as resulting from selective memory search among the wide array of relevant beliefs. Thus motivation helps to determine which self-conceptions will be accessed and which beliefs and what evidence pertaining to causal theories will be accessed.

(c) *Motivated changes of inferential rules*. Motivation affects the evaluation of evidence, so that individuals threatened by some evidence are less likely to believe it. Thus women who are heavy coffee drinkers were found to be particularly reluctant to believe that caffeine causes disease (Kunda, 1987). This reluctance may have been due to the application of particularly stringent inferential rules to the evaluation of the evidence, although to date there is no direct support for this notion. Our model assumes that motivation affects people's willingness to generalize by influencing the threshold required for generalization. Thus larger samples may be required to support generalizations that clash with one's goals.

(d) *Motivated changes of goals*. There is some evidence that when people realize that they are unlikely to obtain their goals, they diminish the importance of these goals to the self. Thus when individuals are outperformed by others on a given task, they come to consider that task as less important to their views of themselves (Tesser & Campbell, 1983). This allows them to maintain positive self-evaluation.

Thus motivation may affect inference by guiding the search among a wide array of potentially relevant beliefs about the world, other people, and the self, and by guiding the application of inferential rules.

2.2. Mechanisms.

We postulate four elements to describe the mechanisms underlying such motivated inferences:

1. *A representation of the self*. This should include *motives* of the self such as staying healthy and *attributes* of the self such as drinking coffee.

2. *A mechanism for evaluating the relevance of a potential conclusion to the motives of the self*. This will be an inference engine for tracing out the consequences of a potential conclusion and determining whether these have any impact on the motives of the self.

3. *Mechanisms for motivated memory search.*. We hypothesize that motivation affects how people retrieve memories and make inferences about their own characteristics and goals and about evidence for and against potential inductive conclusions.

4. *Mechanisms for adjusting the parameters of inference rules*. These will distort the normal inference rules to ensure that inferences favorable to the self are more likely to be made and that unfavorable inferences are less likely to be made.

The interactions of these elements is depicted in figure 1. To begin, inference of a potential conclusion that would result from the application of some inference rule is triggered. The key question is: does the inference rule license the conclusion? In motivated inference this is not simply a matter of seeing whether the available evidence and the inference rule warrant the conclusion, because both the activation of evidence and the parameters of the rule may be influenced by motivation. First the potential relevance of the conclusion to the self is checked. If the conclusion is irrelevant to the self, motivation plays no role in inference and the standard cold inference rule is applied. But if the conclusion has positive consequences for the motivations of the self, then application of the inference rule is enhanced to make it more likely that the inference gets made. Motivated memory search encourages the retrieval of information that will support the conclusion, and evidence thresholds for applying the inference rule are lowered. If, on the other hand, the potential conclusion has negative consequences for the motivations of the self, then application of the rule will be impeded. Sometimes, however, evidence will be so overwhelming that inference will go through anyway, which prompts re-evaluation of the importance to the self of the implicated goals.

3. A COMPUTATIONAL MODEL

The elements just described provide only a sketch of a theory of motivated inference. We need to know a lot more about the structures and processes postulated in Figure 1. The best means currently available for specifying such structures and processes is to develop a detailed computer program that has data structures corresponding to the representations postulated and procedures corresponding to the processes postulated. For a theory of motivated inference, we need an account of the self and of the mechanisms of relevance evaluation, memory

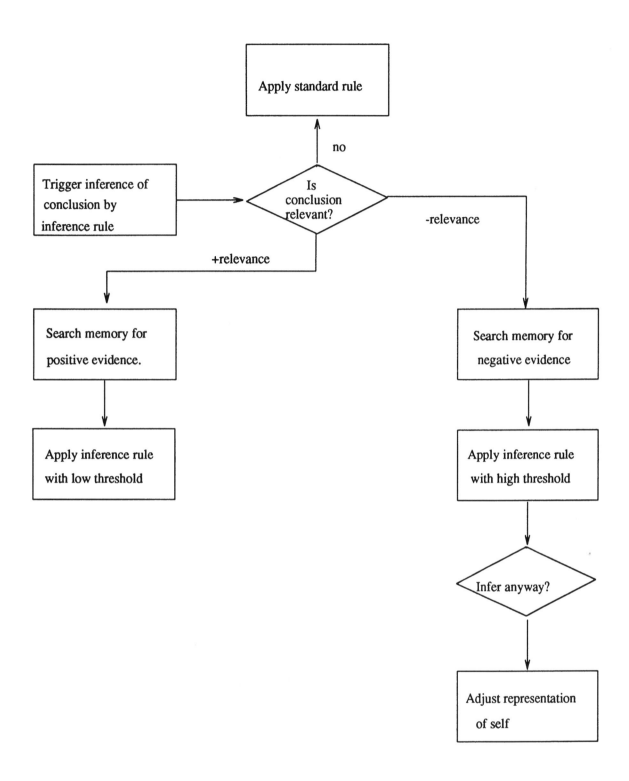

Figure 1: General model of motivated inference.

search, and distorted application of inference rules.

We now propose a computational model that builds on the PI model of problem solving and learning, a more detailed description of which can be found elsewhere (Holland, Holyoak, Nisbett, and Thagard, ch. 4). In PI, problem solving is a process of firing of rules and spreading activation of concepts. During problem solving, the current state of activation triggers various kinds of inductive inference, including generalization, abduction (inference to explanatory hypotheses), and concept formation.

3.1. The Self.

To expand PI into Motiv-PI, our model of motivated inference, we added a representation of the self to PI's set of knowledge structures which currently includes production rules, concepts (schemas), and messages (facts, propositions). Representation of the self has two crucial components: motives and attributes. The attributes of a self can include properties such as being a coffee drinker, a student, or a female. Its motives are its highest level goals, such as being healthy and happy. Figure 2 contains a simplified depiction of a self used in Motiv-PI's simulations. The activation of motives and attributes is variable, to allow differences in the degree to which they matter to the self at different times. For example, your motive to eat may be more active at one time than at another when you have just eaten, and your view of yourself as extraverted might be more active when you are at a party than when you are alone at your desk. Independent of activation, we postulate that motives have different priorities, so that reflection would tell you that being healthy is more important to you than being successful, even if at a particular moment you are overworking yourself to accomplish some career-related goal.

Self: Sandra

Attributes:
  Drinks_coffee:
    Importance: 0.6
    Activation: 0.4
  Female:
    Importance: 0.5
    Activation: 0.6
  Student:
    Importance: 0.1
    Activation: 0.6
  Smokes:
    Importance: 0.7
    Activation: 0.4
  Had_non_working_mother:
    Importance: 0.1
    Activation: 0.4
  Had_close_father:
    Importance: 0.1
    Activation: 0.6

Motives:
  Healthy:
    Priority: 0.9
    Activation: 0.4
  Rich:
    Priority: 0.7
    Activation: 0.8
  Happy:
    Priority: 0.9
    Activation: 0.5
  Successful_career:
    Priority: 0.6
    Activation: 0.7

Figure 2: A sample self.

Induction in Motiv-PI is triggered just as in PI, by the current state of activation of the problem-solving system. For example, the attempt to generalize that all A are B will be triggered by having the concepts of A and B simultaneously active as well as the information that something is both A and B. Providing examples of people who drink coffee and then develop fibrocystic disease would be sufficient to trigger the inference that coffee drinkers get the disease.

756

3.2. Determining Relevance of a Conclusion.

The second element of the theory of motivated inference mentioned in the last section is a means of determining the relevance of a potential conclusion to the self's motives. Motiv-PI does this by using the problem solving apparatus already present in PI. Determining relevance is a special case of problem solving, where the starting conditions of the problem consist of descriptions of the attributes of the self along with the potential conclusion, and the goals of the conclusion are the motives of the self. Normally, the point of attempting to solve a problem is to accomplish all the goals of the problem, but for determining relevance of a conclusion we want only to know which goals, i.e. which motives, have been accomplished. For example, to determine the relevance of the potential conclusion that Sandra has fibrocystic disease to the self Sandra, we set up a new problem whose starting conditions include the hypothetical proposition that she has the disease and whose goals are her motives:

> Problem:
>   Start: (has_fibrocystic_disease (Sandra) projected_true)
>   ... plus other attributes of Sandra.
>   Goals: (is_healthy (Sandra) true)
>   (is_happy (Sandra) true)
>   ... plus other motives of Sandra.

Solving the problem consists of seeing whether any of Sandra's motives could be affected by the hypothetical start. In the case above, the conclusion that Sandra had fibrocystic disease would lead to the consequence that she is not healthy, indicating a negative evaluation of the conclusion. Figure 3 provides an overview of this process. In many cases, a chain of deductions will be necessary to calculate relevance. For example, to assess the motivational relevance of doing well in graduate school, Motiv-PI infers that doing well in graduate school will lead to a good first job, and getting a good first job will help to lead to a successful career, so that doing well in graduate school is positively relevant to having a successful career. Another evaluation, this time of having a stable marriage, is based on inferences that a stable marriage leads one to feel more secure and therefore to be happier. Thus evaluating the relevance of a possible conclusion depends on being able to trace out its consequences. We view it as an advantage of our model that no special mechanism is postulated for doing this, since the normal process of problem solving is used. Just as PI solves problems by simulating the effects of possible actions, Motiv-PI determines relevance to the self by using information stored in concepts and rules to infer the consequences for the self a potential conclusion. The problem solver notes what motives are accomplished by a potential conclusion, yielding a numerical total that takes into account both the priority of the different goals accomplished and their degree of activation.

3.3. Motivated Memory Search.

People possess a broad array of different and sometimes contradictory beliefs about themselves, others, and the world. Motivation can determine which of these beliefs they will retrieve so as to support or hinder potential inductive conclusions. Motiv-PI implements such motivated use of memory very naturally because of the subgoaling mechanism in PI. PI's basic problem solving mechanism is forward chaining, matching production rules such as *If x is sociable, then x is extraverted* against messages such as *Sandra is sociable* to generate conclusions such as *Sandra is extraverted*. But if the system is motivated to show that Sandra is extraverted, it sets that as a goal and chains backwards to activate information about her being sociable. Further subgoaling occurs using an active rule that says that people who go to lots of parties are sociable, leading the system to ask if Sandra goes to lots of parties. Retrieving this information will then make possible the inference that Sandra is sociable and thus is extraverted. If the desired conclusion is reached, the search stops, so the system does not go on to find contradictory information that might imply that the self is introverted. This contradictory information will therefore not be accessed unless it is available to begin with or is activated as a side-effect of the main search. If instead the system were motivated to show that Sandra is introverted, it could do so by activating different rules, such as that shy people are introverted, and different facts, such as episodes where in fact Sandra was shy. Thus motivated memory retrieval occurs using the subgoaling which is an integral part of the problem solving operation of PI.

The same mechanisms govern motivated retrieval of evidence. In motivated generalization, you can be motivated to believe that all A's are B's, in which case you want to retrieve as many A's that are B's as possible. Or, if you are motivated not to form this generalization, then you will want to find examples of A's that are not B's. Motivated retrieval in the former case consists of giving PI's problem solver the goals of finding things that

757

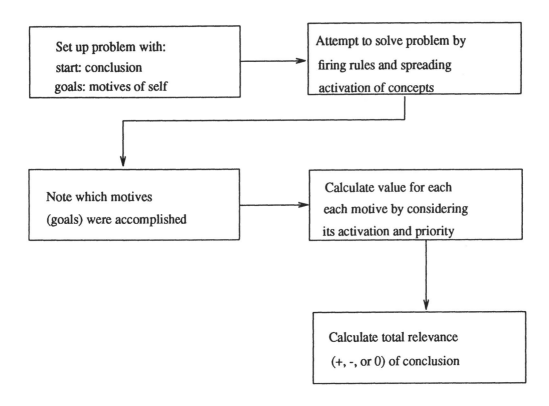

Figure 3: Calculating the relevance of a potential conclusion.

are A's and B's and storing this information away with the the concepts A and B for use as evidence in favor of the generalization. For example, if you want to infer that extraverts are successful, you will do a memory search to construct and retrieve examples of successful extraverts, which may involve the realization that some people whom you had not previously encoded in these terms do in fact fall under both categories.

In Motiv-PI, the extent of the motivated search for evidence is a function of how motivated the self is to form a particular conclusion. The amount of inferences PI's problem solver will employ to try to turn up desired information is a function of the extent to which the potential conclusion is relevant to the self, taking into account both the importance and the degree of activation of the relevant motives. Experimental results suggest that high levels of motivation are required for motivated inference to occur (Kunda, 1987).

3.4. Motivated Generalization.

The calculated motivational relevance of a potential conclusion and motivated memory search can be used by Motiv-PI to distort various inference processes. Here we will concentrate on generalization, since the kinds of inferences studied experimentally by Kunda fall most appropriately under that heading. In PI, generalization is done in accord with a theory developed by Thagard and Nisbett (1982) and tested experimentally by Nisbett, Krantz, Jepson and Kunda (1983). People's willingness to generalize from instances is a function both of the number of instances that provide evidence for the rule to be formed and of background knowledge about variability. You are more prone to generalize that all instances of a new kind of metal burn with a green flame than you are to generalize that all instances of a new kind of bird are blue, since you know that birds are more variable with respect to color than metals are with respect to combustion properties. PI generalizes that all A's are B only when it calculates that a combined measure of the number of instances of A's that are B's and the invariability of A's with respect to B's exceeds a given threshold. (For further discussion of generalization and variability, see Holland, Holyoak, Nisbett, and Thagard, 1986, ch. 8.)

Motiv-PI uses the same considerations when attempting to form generalizations, but motives influence the process in the ways summarized in Figure 4 and described below. A rule "All A are B" is typically relevant to a self if (1) the self has attribute A, and (2) B is motivationally relevant to the self -- the self desires or fears being B. For example, the rule "Women who drink coffee get fibrocystic disease" is only relevant to someone who is female, drinks coffee, and cares about being healthy.

When attempting to form the generalization "all A are B" Motiv-PI first determines whether being B would be relevant to the self. If there is no relevance, then it does generalization just as in PI, with the standard threshold for number of instances and variability. If being B is positively relevant to the self, leading to satisfaction of its motives, then Motiv-PI first attempts to show that the self is A, using the motivated memory search described above. For example, when considering the potential generalization by Sandra that all extraverts are successful, it first determines that being successful is positively relevant to Sandra, and then tries to show that Sandra is extraverted. Since PI simulates the parallelism of numerous rules firing and subgoaling at once, the following subgoaling chains occur simultaneously:

extravert   <- friendly <- has many friends (Sandra)

extravert   <- sociable <- goes to parties (Sandra)

extravert   <- outgoing <- talks to strangers (Sandra)

If B is relevant to the self and the search just described shows that the self is A, then the system fosters generalization that all A's are B's in two ways. First, it attempts to find as many examples of A's that are B's as possible, in order to get above the threshold for the number of instances required for generalization at a given level of variability. In the extravert/success example, the system searches for instances of individuals who are both extraverted and successful by setting itself the subgoals of finding things that fall under these categories. Second, the system adjusts the threshold of the inference rule used in generalization to require fewer instances given the calculated variability.

If being B is a property such as getting breast disease that has negative consequences for the self's motives, then Motiv-PI tries to show that the self is not A. For example, if generalization is triggered by examples of extraverts that failed, Motiv-PI attempts to show that the self is not extraverted. If the self is found nevertheless to be extraverted, then the system attempts in two ways to block generalization that extraverts fail. First, it does a memory search by subgoaling to try to find examples of non-failing extraverts. (PI's normal generalization mechanism blocks inference when such counterexamples are available.) Second, just as positive motivated generalization uses a lower threshold of number of instances and variability, in this negative case a higher threshold is used to impede generalization.

These mechanisms do not, however, guarantee that the undesired generalization will be blocked, since if sufficient evidence is found the generalization will be made nevertheless. In this case, Motiv-PI adjusts the representation of the self to make it less concerned with the bad consequences that derive from the conclusion that all A are B. This form of rationalization consists of deciding that perhaps being B is not so bad after all. In other words, Motive-PI reduces the priority of the motive that B affected. To continue the above example, after concluding that its personality will make it less likely to succeed, the system will decide that success is not all that important to it.

Currently, the variability calculation is not affected by motivation, but conceivably the memory search on which the calculation of variability is based could be substantially affected by motivation. Variability calculations depend on selection of appropriate reference classes. To generalize that all shreebles (a new kind of bird) are blue, PI calculates the background variability of birds with respect to color, but a more sophisticated program would pick reference classes less automatically. It might, for example, be appropriate to consider more specific classes, calculating the variability of tropical birds with respect to primary colors. Perhaps motivation plays a role in searching for reference classes to generate variability estimates that foster or impede generalization. More experimental studies are needed, however, to show whether motivation has an effect on variability calculations and instance retrieval.

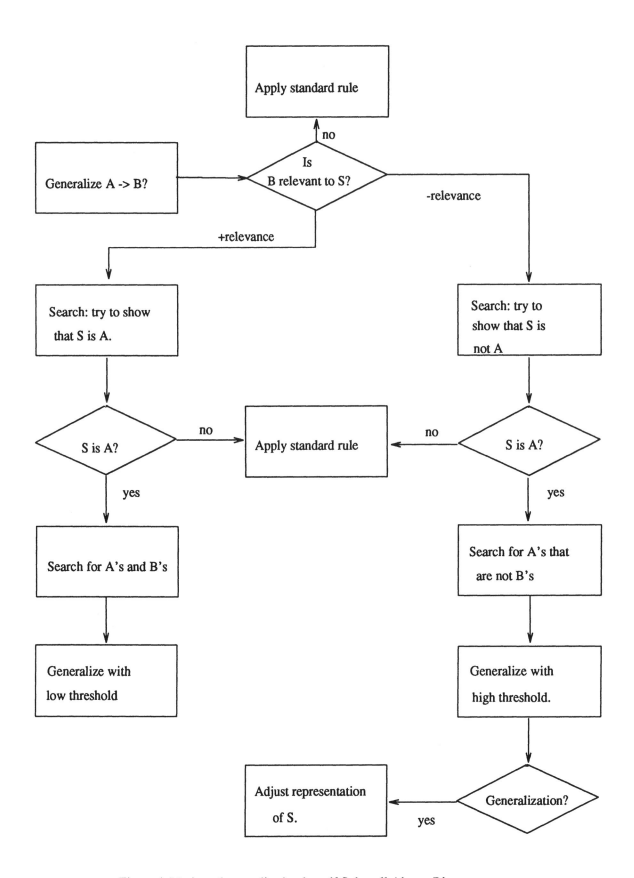

Figure 4: Motivated generalization by self S that all A's are B's.

3.5. Motivated Deduction and Hypothesis Evaluation

Other types of inference besides generalization may also be subject to motivational biases, although these have not yet been studied experimentally or implemented computationally. One might think that deduction is immune from motivational biases, since any conclusion that follows deductively from true premises has to be true. But in any realistic processing system, deduction has to be constrained pragmatically, since any system that made all possible deductive inferences would suffer a combinatorial explosion. In PI, what deductions get made during problem solving is a function of pragmatics such as the problem solving context, particularly the goals to be accomplished. Thus the memory search described above is a kind of motivated deduction.

PI also performs abduction, that is, formation of explanatory hypotheses. The simplest kind offers an explanation of a fact that something has a property B by hypothesizing, on the basis of a rule *All A are B* that it is A. For example, to explain why a friend is exhausted and dissipated, you might abduce that he or she is on drugs, since being on drugs can lead to dissipation. Evaluation of hypotheses formed by abduction is by inference to the best explanation, which PI performs by compiling lists of alternative hypotheses and facts to be explained, and evaluating which of the hypotheses most comprehensively and simply explains the evidence (Thagard, forthcoming, ch. 5). It is easy to see how motivation could distort inference to the best explanation: if you are motivated to accept a hypothesis, then you might be less thorough in searching for alternative hypotheses and more prepared to accept the hypothesis on less evidence. Indeed, it has been shown that the success of a liked person is attributed to the person's ability, whereas the success of a disliked person is attributed to the ease of the task (Regan, Strauss, & Fazio, 1974). For Motiv-PI, alterations could easily be made to bias inference to the best explanation.

3.6. Limitations of Motiv-PI.

Motiv-PI, including representation of the self, calculation of motivational consequences, motivated generalization, and rationalization is implemented in Common LISP and runs in conjunction with PI. A fuller implementation would add features such as the following:

1. A dynamic self, in which the degree of activation of its attributes and motives is updated by the program itself.

2. Motivated variability judgments.

3. Motivated inference to the best explanation.

4. More sophisticated rationalization by adjusting motive importance indirectly through motivated memory search: instead of merely reducing the priority of a motive, the system would try to retrieve evidence that would diminish the inferred importance of the motive.

5. Greater parallelism so that different motivated memory searches could be simulated as occurring at the same time.

4. COMPARISON WITH OTHER VIEWS

4.1. Computational Models of Motivation and Affect

Abelson (1963) proposed an interesting model of sentence evaluation that had some of the features we have discussed in connection with motivated inference. He described a system for cognitive balancing, in which a sentence that enters thought is evaluated and unbalanced sentences are subject to further processing before they are stored. For example, the sentence "My good friend is a murderer" is unbalanced because its subject and predicate get very different evaluations. Here rationalization consists of modifying the subject or predicate in some way to restore balance. Our account can be thought of as a way of achieving balance, or reducing dissonance, between the motives of the self and the inferences it makes, but the empirical and computational work described here are novel in that they shift the focus to the memory and inference processes through which balance is maintained.

We can only briefly mention some other computational studies of affect. Wegman (1985) offers a computational model of Freud's theories of abreaction and repression. Ortony (1986) has proposed a model of event evaluation which is somewhat similar to our account of evaluating the relevance of potential conclusions concerning the self. Dyer (1983) and Hovy (1986) have discussed the relevance of affect in text processing.

761

4.2. Philosophical Applications

Philosophers have paid much attention to phenomena closely related to motivated inference: weakness of will and self-deception (Davidson 1980, Fingarette 1969, Haight 1980, Martin 1985). One central question has been whether these phenomena are possible at all. On a simple view of the mind, it can be difficult to understand how an agent could act other than in its own interests or could deceive itself. However, the structures and processes used in Motiv-PI make it easy to see how weakness of will and self-deception can at least be possible. It is another question, that ought to be answered experimentally, whether people are actually subject to them.

We see self-deception as an extreme case of motivated inference leading to inconsistency. In self-deception, one makes an inference that at some level one knows to be false. If I infer that my finances are solid even though I know that bankruptcy is imminent, then I am guilty of deceiving myself. On some philosophical views, it is hard to see how such a contradiction could exist. But it is perfectly consistent with PI's representations and processes that a system be inconsistent without realizing it. Only a portion of the system's beliefs need be active at any time, so it is very possible to infer a belief that is active but inconsistent with another belief that is not active. According to Audi (1985), in typical cases of self-deception we have not only a motivated inference of a belief that contradicts an existing one, but also removal from consciousness of the original belief, which sounds a lot like Freudian repression. A repression mechanism could easily be added to Motiv-PI. The system could evaluate the relevance to the self of each active message (proposition) and de-activate the negative ones, but this seems much too strong: there is no reason to expect that all information will be active at a given time, so self-deception would seem to be possible without a repression mechanism. Motivated memory retrieval and inference suffice to produce the phenomenon.

On our view, weakness of will can be understood in terms of the difference between priority and activation of motives. Weakness of will occurs when a decision is motivated by desires that are more active than ones to which reflection gives a higher priority. To take an extreme case, consider a cocaine addict fighting a craving for the drug. For physiological reasons, the motive to get some cocaine is far more active than long-term motives such as being healthy, happy, and successful, so that even though the addict would on reflection give higher priority to the latter motives, the desire for cocaine wins out.

5. CONCLUSION

We have described mechanisms for motivated inference that have been worked out in enough detail to be implemented computationally. We close by sketching a practical motive for these investigations. Motivated inference may sometimes be harmless, or even beneficial, despite fostering false conclusions. For example, believing on little evidence that you will succeed at some difficult task may well improve your performance. But motivated inferences can also be dangerous, even life threatening. The smoker who fails to conclude that smoking contributes to cancer may pay for this by incurring the disease. We hope that the investigation of the processes underlying motivated inference will eventually lead to the development of methods for helping people avoid hazardous reasonings from the heart.

6. REFERENCES

Abelson, R. (1963). Computer simulation of "hot" cognition. In S. Tomkins (Ed.) *Computer Simulation of Personality*. New York: John Wiley and Sons. 277-298.

Audi, R. (1985). Self-deception and rationality. In M. Martin (Ed.) *Self-deception and self-understanding*. Lawrence, KS: University Press of Kansas. 169-194.

Davidson, D. (1980). *Essays on action and events*. Oxford: Oxford University Press.

Dyer, M. (1983). The role of affect in narratives. *Cognitive Science, 7*, 211-242.

Fingarette, H. (1969). *Self-deception*. London: Routledge and Kegan Paul.

Greenwald, A. A. (1980). The totalitarian ego: Fabrication and revision of personal history. *American Psychologist*, 35, 603-618.

Haight, M. (1980). *A study of self-deception*. Brighton, Sussex: The Harvester Press.

Hastie, R. (1983). Social inference. *Annual Review of Psychology*, 34, 511-542.

Holland, J., Holyoak, K., Nisbett, R., and Thagard, P. (1986). *Induction: Processes of inference, learning, and discovery*. Cambridge, MA: Bradford Books/MIT Press.

Hovy, E. (1986). Putting affect into text. *Proceedings of the Eighth Annual Conference of the Cognitive Science Society*. Hillsdale, NJ: Erlbaum. 669-675.

Kunda, Z. (1987). Motivation and inference: Self-serving generation and evaluation of causal theories. *Journal of Personality and Social Psychology*. In press.

Kunda, Z. & Santioso, R. (1986). Motivated changes in the self-concept. Unpublished manuscript, Princeton University.

Martin, M. (Ed.). *Self-deception and self-understanding*. Lawrence, KS: University Press of Kansas.

Nisbett, R. E., Krantz, D. H., Jepson, D., and Kunda, Z. (1983). The use of statistical heuristics in everyday inductive reasoning. *Psychological Review, 90*, 339-363.

Ortony, A. (1986). Unpublished manuscript. University of Illinois.

Pascal, B. (1966). *Pensées*. Harmondsworth, Middlesex: Penguin.

Regan, D. T., Strauss, E., & Fazio, R. (1974). Liking and the attribution process. *Journal of Experimental Social Psychology, 10*, 385-392.

Sorrentino, R. M. & Higgins, E. T. (1986). Motivation and cognition: Warming up to synergism. In R.M. Sorrentino and E.T. Higgins (Eds.), *Handbook of motivation and cognition*. New York: Guilford Press.

Tesser, A., & Campbell, J. (1983). Self-definition and self-evaluation maintenance. In J. Suls & A. Greenwald (Eds.), *Social psychological perspectives on the self* (Vol. 2). Hillsdale, NJ: Erlbaum.

Thagard, P., and Holyoak, K. (1985). Discovering the wave theory of sound: Induction in the context of problem solving. *Proceedings of the Ninth International Joint Conference on Artificial Intelligence*. Los Altos: Morgan Kaufmann. 610-612.

Thagard, P., and Nisbett, R. E. (1982). Variability and confirmation. *Philosophical Studies, 42*, 379-394.

Thagard, P. (forthcoming). *Computational philosophy of science*. Cambridge, MA: Bradford Books/MIT Press.

Wegman, C. (1985). *Psychoanalysis and cognitive theory*. Orlando, FL: Academic Press.

Wicklund, R. A. & Brehm, J. W. (1976). *Perspectives on cognitive dissonance*. Hillsdale, NJ: Erlbaum.

A Tale of Two Brains

-or-

The Sinistral Quasimodularity of Language

Thomas G. Bever
The University of Rochester

Caroline Carrithers
Johns Hopkins University

David J. Townsend
Montclair State College

ABSTRACT

Four experiments show that people differ strongly in the extent to which they
depend on linguistic structure during language comprehension.
Structure-dependent people are immediately affected by grammatical variables,
while structure-independent people are less affected by such variables. A
surprising population difference between the two types of people suggests a
genetic and neurological basis for the behavioral difference. All subjects
were right-handed. However, structure-dependent people report no left-handers
in their family, while structure-independent people do report left-handers in
their family. This suggests that the neurological organization for linguistic
ability in right handers with familial left-handedness, is more diffuse than
for right handers with no familial left-handedness. Other facts connect this
to a current hormonal theory of the ontogenesis of hemispheric asymmetries.

It is a common belief that there is a normal way of understanding
sentences, which is essentially the same for everybody. This presupposition
underlies the search for a single mechanism for sentence comprehension. It is
also a common belief that there is a normal neurological configuration for
language, at least among right-handed people. This presupposition underlies
the practice of taking cerebral asymmetries in normals and behavioral syndromes
in aphasics, to reveal the normal function of a universal configuration. The
research reported in this paper suggests that these assumptions are false:
individuals differ markedly in the was they process language, and the
difference is related to a genetic difference in neurological organization for
language.

Most linguistic theories distinguish two aspects of language: The
structural system, which concerns syntactic, semantic and phonological
knowledge; the conceptual system, which includes lexical reference and

764

conceptual knowledge of the world. Normal variation in language behavior can be defined in terms of variation in the degree to which a person depends on each of these systems during behavior. In this discussion, we focus on comprehension, in particular on differentiating those people who depend on structural features of the language from those who depend on more on non-structural knowledge of language.

The extent to which a subject depends on structure might be related to his familial handedness. There are obvious reasons to expect that left-handed people have a different neurological organization for language from right-handers. Accordingly, we have been keeping track of the personal and familial handedness of our subjects in psycholinguistic studies. Over the last decade, we have frequently noticed that 'mixed-background' right-handers, from left-handed families, respond to language stimuli differently from 'pure background' right-handers, from families with no left-handers. In particular, it seemed to us that during comprehension, 'mixed background' right handers are less sensitive to syntactic variables than 'pure background' right handers, but seem to have more facility with tasks involving single words. In the research reported here, we attended directly to handedness background, using it as the basis for isolating a 12-member group of what we thought would be "structure dependent" people from another 12-member group of what we thought would be "structure independent" people. Three experimental studies confirmed the preliminary differentiation that we made between the group of structure-dependent people and the group of structure-independent people, based on handedness background.

The first study used a percept which is affected by clause boundaries, the perceived location of a brief tone presented objectively during a two-clause sentence. In this paradigm, the listener hears the sentence with a tone in it and must be prepared to write the sentence down and report the location of the tone. On critical trials the listener is unexpectedly presented with a written version of the sentence. The listener locates the tone within a 3-4 word long "window" already marked on the sentence as in Figure 1 below. There were 3 kinds of response windows, one with the center on the word before the clause break, one in the clause break and one on the word after the clause break. The correct location for the tone was always in the center of the window. The serial position of the clause break was varied.

All listeners reported the correct tone location better when it was objectively in the break between the clauses, than in either the word preceding or following (Figure 2). Responses to "catch" trials, in which there was no objective tone at all, showed a smaller guessing bias in favor of the between-clause position. When the response patterns are corrected for this guessing bias, the mixed-background listeners showed no overall superiority for the clause break, while the pure-background listeners did (p<.05 by Fisher exact test on subjects, p<.05 by Wilcoxon matched pairs signed ranks test on materials). Apparently, since pure-background subjects are structure-dependent, they have relatively more attention available to listen for the tone between clauses: this suggests that they actively assign each clause a separate representation as they hear it, while mixed-background listeners do not.

We studied the form in which listeners immediately represent a clause with

765

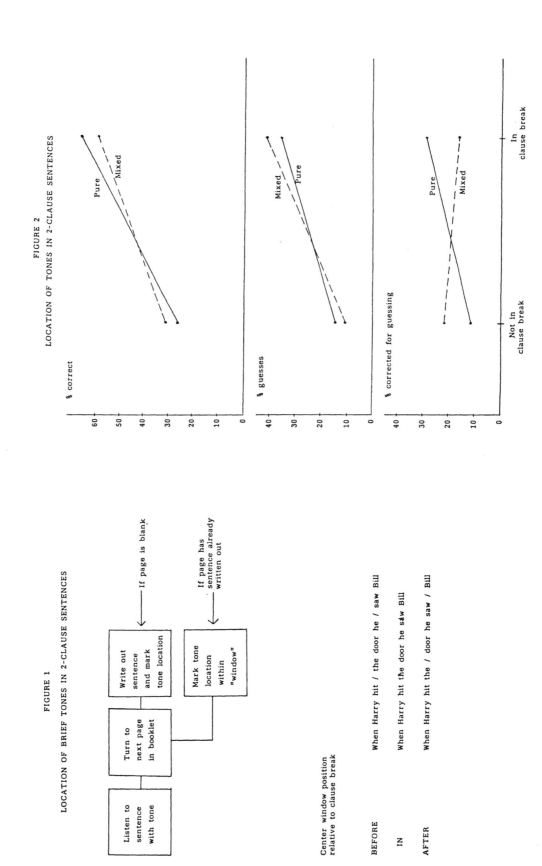

FIGURE 1

LOCATION OF BRIEF TONES IN 2-CLAUSE SENTENCES

FIGURE 2

LOCATION OF TONES IN 2-CLAUSE SENTENCES

a word-recognition technique commonly used in laboratory studies of language processing (Townsend and Bever, 1978). In this paradigm, a listener hears a word sequence, followed by a single probe word: the listener must respond verbally if the probe word was in the original sequence or not. (Table 3 outlines the paradigm). In our study, the critical word sequences were subordinate clauses (introduced by 'if' or 'though'), all spoken with an intonation pattern suggesting that they would continue with a following clause. Among the probe words were verb-particles and adverbs, which could occur in either a late or an early position in each sentence.

This task involves isolating a single word from the representation of a whole clause. We predicted that structure independent subjects would perform relatively well on this task, because, by hypothesis, they make more use of the reference of individual words in their processing. This prediction was confirmed for pure-background subjects: they responded positively to word-probes more slowly than did mixed background listeners (p<.01 By a Fisher exact test) (Figure 4). In English, phrase order is one clue to the thematic structure of a clause (e.g., agents precede the verb unless there are special morphemes indicating the reverse). Accordingly, a structure-dependent listener should retain a sequential representation of a clause, if s/he expects that the clause will be integrated with later material. In contrast, a structure-independent listener should retain a clause in an unordered, conceptual form. Consistent with this difference, pure-background listeners were sensitive to the serial position of the probe word in the fragment, and mixed-background listeners were not (the serial position effect was larger for pure-background subjects, p<.01 on a Fisher exact). Pure-background listeners responded more slowly to late probes than to early probes. This suggests that they scan their representation of the just-heard sentence fragment from beginning to the end, a self-terminating serial scanning pattern that occurs in many non-linguistic probe tasks. The lack of a serial position effect for concept-dependent listeners suggests that their representation and search of the sentence fragment are unordered.

We ran a complementary study with the same subjects and the same type of sentence material, except that the probe was now a 2-4 word phrase - the listener's task was to say whether the phrase was related in meaning to the sentence. (The order of presentation of the three tasks, tone-location, word-probe and phrase-probe was counterbalanced across the groups). In this study, we used both main and subordinate initial clause fragments, as shown in Figure 5. All listeners responded more quickly to main-clause probes than to subordinate-clause probes*. This follows from the general view that a main clause can be immediately processed for meaning while a lower-level representation of a subordinate clause must be retained, so that it can be integrated with the ultimate main clause. The distinction between main and subordinate clauses is structural, which explains why the response time difference is larger for pure-background listeners than for mixed-background listeners (p<.01 by Fisher exact) (Figure 6).

We also carried out a somewhat more naturalistic study of comprehension, which includes reading complete sentences and then answering questions about them. We used a self-paced word-by-word reading paradigm, currently in psycholinguistic fashion. Every time the subject presses a button on a

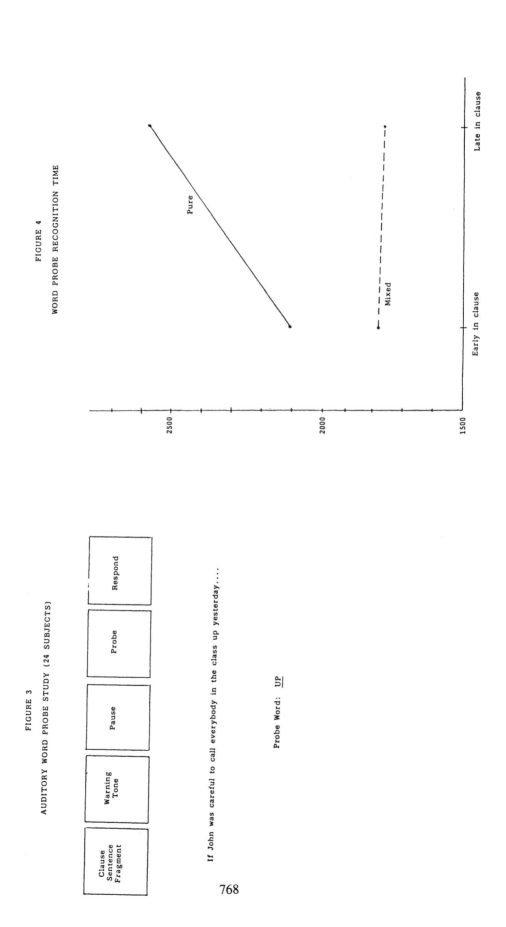

FIGURE 4

WORD PROBE RECOGNITION TIME

FIGURE 3

AUDITORY WORD PROBE STUDY (24 SUBJECTS)

If John was careful to call everybody in the class up yesterday.....

Probe Word: UP

768

computer keyboard, the next word appears in the same location on a video screen, wiping out the previous word. The final word of each sentence was indicated with a punctuation mark, the period. When finished reading the final word, the subject pressed the button again: this elicited a question about the sentence, which the subject answered.

Twenty-four new male subjects of each type of handedness background participated in this study. The results show that pure-background subjects pressed the button for each word 21% slower (483 milliseconds/word) than mixed-background subjects (399 milliseconds/word) (p<.001 by X-square). This finding is consistent with the relative slowness of this kind of person in the word-probe task. On our interpretation, this occurs because forcing a structure-dependent reader to read on a word-by-word basis interferes with the usual comprehension process, in a way that is not true for the structure-independent reader. (We should note that the pure-background readers were slightly better at answering the questions correctly. There was, however, no speed/accuracy trade-off across subjects; r=.04. Overall accuracy was about 80%).

Although they read each word more slowly, pure-background readers are relatively sensitive to structural information within single words as well as word order. We used two types of verbs to explore this. One type of verb was a simple transitive verb such as 'hit, see, love'. These verbs maintain the conventional overlap between grammatical object and thematic object. The other type of verb inverts that relation, such as 'frighten, upset, please': intuitively, the grammatical object of these verbs is the thematic agent of an intransitive form of the verb. The word 'Sam', in "John scared Sam", is the thematic subject of 'scare', in a way that he is not in the sentence "John hit Bill." This fact is reflected differently in different linguistic theories. But, however it is correctly represented, it is a thematic fact about the verb 'scare', not a conceptual fact about the activity of scaring or being scared. We expected thematically-inverse verbs to be comprehended more slowly than simple transitives, just because they violate the conventional relation between syntactis and thematic relations. In fact, final words of sentences with thematically-inverse verbs were read more slowly - as predicted, this effect was much larger for pure-background readers than mixed-background readers (p<.01 by X-square) (Figure 7). (This study also varied other aspects of structure - see Carrithers, 1986).

These results justified our preliminary differentiation of structure-dependent and structure-independent comprehenders. In each of four experiments, the pure-background people depended on aspects of the sentence structure more strongly than the mixed-background people. Consider again the population difference between these two kinds of listeners. Figure 8 lists some population variables that one might think could underly such differences in language processing style. In fact, the subject groups were tightly balanced on each of those variables. (We matched groups closely for SAT scores, using a SAT-yoked design for the first three studies - in the fourth study, the mean verbal SAT scores of the two groups were within 20 points. Right-handedness was determined by a score of at least 98% on a variant of the Oldfield handedness inventory.)

769

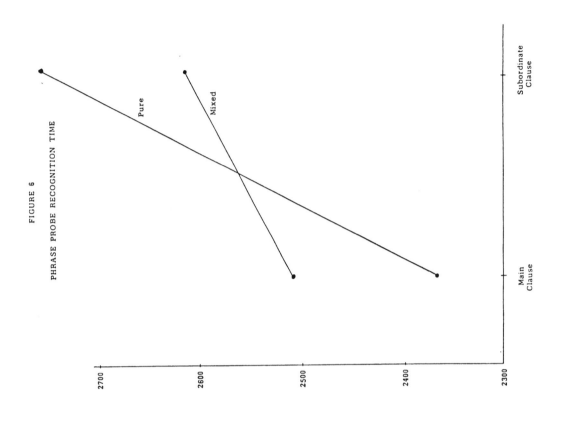

FIGURE 6

PHRASE PROBE RECOGNITION TIME

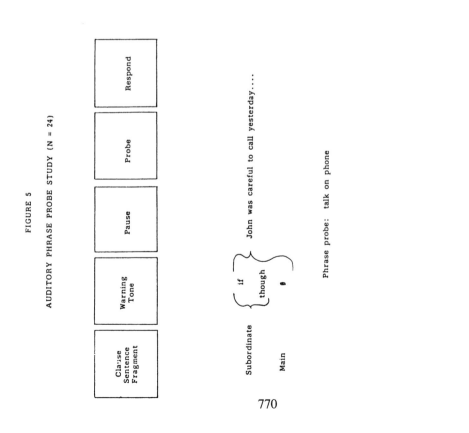

FIGURE 5

AUDITORY PHRASE PROBE STUDY (N = 24)

Phrase probe: talk on phone

770

There was only one consistent population difference between the subjects: The structure-dependent subjects had been chosen as reporting only right-handers in their family envelope, while the structure-independent subjects reported at least one left-hander in their family envelope. The envelope included siblings, parents, uncles or aunts and grandparents. At least one left hander in that envelope predicted that a person was structure independent. (With this envelop as a criterion, about 40% of college undergraduates have left-handers in their family. We counted only full biological relatives; any relative who was reported as "having been left-handed, but forced to become right-handed" was counted as left-handed, as were "ambidextrous" relatives.)

Consider some some morals from this result. First, it suggests that there really is more than one way to initiate understanding of a sentence. Some people rely more on structural representations, some on lexical and conceptual knowledge. The implications of this for interpreting existing psycholinguistic research is troubling: it might seem that experimental results would qualitatively change as a function of subjects' handedness background. Fortunately, we have not found that one type of subject reverses the effects shown by the other type. Rather, the groups are mutually relatively insensitive to the conceptual and structural variables. In general this may have added satistical noise to previous studies that did not control for handedness background, but it will not have caused qualitative eccentricity in all cases.

Our finding also emphasizes the importance of understanding how a person goes about processing language before he becomes aphasic. The individual variability among normals highlights the possibility of a "reduced-efficiency" theory of aphasia. On this theory, damage to a portion of the language neurology results in a reduction of the efficiency of all linguistic behavior. If all systems of linguistic behavior become less efficient in aphasia, then those that were prominent when the patient was normal will still be above behavioral threshold: the patient will appear to have those systems, and appear to have actually lost the other systems. Different configurations of aphasic symptoms will result, simply as a function of normal differences. In this regard, it is interesting to note that concept-dependent listeners have the same relative response pattern as agrammatic aphasics with respect to their relative sensitivity to conceptual information and phrase order.

Finally, we can consider specific hypotheses about why familial left-handedness leads to more reliance on extra-linguistic conceptual knowledge and less reliance on language structure. Luria (1954) pointed out that people with left-handers in their family have a larger chance than normal of having aphasia from a gunshot wound but also have a better chance than normal of recovering from it. This suggests that people with familial sinistrality have a relatively widespread neurological module for language - it is temporarily more vulnerable to a randomly located wound, but has more widespread reserves for recovery. People who have only righthanders in their background may have a tighter, more localized neurological module for language. This difference would offer an interpretation of why people with familial left-handedness are relatively sensitive to conceptual information. If such information is neurologically instantiated throughout the brain, then a larger area for

FIGURE 8

COMPARISON OF
STRUCTURE-DEPENDENT AND CONCEPT-DEPENDENT GROUPS

Age	College	College
Sex - Studies 1-3	6 male, 6 female	6 male, 6 female
Study 4	24 male	24 male
First Language	English	English
VSAT - Studies 1-3	450-700	450-700
		Yoked Design
		(see text)
Study 4	Mean=620	Mean=640
Handedness	Right-handed	Right-handed
Familial Handedness	Right-handed family	At least one left-hander in family

FIGURE 7

FINAL WORD DWELL TIME

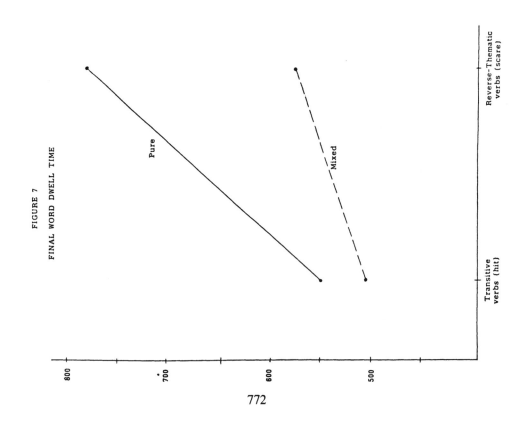

language would allow for more points of contact between language and other kinds of knowledge. Conversely, a small, localized language area would have to depend more on purely linguistic processes. Familial sinistrality may be a phenotypic marker that an individual has a genetic predisposition for a less tighly localized neurological instantiation of language.

It is interesting to integrate this speculation with a recent theory that left-handedness is the expression of particular hormonal events in utero (Geschwind and Galaburda, 1987). They note that left-handers have a much higher incidence of auto-immune disorders than right handers, which indicates special hormonal events in utero. We recently conducted a survey of University of Rochester undergraduates and found that right-handers with left handed families report allergies much more frequently (49%) than pure right handers (15%). It appears possible that familial sinistrality is a marker for people who become neurologically predisposed to acquire left-handedness in utero, but who have not expressed it haptically.

In conclusion, we propose an interpretation that familial right-handers have a beautiful, well-formed small language module, whereas people with left-handers in their family have a correspondingly ugly, mishapen, large module: that is, their language function is neurologically instantiated in a "quasimodule".

References

Bever, T.G. (1983). Cerebral Lateralization, Cognitive Asymmetry and Human Consciousness. In E. Perecman, (Ed.) Cognitive Processing in the Right Hemisphere. New York, NY: Academic Press.

Carrithers, C. (1986). The Special Status of Non-canonical sequences. Ph.D. Dissertation, Columbia University.

Carrithers, C., Townsend, D.J., & Bever, T.G. (In press). Sentence-Processing in Listening and Reading among College and School-age Skilled and Average Readers. In R. Horowitz & S.J. Sammuels, (Eds) Comprehending Spoken and Printed Language. New York, NY: Academic Press.

Geschwind, N., & Galaburda, A.M. (1987). Cerebral Lateralization. Cambridge, MA: MIT Press.

Luria, A.R. (1970). Traumatic Aphasia. The Hague: Mouton.

Townsend, D.J., & Bever, T.G. (1978). Interclausal Relations and Clausal Processing. Journal of Verbal Learning and Verbal Behavior, 17, 509-521.

Computational Demand and Resources in Aphasia[*]

Gary Libben

University of Calgary

This paper presents evidence from a brain-damaged patient's production of English multimorphemic words. It is argued that the patterns of impairment displayed by this patient are best explained as an interaction between the computational demands of English morphology and the limitation of working memory resources resulting from damage to the brain.

The patient's particular disability involved a sensitivity to affixation processes in English word formation. He could not repeat phrases containing unusual morphological sequences such as those found in the expression "no ifs ands or buts". He was also unable to repeat words which contained both a non-neutral prefix (e.g., "un-") and a non-neutral suffix (e.g., "-ity"). Although the patient's morphological deficit could be related to specific types of affixation, his performance was dependent on the overall complexity of the word formation process. It was therefore reasoned that an explanation framed in terms of the representation of linguistic constructs in the brain could not adequately capture the nature of the patient's impairment. The alternative processing explanation accounts for the data and points to a research program in which language pathology research would be brought into closer proximity to the research on language production in normals.

A major goal of the linguistic study of language disorders is to understand the functional architecture of language competence. This paper presents evidence from the study of a brain-damaged patient 's production of English multimorphemic words. It is suggested that the patterns of impairment observed in this patient have significant implications for linguistic and psycholinguistic models of language competence.

Since the development of the doctrine of Phrenology in the late eighteenth century by John Gall, in which a fractionation of the brain was proposed to accommodate a theoretically-motivated fractionation of mind, there has been a countermovement such as that spearheaded by Flourens in the early nineteenth century, in which it was contended that the relevant property of the brain is that it functions as a whole. The debate, of course, is still with us.

The last couple of decades have witnessed a great renaissance of various relatives of Faculty Psychology. Thus, against the background of generative grammar, modularity of mind, human intelligences, and numerous versions of the left-brain, right-brain dichotomy, it is not surprising that putative syndromes such as agrammatism receive much attention in both the aphasiological and linguisitic literature.

The term agrammatism has been used to describe a deficit associated with Broca's aphasia which is characterized by halting, laboured speech and the relative absence of function words and inflectional affixes in speech production. This impairment is often seen in the presence of intact comprehension.

The study of agrammatism has attracted the interest of a number of linguists who have proposed characterizations of the deficit in terms of linguistic theory. Kean (1977, 1980, 1982) has proposed an account of agrammatism which claims that the nature of the deficit can be captured by the formalism of generative phonology (Chomsky and Halle 1968). The approaches taken by Grodzinsky (1983) and Rizzi (1985) have suggested that the deficit can be characterized in terms of Chomsky's (1981) theory of syntax.

[*] This paper developed from a report presented at the Alberta Conference on Language in November 1986. Please address correspondence to: Gary Libben, Department of Linguistics, The University of Calgary, 2500 University Drive N.W., Calgary, Alberta, Canada T2N 1N4

774

Agrammatism has also attracted the attention of investigators who have seen the association between the linguistic deficit termed Broca's aphasia and lesions of the posterior third of the left third frontal convolution (known as Broca's area) as pointing to a significant relationship between brain structure and language structure (Ojemann and Whitaker 1978; Whitaker and Ojemann 1977).

The existence of such a relationship encourages a research program in which units of linguistic theory could be employed in the characterization of the functions of discrete cortical areas.

In this paper, an alternative view of the relationship between linguistic theory and language impairment resulting from damage to the brain is presented. It is argued that it may not be necessary to postulate a modular language-dedicated brain mechanism to account for the relationship between patterns of aphasia and linguistic theory. The observed patterns may well reflect an interaction between general processing factors and the formal properties of language. This view is presented on the basis of evidence obtained from a patient observed in early 1985.

Case 118501: L. J.

History

L. J., a 64 year-old right-handed male, suffered a stroke due to an aneurism of the right middle cerebral artery. The patient was a native speaker of English with some command of conversational French. Immediately following the stroke, he was completely unable to speak, although comprehension seemed intact. His expressive abilities improved greatly within 24 hours of hospitalization, and he was diagnosed as dysarthric rather than dysphasic by the attending neurologists. It should be noted that this diagnosis was largely based on the fact that the cerebral infarction was located in the area of the right pre-central gyrus. Had the infarction been left-hemispheric in a right-handed patient, then a diagnosis of aphasia would have been more likely.

As it turned out, the patient's general language ability had recovered within ten days of the stroke, so that it was comparable to its general pre-morbid state. This general recovery, however, is to be contrasted with a perplexing and persistent difficulty that the patient had with two particular sets of words.

Pluralized Function Words

The first set of words might be termed -- for want of a better descriptor -- pluralized function words. They result from the affixation of plural markers to words such as "the" , "and", "if", etc. Such forms are practically non-existent in samples of English speech. Where they are found, their use is restricted to metalinguistic reference, as in (1)
(1) "A common error is to type two *the's* in a row."
The forms are also found in a highly restricted set of idiomatic expressions, as in (2).
(2) "No ifs ands or buts"
The evidence relating to the patient's performance on such items emerged during a routine examination in which the patient was asked to repeat the phrase in (2). L.J. was unable to repeat the phrase and showed no recognition of it as an idiomatic expression. This appeared not to be an unsystematic performance error. Over a two-week period, the patient was asked to repeat this phrase roughly 50 times. None of his repetitions were correct. Moreover, it appeared that, on each trial, he was hearing the phrase for the first time. An inspection of his erroneous repetitions suggests that his errors reflect an attempt to arrange and/or alter the component words so that some meaning could be attached to the phrase. A representative sample of such attempts are given in Table 1.

Table 1
Oral Repetition of an Idiomatic Expression

TARGET: No ifs ands or buts

REPETITIONS: No ifs and buts
                           But no ifs and buts
                           No ifs and no buts
                           Ends and buts

The inability of L.J. to repeat correctly was revealed not to be limited to either the idiomatic phrase "no ifs ands or buts" or to the task of repetition. In fact, he was unable to correctly repeat any string of pluralized function words. Samples of his repetition performance for random function-word strings are given in Table 2.

Table 2
Oral Repetition of Function-word Strings

TARGET: The's of's and's

REPETITIONS: Ends
                           Of the end
                           And the office [Of's]
                           Ends and buts

L.J.'s inability to repeat pluralizations was restricted to those cases in which the pluralization would be normally blocked by the constraints of English morphology. In contrast to his function word performance, he showed no difficulty repeating series' of concrete nouns such as "dogs, houses, chairs, cards".

As far as the task-specificity of the impairment is concerned, the patient showed all signs of what is popularly termed a 'central impairment'. He manifested an inability to process pluralized function words in any modality. He could not write them to dictation; he was unable to read them aloud, and he showed a 65% error rate when required to copy individual pluralized function words from one sheet of paper to another.

It should be pointed out that L.J.'s impairment was accompanied by virtually unimpaired language production and comprehension. In terms of communicative competence, his linguistic ability was well within the normal range. He did, however, display an articulatory disability which turned out, upon closer inspection, to be limited to a narrowly-defined class of multimorphemic words.

Non-neutral Affixation

In the study of English morphology and phonology, a distinction has been drawn between two types of affixation. The first type may result in sound changes at the point of contact between the stem and the affix and/or a re-assignment of stress internal to the stem. The phenomenon has been termed 'weak' or 'non-neutral' affixation (Kiparsky, 1982) and can be seen in the attachment of the suffix [-ity] to a word such as [legal]. In the resulting form [legality], stress has been moved from the first syllable of the word to the second syllable. The characteristics of non-neutral affixation can also be seen in the attachment of the prefix [in-] to words such as [legal]. In this case, phonetic characteristics of the last segment of the prefix assimilate to the phonetic characteristics of the first segment of the stem. The resulting form is [illegal].

Non-neutral affixation can be contrasted, in English, with neutral affixation in which a strong boundary is said to exist between the affix and the stem. Examples of neutral suffixation are [-ment] attachment (e.g. govern --> government), and [-ness] attachment (e.g. happy --> happiness). The neutral prefixation associated with the attachment of [un-] to adjectives shows no assimilation (e.g. unlikely).

L.J.'s linguistic impairment seemed to include a sensitivity to the distinction between neutral and non-neutral affixation. This sensitivity was related to cumulative effects of non-neutral affixation processes. Specifically, he was unable to repeat, read, or write any word which contained both a non-neutral prefix and a non-neutral suffix. He could not, for example, correctly repeat words such as "inseparable", "illegal", or "irregularity", as is displayed in Table 3.

Table 3
Non-neutral Prefixes and Suffixes

	TARGET	REPETITION
(a)	inseparable	insepudible
(b)	irrefutable	inrefutable
(c)	irregularity	regulalarity
(d)	illegality	legality
(e)	incomparable	incompArable
(f)	irreparable	irrepAIr able

The patient's productions were impaired only in those cases in which both a non-neutral suffix and a non-neutral prefix were present in the target word. He had no difficulty with forms such as "penetrate" or "penetrable". He could not, however, repeat "impenetrable". Similarly, although he could not repeat the word "illegality", he had no difficulty with the words "legal", "illegal", or "legality".

As can be seen in Table 3, the patient also showed a failure to assimilate the [in-] prefix to the stem, producing forms such as "inrefutable". This "unassimilation" occurred in roughly 15% of the erroneous repetitions across forms such as "illegality", "irrefutable", and "irreversible".

In those few cases in which L.J.'s repetitions were partially successful, there was a tendency, as shown in (e) and (f) of Table 3, to improperly assign stress to the multimorphemic word. This was also common in his repetitions of simple non-neutral suffixation. When asked to repeat the word "reputable", where stress is on the first syllable, his pronunciation was typically "repUtable" with stress on the second syllable. Thus, he seemed to be treating non-neutral affixes as though they were neutral.

Again it should be noted that L.J.'s impairment with respect to these words was evident across tasks and across modalities. In fact, the repetition glosses in (a), (b) and (c) of Table 3 correspond exactly to his spelling of these words in a dictation task.

L.J.'s performance in repetition and dictation showed a negative correlation between word length and accuracy. This effect, however, seemed only to be evident in cases of non-neutral affixation. He had no difficulty in the oral or written repetition of other multisyllabic words such as "minimization", "encouragement" etc.

Implications
The linguistic deficit manifested by this particular patient is both puzzling and challenging. The following explanation seemed to best account for the obtained data.
1. The productions of L.J. suggest that , at least in his case, certain morphologically complex items were generated online in the production process for each repetition, reading and writing trial.
2. There is a psychological cost associated with [in-] attachment (and the required assimilation).
3. There is similarly a cost associated with the attachment of non-neutral suffixes which appears to be related to the accompanying stress adjustments.
4. There is a cost associated with the "unblocking" of blocked affixation. This accounts for the patient's difficulty with "no ifs ands or buts".

5. The costs appear to be additive. The notion of additivity is required to account for L.J.'s ability to produce "legal", "legality", "illegal", but not "illegality".

Explanations of the sort given above point to a relationship between the computational resources required to attach non-neutral affixes, and the limitation of those resources caused by the infarction in the non-dominant hemisphere. Put another way, in this patient, online morphology requires unavailable working memory for its execution. A characterization of the deficit in terms of the interaction of working memory as a subject variable and affixation as a linguistic variable, gives rise to the hypothesis that the limitations imposed by working memory would be manifest in other tasks.

As it turned out, the patient performed quite poorly in immediate recall tasks, considering his overall functional ability. He could recall no more than 3 single-digit numbers, and no more than 2 three-digit numbers.

He could recall a maximum of 4 concrete nouns, 3 function words, and a maximum of 2 nonsense words.

There seemed, therefore, to be a considerable amount of converging evidence that an explanation stated in terms of working memory resources and the computational demands of English morphology would turn out to be revealing of properties of English morphology and functional properties of computation in the brain.

It is noteworthy that the patient did not show a dominant hemisphere lesion or major language impairment. One would expect that further investigation of such cases, in addition to the more dramatic cases of aphasia, would bring language pathology research into closer proximity to the normal speech error research and models of speech production of the type suggested by Garrett (1982, 1984).

Finally, it should be emphasized that the approach to linguistic deficits in aphasia which is presented in this paper differs radically from one which views neurolinguistics as the search for the cerebral receptacles of theoretically-motivated constructs. The patient's pattern of impairment maps in a non-trivial manner onto linguistically-motivated characteristics of English morphology. However, it is argued here that the nature of L.J.'s impairment cannot be captured by references to "the representation of morphology in the brain". Rather, the impairment must be captured in terms of a processing account which reflects the interaction between linguistic properties of English morphology and computational properties of the brain.

References

Chomsky, N. (1981). *Lectures on Government and Binding*. Dordrecht: Foris.

Chomsky, N. & Halle, M. (1968). *The sound patterns of English*. New York: Harper and Row

Garrett, M.F. (1982). Production of speech: observations from normal and pathological language use. In A. Ellis (Ed.), *Normality and pathology in cognitive functions*. London: Academic.

Garrett, M.F. (1984). The organization of processing structure for language production: Applications to aphasic speech. In D. Caplan, A.R. Lecours & Smith (Eds.), *Biological perspectives on language*. Cambridge, MA: MIT Press.

Grodzinsky, Y. (1983). The syntactic characterization of agrammatism. *Cognition,* 16, 99-120.

Kean, M.L. (1977). The linguistic interpretation of the aphasias: Broca's Aphasia, an example. *Cognition,* 5, 9-46.

Kean, M.L. (1980). Linguistic representations and the description of language processing. In D. Caplan (Ed.), *Biological Studies of Mental Processes*. Cambridge, MA: MIT Press.

Kean, M.L. (1982). Three perspectives for the analysis of aphasic syndromes. In D. Caplan, M.A. Arbib, & J.C. Marshall (Eds.), *Neural models of language process*. New York: Academic.

Kiparsky, P. (1982). Lexical morphology and phonology. In *Linguistics in the Morning Calm*, edited by the Linguistics society of Korea. Seoul: Hanshin.

Rizzi, L. (1985). Two notes on the linguistic interpretation of Broca's aphasia. In M.L. Kean (Ed.), *Agrammatism*. New York: Academic.

CONSTRAINING INTERACTIVITY:

EVIDENCE FROM ACQUIRED DYSLEXIA[1]

Gordon D. A. Brown

Department of Language and Linguistics
University of Essex

ABSTRACT

It is sometimes claimed that interactive-activation models are too powerful, and that it is difficult to constrain them adequately. I illustrate this problem by showing that the basic interactive-activation architecture has several different possible sources for effects of spelling-to-sound regularity on word naming. I then show how data can constrain the architecture. New data lead to a rather different and more constrained version of the interactive-activation model to account for spelling-to-sound conversion. Analysis of the errors made by patients suffering from acquired surface dyslexia confirms the predictions of the constrained model. It is concluded that the traditional interactive-activation framework must be considerably constrained to account for normal and disturbed word naming.

INTRODUCTION

An early version of the interactive-activation (IA) model (McClelland & Rumelhart 1981; Rumelhart & McClelland, 1982) successfully accounted for contextual effects on letter perception. Since then, the IA framework has been used to account for human performance in a wide variety of domains.

One reason for the popularity of the IA framework is that it provides a general and powerful mechanism for building cognitive models. Some researchers have worried that the resulting models may even be too powerful, and difficult to constrain. In this paper I show that this worry is sometimes justified, for a number of different IA architectures can

---

1 This work was supported by the Economic and Social Research Council (U.K.), reference number C08250011. Reprint requests to: Department of Language and Linguistics, University of Essex, Wivenhoe Park, Colchester CO4 3SQ, England.

predict the "basic findings" in the psychology of spelling-to-sound conversion. Nevertheless, new findings can constrain IA models in this domain, and an appropriately constrained model makes novel predictions that are testable.

A second reason for the popularity of the IA framework is the compatibility of IA models with neural-level modeling techniques. It is not always plausible to interpret IA models as neural nets directly (McClelland, 1985); IA modelers would not always claim that there is just one neuron per node in their IA model. Nevertheless, it is typically assumed that an IA model could easily be cashed out in terms of a more distributed neural network (see Smolensky, 1986). So current IA models often come in between neural modeling and the functionalist approach: much IA modeling is not neural-level because it is not distributed, and it is not functionalist because it is not hardware-independent and because it involves sub-symbolic processing.

The fact that IA models are intended to be cashed out in neural terms means that they should make predictions about the behavior of patients suffering from neurological impairment. That is, IA models and their distributed implementations should not only be able to account for graceful degradation of performance under damage; they should also account for those cases involving severe brain injury where degradation is *not* graceful and leads to quite specific symptom complexes. Progress has already been made in this area, using both local and distributed models (e.g. Cottrell, 1985; Hinton & Sejnowski, 1986; McClelland & Rumelhart, 1986). One aim of the present paper is to present further evidence that an IA model can make novel predictions about the nature of these impairments, and to show that these predictions are upheld. The data can in turn constrain the architecture of the IA model.

BACKGROUND

In this paper I will be concerned with one procedure: the conversion of orthographic representations to phonological representations. This provides us with a classic computational-level mapping problem (Marr, 1982), in which one set of representations (of printed words) must be mapped into another set of representations (of word pronunciations). This particular mapping problem is a difficult one, because the pronunciation of an English word cannot reliably be predicted from its orthography. McClelland and Rumelhart (1981) mention spelling-sound translation as a suitable domain of application for their IA model, and indeed refer to the work of Glushko (1979) as a source of inspiration.

Humans can derive the correct pronunciations of words, even though some words have pronunciations that are not predictable from their spelling. Words will be called *exceptional* or *irregular* here when they contain orthographic segments of at least two letters that are pronounced differently in several other words (see Henderson, 1985, for a discussion of terminology). For example, the word *PINT* has an irregular or exceptional pronunciation compared with its orthographic neighbors such

as *MINT*, *HINT*, *TINT* etc.[2] So the exception word *PINT* may be contrasted
with *PILL*, which has a pronunciation that is regular and consistent (cf.
MILL, *HILL*, *TILL* etc.).

The basic experimental finding is that it takes longer to
prepare pronunciations of exception words like *PINT* than to prepare
pronunciations of consistent words like *PILL* (Glushko, 1979).   This
exception-word effect is more likely to be obtained when the words are
low in frequency and when subjects process the words more slowly
(Seidenberg, Waters, Barnes & Tanenhaus, 1984; Seidenberg, 1985a).
So subjects do sometimes make use of spelling-to-sound correspondence
information in word naming, and this is more likely to happen when
processing is slow and there is more time for phonological
information to become activated (Seidenberg et al., 1984;
Seidenberg, 1985a).

A variety of more detailed findings has been obtained, and many
different models have been put forward to account for the findings (for
recent reviews, see Humphreys & Evett, 1985; Kay, 1985).   Many of the
models are basically IA in orientation, although as most of them have not
been implemented it is not always clear exactly what predictions they
make.   The model that accounts for the widest range of data is that of
Seidenberg and his colleagues (Seidenberg et al., 1984; Seidenberg,
1985a; 1985b; in press; Waters and Seidenberg, 1985).   This model can
account for the basic effects of spelling-to-sound characteristics on
word naming and lexical decision time, and the interactions of such effects
with word frequency and subject speed, within the IA modeling tradition.
Seidenberg (in press) has developed and extended this model to account
for effects of morphological and syllabic structure on lexical
processing.   Sejnowski and Rosenberg (1986) have implemented a
connectionist system, *NETtalk*, which exhibits great success in learning
the spelling-to-sound constraints in English using the back-propagation
algorithm described in Rumelhart, Hinton and Williams (1986) (see also
Rosenberg & Sejnowski, 1986).   So the basic IA framework is apparently
very successful in accounting for a wide variety of sophisticated
experimental data and task performance.   But it may be that this is
because the framework is insufficiently constrained in certain respects,
as we see below.

How do interactive-activation models predict the exception word
effect, whereby words like *PINT* with exceptional pronunciations take
longer to pronounce than matched words like *PILL* with regular consistent
pronunciations? Figure One is similar to the full version of the IA
model set out in McClelland and Rumelhart (1981), although it
differs in that it contains separate lexical levels for orthography and
phonology and does not include a feature level.   Any such a model
can easily be extended to include intermediate levels between words
and letters, representing sub-lexical letter and phoneme clusters (Brown,

---

2   For discussion purposes, we consider just the pronunciation of
terminal trigrams in four-letter words.   Of course, some letter
clusters in an exception word will be pronounced regularly; the model
to be discussed takes account of this.

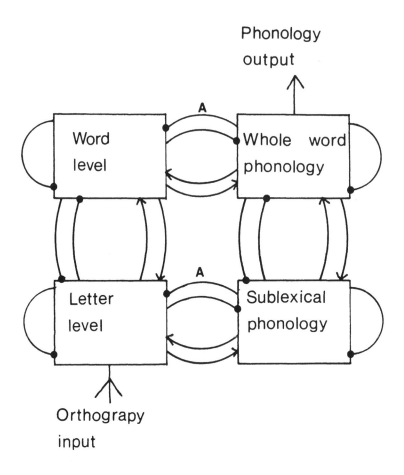

Figure One

1987; Seidenberg, in press). Note that the model has inhibition flowing from words to letters, letters to words, phonology to orthography and orthography to phonology, as well as mutual inhibition within each level[3]. If an IA model contains both orthographic and phonological levels, with connections between them, there are at least two ways in which exception-word effects will be predicted. One way is by inhibition flowing from the phonological to the orthographic levels (pathway A in Figure One). This would operate in the following way. A node for an orthographic segment with more than one possible pronunciation will activate more than one node in the corresponding phonological level. For example, the intermediate-level orthographic node for -INT (as in PINT and MINT) will activate phonology nodes

---

3   Although the letter-to-letter inhibition and the word-to-letter inhibition were both set at zero in the simulation reported by McClelland and Rumelhart (1981). Furthermore, McClelland (1985) reports a problem with letter-to-word inhibition, which is that if three competing letter nodes are all equally active, two of them will produce enough inhibition to cancel out excitation from the remaining candidate.

corresponding to pronunciations /aɪnt/ and /ɪnt/ (not shown on Figure One). These sub-lexical phonology nodes will then cause activation to spread to lexical-level phonology nodes such as /paɪnt/, /mɪnt/, /lɪnt/ etc. If these lexical-level phonology nodes can inhibit lexical-level orthography nodes (pathway A), the result will be that lexical-level nodes for exception words like *PINT* will build up activation more slowly than nodes for consistently-pronounced words like *PILL*. This is mainly because *PINT*'s orthographic neighbors cause inconsistent phonology to be activated, and this inconsistent phonology indirectly inhibits activity in *PINT*'s lexical-orthographic node, via phonology-to-orthography feedback.

But IA models will normally predict exception effects even if there is no feedback from phonological to orthographic levels (Brown, 1987; Seidenberg, in press). A model will still predict exception-word effects if there is *mutual inhibition* within the phonological levels, even if pathway A does not exist, because there will be slower activation of phonology whenever inconsistent phonology is active. An example is the case above where the two different phonological nodes activated at the same level are /aɪnt/ and /ɪnt/. Because of mutual inhibition, both of these will become activated more slowly than they would have on their own. This is because of the reciprocal nature of the inhibition: the most highly activated member in a "winner take all" network will always win eventually (Feldman & Ballard, 1982), but when inhibition is mutual the winner will win more slowly if it has more competition. Just how much more slowly it wins will of course depend on the precise nature of the mutual inhibition function. So, IA models will predict delayed pronunciation of exceptionally-pronounced words even when there is no feedback from phonological to orthographic levels.

Yet another interactive-activation architecture has been put forward to account for exception effects (e.g. Glushko, 1979; Kay & Marcel, 1981; Marcel, 1980). This allows phonology to be activated only as a result of activation within the lexical-level orthographic levels. If there is downward inhibition from phonology to orthography, or if there is mutual inhibition within the phonology levels, exception effects will be predicted. The way this works will depend on whether only whole-word or sublexical phonology is represented, but either case will lead to inhibition for words with orthographic neighbors pronounced differently. Some implementations of this possibility will be equivalent to the architecture in Figure One (see Marcel, 1980 for a detailed discussion).

CONSTRAINING THE ARCHITECTURE

The previous section demonstrated that interactive activation models could account for the exception-word effect in a number of different ways. This suggests that the architecture is underdetermined by the data. However, Brown (1987) has claimed that the reason an exception word like *PINT* takes longer to pronounce than a consistent word like *PILL* is not in fact due to interference coming from the activation of inconsistent phonology associated with *PINT*'s orthographic neighbors. Rather, it is because of the low *frequency* of the spelling-sound correspondence -*INT* -> /aɪnt/ compared with the high-

frequency correspondence -ILL -> /ɪl/.   The contrast can easily   be seen
by considering a word like SOAP.   This   word   is   not   inconsistently
pronounced,   because   there   are   no   orthographic   neighbors
pronounced differently.   SOAP   is   the   only   four-letter English word
ending in -OAP (remember, we are considering just the pronunciations
of the last three letters in four-letter words).   And it turns out
that words like   SOAP   are delayed   in pronunciation just as much as
exception words (PINT) compared to consistent words like PILL, even
though SOAP does not have differently-pronounced orthographic neighbors
to cause interference.   What this strongly suggests is that the
frequency of, not the number of exceptions to, a spelling-to-sound
correspondence in a word determines the speed with which that word is
pronounced by normal adults.   Brown (1987) therefore suggested that the
strength of a link between an orthographic node   and   corresponding
phonology nodes will depend on the frequency of that spelling-
sound correspondence in the language (Seidenberg, in press, makes
the   same   suggestion).   The   implemented   version of the model also
contains   spelling-sound   correspondences at many different levels
(letters, bigrams, trigrams etc).
     What I want to do now is to outline the implications of these data
for the architecture of interactive activation models.   As discussed
above, the standard IA model has between-level inhibition, and
within-level mutual inhibition, both of which predict that there should
be effects of a word's spelling to sound regularity on the time taken to
pronounce that word.   Yet no such effect exists: effects that have
previously been attributed to this variable are in fact due to the
frequency of spelling-to-sound correspondence.   The inhibitory mechanisms
that predict the effect must therefore be   removed.   (I   ignore   the
unattractive alternative possibility that the   inhibition   is simply too
small to be detectable experimentally.)   It   is   a   simple   matter   to
remove   feedback from phonological to orthographic levels.   This
involves removing the relevant excitatory and inhibitory connections
between phonological and orthographic levels from the full version of the
model set out by McClelland and Rumelhart (1981).   But the second
mechanism that predicts exception-word effects is   mutual inhibition
within   the   phonological levels.   Many previous researchers have
attributed exception effects to this source.   It seems undesirable to
remove this mutual inhibition, because of the possible saturation if too
many nodes at a given level can be active at the same time.
     However,   there   are   several ways to preserve the desirable
inhibition within a level without the undesirable side-effect of predicting
non-existing experimental findings.   In a typical IA model a node on a
given level is connected to all the other nodes on the same level. A node
i with activation a_i will sum the inhibitory evidence reaching it, and
(ignoring decay and incoming activation) its activation   at time $(t+\delta t)$
will be given by something of the form:

$$a_i(t+\delta t) = a_i(t) \, (1-n_i(t))$$

where n_i is the summed incoming inhibition (assumed less than 1.0) from other nodes in the same layer, and the node has a resting level of zero. This has the effect that even the node with the highest activation level will be inhibited to some extent by its neighbors, i.e., mutual inhibition. What is needed, however, is a case where the highest-activated node receives no inhibition itself, but inhibits all the other nodes in the level (non-mutual inhibition). Shastri and Feldman (1984) discuss suitable types of unit, with which it is possible for each participating node to receive inhibition dependent on the highest activation of any participating node. This kind of scheme has two quite independent advantages. The first advantage is that the number of connections needed is drastically reduced. In a non-distributed layer with N nodes, the number of connections needed to serve a mutual inhibition process will be a quadratic function of N. But using the Shastri and Feldman "max-calculator" units, the number of connections needed will be only a linear function of N, because there is a master node which receives activation from each node in the layer, and sends the maximum activation it receives back to each node as inhibition. Furthermore, the existence of a separate master node provides a means of control over the within-layer inhibition. This feature is useful for strategic purposes (see Cottrell, 1985).

In our implemented version of such a system, the effect of inhibition is given by the following form of equation, which gives the new activation of a participating node after a cycle of inhibition:

$$a_i(t+\delta t) = a_i(t) \, (1 + \beta[a_i(t) - M(t)])$$

where $M(t)$ is the maximum activation at time t of any node participating in the WTA system, and β is a constant.

In other words, each node is inhibited to an extent that depends on the difference between its activation and the activation of the most active node in the layer. For the most active node itself, of course, this difference is zero, and so there will be no inhibition. This is then a WTA network *par excellence* (because the rich get richer without paying tax on the way).

In terms of the model of phonological processing, this is what is necessary. The within-level inhibition prevents saturation of the network, without slowing down in any way the activation of the winning node. In the currently implemented version of the model, the inhibition works in this way (although the only within-level inhibition is at the lexical-level orthographic and phonological levels in the implemented version of the model, see Brown, 1987). So, according to the constrained model, the reason that *PINT* is named more slowly than *PILL* is because of the low frequency of the spelling-sound correspondence in *PINT*, and not because of interference from *PINT*'s differently-pronounced orthographic neighbors.

We therefore have a resolution in which a different inhibition scheme, which may be independently preferred on the grounds that it requires fewer nodes and allows the possibility of strategic control over inhibition, also provides a better account of the data.

The IA model is therefore considerably constrained. The data suggest that there is no feedback from phonological to orthographic levels, and that the within-level inhibition is not mutual inhibition.

SURFACE DYSLEXIA

We have claimed that the powerful IA architecture needs to be constrained to account for empirical data from normal subjects. But a crucial test of the newly-constrained model is its ability to make novel predictions. We now examine the predictions made by the constrained IA model for the performance of patients suffering from various forms of acquired dyslexia. The syndrome most relevant to the present model is that of surface dyslexia.

Surface dyslexic patients are able to synthesize pronunciations of non-words, but have difficulty in pronouncing many words with exceptional pronunciations. Furthermore, these patients have difficulty in defining homophones (see accounts in Marshall & Newcombe, 1973; Shallice & Warrington, 1980; Coltheart, Masterson, Byng, Prior & Riddoch, 1983; Shallice, Warrington & McCarthy, 1983; Kay & Lesser, 1985; Patterson, Marshall & Coltheart, 1985). These symptoms lead naturally to the suggestion that surface dyslexics are making use of sub-lexical spelling-to-sound correspondence information, and that their access to a semantic lexicon is often via a phonological representation. Most surface dyslexics can pronounce some exception words, especially when they are high-frequency (Bub, Cancelliere & Kertesz, 1985) suggesting that some lexical-level correspondences are preserved. Also they show lexicality effects (Marcel, 1980), suggesting lexical-level involvement (although not all patients show lexicality effects: Shallice et al., 1983; Kay & Lesser, 1985).

Shallice et al. (1983) show that their patient, HTR, is affected by "degrees of irregularity", and can pronounce many "mildly irregular" words correctly. Mildly irregular words are defined as words that contain a spelling-to-sound correspondence that is the second most frequent in the language. Note that although this is a measure of the *relative* frequency of a spelling-sound correspondence, it is likely to be correlated with the *absolute* frequency of that correspondence. It is therefore difficult to tell which of the two factors is causing the effects. The data lead Shallice et al. to conclude that surface dyslexics fall on a continuum according to the size of orthographic units they can translate to phonology. In general, the consensus view is that surface dyslexics are impaired on "irregular" words, where regularity is in some way defined in terms of other words containing the same orthographic segment pronounced differently. In terms of an interactive activation model with spelling-sound links at many different levels, it is reasonable to conclude that surface dyslexics have preserved low-level correspondences but have lost most high-level spelling-to-sound correspondences. Indeed, Shallice et al. give an account very similar to this, and point out that if higher frequency or early-acquired correspondences were more likely to be preserved, surface dyslexia could result. This is because level of correspondence is confounded with frequency of correspondence, because high-level correspondences will occur in fewer words.

What I want to do now is examine the predictions of the newly-constrained IA model for the nature of the errors made by these surface dyslexic patients. The main relevant properties of the constrained version of the model are that (*a*) connections between orthographic and phonological nodes are weighted according to the frequency of that spelling-sound correspondence in the language; (*b*) there is no feedback from phonological to orthographic levels, and (*c*) within-level inhibition is not mutual inhibition.

When higher-level correspondences are abolished, it is reasonable to suppose that the highest frequency correspondences within that level are most likely to be disrupted. Therefore, words that contain low-frequency spelling-to-sound correspondences are most likely to be pronounced incorrectly or not at all, because the correct pronunciation of these words relies on the use of high-level (i.e. lexical or trigram level) correspondences. Most previous models have assumed in contrast that words with exceptional or irregular pronunciations will be susceptible to disruption.

The prediction made by the constrained IA model is, then, that surface/semantic dyslexics will be more likely to make errors on word containing unusual spelling-to-sound correspondences at high levels. The regularity of the word, where regularity is defined (as it normally is) in terms of the number of a word's neighbors that are spelt similarly but pronounced differently, should have no effect.

THE ANALYSIS

Several researchers have examined the prediction that surface dyslexics should make more errors on irregular words. Most have used the lists of regular and irregular words published by Coltheart et al. (1979), and many have published full listings of the words that their patient pronounced wrongly. It is therefore possible to re-analyse these data to determine whether it is in fact the number of differently-pronounced but similarly-spelled words that impairs performance, or whether it is in fact the frequency of the spelling-to-sound correspondence in the word. There are six complete published corpora of errors on the Coltheart et al. words. These are found in Coltheart et al. (1983); Shallice et al. (1983), Kay and Lesser (1985), and Saffran (1985). A number of other papers do include corpora, but these either contain only a subset of the errors made errors (e.g. Margolin, Marcel & Carlson, 1985; Masterson, Coltheart & Meara, 1985) or are based on different sets of regular and irregular words (e.g. Newcombe & Marshall, 1985). We therefore analysed the six complete corpora, although two are from one patient. This patient was tested on two occasions two months apart (Saffran, 1985); on the first occasion 31 errors were made, on the second occasion 17 of those 31 words were misread along with 10 other words. This is similar to the normal overlap between two different patients.[4] One of the corpora came from a "developmental surface dyslexic" ("C.D." in Coltheart et al., 1983); five from acquired surface dyslexics.

For each four-letter and five-letter monosyllabic word in the Coltheart et al. (1979) lists (N=54) we calculated a measure of the frequency of the spelling-to-sound correspondence, and the exceptionality of the spelling-to-sound correspondence. These were calculated by looking at the phonology asociated with each of the trigrams in the word. The exceptionality of each trigram within a word was calculated as the cumulative Kucera & Francis (1967) frequency of all same-length words containing the same trigram in the same position but pronounced differently. The exceptionality of each word was the sum of its trigram exceptionalities. The spelling-sound frequency of each trigram was calculated as the cumulative frequency of all same-length words (including the word in question) containing the same trigram pronounced the same way. The spelling-sound frequency of a word was then obtained by summing the spelling-sound frequencies of the trigrams within that word.

There are therefore two measures for each word, one relating to the frequency of the spelling-sound correspondences contained in the word, and the other relating to the exceptionality of the word (i.e. the number of other words containing the same orthographic segment pronounced differently). And the prediction is that the number of errors made by surface dyslexics will be related to spelling-sound frequency, in contrast to previous claims that the relevant factor will be spelling-sound irregularity.

The 27 of our words classified as irregular by Coltheart et al. had a median exceptionality of 231, and a median spelling-sound frequency of 203. The words classified as regular by Coltheart et al. had a median exceptionality of 54, and a median spelling-sound frequency of 498.

Overall, ignoring the Coltheart et al. classification, there was a clear negative correlation between spelling-sound frequency and error rate: Spearman's Rho = -0.39, t=3.0, p<.01. In contrast, there was no correlation between error rate and exceptionality: Rho = 0.19, t=1.4, p>.10. This clearly supports the prediction made by the constrained IA model discussed above; errors are more likely to be made on words containing infrequent spelling-sound correspondences rather than on words with irregular spelling-sound correspondences.

It could be argued that these correlations result from our own definition of exceptionality, which is based only on high-level spelling-sound correspondences. In fact this is unlikely, because the higher-level correspondences are apparently more susceptible to damage in surface dyslexics. Nevertheless, it could also be argued that four-letter and five-letter words should be analysed separately, in case word length has an independent influence on error rate (over and above the tendency for longer words to contain less frequent spelling-sound correspondences). Therefore, further analysis was carried on on the 12 four-letter monosyllabic words and the 15 five-

---

4 The figures here are based on Saffran's corpus rather than on the figures in the accompanying text.

letter monosyllabic words classified by Coltheart et al. as irregular. The results were as follows.

For five-letter irregular words, spelling-sound frequency correlated significantly and negatively with error rate: Rho = -0.59, $t=2.7$, $p<.02$. Exceptionality did not correlate significantly with error rate: Rho = -0.17, $t=0.6$, $p>.20$. A similar pattern of correlation was observed for the four-letter words, although the correlation between spelling-sound frequency and error rate failed to reach significance: Rho = -0.32, $t=1.1$, $p>.20$. Exceptionality again failed to correlate with error rate: Rho = 0.05, $t=0.2$, $p>.20$.

In combination, these results clearly suggest that surface dyslexics tend to make more errors on words containing infrequent spelling-to-sound correspondences, rather than on words with exceptional spelling-sound patterns. Apparent effects of exceptionality have in fact been due to spelling-sound frequency. This is exactly the pattern of results predicted by the constrained version of the IA model.

It should be noted that we have not controlled for the frequency of purely orthographic regularity in our analysis; this is impossible to do for the words for which error corpora have been reported. This is unlikely to be a major problem, for Brown (1987) found effects of spelling-sound frequency when orthographic regularity was controlled for, and other authors have obtained spelling-sound effects that are not due to orthography. And in reaction-time experiments it is unlikely that orthographic frequency and spelling-sound regularity effects would exactly cancel each other out across a wide range of word frequency and subject decoding speed.

It should also be noted that the spelling-sound frequency of a word is related to the frequency of occurrence of that word, because the frequency of a word contributes to the frequency of the spelling-sound correspondences contained within it. Again, a number of experiments have found effects of spelling-sound characteristics when word frequency is controlled. For example it is clear that surface dyslexics do make more errors on the Coltheart-irregular than on the Coltheart-regular words even though the two sets of words are matched for word frequency. The claim here is just that the effects are really due to the confounded factor of spelling-sound frequency. Indeed, our model interprets effects of word frequency on word naming time as being due to the frequency of spelling-sound correspondences in that word, including the lexical-level spelling-sound correspondence (the strength of which will depend directly on word frequency).

CONCLUSION

We have shown that the full interactive activation framework when applied to the domain of spelling-to-sound conversion is in some respects too powerful, because many different inhibition mechanisms could give rise to delayed processing of words with exceptional pronunciations. Because effects which have previously been seen as exceptionality effects are in fact simple spelling-to-sound frequency effects, the model needs to be constrained. A more constrained IA model is discussed, which has no feedback from phonological to orthographic levels and which also uses a different mechanism for within-level inhibition. This has the dual advantages of giving a better account of the data and requiring fewer within-level inhibitory connections. The constrained model gives rise to novel predictions about the errors made by surface dyslexic patients, and these predictions are confirmed. Thus it is both possible and necessary to constrain models within the interactive-activation framework.

BIBLIOGRAPHY

BROWN, G. D. A. (1987). Resolving inconsistency: A computational model of word naming. *Journal of Memory and Language*, 26, 1-23.

BUB, D., CANCELLIERE, A., & KERTESZ, A. (1985). Whole-word and analytic translation of spelling to sound in a non-semantic reader. In K. E. Patterson, J. C. Marshall and M. Coltheart (Eds.), *Surface Dyslexia: Neuropsychological and cognitive studies of phonological reading*. London: Erlbaum.

COLTHEART, M., BESNER, D., JONASSON, J. T. & DAVELAAR, E. (1979). Phonological encoding in the lexical decision task. *Quarterly Journal of Experimental Psychology*, 31, 489-507.

COLTHEART, M., MASTERSON, J., BYNG, S., PRIOR, M., & RIDDOCH, J. (1985). Surface Dyslexia. *Quarterly Journal of Experimental Psychology*, 35A, 469-495.

COTTRELL, G. (1985). *A connectionist approach to word sense disambiguation*. PhD. thesis, TR 154, University of Rochester.

FELDMAN, J. A., & BALLARD, D. H. (1982). Connectionist models and their properties. *Cognitive Science*, 6, 205-254.

GLUSHKO, R. J. (1979). The organization and activation of orthographic knowledge in reading aloud. *Journal of Experimental Psychology: Human Perception and Performance*, 5, 674-691.

HENDERSON, L. (1985). Issues in the modelling of pronunciation assembly in normal reading. In K. E. Patterson, J. C. Marshall and M. Coltheart (Eds.), *Surface Dyslexia: Neuropsychological and cognitive studies of phonological reading*. London: Erlbaum.

HINTON, G. E., & SEJNOWSKI, T. J. (1986). Learning and relearning in Boltzmann machines. In D. E. Rumelhart and J. L. McClelland (Eds.), *Parallel Distributed Processing Volume 1: Foundations*. Cambridge, Mass: MIT Press.

HUMPHREYS, G. W., & EVETT, L. J. (1985). Are there independent lexical and nonlexical routes in word processing? An evaluation of the dual-route hypothesis. *The Behavioral and Brain Sciences, 8,* 689-740.

KAY, J. (1985). Mechanisms of oral reading: A critical appraisal of cognitive models. In A. W. Ellis, (Ed.), *Progress in the psychology of language,* Vol. 2, London: Erlbaum.

KAY, J., & LESSER, R. (1985). The nature of phonological processing in oral reading: Evidence from surface dyslexia. *Quarterly Journal of Experimental Psychology, 37A,* 39-81.

KAY, J., & MARCEL, A. J. (1981). One process, not two, in reading aloud: Lexical analogies do the work of non-lexical rules. *Quarterly Journal of Experimental Psychology, 33A,* 397-414.

KUCERA, H., & FRANCIS, W. N. (1967). *Computational analysis of present-day American English,* Brown University Press.

MARCEL, A. J. (1980). Surface dyslexia and beginning reading: A revised hypothesis of the pronunciation of print and its impairments. In M. Coltheart, K. E. Patterson & J. C. Marshall, (Eds.), (1980). *Deep Dyslexia.* London: RKP.

MARGOLIN, D. I., MARCEL, A. J., & CARLSON, N. R. (1985). Common mechanisms in dysnomia and post-semantic surface dyslexia: processing deficits and selective attention. In K. E. Patterson, J. C. Marshall and M. Coltheart (Eds.), *Surface Dyslexia: Neuropsychological and cognitive studies of phonological reading.* London: Erlbaum.

MARR, D. (1982). *Vision.* San Francisco: W. H. Freeman & Co.

MARSHALL, J. C., & NEWCOMBE, F. (1973). Patterns of paralexia: a psycholinguistic approach. *Journal of Psycholinguistic Research, 2,* 175-199.

MASTERSON, J., COLTHEART, M., & MEARA, P. (1985). Surface dyslexia in a language without irregularly spelled words. In K. E. Patterson, J. C. Marshall and M. Coltheart (Eds.), *Surface Dyslexia: Neuropsychological and cognitive studies of phonological reading.* London: Erlbaum.

McCLELLAND, J. L. (1985). Putting knowledge in its place: A scheme for programming parallel processing structures on the fly. *Cognitive Science, 9,* 113-146.

McCLELLAND, J. L., & RUMELHART, D. E. (1981). An interactive activation model of context effects in letter perception: Part 1. An account of basic findings. *Psychological Review, 88,* 375-407.

McCLELLAND, J. L., & RUMELHART, D. E. (1986). Amnesia and distributed memory. In J. L. McClelland and D. E. Rumelhart (Eds.),

*Parallel   Distributed   Processing   Volume   2:   Psychological   and
Biological Models*. Cambridge, Mass: MIT Press.

NEWCOMBE, F., & MARSHALL, J. C. (1985).   Reading and writing by letter
     sounds.   In K. E. Patterson, J. C. Marshall and M. Coltheart (Eds.),
     *Surface Dyslexia: Neuropsychological   and   cognitive studies of
     phonological reading*. London: Erlbaum.

PATTERSON, K. E.,   MARSHALL, J. C., &   COLTHEART, M. (1985). (Eds.),
     *Surface Dyslexia: Neuropsychological   and cognitive studies of
     phonological reading*. London: Erlbaum.

ROSENBERG,   C. R., & SEJNOWSKI, T. J.   (1986). The spacing effect on
     NETtalk, a massively-parallel   network.   *Proceedings of the Eighth
     Annual Conference of the Cognitive Science Society*, 72-89.

RUMELHART,   D. E.,   HINTON, G. E., &   WILLIAMS, R. J. (1986).   Learning
     internal representations by error propagation. In D. E. Rumelhart and
     J. L. McClelland (Eds.),   *Parallel Distributed Processing Volume 1:
     Foundations*. Cambridge, Mass: MIT Press.

RUMELHART,   D.   E. & McCLELLAND, J. L. (1982). An interactive activation
     model of   context effects in letter perception: Part 2.   The
     contextual enhancement effect and some tests and extensions of the
     model. *Psychological Review*, 89, 60-94.

SAFFRAN,   E.   M.   (1985).   Lexicalisation and reading performance in
     surface dyslexia. In K. E. Patterson, J. C. Marshall and M. Coltheart
     (Eds.), *Surface Dyslexia: Neuropsychological   and cognitive studies
     of phonological reading*. London: Erlbaum.

SEIDENBERG, M. S. (1985a). The time course of phonological code activation
     in two writing systems. *Cognition*, 19, 1-30.

SEIDENBERG, M. S. (1985b). Constraining models of word recognition.
     Cognition, 20, 169-190.

SEIDENBERG, M. S. (in press). Reading complex words.

SEIDENBERG,   M.   S.,   WATERS, G. S., BARNES, M. A., & TANENHAUS, M. K.
     (1984). When   does   irregular   spelling or pronunciation influence
     word   recognition? *Journal of Verbal Learning and Verbal Behavior*,
     23, 383-404.

SEJNOWSKI,   T. J., & ROSENBERG, C. R. (1986). NETtalk: A parallel network
     that learns to read aloud.   The John Hopkins University, Technical
     Report JHU/EECS-86/01.

SHALLICE,   T.,   & WARRINGTON,   E.   K.   (1980).   Single and multiple
     component central dyslexic syndromes. In M. Coltheart, K. E. Patterson
     & J. C. Marshall, (Eds.), (1980). *Deep Dyslexia*. London: RKP.

SHALLICE, T., WARRINGTON, E. K., & McCARTHY, R. (1983). Reading without semantics. *Quarterly Journal of Experimental Psychology*, 35A, 111-138.

SHASTRI, L., & FELDMAN, J. A. (1984). Semantic networks and neural nets. TR131, University of Rochester.

SMOLENSKY, P. (1986). Neural and conceptual interpretations of PDP models. In J. L. McClelland and D. E. Rumelhart (Eds.), *Parallel Distributed Processing Volume 2: Psychological and Biological Models.* Cambridge, Mass: MIT Press.

WATERS, G. S., & SEIDENBERG, M. S. (1985). Spelling-sound effects in reading: Time course and decision criteria. *Memory & Cognition*, 13, 557-572.

ON THE ROLE OF TIME IN
READER-BASED TEXT COMPREHENSION[1]

Jean-Pierre Corriveau[2]
Department of Computer Science
University of Toronto
Toronto, CANADA
M5S 1A5

Abstract: Information-processing models for comprehension typically regard text as the depository of a *single determinate* meaning, placed *in* it by the writer. Conversely, a *reader-based* approach views meaning as constituted by the interactions between an *individual* and a text. From a computational standpoint, reader-based understanding suggests abandoning models which depend on *a priori* rules of interpretation and limiting the design of an algorithm to the *quantitative* aspects of text comprehension. I propose that the perception of subject matter be viewed as a *race process* where the generation of *bridging inferences* and *expectations* is partly controlled by quantitative factors (such as the delay for memory retrieval) which emphasize the invisible but omnipresent role of *time* during reading.

1 Introduction

In this paper, I am concerned with the *comprehension* of long, unrestricted, *written text* rather than with the design of single-sentence systems. I assume that *subject matter* is what gives a text a certain unity: if we fail to perceive the subject matter of a text, we find it difficult to understand that text (Bransford and Johnson, 1973).

Researchers in text understanding (*e.g.* Dyer, 1983; Graesser and Clark, 1985) typically assume that conceptual structures are constructed during comprehension: each linguistic element must be *connected* to some of the *cognitive constructs* obtained for the text read so far. Intuitively, one can think of each of such constructs as a set of *links* (or *connections*) that specifies which words and clauses are connected one to the other; each link corresponds to a conceptual path of some sort. In this paper, I shall try to avoid representational issues by working at this basic level of conceptual links.

Local coherence allows the reader to perceive successive clauses of a text as a set of related ideas. The connection between two successive clauses may be explicitly stated in the text (*e.g.* using connectives such as *because, therefore, when*). Such explicit inter-clausal connections are not problematic. The difficulty in perceiving coherence instead involves clauses with no explicit connections. In this case, the reader must *bridge* from one clause to another by means of inferences. Researchers generally assume that there is a *correct* inference path to connect two clauses. Correctness is defined with respect to the *a priori* rules of interpretation. Similarly, it is generally hypothesized that the perception of *global coherence* is also rule-based and depends on certain global patterns of organization called *macrostructures* (Phillips, 1985). The few models which tackle the problem of global coherence assume the existence of a small, correct set of macrostructures, which are specified in cognitive schemata. In other words, the possible gists of a text

---

[1] Support from the National Science and Engineering Research Council is gratefully acknowledged. I am indebted to Ed Plantinga and Graeme Hirst for several discussions and valuable comments on this research.

[2] UUCP: jpierre@utai.uucp; CSNet: jpierre@ai.toronto.edu

are, in essence, specified *a priori*: the rules of interpretation of these models come to form an 'understanding algorithm' which defines what it is to correctly understand a text.

The idea that a text *contains a single determinate meaning* has been rejected by the proponents of *reader-based understanding* who consider the meaning of a text to be constituted by the interactions between text and reader (see Holub, 1984). From this standpoint, the algorithms of existing models constitute mere mechanical encodings of sets of rules more or less arbitrarily established by the programmers.

In the context of reader-based understanding, comprehension is taken to proceed from the *private response* of a reader to a text. For example, Gadamer (1976) proposes that the act of interpretation be understood as an interaction between the *horizon* provided by the text and the horizon that the interpreter brings to it. From this perspective, the text acts as a *stabilizing* factor in the *idiosyncratic* interpretation of a reader. From a computational viewpoint, reader-based understanding suggests that we abandon attempts at specifying *a priori* rules of comprehension and, in particular, macrostructures. Instead, the design of an algorithm should be limited to the quantitative (*i.e.* non-qualitative) aspects of comprehension. I investigate these aspects in this paper.

2 From Local to Global Coherence

2.1 Fundamental Hypothesis

By definition, one can establish that A *causes* B only after having inferred that B takes place *after* A in the chronology constructed from the text. Unless one assumes that all causal paths are encoded exhaustively in a knowledge base, in which case comprehension reduces to pattern-matching, it is important to notice that there is an order in which certain inferences are produced: certain types of inferences must be preceded by others. Time displays a similar invisible omnipresence for the problems of anaphora (Hirst, 1981; Stevenson, 1986), and ambiguity (Hirst, 1987). The fundamental hypothesis of my work is that time plays a crucial role during comprehension.

2.2 Time-Constrained Comprehension

2.2.1 Time, Working Memory, and Context

Ultimately, text consists of linguistic elements juxtaposed in a linear sequence. At time t_1, a word $WORD(t_1)$ is input. Most aspects of syntax can be viewed as linguistic devices to establish intra-clausal connections, that is, to connect $WORD(t_1)$ to some of the elements of the cognitive construct existing at t_1 for the clause being currently processed, $CLAUSE(t_1)$.

For simplicity, let me gloss over the syntactic analysis process. A few instants later, at time t_2, once $WORD(t_1)$ has been *integrated* with $CLAUSE(t_1)$, the reader has a new construct for the current clause, $CLAUSE(t_2)$. This construct may contain unresolved references such as pronouns or ambiguous words. Following Stevenson's (1986) work, I assume that the reader immediately tries to connect $CLAUSE(t_2)$ with the constructs he has created for previous clauses, rather than waiting for a complete syntactic representation of the current clause.

Typically, a reader only has a limited number of constructs *in focus*, that is, readily accessible. The other constructs obtained for the text read so far require significantly more time to access, if they can still be accessed at all. This corresponds to the usual *heuristic* distinction between a working memory of limited 'capacity' and another partition of storage, the *background*, which is considerably slower to access.

As the reader advances through the text, the constructs in the working memory change, the focus is constantly modified. Let us denote the set of constructs in focus at time t by $\text{FOCUS}(t)$, and the set of constructs in the background by $\text{BACKGROUND}(t)$. FOCUS and BACKGROUND comprise all the information stored by the comprehender during his reading. Therefore, at time t_2, $\text{CLAUSE}(t_2)$ can only be connected to the elements of $\text{FOCUS}(t_2)$ and $\text{BACKGROUND}(t_2)$. I take *context* to be the union of FOCUS and BACKGROUND at any given point in time. Let us denote the context at time t by $\text{CONTEXT}(t)$.

2.2.2 Bridging Inferences Revisited

At a given point t in time, the reader must connect $\text{CLAUSE}(t)$ with $\text{CONTEXT}(t)$. Ultimately, which connections are made depends on the reader's prior knowledge. Yet I suggest that certain quantitative factors also affect the generation of conceptual connections and thus the perception of subject matter.

Two quantitative factors can be distinguished during the generation of conceptual connections:

- the processing time required for establishing connections.

- the access time of target constructs.

These two factors emphasize a most significant fact about text comprehension, namely, that it is *a real-time process*. The importance of this postulate has been defended by Gigley (1985a, 1985b) in her seminal work in Neurolinguistics from which much of the present research proceeds.

Márkus (1983) makes a crucial observation when he remarks that bridging inferences are generally restricted to what is in focus. For him, the failure to consider the contents of BACKGROUND originates in the *pressure to infer* typically felt by the comprehender who allocates himself but a relatively short amount of time to integrate clauses. Since the cognitive constructs in focus are much faster to access than those in the background, they are the first, and generally, the only ones to be considered. This suggests that the perception of coherence be viewed as a *time-constrained* process: time constitutes the stopping criterion for the generation of conceptual links. More precisely, connecting $\text{CLAUSE}(t)$ to $\text{CONTEXT}(t)$ should be regarded as a *race process* where the *actual* (as opposed to a theoretically *correct*) set of bridging connections is the one available when this race stops. The order in which bridging inferences are created becomes crucial. For example, given a 'hard' text and a short delay between clauses, most temporal connections, but probably only a few implicit causal ones, would be obtained.

2.3 On the Emergence of Global Coherence

Since CLAUSE is necessarily in focus and consists of the same types of links as the contents of the working memory, CLAUSE is taken to be a subset of FOCUS. Thus, each time a conceptual connection is found between CLAUSE and the rest of FOCUS, it is added to FOCUS, that is, to the working memory. Since FOCUS has limited capacity, it may become filled up at a certain point in time. Each time FOCUS is saturated, the addition of a new connection requires that at least one of the existing connections be moved out of focus, that is, either placed in BACKGROUND or suppressed.

There is a fundamental difference between being moved to BACKGROUND and being suppressed: in the first case, the link is still accessible, in the second case, it is lost. Intuitively, we would like the most 'important' or 'relevant' connections of the text to stay in focus, less important ones to be in the background, and trivial ones to be suppressed. Following such an approach, global coherence can simply be defined as the contents of the working memory at a given point in time. The crucial point is that there can be no *a priori* rules of 'importance' or 'relevance'; the selection of what must be moved out of focus must operate at the quantitative level.

Each conceptual connection links two constructs. The connection itself may be direct or consist of an inference path. Since each element in an inference path is only relevant with respect to the whole path, the conceptual connection between two constructs should be thought of as an indivisible unit. In order to *quantitatively* distinguish conceptual connections, I propose that each one be assigned an *activation level* (or equivalently, *energy*): the links with the greatest energy are kept in focus, the others are moved out of focus. Let us investigate the notion of an activation level.

From a quantitative point of view, when a connection A is added to FOCUS, it may affect the energy of a connection B already existing in FOCUS in one of three ways:

1. *independence*: A and B are independent if the addition of A to FOCUS does not affect B's energy.

2. *additivity*: A and B are additive if the addition of A to FOCUS increases B's energy.

3. *adversity*: A and B are adversative if the addition of A to FOCUS decreases B's energy[3].

Each of these quantitative relations is defined over *conceptual* connections which control the perception of coherence, a semantic phenomenon. In other words, how a link affects another one (when one is added to FOCUS and the other is already in the working memory) is not determined quantitatively but *qualitatively and dynamically*, that is, according to the current context and to the semantic knowledge possessed by the comprehender prior to the reading.

The addition of a new connection to FOCUS is not the only factor that affects the energy of the connection in the working memory. More precisely, time also affects activation levels. My hypothesis is two-fold. Firstly, energy is constantly lost as time goes by. This process is known as *constant decaying*. Secondly, the longer a connection stays in FOCUS, the more difficult it becomes to move out of focus. Since a connection can maintain itself in FOCUS only if it keeps a high level of activation, I take a connection's energy, which is a function defined over time, to be inversely proportional to that connection's decay, which is also a function defined over time. Thus, from a qualitative viewpoint, the energy of a link may be seen as a measure of its 'importance' with respect to coherence.

Let us return to the problem of coherence.

Firstly, if the activation level of a connection drops under a specified minimum, that connection is removed from FOCUS and lost. Intuitively, this connection has been *forgotten* (a quantitative phenomenon) or *contradicted* (a semantic phenomenon which, like all others, carries down to the quantitative level). Since it is seldom the case that the exact wording of a text is memorized (Baddeley, 1976, pp.315–316), it seems reasonable to assume that, for example, syntactic links will be quickly forgotten.

Secondly, upon the arrival of a new connection in FOCUS, the working memory constraints will trigger the selection of a (possibly empty) set of connections to be moved out. Several researchers have suggested numerous kinds of constraints for the working memory. Since there is no general agreement, I propose that there be only one constraint for the working memory: its total activation level at any point in time. This assumption has the advantage of being independent of any type of representation, and of not explicitly restricting the number or the size of elements in FOCUS. This unique constraint on the total energy in FOCUS constitutes the criterion that determines how

---

[3]Additivity corresponds to the connectionist notion of an *activation* link, and adversity, to the notion of an *inhibitory* link (see Feldman, 1984).

many and which connections must be moved out of focus. Abstracting from details, connections are moved out so that the constraint is *always* respected. Once it has been selected for removal from FOCUS, a connection can either be transferred to the background or forgotten. Since this decision must be quantitative, admission to BACKGROUND should be controlled by a minimum required energy (which would obviously be greater than the threshold used for forgetting). If a connection is above this threshold, it is transferred to the background; otherwise, it is lost.

I claim that the mechanisms described above adequately cover the construction of global coherence from the local connections established within and between clauses. Indeed, since FOCUS is not restricted to local connections but rather maintains a set of 'important' or 'relevant' connections, each new clause can be immediately linked to both its linear neighbours that would have not yet been moved out of focus, and to the 'important' clauses of the text.

3   On the Perception of Subject Matter

According to the proposed model of comprehension, global coherence is restricted to the conceptual connections triggered by the text. As time goes by, only the 'important' bridges (*i.e.* those that account for global coherence) remain in the working memory, the others being moved to the background or forgotten. From this viewpoint, global coherence strictly proceeds from the text. But subject matter is not bound to the text: it also involves what Gadamer (1976) calls the *horizon of the reader*. Three factors, namely *expectations*, *remindings*, and *interestingness*, account for this horizon. Let us briefly discuss each of these.

3.1   On Expectations and Remindings

A schema is a cluster of 'knowledge', that is, a set of conceptual links. These links are bound together in some way or another so that they come to form a conceptual unit. Thus, when one of the elements of a schema is used to bridge two elements of a text, the whole schema is momentarily accessible. An expectation should be regarded as a connection which 'forces its way' in the working memory when another link of one of the schemata it belongs to is required for bridging. In other words, each time a connection of a schema is used during comprehension, it may 'drag' with it other elements of that schema, which constitute the expectations associated with the bridging connection.

Since an expectation is just another connection in the working memory, it obeys all the rules previously introduced. In particular, it has an activation level and decays with time. As with other elements in FOCUS, its initial energy is determined dynamically according to the current context and to the semantic knowledge possessed by the comprehender prior to the reading. Intuitively, we would like to assume that this initial activation level is relatively low, so that expectations would be among the first things to be moved out of focus when necessary. This low initial energy would also emphasize the need for the reading to *confirm* an expectation soon after its addition to the working memory, that is, before it is forgotten. The confirmation of an expectation consists in the addition to FOCUS of connections that are additive with this expectation. From this point of view, expectations play a most important quantitative role in that they *speed up* the bridging of the clause(s) that confirm(s) them: the presence of an expectation in the working memory constitutes a pre-constructed bridge whose explanatory power is grounded in the schema associated with this expectation. Finally, the selection and addition to FOCUS of an expectation is part of the time-constrained race which delimits the bridging of each clause to its context. In other words, time restricts the number of expectations placed in focus.

It was observed earlier that the few information-processing researchers who address the problem of subject matter postulate an *a priori* set of macrostructures that, in essence, specifies exhaustively the gists that their models can recognize. I reject the idea that one can find a correct and complete set of such macrostructures. However, I do not want to deny that, with experience, each reader *acquires* such macrostructures. But each macrostructure merely provides a set of expectations idiosyncratic to the reader: a macrostructure is viewed as a schema for a particular gist. Thus, the expectations of a reader are not limited to the local level but indeed may concern the global coherence of a text. Macrostructures are not required in the proposed model of comprehension; they do not define what the text is or isn't about, but rather they mostly *facilitate* comprehension.

Schemata provide a mechanism to assemble conceptual links into a cognitive unit. Through the generation of expectations, they allow the reader to place in focus elements which are not required for the perception of global coherence.

Schematic membership is not the only device that can introduce non-bridging (and therefore, non-required) connections in the working memory. *Remindings* (see Schank, 1982) may also place in focus conceptual links which, in this case, have no direct relation to the text. From this observation, a slight distinction can be made between global coherence and subject matter, which extends beyond the connections generated from the text to include such indirect links. Typically, remindings are even more idiosyncratic than expectations and operate at all levels of comprehension: a word, a sentence, or the entire contents of FOCUS at a given point in time may conjure up such remindings. Their effect varies greatly: one may facilitate comprehension, in which case it essentially acts as a schema which sets up expectations, or it may do nothing more than slow down the reader.

3.2 On Interestingness

The proposed model of comprehension operates at the basic level of conceptual connections. One of the characteristics of this model is that when selected to be added to the working memory, a connection is *qualitatively* assigned an initial activation level. This initial energy constitutes a major factor in determining the future role of the connection itself. I now suggest that this initial activation level corresponds to the *interestingness* of the connection. Hidi and Baird (1986) report that the perception of this interestingness is partly idiosyncratic: "Interest occurs only in the interaction of stimulus and person so that one can never stipulate its origin in one to the exclusion of the other". They also remark that "what is central to the response of interest is that a person is compelled to increase intellectual activity to cope with the greater significance of incoming information". What does it mean to "increase intellectual activity"? The answer to this question has mostly to do with time. Upon the addition of a connection to the working memory, its initial energy may affect several parameters of the model. Consider, for example, a 'very interesting' connection. that is, one with a high initial activation level:

1. working memory capacity: the maximum for the total energy of connections in FOCUS could be increased.

2. decay: all decay functions of elements in FOCUS could be 'slowed down'.

3. racing delay: the time allowed to search for bridges could be significantly increased.

The example intuitively corresponds to a comprehender who, after reading something he considers interesting, starts to concentrate (at least temporarily) on subsequent input. Similarly, a reader who gets bored would probably decrease these parameters. Interestingness is necessary for the perception of subject matter. It cannot be grounded solely in the text, and it affects every aspect of comprehension leading to, in effect, a *personalization*, an *appropriation* of the text. And, in the end, this is what *to comprehend* (from *com* + *prehendere* to grasp) means.

4 Conclusion

In this paper, I have investigated some of the quantitative aspects of text comprehension and suggested that they originate in time. Several, if not all, of the quantitative parameters of the proposed model may vary from one individual to the next; they form a first level of explanation for the idiosyncratic nature of reader-based comprehension.

This paper should be regarded as the preliminary specifications of a model of reader-based text comprehension. All aspects of this proposal must be explicited, explained, and illustrated with respect to a particular representational scheme. The ultimate goal of this research is the detailed specifications of an implementation of a model of comprehension which follows a reader-based design philosophy: a text is processed with respect to *a* reader's knowledge base; there is no quest for the knowledge of a 'competent' reader. In other words, the issue is not *what* should be in the knowledge base but rather *how* every element of this knowledge base should be specified and used.

References

[1] Baddeley, Alan (1976) *The Psychology of Memory*, Basic Books, New York.

[2] Bransford, John and Johnson, Marcia (1973) "Considerations of Some Problems of Comprehension", In: Chase, William (ed.), *Visual Information Processing*, Academic Press, New York.

[3] Dyer, Michael G. (1983) *In-Depth Understanding*, MIT Press, Cambridge, MA.

[4] Feldman, Jerome (1984) "Computational Constraints from Biology", *Proceedings of the Sixth Annual Conference of the Cognitive Science Society*, Boulder, Colorado, p.101.

[5] Gadamer, Hans-Georg (1976) *Philosophical Hermeneutics*, translated by David Linge, University of California Press, Berkeley, CA.

[6] Gigley, Helen (1985a) "Computational Neurolinguistics: What Is It All About?", *Proceedings of the Ninth Joint Conference on Artificial Intelligence*, Los Angeles, California, pp.260–266.

[7] Gigley, Helen (1985b) "Grammar Viewed as A Functional Part of a Cognitive System", *Proceedings of the 22nd Anuual Meeting of the Association for Computational Linguistics*, Chicago, Illinois, pp.324–332.

[8] Graesser, Arthur and Clark, Leslie (1985) *Structures and Procedures of Implicit Knowledge*, Ablex Publishing Corporation, Norwood, NJ.

[9] Hidi, Suzanne and Baird, William (1986) "Interestingness—A Neglected Variable in Discourse Processing", *Cognitive Science*, 10:179–194.

[10] Hirst, Graeme (1981) *Anaphora in Natural Language Understanding*, Lecture Notes in Computer Science 119, Springer-Verlag.

[11] Hirst, Graeme (1987) *Semantic Interpretation and the Resolution of Ambiguity*, Studies in Natural Language Processing, Cambridge University Press, Cambridge, England.

[12] Holub, Robert (1984) *Reception Theory—A Critical Introduction*, Methuen, New York.

[13] Márkus, András (1983) "Shifting the Focus of Attention", *Proceedings of the Eighth International Joint Conference on Artificial Intelligence*, Karlsruhe, West Germany, pp.66–68.

[14] Phillips, Martin (1985) *Aspects of Text Structure*, North Holland Linguistic Series, Amsterdam.

[15] Schank, Roger (1982) *Dynamic Memory*, Cambridge University Press, New York.

[16] Stevenson, Rosemary (1986) "The Time Course of Pronoun Comprehension", *Proceedings of the Eighth Annual Conference of the Cognitive Science Society*, Amherst, Massachusetts, pp.102–109.

Causal Reasoning in the Construction
of a Propositional Textbase

Charles R. Fletcher and Charles P. Bloom
University of Minnesota

Abstract

The goal of this research is to unify two different
approaches to the study of text comprehension and
recall. The first of these approaches, exemplified by
the work of Trabasso and his colleagues (Trabasso &
Sperry, 1985; Trabasso & van den Broek, 1985) views
comprehension as a problem solving task in which the
reader must discover a series of causal links that
connect a text's opening to its final outcome. The
second approach, typified by Kintsch and van Dijk
(1978; van Dijk & Kintsch, 1983) emphasizes the
importance of short-term memory as a bottleneck in the
comprehension process. We combine these approaches by
assuming that the most likely causal antecedent to the
next sentence is always held in short-term memory.
Free recall data from three texts are presented in
support of this assumption.

The research reported here represents an attempt to unify
two separate approaches to the study of text comprehension and
recall. The first of these approaches views comprehension as a
problem solving process in which the reader must discover a
sequence of causal links that connect a text's opening to its
final outcome (Black & Bower, 1980; Schank, 1975; Trabasso &
Sperry, 1985; Trabasso & van den Broek, 1985). The second
approach (Kintsch & van Dijk, 1978; Miller & Kintsch, 1980;
Fletcher, 1981, 1986; van Dijk & Kintsch, 1983) emphasizes the
importance of short-term memory as a bottle-neck in the
comprehension process. We will show that a reader's short-term
memory always contains the most likely causal antecedents of the
next sentence. This allows the discovery of a "causal chain"
linking a text's opening to its final outcome within the
constraints imposed by a limited-capacity short-term memory.

Both of the approaches we are considering here have been
used to predict which elements of a text will be recalled best.
In the problem solving approach of Trabasso and his colleagues
clauses are treated as the primary unit of analysis and it has
been demonstrated that: (1) Clauses that lie on the causal chain
that connects a text's opening to its final outcome are recalled
better than otherwise comparable clauses (Black & Bower, 1980;
Trabasso & van den Broek, 1985). (2) The more causal connections
a clause has to the rest of the text, the better it is recalled

(Trabasso & van den Broek, 1985). These results provide clear support the conclusion that the causal structure of a text is an important determinant of how that text will be understood and remembered.

Kintsch and his colleagues (see e.g. Kintsch & van Dijk, 1978; Miller & Kintsch, 1980) have also been successful at predicting which portions of a text will be recalled best. They begin with the following assumptions: (1) The meaning of a text is represented in long-term memory as a network of semantic propositions called a textbase. (2) Texts are processed in cycles, roughly corresponding to sentences. (3) During each cycle, short-term memory can only hold the current sentence plus one to four propositions from earlier in the text. (4) Two propositions must co-occur in short-term memory to be strongly associated in long-term memory. (5) Two propositions must be referentially coherent (i.e., they must refer to the same person, object, or event) to be strongly associated in long-term memory. They then argue that each proposition from a text should be recalled with probability $1-(1-\underline{p})^{\underline{k}}$ where $\underline{p}$ is the probability of recalling a proposition that remains in short-term memory for just one processing cycle and $\underline{k}$ is the number of cycles that a given proposition remains in short-term memory. A procedure called the "leading-edge strategy" is used to predict the contents of short-term memory during each processing cycle and, therefore, the value of $\underline{k}$ for each proposition. This strategy uses formal properties of the underlying textbase to determine which propositions remain in short-term memory.

Clearly, there are important differences between these two approaches. They assume different basic units of analysis (clauses vs. propositions). They assume that the components of a texts are held together by different relations (causal vs. referential). One assumes that two text elements can only be connected if they co-occur in a limited capacity short-term memory, the other assumes that all possible connections are made by the reader. Finally, different mechanisms are assumed to contribute to the recallability of a text element (causal structure vs. time in short-term memory).

We believe that each approach has captured elements of the truth. As an initial step toward a unified model, we will attempt to show: (1) That Trabasso's causal analysis works as well when propositions are taken as the fundamental unit of analysis as it does with clause units. (2) That both referential and causal connections contribute to the coherence of a text. (3) That the propositions most useful for understanding the causal structure of a text are always held in short-term memory. (4) That both the number of processing cycles that a proposition spends in short-term memory and the number of referentially and causally related propositions that co-occur with it in short-term memory influence its recallability.

The success of our approach depends critically on the assumption that readers can identify and hold in short-term memory the propositions that are the most likely causal antecedents of the next sentence they read. Yet it is obvious that Kintsch and van Dijk's (1978) leading-edge strategy does not accomplish this task. In three simple narratives that we analyzed, the leading-edge strategy only allowed 31% of the possible causal connections between propositions to be detected. Two alternative strategies that we examined offer a significant improvement over this figure. The first of these we call the current-state strategy. A reader following this strategy would select the last proposition, or conjunction of propositions, added to the causal chain to retain in short-term memory at the end of each processing cycle. This strategy allows 51% of the causal connections in a text to be detected. The other alternative strategy we wish to consider will be referred to as the current-state plus goal strategy. A reader using this strategy would always retain in short-term memory the current-state in the causal chain (as defined above) as well as the proposition, or conjunction of propositions, describing the current goal in the text. This strategy represents a significant increase in short-term memory load relative to the current-state strategy, essentially doubling the number of propositions that must be held-over from earlier in the text. But it allows 69% of the causal connections in a text to be detected and bears a marked similarity to state-space search models of human problem solving (see e.g. Newell & Simon, 1972). In what follows we will attempt to determine which of these short-term memory allocation strategies (leading-edge, current state, or current-state plus goal) most accurately describes the performance of college student readers.

Method

Subjects

Twenty-four students recruited from the subject pool at the University of Minnesota participated in the study for course credit. All subjects were native speakers of English. Subjects were run in small groups of up to eight people.

Materials

Nine texts were used in the experiment: six fillers and three targets. Each text consisted of ten sentences, and contained four goals hierarchically embedded with one superordinate goal. Test booklets were constructed that contained a page of instructions followed by the nine texts in the following sequence: two filler texts at the beginning, two filler texts randomly distributed among the three target texts, and two filler texts at the end, followed by free recall instructions for each of the five middle texts. Recall of the five texts was in the same order as presented. Each text and each recall was on a separate page.

The propositional structure for each text was derived independently by the two authors using the procedures described in Bovair and Kieras (1985). The causal connections and the propositions included in each causal unit were determined independently by the two authors according to the criteria proposed in Trabasso and Sperry (1985). Any discrepancies were resolved through discussion.

Procedure

The experiment consisted of two self-paced phases. During the first phase, all subjects were instructed to read the nine texts through once at their normal reading speed, paying close attention to the stories because later they would be asked to recall them. In the second phase, subjects were given the titles from the five middle texts on separate pages and instructed to try to write down as much as they could from each text using the exact words if possible.

Results

All recall protocols for the three target texts were scored against their corresponding propositional structures independently by the two authors. A proposition was scored as recalled if any meaning-preserving paraphrase of it was present in the recall protocol. Agreement was 95% and all discrepancies were resolved through discussion.

All analyses were conducted on the three target texts combined (i.e., analyzing for the effect of text and its interactions), as well as independently. But because the effect of text and its interactions accounted for less than one percent of the variance in each of our analyses, we will only present results for the three texts combined.

The first step in attempting to integrate the two approaches to text comprehension and recall is to assess whether or not the causal analysis of text, as proposed by Trabasso and Sperry (1985) can be applied to the proposition as the unit of analysis. Multiple regression analyses were carried out on the probability of recall of each proposition in each story, with the independent variables being whether or not a proposition was on the causal chain (Causal Chain Status), and the number of direct causal connections a proposition had with the other propositions in the story (Causal Connections Possible). Causal Chain Status was a categorical independent variable, with propositions on the causal chain receiving a score of one, and propositions not on the causal chain receiving a score of zero.

Table 1
Proportions of Variance Accounted for by Causal Chain Status and Causal Connections Possible

	R^2	
	Alone	Unique
Full Model = .1934**		
Causal Connections Possible	.1273***	.0016
Causal Chain Status	.1918***	.0661**

* $p<.05$; ** $p<.01$; *** $p<.001$

As can be seen by examination of Table 1, Causal Connections Possible alone, and Causal Chain Status alone each accounted for significant proportions of variance. In addition, Causal Chain Status uniquely accounted for a significant proportion of variance, while Causal Connections Possible failed to account for any significant unique variance. The interaction between the two factors was not significant.

The previous results demonstrate that causal analysis works using the proposition as the unit of analysis. However, these analyses were conducted under the operational assumption that the working memory is of unlimited capacity. A critical assumption of the Kintsch and van Dijk (1978) text processing model is that because of the limited capacity of short-term memory, readers process a text in a number of cycles. During each cycle, a limited number of propositions enter short-term memory and are interrelated with propositions retained from the previous cycle.

The next step in attempting to integrate the two approaches is to test the causal assumptions within the confines of a limited capacity short term memory. To accomplish this, the next set of analyses was conducted to ascertain which of the various short term memory allocation strategies described earlier provides the best fit with the recall data. First, a minimum chi-square criterion was used to find the value of p which produces the best fit between predicted and observed recall probabilities in the equation $Pr(recall) = 1-(1-p)^k$ for each combination of strategy and text. Then separate multiple regression analyses on the probability of recall were computed for each strategy, with the independent variables being the time each proposition was predicted to spend in Short-term memory (Time in STM), computed as $1-(1-p)^k$, the number of direct causal connections a proposition had with the other propositions allowed by their co-occurrence in Short-term memory (Causal Connections Allowed), and the number of referential connections a proposition had with the other propositions allowed by their co-occurrence in Short-term memory (Referential Connections Allowed). The present experiment used sentence boundaries to delimit the number of propositions entering Short-term memory in each cycle. Table 2 presents the proportions of variance accounted for by each model.

Table 2
Proportions of Variance Accounted for by the Different Short Term Memory Allocation Strategies

	R^2	
	Alone	Unique
Current-State: Full Model = .2521***		
Time in STM	.1976***	.0466*
Causal Connections Allowed	.2050***	.0455*
Referential Connections Allowed	.0470*	.0003
Current-State Plus Goal: Full Model = .1506***		
Time in STM	.1099***	.0090
Causal Connections Allowed	.1330***	.0400**
Referential Connections Allowed	.0211	.0105
Leading-Edge: Full Model = .1293***		
Time in STM	.0278*	.0010
Causal Connections Allowed	.1283***	.0921***
Referential Connections Allowed	.0305*	.0006

* $p < .05$; ** $p < .01$; *** $p < .001$

Table 2 shows that although all three full models account for significant amounts of variance, the Current State model accounts for the most. Within the Current State model, all three variables alone account for significant proportions of variance. However, only Time in STM and Causal Connections Allowed account for significant amounts of unique variance. It appears that within the confines of a limited capacity Short-term memory, the use of a strategy based on retaining the last items added to the causal chain provides the best fit with the data.

One result that is somewhat incongruous with previous findings has to do with the influence of referential connections. Trabasso and van den Broek (1985) found that referential connections did not account for any significant variance when compared with causal connections. However, their analyses were based on phrases as the unit of analysis, and on all of the possible connections among those phrases within an unlimited capacity working memory. The present study found referential connections to contribute a significant non-unique amount of variance within the confines of a limited capacity Short-term memory. Subsequent multiple regression analyses carried out on the probability of recall for each proposition in an unlimited capacity Short-term memory, with the number of referential connections and the number of causal connections as the independent variables, replicated the findings of Trabasso and van den Broek (1985). However, a model containing both causal and referential connections accounts for more variance in a limited capacity Short-term memory (R^2=.206), than it does in an unlimited capacity Short-term memory (R^2=.129).

These findings address the question of the manner in which the propositions become interrelated. Trabasso and his colleagues assume these connections to be solely causal. However, the present experimental results seem to suggest that both causal and referential connections are established, with the causal connections being of greater strength.

The final step in integrating these two approaches involved a direct comparison of the variables employed in the structural analyses (e.g., Causal Chain Status and Causal Connections Possible), with the variables employed in the processing analyses (e.g., Time in STM, Causal Connections Allowed, and Referential Connections Allowed).

Table 3
Proportions of Variance Accounted for by Both the Structural Analysis and the Processing Analysis Variables

| | R^2 | |
	Alone	Unique
Structural Analysis Variables	.1934***	.0199
Processing Analysis Variables	.2521***	.0786***

* p<.05; ** p<.01; *** p<.001

Examination of Table 3 reveals that although both the structural and processing analysis variables alone account for significant amounts of variance, the processing analysis variables account for both more variance, as well as a significant amount of unique variance.

Discussion

The results of this experiment can be summarized as follows. First of all, we have shown that the causal analysis suggested by Trabasso and his colleagues (e.g. Trabasso & Sperry, 1985; Trabasso & van den Broek, 1985) can be applied at the level of individual propositions. Next, we have demonstrated that the propositions necessary for building causal chains, as identified by the current-state strategy, are held in short-term memory as a reader progresses through a text. Finally, we have shown that causally significant propositions are recalled best because: (1) they remain in short-term memory longer, and (2) they form more referential and causal links to other propositions. These findings are important because they provide a linkage between two separate, and sometimes competing, approaches to the study of text comprehension and recall.

We are currently extending this research in a number of directions. One of these involves examining of the generality of the current-state strategy. Here we are interested in two issues: (1) is the same strategy used with other genre of texts, and (2) do both good and poor readers employ this strategy? We are particularly interested in the possibility that poor readers might use a more random or idiosyncratic selection strategy. The instructional implications of such a finding are obvious. We are also developing a computer model that uses the current-state strategy to cycle through a text and construct a propositional textbase. Our goal is to combine this comprehension model with a model of retrieval from long-term memory so that we can better understand how these processes interact.

References

Black, J.B., & Bower, G.H. (1980). Story understanding as problem solving. Poetics, 9, 223-250.

Bovair, S., & Kieras, D.E. (1985). A guide to propositional analysis for research on technical prose. In B.K. Britton & J.B. Black (Eds.), Understanding expository text. Hillsdale, NJ: Erlbaum.

van Dijk, T.A., & Kintsch, W. (1983). Strategies of discourse comprehension. New York: Academic Press.

Fletcher, C.R. (1986). Strategies for the allocation of short-term memory during comprehension. Journal of Memory and Language, 25, 43-58.

Fletcher, C.R. (1981). Short-term memory processes in text comprehension. Journal of Verbal Learning and Verbal Behavior, 20, 564-574.

Kintsch, W., & van Dijk, T.A. (1978). Toward a model of text comprehension and production. Psychological Review, 85, 363-394.

Miller, J.R., & Kintsch, W. (1980). Readability and recall of short prose passages: A theoretical analysis. Journal of Experimental Psychology: Human Learning and Memory, 6, 335-354.

Newell, A., & Simon, H.A. (1972). Human problem solving. Englewood Cliffs, NJ: Prentice-Hall.

Schank, R. (1975). The structure of episodes in memory. In D.G. Bobrow & A.M. Collins (Eds.), Representation and understanding: Studies in Cognitive Science. New York: Academic Press.

Trabasso, T., & van den Broek, P. (1985). Causal thinking and the representation of narrative events. Journal of Memory and Language, 24, 612-630.

Trabasso, T., & Sperry, L.L. (1985). Causal relatedness and importance of story events. Journal of Memory and Language, 24, 595-611.

Syntax and the accessibility of antecedents in relation to neurophysiological variation

Wayne Cowart
The Ohio State University

ABSTRACT

Results of a word-by-word reading experiment argue for a specifically syntactic mechanism (N.B., not a discourse mechanism) that assigns antecedents to pronouns such as <u>he</u> and <u>they</u>, even though such assignments are grammatically optional and likely to be revised in many instances by subsequent discourse processes. These results argue for a modular view of mental architecture along the lines of Fodor (1983).

However, this study also draws on certain new proposals concerning possible behaviorally significant variation in the neurophysiological substrates of language processing. Partitioning subjects on certain biological criteria reveals that, while the pattern described above seems to apply to the majority of subjects, there is a large minority that seems to show an importantly different pattern.

On its face, the research reported in this paper is about anaphora. It argues for the existence of an antecedent-assigning mechanism very unlike the powerful discourse-oriented mechanisms that have been evident in much recent research. This mechanism seems to be acutely sensitive to syntactic structure. Apparently, it cannot detect a potential antecedent even in an immediately preceding clause unless there is an intimate syntactic relation between the clauses involved. This seems to occur in spite of the apparent facts that no grammatical principle mandates the coreference assignments the device makes and that many relations formed by this mechanism will probably have to be undone by subsequent discourse processes.

Beyond anaphora, this result bears on important general issues in cognitive science. It supports Fodor's (1983) modular account of mental architecture. Though there are surely discourse mechanisms available that can readily detect any potential antecedent in the preceding clause, the mechanism at work in the present results is somehow unable to access analyses of the context these devices might generate. Thus, the device seems to be "informationally encapsulated."

An important related issue is the question of variation. If the mind is to be regarded as composed of some ensemble of modules (whether or not these conform to Fodor's proposals), the question arises directly whether the character of individual modules, or the manner in which they collaborate, may vary significantly from individual to individual. Modular models of the mind, together with recent work in neurophysiology and psycholinguistics, virtually force the question whether there is significant variation in the logical architecture of the human cognitive system.

Background

Pronominal anaphora within syntactic processing

Much recent linguistic research has suggested that there is an interesting set of syntactic principles bearing on pronominal anaphora (among other phenomena). Within single sentences these principles appear to tightly constrain what pairs of potential antecedents and pronouns must, may or must not be taken to be coreferential (see, for example, Chomsky, 1981, 1986, Reinhart, 1983, Aoun, 1985). Though there are linguists who advocate quite different approaches (Bolinger, 1979, Bosch, 1983, Cornish, 1986), the large body of linguistic work bearing on syntactic aspects of intrasentential pronominal anaphora at least suggests that this area merits some attention in the language processing literature.

Psychological research on pronominal anaphora in adults has generally been concerned almost exclusively with cases where the pronoun and antecedent are in different sentences (see, for example, Hirst & Brill, 1980, Dell, McKoon & Ratcliff, 1983, Tyler and Marslen-Wilson, 1982, and the review in Garnham, 1985, pp. 148-152). Intrasentential relations have sometimes been examined, but usually not in ways that exercise the syntactic principles featured in the linguistic literature. For example, Corbett and Chang (1983) used coordinate structures that function as two separate sentences with respect to the binding theory discussed in Chomsky (1981). Garvey and Caramazza (1974) used main/subordinate clause structures that constitute a more integrated syntactic domain, but their research was concerned with semantic influences on reference relations.

The larger investigation of which the present work is a part is designed, among other things, to explore the role of the syntactic processing system in the assignment of reference relations among pronouns and their various candidate antecedents. In particular, it has examined the possibility that some reference relations (or at least some relations that ultimately get interpreted as reference relations) are assigned by the syntactic processor. Previous experimental results indicate that certain cataphoric instances of they can exert an influence on the syntactic analysis of ambiguous gerund phrases (e.g., flying planes), that the reference relations implicated in this finding are assigned even when they result in a manifestly odd or implausible interpretation, that these relations are blocked when they violate syntactic constraints on reference relations, that these relations are unaffected by alternative antecedents in a preceding sentence, and that effects of these kinds are demonstrable with several experimental paradigms (Cowart &

Cairns, in press, Cowart, 1986a, 1986b).

The work described here extends this line of investigation
to more commonplace instances of pronominal anaphora where the
antecedent precedes the pronoun and where a wider variety of
pronouns can be investigated. The most basic goal of the work
described here was to determine whether a certain variant of the
word-by-word reading procedure can detect any indication that
pronouns (or words following them) are processed differently
according to whether or not an antecedent appears ahead of the
pronoun in the same sentence. A second more theoretically
significant goal was to determine whether any effects of this
kind are sensitive to the syntactic relation between two clauses
where the antecedent is in the first and the pronoun in the
second. The reference-assigning mechanism that appears to be
involved in the cataphoric cases investigated earlier applies, by
hypothesis, to third-person pronouns generally (apart from
reflexives), and thus should be relevant here. If it is, and it
is an essentially syntactic mechanism, it should be sensitive to
syntactically significant variations in clause relations.

Laterality and language processing

There has long been evidence suggesting that the
distribution of language-related functions across and within the
two hemispheres of the brain is subject to some variation.
Though this evidence is difficult to interpret and still the
focus of much controversy, it is nonetheless noteworthy that it
has had virtually no effect on the bulk of sentence processing
research, apart from spotty attempts to control for subject
handedness. This apparently has two causes: 1) it is difficult
to assess dominance, and 2) when it is assessed, there is little
evidence that it has any effects.

Recently, Geschwind and Galaburda (1986) have put forward a
new and very comprehensive theory of cerebral lateralization that
suggests that variability in behavioral lateralization (e.g.,
handedness, ear advantages in speech, etc.) in mature adults is
largely the product of genetically and developmentally induced
differences in the extent, character, and interconnectedness of
the specific neural structures that support particular cognitive
functions. Geschwind and Galaburda argue that there are many
normal asymmetries between the two hemispheres of the brain. For
example, various structures in or near the apparent language
centers in the left hemisphere seem to be typically larger than
homologous structures in the right hemisphere. Geschwind and
Galaburda call this normal pattern of asymmetries "standard
dominance." They also identify a complex, multi-faceted
phenomenon linked to various departures from standard dominance,
i.e., "anomalous dominance." Anomalous dominance is taken to be
not a single alternative pattern but a wide range of differing
dominance patterns that are more or less continuously graded in

the degree to which they depart from the standard pattern. Anomalous dominance is thought to be associated not only with left-handedness but also with a variety of other phenomena, including increased frequency of dyslexia, certain other learning disorders, some special talents, and many immune system disorders, among others. Several further observations are of special relevance here. First, anomalous dominance appears to be considerably more widely distributed than is left-handedness, perhaps affecting 30% to 35% of the population, by Geschwind and Galaburda's estimate. Second, anomalous dominance appears to concentrate in particular families, with frequent evidence of two or more affected individuals among groups of close relatives. Finally, Geschwind and Galaburda suggest that language functions will be affected by anomalous dominance more frequently than will those that determine handedness.

Recently, Bever, Townsend and Carrithers (1986) reported findings suggesting that a fruitful link between Geschwind and Galaburda's work on cerebral lateralization and questions about sentence processing may be possible. Bever, et al., found evidence that some processing phenomena are linked to the presence of left-handers among a subject's biological relatives (i.e., parents, siblings, grandparents, aunts and uncles). For example, in one experiment subjects were asked to indicate whether a probe word heard in isolation shortly after the auditory presentation of a sentence fragment was one of the words in the fragment. Considering only the correct positive responses, subjects who reported no left-handers in their families (hereafter these will be termed 'SD' subjects, for Standard Dominance) were much slower in responding to probes drawn from the latter part of the fragment than they were with words drawn from the earlier part. By contrast, subjects with one or more left-handed relatives ('AD' subjects hereafter) showed no serial order effect whatever; the AD subjects responded equally rapidly to probes drawn from early or late parts of the fragment and they also responded more rapidly overall than the SD subjects. Note that all subjects were themselves strongly right-handed. Bever et al., suggest that the performance of the SD subjects reflects their reliance upon a self-terminating serial search through a linear representation of the utterance just heard. The AD subjects, by contrast, are presumed to treat the task by way of a semantic representation that provides simultaneous access to all parts of the context material.

It is, of course, not at all obvious why processing effects of these kinds should be related to the presence of left-handers in a subject's family. However, categorizing subjects in this way may be regarded simply as a convenient device for separating two populations that differ in the extent to which they exhibit the phenomenon of anomalous dominance. Bever, et al., suggest that in what we are calling AD subjects there is typically a richer interconnection between the language processing system, especially its syntactic component, and the balance of the cognitive system, especially those components involved in semantics and interpretation.

Against this background, the work discussed below was intended to provide a test of the proposals of Bever, et al., via methods and linguistic phenomena different than those they used. Pronoun-antecedent relations are notoriously subject to a great diversity of influences, ranging from stress to syntactic structure to discourse structure. If the phenomena Bever and his colleagues discovered are related to the degree of interconnection between syntactic and semantic modes of processing, anaphoric phenomena should provide a useful body of experimental material. To the degree that the richness of interconnection between the syntactic and semantic (and discourse) processing components varies, this should affect the relative accessibility of various approaches to antecedent-finding.

Experimental Evidence

Kennedy and Murray (1984) provide evidence that a certain variant of the word-by-word reading procedure is much more sensitive to syntactic structure than were earlier forms of this method. One goal of the present experiment was simply to determine whether this revised procedure can detect effects related to the presence or absence of an antecedent for a pronoun. Secondly, the experiment was designed to manipulate the syntactic relation between the clauses bearing antecedent and pronoun to determine whether any simple antecedent effects that might appear are sensitive to this factor. Finally, the experiment was planned to be run on two distinct samples, a group of strongly right-handed SD subjects and an equally strongly right-handed group of AD subjects.

Method

The experimental materials consisted of 24 sets of items similar to (1).

(1) a. Even though the librarians had made an awful lot of noise, she kept on working on her own stuff.
b. Even though the librarian had made an awful lot of noise, she kept on working on her own stuff.
c. The librarians had made an awful lot of noise, but she kept on working on her own stuff.
d. The librarian had made an awful lot of noise, but she kept on working on her own stuff.

Note that the second clauses, including their pronoun subjects, are identical throughout, apart from the coordinating conjunction in the (c) and (d) forms. The subject of each first clause is a lexical NP that provides an acceptable antecedent for the pronoun in the (b) and (d) cases only. The pronouns used included he and she, but they predominated. The two clauses of the (a) and (b) cases are in the relation subordinate-main, while those of the (c) and (d) cases are coordinate.

The experimental design involved three within-subjects factors, Antecedent (No Antecedent, Antecedent Present), Clause Relation (Subordinate, Coordinate) and Word Position (the position of each stimulus word relative to the pronoun in the second clause). These three factors were crossed by a fourth, History (SD vs. AD subjects, those lacking or having left-handed relatives, respectively).

These materials, together with 48 fillers of diverse kinds, were presented to subjects via a minor variant of the cumulative word-by-word procedure discussed by Kennedy and Murray (1984). In this task the subject must press a key to see each succeeding word in the stimulus sentence on a computer display. The interval between key presses is recorded and serves as a crude measure of reading time per word. Unlike other versions of the word-by-word task, each word is presented one space to the right of the word preceding (apart from line breaks) and stays on the screen until the subject presses the key following presentation of the last word. Thus the effect is that of seeing a normally formatted text appear one word at a time. A yes/no question appeared after each sentence presentation and the subject responded via a key press. This response was timed, evaluated and recorded, and the subject was given feedback as to the correctness of the reply. When average response time per word went above 550 msec., the feedback message also urged the subject to respond more rapidly.

In preparation for this work, a survey form was distributed to a large number of students in various undergraduate courses at Ohio State University. This form was derived from Geschwind's variant of the Oldfield inventory. It asked for, among other things, information about the handedness of the respondent's biological relatives. Fifty subjects for this experiment were drawn from a pool of about 430 individuals who completed this form. All were strongly right-handed, with laterality scores (using Geschwind's LS) of 90 to 100. Twenty-four had no left-handed relatives and 26 had one or more such relative.

Results

The results are summarized in Figures 1 and 2. Note that when an antecedent was present, SD subjects responded faster on the pronoun and the three words following it, but only where the clause relation was subordinate/main. By contrast, with AD subjects the antecedent produced faster responses for several

words after the pronoun regardless of the relation between the
two clauses. This pattern seems to be reliable. Variations in
the size of the antecedent effect are best revealed in Figure 2.

The principal statistical analyses covered the first three
words following the pronoun. The limits of this zone were
determined post hoc; it excludes some potentially relevant
contrasts on responses to the pronoun itself and to words
following this zone but seems on the whole to include effects
representative of the overall result. An analysis covering the
span running from the pronoun through the fifth word following
the pronoun produced similar but somewhat weaker results. For
the purposes of this preliminary report, effects and interactions
that do not seem to be theoretically relevant will be ignored.
Extreme response values were reset to +/- 2SD from the subject's
mean.

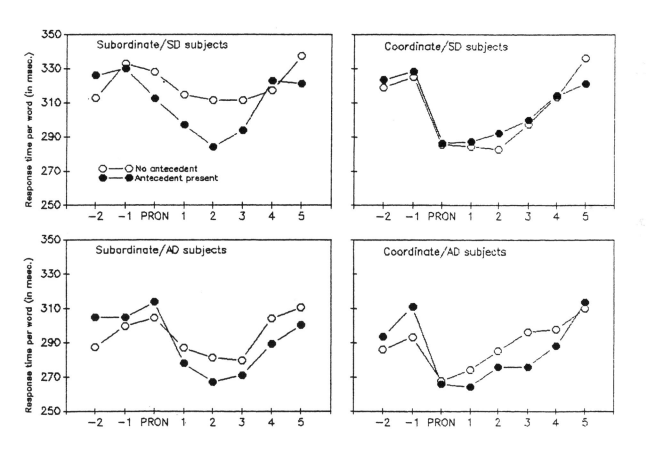

Figure 1. Mean response time per word for SD subjects (upper
panel) and AD subjects as a function of 1) the presence or
absence of an antecedent, 2) the syntactic relation between the
two clauses, and 3) word position relative to the pronoun
('PRON').

The strongest statistical evidence for a contrast between the performance of SD and AD subjects appears when analyses are restricted to just one of these groups at a time. For the SD subjects the Antecedent by Clause Relation interaction is highly significant, $F_1(1,20)=8.89$, $MS_e=1357$, $p<.01$, $F_2(1,22)=8.78$, $MS_e=1470$, $p<.01$, indicating that the apparent contrast between the effects of the Antecedent factor in the two Clause Relation conditions is reliable. In the AD subjects, this same interaction does not approach significance, $F_{1,2}<1$.

On the other hand, the main effect of the Antecedent factor is significant in the results for the AD subjects, $F_1(1,22)=5.64$, $MS_e=1949$, $p<.05$, $F_2(1,22)=4.64$, $MS_e=2092$, $p<.05$, indicating that the antecedent speeded responses generally, without regard to the relation between the clauses. For the SD subjects, this main effect falls well short of significance, $F_1(1,20)=2.42$, $MS_e=1907$, $p>.1$, $F_2(1,22)<1$.

An overall analysis covering results from both subject types produced only inconclusive results. There was an interaction in the by-subjects analysis involving the Antecedent, Clause Relation and History factors, $F_1(1,42)=4.67$, $MS_e=1638$, $p<.05$., $F_2(1,22)=1.2$, NS, as well as a main effect for the Antecedent factor, $F_1(1,42)=7.68$, $MS_e=1929$, $p<.01$, $F_2(1,22)=3.02$, $MS_e=4644$, $p<.1$. The interaction supports the view that the included two-way interaction between the Antecedent and Clause Relation factors is different for SD and AD subjects.

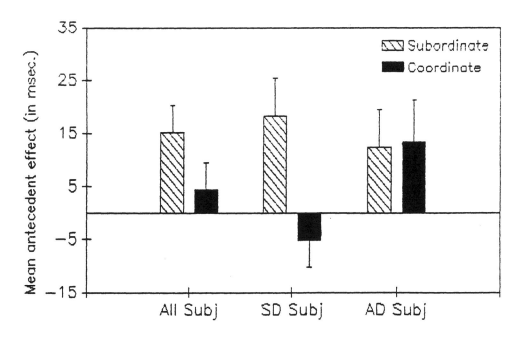

Figure 2. Average antecedent effect (No Antecedent - Antecedent Present) in the two syntax conditions for the three words following the pronoun.

Pilot studies as well as the present experiment suggest that one reliable distinction between SD and AD subjects is that the latter generally respond faster. Though this contrast (the History main effect) is not significant in the by-subjects analysis, it is highly significant in the by-sentences analysis (where it is treated as a within-'subjects' factor), $F_1 (1,42) < 1$, $F_2 (1,22) = 25.7$, $MS_e = 1900$, $p < .001$. Comparing the four SD subject cells at each of eight word positions with the corresponding four AD subject cells shows that the SD subjects were slower in 30 of 32 comparisons, $p < .001$.

These results support two important conclusions. First, there is some syntax-based antecedent-finding mechanism that can influence performance when an antecedent for a pronoun is available in a prior clause that is syntactically integrated with the one bearing the pronoun. Second, effects attributable to such a mechanism are apparent only with subjects who have no left-handers among their close biological relatives.

General Discussion

Pronouns are important from several points of view. Questions about how pronouns are associated with their antecedents define one of the central problems in the theory of discourse processing. These questions bear quite directly on the general organization of the language comprehension system, especially questions about 1) how the diverse kinds of information involved in language comprehension are brought to bear on an incoming utterance, and 2) how the results of diverse analyses are integrated. This in turn can be seen as a special case of the complex of problems in the philosophy of mind that have recently been discussed under the heading of modularity theory (Fodor, 1983).

To properly determine pronoun-antecedent relations, listeners must employ many different kinds of information. Some of the kinds of information used are clearly syntactic, but most are semantic or have to do with discourse structure or knowledge of the world. Modularity theory is consistent with only certain possible accounts of the interface among these various kinds of knowledge. Strictly speaking, the linguistic system is modular in Fodor's sense, so long as there is an informationally-encapsulated parser, regardless of how the syntactic aspects of pronoun-antecedent relations are handled. Nevertheless, there are ways to handle syntactic constraints on pronoun-antecedent relations that would be a serious embarrassment to modularity theory. Suppose that a putatively autonomous syntactic processing system is put in harness with a discourse processing system that, together with various sorts of semantic and discourse analyses, computes c-command relations in the course of assigning antecedents to pronouns. The question would naturally arise as to why other aspects of syntactic analysis might not also be undertaken by this system, thus making the autonomous syntactic processor at least partly redundant. If modularity theory is generally correct, a more consistent outcome would seem to be that an inventory of the capacities of the syntactic processor exhausts the syntactic capacities of the listener, and further, that (conscious reasoning aside) listeners have no capacity to handle syntactic relations apart from what is implemented in the syntactic processing system.

Within this framework, the interface problem for pronouns takes this form: how can the syntactic constraints on pronominal anaphora be implemented without compromising the uniqueness of the various processing subsystems, especially the syntactic processor? Of course, whatever solution is proposed here must respect the fact that for only a relatively small proportion of all pronoun instances will syntactic constraints uniquely and definitively determine an antecedent.

These considerations seem to allow several different ways to organize the interaction between syntactic and discourse processing. One would be for the syntactic processor to add a table to the syntactic representation of each sentence that specifies all possible syntactically acceptable coreference relations within that sentence (cf., Jackendoff, 1972). Another possibility is for the syntactic processor to propose some specific network of coreference relations within each sentence, thus resolving sentence-internal ambiguities. This set of relations is then evaluated by the discourse processor, which has the capacity to revise many of the relations posited by the syntactic processor. The inverse must also be considered; it could be that the syntactic processor makes no assignments of its own, but only evaluates those made by the discourse processor. This would apparently require that there be some mechanism by which it might 'insist' on certain relations, as with reflexives and reciprocals.

The evidence reviewed here suggests that the second of these possibilities is the better model for SD subjects. The large Antecedent effect in the Subordinate condition indicates that something like a reference relation is being assigned, but the extreme sensitivity of this effect to variations in the syntactic relation between the clauses suggests that the mechanism that produces it is essentially syntactic; it seems unlikely that any mechanism that evaluates prospective antecedents in terms of their plausibility or reasonableness in the discourse would be so dramatically sensitive to this sort of syntactic variation. Since these subjects can, presumably, still take the NP in the first clause as the antecedent of the pronoun by later application of discourse processes, these processes seem to be positioned to receive an input from the syntactic processor with some reference relations already specified.

The results for the AD subjects reveal less about the interface between syntactic and discourse processing. The uniformity of the Antecedent effect clearly shows that the mechanism that produces it in these subjects is less sensitive to syntactic structure than is the mechanism controlling the performance of SD subjects. This, however, does not preclude the possibility that some relations are assigned by a syntactic mechanism; it might be that for these subjects the syntax-based assignments are more readily supplemented by those produced by the discourse processor. It does seem clear, however, that a discourse-oriented mode of processing is at least more influential for these subjects than it is for the SD subjects.

The general question about the difference between SD and AD subjects will likely be hard to resolve. Bever, et al., (1986) seem to suggest that, for AD subjects, syntactic and interpretive processing are more intimately integrated, but that these subjects' capacity for syntactic analysis is no less developed than it is in SD subjects. Richer interconnection between syntactic and interpretive modes of analysis simply makes the interpretive modes more salient cognitively and more influential in behavior. Detailed demonstrations of syntactic influences on AD subjects may, however, be difficult to provide.

Though much further research is required, it is clear that the results reported here bear on the two sets of issues raised in the introduction. There does seem to be a syntax-based mechanism for assigning something like a coreference relation. There do seem to be biological differences between subjects that affect the way various modes of language processing are integrated.

Notes

* I am indebted to Tom Bever for a preview of his 1986 Philosophy and Psychology paper (a precursor of Bever, Townsend & Carrithers, 1986), which led directly to the consideration of handedness background in this work, and for further discussions related to these issues. Numbers of others have made valuable contributions to the experimental work described here. These include Deborah Brennan, Heidi Carman, John Dai, Baozhang He, Susan Jasko, Sung-Ae Kim, Julia Sommerkamp, Karen Steensen, and Uma Subramanian. This work was supported in part by a Seed Grant and various small grants from the Office of Research and Graduate Studies of the Ohio State University as well as by various grants from the College of Humanities at OSU.

References

Aoun, J. (1985) *A grammar of anaphora*. Cambridge, MA: MIT Press.

Bever, T. G., Townsend, D. & Carrithers, C. (1986) The quasimodularity of language. Paper presented to the annual meeting of the Psychonomic Society. New Orleans, 1986.

Bolinger, D. (1979) Pronouns in discourse. In T. Givon, *Syntax and semantics Vol 12: Discourse and syntax*. New York and London: Academic Press.

Bosch, P. (1983) *Agreement and Anaphora:A Study of the Role of Pronouns in Syntax and Discourse*. New York: Academic Press.

Chomsky, N. (1981) *Lectures on government and binding*. Dordrecht: Foris.

Chomsky, N. (1986) *Knowledge of Language: Its nature, origin and use*. New York: Praeger.

Corbett, A.T. & Chang, F.R. (1983) Pronoun disambiguation: Accessing potential antecedents. *Memory & Cognition*, *11*, 283-294.

Cornish, F. (1986) *Anaphoric Relations in English and French: A Discourse Perspective*. London: Croom Helm.

Cowart, W. (1986a) Evidence for a strictly sentence-internal antecedent-finding mechanism. In M. van Clay, M. Niepokuj, and V. Nikiforidou (eds.) *Proc. of the 12th Annual Meeting of the Berkeley Linguistics Society*. Berkeley Linguistics Society.

Cowart, W. (1986b) Evidence for structural reference processes. In A. Farley, P. Farley & K.-E. McCullough (eds.), *Proceedings of the 22nd Regional Meeting of the Chicago Linguistics Society*. Chicago: Chicago Linguistics Society.

Cowart, W. & Cairns, H. (in press) Evidence for an anaphoric mechanism within syntactic processing: Some reference relations defy semantic and pragmatic constraints. *Memory and Cognition*.

Dell, G.S., McKoon, G. & Ratcliff, R. (1983) The activation of antecedent information during the processing of anaphoric reference in reading. *Journal of Verbal Learning and Verbal Behavior*, *22*, 121-132.

Fodor, J.A. (1983) <u>Modularity of Mind</u>. Cambridge, MA: MIT Press.

Garnham, A. (1985) <u>Psycholinguistics: Central topics</u>. London &
    New York: Methuen.

Garvey, C. & Caramazza, A. (1974) Implicit causality in verbs.
    <u>Linguistic Inquiry</u>, <u>5</u>, 459-464.

Geschwind, N. & Galaburda, A.M. (1986) <u>Cerebral lateralization</u>.
    Cambridge, MA: MIT Press.

Hirst, W. & Brill, G. (1980) Contextual aspects of pronoun
    assignment. <u>Journal of Verbal Learning and Verbal Behavior</u>,
    <u>19</u>, 168-175.

Jackendoff, R. (1972) <u>Semantic Interpretation in Generative
    Grammar</u>. Cambridge, MA: MIT Press.

Kennedy, A. & Murray, W.S. (1984) Inspection times for words in
    syntactically ambiguous sentences under three presentation
    conditions. <u>Journal of Experimental Psychology: Human
    Perception and Performance</u>, <u>10</u>, 833-849.

Reinhart, T. (1983) <u>Anaphora and semantic interpretation</u>.
    Chicago, IL: Chicago U. Press.

Tyler, L.K. & Marslen-Wilson, W. (1982) Speech comprehension
    processes. In J. Mehler, E.C.T. Walker, M. Garrett,
    <u>Perspectives on mental representation</u>. Hillsdale: Erlbaum.

Inferring Appropriate Responses in Discourse[*]

Melissa P. Chase
The MITRE Corporation
Burlington Road, Bedford, MA 01730

Abstract

This paper discusses how *Scenes*, declarative representations of the intentional and attentional structure of discourse, facilitate the inference of appropriate responses.

1 Introduction

When people engage in conversation, one of the most striking features is the ability of the particpants to infer the meaning of utterances and respond appropriately. Many researchers in discourse processing explain this phenomenon through plan recognition: conversational participants generate and recognize plans to make and understand utterances designed to acheive certain goals. The most successful work along these lines has involved individual utterances [Allen & Perrault 80,Cohen & Perrault 79]. Recently, Grosz and Sidner[Grosz & Sidner 86] have suggested that we require a better understanding of discourse structure in order to extend this work to sequences of utterances in a large discourse.

Grosz and Sidner have proposed a model of discourse structure with three distinct, but interacting components:

[*]This work was supported by MITRE Sponsored Research Project 90780.

1. the structure of the actual sequence of utterances of the discourse

2. a structure of intentions

3. an attentional state

The linguistic structure of the discourse is composed of *discourse segments (DSs)*, which are aggregates of the actual utterances of the discourse. The intentional structure of the discourse consists of *discourse segment purposes (DSPs)*, which specify how the DS's contribute to the overall purpose of the discourse (the DP). Grosz and Sidner have identified two intentional relations that play a crucial role in the structure of a discourse: *dominance* and *satisfaction-precedence*. If an action (physical or linguistic) which satisfies one intention, say DSP1, is intended to partially satisfy another intention, say DSP2, then we say that DSP2 *dominates* DSP1. If DSP1 must be satisfied before DSP2, then we say that DSP1 *satifaction-precedes* DSP2. The attentional structure of the discourse is represented by a collection of *focus spaces*. The attentional state captures the salient objects, properties, and relations at each point of the discourse.

At the MITRE Corporation, we have devised a representation for the intentional and attentional structure of discourse based upon this model, called a *scene*, which serves as the basis for the discourse component of the KING-KONG system (a transportable natural language interface for expert systems). By keeping track of the current intentional and attentional state discourse, KING-KONG is able to reason about the underlying goals and intentions of the user in order provide appropriate responses. In this paper, I would like to briefly describe scenes (for a fuller description, see[Zweben & Chase 87]) and then show how discourse structure contributes to the recognition of the speaker's plans and facilitates appropriate and intelligent response.

2 Scenes

A *scene* is a schema representation, similar in spirit to frames[Minsky 75], scripts[Schank & Abelson 77], and related formalisms[Bobrow & Collins 75].

THE INTENTIONAL STRUCTURE		THE ATTENTIONAL STRUCTURE	
Field	**Description**	**Field**	**Description**
Name	The type of scene	Role–Fillers	The objects filling the roles
Roles	The prominent object classes	Predecessors	The scenes preceding this one
Inferiors	The scenes dominated by this one		in the actual discourse
Superiors	The scene dominating this one	Successors	The scenes following this one
Enables	The post-requisite scenes		in the actual discourse
Enabled–by	The pre-requisite scenes	Focus Cache	The objects available for
Actions	The expert system operations		anaphoric references
	appropriate to this scene	Domain Goal	The current expert system task

Figure 1: The Slots of an Instantiated Scene

A scene defines the potential intentional structure of the discourse of an interaction with the expert system by defining the user's intended actions and their relationships. An *instantiated* scene captures the attentional structure of the discourse by recording which intentions have been satisfied, and what objects and expert system operations were involved.

In a typical expert system, the user wishes to carry out some task which is often decomposed into subtasks. The tasks involve a limited number of operations, which can be performed on a limited number of object classes. Thus, each scene contains information about the possible object classes involved in a task, called *roles*; the potential expert system operations on these roles, called *actions*; and the relations between different tasks, the *inferior*, *superior*, *enables*, and *enabled-by* scenes. Together, these pieces of information permit the computation of the intentional structure of the actual discourse.

When a scene is recognized as the current intentional state, it is instantiated in order to represent the attentional state. Its roles and focus cache are filled with the referents of the objects in the current utterance,

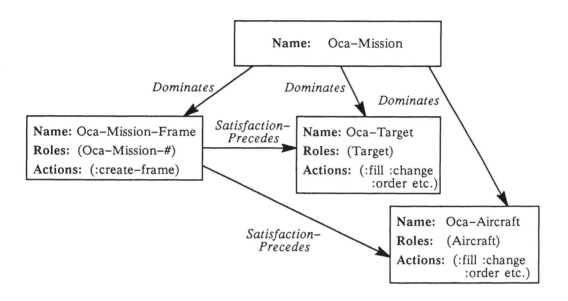

Figure 2: The Intentional Structure

predecessor/successor links are created which model the actual flow of the interaction, and the goal (which expert system task is involved) is recorded. See Figure 1 for a description of the most important information maintained in a scene.

To make this description a little more concrete, consider the following example drawn from the Knowledge-Based Replanning System (KRS) mission planning application. The primary goal of this application is to plan an Offensive Counter Air (OCA) mission. In order to achieve this goal, several choices must be made, such as the target and the type of aircraft. The system is a mixed-initiative expert system, which is capable of fully planning missions, or can guide a user planning a mission by verifying that the user has made appropriate choices and suggesting suitable choices on request. The root of the scene hierarchy corresponds to the DP of the overall discourse–planning an OCA mission. The inferior scenes model the intentions of fulfilling the subtasks of the mission planning task. Each scene has a single role and a number of expert system actions (for making a choice, changing a choice, requesting suitability information, etc.). Figure 2 depicts the intentional structure and Figure 3 the attentional structure generated

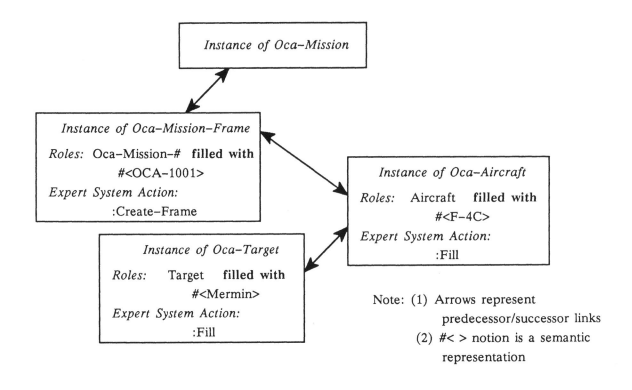

Figure 3: The Attentional State

by the following discourse fragment:

> **User:** Create a mission.
> **Computer:** Displays a new mission template.
> **User:** Send F-4Cs.
> **Computer:** Fills in the aircraft slot of the template with F-4C.
> **User:** Attack Mermin.
> **Computer:** Fills in the target slot of the template with Mermin.

3 Inferring Appropriate Responses

In order to respond appropriately to an utterance, the system must recognize the user's intention in making this utterance within the context of the discourse. This involves recognizing, through use of the intentional and attentional structures of the discourse, both the linguistic act and the domain act intended by the user.

In an expert system interface, the linguistic acts performed by the user are typically requests to inform or requests to act; these requests are made to help satisfy the DP of the interaction, namely, to carry out some expert system task. To respond to these requests, the system must understand the intended meaning of the request, reason about the role of this intention in the overall intentional structure of the discourse, and select an appropriate response based upon the attentional structure of the discourse (the objects and DSPs in focus).

3.1 Recognizing the User's Intention

In many natural language systems, the process of recognizing the intentions of utterances involves describing the user's speech acts as plans and recognizing them[Allen & Perrault 80], [Cohen & Perrault 79], [Sidner & Israel 81], [Sidner 83b], [Litman 86]. In the KING-KONG interface, these acts are described as schemata of patterns to be matched, and the Response Handler (the portion of the system that determines the appropriate response to an utterance) uses the intentional and attentional information in the current scene and linguistic and semantic information from the utterance to recognize the user's plan and respond accordingly. Since each scene encapsulates a collection of objects (the prominent roles of the scene) and the domain actions that manipulate these objects, the amount of inferencing involved in recognizing the user's plan is constrained.

The patterns in the schemata essentially describe the pre-requisites and post-requisites of the plan corresponding to the user's intentions. Each action schema contains the following patterns to be matched:

1. *Scenes* – a list of scenes appropriate for the action, or a list of scenes inappropriate for the action

2. *Speech Act* – the linguistic action of the utterance

3. *Semantic Representation* – the semantic representation of the utterance, either a backend operation and its arguments, or a database relation and its arguments

4. *Scene Roles* – the roles from intentionally related scenes that must already be filled

5. *Effects* – the effects upon the intentional and attentional structure of the discourse, and the backend, of executing this action

When the Scene Controller, the portion of the system that recognizes the intentional and attentional states of the discourse, proposes a scene to the Response Handler, the Response Handler tries to match each action schema against information from the utterance and the current scene. First, the Response Handler checks to see if the scene is appropriate to the action. Then, linguistic features from the utterance are used to deduce the speech act. Next, the system infers the domain action of the request by examining the semantic interpretation of the utterance (determined, in part, by the scene). From the intentional structure encoded in the scene, and the linguistic act and the semantic interpretation of the utterance, the Response Handler infers the intention of the utterance; this includes determining whether the intention is part of the current discourse segment or another one. Finally, the system considers the attentional state contained in the scene (the roles) in order to determine how much of the user's plan has been satisfied. By default, all satisfaction-precedence intentions should have been satisfied, but the schema description permits one to relax this condition. If the match is successful, the Response Handler transmits the effects to the Scene Controller, so the intentional and attentional states can be updated, and issues the appropriate commands to the expert system to execute the specified action.

A couple of examples will clarify how this mechanism allows KING-KONG to provide intelligent responses. The first example illustrates how enabling actions are inferred and executed. The second example demonstrates how an underlying request for information is inferred.

3.2 Example: Implied Actions

In the KRS domain, the choice of target is so central to planning an OCA mission that specifying a target signals the intention of creating a new mission. So, if the user begins a planning session by saying "Make Mermin the target," the system should first create a new mission and then fill in the target slot with Mermin. KING-KONG does indeed carry out this interaction.

The intentional structure of this interaction, captured in the intentional relations encoded in the scene hierarchy is:

> DSP(oca-mission) *dominates* DSP(oca-target)
>
> DSP(oca-mission-frame) *satisfaction-precedes* DSP(oca-target)

These intentional relations describe this formal plan:

> INTEND(user, INTEND(computer, DO(plan-mission))) $\land$
> INTEND(user, INTEND(computer, DO(mission-frame))) $\land$
> INTEND(user, INTEND(computer, DO(fill(target, Mermin)))) $\land$
> BELIEVES(user, GENERATES(plan-mission,
> $\qquad\qquad$ mission-frame $\land$ fill(target, Mermin))) $\land$
> BELIEVES(user, ENABLES(mission-frame, fill(target, Mermin)))

When the Scene Controller proposes the *oca-target* scene, the following action schemata both match:

> *Name:* fill-action
> *Scenes:* (oca-target)
> *Speech Act:* :act
> *Semantic Representation:* (:fill (current-scene prominent-roles))
> *Scene Roles:* (enabling-scene prominent-roles :optional)
> *Effects:* Discourse: DSP(enabling-scene) satisfied $\land$
> $\qquad\qquad\qquad$ DSP(oca-target) satisfied
> $\qquad\qquad$ Backend: fill target in oca-mission frame

and

> *Name:* create-mission-frame-action
> *Scenes:* (oca-target)
> *Speech Act:* :act
> *Semantic Representation:* (:fill (current-scene prominent-roles))
> *Scene Roles:* (enabling-scene prominent-roles :absent)
> *Effects:* Discourse: DSP(enabling-scene) satisfied
> Backend: create oca-mission frame

At this point in the discourse the intention of the utterance "Make Mermin the target" is INTEND(user, INTEND(computer, DO(fill(target, Mermin)))). This intention is recognized because features of the utterance and the state of the discourse represented in the scene match the specifications of both schemata. The domain action is recognized by matching the semantic representation specified in the schemata against the interpretation of the utterance (i.e., to fill the target slot with Mermin). The force of the utterance (a request to act) is deduced by matching the linguistic action of the utterance (an imperative) against the speech act specified in the schemata and determining how far the plan has progressed from the attentional information in the scene hierarchy (what scene is proposed and which intentionally related scenes have had their discourse purposes satisfied). Further, the *fill-action* schema indicates that the satisfaction-precedence intentions of the scene need not be filled (since the prominent-roles of the enabling scene—the scene that *satisfaction-precedes* the current scene—need not have been filled). Since the mission frame has not been created, the schema matches. Similarly, the *create-mission-frame-action* schema specifies that the satisfaction-precedence intention *has not* been satisfied, so this schema also matches. The effects specified in the two schemata ensure that the *create-mission-frame-action* is executed before the *fill-action*.

3.3   Example: Underlying Intentions for Information

Requests for information typically are made to support the satisfaction of the intentional structure of the planning session. Some of these questions directly involve roles and their attributes, and the intention of the request to

832

obtain the information needed to fill a mission frame slot with an appropriate choice.

Other questions are more oblique and do not directly request information about a role. For example, consider the following discourse:

> **User:** Send 4 F-4Cs from Halfort to Mermin.
>
> **Computer:** Displays a new mission template and fills in the slots.
>
> **User:** Leave at 0330 hours.
>
> **Computer:** Fills in the time of departure slot.
>
> **User:** What is an F-4C?
>
> **Computer:** An F-4C is an oca-aircraft.
>
> **User:** Is the mission at night?
>
> **Computer:** Yes. A mission at 0330 hours should not be flown by any of Halfort's aircraft.

The intention underlying the final question reflects a concern about the night-flying abilities of the aircraft.

The intentional structure of this interaction is simply:

$$\text{DSP(oca-mission)} \; \textit{dominates} \; \text{DSP(oca-aircraft)}$$

which represents the plan:

$$\text{INTEND(user, INTEND(computer, DO(plan-mission)))} \; \wedge$$
$$\text{INTEND(user, INTEND(computer, DO(fill(aircraft, X))))} \; \wedge$$
$$\text{BELIEVES(user, GENERATES(plan-mission, fill(aircraft, X)))}$$

The first scene proposed by the Scene Controller is the *oca-aircraft* scene, since that was the previous intentional and attentional state of the discourse. In this context, the following schema matches:

Name: aircraft-inference
Scenes: (oca-aircraft)
Speech Act: :inform
Semantic Representation: (verify(possession(mission-time, night)))
Scene Roles: (:or (current-scene prominent-roles :present)
                      (constraining-scene prominent-roles :present))
Effects: Discourse: DSP(current-scene) satisfied
         Backend: respond to inform request ∧
                  verify role choices consistent with response

The *aircraft-inference* schema specifies that in the context of the *oca-aircraft* scene, requests for information about the mission time are relevant to the discourse purpose of the scene. Thus, the intentional structure of the interaction matches the intention of this request. From the attentional state of the discourse, either the prominent-role of the scene (the aircraft in this case) or the prominent-roles of constraining scenes (scenes that do not *dominate* or *satisfaction-precede* the current scene, because the order in which the scenes are traversed is flexible, but scenes whose intentions involve domain constraints; for example, the choice of airbase and aircraft constrain one another) must have been filled; that is the intentions of the current scene or constraining scenes must have been satisfied. This is the default specification for a request having an underlying intention, and the response it triggers is to answer the surface request for information, and to verify that the role choices specified by the Scene Roles is consistent with the result of the inform act. Hence, the system responds as shown in the above fragment.

This particular schema is extremely specific in its plan; currently we are generalizing this inferential ability. The schema will encode various relationships and attributes that constraint the task represented by a scene (described by the combination of roles and expert system operations), and the effect will be to verify that a particular choice satisfies these constraints, to generate choices and filter them by these constraints, and so on.

4 Conclusion

We have implemented a system based upon the discourse structure model proposed by Grosz and Sidner. In a restricted domain, such as an expert system interface, we have discovered that it is possible to use a representation, called *scenes* to capture the intentional and attentional state of a discourse, and use this information to reason about a user's intentions.

We do not suggest that this approach can solve the general discourse processing problem, but it provides a mechanism for tracking discourse and understanding intentions in a restricted context. To make this system more flexible, we intend to generalize the reponse component, investigate how to acquire new scenes[Mooney & DeJong 85] and reason about misconceptions[Pollack 86] within the scene framework.

References

[Allen & Perrault 80] Allen, J.F. & Perrault, C.R. (1980). Analyzing Intention in Utterances. *Artificial Intelligence*, 15:143–178.

[Bobrow & Collins 75] Bobrow, D. & Collins, A.M. (1975), editor. *Representation and Understanding*. Academic Press, New York.

[Cohen & Perrault 79] Cohen, P. & Perrault, C.R. (1979). Elements of a Plan-Based Theory of Speech Acts. *Cognitive Science*, 3:177–212.

[Grosz & Sidner 86] Grosz, B.J. & C. L. Sidner, C.L. (1986). Attentions, Intentions, and the Structure of Discourse. *Computational Linguistics*, 12:175–204.

[Litman 86] Litman, D.J. (1986). Linguistic Coherence: A Plan-Based Alternative. In *Proceedings of the 24th Annual Meeting of the Association for Computational Linguistics*, pages 215–223, New York.

[Minsky 75]   Minsky, M. (1975). A Framework for Representing Knowledge. In P. H. Winston, editor, *The Psychology of Computer Vision*, McGraw-Hill, New York.

[Mooney & DeJong 85] Mooney, R. & DeJong, G. (1985). Learning Schemata for Natural Language Processing. In *Proceedings of the Ninth International Joint Conference on Artificial Intelligence*, pages 681–687, Los Angeles.

[Pollack 86]   Pollack, M.E. (1986). A Model of Plan Inference that Distinguishes Between the Beliefs of Actors and Observers. In *Proceedings of the 24th Annual Meeting of the Association for Computational Linguistics*, pages 207–214, New York.

[Schank & Abelson 77] Schank, R.C. & Abelson, R.P. (1977). *Scripts, Plans, Goals and Understanding*. Lawrence Erlbaum Associates, Hillsdale, N. J.

[Sidner & Israel 81] Sidner, C.L. & Israel, D.J. (1981). Recognizing Intended Meaning and Speakers' Plans. In *Proceedings of Seventh International Joint Conference on Artificial Intelligence*, pages 203–208, Vancouver, B. C.

[Sidner 83b]   Sidner, C.L. (1983b). What the Speaker Means: The Recognition of Speakers' Plans in Discourse. *International Journal, of Computers and Mathematics*, 9:71–92.

[Zweben & Chase 87] Zweben, M. & Chase, M.P. (1987). Scenes: A Representation of Intentional and Attentional Structure. Submitted to AAAI-87, Seattle, WA.

Using Intentional and Attentional Structure for Anaphor Resolution

Monte Zweben*
MITRE Corporation
Burlington Road, Bedford, MA 01730
(617) 271-7026

12 February 1987

Abstract

This paper describes the *Scenes* knowledge representation that captures the intentional and attentional structure of discourse. Using this information a natural language interface can isolate context and resolve anaphors with focusing heuristics. Further, anaphor resolution can be coordinated with interruptions so that completed digressions are ignored.

1 Introduction

One of the goals of the KING-KONG Expert System Interface developed at the MITRE Corporation is to perform anaphoric resolution using a model of discourse. Grosz and Sidner [Grosz & Sidner 86] claim that any discourse has three main constituents: 1) the structure of the actual sequence of discourse utterances; 2) a structure of intentions; 3) an attentional state. This paper describes *Scenes* [Zweben & Chase 87] which are declarative knowledge representations of the intentional and attentional structure of discourse that facilitate anaphor resolution. Utilizing the attentional structure stored in scenes, anaphors can be resolved. Further, since scenes delineate interruptions, resolution strategies can correctly ignore antecedents that reside in interruptions. This paper describes the discourse model underlying *scenes*, the *scene* mechanism and finally, the anaphor resolution algorithm employed.

2 Intention and Attention

Grosz and Sidner distinguish between the intentional state of discourse and the attentional state. Intentional structure represents the underlying purposes that causally relate the utterances of a coherent discourse. Attentional state, on the other hand, captures the focus of attention in the discourse at any one moment, by recording the salient objects and relationships.

---

*This work was supported by MITRE Sponsored Research Project 90780.

2.1 Intentional Structure

Discourses can be partitioned into segments, each representing some purpose or intention. These discourse segment purposes (DSP's) can be related in special ways. Grosz and Sidner present two kinds of DSP relationships: dominance and satisfaction-precedence. The satisfaction-precedence relation represents one intended action being a pre-requisite of another. The dominance relation states that satisfying one intended action contributes to the satisfaction of another. This relation establishes a hierarchical structure of DSP's representing their dependencies. The intentions that dominate each other depend upon the type of discourse (e.g., general conversation vs. task-oriented dialogue). An expert system interface is primarily concerned with the computer accomplishing tasks. The DSP dominance relation, in an expert system interface, adopts the task-oriented dialogue rule presented here:

$\forall_{i=1,...,n}[Intend(user, Intend(computer, Do(A)))\ \bigwedge$
$Intend(user, Intend(computer, Do(a_i)))\ \bigwedge$
$Believe(user, Generates(A, a_1, a_2, , ..., a_n))]$
$\longleftrightarrow$
$Dominates(Intend(user, Intend(computer, Do(A))),$
$Intend(user, Intend(computer, Do(a_i))))$

A general interpretation of the above is: If the user intends that the expert system executes tasks A and a_i , and the user believes that the performance of task a_i contributes to the performance of task A, then the intention concerning task A dominates the intention of task a_i.

2.2 Attentional Structure

The attentional structure captures the focus of attention in a discourse. It represents the prominent objects and relationships that are dynamically encountered in conversation. The attentional state is modeled by a set of focus spaces and rules for transitioning among them. Focus spaces are paired with their respective discourse purposes to associate intentional state with attentional state. One can view a focus space as the representation of *what* the discourse participant is talking about, while its association with intentional state explains *why*.

3   Knowledge Representation - Scenes

Scenes are knowledge representations of the intentional and attentional states of discourse. They are schema representations [Minsky 75], [Bobrow & Collins 75] , [Schank & Abelson 77] of plans of stereotypical interactions with the expert system. The hierarchical structure of scenes represents the dependencies of the user's intended actions according to the dominance relation defined for intentional structure. Thus, the root of a scene hierarchy represents the overall discourse purpose (DP) of the dialogue, and each remaining scene , in a hierarchy, supports its dominating scene.

In addition to intentional structure, scenes constrain the attentional structure of a discourse by defining the kinds of objects that would be prominent if a scene were active. These object descriptions, called the *roles* of a scene, represent the players participating in the action that the scene represents. Scene recognition is directed by the roles that are observed in the conversation; this process is described in detail later.

When a scene is appropriate (i.e., recognized as the current intentional state), it is instantiated to represent attentional state. Its roles are filled with the referents of the objects in the current clause and other objects already present in the discourse. The preceding scene is linked to the new one, maintaining a predecessor/successor network of scenes modeling the discourse.

Scenes			
The Intentional Structure		**The Attentional Structure**	
Field	Description	Field	Description
Name	The type of scene.	Role–Fillers	The semantic representations of those objects filling the roles.
Roles	The prominent object classes.		
Inferiors	The scenes that this one dominates.	Predecessors	The scenes preceding this one in the actual discourse.
Superior	The scene dominating this one.		
Enables	The post-requisite scenes.	Sucessors	The scenes following this one.
Enables-by	The pre-requisite scenes.	Focus Cache	The objects available for anaphoric reference.
Triggers	The lexical items that are recognized for this scene and their filtering maps.	Expert System Goal	The abstract expert system actions to apply.

Figure 1: The slots of a scene.

An important distinction must be made between a plan and an intentional scene hierarchy. The scenes represent **stereotypical interactions** with an expert system. However, they do not represent the sequence of actions an expert system will take. This information is captured in the expert system's plans, goals and problem-solving strategies. Only data relevant to the user-machine interface is captured by the scene hierarchy.

To clarify the exposition of a scene hierarchy, the following figures present examples of scenes from the KRS mission planning application. The primary goal of this application is to plan a OCA mission task. In order to complete the plan, among other things, a target, an airbase, and a type of aircraft must be chosen. The KRS system is a mixed-initiative system which can fully plan missions or guide a user along using its constraint satisfaction mechanism. The following scenes represent the interactions with a user planning a mission. Figure 2 represents the intentional and attentional structure of the following discourse.

1. Build a mission.

2. Leave from Halfort

3. Send F-4cs.

4. Make Mermin the target.

5. What is the range of an F-4c?

By capturing both intentional and attentional state, our expert system interface demonstrates the ability to perform anaphoric reference. Intentional structure (i.e., scene hierarchies) enables the response handler to perform limited plan recognition [Allen & Perrault 80], [Cohen & Perrault 79], [Sidner & Israel 81], [Sidner 83b], [Litman 86], which isolates context, while attentional structure provides the dynamic information about objects in conversation. The next section describes scene recognition. Subsequently, an extensive demonstration of the system is presented followed by a description of the anaphoric reference algorithm.

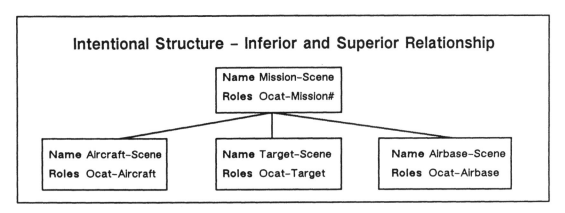

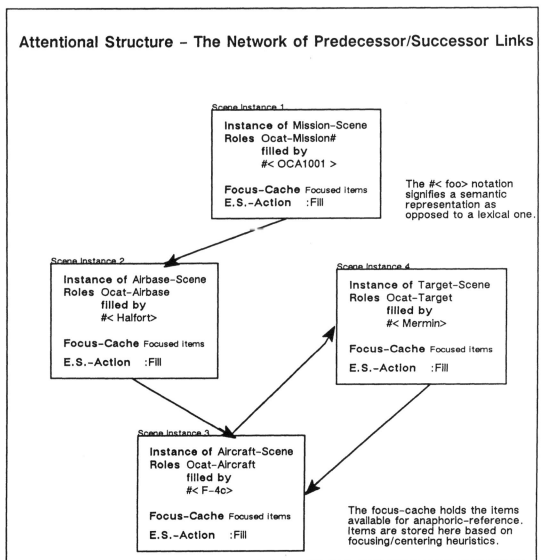

Figure 2: An Example of Intentional and Attentional Structure.

4 Scene Recognition

Scene recognition is the process of determining attentional state. This process is failure driven; a new scene is found if the response mechanism is unable to interpret the input in the current scene. The response handler can either tell the scene controller what scene to move to, or it can instruct the scene controller to use its discourse heuristics to find the new attentional state.

4.1 General Control Flow

Scene recognition is a generate and test process in which heuristics guide the generation of possibilities and roles filter them. When a new scene is required, intentional and attentional structure is used to provide new possibilities. For each scene proposed, the current input clause is tested against the lexical triggers of the scene, which maps the head verb, the arguments and the modifiers of the sentence to the roles of the scene. The goal of this test is to determine whether the referents of the semantic arguments and modifiers in the sentence match the role description specified in the lexical trigger mapping. If all the referents are consistent with the role description, the proposed scene becomes the current scene and is inserted into the predecessor/successor network. Here is an example of an inconsistent match:

Lexical Trigger: Hit [OBJ → Airbase, INSTR → Aircraft]

Input Clause: Hit *the tank* with *an ordnance*.
   Both the object and instrument violate the lexical trigger map.

If the heuristics fail to provide a current scene, the user is asked what context his utterance pertains to (ie. which scene is appropriate) and is then requested to re-phrase his input. In the future, we hope to provide intelligent failure mechanisms with the ability to learn new scenes [Mooney & DeJong 85] and reason about misconceptions [Pollack 86].

4.2 Scene Heuristics

Some of the scene heuristics that currently generate possibilities for scene recognition are:

1. *Intentional Clues* - Choose a context that follows the plan.

   - preceded-by - Try all the post-requisites.

   - precedes - Try all the pre-requisites.

   - superior - Try the more general scene.

   - inferiors - Try the supporting scenes.

   - siblings - Try the scenes at the same intentional level.

   - all-relatives - Try all the scenes that are causally related.

2. *Attentional Clues* - Choose a context that was recently referred to.
   Backtrack through a scene's predecessor/successor network.

3. *Interrupt* - Find a new scene in a different dominance hierarchy.
   Sequence through the contextual lexicon to find a scene that recognizes the main verb of the sentence.

4. *Ask* - Query the user for the current context.
   For each known scene, ask the user whether it is the intended context.

841

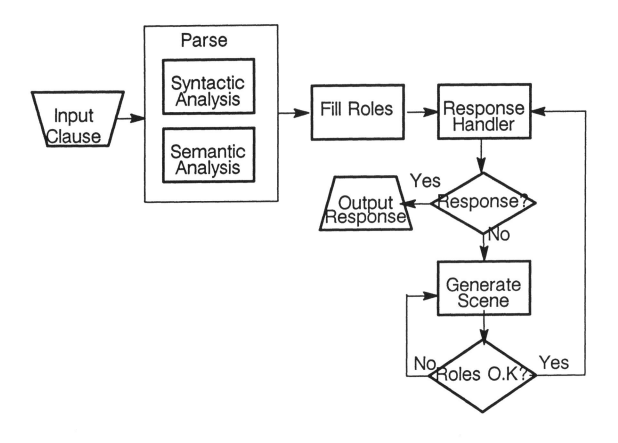

Figure 3: Overall Control Flow

5 Implementation

The entire interface is implemented in Zetalisp on Symbolics lisp machines. Scene instances are represented as flavor instances with instance variables for the fields presented and methods for executing the heuristics. Scene recognition is handled by the scene controller, which is also implemented as a flavor instance to retain dynamic information and to facilitate integration with the expert system's i/o loop. The overall control structure is shown in Figure 3.

6 Demonstration

- Pane 1 is the KRS main window with a command entered to plan a mission.

- In response to the first sentence, KRS creates the mission frame shown in Pane 2 with the appropriate slots filled in. The sentences in Pane 2 demonstrate the ability to respond intelligently and the ability to respond differently to the same utterance in different contexts.

- In pane 3, f-111es are made prominent and then they are referred to with a pronoun.

- Pane 4 is a digression into a refueling context.

- Pane 5 demonstrates that the scene maintained its attentional state with an anphoric reference to the salient F-111e. The conversation ends with a contextual inference that used the context of targets to constrain the weather query.

ATO-LEVEL WINDOW

AIR TASKING ORDER FOR 10FEB87

AUTOPLAN	REPLAN	INTELLIGENCE	DISPLAY	READ/WRITE	SPECIAL-MENU	RESHAPE
PKG		OCA	SSM		AEM	OTHER

(All missions)

> send 4 F-4cs from belfort to nornin

Interaction Pane

02/09/87 14:07:07 MURRE

U.SER:

RADC/COES - MITRE-BEDFORD

KRS ATO-LEVEL WINDOW   AIR TASKING ORDER FOR 10FEB87

AUTOPLAN	REPLAN	INTELLIGENCE	DISPLAY	READY/ABORT	SPECIAL (MENU)	RESHAPE	OTHER
PRG	OCA		SSM		AEW		OTHER

DCA1001   MERMIN   HALFORT   F-4C

OCA1001   2

Autoplan	Switch	Exit	Show Itinerary	Menu

09-FEB-87 14:00

```
AIRBASE:
AIRCRAFT: F-4C
ACNUMBER: 4
ID:
UNIT:
REFUELSVC:
SUPERIOR:
```

TARGET: MERMIN
ORDNANCE:
PD:
TOT:
CALL-SIGN: ~Automatic~
FREQ: ~Automatic~
TRANSPONDER: ~Automatic~

>what is the range of an F-4c

The range of F-4C is 600 MILES

MERMIN is within range of HALFORT's F-4C.
>forget the homebase

>what is the range of an F-4c

The range of F-4C is 600 MILES

MERMIN is within range of F-4C at HALFORT, RASTOR and SPANDEL.

>

OCA1001

Interaction Pane

02/03/87 14:03:41 MURIE   USER:

844

KRS

ATO-LEVEL WINDOW

AIR TASKING ORDER FOR 10FEB87

AUTOPLAN	REPLAN	INTELLIGENCE	DISPLAY	READ/WRITE	SPECIAL-MENU	RESHAPE
PKG	OCA		BSM		AEW	OTHER
▶ OCA1001	MERMIN	HALFORT	F-4C	‡‡	‡‡‡‡	‡‡‡‡‡‡

OCA1001

Autoplan	Switch	Exit	Show Itinerary	Menu	09-FEB-87 14:10	3

```
TARGET: MERMIN                    AIRBASE: HALFORT
ORDNANCE:                         AIRCRAFT: F-4C
PD:                               ACNUMBER: 4
TOT:                              ID:
CALL-SIGN: ~Automatic~            UNIT:
FREQ: ~Automatic~                 REFUELSVC:
TRANSPONDER: ~Automatic~          SUPERIOR:
```

>what is an F-111e

F-111E is a NIGHT-AC
>what airbases have them
Found Referent = F-111E in High-Focus in B<SCENE KONG:OCA1-AC from KONG:NOWHERE 34332513> Content

F-111E is at ALABAR
>leave from halfort

>make kc-135 the refl-aircraft █

OCA1001

Interaction Pane

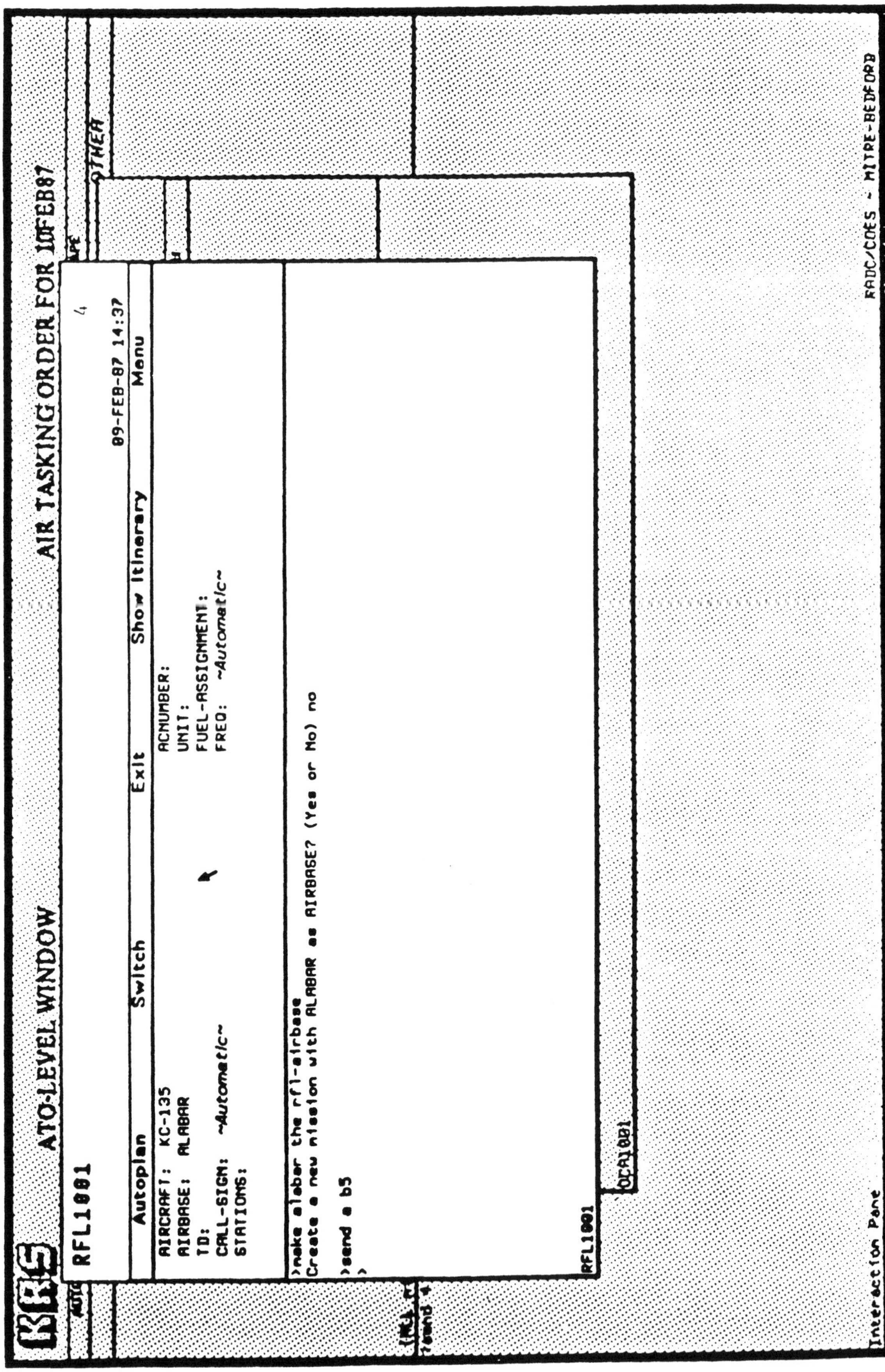

KRS ATO-LEVEL WINDOW

AIR TASKING ORDER FOR 10FEB87

OTHER

RFL1001

| Autoplan | Switch | Exit | Show Itinerary | Menu |

09-FEB-87 14:37

AIRCRAFT: KC-135
AIRBASE: ALABAR
TD:
CALL-SIGN: ~Automatic~
STATIONS:

ACNUMBER:
UNIT:
FUEL-ASSIGNMENT:
FREQ: ~Automatic~

>make alabar the rfl-airbase
Create a new mission with ALABAR as AIRBASE? (Yes or No) no

>send a b5
>

RFL1001

local001

Interaction Pane

02/09/87 14:38:10 MONTE   USER: RADC/COES - MITRE-BEDFORD

846

KRS

ATO-LEVEL WINDOW   AIR TASKING ORDER FOR 10FEB87

RFL1001   OTHER

OCA1001

| Autoplan | Switch | Exit | Show Itinerary | Menu |

                                              09-FEB-87 14:21

TARGET: MERMIN   AIRBASE: HALFORT
ORDNANCE: B5   AIRCRAFT: F-4C
PD: ACNUMBER: 4
TOT: TD:
CALL-SIGN: ~Automatic~ UNIT:
FREQ: ~Automatic~   REFUELSVC:
TRANSPONDER: ~Automatic~ SUPERIOR:

>how fast are they
No Referent found in the B<SCENE OCAT-ORDNANCE from KONG:NOWHERE 34711753> Context
No Referent found in the B<SCENE OCAT-AIRBASE from KONG:NOWHERE 34710322> Context
Found Referent = F-111E in High-Focus in B<SCENE KONG:OCAT-AC from KONG:NOWHERE 34705047> Context

The speed of F-111E is 540 KNOTs
>does the target have sams

Yes for MERMIN.
>what is the weather

The weather at MERMIN is CLOUD-HEIGHT: 9 CLOUD-COVER: 10 VISIBILITY: 1.
>

OCA1001

Interaction Pane

02/09/87 14:23:14 MONTE   USER: RADC/COES - MITRE-BEDFORD

847

7 Anaphors

Pronouns are processed using simple focus heuristics [Sidner 83a], [Reichman 85]. Each scene has an ordered focus cache of semantic representations that are salient. After each sentence is processed, its semantic arguments are pushed onto the current scene's cache. When a pronoun is encountered, the focus cache is searched for a referent. This search requires a semantic analysis that checks whether the cached object makes sense as a substitute for the pronoun. If successful, the object is moved to the front of the list and a message is supplied to the user to inform him of the pronoun resolution. Otherwise, a message that no referent was found in the current scene is provided followed by recursive search in the previous scene's focus cache. Currently, the system does not forget focussed objects and backtracks exhaustively through the scenes until successful. Anaphor resolution is integrated with an interruption component of the scene controller so that completed interruptions do not provide possible antecedents, which is the topic of the next section.

7.1 Interruptions

When the scene controller chooses a scene in a different scene hierarchy, it is changing its expectation of the overall intention of the user. This represents an interruption which is flagged accordingly in the new scene. When the digression is complete and the user returns to the old context, the old context is marked. When searching for antecedents, this region, which represents a full interruption, is skipped. This is shown in the demonstration when the system is asked, "How fast is it ?". The most recent antecedent is the *kc-135* referred to in the interruption. Instead, the interruption is skipped and the *F-111e* is chosen. This demonstrates the utility of scenes and their ability to retain and manage attentional state. Even though the interruption causes a context change, the salient objects (e.g. *F-111e*) are maintained and the interruption is marked, thus enabling the correct anaphor resolution.

7.2 Other Approaches

This approach to anaphoric resolution differs from previous attempts in the following manner: the algorithm is based upon a discourse model that distinguishes intentional and attentional structure. This integrates anaphor resolution with context processing, resulting in the ability to "skip" over interruptions as possible placeholders for antecedents. Most commercial systems simply search for the most recent NP, which often leads to strange resolutions. Further, the use of focus heuristics is not sufficient to provide intelligent resolutions. The most desirable approach is to use focus heuristics on top of a discourse model, which was originally proposed by Grosz and Sidner [Grosz & Sidner 86].

However, the scenes mechanism differs from Grosz and Sidner's model in the representation of attentional state. We maintain a predecessor/successor graph of focus spaces, while Grosz and Sidner use a focus stack. In their model, focus spaces are pushed onto the stack until completed. Anaphors can be resolved with antecedents found by iterating through the focus spaces on the stack until one of two conditions: an interruption is found or the bottom of the stack is encountered. When a focus space is is completed, it is popped off the stack, thereby making its focused objects inaccessible. Hence, completed interruptions are no longer available to provide possible antecedents. If the same intentional state is returned to later, a new focus space is instantiated, which ignores the objects that were prominent in the old space. In the scenes system, completed focus spaces are flagged as such, but they maintain their focus cache. If a discourse participant returns to a closed scene, the cached items in the focus list are still available for reference. While this capability exists in the scene mechanism, the linguistic evidence for this behavior is not conclusive. Nevertheless, users of the natural language system seem to refer to objects that are prominent in completed focus spaces. Grosz and Sidner claim that all these items must be re-introduced to be available for anaphoric reference. Here is an interaction which seems contradictory to this assumption.

User: Are F-4cs and F-111es night aircraft ? *(Aircraft Scene)*

Computer: *Yes*

User: Send an F-4c. *(In a stack model, this focus space would now be popped)*

Computer: *O.K.*

User: Send a B10. *(Ordnance Scene)*

Computer: *F-4cs do not carry B10's.*

User: Send the other one. *(In a stack model, the referent would have been popped)*

The scenes mechanism is able to skip over interruptions without enforcing the stack model. The discourse is represented as a predecessor/successor graph of the scenes used, possibly marked with interruption boundaries. Anaphors are resolved by backtracking through this graph, skipping interruptions. Currently, this graph is never pruned creating a very large structure in lengthy interactions. We recognize the problems of maintaining this graph but the use of the stack model seems too drastic because of examples like the one above. More research is necessary to determine the correct model of attentional state.

8 Conclusion

We have designed a natural language interface that makes extensive use of the **Scene** knowledge representation. Scenes are based upon Grosz and Sidner's model of discourse that distinguishes attentional and intentional structure. This knowledge representation facilitates limited plan recognition which establishes the context of an utterance. Further, it captures the salient objects of a discourse as well as maintaining a model of a user's interaction. Utilizing this information, the interface resolves anaphoric references.

References

[Allen & Perrault 80]   Allen, J.F. & Perrault, C.R. (1980). Analyzing Intention in Utterances. *Artificial Intelligence*, 15:143–178.

[Bobrow & Collins 75]   Bobrow, D. & Collins, A.M. (1975), editor. *Representation and Understanding*. Academic Press, New York.

[Cohen & Perrault 79]   Cohen, P. & Perrault, C.R. (1979). Elements of a Plan-Based Theory of Speech Acts. *Cognitive Science*, 3:177–212.

[Grosz & Sidner 86]   Grosz, B.J. & C. L. Sidner, C.L. (1986). Attentions, Intentions, and the Structure of Discourse. *Computational Linguistics*, 12:175–204.

[Litman 86]   Litman, D.J. (1986). Linguistic Coherence: A Plan-Based Alternative. In *Proceedings of the 24th Annual Meeting of the Association for Computational Linguistics*, pages 215–223, New York.

[Minsky 75]   Minsky, M. (1975). A Framework for Representing Knowledge. In P. H. Winston, editor, *The Psychology of Computer Vision*, McGraw-Hill, New York.

[Mooney & DeJong 85] Mooney, R. & DeJong, G. (1985). Learning Schemata for Natural Language Processing. In *Proceedings of the Ninth International Joint Conference on Artificial Intelligence*, pages 681–687, Los Angeles.

[Pollack 86]   Pollack, M.E. (1986). A Model of Plan Inference that Distinguishes Between the Beliefs of Actors and Observers. In *Proceedings of the 24th Annual Meeting of the Association for Computational Linguistics*, pages 207–214, New York.

[Reichman 85]   Reichman, R. (1985). *Getting Computers to Talk Like You and Me*. The MIT Press, Cambridge, MA.

[Schank & Abelson 77] Schank, R.C. & Abelson, R.P. (1977). *Scripts, Plans, Goals and Understanding*. Lawrence Erlbaum Associates, Hillsdale, N. J.

[Sidner & Israel 81]   Sidner, C.L. & Israel, D.J. (1981). Recognizing Intended Meaning and Speakers' Plans. In *Proceedings of Seventh International Joint Conference on Artificial Intelligence*, pages 203–208, Vancouver, B. C.

[Sidner 83a]   Sidner, C.L. (1983a). Focusing in the Comprehension of Definite Anaphora. In M. Brady and R.C. Berwick, editors, *Computational Models of Discourse*, The MIT Press, Cambridge, MA.

[Sidner 83b]   Sidner, C.L. (1983b). What the Speaker Means: The Recognition of Speakers' Plans in Discourse. *International Journal, of Computers and Mathematics*, 9:71–92.

[Zweben & Chase 87] Zweben, M. & Chase, M.P. (1987). Scenes: A Representation of Intentional and Attentional Structure. Submitted to AAAI-87, Seattle, WA.

850

ACQUIRING SPECIAL CASE SCHEMATA
IN EXPLANATION-BASED LEARNING *

Jude W. Shavlik[†]
Gerald F. DeJong
Artificial Intelligence Research Group
Coordinated Science Laboratory

Brian H. Ross
Department of Psychology

University of Illinois at Urbana-Champaign

ABSTRACT

Much of expertise in problem-solving situations involves rapidly choosing a tightly-constrained schema that is appropriate to the current problem. The paradigm of *explanation-based learning* is being applied to investigate how an intelligent system can acquire these "appropriately general" schemata. While the motivations for producing these specialized schemata are computational, results reported in the psychological literature are corroborated by a fully-implemented computer model. Acquiring these *special case* schemata involves combining schemata from two different classes. One class contains domain-independent problem-solving schemata, while the other class consists of domain-specific knowledge. By analyzing solutions to sample problems, new domain knowledge is produced that often is not easily usable by the problem-solving schemata. Special case schemata result from constraining these general schemata so that a known problem-solving technique is guaranteed to work. This significantly reduces the amount of planning that the problem solver would otherwise need to perform elaborating the general schema in a new problem-solving situation. The model and an application of it in the domain of classical physics are presented.

INTRODUCTION

We are investigating the role of specialized knowledge in schema-based problem solvers. This research illustrates the importance of special-purpose knowledge and demonstrates how the knowledge can be acquired. Furthermore, this specialized level of knowledge is computationally motivated and corroborates a number of findings in the psychological literature.

In a schema-based approach to problem solving, a few general schemata are brought to bear on a problem. Very little searching is performed, therefore the system can only solve those new problems that can easily be made to fit into an existing general schema. We are examining how a computer system can learn new problem-solving schemata for itself. The paradigm we adopt is *explanation-based learning* [DeJong81, DeJong86, Mitchell86], for which there is already some psychological evidence [Ahn87]. In this type of learning, a specific problem solution is generalized into a form that can be later used to solve conceptually similar problems. The generalization process is driven by the *explanation* of why the example solution worked. Extensive knowledge about the domain at hand allows the explanation to be developed and then extended. The resulting schema is quite general. Its generality is limited in part by characteristics of the observed example, but primarily by the system's domain model.

One would expect that such problem-solving schemata should be as general as possible so that they might each cover the broadest class of problems. Indeed, our previous research has been primarily aimed at the acquisition of such maximally general schemata. Recently, for computational reasons, we have adopted an intermediate level of generalization for various schemata. The class of intermediate generality schemata improves the performance of the problem solver by supplying "appropriately general" schemata instead of forcing the system to rely on its maximally general schemata. This results in much improved efficiency at a relatively minor cost in

* This research was partially supported by the National Science Foundation under grant NSF IST 85-11542.
† University of Illinois Cognitive Science/Artificial Intelligence Fellow.

851

generality.

Automatically acquiring schemata of the intermediate level of generality requires that the system's schemata be organized into two classes: a) schemata that represent knowledge of the domain of application and b) schemata that represent general problem-solving knowledge, which apply across many application domains. New schemata are learned by our implemented system as described in [Shavlik87a]. As well as storing the new schema in its general form, the system also stores special cases. These special cases are the result of composing the new, general schema with a small number of problem-solving schemata. A successful composition results in a specialization which is guaranteed to work using the composed problem-solving technique. This frees the problem solver from performing the planning that would otherwise be required to elaborate the general schema to fit the current problem-solving episode. The system can, of course, always resort to its collection of maximally general schema.

An example (discussed further in a later section) that illustrates this idea involves momentum conservation, a fundamental concept in physics. The explanation-based generalization of a sample collision problem leads to a physics formula that describes how external forces change a system's momentum. This general schema is broadly-applicable, but ascertaining that it will lead to the solution of a given problem requires a good deal of work. The constructed special case states that when there are no external forces, momentum is conserved.

Although the motivation for this intermediate level of generalization was computational, the use of this level helps to reconcile our approach with a variety of psychological evidence showing that problem solvers use highly specific schemata [Chase73, Hinsley77, Schoenfeld82, Sweller85]. Much of expertise consists of raidly choosing a tightly-constrained schema appropriate to the current problem. However, the difference between the knowledge of an expert and a novice cannot be explained on the basis of number of schemata alone. The scope and organization of these schemata have been shown in psychological experiments to be qualitatively different [Chi81, Larkin80, Schoenfeld82]. In representing a problem, novices make great use of the specific objects mentioned in the problem statement, while experts first categorize according to the techniques appropriate for solving the problem.

We have found that the intermediate-level schemata generated by our system are similar in scope of applicability to those that human experts appear to possess. For example, the conservation of momentum problem results in a special case schema characterized by the absence of external forces and the specification of a *before* and *after* situation. These features are those cited by experts as the relevant cues for the principle of conservation of momentum (see table 12 of [Chi81]). It should be noted that it was not our explicit intent to model this psychological data. Rather, computational efficiency considerations led to a system that produced results matching this empirical data.

OVERVIEW OF THE MODEL

Our model involves combining two types of schemata: domain-independent problem-solving schemata (e.g., a schema for utilizing a conserved quantity to solve a problem) and schemata that represent domain-specific knowledge (e.g., Newton's laws). People with mature problem-solving backgrounds possess the first type of schema and are told schemata of the second type when introduced to a new domain. Through study they acquire a large collection of schemata that combine aspects of both types, thereby increasing their performance in the domain. Combining general problem-solving techniques with domain specific knowledge produces schemata that, when applied, lead to the rapid solution of new problems.

Figure 1 contains an overview of our model. We assume a known problem-solving schema is used to understand a solution to a specific problem. The explanation-based analysis of the solution may lead to the construction of a new broadly-applicable schema. The generalization process often produces a new schema that, in its fullest form, is not usable by the originally applied problem-solving schema. Constraining the general result so that this problem-solving schema does apply produces a special case. In the special case schema, the constrained schema, its constraints, and the

852

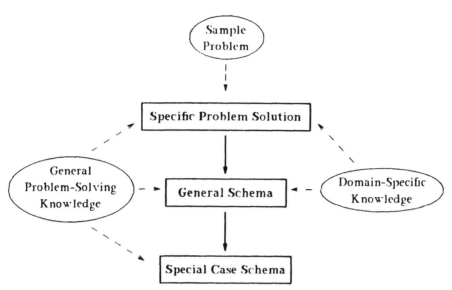

Figure 1. Overview of the Model

original problem-solving schema are packaged together to produce a specialized problem-solving strategy.

Figure 2 shows the relation we propose between a general schema and its special cases. Although not shown in the figure, there can be special cases of the special cases. Retrieval cues directly index the special cases.[1] If the indexed special case is not applicable, the general concept is then accessed. Besides being constructed when the general case is acquired, a new special case may be created whenever the general case is used to solve a later problem.

The next section presents examples of the construction of general and specific case schemata in the domain of classical physics.

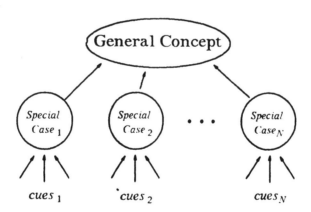

Figure 2. Inter-Schema Organization

---

[1] The method of using these special case cues to select the appropriate schema is not addressed in this paper. Possible indexing techniques include approaches based on *discrimination nets* [Feigenbaum63, Kolodner84, Schank82] and approaches based on *spreading activation* [Anderson83, Quillian68].

AN APPLICATION OF THE MODEL IN CLASSICAL PHYSICS

We have implemented a psychologically-plausible model of the process by which a mathematically-sophisticated student becomes a better problem solver in a new domain. In particular, we have been investigating the transition from novice to expert problem solver in the field of classical physics. Our model, implemented in a computer system named **Physics 101**, assumes the student has an understanding of mathematics through introductory calculus.

There are three main components of our model. The first is a model of how operators are chosen during problem solving [Shavlik86b]. The second explores the processes by which one can understand and generalize solutions to novel problems [Shavlik85]. The third, the topic of this paper, addresses the process of storing learned results so that they can be used to improve subsequent problem solving.

The next sections discuss two sample problems analyzed by **Physics 101**. One involves momentum conservation, and the other, energy conservation. The general schema produced in these two cases, the resulting special cases, and their selection cues are presented.

The schema that makes use of conserved quantities during problem solving is contained in table 1. (Terms beginning with a question mark are universally instantiated variables.) To apply this schema, a formula that is constant with respect to some variable is needed. This formula is instantiated at two different points. If the values of all but one variable at these two points are known, simple algebra can be used to easily find the unknown.

Example 1 - Momentum Conservation

One of the problems presented to **Physics 101** involves a collision among three balls. In this one-dimensional problem (shown in figure 3), there are three balls moving in free space, without the influence of any external forces. Nothing is specified about the forces between the balls. Besides their mutual gravitational attraction, there could, for example, be a long-range electrical interaction and a very complicated interaction during the collision. In the initial state (**state A**) the first ball is moving toward the two stationary ones. Some time later (**state B**) the second and third balls are recoiling from the resulting collision. The task in this problem is to determine the velocity of the first ball after the collision.

Table 1. Conserved Quantity Schema

Preconditions

$$(AND \; (IsaFormula \; ?formula \,)$$
$$(ConstantWithRespectTo \; ?formula \; ?x \,)$$
$$(SpecificPointOf \; ?x_1 \; ?x \,)$$
$$(SpecificPointOf \; ?x_2 \; ?x \,)$$
$$(\neq \; ?x_1 \; ?x_2)$$
$$(= \; ?leftHandSide \; (InstantiatedAt \; ?formula \; ?x_1))$$
$$(= \; ?rightHandSide \; (InstantiatedAt \; ?formula \; ?x_2))$$
$$(= \; ?equation \; (CreateEquation \; ?leftHandSide \; ?rightHandSide \,))$$
$$(AllButOneValueKnown \; ?equation \,) \,)$$

Schema Body

$$(SolveForSingleUnknown \; ?equation \,)$$

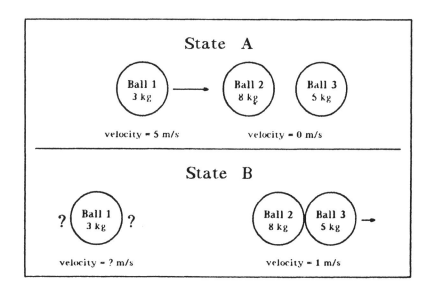

Figure 3. A Three-Body Collision Problem

A teacher's solution to figure 3's problem is analyzed by **Physics 101**. The teacher's solution uses the concept of momentum (*mass × velocity*) conservation to solve the problem. Since this is a conservation law, the time between the two states need not be known.[2] (Another important attribute of momentum conservation is that the properties of the inter-object forces need not be provided.) In verifying the provided solution, the system applies a problem-solving schema in which a constant function is equated at two different points (table 1). In accordance with the explanation-based learning approach, the system's justification of the provided solution is generalized as far as possible while maintaining the veracity of the solution technique. This results in the general schema presented in table 2. (See [Shavlik86a] or [Shavlik87a] for more details on the construction of this schema.)

The explanation-based approach results in a formula that applies to situations significantly different from the sample problem. In addition to not being restricted to problems containing exactly three objects, the newly-acquired formula is not restricted to situations where the external forces are all zero. Instead, an understanding of how the external forces effect momentum is obtained. This process also determines that there is no constraint that restricts this formula to the x-direction. It applies equally well to the y- and z-components of velocity. Hence, the acquired formula is a vector law. The mathematical operations used in the specific solution require, for the solution strategy to be valid, that all objects have non-zero mass and that these masses are constant over time. Finally, the generalization algorithm determines that the inter-object forces need not be known, since they are algebraically cancelled during the derivation of the momentum law.

Notice that the result in table 2 is *not* a conservation law. It describes how the momentum of a system evolves over time. Although this new formula applies to a large class of problems, recognizing its applicability is not easy. The external forces on the system must be summed and a possibly complicated differential equation needs to be solved. Applying this law requires more than counting the number of unknowns in the equation, determining there is only one, and then using simple algebra to find its value.

In order for the originally used problem-solving schema (table 1) to be applicable to this new formula, it must be the case that momentum be constant with respect to time and easily calculable at two different times. This means that the derivative of momentum be zero, which leads to the

---

[2] When the inter-state time is unknown, simply solving the equations of motion resulting from Newton's laws is not possible.

Table 2. The General Momentum Law

Formula

$$\frac{d}{dt} \sum_{i=1}^{n} mass_i \; velocity_{i,?c}(t) \;=\; \sum_{i=1}^{N} force_{external,i,?c}(t)$$

Preconditions

(AND (IsaComponent ?c)
      $\forall\, i \in 1,\ldots,n$ (NOT (ZeroValued $mass_i$)
      $\forall\, i \in 1,\ldots,n$ (IndependentOf $mass_i$ t))

Eliminated Terms

$$\forall\, i \;\forall\, j \neq i \quad force_{i,j,?c}(t)$$

---

requirement that the external forces sum to zero. When this occurs, the momentum of a system can be equated at *any* two distinct states. The special case schema for momentum conservation is contained in table 2sc. Since this is a conservation schema, the time at which each state occurs need not be provided in a problem for this schema to apply.

Table 2sc. The Special-Case Momentum Law

Formula

$$\sum_{i=1}^{n} mass_i \; velocity_{i,?c}(?t_1) \;=\; \sum_{i=1}^{n} mass_i \; velocity_{i,?c}(?t_2)$$

Preconditions

(AND (IsaComponent ?c) (Time $?t_1$) (Time $?t_2$) ($\neq$ $?t_1$ $?t_2$)
      $\forall\, i \in 1,\ldots,n$ (NOT (ZeroValued $mass_i$))
      $\forall\, i \in 1,\ldots,n$ (IndependentOf $mass_i$ t))

Eliminated Terms

$$\forall\, i \;\forall\, j \neq i \quad force_{i,j,?c}(t), \quad ?t_1, \quad ?t_2$$

Special Case Conditions

$$\forall\, i \in 1,\ldots,n \quad force_{external,i,?c} = 0$$

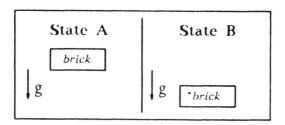

Figure 4. A Falling Brick

Example 2 - Energy Conservation

A second problem (figure 4) presented to **Physics 101** involves a brick falling under the influence of gravity. Again, information at two different states is presented. The mass of the brick, its initial velocity, and its height in the two states are provided. The goal is to find its velocity in the second state. The teacher's solution to this problem uses energy conservation. The kinetic energy ($\frac{1}{2} mass \times velocity^2$) plus the potential energy ($mass \times g \times height$) in the two states is equated. The general law **Physics 101**'s produces by analyzing the sample solution is presented in table 3.

The general energy conservation law applies whenever the total force on an object is known. Notice, though, that a rather complicated vector integral involving the scalar (dot) product of two vectors needs to be computed if this general law is to be used. To use this formula, it is not sufficient to possess knowledge of the values of variables at two different times. A problem solver must also know how the net force depends on position for a continuum of times. In the specific problem there is a *constant* net force (gravity). When the force is constant the problem is greatly simplified. Integrating a constant force leads to a potential energy determined by that constant force multiplied by the object's position. The position only needs to be known at the two distinct times, and not for *all* intervening times. The special case schema for energy conservation is contained in table 3sc. Again, since this is a conservation schema, the time at which each state occurs need not be known.

SIMILARITY-BASED APPROACHES TO LEARNING SPECIAL CASES

A common induction scheme is to posit that learners compare particular instances of a concept (such as specific problems of a problem type) and abstract out those aspects that are common to both problems [Anderson83, Michalski83, Mitchell78, Posner68]. The fact that problem solvers

---

Table 3. The General Energy Law

Formula

$$\frac{d}{dt} \left[\frac{1}{2} mass_{?_i} \; velocity^2_{?_i}(t) - \int \overline{force}_{net, ?_i}(t) \cdot d\overline{position}_{?_i} \right] = 0 \frac{kg \; m^2}{s^3}$$

Preconditions

(AND (Object ?i) (IndependentOf $mass_{?_i}$ t) (NOT (ZeroValued $mass_{?_i}$)))

857

Table 3sc. The Special-Case Energy Law

Formula

$$\frac{1}{2} \; mass_{?i} \; velocity^2{}_{?i} \,(?t_1) + mass_{?i} \; g \; position_{?i,?c}\,(?t_1)$$

$$= \frac{1}{2} \; mass_{?i} \; velocity^2{}_{?i} \,(?t_2) + mass_{?i} \; g \; position_{?i,?c}\,(?t_2)$$

Preconditions

(AND (Object ?i) (Component ?c) (Time ?t$_1$) (Time ?t$_2$) ($\neq$?t$_1$?t$_2$)
(IndependentOf $mass_{?i}$ t) (NOT (ZeroValued $mass_{?i}$)))

Eliminated Terms

?t$_1$, ?t$_2$

Special Case Conditions

$$\overline{force}_{net,?i}(t) = mass_{?i} \; g \; ?\hat{c}$$

---

use highly specific schemata supports such a view, since these schemata would arise whenever two problems from an intermediate level problem type are compared. Although we believe that similarity-based generalization is an important means of learning, especially for novices [Gentner87, Ross84, Ross87], the research described in this paper shows that many of these highly specific schemata can arise from an explanation-based approach. Even some strong proponents of example comparison learning have begun to incorporate some explanation-based ideas in order to account for how much is learned from one example [Anderson87].

Because explanation-based learning requires extensive domain knowledge, it clearly is not appropriate for all learning in a new domain. However, it may be useful even in early learning if the new domain relies heavily upon a domain for which the novice does have substantial knowledge. Because mathematics underlies many other domains, a novice with some mathematical sophistication may be able to make use of explanation-based techniques without extensive knowledge of the new domain.

CONCLUSION

Much of expertise in problem-solving situations involves rapidly choosing a tightly-constrained schema that is appropriate to the current problem. We are applying the paradigm of explanation-based learning to investigate how an intelligent system can acquire these "appropriately general" schemata. While our motivations for producing these specialized schemata are computational, results reported in the psychological literature are corroborated by our fully-implemented computer model.

A major issue in explanation-based learning concerns the *operationality/generality* trade-off [DeJong86, Mitchell86, Segre87]. A schema whose relevance is easy to determine may only be useful in an overly-narrow range of problems. Conversely, a broadly-applicable schema may require extensive work before a problem solver can recognize its appropriateness. Other approaches to selecting the proper level of generality involve pruning easily-reconstructable portions of the explanation structure. Our approach to this problem is to produce as general a schema as possible from the analysis of a specific solution, and then construct a *special case* of this general schema. In constructing the general schema, the original explanation structure is often substantially altered during generalization [Shavlik87a, Shavlik87b]. Augmentation of the explanation is needed in order to generalize such things as the number of entities in a concept or the number of times some action is performed. A special case is produced by constraining a general schema in such a way that its relevance is easily checked. This results in additional features that a situation must possess if the special case is to apply.

Acquiring these special case schemata involves combining schemata from two different classes. One class contains domain-independent problem-solving schemata, while the other class consists of domain-specific knowledge. In our model, learning by analyzing sample problem solutions produces broadly-applicable schemata that, often, are not usable by the originally applied problem-solving schemata. Special case schemata result from constraining these general schemata so that the originally used problem-solving techniques are guaranteed to work. This significantly reduces the amount of planning that the problem solver would otherwise need to perform elaborating the general schema to match a new problem-solving episode.

Besides improving a problem solver's efficiency, special cases also indicate good assumptions to make. For instance, if you do not know what the external forces are, assume they are zero. Physics problems often require one to assume things like "there is no friction", "the string is massless", "the gravity of the moon can be ignored", etc. Problem descriptions given to students contain cues such as these, and students must learn how to take advantage of them. Facts in the initial problem statement suggest possible problem-solving strategies, while any additional requirements of the special case situations indicate good assumptions to make (provided they do not contradict anything else that is known).

This paper demonstrates that these highly-specific schemata can arise in an explanation-based fashion. Explanation-based learning requires extensive knowledge, and seems particularily suited for modelling learning by experts. Although all learning cannot be of this type, explanation-based learning can prove useful even in early learning in a new domain. This can occur if the new domain relies heavily upon another domain in which the novice learner has substantial abilities.

REFERENCES

[Ahn87] W. Ahn, R. J. Mooney, W. F. Brewer and G. F. DeJong, "Schema Acquisition from One Example: Psychological Evidence for Explanation-Based Learning," Technical Report, AI Research Group, Coordinated Science Laboratory, University of Illinois, Urbana, IL, February 1987.

[Anderson83] J. R. Anderson, *The Architecture of Cognition*, Harvard University Press, Cambridge, MA, 1983.

[Anderson87] J. R. Anderson and R. Thompson, "Use of Analogy in a Production System Architecture," in *Similarity and Analogical Reasoning*, S. Vosniadou and A. Ortony (ed.), 1987.

[Chase73] W. G. Chase and H. A. Simon, "Perception in Chess," *Cognitive Psychology 4*, (1973), pp. 55-81.

[Chi81] M. T. Chi, P. J. Feltovich and R. Glaser, "Categorization and Representation of Physics Problems by Experts and Novices," *Cognitive Science 5*, 2 (1981), pp. 121-152.

[DeJong81] G. F. DeJong, "Generalizations Based on Explanations," *Proceedings of the Seventh International Joint Conference on Artificial Intelligence*, Vancouver, B.C., Canada, August 1981, pp. 67-70.

[DeJong86] G. F. DeJong and R. J. Mooney, "Explanation-Based Learning: An Alternative View," *Machine Learning 1*, 2 (April 1986), pp. 145-176.

[Feigenbaum63] E. A. Feigenbaum, "The Simulation of Natural Learning Behavior," in *Computers and Thought*, E. A. Feigenbaum and J. Feldman (ed.), McGraw-Hill, New York, NY, 1963.

859

[Gentner87]   D. Gentner, "Mechanisms of Analogical Learning," in *Similarity and Analogical Reasoning*, S. Vosniadou and A. Ortony (ed.), 1987.

[Hinsley77]   D. A. Hinsley, J. R. Hayes and H. A. Simon, "From Words to Equations: Meaning and Representation on Algebra Word Problems," in *Cognitive Processes in Comprehension*, P. A. Carpenter and M. A. Just (ed.), Lawrence Erlbaum and Associates, Hillsdale, NJ, 1977, pp. 89-105.

[Kolodner84]   J. L. Kolodner, *Retrieval and Organizational Strategies in Conceptual Memory*, Lawrence Erlbaum and Associates, Hillsdale, NJ, 1984.

[Larkin80]   J. H. Larkin, J. McDermott, D. P. Simon and H. A. Simon, "Models of Competence in Solving Physics Problems," *Cognitive Science 4*, 4 (1980), pp. 317-345.

[Michalski83]   R. S. Michalski, "A Theory and Methodology of Inductive Learning," in *Machine Learning: An Artificial Intelligence Approach*, R. S. Michalski, J. G. Carbonell, T. M. Mitchell (ed.), Tioga Publishing Company, Palo Alto, CA, 1983, pp. 83-134.

[Mitchell78]   T. M. Mitchell, "Version Spaces: An Approach to Concept Learning," Ph.D. Thesis, Stanford University, Palo Alto, CA, 1978. (Also appears as Technical Report STAN-CS-78-711, Stanford University)

[Mitchell86]   T. M. Mitchell, R. Keller and S. Kedar-Cabelli, "Explanation-Based Generalization: A Unifying View," *Machine Learning 1*, 1 (January 1986), pp. 47-80.

[Posner68]   M. J. Posner and S. W. Keele, "On the Genesis of Abstract Ideas," *Journal of Experimental Psychology 77*, 3 (July 1968), pp. 353-363.

[Quillian68]   M. R. Quillian, "Semantic Memory," in *Semantic Information Processing*, M. L. Minsky (ed.), MIT Press, Cambridge, MA, 1968.

[Ross84]   B. H. Ross, "Remindings and their Effects in Learning a Cognitive Skill," *Cognitive Psychology 16*, (1984), pp. 371-416.

[Ross87]   B. H. Ross, "Remindings in Learning and Instruction," in *Similarity and Analogical Reasoning*, S. Vosniadou and A. Ortony (ed.), 1987.

[Schank82]   R. C. Schank, *Dynamic Memory*, Cambridge University Press, Cambridge, England, 1982.

[Schoenfeld82] A. H. Schoenfeld and D. Herrmann, "Problem Perception and Knowledge Structure in Expert and Novice Mathematical Problem Solvers," *Journal of Experimental Psychology: Learning, Memory, and Cognition 8*, 5 (1982), pp. 484-494.

[Segre87]   A. M. Segre, "Explanation-Based Learning of Generalized Robot Assembly Tasks," Ph.D. Thesis, Department of Electrical and Computer Engineering, University of Illinois, Urbana, IL, January 1987. (Also appears as UILU-ENG-87-2208, AI Research Group, Coordinated Science Laboratory, University of Illinois at Urbana-Champaign.)

[Shavlik85]   J. W. Shavlik and G. F. DeJong, "Building a Computer Model of Learning Classical Mechanics," *Proceedings of the Seventh Annual Conference of the Cognitive Science Society*, Irvine, CA, August 1985, pp. 351-355.

[Shavlik86a]   J. W. Shavlik and G. F. DeJong, "Computer Understanding and Generalization of Symbolic Mathematical Calculations: A Case Study in Physics Problem Solving," *Proceedings of the 1986 Symposium on Symbolic and Algebraic Computation*, Waterloo, Ontario, Canada, July 1986, pp. 148-153.

[Shavlik86b]   J. W. Shavlik and G. F. DeJong, "A Model of Attention Focussing During Problem Solving," *Proceedings of the Eighth Annual Conference of the Cognitive Science Society*, Amherst, MA, August 1986, pp. 817-822.

[Shavlik87a]   J. W. Shavlik and G. F. DeJong, "Analyzing Variable Cancellations to Generalize Symbolic Mathematical Calculations," *Proceedings of the Third IEEE Conference on Artificial Intelligence Applications*, Orlando, FL, February 1987.

[Shavlik87b]   J. W. Shavlik and G. F. DeJong, "BAGGER: An EBL System that Extends and Generalizes Explanations," Technical Report, AI Research Group, Coordinated Science Laboratory, University of Illinois, Urbana, IL, February 1987.

[Sweller85]   J. Sweller and G. A. Cooper, "The Use of Worked Examples as a Substitute for Problem Solving in Learning Algebra," *Cognition and Instruction 2*, 1 (1985), pp. 59-89.

GENERATING SCRIPTS FROM MEMORY[1]

by

Walter Kintsch & Suzanne M. Mannes

University of Colorado

Abstract

A variation of the Raaijmaker and Shiffrin (1981) retrieval model is proposed to account for typical script generation data. In our model, knowledge is represented as an associative network with propositions as the nodes. A control process which utilizes temporal information in these propositions supplements the probabilistic memory retrieval process to produce ordered retrieval of scriptal events. Simulations are reported which provide a good qualitative fit to data collected from subjects in both script generation and free association tasks. These results support a view of memory as an unorganized knowledge base rather than a stable, organized structure.

1. Scripts as Mental Structures

Scripts are representations of simple stereotyped event sequences. As such, they are a subtype of frames or schemata. In order to be useful, knowledge has to be organized in some way, and scripts, as well as frames, schemata, and semantic nets, provided that organization (e.g. Anderson, 1980; Graesser, 1981; Schank, 1980; Schank & Abelson, 1977). Scripts are claimed to be mental structures, and psychologists set out to demonstrate the psychological reality of these structures and to investigate their properties. Bower, Black, & Turner (1979) observed a high level of agreement when subjects were asked to list the characteristic events that they thought belonged to a script, and their order.

However, the view that scripts are precompiled mental structures was soon challenged from several sides. Computationally, fixed mental structures like scripts turned out to be too inflexible to really serve the purposes for which they were originally designed (van Dijk & Kintsch, 1983; Schank, 1982). If scripts guide retrieval, how close events occur in the script structure should determine the time it takes to retrieve one event, given the other. However, such distance effects have not been observed (Haberlandt & Bingham, 1984; Galambos & Rips, 1982; Bower et al., 1979). At most, it appears that immediately succeeding events are retrieved faster than events farther away (Bower et al.), but that argues more for a local relation like "next", rather than for a larger script structure.

For these reasons, there has been a general move away from considering scripts as fixed mental structures. In the present paper we claim that knowledge is not pre-organized in terms of scripts and schemata, but that such structures are generated from an unorganized associative net in response to a specific task demand in a specific context. Only in this way can the flexibility

---

[1] This research was supported by grant number MH - 15872 from the National Institute of Mental Health. It is publication number 87-3 of the Institue of Cognitive Science at the University of Colorado, Boulder, CO 80309

and context sensitivity that characterize human script use be achieved. Here, we shall show how a behavior that has been taken as prima-facie evidence for the existence of scripts as mental structures can be generated from a knowledge base in which there are no pre-existing global structures like scripts. Specifically, we shall simulate how people go about listing the events which constitute common scriptal activities, such as going to a grocery store.

2. Scripts and Categories

The approach taken here extends earlier work on generating conceptual categories to scripts. Walker & Kintsch (1985) have modelled this process by making two crucial assumptions.

First of all, they assumed that knowledge retrieval obeys the same laws as retrieval from episodic memory. In particular, Walker & Kintsch assumed that the automatic component of the retrieval process can be described by the Raaijmakers & Shiffrin (1981) theory. Given a particular retrieval cue, this model predicts what will be retrieved, and when. On the other hand, the control processes which are necessary to put together an appropriate retrieval cue in the present situation are outside the Raaijmakers & Shiffrin model, and will be discussed below.

Secondly, Walker & Kintsch assumed that the retrieval process operates on an associative net in which categories are not explicitly represented. Of course, there must be information stored with each category member that identifies it as such. In the simplest case this might be an associated IS-A proposition.

A model of script generation will be outlined below which is analogous to the category generation model of Walker & Kintsch, except that it utilizes a different control process to reconstitute exhausted retrieval cues: instead of randomly picking an associated node and adding it to the retrieval cue, local information about what comes next is used. In this way the unproductive search phases which are characteristic of category retrieval are avoided, and an essentially linear retrieval function is obtained. Furthermore, the retrieval process does not slowly peter out, but stops when the end of the chain is reached. Before describing this model in more detail, however, it is necessary to look more closely at how people generate scripts, so that we have a more solid data base with which to compare our model.

3. Generating Scripts: Experimental Results

3.1 Method

Six subjects were asked to think aloud (Ericsson & Simon, 1984) as they pretended to tell a stranger from another culture what typically happens in the following three situations: *going to a restaurant for a meal, going to a grocery store to buy groceries, and going to a doctor's office for a checkup*. These were the same scripts that were used by Walker & Kintsch (1985).

3.2 Results

Each subject's protocol was divided into idea units, roughly corresponding to propositions in the sense of van Dijk & Kintsch (1983).These units were classified as either scriptal events or elaborations. Events were always single propositions, while elaborations were sometimes more complex. For example, the event *make a list* was elaborated with *of what you want to buy*.

There was good agreement among subjects about the events they thought belonged to each of the three scripts. Figure 1 shows the percentage of all responses which were produced by 5 or 6 subjects, 3 or 4 subjects, or by only 1 or 2 subjects. More than half of all responses were given by a majority of the subjects. Almost all items were generated in their natural order; only 1% of all responses were out of order. Figure 2 shows the cumulative number of scriptal units as a function of retrieval time for one of our subjects. The important points to notice are first that the rate at which a script is generated is approximately constant. Secondly, the curve is relatively smooth, in contrast to the severe scalloping observed for category naming tasks.

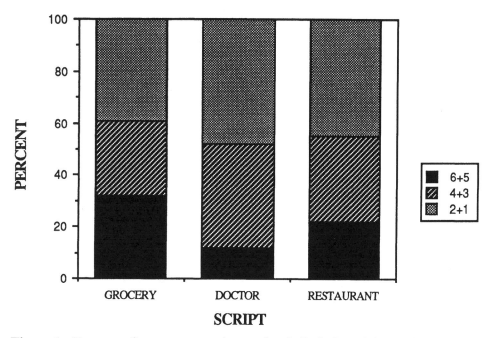

Figure 1. Percent of events agreed upon by 6+5, 4+3, and 2+1 subjects.

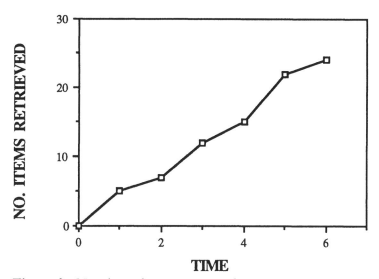

Figure 2. Number of grocery store items retrieved by a single subject
as a function of time (plotted in 30 second intervals).

In listening to these protocols, one is struck by their fluency. Subjects always seem to know what to say next, with little hesitation. There was apparently no need for an extended search for a new retrieval cue. We therefore closely inspected each protocol for possible evidence as to the nature of the retrieval cues that permitted such smooth transition from event to event. In the course of these analyses we arrived at the distinction between scriptal events and elaborations. As already mentioned, events are single-proposition units, expressing an action either by the main actor or by some other participant (checkout-clerk, nurse, waiter) which can

863

stand on their own. That is, events were meaningful by themselves, and did not require reference to some other unit for their interpretation. Elaborations on the other hand, have to be interpreted in the context of some other unit, usually an event. The event *make a list* was elaborated by *of what you want to buy*. Thus, elaborations appear to be dependent on events. Events, however, either follow directly previous events, or are preceded by explicit temporal connectives, such as *then*, or *after that*. A striking asymmetry was noted in this respect: temporal connectives were used by most or all of the subjects in certain places, while they were rare otherwise. For instance, each subject used a temporal marker (connective) between the last item that dealt with preparation for shopping and entering the store and the first item that dealt with the actual shopping. The transition between shopping and checking-out was similarly marked by all subjects, and 5 of the 6 subjects separated checking-out and leaving the store with a temporal marker. Within each of these episodes, in contrast, temporal connectives were rare and used idiosyncratically. Thus, explicit temporal connectives appear to segment the script into separate episodes which are crucial for an understanding of how scripts can be generated from an associative knowledge base. The distribution of these temporal markers is shown in Figure 3.

3.3 A Hypothesis about Retrieval

In Table 1 the events and elaborations for the grocery store script which were produced by our subjects are shown, broken down into episodes as suggested by Figure 3. Items generated by 5 or 6 subjects are shown with asterisks; inferred items (the episode labels) are shown in brackets. The reponses made by one subject are connected by a continuous line, to show this subject's retrieval path.

We are now in a position to state a possible hypothesis about the retrieval cues which control the process of script generation. The answer is simple for elaborations: since all elaborations can be co-ordinated with a specific event, we assume that the events serve as their retrieval cue. For the events themselves, we propose a dual process: some events are retrieved via specific temporal information, while others are retrieved associatively, much as category members are. Our data suggest that goal-directed retrieval on the basis of specific temporal cues, such as X FOLLOWS Y operates at the level of episodes: when enough information within an episode has been generated, the subject does not search for a new retrieval cue by checking random associations, but uses specific temporal information to establish a new episode cue. Thus, once the subject is done with *checking-out*, the knowledge base is searched for a proposition AFTER[CHECK-OUT, $], yielding AFTER[CHECK-OUT, LEAVING], and LEAVING becomes the next retrieval cue. The episode cue itself works much like the category cue in the category naming task: repeated retrieval attempts are made using this cue in conjunction with recently retrieved information. Thus, within-episode retrieval is locally governed, and stops when a certain number of unsuccessful retrieval events have occurred. At that point, a new episode cue is generated in the manner discussed above. Temporal information available in the knowledge base specifies what that cue has to be, and hence a long search is superfluous, giving the script generation process its characteristic smoothness and fluidity. In the next section, this hypothesis about how scripts are generated will be specified further to the point where simulations can be performed. The goal of these simulations will be to determine whether simulated script retrieval is at least qualitatively similar to the human data.

GROCERY STORE

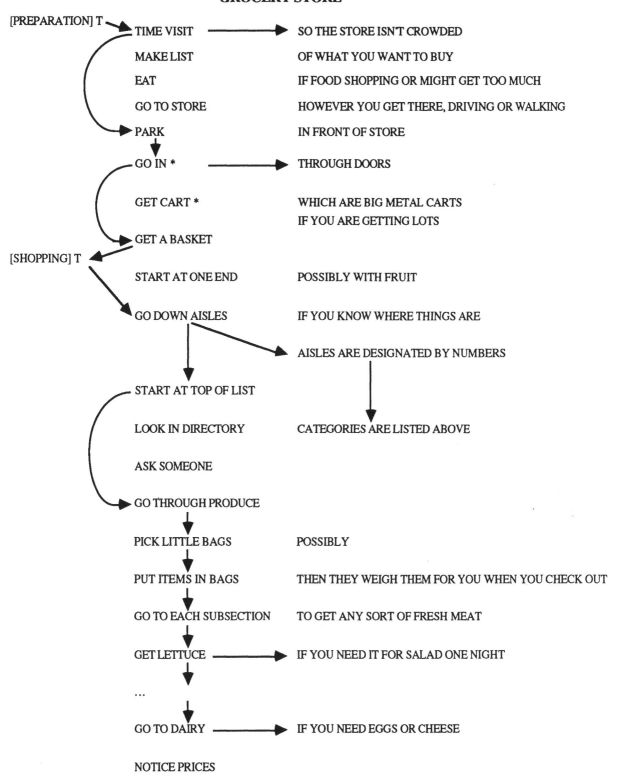

[PREPARATION] T

TIME VISIT	SO THE STORE ISN'T CROWDED
MAKE LIST	OF WHAT YOU WANT TO BUY
EAT	IF FOOD SHOPPING OR MIGHT GET TOO MUCH
GO TO STORE	HOWEVER YOU GET THERE, DRIVING OR WALKING
PARK	IN FRONT OF STORE
GO IN *	THROUGH DOORS
GET CART *	WHICH ARE BIG METAL CARTS IF YOU ARE GETTING LOTS
GET A BASKET	

[SHOPPING] T

START AT ONE END	POSSIBLY WITH FRUIT
GO DOWN AISLES	IF YOU KNOW WHERE THINGS ARE
	AISLES ARE DESIGNATED BY NUMBERS
START AT TOP OF LIST	
LOOK IN DIRECTORY	CATEGORIES ARE LISTED ABOVE
ASK SOMEONE	
GO THROUGH PRODUCE	
PICK LITTLE BAGS	POSSIBLY
PUT ITEMS IN BAGS	THEN THEY WEIGH THEM FOR YOU WHEN YOU CHECK OUT
GO TO EACH SUBSECTION	TO GET ANY SORT OF FRESH MEAT
GET LETTUCE	IF YOU NEED IT FOR SALAD ONE NIGHT
...	
GO TO DAIRY	IF YOU NEED EGGS OR CHEESE
NOTICE PRICES	

Table 1. Items which subjects produced in response to the Grocery Shopping cue. The lines show one subject's path through the network. See text for further explanation.

865

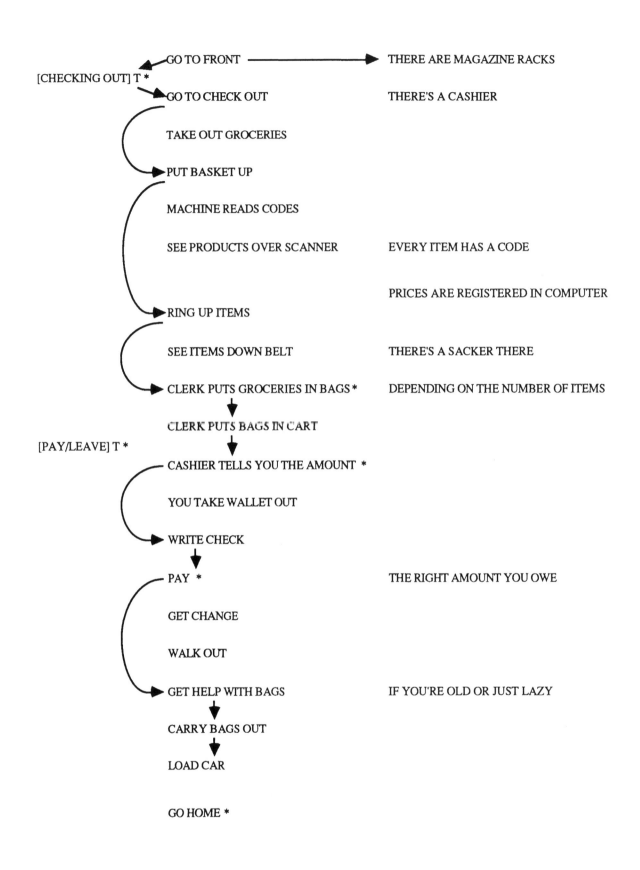

GO TO FRONT ——————————→ THERE ARE MAGAZINE RACKS

[CHECKING OUT] T *

GO TO CHECK OUT   THERE'S A CASHIER

TAKE OUT GROCERIES

PUT BASKET UP

MACHINE READS CODES

SEE PRODUCTS OVER SCANNER   EVERY ITEM HAS A CODE

                                   PRICES ARE REGISTERED IN COMPUTER

RING UP ITEMS

SEE ITEMS DOWN BELT   THERE'S A SACKER THERE

CLERK PUTS GROCERIES IN BAGS *   DEPENDING ON THE NUMBER OF ITEMS

CLERK PUTS BAGS IN CART

[PAY/LEAVE] T *

CASHIER TELLS YOU THE AMOUNT *

YOU TAKE WALLET OUT

WRITE CHECK

PAY *   THE RIGHT AMOUNT YOU OWE

GET CHANGE

WALK OUT

GET HELP WITH BAGS   IF YOU'RE OLD OR JUST LAZY

CARRY BAGS OUT

LOAD CAR

GO HOME *

Table 1. (continued).

4. Generating Scripts: A Simulation

Memory is an associative network with concepts and propositions as nodes. Nodes are related either positively, negatively, or not at all, with connections ranging from -1 to +1. Mathematically, this network will be represented by a matrix, **K**. The rows *(i)* and columns *(j)* of this matrix represent the nodes of the network, and the entries *(s $_{i,j}$)* represent the connection strengths between the nodes.

A retrieval cue consists of one or more nodes of this network. Each retrieval attempt with a particular retrieval cue results in the retrieval of a single node from the network (though not necessarily in an overt response). The model can be concisely statedin terms of the well-known model of memory retrieval proposed by Raaijmakers & Shiffrin (1981). Memory in their model, is represented as an associative net, and the retrieval process which they describe results in the following equation.

$$(1) \qquad Pr\ (j\ /\ i_{1...k}) = \frac{\prod_{i=1}^{k} s_{i,j}}{\sum_{j=1}^{n} \prod_{i=1}^{k} s_{i,j}}$$

Suppose that node *j* has now been retrieved. If it has not already been retrieved previously, it will be output as an overt response. An unsuccessful retrieval attempt occurs when the retrieved node duplicates an earlier retrieval.[2]

The newly retrieved node is added to the retrieval cue, which is now of size *k+1*. If this number exceeds some maximum value *m*, an old element must be dropped. We assume that this will be the most recently added element. Thus, the nature of the retrieval cue changes dynamically in response to local effects, but at the same time retains a stable core.

Free association differs from script generation only in that no temporal cues are used to guide retrieval. *H* is used to retrieve some associate which then takes on the role *H$_1$*. After it has become ineffective, some other associate of *H* takes its place. Thus, we obtain clusters of related associates, with occasional unpredictable jumps to new ones.

4.1 Simulating Grocery Shopping

In order to test the model outlined above, we attempted to simulate the generation of a grocery-shopping script. The goal of this simulation was to account qualitatively for the data described in Section 3. Specifically, we wanted to see whether the simulation yielded an output comparable to the human data in Table 1, and whether the rate with which it was produced was constant, as in Figure 2. For this purpose, an associative knowledge system containing information about grocery shopping had to be constructed first.

4.1.1 The Knowledge Base.
Since merely a qualitative account is attempted here, we need not concern ourselves with all the knowledge people have about grocery shopping. Instead, fifteen typical items from Table 1 were selected more or less randomly and grouped by episodes. In addition, the script heading, the four episode headings as well as the temporal markers between them were included in the

---

[2] In other cases this editing process must be more complex. E.g., in category naming a check needs to be made to determine whether the retrieved node is, in fact, a category member.

knowledge base. Furthermore, for each of the 15 scriptal events plus the 5 headings free associations were obtained from a group of 16 subjects. The two most frequently produced associations for each item were also added to the knowledge base. All the items in the knowledge base thus far are related to *grocery shopping.* To make the simulation more challenging, items unrelated to the grocery script are needed, to see whether or not the retrieval process suffers interference from such items. Therefore, a closely related script was selected, for which three of the four episodes overlapped. Subjects were asked to provide us with a *shoe-store* script. This yielded the three episodes *Entering, Shopping,* and *Leaving.* In addition to the three episode headings, four of the produced scriptal events are associated with both scripts. If the two scripts will interfere with each other, there is certainly opportunity to do so. For each of the *shoe-store* events, two high-frequency associates were also added to the knowledge base. There is, of course, in the knowledge base no distinction between scriptal events, episode headers, or associates: all are just nodes in a network, - we categorize them in this way only on the basis of the protocols subjects generate.

The total matrix thus constructed had 63 rows and columns - a tiny fragment of a human knowledge system. The 3,969 connection strengths $s_{i,j}$ in this matrix were estimated from actual script and free association data. Only rough estimates were made: Whenever a stimulus elicited the same response in 75% of the subjects, a strength value of 1 was used; responses given by fewer subjects were assigned a connection strength of .5; and responses which did not occur in our sample were given a strength of 0. All associations were assumed to be symmetric.

In addition, connections were established on the basis of the script data in the following way. The script headers *grocery shopping* and *shoe store* were connected with a strength of 1 to their respective episode headers and with a value of .5 to the scriptal events within each episode. The episode headers were connected with a strength of 1 to their respective script events. Finally, script events within each episode were connected to each other by a value of .5. However, the strengths of these connections were not necessarily symmetric: *picking up vegetables* and *going through the aisles* are connected both ways, and both are connected to *going to the check-out counter,* but the latter has no strength connecting it to either of the two former nodes. Thus, $s_{(vegetables,aisles)} = s_{(aisles,vegetables)}$, but $s_{(aisles,checkout)} \neq s_{(checkout, aisles)}$. Furthermore, the backwards connections between episode headers and their script headers, as well as between script events and their episode headers, were made less strong (.5) than the forward connections between these elements. Thus, scriptal events may be directional. (It is of course possible that the same may be true for associations in general, but this possibility was not explored here.) The temporal connectives - BEGIN[GROCERY-SHOPPING, ENTER], BEGIN[SHOE-SHOPPING, ENTER], AFTER[ENTER, SHOPPING], AFTER[SHOPPING, CHECK-OUT], AFTER[SHOPPING, LEAVE], AFTER[CHECK-OUT, LEAVE] - which are needed to control the retrieval process in the script generation task are connected by a value of +1 to their second argument, and a value of -1 to their first argument. The decision to use only the values 1 and .5 for connection strengths is both crude and arbitrary, but by restricting ourselves to such rough approximations we have reduced the need for subjective judgments. It should also be noted that precise strength values matter relatively little here: It is the over-all pattern of connections that determines the results.

4.1.2 Retrieval .

Instead of estimating the parameters of the model from the data and trying to fit the data quantitatively, all the parameters were specified a priori.[3] Specifically, we assume that the retrieval cue has three components, two stable (which would be the script header and an episode

---

[3] While it would be possible to estimate the parameters of the retrieval model statistically, a quantitative fit would presume an adequate simulation of the knowledge base, which is beyond our possibilities at present.

header) and the third one variable (the last item retrieved). We also assume L = 3, that is, a retrieval cue is abandoned after three successive retrieval failures. Consider a particular simulation run with this model, then. What we observe is an interplay between the controlled search for a new retrieval cue, and the automatic retrieval process. Given Grocery Shopping, the proposition BEGIN[GROCERY-SHOPPING, $] is used to retrieve ENTER via a pattern matching process. These two nodes then form the first retrieval cue, activating 7 nodes in the system to some extent or other. *Entering Store* is selected. It is produced as a response, and becomes the third component of the retrieval cue. This modified cue now activates four other nodes, and, after 2 failures (the cue retrieves a component of itself), *Getting a Cart* is retrieved. This now replaces *Entering Store* as a component of the retrieval cue, which retrieves, after one failed attempt, a node which entered the network as an elaboration of Getting a Cart: *is fun*. The new retrieval cue activates 5 nodes, 3 of them new ones, but fails anyhow, because the same, old node is retrieved coincidentally three times in a row. Now the control process takes over, again: A memory search is made on the basis of AFTER[ENTER, $], which yields SHOPPING as the new episode cue, and the process once again shifts into its automatic retrieval mode. This interplay continues until the last episode cue is exhausted. In the simulation run under discussion here, a total of 12 nodes is thus retrieved, among them 5 nodes which we have called - on the basis of the data our subjects had given us - scriptal events; the rest were episode headers and elaborations. The data from this simulation run are shown in Table 2. This network looks much like an abbreviated version of Table 1: obviously, what the model does is quite close to what people do when they generate scripts. Most importantly, the retrieval process stayed on track: no intrusions from the shoe-store domain occurred, in spite of the fact that several of the items produced as part of the grocery script were also associated with the former domain. Secondly, the items were produced in the correct order, wherever order mattered, as in the actual data.

Figure 4 shows the results of three independent simulation runs with the grocery script. These simulation results have the same features as the data obtained from the individual subject in the script generation task (Figure 2): the rate at which scriptal events are produced is constant, there is no slowing down towards the end, and the curves are relatively smooth, without the large scallops which characterize category retrieval.

4.1.3 Free Associations.
Figure 5 shows the rate at which free associations to both grocery store and shoe store are produced by the model. To arrive at this figure, a different control process was assumed than for script generation, which does not involves the use of temporal cues. Given the script header, an associated item was produced. These two items then formed the first retrieval cue. As the third item was retrieved it was added to this retrieval cue, but the next item retrieved replaced it, and so on. After three unsuccessful retrieval attempts, this cue was abandoned, and the process was started all over again with the script header as the sole component of the retrieval cue. Once the script header itself lead to three successive retrieval failures, the whole process stopped.

The retrieval functions shown in Figure 5 are negatively accelerated and somewhat scalloped. They look more like a category naming function than a script generation function. Nevertheless, they were generated from the same knowledge base, by the same retrieval process as Figure 4 - only the control process was different. Scripts, categories, and associative structures do not reflect the organization of memory; rather, they are generated from an unorganized knowledge base - an associative net with only local connections between concepts and propositions. Structures need not be in the mind; they may result from particular kind of control processes - that is what this model suggests.

869

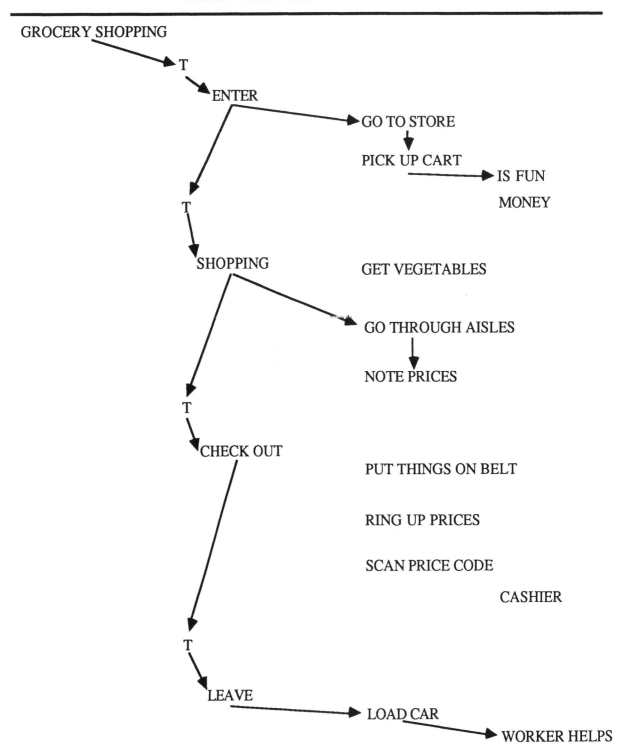

SCRIPT HEADER	TEMPORAL MARKERS AND EPISODE HEADERS	EVENTS	ELABORATIONS
GROCERY SHOPPING			
	T		
	ENTER	GO TO STORE	
		PICK UP CART	IS FUN
			MONEY
	T		
	SHOPPING	GET VEGETABLES	
		GO THROUGH AISLES	
		NOTE PRICES	
	T		
	CHECK OUT	PUT THINGS ON BELT	
		RING UP PRICES	
		SCAN PRICE CODE	
			CASHIER
	T		
	LEAVE	LOAD CAR	
			WORKER HELPS

Table 2. Items which the model retrieved over three runs in response to the Grocery Shopping cue. The lines show the result of one run.

870

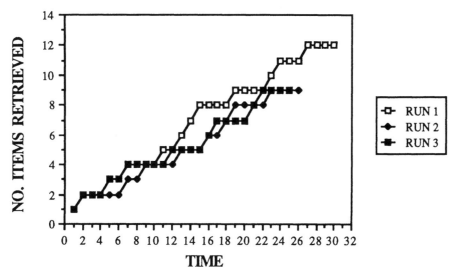

Figure 4. Number of scriptal items retrieved by the model in three runs
as a function of time.

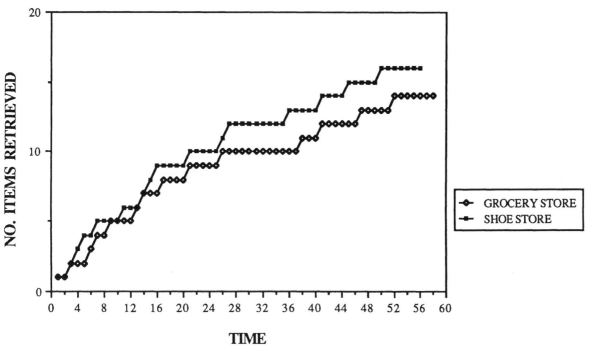

Figure 5. Number of items retrieved by the model in a free
association task as a function of time.

References

Anderson, J. R. (1980). Concepts, propositions, and schemata: What are the cognitive units?
Nebraska Symposium on Motivation, 28, 121-162.

Bower, G. H., Black, J B., & Turner, T. J. (1979). Scripts in memory for text. *Cognitive Psychology, 11,* 177-220.

Ericsson, K. A. & Simon, H. A. (1984). *Protocol analysis: Verbal reports as data.* Cambridge, MA: MIT Press.

Galambos, J. A. & Rips, L. J. (1982). Memory for routines. *Journal of Verbal Learning and Verbal Behavior, 21,* 260-281.

Graesser, A. C. (1981). *Prose comprehension beyond the word.* New York: Springer-Verlag.

Haberlandt, K. & Bingham, G. (1984). The effect of input direction on the processing of script statements. *Journal of Verbal Learning and Verbal Behavior, 23,* 162-177.

Raaijmakers, J. G. & Shiffrin, R. M. (1981). Search of associative memory. *Psychological Review, 88,* 93-134.

Schank, R. C. (1980). Language and memory. *Cognitive Science, 4,* 243-284.

Schank, R. C. (1982). *Dynamic memory.* New York: Cambridge University Press.

Schank, R. C., & Abelson, R. P. (1977). *Scripts, plans, goals, and understanding.* Hillsdale, NJ: Erlbaum.

van Dijk, T. A. & Kintsch, W. (1983). *Strategies of discourse comprehension.* New York: Academic Press.

Walker, W. H. & Kintsch, W. (1985). Automatic and strategic aspects of knowledge retrieval. *Cognitive Science, 9,* 261-283.

Spontaneous Retrieval in a Conceptual Information System

Lisa F. Rau

Artificial Intelligence Branch

Corporate Research and Development

GE Company

Schenectady, NY 12301 USA

Abstract

A traditional paradigm for retrieval from a conceptual knowledge base is to gather up indices or features used to discriminate among or locate items in memory, and then perform a retrieval operation to obtain matching items. These items may then be evaluated for their degree of match against the input. This type of approach to retrieval has some problems. It requires one to look explicitly for items in memory whenever the possibility exists that there might be something of interest there. Also, this approach does not easily tolerate discrepancies or omissions in the input features or indices. In a question-answering system, a user may make incorrect assumptions about the contents of the knowledge base. This makes a tolerant retrieval method even more necessary.

An alternative, two-stage model of conceptual information retrieval is proposed. The first stage is a spontaneous retrieval that operates by a simple marker-passing scheme. It is spontaneous because items are retrieved as a by-product of the input understanding process. The second stage is a graph matching process that filters or evaluates items retrieved by the first stage. This scheme has been implemented in the SCISOR information retrieval system. It is successful in overcoming problems of retrieval failure due to omitted indices, and also facilitates the construction of appropriate responses to a broader range of inputs.

1 Introduction

The System for Conceptual Information Summarization, Organization and Retrieval (SCISOR) is an information retrieval system designed to analyze, answer questions about, and summarize short newspaper stories in natural language. Operating in the domain of corporate takeovers and finance, SCISOR is unique in its approach to retrieval of the complex conceptual events stored in its knowledge base.

Most approaches to conceptual information retrieval can be broken down into the following four phases:

1. **Retrieval decision:** The system comes to a point in its processing when it desires some information from memory.

2. **Retrieval setup:** Indices or features are collected and put into a correct format for retrieval.

3. **Retrieval:** The retrieval process is performed.

4. **Post-processing / Matching:** The outcome of the retrieval process is examined and conditional actions may be taken.

The conceptual information retrieval performed in FRUMP (DeJong, 1979), CYRUS (Kolodner, 1984), COREL (DiBenigno, Cross & DeBessonet, 1986), (which uses the PEARL AI package (Deering, Faletti &Wilensky, 1981), and IPP (Lebowitz, 1983) can all be put into this framework, as could any system that performs deductive information retrieval in the style of Charniak, Riesbeck and McDermott (1980). An exact match restriction may be relaxed after the initial fetch returns a negative result. The model of finding items in memory here is an iterative one of generate and test. This model has certain problems when we consider how it could be used to perform certain desirable functions in an intelligent information retrieval system.

In contrast to this model, in SCISOR, items are retrieved from memory automatically as a result of the instantiation of new input instances. When a user's question is instantiated, potential answers appear in a short-term buffer. When a new story is instantiated, any previously existing context for that story appears in the buffer. This automatic retrieval is implemented with a constrained form of marker-passing. Items spontaneously retrieved in this manner are then run through a more computation intensive matching process.

Three problems with the model of retrieval initially described will be given, followed by a description of the SCISOR system and its solution to these problems.

2 Problem Description

2.1 Find things without looking

One capability the SCISOR system has is to retrieve automatically a user's previously posed but unanswered question when an answer to that question becomes known or refined. For example, consider the following scenario:

User: How much did the ACE company offer to take over ACME?

System: *Figures for the ACE-ACME takeover have not yet been disclosed.*

Intervening time...

System: *BEEP! Figures for the ACE-ACME takeover have just been released. ACE has offered $40 / share for all outstanding shares of ACME.*

In order to provide this capability to a system that performs retrieval as previously described, the system would have to keep a list of unanswered questions present. When new stories were input to the system, it would poll the unanswered questions, asking "does this answer you?". A better approach would be to set up demons on each unanswered question that look for certain input features that might relate to the content of the question. In either case, however, the system is *always looking* for answers to its questions. In SCISOR, if input features happen to relate to an unanswered question in memory, that question is spontaneously brought up for consideration. If the input does not relate to any unanswered questions, nothing happens.

2.2 Retrieval with incorrect or misleading inputs

SCISOR has another capability that would be awkward to implement in the model of retrieval previously described: to find events in memory even when a user's question contains only partial, or even incorrect information. For example, consider the following question, along with some independent potential states of the world that might be true at the time the user asked the question.

- Did ACE food company take over ACME hardware company?

    1. The ACME hardware company took over ACE food company.

    2. The ACE food company took over the BIG-ACME company, which *owns* the ACME hardware company.

    3. The ACE food company made an offer to the ACME hardware company, but has not yet succeeded in taking over the ACME company.

    4. ACE is a conglomerate and not simply a food company.

In each of these cases, the question asked cannot be well answered simply "yes" or "no". Consider what a deductive information retrieval mechanism such as that described in Charniak, *et. al.* (1980). might do to find the answers to the questions above. One possible method would be to retrieve all takeover events in which the company taking over another company was the ACE food company. The episodes found would then be checked to find ones in which ACE took over another company. The resulting events are examined to see whether the object of the takeover was the ACME hardware company. Such a procedure is incapable of detecting any of the scenarios described above without further augmentation.

There are many possible augmentations one could make. For example, one could try switching the arguments, looking up and down an "isa" hierarchy, or looking with all subsets of the set of features. One could also generate plausible indices through reconstruction of what was likely to be present in the situation, as is done in the CYRUS program (Kolodner, 1984). Although this procedure has a certain cognitive appeal, it is not guaranteed to find events present in the memory. In fact, none of the augmentations described is guaranteed to find relevant situations; some are not particularly principled, and all involve substantial additional computation.

In SCISOR, finding partially matching scenarios is a by-product of the retrieval process because the first pass of the two-stage retrieval process SCISOR uses simply finds events with features that are the same as, or similar to, features in the input question. The validity of the relationships between the features is not considered until the second pass of the two-stage process is performed. Thus, any of the scenarios above would be retrieved given the input question as stated. The evaluation mechanism then determines what is the same as the user's question and what the differences are. These differences may then be expressed to the user. Note that when nothing closely matches what the user asked, no events in the system's knowledge base will have enough activation to exceed the threshold, and the system will respond that it doesn't know.

Finally, the SCISOR system can find previous articles stored in memory when a new article is input that deals with the same situation. In the corporate takeover domain, events happen over time. For example, ACE may make an offer to ACME on Monday, and ACME may respond to the offer on Friday. The initial offer should be retrieved from memory when the response to the offer is input, so that this new piece of information can be properly integrated. The way that this could be done with the traditional method of retrieval is similar to the case of retrieving unanswered questions when the answers are input. Checking after each new story to see if it is a

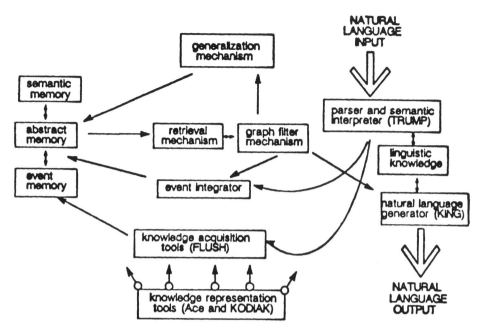

Figure 1: SCISOR System Architecture

continuation of a known story would result in substantial additional overhead. Also, dealing with the features present in a story continuation could have the same problem as dealing with missing or incorrect features in a user's input questions. For example, it would be difficult to find out that ACE was trying to take over ACME if subsequently ACME announced it was trying to take over ACE. Both of these concerns are elegantly addressed with SCISOR's method of retrieval.

This section has described some of the problems with traditional methods of conceptual information retrieval. The next section describes the SCISOR system and in more detail how its approach to retrieval addresses the problems.

3   System Description

3.1   Implementation

SCISOR takes input in natural language, integrates new information in memory, and answers natural language questions in natural language. The natural language input is processed with the TRUMP parser and semantic interpreter (Jacobs, 1986). New events are integrated as a continuation of an ongoing story (if present) by the event integrator, which also stores new events for retrieval. The FLUSH acquisition tools (Besemer, 1986) are aids to the acquisition of vocabulary and phrases in the system's phrasal lexicon. The events in SCISOR are represented using the KODIAK knowledge representation language (Wilensky, 1986), augmented with some scriptal knowledge (Schank & Abelson, 1977) of typical events in the domain. Linguistic knowledge is represented using the Ace linguistic knowledge representation framework (Jacobs & Rau, 1985). Responses to the user are to be generated with the KING (Jacobs, 1985) natural language generator. Figure 1 illustrates the architecture of SCISOR.

The following exchange illustrates the kind of input-output behavior SCISOR was designed to exhibit:

User: Did ACE food company take over the ACME hardware company?

876

System: *Yes they did. The ACE food company completed the $13.2 million friendly takeover last Friday.*

User: What was the sequence of events in the ACE-ACME takeover deal?

System: *Rumors that ACME was to be taken over started May 13, 1985. The stock rose $5 a share to $65. On May 16, the ACE company announced that it had made an offer to ACME valued at $66 a share. The offer was a cash offer to purchase all 2 million outstanding shares of ACME. After a month of deliberation involving some potential conflict of interest matters, the ACE-ACME deal was completed last Friday.*

User: What happened yesterday in the HAL Corporation bid for BIG?

System: *HAL Corporation increased its bid for the BIG company yesterday, up $4 to $70 a share. BIG has not commented on the new offer.*

3.2 Memory Organization

SCISOR manipulates conceptual structures represented in the KODIAK knowledge representation language. KODIAK can be viewed as a hybrid frame and semantic net-based language, similar in spirit to KRYPTON (Brachman, Fikes & Levesque, 1983) or KL-ONE (Brachman & Schmolze, 1985). In SCISOR, knowledge stored is either specific, abstract, or semantic. An example of a specific memory is the memory of the ACE food company acquiring the ACME hardware company. A traditional episodic memory (Tulving, 1972) is composed of specific memories. An example of an abstract memory is the generalization across specific experiences we might have heard about where companies have acquired other companies. Abstract (or generalized episodic) memories are on the border between specific memories and semantic memory. Semantic memory is the memory or knowledge of what "companies" are and what "acquiring" is. Semantic memory is used in understanding and making inferences about the input to the system. Figure 2 illustrates the structure of long-term memory with associated examples.

In addition to this tripartite division, another level of organization is superimposed on memory. Groups of related concepts in episodic or abstract memory are linked together through a common node, called a TAG. Each TAG has a numerical threshold value, currently equal to a fraction of the number of concepts in the episode, currently one-third. Long articles may consist of TAGs that have TAGs as components. Figure 3 illustrates the kind of structure the integrator superimposes on memory.

3.3 The Retrieval Mechanism

Retrieval in SCISOR is a two-stage process. The first stage is a coarse search that finds events in memory likely to be relevant. Relevance is determined by the number of features present in an event in memory related to features in the input. After the most likely candidates have been isolated, a matching process is performed.

The first stage of the retrieval process is performed by a process of priming or constrained spreading activation. As concepts are instantiated in the system, instances of concepts that are related via category membership links are marked (*i.e.*, primed or activated). When a certain subset of the concepts in an event or episode is marked, the entire episode is put into the system's short-term memory buffer. This is the spontaneous retrieval phase. The events that have been spontaneously retrieved can then be run through the match filter to check the nature of the match between the input and the events retrieved. Periodically, all marks are deleted from the system.

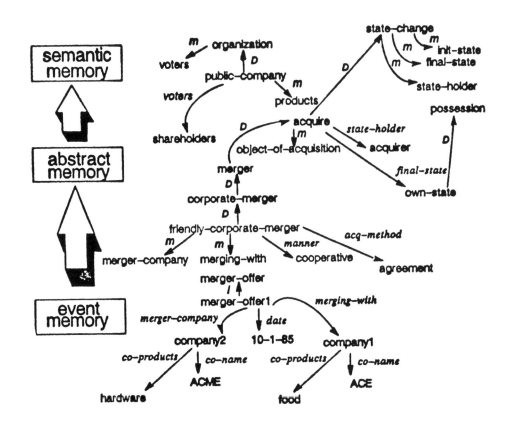

Figure 2: Structure of long-term memory

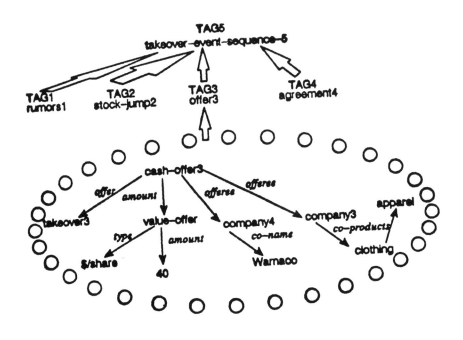

Figure 3: Superimposed Structure

878

Marker-passing "waves" are propagated continuously as new concepts are instantiated. This mechanism retrieves answers to input questions, old unanswered questions to new input answers, and stories given related input stories in exactly the same manner.

In more detail, the retrieval mechanism operates in the following manner:

1. At the time of concept instantiation, related concepts are marked according to the "priming rules" (see next section). Every concept that is marked causes its containing episode or episodes (its TAGs) to have increased activation. The current and threshold levels of activation are simple properties of the TAG.

2. TAGs that have had their values increased check themselves to see whether their current activation levels exceed their threshold.

3. If the TAG threshold has been exceeded, an *intersection* has been detected, and the entire episode is put into the short-term memory buffer.

4. In the current implementation, all concepts in the memory are unmarked after every one or two sentences and after every question are input.

Components of experiences in the input cause other concepts to be marked, but if those other concepts are not components of relevant experiences, the marking does not contribute to retrieval.

The rules that guide how levels of activation or marks are passed through the conceptual hierarchy are given below. These rules were formulated to decrease the possibility of retrieving memories that have limited predictive capacity or relevance to the current situation.

3.3.1 Priming Rules

The effect of the following rules is that all instances of concepts that are components of episodes and are children of the incoming parent of the parent of the incoming instance in the conceptual hierarchy, are marked. Also, all instances of concepts that are components of memories, and are *direct* instances of concepts that are parents of the incoming instance, are marked. This rule prevents everything in the hierarchy from being marked, and limits what is marked to a level of conceptual abstraction supported by the presence of a direct instance at that level.

For example if a user asked "what happened to the ACE company?", the event being asked about is at a fairly general level of conceptual abstraction. Any event more specific than this general "event" instance would be a valid answer. If no event were known, the unanswered question would be stored. In the future, *any* input events involving the ACE company (i.e., offers, rumors) would cause the unanswered question to be activated.

Figure 4 shows a simple example of where components of memories would cross-index each other and when components of memories would not. Double arrows between instances (concepts with numbers following them) signify cross-indexing. Single-headed arrows signify where the instantiation of one concept will activate the concept pointed to, but not vice versa. Note that all concepts as they are instantiated cause any related instances to be marked, which allows any feature in the input to be a potentially useful index key into memory. Also, note that each instance is part of a specific memory that is not shown in Figure 4. For example, "material-company1" may be part of the memory of the unspecified materials company that rumors were circulating about yesterday.

Rules:

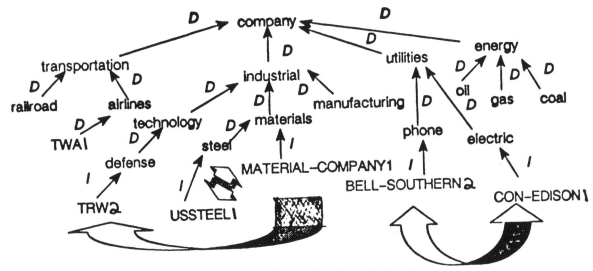

Figure 4: Example of Implicit Indexing

1. Mark only concepts that are components of specific or abstract memories. These concepts are marked with TAGs.

2. Determine the categories to which the incoming instance belongs (Categories-of A).

3. Determine the concepts in the reflexive, transitive closure along category membership links of Categories-of (Categories-of A).

4. Mark the direct instances of concepts in this reflexive, transitive closure. Each marked concept increases the current activation value of each of its TAGs.

5. Check to determine whether the current activation level of the episodes that just had their components marked exceeds the threshold level.

Additional refinements to this algorithm have been made to increase the system's efficiency. One such refinement has been to check the number of episodes containing a certain concept before passing markers to all those episodes. This check can be made easily by maintaining a count of the number of category members in each conceptual category. For example, if the system had read about events involving thousands of companies, this number would be stored at the parent COMPANY node. When this number is very large, the system suppresses the marker-passing operation. Events that are retrieved through the activation of more unique concepts then incorporate the high-frequency concept information. This simple refinement suppresses waves of activation unlikely to cause significant differentiation among events in memory, while still taking all features of the input into consideration.

After a set of events has been spontaneously retrieved, a match filtering operation is performed to ensure that the correct *relationship* exists between concepts within the events.

3.3.2 The Match Filter

The match filter iteratively checks to see that the relationships between marked concepts are correct. For example, consider the example illustrated in Figs. 5 and 6.

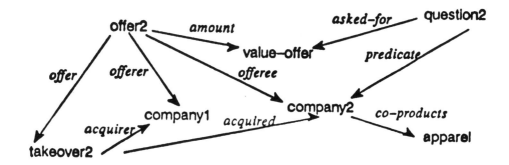

Figure 5: How much are takeover offers for apparel companies?

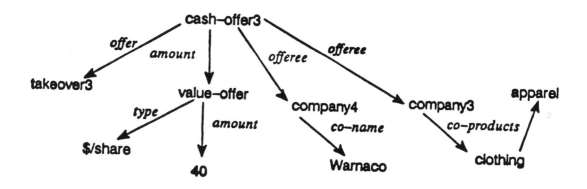

Figure 6: Part of a Story Episode

The story episode contains information about a kind of offer, a CASH-OFFER, made by Warnaco for a clothing company. The question episode requests the value of any offers made by a company in a takeover attempt on an apparel company.

Due to the marker-passing, the instantiation of OFFER in the question episode causes CASH-OFFER to be marked. The match filter begins at a node that is marked only by one node from the input. OFFER2 is such a node, but note that COMPANY2 and COMPANY3 are both marked by COMPANY4 and COMPANY5.

The filter proceeds to check that the OFFER of CASH-OFFER3 (TAKEOVER3) is marked by the OFFER of OFFER2 (TAKEOVER2). This process is repeated for every node in the input. In the case of answering a question, those nodes that do not correspond are added to a list of presuppositions to be expressed to the user. Also, any node that matched but was more general than the input may be expressed to the user. For example, if the user was interested in what pet-food companies were being taken over, and the system only knew about food companies, this generalization is pointed out.

4 Related Research

4.1 Question Answering

In SCISOR, the processes that find the approximate location of an answer to a user's question and the processes that determine what the answer should be are separate. A great deal of work has been done on the second problem, most notably by Lehnert (1978). Determining an appropriate answer to a user's question, given that the context in which the user's question was posed is already known, is a separate process from the initial retrieval of a context in which to search for an answer. This initial retrieval of a context is the spontaneous retrieval this paper describes.

4.2 Conceptual Information Retrieval

Kolodner (1984) has a well-developed theory of conceptual information retrieval. However, the system is not guaranteed to answer correctly. As a cognitive model, its memory failures are understandable and interesting, but when accuracy and reliability of information are important, such a model is not viable. Her theory of retrieval developed can be viewed as one attempt to overcome problems and limitations with existing methods of conceptual information retrieval, such as those described in Charniak, *et. al.*

4.3 Distributed Representation

As Hinton points out (Hinton, 1984) a localist knowledge representation scheme combined with a spreading activation or priming mechanism is hard to distinguish from a distributed representation in terms of functionality. Two main properties of distributed representations are that the knowledge base is contents-addressable and that generalization is automatic. Because the SCISOR knowledge base is represented using the KODIAK localist knowledge representation language and performs retrieval of items in the knowledge base by a constrained form of marker-passing (Charniak, 1983), SCISOR representations form an effective distributed representation, and therefore the structures in the knowledge base are contents-addressable. Thus, the content of input questions or stories "address" events in memory with similar features or content. The automatic generalization that SCISOR also exhibits will be discussed in another paper.

4.4   Cognitive Effects

In systems such as SCISOR that use the same parsing mechanism for both question parsing and story parsing, interesting effects occur. For example, in BORIS (Dyer, 1983), the system would occasionally take as true information present in a user's question not previously known to the system. Although this has been shown to be a cognitively valid effect (Loftus, 1975), and does increase the system's opportunities to learn, it is not a desirable effect for an information retrieval system. In the SCISOR system, a strict difference is maintained between information read in articles, and information present in user's questions in that all information present in questions and not previously known to the system is flagged as potentially unreliable. Thus the system believes nothing it is told, but everything it reads in the paper.

The model of retrieval discussed has a certain cognitive appeal. It exhibits the same kind of spontaneous recall as people do, as is discussed in Schank's work on reminding (Schank, 1982). Also, the answer to a user's question may be retrieved before the user has finished asking the question, an interesting effect also first achieved in Dyer's BORIS system (1983).

5   Summary

SCISOR performs retrieval in a two-stage process. The first step, a spontaneous and coarse search, operates as follows:

1. **New inputs are instantiated**

2. **Marker passing occurs**

3. **Items are spontaneously retrieved**

4. **Items are evaluated**

The second stage is a graph matching process that performs a syntactic matching function on the likely candidates retrieved by the first process. This spontaneous method of retrieval elegantly solves some retrieval problems found in other systems:

1. SCISOR addresses a user's previously unanswered questions if an answer becomes known without always looking for an answer.

2. SCISOR can locate relevant information in response to a user's questions, even when that question contains misleading or partial information.

3. SCISOR retrieves previous events when updates of those events are read. This is done in the same manner as retrieving an unanswered question, and with the same tolerance for partial or contradictory input as in the question-answering case.

6   System Status

SCISOR is implemented in Common Lisp; it is used on VAX computers and Symbolics and SUN workstations. The TRUMP parser and semantic interpreter has not yet been tested with a large grammar or vocabulary but, in these early stages, it has been relatively easy to customize. On the SUN-3 it processes input at the rate of a few seconds per sentence, including the selection of candidate parse and semantic interpretation. The KING natural language generator was implemented in Franz Lisp, and at this writing has not yet been converted to Common Lisp to run with TRUMP.

The system has now been tested with a dozen or so stories stored in the knowledge base. Hundreds of semantic concepts and domain vocabulary are also present. About a dozen questions are answered by the system.

7 Problems

One problem with SCISOR is in the kind of questions it can answer. Currently the SCISOR system is capable of answering only questions about information explicitly stored in the knowledge base. Any information that potentially could be *reconstructed* or *inferred* from information stored in the knowledge base is not available. The line between what is explicit in a story and what can be deduced from that story is not sharp, because some amount of "figuring" must go on to obtain any reasonable understanding of the story. To obtain this understanding, SCISOR computes something similar to a maximally complete inference set (Cullingford, 1986) as the set of information present explicitly in articles and inferred from the context and other world knowledge. Anything in that understanding can be directly retrieved.

For example, SCISOR is able to answer the question "What company was sold for $3 billion?" without pre-indexing a story containing that information by AMOUNT-OF-SALE. However the system cannot answer the question "Which companies have been taken over more times than they have taken over other companies?" for example, because an answer would require counting all the times a company has been taken over and has taken over other companies, comparing these two numbers, and repeating the process for every other company in the knowledge base.

Another problem that ultimately must be addressed is the speed and memory requirements necessary in a viable information system. A simple calculation based on the number of relevant articles per year (10,000) and the average size of the memory requirements per conceptual representation of the articles yields a conservative memory requirement estimate of 25 Mbytes of storage. Such a large knowledge base may effect the speed and accuracy of the retrieval mechanism.

8 Conclusions

SCISOR is an experiment in the usefulness of a spontaneous, marker-passing approach to conceptual information retrieval. In its current implementation, it has demonstrated some promising results. The marker-passing scheme, combined with the knowledge representation used, produces an effective contents-addressable, distributed representation. Retrieval occurs spontaneously when features in the input are related to features of events in memory. SCISOR can find answers to input questions even in light of missing or misleading input information.

The system has not yet been tested on a large number of documents. However so far, the tests that have been performed are quite promising. Before any definitive claims can be made about the ultimate usefulness of this type of system, it must be tested with a large sample of documents in real IR tasks. The next stage of the project will include such tests.

References

[1] D. Besemer. *FLUSH: Beyond the Phrasal Lexicon.* Technical Report 86CRD181, General Electric Corporate Research and Development, 1986.

[2] R. Brachman, R. Fikes, and H. Levesque. Krypton: integrating terminology and assertion. In *Proceedings of the National Conference on Artificial Intelligence*, Washington, D. C., 1983.

[3] R. Brachman and J. Schmolze. An overview of the KL-ONE knowledge representation system. *Cognitive Science*, 9(2), 1985.

[4] E. Charniak. Passing markers: a theory of contextual influence in language comprehension. *Cognitive Science*, 7(3), 1983.

[5] E. Charniak, C. Riesbeck, and D. McDermott. *Artificial Intelligence programming.* Lawrence Erlbaum Associates, Hillsdale, NJ, 1980.

[6] R. E. Cullingford. *Natural Language Processing: A Knowledge-Engineering Approach.* Rowman and Littlefield, Totowa, NJ, 1986.

[7] M. Deering, J. Faletti, and R. Wilensky. PEARL: an efficient language for artificial intelligence programming. In *Proceedings of the Seventh International Joint Conference on Artificial Intelligence*, Vancouver, British Columbia, 1981.

[8] G. DeJong. *Skimming Stories in Real Time: An Experiment in Integrated Understanding.* Research Report 158, Department of Computer Science, Yale University, 1979.

[9] DiBenigno, M.Kathryn, Cross, George R. and DeBessonet, Cary G. *COREL - A Conceptual Retrieval System.* Technical Report CS-86-147, Computer Science Department, Washington State University, 1986.

[10] M.G. Dyer. *In-Depth Understanding.* MIT Press, Cambridge, MA, 1983.

[11] G. Hinton. *Distributed Representations.* Computer Science Technical Report CMU-CS-84-157, Carnegie-Mellon University, 1984.

[12] P. Jacobs. *A knowledge-based approach to language production.* PhD thesis, University of California, Berkeley, 1985. Computer Science Division Report UCB/CSD86/254.

[13] P. Jacobs. Language analysis in not-so-limited domains. In *Proceedings of the Fall Joint Computer Conference*, Dallas, Texas, 1986.

[14] P. Jacobs and L. Rau. Ace: associating language with meaning. In T. O'Shea, editor, *Advances in Artificial Intelligence*, North Holland, Amsterdam, 1985.

[15] J. Kolodner. *Retrieval and Organizational Strategies in Conceptual Memory: A Computer Model.* Lawrence Erlbaum Associates, Hillsdale, NJ, 1984.

[16] M. Lebowitz. Generalization from natural language text. *Cognitive Science*, 7(1), 1983.

[17] W. G. Lehnert. *The Process of Question Answering: Computer Simulation of Cognition*. Lawrence Erlbaum Associates, Hillsdale, NJ, 1978.

[18] E. F. Loftus. Leading questions and the eyewitness report. *Cognitive Psychology*, 7, 1975.

[19] R. C. Schank and R. P. Abelson. *Scripts, Plans, Goals, and Understanding*. Lawrence Erlbaum Associates, Halsted, NJ, 1977.

[20] R.C. Schank. *Dynamic Memory: A Theory of Reminding and Learning in Computers and People*. Cambridge University Press, Cambridge, 1982.

[21] E. Tulving. Episodic and semantic memory. In E. Tulving and W. Donaldson, editors, *Organization and Memory*, Academic Press, New York, 1972.

[22] R. Wilensky. Knowledge Representation - A Critique and a Proposal. In J. Kolodner and C. Riesbeck, editors, *Experience, Memory, and Reasoning*, Lawrence Erlbaum Associates, Hillsdale, NJ, 1986.

A Model of Schema Selection Using Marker Passing and Connectionist Spreading Activation

Hon Wai Chun

Honeywell Bull
300 Concord Road, MA30-895A
Billerica, MA 01821-4186
U.S.A.

Alejandro Mimo

Gold Hill Computers
163 Harvard Street
Cambridge, MA 02139
U.S.A.

ABSTRACT

Schema selection involves determining which pre-stored schema best matches the current input. Traditional serial approaches utilize a match/predict cycle which is heavily dependent upon backtracking. This paper presents a parallel interactive model of schema selection called SAMPAN which is more flexible and adaptive. SAMPAN is a hybrid system that combines marker passing with connectionist spreading activation to provide a highly malleable and general representation for schema selection. This work is motivated by recent success in connectionist schema representations and in natural language marker passing systems. A connectionist schema representation provides many attractive features over traditional schema representations. However, a pure connectionist representation lacks generality; new propositions cannot easily be represented. SAMPAN gets around this problem by using marker passing to perform variable binding on generalized concepts. The SAMPAN system is a constraint satisfaction network with nodes that perform simple pattern matching and input summation. This approach is directly applicable to current schema-based systems.

I. INTRODUCTION

The most difficult task in schema-driven systems[1] is to select a schema that can accurately represent the current input. A typical serial approach to schema selection first extracts cues (in the form of settings, participants, events, etc.) from the input and then uses these cues to help select an appropriate schema. If the selected schema is later shown to be incorrect, the system must backtrack and select another schema. Once a schema is selected, top-down expectation or prediction can be used to assist recognition or natural language understanding. For example, schema or script selection in SAM [SCHA75] involves matching the input with a list of conceptualizations stored in each script. If the input matches, this script will be activated. SAM uses a search list that is dynamically updated to indicate which scripts are most likely to occur based on the conceptualizations or objects seen so far. FRUMP [DEJO79], on the other hand, processes the first paragraph to select a sketchy script. For efficiency, sketchy scripts are matched using a n-way branching discrimination tree.

There are several problems with current serial schema selection techniques [HAVE83]. When a system selects an incorrect schema, it must backtrack, retry another schema, and possibly reinterpret the inputs. For a system with a large number of schemata, this computational overhead may greatly reduce its effectiveness. In addition, there is a "chicken and egg problem." The system cannot make use of schema knowledge while interpreting the input to perform

---

1. Throughout this paper, the terms schemata, frames and scripts are used interchangeably to represent knowledge structures that describe typical situations or objects. The terms massively parallel networks and connectionist networks are also used interchangeably to represent the notion of large parallel networks with many simple processing elements.

schema selection. On the other hand, the system must hypothesize a correct schema before the material it is trying to explain is fully given. There is also a "match or near miss" problem, in which the system must decide whether an adequate match has been made or to continue searching. Although various index and search strategies have been used to improve the selection process, there is still no elegant approach that performs schema selection efficiently. This overall process of backtracking and reinterpreting inputs happens rarely in human perception and interpretation, suggesting that AI's heavy dependence on heuristic search might be inappropriate.

Recently, several connectionist schema representations have been introduced [RUME86, SHAR86] offering solutions to these problems. The parallel nature of connectionist networks avoids the "match or near miss" problem since all schemata are "matched" simultaneously. In addition, in [CHUN85, RUME86], bottom-up schema activation and top-down prediction generation are considered as continuous processes rather than discrete match, predict and backtrack stages. Several schemata may be active at any one time to provide predictions with varying certainty level. Hence, the "chicken and egg problem" disappears since schema knowledge is continuously available during input interpretation. The connectionist spreading activation also provides a default mechanism that is *multivariate distributed* [RUME86] in which defaults are determined by values filling other slots. Charniak [CHAR86a] indicated that the connectionist default mechanism avoided redundancy found in the *square formation* problem where redundant information is generated due to many levels of generality. Connectionist network also provides a "resonance effect" in which activated slots of the same schema reinforce themselves and strengthen the selection of that schema.

However, a pure connectionist representation for schemata lacks generality. For example, a system may have a node that represents "a person went home." However, in order to understand a particular instance, such as "Mary went home," the system must also have this node encoded beforehand. In other words, any possible combination of schema structures with their instantiated bindings must be encoded into the network structure before the network can understand them. Several approaches [MCCL86, TOUR85] have been proposed to solve this problem of representing new propositions or *variable binding*. Although these approaches provide connectionist networks with more flexibility, the set of all possible inputs must still be known beforehand. In addition, they tend to lose the simple structure of connectionist networks which makes them attractive.

SAMPAN maintains the elegance and simplicity of distributed connectionist networks yet solves the variable binding problem by augmenting the system with marker passing capabilities and nodes that represent generalized concepts. The SAMPAN model itself, is influenced by the PDP schema representation [RUME86], the NEXUS dictionary structure [ALTE85], and Wimp's marker passing algorithm [CHAR86]. SAMPAN integrates features found in these recent representational paradigms to construct a model of schema selection that is highly flexible, general and dynamic:

— Connectionist networks are very good at completing patterns from partial information. SAMPAN uses this representation to encode general schema knowledge as well as perform schema activation and prediction generation.

— Marker passing algorithms have proven successful as a method of sharing and locating information. SAMPAN uses a marker passing algorithm similar to that of Wimp [CHAR86] to bind variables.

— In order to represent general episodic events and concepts, SAMPAN uses nodes which are conceptual templates similar to those found in NEXUS's dictionary network.

SAMPAN is relevant to many AI applications, such as natural language understanding (expectation-driven processing), recognition (fill partial patterns with defaults), learning (recognize unexpected events), and run-time plan generation (through expectation). This paper concentrates mainly on the problem of schema selection.

II. OVERVIEW OF SAMPAN

SAMPAN networks are massively parallel constraint satisfaction networks. Nodes represent concept templates and links represent associations between concepts. Communication is in the form of marker passing combined with connectionist value passing. Information passed consists of purely local information. The processing within a SAMPAN network is totally parallel without global procedures.

To demonstrate SAMPAN, a script-driven natural language understanding system is used as an example. In this paper, SAMPAN performs schema selection and generates expectations which can be used by a natural language understanding system. In other words, the SAMPAN network acts as an active schema knowledge base. Input to SAMPAN consists of pre-parsed internal representations. The output of SAMPAN is a stream of expected conceptualizations. The state of the SAMPAN network at any one time represents an emergent schema structure. Currently, SAMPAN is not interfaced to an actual parser. However, the example can still adequately illustrate SAMPAN's performance, since only the pre-parsed input is needed. In the future, we are interested in exploring a tighter integration between the SAMPAN network and a massively parallel parser.

The next section introduces the connectionist schema representation in detail. This is followed by a description of the marker passing used and computations involved. Explanation of the test domain is then given.

III. DISTRIBUTED SCHEMA REPRESENTATION

The SAMPAN network, in terms of its connectionist structure, is adapted from the PDP distributed schema representation [RUME86]. The structure is a constraint satisfaction network of nodes that represents schema slots. A coalition of active nodes represents a schema. In [RUME86], schemata were used to represent different room types (e.g. kitchen, living room, etc.) and slots to represent room descriptors (e.g. ceiling, coffee cup, refrigerator, etc.). In SAMPAN, schemata represent episodic events and slots represent generalized concepts (or event/state concept [ALTE85] such as "a person went to a restaurant"). The representation is distributed since each schema is represented not as a single node but as the activation of a collection of nodes.

Through this connectionist representation, concepts that usually occur together will tend to support each other, while suppressing those that do not. In other words, nodes that are usually active or inactive at the same time will have an activating link between them. In the opposite case, an inhibitory link will be found. In SAMPAN, nodes can be activated either by the parser (analogous to "clamping") or by other nodes through the input links. Links are bi-directional and the weights are symmetrical. Unlike marker passing networks, the links are not labeled and are used solely to modify the information passed through them. For simplicity, the weights are set according to a Bayesian probability measurement [RUME86] (although learning rules can be used as well):

$$w_{ij} = -\ln \frac{P(x_i{=}0 \ \& \ x_j{=}1) \ P(x_i{=}1 \ \& \ x_j{=}0)}{P(x_i{=}1 \ \& \ x_j{=}1) \ P(x_i{=}0 \ \& \ x_j{=}0)}$$

where w_{ij} is the weight of the link between nodes i and j, $P(x_i{=}0 \ \& \ x_j{=}1)$ is the probability that the activation level of node i is 0 and the activation level of node j is 1, and so on.

The result of input activation to this network is a gradual emergence of a coalition of active nodes. This coalition represents a schema structure. This approach is reminiscent of *event concept coherence* presented by Alterman [ALTE85] where concepts are coherent if their positions in the net are *proximal*. In the SAMPAN network, concepts are coherent if they are connected by activating links (links with a positive weight).

There are several representational advantages to this distributed model. Traditional schema structures are often rigid and static, lacking in flexibility and generativity. Although, scripts may have tracks to represent variations or have embedded scripts to represent hierarchical structure, this knowledge is still rigidly defined. In a distributed connectionist representation, a schema is not an explicit representational entity, but an emergent structure based on the events seen so far. The result is a very malleable representation that is highly dynamic and adaptive.

SAMPAN networks are used differently from that of the PDP schema representation. Instead of "clamping" all the inputs at the same time, inputs to SAMPAN networks are fed sequentially. This permits a natural language system which processes sentences sequentially to interpret each sentence in the context of what has been processed before. In addition, instead of representing a schema as the stable coalition of nodes (where all the nodes of a particular schema will eventually be fully activated), SAMPAN allows nodes to decay in the process of schema selection. In this case, a SAMPAN schema is a "dynamic" coalition of nodes that reflects recent inputs and the expectations generated from them. This permits SAMPAN to continuously process input stories without having to first clear network activity after each schema. Otherwise, the network might exhibit a "heat death" phenomenon where too many concepts would be activated or expected.

IV. COMPUTATION OF SAMPAN

This section describes the massively parallel computation involved in schema selection and contextual expectation generation. First, the internal structure of SAMPAN nodes is presented. Next, the pattern matching procedure that is invoked by inputs is introduced. Finally, the relaxation function that is evaluated at each relaxation cycle is described.

Node Structure

In contrast to a connectionist node that contains a memory of its activation level and its internal parameters, a SAMPAN node contains a memory of its local variable bindings augmented with activation levels and a conceptual template. Each node in SAMPAN represents a general concept. Concept templates are in the form:

(*Primitive* (*Role1 Filler1*) (*Role2 Filler2*) ...)

where *Primitive* is a semantic primitive, *Role* is a case structure for *Primitive* and *Filler* is its default value. After each filler, there may be property value pairs to describe the filler. In this paper, the internal representation used for concepts is based upon Schank's [SCHA77] Conceptual Dependency (CD) theory. The choice of internal representation is not crucial to the overall SAMPAN model; other similar representations will also suffice.

For example, a node that represents "a person went to a store" will have the following conceptual template stored:

(PTRANS (ACTOR ?PERSON) (OBJECT ?PERSON) (TO ?STORE))

This will match the input "John went to K-MART":

(PTRANS (ACTOR JOHN-1) (OBJECT JOHN-1) (TO K-MART-1))

Pattern Matching

Pattern matching occurs whenever a node receives direct input from the natural language system (as opposed to inputs from other nodes). When the concept template matches an input, the node creates a mark in the form of a binding list. A SAMPAN mark contains a list of variable bindings augmented with an activation level:

[(*Variable1 Value1 Activation1*) (*Variable2 Value2 Activation2*) ...]

In the above example, the resulting mark will be:

((PERSON JOHN-1 1.0) (STORE K-MART-1 1.0))

Activation is a real number ranging from 0 to 1. This can be considered as a measurement of how certain a particular binding is, based upon the concepts seen so far. The activation level is set to 1.0 if the mark is generated as a result of a direct match with the natural language input. Link will attenuate this activation level as the mark is passed from node to node. This indicates that nodes which are further away from the matched concept will be expected less (or less coherent) than those which are directly connected. Charniak [CHAR86] uses a "zorch" as a rough measure of the "strength" of a mark.

Type constraint checking is also performed during pattern matching. For example, the above template is actually a shorthand notation for:

(PTRANS   (ACTOR ?X TYPE PERSON)
           (OBJECT ?X TYPE PERSON)
           (TO ?Y TYPE STORE))

where PERSON and STORE are object types. An object type hierarchy is needed so that a subclass will match a type constraint requiring its parent class. For example, an input of type PERSON should match an ANIMATE constraint. To avoid storing a type hierarchy in each node, a bit representation of sets [HILL85] is used. Each type is represented as a string of bits. For example, all types (including PERSON) under the set of ANIMATE have their first bit set. When the system tries to match ANIMATE it will only need to check for the first bit. Hierarchical information, in the form of a bit string, is thus passed with the type description. Throughout this paper, only the simplified shorthand notation will be used.

Relaxation Computation

Connectionist relaxation involves a weighted summation of the input activations. In SAMPAN, the relaxation is similar, except the inputs are in the form of activations tagged with variable bindings. When SAMPAN performs weighted summation, only activations with the same variable binding will be summed together. We call this process *marker merging*, i.e. marks with consistent bindings are merged together. In this way, inputs are merged with the binding list stored locally. The net result may be one or more binding lists if inconsistencies are found. Being able to differentiate inconsistent bindings is a feature of SAMPAN that permits multiple instantiation of the same concept template. The following example illustrates this relaxation computation:

NODE #0018:

CONCEPT TEMPLATE:
    (POSS (ACTOR ?PERSON) (OBJECT ?POBJ))
LOCAL BINDINGS:
    ((PERSON MARY-2 0.2) (POBJ BOOK-2 0.1))
DECAY: 0.2
INPUT MARKS:
    ((PERSON MARY-2 0.5))   INPUT WEIGHT: 0.1
    ((PERSON MARY-2 0.4) (POBJ BOOK-2 0.4) (FROM JOHN-1 0.2))
                                      INPUT WEIGHT: 0.2
    ((PERSON JOE-3 0.5) (POBJ CUP-3 0.5))   INPUT WEIGHT: 0.5

LOCAL BINDINGS (AFTER RELAXATION):
    ((PERSON MARY-2 0.29) (POBJ BOOK-2 0.16))
    ((PERSON JOE-3 0.25) (POBJ CUP-3 0.25))

In this example, the local binding of NODE #0018 originally contained only one binding list. After relaxation or *marker merging* with the three input marks, two local binding lists results. This indicates two instantiation of the concept template, namely:

    (POSS (PERSON MARY-2) (POBJ BOOK-2))   and
    (POSS (PERSON JOE-3) (POBJ CUP-3))

In addition, any variable bindings in the input marks that are not needed for the local concept template will be removed. For example the binding (FROM JOHN-1 0.2) in the above example was removed. This keeps the size of the local binding lists manageable and reduces communication overhead. Hence, only bindings required for the local concept template will be passed to other nodes. Even though only a partial binding list is passed, this should not be problematic, since each node receives inputs from many other nodes, the net bindings received collectively fill in the required local variables. Nodes will not have any outputs and hence will not spread marks to other nodes, unless all the required local variables are bound.

In SAMPAN, activated nodes provide contextual expectation. For example, whenever the activation level within a local binding list reaches a particular threshold value, the instantiated concept template can be sent to the natural language system. These expected concepts assist the natural language system in interpreting the inputs. In addition, the predictions provide default values to missing concepts.

V. EXAMPLE APPLICATION

To investigate the use of SAMPAN, a network was constructed to represent five different schemata:

restaurant schema	— general eating out scenario
theater schema	— movie, concert, opera, etc.
subway schema	— subway transportation
shopping schema	— purchasing of merchandise
sport schema	— sport related physical activity

These schemata were chosen to demonstrate the generality and flexibility of SAMPAN schema selection, since they share many common concepts (such as going out, travelling, paying for something, etc.). In addition, these schemata can exhibit many types of interactions. For example, these schemata may occur by themselves, embed with another (e.g. eating in a shopping mall), overlap with another (e.g. buy hotdog while leaving subway), or occur in parallel with another (e.g. going to a restaurant play house).

The network currently contains 70 concept templates and 4830 links. The templates encode general concepts which may be shared by several schemata. The following is a sample of the concept templates encoded into the SAMPAN network:

A person leaves his home.
A person sits on a chair.
A business establishment gives a person an object.
An employee takes a person's money.
A person picks up exercise equipment.
A person is hungry.
A person listens to music.
A person orders food.

The SAMPAN network was constructed by gathering five sets of data from each of ten different human subjects. Each subject was asked to visualize a particular scenario within each of the five schemata. The subjects then checked the list of concept templates and indicated those that were relevant to the scenario. Finally, a program constructed the SAMPAN network based upon the collected statistics.

For experimentation, new stories were constructed and feed to the network. SAMPAN correctly matches concepts it knows about in the stories and generates the appropriate expectations. The following is an example story processed by SAMPAN:

```
; John was hungry.
(IS (ACTOR JOHN-1) (STATE *HUNGER*) (VAL 0))
; John left his home to go to McDonald's.
(PTRANS (ACTOR JOHN-1) (OBJECT JOHN-1) (FROM HOME-1) (TO MCDONALDS-1))
; John told the cashier he wanted a hamburger.
(MTRANS (ACTOR JOHN-1) (OBJECT HAMBURGER-1) (TO CASHIER-1))
; John paid the cashier.
(ATRANS (ACTOR JOHN-1) (OBJECT *MONEY*) (FROM JOHN-1) (TO CASHIER-1))
; The cashier gave John the hamburger.
(ATRANS (ACTOR CASHIER-1) (OBJECT HAMBURGER-1) (FROM CASHIER-1) (TO JOHN-1))
; John sat down.
```

893

(PTRANS (ACTOR JOHN-1) (OBJECT JOHN-1) (TO *CHAIR*))
; John ate his hamburger.
(INGEST (ACTOR JOHN-1) (OBJECT HAMBURGER-1))
; John was happy.
(IS (ACTOR JOHN-1) (STATE *MENTAL*) (VAL 10))
; John threw away some garbage into the garbage can.
(PROPEL (ACTOR JOHN-1) (OBJECT *GARBAGE*) (TO GARBAGE-CAN-1))
; John returned home.
(PTRANS (ACTOR JOHN-1) (OBJECT JOHN-1) (FROM MCDONALDS-1) (TO HOME-1))

For each concept in the story, SAMPAN sends the concept to all the seventy nodes in the network and relaxes the network for one cycle. During the relaxation, pattern matching and marker merging was performed. Figure 1 is a screen printout from a personal computer showing the state of the SAMPAN network when the above story was processed. Each row represents the network activation at a particular cycle. Each column represents a particular node in the network. The darkness of the shading is proportional to the activation level in a node.

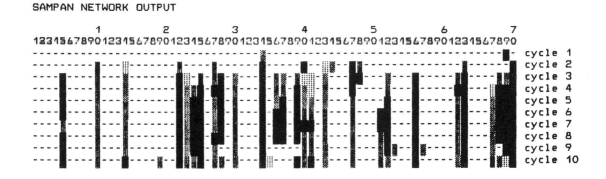

Figure 1. The activation level of the seventy nodes in the example network.

The figure shows that the network formed coalitions of concepts as it processed the story. (Due to the decay factor, some of the nodes gradually died out.) These coalitions represent the emergent schema structure. For example, after SAMPAN processed the third input:

; John told the cashier he wanted a hamburger.
(MTRANS (ACTOR JOHN-1) (OBJECT HAMBURGER-1) (TO CASHIER-1))

activation and bindings were spread to the nodes representing: John having to wait, a cashier giving John a hamburger, John holding a hamburger, John eating a hamburger, John leaving McDonald's to go home, etc. These expected concepts can be used by an expectation-driven natural language system or can help fill in missing concepts not provided by the input.

The example network also handles simple pronoun references. For example, if the concept:

(MTRANS (OBJECT HAMBURGER-1) (TO CASHIER-1))

was used instead of the original third input, the network fills in the ACTOR role to be JOHN-1 through relaxation. However, in complicated stories involving many actors, inference rules need to be used to resolve pronoun references. These inference rules can be encoded in the links as constraints on how variables should be mapped as they are passed from one node to another.

For example, the ACTOR of one concept template might be restricted to be the OBJECT of another.

Summarizing, SAMPAN is able to form coalitions of related concepts that represented the schema structures. The network is flexible enough to accommodate many variations of the same schema as well as novel event sequences. Stories with multiple schemata (embedded or overlapped) are also processed without any difficulty.

SAMPAN was implemented using the MicroAINET language [CHUN86b] written in Golden Common Lisp. MicroAINET is a language to define, construct, and simulate a general class of massively parallel networks. The experiments were performed using 386-based hardware.

VI. RESEARCH DIRECTIONS

This paper represents the first step in providing a massively parallel schema representation. Only the problem of schema selection has been discussed in detail in this paper. We are interested in exploring other issues such as the learning of new structures, the storage of new schema instances, and the integration with existing memory structures [SCHA82]. Another issue that needs to be addressed is the representation of temporal constraints and relationships (we have already done some preliminary work in this area [CHUN86a]). We also hope to integrate SAMPAN with a massively parallel natural language understanding system [LI87]. The SAMPAN marker passing has great potential of providing an inference mechanism and explanation facility within the network (possibly similar to that of Wimp [CHAR86]).

VII. SUMMARY

This paper presented a massively parallel model of schema selection which solved many problems inherent in serial approaches. The selection process is more flexible and adaptive than present methods used. In addition, the resulting schema structure is more malleable and general. SAMPAN has been tested successfully in the domain of natural language understanding.

ACKNOWLEDGEMENTS

The authors would like to thank the members of the Knowledge Engineering Center and Gold Hill Computers for their interest and comments.on this research.

REFERENCES

[ALTE85] Alterman, R., "A Dictionary Based on Concept Coherence," *Artificial Intelligence*, Volume 25 Number 2, February 1985, pp.153-186.

[CHAR86a] Charniak, E., "A Neat theory of Marker Passing," In *Proceedings of AAAI-86*, 1986, pp.584-588.

[CHAR86b] Charniak, E., "Connectionism and Explanation," In *Proceedings of TINLAP 3*, 1986.

[CHUN86a] Chun, H.W., "A Representation for Temporal Sequence and Duration in Massively Parallel Networks: Exploiting Link Interactions," In *Proceedings of AAAI-86*, August 1986.

[CHUN86b] Chun, H.W., "The MicroAINET Language Manual," unpublished draft, 1986.

[CHUN85] Chun, H.W., "Schemata in a Massively Parallel Model of Computation: A Non-Committing Approach to Schema Selection," unpublished Ph.D. thesis proposal, University of Illinois at Urbana-Champaign, 1985.

[DEJO79] DeJong, G., *Skimming Stories in Real Time*, Ph.D. Thesis, Yale University, New Haven, CT, 1979.

[HAVE83] Havens, W.S., "Recognition Mechanisms for Schema-based Knowledge Representations," in N. Cercone (ed.), *Computational Linguistics*, Pergamon Press Limited, 1983.

[HILL85] Hillis, W.D., *The Connection Machine*, MIT Press, Cambridge, 1985.

[LI87] Li, Tangqiu and H.W. Chun, "Massively Parallel Network-based Natural Language Parsing System," In *IEEE-Proceedings of the Second International Conference on Computers and Applications*, Beijing, China, June 1987.

[MCCL86] McClelland, J.L. and A.H. Kawamoto, "Mechanisms of Sentence Processing: Assigning Roles to Constituents," in J.L. McClelland, D.E. Rumelhart (eds.), *Parallel Distributed Processing: Explorations in the Microstructure of Cognition. Vol. 2: Psychological and Biological Models.*, Cambrdige, MA: Bradford Books/ MIT Press, 1986.

[RUME86] Rumelhart, D.E., P. Smolensky, J.L. McClelland, and G.E. Hinton, "Schemata and Sequential Thought Processes in PDP Models," in J.L. McClelland, D.E. Rumelhart (eds.), *Parallel Distributed Processing: Explorations in the Microstructure of Cognition. Vol. 2: Psychological and Biological Models.*, Cambrdige, MA: Bradford Books/ MIT Press, 1986.

[SCHA82] Schank, R., *Dynamic Memory*, Cambridge University Press, 1982.

[SCHA77] Schank, R. and R. Abelson, *Scripts, Plans, Goals and Understanding*, Lawrence Erlbaum Associates, Hillsdale, New Jersey, 1977.

[SCHA75] Schank, R.C. and Yale A.I. Project, "SAM: a story understander," Yale Computer Science Department Research Report 43, New Haven CT., 1975.

[SHAR86] Sharkey, N.E., R.F.E. Sutcliffe and W.R. Wobcke, "Mixing Binary and Continuous Connection Schemes for Knowledge Access," In *Proceedings of AAAI-86*, 1986, pp.262-266.

[TOUR85] Touretzky, D.S. and G.E. Hinton, "Symbols Among the Neurons: Details of a Connectionist Inference Architecture," In *Proceedings of IJCAI-85*, 1985, pp.238-242.

INDIVIDUAL DIFFERENCES IN MECHANICAL ABILITY

Mary Hegarty
Marcel Adam Just

Carnegie-Mellon University

Ian R. Morrison

Interact R & D Corporation

Abstract

People who understand mechanical systems can infer the principles of operation of an unfamiliar device from their knowledge of the device's components and their mechanical interactions. Individuals vary in their ability to make this type of inference. This paper describes studies of performance in psychometric tests of mechanical ability. Based on subjects' retrospective protocols and response patterns, it was possible to identify rules of mechanical reasoning which accounted for the performance of subjects who differ in mechanical ability. The rules are explicitly stated in a simulation model which demonstrates the sufficiency of the rules by producing the kinds of responses observed in the subjects. Three factors are proposed as the sources of individual differences in mechanical ability: (1) ability to correctly identify which attributes of a system are relevant to its mechanical function, (2) ability to use rules consistently, and (3) ability to quantitatively combine information about two or more relevant attributes.

Introduction

We generally associate mechanical ability with a person's understanding of how machines work, the ability to build a machine out of its elementary components, and the ability to determine why a machine is not working. To understand a machine in this way, a person has to be able to identify the elementary components of the machine, know which properties of these elementary components are relevant to their function in the system, and also understand how these elementary components interact to accomplish the machine's function. This paper explores the mental models of individuals with different levels of understanding of machines.

One approach to understanding mechanical ability used by psychometricians is to measure the correlations between tests of mechanical ability and tests of other basic cognitive traits. Studies using this approach suggested that there were several components of mechanical ability, such as general reasoning ability, and knowledge acquired through

---

This work was supported in part by contract number N00014-85-K-0584 from the Office of Naval Research, grant number MH-9617 from the National Institute of Mental Health, and an NSERC post-doctoral fellowship to Ian Morrison. We thank Patricia Carpenter and Carolanne Fisher for their comments and suggestions.

experience with machines (Cronbach, 1984). This suggests that mechanical ability is not a static trait, but can develop as a result of experience.

Our approach includes an analysis of verbal protocols as well as an analysis of the response patterns obtained during the performance on test items. This approach allows us to examine the mental models of different individuals as reflected in which attributes of a mechanical system people consider relevant to its function, their rules relating the attributes to the function, their preferences among different rules, and their methods for combining rules pertaining to different attributes. The resulting models of high and low ability subjects are instantiated as two computer simulation models, whose performance on the test items produces patterns resembling those of human subjects.

We studied performance on pulley problems of the type used in the Bennett test of Mechanical Comprehension (Bennett, 1969). Paper-and-pencil texts such as this have been found to be highly predictive of performance in a number of technical fields such as machine assembly, mechanical repair and vehicle operation (Bennett, 1969, Ghiselli, 1955, Vernon and Parry, 1949). Our focus on pulleys permitted us to construct a large number of pulley problems which systematically varied the number and type of attributes that distinguished the two systems depicted in each problem. The Bennett type of pulley problems were at an appropriate difficulty level for our college-student subjects, allowing measurement of a range of individual differences in performance. Restricting the experiments to pulley problems does not compromise the generality of the research, since previous analyses of the Bennett test (Cronbach, 1984) and our own pilot study have shown that separate scores for different types of items are highly correlated. Thus, our examination of the mechanical ability that deals with pulleys should apply to reasoning about other types of mechanical systems.

Figure 1:   A typical pulley problem.

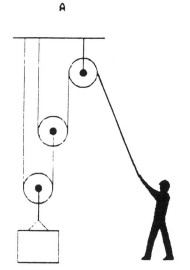
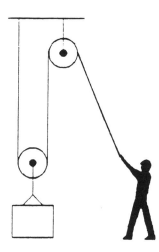

With which pulley sy
does the man have t
with more force to li
the weight?

A
B
If no difference
mark C.

Method

Problems. We analyze performance on 17 pulley system problems, including some items from the Bennett Mechanical Comprehension Test (Bennett, 1969) and other similar items which were constructed especially for this study. All of the items were multiple choice, requiring a selection among three response alternatives. Each problem depicted two pulley systems lifting a weight and asked which pulley system required more force to lift the weight (see Figure 1).

The two pulley systems depicted in each item differed on one or more of the following dimensions: mechanical advantage, weight to be lifted, height (rope length), and pulley diameter. Pulley systems that differed in mechanical advantage, also differed on some other attributes (relevant attributes), which are correlated with mechanical advantage, such as the number of load-bearing ropes and the number of pulleys.

Three types of problems differed in the kinds of attributes that distinguished the two systems depicted in the problem. In one type of problem, the two systems differed only on attributes irrelevant to the mechanical advantage of a pulley system (height or pulley size). In the second type of problem the two pulley systems differed in mechanical advantage, while the weights they lifted were equal. In the third type of problem, both the mechanical advantage and the weights were different for the two systems.

Subjects. The subjects were 43 undergraduate students, 27 students at Carnegie-Mellon University and 16 students at the Community College of Allegheny County. Fourteen of the students had taken two or more courses in physics at college level, while the remainder had taken no college level physics courses.

Procedure. Thirty-eight subjects were administered the test in a group setting, while five other subjects were tested individually and gave verbal protocols while they solved the problems. Two of the five protocol subjects had taken college level physics.

For the purposes of comparing different levels of ability, the data from the 38 subjects who performed the test in a group setting were divided into two groups, a high-scoring group and a low-scoring group, on the basis of their overall scores. A discontinuity in the distribution of scores defined the boundary between the high and low ability subjects. Twenty five subjects solved less than 59% of the problems correctly while thirteen of the remaining subjects scored more than 65% of the problems correctly. The high-scoring group therefore consisted of the top third of the distribution.

Results

General Solution Processes. An analysis of the subjects' verbal protocols suggested the following general account of how they solved the test items. The subjects decided which of the two pulley systems' distinguishing attributes (such as the number of pulleys) were relevant to reducing the effort required to lift the weight. They then compared the two systems using rules which relate these attributes to the amount of effort required.

The repertoire of rules used was inferred from the five subjects who gave verbal protocols. The rules pertain to those attributes that the subjects described as relevant, which were all attributes of the visible components of the systems - either their number, size, or attachments to other components. As Table 1 shows, most of the rules were based on system attributes that are correlated with mechanical advantage. Two of the rules were based on irrelevant attributes (height and pulley size). Two of the rules were quantitative, i.e., they expressed the effort as the ratio of the weight to some some attribute of the pulley system. The remainder of the rules were qualitative. A qualitative rule could state that pulley system with a higher value of some attribute requires less effort or that a system with a lower value the attribute requires less effort.

When two or more of a subject's rules were applicable in a problem, the rule that was used to generate the answer reflected a preference ordering among the rules. The preference ordering among rules implies that even if a subject knows a rule, he will not use it to generate the answer to the problem unless it is the most preferred in the situation.

Table 1: Rules used by the Protocol Subjects

Rule	Number of Subjects who used the Rule.
Qualitative Rules (relevant attributes):	
A pulley system with ... requires less effort	
less weight	5
more pulleys	4
more load-bearing ropes (tensions)	3
more attachments to the ceiling	2
more free pulleys	2
Qualitative Rules (irrelevant attributes):	
A pulley system with ... requires less effort	
larger pulleys	1
less height	1
Quantitative Rules (relevant attributes):	
A pulley system with ... requires less effort.	
less weight per pulley	1
less weight per attachment	1

Individual Differences. The response patterns of a large proportion of subjects could be classified as consistent with the rules observed in the protocols. These response patterns revealed that three factors accounted for individual differences in mechanical ability; (1) ability to discriminate relevant from irrelevant attributes, (2) consistency of rule use and (3) ability to quantitatively combine information about two attributes within a single rule. We will discuss each of these factors in turn.

High-scoring subjects were better able to discriminate relevant from irrelevant attributes of pulley systems. The majority of high-scoring subjects (92%) correctly identified height and pulley size to be irrelevant, while 52% of low-scoring subjects considered height to be relevant and 44% of low-scoring subjects considered pulley size to be relevant. This was reflected in the answers that they chose. High-scoring subjects chose a significantly higher proportion of correct responses (.90) than did low-scoring subjects (.44) in problems that varied the height of the system ($t(36) = 4.00$, $p < .001$). In problems that varied pulley size .98 of high-scoring subjects' responses and .35 of low-scoring subjects' responses were correct ($t(36) = 6.09$, $p < .001$).

High-scoring subjects used rules more consistently in problems that varied mechanical advantage and the rules that they used were more likely to be correct. If consistency is defined as having at least four out of six responses that are consistent with one rule, twelve of the thirteen high-scoring subjects responded consistently. In contrast only eleven of the twenty-five low-scoring subjects responded consistently. Seven of the thirteen high-scoring subjects were classified as using the rule that a system with more load-bearing ropes requires less effort, which gives the correct answer to all of the problems of this type. High-scoring subjects answered a significantly higher proportion (.77) of these problems correctly than did low-scoring subjects (.47), ($t(36) = 4.48$, $p < .001$)

High-scoring subjects also demonstrated the ability to quantitatively combine information about two attributes within a single rule. In problems involving both mechanical advantage and weight differences, the responses of ten of the high-scoring subjects (77%) were consistent with rules expressing a ratio of the weight to some attribute of the system, such as weight per load-bearing strand, attachment, or pulley. The low-scoring subjects, on the other hand, were more likely to base their comparisons of the systems either on weight or on a single attribute of the system, but did not combine the consideration of weight and the system attribute into a single rule. The most common rule used by these subjects was that more effort is required to lift a heavier weight. High-scoring subjects answered a much higher proportion (.62) of these problems correctly than did low-scoring subjects (.33) ($t(36) = 4.03$, $p < .001$).

In summary, high-scoring subjects are better able to identify the attributes relevant to the operation of a pulley system, they are more consistent in their use of rules, and they are more likely to use rules that indicate a quantitative understanding of pulley systems. Not only do the three factors have significant effects on performance, but they are also similarly related to the total scores, as assessed by the following procedure. Each subject was given a score of 1 or 0 on each of the three factors. A score of 1, based on the response pattern on the relevant problems, indicated that the subject had the ability measured by a given factor, while a score of 0 indicated that the subject did not have this ability. Each of the factors had a correlation with the overall score which lay between .49 and .51. Thus the three factors are of approximately comparable importance in predicting an individual's performance. Together the three factors accounted for 38.6% of the variance among the total scores.

A Model of Performance.

In order to specify mechanisms which can underlie the individual differences identified in the experiment, we developed a simulation model, written in Soar (Laird, Newell, and Rosenbloom, in press). The model simulates the performance of one high-scoring subject and one low-scoring subject who gave protocols in the experiment. It simulates the response choices that the subjects gave to the problems, as well as stating the rationale for each choice.

Representational Format. The model operates on a problem description for each of the 17 problems in the experiment. Each problem description contains all the information that is directly available to a human subject through visual inspection. However, not all of the information in the problem description is necessarily used by by the model or by the subject it simulates.

The format of a problem description is a structured description list which consists of identifiers and lists of attributes and values. There are four types of attributes: properties, relations, comparisons, and questions. The simplest type of attribute is a property of a pulley system or component of a pulley system, such as the number of pulleys in the system. The second type of attribute is a relation between two objects. For example, a relation might state that a particular pulley is fixed to the ceiling. The third type of attribute, a comparison, compares two properties or two relations. The fourth type of attribute, a question, contains an attribute with a missing value and states that the value should be obtained. The requirement in each item of the test, namely to compare the relative efforts required to lift the weights with the two depicted pulley systems, is represented as a question about the comparison of the effort attribute.

Production Rules. The simulation model uses a set of productions that can be divided into two subsets, one subset common to all subjects, and a second subset unique to the individual whose solutions were simulated. The common productions control the operators that seek information about the problem and the operators that generate answers to the question posed, express the reasons for producing these answers, and stop the processing when the final answer has been selected. The subject-specific productions determine what information an individual seeks and how he reasons from that information to generate an answer to the problem. These productions reflect the rules that a subject possesses relating attributes of pulley systems to their function (reducing the effort required to lift a weight).

The model can evoke one of two types of operators, elaboration operators and hypothesis operators. When a value in a question is missing, elaboration operators look for information in the problem statement that might be relevant to answering the question. Hypothesis operators suggest values for attributes that are sought by elaboration operators and use these values to suggest tentative answers to the problem. Each suggested answer is accompanied with a reason for this answer. For example, a hypothesis operator may suggest pulley system A requires a greater effort than system B because the weight that system A is lifting is heavier.

The productions choose among elaboration and hypothesis operators on the basis of

902

preferences, expressed in Soar as special data elements. A preference might favor an answer supported by a particular reason. For example a preference might favor an answer based on the amount of weight to be lifted by a system over an answer based on the number of pulleys in a system. Alternatively, a preference might express a response bias. For example a preference might favor a hypothesis operator stating that the efforts required to lift the loads of the two pulley systems are different over an operator stating that the efforts are the same.

Flow of Control. The model proceeds from the problem description and question to its ultimate response by evoking a sequence of operators which derive information from the problem description and suggest answers on the basis of the obtained information (see Figure 2). When the question is first interpreted, an elaboration operator is evoked to seek the information that the question interrogates. The question ("With which pulley system does the man have to pull with more force to lift the weight?") interrogates a comparison of the effort attributes of the two pulley systems. Because there is no information available that allows this comparison to be made directly, additional elaboration operators are evoked to seek other information that might be relevant to the answer. For example, information about the number of pulleys or ceiling attachments in the two pulley systems might be sought at this point. In addition, if the person being modeled has sufficient knowledge to calculate the efforts required by the two pulley systems, a subgoal is generated to calculate the efforts. (The dotted lines in Figure 2 indicate components of the model that are present for subjects with this knowledge). Hypothesis operators use the information obtained by elaboration operators to suggest answers to the question. If no answer is suggested, the model chooses randomly among the possible answers. If only one answer is suggested, it becomes the response of the model for that problem. If more than one answer is suggested, a subgoal is created to resolve the tie. To satisfy the subgoal of resolving the tie, one hypothesis operator may be selected over another as a result of a preference. Otherwise, a random choice is made among the operators.

Modeling individual differences in performance. The three sources of individual differences observed in the experiment are modeled in the simulation in the following ways. To account for the differences among subjects in what they consider to be relevant, the model for a given subject relates the effort required in the case of a particular pulley system to precisely those attributes of the system that the subject considers relevant. That is, the attributes that were considered relevant were in the conditions of the productions embodying the mechanical rules. To account for the differences among subjects in how consistently they use one rule, the model varies or keeps constant its preferences among hypothesis operators across the different problems. If there is a preference for one hypothesis operator over all other hypothesis operators in a situation, the model will always choose the answer and the reason given by that operator in any similar situation. If there is no preference among operators, then the model chooses randomly among applicable operators, producing the same type of inconsistent behavior as observed for low-scoring subjects in the experiment. Finally, to account for the differences among subjects in their ability to quantitatively combine information from two relevant attributes, the model can either contain or not contain productions that suggest values for the effort based on a ratio of the weight of the system to some other relevant attribute.

The model simulated the performance of one high-scoring subject and one low-scoring subject who gave protocols in the experiment. The simulation for both subjects provided the same response and the same explanation of the response as the human subject in 16 of the 17 problems.

Figure 2: Flow of control of the simulation model through a problem.

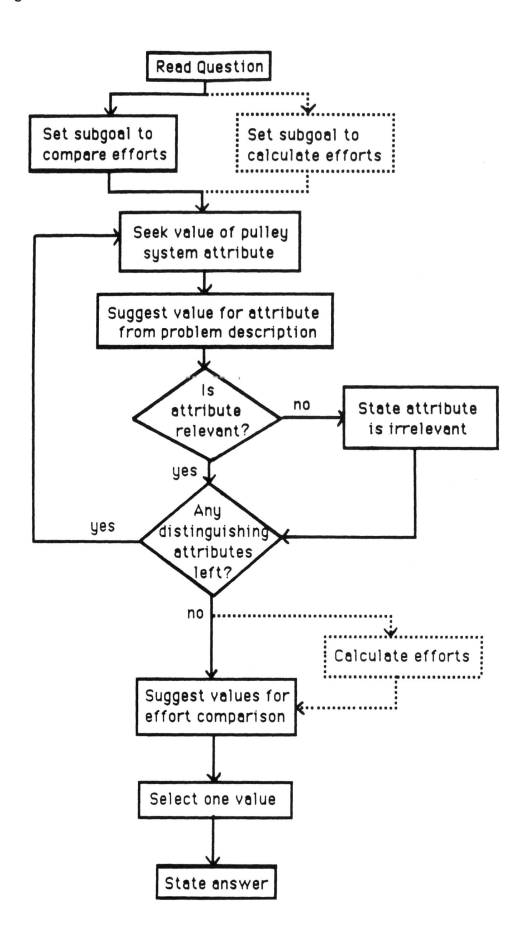

The simulations for the high-scoring and the low-scoring subjects differed in the following ways:

1. The high-scoring simulation produces fewer suggested answers than the low-scoring simulation. This is because the low-scoring model makes suggestions about an irrelevant attribute in addition to making suggestions about a number of relevant attributes. All of the answers suggested by the simulation of the high-scoring subject were based on relevant attributes.

2. The high-scoring simulation calculated numerical values for the the efforts required to lift the loads of the pulley systems by quantitatively combining two attributes: weight and number of pulleys. The low-scoring did not attempt to determine the efforts directly and did not quantitatively combine attributes. When two or more attributes produced the same answer, the low-scoring simulation combined them into a single answer justified by the several explanations. If two or more suggestions produced contradictory answers, these comparisons canceled each other and the low-scoring simulation gave an answer of equality.

3. The high-scoring simulation organized the search for information so that numerical values for the efforts were calculated only when the answer could not be determined by qualitative comparisons. The order of search for information in the low-scoring model was random.

In summary, the model specified mechanisms which can account for the individual differences identified in the experiment. It successfully simulated the performance of a high-scoring and a low-scoring subject indicating the sufficiency of the theoretical proposal. The model suggested that the process of applying rules is similar for high-scoring and low-scoring subjects, but the content of the rules changes with increases in mechanical ability.

Discussion

The research reported in this paper provided both a general model of the processes involved in solving items from tests of mechanical ability and identified sources of individual difference in performance on these tasks. It was found that subjects encoded mechanical systems in terms of attributes of systems that they considered relevant to their function. Comparison of the pulley systems by different subjects was based on rules which expressed a relation between one or more of these attributes and the attribute in question i.e. the effort required to lift the weight with the pulley system.

According to the description of individual differences presented in the paper, low-scoring subjects are characterized as using qualitative rules based on both relevant and irrelevant attributes of pulley systems, and have no clear preferences among their rules so that their responses appear inconsistent with any particular rule. High-scoring subjects, on the other hand, can use quantitative rules when the problem demands the use of these rules, their rules are based on relevant attributes, and they prefer rules based on attributes that are highly correlated with mechanical advantage.

905

A striking feature of the range of solution processes used by subjects with different mechanical ability is their similarity to the developmental stages observed by Siegler (1978, 1981) in his analysis of young children's understanding of a balance beam. The parallel between our findings and developmental findings such as Siegler's suggests the intriguing hypothesis that the processes that underlie the development of mechanical abilities also characterize differences along an individual difference dimension. Our results suggest that mechanical ability should not be thought of as a static trait but as an ability that can develop with increased experience in this domain. This view of mechanical ability is consistent with the dominant view of mechanical ability in the psychometric literature, i.e., that it is a measure of understanding acquired through general exposure to tools and machinery (Cronbach, 1984).

Our results demonstrate that qualitative mental models of pulley systems precede quantitative models. In a qualitative model, attributes of pulley systems are coded by comparison with corresponding attributes of other pulley systems. A qualitative model also includes rules relating attributes of mechanical systems, situations in which it is appropriate to apply these rules, and preferences among these rules. Preferences can resolve conflicts between qualitative rules that are equally applicable in a situation, but there is no simple way in a qualitative model to resolve conflicts between rules with equal preference. In a quantitative model, on the other hand, attributes are given numerical values so that mathematical operations can be applied to these values to resolve conflicts.

The development of understanding of a physical system such as the balance beam or the pulley can be seen as a progression of mental models in which each model elaborates and refines the earlier models, rather than replacing them. The progression from low ability to high ability in mechanical ability involves advancing along a number of different dimensions. One dimension involves adjusting preferences between different rules so that rules based on relevant attributes are preferred to rules based on irrelevant attributes and at higher levels of mechanical ability, preferences among rules based on different correlated attributes correspond to how highly these attributes are correlated with mechanical advantage. A second advance is the progression from a qualitative to a quantitative model of mechanical advantage that enables the subject to quantify the extent to which mechanical advantage can reduce the effort required to lift a weight.

References

Bennett, C. K. (1969). *Bennett Mechanical Comprehension Test.* New York: The

Psychological Corporation.

Cronbach, L. J. (1984). *Essentials of Psychological Testing (4th edition).* New York:

Harper and Row.

Ghiselli, E. E. (1955). The measurement of occupational aptitude. *University of California Publications in Psychology,* *8*(2), 101-216.

Laird, J. E., Newell, A and Rosenbloom, P. S. (in press). Soar: An architecture for general intelligence. *Artificial Intelligence,*

Siegler, R. S. (1978). The origins of scientific reasoning. In R. S. Siegler (Ed.), *Children's Thinking: What Develops?.* Hillsdale, NJ: Erlbaum.

Siegler, R. S. (1981). Developmental sequences within and between concepts. *Monographs of the Society for Research in Child Development, 46*(2), 1-149.

Vernon P. E. and Parry, J. B. (1949). *Personnel Selection in the British Forces.* London: University of London Press.

Dimensionality-Reduction and Constraint in Later Vision

Eric Saund
Department of Brain and Cognitive Sciences and
the Artificial Intelligence Laboratory
Massachusetts Institute of Technology
Cambridge, Massachusetts 02139

Keywords: computational vision, dimensionality-reduction, connectionist networks

Abstract

A computational tool is presented for maintaining and accessing knowledge of certain types of constraint in data: when data samples in an n-dimensional feature space are all constrained to lie on an m-dimensional surface, $m < n$, they can be encoded more concisely and economically in terms of location on the m-dimensional surface than in terms of the n feature coordinates. The recoding of data in this way is called *dimensionality-reduction*. Dimensionality-reduction may prove a useful computational tool relevant to later visual processing. Examples are presented from shape analysis.

1 Introduction

It is commonly understood that vision involves the interaction of incoming data with a priori knowledge about the world. In Early Vision, constraint due to the physics of image formation, rules of imaging geometry, and statistical properties of the world can be analyzed mathematically to support inferences about primitive scene properties based on image measurements. Problems of later vision, however, involving the interpretation of image data in terms of task goals, object models, and meaningful world events, do not offer straightforward mathematical characterizations of the more abstract constraints that give rise to comprehensible images. For example, it is difficult to characterize the constraints on a shape profile that might qualify it to be called a "scissors" shape.

One type of constraint that arises in shape analysis can be cast in terms of surfaces embedded in high-dimensional feature spaces. This paper suggests that the computational tool of *dimensionality-reduction* can be used to maintain and access knowledge of complex constraints of this form.

2 Dimensionality-Reduction

An idealized view of early visual processing is that certain image features are extracted from raw images. For example, figure 1 shows primitive features extracted from an image of a pair of scissors, making explicit the x-location, y-location, and orientations of edges and corners. Slightly more complex features may be constructed by combining primitives, say, by measuring the distances and relative orientations between pairs of corners. We might imagine that one aspect of early visual processing involves the delivery of image descriptions in terms of data points in a huge multidimensional feature space whose coordinate axes are defined by the features measured. The job of later visual processing is to make sense out of the feature descriptions delivered it.

Image feature data can only be meaningfully interpreted by appeal to knowledge of constraints governing the world giving rise to images. These constraints are reflected in the behavior of extracted features for given classes of shapes. For example, features associated with a pair of scissors

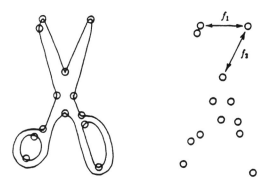

Figure 1. Hypothetical corner primitives extracted from the image of a pair of scissors. These can be combined into more complex features such as the distances between pairs of corners.

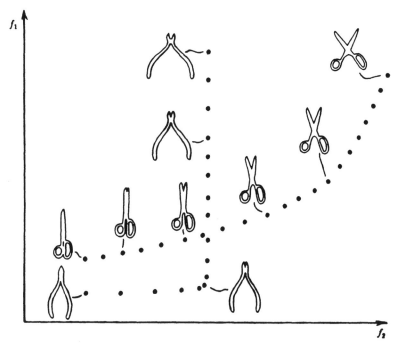

Figure 2. Slice of feature space plotting values of features f_1 and f_2 from figure 1. A scissors shape generates a one-dimensional constraint curve in feature space as the blades open and close. Different objects, such as wire cutters, generate different constraint curves.

generate not a single point, but a path through feature space as the blades rotate about the pivot. The shape class, "pair of scissors," exhibits one degree of freedom in the feature space, and every feature space description of the scissors must be a data point lying somewhere on the constraint curve, just where depending upon how open the blades are (see figure 2). *Dimensionality-reduction* [Krishnaiah and Kanal, 1982; Kohonen, 1984] is the computational mapping between the description of data expressed in terms of its location in a high-dimensional feature space, and in terms of its location on a lower-dimensional *constraint surface*. Figure 3 illustrates. This mapping employs knowledge of the lower-dimensional constraint surface, embedded in the high-dimensional feature

909

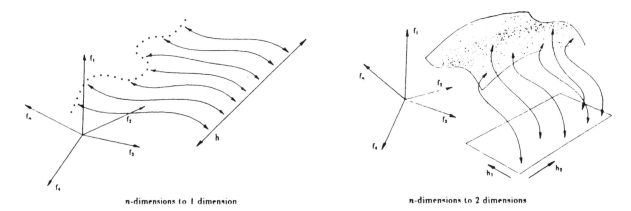

n-dimensions to 1 dimension   n-dimensions to 2 dimensions

Figure 3. Dimensionality-reduction is a mapping between points described in a high-dimensional feature space, and points described in terms of location on a lower-dimensional constraint surface.

space, that is generated by a physical process or class of data. Different objects generate different constraint surfaces in the feature space, as seen in figure 2.

The purpose served by dimensionality-reduction is abstraction and simplification in the description of data. A description consisting of a symbolic token indicating that an input datum lies on the "scissors" constraint surface in the feature space, plus one parameter indicating where it lies on the surface (how open the scissors are), is certainly preferable to a listing of the input's coordinate location along each of the n feature dimensions. In general, a dimensionality-reduced representation can be expected to make explicit descriptive parameters capturing the natural degrees of variability inherent to a class of data, while it factors out redundancy latent among the original measured features.

3 Achieving Dimensionality-Reduction

A black box depiction of a dimensionality-reducer is presented in figure 4. Each such box contains knowledge of one constraint surface, such as, for example, might characterize one class of object shapes. At the bottom of the box enters the description of an image in terms of an n-dimensional feature vector. Out the top emerge n lines, and out the side, m more, where m is the dimensionality of the constraint surface. Each line can represent a bounded real number; for convenience suppose that the feature coordinates are normalized so that all features take values between 0 and 1.

The dimensionality-reducer operates as follows. If the numbers coming out the top of the black box match those coming in the bottom, then it is determined that the data point whose feature vector is given does in fact lie on the constraint surface, and its location on the constraint surface may be read on the m lines coming out the side (the dimensionality-reducer implicitly creates a coordinate system for the constraint surface). If the numbers coming out the top do not match the input feature vector, then it is determined that the data point specified at the input does not lie on the constraint surface.

This black-box dimensionality-reducer may also be used in the opposite direction. That is, an m-dimensional vector specifying a location on the constraint surface may be placed on the side lines as input, and the dimensionality-reducer will then compute, and output at the top, the coordinates of this data point in the n-dimensional feature space.

Several alternative implementations of such a black box dimensionality-reducer are possible.

910

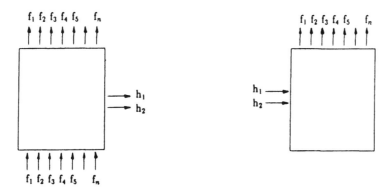

Figure 4. Black box dimensionality-reducer.

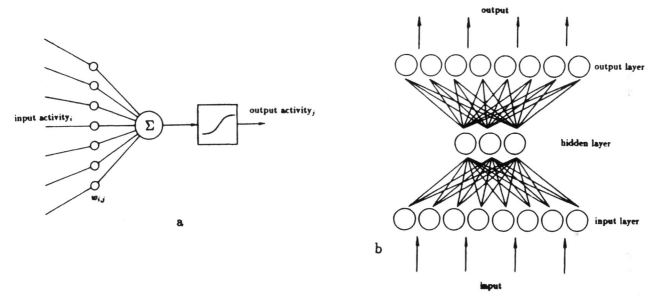

Figure 5. A connectionist network is composed of simple computing elements connected by weighted links, $w_{i,j}$. a. "Activity" in a unit is computed to be the weighted sum of activities in units of the previous layer, passed through a semilinear function mapping the activity to a number between 0 and 1. b. Three layer network.

When it is assumed that the constraint surface is always linear, then dimensionality-reduction amounts to principle components analysis, or factor analysis, and the computation may be expressed as a matrix multiplication [Watanabe, 1965; Tou and Heydorn, 1967; Fukunaga and Koontz, 1970; Kittler and Young, 1973].

A more general implementation, in which the constraint surface may curve to a considerable degree, uses a connectionist network of simple computing elements. Figure 5 illustrates a three-layer network. Each unit takes an activity between 0 and 1. Activity is fixed at the bottom layer as input, then activity for each unit in the middle layer (called the "hidden" layer) is computed as

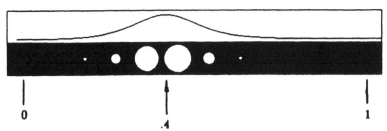

Figure 6. Scalar values between 0 and 1 are represented in sets of units, called *scalar sets*, whose activity takes a characteristic unimodal pattern. Activity of a unit is represented as size of circle. The activity pattern shown in this 12-unit scalar-set represents the number, .4.

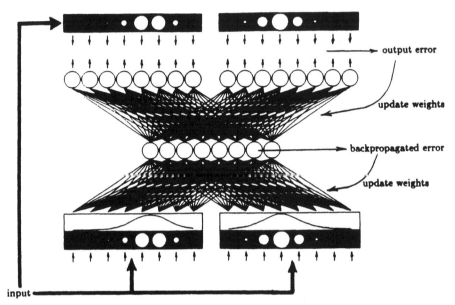

Figure 7. Connectionist dimensionality-reducer. In this case, two scalar-sets are provided at the input and output layers, and the hidden layer is a one-dimensional scalar set, therefore, this network can represent one-dimensional constraint curve in a two-dimensional feature-space. During training period, errors from desired activity are used to train network to reproduce input activity pattern at output.

a semilinear function of the weighted sum of activities on the input units. Output layer activity is computed from the hidden layer activity in a similar way.

Each scalar feature value is represented as the pattern of activity over a set of units, called a *scalar-set*, as illustrated in figure 6. A one-dimensional scalar-set is provided at the input layer and at the output layer for each dimension of the higher-dimensional feature space. The hidden layer is configured as a scalar-set whose dimensionality matches that of the embedded constraint surface. The input, hidden, and output layers of the network correspond to the bottom, side, and top of the black box dimensionality-reducer (see figure 7).

Dimensionality-reducing behavior is achieved by virtue of the link weights between successive layers of the network. These weights are established using the backpropagation training procedure [Rumemlart, Hinton and Williams, 1985; Rumelhart and McClelland, 1986], which furnishes crucial self-organizing properties during the training phase. Training consists of repeated presentation of input activity/desired output activity pairs, where the desired output is defined to be identical to the input activity. At each training trial, activity at the input layer is fixed according to the coordinates of a data sample expressed terms of the high-dimensional feature space. For example, each image of a pair of scissors, with blades open to some degree, generates one training sample, or data point in feature space. Activity propagates through the network, and the output layer activity is compared with that of the input, to create an output error vector. This error is used to incrementally update link weights between the hidden and output layers in such a way as to reduce the error. In addition, the output activity error is backpropagated through the hidden layer/output layer link weights to arrive at an equivalent error in activity at the hidden layer. This essentially amounts to analyzing how each unit at the hidden layer contributed to the error observed at the output layer. The hidden layer error is in turn used to update link weights between the input and hidden layers. In order to achieve patterns of activity at the hidden layer that are interpretable in terms of a location on the lower-dimensional constraint surface, an auxiliary error is introduced at the hidden layer to be added to the backpropagated error. This auxiliary error serves to pressure the input/hidden links into creating patterns of activity at the hidden layer scalar-set taking a characteristic unimodal form representing a location on the constraint surface. A more detailed discussion of the connectionist dimensionality-reducer is presented in [Saund, 1986] and [Saund, 1987].

4 Dimensionality-Reduction in Shape Description

The practical matter of using a dimensionality-reducer involves measuring feature parameters on images and plugging them into appropriate slots in the "black box" input. Each dimensionality-reducer possesses knowledge of legal configurations of features for a given class of data, such as a shape category. Tasks such as object recognition are in principle accomplished by testing feature vectors delivered by early vision against various objects' dimensionality-reducers, amounting to a sort of generalized template matching. Some means must be provided for determining just what measured features to pair with each input line; as the number of measured features increases there occurs a combinatoric explosion of possible feature/input matchings. Therefore, to be utilized by a visual system the dimensionality-reduction tool must be used within a computational shell controlling the mechanics of measuring features, selecting candidate object models, assigning measured feature values to appropriate input lines, evaluating the dimensionality-reducer's applicability to the feature vector, and reading the reduced description off the side output lines.

The choice of features defining a higher-dimensional feature space is important to achieving useful data abstraction through dimensionality-reduction. The example above illustrates that a pair of scissors generates a one-dimensional constraint surface in a feature space derived from simple edge and corner primitives. However, differently shaped scissors give rise to different constraint curves in the feature space because they all create somewhat different configurations of edge and corner features, and a separate dimensionality-reducer would be required for each constraint curve. Where dimensionality-reduction might more realistically come into play is in interpreting feature descriptions at a more abstract level, say, once edge and corner primitives have been grouped into parts. Then, a single dimensionality-reducer might suffice to capture the blade pivoting constraint inherent to all pairs of scissors.

Prior to training a dimensionality-reducer, it is important to select a dimensionality for the constraint surface to match the inherent dimensionality of the data. The connectionist dimensionality-reducer described above provides no means for doing this automatically. However, it is easy to detect whether the constraint surface is of inadequate dimensionality, because under this condition, during training, a network will converge to a state in which it does not correctly map activity at the input layer to (nearly) identical activity at the output layer in terms a unimodal pattern of

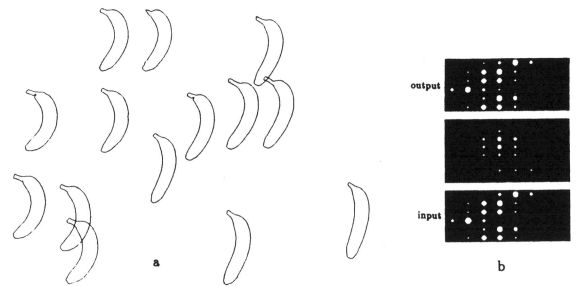

Figure 8. a. Banana shapes arranged according to their locations on a two-dimensional constraint surface found by a connectionist dimensionality-reducer. These were originally described in terms of six simple features such as distances between the ends and lengths and curvatures of various contour segments. The parameters of variability found to pertain among these bananas are roughly the curvature of the lower part of the banana (left/right), and the overall size of the banana (up/down). b. Activity of the dimensionality-reduction network for one banana. Note that output activity matches input activity, and that hidden layer activity is centered around one location in the two-dimensional scalar-set.

activity at the hidden layer.

It is desirable to seek constraint surfaces of low dimensionality for two reasons. First, limits may exist on the tractability of discovering many-dimensional constraint surfaces. The amount of data that must be analyzed in order to establish a constraint surface increases as the power of the surface's dimensionality. In the current computer implementation of the connectionist dimensionality-reducer, the cost in terms of network links and nodes appears to become prohibitive after m becomes three or four. Second, the data simplification afforded by dimensionality-reduction may lose some value when the underlying constraint surface is yet of high dimensionality. For example, it is not certain that much is gained by describing data in terms of a ten-dimensional coordinate system instead of a twenty-dimensional coordinate system. Dimensionality-reduction is perhaps most useful when the constraints operating in a given problem can be decomposed into systems of just a few inherent degrees of freedom each.

Constraint surfaces in multidimensional feature spaces arise in several different ways. The scissors example illustrates a situation in which a single object generates a one-dimensional parameterized class of shapes by virtue of physically constrained motion between parts. Constraint surfaces may also originate in classes of shape objects in which each individual object has a fixed shape, but one that can vary in only certain ways from the shapes of other objects in the class.

Figure 8 shows a set of bananas that were originally described in terms of six properties crudely measured on the banana shapes, such as the distance between the ends and average curvature of various contour segments. By training a connectionist dimensionality-reducer on these data samples, the bananas were found to lie on a two-dimensional constraint surface in the six-dimensional feature space. The organization of this constraint surface is illustrated in the figure; bananas are placed on a plane according to their respective two-dimensional coordinates. Note that banana shapes are organized on the basis of very subtle differences in their geometrical properties.

Although the reduced dimensionality representation concisely encodes the essential parameters of variability among members of the data class falling on a constraint surface, the lower-dimensional coordinate axes do not necessarily align with interpretations of these parameters preferred by human observers. For example, the horizontal and vertical axes of figure 8 roughly correspond to curvature of the lower part of a banana, and banana size, respectively, however, the dimensionality-reduction training procedure run again on the same banana data might rotate these axes an arbitrary amount in the plane.

5 Conclusion

Dimensionality-reduction is a computational tool for exploiting knowledge of constraint latent in a collection of data in order to achieve simpler and more perspicuous descriptions. It is applicable when data samples lie on a lower-dimensional constraint surface embedded in an initial, higher-dimensional, descriptive feature space. Unlike mathematical and physical model based procedures for capturing constraint in early vision, dimensionality-reduction is a quite general concept which holds the possibility for capturing the rather more complex and non-analytic nature of regularities inherent to later visual analysis. The ultimate utility of this tool in computational vision and Cognitive Science rests with the degree to which the visual world exhibits the appropriate type of constraint. This paper presents two rather simple examples of dimensionality-reduction at work in the analysis of visual shape.

References

Fukunaga, K., and Koontz, W.,[1970], "Application of the Karhunen-Loève Expansion to Feature Selection and Ordering," *IEEE Transactions on Computers*, C-19, 311-318.

Kittler, J., and Young, P., [1973], "A New Application to Feature Selection Based on the Karhunen-Loève Expansion," *Pattern Recognition*, 5, 335-352.

Kohonen, T.,[1984], *Self-Organization and Associative Memory*, Springer-Verlag, Berlin.

Krishnaiah and Kanal, eds., [1982], *Handbook on Statistics, Vol. 2: Classification, Pattern Recognition, and Reduction of Dimensionality*, North-Holland.

Rumelhart, D., Hinton, G., and Williams, R., [1985], "Learning Internal Representations by Error Propagation," ICS Report 8506, Institute for Cognitive Science, UC San Diego.

Rumelhart, D., McClelland, J, and the PDP Research Group, [1986], *Parallel Distributed Processing: Explorations in the Microstructure of Cognition*, Bradford Books, Cambridge, Ma.

Saund, E., [1986], "Abstraction and Representation of Continuous Variables in Connectionist Networks," *Proceedings of the Fifth National Conference on Artificial Intelligence*, 638-644.

Saund, E., [1987], "Dimensionalty-Reduction Using Connectionist Networks," AI Memo 941, Massachusetts Institute of Technology.

Tou, J., and Heydorn, R., [1967], "Some Optimal Approaches to Feature Extraction," in Tou, ed., *Computer and Information Sciences II*, Academic Press.

Watanabe, S.,[1965], "Karhunen-Loève Expansion and Factor Analysis," *Trans. 4th Prague Conference on Information Theory*.

Some Causal Models are Deeper than Others[*]

Tom Bylander
Laboratory for Artificial Intelligence Research
Department of Computer and Information Science
The Ohio State University

Abstract

The effort within AI to improve the robustness of expert systems has led to increasing interest in "deep" reasoning, which is representing and reasoning about the knowledge that underlies the compiled knowledge of expert systems. One view is that deep reasoning is the same as causal reasoning. Our aim in this paper is to show that this view is naive, specifically that certain kinds of causal models omit information that is crucial to understanding the causality within a physical situation. Our conclusion is that "deepness" is relative to the phenomena of interest, i.e. whether the representation describes the properties and relationships that mediate interactions among the phenomena and whether the reasoning processes take this information into account.

1. Introduction

Most expert systems depend upon *compiled* representations and reasoning processes. Their representations associate data with conclusions, and their reasoning processes use these associations, but they do not take into account the reasons why the data and conclusions are related. Without this extra knowledge, expert systems will be limited in what explanations they can provide and in reasoning about their own limitations.[**]

Within AI, there has been increasing interest in *deep* reasoning, i.e. representing and reasoning about these "reasons." A number of suggestions have been made that identify deep reasoning with *causal* reasoning. Hart suggests that deep reasoning involves commonsense ideas about causality as well as mathematical modeling (Hart, 1982). Michie suggests that the fundamental laws of the domain constitute deep reasoning (Michie, 1982). A number of programs could be said to perform deep reasoning based on these criteria. Instead of summarizing and comparing these programs, which would probably be confusing rather than enlightening given the plethora of domains and reasoning methods, my strategy is to take one program and compare an explanation of its domain by the program's builders with an explanation produced by the program. The goal of the comparison is to gain insight on the relationship between "causal reasoning" and "deep reasoning."

---

[*]This research is supported by Air Force Office of Scientific Research grant AFOSR$-82-0255$, and grant NIHR GOAOE 82048-02 from the National Institute of Handicapped Research.

[**]This is not a claim that expert systems cannot perform interesting problem solving. Chandrasekaran and Mittal (Chandrasekaran, 1983) have pointed out how an expert system, for a particular reasoning situation, can fully incorporate the appropriate deep knowledge. However, it would not incorporate the deep knowledge for those situations that were not considered in its design.

2. Two Causal Explanations

These two explanations are taken from a paper by Patil, Szolovits, and Schwartz, which describes a program called ABEL (Patil, 1981), one the first programs to perform interesting causal reasoning. The first explanation is by the authors; the second by the ABEL program. The reader is forewarned that these explanations, although they concern the same domain, do not involve exactly the same phenomena.

Explanation #1. "... let us consider the electrolyte and acid-base disturbances that occur with diarrhea, which is the excessive loss of lower gastrointestinal fluid (lower GI loss). The composition of the lower gastrointestinal fluid and plasma fluid are as follows. In comparison with plasma fluid, the lower GI fluid is rich in bicarbonate (HCO_3) and potassium (K) and is deficient in sodium (Na) and chloride (Cl)... The loss of lower GI fluid would result in the loss of corresponding quantities of its constituents (in proportion to the total quantity of fluid loss)... Therefore, an excessive loss of lower GI fluid without proper replacement of fluid and electrolytes would result in a net reduction in the total quantity of fluid in the extracellular compartment (hypovolemia). Because the concentration of K and HCO_3 in lower GI fluid is higher than that in plasma fluid, there is a corresponding reduction in the concentration of K (hypokalemia) and HCO_3 (hypobicarbonatemia) in the extracellular fluid. Finally, as the concentration of Cl and Na in the lower GI fluid is lower than that in plasma fluid, there is an increase in the concentration of Cl (hyperchloremia) and Na (hypernatremia) in the extracellular fluid." (Patil, 1981 - p. 894)

Explanation #2. "Moderate lower GI loss, reduced renal HCO_3 threshold, and normal HCO_3 buffer binding jointly cause no HCO_3 change. The no HCO_3 change causes low extracellular fluid HCO_3, which causes low serum HCO_3. The low serum HCO_3 and low serum pCO_2 jointly cause low serum pH. The low serum pH causes K shift out of cells and causes increased respiration rate. The increased respiration rate causes low serum pCO_2, which causes normal HCO_3 buffer binding. The low serum pCO_2 also causes reduced renal HCO_3 threshold and increased respiration rate causes increased ventilation. The lower GI loss and K shift out of cells jointly cause K loss. The K loss causes low extracellular fluid K, which causes low serum K." (Patil, 1981 - p. 898)

Both of these explanations have a causal story to tell, but in different ways and in different terms. The crucial difference is that the first quote makes use of our *physical* understanding about how the world works. It evokes a physical representation of the body and appeals to our understanding of how physical phenomena behave. The second quote is a different type of a physical explanation. While the second quote causally relates physical states, it does not express any physical relationships that let us understand the causal assertions in terms of some physical principle. Assertions like "low serum pH causes K shift out of cells" im-

917

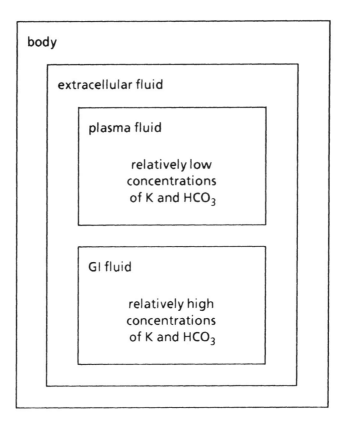

Figure 1: Representation of Patil, Szolovits, and Schwartz's Explanation

plicitly depend on the structure of the human body and how certain parts of the body behave. *With respect to physical phenomena, the first explanation is deep and the second explanation is compiled.*

3. An Analysis of the First Explanation

The first quote builds up the representation displayed in figure 1. (Na and Cl have been omitted for the purposes of this discussion.) The body can be thought of as having a container of extracellular fluid. The extracellular fluid compartment can be decomposed into a plasma fluid compartment and lower GI fluid compartment. Lower GI fluid has certain concentrations of HCO_3 and K, which happen to be greater than in plasma fluid. When the amount of lower GI fluid decreases (as happens in diarrhea), a corresponding amount of HCO_3 and K also decrease. It can be inferred that the total concentration of HCO_3 and K in extracellular fluid also decreases.

This representation lists the parts of the situation: fluid compartments, fluids, HCO_3, and K. It incorporates structural relationships between the parts, e.g., container, composed-of, and concentration, as well as behavioral information about them, e.g., fluid is something that can be contained, and can move. Also a fluid can be composed of other things, including HCO_3 and K in this case. The physical principle that this explanation appeals to is that when a certain amount of

918

fluid moves, the fluid also takes what it is composed of along with it. With a little bit of qualitative (or quantitative) analysis about concentrations, it is not hard to determine how certain concentrations will increase or decrease depending on how fluid moves.

In general, reasoning about physical situations faces two problems: (1) changes in physical structure can change the overall behavior and properties of a situation, and (2) changes in a part's behavior can change the overall behavior and properties of a situation. So to perform deep reasoning about physical phenomena, representations need to express the structure and behavior of physical situations and their constituents, and reasoning processes need to be able to take this information into account. Much of the work in naive physics is aimed at reasoning about physical information such as behavioral properties of components, connections between components, and containment of substances (Hayes, 1985, deKleer, 1984, Forbus, 1984, Bylander, 1985). There has also been research on reasoning about how shape affects behavior (Forbus, 1983, Stanfill, 1983, Shoham, 1985).

4. An Analysis of the Second Explanation

The second quote is a description of the causal network illustrated in figure 2. The physical relationships that supports the causal network is not present in this explanation. For example, one part of the causal network is that loss of GI fluid contributes to low concentration of K in the extracellular fluid. However, this representation does not have structural and behavioral information such as "Extracellular fluid can be decomposed into plasma fluid and GI fluid."

Why is this additional information important? If the program only has causal networks such as in figure 2, the omitted physical information becomes a large set of assumptions that are implicitly encoded into the causal network. The result is that the robustness of the causal network depends on the likelihood that these physical assumptions are true.

For example, suppose that GI fluid in a particular person had a lower concentration of K than plasma fluid, then the causal network would be wrong. Since the causal network does not express where GI fluid sits in the body's structure and that GI fluid normally has a greater concentration of K than plasma fluid, the possibility that this information is wrong cannot be hypothesized and cannot be reasoned about. These are the same characteristics of compiled reasoning that typical expert systems have. Causal networks represent more information about associations between data and conclusions, but because they do not represent physical relationships, causal networks and their reasoning processes are also compiled.[*]

---

[*]Each causal link in ABEL has a "slot" for stating its assumptions. It is unclear what kind of information was being represented by the assumptions, and what reasoning processes could be performed on them. It is conceivable that a causal network could point to the information that supports it, but this additional information would be something different than causal networks.

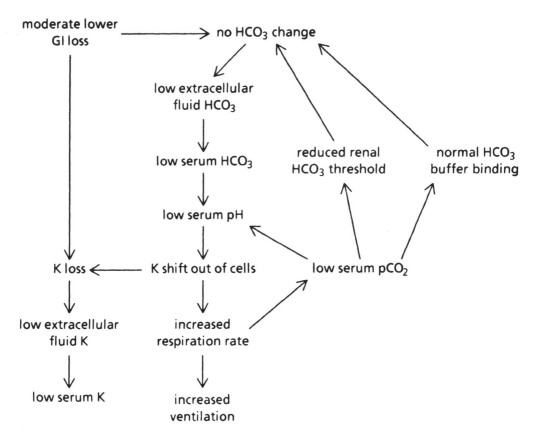

Figure 2: Representation of ABEL's Explanation

5. Some Misconceptions about Deep Reasoning

It might be claimed that representations like figure 1 are no better off than those like figure 2 because the information in figure 1 is a very qualitative representation, while figure 2 could relate physical states in more detail. This leads to the misconception that reasoning at a greater level of detail is "deeper" reasoning. This simply misses the point. Any representation worth considering can describe things at various levels of detail, but without representing physical relationships, certain kinds of reasoning processes can never be applied, no matter the level of detail.

Another misconception is that quantitative reasoning, such as solving or simulating differential equations, is deeper than qualitative reasoning. This is a misconception about the role of quantitative reasoning in reasoning about the world. A quantitative model is used when a situation can be mapped into it, and the results of applying the quantitative process can be interpreted in terms of the situation. To do this, there needs to be an understanding of what the situation is like, when the mapping is applicable, how to apply the mapping, and how to interpret the results. Each of these steps involve representation and reasoning (presumably qualitative) over and above the quantitative model. Quantitative reasoning supplements other reasoning processes; it does not substitute for them.

6. The General Nature of Deep Reasoning

On the basis of these examples, I propose the following definition of "deep":

A representation is "deep" with respect to a class of phenomena iff the representation describes the properties and relationships by which the phenomena interact.

A reasoning strategy is "deep" with respect to a class of phenomena iff the strategy reasons based on how the phenomena interact.

Relative to a certain class of phenomena, deep representations describe the properties and relationships that leads to interaction among these phenomena, and deep reasoning processes operate on this information. Because physical phenomena interact on the basis of physical structure and behavior, there need to be representational primitives whose meaning are structural and behavioral, and reasoning processes that can take this information into account.

References

T. Bylander and B. Chandrasekaran. (1985). Understanding Behavior Using Consolidation. *Proc. Ninth International Joint Conference on Artificial Intelligence.* Los Angeles.

B. Chandrasekaran and S. Mittal. (1983). Deep versus Compiled Knowledge Approaches to Diagnostic Problem-Solving. *International Journal of Man-Machine Studies, 19*, 425-436.

J. de Kleer and J. S. Brown. (1984). A Qualitative Physics Based on Confluences. *Artificial Intelligence, 24*, 7-83.

K. D. Forbus. (1983). Qualitative Reasoning about Space and Motion. In D. Gentner and A. Stevens (Eds.), *Mental Models.* Hillsdale, New Jersey: Lawrence Erlbaum.

K. D. Forbus. (1984). Qualitative Process Theory. *Artificial Intelligence, 24*, 85-168.

P. E. Hart. (1982). Directions for AI in the Eighties. *SIGART, 79*, 11-15.

P. J. Hayes. (1985). The Second Naive Physics Manifesto. In J. Hobbs and R. Moore (Eds.), *Formal Theories of the Commonsense World.* Norwood, New Jersey: Ablex.

D. Michie. (1982). High-road and Low-road Programs. *AI Magazine, 3*(1), 21-22.

R. S. Patil, P. Szolovits, and W. B. Schwartz. (1981). Causal Understanding of Patient Illness in Medical Diagnosis. *Proc. Seventh International Joint Conference on Artificial Intelligence.* Vancouver.

Y. Shoham. (1985). Naive Kinematics: One Aspect of Shape. *Proc. Ninth International Joint Conference on Artificial Intelligence.* Los Angeles.

C. Stanfill. (1983). The Decomposition of a Large Domain: Reasoning about Machines. *Proc. National Conference on Artifical Intelligence.* Washington, D.C..

Applying General Principles to Novel Problems as a Function of Learning History: Learning from Examples vs. Studying General Statements

Catherine A. Clement
Department of Psychology
University of Illinois
Champaign, Illinois 61820

Abstract

This research concerns the effect of learning history for a general principle on the ability to apply the principle to novel situations. Adult subjects learned general problem solving principles under three alternative conditions: (a) abstraction of principles from diverse examples (b) study of explicit general statements of principles and (c) practice in mapping given statements onto examples. The specific aim of this research was to explore how examples given during learning a general principle affect its application to novel problems which do not share "surface" features with the examples.

Results showed that examples did not significantly facilitate application of principles over learning only a given general statement. Moreover, subjects who abstracted principles from examples, although they had abstracted the relevant information, were significantly worse at application than subjects who learned only the general statement or who learned the given statement and examples. These subjects had particular difficulty accessing and selecting the appropriate principle for a problem.

Results suggest that the representation of specific information from examples may interfere with efficiency at matching a principle to a novel problem. Whether such interference occurs may depend on the relationship between the principle and its examples in the memory representation. This relationship may be influenced by the way examples are initially encoded.

The general concern of this research is the ability to apply an abstract concept or general principle to novel situations. It is a common intuition that specific examples are helpful in learning and being able to apply abstract principles such as general scientific principles or general problem solving strategies. Many studies on the effects of examples on the acquisition and use of a principle have focused on "surface similarities" (similarities not strictly related to the principle) between examples and new instances. For example, surface similarities with prior examples provide cues that a principle is relevant to a new situation (e.g. Ross, 1984, 1986; Lewis and Anderson, 1985). Such reliance on surface cues is a not useful if the principle is needed for _novel_ situations "dissimilar" to learning-examples. Other research does suggest that learning-examples can aid application of a principle even when surface similarities with new instances are _absent_ (Gick and Holyoak, 1982; Nitsch, 1977). In these studies subjects who learned a general principle from examples were better at applying the principle to novel instances than those who learned only an abstract description of the principle. However, the strength of these findings is unclear and it is also unclear what factors may allow prior examples to affect application to novel situations (Clement, 1986).

The purpose of the present research is to understand how examples given during learning a general principle affect its application to novel problems which do not share surface features with the examples. Learning histories for a general principle were varied in three ways. Subjects either had to abstract principles from specific examples, study given general descriptions of principles, or study general descriptions and examples.

The contexts in which principles were applied were also varied. Three contexts were used which differed in the extent to which the choice of a principle for a problem was specified for the problem solver. Since each context may demand different cognitive processes the effects of learning history may vary depending on context. In one context the relevant principle for a problem was fully specified for subjects and they only had to map it to the current problem. Such mapping may be described as translating the general terms of the given principle into specific problem elements that generate a solution. In the second context, the set of potentially relevant principles was specified and subjects had to select a principle from the set. Selection may involve exhaustive or terminating tests of the fit between principles in the set and the problem. In the third context no information about the relevant principle was given and subjects had to spontaneously access the principle. Such access may require spontaneously noticing a similarity between an abstract representation of the problem and the features of the relevant principle in LTM. Figure 1 summarizes the processes demanded in each context.

Figure 1. General Description of Processes Required for Application of
Principles in Three Contexts.

I Mapping Only

Translating between the general terms of the principle
in working memory and specific problem elements

II Selection and Mapping

Exhaustive or terminating test mappings of principles
in working memory.

III Spontaneous Access and Mapping

Spontaneous similarity matching between an abstract
representation of the problem and the principle held
in long term memory.

It was speculated that learning histories with examples might affect application of principles for two reasons. First, similarities between the processes of deriving a principle from examples and the processes of applying it new instances may be important. In both situations subjects must translate between a specific and general representation of the principle and must explicitly distinguish between relevant and irrelevant information. Thus, even if the examples and new problems are dissimilar in surface features, the general processes used during learning are similar to those required by application. Processes used during learning may transfer to the task in which the same principle must be applied (Clement, 1986).

Second, learning with examples may affect the representation of a principle in ways that are relevant to application. For example, subjects who represent links with the examples have a concrete model of the principle which they may exploit during application to new instances (even if instances are not surface similar to the model). However, it may be crucial that subjects clearly represent the hierarchical relation between the principle itself and examples. If they inadequately differentiate their description of the principle from the examples, example-specific information may interfere with matching and mapping the principle to novel problems.

EXPERIMENT 1

Two independent groups learned general principles in one of two learning conditions: GS subjects studied given general descriptions of principles and EX subjects abstracted their own general description from diverse examples. After learning, subjects had to use the principles learned to solve novel story problems. The contexts in which they solved the problems varied in the extent to which subjects were informed that a particular principle was relevant.

Method

Subjects

Subjects were 106 undergraduate students.

Learning Materials and Task

Two principles were learned. These were highly abstract and described "survival strategies" used by organisms or organizations to solve problems. For example, according to the "convergence" principle (adapted from Duncker, 1945 and Gick and Holyoak, 1983) "if a strong force cannot be sent along a single path to a target, then weak forces should be sent along many paths simultaneously."

GS subjects (n=45) studied general statements of principles (see Appendix A). Subjects paraphrased the statements from memory and then checked their paraphrase against the given statement.

EX subjects (n=40) studied two or three stories which exemplified each principle (see Appendix B). Subjects had to discover and write a general description of the "survival strategy" common in a set of examples. They also had to illustrate each main point in their description with a part of each example.

After learning, a Recall Task required subjects to describe each principle from memory.

924

Problem Solving Materials and Task

The Problem Solving Task included two Target problems (each soluble with a principle learned) and three Dummy problems (not soluble with either principle). The Target problems described complex "real world" problem situations (see Appendix C). (The Target problem for the convergence principle is a version of Duncker's (1945) "radiation problem").

The task consisted of three phases allowing subjects three passes at the problems. Phase 1 required <u>spontaneous access</u> of principles: subjects were not told that the principles learned were relevant to the problems. At phase 2 subjects had to <u>select</u> one of the principles for problems: subjects were told to figure out which principle applies to which problem. At phase 3 subjects only had to <u>map</u> a specified principle to the appropriate target problem.
Figure 2 summarizes the method for Experiment 1.

Figure 2. Outline of the Procedure for Experiment 1.

<u>Learn Principles</u>

GS Group
- Study given general statement
- Paraphrase from memory
- Correct against original

EX Group
- Discover similar principle in diverse examples
- Write general description
- Illustrate main points

<u>Recall Task</u>
Recall general descriptions

<u>Problem Solving Task</u>
Phase 1- Spontaneous Access and Mapping
Phase 2- Selection and Mapping
Phase 3- Mapping Only

Results and Discussion

Figure 3 shows the proportion of correct solutions to Target problems by each phase. The GS group is significantly better than the EX group at each phase (see figure note).[1] This pattern of between group differences remains when "initial learning" of the principles is taken into account, i.e. when only those subjects who accurately described principles at the recall task are considered. Among these subjects the proportion of correct solutions is higher in both groups but the difference between groups is the same.

Group differences were greatest for access and selection of principles. For mapping, the groups are equivalent for one principle (EX subjects caught up with GS subjects by the end of phase 3). For the other principle, which accounts for group differences in mapping, irrelevant information from examples appeared to lead to an incorrect instantiation of the principle by EX subjects.

925

Figure 3. Experiment 1. The Proportion of Correct Solutions to Target Problems in Each Group[a] by Each Phase. (Phase 1 requires spontaneous access and mapping; Phase 2 requires selection and mapping; Phase 3 only requires mapping).

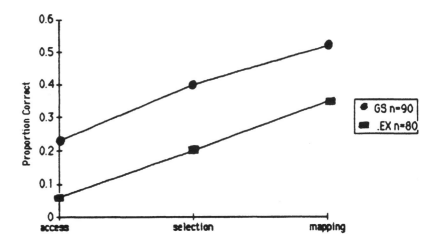

[a]n=the number of responses possible in each group.

Note. Subjects who were correct at an earlier phase, and who did not change to an incorrect response, were included as correct at subsequent phases. At phase 1 statistical analyses considered the proportion of subjects who solved at least one problem, ($p=.01$, fisher exact, two tailed test). At phase 2 and 3 analyses considered the proportion of subjects solving 0, 1 or 2 problems (Chi square $=7.26$, df$=2$, $p \leq .05$ and Chi square$=7.78$, df$=2$, $p \leq .05$, at phases 2 and 3 respectively).

In sum, EX subjects were significantly worse than GS subjects at application of principles to novel problems. These results contrast with the findings of Gick and Holyoak (1983) and provide no evidence that the procedures involved in learning from examples, or the concrete model provided by examples, facilitate application of principles. Moreover, results suggest that the representation of a principle was negatively affected by learning from examples even for those EX subjects who had developed a correct description of a principle. The representation may have failed to clearly differentiate the description of the principle from the examples.

Experiment 2

In Experiment 2, a new group of subjects (PM subjects) were asked to study a given statement of a principle and then practice mapping the statement to examples. This experiment had two aims. First, PM subjects were compared to EX subjects to observe whether processing the examples under the guidance of a given statement would be important. Although PM subjects received the same examples as EX subjects, their learning task might lead to a representation in which the examples and the principle are better differentiated. PM subjects learn a statement of the principle before reading the examples and read the examples only for the purpose of finding elements which instantiate this statement. Thus their encoding of examples may be less thorough and more directed than the encoding by EX subjects. They might be less likely to encode information irrelevant to the principle and more likely to represent the hierarchical relation between the principle and the examples. The second aim of this experiment was to again assess whether examples could facilitate application; PM subjects were also compared to GS subjects from experiment 1. The practice mapping task involved procedures in which subjects had to translate between variables of the general principle and the specific elements of examples. As discussed earlier, such translation processes used during learning a principle may transfer to facilitate its later application.

Method

Subjects

Subjects were 23 undergraduate students. These subjects were compared to a subset of GS and EX subjects in Experiment 1 chosen from the same school as PM subjects.

Design and Procedure

Subjects studied the same general statements of principles used by GS subjects in Experiment 1. Then, given the same examples used previously, they had to find the parts of each example that illustrated the ideas in the principle. (This mapping task had also been given to EX subjects after they had abstracted their general statement.)

After learning, subjects were given the recall and application task used in Experiment 1.

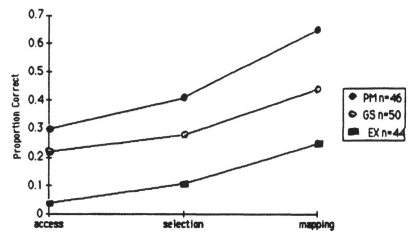

Figure 4. Experiment 2. The Proportion of Correct Solutions to Target Problems in Each Group[a] by Each Phase. (Phase 1 requires spontaneous access and mapping; Phase 2 requires selection and mapping; Phase 3 only requires mapping).

[a]n=the number of responses possible in each group.

Note. Subjects who were correct at an earlier phase, and who did not change to an incorrect response, were included as correct at subsequent phases. At phase 1 and 2 statistical analyses considered the proportion of subjects who solved at least one problem, (p=.01, fisher exact, two tailed test). At phase 3 analyses considered the proportion of subjects solving 0, 1 or 2 problems Chi square =14.19, df=2, p ≤.001).

927

<u>Results and Discussion</u>

Figure 4 shows the proportion correct in each group. Differences between PM and GS subjects are not significant. In contrast, differences with the EX group at each phase are significant. Again results are the same when only subjects who gave good recall descriptions of a principle are considered.

In sum, results suggest that processing the examples under the guidance of the given statement of a principle, rather than having to abstract the principle, allowed better representation and application of the principle. However, contrary to speculations, practice mapping a principle to examples did not significantly improve application over learning only the given statement.

GENERAL DISCUSSION

Subjects who abstracted a principle from examples were poor at application relative to subjects who learned only a general statement of principle even when they had abstracted the relevant information. These subjects were also poor relative to subjects who processed examples under the guidance of the general statement. One account of these findings is that EX subjects may not have adequately differentiated the description of the principle from the examples in their representation of the principle. PM subjects, whose learning task should have lead to a more differentiated representation of the specific and general information, were better at application. For EX subjects their description of the principle may have been represented as part of each example, rather than as a descriptor of a category in which the examples are some of many possible instances. The principle may not have been salient relative to other information. Or the principle may have been represented at a relatively low level of abstraction.

How would a poorly differentiated representation lead to poor application especially when spontaneous access and selection are required? One consequence may be that similarity matching between the problem and the principle is inefficient. The three application contexts may be viewed in terms of a continuum that varies in the extent to which efficient similarity matching is demanded. At one extreme is the context in which principles must be <u>spontaneously accessed.</u> In this context, since the principle is not already available in working memory, subjects may have to automatically notice a similarity between the problem and the principle held in long term memory. Thus, a representation of the principle which is surrounded by specific information from examples should lead to difficulty since this specific information is <u>dissimilar</u> to the target problem when the examples and the problem do not share surface features. (The useful similarity exists at an abstract level of representation of the problem and the principle.) In contrast, a representation of the principle which is not linked to specific examples, or in which the examples and the principle are clearly differentiated, should allow more efficient recognition of similarities between the principle and the problem.

At the other extreme of the continuum is the context in which the relevant principle is already identified and subjects only have to map it. Efficiency at similarity matching should be less of a factor here since the correct principle is already available in working memory. Subjects can work out the correspondence with the problem even if their representation of the principle is poorly differentiated from prior examples. Results suggested that with this

928

decreased demand for efficiency a poorly differentiated representation was less of a problem. EX subjects were equivalent to GS subjects in mapping for one of the two principles. Selection errors are too complex to discuss in the present paper, but results suggest that both efficiency at similarity matching and the intrusion of irrelevant information from the examples lead to poor EX performance.

In sum, results suggest that an abstract representation of a principle, clearly differentiated from specific examples enabled the most ready application of principles to novel problems. Examples may interfere with application if the hierarchical relationship between the example-specific and general information is not clearly represented. A poorly differentiated representation may particularly affect access and selection (rather than mapping) because it may not permit efficient similarity matching. This account is consistent with recent descriptions by Gentner (1987) and Holyoak and Thagard (1986) of access and mapping processes in case-based reasoning.

Results also suggest that the way the examples are initially encoded may affect representation and application. General and particular information may be better differentiated when subjects have prior knowledge of the general principle than when they have the relevant general information only after initial processing of examples. The specific nature of the representation of the principle formed by PM and EX subjects is being explored further in current studies which are varying the similarities between target problems are prior examples.

Even when examples were processed under the guidance of a given statement of a principle application was not significantly better than when only a given statement was learned (however, a trend toward improvement with examples was found). Future research should further explore circumstances in which the processes involved in learning from examples, or the model provided by examples, can be used to facilitate application of a principle to novel instances.

Footnotes

1. In order to get a base solution rate for problems, a control group (n=21), received the Problem Solving Task but received no prior training. GS subjects but not EX subjects gave significantly more correct solutions at phase one than this base rate group.

References

Brown, A.L., & Campione, J.C. (1984). Three faces of transfer: Implications for early competence, individual differences, and instruction. In M. Lamb, A. Brown, & B Rogoff (Eds.), Advances in developmental psychology (Vol.3). Hillsdale, NJ: Erlbaum.

Clement, Catherine (1986). Applying general principles to novel problems as a function of learning history: Abstraction from examples vs. studying general statements. Doctoral Dissertation, Clark University.

Gentner, D. (1987). The mechanisms of analogical learning. In S. Vosniadou & A. Ortony (Eds.) <u>Similarity and analogy in reasoning and learning,</u> in preparation.

Gick, M. L. and Holyoak, K. J. (1983). Schema induction and analogical transfer. <u>Cognitive Psychology, 15,</u> 1-38.

Holyoak, K.J. & Thagard, P. R. (1986). A computational model of analogical problem solving. Unpublished manuscript.

Lewis, M. W. & Anderson, J. R. (1985). Discrimination of operator schemata in problem solving: Learning from examples. <u>Cognitive Psychology, 17,</u> 26-65.

Nitsch, K. E. (1977). Structuring decontextualized forms of knowledge. Doctoral dissertation, Vanderbilt Univesity.

Ross, B. H. (1984). Remindings and their effects in learning a cognitive skill. <u>Cognitive Psychology, 16,</u> 371-416.

Ross, B. H. (1986). This is like that: Object correspondences and remindings and the separation of similarity effects on the access and use of earlier problems. Manuscript submitted for publication.

Appendix A
General Statements Given to Subjects in the GS and PM Groups

Convergence Principle

Often a system[1] must protect against, or destroy, a bad or opposing thing, and it must do this by applying a force in full strength. However, sometimes a force is available but it cannot be applied in full strength along a single path. In this case the strong force should be divided into several weak forces which are applied to the opposing thing simultaneously along several paths. In this way the forces will converge in full strength on the opposing thing.

[1]Subjects are told that a "system" is any kind of organism or organization.

Mimic Principle

Often a system must protect itself against, or destroy, an opposing thing, but there is no force in the system which will allow it to cope with the opposing thing. Unwanted consequences occur when the system and the opposing thing interact.

Sometimes this problem can be solved by finding or making a mimic of the opposing thing, which the system can interact with. The mimic has the same non-destructive features as the opposing thing, but does not have the destructive features.

If the system interacts with the mimic, the system can change, or develop abilities to cope, without having to face destruction. Then, when the system is later confronted with the real opposing thing, it has the abilities to cope with it.

APPENDIX B
Latent Examples Given to Subjects in the LEX and PM Groups

Convergence Examples [1]

The General

A small country was ruled from a strong fortress by a dictator. The fortress was situated in the center of the country, surrounded by farms and villages. Many roads led to the fortress from all different parts of the countryside. A rebel general vowed to capture the fortress. The general knew that an attack by his entire army would capture the fortress. He gathered his army at the head of one of the roads, ready to launch a full-scale direct attack. However, the general then learned that the dictator had planted mines on each of the roads. The mines were set so that small bodies of men could pass over them safely since the dictator needed to move his troops and workers to and from the fortress. However, any large force would detonate the mines. Not only would this blow up the road, but it would also destroy many neighboring villages.

Consequently, the general devised a simple plan. He divided his army into small groups and dispatched different groups to the head of each of the roads which led into the fortress from various sides. When all was ready he gave the signal and each group marched down a different road. Each group continued down its road to the fortress so that the entire army arrived together at the fortress at the same time. In this way, the general captured the fortress.

The Fire

One night a fire broke out in a wood shed full of timber on Mr. Johnson's place. As soon as he saw flames he sounded the alarm and within minutes dozens of neighbors were on the scene. The shed was already burning fiercely, and everyone was afraid that if it wasn't controlled quickly the house would go up next. Fortunately, there was plenty of water available since the shed was right beside a lake.

The men had buckets, but they needed something larger to carry water in. To put the fire out, a large amount of water had to reach the fire at once. The men searched around and found an enormous water trough used for cattle, sitting beside the shed. Several men carried it to the lake and filled it with water. Unfortunately, as they were struggling to bring it to the fire, the big water trough broke. It looked like the house was doomed.

Just then they realized a way to save the house. Everyone took a bucket and filled it with water. Then they waited in a circle surrounding the burning shed. As soon as the last man was prepared, they counted to 3, then all the small buckets of water together were thrown on the fire. The force of all this water at once, dampened the fire right down, and it was quickly brought under control.

[2] These materials are modified versions of stories used by Gick and Holyoak (1983).

Mimic Examples

Boot Camp

Men who sign up for military service do not come to the situation prepared for battle. They must be intensively trained if they are to be effective and survive. A common military training strategy is to put new soldiers in "boot camp". In boot camp, soldiers take part in war-like activities but are not faced with actual bombs, bullets etc. which can kill them.

Thus, soldiers fight one another but stop fighting before serious injury occurs; groups of soldiers take on the roles of enemies and they hunt one another; soldiers attempt to survive alone in the wilderness with no food, though they are rescued before serious harm occurs. Also, fellow soldiers act as enemy interrogators, threatening punishment if certain "secrets" are not revealed; however, real, dangerous punishment is not given.

With these exercises the soldiers develop skills needed for each of these situations which they might find when they go to war.

The Space Vehicle

A space vehicle was being built for a flight to Jupiter. The material used for the space ship had to be able to withstand a gas, called JGAS, which surrounded Jupiter. This was a problem because tests showed that JGAS affected the material usually used for space ships: exposure to the gas caused the ship-material to crumble. Something had to be done.

Engineers working on the ship examined pieces of the crumbled ship-material and discovered that the material was denser and stronger than it was before exposure to the JGAS. It appeared that there were two effects of JGAS: the gas caused the material to crumble, and the gas also changed the density of the material and made it stronger. Apparently the material crumbled before it was made stronger.

The engineers figured out what elements of JGAS had the two different effects on the material. Then they found a different gas. This gas was exactly the same as JGAS, except that it did not contain the element that caused the material to crumble -- this gas was called Jgas-x. The engineers exposed some of the ship-material to Jgas-x. The material did not crumble, though it was made denser and stronger. Then they exposed the same piece of material to JGAS. This time the material withstood the gas and did not crumble.

The engineers decided to treat all of the material to be used in the space vehicle with Jgas-x. Then the material would be able to stand up to the atmosphere on Jupiter which contains JGAS.

The Computer

At a high-technology company a special digital computer has been developed to teach new personnel at the Ajax nuclear powerplant about the operation and repair of the powerplant. The computer has a large, high resolution, color screen which displays its own "Ajax" plant with all its inner workings. When a person is using the computer, the computer will set up a dangerous problem situation that might occur in the real Ajax powerplant; for instance the problem might be that a waste disposal unit is about to leak. The employee must decide how to repair the leak in a limited amount of time. Using voice commands, the person can tell the computer the steps to be taken to fix the unit. Each step being taken to fix the unit and its effects will then be displayed. If the steps are wrong, the "nuclear waste" will leak and the dangerous consequences are graphically demonstrated on the screen. If the steps are correct, the computer will display the unit in a repaired state. This computer method of teaching has some obvious advantages over training personnel at the Ajax powerplant -- there, the personnel (as well as others) might not live long enough to learn the operation of the powerplant.

APPENDIX C
Target Problems Given to All Subjects

Convergence Problem

A doctor was faced with a patient who had a malignant tumor in his stomach. Any operation would kill the patient because of another incurable (though not fatal) illness the patient had. But unless the tumor was destroyed the patient would die. The only possible way to destroy the tumor was to use a kind of ray. If the rays reached the tumor at a sufficiently high intensity, the tumor would be destroyed. Unfortunately, at this intensity the healthy tissue that the rays pass through on the way to the tumor would also be destroyed. At lower intensities the rays are harmless to healthy tissue, but they would not affect the tumor either. The doctor had to discover a way to use the rays to destroy the tumor and at the same time avoid destroying the healthy tissue. Consultants suggested inserting a protected tube into the patient and sending the rays through this tube. The doctor pointed out that this required an incision which was impossible with this patient. Putting a tube through any natural opening in the body was also impossible. What could the doctor do?

Mimic Problem

There are small amphibians called "Zylots" which feed on living, tiny organisms in water; their digestive systems can break down the structure of certain types of organisms. An individual Zylot can readily adapt its digestive system to the available food in the environment, if it is food it can tolerate.

One summer the small fish in a popular man-made lake were dying in large numbers. They were apparently killed by parasites in the lake. The parasite attached itself to the stomach lining of the fish and sucked its blood. At first officials were puzzled because the parasites appeared to be a common variety which were known to be harmless to the fish. However, officials discovered that these parasites were slightly different from the common variety; these had the ability to suck large quantities of blood from a host and then move on to a new host when the original died. These parasites were called Drats. The officials considered using a pesticide to kill the parasites in the lake, but it turned out they were too poor to buy any chemicals. Since they were forced to find a very cheap solution to the problem, the officials felt that killing the Drats may be impossible.

A forest-ranger thought that it might be possible use Zylots to consume Drats; the Zylots might be able to eat Drats and not have their blood sucked. The forest-ranger gathered a population of 1,000,000 Zylots. Unfortunately, when she first took half of the Zylots to a small inlet in the lake things did not work out; the Zylots died. When a Zylot ingested a Drat, the Zylot could not successfully break down the structure of the Drat before having a lot of its blood sucked.

The forest-ranger found a way to solve this problem. One day she brought the rest of the Zylots to the lake and showed officials that they could eat and kill the Drats. How was she able to do this?

DISCOURSE MODELS AND THE ENGLISH TENSE SYSTEM

John Dinsmore
Department of Computer Science
Southern Illinois University at Carbondale

It is shown how Reichenbach's (1947) tense system is a direct consequence of the interaction of some straight-forward and independently motivated semantic rules for the English tense operators (past, present, future, perfect and prospective) and of the knowledge partitioning framework (Dinsmore, 1987), used in representing and structuring discourse knowledge. This account treats the semantics and pragmatics of temporal operators in a coherent and integrated way, and provides additional evidence for the viability of the theory of knowledge partitioning.

1. INTRODUCTION

A central observation in the study of the English verb system is that many tense forms apparently cannot be distinguished truth-conditionally yet differ in some intuitive way pragmatically. A significant and recurrent idea is that the distinctions between the past and the present perfect, for instance, have something to do with the perspective from which an event is viewed. Reichenbach (1947) introduces the term reference time (R) as distinguished from event time (E) and speech time (S) to refer to this temporal perspective. Bull (1960) corrects some mistakes in Reichenbach's system and it is actually Bull's revision with which we will be concerned. The relations among R, E, and S below correspond to the following verb forms.

Past:	$R = E < S$
Past perfect:	$E < R < S$
Past prospective:	$R < S, R < E$
Present:	$R = E = S$
Present perfect:	$E < R = S$
Present prospective:	$R = S < E$
Future:	$S < R = E$
Future perfect:	$S < R, E < R$
Future prospective:	$S < R < E$

There seems to be a concensus, articulated for instance by Taylor (1977), that the pragmatic aspects of the English tense system must be strictly distinguished from the semantic aspects on the basis of truth conditions. In this paper I will oppose this idea and argue that Reichenbach's tense system is a consequence of the semantics of the operators involved, along with the independently motivated underlying discourse model used in representing and structuring discourse knowledge. This is the knowledge partitioning framework developed by Dinsmore (1987). Familiarity with this framework is presupposed in the present paper.

1. A SEMANTICS FOR THE ENGLISH TENSE SYSTEM.

To indicate in the semantics that expression S is true of time t, I will write T(t,S). The following rules are proposed.

(Pres) T(t,pres(S)) iff T(t,S) and t = now.
(Past) T(t,pa(S)) iff T(t,S) and t < now.
(Fut) T(t,fut(S)) iff T(t,S) and t > now.
(Perf) T(t,perf(S)) iff there is a time t' such that T(t',S) and t' < t.
(Pros) T(t,prosp(S)) iff there is a time t' such that T(t',S) and t < t'.

It can be shown that a variety of facts about cooccurence of the tense forms with various adverbials, the truth conditions of complex sentences using non-finite forms of Perfect and Prospective, and the interaction of the tense forms with temporal adverbs can be predicted from these rules (Dinsmore 1981, 1982).

2. THE STRUCTURE OF DISCOURSE MODELS

In knowledge Partitioning the information communicated in a discourse is distributed over a large set of spaces, each of which defines a local domain of reasoning. Spaces are related by contexts, each of which provides a mapping of knowledge belonging to one space into another. A number of logical constraints are imposed on the way spaces are set up and connected.

The following notational conventions have proved helpful in discussing knowledge partitioning: "base" represents the real world. "S | P" means that S is true in space S. "S0 | f [S1]" means that f is a context that maps knowledge of S1 onto knowledge of S0. For instance, to represent George's beliefs, we might use the context base | "George believes that [s1]", then put every proposition P that George believes to be true in s1 as s1 | P. We can always use the mapping to derive base | "George believes that P".

Perhaps the primary implication of this model for natural language discourse processing is that at any point in a discourse some space is in focus. A discourse sentence is rarely assumed to be true in any absolute sense, but contributes knowledge about the current focus space.

3. USING SPACES WITH THE ENGLISH TENSE SYSTEM

Knowledge partitioning in discourse involves the distribution of temporally constrained knowledge over many spaces, each of which could be the focus space at some point in a discourse. It is shown in this section how the tense system serves to refer to temporally bound knowledge from the perspective of different focus spaces.

Within base, the occurance of things like events and states can be expressed by statements of the form base | at(t,P), "P happens at time t". The following describes the semantics of at.

935

(At) For any time t and sentence P, at(t,P) is true iff T(t,P).

Within the knowledge paritioning framework we may derive temporal spaces as follows. First, observe that the environment at(t,___) is entailment preserving, so by the <u>restriction on legal contexts</u>, whenever we have base | at(t,p) we may derive base | at(t,[s]) and s | p, where s is a previously unused space symbol. Second, observe that if at(t,___) is <u>entailment preserving</u>. This means that by the <u>distributive constraint</u> if we have base | at(t,[s]) and base | at(t,p) we should infer s | p. In Dinsmore (in press) the rule involved here is known as <u>space augmentation</u>. A <u>temporal space</u> is any space S with a context of the form S' | at(t,[S]), for some space S' and time t. Like any other space, a temporal space may be the focus space at some point in a discourse.

Reichenbach's tense system follows immediately if we define reference time as follows:

(RefTm)   t is the <u>reference time</u> at a point p in a discourse iff for some space S, S is the focus space at p and base | at(t,[S]).

Reichenbach's observation is that the difference between the reference time and the event time corresponds to the use of perfect, prospective, past and present forms. We can predict this as follows. Suppose that an event E occurs at some past time t2. This can be expressed as base | at(t2,E). Further, suppose that times t1, t3, t4, and t5 are such that

t1 < t2 < t3 < t4 = now < t5.

We define spaces s1 through s5 such that base | at(ti,[si]) for 1 ≤ i ≤ 5. Consider what sentences that mention E would be true in each of these spaces:

For s1: From t1 < t2 and base | at(t2,E) we have base | at(t1,prosp(E)). From this and t1 < now we have base | at(t1,pa(prosp(E))). By space augmentation we have s1 | pa(prosp(E)).

For s2: From t2 < now and base | at(t2,E) we have base | at(t2,pa(E)), and by space augmentation s2 | pa(E).

For s3: From t3 > t2 and base | at(t2,E) we have base | at(t3,perf(E)). Since t3 < now, base | at(t3,pa(perf(E))). By space augmentation, s3 | pa(perf(E)).

For s4: From t4 > t2 and base | at(t2,E) we have base | at(t4,perf(E)) and since t4 = now base | at(t4,pres(perf(E))). By space augmentation, s4 | pres(perf(E)).

For s5: From t5 > t2 and base | at(t2,E) we have base | at(t4,perf(E)). Since t5 > now, base | at(t4,fut(perf(E))). By space augmentation, s5 | fut(perf(E)).

In this way a complete account of reference time emerges. Now we are in a position to see why whe have Perfect and Prospective at all: Without them we would not be able to refer to E while maintaining s1, s3, s4, or s5 as the focus space. A number of implications of this account for the use of the past and the perfect are discussed in Dinsmore (in press).

4.CONCLUSION

This account bridges the gap between the semantics and the pragmatics of the English tense system by showing how the discourse use of English verb forms is related to their semantic interpretations, which may be derived in the usual compositional way. It is well integrated into the knowledge partitioning model of natural language discourse processing.

REFERENCES

Bull, William. 1960. Time, Tense, and the Verb. Berkeley: University of California Press.

Dinsmore, John. 1981. Tense Choice and Time Specification in English. Linguistics 19, 475-494.

Dinsmore, John. 1982. The Semantic Nature of Reichenbach's Tense System. Glossa 16, 216-239.

Dinsmore, John. 1987. Mental Spaces from a Functional Perspective. Cognitive Science 11, 1-21.

Dinsmore, John (in press). The Logic and Functions of the English Past and Perfect. In C. Georgopoulos & R. Ishihara (Eds.), Essays in Honor of Yuki Kuroda. Dordrecht: D. Reidel.

Reichenbach, Hans. 1947. Elements of Symbolic Logic. New York: Macmillan.

Taylor, Barry. 1977. Tense and Continuity. Linguistics and Philosophy 1, 199-220.

The Role of Categories in the Generation of Counterfactuals: A Connectionist Interpretation

Robert M. French and Mark Weaver
Department of Electrical Engineering and Computer Science
University of Michigan
Ann Arbor, Michigan 48109
Tel. (313) 763-5875

Keywords: counterfactuals, norm theory, connectionism, categories

Abstract

This paper proposes that a fairly standard connectionist category mode can provide a mechanism for the generation of counterfactuals -- non-veridical versions of perceived events or objects. A distinction is made between evolved counterfactuals, which generate mental spaces (as proposed by Fauconnier), and fleeting counterfactuals, which do not. This paper explores only the latter in detail. A connection is made with the recently proposed counterfactual theory of Kahneman and Miller; specifically our model shares with theirs a fundamental rule of counterfactual production based on normality. The relationship between counterfactuals and the psychological constructs of "schema with correction" and "goodness" is examined. A computer simulation in support of our model is included.

Introduction

We believe that a picture is emerging in which counterfactuals play a significant role in human cognition; they are not a mere curiosity which may be safely ignored. Humans live in a mental world where actualities are surrounded by possibilities. In other words (as Kahneman and Miller, 1986, have proposed), when people experience an event, they may also experience plausible counterfactual alternatives, and these have a profound effect on their reaction to the actual event.

A simple example is the affective difference between missing an airplane by an hour or by a minute. The outcomes are identical, but the latter case is an order of magnitude more frustrating. The explanation for this centers on the claim that in the latter case, having made the plane is a highly available counterfactual alternative, while in the former case, it is not. Functionally speaking, this affective reaction is appropriate. It "marks" situations that, if repeated, could easily have their outcomes improved the next time.

In keeping with our claim that counterfactuals are not "special purpose" phenomena, we will also propose that their production requires no ad hoc, special purpose machinery. Instead, we suggest that a fairly standard connectionist category model suffices. Furthermore, we propose that the process by which counterfactuals are generated is closely related to the cognitive mechanisms underlying "goodness" or "schema with correction".

Counterfactual Generation vs. Counterfactual Development

The distinction between counterfactual generation and counterfactual development is not a common one, but is nonetheless crucial to our argument. We see the difference between counterfactual generation and counterfactual development as being analogous to the difference between producing an acorn and its subsequent development into an oak tree. For example, Gilles Fauconnier in *Mental Spaces* (1986), is concerned with the counterfactual world (or mental space)

938

to which a counterfactual proposal gives rise. (A mental space is a counterfactual world set up by statements like "If I had a milllon dollars..." or "If I'd left ten minutes earlier....") Similarly, David Lewis (1973) and, more recently, Matthew Ginsberg (1986) examine counterfactuals within the framework of standard and non-standard formal logic. In contrast, our present concern begins and ends with the production of the seed, an initial counterfactual that may (but frequently does not) develop into a counterfactual "world" or "space". We will call these counterfactuals "fleeting counterfactuals".

A concrete example may help to make this clearer. Suppose you are standing at a corner near a puddle talking to a friend. A car passes and swerves slightly to avoid splashing you. You think: "Good thing that car swerved; I would have gotten soaked otherwise" and you go on talking and completely forget the incident. In this case a fleeting counterfactual is produced but then disappears without engendering a mental space. However, this fleeting counterfactual would have served as the "seed" for the development of a mental space had you gone on to reason about the consequences of having been splashed: "I just bought these pants and the dirty water would have stained them...or would it? Anyway, I would have been cold and dirty...probably should stand back from the curb a little farther...".

The important point is the minimal nature of fleeting counterfactuals. They are the result of a low-level process, an automatic by-product of activating a category. This production is *not* under conscious control (you could never *decide* to stop having the possibility of being splashed occur to you when a passing car nears a puddle). Furthermore, the production of fleeting counterfactuals does not divert already active mental processes -- in the example, the ongoing conversation need not have been disrupted. On the other hand, the development of a counterfactual mental space does engage higher level mental processes and does so in direct proportion to the degree of development of a mental space. In the preceding example, if you had gone on to develop a space, it is likely that you also would have missed some of what your companion was saying: "Sorry, I was just thinking about what would have happened had that car hit the puddle and soaked me. What did you just say?"

We believe there is a counterfactual continuum running from fleeting counterfactuals (i.e., no mental space created at all) through the creation of a very small mental space (one or two steps of reasoning within a counterfactual world) to full blown, long-duration counterfactual spaces ("What if I had married Lynn instead of Nancy?").

Normal and Abnormal Events

Kahneman and Miller divide events into normal and abnormal ones. They define an abnormal event as "one that has highly available [counterfactual] alternatives, whether retrieved or constructed" while a normal event is one that "mainly evokes representations that resemble it". Normal events, according to them, do not evoke surprise; abnormal ones do.

The perception of the abnormal feature arises as the result of a (conscious or unconscious) comparison to its normal, expected alternative. Suppose we enter an office in which everything clearly satisfies our expectations except that the desk is upside down. The inverted desk is clearly the "abnormal" aspect of the office, and it generates a counterfactual alternative, specifically the "normal" office. Kahneman and Miller claim that the abnormal aspects of a normal event are the ones that change when we counterfactualize. We claim that the first counterfactualization that occurs "slips" the abnormal feature to one more commonly associated with the prototypical category corresponding to the event.

Representation of categories

In what follows we represent categories as clusters of "feature nodes" with mutually excitatory interconnections. In addition to interconnections, feature nodes also have connections

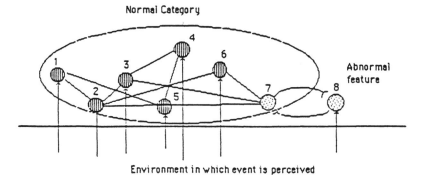

Node 1: pug nose
Node 2: full lips
Node 3: high forehead
Node 4: glasses
Node 5: etc.
Node 6: etc.
Node 7: shoulder-length hair
Node 8: crew cut

Figure 1

through which they receive environmental input. This is not unlike, for example, the schema model of Rumelhart, Smolensky, McClelland, and Hinton (1986). We represent this schematically in Figure 1 for the category "Joe".

To explain initial counterfactual production, we begin by distinguishing two different sources of activation: one external (from the environment via feature detectors) and the other internal (spreading activation from other nodes in the category). Assume that you see your friend Joe, whom you have always known to have shoulder-length hair, with a crew cut. Nodes 1 through 6 represent features that are present and sufficient to allow you to recognize your friend. Node 7 corresponds to the feature "shoulder-length hair" (always active in the past when you saw your friend) while node 8 corresponds to the feature "crew cut" (activated in the present circumstance).

Nodes 1 through 6 receive dual support, both from the environment and from other nodes within the category. Node 8 receives support *only* from the environment, while node 7 receives no external support but clearly has internal support, since your category for "Joe" has always included the feature "shoulder-length hair". Our discussion will focus on these two nodes.

Only in the situation where environmental input violates strong expectations will a strong fleeting counterfactual slip be produced. The situation is summed up in Figure 2 below.

The central role of inhibition

We will focus only on nodes 7 and 8, i.e. the long-hair and crew cut nodes respectively. We may assume that these nodes both receive strong support, the first internally, the second externally. Since people generally perceive the world as it actually is and not merely according to their expectations, we can assume an environmental bias and, consequently, that node 8 receives stronger input than node 7. (Were the opposite to be the case, misperception would occur, which does happen occasionally.) Since crew cuts and long hair are mutually exclusive features, we also assume that the corresponding feature nodes are mutually inhibitory. Initial counterfactual production (i.e., the "slip" to the normal category) would only occur once the inhibitory effect of the "crew-cut" node (8) on the "shoulder-length-hair" node (7) ceased.

The roles of habituation and short-term connection strength

There are essentially two phenomena that would cause the inhibitory effect of node 8 to cease. The most obvious is if the support from the environment simply ceased. The second, less obvious, phenomenon that would cause the activation of Node 7 to dominate that of node 8 is habituation, also called fatigue. In the first case, if environmental input ceases, our model easily demonstrates the slip to the counterfactual alternative. We decided to answer a harder question, namely: Will the slip occur even if environmental support continues unabated? Our model confirmed our intuition that the answer is yes.

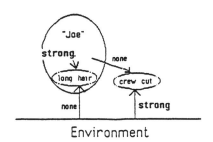

Figure 2.

The notion of neural habituation has been invoked by certain authors to explain a wide range of cognitive phenomena including decreased recall of material immediately following intensive attempts at its memorization (Pomerantz, Kaplan, and Kaplan, 1969), the Necker cube perceptual flip (Feldman,1981), flexibilty in problem solving (Levenick, 1985), etc. We claim that habituation is also an important mechanism in the production of counterfactuals.

In our model, nodes that receive on-going dual support from the category and the environment are less susceptible to the effects of fatigue than singly-supported nodes. In our example, node 8 initially receives support only from the environment. As it fatigues, its inhibiting influence on node 7 decreases. This allows the activity in the latter to rise. Node 7, increasingly active, begins to inhibit the already fatiguing node 8, further contributing to the latter's drop in activity. The activity in node 8 falls even faster, further decreasing its inhibition of node 7, and so on. Near the peak of activity in node 7 (never high enough for perception) we experience the counterfactual alternative "Joe with long hair". (The results of our simulation are consistent with this prediction.)

As we become (quite rapidly) accustomed to "Joe with a crew cut", we no longer experience the counterfactual alternative. The mechanism we propose to account for this is a short-term increase in the connection strength between the category and the new feature node. Short-term potentiation is a well-known phenomenon in neurophysiological investigations of synaptic plasticity (Goddard, 1980) and it also has been proposed as a mechanism to allow activity to briefly persist in a group of previously unassociated neural units, thus supporting the formation of new cell assemblies (Kaplan, 1970). For us, though, the importance of short-term connection strength is that it serves to temporarily strengthen the connection between the category node and the abnormal feature node.

Counterfactuals, goodness, and normality

One of the most significant results of Eleanor Rosch's extensive work with natural categories is the idea of "goodness" (Rosch, 1977). She showed that categories had no definite boundaries and that not all members are equal. Instances closer to the prototype (or most normal instance) are judged as being better than those which are most distant.

We claim that our category model provides a simple, direct coding of goodness by the overall level of activation of the nodes making up the category. Whenever a normal feature is replaced by an abnormal alternative, the normal node acts as an activity sink; it receives activation from other feature nodes but does not generate any in return. Thus the overall level of activity of the category is reduced and the instance is experienced as poorer. When the violation of expectations is strong enough, counterfactualization also occurs.

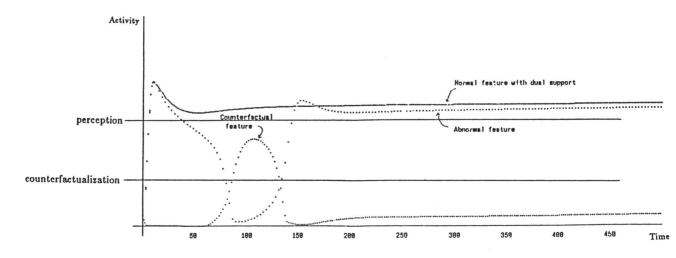

Figure 3 Activity trace showing production of a fleeting counterfactual

Counterfactualization and schema with correction

A time-honored construct in psychology is schema with correction (Woodworth, 1958). The usual context is one of asking people to redraw nonsense figures from memory. Where possible, people remember such figures as familiar figures plus an abnormal feature (eg., a square with a "funny" corner). But this is actually a case of counterfactualization. The "square" with a "funny corner" is, in fact, *not* a square at all. It *would* be a square only if the funny corner were normal. A prerequisite for schema with correction is counterfactualization and identification of an abnormality.

Simulation of the production of a fleeting counterfactual

The feature nodes in our category model do not represent single neural-level elements but rather interconnected groups or assemblies. So, a standard weighted-sum activity function was inappropriate for the simulation. Accordingly, we based our simulation on on a slightly modified mathematical model developed by Stephen Kaplan in 1970. His model was originally designed to describe the time course of activity within a single cell assembly. It allows persistent node activity (due to internal positive feedback) and, significantly, includes fatigue and short term potentiation mechanisms. We created a network of three nodes each of which would behave, in isolation, like Kaplan's cell assembly. We created links between the two nodes in competition. Initially, we did not use changing short-term connection strength between the category and the competing nodes to influence the spread of activity to these nodes. This resulted in a continual alternation of activity peaks between the normal and counterfactual nodes. When we realized that activity spreading *between nodes* was no different than *within* a node, we added a short-term connection strength mechanism to our links. Thereafter the model behaved according to our predictions (Figure 3).

Conclusion

In the world of human cognition, actualities are surrounded by a context of counterfactual possibilities. This context has a powerful effect on people's reactions to objects and events. We have chosen to study what seem to be to us the simplest, most easily isolated counterfactuals which we have called *fleeting counterfactuals*. We have shown that a slightly extended connectionist category model will produce this counterfactual behavior and thus provides the beginnings of an explanatory mechanism for the phenomenon of counterfactuals.

942

Appendix I

Mathematical details of the simulation

The following equations, originally developed by Stephen Kaplan in 1970, model the time course of activity in a neural net. Only very minimal modifications to the original equations were required to adapt them to our own model. (Note: F_t denotes F(t), S_t denotes S(t), etc.)

Sensitivity of connections:

This equation says, in essence, that as fatigue builds, the sensitivity (i.e., the ability to pass activation) of the affected connection drops rapidly. Sensitivity is described by:

$$\alpha_t = \frac{(L_t + S_t)(1 - F_t)^2}{K_\alpha}$$

where:
  L_t and S_t are, respectively, long-term and short-term connection strength;
  F_t is fatigue;
  K_α is a constant designed to ensure that α_t does not exceed 1.

Change in short-term connection strength over time:

$$\Delta S_t = K_{sscale} P_t (1 - S_t)^2 - K_{sdecay} S_t$$

where:
  S_t is short-term connection strength;
  P_t is the activity of the net;
  K_{sscale} and K_{sdecay} are constants.

Change in fatigue over time:

$$\Delta F_t = K_{fscale} P_t (1 - F_t)^2 - K_{fdecay} F_t$$

where:
  F_t is fatigue;
  P_t is the activity of the net;
  K_{fscale} and K_{fdecay} are constants.

Input to the system:

I_t: Input to the nodes under examination was supplied by a square input function. Environmental input to node 8 was weighted somewhat higher (10%) than the category input to node 7, our reason being that environmental support is given more weight than equivalent internal support, in order to improve the probability that misperception of the environment does not occur.

Change in activity over time:

$$\Delta P_t = [P_t + (1 - P_t)I_t](1 - P_t)\alpha_t - [P_t^5 + P_t(1 - P_t)^{10}](1 - \alpha_t)$$

where:
  $[P_t + (1 - P_t)I_t](1 - P_t)\alpha_t$ is the rise component and
  $[P_t^5 + P_t(1 - P_t)^{10}](1 - \alpha_t)$ is the fall component of activity.

For a more detailed explanation of the derivations of these equations, see (Kaplan, 1970).

Bibliography

Fauconnier, G. (1985). Mental Spaces. Cambridge, MA: MIT Press.

Feldman, J.A. (1981). A connectionist model of visual memory. In G.E. Hinton and J.A. Anderson (Eds.), Parallel models of associative memory Hillsdale, NJ:Lawrence Erlbaum Associates.

Ginsberg, M. (1986). Counterfactuals. Artificial Intelligence, 30, 35-79.

Goddard, G. (1980). Components of the memory machine revisited: Hebb revisited. In P.W. Jusczyk and R.M. Klein (Eds.), The nature of thought: Essays in honor of D.O. Hebb. Hillsdale, NJ: Lawrence Erlbaum Associates.

Kahneman, D. and Miller, D.T. (1986). Norm theory: Comparing reality to its alternatives. Psychological Review, 93 (2), 136-153.

Kaplan, S. (1970). The time course of activity in a neural net. Unpublished paper. The University of Michigan.

Levenick, J. (1986). Knowledge representation and intelligent systems. Unpublished doctoral dissertation, University of Michigan, Computer Science Department.

Lewis, D. (1973). Counterfactuals. Cambridge, MA: Harvard University Press.

Pomerantz, J.R., Kaplan, S. and Kaplan, R. (1969). Satiation effects in the perception of single letters. Perception and Psychophysics, 6, 129-132.

Rosch, E. (1977). Principles of categorization. In E. Rosch and B. Lloyd (Eds.), Cognition and Categorization. Hillsdale, NJ: Lawrence Erlbaum Associates.

Rumelhart, D.E., Smolensky, P., McClelland, J.L., and Hinton, G.E. (1986). Schemata and sequential thought. In J.L. McClelland and D.E. Rumelhart, Parallel Distributed Processing. Cambridge, MA: MIT Press.

Woodworth, R. (1958). Dynamics of Behavior. New York: Henry Holt and Co.

Answering Why-Questions:

Test of a Psychological Model of Question Answering

Jonathan M. Golding
Memphis State University

Arthur C. Graesser
Memphis State University

Abstract

We conducted an experimental test of the Graesser and Clark (1985) model of question-answering for why-questions. This model specifies how individuals answer different types of questions by searching through various sources of information after comprehending a text. The sources of information include the passage structure and the generic knowledge structures which are associated with the content words in the query. After these knowledge structures are activated in working memory, search components narrow down a set of relevant answers. A subset of these components were tested in this experiment: (1) an arc search procedure specifying which nodes and arcs within an information source are sampled for answers to a why-question; (2) arc distance, the number of arcs in the representational network that connect the queried node to the answer node; and (3) the intersection of information between the passage structure and the generic knowledge structures associated with the query. After reading short stories, subjects were presented with questions and a number of theoretical answers to each question. Subjects were timed as they judged whether each answer was "Good" (appropriate and relevant) or "Bad" (inappropriate or irrelevant). Results supported the validity of the arc search procedure in that subjects robustly distinguished theoretically good answers from theoretically bad answers to specific questions.

This paper reports an empirical test of Graesser and Clark's (1985) model of question answering in the context of why-questions. This model is an extension of earlier models of question-answering (Graesser & Murachver, 1985; Graesser, Robertson, & Anderson, 1981; Lehnert, 1978), and is developed from a rich data base of question answering protocols and other relevant data (for details about methods, see Graesser & Clark, 1985). The model specifies (1) the major sources of information that are tapped for answers to questions and (2) the processing components that access nodes within the relevant information sources.

945

There are two major information sources for answers to questions: the passage structure and generic knowledge structures (GKSs). The passage structure includes explicit statements and knowledge-based inferences that are needed for establishing coherence between the explicit statements. The generic knowledge structures are either associated with explicit content words in a query or with higher level GKSs which are triggered by patterns of information (e.g, FAIRYTALE). For example, consider the question "Why did the dragon kidnap the maidens?" in the context of narrative passage N. Four information sources would be searched for answers: passage N, the GKS for DRAGON, the GKS for KIDNAP, and the GKS for MAIDEN. The four passages Graesser and Clark (1985) studied contained 25 explicit statements and 100 inferences, on the average. In addition, a typical passage activated 35 GKSs, with approximately 160 statement nodes in each GKS.

Each passage and each GKS were represented in the form of a conceptual graph structure. A structure contains proposition-like statement nodes that are interrelated by a network of categorized, directed arcs. Each statement node is assigned to one of five categories: Event, State, Goal, Action, and Style. Arc categories included: Reason, Outcome, Initiate, Manner, Consequence, Implies, Property, Set Membership, and Referential Pointer. Graesser and Clark (1985, Chapter 2) have defined the statement node and arc categories in detail. For the purpose of the present paper, it is important to note that the conceptual graph is structured according to explicit, quasi-formal constraints.

There are seven major components involved in accessing information sources and converging on a small set of answers relevant to a question. These include (1) working memory, (2) the activation of knowledge structures in working memory, (3) arc search procedures (specifying "legal" arc paths), (4) priorities among knowledge structures in working memory, (5) the intersection between/among knowledge structures, (6) distance between nodes (arc distance), and (7) constraints on knowledge structures. For example, when answering a why-question involving an action (e.g., "Why did the dragon kidnap the maidens?"), specific information would be activated in working memory (the passage structure, and the GKSs for DRAGON, KIDNAP, and MAIDEN). An entry node (E) in working memory would be found which matched the queried statement. An arc search procedure would examine paths of arcs radiating from node E to superordinate goals via forward Reason arcs and backward Manner arcs. Finally, convergence mechanisms (e.g., intersection between structures, priorities, constraints) would converge further on a set of relevant answers.

Testing Components of the Model

The present experiment focuses on three of the model's components for why-questions: arc search procedures, arc distance, and the intersection of information between the passage structure and the GKSs associated with the query. Sixteen subjects read two short stories analyzed by Graesser and Clark (1985). After reading each story, subjects were presented with four why-event questions and four why-action questions on a CRT. Following each question were a number of different answers to that question. Subjects were asked to decide (by pressing a specific key) whether each answer was "good" (appropriate and relevant) or "bad" (inappropriate or irrelevant), based on the story. In addition to recording good/bad judgments, decision time was measured in milliseconds. Each subject made judgments for 256 answers, altogether.

An answer to a specific question was designated as "theoretically good" if the arc search procedure generated the answer. For each good answer there was a legal path of arcs between the queried node and the answer node in the passage structure. Otherwise, an answer was designated as "theoretically bad." Arc distance was defined as the number of arcs (in the passage structure) between the queried node and the answer node. Each answer was also scaled on the number of intersecting GKSs. A GKS was scored as "intersecting" if it contained a node that matched the answer node. For example, there would be a match between the answer node "the daughter cried" and "the person cried" in the GKS for KIDNAP. Based on the Graesser and Clark model, it was predicted that: (1) subjects would distinguish theoretically good from bad answers, (2) subjects would respond more quickly to answers having a greater overlap between passage knowledge and GKSs associated with the query (see also Reder, 1982), and (3) would take longer to respond to answers which were more arcs away from the entry node (see also Anderson, 1983).

Analyses robustly confirmed the validity of the arc search procedures for why-action and why-event questions. Subjects correctly judged good theoretical answers as "good" ($M = .65$ for actions and $M = .62$ for events) more than they judged bad answers as "good" ($M = .20$ for actions and $M = .23$ for events). Multiple regression analyses were performed to assess the impact of several predictor variables on the subjects' likelihood of judging an answer as good. The important predictors included whether the answer was theoretically good or theoretically bad (GOOD/BAD), the number of GKSs with nodes that intersect the

947

answer node (INTERSECT), and the number of arcs between the entry node and the answer node (DISTANCE). A less important predictor was the likelihood that a particular answer presented in the experiment had been generated to its associated question by an independent group of subjects in Graesser and Clark's (1985) question answering task (GENERATE). In addition, each answer was scaled on other predictor variables which were not of direct concern in this study (e.g., number of words in the answer, differences between stories, whether the answer was a state, event, or goal). Among the predictors, there was no problem of co-linearity for either actions (range $r^2 = -.35$ to .34) or events (range $r^2 = -.25$ to .38), except for GENERATE and GOODBAD (action $r^2 = .59$ and event $r^2 = .50$).

The overall regression equations were significant for both answers to why-action questions, $F(7,125) = 18.50$, $p < .01$, $R^2 = .51$ and answers to why-event questions, $F(7, 135) = 12.35$, $p < .01$, $R^2 = .39$. As shown in Table 1, the beta weights indicate that GOOD/BAD was the only significant predictor variable for both of these question types. In fact, GOOD/BAD accounted for most of the predictable variance for both actions ($R^2 = .46$) and events ($R^2 = .34$). It is important to note that the GENERATE predictor was only marginally significant for the queried events. Thus, the theoretical arc search procedure (reflected in the GOOD/BAD variable) is a more robust predictor than simply the number of people who generate a specific answer to a specific question in a question-answering task.

Another set of multiple regression analyses were performed on decision latencies, using the same predictor variables. The overall regression equations significantly predicted latencies for actions, $F(7, 125) = 6.37$, $p < .01$, $R^2 = .26$, and for events, $F(7, 135) = 5.98$, $p < .01$, $R^2 = .24$. The beta-weights in Table 2 indicate that GOOD/BAD was again the only significant predictor. Decision latencies were significantly longer for good answers than bad answers.

The lack of effects for arc distance (DISTANCE) and intersecting GKSs (INTERSECT) led to additional multiple regression analyses. It was felt that the overall speed of a judgment might have determined whether the predictor had an effect on decision latencies. Perhaps the DISTANCE and INTERSECT variables are significant only when decision latencies are comparatively fast; the effect may be masked by other processes when the decision latencies are very long. The analyses were conducted on the four fastest times and four slowest times for each answer, segregating queried actions and queried events. Once again, however, GOOD/BAD was the only consistent significant predictor. Moreover,

948

accuracy was essentially equal for the fast and slow judgments. Regardless of the time taken to make the judgment, the arc search procedure yielded the same degree of accuracy.

Discussion

This experiment investigated three components of Graesser and Clark's (1985) model of question answering: arc search procedures, arc distance, and the intersection between the passage structure and GKSs associated with the query. There was clear support for the arc search component. For both events and actions, subjects robustly distinguished good theoretical answers from bad theoretical answers. Decision latencies were longer for the good theoretical answers than the bad theoretical answers. These results underscore the importance of specifying the legal paths of arcs and nodes while searching for answers during question answering.

The lack of significant effects for other predictor variables awaits further investigation. The lack of an arc distance effect, especially in the fast-slow analyses is perplexing. The use of cohesive texts may have masked this effect. Perhaps the effects of arc distance and passage-GKS intersection will emerge for expository text and narratives which do not fit a clear script or prototypical story format (e.g. Keenan, Baillet, & Brown, 1984).

The absence of an effect for passage-GKS intersection might be a result of the type of question being studied. Why-questions generally probe the explicit statements and comprehension-generated inferences in the passage structure (Graesser & Clark, 1985). Consequently, the GKSs may have had a minimal role in question answering procedures. Other types of questions might highlight the role of GKSs during question answering. For example, how-questions involve GKSs to a greater extent than the passage structure. Perhaps the passage-GKS intersection will be more pronounced when how-questions are analyzed in the context of our narrative passages.

References

Anderson, J.R. (1983). *The architecture of cognition*. Cambridge, MA: Harvard University Press.

Graesser, A. C., & Clark, L. F. (1985). *Structures and procedures of implicit knowledge*. Norwood, NJ: Ablex.

Graesser, A. C., & Murachver, T. (1985). Symbolic procedures of question answering. In A. C. Graesser and J. B. Black (Eds.), *The psychology of questions*. Hillsdale, NJ: Earlbaum.

Graesser, A. C., Robertson, S. P., & Anderson, P. A. (1981). Incorporating inferences in narrative representations: A study of how and why. *Cognitive Psychology*, *13*, 1-26.

Keenan, J. M., Baillet, S. D., & Brown, P. (1984). The effects of causal cohesion on comprehension and memory. *Journal of Verbal Learning and Verbal Behavior*, *23*, 115-126.

Lehnert, W. G. (1978). *The process of question answering.* Hillsdale, NJ: Earlbaum.

Reder, L. M. (1982). Plausibility judgments versus fact retrieval: Alternative strategies for sentence verification. *Psychological Review*, *89*, 250-280.

Table 1

Question Answering Data for Answers to Why-Questions

Regression Coefficients of Predictor Variables[1]:

Probability of Saying Good Answer

	Queried Action	Queried Event
GOOD/BAD	.61***	.48***
INTERSECT	.15**	.04
DISTANCE	-.02	.01
GENERATE	-.02	.17**

Regression Coefficients of Predictor Variables[1]:

Decision Latencies

	Queried Action	Queried Event
GOOD/BAD	.16*	.25***
INTERSECT	.03	.03
DISTANCE	-.04	.10
GENERATE	-.15	-.07

```
*     p < .10
**    p < .05
***   p < .01
```

[1]GOOD/BAD : whether an answer was generated by
             theoretical arc search procedure.
INTERSECT: number of GKSs with a node intersecting the
             answer node in the passage structure.
DISTANCE : number of arcs between queried node and
             answer node in passage structure.
GENERATE : number of subjects who would generate the
             specific answer to the question in
             question-answering protocols.

Observing Machinists' Planning Methods:
Using Goal Interactions to Guide Search

Caroline Hayes

Robotics Institute, Carnegie Mellon University

The following paper describes a model of expert planning behavior, and suggests strategies observed in machinists' behavior that might improve planning performance when applied to other domains. The domain is the design of manufacturing plans for machined parts. The expert machinist uses a planning method which novices do not use. The expert searches for interactions between the problem's goals, then uses the results of the search to guide the construction of a plan that avoids the interactions. Additionally, the expert knows how to divide the problem into two relatively independent subproblems. The subproblems are solved separately and the results merged. A portion of the model is implemented in a program called **Machinist**, *which has successfully created machining plans that were better then those of a machinist with 5 years experience.*

INTRODUCTION

This paper briefly discusses the planning methods observed in expert machinists, and compares behavior of apprentice machinists to that of a program, *Machinist* that implements these planning methods. Also discussed are a number of the methods observed in the machinist's planning behavior that could be applied to other planning domains to improve performance.

Machining is the art of producing metal parts using a variety of power tools to shape metal. It is a highly skilled task requiring 10 to 15 years to become accomplished. Expert machinists are an important resource for almost all manufacturing industries, but there are relatively few highly experienced ones.

The data were gathered from a series of twenty six verbal protocols over a year and a half. Two subjects with more than 15 years experience were studied. Protocols of the machinists behavior with a more detailed analysis are discussed in (Hayes, 1987a). The implementation of the program, a more extensive discussion of it's performance, and a synopsis of work that this research is based on are included in (Hayes, 1987b).

INTERACTIONS MAKE THE PROBLEM DIFFICULT

A major problem that machinists confront in planning is "interactions" between the different "features" that are cut into the part. *Features* are the individual geometric shapes that are cut into a block of metal. *Feature interactions* happen when cutting one collection of features affects the way in which others can be made. The difficulty in making a plan is to find an order in which none of the features interferes too seriously with producing the others.

Most commonly, feature interactions are caused by clamping problems; producing one feature destroys the clamping surfaces needed to grip the piece while cutting another feature. A feature interaction is shown in figure 1. The piece has two features on it: an angle and a slot. If the angle is made first it is difficult to clamp the piece so that the slot can be cut. The right hand vice jaw must press on the angled surface. This is an unstable situation because the part may be forced upward out of the vise. The angle can be said to *interact with* the slot. A simple reordering or the features can avoid this interaction.

THE MACHINIST'S PLANNING METHODS

Examples for this paper were taken from one protocol from a machinist planning how to make the part shown in figure 2.

The protocol showed that the human plans by first scanning the problem specification, (an engineering drawing and notes), and noting cues that indicate problems and feature interactions. Associated with each of these cues, the machinist has a set of restrictions that will help avoid that problem or interaction.

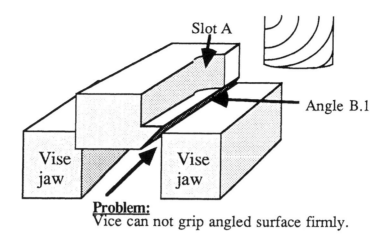

Slot A

Angle B.1

Vise jaw

Vise jaw

Problem:
Vice can not grip angled surface firmly.

Figure 1: *Feature Interaction*: The slot must be cut before the angle.

The features that drew the machinist's attention during the protocol were the large features: C, A, and D, and the large angles: E1, E2, B1 and B2. These features are of particular interest, because they cause the part to have an irregular shape and hence make clamping difficult. When clamping is difficult, the part can be clamped only in restricted ways to produce other features.

Next, the machinist investigated these attention attracting features in more detail. He explored the ways in which each feature restricted and affected others. During this stage he also grouped features into sets that could be made during the same clamping operation. All features in a group could be made with the same side facing up. Grouping makes it easier for the machinist to plan because entire groups are manipulated as single components; all restrictions on any feature in the group apply to the whole group. The strategy effectively reduces the complexity of the planning process. Even so, the machinist eventually found it difficult for keep track of all the restrictions on all the groups at the same time, so he drew a picture which is represented by the graph in figure 3. This is an *interaction graph*.

Whole groups are represented by a single letter in the graph. For instance "C" represents the group consisting of C, R, and the four holes; and "E" represents the group E1, E2, G1, and G2.

The order restrictions represented by this interaction graph are: either shoulder A or group C can go first in either order. Then the angles in group B (B1, B2) can be cut, followed by the E group, followed by D. The numbers beside the letters indicate which side must face up while that feature group is cut. (All the sides in figure 1 are given numbers from 1 - 6). B and E have no numbers because they are angles; the piece must be tilted when they are cut, so no side faces directly up.

Before any of these features can be cut several preliminary steps must be taken. Typically, three orthogonal sides (a group of sides that touch on one corner) must be machined smooth and square to each other before a feature can be cut. This is so that there will be precisely defined sides from which to measure feature positions. If the sides are not accurate, the feature position will not be accurate.

This process of smoothing the sides is known as *squaring.* There are strict rules for squaring that must be followed so that the minimum amount of material is wasted, and accuracy maintained. The squaring rules are dependent only on the characteristics of the starting material. They are independent of the the features to be cut so squaring rules are relatively invariant for all parts regardless of their final shape, while on the other hand, the feature interaction graph always changes from part to part, even when the final shapes are very similar.

The machinist drew a small diagram indicating the order in which he intended to square up the sides. He explicitly drew out only out the first three squaring steps (so he would have three orthogonal sides) and left the others to be decided later. Figure 4 shows the order that he decided on. Steps in the horizontal row (eg. 4, 5, 6) can be done in any order. The numbers indicate which side faces up.

Next, the squaring graph must be merged as efficiently as possible with the interaction graph. The more steps that can be overlapped the better because the final plan will be shorter. Typically. if the same side faces up in two steps, they can be merged, providing that none of the

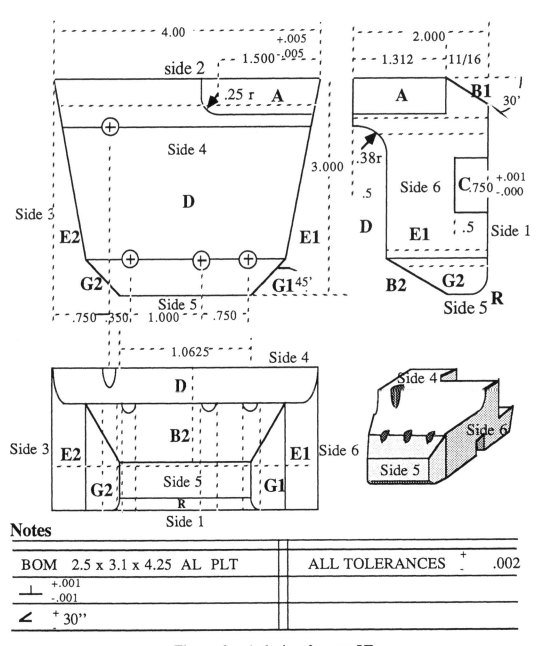

Notes

BOM 2.5 x 3.1 x 4.25 AL PLT	ALL TOLERANCES $\pm$ .002
$\perp$ +.001 / -.001	
$\angle$ $\pm$ 30"	

Figure 2: A design for part XI

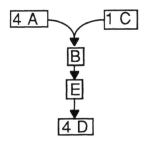

Figure 3: The *Feature Interactions* for PART XI

"before" and "after" constraints of either graph are violated. Figure 5 shows the two graphs merged.

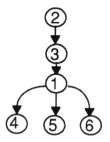

Figure 4: The *Squaring Graph* for Part XI

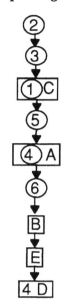

Figure 5: The combination of the *Interaction Graph* and the *Squaring Graph*

Lastly the plan is verified. The machinist looks over the plan and checks that there no interactions were overlooked, and that the part can be clamped properly at each step. If the plan does not pass this test he may have to go back to previous planning stages and replan.

The **Machinist** program, discussed in (Hayes, 1987b) implements these steps, except for the verification step and grouping. Forty nine of it's productions identify feature interactions and other problems. Grouping speeds up the planning process but does not generally affect the quality of the plan. Verification is needed only if interaction is missed, so as long as the program stays within the domain of parts containing interactions that it knows about, verification will be unnecessary.

EVALUATION OF THE PROGRAM AGAINST HUMAN PERFORMANCE

The program's performance was compared with that of four machinists at various experience levels: two second year apprentices, one third year apprentice, and one journeyman with 5 years experience. Each of these subjects was asked to create a machining plan for the same series of three parts. The specifications for the three parts used in the study as well as a few example plans generated by the apprentices, are described in (Hayes, 1987a).

Their resulting plans were judged by two very experienced machinists, each having more than 15 years experience. The program's average performance was better than that of the apprentices or the journeyman. The average scores earned by each machinist or program are shown in figure 6. In fact, Machinist 1 declared the program's plan for Part III to be "Almost the perfect plan. Who ever did this is a man after my own heart." As it turned out, in making the plan for Part III, the program used a heuristic taken from machinist 1. Consequently, the plan coincided with his idea of what was correct and was indeed "after (his) own heart." However, it wasn't made by a man, but a machine.

Performance of Apprentice Machinists and Program

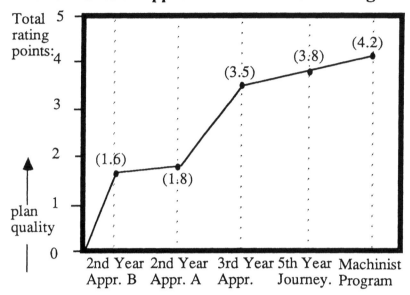

Figure 6: Average scores received by each subject

Judging was done in the following way: for each of the three parts there were five plans: one from each of the four young machinists, and one from the program. All information indicating who (or what) created the plan was removed. Independently, the two experienced machinists ordered each set of five plans from best to worst. The best plans were given a score of 5 and the worst 1.

There was fairly high agreement between the two machinists, despite a few anomalies. The Spearman rank correlation of their judgements is 0.92. Thus, the judgements are quite reliable.

One interesting outcome of this test was that it highlighted the difference between the expert machinists and the novices. The novices with less then two years experience showed no ability to spot feature interactions either before or after planning. Not only could they not foresee the problems, they seemed unable to detect that there was a mistake in the finished plan. It seemed that they lacked the perceptual skill that the expert had for identifying feature interactions. This ability seems to be *key* to the expert machinists' time efficiency and accuracy in planning.

Additionally, the novices with less than five years experience were unable to make efficient plans. It was difficult enough for them to make working plans at all. They used rote methods for many steps. It was not till about 5 years that any of them started to use more flexible methods that allowed more efficiency in the plans.

CONCLUSION

The planning steps for expert machinist are:

1. **Orientation:** Identifies the key features interactions and makes an estimate of the problem difficulty.

2. **Exploration of Feature Interactions:** Groups features and explores the ways in which they interact with others.

3. **Integration of the Feature Interactions with a Squaring Plan:** Merges feature interactions with squaring constraints.

4. **Elaborating and Verifying the Plan:** Checks to see that the plan is correct. If there is an error, he makes a fix or replans.

The **Machinist** program implements this method and produces plans that experts judged to be comparable or better than those of a 5 year journeyman. The plans are very similar in form to those that expert machinists produce. This study also showed that planning methods of experts were very different than those of novices: novices had neither the experts' ability to foresee problems nor their flexibility in planning.

The machinist's method of using patterns to spot problems first and plan around them has potential for improving planning efficiency in other expert domains. Other behaviors seen in machinists behavior that could be applied to other domains are: grouping, to cut down complexity of the plan space; and calculating the varying constraints (features interactions) separately from the invariant constraints (squaring) and latter merging them. This technique simplifies planning by dividing the problem nicely into two relatively independent sub-problems.

ACKNOWLEDGEMENTS

I want to thank Jim Dillinger and Dan McKeel for contributing their machining expertise to this project; Paul Wright, Jaime Carbonell, and Herb Simon, Dick Hayes for their contributions and advice. This research was funded by Cincinnati Milacron and Chrysler.

References

De Groot, A. D. *Thought and Choice in Chess*. The Hauge, Netherlands: Mouton & Co., 1965.

Ericsson, K. A., and H. A. Simon. Verbal Reports as Data. *Psychology Review*, May 1980, *87*(3), 215-251.

Fikes, R. E., P. E. Hart, and N. J. Nilsson. Learning and Executing Generalized Robot Plans. *Artificial Intelligence*, 1972(3), pp. 251-288.

Hammond, Kristian J. CHEF: A Model fo Case-Based Planning. *AAAI-86*, 1986, pp. 267-271.

Hayes, C. C. *Planning in the Machining Domain: Using Goal Interactions to Guide Search*. Master's thesis, Mellon College of Science, Carnegie Mellon University, April 1987.

Hayes, C. C. Using Goal Interactions to Guide Planning. *AAAI*, July 1987.

Hayes-Roth, B.; F. Hayes-Roth. A Cognitive Model of Planning. *Cognitive Science 3*, 1979, pp. 275-310.

Lagoude, Y., J. P. Taang. A Plan Represention Structure for Expert Planning Systems. *Computer-Aided/Intelligent Process Planning*, 1985, *19*, 19-30.

Newell, A, and H. A. Simon. *Human Problem Solving*. Englewood Cliffs, N. J.: Prentice-Hall, Inc., 1972.

Pitrat, Jacques. A Chess Combination Program Which Uses Plans. *Artificial Intelligence 8*, 1977, pp. 275-321.

Sacerdoti, E. D. Planning in a Hierarchy of Abstraction Spaces. *Artificial Intelligence*, 1974, *5*(2), 115-135.

Sussman, Gerald J. *A Computer Model of Skill Acquisition*. New York: American Elsevier Publishing Company, 1975. MIT AI Technical Report TR-297, August 1973.

Wilkins, David. Using Plans in Chess. *IJCAI*, 1979, pp. 960-967.

WHAT MAKES LANGUAGE FORMAL?

Eduard H. Hovy [1]

Information Sciences Institute
of the University of Southern California
4676 Admiralty Way
Marina del Rey, CA 90292-6695
Telephone: 213-822-1511
ARPA address: HOVY@VAXA.ISI.EDU

Topic Area: Natural Language Generation, Text Planning

Abstract

This paper addresses part of the question "how do we say the same thing in different ways in order to communicate non-literal, pragmatic information?". Since the style of the text can communicate much information — it may be stuffy, slangy, prissy — generators that seek to satisfy pragmatic, hearer-related goals in addition to simple informative ones must have rules that control how and when different styles are used. But what is "style"? In this paper, formal and informal language is analyzed to provide stylistic rules that enable a program to produce texts of various levels of formality.

1 Introduction

When we produce language, we tailor our text to the hearer and to the situation. This enables us to communicate more information than is contained in the literal meanings of our words; indeed, the additional information often has a stronger effect on the hearer than the literal content has. This information is carried by both the content and the form of the text. As speakers and hearers, we associate various interpretations of the speaker, his goals, the hearer, and the conversational circumstances, with the various ways of expressing a single underlying fact or idea.

The level of formality of text is one of the strongest carriers of additional information. This level reflects the level of formality of the conversational setting (for instance, a burial or a party) and of the interpersonal distance between the interlocutors. But what does it mean for language to "seem relaxed" or to "be formal"? No single item in the language defines the level of formality; rather, text seems to contain a number of little clues that cumulatively create a certain impression. What are these little clues? Where do they appear in language and how do we decide to use them?

To answer this question, handbooks of writing are of little use: typically, they describe styles in terms of the characteristics of complete paragraphs of text (see, say, [Birk & Birk 65] and [Hill 1892]), which is not useful for a practical, generator-oriented approach. Instead, a functional approach is to describe styles in terms of the decisions a generator has to make: decisions such as sentence content, clause order and content, and word selection.

---

[1] This paper was written while the author was at Yale University Computer Science Department, 2158 Yale Station, New Haven, CT 06520-2158, U.S.A. This work was supported in part by the Advanced Research Projects Agency monitored by the Office of Naval Research under contract N00014-82-K-0149. The work was also supported by AFOSR contract F49620-87-C-0005.

2 Formality

The level of textual formality is probably the pre-eminent stylistic aspect; it comes into play along the whole range of generator decisions (from the initial sentence topic selection and organization down to the final word selection). All language users have rules for making their text more or less formal. The best way to illustrate these rules is to dissect a piece of text:

> *Yesterday, December 7, 1941 — a date which will live in infamy — the United States of America was suddenly and deliberately attacked by naval and air forces of the Empire of Japan.*
>
> *The United States was at peace with that nation and, at the solicitation of Japan, was still in conversation with its Government and its Emperor looking forward to the maintenance of peace in the Pacific.*
>
> *Indeed, one hour after Japanese air squadrons had commenced bombing Oahu, the Japanese Ambassador to the United States and his colleague delivered to the Secretary of State a formal reply to a recent American message. While this reply stated that it seemed useless to continue the existing diplomatic negotiations, it contained no threat or hint of war or armed attack.*
>
> *It will be recorded that the distance of Hawaii from Japan makes it obvious that the attack was deliberately planned many days or even weeks ago. During the intervening time, the Japanese Government has deliberately sought to deceive the United States by false statements and expressions of hope for continued peace.*
>
> ["We Will Gain the Inevitable Triumph — So Help Us God", war address by F.D. Roosevelt to joint session of Congress of the United States, December 8, 1941.]

What characteristics make this address formal? Certainly, one factor is the use of formal verbs and nouns instead of more common ones, such as "solicitation" instead of "request". Another factor is the use of full names and titles instead of their common abbreviations. Accordingly, we replace words and phrases in the address by less formal equivalents ((a) below) and use the common names for entities (b).

The result, however, is definitely not informal. The sentences still seem long and involved. In order to simplify them, we (c) remove conjunctions and multi-predicate phrases, and (d) remove adverbial clauses, or place them toward the ends of sentences. Now, however, the text seems odd; for example, phrases such as "it will be recorded" do not blend with phrases such as "deliberately tried to cheat" (introduced by (a)). To improve this, we (e) eliminate the use of passive voice, and (f) refer to the involved parties — speaker, hearer, and others — directly.

Now some phrases sound flowery and out of place. To simplify, some nominalized verbs can be converted to verbs (g); noun groups can be simplified by dropping redundant adjectives and nouns (h); pronominalization can be increased (i). Finally, a few finishing touches: simplified tenses (j); colloquial phrases (k); complete elision of redundant words where grammatical (l):

> <u>We were</u> $_{(f)}$ *suddenly and deliberately attacked by naval and air forces of* <u>Japan</u> $_{(b)}$ *yesterday, December 7, 1941* $_{(d)}$. *We'll never forget this date* $_{(c,d,k,l)}$.
>
> <u>We were</u> $_{(f)}$ *at peace with* <u>them</u> $_{(i)}$. *[and,]* $_{(c)}$ *At* <u>Japan's request</u> $_{(a,h)}$ <u>we were</u> $_{(f)}$ *still* <u>talking to</u> $_{(a)}$ <u>their</u> $_{(c)}$ *Government.* *[and its Emperor.]* $_{(h)}$ <u>We were</u> $_{(f)}$ *looking forward to* <u>having</u> $_{(a,g)}$ *peace in the Pacific.*

[Indeed,](l) *One hour after Japanese air squadrons [had]*(j) *started*(a) *bombing Oahu, their Ambassador*(a) *[and his colleague]*(l) *gave*(a) *our*(f) *Secretary of State a formal reply to a recent message. [While]*(c) *[This reply said*(a) *that]*(l) *They*(f,i) *thought it was*(e) *useless to continue negotiating*(g). *[there was*(a)*]*(c) *But they*(i) *didn't*(k) *[threaten or]*(h) *talk about*(a) *war. [or armed attack.]*(h)

[Note(e,f) *that]*(l) *The distance of Hawaii from Japan makes it obvious that they*(f) *deliberately planned*(e) *the attack a while*(k) *[or even weeks]*(l) *ago. [In*(a) *the intervening time,]*(d) *The Japanese Government [has]*(j) *deliberately tried*(a) *to cheat*(a) *us*(f) *by [false statements and]*(h) *pretending*(a) *[expressions of hope for continued]*(l) *to hope for peace in the mean time*(k).

3 Rules for Creating Formal Text

A number of texts, ranging from politicians' speeches and writings to discussions with friends, were analyzed in the manner above. The transformation steps were stated as rules that provide criteria by which PAULINE[2] makes appropriate choices at decision points. One of the program's rhetorical goals, the goal controlling formality, takes one of the values *highfalutin, normal, colloquial*. In order to make text more formal, the program examines its options at decision points and applies the strategies paraphrased here:

- **topic inclusion:** to make long sentences, select options that contain causal, temporal, or other relations to other sentence topics

- **topic organization:** to make complex sentences, select options that are subordinated in relative clauses; that conjoin two or more sentence topics; that are juxtaposed into relations and multi-predicate enhancer and mitigator phrases

- **sentence organization:** make sentence seem weighty by including many adverbial clauses; by placing these clauses toward the beginnings of sentences; by building parallel clauses within sentences; by using passive voice; by using more "complex" tenses such as the perfect tenses; by avoiding ellipsis, even though it may be grammatical (such as "Joe got more than Pete [did]", "When [I was] 20 years old, I got married")

- **clause organization:** make weighty, formal clauses, by including many adjectives and adjectival clauses in noun groups; by doubling nouns in noun groups ("Government and Emperor", "statements and expressions"); by including many adverbs and stress words in predicates; by using long, formal phrases; by nominalizing verbs and adverbs ("their flight circled the tree" instead of "they flew round the tree"); by pronominalizing where possible; by not referring directly to the interlocutors or the setting

- **phrase/word choice:** select formal phrases and words; avoid doubtful grammar, slang, and contractions (say "man" rather than "guy" and "cannot" rather than "can't")

In contrast, by following inverted strategies, PAULINE makes its text less formal.

---

[2]PAULINE (Planning And Uttering Language In Natural Environments) is a generator program that produces various texts from a single story representation under various settings that model pragmatic circumstances. PAULINE consists of over 12,000 lines of T, a Scheme-like dialect of LISP developed at Yale University. It generates over 100 variations of a description of an episode that occurred at Yale in April 1986 (see below and [Hovy 87a, 87b]), as well as different versions of texts in two other domains (see [Hovy 86b], [Bain 86], and [Hovy 86a]).

4 Determining Appropriate Levels of Formality

Knowing how to make formal text is not enough. The generator must also know when it is appropriate. Since the level of formality is not actually measurable, it is most apparent only when the level is suddenly changed or is inappropriate. In order to determine the pragmatic effects of formality, then, the important question is: *what does the speaker achieve by altering the level of formality?*

First, if you become less formal, you signal a perceived or desired decrease in the interpersonal distance between yourself and the hearer. In any relationship, the participants maintain a certain distance (say, from intimate to aloof) which is mirrored by a corresponding level of formality. Which interpersonal distance corresponds to which level of formality depends, of course, on social convention and on the interlocutors and their relationship; for example, colloquial or informal language is often used to discuss relatively intimate topics, and more formal language often indicates that you feel, or wish to feel, more distant than the conversation had been implying (perhaps after you are offended or become uncomfortable with the topic). See [Brown & Levinson 78] on the use of formal honorifics and [Kuno 73] and [Harada 76] on Japanese deictic honorifics.

Second, if you alter the level of textual formality, you may perturb the tone or atmosphere of the conversation. Since the conversational atmosphere is also mirrored by textual formality, a serious conversation (a burial speech or a conference talk) requires more formality than an everyday conversation (a report to the family of the day's events). An inappropriate level of formality can affect the hearer's emotion toward you: if you are too informal, you may seem cheeky or irreverent; if you are too distant, you may seem snooty or cold. A large amount of work by sociologists, anthropologists, and psycholinguists describes the characteristics of various settings and the appropriate levels of formality in various cultures (see, for example, [Irvine 79] and [Atkinson 82] on formal events; [Goody 78] and [R. Lakoff 77] on politeness).

Based on these considerations, after PAULINE is given values for the parameters that characterize the conversational setting, the speaker, and the hearer (in boldface), it uses the following rules to activate its rhetorical goal of formality:

1. set the rhetorical goal of formality to

   - *colloquial* when the **depth of acquaintance** is marked *friends*, or when the **relative social status** is marked *equals* in an **atmosphere (tone)** marked *informal*
   - *normal* when the **depth of acquaintance** is marked *acquaintances*
   - *highfalutin* when the **depth of acquaintance** is marked *strangers*

2. then, reset the goal value one step toward *colloquial* if **desired effect on interpersonal distance** is marked *close* or if **tone** is marked *informal*

3. or reset the goal value one step toward *highfalutin* if **desired effect on interpersonal distance** is marked *distant* or if **tone** is marked *formal*

4. and invert the value if **desired effect on hearer's emotion toward speaker** is marked *dislike* or if **desired effect on hearer's emotional state** is marked *angry*

5 The Rules at Work

PAULINE uses these rules to produce the following two texts when it is being *highfalutin* (say, writing for a newspaper) and *colloquial* (say, talking to a friend). (This episode is represented in a property-inheritance network such as described in [Charniak, Riesbeck & McDermott 80] using elements based on Conceptual Dependency (see [Schank 82] and [Schank & Abelson 77]). Approximately 130 representation elements denote the events, actors, locations, props, and their relationships (temporal, intergoal, causal, etc.).)

HIGHFALUTIN	COLLOQUIAL	Decision Type
[IN EARLY APRIL],	[ ] STUDENTS [PUT]	clause position verb formality
A SHANTYTOWN -- NAMED WINNIE MANDELA CITY -- [WAS [ERECTED] BY] [SEVERAL] STUDENTS ON BEINECKE PLAZA,	A SHANTYTOWN, [ ] WINNIE MANDELA CITY, UP [ON BEINECKE PLAZA] [IN EARLY APRIL].	ellipsis mode, verb formality adjective inclusion clause position
[SO THAT]		conjunction
YALE UNIVERSITY WOULD [DIVEST FROM] COMPANIES DOING BUSINESS IN SOUTH AFRICA.	THE STUDENTS WANTED YALE UNIVERSITY TO [PULL THEIR MONEY OUT OF] COMPANIES DOING BUSINESS IN SOUTH AFRICA.	verb formality
[LATER, AT 5:30 AM ON APRIL 14], THE SHANTYTOWN [WAS DESTROYED] BY OFFICIALS; [ALSO, AT THAT TIME,] THE POLICE ARRESTED 76 STUDENTS.	[ ] OFFICIALS [TORE [IT] DOWN] AT 5:30 AM ON APRIL 14, [AND] THE POLICE ARRESTED 76 STUDENTS.	clause position mode, verb formality conjunction
SEVERAL LOCAL POLITICIANS AND FACULTY MEMBERS [EXPRESSED CRITICISM] OF [YALE'S] ACTION.	SEVERAL LOCAL POLITICIANS AND FACULTY MEMBERS [CRITICIZED] THE [ ] ACTION.	verb formality adjective inclusion
[FINALLY], YALE [GAVE] THE STUDENTS [PERMISSION] TO [REASSEMBLE] THE SHANTYTOWN THERE [AND, CONCURRENTLY], THE UNIVERSITY [ANNOUNCED] THAT A COMMISSION WOULD GO TO SOUTH AFRICA IN JULY TO [INVESTIGATE] THE SYSTEM OF APARTHEID.	[LATER,] YALE [ALLOWED] THE STUDENTS TO [PUT [IT] UP] THERE [AGAIN]. [ ] THE UNIVERSITY [SAID] THAT A COMMISSION WOULD GO TO SOUTH AFRICA IN JULY TO [STUDY] THE SYSTEM OF APARTHEID.	word formality verb formality verb formality conjunction verb formality verb formality

6 References

1. Atkinson, J.M., *Understanding Formality: the Categorization and Production of 'Formal' Interaction*, in *British Journal of Sociology*, vol 33 no 1, 1982.

2. Bain, W.M., *A Case-based Reasoning System for Subjective Assessment*, AAAI Proceedings, 1986.

3. Birk, N.P. & Birk, G.B., **Understanding and Using English**, fourth edition, Odyssey Press, New York, 1965.

4. Brown, P. & Levinson, S.C., *Universals in Language Usage: Politeness Phenomena*, in **Questions and Politeness: Strategies in Social Interaction**, Goody E. (ed), Cambridge University Press, Cambridge, 1978.

5. Charniak, E., Riesbeck, C.K. & McDermott, D.V., **Artificial Intelligence Programming**, Lawrence Erlbaum Associates, Hillsdale, 1980.

6. Goody, E. (ed), **Questions and Politeness: Strategies in Social Interaction**, Cambridge University Press, Cambridge 1978.

7. Harada, S.I., *Honorifics*, in **Syntax and Semantics 5: Japanese Generative Grammar**, Shibátani M. (ed), Academic Press, New York, 1976.

8. Hill, A.S., **The Foundations of Rhetoric**, Harper & Brothers, New York, 1892.

9. Hovy, E.H., 1986a, *Some Pragmatic Decision Criteria in Generation*, in **Natural Language Generation: Recent Advances in Artificial Intelligence, Psychology, and Linguistics**, Kempen, G. (ed), Kluwer Academic Publishers, Dordrecht, Boston, 1987.

10. Hovy, E.H., 1986b, *Putting Affect into Text*, Proceedings of the Eighth Conference of the Cognitive Science Society, 1986.

11. Hovy, E.H., 1987a, *Generating Natural Language under Pragmatic Constraints*, in *Journal of Pragmatics*, vol XI no 6, 1987, forthcoming.

12. Hovy, E.H., 1987b, *Interpretation in Generation*, Proceedings of the AAAI conference, 1987.

13. Irvine, J.T., *Formality and Informality of Speech Events*, in *American Anthropologist*, vol 81 no 4, 1979.

14. Kuno, S., **The Structure of the Japanese Language**, Harvard University Press, 1973.

15. Lakoff, R., *Politeness, Pragmatics, and Performatives*, in **Proceedings of the Texas Conference on Performatives, Presuppositions, and Implicatures**, 1977.

16. Schank, R.C., **Dynamic Memory: A Theory of Reminding and Learning in Computers and People**, Cambridge University Press, Cambridge, 1982.

17. Schank, R.C. & Abelson, R.P., **Scripts, Plans, Goals and Understanding**, Lawrence Erlbaum Associates, Hillsdale, 1977.

Grammatical priming of nouns in connected speech

Jola Jakimik
and
Julie Scott

Department of Psychology, University of Wisconsin-Madison

Abstract

On-line processing of inflected spoken words was examined using phoneme-monitoring RT to following targets. Plural and singular nouns followed contexts that required plurals (e.g., A dozen bagels/bagel tumbled ...) or were neutral (e.g., The frozen bagels/bagel tumbled ...). Relative to the neutral contexts, recognition of congruent plural nouns was facilitated, and recognition of incongruent singular nouns was disrupted.

Introduction

The research reported in this paper investigates how inflected words are recognized and understood in spoken sentences. This line of investigation aims to make up for the relative neglect of two related topics in the study of spoken word recognition, the recognition of inflected and derived forms, and the processing of grammatical (syntactic) structure.

Recent research on spoken language processing has focussed on the contribution of preceding context, a focus which has proved to be a fruitful strategy, both theoretically and methodologically. Borrowing from this tradition, the present study examines the contribution of preceding context to the recognition of plural nouns.

The present research uses the on-line task of phoneme-monitoring in the same way that Blank and Foss (1978) used it to examine semantic constraints. They varied the constraints on a critical noun by preceding it with a related verb, adjective, or both. They compared sentences containing semantically related words with control sentences containing no related words. For example, recognition of the word "eye" was measured in the context "The drunk winked his bloodshot ..." and in the context "The drunk concealed his aching ...." Blank and Foss found faster recognition of the critical words when they were preceding by semantically related verbs or adjectives.

Blank and Foss measured recognition of the critical words indirectly, by measuring time to detect phoneme-targets at the beginning of the following words. In the sentences above, the target was /p/ in the word "probably." The reasoning behind this on-line measure is that a response to a word-initial target depends on recognizing that the target sound begins a word, which in turn implies that the end of the previous word has been recognized. Faster recognition of the preceding (critical) word would result in faster detection of the word-initial target phoneme. Delayed recognition of the critical word would delay detection of the target.

The present study examines a local and specific grammatical constraint: the dependency between the initial part of a noun phrase, and the form of the noun. There are some common grammatical constructions in English that require the plural (inflected) form of a noun, and others that do not incorporate this constraint. We refer to this source of constraint as grammatical to emphasize that the constraint is on the form of the word, rather than on its semantic content. Whether it is a purely syntactic constraint is a problem which we will ignore for the present.

There are two possible ways of showing an effect of context on the processing of inflected words. One is to show that the violation of structural constraints causes disruption or slowing of recognition, relative to a condition which obeys the constraints. A second is to show that the availability of additional constraints speeds or facilitates recognition of the inflected forms. The present study looks for both disruptive and facilitative effects. It would not be surprising to find that grammatically incongruous words produced some disruption of sentence processing. It would be more unusual, and therefore interesting, to find that a more constraining grammatical context leads to faster recognition than a less constraining, but nonetheless appropriate context. To our knowledge, no one has demonstrated such a grammatical context effect.

In the present experiment, recognition of the nouns was assessed by measuring phoneme-monitoring RT to targets at the beginning of the words immediately after the critical nouns, as in Blank and Foss' (1978) study. This indirect measure was chosen because this paradigm is most likely to reveal effects on processing of the end of the critical word.

There were four conditions in the present study, which resulted from the combination of two contexts and two forms of the critical noun, plural and singular. In one context, the first part of the critical noun phrase contained a quantifier, number or other word that requires the following noun to be plural; for example, "many," "three," and "various." This context is called the Predicts-plural context. In the second context, the introductory words were ones that could be followed by either a plural or singular noun; for example, "the," "his." This context is referred to as the Neutral context. In each context, the critical noun was either plural or singular. Four versions of one sentence from the experiment are shown below, with the target phoneme underlined.

Neutral context, plural noun:
The frozen bagels tumbled out of the bag when she dropped it.

Predicts-plural context, plural noun:
A dozen bagels tumbled out of the bag when she dropped it.

Neutral context, singular noun:
The frozen bagel tumbled out of the bag when she dropped it.

Predicts-plural context, singular noun:
A dozen bagel tumbled out of the bag when she dropped it.

For plural nouns, a comparison between the Predicts-plural context and the Neutral context asks whether there is facilitation due to greater constraint on the inflected form. For singular nouns, a comparison between the Neutral context and the Predicts-plural context, where a singular noun

966

is inappropriate, reveals whether a violation of this grammatical constraint disrupts on-line processing of the sentence. In this set of comparisons, phoneme-monitoring RTs in the same phonetic context are compared.

Method

Materials

There were 72 critical sentences. Each of the regular plural forms of English was represented. In one-third of the sentences, the critical noun took /s/ in its plural form; in another third, the plural ending was /z/; and in another third it was /Iz/. Within each third, there were equal numbers of one, two, and three syllable nouns. The target phonemes were the six stop (plosive) consonants: /b/, /d/, /g/, /p/, /t/, and /k/.

The critical sentences were randomly assigned to fixed serial positions on a list. Four lists were prepared for recording. Only one version of each sentence occurred on a list, so that each subject would hear only one version of each sentence. Each list contained an equal number of the four versions, as well as equal numbers of the three types of plurals, and of one, two, and three syllable nouns.

In addition to the critical sentences, there were filler sentences of two types. Some contained target phonemes, placed on nouns and adjectives for variety. Other fillers had no targets. There were twenty-eight fillers, for a total of 100 sentences per list. The sentences were divided into 5 blocks of 20 sentences. Each list began with four fillers, and each block began with a filler.

A female speaker recorded the sentences on one channel of a tape. Each sentence trial consisted of the word "Ready" followed by a specification of the target phoneme for that sentence, for example "/b/ as in Bob," then the sentence. Clicks were placed on the second channel, coincident with the stop bursts of the targets. The clicks, which were inaudible to subjects, started the recording of RT by an Apple microcomputer. After 2500 msec, the trial was terminated.

Procedure

Subjects were tested individually. They were given four practice sentences. Subjects were warned to pay close attention to the meaning of the sentences, since they would receive a comprehension test afterwards. The test was a recognition memory test for the fillers.

Subjects

Subjects were students at the University of Wisconsin-Madison. They participated for pay, or for extra course credit. Fifty six subjects were tested.

Results

Only data from subjects who scored 75% or better on the comprehension test were included in the analysis of phoneme-monitoring RTs. This criterion exluded 4 subjects, leaving 13 subjects who heard each tape.

For each subject, a mean RT for each of the four conditions was calculated. For each version of the critical sentences, a mean reaction time was calculated. The subject and item means were analysed in two separate analyses of variance, whose factors were the nature of the preceding context, and the nature of the noun preceding the phoneme target. The table below presents the overall means for the four conditions, averaged over the 72 sentences.

Table 1

Mean Phoneme-monitoring RTs (in msec) to targets after critical nouns.

	Preceding Context	
	Predicts Plural	Neutral
Critical Noun		
Plural	514	558
Singular	552	534

The first row in the table shows that the monitoring time to targets following plural nouns is faster in contexts that predict plurals than in contexts that allow singular and plural nouns. This result indicates that recognition of the inflected form is facilitated in a constraining context, relative to a less constraining context. The second row of the table shows that average monitoring time after singular nouns is slower in contexts that require plural nouns than in neutral contexts. This results indicates that recognition of the nouns is delayed when they violate grammatical constraints.

The main result of the analyses of variance was a significant interaction between the two factors: F_1 (1,51) = 15.31, p < .001; F_2 (1,72) = 17.77, p < .001; min F' (1,115) = 8.22, p < .01. Separate one-way analyses of variance were performed to compare the Neutral contexts and the Predicts-plural contexts. When the critical nouns were plural, the subjects analysis showed a marginally significant effect (F_1 (1,51) = 3.48, p < .07) and the items analysis showed a significant effect[1] (F_2 (1,71) = 12.93, p < .001. When the critical nouns were singular, the subjects analysis showed a significant effect (F_1 (1,51) = 11.76, p < .001) and the items analysis showed a marginal effect (F_2 (1,71) = 3.18, p < .08.

Discussion

The results of the present experiment show two effects of grammatical context on the on-line processing of spoken words, one harmful, the other beneficial. When the preceding context leads the listener to expect the plural form of a noun, and the expectation is violated, processing at that point in the sentence is disrupted, compared either to processing of the singular noun in a neutral context, or to processing of the appropriate plural form in the same constraining context. The latter comparison is between a context-word combination that violates a set of constraints, and a combination that satisfies the same set of constraints. It is analogous to the comparison made in Gurjanov, Lukatela, Lukatela, Savic, and Turvey's (1985) study of inflected forms. Gurjanov et al. compared grammatically correct pairings of possessive adjectives and nouns with incongruous pairings, and found slower recognition of the nouns in incongruous contexts.

In the present study the disruptive effect of violating grammatical constraints can also be seen relative to a neutral context, a comparison that Gurjanov et al., as well as Goodman, McClelland, and Gibbs (1981), were unable to make. This effect might have been stronger except for a problem with the experiment. Whenever there was a grammatical violation, a phoneme target followed. Subjects may have been able to anticipate the occurrence of a target in this condition, counteracting the disruptive effect somewhat.

The second effect of grammatical context was facilitated recognition of inflected words in contexts that predicted plural forms. This conclusion is possible because of the inclusion of an appropriate neutral context. If the contexts are not really neutral, then this novel finding is discredited. An alternative account of the difference between Predicts-plural and Neutral contexts assumes that the neutral contexts actually favor singular noun continuations. Thus, when a plural form follows, it is unexpected. On this account, recognition of plural nouns is slower in the Neutral contexts than in the Predicts-plural contexts, in which no expectation are violated.

The resolution of this problem remains for another experiment. If the neutral contexts are truly neutral, then it should be possible to compare them to different constraining contexts, and also find facilitation. Just as there are demonstratives and quantifiers that require plurals, there are determiners ("a", "an") and demonstratives ("this", "that") that require singulars. An experiment testing this prediction is in progress.

The present experiment provides an interesting demonstration of the effects of grammatical context on word recognition. Its broader implications concern the purpose of various linguistic devices and properties. Agreement phenomena, such as agreement in number and gender between an adjective and noun, are common, and probably serve some purpose other than to frustrate foreign language students. Agreement provides redundancy, and so aids in successful communication. The present results suggest that such grammatical constraints also confer a processing advantage. Ironically, such a processing advantage cannot be measured in a language with full agreement, because there is no neutral baseline.

References

Blank, M. A., & Foss, D. J. (1978). Semantic facilitation and lexical access during sentence processing. Memory & Cognition, 6, 644-652.

Goodman, G. O., McClelland, J. L., & Gibbs, R. W. (1981). The role of syntactic context in word recognition. Memory & Cognition, 9, 580-586.

Gurjanov, M., Lukatela, G., Lukatela, K., Savic, M., & Turvey, M. T. (1985). Grammatical priming of inflected nouns by the gender of possessive adjectives. Journal of Experimental Psychology: Learning, Memory, and Cognition, 11, 692-701.

M.J. GONZALEZ LABRA
Departamento de Psicología Básica
Universidad Nacional de Educación a Distancia
P. Box 50.487
Madrid, Spain
Telephone 4425907

Keywords: inductive/analogical reasoning, problem-solving,
information processing models, intelligence.

Proposal for poster session

ANALOGICAL   REASONING: "Infer-Infer-Compare"   vs
"Infer-Apply-Test" information processing models.

Geometric analogy solution was studied as a function of
systematic variations in the tasks variables used in the
generation and manipulation of item difficulty. Also, the
assumptions on which the task design for componential
analysis are based were tested. The different componential
trials varied in the amount of information presented so as
to include redundant or non redundat terms in solution
trials.

The   experimental   analogies   were   systematically
generated in terms of the number of transformations, types
of dimensional changes, true-false counterparts and zero-two
precueing trials/four terms (redundant)-two terms (non
redundant) solution trials. Analogies with two and three
transformations presented solution times that were closer to
the dimensional change which showed the highest difficulty
level in problems with one transformation. This complexity
graduation did not always correpond to former predictions
formulated in function of the number of transformations.
Attribute salience may probably influence the ease with
which certain dimensions are ⁓encoded and later processed.
These results emphasize the importance of taking in
consideration the information which has to be executed by
the different operations included in processing models of
geometric analogies. Analysis of solution trials with
redundant and non redundant terms suggest that additional
processing mechanisms should be considered if all analogical
terms are included in the design of these second trials. The
results found in the non redundant condition corroborated
Evans' predictions on the processing durations of the
operations executed on the second pair of analogical terms.

Future research should futher explore the circumstances
under which information reprsentation might improve
processing ability in addition to the dimensional salience
effects of tasks variables.

M.J. GONZALEZ LABRA
Departamento de Psicología Básica
Universidad Nacional de Educación a Distancia
P. Box 50.487
Madrid, Spain
Telephone 4425907

ANALOGICAL REASONING: "Infer-Infer-Compare" vs "Infer-Apply-Test" information processing models.

Despite the extensive use of analogy items on intelligence testing and the recent research done on the nature of the processing involved in analogical performance, the principle theoretical and empircal analysis of the information processes necessary for solving these type of problems converge on two main models. These processing models have been described as "infer-infer-compare" and "infer-apply-test" and were postulated by Evans (1968) and Sternberg (1977), respectively. Both models relate process execution and overall performance to the amount of information that must be processed, but the operations included in the procedure for the second part of the problem are substantially different.

Bethell-Fox, Lohman and Snow (1984) suggested that these models could represent different processing strategies: constructive matching for the infer-apply-test model and response elimination for the infer- infer-compare model. Although it seems plausible that these strategies might be used in the solution processing of analogical problems, the identification of Evans 'model with a response elimination strategy is rather questionable. The response elimination notion is Evans' artificial intelligence program is executed when the response alternatives present notorious differences in relation to the problem stem terms, and after the inferential relations have been processed.

The objectives outlined in the present study seek to futher explore the different predictions formulated on the time consuming durations associated with the two alternative processes. In order to analyze the processing durations employed in the executions of each pair of analogical terms, the experimental task was designed following Sternberg's componential method. This method divides the analogical problem in two presentations, each requiring successively less processing demands. Test trials are divided in two parts, with precueing in the first presentation and solution in the second. Precueing conditions may include zero, one, two or three terms and their solution counterparts present the previous information plus the remaining terms necessary for solution. It is assumed that precueing information will be used during this trial and will not be reexecuted during solution trials. Solution tests trials varied the amount of information presented in order to test the former assumption.

Also item information structure research on analogical problems indicate that only two mayor variables -number of elements and transformations- are enough in the prediction of performance data. Recent research on the effects of tasks variables used in the manipulation of item difficulty showed that the different types of transformations should also be considered as an additional source of problem complexity (Whitely Schneider, 1981; Bethell-Fox, Lohman Snow, 1984). Therefore the effects on solution times and error rates of three common types of dimensional changes (rotation, size and color) and their combination in problems with two and three transformations were analyzed.

The analogical problems were generated from seven types of transformations in a true-false presentation format and the geometric figures that constitued the terms were counterbalanced in each type of analogy. Each problema was divided in two test trials: precuening presentations included zero or two terms and solution trials contained all four terms or the last pair . Presentation rate and solution time for each problem was controlled by the subjects on a three field tachistoscope. The sample was constituted by 60 psychology students from the Complutense University of Madrid.

The analysis of solution trials with redundant information (four terms) showed significant longer solution times than the precueing trials. These results could suggest an additional reinspection operation of the previous presented information and its duration would be confounded with the inference/or application processing times. But when solution trials were analyzed in the non redundant condition (second pair of terms) a non significant difference was found between test trials. These results favor Evans'predictions and indicate a significant effect in relation to the type of componential design used in solution trials. This effect should be considered in future interpretations of component durations under the assumption of the differential processing demands required in solution trials.

When the data was analyzed taking in consideration only the number of transformations, it was found that increasing the values of this variable systematically increased time solution and error rates. However, if solution time and errors were analyzed in terms of the different types of transformations the pattern of effects were quite different. Problems with one transformation presented rotation as the highest difficulty dimensional change, followed by changes in size and color. When problems combined two or three transformations the solutime times were closer to the dimensional change that independently presented the highest difficulty level. This finding suggests that averaging solution times in function of the number of transformations is probably canceling the significant effects of the dimensional properties of the terms. This could also imply that the variable number of transformations might be considered as a criterion to facilitate item difficulty manipulation provided that all types of dimensional changes are counterbalanced. However, it is not admissible to infer that the significant differences found in solution times and errors may be explained by the number of times each process has to be executed. The salience of the attributes may probably influence the ease with which certain dimensions are encoded and later processed. Future research in this area should explore the circumstances under which information representation might improve processing ability.

Integrated Learning of Words and their Underlying Concepts[*]

Raymond J. Mooney

Coordinated Science Laboratory
University of Illinois at Urbana-Champaign

Abstract

Models of learning word meanings have generally assumed prior knowledge of the concepts to which the words refer. However, novel natural language text or discourse can often present both unknown concepts and words which refer to these concepts. Also, developmental data suggests that the learning of words and their concepts frequently occurs concurrently instead of concept learning proceeding word learning. This paper presents an integrated computational model for acquiring both word meanings and their underlying concepts concurrently. This model is implemented as a word learning component added to the GENESIS explanation-based schema acquisition system for narrative understanding. A detailed example is described in which GENESIS learns provisional definitions for the words "kidnap", "kidnapper", and "ransom" as well as a kidnapping schema from a single narrative.

Introduction

Previous computational models of the acquisition of word meaning [Berwick83, Granger77, Selfridge82] have assumed existing knowledge of the concept underlying the word to be learned. In these models, word learning is a process of using surrounding context to establish an identification between a new lexical item and a known concept. However, new words are not always encountered as labels for known concepts. When encountering a new concept in natural language text or discourse, it is quite likely that one will also come across unknown words which refer to various aspects of the new concept. A word learning model which requires prior knowledge of the underlying concept will be unable to acquire even provisional meanings for such words.

Developmental studies suggest that the learning of words and their underlying concepts frequently occurs concurrently. Experiments by Gopnik and Meltzoff revealed that children's acquisition of "disappearance" words occurred at about the same time they learned to solve object-permanence tasks involving invisible displacements [Gopnik86]. From this data, they concluded that learning may often involve "*concurrent* cognitive and semantic developments, rather than involving cognitive prerequisites for semantic developments." Bowerman [Bowerman80] and Kuczaj [Kuczaj82] have also used developmental data to argue for an interactive approach to language and concept acquisition.

This paper describes an integrated computational model of the acquisition of word meanings and their underlying concepts. This approach was developed in an attempt to add word learning abilities to the GENESIS explanation-based schema acquisition system [Mooney85]. From a single natural language narrative, the current GENESIS system is able to acquire a new schema as well as provisional meanings for several schema related words.

An Overview of Schema Acquisition in GENESIS

GENESIS [Mooney85] is an explanation-based learning system [DeJong86, Mitchell86] which learns a plan schema from a single instance by determining *why* a particular sequence of actions observed in a specific narrative allowed the actors to achieve their goals. During the understanding process, GENESIS attempts to construct explanations for characters' actions in terms of the goals their actions were meant to achieve. This process involves plan and script-based understanding

---

[*] This research was supported by the Office of Naval Research under grant N-00014-86-K-0309.

974

mechanisms like those employed by previous narrative processing systems [Schank81]. When the system observes that a character has achieved an interesting goal in a novel way, it generalizes the composition of actions the character used to achieve this goal into a new schema. The generalization process (described in [Mooney86]) consists of an analysis of the causal model of the narrative which removes unnecessary details while maintaining the validity of the explanation. The resulting generalized set of actions is then stored as a new schema and used by the system to correctly process narratives which were previously beyond its capabilities.

In [Ahn87], experimental evidence is presented which indicates that, like GENESIS, people can also a acquire a schema by generalizing the explanation of a single narrative. After reading one specific instance of a novel plan, subjects were able to describe the underlying schema in abstract terms, generate a different instance, and correctly answer questions about the general schema.

An Example of Learning Words with their Concepts

This section presents an example of GENESIS' ability to learn word meanings as well as their corresponding concepts from a single example. First the system is given the following kidnapping story. At this point, GENESIS has schemata for threatening, capturing, making bargains, and a number of other actions as well as definitions for many words; however, it does not have a schema for kidnapping-for-ransom nor definitions for the words "kidnap", "kidnapper", or "ransom".

Story 1

Fred is Mary's father and is a millionaire. John approached Mary and pointed a gun at her. She was wearing blue jeans. He told her if she did not get in his car then he would shoot her. He drove her to his hotel and locked her in his room. John called Fred and told him John was holding Mary captive. John told Fred if Fred gave him $250000 at Trenos then John would release Mary. Fred paid him the ransom and the kidnapper released Mary. Valerie is Fred's wife and he told her that someone had kidnapped Mary.

From this single instance, GENESIS learns a general schema for kidnapping-for-ransom (which it calls CaptureBargain based on the two main actions which compose it) as well as preliminary definitions for "kidnapper", "ransom", and "kidnap". A paraphrase of the CaptureBargain schema is shown below:

CaptureBargain(?x97,?a52,?b11,?c4,?y15,?l19)

?b11 is a person. ?c4 is a location. ?x97 is a character. ?b11 is free. ?x97 captures ?b11. ?a52 is a character. ?x97 contacts ?a52 and tells it that ?b11 is ?x97's captive. ?y15 is a valuable. ?x97 wants to have ?y15 more than it wants ?b11 to be ?x97's captive. ?a52 has a positive relationship with ?b11. ?a52 has ?y15. ?x97 and ?a52 carry out a bargain in which ?x97 releases ?b11 and ?a52 gives ?x97 ?y15 at ?l19.

The provisional definition learned for "kidnap" is an action describing an instance of CaptureBargain where the subject is the actor (?x97) and the direct object is the person he captures and then releases in exchange for payment (?b11). The definition conjectured for "kidnapper" is a person filling the actor role of the new schema (?x97) and the definition for "ransom" is a valuable item given to this actor (?y15). These definitions do not exactly match the standard dictionary definitions of these words, but they are reasonable approximations given their use in this one example. The lexical and schematic knowledge acquired from this example enables the system to subsequently explain the following stories as instances of its new CaptureBargain schema.

Story 2

Ted is Alice's husband. A kidnapper took Alice into a room. Bob got $75000 and released Alice.

Story 3

Ted is Alice's husband. John took Alice into a room. Ted paid John the ransom and John released Alice.

Story 4

Steve kidnapped Valerie. Mike was Valerie's father and paid Steve $30000.

Prior to learning, GENESIS could not construct explanations for any of these stories since each one requires both knowledge of the schema and a definition for the appropriate kidnap-related word.

How GENESIS Learns Word Meanings

Since the processes for learning *role labels* such as "kidnapper" and "ransom" and that for learning *schema labels* such as "kidnap" are somewhat different, each of these procedures will be discussed separately.

Learning Role Labels

The procedure used to learn role labels is similar to the technique used by the FOUL-UP system [Granger77] except that it is integrated with the schema learning and schema activation processes. When the parser (a modified version of McDYPAR [Dyer83]) encounters an unknown word when there is an expectation for a noun of some class, a dummy variable is created, annotated with the unknown word, and allowed to fill the expectation. For example, the phrase "Fred paid him the ransom" in Story 1 is parsed into the assertion: Atrans(Person1,?x1,Person3) where ?x1 is marked with the fact that it came from the unknown word "ransom." If an input pattern with an unknown-word variable like ?x1 matches a pattern expected by currently active schema, then a provisional definition for the word is made based on the constraints on the schema role which matches the variable. For example, in the "ransom" case, the previous sentence in Story 1 suggested a Bargain schema between John and Fred and set up expectations for the two proposed actions. Since "Fred paid him the ransom" matches the expected action "Fred gives John the $250000," and "ransom" fills the role of the item whose possession is transferred, an initial definition is made for "ransom" stating that it is a physical object whose possession is transferred during a Bargain.

However, this is not the final definition created for "ransom" since an additional process is performed when a new schema is learned. Each of the sub-actions composing a new schema is checked for roles which are filled by unknown-word variables or which were previously matched to such a variable resulting in an initial definition. In either case, a new definition is created for the unknown word based on the role it fills in the learned schema and the schema constraints on this role. Consequently, when the CaptureBargain schema is subsequently recognized and learned from Story 1, it causes "ransom" to be redefined as a valuable item whose possession is transferred to the actor in a CaptureBargain schema (?y15). The rationale for having *learned* schemata take precedence when defining such words is that a learned schema represents a new situation and therefore new words are assumed more likely to be directly associated with it than with an existing schema like Bargain.

The provisional definition for a role label like "ransom" contains two parts. The first is a set of constraints on the object itself, such as an assertion that it be Valuable. The second part is a suggestion of the schema of which it is a role. The fact that a role label definition can suggest a relevant schema allows GENESIS to use the definition to construct explanations for narratives which it otherwise would not understand. When the word "ransom" is subsequently encountered in Story 3, it suggests that the CaptureBargain schema might be relevant. This schema is then used in a top-down fashion to construct an explanation for the text. Since no other piece of information suggests CaptureBargain, the learned definition for "ransom" is crucial in understanding this story.

However, most words are not role labels like "kidnapper" and "ransom." For example, consider replacing the word "ransom" in Story 1 with the word "moolah." Since the word "moolah" is unknown, GENESIS gives it a definition identical to the one it learned for "ransom." In order to be able to recover from such mistakes, the system monitors the schemata suggested by newly learned words. If a new word subsequently suggests a schema which does not explain any future inputs, the suggestion is removed. Consequently, after receiving a murder-for-inheritance story in which the word "moolah" is used, "moolah" ceases to suggest CaptureBargain.

Learning Schema Labels

Learning meanings for verbs which refer to entire plan schemata is a more difficult task since the relevant context is potentially much broader. A sentence such as "John robbed the store" may be used to introduce a long piece of text elaborating the situation, to succinctly summarize a previous piece of text, or to simply refer to a single action in an even larger plan. A number of heuristics have been developed which allow a reasonable guess to be made regarding the reference of such unknown verbs. The following one is used to resolve the meaning of "kidnap" as used in Story 1.

> If one character informs another that some action occurred and a schema whose actor is the same as this action's was recently acquired from the narrative, and this schema also has roles filled by the speaker and any direct and indirect objects of the action, then assume the speaker is summarizing the event and that the unknown act refers to the new schema.

Specifically, since Fred tells his wife that "someone kidnapped Mary" and both him and Mary were participants in the just completed CaptureBargain schema, GENESIS assumes "kidnap" refers to CaptureBargain (i.e. one can summarize the schema by saying: ?x97 kidnapped ?b11). An appropriate definition is then created for "kidnap" and can be used to help understand Story 4. In this narrative, "Steve kidnapped Valerie" is interpreted as describing an instance of CaptureBargain in which Steve is the actor and Valerie is the victim. The assertions that "Mike is Valerie's father" and "Mike paid Steve $30000" are then understood as parts of the expansion of this instance of CaptureBargain. Another heuristic involves the mentioning of an unknown action followed by an elaboration which describes a novel schema.

Conclusions

Unlike previous approaches to the acquisition of word meanings, the present approach does not assume prior knowledge of the underlying concepts. Learning definitions for new words is integrated with an explanation-based concept learning mechanism. This allows the system to learn concepts and word meanings concurrently, which is a phenomenon which has been observed in developmental studies. However, this paper reports only preliminary work in the area of integrated word learning. There are many problems which still need to be addressed. A few of these are listed below.

(1) The procedure for removing schema suggestions from new definitions is too strict. One counter-example should not eliminate a suggestion and repeated usefulness of a suggestion should make it resistant to elimination.

(2) Morphology of unknown words should be considered. A "kidnapper" is clearly the actor of a "kidnapping."

(3) More and better heuristics are needed for determining whether a word might be a *schema label* and to what schema it might refer.

(4) Only *role labels* and *schema labels* are considered. Many words do not fall into either of these two categories.

(5) Only integration with explanation-based learning is considered. Integration with *similarity-based learning* [Dietterich83] should also be examined.

Nevertheless, the current work demonstrates the feasibility and usefulness of integrating the acquisition of word meanings and concepts. Further research is needed in both AI and psychology to explore the potential symbiotic relationship between language and concept acquisition.

Acknowledgements

The author would like to thank Gerald DeJong for directing this research and Dedre Gentner for pointers to the relevant developmental literature.

References

[Ahn87]   W. Ahn, R. J. Mooney, W. F. Brewer and G. F. DeJong, "Schema Acquisition from One Example: Psychological Evidence for Explanation-Based Learning," Technical Report, AI Research Group, Coordinated Science Laboratory, University of Illinois at Urbana-Champaign, February, 1987.

[Berwick83]   R. C. Berwick, "Learning Word Meanings from Examples," *Proceedings of the Eighth International Joint Conference on Artificial Intelligence*, Karlsruhe, West Germany, August 1983, pp. 459-461.

[Bowerman80]   M. Bowerman, "The Structure and Origin of Semantic Categories in the Language Learning Child," in *Symbols as Sense: New Approaches to the Analysis of Meaning*, M. L. Foster & S. H. Brandes (ed.), Academic Press, New York, 1980, pp. 277-299.

[DeJong86]   G. F. DeJong and R. J. Mooney, "Explanation-Based Learning: An Alternative View," *Machine Learning 1*, 2 (April 1986), pp. 145-176.

[Dietterich83] T. G. Dietterich and R. S. Michalski, "A Comparative Review of Selected Methods for Learning from Examples," in *Machine Learning: An Artificial Intelligence Approach*, R. S. Michalski, J. G. Carbonell and T. M. Mitchell (ed.), Tioga Publishing Company, Palo Alto, CA, 1983, pp. 41-81.

[Dyer83]   M. J. Dyer, *In-Depth Understanding*, MIT Press, Cambridge, MA, 1983.

[Gopnik86]   A. Gopnik and A. N. Meltzoff, "Relations between Semantic and Cognitive Development in the One-Word Stage: The Specificity Hypothesis," *Child Development 57*, (1986), pp. 1040-1053.

[Granger77]   R. H. Granger, "FOUL-UP: A Program that Figures Out Meanings of Words from Context," *Proceedings of the Fifth International Joint Conference on Artificial Intelligence*, Cambridge, MA, August 1977, pp. 172-178.

[Kuczaj82]   S. A. Kuczaj, "Acquisition of Word Meaning in the Context of the Development of the Semantic System," in *Verbal Processes in Children: Progress in Cognitive Developmental Research*, C. J. Brainerd & M. Pressley (ed.), Springer-Verlag, New York, 1982, pp. 95-123.

[Mitchell86]   T. M. Mitchell, R. Keller and S. Kedar-Cabelli, "Explanation-Based Generalization: A Unifying View," *Machine Learning 1*, 1 (January 1986), pp. 47-80.

[Mooney85]   R. J. Mooney and G. F. DeJong, "Learning Schemata for Natural Language Processing," *Proceedings of the Ninth International Joint Conference on Artificial Intelligence*, Los Angeles, CA, August 1985, pp. 681-687.

[Mooney86]   R. J. Mooney and S. W. Bennett, "A Domain Independent Explanation-Based Generalizer," *Proceedings of the National Conference on Artificial Intelligence*, Philadelphia, PA, August 1986, pp. 551-555. (A longer updated version appears as Technical Report UILU-ENG-86-2216, AI Research Group, Coordinated Science Laboratory, University of Illinois at Urbana-Champaign)

[Schank81]   R. C. Schank and C. Riesbeck, *Inside Computer Understanding*, Lawrence Erlbaum and Associates, Hillsdale, NJ, 1981.

[Selfridge82] M. Selfridge, "Inference and Learning in a Computer Model of the Development of Langague Comprehension in a Young Child," in *Strategies for Natural Language Processing*, W. G. Lehnert and M. H. Ringle (ed.), Lawrence Erlbaum and Associates, 1982, pp. 299-326.

INDUCING (NEW) RULES IS DIFFERENT FROM ADJUSTING (OLD) PARAMETERS.

K.Prazdny

Artificial Intelligence Center, FMC Corporation
P.O. Box 580, Santa Clara, CA 95052
prazdny@ai.cel.fmc.com

INTRODUCTION.

Linear (auto)associators capture the structure inherent in a set of patterns as long as the ensemble of example patterns adheres to the linear predictability constraint. Conjunctions and exclusive-disjunctions, however, require that the compound features have a different associative strength than the individual input in isolation. In general, it is hard to specify in advance what order of conjunctions is required to capture the data dependencies. The major motivation behind the development of learning protocols for networks with hidden units (LeCun, 1985, 1986; Hinton, Sejnowski & Ackley, 1984; Rumelhart, Hinton & Williams, 1985) is to discover the relationships in the input automatically.

There is a close relationship between this work and modelling the phenomena of classical conditioning. There, the animal model tries to predict single event (US) while the (auto)associator must, in effect, develop such prediction for each individual vector element. In both cases, the goal is a method of determining which features are predictive of others and distinguishing useful cues from context and background noise. Several contemporary animal learning theories handle conjunctions and disjunctions by assuming that the co-occurence of two stimuli results in some new (external) "resonant" property being signalled by the perceptual system. Other models use "internal resonant" features: new compound features (boolean combinations of existing ones) are introduced into the representation as the result of system's prediction failure (Quinqueton & Sallantin, 1983; Schlimmer & Granger, 1986). Similarly, networks with hidden units can implement an arbitrary input/output mapping if they have the right connections and large enough set of hidden units (Minsky & Papert, 1969). For example, to solve the XOR problem, one can add a unit that detects the conjunction of the two inputs. This amounts essentially to enlarging the dimensionality of the input: from the point of view of the ouput unit the hidden unit is treated as another input unit. In this sense, networks with hidden units can be said to be discovering the "resonant" properties of stimulation (feature combinations predictive of the desired outcome) in a way similar to the classical conditioning models.[1] The number of such potential resonant properties in the general case of a non-linear autoassociator, where the value of a feature is predictable, in general, only from a non-linear combination of values of other set of features, grows exponentially with the vector length, n. Potentially, one needs 2^n-(n+1) "resonant" features for the prediction of a single element (i.e. the domain is the set of all subsets). Hidden units and recurrent connections are, in themselves, of little help, however. One has to find a useful way to use them. To illustrate, McClelland & Rumelhart (1986) attempted to implement a non-linear autoassociator for the "one-same-one" problem using hidden units and recurrent connections. To achieve the required efect they had to train the network with all possible completion patterns. That is, for each pattern to be learned (e.g. 111) they trained the network to associate all of the possible incomplete patterns (11?, 1?1, ?11) with the complete (111) pattern (McClelland & Rumelhart, 1986, p.211). This is, of course, "cheating": they have replaced an autoassociation task with an association task. In general, the enumeration of all possible completions is impractical; a daunting task for even moderately long inputs.

WHAT CAN THE NETWORKS LEARN?

One rather serious disadvantage of the PDP networks with fixed length input[2] is that the information they acquire in the course of their "education" is not cumulative. That is, having been taught in one situation is of absolutely no help in learning in a different but similar situation. There is no possibility of transfering the accumulated knowledge to a new but similar situation, as opposed to a new but similar input. Even the within-situation generalization is apparently difficult to achieve. Confronted with new patterns, the steepest descent weight updating procedures preferentially changes existing, already useful representational features because the error gradient (and thus the weight change) is directly proportional to the current weight magnitude (Sutton, 1986). This protocol thus destroys what has been learned previously. I will illustrate some of these points on the parity problem.

Parity can be defined in two ways for inputs in $\{0, 1\}$: (i) "the sum of inputs is odd", or recursively as (ii) parity(0)=value(0) and parity(n+1)=XOR[parity(n),value(n+1)] where value(n) is the value of the nth input element, and parity(n) is in $\{0, 1\}$. Suppose we have a network with with m input and hidden units, and with a single output unit and teach the parity to the first n inputs (n<m). Does this "education" help in acquiring the parity problem of size n+1 (or n+k), e.g. does the network converge sooner? The answer is that previous experiences with the parity problem on n inputs is of absolutely no help in learning the parity problem of larger (or smaller) size. In fact, in none of our experiments did the network with weights obtained from the previous teaching run even converge (all unused inputs were kept constant at the resting, 0.5, level). This phenomenon is understandable in the view of learning as creating and traversing the "energy landscape" in the weight space: the energy minima are simply at different places! The networks with hidden units do not learn new *concepts*, they are "merely" discovering data dependencies (input/output contingencies) in a particular situation. Similarly, the classical conditioning models and algorithms for learning logical formulas learn "boolean combinations" of antecendent conditions, not the underlying concept.[3]

Suppose that we teach a network the parity problem with 2, 3, 4, ... inputs. What kind of computational mechanism would be required to abstract the concept of parity from such a teaching sequence?[4] An obvious answer may be that one needs another network that sees all the weights and biases, and how they are being modified from one *instance* of the concept to another, and that changes its internal structure so that when a given situation arises it programs the weight distribution of the network appropriately. The problem with this approach is that there is no guarantee of any lawfull relationship between the weight distributions accross the various instances of a given concept. That is, the set of weights for the (n+1) problem may not be predictable from the set of weights for the (n-k) problem. With large enough number of units and connections (weights) relative to the minimum necessary for the given task, i.e. with large degree of freedom, there is a large number of ways in which a given set of input/output specification can be mapped into the weight space, i.e. the mapping is one-to-many. In other words, there is no guarantee that the regularities obvious in one domain will appear in the weight space. In addition, one cannot, in general, know in advance the limit on the number of inputs.[5] A concept can describe an infinite number of situations: it has a generative quality of a rule not captured by the stimulus-response associations of a PDP network.[6]

CONCLUSION.

There are two distinct (not neccessarily related) problems: (I) how do you create/develop a new concept (e.g. parity) as opposed to (II) how do you solve a particular problem (e.g. parity with n inputs). An intelligent agent should (probably) be required to posses the ability to solve (II) using (I). Genuine concept learning, if it can be implemented using connectionist architectures at all, has to be done by a mechanism that monitors the adapting network not limited or constrained by the number of inputs. This evaluative mechanism must be capable of gathering information accross different situation or instances of the same concept. It is this mechanism that would learn new concepts as opposed to adapting to the immediacy of the incoming stimulation. It is doubtfull that this can be done in a connectionist system (of either the weight adjustment or signature table variety) where all knowledge is constrained by a finite structure and there is no distinction between information and control. The ability to construct and manipulate symbolic structures and procedures seems necessary.

REFERENCES.

Chaitin G.J., Randomness and mathematical proof, *Scientific American*, 232, 5, 1975 (May)

LeCun Y., A learning scheme for assymetric threshold networks, *Proceedings Cognitiva-85*, 1985

McClelland J. Rumelhart D., A distributed model of human learning and memory, *Parallel distributed processing*, D. Rumelhart & J. McClelland (eds), 170-215, 1986

Quinqueton J. Sallantin J., Algorithms for learning logical formulas, *Proceedings IJCAI*, 476-478, 1983

Rumelhart D.E. Hinton G. Williams R.J., *Learning internal representations by error propagation*, Tech.Rep. ICS-8506, Institute for Cognitive Science, UCSD, La Jolla

Schlimmer J.C. Granger R.H., Simultaneous configural classical conditioning, *Proceedings Conference Cognitive Society*, 141-153, 1986

Sutton R., Two problems with backpropagation, *Proceedings Conference Cognitive Society*, 823-831, 1986

Notes

[1]Context-sensitive encoding where a code for an element depends on other elements corresponds to creation of compound features or "resonant" properties that are imposed on a system, usually a linear (auto)associator, from the outside. Context-sensitive encoding is performed in an attempt to achieve linear separability that enables the use of a linear system (which has nice and predictable "generalization" properties).

[2]This characteristic shows the close similarity of the PDP networks to the classical pattern recognition work. Input vectors are of fixed length, i.e. the information is coded by position, each position is a different feature, a different (orthogonal) dimension of a (finite dimensional) vector space. Thus, there can be no shift invariancy: 0100 is orthogonal to 0010. Suppose that the input is a histogram. Then <0 16 0 0 0> is more similar to <0 0 16 0 0> than to <0 0 0 0 16> because nearby vector elements encode similar values. In the vector space formalism where the distance is equivalent to the inner product all three are orthogonal (i.e. dissimilar).

[3]How would models relying on explicit rule generation handle the parity concept? They would end up with the exhaustive enumeration of the permissible combinations. E.g., for the 2 input parity (XOR) problem, the system will end up with a compound feature "((1 and 0) or (0 and 1))", and for the 3 input situation it would produce "((1 and 0 and 0) or (0 and 1 and 0) or ......)" or perhaps, given previous experience with the XOR problem, "(((1 and 0) and 0) or ((0 and 1) and 0) ....)" It is relatively easy to envisage a concept formation mechanism operating on such rules that would develop the parity concept defined by (ii).

[4]It may be interesting to know if and how well humans and e.g. the dogs can do this, i.e., if they can develop a genuine parity concept as opposed to a set of responses to a set of situations. Does teaching parity on e.g. 2 and 3 inputs produce generalization, i.e. correct response on e.g. 4 inputs?

[5]There are other, related problems that do not require variable length inputs: parametrized concepts. Consider, for example, the concept of negation (Rumelhart & McClelland, 1986) on a vector with fixed length, N. The concept has, in its simplest form, only one parameter: the position of the negation bit. Is there a connectionist mechanism that can, after it learns concept examples parametrized by the first n inputs, generalize to the remaining (untaught) N-n parameters?

[6]The notion of a concept is probably intimately related to the notion of a program: two strings of different length can be described by the same program. Unfortunately, the theory of program-size complexity (e.g. Chaitin, 1975) while relevant here is not constructive and cannot offer any guidance in the construction of a program from examples.

A Neural Model of Deep Dyslexia

Michael L. Rossen
Department of Psychology
Brown University
Bitnet: Rossen@browncog

Abstract

This paper presents a simulation of the selective deficits and the partial breakdown patterns characteristic of the oral reading performance of deep dyslexics. The most striking symptom of deep dyslexia -- usually considered its defining characteristic -- is the occurrence in oral reading tasks of semantic paralexias: the vocalization of a word semantically related to an isolated, printed target word. The pattern of simulated paralexic errors by the neural model is strongly controlled by the similarity structure of the training set stimuli and, to a lesser extent, the frequency of presentation of stimuli during learning by the model. This result fits well with effects of stimulus type on patterns of paralexic error among deep dyslexics. Further, the model very naturally reproduces the patterns of partial breakdown observed in deep dyslexics, including a slow response time (RT) and within subject variation of response to a particular target word in successive test sessions.

Keywords: neural models, deep dyslexia, similarity structure.

Symptoms and characteristics of deep dyslexia

Loss of a selected subset of cognitive skills often accompanies brain damage. **Deep dyslexia** is an example of a selective symptom-complex resulting from brain damage in which deficits in certain oral reading skills tend to co-occur. The most striking symptom of deep dyslexia -- usually considered its defining characteristic -- is the occurrence in oral reading tasks of **semantic paralexias**: the vocalization of a word semantically related to an isolated, printed target word (e.g., visual target: BEAR; oral response: LION). Deep dyslexics also make visual paralexias, errors where the oral response word graphemically resembles the target word (e.g., visual target: BEAR; oral response: PEAR).

The type of stimulus word used in oral reading tasks strongly affects the likelihood of a paralexic response by a deep dyslexic. Shallice and Warrington (1975) show that nouns of low usage frequency or of low concreteness have a relatively high probability of eliciting a paralexic response, with concreteness having a stronger effect. Syntactic class also has an effect: correct responses are most likely for nouns, with adjectives, verbs and function words (e.g., "is,", "to," "and") causing successively higher rates of paralexic error.

The **selectivity** of deep dyslexia is evident in the stimulus effects just mentioned and from the observation that deep dyslexics can often perform tasks analogous to oral reading, but involving different input and/or output modalities, with little or no evidence of degradation in performance. For example, some deep dyslexics can perform almost perfectly in a picture naming task in which the picture stimuli correspond to word stimuli with which the patients make frequent paralexias during oral reading. (Patterson & Marcel, 1977).

Deep dyslexia involves **patterns of partial breakdown**, patterns of a statistical rather than a deterministic nature. In particular:
(1) Response time (RT) of deep dyslexics in oral reading tasks is slower than for the normal population. In addition, deep dyslexics often make "omissions," in oral reading tasks; that is, they fail to offer any response at all to a printed target stimulus.
(2) Intra-subject response variation. Consider an oral reading task in which the printed target

stimulus "BEAR" is one of several stimuli presented in random order during a single test session. In successive test sessions, a deep dyslexic might correctly read the word "BEAR" aloud, commit a semantic paralexia ("LION"), and a visual paralexia ("PEAR").

(3) Inter-subject response variation. Two deep dyslexics can make different responses to a particular target word, yet have similar statistical likelihoods of semantic and visual paralexias.

Anderson (1983), Kawamoto (1985), and McClelland & Rumelhart (1986) have all employed neural models to examine various characteristics of brain damaged patients. Kawamoto specifically modeled aspects of deep dyslexia. The simulations in this paper concentrate on some characteristics of deep dyslexia not addressed by these previous studies. Simulations one through three show that a fully trained neural model can simulate paralexic responses and reproduce the patterns of partial breakdown just described. Simulations three and four suggest that the stimulus effects outlined above that underscore the selective nature of deep dyslexia can be interpreted in terms of the similarity structure of word representation (see below) and the frequency with which a word is learned by the model.

The Brain-state-in-a-box (BSB) neural network model

The BSB model (Anderson, 1983) employs a training algorithm (1a) and a classification algorithm (1b). The algorithms employ a *vector of idealized neural activities* to represent *information flowing through* the system. Each vector element can take on a continuous range of values between -1 and 1, representing the minimum and maximum activity levels of a neuron. A matrix of idealized *synaptic weights* inter-connecting the neurons represents *information stored within* the system.

$$f^*(\tau) = A\, f(\tau) \qquad \text{(basic computation)}$$

A : matrix of synaptic weights
f : vector of neural activities
τ : scalar discrete time index

$$A(\tau+1) = A(\tau) + \gamma\,[\,f(\tau) - f^*(\tau)\,]\,f^T \qquad \text{(W-H training algorithm)} \qquad (1a)$$

f^T : transpose of f
γ : scalar learning parameter

$$f(\tau+1) = \sigma\,[\,a f^*(\tau) + \beta f(\tau)\,] \qquad \text{(BSB classification algorithm)} \qquad (1b)$$

a : scalar feedback parameter
β : scalar decay parameter
σ : function that limits activities to region [-1,1]

Training procedure. Training consists of modification of a 50 percent connected weight matrix "A" by repeated application of the Widrow-Hoff (W-H) training algorithm (1a) with each of the members of the training stimulus set (see below). Each stimulus is learned an equal number of times, except in simulation three when learning frequency effects are explicitly examined. To eliminate potential recency effects of training, the order of stimulus presentation is randomized and the learning parameter is tapered as training nears completion (Anderson, 1983). To ensure the robustness of training, noisy versions of the training set vectors are used to test the system. The noise is calculated so that, when added to a training vector to create a test stimulus vector, the test stimulus is always within a cone around the training stimulus whose axial angle is half the angle between the training vector and its nearest neighbor in the training set. Training is terminated when the system classifies four sets of noisy training set vectors without error.

The stimuli. A word stimulus is represented as an activity vector that specifies the system's initial state. The classification algorithm (1b) iterates the activity vector through the weight matrix until the vector reaches a corner in activity space, or until until a maximum of 96 iterations is reached. This final activity vector corresponds to a classification of the stimulus vector. In all of the simulations considered here, a 192-dimensional system is used to learn a training set of eight normalized, demeaned stimuli, each pointing to a different corner in activity space. The cosine between two activity vectors defines their *similarity.* Table 1 shows the similarity structure of the training set. Each stimulus vector has a .75 cosine with three *nearest neighbor* stimuli, a .5 cosine with one stimulus and a .375 cosine with the remaining thee stimuli. This idealized similarity structure allows for quantitative analysis of simulated paralexias (see below) as a function of word stimulus similarity.

Performance measures

RT is simulated by the iteration count in the BSB classification algorithm (1b). A final classification by the system is considered an *omission* error if less than 95 percent of the neurons are at their maximum or minimum activity level. If a classification is not an omission, then the system response is defined as the training set vector to which the classification is most similar. If that training set vector was the input stimulus, then the response is considered correct. Otherwise, the response constitutes an erroneous classification, a *simulated paralexia.*

Simulating brain damaged subjects and multiple test sessions

Damage consists of ablation of a randomly chosen, fixed percentage of the synapses of the weight matrix after termination of learning. Different *computer subjects* are simulated using the same trained network with different random ablations. A particular deep dyslexic subject also varies as to which words he/she makes errors on from one test session to the next. To allow this pattern to emerge from the formally deterministic system employed in these simulations, noisy versions of the training set stimuli are employed, as described in the section on training procedure.

Table 1. Similarity structure of training set stimuli.

Training stimulus	Training stimulus							
	1	2	3	4	5	6	7	8
1	1.	.75	.75	.75	.5	.375	.375	.375
2	.75	1.	.75	75	.375	.5	.375	.375
3	.75	.75	1.	.75	.375	.375	.5	.375
4	.75	.75	.75	1.	.375	.375	.375	.5
5	.5	.375	.375	.375	1.	.75	.75	.75
6	.375	.5	.375	.375	.75	1.	.75	.75
7	.375	.375	.5	.375	.75	.75	1.	.75
8	.375	.375	.375	.5	.75	.75	.75	1.

Simulation one: (When) are errors made by a neural model?

Procedure. Ten Computer subjects are derived from the trained system, with damage ranging from five percent to 95 percent synapse ablation. Each computer subject participates in one test session of classifying noisy versions of the training set stimuli.

Results. Ablation of synapses have observable effects on system dynamics as measured by both RT and error occurrence. Figure 1a shows RT , averaged across the test stimuli, plotted against proportion of synapse ablation. RT reaches the system maximum for at least some of the test stimuli after only a 25 percent ablation. Simulated paralexias occur for only a narrow range of ablation levels (Figure 1b). All responses are correct with 45 percent ablation or less, while only omission errors occur with 75 percent or more synapse ablation.

Discussion. The response time data corresponds well with the tendency of deep dyslexics to respond more slowly than normals. The location of the peak for paralexic errors at 65 percent ablation level underscores the robust nature of distributed memory storage. Increasing maximum iteration number does increase the number of paralexias somewhat, but the basic shape of the curve in Figure 1b remains the same. Simulated paralexias have not been observed with the present stimulus set with less than 55 percent ablation. Pilot studies show that higher or lower cosines among nearest neighbors leads to more or less simulated paralexias, respectively, given a fixed percentage of ablated synapses. No simulated paralexias could be observed at any damage level, given present operational definitions, for stimulus sets in which nearest neighbors had a cosine of only .25. Also, additional training decreases the number of errors for a given amount of damage, but the effect was negligible.

Simulation two: Within subject variability of error

Procedure. This simulation tests the ability of a computer subject with simulated brain damage to give variable classifications to the same test stimulus in successive test sessions. One

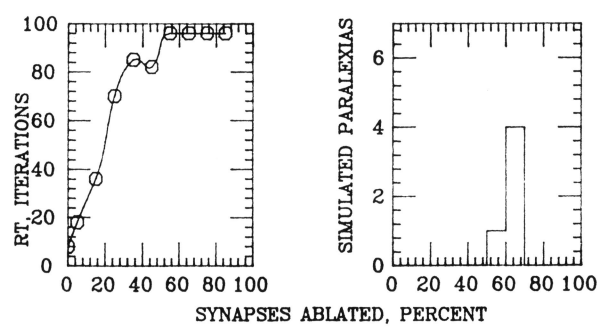

Figure 1a   Figure 1b

Table 2. Simulation 2: Simulated paralexias by one computer subject in 24 test sessions.

Input	Output corner								
stimulus	1	2	3	4	5	6	7	8	paralexias
4	0	15	1	(3)	0	0	0	0	16
5	0	0	0	0	(12)	0	12	0	12

computer subject with 65 percent synapse ablation performs 24 test sessions of identifying noisy versions of the training set stimuli.

Results. Table 2 shows the number and type of simulated paralexias for each test stimulus (omission errors are not included). No errors occur for 6 of the stimuli. 16 simulated paralexias are made on stimulus four; 15 of them converge to stimulus two and one of them to stimulus three. 12 simulated paralexias are made with stimulus five, all to stimulus seven.

Discussion. Simulated paralexic error patterns are variable over the sessions, conforming to the error patterns reported in the literature for deep dyslexic patients. Of the training set members with which simulated paralexias are made, neither of them always produced errors with the system, and one of them produced two different types of paralexias.

Simulation three: Error patterns across computer subjects

Procedure. This simulation examines the effect of similarity between activity vectors on the pattern of simulated paralexias by computer subjects. Because of the the symmetric similarity structure of the training set, paralexic errors should be equally likely for each stimulus. On the other hand, if similarity structure is important, then paralexic errors should be more likely to be classified as near neighbor stimuli. 25 computer subjects each with 65 percent synapse ablation perform one test session of identifying the training set stimuli with no noise added.

Results. Table 3 shows the distribution of simulated paralexias from and to each of the training set stimuli. The most prominent characteristic of these results is that errors only correspond to nearest neighbor stimuli. Within sets of nearest neighbors (stimuli one though four and stimuli five through eight, as can be seen by consulting Table 1), distribution of errors from and to each of the stimuli is almost flat, with the exception of stimulus one to which more errors tended and from which less errors were made.

Discussion. The absence of simulated paralexias to anything but a nearest neighbor stimulus can be interpreted as a tendency for the system to make semantic (as opposed to random) errors when damage is in the form of random synaptic ablation. This prediction assumes that semantically related words are represented in memory as highly correlated patterns of activation. Support for this assumption can be found in the work of Anderson (1983) and Kawamoto (1985). One might speculate further that nouns and adjectives (as opposed to verbs), and words with concrete meanings are represented by activity patterns relatively isolated from the activity patterns of nearest neighbor words. This would serve as an explanation for the syntactic class and concreteness effects found in deep dyslexic oral reading performance.

987

Table 3. Simulation 3: Classification of test stimuli by 25 computer subjects.

Input stimulus	Output corner								paralexias
	1	2	3	4	5	6	7	8	
1	20	2	1	2	0	0	0	0	5
2	7	13	2	3	0	0	0	0	12
3	3	2	17	3	0	0	0	0	8
4	3	3	2	17	0	0	0	0	8
5	0	0	0	0	15	1	5	4	10
6	0	0	0	0	5	16	2	2	9
7	0	0	0	0	4	2	15	4	10
8	0	0	0	0	2	5	3	15	10
Total paralexias	13	7	5	8	11	8	10	10	72

Simulation four: The effect of variable frequency of stimulus presentation during training.

Procedure. This simulation is concerned with the effects of variable frequency of stimulus presentation during learning on the pattern of simulated paralexias by computer subjects. Toward this end, stimulus one is learned by the system three times as frequently as any of the other stimuli. High frequency of presentation for a particular stimulus causes that stimulus to be learned more quickly than other stimuli. Though this effect is muted somewhat by the error correction nature of the W-H learning algorithm, it is still possible that the high presentation frequency of stimulus one during learning renders it more resistant to error within a damaged system.

A side effect of the skewed presentation frequency is that the system requires relatively more training to learn all the stimuli. As was mentioned in simulation one, one of the results of extra training is a slight increase in stability with respect to damage. Thus, a 75 percent ablation level is used to maximize the number of simulated paralexias. Data is gathered from 100 computer subjects that each perform one test session identifying the training set stimuli with no noise added.

Results. Table 4 shows the distribution of classification errors from and to each of the training set stimuli. A strong tendency exists for simulated paralexias to converge to stimulus one, the stimulus trained with a high frequency of presentation. Conversely, the computer subjects are relatively unlikely to make a simulated paralexia with stimulus one as input.

As in simulation three, no errors at all are made to any but nearest neighbor stimuli. The total number of classification errors made on stimuli five through eight is virtually equal to the number made with stimuli one through four. No stimulus within the set of stimuli five through eight is markedly more or less likely to cause a simulated paralexia.

Table 4. Simulation 4: Classification of test stimuli by 100 computer subjects. Learning frequency of stimulus one was triple that of the others.

Input stimulus	Output corner								paralexias
	1	2	3	4	5	6	7	8	
1	73	10	8	9	0	0	0	0	27
2	20	49	11	20	0	0	0	0	51
3	34	14	44	8	0	0	0	0	56
4	22	23	7	48	0	0	0	0	52
5	0	0	0	0	60	11	18	11	40
6	0	0	0	0	11	56	15	18	44
7	0	0	0	0	18	16	52	14	48
8	0	0	0	0	18	28	10	44	54
Total paralexias	76	47	26	37	47	55	42	43	373

Discussion. Deep dyslexics tend to make more paralexias with low frequency words; however, this effect is often overshadowed by word concreteness effects. If word concreteness can in part be interpreted as the degree of isolation of a word's activity pattern, then the results of this simulation are in accordance with both of these findings. Moreover, in a pilot study identical to this simulation except that only stimuli one through five were learned by the system, no paralexic errors at all were made with stimulus five, even though stimulus one had a higher learning presentation rate. This model makes the further prediction that when semantic paralexias occur, the paralexia is likely to be of higher frequency in the language than the target word.

References

Anderson, J. A. (1983). Cognitive and psychological computation with neural models. *IEEE transactions on systems, man, and cybernetics, SMC-13*(5), 799-815.

Kawamoto, A. H. (1985). Dynamic processes in the (re)solution of lexical ambiguity. Unpublished doctoral dissertation, Providence: Brown University.

McClelland, J. L., & Rumelhart, D. E. (1986). Amnesia and distributed memory. In J. L. McClelland & D. E. Rumelhart (Eds.), *Parallel Distributed Processing: Explorations in the Microstructure of Cognition: Vol 2. Psychological and Biological Models* (pp. 503-527). Cambridge, MA: Bradford.

Patterson, K. E., and Marcel, A. J. (1977). Aphasia, dyslexia and phonological coding of written words. *Quarterly Journal of Experimental Psychology, 29*, 307-318.

Shallice, T., & Warrington, E. K. (1975). Word recognition in a phonemic dyslexic patient. *Quarterly journal of experimental psychology, 27*, 187-199.

Planning Principles Specific to Mutual Goals

Colleen M. Seifert
Institute for Cognitive Science - UCSD
and Navy Personnel Research and Development Center

Abstract

A theory of planning should provide a model of how planning knowledge might be learned and stored in memory so as to be available and utilized in appropriate situations. This paper presents a content theory of the planning strategies and constraints on planning specific to joint planning situations. The categorization helps to explain what information is relevant to this general class of planning problems, namely goal pursuit situations where goals cannot be satisfied without the participation of another planner. The taxonomy of planning principles presented outlines the common problems in mutual goal pursuit situations, and provides strategies for resolving the problematic interactions. The principles apply to a variety of types of mutual goal pursuit arrangements such as business partners, a political coalition, or social relationships.

Various types of planning principles have been proposed to account for constraints on the planning process based upon the type of goal pursuit involved; for example, the characterization of goal interactions in terms of whether the goals involve one planner or more than one, and whether the relationship between the planners is positive (concord) or negative (competition) has been proposed by Wilensky (1983). *Mutual goal pursuits* pose a unique class of planning problems having to do with the involvement of another planner in joint operations. Wilensky defined the positive goal relationship between two actors as "goal concord," where a planner has the same goal as someone else, resulting in the plan to accomplish the goal together. The goals of different planners can be mutually beneficial; such a shared commitment to a goal is termed an alliance (Wilensky, 1983). However, in this paper I explore the planning situation where the mutuality of goals involves not simply having the same goals (e.g., both parties want to be in New York), but rather that the mutual goals are *joint* goals -- the goals can not be satisfied without the active participation of the partner in the goal, and they involve a particular partner rather than any agent who happens to share a goal. In a mutual goal pursuit, two or more partners pursue shared goals with shared responsibility, forming the basis for a relationship between the planners defined by the content of the mutual goal (or goals) involved. This definition of mutual goal pursuit therefore includes an important class of planning situations: namely, the relationships people form to satisfy their goals, such as marriages, collaborations, and business partnerships.

The planning theory presented here is a type of memory-based planning, where appropriate plans and planning heuristics are retrieved from memory based upon the features of the planning situation. Relevant planning strategies retrieved from memory aid the planner in decisions about the mutual goal pursuit in relation to other goals the planner is pursuing. Characterizing the planning constraints for different types of planning situations provides a means to access general knowledge about planning strategies from many different contexts, and this knowledge can be used to prevent planning errors from recurring in similar situations (Schank, 1982; Dyer, 1983). Each principle describes a constraint on planning in terms of a causal vocabulary of goals, plans and their context, and provides inferences applicable to the situation as well as possible repair strategies for the problem. Examples of problems evident in mutual goal pursuit (MGP) situations were analyzed to determine the interactions and strategies that arise which affect such situations. Basically, the job of categorizing the planning strategies involved in joint planning situations requires specifying the types of things that can go wrong, and characterizing the many factors that may affect MGP arrangements. These factors break down into three categories:

- assignment of responsibility for pursuit

- amount of effort towards pursuit from either partner

- differences in criterion of satisfaction for partners

The importance of characterizing the planning strategies involved in MGP is that they provide information about people's behavior in mutual goal situations. From the specification of planning strategies in MGP, it will be possible to explain problems by identifying the planning principles operating in them and perhaps their misuse, and it will be possible to predict which strategies may be useful in particular problem situations. The MGP strategies proposed in the next sections allow for planning within the constraints of the factors unique to MGP situations. In addition to these strategies, general plans to motivate another person to work on a goal, such as the power to remind them of the commitment they made to the goal pursuit, provide common sense solutions based upon the problem and the variables within the MGP.

Assignment of Responsibility for Pursuit

A main source of problems in MGP is the assignment of responsibility. Since the mutual goal pursuit situation is characterized by the joint effort of two planners, it requires coordination beyond that required of two actors who happen to have the same goal at the same time. When an agent is hired to adopt your goal (i.e., a housekeeper who will also have the goal of keeping your house clean), responsibility for pursuit is clear. The agent relationship is invoked simply to pass responsibility to the agent. However, MGP is much more complicated: not only are there joint goals which both partners want achieved, such as keeping the shared household clean, but there are mutual goals where the successful attainment of the goals *requires* the efforts of both partners (e.g., communication). The possible ways for the actors to fail to satisfy their responsibility for goals are correspondingly complicated. Further, the two partners may not agree upon the coordination of responsibilities for the mutual goal pursuits. Thus, MGP situations must settle the question of the *assignment* of responsibility among partners as well as *execution* of plans for which each partner is responsible.

Agree on assignment of responsibilities. If the responsibility for a particular goal or plan is not clearly defined, several outcomes are possible:

- Both planners can independently assume responsibility and pursue the goal. This will satisfy the goal, but will result in a waste of resources, since the goal will be satisfied twice, or twice the necessary effort will have been expended.

- One planner can assume responsibility and pursue the goal while the other does not. This will satisfy the goal, but will result in a long term commitment to taking care of the goal without the advantage of sharing the goal pursuit with the partner who benefits from it.

- Both partners assume their coplanner is responsible for the goal, therefore neither partner pursues it. This results in goal failure, which may be further complicated by a difficulty in quickly detecting this problem.

The repair for these bugs is to coordinate responsibility. It is not always easy to do this since multiple goals may be involved, but some long term arrangement is essential, as the recurrence of goals will make errors in assignment very costly.

Separate responsibility. How responsibility is assigned may result in other MGP problems:

- Both partners pursuing a plan may impede plan execution. This is represented by the adage *too many cooks spoil the broth*: separate actors may undo or adversely affect each others' goal-directed actions.

- A plan may require more than one actor; if so, the actions in service of the plan must be coordinated to avoid unplanned interactions of each partner's efforts while accounting for who is responsible for what parts of the plan.

- A planner assigned responsibility should be allowed to function independently. Once assignment of responsibility has been made, interference or assistance on the part of the other planner may be problematic. "Kibitzing" is greatly resented once responsibility is undertaken by one planner.

Communicate plan contents to coplanner. At times, the particular plans or steps involved in pursuing a goal may require informing the coplanner of the contents of the plan. Failure to do so may result in the following problem configurations:

- One partner inadvertently undoes the plan steps already executed by the other partner.

- One partner misunderstands the other's actions as failing to pursue the goal.

- One partner believes a deceptive plan that is intended by the other to foil outsiders. The coplanner must be informed of deceptive tactics.

- Coplanners must be informed of changes in plans; otherwise, they may take further action based upon presumed outcomes that will not occur. This involves "counting your chickens before they have hatched" through the lack of courtesy of the coplanner to inform.

- Coplanners must communicate planning decisions involving shared resources. This is illustrated by the story "Gift of the Magi."

- While communication is necessary in all of these cases, there is a caveat: overinformation may be the equivalent of forcing the coplanner to perform the goal pursuit themselves.

Optimize planning choices over the mutual pursuit

- Use the individual planner's abilities to their advantage. While each partner may be capable of handling each goal pursuit, the optimal arrangement for the MGP is to place the individual best suited for the goal in the position of responsibility.

992

- Use the mutual resources to the best advantage. Since resources are joint and are involved in many plans, coordination of resources will be necessary.

- Take advantage of joint planning ability. For some planning needs, the partnership will provide an advantage in ability to plan for complicated needs. This is captured by the adage, "two heads are better than one."

Amount of Effort towards Pursuit from Either Partner

A second major source of problems in MGP situations involves the expenditure of effort towards mutual goals. In a mutual goal situation, where effort is required of both partners, a variety of responses are possible. The two partners might not be willing to put forth the same amount of effort to satisfy the goal: one or the other partner may expend too little or too much effort on the goal pursuit, according to the judgment of the other partner. This conflict is due to the individual's goal structure, which includes the MGP, competing with the MGP's needs for effort from the planner. Because of the meta-plan to "minimize effort in goal pursuit," it is always advantageous to find optimal arrangements in goal pursuit to avoid wasting resources, noted by Zipf (1949) as the principle of least effort. The decision about the amount of effort to put forth towards a goal is particularly important when it comes to recurring goals. When the satisfaction of a goal is going to be periodically required, then attention towards minimizing the expenditure of effort towards the goal is more important than when pursuing a goal that is only satisfied once. The resolution of the conflict in amount of effort expended towards goals results in these patterns of MGP interactions and strategies for their solution:

- **Pull your own weight.** When you are in a mutual goal pursuit, you have equal responsibility with your partner for the goal pursuit; therefore, you should expend equal effort in the goal pursuit. Abandon this strategy if your partner fails to make effort in goal pursuit.

- **Optimize effort.** When you are in a mutual goal pursuit, and a recurring goal makes optimizing effort important, make little effort in the pursuit and your co-planner will be forced to satisfy the goal. Abandon if this strategy threatens the mutual goal (no one serves the goal) or threatens the basis of the mutual goal (the partnership agreement).

- **Playing hard to get.** When you are in a mutual goal pursuit, and your co-planner fails to pursue a mutual goal, abandon the goal until the co-planner carries out pursuit. Abandon if the basis of the mutual goal (the partnership agreement) is in question.

- **Expend effort only when detectable.** No "credit" will accrue for work on a MGP if the effort required is not observed by the coplanner. Coast along without effort if the effort will not be perceived, and draw attention to effort when it is made.

- **Expend effort only when required.** Perfunctory performance on recurring goals is adequate unless circumstances require a best effort. An example is going through the motions instead of hustling during practice drills in sports events.

- **Fill the gap.** Let your partner pursue the goal, and match your effort to the difference between your partner's effort and the required level. Since your effort is required for successful achievement, you contribute without bearing heavy burden.

- **Tit for tat.** Match your effort to your partner's. If it is sufficient, you can try to adjust the amount to avoid wasting effort. If it is insufficient, it will convince both partners that more effort is required from both.

Some general strategies for goal pursuit are of particular importance in mutual goal pursuit situations. These include:

- **Juggle effort towards threatened goals.** Work hardest on those goals that threaten to fail imminently; coast on those that are currently satisfactory. This refers to the carnival trick of keeping a set of dishes spinning on individual rods -- adding spin must be done to the plate that needs it most first, leaving the others to struggle on until their turn.

- **Prioritize goals for extra effort.** Place more effort into those goals considered important. Less important goals can suffer from little effort better than important ones.

- **Cut your losses.** Minimize effort towards failing *and* unimportant goals. This is "not throwing good money after bad."

Differences in Criterion of Satisfaction for Partners

Another source of MGP problems lies in differences in the criterion of satisfaction for the different partners. If both partners agree about the amount of effort required to satisfy a goal, then the problem of optimizing the planners' efforts to achieve that level is relatively straightforward. However, the two partners in the MGP may not agree on the nature of the mutual goal: X may think that particular plans or actions he performs are in service of the goal pursuit, but Y may think the efforts do not lead to goal satisfaction. In addition, the two parties might not agree on the importance of the goal; for example, one partner might feel that financial support is most important, and pursue it over the emotional support the other prefers. Finally, the problem is greatly complicated by the nature of the goals commonly pursued in mutual goal situations. For example, suppose the mutual goal is emotional support; the level of satisfaction required for one individual may not be the same as that required for the other. Because of the "fair play" notion of effort in mutual goal pursuits, the partner more easily satisfied will complain about the extra effort required to satisfy the more demanding partner. The principle of fair exchange (Blau, 1964) forms the basis for a standard of what can be expected of a partner. This is a frequent problem in MGP situations, and the following strategies propose some responses:

- **Adopt your partner's satisfaction criteria.** When your partner's criterion requires more effort than yours, accept the additional demands and work to satisfy them if possible.

- **Attend to your partner's priority goals.** Expend the effort demanded on those goals deemed most important by your partner; slack off on the rest. For example, if being a good economic provider is most important to your partner, decide to work late rather than spend the evening at home.

- **Increase your criterion on goals important to you.** Make up the difference in effort that you expend on a demanding partner by increasing your criteria on goals more important to you.

- **Quid pro quo.** Make a clear exchange of needs. Offer to satisfy your partner's particular satisfaction criteria in exchange for the satisfaction of one of yours.

The MGP requires planners to make decisions not only about their individual planning efforts, but also to influence another planner's efforts in order to achieve the goal. Even when there are problems in the MGP, and one partner has to plan to get the other to live up to his side of the agreement, this does not mean that the relationship is competitive. You are not out to cause your partner to fail; you are out to influence him into cooperating. Therefore, even when you are "counterplanning" against him, the end goal is to enlist his cooperation. This puts definite constraints on what plans can be used to influence the other planner; for example, avoiding strategies that might produce negative responses which may linger after the issue is resolved. Resolving problems in MGP situations requires strategies that elicit increased cooperation from the partner while minimizing your own effort.

Conclusion

The purpose of the theory of planning is to account for the choices a planner makes in attempting to reach a goal, and to provide a model of how planning knowledge might be learned and stored in memory so as to be available and utilized in appropriate situations. The above categorization outlines the common problems in MGP, and provides strategies for resolving the problematic interactions. The principles are general enough to apply as well to MGP with any number of actors, and to a variety of types of MGP arrangements. A taxonomy of constraints specific to MGP planning situations provides and understanding of what features of the planning environment are relevant to plan selection and planning failures, and therefore serve as specifications for planning rules in computer models of planners. The content theory presented advances an understanding about what information is relevant to planning situations and broadens the scope of problems that memory-based planning theories can account for.

References

Blau, P. M. (1964). Change and power in social life. New York: Wiley.

Dyer, M. G. (1983). *In-depth understanding: A computer model of integrated processing for narrative comprehension.* Cambridge, MA: MIT Press.

Schank, R. C. (1982). *Dynamic memory: A theory of reminding and learning in computers and people.* New York: Cambridge University Press.

Wilensky, R. (1983). *Planning and understanding.* Reading, MA: Addison-Wesley.

Zipf, G. K. (1949). Human behavior and the principle of least effort. Cambridge, MA: Addison-Wesley.

Finding Creative Solutions in Adversarial Impasses*

Katia Sycara

School of Information and Computer Science
Georgia Institute of Technology
Atlanta, Georgia 30332
Telephone: (404) 894-5550
Computer address: katia@gatech (CSNET)

ABSTRACT

This paper presents a method for generating creative solutions to resolve adversarial impasses that makes use of memory structures based on goal interactions and blame attribution. These knowledge structures are called Situational Assessment Packets (SAPs). SAPs contain general strategies for satisfying multiple conflicting goals either totally or partially. These resolution strategies are evaluated for applicability to a situation by considering interactions of the goals of the problem solver and the goals of the interacting agents. This work is part of the PERSUADER, a computer program that functions as a third party problem solver (mediator) in hypothetical labor negotiations.

May 11, 1987

Poster session
Key Words: Knowledge structures, Adversarial Resoning, Multi-agent Planning,

996

Finding Creative Solutions in Adversarial Impasses*

Katia Sycara

School of Information and Computer Science
Georgia Institute of Technology
Atlanta, Georgia 30332
Telephone: (404) 894-5550
Computer address: katia@gatech (CSNET)

1. Introduction

Event 1: The Blackhound union's demands in contract negotiations are a 20 percent wage increase, and 6 percent increase in pensions. The company's counterproposal is 3 percent wage increase, and 1 percent increase in pensions. After an incremental shifting of positions, the parties agree to the following proposal of a mediator: 8 percent wage increase, and 4 percent increase in pensions.

Event 2: During contract negotiations, Southern Airlines presents its employees with the ultimatum that if they don't take wage cuts of 8%, the company which has become non-competitive, will go bankrupt. The employees protest and a mediator is called in. The mediator finds out that Southern Airlines has been loosing money because of mismanagement in an industry where other airlines are making money. She proposes that the employees accept 6% wage cuts and that the company have employee representatives sit on the board of directors.

Events 1 and 2 deal with adversarial situations. In adversarial situations the goals of two or more disputants are in conflict. Both Events illustrate a resolution by compromise of the conflicting goals. There is a main difference, however. The solution in Event 1 is a perturbation of the values of the adversaries' original goals. In Event 2, on the other hand, new elements not predicted by the input (employee representation on the board of directors) also enter the solution.

While an event such as Event 1 is fairly typical and therefore requires fairly shallow reasoning, Event 2 is atypical (companies in a prosperous industry do not usually lose money) and requires creative problem solving. In cases where the situation is novel, a problem solver has to take into account, not only the apparent goals of the agents, but also higher level goals and their interactions. In general, cases that require creative problem solving are those that violate expectations about (a) prevailing practice, namely what the situation of similar agents is (b) role themes of the interacting agents, (c) beliefs about the rationality of the agents,** (d) beliefs about the temporal continuation of a state.

We propose a high level knowledge structure, called a Situational Assessment Packet (SAP) that captures the abstract structure of an atypical problem solving situation involving interacting agents in terms of expectation violations. SAPs are like MOPs (Schank, 1982) in that they are organizing memory structures that store generalized information. They are also like TOPs (Schank, 1982) in that they organize expectations and explanations of expectation failures in a domain-independent manner. While Hammond's TOPs (1986) generally involve goal interactions associated with a *single agent*, SAPs involve interactions of goals in a *multiagent* situation. The inclusion of partial goal fulfillment strategies is another novel feature of SAPs,

---

* This work was supported by ARO Grant No. DAAG 29-85-K-0023
** By rationality we mean an agent's reluctance to follow a course of action that will result in loss of benefits.

997

and one that further differentiates SAPs from planning TOPs (Hammond, 1986), and from Wilensky's (1983) work on planning. SAPs, unlike TAUs (Dyer, 1983) record the *reason* for the failure of the outcome of a situation. Because SAPs include blame attribution information, they provide a problem solver with *predictions* about the agents' subsequent behavior, something that neither TAUs nor TOPs allow.

The ability to store cross-contextual episodes/cases makes SAPs very powerful mechanisms since, what is learned in one planning situation can be used in planning for a subsequent situation that fits the same SAP. SAPs are accessed when the case under consideration is atypical. They contain an abstracted structure representing situation-outcome patterns in terms of: (1) a problem solving situation, (2) expectations associated with the situation, (3) the reason the expectation is violated (4) who/what is responsible for the violation, (5) how a third party problem solver can find an equitable solution, and (6) how to justify the solution. In addition, SAPs warn the problem solver about potential failures, so that these failures can be avoided.

SAPs are intended to prepackage knowledge about situations in which psychological considerations affect the acceptability of solutions or arguments to influence the acceptance of solutions. Psychological validity applies only to individual SAPs because SAPs are a general mechanism for introducing departures from a model based on rational agents. SAP MISMANAGEMENT, that captures the situation in Event 2, embodies equity theory Walser (1978) and the principle of distributive justice which hold that human agents will seek equity and/or proportionality in payoffs even at some sacrifice to themselves.* SAP MISPLACED-LOYALTY employs Festinger's (1957) cognitive dissonance theory to direct arguments at loyalty (attraction) to a party in order to weaken attraction to that party's position. Other SAPs function in similar ways to customize rational solutions to the known vagaries of human agents.

1.1. Situational Assessment Packets (SAPs)

In adversarial situations a problem solver has to come up with a solution that satisfies multiple goals simultaneously. Describing a problem in terms of goals and plans alone forces the problem solver in considering separate plans for the satisfaction of each goal. This deprives the problem solver of the opportunity to access "analogous" plans useful for similarity between the abstract interelations of the goals and plans. For example, such a planner would not be able to recognize the similarity of Event 2 to the following event:

> Event 3: Tom and Jerry are project partners in a computer science class. Tom finishes his part of the project in time, but Jerry starts his part the night before the project is due and does a lousy job. The teacher assigns separate grades to the two parts of the project, though this is not his customary practice.

Event 3 is an example of the SAP MISMANAGEMENT, shown below, since Jerry mismanages time with the result of jeopardizing the success of the joint project.

SAP MISMANAGEMENT

recognition criteria:

> (a) x and y have a non-competitive high level goal G
>
> (b) x mismanages some resource that is an enablement condition C for the achievement of G.
>
> (c) G is in danger of failing

solution:

> an equitable solution to prevent the failure of G is to have x, the guilty party, bear the brunt of the recovery cost

---

* This has been borne out in real situations, as for example the recent (1985) Eastern Airlines settlement with the International Pilots' Association.

justification:

> appeal to theme of fairness and add that if y does not perceive the solution as just, then y will not cooperate and thus G will fail (which certainly x does not want).

SAP MISMANAGEMENT, for example, organizes cases such as Events 2 and 3, cases where an officer of an organization (e.g., a union, a church) has mismanaged funds, or where a military leader has mismanaged his part of a campaign.

Another SAP is MISPLACED-LOYALTY that captures the dilemma of an actor caused by his loyalty to another actor with whom the first is bound through a thematic relationship. A third party problem solver is asked to give advice regarding this dilemma. Consider the following events:

> Event 4: A local union gives high priority to pensions although the majority of its members are young with the result of making agreement during contract negotiations impossible. The mediator that handles the case finds out that this union attitude is due to its desire to follow the guidelines of the international union whose program mandates as a negotiation goal high pension increases.

> Event 5: Susan is torn between her love for John and wish to marry him and her family's opposition because of disapproval of his lifestyle and political views. A friend tells her that if her family really cared for her they would respect her wishes.

SAP MISPLACED-LOYALTY

recognition criteria:

> (a) x goes against his interest in order to conform to third party's wishes

> (b) this endangers an achievement goal G of x

> (c) third party is blamed

solution:

> suggestion to look after own interest

justification:

> point out contradictory attitude of third party towards x

Other SAPs we have identified include UNFORESEEN-DISASTER (external chance events force the agents into unpleasant situations), LAME-DUCK (an agent has the title of an office but not the authority), and DETRIMENTAL-PREROGATIVES (a prerogative is no longer advantageuos to the grantor because of changing circumstances). A full presentation of SAPs can be found in (Sycara, 1987).

SAPs are recognized based on (1) expectations that the problem solving situation violates, and (2) the cause of the violation (who is to blame). SAP recognition rules are associated with each expectation violation category (presented in the introduction). If a given situation indicates the possibility of a failure, preventive advice from the appropriate SAP can help the reasoner avoid the failure.

1.2. Causal Structure of SAPs

The causal structure of a SAP is a graph whose nodes represent goals, states and actions of the agents and are connected via links that indicate relations between the goals and states. The abstract causal structure of SAP MISPLACED-LOYALTY is illustrated below:

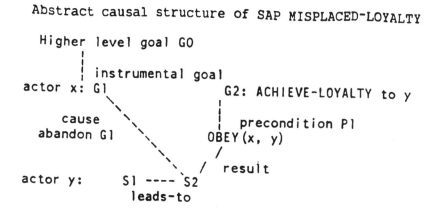

Figure 1

G1 is instrumental to the higher level goal G0 of actor x. S1 causes S2. Actor y's state S2 is a cause for the abandonment of G1 by actor x. OBEY(x,y) is a precondition for the satisfaction of G2. While the content of particular goals of different episodes within the above SAP is different, the interactions that occur between the goals in these situations are similar. They all involve an actor's desire to satisfy a primary goal G1, as well as a secondary goal, G2 which is ACHIEVE-LOYALTY to y. G1 is in danger of failing because of belief states S1 and S2 of y. Thus, Events 4 and 5 can be handled through the use of similar strategies for dealing with the particular goal and state interaction. Figure 2 represents graphically how Event 5 fits into the SAP MISPLACED-LOYALTY.

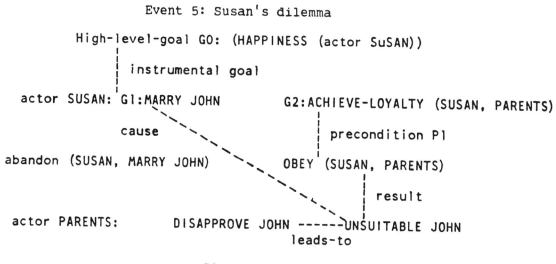

Figure 2

1.3. Using SAPs to guide problem solving

One advantage of using SAPs in problem solving is that they contain domain-independent planning strategies that depend only on the causal structure of a SAP. Thus, a planner does not have to consider all possible plans for the achievement of a goal, but only the ones suggested by the strategies in the SAP. For example, the possible general strategies in MISPLACED-LOYALTY include: (1) fulfill both G1 and G2, (2) abandon G2, (3) partially fulfill G2, (4) abandon G1, (5) partially fulfill G1. To achieve both goals G1 and G2 of x, for example, more specific strategies are (a) cause y to change his belief state S1, (b) wait for y to change his belief state S1 An example of an argument to change S1 in Event 5 might be that John's

1000

political views are not so undesirable since politician z's (whom Susan's parents respect) son also has them and z does not seem to disapprove of him.

The interactions of the goals of the agents with the goals of the problem solver serve as meta-planning knowledge. The problem solver will suggest one strategy over another depending on how his goals interact with those of the disputants. For example, whether the strategy "Abandon G2" will be suggested depends on whether the ACHIEVE-LOYALTY goal G2 interferes with G3, a goal of the third party problem solver (in conflict resolution G3 is finding an acceptable compromise). In Event 4, the union's high priority to pensions interferes with the mediator's goal of finding a mutually acceptable settlement. If the union abandons its loyalty goal to the international and lowers the priority it attaches to pensions, the search for an acceptable solution will be facilitated. Hence, the plan "assert own priorities" that is instrumental to strategy "Abandon G2" will be suggested by the mediator.

SAPs also give a problem solver the ability, absent from present planning work, to include solutions that effect *partial goal satisfaction*. This could be helpful when, for example, there was no plan that could be used for total goal satisfaction.

1.3.1. Indexing in SAPs

SAPs organize generalized episodes (Kolodner et al., 1985) or MOPs (Schank, 1982) as well as single cases. Once a solution has been discovered for a particular problem, and the case has been appropriately indexed inside a SAP, it is available through reminding when another case sharing the same abstract goal/plan/state/condition interrelations is being processed. This reminding focuses the attention of the problem solver to a solution that might be directly applicable in the current problem.

To index episodes and planning advice under SAPs, the indices have to be such that they afford efficient retrieval of strategies and plans. To do this we index plans (strategies) under the preconditions they satisfy or the effects that need to be achieved. Under the plans either generalized or simple episodes can be indexed.

Figure 3 shows part of the indexing structure for SAP MISPLACED-LOYALTY. The upper part of this figure is the same as Figure 1 and depicts the causal structure of MISPLACED-LOYALTY. In addition to the goals and states the preconditions for the accomplishment of the goals are depicted.

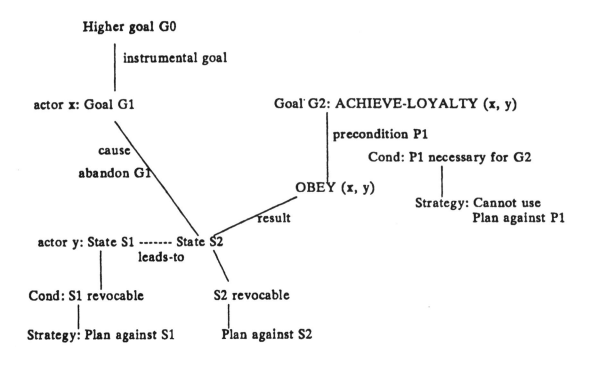

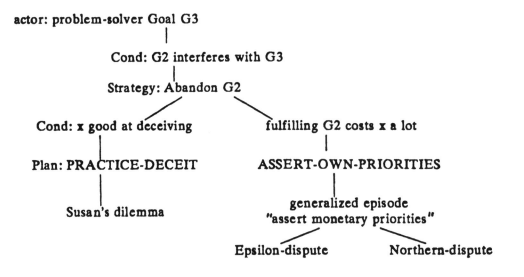

Figure 3

At the leaves of a SAP are pointers to generalized and individual episodes. If a generalized episode is reached, additional indices are traversed to access a particular experience within the generalized episode. For example, the two cases shown organized under the generalized episode "assert monetary priorities" are the Epsilon dispute and Northern dispute. Both disputes are indexed in this location in memory since they used the same plan.

1.4. Summary

In this paper I have presented a class of abstract knowledge structures, called SAPs, which represent the causal structure of atypical problem solving situations involving many interacting agents. SAPs perform the following major functions in problem solving: (1) provide the basis for finding an appropriate solution, (2) act as a source of preventive and recovery advice, (3) provide justifications for the proposed solution, (4) provide remindings across different domains.

1.5. References

Dyer, M.G. *In-Depth Understanding: A Computer Model of Integrated Processing for Narrative Comprehension*, The MIT Press, Cambridge, Mass., 1983.

Festinger, L.A., *A theory of cognitive dissonance.* Evanston, Illinois: Row Peterson & Co., 1957.

Hammond, K. "CHEF: A Model of Case-based Planning", *AAAI-86*, pp. 267-271, Philadelphia, Pa., 1986.

Kolodner, J.L. Simpson, R.L., and Sycara-Cyranski, K. "A Process Model of Case-Based Reasoning in Problem Solving", *IJCAI-85*, pp. 284-290, Los Angeles, Ca. 1985.

Schank, R.C. *Dynamic Memory: A Theory of Reminding and Learning in Computers and People*, Cambridge University Press, Cambridge, Mass., 1982.

Sycara, K. "Adversarial Reasoning in Conflict Resolution", Ph.D. Thesis, School of ICS, Georgia Institute of Technology, Atlanta, Ga., 1987.

Walser, E., Walser, G.W., and Berscheid, E., *Equity: theory and research.* Boston: Allyn & Bacon, Inc., 1978.

Wilensky, R. *Planning and Understanding: A Computational Approach to Human Reasoning*, Addison-Wesley Publishing Company, Reading, Mass., 1983.

ACADEMIC PRESS